Life

The Science of Biology

Sixth Edition

Life

Sixth Edition

The Science of Biology

William K. Purves
Emeritus, Harvey Mudd College
Claremont, California

David Sadava
The Claremont Colleges
Claremont, California

Gordon H. Orians
Emeritus, The University of Washington
Seattle, Washington

H. Craig Heller
Stanford University
Stanford, California

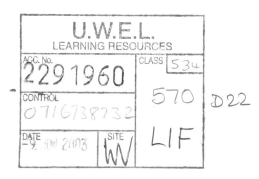
 Sinauer Associates, Inc.

 W. H. Freeman and Company

The Cover

Giraffes (*Giraffa camelopardalis*) near Samburu, Kenya.
Photograph © BIOS/Peter Arnold, Inc.

The Opening Page

Soap yucca (*Yucca elata*), White Sands National Monument, New Mexico.
Photograph © David Woodfall/DRK PHOTO.

The Title Page

The endangered Florida panther (*Felis concolor coryi*).
Photograph © Thomas Kitchin/Tom Stack & Associates.

Life: The Science of Biology, Sixth Edition

Copyright © 2001 by Sinauer Associates, Inc. All rights reserved. This
book may not be reproduced in whole or in part without permission.

Address editorial correspondence to:
Sinauer Associates, Inc., 23 Plumtree Road, Sunderland, Massachusetts 01375 U.S.A.
www.sinauer.com

Email: publish@sinauer.com

Address orders to:
VHPS/W. H. Freeman & Co. Order Department, 16365 James Madison Highway,
U.S. Route 15, Gordonsville, VA 22942 U.S.A.
www.whfreeman.com

Examination copy information: 1-800-446-8923
Orders: 1-888-330-8477

Library of Congress Cataloging-in-Publication Data

Life, the science of biology / William K. Purves...[et al.].--6th ed.
      p. cm.
      Includes index.
      ISBN 0-7167-3873-2 (hardcover) – ISBN 0-7167-4348-5 (Volume 1) –
      ISBN 0-7167-4349-3 (Volume 2) – ISBN 0-7167-4350-7 (Volume 3)
      1. Biology I. Purves, William K. (William Kirkwood), 1934–

QH308.2 .L565 2000
570--dc21   00-048235

Printed in U.S.A.

Third Printing 2002 Courier Companies Inc.

This book is dedicated to the memory of Angeline Douvas

About the Authors

Gordon Orians   Craig Heller   Bill Purves   David Sadava

William K. Purves is Professor Emeritus of Biology as well as founder and former chair of the Department of Biology at Harvey Mudd College in Claremont, California. He received his Ph.D. from Yale University in 1959 under Arthur Galston. A fellow of the American Association for the Advancement of Science, Professor Purves has served as head of the Life Sciences Group at the University of Connecticut and as chair of the Department of Biological Sciences, University of California, Santa Barbara, where he won the Harold J. Plous Award for teaching excellence. His research interests focused on the chemical and physical regulation of plant growth and flowering. Professor Purves elected early retirement in 1995, after teaching introductory biology for 34 consecutive years, in order to turn his skills to writing and producing multimedia for introductory biology students. That year, he was awarded the Henry T. Mudd Prize as an outstanding member of the Harvey Mudd faculty or administration.

David Sadava is now responsible for *Life*'s chapters on the cell (2–8), in addition to the chapters on genetics and heredity that he assumed in the previous edition. He is the Pritzker Family Foundation Professor of Biology at Claremont McKenna, Pitzer, and Scripps, three of the Claremont Colleges. Professor Sadava received his Ph.D. from the University of California, San Diego in 1972, and has been at Claremont ever since. The author of textbooks on cell biology and on plants, genes, and agriculture, Professor Sadava has done research in many areas of cell biology and biochemistry, ranging from developmental biology, to human diseases, to pharmacology. His current research concerns human lung cancer and its resistance to chemotherapy. Vir-

tually all of the research articles he has published have undergraduates as coauthors. Professor Sadava has taught a variety of courses to both majors and nonmajors, including introductory biology, cell biology, genetics, molecular biology, and biochemistry, and he recently developed a new course on the biology of cancer. For the last 15 years, Professor Sadava has been a visiting professor in the Department of Molecular, Cellular, and Developmental Biology at the University of Colorado, Boulder, and is currently a visiting scientist at the City of Hope Medical Center.

Gordon H. Orians is Professor Emeritus of Zoology at the University of Washington. He received his Ph.D. from the University of California, Berkeley in 1960 under Frank Pitelka. Professor Orians has been elected to the National Academy of Sciences and the American Academy of Arts and Sciences, and is a Foreign Fellow of the Royal Netherlands Academy of Arts and Sciences. He was President of the Organization for Tropical Studies, 1988–1994, and President of the Ecological Society of America, 1995–1996. He is chair of The Board on Environmental Studies and Toxicology of the National Research Council and a member of the board of directors of World Wildlife Fund–US. He is a recipient of the Distinguished Service Award of the American Institute of Biological Sciences. Professor Orians is a leading authority in ecology, conservation biology, and evolution, with research experience in behavioral ecology, plant–herbivore interactions, community structure, the biology of rare species, and environmental policy. He elected early retirement to be able to devote more time to writing and environmental policy activities.

H. Craig Heller is the Lorry Lokey/Business Wire Professor of Biological Sciences and Human Biology at Stanford University. He has served as Director of the popular interdisciplinary undergraduate program in Human Biology and is now Chairman of Biological Sciences. Professor Heller received his Ph.D. from Yale University in 1970 and did postdoctoral work at Scripps Institute of Oceanography on how the brain regulates body temperature of mammals. His current research focuses on the neurobiology of sleep and circadian rhythms. Professor Heller has done research on a great variety of animals ranging from hibernating squirrels to exercising athletes. He teaches courses on animal and human physiology and neurobiology.

Preface

Biologists' understanding of the living world is growing explosively. This isn't the world that the four authors of this book were born into. We never dreamed, as we began our research careers as freshly minted Ph.D.'s, that our science could move so rapidly. Biology has now entered the post-genomic era, allowing biologists and biomedical scientists to tackle once-unapproachable challenges. We are also at the threshold of some experiments that raise ethical concerns so great that we must stand back and participate with others in determining what is right to do and what is not.

The enormous growth and changes in biology create a special challenge for textbook authors. How can a biology textbook provide the basics, keep up with the exciting new discoveries, and not become overwhelming. The increasing bulk of textbooks is of great concern to authors as well as to instructors and their students, who blanch at the prospect of too many pages, too many term papers, and too little sleep. Some reconsideration of what is essential and how that is best presented needs to be made if the proliferation of facts is not to obscure the fundamental principles.

Our major goals were brevity, emphasis on experiments, and better ways to help students learn

In writing the Sixth Edition of *Life*, we committed ourselves to reversing the pattern of ever increasing page lengths in new editions. We wanted a shorter book that brings the subject into sharper focus. We tried to achieve this by judicious reduction of detail, by more concise writing, and by more use of figures as primary teaching sources. It worked! Our efforts were successful. This edition is 200 pages shorter than its predecessor, yet it covers much exciting new material.

While working to tighten and shorten the text, we were also determined to retain and even increase our emphasis on *how* we know things, rather than just *what* we know. To that end, the Sixth Edition inaugurates 72 specially formatted figures that show how experiments, field observations, and comparative methods help biologists formulate and test hypotheses (the figure at right is an example). Another 26 figures highlight some of the many field and laboratory methods created to do this research. These Experiment and Research Methods illustrations are listed on the endpapers at the back of the book.

In the Fifth Edition, we introduced "balloon captions" that guide the reader through the illustrations (rather than having to wade through lengthy captions). This feature was widely applauded and we have worked to refine the balloons' effectiveness. In response to suggestions from users

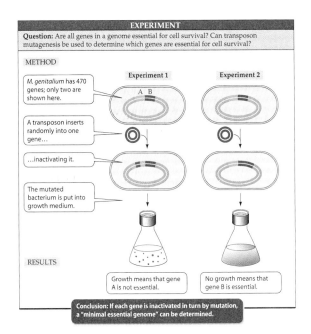

13.22 Using Transposon Mutagenesis to Determine the Minimal Genome
By inactivating genes one by one, scientists can determine which ones are essential for the cell's survival.

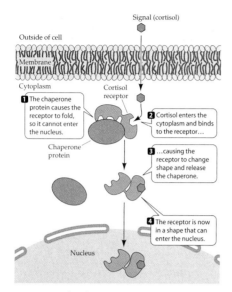

15.9 A Cytoplasmic Receptor
The receptor for cortisol is bound to a chaperone protein. Binding of the signal (which diffuses directly through the membrane) releases the chaperone and allows the receptor protein to enter the cell's nucleus, where it functions as a transcription factor.

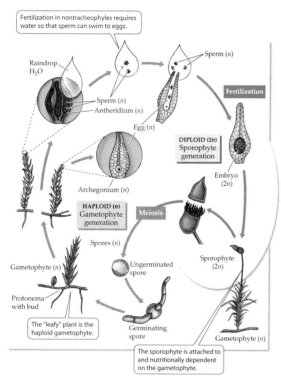

28.3 A Nontracheophyte Life Cycle
The life cycle of nontracheophytes, illustrated here by a moss, is dependent on an external source of liquid water. The visible green structure of nontracheophytes is the gametophyte; in nontracheophyte plants, the "leafy" structures are sporophytes.

of the Fifth Edition, in the Sixth Edition we now number many of the balloons, emphasizing the flow of the figure and making the sequence easier to follow (Figure 15.9 at left is an example).

This edition is accompanied by a comprehensive website, www.thelifewire.com (and an optional CD-ROM that contains the same material) that reinforces the content of every chapter. A key component of the website is a combination of animated tutorials and activities for each chapter, all of which include self-quizzes. Within each book chapter, this [icon] icon refers students to a tutorial or an activity. An index of the icons begins in the front endpapers of the book. Figure 28.3 (left, below) shows a typical web icon placement.

As part of the ongoing challenge of keeping the writing and illustrations as clear as possible, we frequently employ bulleted lists. We think these lists will help students sort through what is, even after pruning, a daunting amount of material. And we have continued to provide plenty of interim summaries and bridges that link passages of text.

In all the introductory textbooks, the chapters end with summaries. In ours, we have organized the material within the chapter's main headings. In most cases, we tie key concepts to the figure (or figures) that illustrate it. For visual learners, this provides an efficient mode of reviewing the chapter.

From our many decades in the classroom, we know how important it is to motivate students. Each chapter begins with a brief description of some event, phenomenon, or idea that we hope will engage the reader while conveying a sense of the significance and purpose of the chapter's subject.

Evolution Continues to be the Dominant Theme

Evolution continues to be the most important of the themes that link our chapters and provide continuity. As we have written the various editions of the book, however, the emergence of *genomics* as a new paradigm in the late twentieth century has developed, revolutionizing most areas of biology. In this new century, understanding the workings of the genome is of paramount importance in almost any biological discussion.

In this edition, we have moved further toward updating the evolutionary theme to encompass the postgenomic era. Just two examples are the addition of a section on genomic evolution to our coverage of molecular evolution, and a section on "evo/devo" in the chapter on molecular biology of development. In addition, the chapters on the diversity of life reflect the vast changes in our understanding of systematics and phylogenetic relationships thanks to the genomic perspective.

In fact, each chapter of the book has undergone important changes.

The Seven Parts: Content, Changes, and Themes

In Part One, The Cell, the emphasis in the discussions of biological molecules and thermodynamics has shifted more decisively toward biological aspects and away from pure chemistry. We have made our discussions of enzymes, cell respiration, and photosynthesis less detailed and more focused on the biological applications.

A major addition to Part Two, Information and Heredity, is a new chapter (Chapter 15) on cell signaling and communication, introduced at a place where the students have the necessary grounding in cell biology and molecular genetics. That chapter leads logically into an updated chapter (Chapter 16) on the molecular biology of development, which includes a new section on the intersection of evolutionary and developmental biology—"evo-devo" in the modern jargon. Several chapters incorporate the exciting new work in genomics of prokaryotes, humans, and other eukaryotes.

We have updated all the chapters in Part Three, Evolutionary Processes. In particular, Chapter 24 ("Molecular and Genomic Evolution") reflects the rapid advances in this exciting field. The section on genomic evolution (on pages 446–447) is brand new and includes Figure 24.9 (shown at right).

Part Four, The Evolution of Diversity, now reflects some exciting changes. The chapter on the protists—which can no longer be treated as a single "kingdom"—reflects the continuing uncertainty over the origin and early diversification of eukaryotes. The equally great uncertainty over prokaryote phylogeny, as we deal with the implications of extensive lateral transfer of genes, is evident in the chapter on prokaryote phyla.

We have extended the coverage of the evolution and diversity of plants to two chapters, and that of the animals to three. Recent findings stemming largely from molecular research have led to modifications of the phylogenies of angiosperms and of the animal kingdom. These changes are reflected in the many simplified "trees" that give a broad overview of systematic relationships. Key evolutionary events that separate and unite the different groups are highlighted with red "hot spots" (see Figure 33.1 at right).

We have rearranged Part Five, "The Biology of Flowering Plants," to allow Chapter 39 ("Plant Responses to Environmental Challenges") to serve as a capstone to the whole part, drawing together some of the major threads. We have added sections on hormones and photoreceptors discovered in recent years, and on their signal transduction pathways. The opening chapter (Chapter 34) on "The Flowering Plant Body" has an increased emphasis on meristems.

Part Six, The Biology of Animals, continues to be a broad, comparative treatment of animal physiology with an emphasis on mechanisms of control and regulation. Much new material has been added, including a major revision of Animal

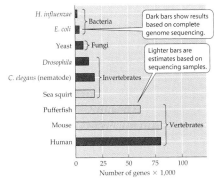

24.9 Complex Organisms Have More Genes than Simpler Organisms
Genome sizes have been measured or estimated in a variety of organisms, ranging from single-celled prokaryotes to vertebrates.

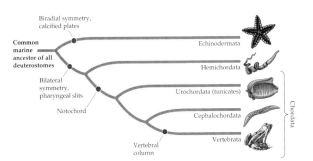

33.1 A Probable Deuterostomate Phylogeny
There are fewer major lineages and many fewer species of deuterostomes than of protostomes.

Development (Chapter 43) to complement and extend the earlier Chapter 16 (Development: Differential Gene Expression). Some other new topics are the role of melatonin in photoperiodism, the role of leptin in the control of food intake, and the discovery in fruit flies of a gene that controls male mating behavior. The extensive coverage of the fast moving field of neurobiology has been substantially updated.

Throughout Part Seven, Ecology and Biogeography, we have added examples of experimental approaches to understanding the dynamics of ecological systems. Some of the examples illustrate the use of experimental and comparative methods. As before, we conclude the book with a chapter on conservation biology (Chapter 58), emphasizing the use of scientific principles to help preserve Earth's vast biological diversity.

There Are Many People to Thank

The reviewing process for *Life*, once a single pass at the stage of draft manuscript, has become an ongoing phenomenon. When the Fifth Edition was still young, we received critiques that influenced our work on this Sixth Edition. The two most penetrating ones came from Zach Gertz, then an undergraduate at Harvard, and Joseph Vanable, a veteran introductory biology professor at Purdue.

Next, still during the Fifth Edition run, 18 instructors recorded their suggestions for improvements in *Life* while teaching from the book. We call these reviews Diary Reviews. The third stage was the Manuscript Reviews. Seventy-three dedicated teachers and researchers read the first-draft chapters and gave us significant and cogent advice. Still another stage has been added to the process and it turned out to be invaluable. We are indebted to 16 Accuracy Reviewers, colleagues who carefully reviewed the almost final page proofs of each chapter to spot lingering errors or imprecisions in the text and art that inevitably escape our weary eyes. Finally, we appreciate the advice given by several experts who reviewed the animations and activities that our publishers developed for the student Web Site/CD-ROM that accompanies this edition of *Life*. We thank all these reviewers and hope this new edition measures up to their expectations. They are listed after this Preface.

J/B Woolsey Associates has again worked closely with each of us to improve an already excellent art program. They helped to refine the very successful "balloon captions" that were introduced in the Fifth Edition. With their creative input we introduced the Experiment and Research Method illustrations found throughout the text.

James Funston joined us again as the developmental editor for the Sixth Edition. As always, James enforced a rigorous standard for clear writing and illustrating. And he contributed significantly to the process of shortening the book. Norma Roche also suggested cuts, and provided incisive copy editing from beginning to end. Her many astute queries often led to rewrites that enhanced the clarity of the presentation. From first draft to final pages, Susan McGlew was tireless in arranging for expert academic reviews of all of the chapters. Since the First Edition, we have profited immeasurably from the work of Carol Wigg, who again coordinated the pre-production process, including illustration editing and copy editing. She wrote many figure captions, suggested several of the chapter-opening stories, orchestrated the flow of the text and art, kept us mostly on schedule, enforced—sometimes with her red pen—the mandate to be concise, and what's more, did it all with good humor, even under pressure. David McIntyre, photo researcher, found many wonderful new photographs to enhance the learning experience and enliven the appearance of the book as a whole.

We again wish to thank the dedicated professionals in W. H. Freeman's marketing and sales group. Their enthusiasm has helped bring *Life* to a wider audience with each edition. We appreciate their continuing support and valuable input on ways to improve the book. A large share of *Life's* success is due to their efforts in this publishing partnership.

We have always respected Sinauer Associates for their outstanding list of biology books at all levels and we have enjoyed having them lead and assist us through yet another edition. Andy Sinauer has been the guiding spirit behind the development of *Life* since two of us first began to write the First Edition. Andy never ceases helping his authors to achieve our goals, while remaining gentle but firm about his agendas. It has been a very satisfying experience for us to work with him yet again, and we look forward to a continuing association.

Bill Purves   David Sadava   Gordon Orians   Craig Heller

November, 2000

Reviewers for the Sixth Edition

Diary Reviewers

Carla Barnwell, University of Illinois

Greg Beaulieu, University of Victoria

Gordon Fain, University of California, Los Angeles

Ruth Finkelstein, University of California, Santa Barbara

Steve Fisher, University of California, Santa Barbara

Alice Jacklet, SUNY, Albany

Clare Hasenkampf, University of Toronto, Scarborough

Werner Heim, Colorado College

David Hershey, Hyattsville, MD

Hans-Willi Honegger, Vanderbilt University

Durrell Kapan, University of Texas, Austin

Cheryl Kerfeld, University of California, Los Angeles

Michael Martin, University of Michigan, Ann Arbor

Murray Nabors, Colorado State University

Ronald Poole, McGill University

Nancy Sanders, Truman State University

Susan Smith, Massasoit Community College

Raymond White, City College of San Francisco

Manuscript Reviewers

John Alcock, Arizona State University

Allen V. Barker, University of Massachusetts, Amherst

Andrew R. Blaustein, Oregon State University

Richard Brusca, University of Arizona

Matthew Buechner, University of Kansas

Warren Burggren, University of North Texas

Jung Choi, Georgia Institute of Technology

Andrew Clark, Pennsylvania State University

Carla D'Antonio, University of California, Berkeley

Alan de Queiroz, University of Colorado

Michael Denbow, Virginia Tech

Susan Dunford, University of Cincinnati

William Eickmeier, Vanderbilt University

John Endler, University of California, Santa Barbara

Gordon L. Fain, University of California, Los Angeles

Stu Feinstein, University of California, Santa Barbara

Danilo Fernando, SUNY, Syracuse

Steve Fisher, University of California, Santa Barbara

Doug Futuyma, SUNY, Stony Brook

Scott Gilbert, Swarthmore College

Janice Glime, Michigan Technological University

Elizabeth Godrick, Boston University

Robert Goodman, University of Wisconsin, Madison

Nancy Guild, University of Colorado

Jessica Gurevitch, SUNY, Stony Brook

Jeff Hardin, University of Wisconsin, Madison

Joseph Heilig, University of Colorado

David Hershey, Hyattsville, MD

Mark Johnston, Dalhousie University

Walter Judd, University of Florida

Thomas Kane, University of Cincinnati

Laura Katz, Smith College

Elizabeth Kellogg, University of Missouri, St. Louis

Peter Krell, University of Guelph

Thomas Kursar, University of Utah

Wayne Maddison, University of Arizona

William Manning, University of Massachusetts, Amherst

Michael Marcotrigiano, Smith College

Lloyd Matsumoto, Rhode Island College

Stu Matz, The Evergreen State College

D. Jeffrey Meldrum, Idaho State University

Mike Millay, Ohio University (Southern Campus)

David Mindell, University of Michigan, Ann Arbor

Deborah Mowshowitz, Columbia University

Laura Olsen, University of Michigan, Ann Arbor

Guillermo Orti, University of Nebraska

Constance Parks, University of Massachusetts, Amherst

Jane Phillips, University of Minnesota

Ronald Poole, McGill University

Warren Porter, University of Wisconsin, Madison

Thomas Poulson, University of Illinois, Chicago

Loren Rieseberg, Indiana University

Ian Ross, University of California, Santa Barbara

Nancy Sanders, Truman State University

Paul Schroeder, Washington State University

Jim Shinkle, Trinity University

Mitchell Sogin, Marine Biological Laboratory, Woods Hole

Wayne Sousa, University of California, Berkeley

Charles Staben, University of Kentucky

James Staley, University of Washington

Steve Stanley, The Johns Hopkins University

Barbara Stebbins-Boaz, Willamette University

Antony Stretton, University of Wisconsin, Madison

Steven Swoap, Williams College

Gerald Thrush, California State University, San Bernardino

Richard Tolman, Brigham Young University

Mary Tyler, University of Maine

Michael Wade, Indiana University

Bruce Walsh, University of Arizona

Steven Wasserman, University of California, San Diego

Alex Weir, SUNY, Syracuse

Mary Williams, Harvey Mudd College

Jonathan Wright, Pomona College

Accuracy Reviewers

Andrew Clark, Pennsylvania State University

Joanne Ellzey, University of Texas, El Paso

Tejendra Gill, University of Houston, University Park

Paul Goldstein, University of Texas, El Paso

Laura Katz, Smith College

Hans Landel, North Seattle Community College

Sandy Ligon, University of New Mexico

Peter Lortz, North Seattle Community College

Roger Lumb, Western Carolina University

Coleman McCleneghan, Appalachian State University

Janie Milner, Santa Fe Community College

Zack Murrell, Appalachian State University

Ben Normark, University of Massachusetts, Amherst

Mike Silva, El Paso Community College

Phillip Snider, University of Houston, University Park

Steven Wasserman, University of California, San Diego

Media Reviewers

Karen Bernd, Davidson College

Mark Browning, Purdue University

William Eldred, Boston University

Joanne Ellzey, University of Texas, El Paso

Randall Johnson, University of California, San Diego

Coleman McCleneghan, Appalachian State University

Melissa Michael, University of Illinois

Tom Pitzer, Florida International University

Kenneth Robinson, Purdue University

To the Student

Welcome to the study of life! In our student days—and ever since—we have enjoyed studying the fascinating and fast-changing field of biology, and we hope that you will, too.

Getting the Most Out of the Book

There are a few things you can do to help you get the most from this book and from your course. For openers, read the book actively—don't just read passively, but do things that force you to think as you read. If we pose questions, stop and think about them. Ask questions of the text as you go. Do you understand what is being said? Does it relate to something you already know? Is it supported by experimental or other evidence? Does that evidence convince you? How does this passage fit into the chapter as a whole? Annotate the book—write down comments in the margins about things you don't understand, or about how one part relates to another, or even when you find an idea particularly interesting. People remember things they think about much better than they remember things they have read passively. Highlighting is passive; copying is drudge work; questioning and commenting are active and well worthwhile.

"Read" the illustrations actively too. You will find the balloon captions in the illustrations especially useful—they are there to guide you through the complexities of some topics and to highlight the major points.

The chapter summaries will help you quickly review the high points of what you have read. A summary identifies particular illustrations that you should study to help organize the material in your mind. Add concepts and details to the framework by reviewing the text. A way to review the material in slightly more detail after reading the chapter is to go back and look at the boldfaced terms. You can use the boldfaced terms to pose questions—and see if you can answer those questions. The boldfacing will probably be more useful on a second reading than on the first.

Use the "For Discussion" questions at the end of each chapter. These questions are usually open-ended and are intended to cause you to reflect on the material.

The glossary and the index can help you a great deal. When you are uncertain of the meaning of a term, check the glossary first—there are more than 1,500 definitions in it. If you don't find a term in the glossary, or if you want a more thorough discussion of the term, use the index to find where it's discussed.

The Web Site

Use the student Web Site/CD-ROM to help you understand some of the more detailed material and to help you sort out the information we have laid before you. An illustrated guide to the learning resources found on the Web Site/CD-ROM is in the front of this book. Pay particular attention to the activities and animated tutorials on key concepts, and to the self-quizzes. The self-quizzes provide extensive feedback for each correct and incorrect answer, and include hot-linked references to text pages. If you'd like to pursue some topics in greater detail, you'll find a chapter-by-chapter annotated list of suggested readings. We have tried to choose readings from books and magazines, especially *Scientific American*, that should be available in your college library.

What If the Going Gets Tough?

Most students occasionally have difficulty in courses, including biology courses. If you find that you are slipping behind in the course, or if a particular topic is giving you an unreasonable amount of trouble, here are some useful steps you might take. First, the basics: attend class, take careful lecture notes, and read the textbook assignments. Second, note that one of the most important roles of studying is to discover what you *don't* know, so that you can do something about it. Use the index, the glossary, the chapter summaries, and the text itself to try to answer any questions you have and to help you organize the material. Make a habit of looking over your lecture notes within 24 hours of when you take them—find out right away what points are unclear, and get them straightened out in your mind. The web site can help by providing a different perspective.

If none of these self-help remedies does the trick, get help! Other students are often a good source of help, because they are dealing with the material at the same level as you are. Study groups can be very useful, as long as the participants are all committed to learning the material. Tutors are almost always helpful, as are faculty members. The main thing is to *get help when you need it*. It is not a good idea to be strong and silent and drift into a low grade.

But don't make the grade the point of this or any other course. You are in college to learn, to pursue interesting subjects, and to enjoy the subjects you are pursuing. We hope you'll enjoy the pursuit of biology.

Bill Purves   David Sadava   Gordon Orians   Craig Heller

Life's Supplements

For the Student

Web Site/CD-ROM

Student Web Site at www.thelifewire.com
Life 6.0 CD-ROM (optionally bundled with the text)

The Web Site and CD-ROM each support the entire text, offering:

▸ Over 65 **Animated Tutorials** clarifying key topics from the text

▸ **Activities**, including flashcards for key terms and concepts, and drag-and-drop exercises

▸ **Self-quizzes** with extensive feedback, references to the Study Guide, and hot-linked references to *Life: The Science of Biology*, Sixth Edition

▸ **Glossary** of key terms and concepts

▸ **End-of-chapter Online Quizzes** (see "Online Quizzing" under "For the Instructor")

▸ **Lifelines**

Study Skills (Jerry Waldvogel, *Clemson University*) provides class-tested practical advice on time management, test-taking, note-taking, and how to read the textbook

Math for Life (Dany Adams, *Smith College*) helps students learn or reacquire basic quantitative skills

▸ **Suggested Readings** for further study
Order ISBN 0-7167-3874-0, *Life 6.0* CD-ROM, or ISBN 0-7167-3875-9, Text/CD-ROM bundle

Study Guide

Christine Minor, *Clemson University*, Edward M. Dzialowski and Warren W. Burggren, *University of North Texas*, Lindsay Goodloe, *Cornell University*, and Nancy Guild, *University of Colorado at Boulder*.

For each chapter of the text, the study guide offers clearly defined learning objectives, summaries of key concepts, references to *Life* and to the student *Web/CD-ROM*, and review and exam-style self-test questions with answers and explanations.
Order ISBN 0-7167-3951-8

Lecture Notebook

This new tool presents black and white reproductions of all the Sixth Edition's line art and tables (more than 1000 images, with labels). The *Notebook* provides ample ruled spaces for note-taking.
Order ISBN 0-7167-4449-X

For the Instructor

Instructor's Teaching Kit

This **new** comprehensive teaching tool (in a three-ring binder) combines:

1. Instructor's Manual

Erica Bergquist, *Holyoke Community College*

The Manual includes:

▸ Chapter overviews

▸ Chapter outlines

▸ A "What's New" guide to the Sixth Edition

▸ All the bold-faced key terms from the text

▸ Key concepts and facts for each chapter

▸ Overviews of the animated tutorials from the Student Web Site/CD-ROM

▸ Custom lab ordering information (see "Custom Labs")

2. Enriched Lecture Notes, with diagrams
Charles Herr, *Eastern Washington University*

3. A PowerPoint® Thumbnail Guide to the PowerPoint® presentations on the Instructor's CD-ROM

Test Bank

Charles Herr, *Eastern Washington University*

The test bank, available in both computerized and printed formats, offers more than 4000 multiple-choice and sentence-completion questions.

The easy-to-use computerized test bank on CD-ROM includes Windows and Macintosh versions in a format that lets instructors add, edit, and resequence questions to suit their needs. From this same CD-ROM, instructors can access *Diploma* Online Testing from the Brownstone Research Group. *Diploma* allows instructors to easily create and administer secure exams over a network and over the Internet, with questions that incorporate multimedia and interactive exercises. More information about *Diploma* is available at http://www.brownstone.net

Online Quizzing

The online quizzing function is accessed via the Student Web Site at www.thelifewire.com. Using Question Mark's *Perception*, instructors can easily and securely quiz students online using multiple-choice questions for each text chapter and its media resources.

Instructor's Resource CD-ROM

The Instructor's Resource CD-ROM employs **Presentation Manager** and includes:

▶ All four-color line art and tables from the text (more than 1000 images), resized and reformatted to maximize large-hall projection

▶ More than 1500 photographic images, including electron micrographs, from the Biological Photo Service collection—all keyed to *Life* chapters

▶ More than 60 animations from the Student Web Site/CD-ROM

▶ Exceptional video microscopy from Jeremy Pickett-Heaps and others

▶ Chapter outlines and lecture notes from the Instructor's Teaching Kit in editable Microsoft® Word documents

PowerPoint® Presentations

The PowerPoint® slide set for *Life* follows the chapter summaries provided in the Instructor's Teaching Kit and can be used directly or customized. Each slide incorporates a figure from *Life*.

PowerPoint® Tutorials
QuickTime™ movies demonstrate how to use PowerPoint®.

Classroom Management

As a service for adopters using WebCT, we will provide a fully-loaded WebCourselet, including the instructor and student resources for this text. The files can then be customized to fit your specific course needs, or can be used as is. Course outlines, pre-built quizzes, activities, and a whole array of materials are included, eliminating hours of work for instructors interested in creating WebCT courses. For more information and a demo of the WebCourselet for this text, please visit our Web Site (http://bfwpub.com/mediaroom/Index.html) and click "WebCT".

Overhead Transparencies

The transparency set includes all four-color line art and tables from the text (more than 1000 images) in a convenient three-ring binder. Balloon captions (and some labels) are deleted to enhance projection and allow for classroom quizzing. Labels and images have been resized for maximum readability.

Slide Set

The slide set includes selected four-color figures from the text. Labels and images have been resized for maximum readability.

Laboratory Manuals

Biology in the Laboratory, Third Edition

Doris Helms, Robert Kosinski, and John Cummings, *all of Clemson University*

The revised edition of this popular lab manual, which includes a CD-ROM, is available to accompany the Sixth Edition of *Life*.
Order ISBN 0-7167-3146-0

Laboratory Outlines in Biology VI

Peter Abramoff and Robert G. Thomson, *Marquette University*
Order ISBN 0-7167-2633-5

The following manuals are available in a bound volume or as separates:
Anatomy and Dissection of the Rat, Third Edition
Warren F. Walker, Jr., *Oberlin College*, and Dominique Homberger, *Louisiana State University*
Order ISBN 0-7167-2635-1
Anatomy and Dissection of the Fetal Pig, Fifth Edition
Warren F. Walker, Jr., *Oberlin College*, and Dominique Homberger, *Louisiana State University*
Order ISBN 0-7167-2637-8
Anatomy and Dissection of the Frog, Second Edition
Warren F. Walker, Jr., *Oberlin College*
Order ISBN 0-7167-2636-X

Custom Labs

Custom Publishing for Laboratory Manuals at www.custompub.whfreeman.com

With this custom publishing option, instructors can build and order customized lab manuals in just minutes, choosing material from Freeman's acclaimed biology laboratory manuals—lab-tested experiments that have been used successfully by hundreds of thousands of students. Instructors determine the manual's content (with the option to incorporate their own material or blank pages), table of contents or index styles, and cover design, and submit the order. A streamlined production process provides a quick turnaround to meet crucial deadlines.

Contents in Brief

Contents

Part One
THE CELL

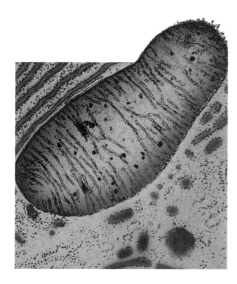

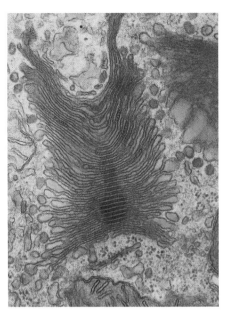

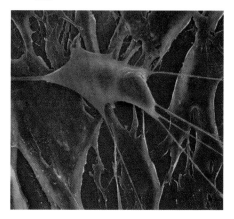

Part Two

INFORMATION AND HEREDITY

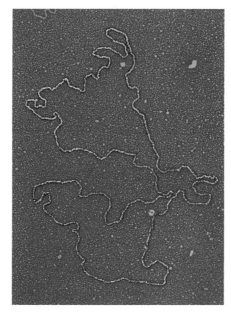

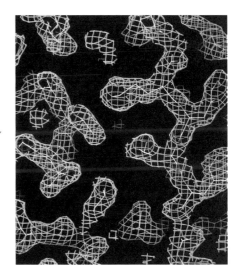

Part Three
EVOLUTIONARY PROCESSES

Part Five
THE BIOLOGY OF FLOWERING PLANTS

Part Six
THE BIOLOGY OF ANIMALS

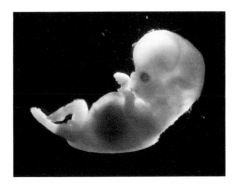

Part Seven
ECOLOGY AND BIOGEOGRAPHY

1 An Evolutionary Framework for Biology

AT MIDNIGHT ON DECEMBER 31, 1999, MAS-sive displays of fireworks exploded in many places on Earth as people celebrated a new millennium—the passage from one thousand-year time frame into the next—and the advent of the year 2000. One such millennial display took place above the Egyptian pyramids.

We are impressed with the size of the pyramids, how difficult it must have been to build them, and how ancient they are. The oldest of these awe-inspiring monuments to human achievement was built more than 4,000 years ago; in the human experience, this makes the Egyptian pyramids very, very old. Yet from the perspective of the age of Earth and the time over which life has been evolving, the pyramids are extremely young. Indeed, if the history of Earth is visualized as a 30-day month, recorded human history—the dawn of which coincides roughly with the construction of the earliest pyramids—is confined to the last *30 seconds* of the final day of the month (Figure 1.1).

The development of modern biology depended on the recognition that an immense length of time was available for life to arise and evolve its current richness. But for most of human history, people had no reason to suspect that Earth was so old. Until the discovery of radioactive decay at the beginning of the twentieth century, no methods existed to date prehistoric events. By the middle of the nineteenth century, however, studies of rocks and the fossils they contained had convinced geologists that Earth was much older than had generally been believed. Darwin could not have conceived his theory of evolution by natural selection had he not understood that Earth was very ancient.

In this chapter we review the events leading to the acceptance of the fact that life on Earth has evolved over several billion years. We then summarize how evolutionary mechanisms adapt organisms to their environments, and we review the major milestones in the evolution of life on Earth. Finally, we briefly describe how scientists generate new knowledge, how they develop and test hypotheses, and how that knowledge can be used to inform public policy.

A Celebration of Time
One millennial fireworks display celebrating the year 2000 took place over the ancient pyramids of Egypt, structures that represent more than 4,000 years of human history but an infinitesimal portion of Earth's geologic history.

Organisms Have Changed over Billions of Years

Long before the mechanisms of biological evolution were understood, some people realized that organisms had changed over time and that living organisms had evolved from organisms no longer alive on Earth. In the 1760s, the French naturalist Count George-Louis Leclerc de Buffon (1707–1788) wrote his *Natural History of Animals*, which contained a clear statement of the possibility of evolution. Buffon originally believed that each species had been divinely created for a particular way of life, but as he studied animal anatomy, doubts arose. He observed that the limb bones of all mammals, no matter what their way of life, were re-

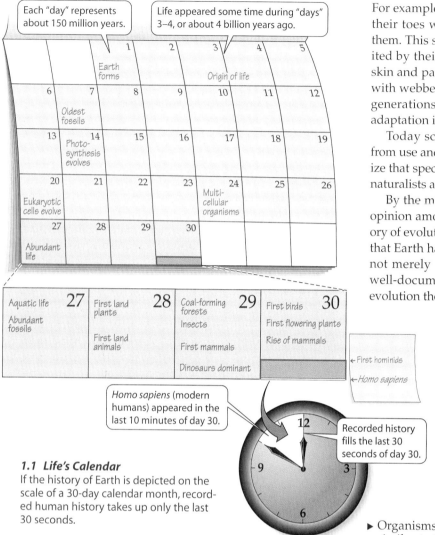

1.1 Life's Calendar
If the history of Earth is depicted on the scale of a 30-day calendar month, recorded human history takes up only the last 30 seconds.

For example, Lamarck suggested that aquatic birds extend their toes while swimming, stretching the skin between them. This stretched condition, he thought, could be inherited by their offspring, which would in turn stretch their skin and pass this condition along to their offspring; birds with webbed feet would thereby evolve over a number of generations. Lamarck explained many other examples of adaptation in a similar way.

Today scientists do not believe that changes resulting from use and disuse can be inherited. But Lamarck did realize that species change with time. And after Lamarck, other naturalists and scientists speculated along similar lines.

By the middle of the nineteenth century, the climate of opinion among many scholars was receptive to a new theory of evolutionary processes. By then geologists had shown that Earth had existed and changed over millions of years, not merely a few thousand years. The presentation of a well-documented and thoroughly scientific argument for evolution then triggered a transformation of biology.

The theory of evolution by natural selection was proposed independently by Charles Darwin and Alfred Russel Wallace in 1858. We will discuss evolutionary theory in detail in Chapter 21, but its essential features are easy to understand. The theory rests on two facts and one inference drawn from them. The two facts are:

▶ The reproductive rates of all organisms, even slowly reproducing ones, are sufficiently high that populations would quickly become enormous if mortality rates did not balance reproductive rates.

▶ Organisms of all types are variable, and offspring are similar to their parents because they inherit their features from them.

The inference is:

▶ The differences among individuals influence how well those individuals survive and reproduce. Traits that increase the probability that their bearers will survive and reproduce are more likely to be passed on to their offspring and to their offspring's offspring.

Darwin called the differential survival and reproductive success of individuals **natural selection**. The remarkable features of all organisms have evolved under the influence of natural selection. Indeed, *the ability to evolve by means of natural selection clearly separates life from nonlife.*

Biology began a major conceptual shift a little more than a century ago with the general acceptance of long-term evolutionary change and the recognition that differential survival and reproductive success is the primary process that adapts organisms to their environments. The shift has taken a long time because it required abandoning many components of an earlier worldview. The pre-Darwinian view held that the world was young, and that organisms had been created in their current forms. In the Darwinian view,

markably similar in many details (Figure 1.2). Buffon also noticed that the legs of certain mammals, such as pigs, have toes that never touch the ground and appear to be of no use. He found it difficult to explain the presence of these seemingly useless small toes by special creation.

Both of these troubling facts could be explained if mammals had not been specially created in their present forms, but had been modified over time from an ancestor that was common to all mammals. Buffon suggested that the limb bones of mammals might all be similar, and that the functionless toes of pigs might be inherited from ancestors with fully formed and functional toes. Buffon's idea was an early statement of evolution (descent with modification), although he did not attempt to explain how such changes took place.

Buffon's student Jean Baptiste de Lamarck (1744–1829) was the first person to propose a mechanism of evolutionary change. Lamarck suggested that lineages of organisms may change gradually over many generations as offspring inherit structures that have become larger and more highly developed as a result of continued use or, conversely, have become smaller and less developed as a result of disuse.

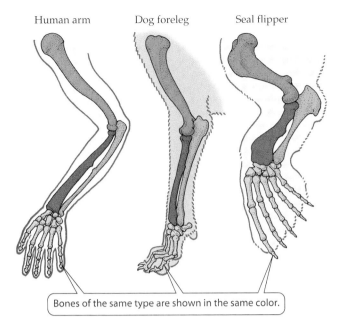

Human arm   Dog foreleg   Seal flipper

Bones of the same type are shown in the same color.

1.2 Mammals Have Similar Limbs
Mammalian forelimbs have different purposes, but the number and types of their bones are similar, indicating that they have been modified over time from a common ancestor.

the world is ancient, and both Earth and its inhabitants have been continually changing. In the Darwinian view of the world, organisms evolved their particular features because individuals with those features survived and reproduced better than individuals with different features.

Adopting this new view of the world means accepting not only the processes of evolution, but also the view that the living world is constantly evolving, and that evolutionary change occurs without any "goals." The idea that evolution is not directed toward a final goal or state has been more difficult for many people to accept than the process of evolution itself. But even though evolution has no goals, evolutionary processes have resulted in a series of profound changes—milestones—over the nearly 4 billion years life has existed on Earth.

Evolutionary Milestones

The following overview of the major milestones in the evolution of life provides both a framework for presenting the characteristics of life that will be described in this book and an overview of how those characteristics evolved during the history of life on Earth.

Life arises from nonlife

All matter, living and nonliving, is made up of chemicals. The smallest chemical units are atoms, which bond together into molecules; the properties of those molecules are the subject of Chapter 2. The processes leading to life began nearly 4 billion years ago with interactions among small molecules that stored useful information.

The information stored in these simple molecules eventually resulted in the synthesis of larger molecules with complex but relatively stable shapes. Because they were both complex and stable, these units could participate in increasing numbers and kinds of chemical reactions. Some of these large molecules—carbohydrates, lipids, proteins, and nucleic acids—are found in all living systems and perform similar functions. The properties of these complex molecules are the subject of Chapter 3.

Cells form from molecules

About 3.8 billion years ago, interacting systems of molecules came to be enclosed in compartments surrounded by membranes. Within these membrane-enclosed units, or **cells**, control was exerted over the entrance, retention, and exit of molecules, as well as the chemical reactions taking place within the cell. Cells and membranes are the subjects of Chapters 4 and 5.

Cells are so effective at capturing energy and replicating themselves—two fundamental characteristics of life—that since the time they evolved, they have been the unit on which all life has been built. Experiments by the French chemist and microbiologist Louis Pasteur and others during the nineteenth century convinced most scientists that, under present conditions on Earth, cells do not arise from noncellular material, but must come from other cells.

For 2 billion years, cells were tiny packages of molecules each enclosed in a single membrane. These **prokaryotic cells** lived autonomous lives, each separate from the other. They were confined to the oceans, where they were shielded from lethal ultraviolet sunlight. Some prokaryotes living today may be similar to these early cells (Figure 1.3).

1.3 Early Life May Have Resembled These Cells
"Rock-eating" bacteria, appearing red in this artificially colored micrograph, were discovered in pools of water trapped between layers of rock more than 1,000 meters below Earth's surface. Deriving chemical nutrients from the rocks and living in an environment devoid of oxygen, they may resemble some of the earliest prokaryotic cells.

To maintain themselves, to grow, and to reproduce, these early prokaryotes, like all cells that have subsequently evolved, obtained raw materials and energy from their environment, using these as building blocks to synthesize larger, carbon-containing molecules. The energy contained in these large molecules powered the chemical reactions necessary for the life of the cell. These conversions of matter and energy are called **metabolism**.

All organisms can be viewed as devices to capture, process, and convert matter and energy from one form to another; these conversions are the subjects of Chapters 6 and 7. *A major theme in the evolution of life is the development of increasingly diverse ways of capturing external energy and using it to drive biologically useful reactions.*

Photosynthesis changes Earth's environment

About 2.5 billion years ago, some organisms evolved the ability to use the energy of sunlight to power their metabolism. Although they still took raw materials from the environment, the energy they used to metabolize these materials came directly from the sun. Early photosynthetic cells were probably similar to present-day prokaryotes called cyanobacteria (Figure 1.4). The energy-capturing process they used—**photosynthesis**—is the basis of nearly all life on Earth today; it is explained in detail in Chapter 8. It used new metabolic reactions that exploited an abundant source of energy (sunlight), and generated a new waste product (oxygen) that radically changed Earth's atmosphere.

The ability to perform photosynthetic reactions probably accumulated gradually during the first billion years or so of evolution, but once this ability had evolved, its effects were dramatic. Photosynthetic prokaryotes became so abundant that they released vast quantities of oxygen gas (O_2) into the atmosphere. The presence of oxygen opened up new avenues of evolution. Metabolic reactions that use O_2, called **aerobic metabolism**, came to be used by most organisms on Earth. The oxygen in the air we breathe today would not exist without photosynthesis.

Over a much longer time, the vast quantities of oxygen liberated by photosynthesis had another effect. Formed from O_2, ozone (O_3) began to accumulate in the upper atmosphere. The ozone slowly formed a dense layer that acted as a shield, intercepting much of the sun's deadly ultraviolet radiation. Eventually (although only within the last 800 million years of evolution), the presence of this shield allowed organisms to leave the protection of the oceans and establish new lifestyles on Earth's land surfaces.

Sex enhances adaptation

The earliest unicellular organisms reproduced by doubling their hereditary (genetic) material and then dividing it into two new cells, a process known as mitosis. The resulting progeny cells were identical to each other and to the parent. That is, they were clones. But **sexual reproduction**—the combining of genes from two cells in one cell—appeared early during the evolution of life. Sexual reproduction is advantageous because an organism that combines its genetic information with information from another individual produces offspring that are more variable. *Reproduction with variation is a major characteristic of life.*

Variation allows organisms to adapt to a changing environment. **Adaptation** to environmental change is one of life's most distinctive features. An organism is adapted to a given environment when it possesses inherited features that enhance its survival and ability to reproduce in that environment. Because environments are constantly changing, organisms that produce variable offspring have an advantage over those that produce genetically identical "clones," because they are more likely to produce some offspring better adapted to the environment in which they find themselves.

Eukaryotes are "cells within cells"

As the ages passed, some prokaryotic cells became large enough to attack, engulf, and digest smaller cells, becoming the first predators. Usually the smaller cells were destroyed within the predators' cells. But some of these smaller cells survived and became permanently integrated into the operation of their hosts' cells. In this manner, cells with complex internal compartments arose. We call these cells **eukaryotic cells**. Their appearance slightly more than 1.5 billion years ago opened more new evolutionary opportunities.

Prokaryotic cells—the Bacteria and Archaea—have no membrane-enclosed compartments. Eukaryotic cells, on the

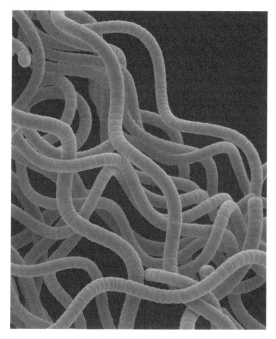

1.4 Oxygen Produced by Prokaryotes Changed Earth's Atmosphere
These modern cyanobacteria are probably very similar to early photosynthetic prokaryotes.

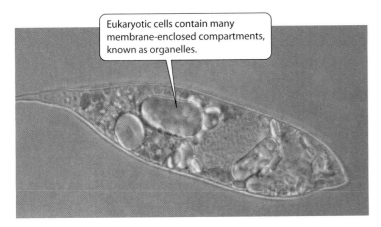

Eukaryotic cells contain many membrane-enclosed compartments, known as organelles.

1.5 Multiple Compartments Characterize Eukaryotic Cells
The nucleus and other specialized organelles probably evolved from small prokaryotes that were ingested by a larger prokaryotic cell. This is a photograph of a single-celled eukaryotic organism known as a protist.

other hand, are filled with membrane-enclosed compartments. In eukaryotic cells, genetic material—genes and chromosomes—became contained within a discrete nucleus and became increasingly complex. Other compartments became specialized for other purposes, such as photosynthesis. We refer to these specialized compartments as **organelles** (Figure 1.5).

Multicellularity permits specialization of cells

Until slightly more than 1 billion years ago, only single-celled organisms existed. Two key developments made the evolution of multicellular organisms—organisms consisting of more than one cell—possible. One was the ability of a cell to change its structure and functioning to meet the challenges of a changing environment. This was accomplished when prokaryotes evolved the ability to change from rapidly growing cells into resting cells called **spores** that could survive harsh environmental conditions. The second development allowed cells to stick together in a "clump" after they divided, forming a multicellular organism.

Once organisms could be composed of many cells, it became possible for the cells to specialize. Certain cells, for example, could be specialized to perform photosynthesis. Other cells might become specialized to transport chemical materials such as oxygen from one part of an organism to another. Very early in the evolution of multicellular life, certain cells began to be specialized for sex—the passage of new genetic information from one generation to the next.

With the presence of specialized sex cells, genetic transmission became more complicated. Simple nuclear division—mitosis—was and is sufficient for the needs of most cells. But among the sex cells, or gametes, a whole new method of nuclear division—meiosis—evolved. Meiosis allows gametes to combine and rearrange the genetic infor-

mation from two distinct parent organisms into a genetic package that contains elements of both parent cells but is different from either. The recombinational possibilities generated by meiosis had great impact on variability and adaptation and on the speed at which evolution could occur.

Mitosis and meiosis are covered in detail in Chapter 9.

Controlling internal environments becomes more complicated

The pace of evolution, quickened by the emergence of sex and multicellular life, was also heightened by changes in Earth's atmosphere that allowed life to move out of the oceans and exploit environments on land. Photosynthetic green plants colonized the land, providing a rich source of energy for a vast array of organisms that consumed them. But whether it is made up of one cell or many, an organism must respond appropriately to its external environment. Life on land presented a new set of environmental challenges.

In any environment, external conditions can change rapidly and unpredictably in ways that are beyond an organism's control. An organism can remain healthy only if its internal environment remains within a given range of physical and chemical conditions. Organisms maintain relatively constant internal environments by making metabolic adjustments to changes in external and internal conditions such as temperature, the presence or absence of sunlight, the presence or absence of specific chemicals, the need for nutrients (food) and water, or the presence of foreign agents inside their bodies. Maintenance of a relatively stable internal condition—such as a constant human body temperature despite variation in the temperature of the surrounding environment—is called **homeostasis**. *A major theme in the evolution of life is the development of increasingly complicated systems for maintaining homeostasis.*

Multicellular organisms undergo regulated growth

Multicellular organisms cannot achieve their adult shapes or function effectively unless their growth is carefully regulated. Uncontrolled growth—one example of which is cancer—ultimately destroys life. *A vital characteristic of living organisms is regulated growth.* Achieving a functional multicellular organism requires a sequence of events leading from a single cell to a multicellular adult. This process is called **development**.

The adjustments that organisms make to maintain constant internal conditions are usually minor; they are not obvious, because nothing appears to change. However, at some time during their lives, many organisms respond to changing conditions not by maintaining their status, but by undergoing major cellular and molecular reorganization. An early form of such developmental reorganization was the prokaryotic spores that were generated in response to environmental stresses. A striking example that evolved much later is **metamorphosis**, seen in many modern in-

1.6 Organisms May Change Dramatically During Their Lives
The caterpillar, pupa, and adult are all stages in the life cycle of a monarch butterfly. The transition from one stage to another is triggered by internal signals.

sects, such as butterflies. In response to internal chemical signals, a caterpillar changes into a pupa and then into an adult butterfly (Figure 1.6).

The activation of gene-based information within cells and the exchange of signal information among cells produce the well-timed events that are required for the transition to the adult form. Genes control the metabolic processes necessary for life. The nature of the genetic material that controls these lifelong events has been understood only within the twentieth century; it is the story to which much of Part Two of this book is devoted.

Altering the timing of development can produce striking changes. Just a few genes can control processes that result in dramatically different adult organisms. Chimpanzees and humans share more than 98 percent of their genes, but the differences between the two in form and in behavioral abilities—most notably speech—are dramatic (Figure 1.7). When we realize how little information it sometimes takes to create major transformations, the still mysterious process of **speciation** becomes a little less of a mystery.

Speciation produces the diversity of life

All organisms on Earth today are the descendants of a kind of unicellular organism that lived almost 4 billion years ago. The preceding pages described the major evolutionary events that have led to more complex living organisms. The course of this evolution has been accompanied by the storage of larger and larger quantities of information and increasingly complex mechanisms for using it. But if that were the entire story, only one kind of organism might exist on Earth today. Instead, Earth is populated by many millions of kinds of organisms that do not interbreed with one another. We call these genetically independent groups of organisms **species**.

As long as individuals within a population mate at random and reproduce, structural and functional changes may occur, but only one species will exist. However, if a population becomes divided into two or more groups, and individuals can mate only with individuals in their own group, differences may accumulate with time, and the groups may evolve into different species.

The splitting of groups of organisms into separate species has resulted in the great variety of life found on Earth today, as described in Chapter 20. How species form is explained in Chapter 22. From a single ancestor, many species may arise as a result of the repeated splitting of populations. How biologists determine which species have descended from a particular ancestor is discussed in Chapter 23.

1.7 Genetically Similar Yet Very Different
By looking at the two, you might be surprised to learn that chimpanzees and humans share more than 98 percent of their genes.

(a)

(b)

(c)   (d)

1.8 Adaptations to the Environment
(*a*) The long, pointed wings of the peregrine falcon allow it to accelerate rapidly as it dives on its prey. (*b*) The action of a hummingbird's wings allows it to hover in front of a flower while it extracts nectar. (*c*) In a water-limited environment, this saguaro cactus stores water in its fleshy trunk. Its roots spread broadly to extract water immediately after it rains. (*d*) The aboveground root system of mangroves is an adaptation that allows these plants to thrive while inundated by salt water—an environment that would kill most terrestrial plants.

Sometimes humans refer to species as "primitive" or "advanced." These and similar terms, such as "lower" and "higher," are best avoided because they imply that some organisms function better than others. In this book, we use the terms "ancestral" and "derived" to distinguish characteristics that appeared earlier from those that appeared later in the evolution of life.

It is important to recognize that *all* living organisms are successfully adapted to their environments. The wings that allow a bird to fly and the structures that allow green plants to survive in environments where water is either scarce or overabundant are examples of the rich array of adaptations found among organisms (Figure 1.8).

The Hierarchy of Life

Biologists study life in two complementary ways:

▶ They study structures and processes ranging from the simple to the complex and from the small to the large.

▶ They study the patterns of life's evolution over billions of years to determine how evolutionary processes have resulted in lineages of organisms that can be traced back to recent and distant ancestors.

These two themes of biological investigation help us synthesize the hierarchical relationships among organisms and the role of these relationships in space and time. We first describe the hierarchy of interactions among the units of biology from the smallest to the largest—from cells to the biosphere. Then we turn to the hierarchy of evolutionary relationships among organisms.

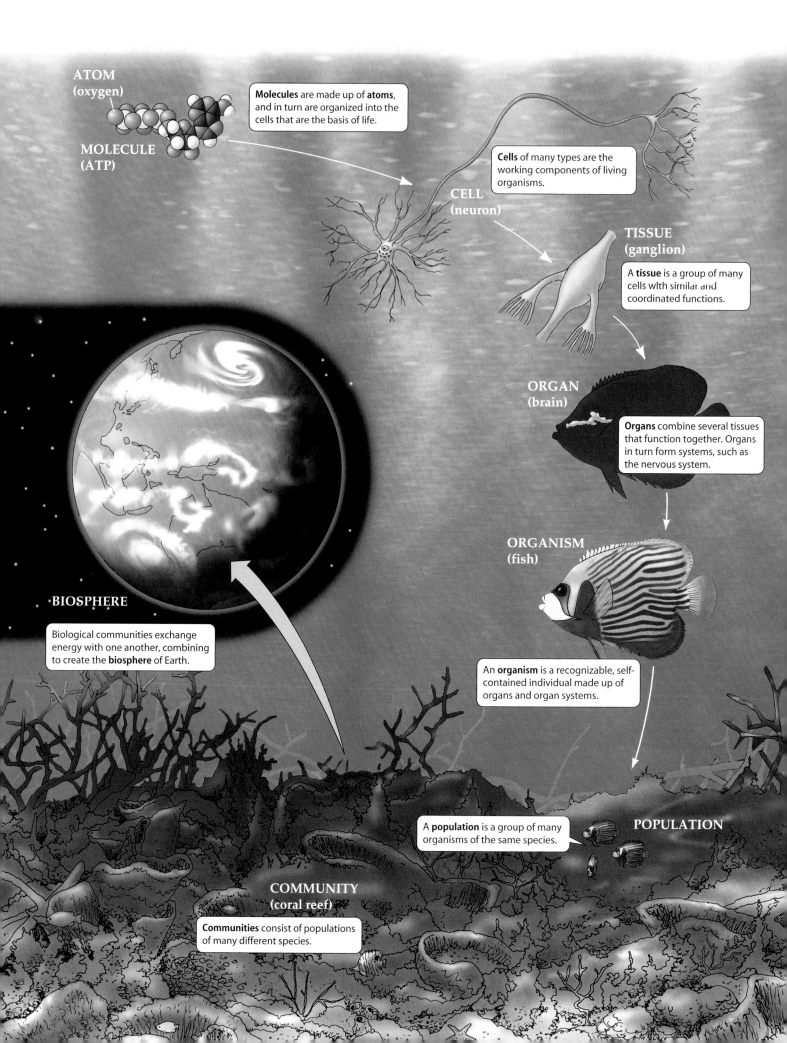

ATOM
(oxygen)

Molecules are made up of atoms, and in turn are organized into the cells that are the basis of life.

MOLECULE
(ATP)

Cells of many types are the working components of living organisms.

CELL
(neuron)

TISSUE
(ganglion)

A tissue is a group of many cells with similar and coordinated functions.

ORGAN
(brain)

Organs combine several tissues that function together. Organs in turn form systems, such as the nervous system.

ORGANISM
(fish)

BIOSPHERE

Biological communities exchange energy with one another, combining to create the biosphere of Earth.

An organism is a recognizable, self-contained individual made up of organs and organ systems.

A population is a group of many organisms of the same species.

POPULATION

COMMUNITY
(coral reef)

Communities consist of populations of many different species.

◄ 1.9 The Hierarchy of Life
The individual organism is the central unit of study in biology, but understanding it requires a knowledge of many levels of biological organization both above and below it. At each higher level, additional and more complex properties and functions emerge.

Biologists study life at different levels

Biology can be visualized as a hierarchy in which the units, from the smallest to the largest, include atoms, molecules, cells, tissues, organs, organisms, populations, and communities (Figure 1.9).

The organism is the central unit of study in biology. Parts Five and Six of this book discuss organismal biology in detail. But to understand organisms, biologists must study life at all its levels of organization. Biologists study molecules, chemical reactions, and cells to understand the operations of tissues and organs. They study organs and organ systems to determine how organisms function and maintain internal homeostasis. At higher levels in the hierarchy, biologists study how organisms interact with one another to form social systems, populations, ecological communities, and biomes, which are the subjects of Part Seven of this book.

Each level of biological organization has properties, called **emergent properties**, that are not found at lower levels. For example, cells and multicellular organisms have characteristics and carry out processes that are not found in the molecules of which they are composed.

Emergent properties arise in two ways. First, many *emergent properties of systems result from interactions among their parts*. For example, at the organismal level, developmental interactions of cells result in a multicellular organism whose adult features are vastly richer than those of the single cell from which it grew. Other examples of properties that emerge through complex interactions are memory and emotions. In the human brain, these properties result from interactions among the brain's 10^{12} (trillion) cells with their 10^{15} (quadrillion) connections. No single cell, or even small group of cells, possesses them.

Second, *emergent properties arise because aggregations have collective properties* that their individual units lack. For example, individuals are born and they die; they have a life span. An individual does not have a birth rate or a death rate, but a population (composed of many individuals) does. Birth and death rates are emergent properties of a population. Evolution is an emergent property of populations that depends on variation in birth and death rates, which emerges from the different life spans and reproductive success of individuals in the various populations.

Emergent properties do not violate the principles that operate at lower levels of organization. However, emergent properties usually cannot be detected, predicted, or even suspected by studying lower levels. Biologists could never discover the existence of human emotions by studying single nerve cells, even though they may eventually be able to explain it in terms of interactions among many nerve cells.

Biological diversity is organized hierarchically

As many as 30 million species of organisms inhabit Earth today. Many times that number lived in the past but are now extinct. If we go back four billion years, to the origin of life, all organisms are believed to be descended from a single *common ancestor*. The concept of a common ancestor is crucial to modern methods of classifying organisms. *Organisms are grouped in ways that attempt to define their evolutionary relationships, or how recently the different members of the group shared a common ancestor.*

To determine evolutionary relationships, biologists assemble facts from a variety of sources. Fossils tell us where and when ancestral organisms lived and what they looked like. The physical structures different organisms share—toes among mammals, for example—can be an indication of how closely related they are. But a modern "revolution" in classification has emerged because technologies developed in the past 30 years now allow us to compare the genomes of organisms: We can actually determine how many genes different species share. The more genes species have in common, the more recently they probably shared a common ancestor.

Because no fossil evidence for the earliest forms of life remains, the decision to divide all living organisms into three major **domains**—the deepest divisions in the evolutionary history of life—is based primarily on molecular evidence (Figure 1.10). Although new evidence is constantly being brought to light, it seems clear that organisms belonging to a particular domain have been evolving separately from organisms in the other two domains for more than a billion years.

Organisms in the domains **Archaea** and **Bacteria** are prokaryotes—single cells that lack a nucleus and the other internal compartments found in the Eukarya. Archaea and Bacteria differ so fundamentally from each other in the chemical reactions by which they function and in the products they produce that they are believed to have separated into distinct evolutionary lineages very early during the evolution of life. These domains are covered in Chapter 26.

Members of the third domain have eukaryotic cells containing nuclei and complex cellular compartments called organelles. The **Eukarya** are divided into four groups—the protists and the classical kingdoms Plantae, Fungi, and Animalia (see Figure 1.10). Protists, the subject of Chapter 27, are mostly single-celled organisms. The remaining three kingdoms, whose members are all multicellular, are believed to have arisen from ancestral protists.

Some bacteria, some protists, and most members of the kingdom Plantae (plants) convert light energy to chemical energy by photosynthesis. The biological molecules that they produce are the primary food for nearly all other living organisms. The Plantae are covered in Chapters 28 and 29.

The Fungi, the subject of Chapter 30, include molds, mushrooms, yeasts, and other similar organisms, all of

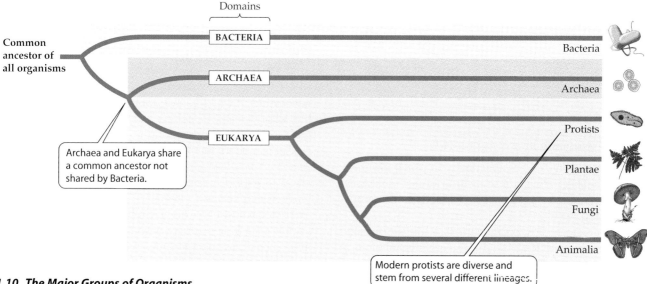

Domains

BACTERIA

Bacteria

ARCHAEA

Archaea

EUKARYA

Protists

Common ancestor of all organisms

Archaea and Eukarya share a common ancestor not shared by Bacteria.

Plantae

Fungi

Animalia

Modern protists are diverse and stem from several different lineages.

1.10 The Major Groups of Organisms
The classification system used in this book divides Earth's organisms into three domains. The domain Eukarya contains numerous groups of unicellular and multicellular organisms. This "tree" diagram gives information on evolutionary relationships among the groups, as described in Chapter 23.

which are **heterotrophs**: They require a food source of energy-rich molecules synthesized by other organisms. Fungi absorb food substances from their surroundings and break them down (digest them) within their cells. They are important as decomposers of the dead bodies of other organisms.

Members of the kingdom Animalia (animals) are also heterotrophs. These organisms ingest their food source, digest the food outside their cells, and then absorb the products. Animals get their raw materials and energy by eating other forms of life. Perhaps because we are animals ourselves, we are often drawn to study members of this kingdom, which is covered in Chapters 31, 32, and 33.

The biological classification system used today has many hierarchical levels in addition to the ones shown in Figure 1.10. We will discuss the principal levels in Chapter 23. But to understand some of the terms we will use in the intervening chapters, you need to know that each species of organism is identified by two names. The first identifies the **genus**—a group of species that share a recent common ancestor—of which the species is a member. The second name is the species name. To avoid confusion, a particular combination of two names is assigned to only a single species. For example, the scientific name of the modern human species is *Homo sapiens*.

Asking and Answering "How?" and "Why?"

Because biology is an evolutionary science, biological processes and products can be viewed from two different but complementary perspectives. Biologists ask, and try to answer, functional questions: How does it work? They also

ask, and try to answer, adaptive questions: Why has it evolved to work that way?

Suppose, for example, that some marine biologists walking on mudflats in the Bay of Fundy, Nova Scotia, Canada, observe many amphipods (tiny relatives of shrimps and lobsters) crawling on the surface of the mud (Figure 1.11). Two obvious questions they might ask are

▶ *How* do these animals crawl?
▶ *Why* do they crawl?

To answer the "how" question, the scientists would investigate the molecular mechanisms underlying muscular contraction, nerve and muscle interactions, and the receipt of stimuli by the amphipods' brains. To answer the "why" question, they would attempt to determine why crawling on the mud is adaptive—that is, why it improves the survival and reproductive success of amphipods.

Is either of these two types of questions more basic or important than the other? Is any one of the answers more fundamental or more important than the other? Not really. The richness of possible answers to apparently simple questions makes biology a complex field, but also an exciting one. Whether we're talking about molecules bonding, cells dividing, blood flowing, amphipods crawling, or forests growing, we are constantly posing both how and why questions. To answer these questions, scientists generate hypotheses that can be tested.

Hypothesis testing guides scientific research

The most important motivator of most biologists is curiosity. People are fascinated by the richness and diversity of life, and they want to learn more about organisms and how they function and interact with one another. Curiosity is probably an adaptive trait. Humans who were motivated to learn about their surroundings are likely to have survived and reproduced better, on average, than their less curious relatives. We hope this book will help you share in the ex-

1.11 An Amphipod from the Mud Flats
Scientists studied this tiny crustacean (whose actual size of approximately 1 centimeter is shown by the scale bar) in an attempt to see whether its behavior changes when it is infected by a parasitic worm. The female of this amphipod species is at the top; the lower specimen is a male.

citement biologists feel as they develop and test hypotheses. There are vast numbers of how and why questions for which we do not have answers, and new discoveries usually engender questions no one thought to ask before. Perhaps *your* curiosity will lead to an important new idea.

Underlying all scientific research is the **hypothetico-deductive (H-D) approach** by which scientists ask questions and test answers. The H-D approach allows scientists to modify and correct their beliefs as new observations and information become available. The method has five stages:

▶ Making observations.
▶ Asking questions.
▶ Forming **hypotheses,** or tentative answers to the questions.
▶ Making predictions based on the hypotheses.
▶ Testing the predictions by making additional observations or conducting experiments.

The data gained may support or contradict the predictions being tested. If the data support the hypothesis, it is subjected to still more predictions and tests. If they continue to support it, confidence in its correctness increases, and the hypothesis comes to be considered a **theory**. If the data do not support the hypothesis, it is abandoned or modified in accordance with the new information. Then new predictions are made, and more tests are conducted.

Applying the hypothetico-deductive method

The way in which marine biologists answered the question "Why do amphipods crawl on the surface of the mud rather than staying hidden within?" illustrates the H-D approach. As we saw above, the biologists observed something occurring in nature and formulated a question about it. To begin answering the question, they assembled available information on amphipods and the species that eat them.

They learned that during July and August of each year, thousands of sandpipers assemble for four to six weeks on the mudflats of the Bay of Fundy, during their southward migration from their Arctic breeding grounds to their wintering areas in South America (Figure 1.12). On these mud-

1.12 Sandpipers Feed on Amphipods
Migrating sandpipers crowd the exposed tidal flats in search of food. By consuming infected amphipods, the sandpipers also become infected, serving as hosts and allowing the parasitic worm to complete its life cycle.

flats, which are exposed twice daily by the tides, they feed vigorously, putting on fat to fuel their next long flight. Amphipods living in the mud form about 85 percent of the diet of the sandpipers. Each bird may consume as many as 20,000 amphipods per day!

Previous observations had shown that a nematode (roundworm) parasitizes both the amphipods and the sandpipers. To complete its life cycle, the nematode must develop within both a sandpiper and an amphipod. The nematodes mature within the sandpipers' digestive tracts, mate, and release their eggs into the environment in the birds' feces. Small larvae hatch from the eggs and search for, find, and enter amphipods, where they grow through several larval stages. Sandpipers are reinfected when they eat parasitized amphipods.

GENERATING A HYPOTHESIS AND PREDICTIONS. Based on the available information, biologists generated the following hypothesis: *Nematodes alter the behavior of their amphipod hosts in a way that increases the chance that the worms will be*

1.13 Collecting Field Data
Amphipods are collected from the mud to be tested for infection by parasites. Some of these crustaceans will be used in laboratory experiments.

passed on to sandpiper hosts. From this general hypothesis they generated two specific predictions.

▶ First, they predicted that amphipods infected by nematodes would increase their activity on the surface of the mud during daylight hours, when the sandpipers hunted by sight, but not at night, when the sandpipers fed less and captured prey by probing into the mud.

▶ Second, they predicted that only amphipods with late-stage nematode larvae—the only stage that can infect sandpipers—would have their behavior manipulated by the nematodes.

For each hypothesis proposing an effect, there is a corresponding **null hypothesis**, which asserts that the proposed effect is absent. For the hypothesis we have just stated, the null hypothesis is that nematodes have no influence on the behavior of their amphipod hosts. The alternative predictions that would support the null hypothesis are (1) that infected amphipods show no increase their activity either during the day or at night and (2) that all larval stages affect their hosts in the same manner. It is important in hypothesis testing to generate and test as many alternate hypotheses and predictions as possible.

TESTING PREDICTIONS. Investigators collected amphipods in the field, taking them from the surface and from within the mud, during the day and at night (Figure 1.13). They found that during the day, amphipods crawling on the surface were much more likely to be infected with nematodes than were amphipods collected from within the mud. At night, however, there was no difference between the proportion of infected amphipods on the surface and those burrowing within the mud. This evidence supported the first prediction.

The field collections also showed that a higher proportion of the amphipods collected on the surface than of those collected from within the mud were parasitized by late-stage nematode larvae. However, amphipods crawling on the surface were no more likely to be infected by early-stage nematode larvae than were amphipods collected from the mud. These findings supported the *second* prediction.

To test the prediction that nematode larvae are more likely to affect amphipod behavior once they become infective, biologists performed laboratory experiments. They artificially infected amphipods with nematode eggs they obtained from sandpipers collected in the field. The infected amphipods established themselves in mud in laboratory containers.

By examining infected amphipods, investigators determined that it took about 13 days for the nematode larvae to reach the late, infective stage. By monitoring the behavior of the amphipods in the test tubes, the researchers determined that the amphipods were more likely to expose themselves on the surface of the mud once the parasites had reached the infective stage (Figure 1.14). This finding supported the second prediction.

Thus a combination of field and laboratory experiments, observation, and prior knowledge all supported the hypothesis that nematodes manipulate the behavior of their amphipod hosts in a way that decreases the survival of the amphipods, but increases the survival of the nematodes.

As is common practice in all the sciences, the researchers gathered all their data and collected them in a report, which they submitted to a scientific journal. Once such a report is published,* other scientists can evaluate the data, make their own observations, and formulate new ideas and experiments.

Experiments are powerful tools

The key feature of **experimentation** is the control of most factors so that the influence of a single factor can be seen clearly. In the laboratory experiments with amphipods, all individuals were raised under the same conditions. As a result, the nematodes reached the infective stage at about the same time in all of the infected amphipods.

Both laboratory and field experiments have their strengths and weaknesses. The advantage of working in a laboratory is that control of environmental factors is more

---

*In the case illustrated here, the data on amphipod behavior were published in the journal *Behavioral Ecology*, Volume 10, Number 4 (1998). D. McCurdy et al., "Evidence that the parasitic nematode *Skrjabinoclava* manipulates host *Corephium* behavior to increase transmission to the sandpiper, *Calidris pusilla*."

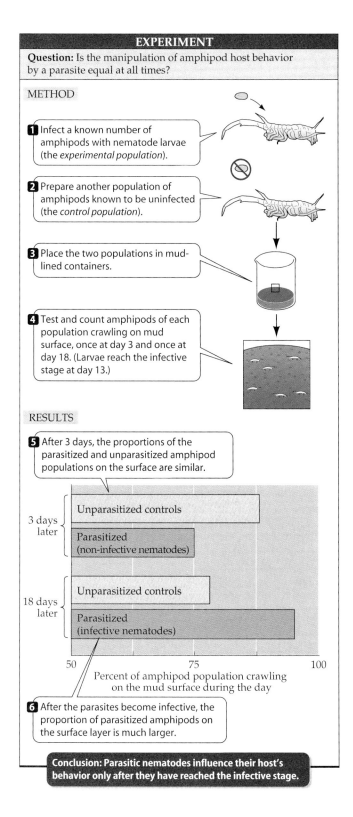

EXPERIMENT

Question: Is the manipulation of amphipod host behavior by a parasite equal at all times?

METHOD

1 Infect a known number of amphipods with nematode larvae (the *experimental population*).

2 Prepare another population of amphipods known to be uninfected (the *control population*).

3 Place the two populations in mud-lined containers.

4 Test and count amphipods of each population crawling on mud surface, once at day 3 and once at day 18. (Larvae reach the infective stage at day 13.)

RESULTS

5 After 3 days, the proportions of the parasitized and unparasitized amphipod populations on the surface are similar.

3 days later
- Unparasitized controls
- Parasitized (non-infective nematodes)

18 days later
- Unparasitized controls
- Parasitized (infective nematodes)

50   75   100
Percent of amphipod population crawling on the mud surface during the day

6 After the parasites become infective, the proportion of parasitized amphipods on the surface layer is much larger.

Conclusion: Parasitic nematodes influence their host's behavior only after they have reached the infective stage.

1.14 An Experiment Demonstrates that Parasites Influence Amphipod Behavior
Amphipods are more likely to crawl on the surface of the mud, exposing themselves to being captured by sandpipers, when their parasitic nematodes have reached the stage at which they can infect a sandpiper.

and field experiments are needed to test most hypotheses about what organisms do.

A single piece of supporting evidence rarely leads to widespread acceptance of a hypothesis. Similarly, a single contrary result rarely leads to abandonment of a hypothesis. Results that do not support the hypothesis being tested can be obtained for many reasons, only one of which is that the hypothesis is wrong. Incorrect predictions may have been made from a correct hypothesis. A negative finding can also result from poor experimental design, or because an inappropriate organism was chosen for the test. For example, a species of sandpiper that fed only by probing in the mud for its prey would have been an unsuitable subject for testing the hypothesis that nematodes alter their hosts in a way to make them more visible to predators.

Accepted scientific theories are based on many kinds of evidence

A general textbook like this one presents hypotheses and theories that have been extensively tested, using a variety of methods, and are generally accepted. When possible, we illustrate hypotheses and theories with observations and experiments that support them, but we cannot, because of space constraints, detail all the evidence. Remember as you read that statements of biological "fact" are mixtures of observations, predictions, and interpretations.

No amount of observation could possibly substitute for experimentation. However, this does not mean that scientists are insensitive to the welfare of the organisms with which they work. Most scientists who work with animals are continually alert to finding ways of getting answers that use the smallest number of experimental subjects and that cause the subjects the least pain and suffering.

Not all forms of inquiry are scientific

If you understand the methods of science, you can distinguish science from non-science. Recently some people have claimed that "creation science," sometimes called "scientific creationism," is a legitimate science that deserves to be taught in schools together with the evolutionary view of the world presented in this book. In spite of these claims, creation science is not science.

Science begins with observations and the formulation of hypotheses that can be tested and that will be rejected if significant contrary evidence is found. Creation science begins with the assertions, derived from religious texts, that Earth is only a few thousand years old and that all species of organisms were created in approximately their present forms. These assertions are not presented as a hypothesis

complete. Field experiments are more difficult because it is usually impossible to control more than a small number of environmental factors. But field experiments have one important advantage: Their results are more readily applicable to what happens where the organisms actually live and evolve. Just because an organism does something in the laboratory does not mean that it behaves the same way in nature. Because biologists usually wish to explain nature, not processes in the laboratory, combinations of laboratory

from which testable predictions can be derived. Advocates of creation science assume their assertions to be true and that no tests are needed, nor are they willing to accept any evidence that refutes them.

In this chapter we have outlined the hypotheses that Earth is about 4 billion years old, that today's living organisms evolved from single-celled ancestors, and that many organisms dramatically different from those we see today lived on Earth in the remote past. The rest of this book will provide evidence supporting this scenario. To reject this view of Earth's history, a person must reject not only evolutionary biology, but also modern geology, astronomy, chemistry, and physics. All of this extensive scientific evidence is rejected or misinterpreted by proponents of "creation science" in favor of their particular religious beliefs.

Evidence gathered by scientific procedures does not diminish the value of religious accounts of creation. Religious beliefs are based on faith—not on falsifiable hypotheses, as science is. They serve different purposes, giving meaning and spiritual guidance to human lives. They form the basis for establishing values—something science cannot do. The legitimacy and value of both religion and science is undermined when a religious belief is presented as scientific evidence.

Biology and Public Policy

During the Second World War and immediately thereafter, the physical sciences were highly influential in shaping public policy in the industrialized world. Since then, the biological sciences have assumed increasing importance. One reason is the discovery of the genetic code and the ability to manipulate the genetic constitution of organisms. These developments have opened vast new possibilities for improvements in the control of human diseases and agricultural productivity. At the same time, these capabilities have raised important ethical and policy issues. How much, and in what ways, should we tinker with the genetics of people and other species? Does it matter whether organisms are changed by traditional breeding experiments or by gene transfers? How safe are genetically modified organisms in the environment and in human foods?

Another reason for the importance of the biological sciences is the vastly increased human population. Our use of renewable and nonrenewable natural resources is stressing the ability of the environment to produce the goods and services upon which society depends. Human activities are causing the extinction of a large number of species and are resulting in the spread of new human diseases and the resurgence of old ones. Biological knowledge is vital for determining the causes of these changes and for devising wise policies to deal with them.

Therefore, biologists are increasingly called upon to advise governmental agencies concerning the laws, rules, and regulations by which society deals with the increasing number of problems and challenges that have at least a par-

tial biological basis. We will discuss these issues in many chapters of this book. You will see how the use of biological information can contribute to the establishment and implementation of wise public policies.

Chapter Summary

▶ If the history of Earth were a month with 30 days, recorded human history would occupy only the last 30 seconds. **Review Figure 1.1**

Organisms Have Changed over Billions of Years

▶ Evolution is the theme that unites all of biology. The idea of, and evidence for, evolution existed before Darwin. **Review Figure 1.2**

▶ The theory of evolution by natural selection rests on two simple observations and one inference from them.

Evolutionary Milestones

▶ Life arose from nonlife about 3.8 billion years ago when interacting systems of molecules became enclosed in membranes to form cells.

▶ All living organisms contain the same types of large molecules—carbohydrates, lipids, proteins, and nucleic acids.

▶ All organisms consist of cells, and all cells come from pre-existing cells. Life no longer arises from nonlife.

▶ A major theme in the evolution of life is the development of increasingly diverse ways of capturing external energy and using it to drive biologically useful reactions.

▶ Photosynthetic single-celled organisms released large amounts of oxygen into Earth's atmosphere, making possible the oxygen-based metabolism of large cells and, eventually, multicellular organisms.

▶ Reproduction with variation is a major characteristic of life. The evolution of sexual reproduction enhanced the ability of organisms to adapt to changing environments.

▶ Complex eukaryotic cells evolved when some large prokaryotes engulfed smaller ones. Eukaryotic cells evolved the ability to "stick together" after they divided, forming multicellular organisms. The individual cells of multicellular organisms became modified for specific functions within the organism.

▶ A major theme in the evolution of life is the development of increasingly complicated systems for responding to changes in the internal and external environments and for maintaining homeostasis.

▶ Regulated growth is a vital characteristic of life.

▶ Speciation resulted in the millions of species living on Earth today.

▶ Adaptation to environmental change is one of life's most distinctive features and is the result of evolution by natural selection.

The Hierarchy of Life

▶ Biology is organized into a hierarchy of levels from molecules to the biosphere. Each level has emergent properties that are not found at lower levels. **Review Figure 1.9**

▶ Species are classified into three domains: Archaea, Bacteria, and Eukarya. The domains Archaea and Bacteria consist of prokaryotic cells. The domain Eukarya contains the protists and the kingdoms Plantae, Fungi, and Animalia, all of which have eukaryotic cells. **Review Figure 1.10**

Asking and Answering "How?" and "Why?"

▶ Biologists ask two kinds of questions. "How" questions ask how organisms work. "Why" questions ask why they evolved to work that way.

▶ Both how and why questions are usually answered using a hypothetico-deductive (H-D) approach. Hypotheses are tentative answers to questions. Predictions are made on the basis of a hypothesis. The predictions are tested by observations and experiments, the results of which may support or refute the hypothesis. **Review Figure 1.14**

▶ Science is based on the formulation of testable hypotheses that can be rejected in light of contrary evidence. The acceptance on faith of already refuted, untested, or untestable assumptions is not science.

Biology and Public Policy

▶ Biologists are often called upon to advise governmental agencies on the solution of important problems that have a biological component.

For Discussion

1. According to the theory of evolution by natural selection, a species evolves certain features because they improve the chances that its members will survive and reproduce. There is no evidence, however, that evolutionary mechanisms have foresight or that organisms can anticipate future conditions. What, then, do biologists mean when they say, for example, that wings are "for flying"?

2. Why is it so important in science that we design and perform tests capable of rejecting a hypothesis?

3. One hypothesis about the manipulation of a host's behavior by a parasite was discussed in this chapter, and some tests of that hypothesis were described. Suggest some other hypotheses about the ways in which parasites might change the behavior and physiology of their hosts. Develop some critical tests for one of these alternatives. What are the appropriate associated null hypotheses?

4. Some philosophers and scientists believe that it is impossible to prove any scientific hypothesis—that we can only fail to find a reason to reject it. Evaluate this view. Can you think of reasons why we can be more certain about rejecting a hypothesis than about accepting it?

5. Discuss one current environmental problem whose solution requires the use of biological knowledge. How well is biology being used? What factors prevent scientific data from playing a more important role in finding a solution to the problem?

Part One

THE CELL

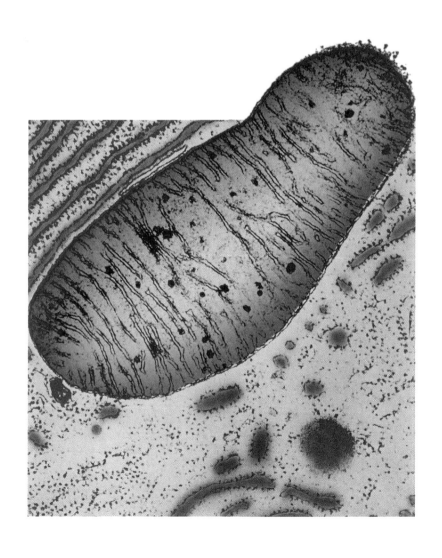

2 Small Molecules: Structure and Behavior

IN RECENT YEARS, SOME STARTLING DISCOVeries have shown that life can exist in places we never dreamed possible. There are organisms living in hot springs at temperatures above the boiling point of water, beneath the frozen Antarctic ice, 2 miles below Earth's surface, 3 miles below the surface of the sea, in extremely acid environments, in extremely salty conditions, and even inside nuclear reactors. Such findings have rekindled interest in astrobiology, the science of and search for life outside Earth.

The one absolute requirement for life is water. Without water to act as a solvent for biochemicals, to receive wastes, to absorb heat, and to participate directly in chemical reactions, life would not exist as we know it. With strong recent evidence that there was once flowing water on Mars, and that Europa (one of Jupiter's moons) may have a thin crust of ice with liquid water below it, there is great excitement about the possibility of life on nearby extraterrestrial bodies.

But what form would this life take? A major discovery of biology is that living things are composed of the same types of chemical elements as the vast nonliving portion of the universe. This *mechanistic* view—that life is chemically based and obeys universal physicochemical laws—is a relatively recent one in human history. The concept of a "vital force" responsible for life, different from the forces found in physics and chemistry, was common in Western culture until the nineteenth century, and many people still assume such a force exists. However, most scientists adhere to a mechanistic view of life.

Before describing how chemical elements are arranged in living creatures, we examine some fundamental chemical concepts. The first part of this chapter will address the constituents of matter: atoms. We examine their variety, their properties, and their capacity to combine with other atoms. Then we consider how matter changes. In addition to changes in state (solid to liquid to gas), substances undergo changes that transform both their composition and their characteristic properties. Then we return to a consideration of the structure and properties of water and its relationship to acids and bases. We close with a consideration of characteristic groups of atoms that contribute specific properties to larger molecules of which they are part, and which will be the subject of Chapter 3.

Atoms: The Constituents of Matter

More than a trillion (10^{12}) atoms could fit over the period at the end of this sentence. Each atom consists of a dense, positively charged nucleus, around which one or more negatively charged electrons move. The nucleus contains one or more protons and may contain one or more neutrons. Atoms and their component particles have **mass**, a property of all matter. Mass measures the quantity of matter present; the greater the mass, the greater the quantity of matter.

The mass of a proton serves as a standard unit of measure: the *atomic mass unit* (amu), or *dalton* (named after the English chemist John Dalton). A single proton or neutron has a mass of about 1 dalton, which is 1.7×10^{-24} grams (0.0000000000000000000000017 g). The mass of an electron is 9×10^{-28} g (0.0005 dalton). Because the mass of an electron is so much less than the mass of a proton or a neutron, the contribution of electrons to the mass of an atom can usually be ignored.

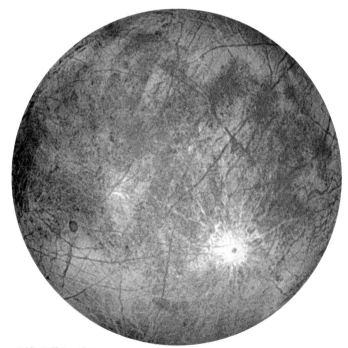

Life Off Earth?
Orbiting 400,000 miles above the giant planet Jupiter, Europa has a surface of water ice, possibly covering a slushy ocean. Where there is water, there could be, or could have been, life.

Even the smallest atoms, such as helium, have measurable mass:

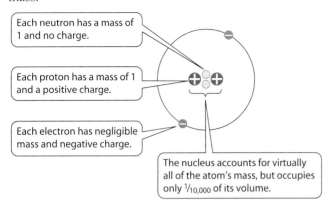

Each neutron has a mass of 1 and no charge.

Each proton has a mass of 1 and a positive charge.

Each electron has negligible mass and negative charge.

The nucleus accounts for virtually all of the atom's mass, but occupies only $1/10,000$ of its volume.

The positive electric charge of a proton is defined as a unit of charge. An electron has a negative charge equal and opposite to that of a proton. Thus the charge of a proton is +1 unit, and that of an electron is –1 unit. Unlike charges (+/–) attract each other; like charges (+/+ or –/–) repel each other. The neutron, as its name suggests, is electrically neutral, so its charge is 0 unit. When the number of protons in an atom equals the number of electrons, the atom is electrically neutral. An atom with more or fewer electrons than protons has an electric charge and is called an ion; we will discuss ions in detail later in the chapter.

An element is made up of only one kind of atom

An **element** is a pure substance that contains only one type of atom. The element hydrogen consists only of hydrogen atoms; the element iron consists only of iron atoms. The atoms of each element have certain characteristics or properties that distinguish them from the atoms of other elements. The more than 100 elements found in the universe are arranged in the periodic table (Figure 2.1). These elements are not found in equal amounts. Earth's crust is half oxygen; 28% silicon; 8% aluminum; 3–5% each of sodium, magnesium, potassium, calcium, and iron; and much smaller amounts of the other elements.

About 98% of the mass of every living organism (bacterium, turnip, or human) is composed of just six elements: carbon, hydrogen, nitrogen, oxygen, phosphorus, and sulfur. Other elements are present in small amounts. The chemistry of the six major elements will be our primary concern here, but the others are not unimportant. Sodium and potassium, for example, are essential for nerves to function; calcium can act as a biological signal; iodine is a component of a vital hormone; and plants need molybdenum in order to incorporate nitrogen into biologically useful substances.

2.1 The Periodic Table
The periodic table groups the elements according to their physical and chemical properties.

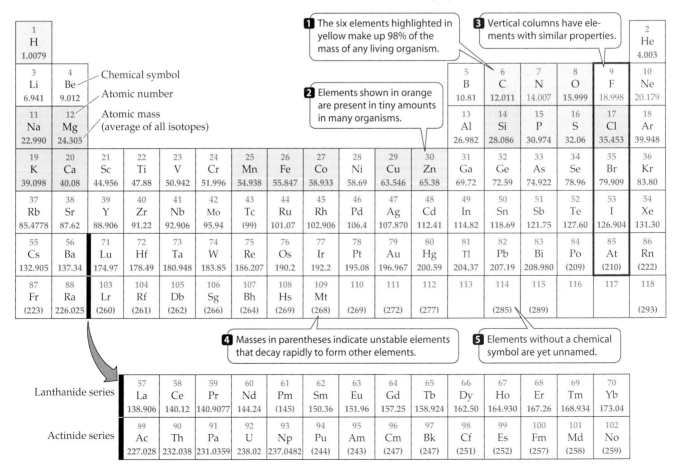

The number of protons identifies the element

An element is distinguished from other elements by the number of protons in each of its atoms. This number, which does not change, is called the **atomic number**. An atom of hydrogen contains 1 proton, a helium atom has 2 protons, carbon has 6 protons, and plutonium has 94 protons. The atomic numbers of these elements are thus 1, 2, 6, and 94, respectively.

Every element except hydrogen has one or more neutrons in its nucleus. The **mass number** of an atom equals the total number of protons and neutrons in its nucleus. Because the mass of an electron is infinitesimal compared with that of a neutron or proton, electrons are ignored in calculating the mass number. The nucleus of a helium atom contains 2 protons and 2 neutrons; oxygen has 8 protons and 8 neutrons. Helium, therefore, has a mass number of 4 and oxygen a mass number of 16. The mass number may be thought of as the mass of the atom in daltons.

Each element has its own one- or two-letter chemical symbol. For example, H stands for hydrogen, He for helium, and O for oxygen. Some symbols come from other languages: Fe (from the Latin *ferrum*) stands for iron, Na (Latin *natrium*) for sodium, and W (German *Wolfram*) for tungsten. The periodic table (see Figure 2.1) gives the symbols for the 92 natural elements, as well as showing 26 elements (elements 93–118) that have been synthesized in laboratories but have not been found in nature.

In text, the atomic number and mass number of an element are written to the left of the element's symbol:

Mass number **12**
Atomic number **6** C Symbol of element

Thus, hydrogen, carbon, and oxygen are written as $^{1}_{1}H$, $^{12}_{6}C$, and $^{16}_{8}O$, respectively.

Isotopes differ in number of neutrons

We have been speaking of elements as if each had only one atomic form, but this is not true. **Isotopes** of the same element all have the same number of protons, but differ in the number of neutrons in the atomic nucleus (Figure 2.2).

In nature, many elements exist as several isotopes. For example, the natural isotopes of carbon are ^{12}C, ^{13}C, and ^{14}C. Unlike the isotopes of hydrogen, which have special names (see Figure 2.2), the isotopess of most elements do not have distinct names. Rather, they are written in the form shown above and are referred to as carbon-12, carbon-13, and carbon-14, respectively. Most carbon atoms are ^{12}C, about 1.1 percent are ^{13}C, and a tiny fraction are ^{14}C. An element's **atomic mass**, or **atomic weight**,* is the average of the mass numbers of a representative sample of atoms of the element, with all isotopes in their normally occurring

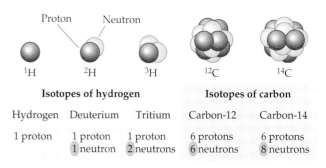

2.2 Isotopes Have Different Numbers of Neutrons
Deuterium and tritium are rare isotopes of hydrogen. Unlike these two isotopes, isotopes of other elements do not have have distinct names. Carbon-12 is the most common isotope of carbon; carbon-14 is a rare form.

proportions. The atomic weight of carbon is thus calculated to be 12.011.

Some isotopes, called **radioisotopes**, are unstable and spontaneously give off energy as α (alpha), β (beta), or γ (gamma) radiation from the atomic nucleus. Such radioactive decay transforms the original atom into another atom, usually of another element. For example, carbon-14 loses a beta particle (actually an electron) to form nitrogen-14. Biologists and physicians can incorporate radioisotopes into molecules and use the emitted radiation as a tag to locate those molecules or to identify changes that the molecules undergo inside the body (Figure 2.3). Three radioisotopes commonly used in this way are ^{3}H (tritium), ^{14}C (carbon-14), and ^{32}P (phosphorus-32). In addition to these applications, radioisotopes can be used to date fossils (see Chapter 20).

Although radioisotopes are useful for experiments and in medicine, even low doses of their radiation have the potential to damage molecules and cells. Gamma radiation from cobalt-60 (^{60}Co) is used medically to damage or kill rapidly dividing cancer cells.

Electron behavior determines chemical bonding

When considering atoms, biologists are concerned primarily with electrons because the behavior of electrons explains how chemical changes occur in living cells. These changes, called **chemical reactions** or just **reactions**, are changes in the atomic composition of substances. The characteristic number of electrons in each atom of an element determines how its atoms react with other atoms. All chemical reactions involve changes in the relationships of electrons with one another.

The location of a given electron in an atom at any given time is impossible to determine. We can only describe a volume of space within the atom where the electron is likely to be. The region of space where the electron is found at least

---

*The concepts of "weight" and "mass" are not identical. Weight is the measure of the Earth's gravitational attraction for mass; on another planet, the same quantity of mass would have a different weight. On Earth, however, the term "weight" is often used as a measure of mass, and in biology one encounters the terms "weight" and "atomic weight" more frequently than "mass" and "atomic mass." Therefore, we will use "weight" for the remainder of this book.

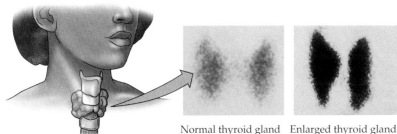

Normal thyroid gland   Enlarged thyroid gland

2.3 A Radioisotope Used in Medicine
The thyroid gland takes up iodine and uses it in the synthesis of thyroid hormone. A patient suspected of having thyroid disease is injected with radioactive iodine, which allows the thyroid gland to be visualized by a scanning device.

90 percent of the time is the electron's **orbital** (Figure 2.4). In an atom, a given orbital can be occupied by at most two electrons. Thus any atom larger than helium (atomic number 2) must have electrons in two or more orbitals. As Figure 2.4 shows, the different orbitals have characteristic forms and orientations in space.

The orbitals in turn constitute a series of **electron shells**, or energy levels, around the nucleus (Figure 2.5). The first, or innermost, electron shell consists of only one orbital, called an *s* orbital. Hydrogen ($_1$H) has one electron in its first shell; helium ($_2$He) has two. All other elements have two first-shell electrons, as well as electrons in other shells.

The second shell is made up of four orbitals (an *s* orbital and three *p* orbitals) and hence can hold up to eight electrons. The *s* orbitals fill with electrons first, and their electrons have the lowest energy. Subsequent shells have different numbers of orbitals, but the outermost shells usually hold only eight electrons.

In any atom, the outermost electron shell determines how the atom combines with other atoms; that is, how an atom behaves chemically. When an outermost shell consisting of four orbitals contains eight electrons, there are no unpaired electrons (see Figure 2.5). Such an atom is stable and will not react with other atoms. Examples of chemically inert elements are helium, neon, and argon.

The atoms of chemically reactive elements seek to attain the stable condition of having no unpaired electrons in their outer shells. They attain this stability by sharing electrons with other atoms, or by gaining or losing one or more electrons from their outermost shells. When they share electrons, atoms are bonded together. Such bonds create stable associations of atoms called molecules.

A **molecule** can be defined as two or more atoms linked by chemical bonds. The tendency of atoms in stable molecules to have eight electrons in their outermost shell is known as the *octet rule*. Many atoms in biologically important molecules—for example, carbon (C) and nitrogen (N)—follow the octet rule. However, some biologically important atoms are exceptions to the rule. Hydrogen (H) is an obvious exception, attaining stability when only two electrons occupy its single shell.

Chemical Bonds: Linking Atoms Together

A **chemical bond** is an attractive force that links two atoms to form a molecule. There are several kinds of chemical bonds (Table 2.1). In this section, we first discuss covalent bonds, the strong bonds that result from the sharing of electrons. Then we examine other kinds of interactions, including hydrogen bonds, that are weaker than covalent bonds but enormously important to biology. Finally, we consider ionic bonding, which results as a consequence of the loss or gain of electrons by atoms.

Covalent bonds consist of shared pairs of electrons

When two atoms attain stable electron numbers in their outer shells by sharing one or more pairs of electrons, a **covalent bond** forms. Consider two hydrogen atoms in close proximity, each with a single unpaired electron in the outer shell. Each positively charged nucleus exerts some attraction on the other atom's un-

y

z

The two electrons closest to the nucleus move in a spherical *s* orbital.

x

s Orbital

p$_x$ Orbital   *p*$_y$ Orbital   *p*$_z$ Orbital   All *p* orbitals full

Two electrons form a dumbell-shaped *x*-axis (*p*$_x$) orbital...

...two more fill the *p*$_y$ orbital...

...and two fill the *p*$_z$ orbital.

Six electrons fill all three *p* orbitals.

2.4 Electron Orbitals
Each orbital holds a maximum of two electrons. The *s* orbitals have a lower energy level and fill with electrons before the *p* orbitals do.

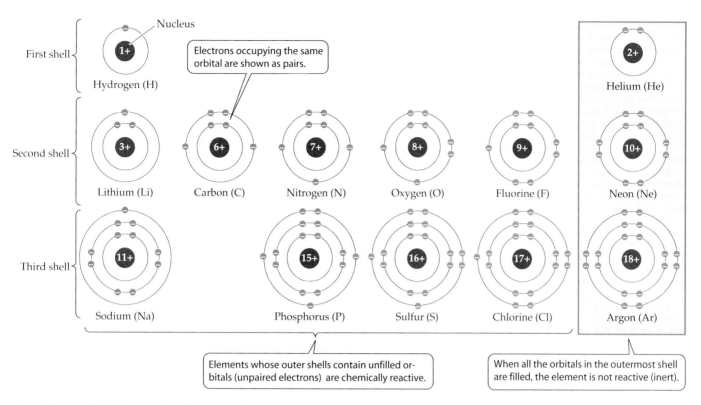

Elements whose outer shells contain unfilled orbitals (unpaired electrons) are chemically reactive.

When all the orbitals in the outermost shell are filled, the element is not reactive (inert).

2.5 Electron Shells Determine the Reactivity of Atoms

Each orbital holds a maximum of two electrons, and each shell can hold a specific maximum number of electrons. Each shell must be filled before electrons move into the next shell. The energy level of electrons is higher in shells farther from the nucleus. An atom with unpaired electrons in its outermost shell may react (bond) with other atoms.

paired electron, but this attraction is balanced by each electron's attraction to its own nucleus. So the two unpaired electrons become shared by both atoms, filling the outer shells of both of them (Figure 2.6).

A carbon atom has a total of six electrons; two electrons fill its inner shell and four are in its outer shell. Because the outer shell can hold up to eight electrons, this atom can share electrons with up to four other atoms. Thus it can

2.1 Chemical Bonds and Interactions

NAME	BASIS OF INTERACTION	STRUCTURE	BOND ENERGY[a] (KCAL/MOL)
Covalent bond	Sharing of electron pairs		50–110
Hydrogen bond	Sharing of H atom		3–7
Ionic interaction	Attraction of opposite charges		3–7
van der Waals interaction	Interaction of electron clouds		1
Hydrophobic interaction	Interaction of nonpolar substances		1–2

[a]Bond energy is the amount of energy needed to separate two bonded or interacting atoms under physiological conditions.

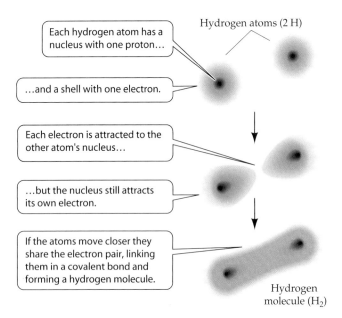

Each hydrogen atom has a nucleus with one proton…

…and a shell with one electron.

Each electron is attracted to the other atom's nucleus…

…but the nucleus still attracts its own electron.

If the atoms move closer they share the electron pair, linking them in a covalent bond and forming a hydrogen molecule.

Hydrogen atoms (2 H)

Hydrogen molecule (H$_2$)

2.6 Electrons Are Shared in Covalent Bonds
Two hydrogen atoms combine to form a hydrogen molecule. Each electron is attracted to both protons. A covalent bond forms when the electron orbitals of the two atoms overlap.

form four covalent bonds. When an atom of carbon reacts with four hydrogen atoms, a substance called methane (CH$_4$) forms (Figure 2.7a and b). Thanks to electron sharing, the outer shell of methane's carbon atom is filled with eight electrons, and the outer shell of each hydrogen atom is also filled. Thus four covalent bonds—each consisting of a shared pair of electrons—hold methane together. Table 2.2 shows the covalent bonding capacities of some biologically significant elements.

ORIENTATION OF COVALENT BONDS. Covalent bonds are very strong. The thermal energy that biological molecules ordinarily have at body temperature is less than 1 percent of that needed to break covalent bonds. So biological molecules, most of which are put together with covalent bonds, are quite stable. A second property of covalent bonds is that, for a given pair of atoms, they are the same in length, angle, and direction, regardless of the larger molecule of which the particular bond is a part. The four filled orbitals around the carbon nucleus of methane, for example, distribute themselves in space so that the bonded hydrogens are directed to the corners of a regular tetrahedron with carbon in the center (Figure 2.7c). This three-dimensional structure of carbon and hydrogen is the same in complicat-

2.7 Covalent Bonding with Carbon
Different representations of covalent bond formation in methane (CH$_4$). (a) Diagram illustrating the filling and stabilizing of the outer electron shells in carbon and hydrogen atoms. (b) Two common ways of representing bonds. (c) The spatial orientation of methane's bonds, represented in two ways.

2.2	**Covalent Bonding Capabilities of Some Biologically Important Elements**	
ELEMENT		USUAL NUMBER OF COVALENT BONDS
Hydrogen (H)		1
Oxygen (O)		2
Sulfur (S)		2
Nitrogen (N)		3
Carbon (C)		4
Phosphorus (P)		5

ed proteins as it is in the simple methane molecule. It makes the prediction of biological structure possible.

Although the orientation of orbitals and the shapes of molecules differ depending on the kinds of atoms involved and how they are linked together, it is essential to remember that all molecules occupy space and have three-dimensional shapes. The shapes of molecules contribute to their biological functions, as we will see in Chapter 3.

MULTIPLE COVALENT BONDS. A covalent bond is represented by a line between the chemical symbols for the atoms. A bond in which a single pair of electrons is shared is called a

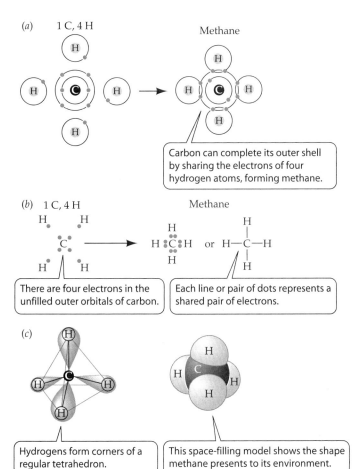

(a)   1 C, 4 H   Methane

Carbon can complete its outer shell by sharing the electrons of four hydrogen atoms, forming methane.

(b)   1 C, 4 H   Methane

There are four electrons in the unfilled outer orbitals of carbon.

Each line or pair of dots represents a shared pair of electrons.

(c)

Hydrogens form corners of a regular tetrahedron.

This space-filling model shows the shape methane presents to its environment.

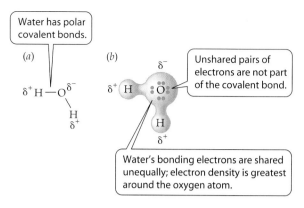

2.8 The Polar Covalent Bond in the Water Molecule
(a) A covalent bond between atoms with different electronegativities is a polar covalent bond, and has partial (δ) charges at the ends. (b) In water, the electrons are displaced toward the oxygen atom and away from the hydrogen atoms.

single bond (for example, H—H, C—H). When four electrons (two pairs) are shared, the link is called a double bond (C=C). In the gas ethylene ($H_2C=CH_2$), two carbon atoms share two pairs of electrons. Triple bonds (six shared electrons) are rare, but there is one in nitrogen gas (N≡N), the chief component of the air we breathe. In the covalent bonds in these five examples, the electrons are shared more or less equally between the nuclei; consequently, all regions of the bonds are identical.

UNEQUAL SHARING OF ELECTRONS. If two atoms of the same element are covalently bonded, there is an equal sharing of the pair(s) of electrons in the outer shell. However, when the two atoms are of different elements, the sharing is not necessarily equal. One nucleus may exert a greater attractive force on the electron pair than the other nucleus, and so the pair tends to be closer to that atom.

The attractive force that an atom exerts on electrons is its **electronegativity**. It depends on how many positive charges a nucleus has (nuclei with more protons are more positive and thus more attractive to electrons) and how far away the electrons are from the nucleus (closer means more electronegativity). *The closer two atoms are in electronegativity, the more equal their sharing of electrons will be.*

Table 2.3 shows the electronegativities of some elements important in biological systems. Looking at the table, it is

2.3	**Some Electronegativities**
ELEMENT	**ELECTRONEGATIVITY**
Oxygen (O)	3.5
Chlorine (Cl)	3.1
Nitrogen (N)	3.0
Carbon (C)	2.5
Phosphorus (P)	2.1
Hydrogen (H)	2.1
Sodium (Na)	0.9
Potassium (K)	0.8

obvious that two oxygen atoms, both with electronegativity of 3.5, will share electrons equally in a covalent bond. So will two hydrogen atoms (both with 2.1). But when hydrogen bonds with oxygen to form water, the electrons involved are *unequally* shared: They tend to be nearer to the oxygen nucleus because it is the more electronegative of the two. The result is called a *polar covalent bond* (Figure 2.8).

Because of this unequal sharing of electrons, the oxygen end of the hydrogen–oxygen bond has a slightly negative charge (symbolized δ⁻ and spoken as "delta negative," meaning a partial unit of charge), and the hydrogen end is slightly positive (δ⁺). The bond is polar because these opposite charges are separated at the two ends of the bond. The partial charges that result from polar covalent bonds produce **polar molecules** or polar regions of large molecules. Polar bonds greatly influence the interactions between molecules that contain them.

Hydrogen bonds may form between molecules

In liquid water, the negatively charged oxygen (δ⁻) atom of one water molecule is attracted to the positively charged hydrogen (δ⁺) atoms of another water molecule. (Remember, negative charges attract positive charges.) The bond resulting from this attraction is called a **hydrogen bond**.

Hydrogen bonds are not restricted to water molecules. They may form between an electronegative atom and a hydrogen covalently bonded to a different electronegative atom (Figure 2.9).

A hydrogen bond is a weak bond; it has about one-tenth (10%) of the strength of a covalent bond between a hydrogen atom and an oxygen atom (see Table 2.1). However, where many hydrogen bonds form, they have considerable strength and greatly influence the structure and properties of substances. Later in this chapter we'll see how hydrogen bonding in water contributes to many of the properties that make water significant for living systems. Hydrogen bonds also play important roles in determining and maintaining the three-dimensional shapes of giant molecules such as DNA and proteins (see Chapter 3).

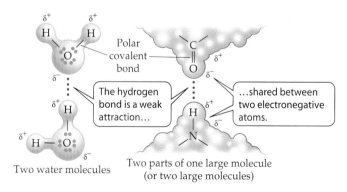

2.9 Hydrogen Bonds Can Form Between or within Molecules
Hydrogen bonds can form between two molecules or, if a molecule is large, between two different parts of the same molecule. Covalent and polar covalent bonds, on the other hand, are always found within molecules.

Ions form bonds by electrical attraction

When one interacting atom is much more electronegative than the other, a complete transfer of one or more electrons may take place. Consider sodium (electronegativity 0.9) and chlorine (3.1). A sodium atom has only one electron in its outermost shell; this condition is unstable. A chlorine atom has seven electrons in its outer shell—another unstable condition. Since the electronegativities of these elements are so different, any electrons involved in bonding will tend to be much nearer to the chlorine nucleus—so near, in fact, that there is a complete transfer of the electron from one element to the other (Figure 2.10). This reaction between sodium and chlorine makes both atoms more stable. The result is two ions. **Ions** are electrically charged particles that form when atoms gain or lose one or more electrons.

▶ The sodium ion (Na^+) has a +1 unit charge because it has one less electron than it has protons. The outermost electron shell of the sodium ion is full, with eight electrons, so the ion is stable. Positively charged ions are called **cations**.

▶ The chloride ion (Cl^-) has a –1 unit charge because it has one more electron than it has protons. This additional electron gives Cl^- an outer shell with a stable load of eight electrons. Negatively charged ions are called **anions**.

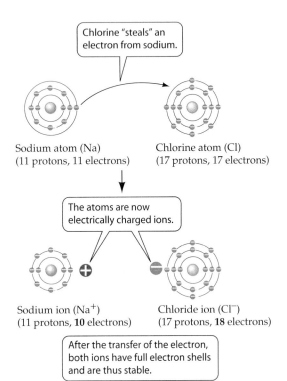

Sodium atom (Na)
(11 protons, 11 electrons)

Chlorine atom (Cl)
(17 protons, 17 electrons)

Sodium ion (Na^+)
(11 protons, **10** electrons)

Chloride ion (Cl^-)
(17 protons, **18** electrons)

2.10 Formation of Sodium and Chloride Ions
When a sodium atom reacts with a chlorine atom, the more electronegative chlorine acquires a more stable, filled outer shell by obtaining an electron from the sodium. In so doing, the chlorine atom becomes a negatively charged chloride ion (Cl^-). The sodium atom, upon losing the electron, becomes a positively charged sodium ion (Na^+).

Some elements form ions with multiple charges by losing or gaining more than one electron. Examples are Ca^{2+} (calcium ion, created from a calcium atom that has lost two electrons) and Mg^{2+} (magnesium ion). Two biologically important elements each yield more than one stable ion: Iron yields Fe^{2+} (ferrous ion) and Fe^{3+} (ferric ion), and copper yields Cu^+ (cuprous ion) and Cu^{2+} (cupric ion). Groups of covalently bonded atoms that carry an electric charge are called **complex ions**; examples include NH_4^+ (ammonium ion), SO_4^{2-} (sulfate ion), and PO_4^{3-} (phosphate ion).

The charge from an ion radiates from it in all directions. Once they form, ions are usually stable, and no more electrons are lost or gained. Ions can form stable bonds, resulting in stable solid compounds such as sodium chloride (NaCl) and potassium phosphate (K_3PO_4).

Ionic bonds are bonds formed by electrical attractions between ions bearing opposite charges. In sodium chloride—familiar to us as table salt—cations and anions are held together by ionic bonds. In solids, the ionic bonds are strong because the ions are close together. However, when ions are dispersed in water, the distance between them can be large; the strength of their attraction is thus greatly reduced. Under the conditions that exist in the cell, an ionic attraction is less than one-tenth as strong as a covalent bond that shares electrons equally (see Table 2.1).

Not surprisingly, ions with one or more units of charge can interact with polar molecules as well as with other ions. Such interaction results when table salt, or any other ionic solid, dissolves in water: "Shells" of water molecules surround the individual ions, separating them (Figure 2.11). The hydrogen bond that we described earlier is a type of ionic bond, because it is formed by electrical attractions. However, it is weaker than most ionic bonds because the hydrogen bond is formed by partial charges (δ^+ and δ^-) rather than by whole-unit charges (+1 unit, –1 unit).

Polar and nonpolar substances interact best among themselves

"Like attracts like" is an old saying, and nowhere is it more true than in polar and nonpolar molecules, which tend to interact with their own kind. Just as water molecules interact with one another through their polarity-induced hydrogen bonds, any molecule that is itself polar will interact with other polar molecules by weak (δ^+ to δ^-) attractions in hydrogen bonds. If a polar molecule interacts with water in this way, it is called *hydrophilic* ("water-loving").

What about nonpolar molecules? For example, carbon (electronegativity 2.5) forms nonpolar bonds with hydrogen (electronegativity 2.1). The resulting *hydrocarbon* molecule—ethane—is nonpolar (Figure 2.12), and in water it will tend to aggregate with other nonpolar molecules rather than with polar water. Such molecules are called *hydrophobic* ("water-hating"), and the interactions between them are hydrophobic interactions. It is important to realize that hydrophobic substances do not really "hate" water; they can form weak interactions with it (recall that the electronegativities of carbon and hydrogen are not exactly the same).

But these interactions are far weaker than the hydrogen bonds between the water molecules, and so the nonpolar substances keep to themselves.

These weak interactions between nonpolar substances are enhanced by **van der Waals forces**, which occur when two atoms are in close proximity. These forces result from random variations in the electron distribution in one mole-

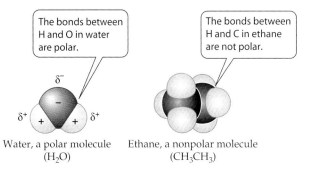

The bonds between H and O in water are polar.

The bonds between H and C in ethane are not polar.

δ^-

δ^+ + + δ^+

Water, a polar molecule (H_2O)

Ethane, a nonpolar molecule (CH_3CH_3)

2.12 Polar and Nonpolar Molecules
Because the hydrocarbon ethane is nonpolar, it does not interact with water, but tends to interact with other nonpolar substances.

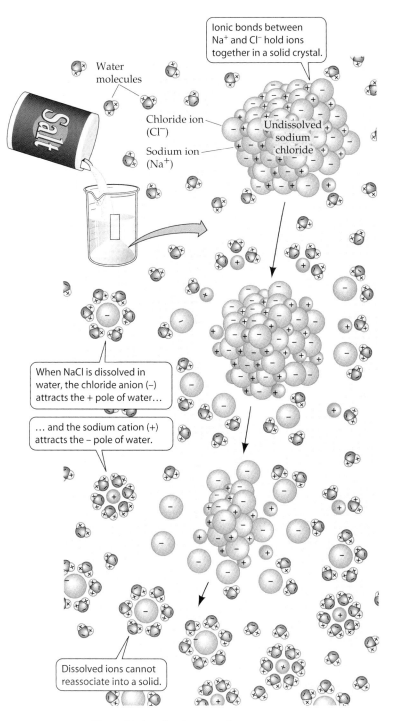

Ionic bonds between Na^+ and Cl^- hold ions together in a solid crystal.

Water molecules

Chloride ion (Cl^-)

Undissolved sodium chloride

Sodium ion (Na^+)

When NaCl is dissolved in water, the chloride anion (–) attracts the + pole of water…

… and the sodium cation (+) attracts the – pole of water.

Dissolved ions cannot reassociate into a solid.

2.11 Water Molecules Surround Ions
When an ionic solid dissolves in water, polar water molecules cluster around cations or anions, blocking their reassociation into a solid and forming a solution.

cule, which create an opposite charge distribution in the adjacent molecule. The result is a brief, weak attraction. Although each such interaction is brief and weak at any one site, the summation of many such interactions over the entire span of a large nonpolar molecule can produce substantial attraction. van der Waals forces are important in maintaining the structures of many biologically important substances.

Chemical Reactions: Atoms Change Partners

A **chemical reaction** occurs when atoms combine or change bonding partners. Consider the combustion reaction that takes place in the flame of a propane stove. When propane (C_3H_8) reacts with oxygen gas (O_2), the carbon atoms become bonded to oxygen atoms instead of to hydrogen atoms, and the hydrogen atoms become bonded to oxygen instead of carbon (Figure 2.13). As the covalently bonded atoms change partners, the composition of the matter changes, and propane and oxygen gas become carbon dioxide and water. This chemical reaction can be represented by the balanced equation

$$C_3H_8 + 5\,O_2 \rightarrow 3\,CO_2 + 4\,H_2O$$

In this equation, the propane and oxygen are the **reactants**, and the carbon dioxide and water are the **products**. In this case, the reaction is complete: All the propane and oxygen are used up in forming the two products. The arrow symbolizes the chemical reaction. The numbers preceding the molecular formulas balance the equation and indicate how many molecules are used or are produced.

In this and all other chemical reactions, matter is neither created nor destroyed. The total number of carbons on the left equals the total number on the right. However, there is another product of this reaction: energy. The heat of the stove's flame and its blue light reveal that the reaction of propane and oxygen releases a great deal of energy. **Energy** is defined as the capacity to do work, but on a more intuitive level, it can be thought of as the capacity for change. Chemical reactions do not create or destroy energy, but *changes* in energy usually accompany chemical reactions.

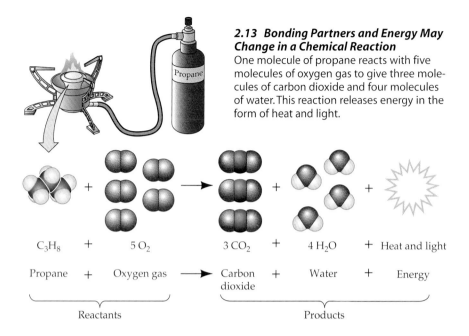

2.13 Bonding Partners and Energy May Change in a Chemical Reaction
One molecule of propane reacts with five molecules of oxygen gas to give three molecules of carbon dioxide and four molecules of water. This reaction releases energy in the form of heat and light.

$$C_3H_8 \quad + \quad 5\,O_2 \quad \longrightarrow \quad 3\,CO_2 \quad + \quad 4\,H_2O \quad + \quad \text{Heat and light}$$

Propane   +   Oxygen gas   →   Carbon dioxide   +   Water   +   Energy

Reactants   Products

In the reaction between propane and oxygen, the energy that was released as heat and light was already present in the reactants in another form, called *potential energy.* In some chemical reactions, energy must be supplied from the environment (for example, some substances will react only after being heated), and some of this supplied energy becomes stored as potential chemical energy in the bonds formed in the products.

We can measure the energy associated with chemical reactions using a unit called a **calorie (cal)**. A calorie* is the amount of heat energy needed to raise the temperature of 1 gram of pure water from 14.5°C to 15.5°C. Another unit of energy that is increasingly used is the **joule (J)**. When you compare data on energy, always compare joules to joules and calories to calories. The two units can be interconverted: 1 J = 0.239 cal, and 1 cal = 4.184 J. Thus, for example, 486 cal = 2,033 J, or 2.033 kJ. Although defined in terms of heat, the calorie and the joule are measures of any form of energy—mechanical, electric, or chemical.

Within living cells, chemical reactions called *oxidation–reduction reactions* take place. These biological reactions have much in common with the combustion of propane. The fuel is different (the sugar glucose, rather than propane), and the reactions proceed by many intermediate steps that permit the energy released from the glucose to be harvested and put to use by the cell. But the products are the same: carbon dioxide and water.

We will present and discuss energy changes, oxidation–reduction reactions, and several other types of chemical reactions that are prevalent in living systems in the chapters that follow.

*The nutritionist's or dieter's Calorie, with a capital C, is what biologists call a kilocalorie (kcal) and is equal to 1,000 heat-energy calories.

Water: Structure and Properties

Water, like all other matter, can exist in three states: solid (ice), liquid, and gas (vapor) (Figure 2.14). Liquid water is the medium in which life originated on Earth more than 3.8 billion years ago, and it is in water that life evolved for its first billion years. Today, water covers three-fourths of Earth's surface, and the bodies of all active organisms contain between 45 and 95 percent water.

No organism can remain biologically active without water. Within cells, water participates directly in many chemical reactions, and it is the medium (or solvent) in which most biological reactions take place. In this section we will consider the structure and interactions of water molecules, exploring how these generate properties essential to life.

Water has a unique structure and special properties

Each water molecule is composed of one oxygen atom bonded to two hydrogen atoms (H_2O). In the molecule, the four pairs of electrons in the outer shell of oxygen repel each other, producing a tetrahedral shape:

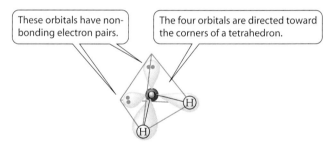

These orbitals have non-bonding electron pairs.

The four orbitals are directed toward the corners of a tetrahedron.

The shape of the water molecule, its polar nature, and its capacity to form hydrogen bonds give water its unusual properties. For example, ice floats, and compared with other liquids, water is an excellent solvent, making it an ideal medium for biochemical reactions. Water is both cohesive (sticking to itself) and adhesive (sticking to other things). And the energy changes that accompany its transitions from solid to liquid to gas are significant in living systems.

ICE FLOATS.   In its solid state (ice), water is held by its hydrogen bonds in a rigid, crystalline structure in which each water molecule is hydrogen-bonded to four others (Figure 2.15a). Although these molecules are held firmly in place, they are not as tightly packed as they are in liquid water (Figure 2.15b). In other words, *solid water is less dense than liquid water*, which is why ice floats in water.

If ice sank in water, as almost all other solids do in their corresponding liquids, ponds and lakes would freeze from

2.14 Water: Solid and Liquid
Solid water from a glacier floats in its liquid form. The clouds are also water, but not in its gaseous phase: They are composed of fine drops of liquid water.

the bottom up, becoming solid blocks of ice in winter and killing most of the organisms living in them. Once the whole pond had frozen, its temperature could drop well below the freezing point of water. However, because ice floats, it forms a protective insulating layer on the top of the pond, reducing heat flow to the cold air above. Thus fish, plants, and other organisms in the pond are not subjected to temperatures lower than 0°C, the freezing point of pure water.

MELTING AND FREEZING.   Compared with other nonmetallic substances of the same size, molecular ice requires a great

deal of heat energy to melt. Melting 1 mole (a standard quantity—6.02×10^{23}; see page 28) of water molecules requires the addition of 5.9 kJ of energy. This value is high because more than a mole of hydrogen bonds must be broken for 1 mole of water to change from solid to liquid. In the opposite process, freezing, a great deal of energy must be lost for water to transform from liquid to solid. These properties help make water a moderator of temperature changes.

HEAT AND COOLING.   Another property of water that moderates temperature is the high heat capacity of liquid water. The **specific heat** of a substance is the amount of heat energy required to raise the temperature of 1 gram of that substance by 1°C. Raising the temperature of liquid water takes a relatively large amount of heat because much of the heat energy is used to break the hydrogen bonds that hold the liquid together. Compared with other small molecules that are liquids, water has a high specific heat. This phenomenon contributes to the surprising constancy of the temperature of the oceans and other large bodies of water through the seasons of the year. The temperature changes of coastal land masses are also moderated by large bodies of water. Indeed, water helps minimize variations in atmospheric temperature throughout the planet.

EVAPORATION AND COOLING.   Water also has a high **heat of vaporization**, which means that a lot of heat is required to change water from its liquid state to its gaseous state (the process of evaporation). This heat is absorbed from the environment in contact with the water. Once again, much of the heat energy is used to break hydrogen bonds. Evaporation thus has a cooling effect on the environment— whether a leaf, a forest, or an entire land mass. This effect

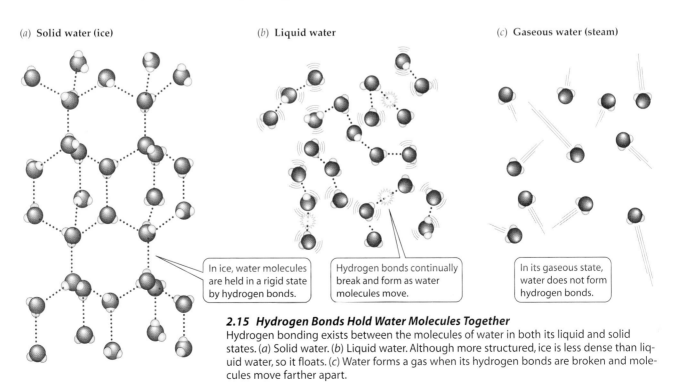

(*a*) **Solid water (ice)**   (*b*) **Liquid water**   (*c*) **Gaseous water (steam)**

In ice, water molecules are held in a rigid state by hydrogen bonds.

Hydrogen bonds continually break and form as water molecules move.

In its gaseous state, water does not form hydrogen bonds.

2.15 Hydrogen Bonds Hold Water Molecules Together
Hydrogen bonding exists between the molecules of water in both its liquid and solid states. (*a*) Solid water. (*b*) Liquid water. Although more structured, ice is less dense than liquid water, so it floats. (*c*) Water forms a gas when its hydrogen bonds are broken and molecules move farther apart.

explains why sweating cools the human body: As sweat evaporates off the skin, it uses up some of the adjacent body heat.

COHESION AND SURFACE TENSION. In liquid water, the molecules are free to move about. The hydrogen bonds between the water molecules continually form and break. In other words, liquid water has a dynamic structure. On average, every water molecule forms 3.4 hydrogen bonds with other water molecules. This number represents fewer bonds than exist in ice, but it is still a high number.

These hydrogen bonds explain the cohesive strength of liquid water. The **cohesive strength** of water is what permits narrow columns of water to stretch from the roots to the leaves of trees more than 100 meters high. When water evaporates from leaves, the entire column moves upward in response to the pull of the molecules at the top.

Water also has a high **surface tension**, which means that the surface of liquid water exposed to the air is difficult to puncture. The water molecules in this surface layer are hydrogen-bonded to other water molecules below. The surface tension of water permits a container to be filled slightly above its rim without overflowing, and it permits small animals to walk on the surface of water (Figure 2.16).

Most biological substances are dissolved in water

A **solution** is produced when a substance is dissolved in water (an aqueous solution) or another liquid. Many of the important molecules in biological systems are polar, and therefore are soluble in water. Much of biochemistry takes place in an aqueous solution.

One branch of the study of solutions is *qualitative analysis*, which deals with substances dissolved in a solvent (in this case, water) and the chemical reactions that occur there. Qualitative analysis is the subject of much of the next few chapters.

Solutions can also be studied by *quantitative analysis*, in which concentrations—the amount of substance in a given amount of solution—are measured. What follows is a brief introduction to some of the quantitative chemical terms you will see in this text.

▶ A *molecular formula* uses chemical symbols to identify the different atoms in a compound, and subscript numbers to show how many of each type of atoms are present. Thus, the formula for sucrose—table sugar—is $C_{12}H_{22}O_{11}$.

▶ Each compound has a **molecular weight** (**molecular mass**) that is the sum of the atomic weights of all atoms in the molecule. Looking at the periodic table in Figure 2.1, you can calculate the molecular weight of table sugar to be approximately 342. Molecular weights are usually related to the molecule's size (Figure 2.17).

▶ A **mole** is the amount of an ion or compound in grams whose weight is numerically equal to its molecular weight. So one mole of sugar weighs 342 grams.

One aim of quantitative analysis is to study the behaviors of precise numbers of molecules in solution. But it is

2.16 Surface Tension
Water striders "skate" along, supported by the surface tension of the water that is their home.

not possible to count molecules directly. Instead, chemists use a constant that relates the *weight* of any substance to the *number* of molecules of that substance. This constant is called *Avogadro's number*, which is 6.02×10^{23} *molecules per mole*. It allows chemists to work with moles of substances (which can be weighed out in the laboratory) instead of actual molecules. The mole concept is analogous to the concept of a dozen: We buy a dozen eggs or a dozen doughnuts, knowing that we will get 12 of whichever we buy.

In the same way, chemists can dissolve a mole of sugar in water to make 1 liter, knowing that the mole contains 6.02×10^{23} individual sugar molecules. This solution—1 mole of a substance dissolved in water to make 1 liter—is called a 1 molar (1 *M*) solution.

The many molecules that dissolve in water in living tissues are not present at anything close to a 1 molar concentration. Most are in the micromolar (millionths of a mole; μ*M*) to millimolar (thousandths of a mole; m*M*) range. Some, such as hormones, are far less concentrated than this.

While these abbreviations seem to indicate very low concentrations, remember that even a 1 μ*M* solution has 6.02×10^{17} molecules of the solute per liter.

Acids, Bases, and the pH Scale

Some substances dissolve in water and release hydrogen ions (H^+), which are actually single, positively charged protons. These tiny bits of charged matter can attach to other molecules, and in doing so, change their properties. In this section, we examine the properties of substances that release H^+ (called acids) and attach to H^+ (called bases). We will distinguish strong and weak acids and bases, and provide a quantitative means for stating the concentration of H^+ in solutions: the pH scale.

Acids donate H⁺, bases accept H⁺

If hydrochloric acid (HCl) is added to water, it dissolves and ionizes, releasing the ions H^+ and Cl^-:

$$HCl \rightarrow H^+ + Cl^-$$

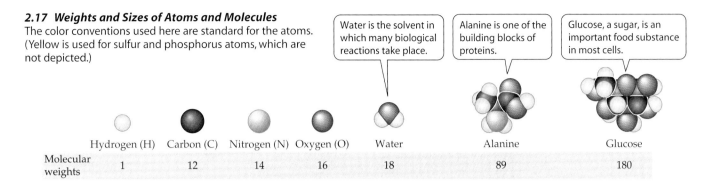

2.17 Weights and Sizes of Atoms and Molecules
The color conventions used here are standard for the atoms. (Yellow is used for sulfur and phosphorus atoms, which are not depicted.)

Water is the solvent in which many biological reactions take place.

Alanine is one of the building blocks of proteins.

Glucose, a sugar, is an important food substance in most cells.

	Hydrogen (H)	Carbon (C)	Nitrogen (N)	Oxygen (O)	Water	Alanine	Glucose
Molecular weights	1	12	14	16	18	89	180

Because its H^+ concentration has increased, such a solution is *acidic*. Just like the burning reaction of propane and oxygen (see Figure 2.13), the dissolution of HCl to form its ions is a complete reaction. HCl is therefore called a *strong acid*.

An **acid** *releases* H^+ ions in solution. HCl is an acid, as is H_2SO_4 (sulfuric acid). One molecule of sulfuric acid may ionize to yield two H^+ and one SO_4^{2-}. Biological compounds that contain —COOH (the carboxyl group; see Figure 2.20) are also acids (such as acetic acid and pyruvic acid), because

$$—COOH \rightarrow —COO^- + H^+$$

Not all acids dissolve fully in water. For example, if acetic acid is added to water, at the end of the reaction, there are not just the two ions, but some of the original acid as well. Because the reaction is not complete, acetic acid is a weak acid.

Bases *accept* H^+. Like acids, there are strong and weak bases. If NaOH (sodium hydroxide) is added to water, the NaOH dissolves and ionizes, releasing OH^- and Na^+ ions:

$$NaOH \rightarrow Na^+ + OH^-$$

Because the concentration of OH^- increases, such a solution is *basic*, and because this reaction is complete, NaOH is a strong base.

Weak bases include the bicarbonate ion (HCO_3^-), which can accept a H^+ ion and become carbonic acid (H_2CO_3), and ammonia (NH_3), which can accept a H^+ and become an ammonium ion (NH_4^+). Amino groups in biological molecules can also accept protons, acting as bases:

$$—NH_2 + H^+ \rightarrow —NH_3^+$$

When acetic acid is dissolved in water, two reactions happen. First, acetic acid forms its ions:

$$CH_3COOH \rightarrow CH_3COO^- + H^+$$

Then, once ions are formed, they re-form acetic acid:

$$CH_3COO^- + H^+ \rightarrow CH_3COOH$$

This pair of reactions is reversible. The formula for a reversible reaction can be written with two arrows:

$$CH_3COOH \rightleftharpoons CH_3COO^- + H^+$$

A **reversible reaction** can proceed in either direction—left to right or right to left—depending on the relative starting concentrations of the reactants and products.

In principle, *all* chemical reactions are reversible. In terms of acids and bases, there are two types of reactions, depending on the extent of reversibility:

▸ Ionization of strong acids and bases is virtually irreversible.
▸ Ionization of weak acids and bases is somewhat reversible.

Many of the acid and base groups on large molecules in biological systems are weak.

Water is a weak acid

The water molecule has a slight but significant tendency to ionize into a hydroxide ion (OH^-) and a hydrogen ion (H^+). Actually, *two* water molecules participate in this ionization. One of the two molecules "captures" a hydrogen ion from the other, forming a hydroxide ion and a hydronium ion:

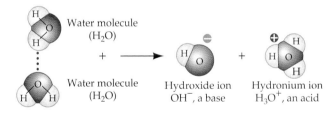

Water molecule (H_2O)

Water molecule (H_2O)

Hydroxide ion OH^-, a base

Hydronium ion H_3O^+, an acid

The hydronium ion is in effect a hydrogen ion bound to a water molecule. For simplicity, biochemists tend to use a modified representation of the ionization of water:

$$H_2O \rightarrow H^+ + OH^-$$

The ionization of water is very important for all living creatures. This fact may seem surprising, since only about one water molecule in 500 million is ionized at any given time. But we are less surprised if we focus on the abundance of water in living systems and the reactive nature of the H^+ produced by ionization.

pH is the measure of hydrogen ion concentration

The terms "acid*ic*" and "bas*ic*" refer only to *solutions*. How acidic or basic a solution is depends on the relative concentrations of H^+ and OH^- ions in it. "Acid" and "base" refer to *compounds* and *ions*. A compound or ion that is an acid can donate H^+; one that is a base can accept H^+.

How do we specify how acidic or basic a solution is? First, let's look at the H^+ concentrations of a few contrasting solutions. In 1 liter of pure water, the H^+ concentration is 10^{-7} M. In 1 M hydrochloric acid, the H^+ concentration is 1 M; and in 1 M sodium hydroxide, the H^+ concentration is 10^{-14} M. Because its values range so widely, the H^+ concentration itself is an inconvenient quantity to measure. It is easier to work with the logarithm of the concentration, because logarithms compress this range.

We indicate how acidic or basic a solution is by its pH ("*potential of Hydrogen*"). The **pH value** is defined as the negative logarithm of the hydrogen ion concentration in moles per liter (molar concentration). In chemical notation, molar concentration is often indicated by putting square brackets around the symbol for a substance; thus $[H^+]$ stands for the molar concentration of H^+. The equation for pH is

$$pH = -\log_{10}[H^+]$$

Since the H^+ concentration of pure water is 10^{-7} M, its pH is $-\log(10^{-7}) = -(-7)$, or 7. A smaller negative logarithm means a larger number. In practical terms, a lower pH means a higher H^+ concentration, or greater acidity. In 1 M HCl, the H^+ concentration is 1 M, so the pH is the negative logarithm of 1 ($-\log 10^0$), or 0. The pH of 1 M NaOH is the negative logarithm of 10^{-14}, or 14.

A solution with a pH of less than 7 is acidic—it contains more H^+ ions than OH^- ions. A solution with a pH of 7 is neutral, and a solution with a pH value greater than 7 is basic. Figure 2.18 shows the pH values of some common substances.

Buffers minimize pH change

An organism must control the pH of the separate compartments within its cells. Animals must also control the pH of their blood. The normal pH of human blood is 7.4, and deviations of even a few tenths of a pH unit can be fatal. The control of pH is made possible in part by buffers—chemical systems that maintain a relatively constant pH even when substantial amounts of acid or base are added.

A **buffer** is a mixture of a weak acid and its corresponding base—for example, carbonic acid (H_2CO_3) and bicarbonate ions (HCO_3^-). If acid is added to this buffer, not all the H^+ ions from that acid stay in solution. Instead, many of them combine with the bicarbonate ions to produce more carbonic acid. This reaction uses up some of the H^+ ions in the solution and decreases the acidifying effect of the added acid:

$$HCO_3^- + H^+ \rightleftharpoons H_2CO_3$$

If a base is added, the reaction essentially reverses. Some of the carbonic acid ionizes to produce bicarbonate ions and more H^+, which counteracts some of the added base. In this way, the buffer minimizes the effects of an added acid or base on pH. A given amount of acid or base causes a

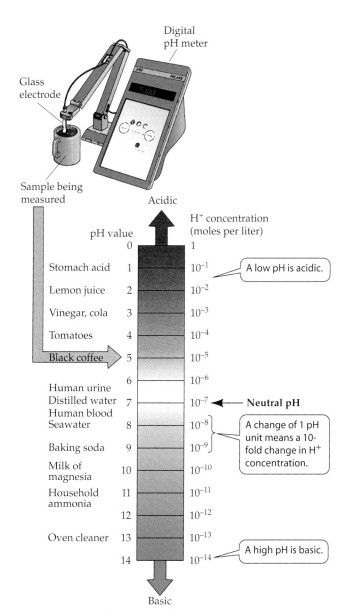

2.18 pH Values of Some Familiar Substances
An electronic instrument similar to the one drawn at the top of the figure is used to measure the pH of a solution.

smaller change in pH in a buffered solution than in an unbuffered one (Figure 2.19).

Buffers illustrate an important chemical principle in reversible reactions called the *law of mass action*. Addition of a component on one side of a reversible system drives the reaction in the direction that uses up that compound. In this case, addition of an acid drives the reaction in one direction; addition of a base drives it in the other.

The Properties of Molecules

Some molecules are small, such as H_2 and CH_4. Others are larger, such as a molecule of table sugar (sucrose), which has 45 atoms. Still other molecules, such as proteins, are gigantic, sometimes containing tens of thousands of atoms bonded together in specific ways.

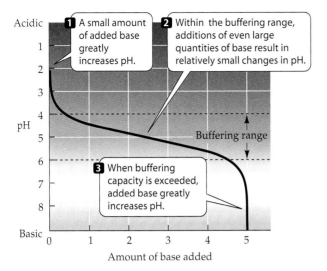

2.19 Buffers Minimize Changes in pH
With increasing amounts of added base, the overall slope of a graph of pH is downward. In the buffering range, however, the slope is shallow. At high and low values of pH, where the buffer is ineffective, the slopes are much steeper.

Whether large, medium, or small, most of the molecules in living systems contain carbon atoms and are thus referred to as **organic molecules**. Most organic molecules include hydrogen and oxygen atoms as well as carbon, and many also include nitrogen and phosphorus.

All molecules have a specific three-dimensional shape. For example, the orientation of the bonding orbitals around the carbon atom gives the methane molecule (CH_4) the shape of a regular tetrahedron (see Figure 2.7c). In carbon dioxide (CO_2), the three atoms are in line. Larger molecules have complex shapes that result from the numbers and kinds of atoms present and the ways in which they are linked together. Some large molecules have compact, ball-like shapes. Others are long, thin, ropelike structures. Their shapes relate to the roles these molecules play in living cells.

In addition to size and shape, molecules have certain properties that characterize them and determine their biological roles. Chemists use the characteristics of composition, structure (three-dimensional shape), reactivity, and solubility to distinguish a sample of one pure molecule from another. That certain groups of atoms are found together in a variety of different molecules simplifies our understanding of the reactions that molecules undergo in living cells.

Functional groups give specific properties to molecules

Functional groups are groups of atoms that make up part of a larger molecule and have particular chemical properties (shape, polarity, reactivity, solubility). The same functional group may be part of very different molecules. You will encounter several functional groups in your study of biology (Figure 2.20).

2.20 Some Functional Groups Important to Living Systems
These functional groups (highlighted in white boxes) are the most common ones found in biologically important molecules. R represents the "remainder" of the molecule, which may be any of a large number of carbon skeletons or other chemical group.

An important kind of biological molecule containing functional groups is the *amino acids*, which have both a carboxyl group and an amino group attached to the same carbon atom, the α (alpha) carbon. Also attached to the α carbon atom are a hydrogen atom and a side chain, designated by the letter R:

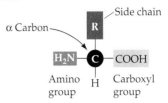

Functional group	Class of compounds	Structural formula	Example
Hydroxyl —OH	Alcohols	R—OH	Ethanol
Aldehyde —CHO	Aldehydes	R—C(=O)H	Acetaldehyde
Keto CO	Ketones	R—C(=O)—R	Acetone
Carboxyl —COOH	Carboxylic acids	R—C(=O)OH	Acetic acid
Amino —NH₂	Amines	R—NH₂	Methylamine
Phosphate —OPO₃²⁻	Organic phosphates	R—O—P(=O)(O⁻)—O⁻	3-Phosphoglyceric acid
Sulfhydryl —SH	Thiols	R—SH	Mercaptoethanol

Different side chains have different chemical compositions, structures, and properties. Each of the 20 amino acids found in proteins has a different side chain that gives it its distinctive chemical properties, as we'll see in Chapter 3. Because they possess both carboxyl and amino groups, amino acids are simultaneously acids and bases. At the pH values commonly found in cells, both the carboxyl and the amino groups are ionized: The carboxyl group has lost a proton, and the amino group has gained one.

Isomers have different arrangements of the same atoms

Isomers are molecules that have the same chemical formula but different arrangements of the atoms. (The prefix "iso-" means "same" and is encountered in many biological terms.) Of the different kinds of isomers, we will consider two: structural isomers and optical isomers.

Structural isomers differ in how their atoms are joined together. Consider two simple molecules, each composed of 4 carbon and 10 hydrogen atoms bonded covalently, with the formula C_4H_{10}. These atoms can be linked together in two different ways, resulting in two forms of the molecule:

$$\begin{array}{c} \quad\ \ \text{H}\quad\text{H} \\ \quad\ \ | \quad\ | \\ \text{H}_3\text{C}-\text{C}-\text{C}-\text{CH}_3 \\ \quad\ \ | \quad\ | \\ \quad\ \ \text{H}\quad\text{H} \\ \text{Butane} \end{array} \qquad \begin{array}{c} \text{CH}_3 \\ | \\ \text{H}_3\text{C}-\text{C}-\text{CH}_3 \\ | \\ \text{H} \\ \text{Isobutane} \end{array}$$

The different bonding relationships of butane and isobutane are distinguished in structural formulas, and the compounds have different chemical properties.

Many molecules of biological importance, particularly the sugars and amino acids, have **optical isomers**. Optical isomers occur whenever a carbon atom has four *different* atoms or groups attached to it. This pattern allows two different ways of making the attachments, each the mirror image of the other (Figure 2.21). Such a carbon atom is an *asymmetric carbon*, and the pair of compounds are optical isomers of each other. Your right and left hands are optical isomers. Just as a glove is specific for a particular hand, some biochemical molecules can interact with one optical isomer of a compound, but are unable to "fit" the other.

The α carbon in an amino acid is an asymmetric carbon because it is bonded to four different functional groups. Therefore, amino acids exist in two isomeric forms, called D-amino acids and L-amino acids. "D" and "L" are abbreviations for the Latin terms for right (*dextro*) and left (*levo*), respectively. Only L-amino acids are commonly found in most organisms.

Between the small molecules we have discussed in this chapter and the world of the living cell stands another level, that of the macromolecules. These huge molecules—the proteins, lipids, carbohydrates, and nucleic acids—are the subject of the next chapter.

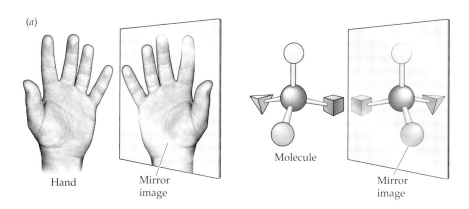

(a)

Hand   Mirror image   Molecule   Mirror image

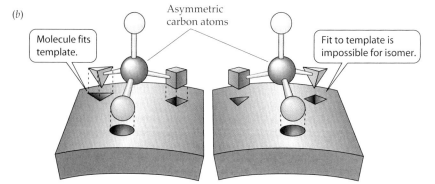

(b)

Asymmetric carbon atoms

Molecule fits template.   Fit to template is impossible for isomer.

2.21 Optical Isomers
(a) Optical isomers are mirror images of each other. (b) Molecular optical isomers result when four different groups are attached to a single carbon atom. (c) If a template is laid out to match the groups on one carbon atom, the groups on the mirror-image isomer cannot be rotated to fit the same template.

Chapter Summary

Atoms: The Constituents of Matter

▶ Matter is composed of atoms. Each atom consists of a positively charged nucleus of protons and neutrons, surrounded by electrons bearing negative charges.

▶ There are many elements in nature, but only a few of them make up the bulk of living systems. **Review Figure 2.1**

▶ Isotopes of an element differ in their numbers of neutrons. Some isotopes are radioactive, emitting radiation as they decay. **Review Figure 2.2**

▶ Electrons are distributed in shells consisting of orbitals. Each orbital contains a maximum of two electrons. **Review Figures 2.4, 2.5**

▶ In losing, gaining, or sharing electrons to become more stable, an atom can combine with other atoms to form molecules. **Review Table 2.1**

Chemical Bonds: Linking Atoms Together

▶ Covalent bonds are strong bonds formed when two atomic nuclei share one or more pairs of electrons. Covalent bonds have spatial orientations that give molecules three-dimensional shapes. **Review Figures 2.6, 2.7, Table 2.2**

▶ Nonpolar covalent bonds are formed when the electronegativities of two atoms are approximately equal. When atoms with strong electronegativity (such as oxygen) bond to atoms with weaker electronegativity (such as hydrogen), a polar covalent bond is formed, in which one end is δ^+ and the other is δ^-. **Review Figure 2.8, Table 2.3**

▶ Hydrogen bonds are weak electrical attractions that form between a δ^+ hydrogen atom in one molecule and a δ^- nitrogen or oxygen atom in another molecule or in another part of a large molecule. Hydrogen bonds are abundant in water. **Review Figure 2.9**

▶ Ions are electrically charged bodies that form when an atom gains or loses one or more electrons. Ionic bonds are electrical attractions between oppositely charged ions. Ionic bonds are strong in solids, but weaker when the ions are separated from one another in solution. **Review Figures 2.10, 2.11**

▶ Nonpolar molecules do not interact directly with polar substances, including water. Nonpolar molecules are attracted to each other by very weak bonds called van der Waals forces. **Review Figure 2.12**

Chemical Reactions: Atoms Change Partners

▶ In chemical reactions, substances change their atomic compositions and properties. Energy is released in some reactions, whereas in others energy must be provided. Neither matter nor energy is created or destroyed in a chemical reaction, but both change form.

▶ Combustion reactions are oxidation–reduction reactions in which a fuel is converted to carbon dioxide and water, while energy is released as heat and light. In living cells, combustion reactions take place in multiple steps so that the released energy can be harvested for cellular activities. **Review Figure 2.13**

Water: Structure and Properties

▶ Water's molecular structure and its capacity to form hydrogen bonds give it unusual properties that are significant for life. Water is an excellent solvent; solid water floats in liquid water; and water gains or loses a great deal of heat when it changes its state, a property that moderates environmental temperature changes. **Review Figure 2.15**

▶ The cohesion of water molecules permits liquid water to rise to great heights in narrow columns and produces a high surface tension. Water's high heat of vaporization assures effective cooling when water evaporates.

▶ Solutions are produced when substances dissolve in water. The concentration of a solution is the amount of a given substance in a given amount of solution. Most biological substances are dissolved in water at very low concentrations.

Acids, Bases, and the pH Scale

▶ Acids are substances that donate hydrogen ions (H^+). Bases are substances that accept hydrogen ions.

▶ The pH of a solution is the negative logarithm of the hydrogen ion concentration. Values lower than pH 7 indicate an acidic solution; values above pH 7 indicate a basic solution. **Review Figure 2.18**

▶ Buffers are systems of weak acids and bases that limit the change in pH when hydrogen ions are added or removed. **Review Figure 2.19**

The Properties of Molecules

▶ Molecules vary in size, shape, reactivity, solubility, and other chemical properties.

▶ Functional groups make up part of a larger molecule and have particular chemical properties. The consistent chemical behavior of functional groups helps us understand the properties of the molecules that contain them. **Review Figure 2.20**

▶ Structural and optical isomers have the same kinds and numbers of atoms, but differ in their structures and properties. **Review Figure 2.21**

For Discussion

1. Would you expect the elemental composition of Earth's crust to be the same as that of the human body? How could you find out?

2. Lithium (Li) is the element with atomic number 3. Draw the electronic structures of the Li atom and of the Li^+ ion.

3. Draw the structure of a pair of water molecules held together by a hydrogen bond. Your drawing should indicate the covalent bonds.

4. The molecular weight of sodium chloride (NaCl) is 58.45. How many grams of NaCl are there in 1 liter of a 0.1 M NaCl solution? How many in 0.5 liter of a 0.25 M NaCl solution?

5. The side chain of the amino acid glycine is simply a hydrogen atom (—H). Are there two optical isomers of glycine? Explain.

 Self-quizzes and Supplemental Readings for each chapter are on the Student Web Site/CD-ROM.

3 Macromolecules: Their Chemistry and Biology

A SPIDER WEB IS AN AMAZING STRUCTURE. NOT only is it beautiful to look at, but it is an architectural wonder that serves as the spider's home, its mating ground, and its means of hunting and capturing food. Consider a fly that happens to intersect with a spider web. The fibers of the web must slow down the fly, but they cannot break, so they must stretch in order to dissipate the energy of the fly's movement. On the other hand, the fibers holding the web together cannot stretch too much, because they must be strong enough to hold the entire structure in place and not let the web wobble out of control. Web fibers are far thinner than human hair, yet they are stronger than steel and, in some cases, more elastic than nylon. In fact, spider silk may be as strong as Kevlar, a synthetic substance used to make bulletproof vests and the cords attached to parachutes.

Spider silk is composed of slight variations on a single type of huge molecule—a macromolecule—called a protein. The many types of proteins in biological systems are composed of different amounts of the 20 molecules known as amino acids, and spider silks have their own unique selections of these molecules. The silk protein that stretches contains amino acids that allow it to curl into a spiral, and when these spirals associate into silk fibers, they can slip along each other to change the fiber's length. The strong fibers, in contrast, are made up of amino acids that fold the individual proteins into flat sheets, with ratchets that fit parallel sheets together (much like Lego blocks), so that the fibers are hard to pull apart. The relationship between chemical structure and biological function is the theme of this chapter and many of the succeeding ones in this section.

The four major types of biological macromolecules—proteins, carbohydrates, lipids, and nucleic acids—are composed of building blocks called monomers. In the case of proteins like spider silk, the monomers are amino acids; carbohydrate monomers are sugars, and nucleic acid monomers are nucleotides. Some lipids are composed of a small molecule, glycerol, covalently bonded to larger fatty acids. Lipids interact to form huge macromolecular aggregates, such as the membranes that surround cells.

The four kinds of large molecules are made the same way in all living things, and are present in roughly the same proportions in all organisms (Figure 3.1). Although an apple tree is obviously different from a person, their basic chemistry is the same, demonstrating the unity of life. A protein that has a certain role in the apple probably has a similar role in the human. One important advantage of biochemical unity is that organisms can eat one another. When you eat an apple, the molecules you take in include carbohydrates, lipids, and proteins that can be re-fashioned into the special varieties of those molecules used by humans.

Macromolecules: Giant Polymers

Macromolecules are giant **polymers** (poly-, "many"; -mer, "unit") constructed by the covalent linking of smaller molecules called **monomers** (Table 3.1). These monomers may or may not be identical, but they always have similar chemical structures. Molecules with molecular weights exceeding 1,000 are usually considered macromolecules, and the proteins, polysaccharides (large carbohydrates), and nucleic acids of living systems certainly fall into this category.

Each type of macromolecule performs some combination of a diversity of functions: energy storage, structural

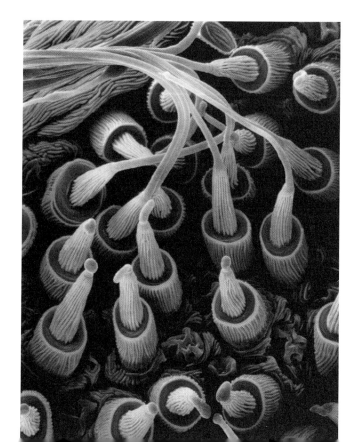

A Complex Macromolecule
Spider silk (purple) being spun into web material from a gland by the shiny black spider, *Castercantha*.

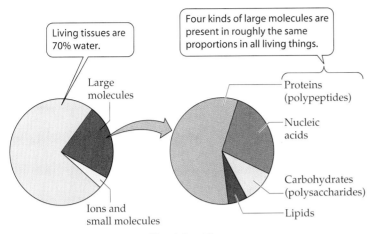

3.1 Substances Found in Living Tissues
The substances shown here make up the nonmineral components of living tissue (bone would be an example of a "mineral tissue"). Most tissues are at least 70 percent water.

support, protection, catalysis, transport, defense, regulation, movement, and heredity. These roles are not necessarily exclusive. For example, both carbohydrates and proteins can play structural roles, supporting and protecting tissues and organisms. However, only nucleic acids specialize in information storage and function as hereditary material, carrying both species and individual traits from generation to generation.

The functions of macromolecules are directly related to their shapes and to the chemical properties of their monomers. Some macromolecules, such as catalytic and defensive proteins, fold into compact spherical forms with surface features that make them water-soluble and capable of intimate interaction with other molecules. Other proteins and carbohydrates form long, fibrous systems that provide strength and rigidity to cells and organisms. Still other long, thin assemblies of proteins can contract and cause movement.

Because macromolecules are so large, they contain many different functional groups (see Figure 2.20). For example, a large protein may contain hydrophobic, polar, and charged functional groups that give specific properties to local sites on a macromolecule. As we will see, this diversity of properties determines the shapes of macromolecules and their interactions with both other macromolecules and smaller molecules.

3.1 The Building Blocks of Organisms

MONOMER	SIMPLE POLYMER	COMPLEX POLYMER (MACROMOLECULE)
Amino acid	Peptide or oligopeptide	Polypeptide (protein)
Nucleotide	Oligonucleotide	Nucleic acid
Monosaccharide (sugar)	Oligosaccharide	Polysaccharide (carbohydrate)

Condensation Reactions

The polymers of living things are constructed from monomers by a series of reactions called **condensation reactions** or *dehydration reactions* (both words refer to the loss of water). Condensation reactions result in covalently bonded monomers (Figure 3.2*a*). The condensation reactions that produce the different kinds of macromolecules differ in detail, but in all cases, polymers will form only if energy is added to the system. In living systems, specific energy-rich molecules supply this energy.

The reverse of a condensation reaction is a **hydrolysis reaction** (hydro-, "water"; -lysis, "break"). These reactions digest polymers and produce monomers. Water reacts with the bonds that link the polymer together, and the products are free monomers. The elements (H and O) of H_2O become part of the products (Figure 3.2*b*). Like condensation reactions, hydrolysis requires the addition of energy.

We begin our study of biological macromolecules with a very diverse group of polymers, the proteins.

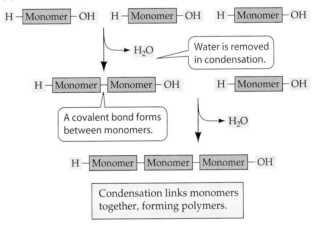

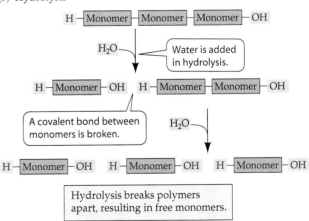

3.2 Condensation and Hydrolysis of Polymers
(*a*) A condensation reaction links monomers into polymers. (*b*) A hydrolysis reaction digests polymers into individual monomers. In living tissues, these reactions do not occur spontaneously, but require added energy.

Proteins: Polymers of Amino Acids

Proteins are involved in structural support, protection, catalysis, transport, defense, regulation, and movement. Among the functions of macromolecules listed earlier, only energy storage and heredity are not usually performed by proteins.

Of particular importance are proteins called **enzymes** that increase the rates of chemical reactions in cells, a function known as *catalysis*. In general, each chemical reaction requires a different enzyme, because proteins show great specificity for the smaller molecules with which they interact.

Proteins range in size from small ones such as the RNA-digesting enzyme ribonuclease A, which has a molecular weight of 5,733 and 51 amino acid residues, to huge molecules such as the cholesterol transport protein apolipoprotein B, which has a molecular weight of 513,000 and 4,636 amino acid residues. (The word "residue" refers to a monomer when it is part of a polymer.) Each of these proteins consists of a single chain of amino acids (a *polypeptide chain*) folded into a specific three-dimensional shape that is required for protein function.

Some proteins have more than one polypeptide chain. For example, the oxygen-carrying protein hemoglobin has four chains that are folded separately and associate together to make the functional protein. As we will see later in this book, there are many such "multi-protein machines," composed of dozens of interacting polypeptides.

Each of these proteins has a characteristic amino acid composition. But not every protein contains all kinds of amino acids, nor an equal number of different ones. The diversity in amino acid content and sequence is the source of the diversity in protein structures and functions. In some cases, additional chemical structures called **prosthetic groups** may be attached covalently to the protein. These groups include carbohydrates, lipids, phosphate groups, the iron-containing heme group that binds to hemoglobin, and metal ions such as copper and zinc. Prosthetic groups are discussed further in Chapter 6.

The next several chapters will describe the many functions of proteins. To understand them, we must first explore protein structure. First, we will examine the properties of the amino acids and how they link to form proteins. Then we will systematically examine protein structure and look at how a linear chain of amino acids is consistently folded into a compact three-dimensional shape. Finally, we will see how this structure provides a specific physical and chemical environment for other molecules that can interact with the protein.

Proteins are composed of amino acids

The 20 amino acids commonly found in proteins have a wide variety of properties. In Chapter 2, we looked at the structure of amino acids and identified four different groups attached to a central (α) carbon atom: a hydrogen atom, an amino group (NH_3^+), a carboxyl group (COO^-), and a **side chain**, or **R group**. The R groups (R stands for

"remainder" or "residue") of amino acids are important in determining the three-dimensional structure and function of the macromolecule. They are highlighted in white in Table 3.2.

As Table 3.2 shows, amino acids are grouped and distinguished by their side chains. Some side chains are electrically charged (+1, –1), while others are polar ($\delta+$, $\delta-$) or nonpolar and hydrophobic.

▶ The five amino acids that have *electrically charged* side chains attract water and oppositely charged ions of all sorts.

▶ The five amino acids that have *polar* side chains tend to form weak hydrogen bonds with water and with other polar or charged substances.

▶ Seven amino acids have side chains that are *nonpolar* hydrocarbons or very slightly modified hydrocarbons. In the watery environment of the cell, the hydrophobic side chains may cluster together.

▶ Three amino acids—cysteine, glycine, and proline—are special cases, although their R groups are generally hydrophobic.

The cysteine side chain, which has a terminal —SH group, can react with another cysteine side chain to form a covalent bond called a **disulfide bridge** (—S—S—) (Figure 3.3). Disulfide bridges help determine how a protein chain folds. When cysteine is not part of a disulfide bridge, its side chain is hydrophobic.

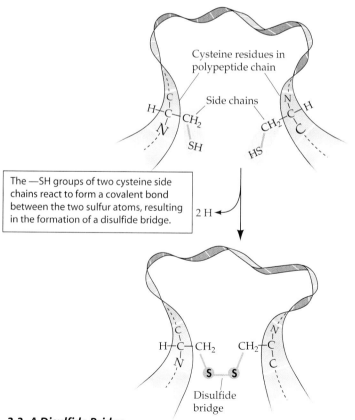

The —SH groups of two cysteine side chains react to form a covalent bond between the two sulfur atoms, resulting in the formation of a disulfide bridge.

3.3 A Disulfide Bridge
Disulfide bridges (—S—S—) are important in maintaining the proper three-dimensional shapes of some protein molecules.

3.2 *Twenty Amino Acids Found in Proteins*

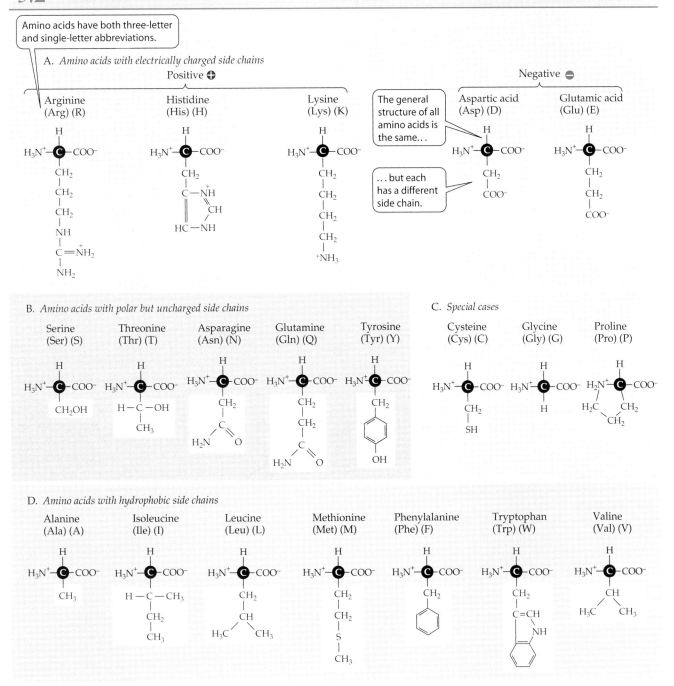

Amino acids have both three-letter and single-letter abbreviations.

A. *Amino acids with electrically charged side chains*

Positive ⊕

Arginine (Arg) (R)   Histidine (His) (H)   Lysine (Lys) (K)

The general structure of all amino acids is the same…

…but each has a different side chain.

Negative ⊖

Aspartic acid (Asp) (D)   Glutamic acid (Glu) (E)

B. *Amino acids with polar but uncharged side chains*

Serine (Ser) (S)   Threonine (Thr) (T)   Asparagine (Asn) (N)   Glutamine (Gln) (Q)   Tyrosine (Tyr) (Y)

C. *Special cases*

Cysteine (Cys) (C)   Glycine (Gly) (G)   Proline (Pro) (P)

D. *Amino acids with hydrophobic side chains*

Alanine (Ala) (A)   Isoleucine (Ile) (I)   Leucine (Leu) (L)   Methionine (Met) (M)   Phenylalanine (Phe) (F)   Tryptophan (Trp) (W)   Valine (Val) (V)

The glycine side chain consists of a single hydrogen atom and is small enough to fit into tight corners in the interior of a protein molecule, where a larger side chain could not fit.

Proline differs from other amino acids because it possesses a modified amino group lacking a hydrogen on its nitrogen, which limits its hydrogen bonding ability. Also, the ring system of proline limits rotation about its α carbon, so proline is often found at bends or loops in a protein.

Peptide linkages covalently bond amino acids together

When amino acids polymerize, the carboxyl group of one amino acid reacts with the amino group of another, undergoing a condensation reaction that forms a **peptide linkage**. Figure 3.4 gives a simplified description of this reaction. (In reality, other molecules must activate the amino acids in order for this reaction to proceed, and there are intermediate steps in the process. We will examine these in Chapter 12).

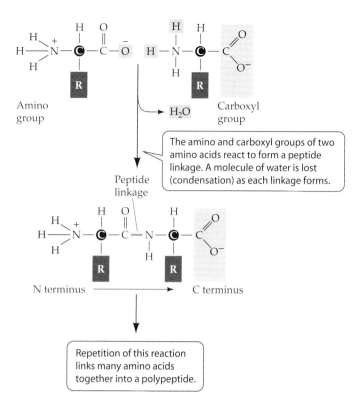

3.4 Formation of Peptide Linkages
In living things, the reaction leading to a peptide linkage has many intermediate steps, but the reactants and products are the same as those shown in this simplified diagram.

Just as a sentence begins with a capital letter and ends with a period, polypeptide chains have a linear order. The chemical "capital letter" marking the beginning of a polypeptide chain is the amino group of the first amino acid in the chain and is known as the *N terminus*. The chemical punctuation mark for the end of the chain is the carboxyl group of the last amino acid (the *C terminus*).

All the other amino and carboxyl groups (except those in side chains) are involved in peptide bond formation, so they do not exist in the chain as "free," intact groups. Biochemists refer to the "N → C," or "amino-to-carboxyl" orientation of polypeptides.

The peptide linkage has two characteristics that are important in the three-dimensional structure of proteins. First, in many single covalent bonds, the groups on either side of the bonds are free to rotate in space. This is not so with the C—N peptide bond. The adjacent atoms (the α carbons of the two adjacent amino acids) are not free to rotate because of the partial double-bond character of the peptide bond. Chemists will realize that this is due to the resonance between the strong electronegativity of the oxygen bound to the carbon and the weak electronegativity of the hydrogen bound to the nitrogen. This characteristic limits the folding of the polypeptide.

Second, the oxygen bound to the carbon carries a slight negative charge (δ–), whereas the hydrogen bound to the nitrogen is slightly positive (δ+). This asymmetry of charge favors hydrogen bonding within the protein molecule itself and with other molecules, contributing to both the structure and the function of many proteins.

The primary structure of a protein is its amino acid sequence

There are four levels of protein structure, called primary, secondary, tertiary, and quaternary. The precise sequence of amino acids in a polypeptide constitutes the **primary structure** of a protein (Figure 3.5*a*). The *peptide backbone* of this primary structure consists of a repeating sequence of three atoms (—N—C—C—): the N from the amino group, the α carbon, and the C from the carboxyl group of each amino acid.

Scientists have deduced the primary structure of many proteins, and use the single-letter abbreviations for amino acids (see Table 3.2) to record the sequence. Here, for example, are the first 25 amino acids (out of a total of 457) for the protein hexokinase, from baker's yeast:

AASXDXSLVEVHXXVFIVPPXILQA

The theoretical number of different proteins is enormous. Since there are 20 different amino acids, there are $20 \times 20 = 400$ distinct dipeptides (two linked amino acids), and $20 \times 20 \times 20 = 8{,}000$ different tripeptides (three linked amino acids). Imagine this process of multiplying by 20 extended to a protein made up of 100 amino acids (which is considered a small protein). There could be 20^{100} such small proteins, each with its own distinctive primary structure. How large is the number 20^{100}? There aren't that many electrons in the entire universe!

At the higher levels of protein structure, local coiling and folding give the molecule its final functional shape, but all of these levels derive from the primary structure—that is, which amino acids are at which locations on the polypeptide chain. The properties associated with a precise sequence of amino acids determine how the protein can twist and fold, thus adopting a specific stable structure that distinguishes it from every other protein.

The secondary structure of a protein requires hydrogen bonding

Although the primary structure of each protein is unique, the secondary structures of many different proteins may be quite similar. A protein's **secondary structure** consists of regular, repeated patterns in different regions of a polypeptide chain. There are two basic types of secondary structure, both of them determined by hydrogen bonding between the amino acid residues that make up the primary structure.

THE α HELIX. The α (alpha) **helix** is a right-handed coil that is "threaded" in the same direction as a standard wood screw (Figure 3.5*b*). The R groups extend outward from the peptide backbone of the helix. The coiling results from hydrogen bonds between the slightly positive hydrogen of the N—H of one amino acid residue and the slightly negative oxygen of the C=O of another. When this pattern of hydrogen bonding is established repeatedly over a segment of the protein, it stabilizes the coil, resulting in an α helix. Amino acids with large R groups that distort the coil or

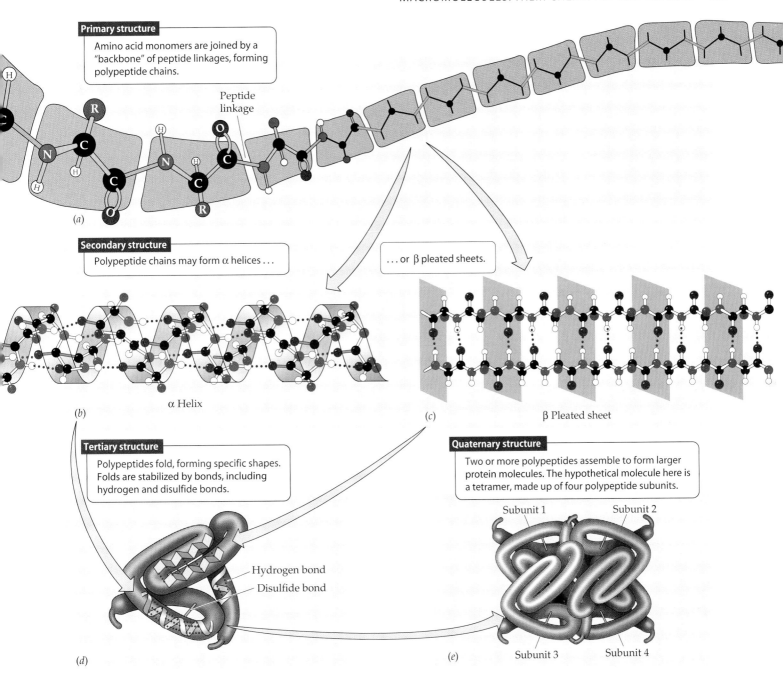

Primary structure

Amino acid monomers are joined by a "backbone" of peptide linkages, forming polypeptide chains.

Peptide linkage

(a)

Secondary structure

Polypeptide chains may form α helices ...

... or β pleated sheets.

α Helix

(b)

β Pleated sheet

(c)

Tertiary structure

Polypeptides fold, forming specific shapes. Folds are stabilized by bonds, including hydrogen and disulfide bonds.

Hydrogen bond
Disulfide bond

(d)

Quaternary structure

Two or more polypeptides assemble to form larger protein molecules. The hypothetical molecule here is a tetramer, made up of four polypeptide subunits.

Subunit 1   Subunit 2

Subunit 3   Subunit 4

(e)

3.5 The Four Levels of Protein Structure
Secondary, tertiary, and quaternary structure all arise from the primary structure of the protein.

otherwise prevent the formation of the necessary hydrogen bonds will keep the α helix from forming.

Alpha-helical secondary structure is particularly evident in the insoluble fibrous structural proteins called keratins, which make up hair, hooves, and feathers. Hair can be stretched because stretching requires that only the hydrogen bonds of the α helix, not the covalent bonds, be broken; when the tension on the hair is released, both the hydrogen bonds and the helix re-form.

THE β PLEATED SHEET.   The β (beta) pleated sheet is formed from two or more polypeptide chains that are almost com-

pletely extended and lying next to one another. The sheet is stabilized by hydrogen bonds between the N—H groups on one chain and the C=O groups on the other (Figure 3.5c). A β pleated sheet may form between separate polypeptide chains, as in spider silk, or between different regions of the same polypeptide that is bent back on itself. Many proteins contain regions of both α helix and β pleated sheet in the same polypeptide chain.

The tertiary structure of a protein is formed by bending and folding

In many proteins, the polypeptide chain is bent at specific sites and folded back and forth, resulting in the **tertiary structure** of the protein (Figure 3.5d). Although the α helices and β pleated sheets contribute to the tertiary struc-

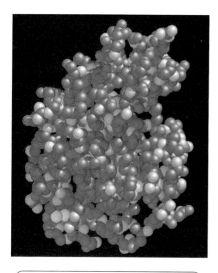

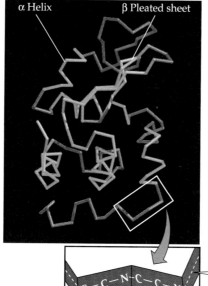

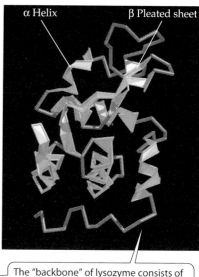

α Helix   β Pleated sheet

α Helix   β Pleated sheet

A realistic depiction of lysozyme shows dense packing of its atoms.

C—C—N—C—C—N

The "backbone" of lysozyme consists of repeating N—C—C units of amino acids.

3.6 Three Representations of Lysozyme
Different molecular representations of a protein emphasize different aspects of its tertiary structure. These three representations of lysozyme are similarly oriented.

ture, only parts of the macromolecule usually have these secondary structures, and large regions consist of structures unique to a particular protein.

While hydrogen bonding is responsible for secondary structure, the interactions between R groups determine tertiary structure. We described the various strong and weak interactions between atoms in Chapter 2 (see Table 2.1). Many of these interactions are involved in determining tertiary structure:

▶ Covalent *disulfide bridges* can form between specific cysteine residues (see Figure 3.3), holding a folded polypeptide in place.
▶ *Hydrophobic side chains* can aggregate together in the interior of the protein, away from water, folding the polypeptide in the process.
▶ *Van der Waals forces* can stabilize the close interactions between the hydrophobic residues.
▶ *Ionic interactions* can occur between positively and negatively charged side chains buried deep within a protein, away from water, forming a *salt bridge*.

A complete description of a protein's tertiary structure specifies the location of every atom in the molecule in three-dimensional space, in relation to all the other atoms. The tertiary structure of the protein lysozyme is represented in Figure 3.6.

Bear in mind that both tertiary structure and secondary structure derive from the protein's primary structure. If lysozyme is heated slowly, the heat energy will disrupt only the weak interactions and cause only the tertiary structure to break down. But the protein will return to its normal tertiary structure when it cools, demonstrating that all the

information needed to specify the unique shape of a protein is contained in its primary structure.

The quaternary structure of a protein consists of subunits

As mentioned earlier, many functional proteins have two or more polypeptide chains, called **subunits**, each of them folded into its own unique tertiary structure. The protein's **quaternary structure** results from the ways in which these multiple polypeptide subunits bind together and interact.

Quaternary structure is illustrated by hemoglobin (Figure 3.7). Hydrophobic interactions, van der Waals forces, hydrogen bonds, and ionic bonds all help hold the four subunits together to form the hemoglobin molecule. The function of hemoglobin is to carry oxygen in red blood cells. As hemoglobin binds one O_2 molecule, the four subunits shift their relative positions slightly, changing the quaternary structure. Ionic bonds are broken, exposing buried side chains that enhance the binding of additional O_2 molecules. The structure changes again when hemoglobin releases its oxygen molecules to the cells of the body.

The surfaces of proteins have specific shapes

Small molecules in a solution are in constant motion. They vibrate, rotate, and move from place to place like corn in a popper. If two of them collide in the right circumstances, a chemical reaction can occur. The specific shapes of proteins allow them to bind noncovalently with other molecules, which in turn allows other important biological events to occur. For example:

▶ Two adjacent cells can stick together because proteins protruding from each of the cells interact with each other (see Chapter 5).
▶ A substance can enter a cell by binding to a carrier protein in the cell surface membrane (see Chapter 5).

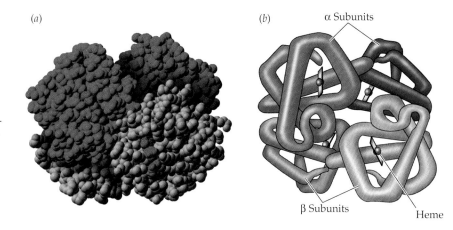

(a)   (b)   α Subunits

3.7 Quaternary Structure of a Protein
Hemoglobin consists of four folded polypeptide subunits that assemble themselves into the quaternary structure shown here. In these two graphic representations, each type of subunit is a different color. The heme groups contain iron and are the oxygen-carrying sites.

β Subunits   Heme

▶ A chemical reaction can be speeded up when an enzyme protein binds to one of the reactants (see Chapter 6).

▶ A multi-protein "machine," DNA polymerase, can catalyze the replication of DNA (see Chapter 11).

▶ Another multi-protein "machine," the ribosome, can synthesize proteins (see Chapter 12).

▶ Proteins on a cell's outer surface can bind to chemical signals such as hormones (see Chapter 15).

▶ Defensive proteins called antibodies can recognize the shape of a virus coat and bind to it (see Chapter 19).

When a small molecule collides with and binds to a much larger protein, it is like a baseball being caught by a catcher: The catcher's mitt has a shape that binds to the ball and fits around it. A hockey puck or a ping-pong ball would not fit the baseball mitt. Thus, the binding of a small molecule to a protein involves a general interaction between two three-dimensional objects that becomes more specific after initial binding. When two large polypeptide chains bind to each other, the interactions are more complicated because extensive surfaces of each macromolecule must come into contact, but the principle is the same.

Biological specificity depends not just on the shape of a protein, but also on the surface chemical groups that it presents to a substance attempting to bind to it (Figure 3.8). The groups on the surface are the R groups of the exposed amino acids, and are therefore a property of the protein's primary structure.

Look again at the structures of the 20 amino acids in Table 3.2, noting the properties of the R groups. Exposed hydrophobic groups will bind to similarly nonpolar groups in the substance with which the protein interacts (often called the *ligand*). Charged R groups will bind to oppositely charged groups on the ligand. Polar R groups containing a hydroxyl (—OH) group can form a hydrogen bond with an incoming ligand. These three types of interactions—hydrophobic, ionic, and hydrogen bonding—are weak by themselves, but strong when all of them act together. So the exposure of appropriate amino acid R groups on the protein surface allows specific binding of a ligand to occur.

Protein shapes are sensitive to the environment

The three-dimensional structure of a protein determines what it binds and, therefore, its function. The primary structure of a protein constrains its secondary, tertiary, and (if subunits exist) quaternary structures. A major effort in

biochemistry is to try to predict three-dimensional protein structure from amino acid sequence. For some short sequences, this is relatively straightforward: For example, certain amino acid sequences will fold into a β pleated sheet. But for large polypeptide chains, the multitude of potential interactions make structure prediction a problem, approachable only by computer. Indeed, a whole new field of computational biochemistry has emerged to tackle the challenge of structure prediction.

Knowing the exact shape of a protein and what can bind to it is important not only in understanding basic biology, but in applied fields such as medicine as well. For example, the three-dimensional structure of a protease, a protein es-

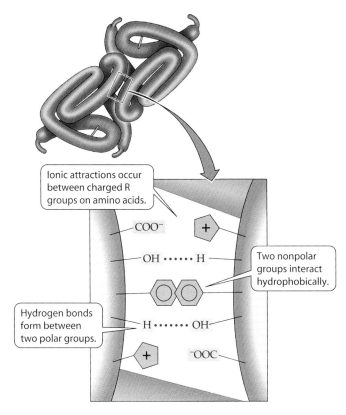

Ionic attractions occur between charged R groups on amino acids.

COO⁻   +

OH ••••• H   Two nonpolar groups interact hydrophobically.

Hydrogen bonds form between two polar groups.   H ••••• OH

+   ⁻OOC

3.8 Noncovalent Interactions between Polypeptides and Other Molecules
Noncovalent interactions allow a protein to bind tightly to another molecule with specific properties, or allow regions within a protein to interact with one another.

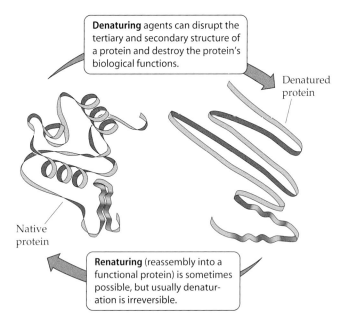

Denaturing agents can disrupt the tertiary and secondary structure of a protein and destroy the protein's biological functions.

Denatured protein

Native protein

Renaturing (reassembly into a functional protein) is sometimes possible, but usually denaturation is irreversible.

3.9 Denaturation Is the Loss of Tertiary Protein Structure and Function
Agents that can cause denaturation include high temperatures and certain chemicals.

sential for the replication of HIV—the virus that causes AIDS—was first determined in this way. Then specific inhibitors were designed to interact with its surface. These protease inhibitors have prolonged the lives of countless people living with HIV.

Because it is determined by weak forces, protein shape is sensitive to environmental conditions that would not break covalent bonds but do upset weaker noncovalent interactions. Elevated temperatures, pH changes, or altered salt concentrations can cause a protein to adopt a different, biologically inactive tertiary structure. Increases in temperature cause more rapid molecular movements and thus can break hydrogen bonds and hydrophobic interactions. Alterations in pH can change the pattern of ionization of carboxyl and amino groups in the R groups of amino acids, thus disrupting the pattern of ionic attractions and repulsions that contributes to normal tertiary structure.

The loss of normal tertiary structure is called **denaturation**, and it is always accompanied by a loss of the normal biological function of the protein (Figure 3.9). Denaturation can be caused by heat or by high concentrations of polar substances such as urea, which disrupt the hydrogen bonding that is crucial to protein structure. Nonpolar solvents may also disrupt normal structure.

Usually denaturation is irreversible, because amino acids that were buried may now be exposed and vice versa, causing a new structure to form or different molecules to bind to the protein. Boiling an egg denatures its proteins and is, as you know, not reversible. However, as we saw earlier, denaturation is often reversible in the laboratory, especially if it was caused originally by disruption of weak forces. If the denaturing chemicals are removed, the protein returns to its "native" shape and normal function.

Chaperonins help shape proteins

There are two occasions when a polypeptide chain is in danger of binding the wrong ligand. First, following denaturation, hydrophobic R groups, previously on the inside of the protein away from water, become exposed on the surface. Since these groups can interact with similar groups on other molecules, the denatured proteins may aggregate and become insoluble, losing their function. Second, when a protein has just been synthesized and has not yet folded completely, it could present a surface that binds the wrong molecule.

Living systems limit inappropriate protein interactions by making a class of proteins called, appropriately, **chaperonins** (recall the chaperones—usually teachers—at school dances who try to prevent "inappropriate interactions" among the students). Chaperonins were first identified in fruit flies as "heat shock" proteins, which prevented denaturing proteins from clumping together when the flies' temperature was raised.

Some chaperonins work by trapping proteins in danger of inappropriate binding inside a molecular "cage" (Figure 3.10). This cage is composed of identical subunits, and is itself a good example of quaternary protein structure. Inside the cage, the targeted protein folds into the right shape, and then is released at the appropriate time and place.

3.10 Chaperonins Protect Proteins from Inappropriate Binding
Chaperonins surround new or denatured proteins and prevent them from binding to the wrong ligand.

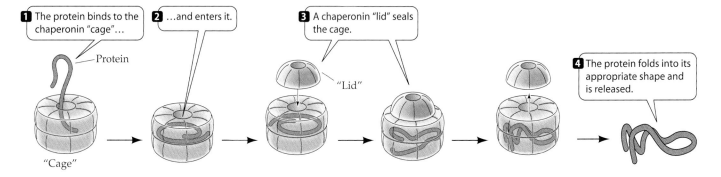

1 The protein binds to the chaperonin "cage"...

Protein

"Cage"

2 ...and enters it.

3 A chaperonin "lid" seals the cage.

"Lid"

4 The protein folds into its appropriate shape and is released.

Carbohydrates: Sugars and Sugar Polymers

Carbohydrates are a diverse group of compounds containing primarily carbon atoms flanked by hydrogen and hydroxyl groups (H—C—OH). They have two major biochemical roles:

▶ They act as a source of energy that can be released in a form usable by body tissues. This energy is stored in strong C—C and C=O covalent bonds.
▶ They serve as carbon skeletons that can be rearranged to form other molecules important for biological structures and functions.

Some carbohydrates are relatively small, with molecular weights less than 100. Others are true macromolecules, with molecular weights in the hundreds of thousands.

There are four categories of biologically important carbohydrates, which we will discuss in turn:

▶ *Monosaccharides* (mono-, "one"; saccharide, "sugar"), such as glucose, ribose, or fructose, are simple sugars and are the monomers out of which the larger forms are constructed.
▶ *Disaccharides* (di-, "two") consist of two monosaccharides.
▶ *Oligosaccharides* (oligo-, "several") have several monosaccharides (3 to 20).
▶ *Polysaccharides* (poly-, "many"), such as starch, glycogen, and cellulose, are large polymers composed of hundreds of thousands of monosaccharide units.

The relative proportions of carbon, hydrogen, and oxygen indicated by the general formula for carbohydrates, CH_2O (i.e., the proportions of these atoms are 1:2:1), apply to monosaccharides. In disaccharides, oligosaccharides, and polysaccharides, these proportions differ slightly from the general formula because two hydrogens and an oxygen are lost during the condensation reactions that form them.

Monosaccharides are simple sugars

Green plants produce monosaccharides through photosynthesis, and animals acquire them directly or indirectly from plants. All living cells contain the monosaccharide glucose. Cells use glucose as an energy source, breaking it down through a series of hydrolysis reactions that release stored energy and produce water and carbon dioxide.

Glucose exists in two forms, the straight chain and the ring; the ring structure predominates in more than 99 percent of circumstances. There are also two forms of the ring structure (α-glucose and β-glucose), which differ only in the placement of the —H and —OH attached to carbon 1 (Figure 3.11). The α and β forms interconvert and exist in equilibrium when dissolved in water.

Different monosaccharides contain different numbers of carbons. (The standard convention for numbering carbons shown in Figure 3.11 is used throughout this book.) Most of the monosaccharides found in living systems belong to the D series of optical isomers (see Chapter 2). But some monosaccharides are *structural* isomers, which have the same kinds and numbers of atoms, but arranged differently by bonding. For example, the *hexoses* (hex-, "six"), a group of structural isomers, all have the formula $C_6H_{12}O_6$. Included among the hexoses are glucose, fructose (so named because it was first found in fruits), mannose, and galactose (Figure 3.12).

Pentoses (pent-, "five") are five-carbon sugars. Some pentoses are found primarily in the cell walls of plants. Two pentoses are of particular biological importance: Ribose and deoxyribose form part of the backbones of the nucleic acids RNA and DNA, respectively. These two pentoses are not isomers; rather, one oxygen atom is missing from carbon 2 in deoxyribose (de-, "absent") (see Figure 3.12). As we will see in Chapter 12, the absence of this oxygen atom has important consequences for the functional distinction of RNA and DNA.

3.11 Glucose: From One Form to the Other

All glucose molecules have the formula $C_6H_{12}O_6$, but their structures vary. When dissolved in water, the α and β "ring" forms of glucose interconvert. The dark line at the bottom of each ring indicates that that edge of the molecule extends toward you; the upper, lighter edge extends back into the page.

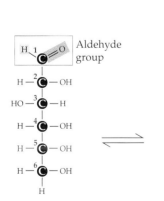

Straight-chain form | Intermediate form | α-Glucose | β-Glucose

The straight-chain form of glucose has an aldehyde group at carbon 1.

A reaction between this aldehyde group and the hydroxyl group at carbon 5 gives rise to a ring form.

Depending on the orientation of the aldehyde group when the ring closes, either of two rapidly and spontaneously interconverting molecules—α-glucose and β-glucose—forms.

Three-carbon sugar

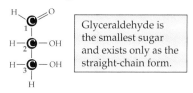

Glyceraldehyde is the smallest sugar and exists only as the straight-chain form.

Five-carbon sugars

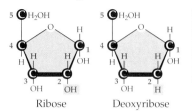

Ribose   Deoxyribose

Ribose and deoxyribose each have five carbons, but very different chemical properties and biological roles.

Six-carbon sugars

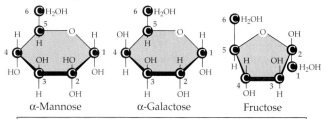

α-Mannose   α-Galactose   Fructose

These hexoses are isomers. All have the formula $C_6H_{12}O_6$, but each has distinct chemical properties and biological roles.

3.12 Monosaccharides Are Simple Sugars
Monosaccharides are made up of varying numbers of carbons. Some are structural isomers, which have the same number of carbons, but arranged differently. Fructose, for example, is a hexose but forms a five-sided ring like the pentoses.

Glycosidic linkages bond monosaccharides together

Monosaccharides are covalently bonded together by condensation reactions that form *glycosidic linkages*. Such a linkage between two monosaccharides forms a disaccharide. For example, a molecule of sucrose (table sugar) is formed from a glucose and a fructose molecule, while lactose (milk sugar) contains glucose and galactose.

The disaccharide maltose contains two glucose molecules, but it is not the only disaccharide that can be made from two glucoses. When glucose molecules form glycosidic linkages, the disaccharide product will be one of two types: α-linked or β-linked, depending on whether the molecule that bonds by its carbon 1 is α-glucose or β-glucose (see Figure 3.11). An α linkage with carbon 4 of a second glucose molecule gives maltose, whereas a β linkage gives cellobiose (Figure 3.13).

Maltose and cellobiose are disaccharide isomers, both having the formula $C_{12}H_{22}O_{11}$. However, they are different compounds with different properties. They undergo different chemical reactions and are recognized by different enzymes. For example, maltose can be hydrolyzed to its monosaccharides in the human body, whereas cellobiose cannot. Certain microorganisms have the chemistry to break down cellobiose.

Oligosaccharides contain several monosaccharides linked by glycosidic linkages at various sites. Many oligosaccharides have additional functional groups, which give them special properties. Oligosaccharides are often covalently bonded to proteins and lipids on the outer cell surface, where they serve as cell recognition signals. The human blood groups (such as ABO) get their specificity from oligosaccharide chains.

Maltose is produced when an α-1,4 glycosidic linkage forms between two glucose molecules. The hydroxyl group on carbon 1 of one glucose in the α (down) position reacts with the hydroxyl group on carbon 4 of the other glucose.

In **cellobiose**, two glucoses are linked by a β-1,4 glycosidic linkage.

3.13 Disaccharides Are Formed by Glycosidic Linkages
Glycosidic linkages between two monosaccharides create many different disaccharides. Which disaccharide is formed depends on which monosaccharides are linked, and on the site (which carbon atom is linked) and form (α or β) of the linkage.

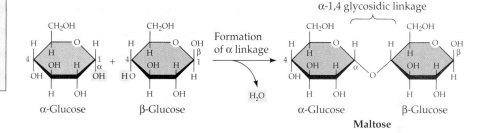

Polysaccharides serve as energy stores or structural materials

Polysaccharides are giant chains of monosaccharides connected by glycosidic linkages. *Starch* is a polysaccharide of glucose with glycosidic linkages in the α-orientation. *Cellulose*, too, is a giant polysaccharide made up solely of glucose, but its individual monosaccharides are connected by β linkages (Figure 3.14*a*). Cellulose is the predominant component of plant cell walls, and is by far the most abundant organic compound on Earth. Both starch and cellulose are composed of nothing but glucose, but their very different chemical and physical properties give them distinct biological functions.

Starch can be more or less easily degraded by the actions of chemicals or enzymes. Cellulose, however, is chemically more stable because of its β-glycosidic linkages. Thus starch

(*a*) **Molecular structure**

Cellulose

Starch and glycogen

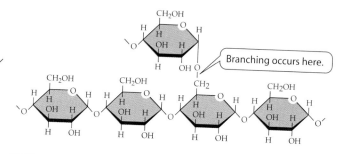

Branching occurs here.

Hydrogen bonding to other cellulose molecules can occur at these points.

Cellulose is an unbranched polymer of glucose with β-1,4 glycosidic linkages that are chemically very stable.

Glycogen and starch are polymers of glucose with α-1,4 glycosidic linkages. α-1,6 glycosidic linkages produce branching at carbon 6.

(*b*) **Macromolecular structure**

Linear (cellulose)

Branched (starch)

Highly branched (glycogen)

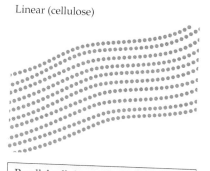

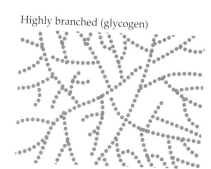

Parallel cellulose molecules hydrogen-bond to form long thin fibrils.

Branching limits the number of hydrogen bonds that can form in starch molecules, making starch less compact than cellulose.

The high amount of branching in glycogen makes its solid deposits less compact than starch.

(*c*) **Polysaccharides in cells**

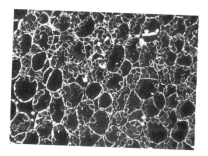

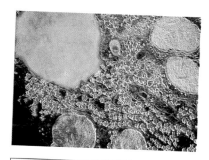

Layers of cellulose fibrils, as seen in this scanning electron micrograph, give plant cell walls great strength.

Dyed red in this micrograph, starch deposits have a large granular shape within cells.

Colored pink in this electron micrograph of human liver cells, glycogen deposits have a small granular shape.

3.14 Representative Polysaccharides
Cellulose, starch, and glycogen demonstrate different levels of branching and compaction in polysaccharides.

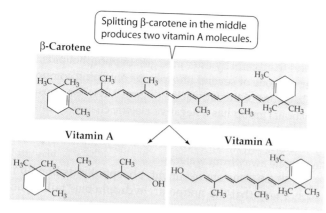

Splitting β-carotene in the middle produces two vitamin A molecules.

β-Carotene

Vitamin A   Vitamin A

3.23 β-Carotene is the Source of Vitamin A
The carotenoid β-carotene is symmetrical around its central double bond; when split, β-carotene becomes two vitamin A molecules. The simplified structural formula used here is standard chemical shorthand for large organic molecules with many carbon atoms. Structural formulas are simplified by omitting the C's (for carbon atoms) at the intersections of the lines representing covalent bonds. H's to fill all the available bonding sites on each C are assumed.

two vitamin A molecules (Figure 3.23), from which we make the pigment rhodopsin, which is required for vision. Carotenoids are responsible for the colors of carrots, tomatoes, pumpkins, egg yolks, and butter.

STEROIDS ARE SIGNAL MOLECULES. The **steroids** are a family of organic compounds whose multiple rings share carbons (Figure 3.24). The steroid cholesterol is an important constituent of membranes. Other steroids function as hormones, chemical signals that carry messages from one part of the body to another. Testosterone and the estrogens are steroid hormones that regulate sexual development in vertebrates. Cortisol and related hormones play many regulatory roles in the digestion of carbohydrates and proteins, in the maintenance of salt balance and water balance, and in sexual development.

Cholesterol is synthesized in the liver and is the starting material for making testosterone and other steroid hormones, as well as the bile salts that help break down dietary fats so that they can be digested. Cholesterol is absorbed from foods such as milk, butter, and animal fats. An

excess of cholesterol in the blood can lead to its deposition (along with other substances) in the arteries, a condition that may lead to arteriosclerosis and heart attack.

Some lipids are vitamins

Vitamins are small organic molecules that are not synthesized in the body and must be acquired from dietary sources, and whose deficiencies lead to defined diseases.

Vitamin A is formed from the β-carotene found in green and yellow vegetables (see Figure 3.23). In humans, a deficiency of vitamin A leads to dry skin, eyes, and internal body surfaces; retarded growth and development; and night blindness, which is a diagnostic symptom for the deficiency. Vitamin D regulates the absorption of calcium from the intestines. It is necessary for the proper deposition of calcium in bones; a deficiency of vitamin D can lead to rickets, a bone-softening disease.

Vitamin E seems to protect cells from damaging effects of oxidation–reduction reactions. For example, it has an important role in preventing unhealthy changes in the double bonds in the unsaturated fatty acids of membrane phospholipids. Commercially, vitamin E is added to some foods to slow spoilage. Vitamin K is found in green leafy plants and is also synthesized by bacteria normally present in the human intestine. This vitamin is essential to the formation of blood clots. Predictably, a deficiency of vitamin K leads to slower clot formation and potentially fatal bleeding from a wound.

Wax coatings repel water

The sheen on human hair is not there only for cosmetic purposes. Glands in the skin secrete a waxy coating that repels water and keeps the hair pliable. Birds that live near water have a similar waxy coating on their feathers. The shiny leaves of holly plants, familiar during winter holidays, also have a waxy coating. Finally, bees make their honeycombs out of wax. All waxes have the same basic structure: They are formed by an ester linkage between a saturated, long-chain fatty acid and a saturated, long-chain

3.24 All Steroids Have the Same Ring Structure
The steroids shown, all important in vertebrates, are composed of carbon and hydrogen and are highly hydrophobic. However, small chemical variations, such as the presence or absence of a methyl or hydroxyl group, can produce enormous functional differences.

Cholesterol is a constituent of membranes and is the source of steroid hormones.

Vitamin D₂ can be produced in the skin by the action of light on a cholesterol derivative.

Cortisol is a hormone secreted by the adrenal glands.

Testosterone is a male sex hormone.

alcohol. The result is a very long molecule, with 40–60 CH_2 groups. For example, here is the structure of beeswax:

$$\underbrace{CH_3 - (CH_2)_{14} - \overset{\overset{\textstyle O}{\|}}{C}}_{\text{Fatty acid}} - \underbrace{O - CH_2 - (CH_2)_{28} - CH_3}_{\text{Alcohol}}$$

This highly nonpolar structure accounts for the impermeability of wax to water.

The Interactions of Macromolecules

We have treated the classes of macromolecules as if each were separate from the others. In cells, however, certain macromolecules of different classes may be covalently bonded to one another. Proteins with attached oligosaccharides are called *glycoproteins* (glyco-, "sugar"). The specific oligosaccharide chain attached can determine where within the cell a newly synthesized protein will reside. Other carbohydrate chains covalently bond to lipids, resulting in *glycolipids*, which reside in the cell surface membrane, with the carbohydrate chain extending out into the cell's environment. The carbohydrates that determine a person's blood type (A, B, AB, or O) are attached to either proteins or lipids sticking out from the surfaces of red blood cells.

We have already mentioned the fact that proteins can bind noncovalently to other proteins in quaternary structures. But proteins can bind noncovalently to the other types of macromolecules as well. For example, there are hundreds of different proteins that recognize and bind to DNA, regulating its function. Other proteins, in combination with cholesterol and other lipids, form lipoproteins. Some lipoproteins serve as carrier proteins, which make it possible to move very hydrophobic lipids such as cholesterol through water-rich environments such as the blood.

Summary

Macromolecules: Giant Polymers

▶ Macromolecules are constructed by the formation of covalent bonds between smaller molecules called monomers. Macromolecules include polysaccharides, proteins, and nucleic acids. **Review Figure 3.1 and Table 3.1**

▶ Macromolecules have specific, characteristic three-dimensional shapes that depend on the structures, properties, and sequence of their monomers. Different functional groups give local sites on macromolecules specific properties that are important for their biological functioning and their interactions with other macromolecules.

Condensation Reactions

▶ Monomers are joined by condensation reactions, which release a molecule of water for each bond formed. Hydrolysis reactions use water to break polymers into monomers. **Review Figure 3.2**

Proteins: Polymers of Amino Acids

▶ The functions of proteins include support, protection, catalysis, transport, defense, regulation, and movement. Protein function sometimes requires an attached prosthetic group.

▶ There are 20 amino acids found in proteins. Each amino acid consists of an amino group, a carboxyl group, a hydrogen, and a side chain bonded to the α carbon atom. **Review Table 3.2**

▶ The side chains of amino acids may be charged, polar, or hydrophobic; there are also "special cases," such as the —SH groups, which can form disulfide bridges. The side chains give different properties to each of the amino acids. **Review Table 3.2 and Figure 3.3**

▶ Amino acids are covalently bonded together by peptide linkages, which form by condensation reactions between the carboxyl and amino groups. **Review Figure 3.4**

▶ The polypeptide chains of proteins are folded into specific three-dimensional shapes. Four levels of structure are possible: primary, secondary, tertiary, and quaternary.

▶ The primary structure of a protein is the sequence of amino acids bonded by peptide linkages. This primary structure determines both the higher levels of structure and protein function. **Review Figure 3.5a**

▶ Secondary structures of proteins, such as α helices and β pleated sheets, are maintained by hydrogen bonds between atoms of the amino acid residues. **Review Figure 3.5b,c**

▶ The tertiary structure of a protein is generated by bending and folding of the polypeptide chain. **Review Figures 3.5d, 3.6**

▶ The quaternary structure of a protein is the arrangement of polypeptides in a single functional unit consisting of more than one polypeptide subunit. **Review Figures 3.5e, 3.7**

▶ Weak chemical interactions are important in the binding of proteins to other molecules. **Review Figure 3.8**

▶ Proteins denatured by heat, acid, or certain chemicals lose their tertiary and secondary structure as well as their biological function. Renaturation is not always possible. **Review Figure 3.9**

▶ Chaperonins assist protein folding by preventing binding to inappropriate ligands. **Review Figure 3.10**

Carbohydrates: Sugars and Sugar Polymers

▶ All carbohydrates contain carbon bonded to H and OH groups.

▶ Hexoses are monosaccharides that contain six carbon atoms. Examples of hexoses include glucose, galactose, and fructose, which can exist as chains or rings. **Review Figures 3.11, 3.12**

▶ The pentoses are five-carbon monosaccharides. Two pentoses, ribose and deoxyribose, are components of the nucleic acids RNA and DNA, respectively. **Review Figure 3.12**

▶ Glycosidic linkages may have either α or β orientation in space. They covalently link monosaccharides into larger units such as disaccharides (for example, cellobiose), oligosaccharides, and polysaccharides. **Review Figures 3.13, 3.14**

▶ Cellulose, a very stable glucose polymer, is the principal component of the cell walls of plants. It is formed by glucose units linked together by β-glycosidic linkages between carbons 1 and 4. **Review Figure 3.14**

▶ Starches, less dense and less stable than cellulose, store energy in plants. Starches are formed by α-glycosidic linkages between carbons 1 and 4 and are distinguished by the amount of branching that occurs through glycosidic bond formation at carbon 6. **Review Figure 3.14**

▶ Glycogen contains α-1,4 glycosidic linkages and is highly branched. Glycogen stores energy in animal livers and muscles. **Review Figure 3.14**

▶ Chemically modified monosaccharides include the sugar phosphates and amino sugars. A derivative of the amino sugar glucosamine polymerizes to form the polysaccharide

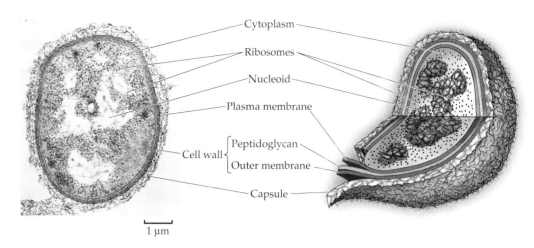

4.4 A Prokaryotic Cell
The bacterium *Pseudomonas aeruginosa* illustrates typical prokaryotic cell structures. The electron micrograph on the left is magnified about 80,000 times. Note the existence of several protective structures external to the plasma membrane.

1 μm

All cells are surrounded by a plasma membrane

A **plasma membrane** separates each cell from its environment, creating a segregated (but not isolated) compartment. The plasma membrane is composed of a phospholipid bilayer, with the hydrophilic ends of the lipids facing the cell's aqueous interior on one side and the extracellular environment on the other (see Figure 3.22). Proteins are embedded in the lipids. In many cases, the proteins protrude into the cytoplasm and into the extracellular environment. We will devote most of the next chapter to the structure and functions of the plasma membrane, but summarize its roles here:

▶ The plasma membrane acts as a *selectively permeable barrier*, preventing some substances from crossing while permiting other substances to enter and leave the cell.
▶ As the cell's boundary with the outside environment, the plasma membrane is important in *communicating with adjacent cells and receiving extracellular signals*. We will describe this function in Chapter 15.
▶ The plasma membrane allows the cell to maintain a more or less *constant internal environment*. A self-maintaining, constant internal environment is a key characteristic of life and will be discussed in detail in Chapter 40.

Cells show two organizational patterns

Once the microscope was applied to biological samples, it soon became apparent that there are two types of cell structures in the living world.

Prokaryotic cell organization is characteristic of the domain Bacteria and Archaea. Organisms in these domains are called *prokaryotes*. Their cells do not have membrane-enclosed internal compartments.

Eukaryotic cell organization is found in the domain Eukarya, which includes the protists, plants, fungi, and animals. The genetic material (DNA) of eukaryotic cells is contained in a special membrane-enclosed compartment called the nucleus. Eukaryotic cells also contain other membrane-enclosed compartments in which specific chemical reactions take place. Organisms with this type of cell are known as *eukaryotes*.

Both prokaryotes and eukaryotes have prospered for many hundreds of millions of years of evolution, and both are great success stories. Let's look first at prokaryotic cells.

Prokaryotic Cells

Prokaryotes can live off more different and diverse energy sources than any other living creatures, and they inhabit greater environmental extremes, such as very hot springs and very salty water. The vast diversity within the prokaryotic domains is the subject of Chapter 26.

Prokaryotic cells are generally smaller than eukaryotic cells, ranging from 0.25×1.2 μm to 1.5×4 μm. So they are generally visible by light microscopy, although their substructures are visible only by electron microscopy. Each prokaryote is a single cell, but many types of prokaryotes are usually seen in chains, small clusters, or even clusters containing hundreds of individuals.

In this section, we will first consider the features that cells in the domains Bacteria and Archaea have in common. Then we will examine structural features that are found in some, but not all, prokaryotes.

All prokaryotic cells share certain features

All prokaryotic cells have the same basic structure (Figure 4.4):

▶ The plasma membrane encloses the cell, regulating the traffic of materials into and out of the cell and separating it from its environment.
▶ A region called the **nucleoid** contains the hereditary material (DNA) of the cell.

The rest of the material enclosed in the plasma membrane is called the **cytoplasm**. Cytoplasm is composed of two parts: the liquid cytosol, and insoluble suspended particles, including ribosomes.

▶ The **cytosol** consists mostly of water that contains dissolved ions, small molecules, and soluble macromolecules such as proteins.
▶ **Ribosomes** are granules about 25 nm in diameter that are sites of protein synthesis.

Although structurally less complicated than eukaryotic cells, prokaryotic cells are functionally complex, carrying out thousands of biochemical transformations.

Some prokaryotic cells have specialized features

Many prokaryotic cells have at least a few structural complexities. For example, most prokaryotes have a **cell wall** lo-

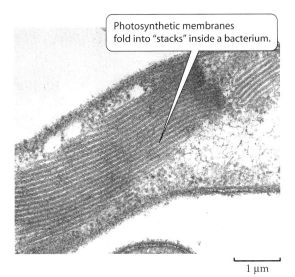

Photosynthetic membranes fold into "stacks" inside a bacterium.

1 μm

4.5 Some Prokaryotes Have Internal Membrane Systems
The presence of internal membranes contradicts the notion that prokaryotes are nothing more than tiny bags of molecules. These photosynthetic membranes contain compounds needed for photosynthesis.

cated outside the plasma membrane (see Figure 4.4). The rigidity of the cell wall supports the cell and determines its shape. The cell walls of most bacteria, but not archaea, contain *peptidoglycan*, a polymer of amino sugars, cross-linked by covalent bonds to form a single giant molecule around the entire cell. In some bacteria, another layer—the *outer membrane* (a polysaccharide-rich phospholipid membrane)—encloses the cell wall. Unlike the plasma membrane, this outer membrane is not a major permeability barrier, and some of its polysaccharides are disease-causing toxins.

Enclosing the cell wall and outer membrane in some bacteria is a layer of slime, composed mostly of polysaccharides and referred to as a **capsule**. The capsules of some bacteria may protect them from attack by white blood cells in the animals they infect. The capsule helps keep the cell from drying out, and sometimes it traps other cells for the bacterium to attack. Many prokaryotes produce no capsule, and those that do have capsules can survive even if they lose them, so the capsule is not essential to cell life.

Some groups of bacteria—the cyanobacteria and some others—carry on photosynthesis. In *photosynthesis*, the energy of sunlight is converted to chemical energy that can be used for a variety of energy-requiring reactions, such as the synthesis of cellular proteins and DNA. In these photosynthetic bacteria, the plasma membrane folds into the cytoplasm to form an internal membrane system that contains bacterial chlorophyll and other compounds needed for photosynthesis (Figure 4.5).

Other groups of prokaryotes possess different types of membranous structures called **mesosomes**, which may function in cell division or in various energy-releasing reactions. Like the photosynthetic membrane systems, mesosomes are formed by infolding of the plasma membrane. They remain attached to the plasma membrane and never form the free-floating, separate membranous organelles that are characteristic of eukaryotic cells.

Some prokaryotes swim by using appendages called **flagella** (Figure 4.6*a*). A single flagellum, made of a protein called flagellin, looks at times like a tiny corkscrew. It spins on its axis like a propeller, driving the cell along. Ring structures anchor the flagellum to the plasma membrane and, in some bacteria, to the outer membrane of the cell wall (Figure 4.6*b*). We know that the flagella cause the motion of the cell because if they are removed, the cell cannot move.

Pili project from the surface of some groups of bacteria (Figure 4.6*c*). Shorter than flagella, these threadlike structures help bacteria adhere to one another during mating, as well as to animal cells for protection and food.

Eukaryotic Cells

Animals, plants, fungi, and protists have cells that are usually larger and structurally more complex than those of the prokaryotes (Figure 4.7). To get a sense of the most promi-

4.6 Prokaryotic Projections
Surface projections such as these bacterial flagella (*a*, *b*) and pili (*c*) contribute to movement, to adhesion, and to the complexity of prokaryotic cells.

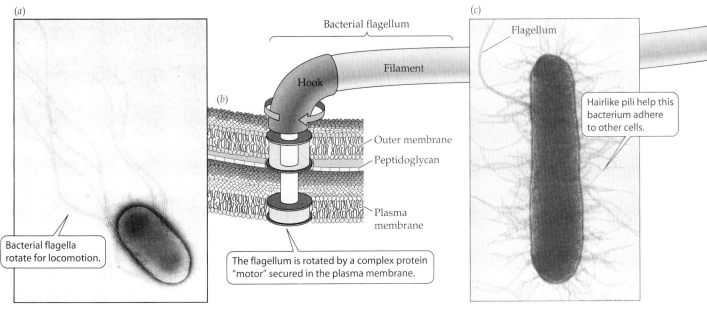

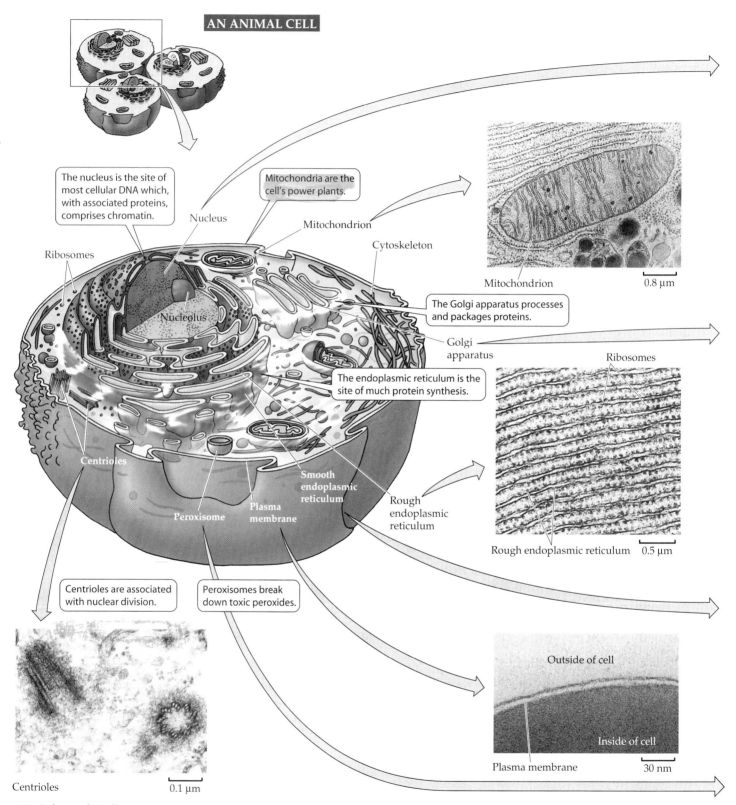

AN ANIMAL CELL

The nucleus is the site of most cellular DNA which, with associated proteins, comprises chromatin.

Mitochondria are the cell's power plants.

Nucleus

Mitochondrion

Cytoskeleton

Ribosomes

Nucleolus

Mitochondrion

0.8 μm

The Golgi apparatus processes and packages proteins.

Golgi apparatus

Ribosomes

The endoplasmic reticulum is the site of much protein synthesis.

Centrioles

Smooth endoplasmic reticulum

Peroxisome

Plasma membrane

Rough endoplasmic reticulum

Rough endoplasmic reticulum   0.5 μm

Centrioles are associated with nuclear division.

Peroxisomes break down toxic peroxides.

Outside of cell

Inside of cell

Plasma membrane

30 nm

Centrioles   0.1 μm

4.7 Eukaryotic Cells

In electron micrographs, many plant cell organelles are nearly identical in form to those observed in animal cells. Cellular structures unique to plant cells include the cell wall and the chloroplasts. Animal cells contain centrioles, which are not found in plant cells.

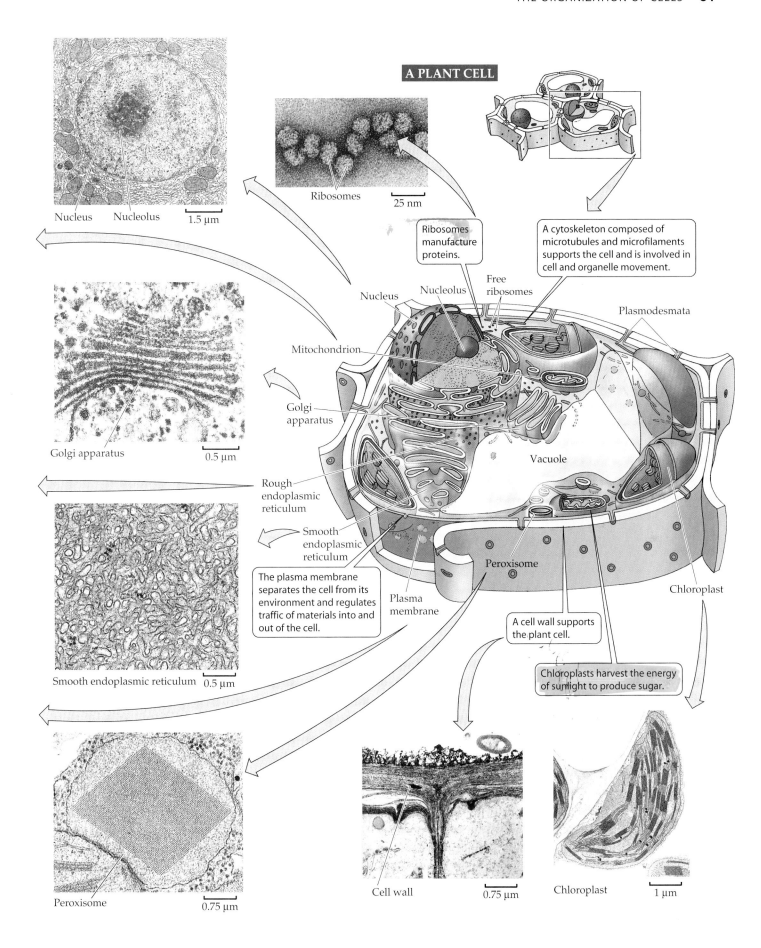

Nucleus Nucleolus
1.5 μm

Ribosomes
25 nm

A PLANT CELL

Golgi apparatus
0.5 μm

Smooth endoplasmic reticulum 0.5 μm

Peroxisome
0.75 μm

Cell wall 0.75 μm

Chloroplast 1 μm

Ribosomes manufacture proteins.

A cytoskeleton composed of microtubules and microfilaments supports the cell and is involved in cell and organelle movement.

Nucleus

Nucleolus

Free ribosomes

Plasmodesmata

Mitochondrion

Golgi apparatus

Vacuole

Rough endoplasmic reticulum

Smooth endoplasmic reticulum

Peroxisome

Chloroplast

The plasma membrane separates the cell from its environment and regulates traffic of materials into and out of the cell.

Plasma membrane

A cell wall supports the plant cell.

Chloroplasts harvest the energy of sunlight to produce sugar.

nent differences, compare the eukaryotic plant and animal cells on the preceding two pages with the prokaryotic cell in Figure 4.4.

Eukaryotic cells generally have dimensions ten times greater than those of prokaryotes; for example, the spherical yeast cell has a diameter of 8 μm. Like prokaryotic cells, eukaryotic cells have a plasma membrane, cytoplasm, and ribosomes. But added on to this basic organization are two elements not found in prokaryotes:

▶ An internal **cytoskeleton** that maintains cell shape and moves materials
▶ **Membranous compartments** in the cytoplasm whose interiors are separated from the cytosol by a membrane

Compartmentalization is the key to eukaryotic cell function

Recall that prokaryotic cells are surrounded by a plasma membrane that regulates molecular traffic into and out of the cell. In addition, eukaryotic cells have "cells within cells"—interior compartments surrounded by membranes that regulate what enters or leaves that compartment. The membranes ensure that conditions inside the compartment are different from those in the surrounding cytoplasm.

Some of the compartments are like little factories that make specific products. Others are like power plants that take in energy in one form and convert it to a more useful form. These membranous compartments, as well as other structures (such as ribosomes) that lack membranes but possess distinctive shapes and functions, are called **organelles** (see Figure 4.7). Each of these organelles has specific roles in its particular cell. These roles are defined by chemical reactions.

4.8 *The Nucleus Is Enclosed by a Double Membrane*
The electron micrograph shows the nucleus of a nondividing animal cell. The double-membraned nuclear envelope, nucleolus, nuclear lamina, and nuclear pores are common features of all cell nuclei.

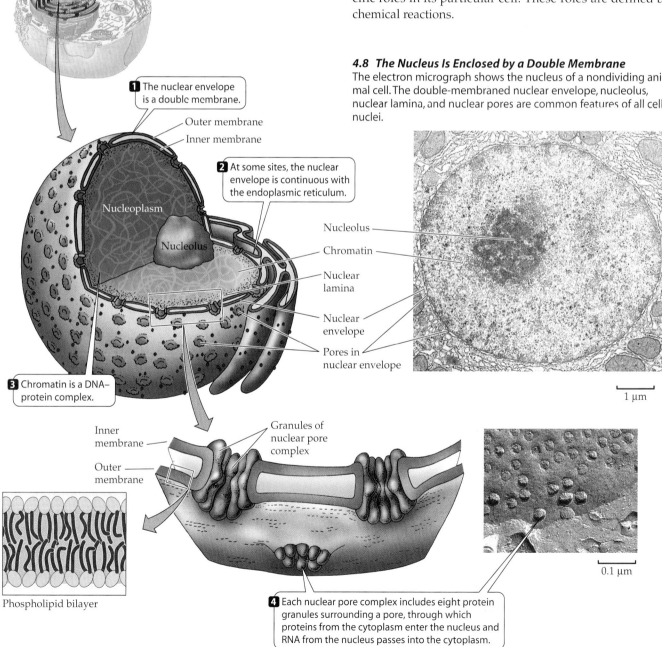

1 The nuclear envelope is a double membrane.

Outer membrane
Inner membrane

2 At some sites, the nuclear envelope is continuous with the endoplasmic reticulum.

Nucleoplasm

Nucleolus

3 Chromatin is a DNA–protein complex.

Nucleolus
Chromatin
Nuclear lamina
Nuclear envelope
Pores in nuclear envelope

1 μm

Inner membrane
Outer membrane

Granules of nuclear pore complex

Phospholipid bilayer

0.1 μm

4 Each nuclear pore complex includes eight protein granules surrounding a pore, through which proteins from the cytoplasm enter the nucleus and RNA from the nucleus passes into the cytoplasm.

▶ The **nucleus** contains most of the cell's genetic material (DNA). It determines the expression of this material as cell functions and its duplication when the cell reproduces.

▶ The **mitochondrion** is a power plant and industrial park, where energy stored in the bonds of carbohydrates is converted to a form more useful to the cell and certain essential biochemical conversions of amino acids and fatty acids occur.

▶ The **endoplasmic reticulum** and **Golgi apparatus** make up a compartment where proteins are packaged and sent to appropriate locations in the cell.

▶ The **lysosome** and **vacuole** are cellular digestive systems, where large molecules are hydrolyzed into usable monomers.

▶ The **chloroplast** performs photosynthesis.

All of these organelles have unique chemical compositions and functions. The membrane surrounding each does two essential things: First, it keeps the organelle's molecules away from other molecules in the cell with which they might react inappropriately. Second, it acts as a traffic regulator, letting important raw materials into the organelle and releasing its products to the cytoplasm.

Organelles that Process Information

Living things depend on accurate, appropriate information—internal signals, environmental cues, and stored instructions—to respond appropriately to changing conditions and maintain a constant internal environment. In the cell, information is *stored* as the sequence of nucleotides in DNA molecules. Most DNA in eukaryotic cells resides in the nucleus. Information is *translated* from the language of DNA into the language of proteins at the ribosomes. This process is described in detail in Chapter 12.

The nucleus stores most of the cell's DNA

The single nucleus is usually the largest organelle in a cell (Figure 4.8; see also Figure 4.7). The nucleus of most animal cells is approximately 5 μm in diameter—substantially larger than most entire prokaryotic cells. The nucleus has several roles in the cell:

▶ The nucleus is the site of DNA duplication to support cell reproduction.

▶ The nucleus is the site of DNA control of cellular activities.

▶ A region within the nucleus, called the **nucleolus**, begins the assembly of ribosomes from specific proteins and RNA.

The nucleus is surrounded by *two* membranes, which together form the **nuclear envelope**. The two membranes of the nuclear envelope are separated by only a few tens of nanometers and are perforated by **nuclear pores** approximately 9 nm in diameter, which connect the interior of the nucleus with the cytoplasm. At these pores, the outer membrane of the nuclear envelope is continuous with the inner membrane. Each pore is surrounded by a *pore complex*: eight large protein granules arranged in an octagon where the inner and outer membranes merge. RNA and proteins pass through these pores to enter or leave the nucleus.

At certain sites, the outer membrane of the nuclear envelope folds outward into the cytoplasm and is continuous with the membrane of another organelle, the endoplasmic reticulum (discussed later in the chapter).

Inside the nucleus, DNA combines with proteins to form a fibrous complex called **chromatin**. These are exceedingly long, thin, entangled threads that, prior to cell division, condense to form readily visible objects called **chromosomes** (Figure 4.9).

Surrounding the chromatin are water and dissolved substances collectively referred to as the **nucleoplasm**. Within the nucleoplasm, a network of apparently structural proteins called the *nuclear matrix* organizes the chromatin. At the periphery of the nucleus, the chromatin attaches to a protein meshwork, called the **nuclear lamina**, which is formed by the polymerization of proteins called *lamins* into filaments (Figure 4.10). The nuclear lamina maintains the

4.9 Chromatin and Chromosomes
(*a*) When a cell is not dividing, the nuclear DNA and proteins are aggregated as chromatin, which is dispersed throughout the nucleus. (*b*) The chromatin in a dividing cell is packed into dense bodies called chromosomes.

(*a*)

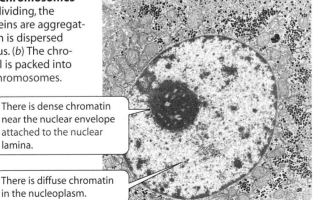

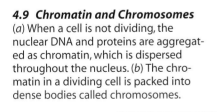

There is dense chromatin near the nuclear envelope attached to the nuclear lamina.

There is diffuse chromatin in the nucleoplasm.

1 μm

(*b*)

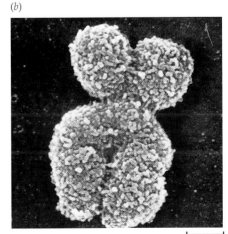

0.5 μm

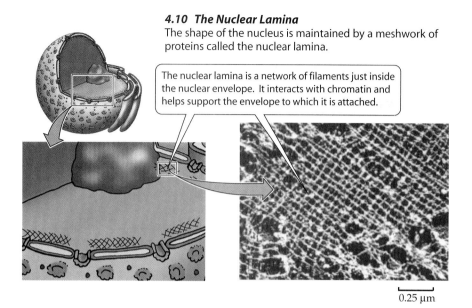

4.10 The Nuclear Lamina
The shape of the nucleus is maintained by a meshwork of proteins called the nuclear lamina.

The nuclear lamina is a network of filaments just inside the nuclear envelope. It interacts with chromatin and helps support the envelope to which it is attached.

0.25 μm

shape of the nucleus by its attachment to both chromatin and the nuclear envelope.

During most of the life cycle of the cell, the nuclear envelope is a stable structure. When the cell divides, however, the nuclear envelope fragments into pieces of membrane with attached pore complexes. The envelope re-forms when distribution of the duplicated DNA to the daughter cells is completed.

Ribosomes are the sites of protein synthesis

In both eukaryotic and prokaryotic cells, proteins are synthesized on thousands of ribosomes. Ribosomes are tiny granules found in three places in almost all eukaryotic cells: free in the cytoplasm, attached to the surface of endoplasmic reticulum (as will be described later in this chapter), and inside the mitochondria, where energy is processed. Ribosomes are also found in chloroplasts, the photosynthetic organelles of plant cells. In each of these locations, the ribosomes provide the sites where proteins are synthesized under the direction of nucleic acids. Although they seem small in comparison to the cell in which they are contained, ribosomes are huge machines composed of several dozen kinds of molecules.

The ribosomes of prokaryotes and of eukaryotes are similar in that both consist of two different-sized subunits. Eukaryotic ribosomes are somewhat larger, but the structure of prokaryotic ribosomes is better understood. Chemically, ribosomes consist of a special type of RNA, called *ribosomal RNA*, to which more than 50 different protein molecules are noncovalently bound.

The Endomembrane System

Much of the volume of some eukaryotic cells is taken up by an extensive **endomembrane system**. This system includes two main components, the endoplasmic reticulum and the Golgi apparatus. Continuities between the nuclear envelope and the endomembrane system are visible by electron microscopy. Tiny vesicles appear to shuttle between the various components of the endomembrane system. This system has various structures, but all of them are essentially compartments, closed off by their membranes from the cytoplasm.

In this section, we will examine the functional significance of these compartments, and show that materials synthesized in the endoplasmic reticulum can be transferred to another organelle, the Golgi apparatus, for further processing, storage, or transport. We will also describe the role of the lysosome in cell digestion.

The endoplasmic reticulum is a complex factory

Electron micrographs reveal a network of interconnected membranes branching throughout the cytoplasm, forming tubes and flattened sacs. These membranes are collectively called the **endoplasmic reticulum**, or **ER**. The interior compartment of the ER, referred to as the *lumen*, is separate and distinct from the surrounding cytoplasm (Figure 4.11). The surface area of the ER can occupy up to 15 percent of the entire interior volume of the cell, and its foldings result in a surface area many times greater than that of the plasma membrane. At certain sites, the ER is continuous with the outer membrane of the nuclear envelope.

Parts of the ER are liberally sprinkled with ribosomes, which are temporarily attached to the outer faces of the flattened sacs. Because of their appearance in the electron microscope, these regions are called **rough ER**, or **RER**. RER has two roles:

▸ As a compartment, it segregates certain newly synthesized proteins away from the cytoplasm and transports them to other locations in the cell.
▸ While inside the RER, proteins can be chemically modified so as to alter their function and intracellular destination.

The attached ribosomes are sites for the synthesis of proteins that function outside the cytosol—that is, proteins that are to be exported from the cell, incorporated into membranes, or moved into organelles of the endomembrane system. These proteins enter the lumen of the ER as they are synthesized. Once in the lumen of the ER, these proteins undergo several changes, including the formation of disulfide bridges and folding into their tertiary structures (see Figure 3.5). Proteins gain carbohydrate groups in the RER, thus becoming glycoproteins. The carbohydrate groups are part of an "addressing" system that ensures that the right proteins are directed to the right parts of the cell.

Some parts of the endoplasmic reticulum, called the **smooth ER** or **SER**, are more tubular (less like flattened sacs) and lack ribosomes (see Figure 4.11). Within the lumen of the SER, proteins that have been synthesized on the RER are chemically modified. In addition, the SER has two other important roles:

▶ It is responsible for chemically modifying small molecules taken in by the cell. This is especially true for drugs and pesticides.
▶ It is the site for the hydrolysis of glycogen and the synthesis of steroids.

Cells that synthesize a lot of protein for export are usually packed with ER. Examples include glandular cells that secrete digestive enzymes and plasma cells that secrete antibodies. In contrast, cells that carry out less protein synthesis (such as storage cells) contain less ER.

The Golgi apparatus stores, modifies, and packages proteins

In 1898, the Italian microscopist Camillo Golgi discovered a delicate structure in nerve cells, which came to be known as the **Golgi apparatus**. Because of the resolution limits of light microscopy, and because the staining techniques of the time often failed to reveal the structure, many biologists regarded it as a product of Golgi's imagination. In the late 1950s, however, the electron microscope showed clearly that the Golgi apparatus does exist—and not just in nerve cells, but in most eukaryotic cells.

The exact appearance of the Golgi apparatus varies from species to species, but it always consists of flattened membranous sacs called *cisternae* and small membrane-enclosed *vesicles*. The cisternae appear to be lying together like a stack of saucers (Figure 4.12*a*).The entire apparatus is about 1 μm long.

The Golgi apparatus has several roles:

▶ It receives proteins from the ER and chemically modifies them.
▶ Proteins within the Golgi apparatus are concentrated, packaged, and sorted before being sent to their cellular or extracellular destinations.
▶ The Golgi apparatus is where some polysaccharides for the plant cell wall are synthesized.

In the cells of plants, protists, fungi, and many invertebrate animals, the stacks of cisternae are individual units scattered throughout the cytoplasm. In vertebrate cells, a few such stacks usually form a larger, single, more complex Golgi apparatus.

The Golgi appears to have three functionally distinct parts: a bottom, a middle, and a top. The bottom cisternae, constituting the *cis* region of the Golgi apparatus, lie nearest to the nucleus or a patch of RER (see Figure 4.12). The top cisternae, constituting the *trans* region, lie closest to the surface of the cell. The cisternae in the middle make up the *medial* region of the complex. These three parts of the Golgi apparatus contain different enzymes and perform different functions.

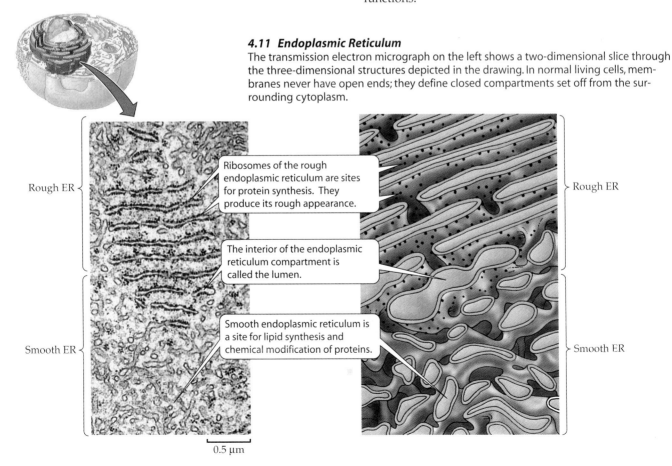

4.11 Endoplasmic Reticulum
The transmission electron micrograph on the left shows a two-dimensional slice through the three-dimensional structures depicted in the drawing. In normal living cells, membranes never have open ends; they define closed compartments set off from the surrounding cytoplasm.

Rough ER

Ribosomes of the rough endoplasmic reticulum are sites for protein synthesis. They produce its rough appearance.

The interior of the endoplasmic reticulum compartment is called the lumen.

Smooth endoplasmic reticulum is a site for lipid synthesis and chemical modification of proteins.

Smooth ER

Rough ER

Smooth ER

0.5 μm

The Golgi apparatus receives proteins from the ER, packages them, and sends them on their way. The chemical modifications made to proteins within the Golgi apparatus generally "tag" them to their proper destinations—a process we will describe further in Chapter 12. So in some sense the Golgi apparatus is a "post office" for the cell.

Since there is often no direct membrane continuity between ER and Golgi apparatus, how does a protein get from one organelle to the other? The protein could simply leave the ER, travel across the cytoplasm, and enter the Golgi apparatus. But this would expose the protein to interactions with other molecules in the cytoplasm. On the other hand, segregation from the cytoplasm could be maintained if a piece of the ER could "bud off," forming a vesicle that contains the target protein—and this is in fact exactly what happens. The protein makes the passage from ER to Golgi apparatus safely enclosed in the vesicle. Once it arrives, the vesicle fuses with the membrane of the Golgi apparatus, releasing its cargo .

Vesicles form from the rough ER, move through the cytoplasm, and fuse with the *cis* region of the Golgi appara-

tus, where their contents are released into the lumen of the Golgi. Other small vesicles may move between the cisternae, transporting proteins. Associated with the cisternae, particularly those toward the *trans* region, are tiny vesicles that pinch off and move to other cisternae or away from the Golgi (see Figure 4.12b).

The membranes of two vesicles can sometimes make contact with each other and fuse, resulting in a larger vesicle and a mixing of the contents. Vesicles may also fuse with other organelles, or with the plasma membrane, where they release their contents to the outside of the cell. The formation, transport, and fusing behavior of vesicles is essential to the function of the Golgi apparatus. Structurally, *vesicles are the transport vehicles into and out of the Golgi apparatus and to the ultimate destinations of the proteins.*

Lysosomes contain digestive enzymes

Originating in part from the Golgi apparatus are organelles called **lysosomes**. They contain digestive enzymes, and they are the sites of hydrolysis of macromolecules—proteins, polysaccharides, nucleic acids, and lipids—to their monomers (see Figure 3.2). Lysosomes are about 1 μm in diameter, are surrounded by a single membrane, and have a densely staining, featureless interior (Figure 4.13a). There may be dozens of lysosomes in a cell, depending on its needs.

Lysosomes are sites for the breakdown of food and foreign objects taken up by the cell. How do these materials get into the cell in the first place? In a process called *phagocytosis* (phago-, "eating"; cytosis, "cellular"), a pocket forms in the plasma membrane and eventually deepens and en-

(a)

Nucleus ER GA

0.25 μm

4.12 The Golgi Apparatus
(a) The Golgi apparatus appears as stacked disks in an electron micrograph. (b) The Golgi apparatus modifies proteins from the ER and "targets" them to the correct addresses.

(b)

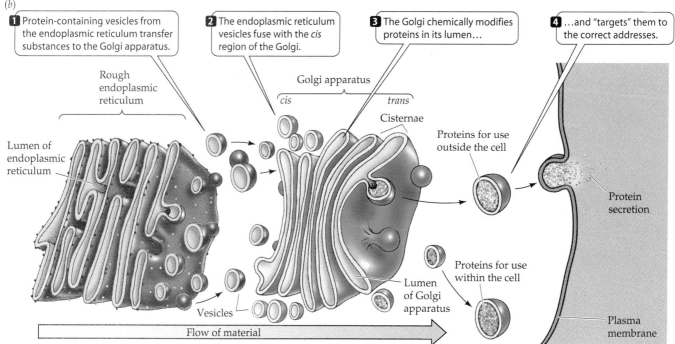

1 Protein-containing vesicles from the endoplasmic reticulum transfer substances to the Golgi apparatus.

2 The endoplasmic reticulum vesicles fuse with the *cis* region of the Golgi.

3 The Golgi chemically modifies proteins in its lumen…

4 …and "targets" them to the correct addresses.

Rough endoplasmic reticulum

Golgi apparatus

cis

trans

Cisternae

Lumen of endoplasmic reticulum

Proteins for use outside the cell

Proteins for use within the cell

Lumen of Golgi apparatus

Vesicles

Flow of material

Protein secretion

Plasma membrane

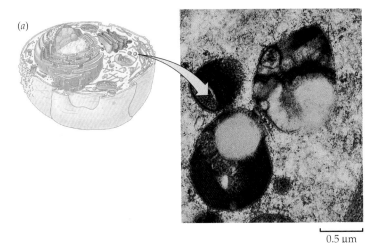

(a)

4.13 Lysosomes Isolate Digestive Enzymes from the Cytoplasm
(a) In this electron micrograph of a rat cell, the darkly stained organelles are secondary lysosomes in which digestion is taking place. (b) The origin and action of lysosomes and lysosomal digestion.

0.5 μm

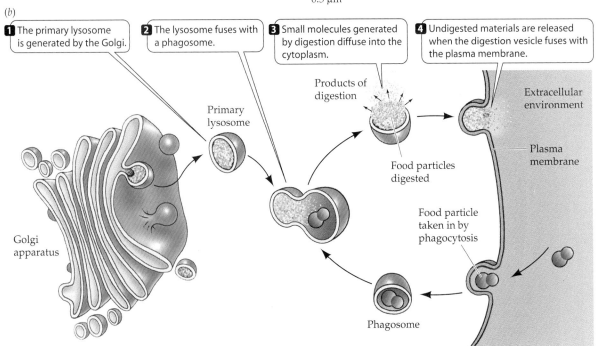

(b)

1 The primary lysosome is generated by the Golgi.

2 The lysosome fuses with a phagosome.

3 Small molecules generated by digestion diffuse into the cytoplasm.

4 Undigested materials are released when the digestion vesicle fuses with the plasma membrane.

Products of digestion

Extracellular environment

Primary lysosome

Plasma membrane

Food particles digested

Food particle taken in by phagocytosis

Golgi apparatus

Phagosome

closes material from outside the cell. This pocket becomes a small vesicle and breaks free of the plasma membrane to move into the cytoplasm as a *phagosome* containing food or other material (Figure 4.13*b*). The phagosome fuses with a *primary lysosome* to form a *secondary lysosome*, where digestion occurs.

The effect of this fusion is rather like releasing hungry foxes into a chicken coop. The enzymes in the secondary lysosome quickly hydrolyze the food particles. These reactions are enhanced by the mild acidity of the lysosome's interior, where the pH is lower than in the surrounding cytoplasm. The products of digestion exit through the membrane of the lysosome, providing fuel molecules and raw materials for other cell processes. The "used" secondary lysosome containing undigested particles then moves to the plasma membrane, fuses with it, and releases the undigested contents to the environment.

Lysosomes are also where the cell digests its own material in a process called *autophagy*. Autophagy is an ongoing process, in which macromolecules such as proteins are en-

gulfed by lysosomes and hydrolyzed to amino acids, which pass out of the lysosome through its membrane into the cytoplasm for reuse.

Plant cells do not appear to contain lysosomes, but the central vacuole of a plant cell may function in an equivalent capacity because it, like lysosomes, contains many digestive enzymes.

Organelles that Process Energy

A cell uses energy to transform raw materials into cell-specific materials that it can use for activities such as growth, reproduction, and movement. Energy is transformed from one form to another in mitochondria (found in all eukaryotic cells) and in chloroplasts (found in eukaryotic cells that harvest energy from sunlight). In contrast, energy transformations in prokaryotic cells are associated with enzymes attached to the inner surface of the plasma membrane or extensions of the plasma membrane that protrude into the cytoplasm.

Mitochondria are energy transformers

In eukaryotic cells, the utilization of food molecules such as glucose begins in the cytosol. The fuel molecules that result from partial degradation of this food enter the **mitochondria** (singular mitochondrion), whose primary function is to *convert the potential chemical energy of fuel molecules into a form that the cell can use*: the energy-rich molecule called *ATP*, or *adenosine triphosphate*. ATP is not a long-term energy storage form, but rather a kind of energy currency. Its role in the cell is analogous to the role of paper money in an economy. Chemically, ATP can participate in a great number of different cellular reactions and processes that require energy. In the mitochondria, the production of ATP using fuel molecules and O_2 is called *cellular respiration*.

Typical mitochondria are small—somewhat less than 1.5 µm in diameter and 2–8 µm in length—about the size of many bacteria. Mitochondria are visible with a light microscope, but almost nothing was known of their precise structure until they were examined with the electron microscope. Electron micrographs revealed that mitochondria have two membranes. The *outer membrane* is smooth and protective, and it offers little resistance to the movement of substances into and out of the mitochondrion. Immediately inside the outer mitochondrial membrane is an *inner membrane*, which folds inward in many places, giving it a much greater surface area than that of the outer membrane (Figure 4.14). These folds tend to be quite regular, giving rise to shelflike structures called **cristae**.

The inner mitochondrial membrane contains many large protein molecules that participate in cellular respiration and the production of ATP. The inner membrane exerts much more control over what enters and leaves the mitochondrion than does the outer membrane. The region enclosed by the inner membrane is referred to as the **mitochondrial matrix**. In addition to many proteins, the matrix contains some ribosomes and DNA that are used to make some of the proteins needed for cellular respiration.

The number of mitochondria per cell ranges from one contorted giant in some unicellular protists to a few hundred thousand in large egg cells. An average human liver cell contains more than a thousand mitochondria. Cells that require the most chemical energy tend to have the most mitochondria per unit of volume. In Chapter 7 we will see how the different parts of the mitochondrion work together in cellular respiration.

Plastids photosynthesize or store materials

One class of organelles—the **plastids**—is produced only in plants and certain protists. There are several types of plastids, with different functions.

CHLOROPLASTS. The most familiar of the plastids is the **chloroplast**, which contains the green pigment chlorophyll and is the site of photosynthesis (Figure 4.15). In photosynthesis, light energy is converted into the chemical energy of bonds between atoms. The molecules formed in photosynthesis provide food for plants themselves and for other

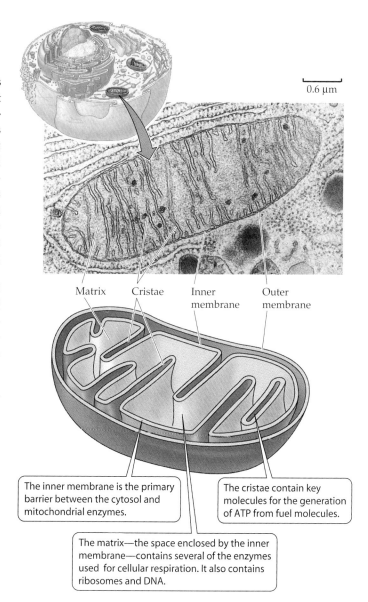

0.6 µm

Matrix | Cristae | Inner membrane | Outer membrane

The inner membrane is the primary barrier between the cytosol and mitochondrial enzymes.

The cristae contain key molecules for the generation of ATP from fuel molecules.

The matrix—the space enclosed by the inner membrane—contains several of the enzymes used for cellular respiration. It also contains ribosomes and DNA.

4.14 A Mitochondrion Converts Energy from Fuel Molecules into ATP
The electron micrograph is a two-dimensional slice through a three-dimensional reality. As the drawing emphasizes, the cristae are extensions of the inner mitochondrial membrane.

organisms that eat plants. Directly or indirectly, photosynthesis is the energy source for most of the living world.

Chloroplasts are quite variable in size and shape (Figure 4.16a,b). Like the mitochondrion, the chloroplast is surrounded by two membranes. Arising from the inner membrane is a series of discrete internal membranes whose structure and arrangement vary from one group of photosynthetic organisms to another. Here we concentrate on the chloroplasts of the flowering plants. Even these show some variation, but the pattern shown in Figure 4.15 is typical.

As seen in electron micrographs, the internal membranes of chloroplasts look like stacks of flat, hollow pita bread. These stacks, called **grana** (singular granum), consist of a series of flat, closely packed, circular compartments called **thylakoids**. In addition to phospholipids and proteins, the membranes of the thylakoids contain molecules

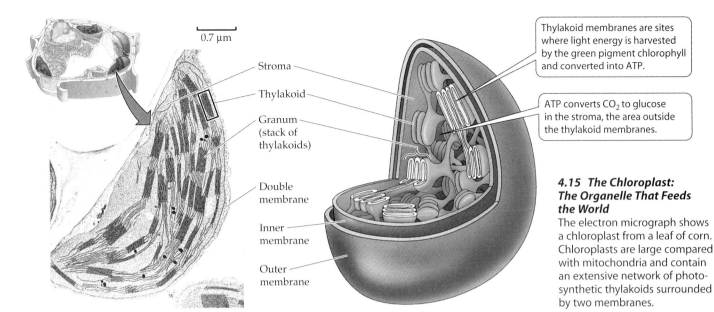

0.7 µm

Stroma
Thylakoid
Granum (stack of thylakoids)
Double membrane
Inner membrane
Outer membrane

Thylakoid membranes are sites where light energy is harvested by the green pigment chlorophyll and converted into ATP.

ATP converts CO_2 to glucose in the stroma, the area outside the thylakoid membranes.

4.15 The Chloroplast: The Organelle That Feeds the World
The electron micrograph shows a chloroplast from a leaf of corn. Chloroplasts are large compared with mitochondria and contain an extensive network of photosynthetic thylakoids surrounded by two membranes.

of the green pigment chlorophyll and the yellow-orange carotenoids. These two pigment families harvest light for photosynthesis. Thylakoids of one granum may be connected to those of other grana, making the interior of the chloroplast a highly developed network of membranes, much like the ER.

The fluid in which the grana are suspended is referred to as **stroma**. Like the mitochondrial matrix, the chloroplast stroma contains ribosomes and DNA, and these are used to synthesize some, but not all, of the proteins that make up the chloroplast.

Animal cells do not *produce* chloroplasts, but some do *contain* functional chloroplasts. These are either taken up as free chloroplasts derived from the partial digestion of green plants, or contained within unicellular algae that live within the animal's tissues. The green color of some corals and sea anemones results from chloroplasts in algae that live within those animals (Figure 4.16c). The animals derive some of

their nutrition from the photosynthesis that their chloroplast-containing "guests" carry out. Such an intimate relationship between two different organisms is called *symbiosis*.

OTHER TYPES OF PLASTIDS. The red color of a flower or a ripe tomato results from the presence of legions of plastids called **chromoplasts** (Figure 4.17a). Just as chloroplasts derive their color from the pigment chlorophyll, chromoplasts are red, orange, or yellow depending on the kinds of carotenoid pigments present. The chromoplasts have no known chemical function in the cell, but the colors they give to some petals and fruits probably help attract animals that assist in pollination or seed dispersal. (On the other hand, carrot roots gain no apparent advantage from being orange.)

4.16 Being Green
(a) In green plants, chloroplasts are concentrated in the leaf cells.
(b) Green algae are photosynthetic and filled with chloroplasts.
(c) No animal species produces its own chloroplasts, but in this symbiotic arrangement, unicellular green algae nourish a giant sea anemone.

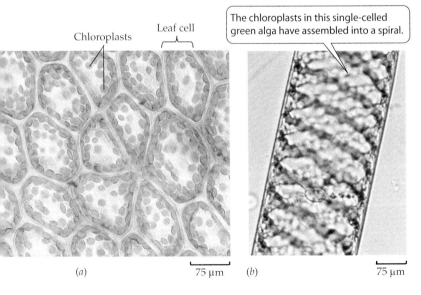

Chloroplasts
Leaf cell

The chloroplasts in this single-celled green alga have assembled into a spiral.

(a)   75 µm   (b)   75 µm

Chloroplast-filled green algae live in the tissues of this sea anemone.

(c)

(a)

4.17 Chromoplasts and Leucoplasts
(a) Colorful pigments stored in the chromoplasts of flowers like this begonia may help attract pollinating insects. (b) Leucoplasts in the cells of a potato are filled with white starch grains.

5 µm

Leucoplast

Starch grains

1 µm

Other plastids, called **leucoplasts**, are storage depots for starch and fats (Figure 4.17b).

Mitochondria and chloroplasts may have an endosymbiotic origin

Chloroplasts and mitochondria are about the size of prokaryotic cells. They contain DNA and have ribosomes that are similar to prokaryotic ribosomes, and they reproduce and divide within the cell to produce additional mitochondria and chloroplasts.

But, these organelles, even though they have the genetic material and protein synthesis machinery needed to make some of their own components, are not independent of control by the nucleus. The vast majority of their proteins are encoded by nuclear DNA, made in the cytoplasm, and imported into the organelle. These observations have led to speculation on the origin of these organelles. One proposal for this origin is the **endosymbiosis theory** of the origin of mitochondria and chloroplasts, which envisions the following scenario.

About 2 billion years ago, only prokaryotes inhabited Earth. Some of them absorbed their food directly from the environment. Others were photosynthetic. Still others fed on smaller prokaryotes by engulfing them (Figure 4.18).

Suppose that a small, photosynthetic prokaryote was *in*gested by a larger one, but was not *di*gested. Instead, it survived trapped within a vesicle in the cytoplasm of the larger cell. The smaller, ingested prokaryote divided at about the same rate as the larger one, so successive generations of the larger cell also contained the offspring of the smaller one. We call this phenomenon *endosymbiosis* (endo-, "within"; symbiosis, "living together"); it is comparable to the algae that live within sea anemones (see Figure 4.16c).

According to this scenario, endosymbiosis provided benefits for both organisms. The larger cell obtained the photosynthetic products from the smaller cell, and the smaller cell was protected by the larger one. The smaller cell gradually lost much of its DNA to the nucleus, resulting in the modern chloroplast.

Much circumstantial evidence favors the endosymbiosis theory. Chloroplast DNA sequences are more like certain prokaryotic sequences than like any plant DNA. Moreover, on an evolutionary time scale of millions of years, there is evidence for DNA moving between organelles in a cell. Finally, there are many biochemical similarities between chloroplasts and modern bacteria.

Similar evidence and arguments also support the proposition that mitochondria are the descendants of respiring prokaryotes engulfed by larger prokaryotes. The benefits of this endosymbiotic relationship might have been due to the capacity of the engulfed prokaryote to detoxify molecular oxygen (O_2), which was increasing in Earth's atmosphere because of photosynthesis.

However, mitochondria and chloroplasts are not enough to turn a prokaryote into a eukaryote. The endosymbiosis theory is still incomplete. For example, the origins of the nuclear envelope and other important structures—including those responsible for nuclear division—still need to be understood. We discuss further aspects of the origin of the eukaryotic cell in Chapter 27.

Membrane of larger cell

Double membranes may have originated when one cell engulfed another.

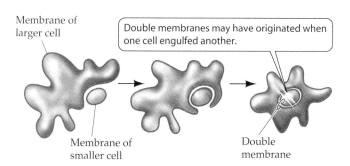

Membrane of smaller cell

Double membrane

4.18 The Endosymbiosis Theory
The double membrane that encloses mitochondria and chloroplasts may have arisen from two different sources: the outer membrane from the engulfing cell's plasma membrane and the inner membrane from the engulfed cell's plasma membrane.

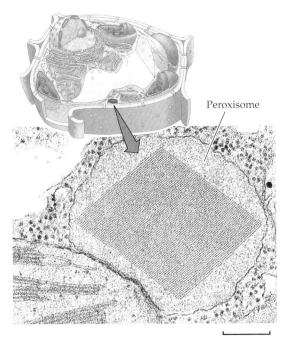

4.19 A Peroxisome
A diamond-shaped crystal, composed of an enzyme, almost entirely fills this rounded peroxisome in a leaf cell. The enzyme catalyzes one of the reactions fulfilling the special function of the peroxisome.

Other Organelles

In addition to the information-processing organelles (nucleus and ribosomes), the energy-processing organelles (mitochondria and chloroplasts), and the organelles of the endomembrane system (endoplasmic reticulum, Golgi apparatus, and lysosomes), there are two other kinds of membrane-enclosed organelles: peroxisomes and vacuoles. Both are surrounded by a single membrane.

Peroxisomes house specialized chemical reactions

Peroxisomes are small organelles—0.2 to 1.7 μm in diameter. They have a single membrane and a granular interior (Figure 4.19). Peroxisomes are found at one time or another in at least some of the cells of almost every eukaryotic species. Peroxisomes are organelles within which toxic peroxides (such as hydrogen peroxide, H_2O_2) are formed as unavoidable side products of chemical reactions. Subsequently, the peroxides are safely broken down within the peroxisomes without mixing with other parts of the cell.

A structurally similar organelle, the **glyoxysome**, is found only in plants. Glyoxysomes, which are most prominent in young plants, are the sites where stored lipids are converted into carbohydrates for transport to growing cells.

Vacuoles are filled with water and soluble substances

Many eukaryotic cells, but particularly those of plants and protists, contain membrane-enclosed organelles that look empty under the electron microscope. These organelles are called **vacuoles** (Figure 4.20). They are not actually empty; rather, they are filled with aqueous solutions that contain many dissolved substances.

Despite their structural simplicity, vacuoles have a variety of functions. For example, like animals and other organisms, plant cells produce a number of toxic by-products and waste materials. Animals have specialized excretory mechanisms for getting rid of such wastes, but plants do not. Although plants can secrete some wastes to their environment, many are simply stored within vacuoles. And since they are poisonous or distasteful, these stored materials deter some animals from eating the plants. Thus stored wastes may contribute to plant survival.

In many plant cells, enormous vacuoles take up more than 90 percent of the cell volume and grow as the cell grows. But vacuoles are by no means a waste of space, for the dissolved substances in the vacuole, working together with the vacuolar membrane, provide the *turgor*, or stiffness, of the cell, which in turn provides support for the structure of nonwoody plants. The presence of the dissolved substances causes water to enter the vacuole, making it tend to swell like a balloon. Plant cells have a rigid cell wall, which acts like a box, resisting the swelling of the vacuole but providing strength in the process.

Vacuoles even play a role in the sex life of plants. Some pigments (especially blue and pink ones) in petals and fruits are contained in vacuoles. These pigments—the anthocyanins—are visual cues that encourage animals to visit flowers and thus aid in pollination, or to eat fruits and thus aid in seed dispersal.

Food vacuoles are found in some simple and evolutionarily ancient groups of organisms: single-celled protists and simple multicellular organisms such as sponges. In these organisms, the cells engulf food particles by phagocytosis, generating a food vacuole. Fusion of this vacuole with a

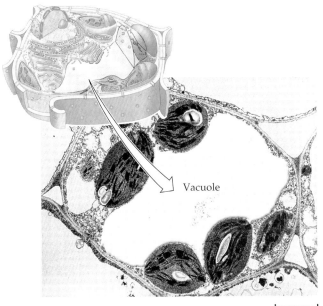

4.20 Vacuoles in Plant Cells Are Usually Large
The large central vacuole in this cell is typical of mature plant cells. Smaller vacuoles are visible toward each end of the cell.

4.21 The Cytoskeleton

Three highly visible and important structural components of the cytoskeleton are shown in detail. These structures maintain and reinforce cell shape, and contribute to cell movement.

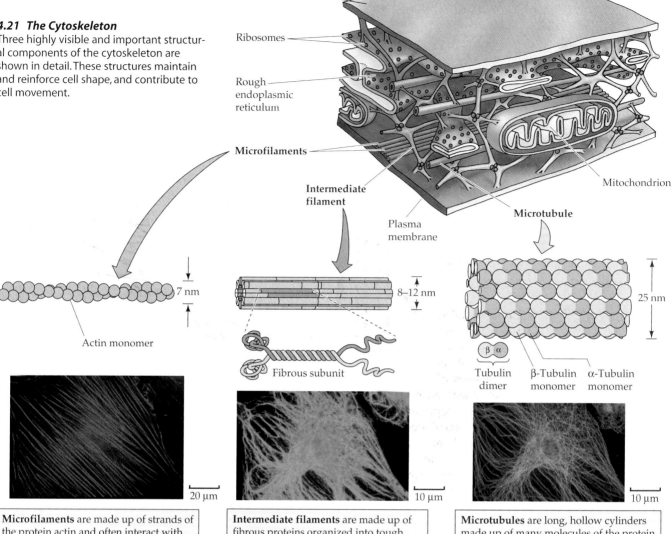

Microfilaments are made up of strands of the protein actin and often interact with strands of other proteins. They change cell shape and drive cellular motion, including contraction, cytoplasmic streaming, and the "pinched" shape changes that occur during cell division. Microfilaments and myosin strands together drive muscle action.

Intermediate filaments are made up of fibrous proteins organized into tough, ropelike assemblages that stabilize a cell's structure and help maintain its shape. Some intermediate filaments hold neighboring cells together. Others make up the nuclear lamina.

Microtubules are long, hollow cylinders made up of many molecules of the protein tubulin. Tubulin consists of two subunits, α-tubulin and β-tubulin. Microtubules lengthen or shorten by adding or subtracting tubulin dimers. Microtubule shortening moves chromosomes. Interactions between microtubules drive the movement of cells. Microtubules serve as "tracks" for the movement of vesicles.

lysosome results in digestion, and small molecules leave the vacuole and enter the cytoplasm for use or distribution to other organelles.

Many freshwater protists have a highly specialized *contractile vacuole*. Its function is to rid the cell of the excess water that rushes in because of the imbalance in salt concentration between the relatively salty interior of the cell and its freshwater environment. The contractile vacuole enlarges as water enters, then abruptly contracts, forcing the water out of the cell through a special pore structure.

The Cytoskeleton

In addition to the many membrane-enclosed organelles, the eukaryotic cytoplasm has a set of long, thin fibers called the **cytoskeleton**, which fills at least three important roles:

▶ It maintains cell shape and support.
▶ It provides for various types of cell movement.
▶ Some of its fibers act as tracks or supports for "motor proteins," which help the cell move or move things within the cell.

In the discussion that follows, we'll look at three components of the cytoskeleton: microfilaments, intermediate filaments, and microtubules (Figure 4.21).

Microfilaments function in support and movement

Microfilaments can exist as single filaments, in bundles, or in networks. They are about 7 nm in diameter and several μm long. They are assembled from **actin**, a protein that exists in several forms and has many functions among members of the animal phyla. The actin found in microfilaments (which are also known as actin filaments) is extensively folded and has distinct "head" and "tail" sites. These sites interact with similar actin molecules to assemble into a long chain (see Figure 4.21). Two of these chains interact to form the double helix of a microfilament. The polymerization of actin into microfilaments is reversible, and they can disappear from cells, breaking down into units of free actin.

Microfilaments have two major roles:

▶ They help the entire cell or parts of the cell to contract.
▶ They stabilize cell shape.

In muscle cells, actin fibers are associated with another protein called *myosin*, and their interactions account for the contraction of muscles. In nonmuscle cells, actin fibers are associated with localized changes of shape in cells. For example, microfilaments are involved in a flowing movement of the cytoplasm called *cytoplasmic streaming*, in movements of specific organelles and particles within cells, and in the "pinching" contractions that divide an animal cell into two daughter cells. Microfilaments are also involved in the formation of cellular extensions, called *pseudopodia* (pseudo-, "false;" podia, "feet"), that enable cells to move (Figure 4.22).

In some cell types, microfilaments form a meshwork just inside the plasma membrane. Actin-binding proteins then cross-link the microtubules to form a rigid structure that supports the

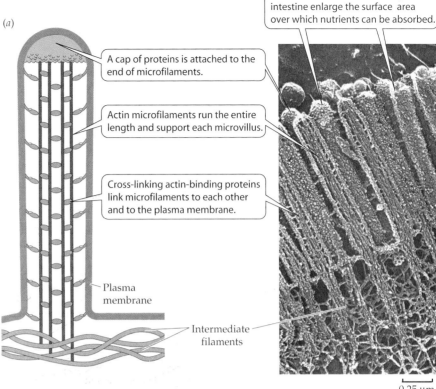

(a)

A cap of proteins is attached to the end of microfilaments.

Actin microfilaments run the entire length and support each microvillus.

Cross-linking actin-binding proteins link microfilaments to each other and to the plasma membrane.

Plasma membrane

Intermediate filaments

The microvilli of the cells lining the intestine enlarge the surface area over which nutrients can be absorbed.

0.25 μm

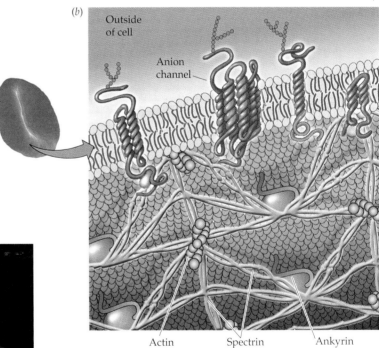

(b)

Outside of cell

Anion channel

Actin   Spectrin   Ankyrin

4.23 Microfilaments for Support
(a) Microfilaments form the backbone of the microvilli that increase the surface area of some cells, such as intestinal cells that absorb nutrients. (b) Actin microfilaments, along with ankyrin and spectrin proteins, support the "doughnut" shape of red blood cells.

20 μm

4.22 Microfilaments for Motion
The green-stained microfilaments in these cells provide a way for the cell to move.

cell. For example, microfilaments support the tiny *microvilli* that line the intestine, giving it a larger surface area through which to absorb nutrients. Such a "submembrane skeleton" also helps keep the red blood cell in its familiar doughnut shape (Figure 4.23).

Intermediate filaments are tough supporting elements

Intermediate filaments (see Figure 4.21) are found only in multicellular organisms. Although there are at least five distinct types of intermediate filaments, all share the same general structure and are composed of fibrous proteins of the keratin family, similar to the protein that makes up hair and fingernails. In cells, these proteins are organized into tough, ropelike assemblages 8 to 12 nm in diameter.

Intermediate filaments have two major structural functions:

▶ They stabilize cell structure.
▶ They resist tension.

In some cells, intermediate filaments radiate from the nuclear envelope and may maintain the positions of the nucleus and other organelles in the cell. The lamins of the nuclear lamina are intermediate filaments. Other kinds of intermediate filaments help hold a complex apparatus of microfilaments in place in muscle cells. Still other kinds stabilize and help maintain rigidity in surface tissues by connecting "spot welds" called *desmosomes* between adjacent cells (see Figure 5.6*b*).

Microtubules are long and hollow

Microtubules are long, hollow, unbranched cylinders about 25 nm in diameter and up to several micrometers long. Assembled from molecules of the protein **tubulin**, microtubules have two roles:

▶ They form a rigid internal skeleton for some cells, especially at cell extensions.
▶ They act as a framework on which motor proteins can move structures in the cell.

Tubulin is a dimer made up of two polypeptide monomers, called α-tubulin and β-tubulin. Thirteen rows, of tubulin dimers surround the central cavity of the microtubule (see Figure 4.21). The two ends of a microtubule are different. One end is designated the + end, the other the – end. Tubulin dimers can be added or subtracted mainly at the + end, lengthening or shortening the microtubule. This capacity to change length rapidly makes microtubules dynamic structures.

This dynamic property is seen in animal cells, where microtubules are often found in parts of the cell that are changing shape. Many microtubules radiate from a region of the cell called the *microtubule organizing center*. Tubule polymerization results in a rigid cell, and tubule depolymerization leads to a collapse of this rigid structure. In plants, microtubules help control the arrangement of the cellulose fibers of the cell wall. Electron micrographs of plants frequently show microtubules lying just inside the plasma membrane of cells that are forming or extending their cell walls. Experimental alteration of the orientation of these microtubules leads to a similar change in the cell wall, and a new shape for the cell. In many cells, microtubules serve as tracks for *motor proteins*, specialized molecules that use energy to change their shape and move. Motor proteins bond to and move along the microtubules, carrying materials from one part of the cell to another. Microtubules are also essential in distributing chromosomes to daughter cells during cell division. And they are intimately associated with movable cell appendages: the flagella and cilia.

Microtubules power cilia and flagella

Many eukaryotic cells possess flagella and/or cilia. These whiplike organelles push or pull the cell through its aque-

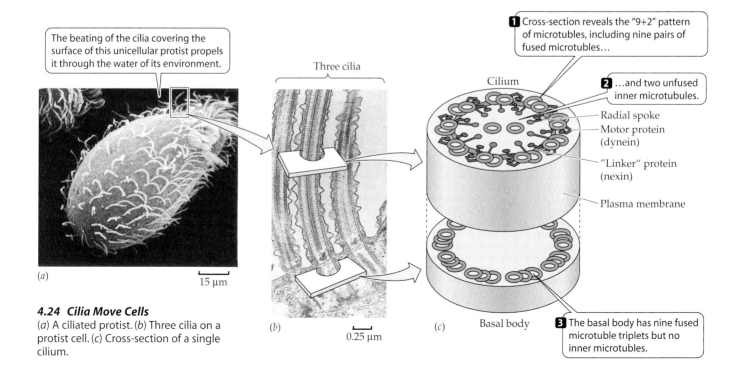

4.24 Cilia Move Cells
(*a*) A ciliated protist. (*b*) Three cilia on a protist cell. (*c*) Cross-section of a single cilium.

The beating of the cilia covering the surface of this unicellular protist propels it through the water of its environment.

(*a*)   15 µm

Three cilia

(*b*)   0.25 µm

Cilium

1 Cross-section reveals the "9+2" pattern of microtubles, including nine pairs of fused microtubules...

2 ...and two unfused inner microtubules.

Radial spoke
Motor protein (dynein)
"Linker" protein (nexin)
Plasma membrane

(*c*)   Basal body

3 The basal body has nine fused microtuble triplets but no inner microtubles.

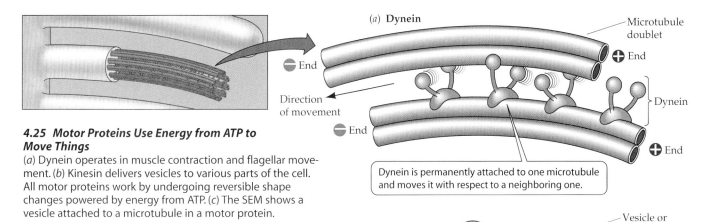

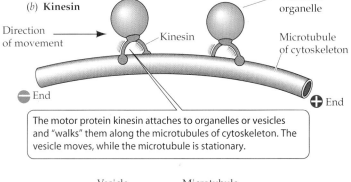

(a) **Dynein**

Microtubule doublet

⊖ End

Direction of movement

⊖ End

⊕ End

Dynein

⊕ End

> Dynein is permanently attached to one microtubule and moves it with respect to a neighboring one.

4.25 Motor Proteins Use Energy from ATP to Move Things
(a) Dynein operates in muscle contraction and flagellar movement. (b) Kinesin delivers vesicles to various parts of the cell. All motor proteins work by undergoing reversible shape changes powered by energy from ATP. (c) The SEM shows a vesicle attached to a microtubule in a motor protein.

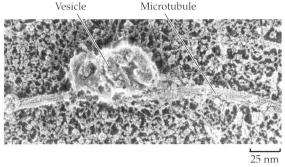

(b) **Kinesin**

Vesicle or organelle

Direction of movement

Kinesin

Microtubule of cytoskeleton

⊖ End

⊕ End

> The motor protein kinesin attaches to organelles or vesicles and "walks" them along the microtubules of cytoskeleton. The vesicle moves, while the microtubule is stationary.

Vesicle   Microtubule

25 nm

ous environment, or they may move surrounding liquid over the surface of the cell (Figure 4.24a). Cilia and eukaryotic (but not prokaryotic*) flagella are both assembled from specialized microtubules and have identical internal structures, but they differ in their relative lengths and their patterns of beating:

▶ **Flagella** are longer than cilia and are usually found singly or in pairs. Waves of bending propagate from one end of a flagellum to the other in snakelike undulation.

▶ **Cilia** are shorter than flagella and are usually present in great numbers. They beat stiffly in one direction and recover flexibly in the other direction (like a swimmer's arm), so that the recovery stroke does not undo the work of the power stroke.

Observed by electron microscopy in cross section, a typical cilium or eukaryotic flagellum is surrounded by the plasma membrane and contains a "9 + 2" array of microtubules. As Figure 4.24b shows, nine fused pairs of microtubules—called **doublets**—form an outer cylinder, and one pair of unfused microtubules runs up the center. A spoke radiates from one microtuble of each pair and connects the doublet to the center of the structure.

In the cytoplasm at the base of every eukaryotic flagellum or cilium is an organelle called a **basal body**. The nine microtubule doublets extend into the basal body. In the basal body, each doublet is accompanied by another microtubule, making nine sets of *three* microtubules. The central, unfused microtubules do not extend into the basal body.

The microtuble doublets of cilia and flagella are linked by proteins. The motion of cilia and flagella results from the sliding of the microtubules past each other, driven by a motor protein called **dynein**, which can undergo changes

*Some prokaryotes have flagella, as we saw earlier, but prokaryotic flagella lack microtubules and dynein. The flagella of prokaryotes are neither structurally nor evolutionarily related to those of eukaryotes. The prokaryotic flagellum is assembled from a protein called *flagellin*, and it has a much simpler structure and a smaller diameter than a single eukaryotic microtubule. And whereas eukaryotic flagella beat in a wavelike motion, prokaryotic flagella rotate (see Figure 4.7).

in its shape driven by energy from ATP. Dynein molecules attached to one microtubule bind to a neighboring microtubule. As the dynein molecules change shape, they move the microtubule past its neighbor (Figure 4.25a). Blocking the motor action of dynein is the idea behind a new class of spermicides used for contraception: Because these spermicides inhibit dynein, the sperm cannot swim toward the egg, and fertilization cannot occur.

Dynein and another motor protein, **kinesin**, are responsible for carrying protein-laden vesicles from one part of the cell to another. Recall that microtubules have a + end and a − end. Dynein binds to a microtubule and moves attached vesicles and other organelles toward the − end, while kinesin moves them toward the + end (Figure 4.25b).

Centrioles are almost identical to basal bodies. Centrioles are found in all eukaryotes except the flowering plants, pine trees and their relatives, and some protists. Under the light microscope, a centriole looks like a small, featureless particle, but the electron microscope reveals that it is made up of a precise bundle of microtubules, arranged as nine sets of three fused microtubules each (Figure 4.26). Centri-

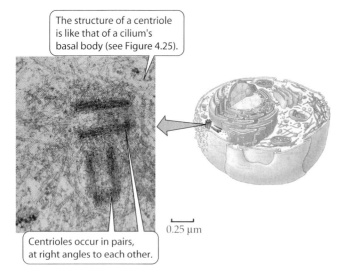

The structure of a centriole is like that of a cilium's basal body (see Figure 4.25).

0.25 μm

Centrioles occur in pairs, at right angles to each other.

4.26 Centrioles Contain Triplets of Microtubules
Centrioles are found in the microtubule organizing center, a region near the nucleus. The electron micrograph shows a pair of centrioles at right angles to each other.

oles lie in the microtubule organizing center in cells that are about to undergo division. As you will see in Chapter 9, they are involved in the formation of the mitotic spindle, to which the chromosomes attach.

Extracellular Structures

Although the plasma membrane is the functional barrier between the inside and outside of a cell, many structures outside the plasma membrane are produced by cells, secreted to the outside, and play essential roles in protecting, supporting, or attaching cells. These structures are said to be **extracellular** because they are outside the plasma membrane. The peptidoglycan cell wall of bacteria is such an extracellular structure. In eukaryotes, other extracellular structures play the same roles: in plants, the cellulose cell wall, and in multicellular animals, the extracellular matrix found between cells. Both of these structures are made up of a prominent fibrous macromolecule embedded in a jelly-like medium.

The plant cell wall consists largely of cellulose

The **cell wall** of plant cells is a semirigid structure outside the plasma membrane (Figure 4.27). It consists of cellulose fibers embedded in other complex polysaccharides and proteins. The cell wall has two major roles in plants:

▶ It provides support for the cell and limits its volume by remaining rigid.
▶ It acts as a barrier to infections by fungi and other organisms that can cause plant diseases.

Because of their thick cell walls, plant cells viewed under a light microscope appear to be entirely isolated from each other. But electron microscopy reveals that this is not the

case. The cytoplasm of adjacent plant cells is connected by numerous plasma membrane-lined channels, called *plasmodesmata*, that are about 20 to 40 nm in diameter and extend through the walls of adjoining cells (see Figure 4.27). These connections permit the diffusion of water, ions, small molecules, and RNA and proteins between connected cells. Such diffusion ensures that the cells of a plant have uniform concentrations of these substances.

Animal cells have elaborate extracellular matrices

The cells of multicellular animals lack the semirigid cell wall that is characteristic of plant cells, but many animal cells are surrounded by, or are in contact with, an **extracellular matrix**. This matrix is composed of fibrous proteins such as collagen (the most abundant protein in mammals) and glycoproteins (Figure 4.28). These proteins, as well as other substances particular to certain body tissues, are secreted by cells that are present in or near the matrix. In the human body, some tissues, such as those in the brain, have very little extracellular matrix; other tissues, such as bone and cartilage, have large amounts of extracellular matrix. The functions of the extracellular matrix are many:

▶ It holds cells together in tissues.
▶ It contributes to the physical properties of cartilage, skin, and other tissues.
▶ It helps filter materials passing between different tissues.
▶ It helps orient cell movements during embryonic development and during tissue repair.
▶ It plays a role in chemical signaling from one cell to another.

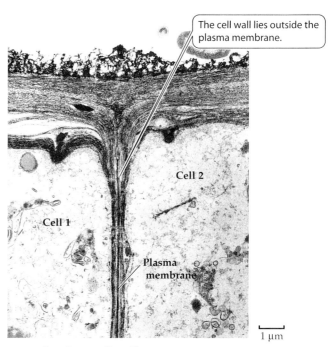

The cell wall lies outside the plasma membrane.

Cell 2

Cell 1

Plasma membrane

1 μm

4.27 The Plant Cell Wall
The semirigid cell wall provides support for plant cells.

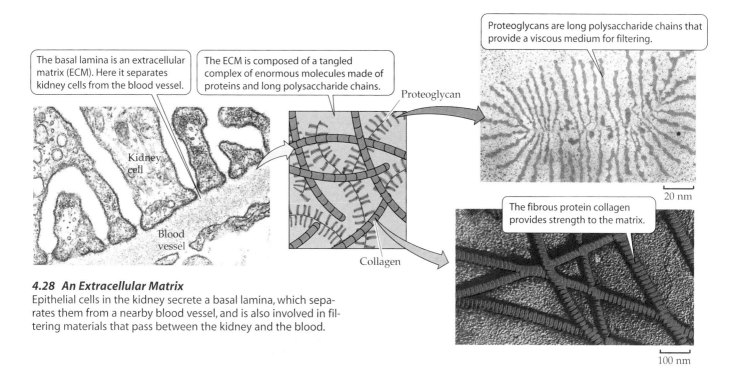

The basal lamina is an extracellular matrix (ECM). Here it separates kidney cells from the blood vessel.

The ECM is composed of a tangled complex of enormous molecules made of proteins and long polysaccharide chains.

Proteoglycans are long polysaccharide chains that provide a viscous medium for filtering.

Proteoglycan

Kidney cell

Blood vessel

Collagen

The fibrous protein collagen provides strength to the matrix.

20 nm

100 nm

4.28 An Extracellular Matrix
Epithelial cells in the kidney secrete a basal lamina, which separates them from a nearby blood vessel, and is also involved in filtering materials that pass between the kidney and the blood.

The cells embedded in bone and cartilage, for example, secrete and maintain the extracellular material that makes up these structures. Bone cells are embedded in an extracellular matrix that consists primarily of collagen and calcium phosphate. This matrix gives bone its familiar rigidity. Epithelial cells, which line body cavities, lie together as a sheet spread over a **basal lamina**, or basement membrane, a form of extracellular matrix (see Figure 4.28).

Some extracellular matrices are made up, in part, of an enormous **proteoglycan**. A single molecule of this proteoglycan consists of many hundreds of polysaccharides covalently attached to about a hundred proteins, all of which are attached to one enormous polysaccharide. The molecular weight of this proteoglycan can exceed 100 million; the molecule takes up as much space as an entire prokaryotic cell.

Chapter Summary

The Cell: The Basic Unit of Life

▶ All cells come from preexisting cells and have certain processes, types of molecules, and structures in common.

▶ To maintain adequate exchanges with its environment, a cell's surface area must be large compared with its volume. **Review Figure 4.2**

▶ Microscopes are needed to visualize cells. Because of their greater resolving power, electron microscopes allow observation of greater detail than can be seen with light microscopes. **Review Figure 4.3**

▶ Prokaryotic cell organization is characteristic of the kingdoms Eubacteria and Archaebacteria. Prokaryotic cells lack internal compartments. **Review Figure 4.4**

▶ Eukaryotic cell organization is characteristic of cells in the other four kingdoms. Eukaryotic cells have many membrane-enclosed compartments, including a nucleus that contains DNA. **Review Figure 4.7**

Prokaryotic Cells

▶ All prokaryotic cells have a plasma membrane, a nucleoid region with DNA, and a cytoplasm that contains ribosomes, dissolved enzymes, water, and small molecules. Some prokaryotes have additional protective structures: cell wall, outer membrane, and capsule. Some prokaryotes contain photosynthetic membranes, and some have mesosomes. **Review Figures 4.4, 4.5**

▶ Projecting from the surface of some prokaryotes are rotating flagella, which move the cells from place to place. Pili are projections by which prokaryotic cells attach to one another or to environmental surfaces. **Review Figure 4.6**

Eukaryotic Cells

▶ Like prokaryotic cells, eukaryotic cells have a plasma membrane, cytoplasm, and ribosomes. However, eukaryotic cells are larger and contain many membrane-enclosed organelles. **Review Figure 4.7**

▶ The membranes that envelop organelles in the eukaryotic cell are partial barriers, ensuring that the chemical composition of the interior of the organelle differs from that of the surrounding cytoplasm.

Organelles that Process Information

▶ The nucleus is usually the largest organelle in a cell. It is surrounded by a double membrane, the nuclear envelope, which disassembles during cell division. Within the nucleus, the nucleolus is the source of the ribosomes found in the cytoplasm. **Review Figure 4.8**

▶ Nuclear pores have complex structures that govern what enters and leaves the nucleus. **Review Figure 4.8**

▶ The nucleus contains most of the cell's DNA, which associates with protein to form chromatin. Chromatin is diffuse throughout the nucleus until just before cell division, when it condenses to form chromosomes. **Review Figure 4.9**

The Endomembrane System

▶ The endomembrane system is made up of a series of interrelated membranes and compartments.

▶ The rough endoplasmic reticulum has attached ribosomes that synthesize proteins. The smooth endoplasmic reticulum lacks ribosomes and is associated with the synthesis of lipids. **Review Figures 4.7, 4.11**

▶ The Golgi apparatus adds signal molecules to proteins, directing them to their proper destinations. It receives materials from the rough ER by means of vesicles that fuse with the *cis* region of the Golgi. **Review Figures 4.7, 4.12, 4.13**

▶ Vesicles originating from the *trans* region of the Golgi contain proteins for different cellular locations. Some of these vesicles fuse with the plasma membrane and release their contents outside the cell. **Review Figure 4.12**

▶ Lysosomes contain many digestive enzymes. Lysosomes fuse with the phagosomes produced by phagocytosis to form secondary lysosomes, in which engulfed materials are digested. Undigested materials are secreted from the cell when the secondary lysosome fuses with the plasma membrane. **Review Figure 4.13**

Organelles that Process Energy

▶ Mitochondria are enclosed by an outer membrane and an inner membrane that folds inward to form cristae. Mitochondria contain the proteins needed for cellular respiration and the generation of ATP. **Review Figure 4.14**

▶ All eukaryotic cells contain mitochondria. Green plant cells also contain chloroplasts These organelles are enclosed by double membranes and contain an internal system of thylakoids organized as grana. **Review Figures 4.7, 4.16**

▶ Thylakoids within chloroplasts contain the chlorophyll and proteins that harvest light energy for photosynthesis. **Review Figure 4.16**

▶ Both mitochondria and chloroplasts contain their own DNA and ribosomes and are capable of making some of their own proteins.

▶ The endosymbiosis theory of the evolutionary origin of mitochondria and chloroplasts states that these organelles originated when larger prokaryotes engulfed, but did not digest, smaller prokaryotes. Mutual benefits permitted this symbiotic relationship to be maintained and to evolve into the eukaryotic organelles observed today. **Review Figure 4.18**

Other Organelles Enclosed by Membranes

▶ Peroxisomes and glyoxysomes contain special enzymes and carry out specialized chemical reactions inside the cell.

▶ Vacuoles are prominent in many plant cells and consist of a membrane-enclosed compartment full of water and dissolved substances. By taking in water, vacuoles enlarge and provide the pressure needed to stretch the cell wall and provide structural support for the plant.

The Cytoskeleton

▶ The cytoskeleton within the cytoplasm of eukaryotic cells provides shape, strength, and movement. It consists of three interacting types of protein fibers. **Review Figure 4.21**

▶ Microfilaments consist of two chains of actin units that together form a double helix. Microfilaments strengthen cellular structures and provide the movement in animal cell division, cytoplasmic streaming, and pseudopod extension. Microfilaments may be found as individual fibers, bundles of fibers, or networks of fibers joined by linking proteins. **Review Figures 4.21, 4.23**

▶ Intermediate filaments are formed of keratins and are organized into tough, ropelike structures that add strength to cell attachments in multicellular organisms. **Review Figure 4.21**

▶ Microtubules are composed of dimers of the protein tubulin. They can lengthen and shorten by adding and losing tubulin dimers. They are involved in the structure and function of cilia and flagella, both of which have a characteristic 9 + 2 pattern of microtubules. **Review Figures 4.21, 4.24**

▶ The movements of cilia and flagella are due to the binding of the motor protein dynein to the microtubules. Microtubules also bind motor proteins, including kinesin and dynein, that move organelles through the cell. **Review Figure 4.25**

▶ Centrioles, made up of triplets of microtubules, are involved in the distribution of chromosomes during nuclear division. **Review Figure 4.26**

Extracellular Structures

▶ Materials external to the plasma membrane provide protection, support, and attachment for cells in multicellular systems.

▶ The cell wall of plants consists principally of cellulose. It is pierced by plasmodesmata that join the cytoplasm of adjacent cells. **Review Figure 4.27**

▶ In multicellular animals, the extracellular matrix consists of different kinds of proteins, including proteoglycan. In bone and cartilage, the protein collagen predominates. **Review Figure 4.28**

For Discussion

1. Which organelles and other structures are found in both plant and animal cells? Which are found in plant but not animal cells? In animal but not plant cells? Discuss these differences in relation to the activities of plants and animals.

2. Through how many membranes would a molecule have to pass in going from the interior of a chloroplast to the interior of a mitochondrion? From the interior of a lysosome to the outside of a cell? From one ribosome to another?

3. How does the possession of double membranes by chloroplasts and mitochondria relate to the endosymbiosis theory of the origins of these organelles? What other evidence supports the theory?

4. What kinds of cells and subcellular structures would you choose to examine by transmission electron microscopy? By scanning electron microscopy? By light microscopy? What are the advantages and disadvantages of each of these modes of microscopy?

5 *Cellular Membranes*

The nonp
"tails" int
interior o

The char
"head" p
polar wa

5.2 A P
The eigl
cross se

other n
cholest
lestero
molecu
saturat
lestero
Figure
Ch
fluidit
positic
memb
brane
Since i
reduce
organi
this pi
positic
satura
Such o
nating

Mem
All b
plasn
phosp
meml
chonc

5.3 N
Freez
This n
then :

"No sweat" may describe your reaction to a course with a light workload. But it certainly does not apply to a professional athlete—or to anyone else—who is engaging in vigorous activity. The harder we work physically, the hotter we get, and soon we start to sweat. Sweating is a way to reduce body heat by using the excess heat to evaporate water. At peak activity, we lose as much as 2 liters of water in an hour.

The sweat glands lie just below the surface of the skin. They are essentially tubes bathed in extracellular fluid. When stimulated by physical activity or other signals, these tubes fill with water and dissolved solutes. To get from the extracellular fluid into the tubes, water must pass into and through the cells that line the tube.

A hallmark of living cells is their ability to regulate what enters and leaves their cytoplasm. This is a function of the plasma membrane, which is composed of a hydrophobic lipid bilayer with associated proteins. When a person engages in normal activities, the membranes of the cells lining their sweat glands do not allow much water to enter or leave. But when the same person exercises, special pore proteins in the membrane, called aquaporins, open and allow water from the extracellular fluid to pass through the cells into a tube that leads to the surface of the skin.

Membranes are dynamic structures whose components move, change, and perform vital physiological roles as they allow cells to interact with other cells and molecules in the environment. We describe the structural aspects of these interactions here. Membranes also regulate ionic and molecular traffic into and out of the cell. This selective permeability, which we describe in this chapter, is an important characteristic of life. Later, we will see it in action in such diverse situations as the transduction of light energy into chemical energy in the chloroplast and the retention of water and ions in the mammalian kidney.

Membrane Composition and Structure

The chemical makeup, physical organization, and functioning of a biological membrane depend on three classes of biochemical compounds: lipids, proteins, and carbohy-

Sweating: A Regulated Membrane Activity
Tennis star Venus Williams, shown here winning a gold medal at the 2000 Olympic Games, can lose up to 2 liters of water in an hour by sweating. The excess body heat generated by her physical activity is used to evaporate the sweat, helping to keep her body temperature at normal levels.

drates (Figure 5.1). The lipids establish the physical integrity of the membrane and create an effective barrier to the rapid passage of hydrophilic materials such as water and ions. In addition, the phospholipid bilayer serves as a lipid "lake" in which a variety of proteins "float." This general design is known as the *fluid mosaic model* of the membrane. Membrane proteins embedded in the phospholipid bilayer have a number of functions, including moving materials through the membrane and receiving chemical signals from the cell's external environment.

Like some proteins, carbohydrates—the third class of compounds important in membranes—are crucial in recognizing specific molecules. The carbohydrates attach either to lipid or to protein molecules on the outside of the plasma membrane, where they protrude into the environment, away from the cell.

Lipids constitute the bulk of a membrane

Nearly all of the lipids in biological membranes are phospholipids. Recall from Chapter 2 that some compounds are hydrophilic ("water-loving") and others are hydrophobic

("water-h
hydrophi
nonpolar
hydropho
rials, but
drophilic
the phosp
drophilic
As a c
phospho
layer, wit
each othe
environn
branes w
tory. Bot
ous shee
acids to
small ho
neously.
cle fusio

Ou

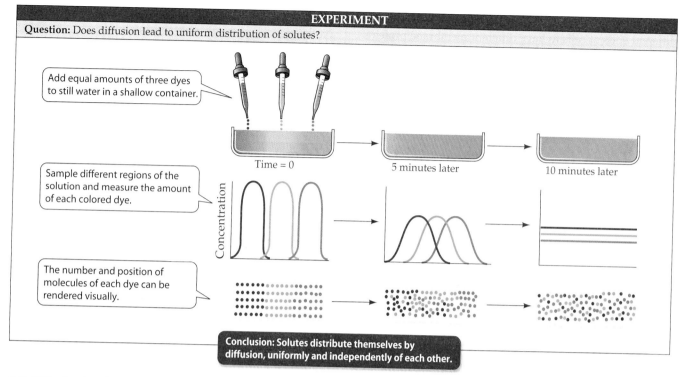

EXPERIMENT

Question: Does diffusion lead to uniform distribution of solutes?

Add equal amounts of three dyes to still water in a shallow container.

Time = 0   5 minutes later   10 minutes later

Sample different regions of the solution and measure the amount of each colored dye.

Concentration

The number and position of molecules of each dye can be rendered visually.

Conclusion: Solutes distribute themselves by diffusion, uniformly and independently of each other.

5.7 Diffusion Leads to Uniform Distribution of Solutes

Diffusion is the net movement of a solute from regions of greater concentration to regions of lesser concentration. The speed of diffusion varies with the substances involved, but the process continues until the solution reaches equilibrium.

diffuse from one compartment to the other until their concentrations are equal on both sides of the membrane. Molecules to which the membrane is impermeable remain in separate compartments, and their concentrations remain different on the two sides of the membrane. Equilibrium is reached when the concentrations of the diffusing substance are identical on both sides of the permeable membrane. Individual molecules are still passing through the membrane when equilibrium is established, but equal numbers of molecules are moving in each direction, so there is no net change in concentration.

Simple diffusion takes place through the membrane bilayer

In **simple diffusion**, small molecules pass through the lipid bilayer of the membrane. The more lipid-soluble the molecule, the more rapidly it diffuses through the bilayer. This statement holds true over a wide range of molecular weights. Only water and the smallest of molecules seem to deviate from this rule, passing through bilayers much more rapidly than their lipid solubilities would predict.

Charged and/or polar molecules such as amino acids, sugars, and ions do not pass readily through a membrane, for two reasons. First, cells are made up of, and exist in, water, and polar or charged substances form many hydrogen bonds with water, preventing their "escape" to the membrane. Second, the interior of the membrane is hy-

drophobic, and hydrophilic substances tend to be excluded from it. On the other hand, a molecule that is itself hydrophobic, and hence soluble in lipids, enters the membrane readily and is thus able to pass through it.

Osmosis is the diffusion of water across membranes

Water molecules are abundant enough and small enough that they move through membranes by a diffusion process called **osmosis**. This completely passive process uses no metabolic energy and can be understood in terms of the concentrations of solutions. Osmosis depends on the *number* of solute particles present—not the kind of particles. We will describe osmosis using red blood cells and plant cells as examples.

Red blood cells are normally suspended in a fluid called *plasma*, which contains salts, proteins, and other solutes. If a drop of blood is examined under the light microscope, the red cells are seen to have their characteristic donut shape. If pure water is added to the drop of blood, the cells quickly swell and burst (Figure 5.8a). Similarly, if slightly wilted lettuce is put in pure water, it soon becomes crisp; by weighing it before and after, we can show that it has taken up water (Figure 5.8b).

If, on the other hand, red blood cells or crisp lettuce leaves are placed in a relatively concentrated solution of salt or sugar, the leaves become limp (wilt) and the red blood cells pucker and shrink. From analyses of such observations, we know that the difference in solute concentrations is the principal factor that determines whether water will move from the surrounding environment into cells, or out of cells into the environment.

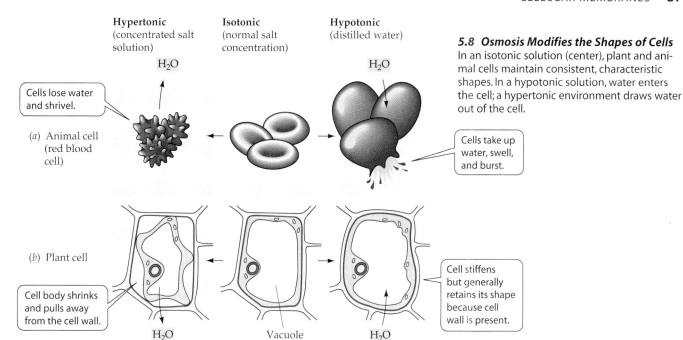

5.8 Osmosis Modifies the Shapes of Cells
In an isotonic solution (center), plant and animal cells maintain consistent, characteristic shapes. In a hypotonic solution, water enters the cell; a hypertonic environment draws water out of the cell.

Hypertonic
(concentrated salt solution)

Isotonic
(normal salt concentration)

Hypotonic
(distilled water)

H_2O

H_2O

Cells lose water and shrivel.

(a) Animal cell (red blood cell)

Cells take up water, swell, and burst.

(b) Plant cell

Cell body shrinks and pulls away from the cell wall.

Cell stiffens but generally retains its shape because cell wall is present.

H_2O

Vacuole

H_2O

Other things being equal, if two different solutions are separated by a membrane that allows water, but not solutes, to pass through, water molecules will move across the membrane toward the solution with a higher solute concentration. In other words, water will diffuse from a region of *its higher concentration* (lower concentration of solutes) to a region of *its lower concentration* (higher concentration of solutes).

Three terms are used to compare the solute concentrations of two solutions separated by a membrane:

▶ *Isotonic solutions* have equal total solute concentrations.
▶ A *hypertonic solution* has a higher total solute concentration than the other solution with which it is being compared.
▶ A *hypotonic* solution has a lower total solute concentration than the other solution with which it is being compared.

Water moves from a hypotonic solution across a membrane to a hypertonic solution.

When we say that "water moves," bear in mind that we are referring to the *net* movement of water. Since it is so abundant, water is constantly moving across the plasma membrane into and out of cells. Whether the overall movement is greater in one direction or the other is what concerns us here.

The concentration of solutes in the environment determines the direction of osmosis in all animal cells. A red blood cell takes up water from a solution that is hypotonic to the cell's contents. The cell bursts because its plasma membrane cannot withstand the swelling of the cell (see Figure 5.8a). The integrity of red blood cells (and other blood cells) is absolutely dependent on the maintenance of a constant solute concentration in the plasma in which they are suspended: The plasma must be isotonic with the cells if the cells are not to burst or shrink.

In contrast to animal cells, the cells of plants, archaea, bacteria, fungi, and some protists have cell walls that limit the volume of the cells and keep them from bursting. Cells with sturdy cell walls take up a limited amount of water and, in so doing, build up internal pressure against the cell wall that prevents further water from entering. This pressure within the cell, called *turgor pressure*, is the driving force for the enlargement of plant cells—it is a normal and essential component of plant development.

Diffusion may be aided by channel proteins

As we saw earlier, polar substances such as amino acids and sugars and charged substances such as ions do not diffuse across membranes. Instead, they cross the hydrophobic lipid barrier through protein-lined channels in a process called **facilitated diffusion**. Integral membrane proteins form these channels (Figure 5.9), which are lined with polar amino acids and water on the inside (to bind to the polar or charged substance and allow it to pass) and nonpolar amino acids on the outside (to allow the protein channel to insert itself into the lipid bilayer).

The best-studied protein channels are the *ion channels*. As you will see, the movement of ions into and out of cells is important in many biological processes, ranging from the electrical activity of the nervous system to the opening of pores in leaves that allow gas exchange with the environment. Hundreds of these channels have been identified, and all show the basic structure of a water-lined pore that just fits the ion that moves through it.

Ion channels are *gated*: they can be closed to ion passage, or open. A gated channel opens when something happens to change the shape of the protein. Depending on the channel, this stimulus can range from the binding of a chemical signal to an electrical charge caused by an imbalance of ions. Once the channel opens, millions of ions can rush through it

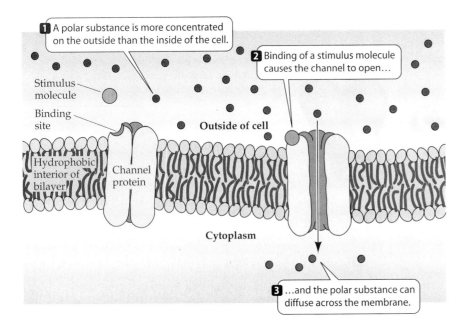

1 A polar substance is more concentrated on the outside than the inside of the cell.

Stimulus molecule

Binding site

2 Binding of a stimulus molecule causes the channel to open…

Outside of cell

Hydrophobic interior of bilayer

Channel protein

Cytoplasm

3 …and the polar substance can diffuse across the membrane.

5.9 A Gated Channel Protein Opens in Response to a Stimulus
The membrane protein changes its three-dimensional shape when the stimulus binds.

favoring glucose entry, with a higher concentration outside the cell (in blood capillaries or the intestine) than inside.

Transport by carrier proteins is different from simple diffusion. In both processes, the *rate* of movement depends on the concentration gradient across the membrane. However, in facilitated diffusion, a point is reached at which further increases in the concentration gradient are not accompanied by an increased rate of diffusion. At this point, the facilitated diffusion system is said to be *saturated*. Because there are only a limited number of carrier protein molecules per unit of membrane area, the rate of movement reaches a maximum when all the carrier molecules are fully loaded with solute molecules. In other words, when the differences in solute concentration across the membrane are sufficiently high, not enough carrier molecules are free at a given moment to handle all the solute molecules.

per second. How fast this happens, and in which direction (into or out of the cell), depends on the concentration gradient of the ion between the cytoplasm and the exterior environment of the cell. For example, if the concentration of potassium ion is much higher outside of the cell than inside, potassium will enter the cell through a potassium channel by diffusion; if it is higher inside the cell, potassium ion will diffuse out of the cell.

As we mentioned, water crosses the plasma membrane at a rate far in excess of expectations, given its polarity. One way that water can do this is by hydrating ions as they pass through ion channels. Up to 12 water molecules may coat an ion as it traverses a channel. Another way that water enters cells rapidly is through water channels called *aquaporins*. Membrane proteins that allow water to pass through them have been characterized in many cells, from the plant vacuole, where they are important in maintaining turgor, to the mammalian kidney, where they act in retaining water that would otherwise be lost through urine.

Carrier proteins aid diffusion by binding substances

Another kind of facilitated diffusion involves not just the opening of a channel, but the actual binding of the transported substance to a membrane protein. These proteins are called *carriers*, and, like channel proteins, they allow diffusion both into and out of the cell. They are used to transport polar molecules such as sugars and amino acids.

Glucose, for example, is the major energy source for most mammalian cells, and those cells have a carrier protein called the glucose transporter that facilitates the uptake of glucose (Figure 5.10). Since glucose is rapidly altered as soon as it gets into a cell, there is almost always a strong concentration gradient

Active Transport

In many biological situations, an ion or molecule must be moved across a membrane from a region of lower concentration to a region of higher concentration. In these cases,

5.10 A Carrier Protein Facilitates Diffusion
The carrier protein allows glucose to enter the cell at a faster rate than would be possible by simple diffusion across the membrane barrier.

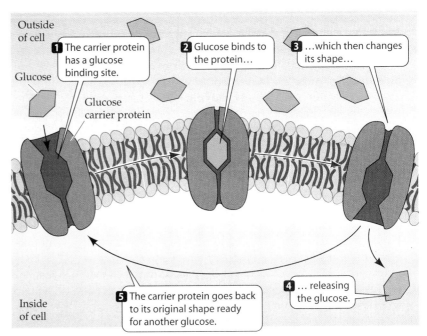

Outside of cell

Glucose

1 The carrier protein has a glucose binding site.

2 Glucose binds to the protein…

3 …which then changes its shape…

Glucose carrier protein

4 … releasing the glucose.

5 The carrier protein goes back to its original shape ready for another glucose.

Inside of cell

5.1 *Membrane Transport Mechanisms*

	SIMPLE DIFFUSION	FACILITATED DIFFUSION	ACTIVE TRANSPORT
Direction	With concentration gradient	With concentration gradient	Against concentration gradient
Energy source	Concentration gradient	Concentration gradient	ATP hydrolysis (primary)
Membrane protein required?	No	Yes	Yes
Specificity	Not specific	Specific	Specific

the substance cannot not rush into or out of cells by diffusion. The movement of a substance across a biological membrane *against* a concentration gradient—called **active transport**—requires the expenditure of energy. The differences between diffusion and active transport are summarized in Table 5.1.

Active transport is directional

Three types of proteins are involved in active transport (Figure 5.11):

▶ **Uniport** transporters move a single solute in one direction. For example, a Ca^{2+}-binding protein found in the plasma membrane and endoplasmic reticulum membranes of many cells actively transports this ion to regions of higher concentration either outside the cell or inside the ER.

▶ **Symport** transporters move two solutes in the same direction. For example, the uptake of amino acids from the intestine into the cells that line it requires the simultaneous binding of Na^+ and amino acid to the same carrier protein.

▶ **Antiport** transporters move two solutes in opposite directions, one into the cell and the other out of the cell. For example, many cells have an "Na^+–K^+ pump" that moves Na^+ out of the cell and K^+ into it.

Primary and secondary active transport rely on different energy sources

There are two basic types of active transport processes. The first, **primary active transport**, requires the direct participation of ATP. Energy released by the hydrolysis of ATP drives the movement of specific ions against a concentration gradient. For example, if we compare the concentrations of potassium ions (K^+) and sodium ions (Na^+) inside a nerve cell and in the fluid bathing the nerve (Table 5.2), we can see that the K^+ concentration is much higher inside the cell, whereas the Na^+ concentration is much higher outside. Nevertheless, a protein in the nerve cells continues to pump Na^+ out and K^+ in, against these concentration gradients, ensuring that the gradients are maintained. This *sodium–potassium pump* is found in all animal cells and is an integral membrane glycoprotein. It breaks down a molecule of ATP to ADP and phosphate (P_i), and uses the energy released to bring two K^+ ions into the cell and export three Na^+ ions (Figure 5.12). The Na^+–K^+ pump is thus an antiport transport system.

Only cations are transported directly by pumps in primary active transport. Other solutes are transported by **secondary active transport**. This form of active transport does not use ATP directly; rather, the transport of the solute is tightly coupled to an ion concentration gradient established by primary active transport. The movement of the solute *against* its concentration gradient is accomplished using energy "regained" by letting ions move across the membrane *with* their concentration gradient.

For example, energy from ATP is used in primary active transport to establish concentration gradients of potassium and sodium ions; then the passive diffusion of some sodium ions in the opposite direction provides energy for the secondary active transport of the sugar glucose (Figure 5.13). Other secondary active transporters aid in the uptake

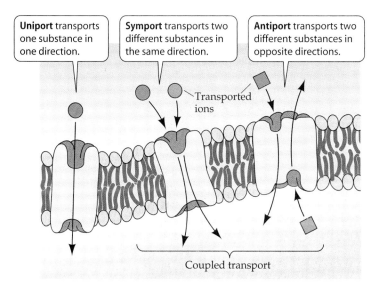

Uniport transports one substance in one direction.

Symport transports two different substances in the same direction.

Antiport transports two different substances in opposite directions.

Transported ions

Coupled transport

5.11 Proteins for Active Transport
Note that in each of the three cases, transport is directional.

5.2 *Concentration of Major Ions Inside and Outside the Nerve Cell of a Squid*

	CONCENTRATION (MOLAR)	
ION	INSIDE	OUTSIDE
K^+	0.400	0.020
Na^+	0.050	0.440
Cl^-	0.120	0.560

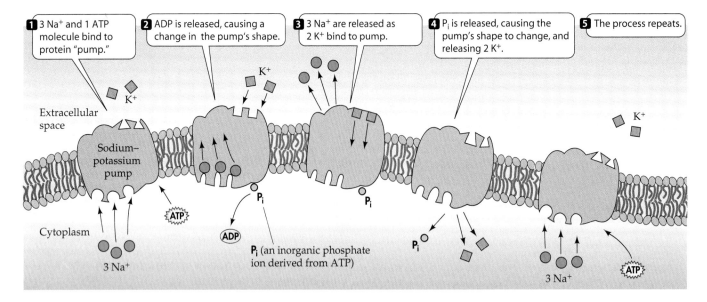

1 3 Na$^+$ and 1 ATP molecule bind to protein "pump."

2 ADP is released, causing a change in the pump's shape.

3 3 Na$^+$ are released as 2 K$^+$ bind to pump.

4 P$_i$ is released, causing the pump's shape to change, and releasing 2 K$^+$.

5 The process repeats.

K$^+$

K$^+$

K$^+$

Extracellular space

Sodium–potassium pump

ATP

ADP

Cytoplasm

3 Na$^+$

P$_i$

P$_i$

P$_i$ (an inorganic phosphate ion derived from ATP)

P$_i$

3 Na$^+$

ATP

5.12 Primary Active Transport: The Na$^+$–K$^+$ Pump
In active transport, energy is used to move a solute against its concentration gradient. Even though the Na$^+$ concentration is higher outside the cell and the K$^+$ concentration is higher inside the cell, for each molecule of ATP used, two K$^+$ are pumped *into* the cell and three Na$^+$ are pumped *out of* the cell.

of amino acids and other sugars, which are essential raw materials for cell maintenance and growth. Both types of coupled transport proteins—symports and antiports—are used for secondary active transport.

Endocytosis and Exocytosis

Macromolecules such as proteins, polysaccharides, and nucleic acids are simply too large and too charged or polar to pass through membranes. This is a fortunate property. Think of the consequences if these molecules could diffuse out of cells: A red blood cell would not retain its hemoglobin! On the other hand, cells must sometimes take up or secrete intact large molecules. As we saw in Chapter 4, this occurs by means of vesicles that either pinch off from the plasma mem-

brane and enter the cell (endocytosis) or fuse with the plasma membrane and release their contents (exocytosis).

Macromolecules and particles enter the cell by endocytosis

Endocytosis is a general term for a group of processes that bring macromolecules, large particles, small molecules, and even small cells into the eukaryotic cell (Figure 5.14*a*). There are three types of endocytosis: phagocytosis, pinocytosis, and receptor-mediated endocytosis. In all three, the plasma membrane invaginates (folds inward) around materials from the environment, forming a small pocket. The pocket deepens, forming a vesicle. This vesicle separates from the surface of the cell and migrates with its contents to the cell's interior.

5.13 Secondary Active Transport
The sodium ion concentration gradient established by primary active transport (right) powers the secondary active transport of glucose (left). The movement of glucose across the membrane against its concentration gradient is coupled by a symport protein to the movement of Na$^+$ into the cell.

Secondary active transport
Sodium ions, moving with the concentration gradient established by the sodium–potassium pump, drive transport of glucose against its concentration gradient.

Primary active transport
The sodium–potassium pump moves sodium ions, using the energy of ATP hydrolysis to establish a concentration gradient of Na$^+$.

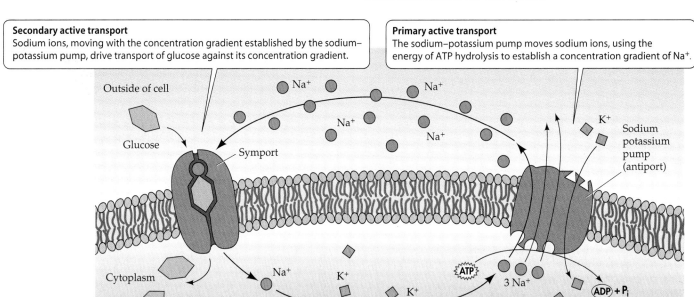

Outside of cell

Na$^+$

Na$^+$

Na$^+$

Na$^+$

K$^+$

Glucose

Symport

Sodium potassium pump (antiport)

Cytoplasm

Na$^+$

K$^+$

ATP

3 Na$^+$

K$^+$

ADP + P$_i$

K$^+$

K$^+$

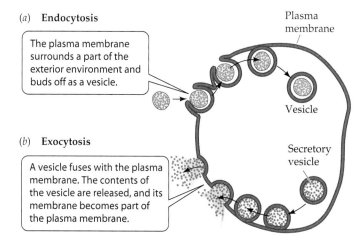

(a) **Endocytosis**

> The plasma membrane surrounds a part of the exterior environment and buds off as a vesicle.

Plasma membrane

Vesicle

(b) **Exocytosis**

> A vesicle fuses with the plasma membrane. The contents of the vesicle are released, and its membrane becomes part of the plasma membrane.

Secretory vesicle

5.14 Endocytosis and Exocytosis
Endocytosis and exocytosis are used by all eukaryotic cells to take up substances from and release substances to the outside environment.

In **phagocytosis**, part of the plasma membrane engulfs fairly large particles or even entire cells. Phagocytosis is used as a cellular feeding process by unicellular protists and by some white blood cells that defend the body against foreign cells and substances. The food vacuole or phagosome formed usually fuses with a lysosome, and its contents are digested (see Figure 4.13*b*).

In **pinocytosis** ("cellular drinking"), vesicles also form. However, these vesicles are smaller, and the process operates to bring in small dissolved substances or fluids. It is relatively nonspecific as to what it brings into the cell. For example, pinocytosis goes on constantly in the endothelium, the single layer of cells that separates a tiny blood capillary from its surrounding tissue (Figure 5.15), and is a way for the cells to rapidly acquire the fluids of the blood.

In **receptor-mediated endocytosis**, specific reactions at the cell surface trigger the uptake of specific materials. Let's take a closer look at this process.

Receptor-mediated endocytosis is highly specific

Receptor-mediated endocytosis is used by animal cells to capture specific macromolecules from the cell's environment. The uptake process is similar to nonspecific endocytosis, as already described. However, in receptor-mediated endocytosis, receptor proteins at particular sites on the outer surface of the plasma membrane bind to specific substances in the environment outside the cell. These sites are called **coated pits** because they form a slight depression in the plasma membrane whose cytoplasmic surface is coated by fibrous proteins, such as *clathrin*.

When a receptor protein binds to its specific macromolecule outside the cell, its coated pit invaginates and forms a **coated vesicle** around the bound macromolecule. Strengthened and stabilized by clathrin molecules, this vesicle carries the macromolecule into the cell (Figure 5.16). Once inside, the vesicle loses its clathrin coat and may fuse with a lysosome, where the engulfed material is processed and released into the cytoplasm. Because of its specificity for particular macromolecules, receptor-mediated endocytosis is a rapid and efficient method of taking up what may be minor constituents of the cell's environment.

Receptor-mediated endocytosis is the method by which cholesterol is taken up by most mammalian cells. Water-insoluble cholesterol is synthesized in the liver and transported in the blood attached to a protein, forming a lipoprotein called *low-density lipoprotein*, or *LDL*. The uptake of cholesterol begins with the binding of LDL to specific receptor proteins in coated pits. After being engulfed by endocytosis, the LDL particle is freed from the receptors. The receptors segregate to a region of the vesicle that buds off to form a new vesicle, which is recycled to the plasma membrane. The freed LDL particle remains in the original vesicle, which fuses with a lysosome in which the LDL is digested and the cholesterol made available for cell use. Persons with the inherited disease *hypercholesterolemia* (*-emia*, "blood") have dangerously high levels of cholesterol in their blood because of a deficient receptor for LDL.

Exocytosis moves materials out of the cell

Exocytosis is the process by which materials packaged in vesicles are secreted from a cell when the vesicle membrane fuses with the plasma membrane (see Figure 5.14*b*). The initial event in this process is the binding of a membrane protein protruding from the cytoplasmic side of the vesicle with a membrane protein on the cytoplasmic side of the target site on the plasma membrane. The phospholipid regions of the two membranes merge, and an opening to the outside of the cell develops. The contents of the vesicle are released to the environment, and the vesicle membrane is smoothly incorporated into the plasma membrane.

In Chapter 4, we encountered exocytosis as the last step in the processing of material engulfed by phagocytosis: the secretion of indigestible materials to the environment. Exocytosis is also important in the secretion of many different sub-

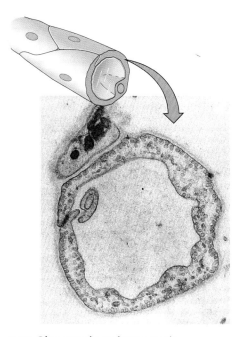

5.15 Pinocytosis and Exocytosis
The single endothelial cell that surrounds a blood capillary uses pinocytosis and exocytosis to transport substances between the blood and the surrounding tissue.

❶ The protein clathrin coats the cytoplasmic side of the plasma membrane at a coated pit.

❹ The endocytosed contents are surrounded by a clathrin-coated vesicle.

Specific substance binding to receptor proteins

Cytoplasm

Coated pit

Clathrin coat

0.1 μm

5.16 Formation of a Coated Vesicle
In receptor-mediated endocytosis, the receptor proteins in a coated pit bind specific macromolecules, which are then carried into the cell by the coated vesicle.

stances, including digestive enzymes from the pancreas, neurotransmitters from nerve cells, and materials for the construction of the plant cell wall.

Membranes Are Not Simply Barriers

We have discussed several functions of membranes—the compartmentalization of cells, the regulation of traffic between compartments, and the movement of materials into and out of cells—but there are more. In Chapter 4, we described how the membrane of the rough endoplasmic reticulum serves as a site for ribosome attachment. Newly formed proteins are passed from the ribosomes through the membrane and into the interior of the ER for modification and delivery to other parts of the cell. On the other hand, the membranes of nerve cells, muscle cells, some eggs, and other cells are electrically excitable. In nerve cells, the plasma membrane is the conductor of the nerve impulse from one end of the cell to the other.

Numerous other biological activities and properties discussed in the chapters to follow are associated with membranes. We review three of these here.

INFORMATION PROCESSING. As we have seen, the plasma membranes at cell surfaces and the membranes within cells may have protruding integral membrane proteins or attached carbohydrates that can bind to specific substances in the environment. The binding of a specific substance can serve as a signal to initiate, modify, or turn off a cell function (Figure 5.17a).

In this type of information processing, specificity in binding is essential. We have already seen the role of a specific receptor protein in the endocytosis of LDL and its cargo of cholesterol (see Figure 5.17). Another example is the binding of a hormone, such as insulin, to specific receptors on a target cell, such as a liver cell, to elicit a response—in this case, the uptake of glucose. There are many other examples, which we will discuss in Chapter 15.

ENERGY TRANSFORMATION. In a variety of cells, the membranes of organelles are specialized for processing energy (Figure 5.17b). For example, the inner mitochondrial membrane helps convert the energy of fuel molecules to the energy in ATP, and the thylakoid membranes of chloroplasts participate in the conversion of light energy to the energy of chemical bonds. The two characteristics of membranes that enable them to participate in these processes are their structural organization and their separation of electric charges.

ORGANIZING CHEMICAL REACTIONS. Many processes in cells depend on a series of enzyme-catalyzed reactions in which the products of one reaction serve as the reactants for the next. For such a reaction to occur, all the necessary molecules must come together. In a solution, the reactants and enzymes are all randomly distributed, and collisions among them are random. For this reason, a complete series of chemical reactions in solution may occur very slowly. However, if the different enzymes are bound to a membrane in sequential order, the product of one reaction can be released close to the enzyme for the next reaction. With such an "assembly line," reactions proceed more rapidly and efficiently (Figure 5.17c).

Membranes Are Dynamic

As we have seen in this chapter, membranes participate in numerous physiological and biochemical processes. Membranes are dynamic in another sense as well: They are constantly forming, transforming from one type to another, fusing with one another, and breaking down.

In eukaryotes, phospholipids are synthesized on the surface of the smooth endoplasmic reticulum and rapidly distributed to membranes throughout the cell as vesicles form from the ER, move away, and fuse with other organelles. Membrane proteins are inserted into the rough endoplasmic reticulum as they form on ribosomes. Functioning membranes also move about within eukaryotic cells. For example, portions of the rough ER bud away from the ER and join the *cis* faces of the Golgi apparatus (see Chapter 4). Rapidly—often in less than an hour—these segments of membrane

(a) **Information processing**

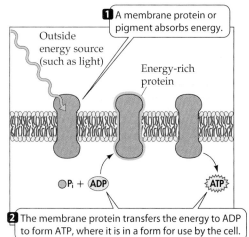

Signal molecule

Signal binding site

Outside of cell

1 Signal binding induces a change in the receptor protein...

Cytoplasm

2 ...causing some effect inside the cell.

(b) **Energy transformation**

1 A membrane protein or pigment absorbs energy.

Outside energy source (such as light)

Energy-rich protein

P_i + ADP

ATP

2 The membrane protein transfers the energy to ADP to form ATP, where it is in a form for use by the cell.

5.17 More Membrane Functions
(a) Membrane proteins conduct signals from outside the cell that trigger changes inside the cell. (b) The membranes of organelles such as mitochondria and chloroplasts are specialized for the transformation of energy. (c) When a series of biochemical reactions must take place in sequence, the membrane can sometimes arrange the enzymes in an "assembly line" to ensure that the reactions occur in proximity to each other.

(c) **Organizing chemical reactions**

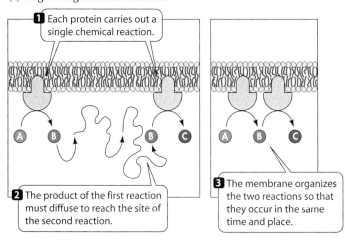

1 Each protein carries out a single chemical reaction.

A → B

B → C

2 The product of the first reaction must diffuse to reach the site of the second reaction.

A → B → C

3 The membrane organizes the two reactions so that they occur in the same time and place.

find themselves in the *trans* regions of the Golgi, from which they bud away to join the plasma membrane (Figure 5.18)

During this journey, changes in the membrane's proteins and phospholipids occur. Membrane from vesicles is constantly merging with the plasma membrane by exocytosis, but this process is largely balanced by the removal of membrane in endocytosis, affording a recovery path by which internal membranes are replenished. In sum, there is a steady flux of membranes and membrane components in cells.

Because all membranes appear similar under the electron microscope, and because they interconvert readily, we might expect all subcellular membranes to be chemically identical. However, that is not the case, for there are major chemical differences among the membranes of even a single cell. Membranes are changed chemically when they form parts of certain organelles. In the Golgi apparatus, for example, the membranes of the *cis* face closely resemble those of the endoplasmic reticulum in chemical composition, but the *trans*-face membranes are more similar to the plasma membrane. As a vesicle is formed, the mix of pro-

5.18 Dynamic Continuity of Cellular Membranes
Membranes continually form, move, and fuse in cells.

1 New stretches of membrane may be generated at certain locations, such as the endoplasmic reticulum.

2 Vesicles budding from the *trans* region of the Golgi apparatus are also membrane-enclosed.

3 The vesicles may remain inside the cell as organelles, such as lysosomes...

4 ...or they may fuse with the plasma membrane, delivering their contents to the exterior of the cell (exocytosis) and adding their membranes to the plasma membrane.

5 Membrane is subtracted from the plasma membrane by endocytosis.

Nuclear envelope

Endoplasmic reticulum

Secretory vesicle

Exocytosis

Plasma membrane

cis region of Golgi apparatus

Lysosome

Endocytosis

trans region of Golgi apparatus

Outside of cell

teins and lipids in its membrane is selected, just as its internal contents are selected, to correspond with the vesicle's target membrane.

Ceaselessly moving, functioning, changing their composition and roles, biological membranes are central to life.

Chapter Summary

Membrane Composition and Structure

▶ Biological membranes consist of lipids, proteins, and carbohydrates. The fluid mosaic model of membrane structure describes a phospholipid bilayer in which membrane proteins can move about laterally within the membrane. **Review Figures 5.1, 5.2**

▶ Integral membrane proteins are at least partially inserted into the phospholipid bilayer. Peripheral proteins attach to the surface of the bilayer by ionic bonds. **Review Figure 5.1**

▶ The two surfaces of a membrane may have different properties because of their different phospholipid composition, exposed domains of integral membrane proteins, and their peripheral membrane proteins. **Review Figures 5.1, 5.2**

▶ Carbohydrates attached to proteins or phospholipids project from the external surface of the plasma membrane and function as recognition signals for interactions between cells. **Review Figure 5.1**

Cell Adhesion

▶ In an organism or tissue, cells recognize and bind to each other by means of membrane proteins that protrude from the cell surface. **Review Figure 5.5**

▶ Tight junctions prevent the passage of molecules through the space around cells, and they define functional regions of the plasma membrane by restricting the migration of membrane proteins uniformly over the cell surface. Desmosomes allow cells to adhere strongly to one another. Gap junctions provide channels for chemical and electrical communication between adjacent cells. **Review Figure 5.6**

Passive Processes of Membrane Transport

▶ Substances can diffuse passively across a membrane by three processes: unaided diffusion through the phospholipid bilayer, facilitated diffusion through protein channels, or facilitated diffusion by means of a carrier protein. **Review Table 5.1**

▶ Solutes diffuse across a membrane from a region with a greater solute concentration to a region with a lesser solute concentration. Equilibrium is reached when the concentrations of a given solute are identical on both sides of the membrane. **Review Figure 5.7**

▶ The rate of simple diffusion of a solute across a membrane is directly proportional to the concentration gradient across the membrane. An important factor in simple diffusion across a membrane is the lipid solubility of the solute.

▶ In osmosis, water diffuses from regions of higher water concentration to regions of lower water concentration.

▶ In hypotonic solutions, cells tend to take up water, while in hypertonic solutions, cells tend to lose water. Animal cells must remain isotonic to the environment to prevent destructive loss or gain of water. **Review Figure 5.8***a*

▶ The cell walls of plants and some other organisms prevent the cells from bursting under hypotonic conditions. The turgor pressure that develops under these conditions keeps plants upright and stretches the cell wall during plant cell growth. **Review Figure 5.8***b*

▶ Channel proteins and carrier proteins function in facilitated diffusion. **Review Figures 5.9, 5.10**

▶ The rate of carrier-mediated facilitated diffusion reaches a maximum when a solute concentration is reached that saturates the carrier proteins so that no increase in rate is observed with further increases in solute concentration.

Active Transport

▶ Active transport requires the use of energy to move substances across a membrane against a concentration gradient. **Review Table 5.1**

▶ Active transport proteins may be uniports, symports, or antiports. **Review Figure 5.11**

▶ In primary active transport, energy from the hydrolysis of ATP is used to move ions into or out of cells against their concentration gradients. **Review Figure 5.12**

▶ Secondary active transport couples the passive movement of one solute with its concentration gradient to the movement of another solute against its concentration gradient. Energy from ATP is used indirectly to establish the concentration gradient that results in the movement of the first solute. **Review Figure 5.13**

Endocytosis and Exocytosis

▶ Endocytosis transports macromolecules, large particles, and small cells into eukaryotic cells by means of engulfment by and vesicle formation from the plasma membrane. Phagocytosis and pinocytosis are both nonspecific types of endocytosis. **Review Figures 5.14, 5.15**

▶ In receptor-mediated endocytosis, a specific membrane receptor binds to a particular macromolecule. **Review Figure 5.16**

▶ In exocytosis, materials in vesicles are secreted from the cell when the vesicles fuse with the plasma membrane. **Review Figure 5.14**

Membranes Are Not Simply Barriers

▶ Membranes function as sites for recognition and initial processing of extracellular signals, for energy transformations, and for organizing chemical reactions. **Review Figure 5.17**

Membranes Are Dynamic

▶ Although not all cellular membranes are identical, ordered modifications in membrane composition accompany the conversions of one type of membrane into another type. **Review Figure 5.18**

For Discussion

1. In Chapter 47, we will see that the functioning of muscles requires calcium ions to be pumped into a subcellular compartment against a calcium concentration gradient. What types of molecules are required for this to happen?

2. Some algae have complex glassy structures in their cell walls. These structures form within the Golgi apparatus. How do these structures reach the cell wall without having to pass through a membrane?

3. Organisms that live in fresh water are almost always hypertonic to their environment. In what way is this a serious problem? How do some organisms cope with this problem?

4. Contrast nonspecific endocytosis and receptor-mediated endocytosis with respect to mechanism and to performance.

6 Energy, Enzymes, and Metabolism

THE 5-YEAR-OLD BOY PRESENTED BOSTON SURgeon Joseph Upton with a problem. Dr. Upton had sewn the boy's ear back on after a dog had bitten it off, but after 4 days, blood flow to and from the ear was blocked. If blood flow was not restored quickly, the reattachment would fail. To open up the blood vessels, Dr. Upton tried an old technique: He applied 24 leeches to the wound. The leeches used their sucking mouthparts to attach to the boy's ear and drank his blood. In the process, they released a molecule called hirudin (after the scientific name of the leech, *Hirudo medicinalis*) into the boy's blood. A potent anti-clotting agent, the hirudin acted slowly but consistently over 24 hours to clear the obstructed blood vessels, and the boy's ear was saved.

The medical use of leeches to prevent blood clotting goes back thousands of years. One of the most powerful anti-clotting agents known, hirudin is a small protein of 65 amino acids that folds into a specific shape that allows it to bind tightly to thrombin, a protein present in human blood. In the absence of hirudin, thrombin has a three-dimensional structure that allows it to bind to fibrinogen (yet another blood plasma protein). When this binding occurs, a peptide bond between two of the amino acids in fibrinogen is broken, forming fibrin—the protein that forms blood clots. If hirudin binds to thrombin, thrombin cannot act on fibrinogen, and blood clots do not form.

In chemical terms, thrombin is an enzyme, or biological catalyst, for fibrin formation. The hydrolysis of peptide bonds to form fibrin would happen whether thrombin was there or not; it's just that it would happen much more slowly, certainly too slowly to have any benefit in the lifetime of the organism! Thousands of such reactions go on all the time in every organism, each reaction catalyzed by a specific protein with a particular three-dimensional structure. Taken together, these reactions make up metabolism, which is the sum total of all of the chemical conversions in a cell.

Many metabolic reactions can be classified as either (1) building up complexity in the cell, using energy to do so; or (2) breaking down complex substances into simpler ones, releasing energy in the process.

Biomedical Medicine from a Natural Source
Leeches are the source of hirudin, a molecule that prevents blood coagulation by inhibiting the action of an enzyme, thrombin, in mammalian blood.

This chapter is concerned with energy and enzymes. Without them, neither we nor any other organism would be able to function. Indeed, when an enzyme is inactivated, either by the binding of an inhibitor such as hirudin that keeps the enzyme from binding to its target, or by some error leading to an alteration in its three-dimensional structure, its function is destroyed. This can have dire consequences: What if we had hirudin in our blood all the time?

Before considering how enzymes perform their molecular wizardry, let us consider the general principles of energy in biological systems.

Energy and Energy Conversions

Physicists define *energy* as the capacity to do work, which occurs when a force operates on an object over a distance. In biochemistry, energy represents the capacity for change. All living things must obtain energy from the environment—no cell manufactures energy. Indeed, one of the fundamental physical laws is that energy can neither be created nor destroyed. However, energy can be transformed from one kind into another. Energy transformations are linked to the chemical transformations that occur in cells. *Metabolism* is the total chemical activity of a living organism; at any instant, metabolism consists of thousands of individual chemical reactions.

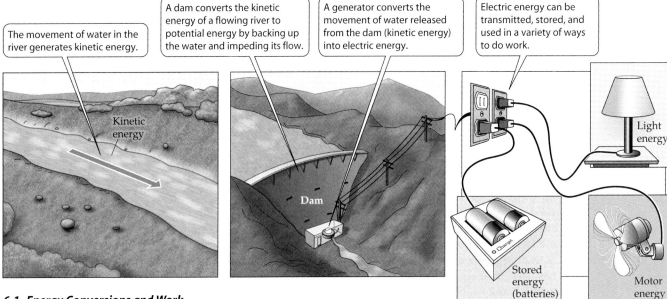

The movement of water in the river generates kinetic energy.

A dam converts the kinetic energy of a flowing river to potential energy by backing up the water and impeding its flow.

A generator converts the movement of water released from the dam (kinetic energy) into electric energy.

Electric energy can be transmitted, stored, and used in a variety of ways to do work.

Kinetic energy

Dam

Light energy

Stored energy (batteries)

Motor energy

6.1 Energy Conversions and Work
The kinetic energy of a flowing river can be converted to potential energy by a dam. Release of water from the dam converts the potential energy back into kinetic energy, which a generator can convert into electric energy.

Energy changes are related to changes in matter

Energy comes in many forms, such as chemical energy, light energy, and mechanical energy. But all forms of energy can be considered as one of two basic types:

▶ **Kinetic energy** is the energy of movement. This type of energy does work that alters the state or motion of matter. It can exist in the form of heat, light, electric, and mechanical energy, among others.
▶ **Potential energy** is the energy of state or position—that is, stored energy. It can be stored in chemical bonds, as a concentration gradient, and as electric potential, among other ways.

Water stored behind a dam has potential energy. When the water is released from the dam, some of this potential energy is converted into kinetic energy (Figure 6.1). Likewise, fatty acids, with their many C—C and C—H bonds, store chemical energy, which can be released to do biochemical work. In Chapter 5, we saw an example of the potential energy of a concentration gradient in secondary active transport, in which the gradient of one substance (Na^+) across a plasma membrane powers the transport of another (glucose) (see Figure 5.13).

In all cells of all organisms, two types of metabolic reactions occur:

▶ **Anabolic reactions** link together simple molecules to form more complex molecules. The synthesis of a protein from amino acids is anabolic. Anabolic reactions store energy in the chemical bonds that are formed.
▶ **Catabolic reactions** (catabolism) break down complex molecules into simpler ones and release stored energy.

Catabolic and anabolic reactions are often linked. The energy released in catabolic reactions is used to do biological work and drive anabolic reactions (Figure 6.2). The energy needed during anabolism to form the peptide bonds that link amino acids together into proteins comes from catabolism.

Cellular activities such as growth, motion, and active transport of ions across a membrane all require energy, and

6.2 Biological Energy Transformations
Cavorting lionesses convert chemical energy, obtained from the prey they have eaten, into a burst of kinetic energy of motion. Their prey obtained chemical energy by consuming plants. The plants trapped light energy and produced the prey's food by photosynthesis.

6.7 Using /
Fireflies con\
energy, emit
to mate. Ver

ganic pho:
well as fre

Two impc

▶ It is exe
energy
temper
of livin

▶ The eq
is, tow;
there i:

What ch
leased b
phosph;
AMP—t
charged
of free (
phates
the adc
diphos]
charged
added.
compre
energy
when t
make ;

ATP c

As we
and y
the fo

*The
kJ/mc
and p
that d

(a)

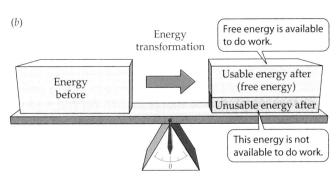

Energy
transformation

Energy
before

Energy
after

A measuring de-
vice indicates that
the total energy
does not change.

0

The First Law of Thermodynamics. The total
amount of energy before a transformation
equals the total amount after a transformation.
No new energy is created, and no energy is lost.

(b)

Energy
transformation

Free energy is available
to do work.

Energy
before

Usable energy after
(free energy)

Unusable energy after

This energy is not
available to do work.

0

The Second Law of Thermodynamics. Although a transformation
does not change the total amount of energy within a closed system,
after any transformation the amount of free energy available to do
work is always less than the original amount of energy.

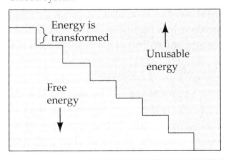

Closed system

Energy is
transformed

Unusable
energy

Free
energy

Another statement of the Second Law is that in a closed system,
with repeated energy transformations, free energy decreases and
unusable energy increases—a phenomenon known as **entropy**.

6.3 The Laws of Thermodynamics
(a) The first law is that energy cannot be created or destroyed.
(b) The second law is that during energy transformations, free
energy is lost.

none of them would proceed without a source of energy. In
the discussion that follows, you will discover the physical
laws that govern all energy transformations, identify the
energy available to do work, and consider the direction of
energy flow.

The first law: Energy is neither created nor destroyed

Energy can be converted from one form to another. For example, by striking a match, you convert potential chemical
energy to light and heat. In any conversion of energy from
one form to another (chemical to light, mechanical to electric), energy is neither created nor destroyed. This is the
first law of thermodynamics (Figure 6.3a).

The first law applies to the universe as a whole or to
any closed system within the universe. By "system" we
mean any part of the universe containing specified matter
and energy. A *closed system* is one that is not exchanging
energy with its surroundings. For example, a thermos bot-
tle does not gain or lose heat, and so the material inside it
is a closed system (Figure 6.4a).

Open systems, such as living cells, exchange matter and
energy with their surroundings (Figure 6.4b). Does this
mean that cells disobey the first law, or that the first law
does not apply to living organisms? Not at all. It means
that an open system is merely one part of a larger closed
system and receives energy from other parts of that larger
system.

The first law tells us that in any interconversion of the
forms of energy, the total energy before and after the conversion is the same. As you will see in the next two chapters, potential energy in the chemical bonds of carbohydrates and lipids can be converted to potential energy in
ATP. This energy can then be used to produce potential energy in the concentration gradients established by active
transport, which can be converted to kinetic energy and
used to do mechanical work, such as muscle contraction.

The second law: Not all energy can be used, and disorder tends to increase

The **second law of thermodynamics** states that, although
energy cannot be created or destroyed, when energy is
converted from one form to another, some of the energy
becomes unavailable to do work (Figure 6.3b). In other
words, no physical process or chemical reaction is 100 percent efficient, and not all the energy released can be converted to work. Some energy is lost to a form associated
with disorder. The second law applies to all energy transformations, but we will focus here on chemical reactions in
living systems.

NOT ALL ENERGY CAN BE USED. In any system, the total energy includes the *usable* energy that can do work and the *unusable* energy that is lost to disorder:

total energy = usable energy + unusable energy

In biological systems, the total energy is called **enthalpy**
(*H*). The usable energy that can do work is called **free energy** (*G*). Free energy is what cells require for all the chemical reactions of cell growth, cell division, and the maintenance of cell health. The unusable energy is represented by
entropy (*S*), which is the disorder of the system, multiplied

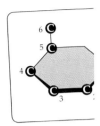

The pathways of glycolysis and cellular respiration (pyruvate oxidation, citric acid cycle and respiratory chain) are represented by a "road map" of symbols that guide you to better understand the relationship of these pathways in the illustrations that follow.

7.7 Glycolysis Converts Glucose to Pyruvate
Ten enzymes, starting with hexokinase, catalyze ten reactions in turn. Along the way, ATP is produced (reactions 7 and 10), and two NAD^+ are reduced to two $NADH + H^+$ (reaction 6).

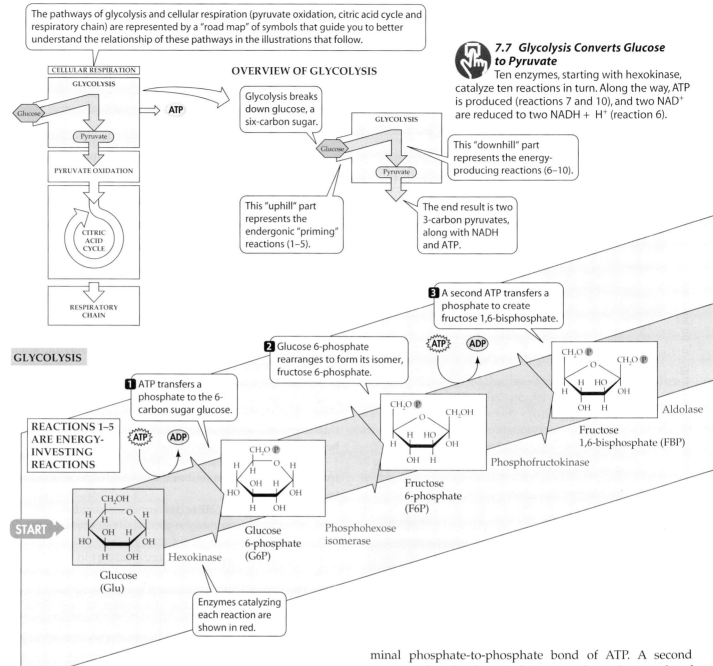

OVERVIEW OF GLYCOLYSIS

Glycolysis breaks down glucose, a six-carbon sugar.

This "downhill" part represents the energy-producing reactions (6–10).

This "uphill" part represents the endergonic "priming" reactions (1–5).

The end result is two 3-carbon pyruvates, along with NADH and ATP.

GLYCOLYSIS

3 A second ATP transfers a phosphate to create fructose 1,6-bisphosphate.

2 Glucose 6-phosphate rearranges to form its isomer, fructose 6-phosphate.

1 ATP transfers a phosphate to the 6-carbon sugar glucose.

REACTIONS 1–5 ARE ENERGY-INVESTING REACTIONS

Glucose (Glu)

Hexokinase

Glucose 6-phosphate (G6P)

Phosphohexose isomerase

Fructose 6-phosphate (F6P)

Phosphofructokinase

Fructose 1,6-bisphosphate (FBP)

Aldolase

Enzymes catalyzing each reaction are shown in red.

6.6 Concentra
No matter what phosphate are d there will always cent glucose 1-p

molecules are ance between *chemical equili*

Chemical ec

Every chemi not necessari tants presen Each reactio equilibrium the reaction principle of e

Every livi converted ir we start out phate that 1 molar conce tained unde pH 7). As th tration of th 0.019 *M*, wl to 0.001 *M* reached at the reverse 1-phospha action.

At equi actant rati has gone right," as spontane the experi tion is des

glt

The ch lated dir ward cor

molecules of ADP, with a rearrangement in between. More than 20 kcal (83.6 kJ/mol) of free energy is stored in ATP for every mole of BPG broken down. Finally, we are left with two moles of pyruvate for every mole of glucose that entered glycolysis.

SUBSTRATE-LEVEL PHOSPHORYLATION. The enzyme-catalyzed transfer of phosphate groups from donor molecules to ADP molecules (reaction 7) is called **substrate-level phosphorylation**. This process is driven by energy obtained from oxidation. For example, when G3P reacts with a phosphate group (P_i) and NAD^+, becoming BPG, an aldehyde is oxidized to a carboxylic acid, with NAD^+ acting as the oxidizing agent. The oxidation provides so much energy that the newly added phosphate group is linked to the rest of the molecule by a bond that has even more energy than the ter-

minal phosphate-to-phosphate bond of ATP. A second enzyme, phosphoglycerate kinase, catalyzes the transfer of this phosphate group from BPG to ADP in reaction 7, forming ATP. Both reactions are exergonic, even though a substantial amount of energy is consumed in the formation of ATP.

GLYCOLYSIS MAY BE FOLLOWED BY FERMENTATION. A review of the glycolytic pathway shows that at the beginning of glycolysis, two molecules of ATP are used per molecule of glucose, but that ultimately four molecules of ATP are produced (two for each of the two BPG molecules)—a net gain of two ATP molecules and two $NADH + H^+$.

When fermentation follows glycolysis, the total usable energy yield is just these two ATP molecules per glucose molecule. Under anaerobic conditions, the $NADH + H^+$ is rapidly recycled to NAD^+ by the reduction of pyruvate. The NAD^+ is then available for the glycolytic reaction catalyzed by the enzyme triose phosphate dehydrogenase (reaction 6 in Figure 7.7).

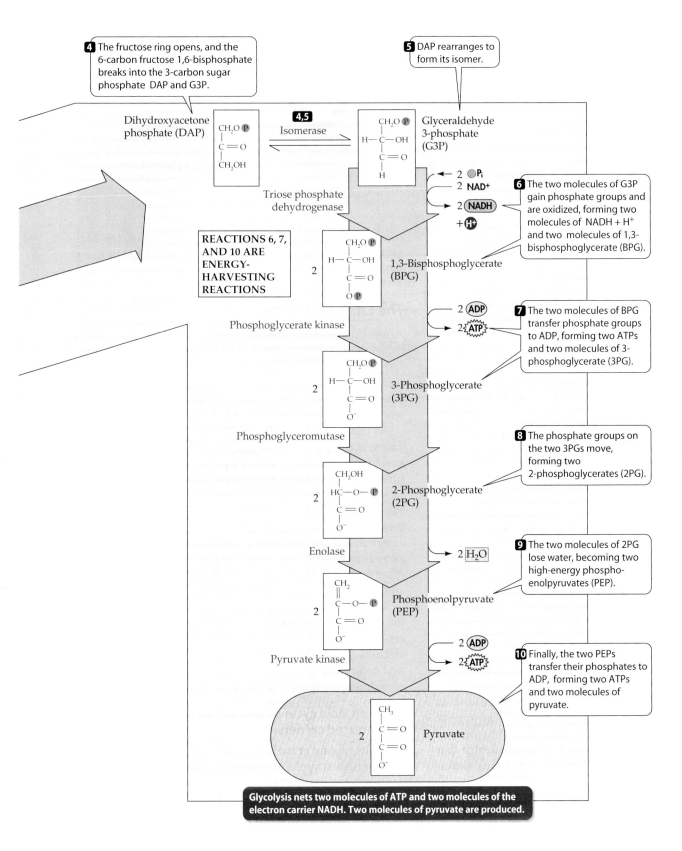

4 The fructose ring opens, and the 6-carbon fructose 1,6-bisphosphate breaks into the 3-carbon sugar phosphate DAP and G3P.

5 DAP rearranges to form its isomer.

Dihydroxyacetone phosphate (DAP)

4,5 Isomerase

Glyceraldehyde 3-phosphate (G3P)

Triose phosphate dehydrogenase

REACTIONS 6, 7, AND 10 ARE ENERGY-HARVESTING REACTIONS

2 P_i
2 NAD+
2 NADH
+ H+

6 The two molecules of G3P gain phosphate groups and are oxidized, forming two molecules of NADH + H+ and two molecules of 1,3-bisphosphoglycerate (BPG).

2 1,3-Bisphosphoglycerate (BPG)

Phosphoglycerate kinase

2 ADP
2 ATP

7 The two molecules of BPG transfer phosphate groups to ADP, forming two ATPs and two molecules of 3-phosphoglycerate (3PG).

2 3-Phosphoglycerate (3PG)

Phosphoglyceromutase

8 The phosphate groups on the two 3PGs move, forming two 2-phosphoglycerates (2PG).

2 2-Phosphoglycerate (2PG)

Enolase

2 H_2O

9 The two molecules of 2PG lose water, becoming two high-energy phosphoenolpyruvates (PEP).

2 Phosphoenolpyruvate (PEP)

Pyruvate kinase

2 ADP
2 ATP

10 Finally, the two PEPs transfer their phosphates to ADP, forming two ATPs and two molecules of pyruvate.

2 Pyruvate

Glycolysis nets two molecules of ATP and two molecules of the electron carrier NADH. Two molecules of pyruvate are produced.

On the other hand, in the presence of oxygen, eukaryotes and some bacteria reap far more energy by completely oxidizing pyruvate and by oxidizing NADH + H+ through the respiratory chain, as we will see in the sections that follow. In eukaryotes, these reactions take place in the mitochondria.

8 Photosynthesis: Energy from the Sun

FOR SEVERAL DECADES, CORN GROWERS IN THE United States competed to see who could coax the highest yield of grain from their acreage. After rising rapidly in the first half of the twentieth century, yields continued to increase, albeit somewhat more slowly. But the trend was clearly up—until the last decade of the century. From 1990 on, crop yields per acre have leveled off for corn, rice, and wheat—three grains which together supply over half the human race with food.

Although overall food production continues to rise as more land is put into production and the environment of the crops is more intensively manipulated with fertilizers and pesticides, the increase of the human population is wiping out any per capita gains. Per person food production has not improved much since 1960–1980, the peak of the so-called "Green Revolution" in agriculture. This was a period when new genetic strains and more intensive environmental management combined to more than double crop yields.

To coax crop plants to grow more and produce more on the available land, scientists are now focusing on photosynthesis, the biochemical process by which plants turn sunlight into carbohydrates, sugars, and starch. Photosynthesis is the very basis of life on Earth.

The basic transformation of photosynthesis—the conversion of solar energy into chemical energy—is a familiar example of the laws of thermodynamics. As the first law tells us, when the form of energy changes from sunlight to plant, no energy is lost. However, as the second law states, the conversion is relatively inefficient, with only about 4 percent of the incident solar energy ending up in chemical bonds. Moreover, the use of solar energy initially captured as ATP and reduced electron carriers to reduce carbon dioxide to sugars is also inefficient.

How can these efficiencies be improved? An important first step is a thorough understanding of photosynthesis. The process of photosynthesis can be neatly broken down into two steps. The first step is the conversion of energy from light to chemical bonds in reduced electron carriers and ATP. In the second step, these two sources of chemical energy are used to drive the synthesis of carbohydrates from carbon dioxide. In this chapter, we will examine these two processes, and show how they are related to each other and to plant growth.

Identifying Photosynthetic Reactants and Products

By the beginning of the nineteenth century, scientists understood the broad outlines of photosynthesis. It was known to use three principal ingredients—water, carbon dioxide (CO_2), and light—and to produce not only carbohydrate but also oxygen gas (O_2). Scientists had learned that:

► The water for photosynthesis in land plants comes primarily from the soil and must travel from the roots to the leaves.
► Carbon dioxide is taken in, and water and O_2 are released, through tiny openings in leaves, called *stomata* (singular *stoma*) (Figure 8.1)
► Light is absolutely necessary for the production of oxygen and carbohydrate.

Primary Producers
Powered by sunlight, corn plants (*Zea mays*) convert atmospheric CO_2 and water into an energy source (food) for humans and animals.

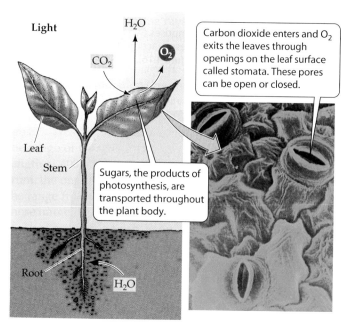

8.1 Ingredients for Photosynthesis
A typical terrestrial plant uses light from the sun, water from the soil, and carbon dioxide from the atmosphere to form organic compounds by photosynthesis.

By 1804, scientists could summarize photosynthesis as follows:

carbon dioxide + water + light energy → sugar + oxygen

which turns into an equation that is the *reverse* of the overall equation for cellular respiration given in Chapter 7:

$$6 CO_2 + 6 H_2O \rightarrow C_6H_{12}O_6 + 6 O_2$$

Although correct, these statements say nothing about the details of the process. What roles does light play? How do the carbons become linked? And does the oxygen gas come from the CO_2 or from the H_2O?

Almost a century and a half passed before the source of the O_2 released during photosynthesis was determined. Its identification was one of the first uses of an isotopic tracer in biological research. In these experiments, two groups of green plants were allowed to carry on photosynthesis. Plants in the first group were supplied with water containing the heavy-oxygen isotope ^{18}O and with CO_2 containing only the common oxygen isotope ^{16}O; plants in the second group were supplied with CO_2 labeled with ^{18}O and water containing only ^{16}O.

When oxygen gas was collected from each group of plants and analyzed, it was found that O_2 containing ^{18}O was produced in abundance by the plants that had been given ^{18}O-labeled water, but not by the plants given labeled CO_2. These results showed that all the oxygen gas produced during photosynthesis comes from water (Figure 8.2). This discovery is reflected in a revised balanced equation:

$$6 CO_2 + 12 H_2O \rightarrow C_6H_{12}O_6 + 6 O_2 + 6 H_2O$$

Water appears on both sides of the equation because water is both used as a reactant (the twelve molecules on the left) and released as a product (the six new ones on the right). In this revised equation, there are now sufficient water molecules to account for all the oxygen gas produced.

The photosynthetic production of oxygen by green plants is an important source of atmospheric oxygen, which most organisms—including plants themselves—require in order to complete their respiratory chains and obtain the energy for life.

The Two Pathways of Photosynthesis: An Overview

The overall photosynthetic reaction takes place in the chloroplasts of photosynthetic cells, which in most plants are found in the leaves. But photosynthesis does not proceed in a single step. In fact, in all of chemistry, no such complex reaction is accomplished in a single step. Rather, a series of simpler steps is required.

By the middle of the twentieth century, it was clear that photosynthesis consists of many reactions that can be divided into two pathways:

▶ The first pathway, called the *light reactions*, is driven by light energy. It produces ATP and a reduced electron carrier ($NADPH + H^+$).

▶ The second pathway, called the *Calvin–Benson cycle*, does not use light directly. It uses ATP, $NADPH + H^+$, and CO_2 to produce sugar.

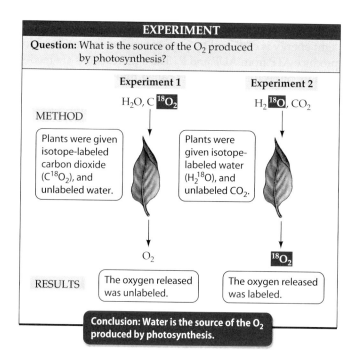

8.2 Water Is the Source of the Oxygen Produced by Photosynthesis
Because only plants given isotope-labeled water released labeled O_2, this experiment showed that water is the source of the oxygen released during photosynthesis.

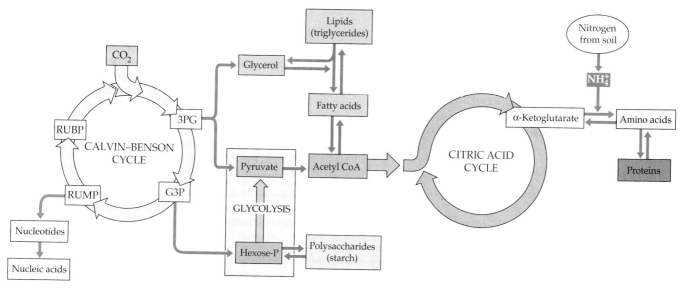

8.22 Metabolic Interactions in a Plant Cell
Note the relationships among the Calvin–Benson cycle, the citric acid cycle, and glycolysis.

▶ 3PG from the Calvin–Benson cycle can be converted to pyruvate, the end product of glycolysis.

▶ G3P from the Calvin–Benson cycle can be converted to hexose phosphates (such as glucose 1-phosphate), which can enter glycolysis.

In both cases, the result is the catabolic breakdown of Calvin–Benson cycle products to CO_2, with the associated synthesis of ATP. Once the carbon skeletons from the Calvin–Benson cycle enter the "central switching yard" of glycolysis and the citric acid cycle, they can be used anabolically to make lipids, proteins, and other carbohydrates (see Figure 7.19).

Energy flows from sunlight to reduced carbon in photosynthesis to ATP in respiration. Energy can also be stored in the bonds of macromolecules such as polysaccharides, lipids, and proteins. For a plant to grow, energy storage (as body structures) must exceed energy release; that is, overall photosynthesis to fixed carbon must exceed respiration. This is the aim of the farmers growing corn whom we described in the opening of this chapter. And it is the basis of the ecological food chain, as we will see in later chapters.

Chapter Summary

▶ Life on Earth depends on the absorption of light energy from the sun.

▶ In plants, photosynthesis takes place in chloroplasts.

Identifying Photosynthetic Reactants and Products

▶ Photosynthesizing plants take in CO_2, water, and light energy, producing O_2 and carbohydrate. The overall reaction is

$$6\,CO_2 + 12\,H_2O + light \rightarrow C_6H_{12}O_6 + 6\,O_2 + 6\,H_2O$$

The oxygen atoms in O_2 come from water, not from CO_2. **Review Figures 8.1, 8.2**

The Two Pathways of Photosynthesis: An Overview

▶ In the light reactions of photosynthesis, electron flow and photophosphorylation produce ATP and reduce $NADP^+$ to $NADPH + H^+$. **Review Figure 8.3**

▶ ATP and $NADPH + H^+$ are needed for the reactions that fix and reduce CO_2 in the Calvin–Benson cycle, forming sugars. **Review Figure 8.3**

Properties of Light and Pigments

▶ Light energy comes in packets called photons, but it also has wavelike properties. **Review Figure 8.4**

▶ Pigments absorb light in the visible spectrum. **Review Figure 8.5**

▶ Absorption of a photon puts a pigment molecule in an excited state that has more energy than its ground state. **Review Figure 8.6**

▶ Each compound has a characteristic absorption spectrum. An action spectrum reveals the biological effectiveness of different wavelengths of light. **Review Figures 8.7, 8.8**

▶ Chlorophylls and accessory pigments form antenna systems for absorption of light energy. **Review Figures 8.7, 8.9, 8.11**

Light Reactions: Light Absorption

▶ An excited pigment molecule may lose its energy by fluorescence, or by transferring it to another pigment molecule. **Review Figures 8.10, 8.11**

Electron Flow, Photophosphorylation, and Reductions

▶ Noncyclic electron flow uses two photosystems (I and II), producing ATP, $NADPH + H^+$, and O_2. Photosystem II uses P_{680} chlorophyll, from which light-excited electrons are passed to a redox chain that drives chemiosmotic ATP production. Light-driven oxidation of water releases O_2 and passes electrons from water to the P_{680} chlorophyll. Photosystem I passes electrons from P_{700} chlorophyll to another redox chain and then to $NADP^+$, forming $NADPH + H^+$. **Review Figure 8.12**

▶ Cyclic electron flow uses P_{700} chlorophyll and produces only ATP. Its operation maintains the proper balance of ATP and $NADPH + H^+$ in the chloroplast. **Review Figure 8.13**

▶ Chemiosmosis is the source of ATP in photophosphorylation. Electron transport pumps protons from the stroma into the thylakoids, establishing a proton-motive force. Diffusion of the protons back to the stroma via ATP synthase channels drives ATP formation from ADP and P_i. **Review Figure 8.14**

▶ Photosynthesis probably originated in anaerobic bacteria that used H_2S as a source of electrons instead of H_2O. Oxygen production by bacteria was an important event in the evolution of eukaryotes.

Making Sugar from CO_2: The Calvin–Benson Cycle

▶ The Calvin–Benson cycle makes sugar from CO_2. This pathway was elucidated through the use of radioactive tracers. **Review Figure 8.15**

▶ The Calvin–Benson cycle consists of three phases: fixation of CO_2, reduction and carbohydrate production, and regeneration of RuBP. RuBP is the initial CO_2 acceptor, and 3PG is the first stable product of CO_2 fixation. The enzyme rubisco catalyzes the reaction of CO_2 and RuBP to form 3PG. **Review Figures 8.16, 8.17**

Photorespiration and Its Evolutionary Consequences

▶ The enzyme rubisco can catalyze a reaction between O_2 and RuBP in addition to the reaction between CO_2 and RuBP. This consumption of O_2 is called photorespiration and significantly reduces the efficiency of photosynthesis. The reactions that constitute photorespiration are distributed over three organelles: chloroplasts, peroxisomes, and mitochondria. **Review Figure 8.18**

▶ At high temperatures and low CO_2 concentrations, the oxygenase function of rubisco is favored.

▶ C_4 plants bypass photorespiration with special chemical reactions and specialized leaf anatomy. In C_4 plants, PEP carboxylase in mesophyll chloroplasts initially fixes CO_2 in four-carbon acids, which then diffuse into bundle sheath cells, where their decarboxylation produces locally high concentrations of CO_2. **Review Figures 8.19, 8.20**

▶ CAM plants operate much like C_4 plants, but their initial CO_2 fixation by PEP carboxylase is temporally separated

from the Calvin–Benson cycle, rather than spatially separated as in C_4 plants. **Review Figure 8.21**

Metabolic Pathways in Plants

▶ Plants respire both in the light and in the dark, but photosynthesize only in the light. To survive, a plant must photosynthesize more than it respires, giving it a net gain of reduced energy-rich compounds.

▶ Photosynthesis and respiration are linked through the Calvin–Benson cycle, the citric acid cycle, and glycolysis. **Review Figure 8.22**

For Discussion

1. Both electron flow and the Calvin–Benson cycle stop in the dark. Which specific reaction stops first? Which stops next? Continue answering the question "Which stops next?" until you have explained why both pathways have stopped.

2. In what principal ways are the reactions of electron flow in photosynthesis similar to the respiratory chain and oxidative phosphorylation discussed in Chapter 7? Differentiate between cyclic and noncyclic electron flow in terms of (1) the products and (2) the source of electrons for the reduction of oxidized chlorophyll.

3. The development of what two experimental techniques made it possible to elucidate the Calvin–Benson cycle? How were these techniques used in the investigation?

4. If water labeled with ^{18}O is added to a suspension of photosynthesizing chloroplasts, which of the following compounds will first become labeled with ^{18}O: ATP, NADPH, O_2, or 3PG? If water labeled with ^3H is added to a suspension of photosynthesizing chloroplasts, which of the same compounds will first become radioactive? If CO_2 labeled with ^{14}C is added to a suspension of photosynthesizing chloroplasts, which of those compounds will first become radioactive?

Part Two

INFORMATION
AND HEREDITY

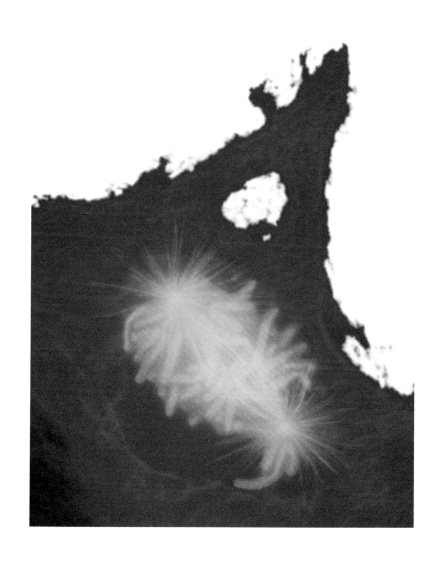

9 Chromosomes, the Cell Cycle, and Cell Division

IN 1951, 31-YEAR-OLD HENRIETTA LACKS EN-tered Johns Hopkins Hospital to be treated for a cancerous tumor. Although she died a few months later, her tumor cells are alive today. Scientists found that, given adequate nourishment, cancerous cells from the tumor reproduced themselves indefinitely in a laboratory dish. These "HeLa cells" became a test-tube model for studies of human cell biology and biochemistry. Over the past half-century, tens of thousands of research articles have been published using information obtained from Henrietta's cells. But are these "immortal" cells really a good model for human biology?

In one sense, they are. Most multicellular organisms come from a single cell: the fertilized egg. This cell reproduces itself to make two cells, these in turn divide to become four cells, and so on until all the cells of a new organism have been produced. An organism is not just a ball of many cells, however; the cells must specialize into tissues and organs, each with specific roles to perform. This process of specialization, or differentiation, is a subject we will return to in later chapters of Part Two.

In normal tissues, cell reproduction ("births") is offset by cell loss ("deaths"). We know cell death is important from careful studies of a tiny worm, in which 1,090 cells are produced from the fertilized egg and exactly 131 of them die before the worm is born. If they do not die, the worm's organs are severely malformed. Another example occurs in the mammalian brain. Young mice, for instance, lose hundreds of thousands of brain cells each day; if these cells do not die, the mouse's overcrowded brain simply does not work.

A cell's death is often programmed into its genetic message; normal cells "sacrifice" themselves for the greater good of the organism. Once an organism reaches its adult size, it stays that way through a combination of cell division and programmed cell death. Like most cancerous cells, Henrietta Lacks's tumor cells keep growing because they have a genetic imbalance that heavily favors cell reproduction over cell death.

Unicellular organisms use cell division primarily to reproduce themselves, whereas in multicellular organisms cell division also plays important roles in the growth and repair of tissues (Figure 9.1) In this chapter, we first de-scribe how prokaryotic cells produce two new organisms from the original single-celled organism. Then we describe two types of cell and nuclear division—mitosis and meio-sis—and relate these two modes of cell division to asexual and sexual reproduction in eukaryotic organisms. Finally, to balance our discussion of cell "birth" through division, we will describe the important process of programmed cell death, also known as apoptosis.

Systems of Cell Reproduction

In order for any cell to divide, four events must occur:

▶ There must be a *reproductive signal*. This signal, which may come either from inside or outside the cell, initiates the cellular reproductive events.
▶ *Replication* of DNA, the genetic material, and other vital cell components must occur so that each of the two new cells will have complete cell functions.
▶ The cell must *distribute* (segregate) the replicated DNA to each of the two new cells.
▶ The cell membrane (and the cell wall, in organisms that have one) must grow to separate the two new cells in a process called *cytokinesis*.

Prokaryotes divide by fission

In prokaryotes, cell division often means reproduction of the entire single-celled organism. The cell grows in size, replicates its DNA, and then essentially divides into two new cells—a process called *fission.*

HeLa Cells: More Births Than Deaths
These cells have been cultured in a laboratory since 1951. They are the source of much data relating the reproduction of human cells.

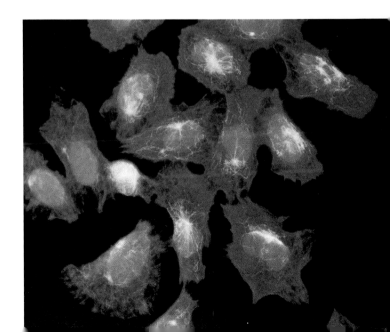

(a) Cell division contributes to the growth of this root tissue.

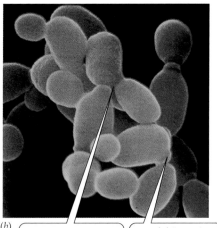

(b) Yeast cells divide by budding. This one has nearly divided…

…and this one is beginning to bud.

(c) Cell division contributes to the regeneration of a lizard's tail.

9.1 Important Consequences of Cell Division
Cell division is the basis for growth, reproduction, and regeneration.

REPRODUCTIVE SIGNALS.   The reproductive rates of many prokaryotes respond to conditions in the environment. The bacterium *Escherichia coli*, a species that is commonly used in genetic studies, is a "cell division machine" that essentially divides continuously. Typically, cell division takes 40 minutes at 37°C. But if there are abundant sources of carbohydrates and salts available, the division cycle speeds up so that cells may divide in 20 minutes. Another bacterium, *Bacillus subtilis*, stops dividing under adverse nutritional conditions, then resumes dividing when things improve. These observations suggest that the initiation of cell division in prokaryotes is under the control of metabolic intermediates, such as carbohydrates, in the environment.

REPLICATION OF DNA.   A **chromosome**, as we saw in Chapter 4, is a DNA molecule containing genetic information. When a cell divides, its chromosomes must be copied, or *replicated*, and each of the two resulting copies must find its way into one of the two new cells.

Most prokaryotes have only one chromosome, a single long DNA molecule with proteins bound to it. In the bacterium *E. coli*, the DNA is a circular molecule about 1.6 million nm (1.6 mm) in circumference. The bacterium itself is only about 1 μm (1,000 nm) in diameter and about 4 μm long. Thus the long thread of DNA, which could form a circle over 100 times larger if fully expanded, is packed into a very small space. So it is not surprising that the molecule usually appears in electron micrographs as a hopeless tangle of fibers (Figure 9.2). The DNA molecule accomplishes some packing by folding in on itself, and positively charged (basic) proteins bound to negatively charged (acidic) DNA contribute to this packing. Circular chromosomes appear to be characteristic of all prokaryotes, as well as some viruses, and are also found in the chloroplasts and mitochondria of eukaryotic cells.

Functionally, the prokaryotic DNA molecule has two regions that are important for cell reproduction:

▶ *Ori* is the origin of replication, where replication of the circle starts.

▶ *Ter* is the terminus of replication, where it ends.

The process of chromosome replication occurs as the DNA is threaded through a "replication complex" of proteins at the center of the cell.

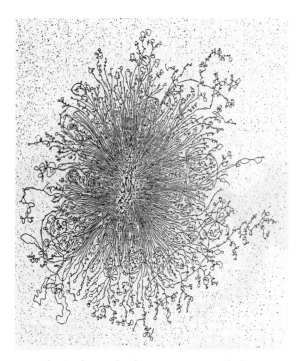

9.2 The Prokaryotic Chromosome Is a Circle
These long, looping fibers of DNA from a cell of the bacterium *Escherichia coli* are all part of one continuous circular chromosome.

DISTRIBUTION OF DNA.   DNA replication actively drives the parceling out of the two new DNA molecules to the new cells. The first region to be replicated is *ori*. The two *ori* regions are attached to the plasma membrane, and they separate as the new chromosome forms and new plasma membrane forms between them (Figure 9.3). By the end of replication, there are two chromosomes, one at either end of the bacterial cell.

CYTOKINESIS. Cell partition, or **cytokinesis**, begins 20 minutes after chromosome duplication is finished. The first event of cytokinesis is a pinching in of the plasma membrane to form a ring similar to a purse string. Fibers composed of a protein similar to eukaryotic tubulin (which makes up microtubules) are major components of this ring. As the membrane pinches in, new cell wall materials are synthesized, which finally separate the two cells.

Eukaryotic nuclei divide by mitosis or meiosis

Cell reproduction in eukaryotes also involves reproductive signals, DNA replication, segregation, and cytokinesis. But, as you might expect, events in eukaryotes are somewhat more complex.

First, unlike prokaryotes, eukaryotic cells do not constantly divide whenever environmental conditions are adequate. In fact, eukaryotic cells that have *differentiated* (become specialized) seldom divide. So the signals for cell division are related not to the physiology of the single cell, but to the needs of the entire organism. Second, instead of a single chromosome, eukaryotes usually have many (humans have 46), so the processes of replication and segregation, while basically the same as in prokaryotes, are more intricate. Third, eukaryotic cells have a distinct nucleus, which has to be replicated and then divided into two new nuclei. Finally, cytokinesis is different in plant cells (which have a cell wall) than in animal cells (which do not).

Mitosis is a nuclear division mechanism that operates in most types of cells. Mitosis sorts the genetic material into two new nuclei and ensures that both contain exactly the same genetic information. A second mechanism of nuclear division, **meiosis**, occurs in the gametes—those cells that will contribute to the reproduction of a new organism. Meiosis generates diversity by shuffling the genetic material, resulting in new gene combinations. It plays a key role in sexual life cycles.

The duplication of a eukaryotic cell typically consists of three steps:

▶ The replication of the genetic material within the nucleus
▶ The packaging and separation of the genetic material into two new nuclei
▶ The division of the cytoplasm

What determines whether a cell will divide? How does mitosis lead to identical cells, and meiosis to diversity? Why do we need both identical copies and diverse cells? Why do most eukaryotic organisms reproduce sexually? In the pages that follow, we will describe the details of mitosis, meiosis, and interphase, as well as their consequences for heredity, development, and evolution.

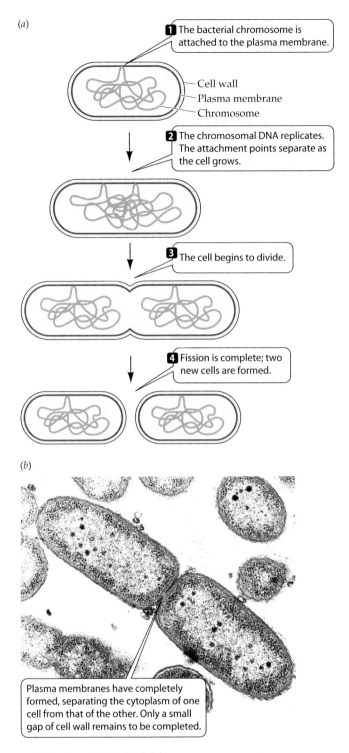

(a)

1. The bacterial chromosome is attached to the plasma membrane.

Cell wall
Plasma membrane
Chromosome

2. The chromosomal DNA replicates. The attachment points separate as the cell grows.

3. The cell begins to divide.

4. Fission is complete; two new cells are formed.

(b)

Plasma membranes have completely formed, separating the cytoplasm of one cell from that of the other. Only a small gap of cell wall remains to be completed.

9.3 Prokaryotic Cell Division
(*a*) The steps of cell division in prokaryotes. (*b*) These two cells of the bacterium *Pseudomonas aeruginosa* have almost completed fission. Each cell contains a complete chromosome, visible as the nucleoid in the center of the cell.

Interphase and the Control of Cell Division

Between divisions of the cytoplasm—that is, for most of its life—a eukaryotic cell is in a condition called **interphase**. A cell lives and functions until it divides or dies—or, if it is a

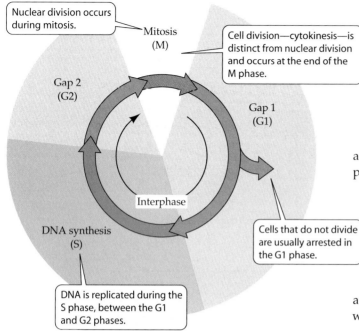

Nuclear division occurs during mitosis.

Mitosis (M)

Cell division—cytokinesis—is distinct from nuclear division and occurs at the end of the M phase.

Gap 2 (G2)

Gap 1 (G1)

Interphase

DNA synthesis (S)

Cells that do not divide are usually arrested in the G1 phase.

DNA is replicated during the S phase, between the G1 and G2 phases.

9.4 The Eukaryotic Cell Cycle

The cell cycle consists of a mitotic (M) phase, during which first nuclear division (mitosis) and then cell division (cytokinesis) take place. The M phase is followed by a long period of growth known as interphase. Interphase has three subphases (G1, S, and G2) in cells that divide.

sex cell, until it fuses with another sex cell. Some types of cells, such as red blood cells, muscle cells, and nerve cells, lose the capacity to divide as they mature. Other cell types, such as cortical cells in plant stems, divide only rarely. Most cells, however, have some probability of dividing, and some are specialized for rapid division. For most types of cells, we may speak of a **cell cycle** that has two phases: mitosis and interphase.

A given cell lives for one turn of the cell cycle and then becomes two cells. The cell cycle, when repeated again and again, is a constant source of new cells. However, even in tissues engaged in rapid growth, cells spend most of their time in interphase. Examination of any collection of dividing cells, such as the tip of a root or a slice of liver, will reveal that most of the cells are in interphase most of the time; only a small percentage of the cells will be in mitosis at any given moment. We can confirm this fact by watching a single cell through its entire cycle.

In this section, we will describe the cell cycle events that occur during interphase, especially the "decision" to enter mitosis.

Interphase consists of three subphases, identified as G1, S, and G2. The cell's DNA replicates during the **S phase** (the S stands for *s*ynthesis). The period between the end of mitosis and the onset of the S phase is called **G1**, or Gap 1. Another gap—**G2**—separates the end of the S phase and the beginning of mitosis, when nuclear and cytoplasmic division take place and two new cells are formed. Mitosis and cytokinesis are referred to as the **M phase** of the cell cycle (Figure 9.4).

The process of DNA replication is a major topic by itself, and will be discussed in Chapter 11. But its result is, at the end of S phase, where there was formerly one chromosome, there are now two, joined together and awaiting segregation into two new cells by mitosis or meiosis.

Although one key event—DNA replication—dominates and defines the S phase, important cell cycle processes take place in the gap phases as well. G1 is quite variable in length in different cell types. Some rapidly dividing embryonic cells dispense with it entirely, while other cells may remain in G1 for weeks or even years. The biochemical hallmark of a G1 cell is that it is preparing for the S phase. It is at the G1-to-S transition that the commitment to enter another cell cycle is made. During G2, the cell makes preparations for mitosis—for example, synthesizing components of the microtubules that will form the spindle.

Cyclins and other proteins signal events in the cell cycle

How are appropriate decisions to enter the S or M phases made? These transitions—from G1 to S and from G2 to M—depend on the activation of a protein called **cyclin-dependent kinase**, or **Cdk**. A *kinase* is an enzyme that catalyzes the transfer of a phosphate group from ATP to another molecule; this phosphate transfer is called *phosphorylation*. Cdk is a kinase that can catalyze the phosphorylation of certain amino acids in proteins.* Activated Cdk's are important in initiating the steps of the cell cycle.

The discovery that Cdk's induce cell division is a beautiful example of research on different organisms and different cell types converging on a single mechanism. One group of scientists was studying immature sea urchin eggs, trying to find out how they are stimulated to divide and form mature eggs. A protein called *maturation promoting factor* was purified from the maturing eggs, which by itself prodded the immature eggs into division. At the same time, other scientists studying the cell cycle in yeast, a single-celled eukaryote, found a strain that was stalled at the G1–S boundary because it lacked a Cdk. This yeast Cdk was very similar to the sea urchin's maturation promoting factor. Similar Cdk's were soon found to control the G1–S transition in many other organisms, including humans.

But Cdk's are not active by themselves. They must be bound to a second type of protein, called **cyclin**. This binding—an example of allosteric interaction—causes the Cdk to alter its shape and exposes its active site. It is the cyclin-Cdk *complex* that acts as a protein kinase and triggers the transition from G1 to S phase. Then the cyclin breaks down and the Cdk becomes inactive (Figure 9.5).

*Phosphorylation changes the three-dimensional structure of the targeted protein, sometimes simultaneously changing that protein's function. This important biochemical process is discussed further in Chapters 12 and 15.

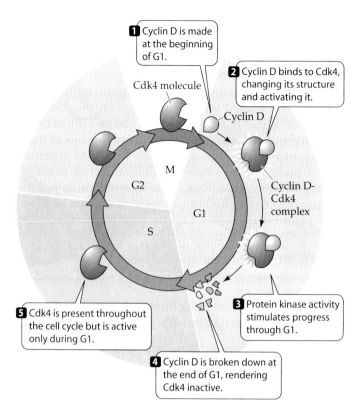

1 Cyclin D is made at the beginning of G1.

Cdk4 molecule

2 Cyclin D binds to Cdk4, changing its structure and activating it.

Cyclin D

Cyclin D–Cdk4 complex

M

G2

G1

S

5 Cdk4 is present throughout the cell cycle but is active only during G1.

3 Protein kinase activity stimulates progress through G1.

4 Cyclin D is broken down at the end of G1, rendering Cdk4 inactive.

9.5 Cyclin-Dependent Kinase and Cyclin Trigger Decisions in the Cell Cycle
A human cell makes the decision to enter the cell cycle during G1, when cyclin D binds to a cyclin-dependent kinase (Cdk4). There are four such cyclin-Cdk controls during the typical cell cycle in humans.

Several different cyclin-Cdk combinations act at various stages of the mammalian cell cycle:

► Cyclin D-Cdk4 acts during the middle of G1. This is the *restriction point*, a key decision point beyond which the rest of the cell cycle is normally inevitable (see Figure 9.5).
► Cyclin E-Cdk2 acts at the G1–S boundary, initiating DNA replication.
► Cyclin A-Cdk2 acts during S, and also stimulates DNA replication.
► Cyclin B-Cdk1 acts at the G2–M boundary, initiating the transition to chromosome condensation and mitosis.

The cyclin-Cdk complexes act as *checkpoints*, points at which cell cycle progress can be monitored to determine if the next step can be taken. For example, if DNA is damaged by radiation during G1, a protein called p21 is made. (The p stands for "protein," and the 21 stands for its molecular weight—about 21,000 daltons.) The p21 protein then binds to the two G1 Cdk's, preventing their activation by cyclins. So the cell cycle stops while repairs are made to DNA. The p21 protein itself is targeted for degradation, so that it breaks down after the DNA is repaired, allowing cyclins to bind to the Cdk's and the cell cycle to proceed.

What molecules do cyclin-Cdk complexes target for phosphorylation? Some important targets are known. For example, the cyclin B-Cdk1 complex catalyzes the phosphorylation of target proteins that then bind to DNA and initiate chromosome condensation. Phosphorylation of other target proteins results in the disaggregation of the nuclear envelope early in mitosis.

Because cancer results from inappropriate cell division, it is not surprising that the cyclin-Cdk controls are disrupted in cancer cells. For example, some fast-growing breast cancers have too much cyclin D, which overstimulates Cdk4 and cell division. As we will describe in Chapter 18, a major protein in normal cells that prevents them from dividing is p53, which leads to inhibition of Cdk's. More than half of all human cancers contain defective p53, resulting in the absence of cell cycle controls.

Growth factors can stimulate cells to divide

Cyclin-Cdk complexes provide an *internal* control for progress through the cell cycle. But there are situations in the body in which cells that are slowly cycling, or not cycling at all, must be stimulated to divide through *external* controls, called **growth factors**. When you cut yourself and bleed, specialized cell fragments called platelets gather at the wound and help initiate blood clotting. The platelets also produce and release a protein, called *platelet-derived growth factor*, that diffuses to the adjacent cells in the skin and stimulates them to divide and heal the wound.

Other growth factors include *interleukins*, which are made by one type of white blood cell and promote cell division in other cells that are essential for the body's immune system defenses. *Erythropoietin*, made by the kidney, stimulates the division of bone marrow cells and the production of red blood cells. In addition, many hormones promote division in specific cell types.

We will describe the physiological roles of these external mitotic inducers in later chapters, but all growth factors act in a similar way. They bind to their target cells via specialized receptor proteins on the target cell surface. This specific binding triggers events within the target cell that initiate a cell division cycle. Cancer cells often cycle inappropriately because they make their own growth factors, or because they no longer require growth factors to start cycling.

Eukaryotic Chromosomes

Most human cells other than eggs and sperm contain two full sets of genetic information, one from the mother and the other from the father. As in prokaryotes, this genetic information consists of molecules of DNA packaged as chromosomes. However, unlike prokaryotes, eukaryotes have more than one chromosome, and during interphase these chromosomes reside within a membrane-enclosed organelle, the nucleus.

The basic unit of the eukaryotic chromosome is a gigantic, linear, double-stranded molecule of DNA complexed

Fungus (*Rhizopus oligosporus*)

Fern (*Osmunda cinnamomcea*)

Elephant (*Loxodonta africana*)

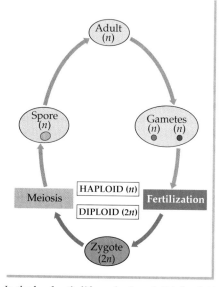

In the **haplontic life cycle**, the adult is haploid and the zygote is the only diploid stage.

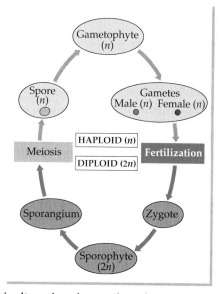

In **alternation of generations**, there are both haploid and diploid adult stages.

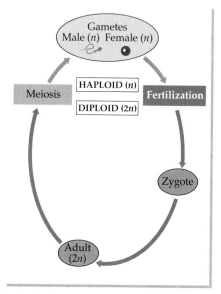

In the **diplontic life cycle**, the adult is diploid and the gametes are the only haploid stage.

9.12 Fertilization and Meiosis Alternate in Sexual Reproduction

In sexual reproduction, haploid (*n*; yellow) cells or organisms alternate with diploid (2*n*; blue) cells or organisms.

mologs) of a homologous pair bear corresponding, though generally not identical, types of genetic information.

Haploid cells contain only one homolog from each pair of chromosomes. The number of chromosomes in such a single set is denoted by *n*. When haploid gametes fuse in fertilization, the resulting zygote has two homologs of each type. It is thus said to be **diploid**, denoted 2*n*.

As you can see in Figure 9.12, sexual life cycles exhibit different patterns of development after zygote formation. In *haplontic* organisms, such as protists and many fungi, the mature organism is haploid. The zygote undergoes a reduction division—meiosis—to produce haploid cells, or *spores*. These spores then form the new organism by mitosis of haploid cells, which may be single-celled or multicellular. Gametes are then produced by this organism by mitosis. So in haplontic organisms, the zygote is the only diploid cell in the life cycle.

At the other extreme are *diplontic* organisms, which include animals and some plants. Here, the gametes are the only haploid cells, and the organism itself is diploid. Gametes are formed by meiosis, and the formation of the organism involves mitosis of diploid cells.

In the middle are organisms that have an alternation of haploid and diploid generations. Most plants fall into this category. Here, the zygote divides by mitosis of diploid cells into a diploid organism. Meiosis does not give rise to gametes, but instead to haploid spores that divide by mitosis to form an alternate, haploid life stage. It is this haploid organism that forms gametes by mitosis, and after fertilization the cycle begins anew. We will look at all of these life cycles in greater detail in subsequent chapters.

The essence of sexual reproduction is the random selection of half of a parent's diploid chromosome set to make a haploid gamete, followed by the fusion of two such haploid gametes to produce a diploid cell that contains genetic information from both gametes. Both of these steps contribute to a shuffling of genetic information in the population, so no two individuals have exactly the same genetic constitution. The diversity provided by sexual reproduction opens up enormous opportunities for evolution.

Centromeres occupy characteristic positions (arrows) on homologous chromosomes.

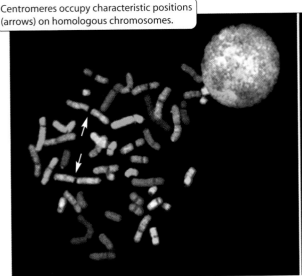

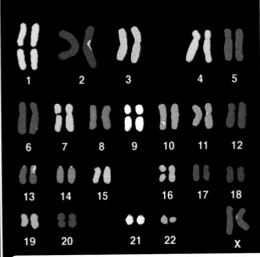

The karyotype shows 23 pairs of chromosomes, including the sex chromosomes. This female's sex chromosomes are X and X; a male would have X and Y chromosomes.

9.13 Human Cells Have 46 Chromosomes
Chromosomes from a human cell are shown in metaphase of mitosis. In this "chromosome painting" technique, each homologous pair shares a distinctive color. The multicolored globe is an interphase nucleus. The karyotype on the right is produced by computerized analysis of the image on the left.

The number, shapes, and sizes of the metaphase chromosomes constitute the karyotype

When nuclei are in metaphase of mitosis, it is often possible to count and characterize the individual chromosomes. This is a relatively simple process in some organisms, thanks to techniques that can capture cells in metaphase and spread out the chromosomes. A photograph of the entire set of chromosomes can then be made, and the images of the individual chromosomes can be placed in an orderly arrangement. Such a rearranged photograph reveals the

9.1	Numbers of Pairs of Chromosomes in Some Plant and Animal Species	
COMMON NAME	**SPECIES**	**NUMBER OF CHROMOSOME PAIRS**
Mosquito	*Culex pipiens*	3
Housefly	*Musca domestica*	6
Toad	*Bufo americanus*	11
Rice	*Oryza sativa*	12
Frog	*Rana pipiens*	13
Alligator	*Alligator mississippiensis*	16
Rhesus monkey	*Macaca mulatta*	21
Wheat	*Triticum aestivum*	21
Human	*Homo sapiens*	23
Potato	*Solanum tuberosum*	24
Donkey	*Equus asinus*	31
Horse	*Equus caballus*	32
Dog	*Canis familiaris*	39
Carp	*Cyprinus carpio*	52

number, shapes, and sizes of chromosomes in a cell, which together constitute its **karyotype** (Figure 9.13).

Individual chromosomes can be recognized by their lengths, the positions of their centromeres, and characteristic banding when they are stained and observed at high magnification. When the cell is diploid, the karyotype consists of homologous pairs of chromosomes—23 pairs for a total of 46 chromosomes in humans, and greater or smaller numbers of pairs in other diploid species. There is no simple relationship between the size of an organism and its chromosome number (Table 9.1).

Meiosis: A Pair of Nuclear Divisions

Meiosis consists of two nuclear divisions that reduce the number of chromosomes to the haploid number in preparation for sexual reproduction. Although the *nucleus divides twice* during meiosis, the *DNA is replicated only once*. To understand the process of meiosis and its specific details, it is useful to keep in mind the overall functions of meiosis:

▶ To reduce the chromosome number from diploid to haploid.
▶ To ensure that each of the haploid products has a complete set of chromosomes.
▶ To promote genetic diversity among the products.

Two unique features characterize the first meiotic division, **meiosis I**. The first is that homologous chromosomes pair along their entire lengths. This process, called **synapsis**, lasts from prophase to the end of metaphase. The second is that after this metaphase, the homologous chromosomes separate. The individual chromosomes, each consisting of two joined chromatids, remain intact until the end of the metaphase of **meiosis II**, the second meiotic division. In the discussion that follows, you can refer to Figure 9.14 to help you visualize each step.

MEIOSIS I

Middle Prophase I

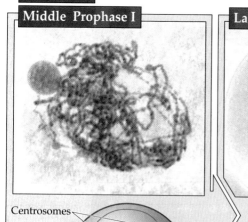

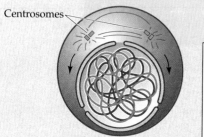

Centrosomes

The chromatin begins to condense following interphase.

Later Prophase I

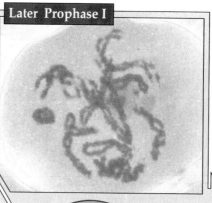

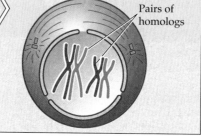

Pairs of homologs

Synapsis aligns homologs, and chromosomes condense. Homologs are shown in different colors indicating those coming from each parent. In reality, their differences are very small, usually comprising different alleles of some genes.

Late Prophase I–Prometaphase

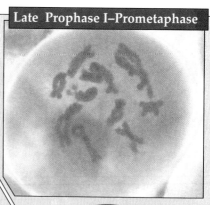

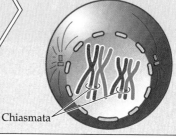

Chiasmata

The chromosomes continue to coil and shorten. Crossing-over at chiasmata results in an exchange of genetic material. In prometaphase the nuclear envelope breaks down.

MEIOSIS II

Prophase II

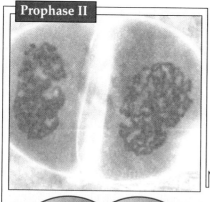

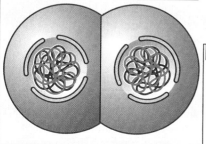

The chromosomes condense again, following a brief interphase (interkinesis) in which DNA does not replicate.

Metaphase II

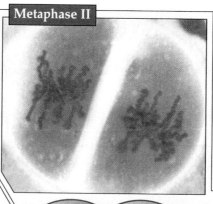

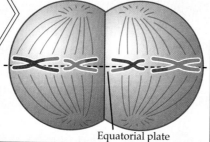

Equatorial plate

Kinetochores of the paired chromatids line up across the equatorial plates of each cell.

Anaphase II

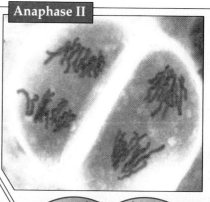

The chromatids finally separate, becoming chromosomes in their own right, and are pulled to opposite poles. Because of crossing over in prophase I, each new cell will have a different genetic makeup.

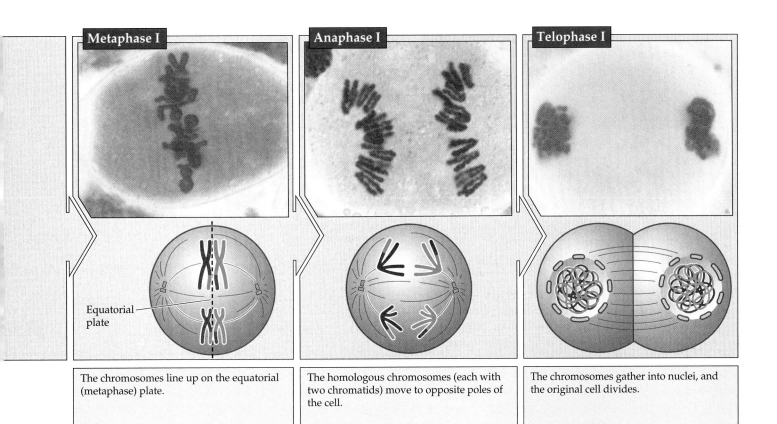

Metaphase I

Equatorial
plate

The chromosomes line up on the equatorial (metaphase) plate.

Anaphase I

The homologous chromosomes (each with two chromatids) move to opposite poles of the cell.

Telophase I

The chromosomes gather into nuclei, and the original cell divides.

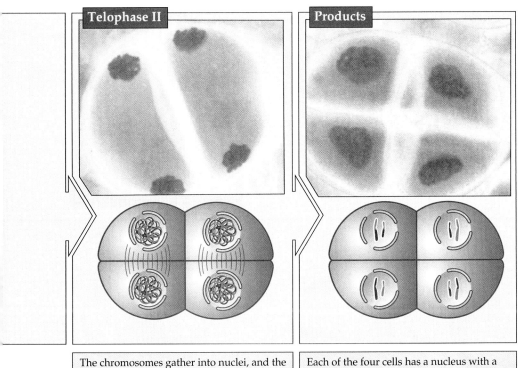

Telophase II

The chromosomes gather into nuclei, and the cells divide.

Products

Each of the four cells has a nucleus with a haploid number of chromosomes.

9.14 Meiosis

In meiosis, two sets of chromosomes are divided among four nuclei, each of which then has half as many chromosomes as the original cell. These four haploid cells are the result of two successive nuclear divisions. The photomicrographs shown here are of meiosis in the male reproductive organ of a lily. As in Figure 9.8, the diagrams show corresponding phases in an animal.

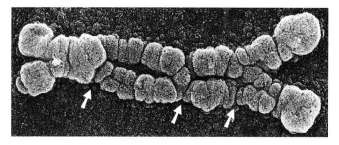

9.15 Chiasmata: Evidence of Exchange between Chromatids
Chiasmata are visible near the middle of some chromatids from a desert locust in this scanning electron micrograph, and near the ends of others. Three chiasmata are indicated with arrows.

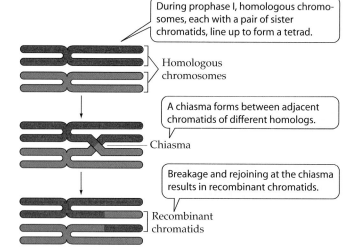

During prophase I, homologous chromosomes, each with a pair of sister chromatids, line up to form a tetrad.

Homologous chromosomes

A chiasma forms between adjacent chromatids of different homologs.

Chiasma

Breakage and rejoining at the chiasma results in recombinant chromatids.

Recombinant chromatids

9.16 Crossing Over Forms Genetically Diverse Chromosomes
The exchange of genetic material by crossing over may result in new combinations of genetic information on the recombinant chromosomes.

The first meiotic division reduces the chromosome number

Like mitosis, meiosis I is preceded by an interphase with an S phase during which each chromosome is replicated. As a result, each chromosome consists of two sister chromatids.

Meiosis I begins with a long prophase I (the first three frames of Figure 9.14), during which the chromosomes change markedly. A key change is that homologous chromosomes join together, or *synapse*. By the time they can be clearly seen under light microscope, the two homologs are already tightly joined. This joining begins at the centromeres and is mediated by a recognition of homologous DNA sequences on homologous chromosomes. In addition, a special group of proteins may form a scaffold called the *synaptonemal complex* that runs lengthwise along the homologous chromosomes and appears to join them together. The

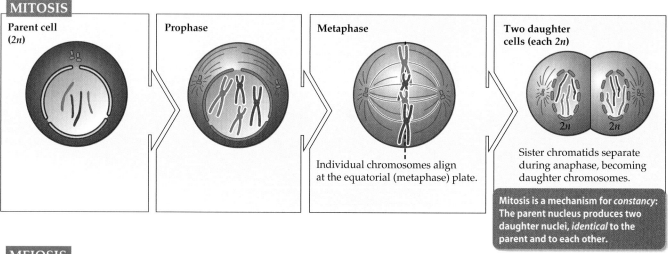

MITOSIS

Parent cell (2n)

Prophase

Metaphase

Individual chromosomes align at the equatorial (metaphase) plate.

Two daughter cells (each 2n)

Sister chromatids separate during anaphase, becoming daughter chromosomes.

Mitosis is a mechanism for *constancy*: The parent nucleus produces two daughter nuclei, *identical* to the parent and to each other.

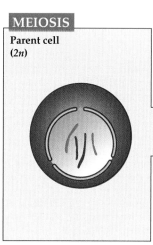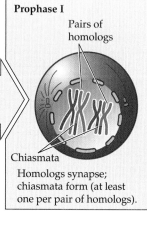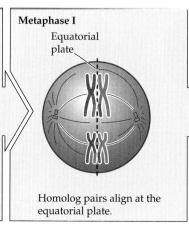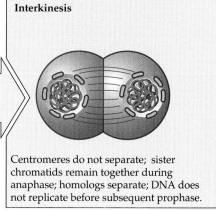

MEIOSIS

Parent cell (2n)

Prophase I

Pairs of homologs

Chiasmata

Homologs synapse; chiasmata form (at least one per pair of homologs).

Metaphase I

Equatorial plate

Homolog pairs align at the equatorial plate.

Interkinesis

Centromeres do not separate; sister chromatids remain together during anaphase; homologs separate; DNA does not replicate before subsequent prophase.

four chromatids of each pair of homologous chromosomes form a *tetrad*, or *bivalent*. To summarize: A tetrad is four chromatids, two each from two homologous chromosomes. For example, there are 46 chromosomes in a human diploid cell, so there are 23 homologous pairs of chromosomes, each with two chromatids, for a total of 92 chromatids during prophase I. In other words, there are 23 tetrads, each containing two homologous chromosomes and four chromatids.

Throughout prophase I and metaphase I, the chromatin continues to coil and compact progressively, so the chromosomes appear ever thicker. At a certain point, the homologous chromosomes seem to *repel* each other, especially near the centromeres, but they are held together by physical attachments. Regions having these attachments take on an X-shaped appearance and are called **chiasmata** (from the Greek word "chiasma," meaning "cross"; Figure 9.15). A chiasma reflects an exchange of material between chromatids on homologous chromosomes—what geneticists call *crossing over* (Figure 9.16). The chromosomes begin exchanging material shortly after synapsis begins, but the chiasmata do not become visible until later, when the homologs are repelling each other.

Crossing over increases the genetic variation among the products of meiosis. We will have a great deal to say about crossing over and its genetic consequences in the coming chapters.

There seems to be plenty of time for the complicated events of prophase I to occur. Whereas mitotic prophase is usually measured in minutes, and all of mitosis seldom takes more than an hour or two, meiosis can take much longer. In human males, the cells in the testis that undergo meiosis take about a week for prophase I and about a month for the entire meiotic cycle. In the cells that will become eggs, prophase I begins long before a woman's birth, during early fetal development, and ends as much as decades later, during the monthly ovarian cycle.

Prophase I is followed by prometaphase I (not pictured in Figure 9.14), during which the nuclear envelope and the nucleoli disappear. A spindle forms, and microtubules become attached to the kinetochores of the chromosomes. In meiosis I, there is only *one kinetochore per chromosome*, not one per chromatid as in mitosis. Thus the entire chromosome, consisting of two chromatids, will migrate to one pole in the meiotic cell.

By metaphase I, all the chromosomal kinetochores have become connected to polar microtubules, and all the chromosomes have moved to the equatorial plate. Until this point, they have been held together by chiasmata.

The homologous chromosomes separate in anaphase I, when *individual chromosomes*, each still consisting of two chromatids, are pulled to the poles, with one homolog of a pair going to one pole and the other homolog going to the opposite pole. (Note that this process differs from the separation of *chromatids* during mitotic anaphase.) Each of the two daughter nuclei from this division is haploid; that is, it contains only one set of chromosomes, not the two sets that were present in the original diploid nucleus. However, because they consist of two chromatids rather than just one, each of these chromosomes has twice the mass that a chromosome at the end of a mitotic division has.

In some species, but not in others, there is a telophase I, with the reappearance of nuclear envelopes and so forth. When there is a telophase I, it is followed by an interphase, called **interkinesis**, similar to the mitotic interphase. During interkinesis the chromatin is partially uncoiled; however, there is no replication of the genetic material, because each chromosome already consists of two chromatids. Furthermore, the sister chromatids in interkinesis are generally not genetically identical, because crossing over in prophase I has reshuffled genetic material between maternal and paternal chromosomes.

The second meiotic division separates the chromatids

Meiosis II is similar to mitosis in many ways. In each nucleus produced by meiosis I, the chromosomes line up at equatorial plates in metaphase II, the chromatids—each of which has a centromere—separate, and new daughter chromosomes move to the poles in anaphase II.

The three major differences between meiosis II and mitosis are:

▶ DNA replicates before mitosis, but not before meiosis II.
▶ In mitosis, the sister chromatids that make up a given chromosome are identical; in meiosis II, they differ over part of their length if they participated in crossing over during prophase of meiosis I.
▶ The number of chromosomes on the equatorial plate of each of the two nuclei in meiosis II is half the number in the single mitotic nucleus.

Figure 9.17 compares mitosis and meiosis. The result of meiosis is four nuclei; each nucleus is haploid and has a single set of chromosomes that differs from other such sets

9.17 Mitosis and Meiosis: A Comparison
Meiosis differs from mitosis by synapsis and by the failure of the centromeres to separate at the end of metaphase I.

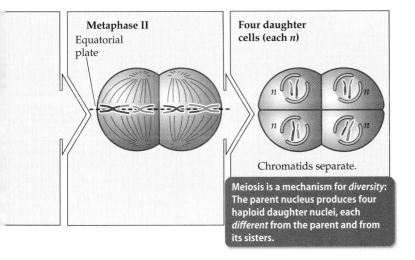

Metaphase II
Equatorial plate

Four daughter cells (each *n*)

Chromatids separate.

Meiosis is a mechanism for *diversity*: The parent nucleus produces four haploid daughter nuclei, each *different* from the parent and from its sisters.

in its exact genetic composition. The differences, to repeat a very important point, result from crossing over during prophase I and from the segregation of homologous chromosomes during anaphase I.

Meiosis leads to genetic diversity

What are the consequences of the synapsis and separation of homologous chromosomes during meiosis? In mitosis, each chromosome behaves independently of its homolog; its two chromatids are sent to opposite poles at anaphase. If we start a mitotic division with x chromosomes, we end up with x chromosomes in each daughter nucleus, and each chromosome consists of one chromatid. In meiosis, things are very different.

In meiosis, synapsis organizes things so that chromosomes of maternal origin pair with their paternal homologs. Then their separation during meiotic anaphase I ensures that each pole receives one member of each homologous pair. (Remember that each chromosome still consists of two chromatids.) For example, at the end of meiosis I in humans, each daughter nucleus contains 23 of the original 46 chromosomes. In this way, the chromosome number is decreased from diploid to haploid. Furthermore, meiosis I guarantees that each daughter nucleus gets one full set of chromosomes, for it must have one of each homologous pair.

The products of meiosis I are genetically diverse for two reasons. First, synapsis during prophase I allows the maternal chromosome to interact with the paternal one; after crossing over, the recombinant chromatids contain some genetic material from each chromosome. Second, which member of a homologous pair goes to which daughter cell at anaphase I is a matter of pure chance. For example, if there are two pairs of chromosomes in the diploid parent nucleus, a particular daughter nucleus could get paternal chromosome 1 and maternal chromosome 2, or paternal 2 and maternal 1, or both maternals, or both paternals. It all depends on the way in which the homologous pairs line up at metaphase I.

Note that of the four possible chromosome combinations just described, two produce daughter nuclei that are the same as one of the parental types (except for any material exchanged by crossing over). The greater the number of chromosomes, the less probable that the original parental combinations will be reestablished, and the greater the potential for genetic diversity. Most species of diploid organisms do, indeed, have more than two pairs. In humans, with 23 chromosome pairs, 2^{23} different combinations can be produced.

Meiotic Errors

A pair of homologous chromosomes may fail to separate during meiosis I, or sister chromatids may fail to separate during meiosis II or during mitosis. This phenomenon is called **nondisjunction**, and it results in the production of aneuploid cells (Figure 9.18). **Aneuploidy** is a condition in

which one or more chromosomes or pieces of chromosomes are either lacking or present in excess.

Aneuploidy can give rise to genetic abnormalities

One reason for nondisjunction may be a lack of chiasmata. Recall that these structures, formed during prophase I, hold the two homologous chromosomes together into metaphase I. This ensures that one homolog will face one pole and the other homolog the other pole. Without this "glue," the two homologs may line up randomly at metaphase I, just like chromosomes during mitosis, and there is a 50 percent chance that both will go to the same pole.

If, for example, the chromosome 21 pair fails to separate during the formation of a human egg (and thus both mem-

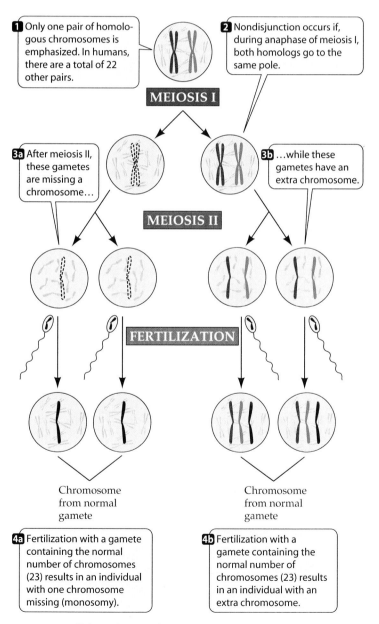

1 Only one pair of homologous chromosomes is emphasized. In humans, there are a total of 22 other pairs.

2 Nondisjunction occurs if, during anaphase of meiosis I, both homologs go to the same pole.

MEIOSIS I

3a After meiosis II, these gametes are missing a chromosome…

3b …while these gametes have an extra chromosome.

MEIOSIS II

FERTILIZATION

Chromosome from normal gamete

Chromosome from normal gamete

4a Fertilization with a gamete containing the normal number of chromosomes (23) results in an individual with one chromosome missing (monosomy).

4b Fertilization with a gamete containing the normal number of chromosomes (23) results in an individual with an extra chromosome.

9.18 Nondisjunction Leads to Aneuploidy
Nondisjunction occurs if homologous chromosomes fail to separate during meiosis I. The result is aneuploidy: One or more chromosomes are either lacking or present in excess.

bers of the pair go to one pole during anaphase I), the resulting egg will contain either two of chromosome 21 or none at all. If an egg with two of these chromosomes is fertilized by a normal sperm, the resulting zygote will have three copies of the chromosome: It will be **trisomic** for chromosome 21. A child with an extra chromosome 21 demonstrates the symptoms of *Down syndrome:* impaired intelligence; characteristic abnormalities of the hands, tongue, and eyelids; and an increased susceptibility to cardiac abnormalities and diseases such as leukemia.

Other abnormal events can also lead to aneuploidy. In a process called *translocation,* a piece of a chromosome may break away and become attached to another chromosome. For example, a particular large part of one chromosome 21 may be translocated to another chromosome. Individuals who inherit this translocated piece along with two normal chromosomes 21 will have Down syndrome.

Trisomies (and the corresponding monosomies) are surprisingly common in human zygotes, but most of the embryos that develop from such zygotes do not survive to birth. Trisomies for chromosomes 13, 15, and 18 greatly reduce the probability that an embryo will survive to birth, and virtually all infants who are born with such trisomies die before the age of 1 year. Trisomies and monosomies for other chromosomes are lethal to the embryo. At least one-fifth of all recognized pregnancies spontaneously terminate during the first two months, largely because of such trisomies and monosomies. (The actual proportion of spontaneously terminated pregnancies is certainly higher, because the earliest ones often go unrecognized.)

Polyploids can have difficulty in cell division

Both diploid and haploid nuclei divide by mitosis. Multicellular diploid and multicellular haploid individuals develop from single-celled beginnings by mitotic divisions. Likewise, mitosis may proceed in diploid organisms even when a chromosome from one of the haploid sets is missing or when there is an extra copy of one of the chromosomes (as in Down syndrome).

Under some circumstances, triploid ($3n$), tetraploid ($4n$), and higher-order polyploid nuclei may form. Each of these *ploidy levels* represents an increase in the number of complete sets of chromosomes present. If, by accident, the nucleus has one or more extra full sets of chromosomes—that is, if it is triploid, tetraploid, or of still higher ploidy—this abnormally high ploidy in itself does not prevent mitosis. In mitosis, each chromosome behaves independently of the others.

In meiosis, by contrast, chromosomes synapse to begin division. If even one chromosome has no homolog, anaphase I cannot send representatives of that chromosome to both poles. A diploid nucleus can undergo normal meiosis; a haploid one cannot. A tetraploid nucleus has an even number of each kind of chromosome, so each chromosome can pair with its homolog. But a triploid nucleus cannot undergo normal meiosis, because one-third of the chromosomes would lack partners.

This limitation has important consequences for the fertility of triploid, tetraploid, and other chromosomally unusual organisms that may be produced by plant breeding or by natural accidents. Modern bread wheat plants are hexaploids, the result of the accidental crossing of three different grasses, each having its own diploid set of 14 chromosomes.

Cell Death

As we mentioned at the start of this chapter, an essential role of cell division in complex eukaryotes is to replace cells that die. In humans, billions of cells die each day, mainly in the blood and the epithelia lining organs such as the intestine. Cells die in one of two ways. The first, **necrosis**, occurs when cells either are damaged by poisons or are starved of essential nutrients. These cells usually swell up and burst, releasing their contents into the extracellular environment. This often results in inflammation (see Chapter 19). The scab that forms around a wound is a familiar example of necrotic tissue.

More typically, cell death in an organism is due to **apoptosis** (from the Greek word meaning "falling off"). Apoptosis is a prescribed series of events that constitute genetically programmed cell death. These two ways for cells to die are compared in Table 9.2.

9.2 **Two Different Ways for Cells to Die**

	NECROSIS	APOPTOSIS
Stimuli	Low O_2, toxins, ATP depletion, damage	Specific, genetically programmed physiological signals
ATP required	No	Yes
Cellular pattern	Swelling, organelle disruption, tissue death	Chromatin condensation, membrane blebbing, single-cell death
DNA breakdown	Random fragments	Nucleosome-sized fragments
Plasma membrane	Burst	Blebbed (see Figure 9.19*b*)
Fate of dead cells	Ingested by phagocytes	Ingested by neighboring cells
Reaction in tissue	Inflammation	No inflammation

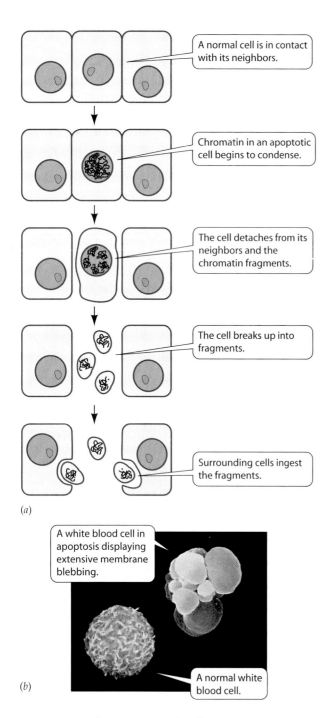

(a)

(b)

A normal cell is in contact with its neighbors.

Chromatin in an apoptotic cell begins to condense.

The cell detaches from its neighbors and the chromatin fragments.

The cell breaks up into fragments.

Surrounding cells ingest the fragments.

A white blood cell in apoptosis displaying extensive membrane blebbing.

A normal white blood cell.

9.19 Apoptosis: Programmed Cell Death
Many cells are genetically programmed to "self-destruct" when they are no longer needed, or when they have lived long enough to accumulate a burden of DNA damage that might harm the organism.

Why would a cell initiate apoptosis, which is essentially "cell suicide"? One reason is that the cell in question is no longer needed by the organism. For example, before birth, a human fetus has weblike hands, with connective tissue between the fingers. As development proceeds, this unneeded tissue disappears as its cells undergo apoptosis (see Figure 16.10).

A second reason for apoptosis is that the longer cells live, the more prone they are to damage that could lead to cancer. This is especially true of cells in the blood and intestine, which are exposed to high levels of toxic substances. In these cases, cells "sacrifice their lives for the good of the organism." Such cells normally die after only days or weeks.

Like the cell division cycle, the cell death cycle has signals controlling its progress. These include the lack of a mitotic signal, such as a growth factor, and recognition of DNA damage. As we will see in Chapter 17, many of the drugs used to treat diseases of cell proliferation such as cancer work via these signals.

The events of apoptosis are very similar in most organisms (Figure 9.19). The cell becomes isolated from its neighbors, chops up its chromatin into nucleosome-sized pieces, and then fragments itself. In a remarkable example of the economy of nature, the surrounding living cells usually ingest the remains of the dead cell. The genetic signals that lead to apoptosis are also common to many organisms.

Chapter Summary

▶ Cell division is necessary for reproduction, growth, and repair of an organism. **Review Figure 9.1**

Systems of Cell Reproduction

▶ Cell division must be initiated by a reproductive signal. Cell division consists of three steps: replication of the genetic material (DNA), partitioning of the two DNA molecules to separate portions of the cell, and division of the cytoplasm.

▶ In prokaryotes, cellular DNA is a single molecule, or chromosome. Prokaryotes reproduce by cell fission. **Review Figure 9.3**

▶ In eukaryotes, nuclei divide by either mitosis or meiosis.

Interphase and the Control of Cell Division

▶ The mitotic cell cycle has two main phases: interphase (during which cells are not dividing) and mitosis (when cells divide).

▶ During most of the cell cycle the cell is in interphase, which is divided into three subphases: S, G1, and G2. DNA is replicated during S phase. **Review Figure 9.4**

▶ Cyclin-Cdk complexes regulate the passage of cells from G1 into S phase and from G2 into M phase. **Review Figure 9.5**

▶ In addition to the internal cyclin-Cdk complexes, controls external to the cell, such as growth factors and hormones, can also stimulate the cell to begin a division cycle.

Eukaryotic Chromosomes

▶ Chromosomes contain DNA and proteins. At mitosis, chromosomes initially appear to be double because two sister chromatids are held together at the centromere. Each sister chromatid consists of one double-stranded DNA molecule complexed with proteins and referred to as chromatin. **Review Figure 9.6**

▶ During interphase, the DNA in chromatin is wound around cores of histones to form nucleosomes. DNA folds over and over again, packing itself within the nucleus. When mitotic chromosomes form, it folds even more. **Review Figure 9.7**

Mitosis: Distributing Exact Copies of Genetic Information

▶ After DNA is replicated during S phase, the first sign of mitosis is the separation of centrosomes, which initiate microtubule formation for the spindle. **Review Figure 9.9**

▶ Mitosis can be divided into several phases, called prophase, prometaphase, metaphase, anaphase, and telophase. **Review Figure 9.8**

▶ During prophase, the chromosomes condense and appear as paired chromatids.

▶ During prometaphase, the chromosomes move toward the middle of the spindle. In metaphase, they gather at the middle of the cell with their centromeres on the equatorial plate. At the end of metaphase, the centromeres holding the chromatid pairs together separate, and during anaphase each member of the pair, now called a daughter chromosome, migrates to its pole along the microtubule track.

▶ During telophase, the chromosomes become less condensed. The nuclear envelopes and nucleoli re-form, thus producing two nuclei whose chromosomes are identical to each other and to those of the cell that began the cycle. **Review Figure 9.8**

Cytokinesis: The Division of the Cytoplasm

▶ Nuclear division is usually followed by cytokinesis. Animal cell cytoplasm usually divides by a furrowing of the plasma membrane, caused by the contraction of cytoplasmic microfilaments. In plant cells, cytokinesis is accomplished by vesicle fusion and the synthesis of new cell wall material. **Review Figure 9.10**

Reproduction: Sexual and Asexual

▶ The cell cycle can repeat itself many times, forming a clone of genetically identical cells.

▶ Asexual reproduction produces a new organism that is genetically identical to the parent. Any genetic variety is the result of mutations.

▶ In sexual reproduction, two haploid gametes—one from each parent—unite in fertilization to form a genetically unique, diploid zygote. **Review Figure 9.12**

▶ In sexually reproducing organisms, certain cells in the adult undergo meiosis, a process by which a diploid cell produces haploid gametes. Each gamete contains a random mix of one of each pair of homologous chromosomes from the parent.

▶ The number, shapes, and sizes of the chromosomes constitute the karyotype of an organism. **Review Figure 9.13**

Meiosis: A Pair of Nuclear Divisions

▶ Meiosis reduces the chromosome number from diploid to haploid and ensures that each haploid cell contains one member of each chromosome pair. It consists of two nuclear divisions. **Review Figure 9.14**

▶ During prophase I of the first meiotic division, homologous chromosomes pair up with each other, and material may be exchanged by crossing over between nonsister chromatids of two adjacent homologs. In metaphase I, the paired homologs gather at the equatorial plate. Each chromosome has only one kinetochore and associates with polar microtubules for one pole. In anaphase I, entire chromosomes, each with two chromatids, migrate to the poles. By the end of meiosis I, there are two nuclei, each with the haploid number of chromosomes with two sister chromatids. **Review Figures 9.14, 9.16**

▶ In meiosis II, the sister chromatids separate. No DNA replication precedes this division, which in other aspects is similar to mitosis. The result of meiosis is four cells, each with a haploid chromosome content. **Review Figures 9.14, 9.17**

▶ Both crossing over during prophase I and the random selection of which homolog of a pair migrates to which pole during anaphase I ensure that the genetic composition of each haploid gamete is different from that of the parent and from that of the other gametes. The more chromosome pairs there are in a diploid cell, the greater the diversity of chromosome combinations generated by meiosis.

Meiotic Errors

▶ In nondisjunction, one member of a homologous pair of chromosomes fails to separate from the other, and both go to the same pole. This event leads to one gamete with an extra chromosome and another other lacking that chromosome. Fertilization with a normal haploid gamete results in aneuploidy and genetic abnormalities that are invariably harmful or lethal to the organism. **Review Figure 9.18**

Cell Death

▶ Cells may die by necrosis or may self-destruct by apoptosis, a genetically programmed series of events that includes the detachment of the cell from its neighbors and the fragmentation of its nuclear DNA. **Review Figure 9.19**

Applying Concepts

1. Compare chromatids and chromosomes. At what stages during mitosis and meiosis are chromatids present?

2. Strains of organisms unable to carry out certain functions in the cell cycle have been invaluable to scientists to determine what happens in the normal cell cycle. Describe the cell cycle in cells

   a lacking the G1 cyclin.

   b. lacking the mitotic spindle.

   c. lacking the microfilaments involved in plasma membrane contraction.

3. Compare the sequence of events in the mitotic cell cycle with the sequence in programmed cell death.

4. The potato plant has 24 pairs of chromosomes. What is the number of

   a. chromatids in a cell at prophase of mitosis?

   b. chromosomes in a cell at anaphase of mitosis?

   c. chromatids in a cell at metaphase I of meiosis?

   d. chromatids in a cell at prophase II of meiosis?

10 Genetics: Mendel and Beyond

✡ IN THE MIDDLE EASTERN DESERT 1,800 YEARS ago, a rabbi faced a serious dilemma. A Jewish woman had given birth to a son. As required by laws first set down by God's commandment to Abraham almost 2,000 years previously and reiterated later by Moses, the mother brought her 8-day-old son to the rabbi for ritual penile circumcision. The rabbi knew that the woman's two previous sons had bled to death when their foreskins were cut. Yet the Biblical commandment remained: Unless he was circumcised, the boy could not be counted among those with whom God had made His solemn covenant. After consultation with other rabbis, it was decided to exempt this, the third son.

Almost one thousand years later, in the twelfth century, the physician and biblical commentator Moses Maimonides reviewed this and numerous other cases in the rabbinical literature, and stated that in such instances the third son should not be circumcised. Furthermore, the ban should apply whether the son was "from her first husband or from her second husband." The bleeding disorder, he reasoned, was clearly carried by the mother and passed on to her sons.

Knowing nothing of our modern vision of genetics, these rabbis linked a human disease (which turns out to be hemophilia A) to a pattern of inheritance (which we know as sex linkage). Only in the past several decades have the precise biochemical nature of hemophilia A and its genetic determination been worked out.

How do we account for, and predict, such patterns of inheritance? In this chapter, we will discuss how the units of inheritance, called genes, are transmitted from generation to generation of plants and animals, and show how many of the rules that govern genetics can be explained by the behavior of chromosomes during meiosis. We will also describe the interactions of genes with one another and with the environment, and the consequences of the fact that genes occupy specific positions on chromosomes.

An Ancient Ritual
A male infant undergoes ritual circumcision in accordance with Jewish laws. Sons of Jewish mothers who carry the gene for hemophilia may be exempt from the ritual.

The Foundations of Genetics

Much of the early study of biological inheritance was done with plants and animals of economic importance. Records show that people were deliberately cross-breeding date palm trees and horses as early as 5,000 years ago. By the early 1800s, plant breeding was widespread, especially with ornamental flowers such as tulips. Half a century later, in 1866, Gregor Mendel used the knowledge of plant reproduction to design and conduct experiments on inheritance. Although his published results were neglected by scientists for 40 years, they ultimately became the foundation for the science of genetics.

Plant breeders showed that both parents contribute equally to inheritance

Plants are easily grown in large quantities, many produce large numbers of offspring (in the form of seeds), and many have relatively short generation times. In most plant species, the same individuals have both male and female reproductive organs, permitting each plant to reproduce as a male, as a female, or as both. Best of all, it is often easy to control which individuals mate (Figure 10.1).

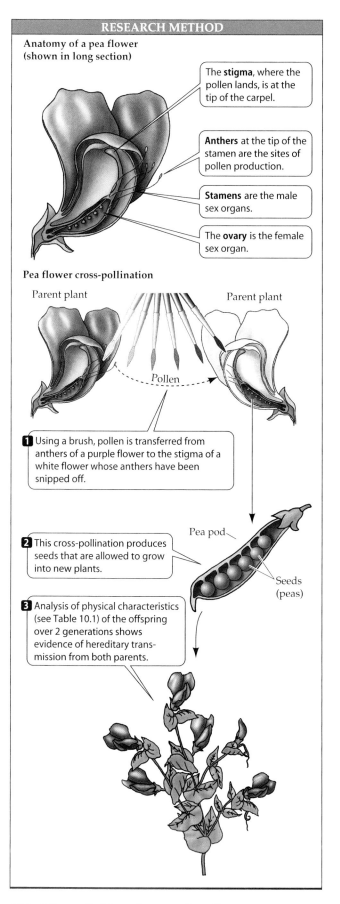

RESEARCH METHOD

Anatomy of a pea flower
(shown in long section)

The **stigma**, where the pollen lands, is at the tip of the carpel.

Anthers at the tip of the stamen are the sites of pollen production.

Stamens are the male sex organs.

The **ovary** is the female sex organ.

Pea flower cross-pollination

Parent plant

Parent plant

Pollen

1 Using a brush, pollen is transferred from anthers of a purple flower to the stigma of a white flower whose anthers have been snipped off.

Pea pod

2 This cross-pollination produces seeds that are allowed to grow into new plants.

Seeds (peas)

3 Analysis of physical characteristics (see Table 10.1) of the offspring over 2 generations shows evidence of hereditary transmission from both parents.

10.1 A Controlled Cross between Two Plants
Plants were widely used in early genetic studies because it is easy to control which individuals mate with which. Mendel used the pea plant, *Pisum sativum*, in many of his experiments.

Some discoveries that Mendel found useful in his studies had been made in the late eighteenth century by a German botanist, Josef Gottlieb Kölreuter. Kölreuter studied the offspring of reciprocal crosses between plants and showed that the two parents contributed equally to the characteristics inherited by their offspring. In a **reciprocal cross**, plants are crossed with (mated with) each other in opposite directions. For example, in one cross, males that have white flowers are mated with females that have red flowers, while in a complementary cross, red-flowered males and white-flowered females are the parents. In Kölreuter's experience, such reciprocal crosses always gave identical results.

Although the concept of equal parental contributions was an important discovery, the nature of what exactly the parents were contributing—the units of inheritance—remained unknown. Laws of inheritance proposed at the time favored the concept of *blending*. If a plant that had one form of a characteristic (say, red flowers) was crossed with one that had a different form of that characteristic (blue flowers), the offspring would be a blended combination of the two parents (purple flowers).

According to the blending concept, it was thought that once heritable elements were combined, they could not be separated again (like combined inks). The red and blue genetic determinants were thought to be forever blended into the new purple one. Then, about a century after Kölreuter completed his work, Mendel began his.

Mendel's discoveries were overlooked for decades

Gregor Mendel was an Austrian monk, not an academic scientist, but he was qualified to undertake scientific investigations. Although in 1850 he had failed an examination for a teaching certificate in natural science, he later undertook intensive studies in physics, chemistry, mathematics, and various aspects of biology at the University of Vienna. His work in physics and mathematics probably led him to apply experimental and quantitative methods to the study of heredity—and these were the key ingredients in his success.

Mendel worked out the basic principles of inheritance in plants over a period of about 9 years. His work culminated in a public lecture in 1865 and a detailed written account published in 1866. Mendel's paper appeared in a journal that was received by 120 libraries, and he sent reprinted copies (of which he had obtained 40) to several distinguished scholars. However, his theory was not accepted. In fact, it was ignored.

The chief difficulty was that the most prominent biologists of Mendel's time were not in the habit of thinking in mathematical terms, even the simple terms used by Mendel. Even Charles Darwin, whose theory of evolution by natural selection depended on genetic variation among individuals, failed to understand the significance of Mendel's findings. In fact, Darwin performed breeding experiments like Mendel's on snapdragons and got data similar to Mendel's, but he missed the point, still relying on the concept of blending. In addition, Mendel had little credibility as a biologist; in-

10.1 Mendel's Results from Monohybrid Crosses

DOMINANT × RECESSIVE	DOMINANT	RECESSIVE	TOTAL	RATIO
Spherical seeds × Wrinkled seeds	5,474	1,850	7,324	2.96:1
Yellow seeds × Green seeds	6,022	2,001	8,023	3.01:1
Purple flowers × White flowers	705	224	929	3.15:1
Inflated pods × Constricted pods	882	299	1,181	2.95:1
Green pods × Yellow pods	428	152	580	2.82:1
Axial flowers × Terminal flowers	651	207	858	3.14:1
Tall stems × Dwarf stems (1 m) (0.3 m)	787	277	1,064	2.84:1

deed, his lowest grades were in biology! Whatever the reasons, Mendel's pioneering paper had no discernible influence on the scientific world for more than 30 years.

Then, in 1900, Mendel's discoveries burst into prominence as a result of independent experiments by three plant geneticists: the Dutch Hugo de Vries, the German Karl Correns, and the Austrian Erich von Tschermak. Each of these scientists carried out crossing experiments and obtained quantitative data about the progeny; each published his principal findings in 1900; each cited Mendel's 1866 paper. By that time, meiosis had been observed and described. At last the time was ripe for biologists to appreciate the significance of what these four geneticists had discovered.

Mendel's Experiments and the Laws of Inheritance

That Mendel was able to make his discoveries before the discovery of meiosis was due in part to the methods of experimentation he used. Mendel's work is a fine example of preparation, execution, and interpretation. Let's see how he approached each of these steps.

Mendel devised a careful research plan

Mendel chose the garden pea for his studies because of its ease of cultivation, the feasibility of controlled pollination (see Figure 10.1), and the availability of varieties with differing traits. He controlled pollination, and thus fertilization, of his parent plants by manually moving pollen from one plant to another. Thus he knew the parentage of the offspring in his experiments. If untouched, the pea plants Mendel studied naturally *self-pollinate*—that is, the female organ of each flower receives pollen from the male organs of the same flowers—and he made use of this natural phenomenon in some of his experiments.

Mendel began by examining different varieties of peas in a search for heritable characters and traits suitable for study. A **character** is a feature, such as flower color; a **trait** is a particular form of a character, such as white flowers. A **heritable** character trait is one that is passed from parent to offspring. Mendel looked for characters that had well-defined, contrasting alternative traits, such as purple flowers versus white flowers, that were true-breeding.

To be considered **true-breeding**, the observed trait must be the only form present for many generations. In other words, peas with white flowers, when crossed with one another, would have to give rise only to progeny with white flowers for many generations; tall plants bred to tall plants would have to produce only tall progeny.

Mendel isolated each of his true-breeding strains by repeated inbreeding (done by crossing of sibling plants that were seemingly identical, or allowing individuals to self-pollinate) and selection. In most of his work, Mendel concentrated on the seven pairs of contrasting traits shown in Table 10.1. Before performing any given cross, he made sure that each potential parent was from a true-breeding strain—an essential point in his analysis of his experimental results.

10.2 Contrasting Traits
In Experiment 1, Mendel studied the inheritance of seed shape. We know today that the wrinkled seeds possess an abnormal form of starch. Contrast their appearance with that of the spherical seeds below.

Mendel then collected pollen from one parental strain and placed it onto the stigma (female organ) of flowers of the other strain. The plants providing and receiving the pollen were the **parental generation**, designated **P**. In due course, seeds formed and were planted. The resulting new plants constituted the **first filial generation, F₁**. Mendel and his assistants examined each F₁ plant to see which traits it bore and then recorded the number of F₁ plants expressing each trait. In some experiments the F₁ plants were allowed to self-pollinate and produce a **second filial generation**, or **F₂**. Again, each F₂ plant was characterized and counted.

In sum, Mendel devised a well-organized plan of research, pursued it faithfully and carefully, recorded great amounts of quantitative data, and analyzed the numbers he recorded to explain the relative proportions of the different kinds of progeny. His 1866 paper stands to this day as a model of clarity. His results and the conclusions to which they led are the subject of the next few sections.

Mendel's Experiment 1 examined a monohybrid cross

"Experiment 1" in Mendel's paper involved a **monohybrid cross**—one in which each parent pea plant was true-breeding for a given character, but in this case each displayed a different form of that character (a different trait). He took pollen from plants of a true-breeding strain with wrinkled seeds and placed it on the stigmas of flowers of a true-breeding, spherical-seeded strain (Figure 10.2). He also performed the reciprocal cross, placing pollen from the spherical-seeded strain on the stigmas of flowers of the wrinkled-seeded strain.

In both cases, all the F₁ seeds that were produced were spherical—it was as if the wrinkled trait had disappeared completely. The following spring Mendel grew 253 F₁ plants from these spherical seeds, each of which was allowed to self-pollinate—this was the monohybrid cross—to produce F₂ seeds. In all, there were 7,324 F₂ seeds, of which 5,474 were spherical and 1,850 wrinkled (Figure 10.3).

Mendel concluded that the spherical seed trait was **dominant**: In the F₁ generation, it was always expressed rather

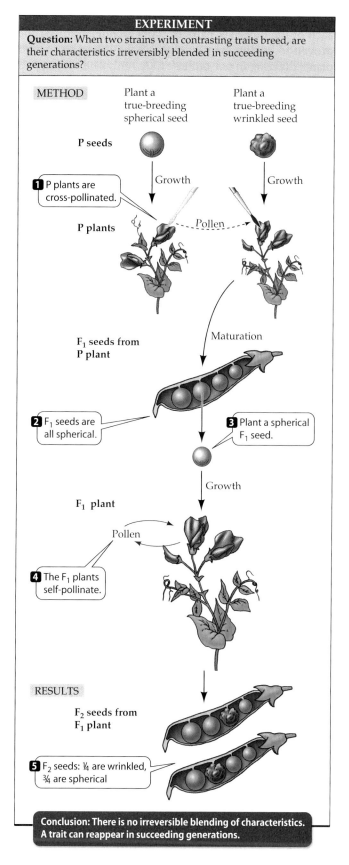

10.3 Mendel's Experiment 1
The pattern Mendel observed in the F₂ generation—¼ of the seeds wrinkled, ¾ spherical—was the same no matter which variety contributed the pollen in the parental generation.

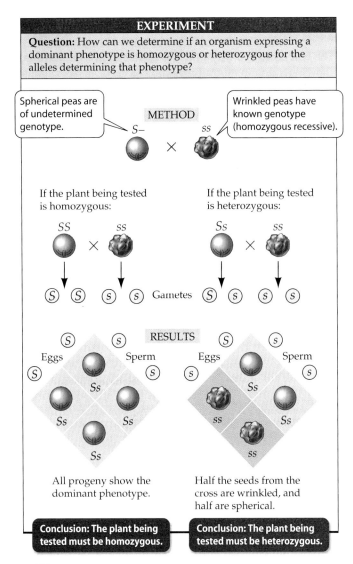

EXPERIMENT

Question: How can we determine if an organism expressing a dominant phenotype is homozygous or heterozygous for the alleles determining that phenotype?

METHOD

Spherical peas are of undetermined genotype.

Wrinkled peas have known genotype (homozygous recessive).

S–   ×   *ss*

If the plant being tested is homozygous:

SS   ×   *ss*

If the plant being tested is heterozygous:

Ss   ×   *ss*

Gametes

RESULTS

Eggs   Sperm

Ss
Ss   *Ss*
Ss

All progeny show the dominant phenotype.

Eggs   Sperm

Ss
ss   *Ss*
ss

Half the seeds from the cross are wrinkled, and half are spherical.

Conclusion: The plant being tested must be homozygous.

Conclusion: The plant being tested must be heterozygous.

10.6 Homozygous or Heterozygous?
A plant with a dominant phenotype may be homozygous or heterozygous. Its genotype can be determined by making a test cross, which involves crossing it with a homozygous recessive plant and observing the phenotypes of the progeny produced.

we do not yet know the identity of the second allele. There are two possible results:

▶ If the individual being tested is homozygous dominant (*SS*), all offspring of the test cross will be *Ss* and show the dominant trait (spherical seeds).

▶ If the individual being tested is heterozygous (*Ss*), then approximately half of the offspring of the test cross will show the dominant trait (*Ss*), but the other half will be homozygous for, and will show, the recessive trait (*ss*) (Figure 10.6).

These were exactly the results that Mendel obtained; thus Mendel's model accurately predicts the results of such test crosses.

Mendel's second law says that alleles of different genes assort independently

What happens if two parents that differ at two or more loci are crossed? Consider an organism heterozygous for two genes, *Ss* and *Yy*, in which *S* and *Y* came from its mother and *s* and *y* came from its father. When this organism makes gametes, do the alleles of maternal origin (*S* and *Y*) go together to one gamete and those of paternal origin (*s* and *y*) to another gamete? Or can a single gamete receive one maternal and one paternal allele, *S* and *y* (or *s* and *Y*)? To answer these questions, Mendel performed a series of **dihybrid crosses**, crosses made between parents that are identical double heterozygotes.

In these experiments, Mendel began with peas that differed for two characters of the seeds: seed shape and seed color. One true-breeding strain produced only spherical, yellow seeds (*SSYY*) and the other strain produced only wrinkled, green ones (*ssyy*). A cross between these two strains produced an F_1 generation in which all the plants were *SsYy*. Because the *S* and *Y* alleles are dominant, these F_1 seeds were all yellow and spherical.

Mendel continued this experiment to the next generation—the dihybrid cross. There are two ways in which these doubly heterozygous plants might produce gametes, as Mendel saw it. (Remember that he had never heard of chromosomes or meiosis.)

First, if the alleles maintain the associations they had in the original parents (that is, if they are *linked*), then the F_1 plants should produce two types of gametes (*SY* and *sy*), and the F_2 progeny resulting from self-pollination of the F_1 plants should consist of three times as many plants bearing spherical, yellow seeds as ones with wrinkled, green seeds. Were such results to be obtained, there might be no reason to suppose that seed shape and seed color were regulated by two different genes, because spherical seeds would always be yellow, and wrinkled ones always green.

The second possibility is that the segregation of *S* from *s* is independent of the segregation of *Y* from *y* during the production of gametes (that is, that they are *unlinked*). In this case, four kinds of gametes should be produced, in equal numbers: *SY*, *Sy*, *sY*, and *sy*. When these gametes combine at random, they should produce an F_2 of nine different genotypes. The progeny could have any of three possible genotypes for shape (*SS*, *Ss*, or *ss*) and any of three possible genotypes for color (*YY*, *Yy*, or *yy*). The combined nine genotypes should produce just four phenotypes (spherical yellow, spherical green, wrinkled yellow, wrinkled green). By using a Punnett square, we can show that these four phenotypes would be expected to occur in a ratio of 9:3:3:1. (Figure 10.7).

Mendel's dihybrid crosses produced the results predicted by the second possibility. Four different phenotypes appeared in the F_2 in a ratio of about 9:3:3:1. The parental traits appeared in new combinations of two of the phenotypic classes (spherical green and wrinkled yellow). Such new combinations are called **recombinant phenotypes**.

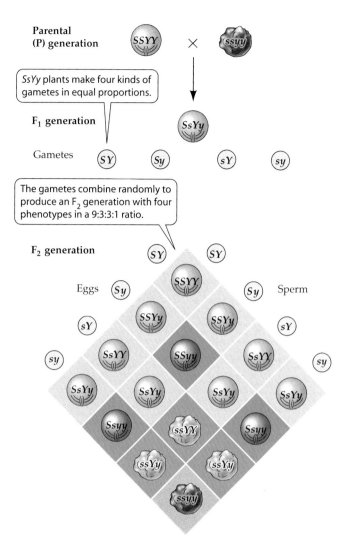

Parental (P) generation

SsYy plants make four kinds of gametes in equal proportions.

F₁ generation

Gametes

The gametes combine randomly to produce an F₂ generation with four phenotypes in a 9:3:3:1 ratio.

F₂ generation

Eggs   Sperm

10.7 Independent Assortment
The 16 possible combinations of gametes result in 9 different genotypes. Because *S* and *Y* are dominant over *s* and *y*, respectively, the 9 genotypes determine 4 phenotypes in the ratio of 9:3:3:1.

These results led Mendel to the formulation of what is now known as Mendel's second law: *Alleles of different genes assort independently of one another during gamete formation.* This **law of independent assortment** is not as universal as the law of segregation, because it applies to genes that lie on separate chromosomes but not necessarily to those that lie on the same chromosome. However, it is correct to say that *chromosomes* segregate independently during the formation of gametes, and so do any two genes on separate chromosome pairs (Figure 10.8).

Punnett squares or probability calculations: A choice of methods

Many people find it easiest to solve genetics problems using probability calculations, perhaps because the principle is familiar. When we flip a coin, for example, we expect that it has an equal probability of landing "heads" or "tails." For a given toss of a fair coin, the probability of heads is independent of what happened in all the previous tosses. A run of ten straight heads implies nothing about

the next toss. No "law of averages" increases the likelihood that the next toss will come up tails, and no "momentum" makes an eleventh occurrence of heads any more likely. On the eleventh toss, the odds are still 50:50.

The basic conventions of probability are simple: If an event is absolutely certain to happen, its probability is 1. If it cannot happen, its probability is 0. Otherwise, its probability lies between 0 and 1. A coin toss results in heads approximately half the time, and the probability of heads is ½—as is the probability of tails.

MULTIPLYING PROBABILITIES. If two coins (a penny and a dime, say) are tossed, each acts independently of the other. What, then, is the probability of both coins coming up heads? Half the time, the penny comes up heads; of that fraction, half the time the dime also comes up heads. Therefore, the joint probability of two heads is half of one-half, or ½ × ½ = ¼. To find the joint probability of independent events, then, the

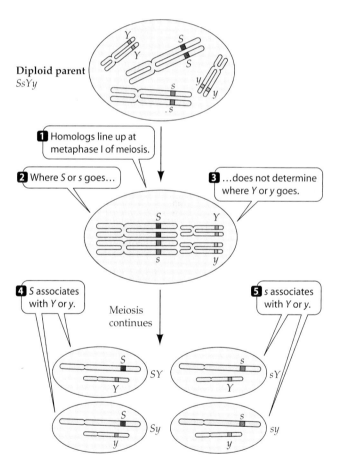

Diploid parent
SsYy

1 Homologs line up at metaphase I of meiosis.

2 Where *S* or *s* goes... **3** ...does not determine where *Y* or *y* goes.

4 *S* associates with *Y* or *y*.   Meiosis continues   **5** *s* associates with *Y* or *y*.

Four haploid gametes
SY, Sy, sY, sy

10.8 Meiosis Accounts for Independent Assortment of Alleles
We now know that alleles are segregated independently during metaphase I of meiosis. Thus a parent of genotype *SsYy* can form gametes with four different genotypes; which ones actually form is a matter of chance.

10.9 Joint Probabilities of Independent Events

Like two tosses of a coin, the segregation of each allele into a sperm or an egg is an independent event. The probability of any given combination of alleles from a sperm and an egg is obtained by multiplying the probabilities of each event; this is the probability of producing a homozygote. Since a heterozygote can be formed in two ways, the two probabilities are added together.

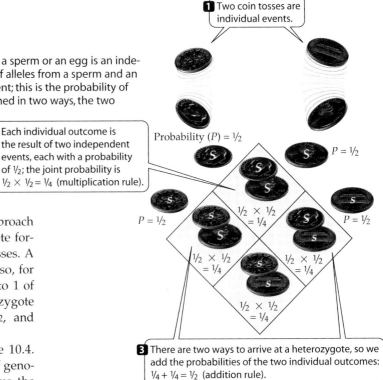

1 Two coin tosses are individual events.

2 Each individual outcome is the result of two independent events, each with a probability of ½; the joint probability is ½ × ½ = ¼ (multiplication rule).

Probability (P) = ½

P = ½   P = ½

P = ½   P = ½

½ × ½ = ¼

½ × ½ = ¼   ½ × ½ = ¼

½ × ½ = ¼

3 There are two ways to arrive at a heterozygote, so we add the probabilities of the two individual outcomes: ¼ + ¼ = ½ (addition rule).

general rule is to multiply the probabilities of the individual events (Figure 10.9).

THE MONOHYBRID CROSS.   To apply a probabilistic approach to genetics problems, we need only deal with gamete formation and random fertilization instead of coin tosses. A homozygote can produce only one type of gamete, so, for example, an SS individual has a probability equal to 1 of producing gametes with the genotype S. The heterozygote Ss produces S gametes with a probability of ½, and s gametes with a probability of ½.

Consider the F_2 progeny of the cross in Figure 10.4. They are obtained by self-pollination of F_1 plants of genotype Ss. The probability that an F_2 plant will have the genotype SS must be ½ × ½ = ¼ because there is a 50:50 chance that the sperm will have the genotype S, and this chance is independent of the 50:50 chance that the egg will have the genotype S. Similarly, the probability of ss offspring is ½ × ½ = ¼.

ADDING PROBABILITIES.   The probability of an F_2 plant getting S from the sperm and s from the egg is also ¼, but remember that the same genotype can also result from s in the sperm and S in the egg, with a probability of ¼. The probability of an event that can occur in two or more different ways is the sum of the individual probabilities of those ways. Thus the probability that an F_2 plant will be a heterozygote is equal to the sum of the probabilities of each way of forming a heterozygote: ¼ + ¼ = ½ (see Figure 10.9). The three genotypes are therefore expected in the ratio ¼ SS:½ Ss:¼ ss—hence the 1:2:1 ratio of genotypes and the 3:1 ratio of phenotypes seen in Figure 10.4.

THE DIHYBRID CROSS.   If F_1 plants heterozygous for two independent characters self-pollinate, the resulting F_2 plants express four different phenotypes. The proportions of these

phenotypes are easily determined by probabilities. Let's see how this works for the experiment shown in Figure 10.7.

The probability that a seed will be spherical is ¾, as we have just seen. By the same reasoning, the probability that a seed will be yellow is also ¾. The two characters are determined by separate genes and are independent of each other, so the joint probability that a seed will be both spherical and yellow is ¾ × ¾ = ⁹⁄₁₆. For the wrinkled, yellow members of the F_2 generation, the probability of being yellow is again ¾; the probability of being wrinkled is ½ × ½ = ¼. The joint probability that a seed will be both wrinkled and yellow, then, is ¾ × ¼ = ³⁄₁₆. The same probability applies, for similar reasons, to the spherical, green F_2 seeds. Finally, the probability that F_2 seeds will be both wrinkled and green must be ¼ × ¼ = ¹⁄₁₆. Looking at all four phenotypes, we see they are expected in the ratio of 9:3:3:1.

10.10 Pedigree Analysis and Dominant Inheritance

A human pedigree showing dominant inheritance. This family carries the allele for Huntington's disease. Everyone who inherits this allele is affected.

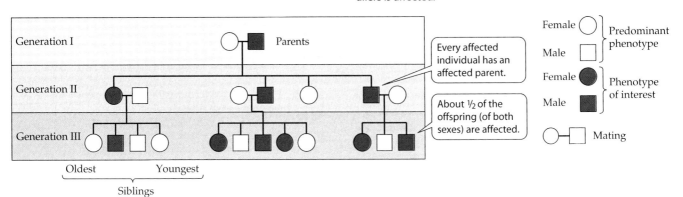

Generation I   Parents

Generation II

Generation III

Oldest   Youngest

Siblings

Every affected individual has an affected parent.

About ½ of the offspring (of both sexes) are affected.

Female ◯ Predominant
Male ☐   phenotype

Female ● Phenotype
Male ■   of interest

◯—☐ Mating

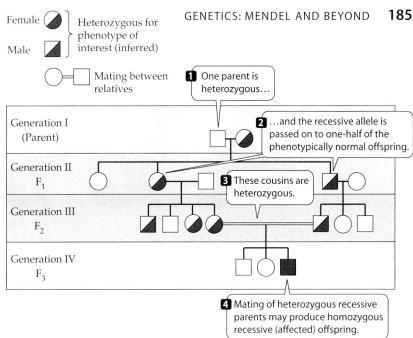

10.11 *Recessive Inheritance*
This family carries the allele for albinism, a recessive trait. In an affected individual, the trait must be inherited from two heterozygous parents or (rarely) from one homozygous and one heterozygous parent. In this case the heterozygous parents are cousins, but the same result could occur if the parents were unrelated but heterozygous.

Female — Heterozygous for phenotype of interest (inferred)

Male

⬡—▭ Mating between relatives

1 One parent is heterozygous…

Generation I (Parent)

2 …and the recessive allele is passed on to one-half of the phenotypically normal offspring.

Generation II F₁

3 These cousins are heterozygous.

Generation III F₂

Generation IV F₃

4 Mating of heterozygous recessive parents may produce homozygous recessive (affected) offspring.

Probability calculations and Punnett squares give the same results. Learn to do genetics problems both ways, and then decide which method you prefer.

Mendel's laws can be observed in human pedigrees

A few years after Mendel's work was uncovered by plant breeders, Mendelian inheritance was found in humans. By now, patterns of over 2,500 inherited human characteristics have been described.

Mendel worked out the rules of inheritance by performing many planned crosses and counting many offspring. Neither of these approaches is possible with humans. So human geneticists rely on **pedigrees**, family trees that show the segregation of phenotypes (and alleles) in several generations of related individuals.

Because humans have such small numbers of offspring, human pedigrees do not show the clear proportions of offspring that Mendel saw in his pea plants (see Table 10.1). For example, when two people marry who are both heterozygous for a recessive allele (say, *Aa*), there will be, for each of their children, a 25 percent probability that the child will be homozygous recessive (*aa*). Over many such marriages, one-fourth of all the children will be homozygous recessive (*aa*). But what about a single marriage? In human families, while the odds for each child remain the same, small numbers of children mitigate against getting the exact one-fourth proportion. So, in a family with two children, both could easily be *aa* (or *Aa* or *AA*).

To deal with this ambiguity, human geneticists assume that any allele that causes an abnormal phenotype is rare in the population. This means that in a given family with the rare allele (say, one parent is *Aa*), it is highly unlikely that an outsider marrying into the family will have that same rare allele (the outsider is most likely *AA*).

Human geneticists may wish to know whether a particular rare allele is dominant or recessive. Figure 10.10 depicts a pedigree showing the pattern of inheritance of a rare dominant phenotype. The following are the key features to look for in such a pedigree:

▶ Every affected person has an affected parent.
▶ About half of the offspring of an affected person are also affected.
▶ The phenotype occurs equally in both sexes.

Compare this pattern with Figure 10.11, which shows the pattern of inheritance of a rare recessive phenotype:

▶ Affected people usually have parents who are both not affected.
▶ About one-quarter of the children of unaffected parents can be affected.
▶ The phenotype occurs equally in both sexes.

In pedigrees showing recessive inheritance, it is not uncommon to find a marriage of two relatives. This observation is a result of the rarity of phenotypically abnormal alleles. For two phenotypically normal parents to have an affected child (*aa*), the parents must both be heterozygous (*Aa*). If the *a* allele is rare in the general population, the chance of two people marrying who are both carrying the same rare allele is quite low. On the other hand, if the particular recessive allele is present in a family, two cousins might share it (see Figure 10.10). This is why studies on populations isolated either culturally (by religion, as with the Amish in the United States) or geographically (as on islands) have been so valuable to human geneticists. People in these groups tend either to have large families, or marry among themselves, or both.

Because the major use of pedigree analysis is the clinical evaluation and counseling of patients with inherited abnormalities, a single pair of alleles is usually followed. However, just as pedigree analysis shows the segregation of alleles, it also can show independent assortment if two different allele pairs are considered.

Alleles and Their Interactions

Let's move on to the extensions of Mendelian genetics that have been developed by other researchers, mostly in the early part of the twentieth century. Decades after Mendel's work, others discovered that his hereditary particles—genes—are chemical entities—DNA sequences—that are

Possible genotypes	CC, Cc^{ch}, Cc^h, Cc	c^{ch}, c^{ch}	c^{ch}c^h, c^{ch}c	c^{h}c^h, c^{h}c	cc
Phenotype	Dark gray	Chinchilla	Light gray	Himalayan	Albino

10.12 Inheritance of Coat Color in Rabbits
There are four alleles of the gene for coat color in rabbits. Different combinations of two alleles give different colors.

usually expressed as proteins. Accordingly, the different alleles of a gene are slightly different sequences of DNA at the same locus, which result in slightly different protein products. In the next chapter we'll see the molecular basis for the distinctions between alleles. In this section we deal with how alleles relate to one another, some of their general properties, and how they arise.

In many cases, alleles do not show simple relationships between dominance and recessiveness. In others, a single allele may have multiple phenotypic effects when it is expressed. Existing alleles can form new alleles by mutation, so there can be many alleles for a single character.

New alleles arise by mutation

Different alleles exist because any gene is subject to *mutation*, which occurs when a gene is changed to a *stable, heritable* new form. In other words, an allele can mutate to become a different allele. Mutation, which will be discussed in detail in Chapter 12, is a random process; different copies of the same gene may be changed in different ways, depending on how and where the DNA sequence changes.

One particular allele of a gene may be defined as the **wild type**, or standard, because it is present in most individuals in nature and gives rise to an expected trait or phenotype. Other alleles of that same gene, often called **mutant** alleles, may produce a different phenotype. The wild-type and mutant alleles reside at the same locus and are inherited according to the rules set forth by Mendel. A genetic locus with a wild-type allele that is present less than 99 percent of the time (the rest of the alleles being mutant) is said to be **polymorphic** (from the Greek *poly*, "many," and *morph*, "form").

Many genes have multiple alleles

Because of random mutations, a group of individuals may have more than two alleles of a given gene. (Any one individual has only two alleles, of course—one from its mother and one from its father.) In fact, there are many examples of such multiple alleles.

Coat color in rabbits is determined by one gene with four alleles. There is a dominance hierarchy in the gene combinations:

$$C > c^{ch} > c^h > c$$

Any rabbit with the C allele (along with any of the four) is gray, and a rabbit that is cc is albino. The intermediate colors result from the different allelic combinations shown in Figure 10.12.

Multiple alleles increase the number of possible phenotypes. In Mendel's monohybrid cross, there was just one pair of alleles (Ss) and two possible phenotypes (resulting from SS or Ss and ss). The four alleles of the rabbit coat color gene produce five phenotypes.

Dominance is usually not complete

In the single-pair alleles studied by Mendel, dominance is complete when an individual is heterozygous. That is, an Ss individual will express the S phenotype. However, many genes have alleles that are not dominant or recessive to one another. Instead, the heterozygotes show an intermediate phenotype—at first glance like that predicted by the old blending theory of inheritance. For example, if a true-breeding red snapdragon is crossed with a true-breeding white one, all the F$_1$ flowers are pink. That this phenomenon can still be explained in terms of Mendelian genetics, rather than blending, is readily demonstrated by a further cross.

According to the blending theory, if one of the pink F$_1$ snapdragons is crossed with a true-breeding white one, all the offspring should be a still lighter pink. In fact, approximately $^1/_2$ of the offspring are white, and $^1/_2$ are the same shade of pink as the original F$_1$. When the F$_1$ pink snapdragons are allowed to self-pollinate, the resulting F$_2$ plants are distributed in a ratio of 1 red:2 pink:1 white (Figure 10.13). Clearly the hereditary particles—the genes—have not blended; they are readily sorted out in the F$_2$.

We can understand these results in terms of the Mendelian model. When a heterozygous phenotype is intermediate, as in the snapdragon example, the gene is said to be governed by **incomplete dominance**. All we need to do in cases like this is recognize that the heterozygotes show a phenotype intermediate between those of the two homozygotes.

We can also understand incomplete dominance in molecular terms. Remember that genes code for specific proteins, many of which are enzymes. Different alleles at a locus code for alternative forms of a protein. When the protein is an enzyme, the different forms often have different degrees of catalytic activity. In the snapdragon example, one allele codes for an enzyme that catalyzes a reaction leading to the forma-

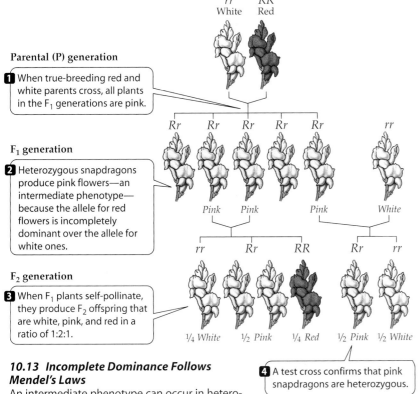

Parental (P) generation

1 When true-breeding red and white parents cross, all plants in the F₁ generations are pink.

F₁ generation

2 Heterozygous snapdragons produce pink flowers—an intermediate phenotype—because the allele for red flowers is incompletely dominant over the allele for white ones.

F₂ generation

3 When F₁ plants self-pollinate, they produce F₂ offspring that are white, pink, and red in a ratio of 1:2:1.

4 A test cross confirms that pink snapdragons are heterozygous.

10.13 Incomplete Dominance Follows Mendel's Laws
An intermediate phenotype can occur in heterozygotes when neither allele is dominant. The phenotype (here, pink flowers) may give the appearance of a blended trait, but dominant and recessive traits reappear in their original forms in succeeding generations, as predicted by Mendel's laws.

tion of a red pigment in the flowers. The alternative allele codes for an altered enzyme that lacks catalytic activity for pigment production. Plants homozygous for this alternative allele cannot synthesize red pigment, and their flowers are white. Heterozygous plants, with only one allele for the functional enzyme, produce just enough red pigment that their flowers are pink.

There are more examples of incomplete dominance than of complete dominance in nature. Thus an unusual feature of Mendel's report is that all seven of the examples he described (see Table 10.1) are characterized by complete dominance. For dominance to be complete, a single copy of the dominant allele must produce enough of its protein product to give the maximum phenotypic response. For example, just one copy of the dominant allele T at one of the loci studied by Mendel leads to the production of enough of a growth-promoting chemical that Tt heterozygotes are as tall as homozygous dominant plants (TT)—the second copy of T causes no further growth of the stem. Homozygous recessive plants (tt) are much shorter because the allele t does not lead to the production of the growth promoter.

In codominance, both alleles are expressed

Sometimes two alleles at a locus produce two different phenotypes that both appear in heterozygotes. An example of this phenomenon, called **codominance**, is seen in the ABO blood group system in humans.

Early attempts at blood transfusion—made before blood types were understood—frequently killed the patient. Around 1900, however, the Austrian scientist Karl Landsteiner mixed blood cells and serum (blood from which cells have been removed) from different individuals. He found that only certain combinations of blood are compatible. In other combinations, the red blood cells of one individual form clumps because of the presence in the other individual's serum of specific proteins, called *antibodies*, that react with foreign, or "nonself," cells. Proteins on nonself cells, called *antigens*, prompt the synthesis of antibodies. This discovery led to our ability to administer compatible blood transfusions that do not kill the recipient.

Blood compatibility is determined by a set of three alleles (I^A, I^B, and I^O) at one locus, which determines certain proteins (antigens) on the surface of red blood cells. Different combinations of these alleles in different people produce four different blood types, or phenotypes: A, B, AB, and O (Figure 10.14).

Some alleles have multiple phenotypic effects

When a single allele has more than one distinguishable phenotypic effect, we say that the allele is **pleiotropic**. A

Blood type of cells	Genotype	Antibodies made by body	Reaction to added antibodies	
			Anti-A	Anti-B
A	I^AI^A or I^AI^O	Anti-B		
B	I^BI^B or I^BI^O	Anti-A		
AB	I^AI^B	Neither anti-A nor anti-B		
O	I^OI^O	Both anti-A and anti-B		

Red blood cells that do not react with antibody remain evenly dispersed.

Red blood cells that react with antibody clump together (speckled appearance).

10.14 ABO Blood Reactions Are Important in Transfusions
Cells of blood types A, B, AB, and O were mixed with anti-A or anti-B antibodies. As you look down the columns, note that each of the types, when mixed separately with anti-A and with anti-B, gives a unique pair of results; this is the basic method by which blood is typed. A person with type O blood is a good blood donor because O cells do not provoke or react with either anti-A or anti-B antibodies. A person with type AB blood is a good recipient, since neither type of antibody is made.

familiar example of pleiotropy involves the allele responsible for the coloration pattern (light body, darker extremities) of Siamese cats, discussed later in this chapter. The same allele is also responsible for the characteristic crossed eyes of Siamese cats. Although these effects appear to be unrelated, both result from the same protein produced under the influence of the allele.

Gene Interactions

Thus far we have treated the phenotype of an organism, with respect to a given character, as a simple result of its genotype, and we have implied that a single trait results from the alleles of a single gene. In fact, several genes may interact to determine a trait's phenotype. For example, height in people is determined by the actions of many genes, such as those that determine bone growth, hormone concentrations, and other aspects of development. Sometimes several genes act *additively*, so that the phenotype can be predicted by how many of these genes are active. To complicate things further, the physical environment may interact with the genetic constitution of an individual in determining the phenotype. Height in people, for example, is not determined only by their genes. Nutrition is just one environmental factor that undoubtedly has a strong influence on height.

Some genes alter the effects of other genes

Epistasis occurs when the phenotypic expression of one gene is affected by another gene. For example, several genes determine coat color in mice. The wild-type color is agouti, a grayish pattern resulting from bands on the individual hairs. The dominant allele *B* determines that the hairs will have bands and thus that the color will be agouti, whereas the homozygous recessive genotype *bb* results in unbanded hairs, and the mouse appears black. On another chromosome, a second locus affects an early step in the formation of hair pigments. The dominant allele *A* at this locus allows normal color development, but *aa* blocks all pigment production. Thus, *aa* mice are all-white albinos, irrespective of their genotype at the *B* locus (Figure 10.15).

If a mouse with genotype *AABB* (and thus the agouti phenotype) is crossed with an albino of genotype *aabb*, the F$_1$ is *AaBb* and has the agouti phenotype. If the F$_1$ mice are crossed with each other to produce an F$_2$ generation, then epistasis will result in an expected phenotypic ratio of 9 agouti:3 black:4 albino. (Can you show why? The underlying ratio is the usual 9:3:3:1 for a dihybrid cross with unlinked genes, but look closely at each genotype and watch out for epistasis.)

In another form of epistasis, two genes are mutually dependent: The expression of each depends on the alleles of the other. The epistatic action of such *complementary genes* may be explained as follows: Suppose gene *A* codes for enzyme A in the metabolic pathway for purple pigment in flowers, and gene *B* codes for enzyme B:

| colorless precursor | $\xrightarrow{\text{enzyme A}}$ | colorless intermediate | $\xrightarrow{\text{enzyme B}}$ | purple pigment |

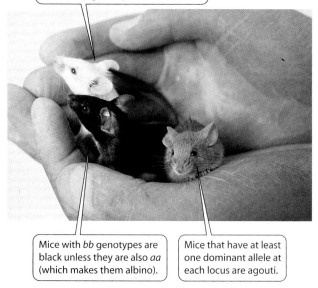

Mice with genotype *aa* are albino regardless of their genotype for the other locus, because the *aa* genotype blocks all pigment production.

Mice with *bb* genotypes are black unless they are also *aa* (which makes them albino).

Mice that have at least one dominant allele at each locus are agouti.

10.15 Genes May Interact Epistatically
Epistasis occurs when one gene alters the phenotypic effect of another gene. In these mice, the presence of the recessive genotype (*aa*) at one locus blocks pigment production, producing an albino mouse no matter what the genotype is at the second locus.

In order for the pigment to be produced, both reactions must take place. The recessive alleles *a* and *b* code for nonfunctional enzymes. If a plant is homozygous for either *a* or *b*, the corresponding reaction will not occur, no purple pigment will form, and the flowers will be white.

Hybrid vigor results from new gene combinations and interactions

If Mendel's paper was the most important event in genetics in the nineteenth century, perhaps an equally important paper in applied genetics was published early in the twentieth century by G. H. Shull, entitled "The composition of a field of maize". For centuries, it has been known that if one takes two pure, homozygous genetic strains of a plant or animal, and crosses them, the result is offspring that are phenotypically much stronger, larger, and in general more "vigorous" than either of the parents (Figure 10.16).

Conversely, avoidance of *inbreeding* (mating between close relatives) is a time-honored tradition among farmers growing crops and in human societies (where it is called incest). The reason for this is that close relatives tend to have the same recessive alleles, some of which may be harmful, as we saw in our discussion of human pedigrees above.

Shull crossed two of the thousands of existing varieties of corn (maize). Both varieties produced about 20 bushels of corn per acre. But when he crossed them, the yield of their offspring was an astonishing 80 bushels per acre. This phenomenon is known as **heterosis** (short for heterozygosis), or **hybrid vigor**. The cultivation of hybrid corn spread rapidly in the United States and all over the world, quadrupling grain production. The practice of hybridization has spread to many other crops and animals used in agriculture.

10.16 Hybrid Vigor in Corn
The heterozygous F₁ offspring is larger and stronger than either homozygous parent.

The actual mechanism by which hybrid vigor works is not known. A widely accepted hypothesis is *overdominance*, a situation in which the heterozygous condition in certain important genes is superior to either homozygote.

Polygenes mediate quantitative inheritance

Individual heritable characters are often found to be controlled by groups of several genes, called **polygenes**, of which each allele intensifies or diminishes the observed phenotype. As a result, variation in such characters is **continuous**, or *quantitative*) rather than, as in the examples we have been considering, **discontinuous** (or *discrete*). Many characters that vary continuously—such as height and other aspects of size, or skin color—are under polygenic control. The polygenes affecting a particular quantitative character are commonly located on many different chromosomes.

Humans differ with respect to the amount of a dark pigment, *melanin*, in their skin (Figure 10.17). There is great variation in the amount of melanin among different people, but much of this variation is determined by alleles at many loci. No alleles at these loci demonstrate dominance. Of course, skin color is not entirely determined by the genotype, since exposure to sunlight in light-skinned people can cause the production of more melanin (that is, a suntan).

The environment affects gene action

The phenotype of an individual does not result from its genotype alone. Genotype and environment interact to determine the phenotype of an organism. Environmental variables such as light, temperature, and nutrition can affect the translation of a genotype into a phenotype. A familiar example involves the Siamese cat. This handsome animal normally has darker fur on its ears, nose, paws, and tail than on the rest of its body. These darkened extremities normally have a lower temperature than the rest of the body.

A few simple experiments show that the Siamese cat has a genotype that results in dark fur, but only at temperatures below the general body temperature. If some dark fur is removed from the tail and the cat is kept at higher than usual

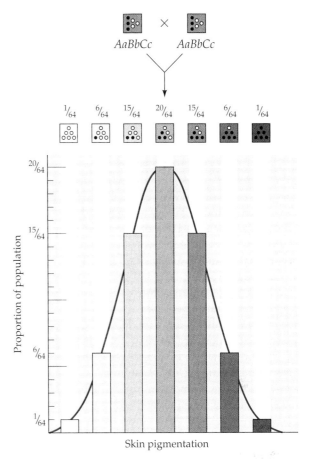

10.17 Polygenes Determine Human Skin Pigmentation
A model of polygenic inheritance based on three genes. Alleles A, B, and C contribute melanin to the skin, but alleles a, b, and c do not. The more A, B, and C alleles an individual possesses, the darker that person's skin will be. If both members of a couple have intermediate pigmentation (in this example, *AaBbCc*), it is unlikely (but not impossible) that their children will have either very light or very dark skin. The actual number of genes involved is much higher.

temperatures, the new fur that grows in is light. Conversely, removal of light fur from the back, followed by local chilling of the area, causes the spot to fill in with dark fur.

It is sometimes possible to determine the proportion of individuals in a group with a given genotype that actually show the expected phenotype. This proportion is called the **penetrance** of the genotype. The environment may also affect the **expressivity** of the genotype—that is, the degree to which it is expressed in an individual. For an example of environmental effects on expressivity, consider how Siamese cats kept indoors or outdoors in different climates might look.

Uncertainty over how much of the phenotypic variation we observe is due to the environment and how much to the effects of polygenes complicates the analysis of quantitative inheritance. A useful approach that avoids this difficulty is to study identical twins, which develop from the same fertilized egg. Since such twins are genetically identical, any differences between them can be attributed to environmental effects.

Genes and Chromosomes

The recognition that genes occupy characteristic positions on chromosomes and thus are segregated by meiosis enabled Mendel's successors to provide a physical explanation for his model of inheritance. It soon became apparent that the association of genes with chromosomes has other genetic consequences as well.

In this section we will address the following questions: What is the pattern of inheritance of genes that occupy nearby loci on the same chromosome? How do we determine the order of genes on a chromosome—and the distances between them? Why were all the carriers of hemophilia in Queen Victoria's family women, and why were all of her descendants who had hemophilia men?

The answers to these and many other genetic questions were worked out in studies of the fruit fly *Drosophila melanogaster* (Figure 10.18). Its small size, its ease of cultivation, and its short generation time made this animal an attractive experimental subject. Beginning in 1909, Thomas Hunt Morgan and his students established *Drosophila* as a highly useful laboratory organism in Columbia University's famous "fly room," where they discovered the phenomena described in this section. *Drosophila* remains extremely important in studies of chromosome structure, population genetics, the genetics of development, and the genetics of behavior.

Genes on the same chromosome are linked

In the immediate aftermath of the rediscovery of Mendel's laws, the second law—independent assortment—was considered to be generally applicable. However, some investigators, including R. C. Punnett (the inventor of the Punnett square), began to observe strange deviations from the expected 9:3:3:1 ratio in some dihybrid crosses. T. H. Morgan, too, obtained data not in accord with Mendelian ratios, and specifically not in accord with the law of independent assortment.

Morgan crossed *Drosophila* of two known genotypes, *BbVgvg × bbvgvg*, in which *B*, the wild type (gray body), is dominant over *b* (black body), and *Vg* (wild-type wing) is dominant over *vg* (*vestigial*, a very small wing). (Do you recognize this type of cross? It is a test cross for the two gene pairs—see Figure 10.6.) Morgan expected to see four phenotypes in a ratio of 1:1:1:1, but this was not what he observed. The body color gene and the wing size gene were not assorting independently; rather, they were for the most part inherited together (Figure 10.19).

These results became understandable to Morgan when he assumed that the two loci are *on the same chromosome*—

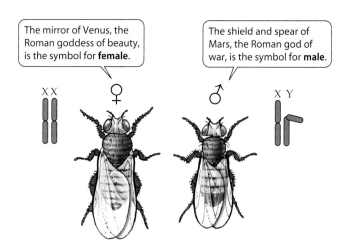

10.18 Drosophila melanogaster, *the Star of Morgan's Fly Room*
The fruit fly (whose Latin name means "vinegar-loving, dark-bodied fly") has a short generation time—a major reason for its widespread use as a laboratory organism in genetics experiments.

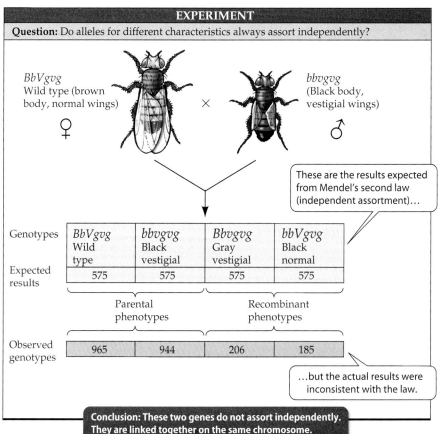

10.19 Alleles That Do Not Assort Independently
Morgan's studies showed that the genes for body color and wing size in *Drosophila* are linked, so their alleles do not assort independently. Linkage accounts for the departure of the phenotype ratios observed from the results predicted by Mendel's laws.

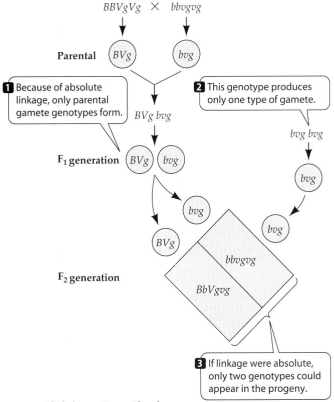

$BBVgVg \times bbvgvg$

Parental

BVg bvg

1 Because of absolute linkage, only parental gamete genotypes form.

2 This genotype produces only one type of gamete.

$BVg\ bvg$

$bvg\ bvg$

F₁ generation BVg bvg

bvg

bvg

bvg

BVg

F₂ generation

$bbvgvg$

$BbVgvg$

3 If linkage were absolute, only two genotypes could appear in the progeny.

10.20 If Linkage Were Absolute
If two genes are absolutely linked on the same chromosome, all the F₂ offspring from a dihybrid test cross would have parental genotypes. If the genes in Morgan's experiment had been absolutely linked, they would have been inherited as if they were a single gene.

that is, that they are **linked**. After all, since the number of genes in a cell far exceeds the number of chromosomes, each chromosome must contain many genes. The full set of loci on a given chromosome constitutes a *linkage group*. The number of linkage groups in a species equals the number of homologous chromosome pairs.

Suppose, now, that the *Bb* and *Vgvg* loci are indeed located on the same chromosome. If we assume that the linkage is absolute, we expect to see just *two* types of progeny from Morgan's test cross (Figure 10.20). These two would resemble the original (grand)parents. However, this is not always the case.

Genes can be exchanged between chromatids

Absolute linkage is extremely rare. If linkage were absolute, Mendel's second law (independent assortment of alleles at different loci) would apply only to loci on different chromosomes. What actually happens is more complex, and therefore more interesting. The chromosome is not unbreakable, so **recombination** of genes can occur. Genes at different loci on the same chromosome do sometimes separate from one another during meiosis.

Genes may recombine when two homologous chromosomes physically exchange corresponding segments during prophase I of meiosis—that is, by crossing over (see Figure 9.16). In other words, recombination may occur at a chiasma when homologous chromosomes are paired up during meiosis (Figure 10.21).

Recall that the DNA has been replicated by this stage, and that each chromosome consists of two chromatids. The exchange event involves only two of the four chromatids, one from each member of the chromosome pair. The chiasma can occur at any point along the length of the chromosome. The chromosome sections involved are exchanged reciprocally, so both chromatids involved in crossing over become recombinant (that is, each chromatid ends up with genes from both parents).

When crossing over takes place between two linked genes, not all progeny of a cross will have the parental types. Instead, recombinant offspring appear as well, and

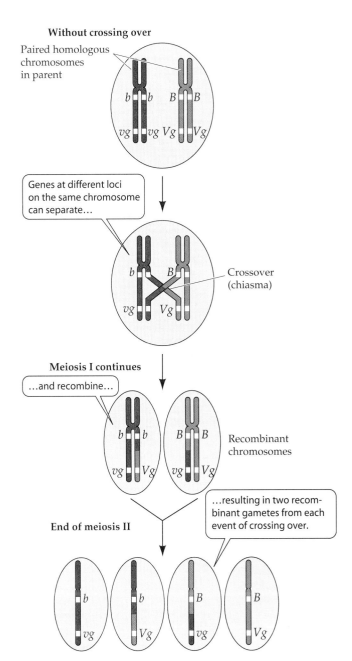

Without crossing over

Paired homologous chromosomes in parent

b b B B

vg vg Vg Vg

Genes at different loci on the same chromosome can separate…

b B

vg Vg

Crossover (chiasma)

Meiosis I continues

…and recombine…

b b B B

vg Vg vg Vg

Recombinant chromosomes

…resulting in two recombinant gametes from each event of crossing over.

End of meiosis II

b b B B

vg Vg vg Vg

10.21 Crossing Over Results in Genetic Recombination
Genes at different loci on the same chromosome can separate from one another and recombine by crossing over. Recombination occurs at a chiasma during prophase I of meiosis.

10.22 Recombinant Frequencies
The frequency of recombinant offspring (those with a phenotype different than either parent's) can be calculated. Recombinant frequencies will be larger for loci that are far apart than for those that are close together on the chromosome.

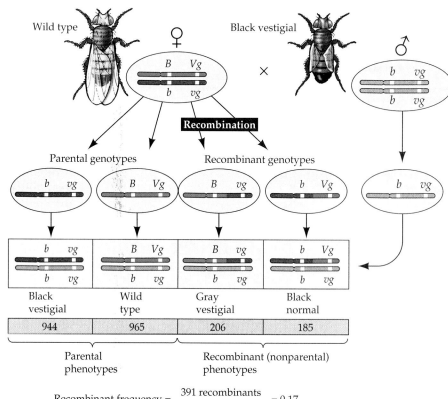

Recombinant frequency = $\dfrac{391\ \text{recombinants}}{2{,}300\ \text{total offspring}}$ = 0.17

they appear in repeatable proportions called **recombinant frequencies**, which equal the number of recombinant progeny divided by the total number of progeny (Figure 10.22). Recombinant frequencies will be greater for loci that are far apart on the chromosome than for loci that are closer together, because a chiasma is more likely to cut between genes that are far apart than genes that are close together.

Geneticists make maps of eukaryotic chromosomes

If two loci are very close together on a chromosome, the odds for crossing over between them are small. In contrast, if two loci are far apart, crossing over could occur between them at many points. In 1911, Alfred Sturtevant, then an undergraduate student in T. H. Morgan's fly room, realized how that simple insight could be used to show where different genes lie on the chromosome in relation to one another. He suggested that the farther apart two genes are on a chromosome, the greater the likelihood that they will separate and recombine in meiosis.

The Morgan group had determined recombinant frequencies for many pairs of linked genes. Sturtevant used these recombinant frequencies to create genetic maps that indicated the arrangement of genes along the chromosome (Figure 10.23). Ever since Sturtevant demonstrated this important point, geneticists have mapped the chromosomes of eukaryotes, prokaryotes, and viruses, assigning distances between genes in **map units**. A map unit corresponds to a recombinant frequency of 0.01; it is also referred to as a centimorgan (cM), in honor of the founder of the fly room. You, too, can work out a genetic map (Figure 10.24).

Sex Determination and Sex-Linked Inheritance

In Kölreuter's experience, and later in Mendel's, reciprocal crosses apparently always gave identical results. The reason is that in diploid organisms, chromosomes come in pairs. One member of each chromosome pair derives from each parent; it does not matter, in general, whether a dominant allele was contributed by the mother or by the father. But sometimes the parental origin of a chromosome does matter. To understand the types of inheritance in which paren-

tal origin is important, we must consider the ways in which sex is determined in different species.

Sex is determined in different ways in different species

In corn, a plant much studied by geneticists, every diploid adult has both male and female reproductive structures. The two types of tissue are genetically identical, just as roots and leaves are genetically identical. Plants such as maize and Mendel's pea plants, and animals such as earthworms, which produce both male and female gametes in the same organism, are said to be *monoecious* (from the Greek for "one house"). Other plants, such as date palms and oak trees, and most animals are *dioecious* ("two houses"), meaning that some individuals can produce only male gametes and the others can produce only female gametes. In other words, dioecious organisms have two sexes.

In most dioecious organisms, sex is determined by differences in the chromosomes, but such determination operates in different ways in different groups of organisms. For example, the sex of a honeybee depends on whether it develops from a fertilized or an unfertilized egg. A fertilized egg is diploid and gives rise to a female bee—either a worker or a queen, depending on the diet during larval life (again, note how the environment affects the phenotype). An unfertilized egg is haploid and gives rise to a male drone:

Diploid worker

Diploid queen

Haploid drone

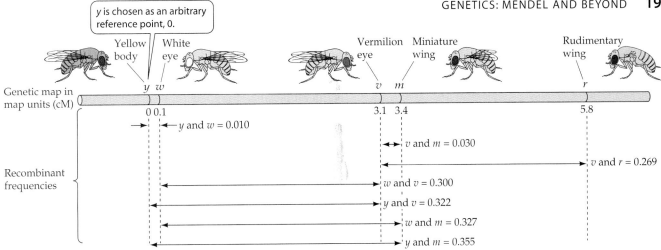

y is chosen as an arbitrary reference point, 0.

10.23 Steps Toward a Genetic Map
Because the chance of a recombinant genotype occurring increases the farther apart two loci fall on a chromosome, Sturtevant was able to derive this partial map of a *Drosophila* chromosome from the Morgan group's data on the recombinant frequencies of five recessive traits. He assigned an arbitrary unit of distance—the map unit, or centimorgan (cM)—equivalent to a recombinant frequency of 0.01.

1 At the outset, we have no idea of the individual distances, and there are several possible sequences (a-b-c, a-c-b, b-a-c).

We make a cross *AABB* × *aabb*, and obtain an F₁ generation with a genotype *AaBb*. We test cross these *AaBb* individuals with *aabb*. Here are the genotypes of the first 1,000 progeny:

450 *AaBb*, 450 *aabb*, 50 *AaBb*, and 50 *aaBb*.

2 How far apart are the *a* and *b* genes? Well, what is the recombinant frequency? Which are the recombinant types, and which are the parental types?

Recombinant frequency (*a* to *b*) = (50 + 50)/1,000 = 0.1
So the map distance is
Map distance = 100 × recombinant frequency = 100 × 0.1 = 10 cM

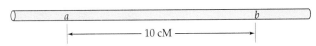

3 Now we make a cross *AACC* × *aacc*, obtain an F₁ generation, and test cross it, obtaining:

460 *AaCc*, 460 *aacc*, 40 *Aacc*, and 40 *aaCc*.

How far apart are the *a* and *c* genes?

Recombinant frequency (*a* to *c*) = (40 + 40)/1,000 = 0.08
Map distance = 100 × recombinant frequency = 100 × 0.08 = 8 cM

10.24 Map These Genes
We want to determine the order of three loci (*a*, *b*, and *c*) on a chromosome, as well as the map distances (in cM) between them. How do we determine a map distance?

4 How far apart are the *b* and *c* genes?

We make a cross *BBCC* × *bbcc*, obtain an F₁ generation, and test cross it, obtaining:

490 *BbCc*, 490 *bbcc*, 10 *Bbcc*, and 10 *bbCc*.

Determine the map distance between *b* and *c*.

Recombinant frequency (*b* to *c*) = (10 + 10)/1,000 = 0.02
Map distance = 100 × recombinant frequency = 100 × 0.02 = 2 cM

5 Which of the three genes is between the other two?
Because *a* and *b* are the farthest apart, *c* must be between them.

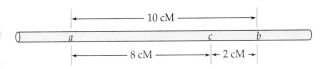

These numbers add up perfectly, but in most real cases they don't add up perfectly because of multiple crossovers.

In many other animals, including humans, sex is determined by a single **sex chromosome** (or by a pair of them). Both males and females have two copies of each of the rest of the chromosomes, which are called **autosomes**.

Female grasshoppers, for example, have two **X chromosomes**, whereas males have only one. Female grasshoppers are described as being XX (ignoring the autosomes) and males as XO (pronounced "ex-oh"):

Females form eggs that contain one copy of each autosome and one X chromosome. Males form approximately equal amounts of two types of sperm: One type contains one copy of each autosome and one X chromosome; the other type contains only autosomes. When an X-bearing sperm fertilizes an egg, the zygote is XX, and develops into a female. When a sperm without an X fertilizes an egg, the zygote is XO, and develops into a male. This chromosomal mechanism ensures that the two sexes are produced in approximately equal numbers.

As in grasshoppers, female mammals have two X chromosomes and males have one. However, male mammals also have a sex chromosome that is not found in females: the **Y chromosome**. Females may be represented as XX and males as XY:

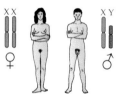

Males produce two kinds of gametes: Each has a complete set of autosomes, but half the gametes carry an X chromosome and the other half carry a Y. When an X-bearing sperm fertilizes an egg, the resulting XX zygote is female; when a Y-bearing sperm fertilizes an egg, the resulting XY zygote is male.

Some subtle but important differences show up clearly in mammals with abnormal sex chromosome constitutions. These conditions, which result from nondisjunctions, as described in Chapter 9, tell us something about the functions of the X and Y chromosomes. In humans, XO individuals sometimes appear. Human XO individuals are females who are physically moderately abnormal but mentally normal; usually they are also sterile. The XO condition in humans is called *Turner syndrome*. It is the only known case in which a human can survive with only one member of a chromosome pair (here, the XY pair), although most XO conceptions terminate spontaneously early in development. XXY individuals also occur; this condition is known as *Klinefelter syndrome*. People with this genotype are sometimes taller than average, always sterile, and always male.

The X and Y chromosomes have different functions

The gene that determines maleness was identified through observations of people with chromosomal abnormalities. For example, some XY individuals who are phenotypically women have been identified and studied; in these people, a small portion of the Y chromosome was missing. In other cases, men who were genetically XX had a small piece of the Y chromosome present, attached to another chromosome. The missing and present Y fragment in these two examples, respectively, contained the maleness-determining gene, which was named *SRY* (*sex-determining region on the Y chromosome*).

The *SRY* gene codes for a protein involved in *primary sex determination*—that is, the determination of the kinds of gametes that will be produced gametes and the organs that will make them. In the presence of functional SRY protein, the embryo develops sperm-producing testes. If SRY protein is absent, the primary sex determination is female: ovaries and eggs develop. In this case, a gene on the X chromosome called *DAX1* produces an anti-testis factor. So the role of *SRY* in a male is to inhibit the maleness inhibitor made by *DAX1*.

Primary sex determination is not the same as *secondary sex determination*, which results in the outward manifestations of maleness and femaleness (body type, breast development, body hair, and voice). These outward characteristics are not determined directly by the presence or absence of the Y chromosome. Rather, they are determined by the actions of *hormones*, such as testosterone and estrogen.

The Y chromosome functions differently in *Drosophila melanogaster*. Superficially, *Drosophila* follows the same pattern of sex determination as mammals—females are XX and males are XY. However, XO individuals are males (rather than females as in mammals) and almost always are indistinguishable from normal XY males except that they are sterile. XXY *Drosophila* are normal, fertile females:

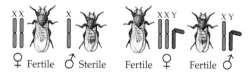

Thus, in *Drosophila*, sex is determined strictly by the ratio of X chromosomes to autosome sets. If there is one X chromosome for each set of autosomes, the individual is a female; if there is only one X chromosome for the two sets of autosomes, the individual is a male. The Y chromosome plays no sex-determining role in *Drosophila*, but it is needed for male fertility.

In birds, moths, and butterflies, males are XX and females are XY. To avoid confusion, these forms are usually expressed as ZZ (male) and ZW (female):

In these organisms, the female produces two types of gametes. Thus the egg determines the sex of the offspring, rather than the sperm, as in humans and fruit flies.

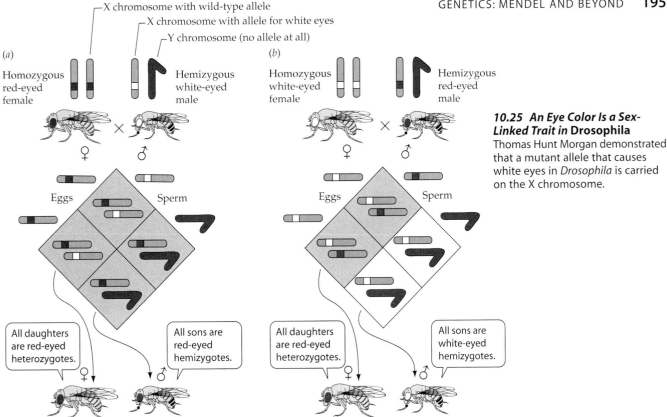

10.25 An Eye Color Is a Sex-Linked Trait in Drosophila
Thomas Hunt Morgan demonstrated that a mutant allele that causes white eyes in *Drosophila* is carried on the X chromosome.

Genes on sex chromosomes are inherited in special ways

In *Drosophila* and in humans, the Y chromosome carries few known genes, but a substantial number of genes affecting a great variety of characters are carried on the X chromosome. The result of this arrangement is a deviation from the usual Mendelian ratios for the inheritance of genes located on the X chromosome. Any such gene is present in two copies in females, but in only one copy in males. Therefore, females may be heterozygous for genes that are on the X chromosome, but males will always be **hemizygous** for genes on the X chromosome—they will have only one of each, and it will be expressed.

Kölreuter's historic reciprocal crosses, mentioned at the beginning of this chapter, always gave the same outcome regardless of which parent displayed which trait. However, reciprocal crosses do not give identical results for characters whose genes are carried on the sex chromosomes. This is a sharp deviation from the rules governing the inheritance of alleles on autosomes.

The first and still one of the best examples of **sex-linked inheritance**—inheritance of characters governed by loci on the sex chromosomes—is that of eye color in *Drosophila*. The wild-type eye color of these flies is red. In 1910, Morgon discovered a mutation that causes white eyes. He experimented by crossing flies of the wild-type and mutant phenotypes. His results demonstrated that the eye color locus is on the X chromosome.

When homozygous red-eyed females were crossed with (hemizygous) white-eyed males, all the sons and daughters had red eyes, because red is dominant over white, and all the progeny had inherited a wild-type X chromosome from

their mothers (Figure 10.25*a*). However, in the reciprocal cross, in which a white-eyed female was mated with a red-eyed male, all the sons were white-eyed and all the daughters red-eyed (Figure 10.25*b*).

The sons from the reciprocal cross inherited their only X chromosome from their white-eyed mother; the Y chromosome they inherited from their father does not carry the eye color locus. The daughters, on the other hand, got an X chromosome with the white allele from their mother and an X chromosome bearing the red allele from their father; they were therefore red-eyed heterozygotes.

When Morgan mated heterozygous females with red-eyed males, he observed that half their sons had white eyes, but all their daughters had red eyes. Thus, in this case, eye color was carried on the X chromosome and not on the Y.

Human beings display many sex-linked characters

The human X chromosome carries thousands of genes. The alleles at these loci follow the same pattern of inheritance as those for white eyes in *Drosophila*. One human X chromosome gene, for example, has a mutant recessive allele that leads to red-green color blindness, a hereditary disorder. Red-green color blindness appears in individuals who are homozygous or hemizygous for the mutant allele.

Pedigree analysis of X-linked recessive phenotypes (Figure 10.26) reveals the following patterns:

▶ The phenotype appears much more often in males than in females, because only one copy of the rare allele is needed for its expression in males, while two copies must be present in females.

▶ A male with the mutation can pass it on only to his daughters; all his sons get his Y chromosome.

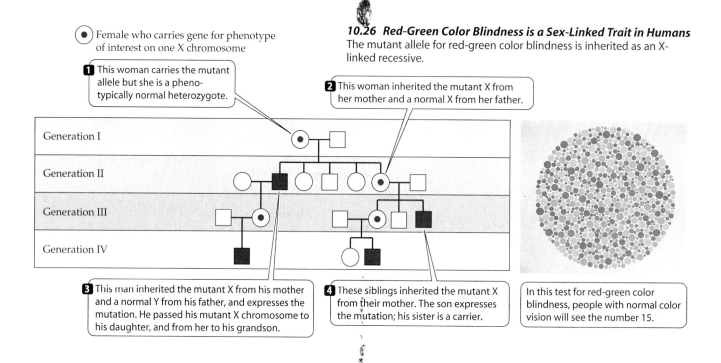

Female who carries gene for phenotype of interest on one X chromosome

1 This woman carries the mutant allele but she is a phenotypically normal heterozygote.

2 This woman inherited the mutant X from her mother and a normal X from her father.

10.26 Red-Green Color Blindness is a Sex-Linked Trait in Humans
The mutant allele for red-green color blindness is inherited as an X-linked recessive.

Generation I

Generation II

Generation III

Generation IV

3 This man inherited the mutant X from his mother and a normal Y from his father, and expresses the mutation. He passed his mutant X chromosome to his daughter, and from her to his grandson.

4 These siblings inherited the mutant X from their mother. The son expresses the mutation; his sister is a carrier.

In this test for red-green color blindness, people with normal color vision will see the number 15.

▶ Daughters who receive one mutant X chromosome are heterozygous *carriers*. They are phenotypically normal, but they can can pass the mutant X to both sons and daughters (only half of the time; half of their X chromosomes carry the normal allele).

▶ The mutant phenotype can skip a generation if the mutation passes from a male to his daughter (phenotypically normal) to her son.

As we will see in later chapters, there are several important human diseases that are inherited as X-linked recessives, including the most common forms of muscular dystrophy and hemophilia. England's Queen Victoria was a heterozygous carrier of hemophilia A, the bleeding disorder mentioned at the beginning of this chapter. She passed it on to some of her male offspring and thereby to several of the royal familes of Europe.

Human mutations inherited as X-linked dominants are rarer than recessives, because dominants appear in every generation, and because people carrying the harmful mutations, even as heterozygotes, often fail to survive and/or reproduce. (Look at the four points above and try to determine what would happen if the mutation were dominant.)

The small human Y chromosome carries only about 20 known genes. Among them are the maleness determinants, whose existence was suggested by the phenotypes of the XO and XXY individuals described on page 20. Y-linked alleles are passed from father to son. (You can verify this with a Punnett square.)

Non-Nuclear Inheritance

The nucleus is not the only organelle in a eukaryotic cell that carries genetic material. As we described in Chapter 4, mitochondria and plastids, which may have arisen from

bacteria that colonized other cells, contain small numbers of genes. For example, in humans, there are about 60,000 genes in the nucleus and 37 in mitochondria. Plastid genomes are about five times larger than those of mitochondria. In any case, the organelle genes include several that are important for organelle assembly and function, so it is not surprising that mutations of these genes have profound effects on the organism.

The inheritance of organelle genes differs from that of nuclear genes for several reasons. First, mitochondria and plastids are apparently passed on from the mother only. As you will see in later chapters, eggs contain abundant cytoplasm and organelles, but the only part of the sperm that survives to take part in the union of haploid gametes is the nucleus. So you have inherited your mother's mitochondria (with their genes), but not your father's.

Second, there may be hundreds of mitochondria and/or plastids in a cell. So a cell is not diploid for organelle genes; rather, it is highly polyploid. A third factor is that organelle genes tend to mutate at much faster rates than nuclear genes, so there are multiple alleles of organelle genes.

The phenotypes of mutations in the DNA of organelles reflect the organelles' roles. For example, some plastid mutations affect proteins that assemble chlorophyll molecules into the photosystem reaction centers (see Figure 8.11), and result in a phenotype that is essentially a white instead of a green tissue. Mitochondrial mutations that affect one of the complexes in the electron transport chain result in less ATP production. They have especially noticeable effects in tissues with a high energy requirement, such as the nervous system, muscles, and kidneys. In 1995, Greg Lemond, a professional cyclist who had won the famous Tour de France three times, was forced to retire because of muscle weakness suspected to be caused by a mitochondrial mutation.

Chapter Summary

The Foundations of Genetics

▶ Plant breeders can control which plants mate. Although it has long been known that both parent plants contribute equally to the character traits of their offspring, before Mendel's time it was believed that, once they were brought together, the units of inheritance blended and could never be separated. **Review Figure 10.1**

▶ Although Gregor Mendel's work was meticulous and well documented, his discoveries, reported in the 1860s, lay dormant until decades later, when others rediscovered them.

Mendel's Experiments and Laws of Inheritance

▶ Mendel used garden pea plants for his studies because they were easily cultivated and crossed, and because they showed numerous characters (such as seed shape) with clearly different traits (spherical or wrinkled). **Review Table 10.1**

▶ In a monohybrid cross, the offspring showed only one of the two traits. Mendel proposed that the trait observed in the first generation (F_1) was dominant and the other was recessive. **Review Table 10.1**

▶ When the F_1 offspring were self-pollinated, the resulting F_2 generation showed a 3:1 phenotypic ratio, with the recessive phenotype present in one-fourth of the offspring. This reappearance of the recessive phenotype refuted the blending hypothesis. **Review Figure 10.3**

▶ Because some alleles are dominant and some are recessive, the same phenotype can result from different genotypes. Homozygous genotypes have two copies of the same allele; heterozygous genotypes have two different alleles. Heterozygous genotypes yield phenotypes that show the dominant trait.

▶ On the basis of many crosses using different characters, Mendel proposed his first law: that the units of inheritance (now known as genes) are particulate, that there are two copies (alleles) of each gene in every parent, and that during gamete formation the two alleles for a character segregate from each other. **Review Figure 10.4**

▶ Geneticists who followed Mendel showed that genes are carried on chromosomes and that alleles are segregated during meiosis I. **Review Figure 10.5**

▶ Using a test cross, Mendel was able to determine whether a plant showing the dominant phenotype was homozygous or heterozygous. The appearance of the recessive phenotype in half of the offspring of such a cross indicates that the parent is heterozygous. **Review Figure 10.6**

▶ From studies of the simultaneous inheritance of two characters, Mendel concluded that alleles of different genes assort independently. **Review Figures 10.7, 10.8**

▶ We can predict the results of hybrid crosses either by using a Punnett square or by calculating probabilities. To determine the joint probability of independent events, we multiply the individual probabilities. To determine the probability of an event that can occur in two or more different ways, we add the individual probabilities. **Review Figure 10.9**

▶ That humans exhibit Mendelian inheritance can be inferred by the analysis of pedigrees. **Review Figures 10.10, 10.11**

Alleles and Their Interactions

▶ New alleles arise by mutation, and many genes have multiple alleles. **Review Figure 10.12**

▶ Dominance is usually not complete, since both alleles in a heterozygous organism may be expressed in the phenotype. **Review Figures 10.13, 10.14**

Gene Interactions

▶ In epistasis, the products of different genes interact to produce a phenotype. **Review Figure 10.15**

▶ In some cases, the phenotype is the result of the additive effects of several genes (polygenes), and inheritance is quantitative. **Review Figure 10.17**

▶ Environmental variables such as temperature, nutrition, and light affect gene action.

Genes and Chromosomes

▶ Each chromosome carries many genes. Genes located on the same chromosome are said to be linked, and they are often inherited together. **Review Figures 10.19, 10.20**

▶ Linked genes recombine by crossing over in prophase I of meiosis. The result is recombinant gametes, which have new combinations of linked genes because of the exchange. **Review Figures 10.21, 10.22**

▶ The distance between genes on a chromosome is proportional to the frequency of crossing over between them. Genetic maps are based on recombinant frequencies. **Review Figures 10.23, 10.24**

Sex Determination and Sex-Linked Inheritance

▶ Sex chromosomes carry genes that determine whether male or female gametes are produced. The specific functions of X and Y chromosomes differ among species.

▶ In fruit flies and mammals, the X chromosome carries many genes, but its homolog, the Y chromosome, has only a few. Males have only one allele for most X-linked genes, so rare alleles show up phenotypically more often in males than in females. **Review Figures 10.25, 10.26**

Non-Nuclear Inheritance

▶ Cytoplasmic organelles such as plastids and mitochondria contain some heritable genes.

▶ Organelle genes are generally inherited by way of the egg (maternal inheritance), because male gametes contribute only their nucleus to the zygote at fertilization.

Some Genetics Problems

1. Using the Punnett squares below, show that for typical dominant and recessive autosomal traits, it does not matter which parent contributes the dominant allele and which the recessive allele. Cross true-breeding tall plants (*TT*) with true-breeding dwarf plants (*tt*).

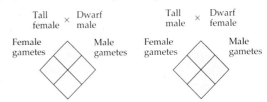

2. The accompanying photograph shows the shells of 15 bay scallops, *Argopecten irradians*. These scallops are hermaphroditic; that is, a single individual can reproduce sexually by self-fertilization, as did the pea plants of the F_1 generation in Mendel's experiments. Three color schemes are evident: yellow, orange, and black and white. The color-determining gene has three alleles. The top row shows a yellow scallop and a representative sample of its offspring, the middle row shows a black-and-white scallop and its offspring, and the bottom row shows an orange scallop and its offspring. Assign a suitable symbol to each of the three alleles participating in color determination;

then determine the genotype of each of the three parent individuals and explain what you can about the genotypes of the different offspring. Explain your results carefully.

3. Show diagrammatically what occurs when the F₁ offspring of the cross in Question 1 self-pollinate.

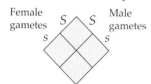

4. A new student of genetics suspects that a particular recessive trait in fruit flies (dumpy wings, which are somewhat smaller and more bell-shaped than the wild type) is sex-linked. A single mating between a fly having dumpy wings (*dp*; female) and a fly with wild-type wings (*Dp*; male) produces three dumpy-winged females and two wild-type males. On the basis of these data, is the trait sex-linked or autosomal? What were the genotypes of the parents? Explain how these conclusions can be reached on the basis of so few offspring.

5. The sex of fishes is determined by the same XY system as in humans. An allele at one locus on the Y chromosome of the fish *Lebistes* causes a pigmented spot to appear on the dorsal fin. A male fish that has a spotted dorsal fin is mated with a female fish that has an unspotted fin. Describe the phenotypes of the F₁ and the F₂ generations from this cross.

6. In *Drosophila melanogaster*, the recessive allele *p*, when homozygous, determines pink eyes. *Pp* or *PP* results in wild-type eye color. Another gene, on another chromosome, has a recessive allele, *sw*, that produces short wings when homozygous. Consider a cross between females of genotype *PPSwSw* and males of genotype *ppswsw*. Describe the phenotypes and genotypes of the F₁ generation and of the F₂ generation produced by allowing the F₁ progeny to mate with one another.

7. On the same chromosome of *Drosophila melanogaster* that carries the *p* (pink eyes) locus, there is another locus that affects the wings. Homozygous recessives, *byby*, have blistery wings, while the dominant allele *By* produces wild-type wings. The *P* and *By* loci are very close together on the chromosome; that is, they are tightly linked. In answering the following questions, assume that no crossing over occurs.

a. For the cross *PPByBy* x *ppbyby*, give the phenotypes and genotypes of the F₁ generation and of the F₂ generation produced by interbreeding of the F₁ progeny.

b. For the cross *PPbyby* x *ppByBy*, give the phenotypes and genotypes of the F₁ and F₂ generations.

c. For the cross in Question 7*b*, what further phenotype(s) would appear in the F₂ generation if crossing over occurred?

d. Draw a nucleus undergoing meiosis, at the stage in which the crossing over in Question 7*c* occurred. In which generation (P, F₁, or F₂) did this crossing over take place?

8. Consider the following cross of *Drosophila melanogaster* with alleles as described in Question 6. Males with genotype *Ppswsw* are crossed with females of genotype *ppSwsw*. Describe the phenotypes and genotypes of the F₁ generation.

9. In the Andalusian fowl, a single pair of alleles controls the color of the feathers. Three colors are observed: blue, black, and splashed white. Crosses among these three types yield the following results:

PARENTS	PROGENY
Black × blue	Blue and black (1:1)
Black × splashed white	Blue
Blue × splashed white	Blue and splashed white (1:1)
Black × black	Black
Splashed white × splashed white	Splashed white

a. What progeny would result from the cross blue × blue?

b. If you wanted to sell eggs, all of which would yield blue fowl, how should you proceed?

10. In *Drosophila melanogaster*, white (*w*), eosin (*w^e*), and wild-type red (*w^+*) are multiple alleles of a single locus for eye color. This locus is on the X chromosome. A female that has eosin (pale orange) eyes is crossed with a male that has wild-type eyes. All the female progeny are red-eyed; half the male offspring have eosin eyes, and half have white eyes.

a. What is the order of dominance of these alleles?

b. What are the genotypes of the parents and progeny?

11. Color blindness is a recessive trait. Two people with normal color vision have two sons, one color-blind and one with normal color vision. If the couple also has daughters, what proportion of them will have normal color vision? Explain.

12. A mouse with an agouti coat is mated with an albino mouse of genotype *aabb*. Half of the offspring are albino, one-fourth are black, and one-fourth are agouti. What are the genotypes of the agouti parents and of the various kinds of offspring? (*Hint*: See the section on epistasis.)

13. The disease Leber's optic neuropathy is caused by a mutation in a gene carried on mitochondrial DNA. What would be the result in their first child if a man with this disease married a woman who did not have the disease? What would be the result if the wife had the disease and the husband did not?

11 DNA and Its Role in Heredity

THE IMAGE OF THE DNA DOUBLE HELIX IS ONE of the great secular icons to emerge in the last half of the twentieth century. Its elegance and simplicity make it instantly recognizable by the general public. The story of how scientists determined that the gene envisioned by Mendel is made of DNA is one of the epics of experimental biology. These studies opened up an entirely new field of natural science: molecular biology, which is concerned with DNA and its expression in cells.

The representation of DNA shown below is the familiar double helix, but with an added chemical shown in green. The green molecule is benzpyrene, one of the toxic chemicals emitted in tobacco smoke. This extra chemical entity has dire consequences. The regular, twisted structure of DNA is just the right shape to allow benzpyrene to wedge into the groove of the helix. A covalent bond forms between the benzpyrene and a DNA monomer, causing a major problem when DNA is expressed and replicated. Ultimately, there are irreversible changes in the DNA, and these changes are passed on to the two daughter cells after a cell division cycle. This damage to the DNA is the key event that begins cancer—in the case of benzpyrene, usually lung cancer.

In this and the next several chapters, we focus on the structure, replication, and function of DNA. As you will see, the structure of DNA determines its functions. This chapter first describes the key experiments that led to the determination that the genetic material is DNA. Then the structure and replication of the molecule are described. Finally, we present two practical applications that have arisen from our knowledge of DNA replication: DNA sequencing, and the polymerase chain reaction.

DNA: The Genetic Material

During the first half of the twentieth century, the hereditary material was generally assumed to be a protein. The impressive chemical diversity of proteins made this assumption seem reasonable. In addition, some proteins—notably enzymes and antibodies—showed great specificity. Nucleic acids, by contrast, were known to have only a few components and seemed too simple to carry the complex information expected in the genetic material.

Circumstantial evidence, however, pointed to DNA. It was in the right place, since it was an important component of the nucleus and chromosomes, which were known to carry genes. And it was present in the right amounts. During the 1920s, a dye was developed that bound specifically to DNA and turned red in direct proportion to the amount of DNA present. When different cells were stained with this dye and their color intensity measured, each species appeared to have its own specific nuclear DNA content. Furthermore, the quantity in somatic cells was twice that in eggs or sperm—as might be expected for diploid and haploid cells, respectively. These two observations were consistent with DNA as the genetic material.

But circumstantial evidence is not a scientific demonstration of cause and effect. After all, proteins are also present in nuclei. The convincing demonstration that DNA is the genetic material came from two lines of experiments, one on bacteria and the other on bacterial viruses.

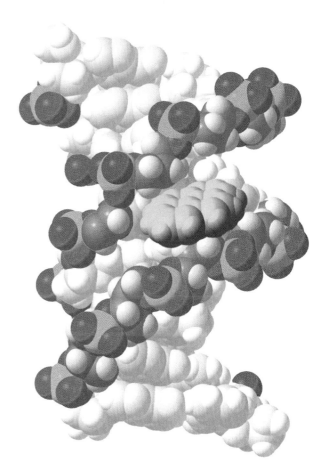

The Double Helix of DNA
A computer-generated model of DNA. The molecule in green is benzpyrene, a major cancer-causing component of tobacco smoke. The "backbone" of the DNA molecule is visible as a chain of sugars (gray) and phosphate groups (red and orange).

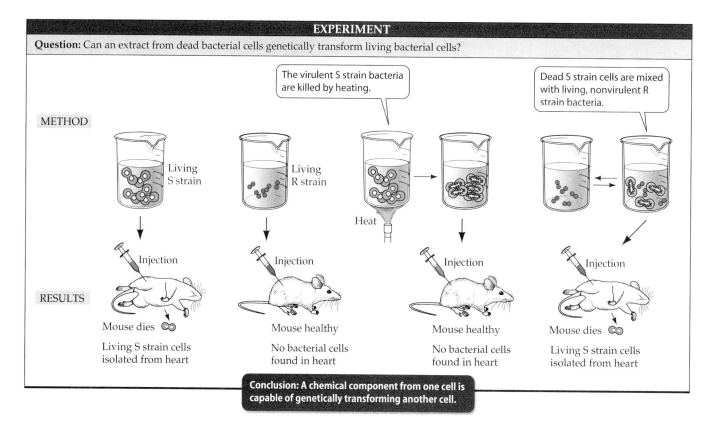

EXPERIMENT

Question: Can an extract from dead bacterial cells genetically transform living bacterial cells?

METHOD

The virulent S strain bacteria are killed by heating.

Dead S strain cells are mixed with living, nonvirulent R strain bacteria.

Living S strain

Living R strain

Heat

RESULTS

Injection

Injection

Injection

Injection

Mouse dies

Living S strain cells isolated from heart

Mouse healthy

No bacterial cells found in heart

Mouse healthy

No bacterial cells found in heart

Mouse dies

Living S strain cells isolated from heart

Conclusion: A chemical component from one cell is capable of genetically transforming another cell.

11.1 Genetic Transformation of Nonvirulent R Pneumococci
Frederick Griffith's experiments demonstrated that something in the virulent S strain could transform nonvirulent R strain bacteria into a lethal form, even when the S strain bacteria had been killed by high temperatures.

DNA from one type of bacterium genetically transforms another type

The history of biology is filled with incidents in which research on one specific topic has—with or without answering the question originally asked—contributed richly to another, apparently unrelated area. Such a case of "serendipity" is the work of Frederick Griffith, an English physician.

In the 1920s, Griffith was studying the disease-causing behavior of the bacterium *Streptococcus pneumoniae,* or pneumococcus, one of the agents that causes pneumonia in humans. He was trying to develop a vaccine against this devastating illness (antibiotics had not yet been discovered). He was working with two strains of pneumococcus. A bacterial *strain* is a population of cells descended from a single parent cell; strains differ in one or more inherited characteristics. Griffith's strains were designated S and R because, when grown in the laboratory, one produces shiny, smooth (S) colonies, and the other produces colonies that look rough (R).

When the S strain was injected into mice, the mice died within a day, and the hearts of the dead mice were found to be teeming with the deadly bacteria. When the R strain was injected, the mice did not become diseased. In other words, the S strain is *virulent* (disease-causing) and the R strain is *nonvirulent*. The virulence of the S strain is caused by a

polysaccharide capsule that protects the bacterium from the immune defense mechanisms of the host. The R strain lacks this capsule, so the R strain cells can be inactivated by the defenses of a mouse.

With the hope of developing a vaccine against pneumonia, Griffith inoculated some mice with heat-killed S pneumococci. These heat-killed bacteria did not produce infection. However, when Griffith inoculated other mice with a mixture of living R bacteria and heat-killed S bacteria, to his astonishment, the mice died of pneumonia. When he examined blood from the hearts of these mice, he found it full of living bacteria—many of them with characteristics of the virulent S strain! Griffith concluded that, in the presence of the dead S pneumococci, some of the living R pneumococci had been transformed into virulent S-strain organisms (Figure 11.1).

We now call the phenomenon of the genetic alteration of an organism **transformation**. In terms of Griffith's observations, one could say that transformation is the uptake of information from the environment. As we'll see, today's definition of transformation is more precise. For now, note that living R pneumococci had gained a trait—virulence—from something in their environment.

Did this transformation of the bacteria depend on something the mouse did? No. It was shown that simply incubating living R and heat-killed S bacteria together in a test tube yielded the same transformation. Next it was discovered that a cell-free extract of heat-killed S cells also transformed R cells. (A cell-free extract contains all the contents of ruptured cells, but no intact cells.) This result demonstrated that some substance—called at the time a chemical transforming principle—from the dead S pneumococci could cause a heritable change in the affected R cells. From these observations, some scientists concluded that this transforming principle carried heritable information, and thus was the genetic material.

The transforming principle is DNA

The identification of the transforming principle was a crucial step in the history of biology, accomplished over a period of several years by Oswald T. Avery and his colleagues at what is now Rockefeller University. They treated samples of the transforming principle in a variety of ways to destroy different types of substances—proteins, nucleic acids, carbohydrates, and lipids—and tested the treated samples to see if they had retained transforming activity.

The answer was always the same: If the DNA in the sample was destroyed, transforming activity was lost; everything else was dispensable. As a final step, Avery, with Colin MacLeod and Maclyn McCarty, isolated virtually pure DNA from a sample of pneumococcal transforming principle and showed that it caused bacterial transformation.

The work of Avery, MacLeod, and McCarty, published in 1944, was a milestone in establishing that DNA is the genetic material in cells. However, it had little impact at the time, for two reasons. First, most scientists did not believe that DNA was chemically complex enough to be the hereditary material, especially given the great chemical complexity of proteins. Second, and perhaps more important, it was not yet obvious that bacteria even had genes; bacterial genetics was still to be elucidated (see Chapter 13).

Viral replication experiments confirm that DNA is the genetic material

A report published in 1952 by Alfred D. Hershey and Martha Chase of the Carnegie Laboratory of Genetics had a much greater immediate impact than did Avery's 1944 paper. The Hershey–Chase experiment was carried out with a virus that infects bacteria. This virus, called T2 bacteriophage, consists of little more than a DNA core packed inside a protein coat (Figure 11.2*a*). The virus is thus made of the two materials that were, at the time, the leading candidates for the genetic material.

When a T2 bacteriophage attacks a bacterium, part (but not all) of the virus enters the bacterial cell. Hershey and Chase set out to determine which part of the virus—protein or DNA—enters the bacterial cell. To trace these two components during the life cycle of the virus (Figure 11.2*b*), Hershey and Chase labeled each with a specific radioactive tracer.

All proteins contain some sulfur (in the amino acids cysteine and methionine), an element not present in DNA, and sulfur has a radioactive isotope, ^{35}S. The deoxyribose–phosphate "backbone" of DNA, on the other hand, is rich in phosphorus (see Chapter 3), an element that is *not* present in most proteins—and phosphorus also has a radioactive isotope, ^{32}P. Hershey and Chase grew one batch of T2 in a bacterial culture in the presence of ^{32}P, so that all the viral DNA was labeled with ^{32}P. Similarly, all the proteins of another batch of T2 were labeled with ^{35}S.

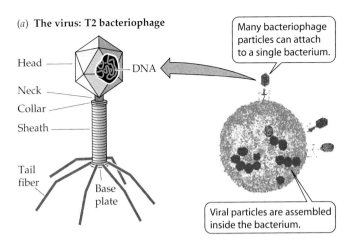

(a) **The virus: T2 bacteriophage**

Head
DNA
Neck
Collar
Sheath
Tail fiber
Base plate

Many bacteriophage particles can attach to a single bacterium.

Viral particles are assembled inside the bacterium.

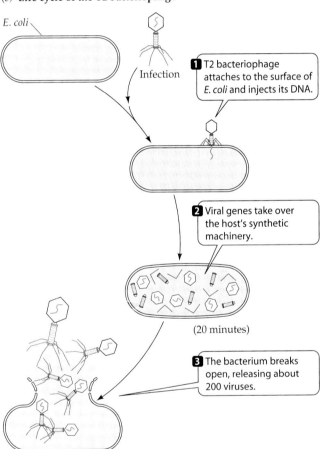

(b) **Life cycle of the T2 bacteriophage**

E. coli

Infection

1 T2 bacteriophage attaches to the surface of *E. coli* and injects its DNA.

2 Viral genes take over the host's synthetic machinery.

(20 minutes)

3 The bacterium breaks open, releasing about 200 viruses.

11.2 T2 and the Bacteriophage Reproduction Cycle
(a) The external structures of the bacteriophage T2 consist entirely of protein. This cutaway view shows a strand of DNA within the head. (b) T2 is parasitic on *E. coli*, depending on the bacterium to produce new viruses.

In separate experiments, Hershey and Chase combined radioactive viruses containing either ^{32}P or ^{35}S with bacteria. After a few minutes, they agitated the mixtures vigorously in a kitchen blender, which (without bursting the bacteria) stripped away the parts of the virus that had not penetrated the bacteria. Then, using a centrifuge, Hershey and Chase separated the bacteria from the rest of the mate-

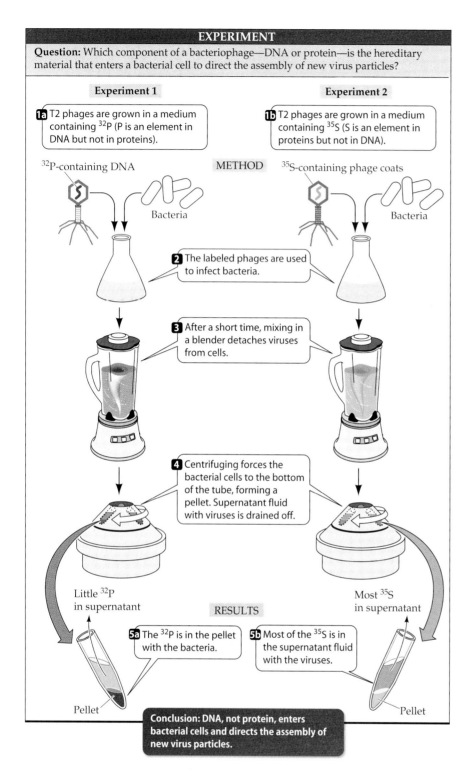

EXPERIMENT

Question: Which component of a bacteriophage—DNA or protein—is the hereditary material that enters a bacterial cell to direct the assembly of new virus particles?

Experiment 1

1a T2 phages are grown in a medium containing ^{32}P (P is an element in DNA but not in proteins).

^{32}P-containing DNA

METHOD

Bacteria

Experiment 2

1b T2 phages are grown in a medium containing ^{35}S (S is an element in proteins but not in DNA).

^{35}S-containing phage coats

Bacteria

2 The labeled phages are used to infect bacteria.

3 After a short time, mixing in a blender detaches viruses from cells.

4 Centrifuging forces the bacterial cells to the bottom of the tube, forming a pellet. Supernatant fluid with viruses is drained off.

Little ^{32}P in supernatant

Most ^{35}S in supernatant

RESULTS

5a The ^{32}P is in the pellet with the bacteria.

5b Most of the ^{35}S is in the supernatant fluid with the viruses.

Pellet

Pellet

Conclusion: DNA, not protein, enters bacterial cells and directs the assembly of new virus particles.

11.3 The Hershey–Chase Experiment
Because only DNA entered the bacterial cell during infection, the experiment demonstrated that DNA, not protein, is the hereditary material.

^{35}S, but about one-third of the original ^{32}P—and thus, presumably, one-third of the original DNA. Because DNA was carried over in the virus from generation to generation but protein was not, a logical conclusion was that the hereditary information of the virus is contained in the DNA.

The Hershey–Chase experiment convinced most scientists that DNA is the carrier of hereditary information. By this time, other researchers had identified mutations—and therefore genes—in viruses and bacteria.

The Structure of DNA

Once scientists agreed that the genetic material is DNA, they wanted to learn its precise chemical structure. In the structure of DNA, they hoped to find the answers to two questions: how DNA is replicated between nuclear divisions, and how it causes the synthesis of specific proteins. Both expectations were fulfilled. X-ray crystallography studies provided the first clues about the dimensions of DNA and hinted that it had a helical form. Dimensionally accurate models built by James Watson and Francis Crick completed the picture.

X-ray crystallography provided clues to DNA structure

The structure of DNA was deciphered only after many types of experimental evidence and theoretical considerations were combined. The most crucial evidence was obtained by X-ray crystallography (Figure 11.4). The positions of atoms in a crystalline substance can be inferred from the pattern of diffraction of X-rays passed through it, but even today this is not an easy task when the substance is of enormous molecular weight.

In the early 1950s, even a highly talented X-ray crystallographer could (and did) look at the best available images from DNA preparations and fail to see what they meant. Nonetheless, the attempt to characterize DNA would have been impossible without the crystallographs prepared by

rial. They found that most of the ^{35}S (and thus the protein) had separated from the bacteria, and that most of the ^{32}P (the DNA) had stayed with the bacteria. These results suggested that the DNA was transferred to the bacteria, whereas the protein remained outside (Figure 11.3).

Hershey and Chase then performed similar but longer experiments, allowing a progeny generation of viruses to be collected. The resulting T2 progeny (the "offspring" of the original viruses) contained almost none of the original

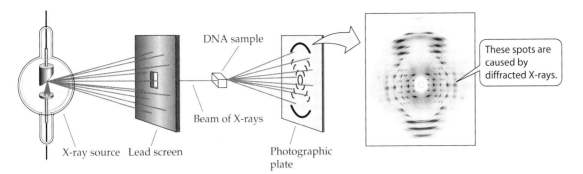

11.4 X-Ray Crystallography Revealed the Basic Helical Nature of DNA Structure
The positions of atoms in DNA can be inferred by the pattern of diffraction of X-rays passed through it, although the task requires tremendous skill.

the English chemist Rosalind Franklin. Franklin's work, in turn, depended on the success of the English biophysicist Maurice Wilkins, who prepared very uniformly oriented DNA fibers, which provided samples for diffraction that were far better than previous ones.

The chemical composition of DNA was known

The chemical composition of DNA also provided important clues about its structure. Biochemists knew that DNA was a polymer of nucleotides. Each nucleotide of DNA consists of a molecule of the sugar deoxyribose, a phosphate group, and a nitrogen-containing base (see Figures 3.16 and 3.17). The only differences among the four nucleotides of DNA are their nitrogenous bases: the purines **adenine** (**A**) and **guanine** (**G**), and the pyrimidines **cytosine** (**C**) and **thymine** (**T**).

In 1950, Erwin Chargaff at Columbia University reported some observations of major importance. He and his colleagues found that DNA from many different species— and from different sources within a single organism—exhibits certain regularities. In almost all DNA the following rule holds: The amount of adenine equals the amount of thymine, and the amount of guanine equals the amount of cytosine (Figure 11.5). As a result, the total abundance of purines equals the total abundance of pyrimidines. The structure of DNA could not have been worked out without this information, yet its significance was overlooked for at least three years. Interestingly, the ratio of A + T to G + C

varies widely among different organisms (Table 11.1). This observation reinforced the importance of Chargaff's rule.

Watson and Crick described the double helix

The solution to the puzzle of the structure of DNA was accelerated by the technique of model building: assembling three-dimensional representations of possible molecular structures using known relative molecular dimensions and known bond angles. This technique, originally exploited in structural studies by the American chemist Linus Pauling, was used by the English physicist Francis Crick and the American geneticist James D. Watson, then both at the Cavendish Laboratory of Cambridge University.

Watson and Crick attempted to combine all that had been learned so far about DNA structure into a single coherent model. The crystallographers' results (see Figure 11.4) convinced Watson and Crick that the DNA molecule is **helical** (cylindrically spiral) and provided the values of certain distances within the helix. The results of density measurements and previous model building suggested that there are two polynucleotide chains in the molecule. The modeling studies had also led to the conclusion that the two chains in DNA run in opposite directions—that is, that they are *antiparallel*. (We'll clarify this point in the next section.)

Crick and Watson built several models. Late in February of 1953, they built the one that established the general structure of DNA. There have been minor amendments to their first published structure, but the principal features remain unchanged.

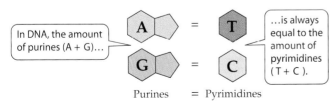

11.5 Chargaff's Rule
The total abundances of purines and pyrimidines are equal in DNA.

	AMOUNT OF BASE (PERCENTAGE OF TOTAL DNA)			
DNA ORIGIN	A	T	G	C
Human (*Homo sapiens*)	31.0	31.5	19.1	18.4
Corn (*Zea mays*)	25.6	25.3	24.5	24.6
Fruit fly (*Drosophila melanogaster*)	27.3	27.6	22.5	22.5
Bacterium (*Escherichia coli*)	26.1	23.9	24.9	25.1

11.1 **Percentages of Bases in the DNA of Some Well-Studied Species**

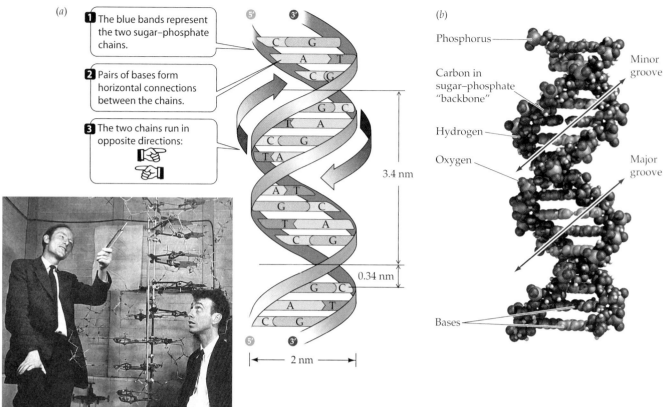

(a)

1 The blue bands represent the two sugar–phosphate chains.

2 Pairs of bases form horizontal connections between the chains.

3 The two chains run in opposite directions:

3.4 nm

0.34 nm

2 nm

(b)

Phosphorus

Carbon in sugar–phosphate "backbone"

Hydrogen

Oxygen

Bases

Minor groove

Major groove

11.6 DNA Is a Double Helix
(a) Watson and Crick proposed that DNA is a double helical molecule. *(b)* Biochemists can now pinpoint the position of every atom in a DNA macromolecule. To see that the essential features of the original Watson–Crick model have been verified, follow with your eyes the double helical ribbons of sugar–phosphate groups and note the horizontal rungs of the bases (see also Figure 3.18).

Four key features define DNA structure

Four features summarize the molecular architecture of DNA. The DNA molecule is

▶ a double-stranded helix,
▶ of uniform diameter,
▶ right-handed (that is, it twists to the right, as do the threads on most screws), and
▶ antiparallel (the two strands run in opposite directions).

The sugar–phosphate backbones of the polynucleotide chains coil around the outside of the helix, and the nitrogenous bases point toward the center (Figure 11.6).

The two chains are held together by hydrogen bonding between specifically paired bases. Consistent with Chargaff's rule, adenine (A) pairs with thymine (T) by forming two hydrogen bonds, and guanine (G) pairs with cytosine (C) by forming three hydrogen bonds (Figure 11.7). Every base pair consists of one purine (A or G) and one pyrimidine (C or T). This pattern is known as **complementary base pairing**. Because the AT and GC pairs, like rungs of a ladder, are of equal length and fit identically into the double helix, the diameter of the helix is uniform. The base pairs are flat, and their stacking in the center of the molecule is stabilized by hydrophobic interactions (see Chapter 2), contributing to the overall stability of the double helix.

What does it mean to say that the two DNA strands run in opposite directions? The direction of a polynucleotide can be defined by looking at the linkages (called phosphodiester bonds) between adjacent nucleotides. In the sugar–phosphate backbone of DNA, the phosphate groups connect to the 3′ carbon of one deoxyribose molecule and the 5′ carbon of the next, linking successive sugars together (see Figure 11.7). The prime (′) designates the position of a carbon atom in the five-carbon sugar deoxyribose.

Thus the two ends of a polynucleotide differ. Polynucleotides have a free (not connected to another nucleotide) 5′ phosphate group ($—OPO_3^-$) at one end, called the **5′ end**; a free 3′ hydroxyl group ($—OH$) is at the other, the **3′ end**. The 5′ end of one strand in a DNA double helix is paired with the 3′ end of the other strand, and vice versa; in other words, the strands run in opposite directions .

The double helical structure of DNA is essential to its function

The genetic material must perform four important functions, and the DNA molecule modeled by Watson and Crick was elegantly suited to three of them.

▶ *The genetic material should be able to store an organism's genetic information.* With its millions of nucleotides in a sequence that differs in every species and every individual, DNA fits this role nicely.

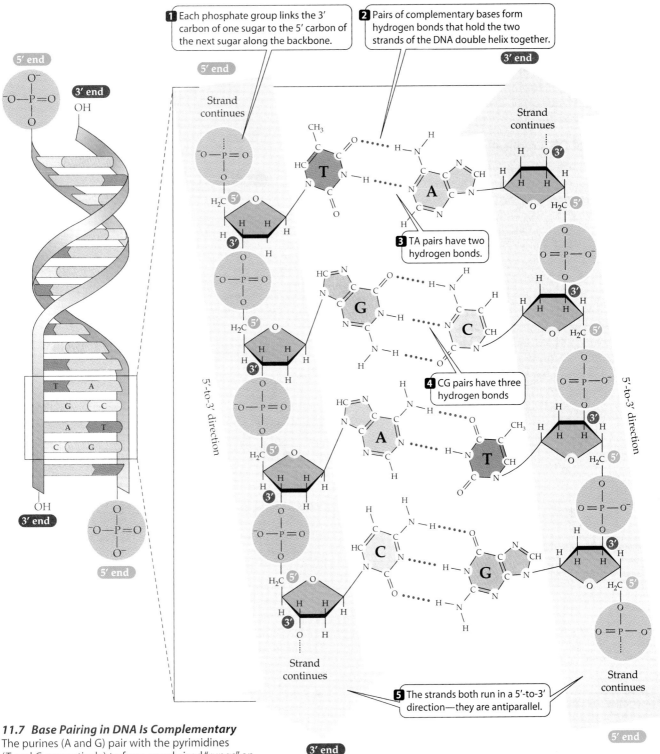

1 Each phosphate group links the 3' carbon of one sugar to the 5' carbon of the next sugar along the backbone.

2 Pairs of complementary bases form hydrogen bonds that hold the two strands of the DNA double helix together.

3 TA pairs have two hydrogen bonds.

4 CG pairs have three hydrogen bonds

5 The strands both run in a 5'-to-3' direction—they are antiparallel.

5' end

3' end

OH

3' end

5' end

Strand continues

5'-to-3' direction

3' end

OH

3' end

5' end

5' end

3' end

Strand continues

5'-to-3' direction

Strand continues

Strand continues

5' end

11.7 Base Pairing in DNA Is Complementary
The purines (A and G) pair with the pyrimidines (T and C, respectively) to form equal-sized "rungs" on a "ladder" (the sugar–phosphate backbones). The ladder is twisted into a double helical structure.

▶ *The genetic material must be susceptible to mutation, or permanent changes in its information.* For DNA, mutations might be simple changes in the linear sequence of nucleotide pairs.

▶ *The genetic material must be precisely replicated in the cell division cycle.* Replication could be accomplished by

complementary base pairing, A with T and G with C. In the original publication of their findings in the journal *Nature* in 1953, Watson and Crick coyly pointed out, "It has not escaped our notice that the specific pairing we have postulated immediately suggests a possible copying mechanism for the genetic material."

▶ *The genetic material must be expressed as the phenotype.* This function is not inherent in the structure of DNA; however, as we show in the next chapter, it also turns out to be well served by DNA.

DNA Replication

Watson and Crick's model for DNA replication was soon confirmed. First, experiments showed DNA replicated from template strands in a test tube containing simple substrates and an enzyme. Then an elegant experiment showed that each of the two strands of the double helix serves as a template for a new strand.

Three modes of DNA replication appeared possible

Just three years after Watson and Crick published their paper in *Nature*, their prediction that the DNA molecule contains the information needed for its own replication was demonstrated by the work of Arthur Kornberg, then at Washington University in St. Louis. Kornberg showed that DNA can replicate in a test tube with no cells present. The only requirements are DNA, a specific enzyme (which he obtained from bacteria and called **DNA polymerase**), and a mixture of four precursors: the deoxyribonucleoside triphosphates dATP, dCTP, dGTP, and dTTP. If any one of the four deoxyribonucleoside triphosphates is omitted from the reaction mixture, DNA does not replicate.

Somehow, the DNA itself serves as a template for the reaction—a guide to the exact placement of nucleotides in the

Original DNA   After one round of replication

(a)

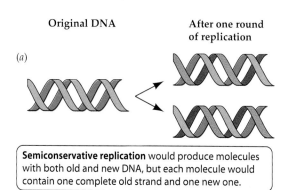

Semiconservative replication would produce molecules with both old and new DNA, but each molecule would contain one complete old strand and one new one.

(b)

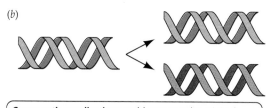

Conservative replication would preserve the original molecule and generate an entirely new molecule.

(c)

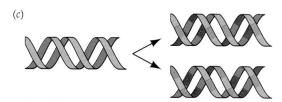

Dispersive replication would produce two molecules with old and new DNA interspersed along each strand.

11.8 Three Models for DNA Replication
In each model, original DNA is shown in blue and newly synthesized DNA in red.

RESEARCH METHOD

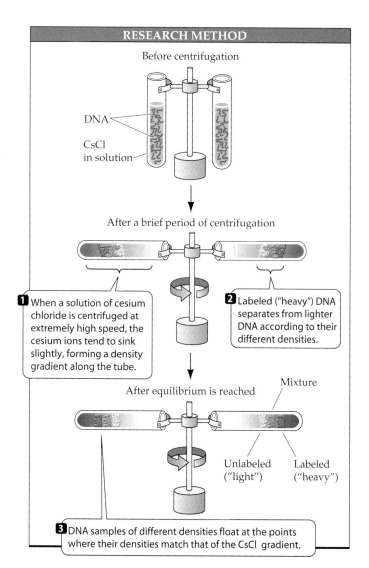

Before centrifugation

DNA

CsCl in solution

After a brief period of centrifugation

1 When a solution of cesium chloride is centrifuged at extremely high speed, the cesium ions tend to sink slightly, forming a density gradient along the tube.

2 Labeled ("heavy") DNA separates from lighter DNA according to their different densities.

After equilibrium is reached   Mixture

Unlabeled ("light")   Labeled ("heavy")

3 DNA samples of different densities float at the points where their densities match that of the CsCl gradient.

11.9 Density Gradient Centrifugation
Labeled ("heavy") DNA will separate from lighter DNA in the density gradient formed by a cesium chloride solution.

new strand. Where there is a T in the template, there must be an A in the new strand, and so forth. How does DNA perform the template function; that is, how exactly does the molecule replicate?

There were three possible replication patterns that would result in complementary base pairing:

▸ *Semiconservative replication*, in which each parent strand serves as a template for a new strand and the two new DNA's each have one old and one new strand (Figure 11.8a)

▸ *Conservative replication*, in which the original double helix serves as a template for, but does not contribute to, the new double helix (Figure 11.8b)

▸ *Dispersive replication*, in which fragments of the original DNA molecule serve as templates for assembling two molecules, each containing old and new parts, perhaps at random (Figure 11.8c)

Watson and Crick's original paper suggested that DNA replication was semiconservative, but Kornberg's experiment did not provide a basis for choosing among these three models.

Meselson and Stahl demonstrated that DNA replication is semiconservative

A clever experiment conducted by Matthew Meselson and Franklin Stahl convinced the scientific community that semiconservative replication is the correct model. Working at the California Institute of Technology in 1957, they devised a simple way to distinguish old strands of DNA from new ones: density labeling.

The key to their experiment was the use of a "heavy" isotope of nitrogen. Heavy nitrogen (^{15}N) is a rare, nonradioactive isotope that makes molecules containing it more dense than chemically identical molecules containing the common isotope, ^{14}N. To distinguish DNA of different densities (that is, DNA containing ^{15}N versus DNA containing ^{14}N), Meselson, Stahl, and Jerome Vinograd invented a new centrifugation procedure using a cesium chloride (CsCl) solution.

Spinning solutions or suspensions at high speed in a centrifuge causes the solutes or particles to separate, and they form a gradient according to their density. A concentrated solution of CsCl has a density very close to that of DNA. At high gravitational forces, cesium ions sediment out of the solution to some extent, establishing a gradient from low density at the top of the centrifuge tube to high density at the bottom. When a DNA sample is dissolved in CsCl and centrifuged at about 100,000 times the force of gravity, the DNA gathers in a band at a position in the tube where the density of the CsCl solution equals its own density (Figure 11.9).

After developing this method of distinguishing DNA densities, Meselson and Stahl grew a culture of the bacterium *Escherichia coli* for 17 generations in a medium in which the nitrogen source (ammonium chloride, NH$_4$Cl) was made with ^{15}N instead of ^{14}N. As a result, all the DNA in the bacteria was "heavy." They grew another culture in a medium with ^{14}N, and extracted DNA from both cultures. When the extracts were combined and centrifuged with CsCl, two separate DNA bands formed, showing that this method could distinguish DNA samples of slightly different densities.

Next, Meselson and Stahl grew another *E. coli* culture on ^{15}N medium, then *transferred* it to normal ^{14}N medium and allowed the bacteria to continue growing (Figure 11.10). Under the conditions they used, *E. coli* replicates its DNA every 20 minutes. Meselson and Stahl collected some of the

11.10 The Meselson–Stahl Experiment
Density gradient centrifugation revealed a pattern that supports the semiconservative model of DNA replication.

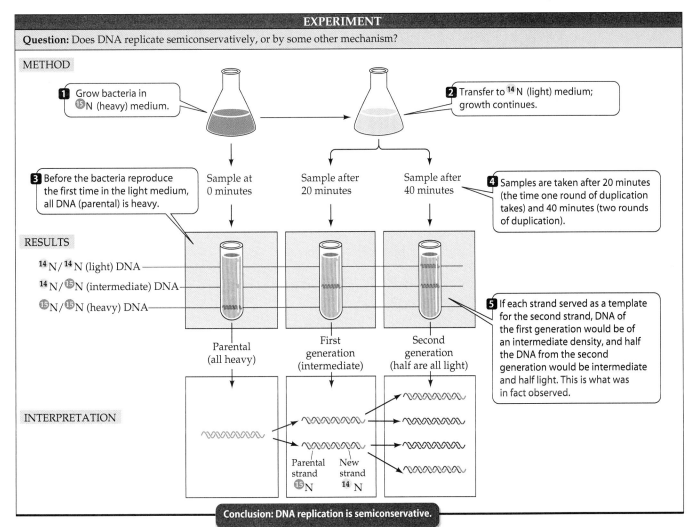

EXPERIMENT

Question: Does DNA replicate semiconservatively, or by some other mechanism?

METHOD

1 Grow bacteria in ^{15}N (heavy) medium.

2 Transfer to ^{14}N (light) medium; growth continues.

3 Before the bacteria reproduce the first time in the light medium, all DNA (parental) is heavy.

Sample at 0 minutes

Sample after 20 minutes

Sample after 40 minutes

4 Samples are taken after 20 minutes (the time one round of duplication takes) and 40 minutes (two rounds of duplication).

RESULTS

^{14}N/^{14}N (light) DNA

^{14}N/^{15}N (intermediate) DNA

^{15}N/^{15}N (heavy) DNA

Parental (all heavy)

First generation (intermediate)

Second generation (half are all light)

5 If each strand served as a template for the second strand, DNA of the first generation would be of an intermediate density, and half the DNA from the second generation would be intermediate and half light. This is what was in fact observed.

INTERPRETATION

Parental strand ^{15}N   New strand ^{14}N

Conclusion: DNA replication is semiconservative.

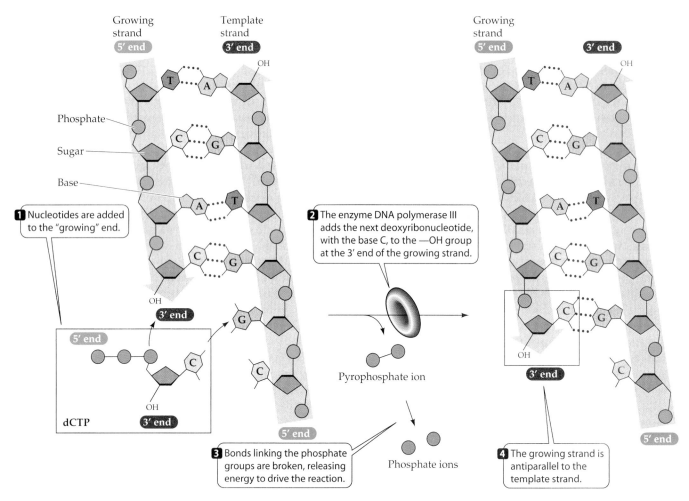

11.11 Each New DNA Strand Grows from its 5′ End to its 3′ End
A DNA strand, with its 3′ end at the top and its 5′ end at the bottom, is the template for the synthesis of the complementary strand at the left.

bacteria after each division and extracted DNA from the samples.

Meselson and Stahl observed that the DNA banding in the density gradient was different in each bacterial generation. At the time of the transfer, the DNA was uniformly labeled with ^{15}N, and hence was relatively dense. After one generation, when the DNA had been duplicated once, all the DNA was of an intermediate density. After two generations, there were two equally large DNA bands: one of low density and one of intermediate density. In samples from subsequent generations, the proportion of low-density DNA increased steadily.

The results of this experiment can be explained by the semiconservative model of DNA replication. The high-density DNA had two ^{15}N strands; the intermediate-density DNA had one ^{15}N and one ^{14}N strand; and the low-density DNA had two ^{14}N strands. In the first round of DNA replication, the strands of the double helix—both heavy with ^{15}N—separated. While separated, each strand

acted as the template for a second strand, which contained only ^{14}N and hence was less dense. Each double helix then consisted of one ^{15}N and one ^{14}N strand and was of intermediate density. In the second replication, the ^{14}N-containing strands directed the synthesis of partners with ^{14}N, creating low-density DNA, and the ^{15}N strands got new ^{14}N partners (see Figure 11.10).

The crucial observation demonstrating the semiconservative model was that intermediate-density DNA (^{15}N–^{14}N) appeared in the first generation and continued to appear in subsequent generations. With the other models, the results would have been quite different. In the conservative model, the first generation would have had both high-density DNA (^{15}N–^{15}N) and low density DNA (^{14}N–^{14}N), but no intermediate DNA. In the dispersive model, the density of the new DNA would have been between low and high, and not exactly intermediate.

Soon after Meselson and Stahl published their work, other scientists showed that semiconservative replication occurred in the DNAs of eukaryotic plant and animal cells. Using labeled DNA, they even demonstrated that chromatids appeared to replicate semiconservatively, providing the first evidence that a chromatid is a single molecule of double-helical DNA.

The Mechanism of DNA Replication

But how does DNA get replicated semiconservatively? There are four requirements for this process:

▶ DNA must act as a *template* for complementary base pairing.
▶ The four *deoxyribonucleoside triphosphates*, dATP, dGTP, dCTP, and dTTP, must be present.
▶ A DNA polymerase *enzyme* is needed to bring the substrates to the template and catalyze the reactions.
▶ A source of chemical *energy* is needed to drive this highly endergonic synthesis reaction.

DNA replication takes place in two steps:

▶ The DNA is locally *denatured* (unwound) to separate the two template strands and make them available for base pairing.
▶ The new *nucleotides are linked* by covalent bonding to each growing strand in a sequence determined by complementary base pairing.

A key observation of virtually all DNA replication is that nucleotides are always added to the growing strand at the 3′ end—the end at which the DNA strand has a free hydroxyl (—OH) group on the 3′ carbon of its terminal deoxyribose (Figure 11.11). The three phosphate groups in a deoxyribonucleoside triphosphate are attached to the 5′ position of the sugar (see Figure 11.7). So when a new nucleotide is added to DNA, it can attach only to the 3′ end.

When DNA polymerase brings a new deoxyribonucleoside triphosphate to the 3′ end of a growing chain, the free hydroxyl group on the chain reacts with one of the substrate's phosphate groups. As this happens, the bond linking the terminal two phosphate groups to the rest of the deoxyribonucleoside triphosphate breaks, and thereby releases energy for this reaction. The resulting pyrophosphate ion, consisting of the two terminal phosphate groups, also breaks, forming two phosphate ions and in the process releasing additional free energy. The phosphate group still on the nucleotide becomes part of the sugar–phosphate backbone of the growing DNA molecule.

DNA is threaded through a replication complex

DNA is replicated through the interaction of the template DNA with a huge protein complex that catalyzes the reactions. All chromosomes have at least one sequence of nucleotides, called the **origin of replication**, that is recognized by this replication complex. DNA replicates in both directions from the origin, forming two **replication forks**. Both of the separated strands of the parent molecule act as templates, and the formation of the new strands is guided by complementary base pairing.

Until recently, DNA replication was depicted as a locomotive (the replication complex) moving along a railroad track (the DNA) (Figure 11.12*a*). The current view is that this is not so. Instead, the replication complex is stationary,

attached to nuclear structures, and it is the DNA that moves, essentially threading through the complex as single strands and emerging as double strands (Figure 11.12*b*). During S phase in eukaryotes, there are about 100 replication complexes, and each of them contains as many as 300 individual replication forks.

All replication complexes contain several proteins with different roles. We will describe these proteins as we examine the steps of the process:

▶ *DNA helicase* opens up the double helix.
▶ *Single-strand binding proteins* keep the two strands separated.
▶ *RNA primase* makes the primer strand needed to get replication under way.
▶ *DNA polymerase* adds nucleotides to the primer that are complementary to the template, proofreads the DNA, and repairs it.
▶ *DNA ligase* seals up breaks in the sugar–phosphate backbone.

Small, circular DNA's replicate from a single origin, while large, linear DNA's have many origins

The key event at the origin of replication is the localized unwinding (denaturation) of DNA. There are several forces that hold the two strands together, including hydrogen bonding and the hydrophobic stacking of bases. An enzyme, **DNA helicase**, uses energy from ATP hydrolysis to unwind the DNA, and special proteins bind to the unwound strands to keep them that way, preventing them from reassociating into a double helix. This makes the two template strands available for complementary base pairing.

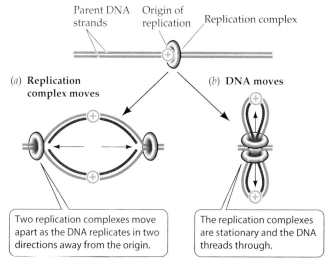

11.12 Two Views of DNA Replication
(*a*) It was once thought that the replication complex moved along DNA. (*b*) Newer evidence suggests that the DNA is threaded through the stationary complex.

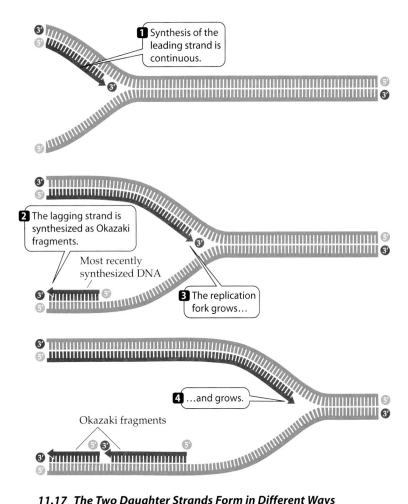

11.17 The Two Daughter Strands Form in Different Ways
As the original DNA unwinds, both daughter strands are synthesized in the 5'-to-3' direction, although their template strands are antiparallel. The leading strand grows continuously forward, but the lagging strand grows in shorter, backward stretches called Okazaki fragments. Eukaryotic Okazaki fragments are hundreds of nucleotides long, with gaps between them.

bond between the adjacent Okazaki fragments is missing (Figure 11.18). Another enzyme, **DNA ligase**, catalyzes the formation of that bond, linking the fragments and making the lagging strand whole.

Working together, DNA helicase, the two DNA polymerases, RNA primase, DNA ligase, and the other proteins do the complex job of DNA synthesis with a speed and accuracy that are almost unimaginable. In *E. coli*, the replication complex makes new DNA at a rate in excess of 1,000 base pairs per second, committing errors in fewer than one base in 10^6; or one in a million.

DNA Proofreading and Repair

DNA must be faithfully replicated and maintained. The price of failure can be great; the transmission of genetic information is at stake, as is the functioning and even the life of a cell or multicellular organism. Yet the replication of DNA is *not* perfectly accurate, and the DNA of nondividing

cells is subject to damage by environmental agents. In the face of these threats, how has life gone on so long?

The preservers of life are DNA repair mechanisms. DNA polymerases initially make a significant number of mistakes in assembling polynucleotide strands. In *E. coli*, for example, the observed error rate every 10^6 bases replicated would result in flaws in approximately one out of every three genes each time the cell divided. In humans, about 1,000 genes in every cell would be affected each time the cell divided.

11.18 The Lagging Strand Story
DNA polymerase I and DNA ligase join DNA polymerase III to complete the complex task of synthesizing the lagging strand.

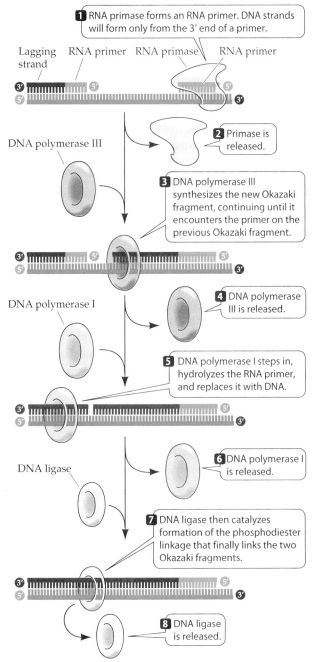

Fortunately, our cells normally have at least three DNA repair mechanisms at their disposal:

▶ A *proofreading mechanism* corrects errors as DNA polymerase makes them.

▶ A *mismatch repair mechanism* scans DNA after it has been made and corrects any base-pairing mismatches.

▶ An *excision repair mechanism* removes abnormal bases that have formed because of chemical damage and replaces them with functional bases.

Proofreading and repair mechanisms ensure that DNA replication is accurate

After introducing a new nucleotide into a growing polynucleotide strand, the DNA polymerases perform a **proofreading** function (Figure 11.19*a*). When a DNA polymerase recognizes a mispairing of bases, it removes the improperly introduced nucleotide and tries again. (DNA helicase, DNA ligase, and other proteins of the replication complex also play roles in this key mechanism.) The polymerase is usually successful in inserting the correct monomer the second time around. As a result of this proofreading mechanism, the overall error rate for DNA replication is greatly lowered: Starting from an initial error rate of 1 base in every 10^6, the final error rate is only about 1 base in every 10^9 bases replicated.

After DNA has been replicated and during genetic recombination, a second mechanism surveys the newly replicated molecule and looks for mismatched base pairs (Figure 11.19*b*). For example, this **mismatch repair** system might detect an AC base pair instead of an AT pair. Since both AT and GC pairs obey the base-pairing rules, how does the repair mechanism "know" whether the AC pair should be repaired by removing the C and replacing it with T, for instance, or by removing the A and replacing it with G?

The repair mechanism can detect the "wrong" base because a newly synthesized DNA strand is chemically modified some time after replication. In eukaryotes, methyl groups ($-CH_3$) are added to some cytosines to form 5-methylcytosine. In prokaryotes, guanine is methylated. Right after replication, methylation has not yet occurred, so the newly replicated strand is "marked" by being unmethylated, as the one in which errors should be corrected. When mismatch repair fails, DNA sequences are altered. One form of colon cancer arises in part from a failure of mismatch repair.

DNA molecules can also be damaged during the life of a cell (e.g., when it is in G1). Some cells live and play important roles in the organism for many years, even though their DNA is constantly at risk from hazards such as high-energy radiation, chemicals that induce mutations, and random spontaneous chemical reactions. Cells owe their lives

11.19 DNA Repair Mechanisms
The proteins of DNA replication also play roles in the life-preserving DNA repair mechanisms, helping to ensure the exact replication of template DNA.

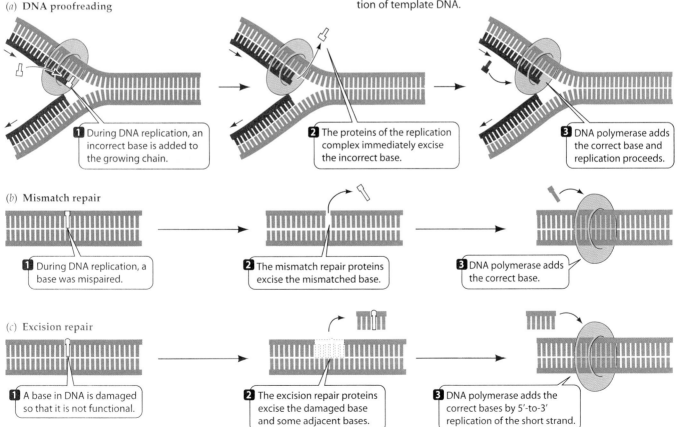

(*a*) **DNA proofreading**

1 During DNA replication, an incorrect base is added to the growing chain.

2 The proteins of the replication complex immediately excise the incorrect base.

3 DNA polymerase adds the correct base and replication proceeds.

(*b*) **Mismatch repair**

1 During DNA replication, a base was mispaired.

2 The mismatch repair proteins excise the mismatched base.

3 DNA polymerase adds the correct base.

(*c*) **Excision repair**

1 A base in DNA is damaged so that it is not functional.

2 The excision repair proteins excise the damaged base and some adjacent bases.

3 DNA polymerase adds the correct bases by 5'-to-3' replication of the short strand.

to DNA repair mechanisms. For example, in **excision repair**, certain enzymes "inspect" the cell's DNA (Figure 11.19c). When they find mispaired bases, chemically modified bases, or points at which one strand has more bases than the other (with the result that one or more bases of one strand form an unpaired loop), these enzymes cut the defective strand. Another enzyme cuts away the bases adjacent to and including the offending base, and DNA polymerase and DNA ligase synthesize and seal up a new (usually correct) piece to replace the excised one.

Our dependence on this repair mechanism is underscored by our susceptibility to various diseases that arise from excision repair defects. One example is the skin disease xeroderma pigmentosum. People with this disease lack a mechanism that normally repairs damage caused by the ultraviolet radiation in sunlight. Without this mechanism, a person exposed to sunlight develops skin cancer.

DNA repair requires energy

What does it cost the cell to keep its DNA accurate and ensure that it replicates properly? At first glance, you might expect DNA polymerization to be fairly "neutral" energetically: Adding a new monomer to the chain requires the formation of a new phosphodiester bond, but is supported by the hydrolysis of one of the high-energy bonds in the deoxyribonucleoside triphosphate (see Figure 11.11). Overall, however, this reaction is slightly endergonic. But help is available in the form of the pyrophosphate ion released in the polymerization reaction. The enzyme pyrophosphatase cleaves the high-energy bond in the pyrophosphate. Coupling this reaction to the polymerization gives it a big boost.

Noncovalent bonds also play a major role in favoring DNA polymerization. Hydrogen bonds form between the complementarily paired bases, and other weak interactions form as the bases stack in the middle of the double helix. These bonds and interactions stabilize the DNA molecule and help drive the polymerization reaction. Thus DNA synthesis itself does not take a tremendous toll in energy.

DNA repair processes, however, are far from cheap energetically. Some are very inefficient. Nonetheless, the cell deploys many DNA repair mechanisms, some overlapping in function with others. Why? Perhaps because the cell simply can't afford to leave its genetic information unprotected, regardless of the cost.

Practical Applications of DNA Replication

The principles that underlie DNA replication in cells have been applied to develop two laboratory techniques that have been vital in analyzing genes and genomes. First, the nucleotide sequence of a DNA molecule can be determined, and second, short DNA sequences can be copied using the polymerase chain reaction technique.

The nucleotide sequence of DNA can be determined

As we saw earlier in this chapter, the deoxyribonucleoside triphosphates (dNTP's) that are the normal substrates for DNA replication contain the sugar 2-deoxyribose. If instead of this, the sugar used is 2,3-dideoxyribose, the resulting nucleoside triphosphates (ddNTP's) will still be picked up by DNA polymerase and added to a growing DNA chain. However, because ddNTP's lack a hydroxyl group at the 3' position, the next nucleotide cannot be added (Figure 11.20a). Thus synthesis stops at the place where ddNTP is incorporated into the growing end of a DNA strand.

In the sequencing technique, a molecule of DNA (usually no more than 500 base pairs long) is denatured. The resulting single-strands of DNA are mixed with:

▸ DNA polymerase to synthesize the complementary strand,
▸ short primers appropriate for that sequence,
▸ the four dNTPs (dATP, dGTP, dCTP, and dTTP), and
▸ small amounts of the four ddNTPs, each with a fluorescent "tag" that emits a different color of light.

The reaction mixture soon contains mostly a DNA mixture of the template DNA strands and shorter, new complementary strands. The latter, each ending with a ddNTP, are of varying lengths. For example, each time a T is reached on the template strand, the growing complementary strand adds either dATP or ddATP. If dATP is added, the strand continues to grow. If ddATP is added, chain growth stops.

After DNA replication has been allowed to proceed for a while in a test tube, the numerous short fragments are denatured from their templates and separated by electrophoresis (see Figure 17.2). This technique measures the length of the DNA fragments, and can detect differences in fragment length as short as one base in 500. During the electrophoresis run, the fragments pass through a laser beam that excites the fluorescent tags. The light emitted is then detected, and the resulting information—that is, which ddNTP is at the end of a strand of which length—is fed into a computer, which processes it and prints out the sequence (Figure 11.20b). The computer can also be programmed to analyze the sequence, and these analyses have formed the basis of the new science of genomics, as we will describe in Chapters 13, 14, and 18.

The polymerase chain reaction makes multiple copies of DNA

Since DNA can be replicated in the test tube, using enzymes from *E. coli* and simple substrates, it is possible to make quantities of a single DNA sequence. The **polymerase chain reaction** (**PCR**) technique is an extension of this early work, which essentially automates the process and makes it much more efficient. PCR is not very complicated: A short

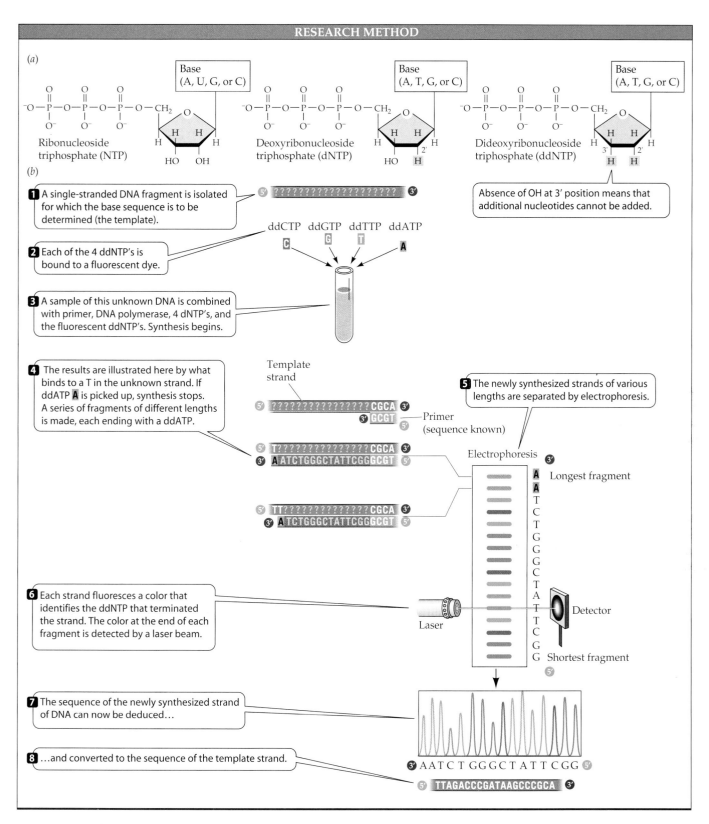

RESEARCH METHOD

(a)

Ribonucleoside triphosphate (NTP)

Base (A, U, G, or C)

Deoxyribonucleoside triphosphate (dNTP)

Base (A, T, G, or C)

Dideoxyribonucleoside triphosphate (ddNTP)

Base (A, T, G, or C)

Absence of OH at 3′ position means that additional nucleotides cannot be added.

(b)

1 A single-stranded DNA fragment is isolated for which the base sequence is to be determined (the template).

5′ ??????????????????? 3′

2 Each of the 4 ddNTP's is bound to a fluorescent dye.

ddCTP   ddGTP   ddTTP   ddATP
C   G   T   A

3 A sample of this unknown DNA is combined with primer, DNA polymerase, 4 dNTP's, and the fluorescent ddNTP's. Synthesis begins.

4 The results are illustrated here by what binds to a T in the unknown strand. If ddATP **A** is picked up, synthesis stops. A series of fragments of different lengths is made, each ending with a ddATP.

Template strand

5′ ?????????????????CGCA 3′
                    3′ GCGT 5′
Primer (sequence known)

5′ T?????????????????CGCA 3′
3′ AATCTGGGCTATTCGGGCGT 5′

5′ TT?????????????????CGCA 3′
3′ ATCTGGGCTATTCGGGCGT 5′

5 The newly synthesized strands of various lengths are separated by electrophoresis.

Electrophoresis

3′
A Longest fragment
A
T
C
T
G
G
G
C
T
A
T
T
C
G
G Shortest fragment
5′

Laser   Detector

6 Each strand fluoresces a color that identifies the ddNTP that terminated the strand. The color at the end of each fragment is detected by a laser beam.

7 The sequence of the newly synthesized strand of DNA can now be deduced…

8 …and converted to the sequence of the template strand.

3′ AATC T GGGCT A T T C G G 5′

5′ TTAGACCCGATAAGCCCGCA 3′

11.20 Sequencing DNA
(a) The normal substrates for DNA replication are dNTP's. The slightly different structure of ddNTP's can cause DNA synthesis to stop. (b) When labeled ddNTP's are incorporated into a mixture containing a single-stranded DNA template of unknown sequence, the result is an electrophoresis of fragments of varying lengths.

RESEARCH METHOD

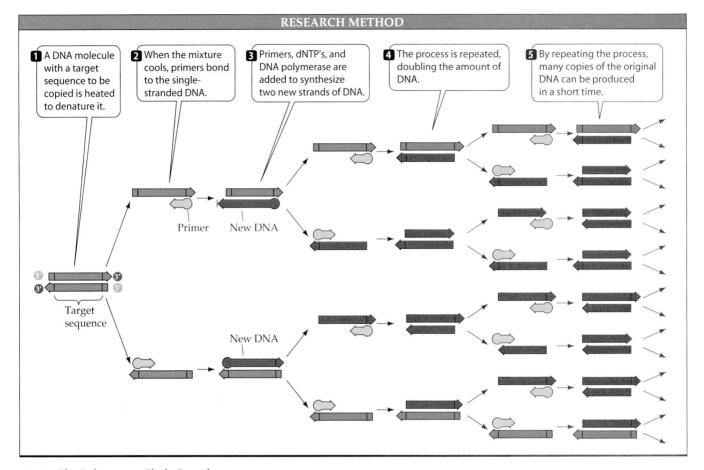

11.21 The Polymerase Chain Reaction
The steps in this cyclic process are repeated many times to produce multiple copies
of a DNA sequence.

region of DNA is copied many times in the test tube by DNA polymerase.

PCR is a cyclic process in which the following sequence of steps is repeated over and over again (Figure 11.21). It begins with the same first two steps as DNA sequencing:

▶ Double-stranded DNA is denatured by heat into single strands.
▶ Short primers for DNA replication are added to the mixture.
▶ DNA polymerase catalyzes the production of complementary new strands.

A single cycle, taking a few minutes, doubles the amount of DNA and leaves the new DNA in the double-stranded state. Repeating the cycle many times can theoretically lead to a geometric increase in the number of copies of the DNA sequence.

The PCR technique requires that the base sequences at the 3′ end of each strand of the target DNA be known so that a complementary primer, usually 15 to 20 bases long, can be made in the laboratory. Because of the uniqueness of DNA sequences, usually two primers of this length will bind to only one region of DNA in an organism's genome.

This specificity in the face of the incredible diversity of target DNA is a key to the power of PCR.

One potential problem with PCR involves its temperature requirements. To denature the DNA during each cycle, it must be heated to more than 90°C—a temperature that destroys most DNA polymerases. Then it must be cooled to about 55°C to allow the primer to hydrogen-bond to the single strands of template DNA. The PCR method would not be practical if new polymerase had to be added during each cycle after denaturation—an expensive and laborious proposition.

This problem was solved by nature: In the hot springs at Yellowstone National Park, as well as other locations, there live bacteria called, appropriately, *Thermus aquaticus*. The means by which these organisms survive temperatures up to 95°C was investigated by bacteriologist Thomas Brock and his colleagues. They discovered that *T. aquaticus* has an entire metabolic machinery that is heat-resistant, including DNA polymerase that does not denature at this high temperature.

Scientists pondering the problem of amplifying DNA by PCR read Brock's basic research articles and got a clever idea: Why not use *T. aquaticus* DNA polymerase in the PCR reaction? It would not be denatured, and thus would not have to be added during each cycle. The idea worked, and earned biochemist Kerry Mullis a Nobel prize. PCR has had an enormous impact on genetic research. Some of its most striking applications will be described in Chapters 13 through 17.

Chapter Summary

DNA: The Genetic Material

▶ In addition to circumstantial evidence (the location and quantity of DNA in the cell), two experiments provided a convincing demonstration that DNA is the genetic material.

▶ In one experiment, DNA from a virulent strain of pneumococcus bacteria genetically transformed nonvirulent bacteria into virulent bacteria. **Review Figure 11.1**

▶ In a second set of experiments, labeled viruses were incubated with host bacteria. Labeled viral DNA entered the host cells, where it produced hundreds of new viruses bearing the label. **Review Figures 11.2, 11.3**

The Structure of DNA

▶ X-ray crystallography showed that the DNA molecule is a helix. **Review Figure 11.4**

▶ DNA is composed of nucleotides, each containing one of four bases—adenine, cytosine, thymine, or guanine. Biochemical analysis revealed that the amount of adenine equals the amount of thymine and the amount of guanine equals the amount of cytosine. **Review Figure 11.5**

▶ Putting the accumulated data together, Watson and Crick built a model of the DNA molecule. They proposed that DNA is a double-stranded helix in which the strands are antiparallel and the bases are held together by hydrogen bonding. This model accounts for the genetic information, mutation, and replication functions of DNA. **Review Figures 11.6, 11.7**

DNA Replication

▶ Three possible models for DNA replication were hypothesized: semiconservative, conservative, and dispersive. **Review Figure 11.8**

▶ An experiment by Meselson and Stahl proved the replication of DNA to be semiconservative. Each parent strand acts as a template for the synthesis of a new strand; thus the two replicated DNA helices contain one parent strand and one newly synthesized strand each. **Review Figures 11.9, 11.10**

The Mechanism of DNA Replication

▶ In DNA replication, the enzyme DNA polymerase catalyzes the addition of nucleotides to the 3′ end of each strand. Nucleotides are added by complementary base pairing with the template strand of DNA. The substrates are deoxyribonucleoside triphosphates, which are hydrolyzed as they are added to the growing chain, releasing energy that fuels the synthesis of DNA. **Review Figure 11.11**

▶ The DNA replication complex is in a fixed location and DNA is threaded through it for replication. **Review Figure 11.12**

▶ Many proteins assist in DNA replication. DNA helicases unwind the double helix, and the template strands are stabilized by other proteins.

▶ Prokaryotes have a single origin of replication; eukaryotes have many. Replication in both cases proceeds in both directions from an origin of replication. **Review Figure 11.13**

▶ An RNA primase catalyzes the synthesis of short RNA primers, to which nucleotides are added as the chain grows. **Review Figure 11.15**

▶ Through the action of DNA polymerase, the leading strand grows continuously in the 5′-to-3′ direction until the replication of that section of DNA has been completed. Then the RNA primer is degraded and DNA is added in its place.

▶ On the lagging strand, which grows in the other direction, DNA is still made in the 5′-to-3′ direction (away from the replication fork). But synthesis of the lagging strand is discontinuous: The DNA is added as short fragments to primers, then the polymerase skips past the 5′ end to make the next fragment. **Review Figures 11.16, 11.17, 11.18**

DNA Proofreading and Repair

▶ The machinery of DNA replication makes about one error in 10^6 nucleotides bases added. These errors are repaired by three different mechanisms: proofreading, mismatch repair, and excision repair. DNA repair mechanisms lower the overall error rate of replication to about one base in 10^9. **Review Figure 11.19**

▶ Although energetically costly and somewhat redundant, DNA repair is crucial to the survival of the cell.

Practical Applications of DNA Replication

▶ The principles of DNA replication can be used to determine the nucleotide sequence of DNA. **Review Figure 11.20**

▶ The polymerase chain reaction technique uses DNA polymerases to repeatedly replicate DNA in the test tube. **Review Figure 11.21**

For Discussion

1. Outline a series of experiments using radioactive isotopes to show that bacterial DNA and not protein enters the host cell and is responsible for bacterial transformation.

2. Suppose that Meselson and Stahl had continued their experiment on DNA replication for another ten bacterial generations. Would there still have been any $^{14}N-^{15}N$ hybrid DNA present? Would it still have appeared in the centrifuge tube? Explain.

3. If DNA replication were conservative rather than semiconservative, what results would Meselson and Stahl have observed? Diagram the results using the conventions of Figure 11.10.

4. Using the following information, calculate the number of origins of DNA replication on a human chromosome: DNA polymerase adds nucleotides at 3,000 base pairs per minute in one direction; replication is bidirectional; the S phase lasts 300 minutes; there are 120 million base pairs per chromosome. With a typical chromosome 3 cm long, how many origins are there per micrometer?

5. The drug dideoxycytidine (used to treat certain viral infections) is a nucleoside made with 2′-3′-dideoxyribose. This sugar lacks —OH groups at both the 2′ and the 3′ positions. Explain why this drug would stop the growth of a DNA chain if it was added to DNA.

mutant strains could no longer grow on minimal medium, but needed additional nutrients that they could not make on their own. The scientists proposed that these *auxotrophs* ("increased eaters") must have suffered mutations in genes that coded for enzymes necessary for the synthesis of the nutrients they now needed to ingest. For each auxotrophic strain, Beadle and Tatum were able to find a single compound that, when added to the minimal medium, supported the growth of that strain. This result supported the idea that mutations have simple effects, and that each mutation causes a defect in only one enzyme in a metabolic pathway.

One group of auxotrophs, for example, could grow on minimal medium supplemented with the amino acid arginine. (Wild-type *Neurospora* makes arginine by itself.) These mutant strains were classified as *arg* mutants. Were their mutations different alleles of the same gene, or were they in different genes, each coding for an enzyme along the biochemical pathway for arginine synthesis? Mapping studies established that some of the *arg* mutations were at different loci, or on different chromosomes, and so were not alleles. Beadle and Tatum concluded that different genes could participate in governing a single biosynthetic pathway—in this case, the pathway leading to arginine synthesis.

By growing different *arg* mutants in the presence of various compounds suspected to be intermediates in the synthetic metabolic pathway for arginine, Beadle and Tatum were able to classify each mutation as affecting one enzyme or another, and to order the compounds on the pathway (see Figure 12.1). Then they broke open the wild-type and mutant cells and examined them for enzyme activities. The results confirmed their hypothesis: Each mutant strain was indeed missing a single active enzyme in the pathway.

If a genetic mutation results in an abnormal or missing enzyme, then the wild-type gene must code for the normal enzyme. This conclusion led Beadle and Tatum to formulate the one-gene, one-enzyme hypothesis. According to this hypothesis, the *function of a gene is to control the production of a single, specific enzyme.* This proposal strongly influenced the subsequent development of the sciences of genetics and molecular biology.

The English physician Archibald Garrod, who studied the inherited disease alkaptonuria, had made this proposal 40 years before. He linked the biochemical phenotype of the disease to an abnormal gene and a missing enzyme. There are hundreds of examples of such hereditary diseases, and we will return to them in Chapter 18.

The gene–enzyme relationship requires modification when we consider that many enzymes are composed of more than one polypeptide chain, or subunit (that is, they have a quaternary structure). In this case, each polypeptide chain is specified by its own separate gene. Thus, it is more correct to speak of a **one-gene, one-polypeptide hypothesis**: *The function of a gene is to control the production of a single, specific polypeptide.*

Much later, it was discovered that some genes code for forms of RNA that do not become translated into polypep-

tides, and that still other genes are involved in controlling which other DNA sequences are expressed. While these discoveries have overthrown the idea that all genes code for proteins, they did not invalidate the relationship between genes and polypeptides.

DNA, RNA, and the Flow of Information

Now let us turn our attention to *how* a gene expresses itself as a polypeptide. This expression occurs in two steps. The first step, transcription, copies the information of a DNA sequence (the gene) into corresponding information in an RNA sequence. The second step, translation, converts this RNA information into an appropriate amino acid sequence in a polypeptide.

RNA differs from DNA

To understand the transcription and translation of genetic information, you need to understand the structure of RNA. **RNA (ribonucleic acid)** is a polynucleotide similar to DNA (see Figure 3.17), but it differs from DNA in three ways:

▶ RNA generally consists of only one polynucleotide strand [thus Chargaff's rule, G = C and A = T (see Figure 11.5), is usually true for DNA and not for RNA].

▶ The sugar molecule found in RNA is ribose, rather than the deoxyribose found in DNA.

▶ Although three of the nitrogenous bases (adenine, guanine, and cytosine) in RNA are identical to those in DNA, the fourth base in RNA is uracil (U), which is similar to thymine but lacks the methyl ($-CH_3$) group.

Thymine   Uracil

RNA can base-pair with single-stranded DNA, and this pairing obeys the same hydrogen-bonding rules as in DNA, except that adenine pairs with uracil instead of thymine. RNA can also fold over and base-pair with itself, as we will see with tRNA later in this chapter.

Information flows in one direction when genes are expressed

Francis Crick (of the Watson–Crick model) proposed what he called the **central dogma** of molecular biology. The central dogma is, simply, that DNA codes for the production of RNA (transcription), RNA codes for the production of protein (translation), and protein does *not* code for the production of protein, RNA, or DNA (Figure 12.2a). In Crick's words, "once 'information' has passed into protein *it cannot get out again.*"

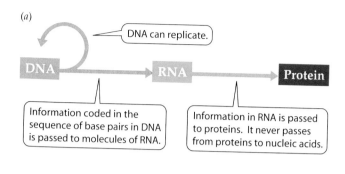

(a)

DNA can replicate.

DNA → RNA → Protein

Information coded in the sequence of base pairs in DNA is passed to molecules of RNA.

Information in RNA is passed to proteins. It never passes from proteins to nucleic acids.

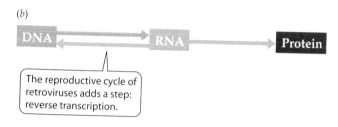

(b)

DNA ⇄ RNA → Protein

The reproductive cycle of retroviruses adds a step: reverse transcription.

12.2 The Central Dogma
(a) Information flows from DNA to proteins, as indicated by the arrows. In certain viruses, RNA can replicate to RNA. (b) The reproductive cycle of retroviruses adds a step, reverse transcription, to the central dogma.

The central dogma posed two questions:

▶ How does genetic information get from the nucleus to the cytoplasm? (As you know, most of the DNA of a eukaryotic cell is confined to the nucleus, but proteins are synthesized in the cytoplasm.)

▶ What is the relationship between a specific nucleotide sequence (in DNA) and a specific amino acid sequence (in a protein)?

To answer the first question, Crick and his colleagues developed the messenger hypothesis, according to which an RNA molecule forms as a complementary copy of one DNA strand of a particular gene. The process by which this RNA forms is called **transcription**. If each such RNA molecule contains the information from a gene, there should be as many different kinds of RNA molecules as there are genes. This **messenger RNA**, or **mRNA**, then travels from the nucleus to the cytoplasm, where it serves as a template for the synthesis of proteins.

To answer the second question, Crick proposed the adapter hypothesis: there must be an adapter molecule that can bind a specific amino acid at one end and recognize a sequence of nucleotides with another region. In due course, these adapters, called **transfer RNA**, or **tRNA**, were identified. Because they recognize the genetic message of mRNA and simultaneously carry specific amino acids, tRNA's can translate the language of DNA into the language of proteins. The tRNA adapters line up on the mRNA so that the amino acids are in the proper sequence for a growing polypeptide chain—a process called **translation** (Figure 12.3).

Summarizing the main features of the central dogma, the messenger hypothesis, and the adapter hypothesis, we may say that *a given gene is transcribed to produce a messenger RNA (mRNA) complementary to one of the DNA strands*, and that *transfer RNA (tRNA) molecules translate the sequence of bases in the mRNA into the appropriate sequence of amino acids.*

RNA viruses modify the central dogma

According to the central dogma, DNA codes for RNA, and RNA codes for protein. All cellular organisms have DNA as their hereditary material. Only among viruses (and certain DNA sequences) are variations on the central dogma found.

Many viruses, such as the tobacco mosaic virus, influenza virus, and poliovirus, have RNA rather than DNA as their genetic material. With its nucleotide sequence, RNA could potentially act as an information carrier and be expressed as proteins. But since RNA is usually single-stranded, its replication is a problem. The viruses generally solve this problem by transcribing from RNA to RNA, making an RNA strand that is complementary to the genome. This "opposite" strand is used to make to make more copies of the genome by transcription.

HIV and certain tumor viruses also have RNA as their genome, but do not replicate it as RNA-to-RNA. Instead, after infecting a host cell, they make a DNA copy of their genome, and use it to make more RNA. This RNA is then used both as genomes for more copies of the virus and as mRNA (see Figure 12.2b). Synthesis of DNA from RNA is called reverse transcription. Not surprisingly, such viruses are called *retroviruses*. We will examine both types of RNA viruses in detail in the next chapter.

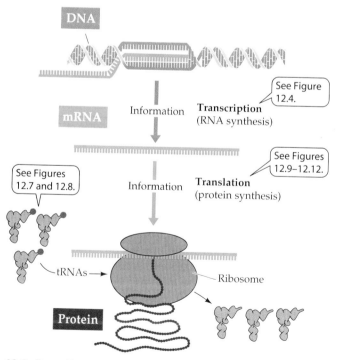

DNA

Information   **Transcription**
(RNA synthesis)

See Figure 12.4.

mRNA

Information   **Translation**
(protein synthesis)

See Figures 12.9–12.12.

See Figures 12.7 and 12.8.

tRNAs

Ribosome

Protein

12.3 From Gene to Protein
This figure summarizes the processes of gene expression in prokaryotes. In eukaryotes, the processes are somewhat more complex.

and water. And, as we saw at the opening of this chapter, infectious diseases caused by prokaryotes and viruses continue to challenge humankind.

Viruses: Reproduction and Recombination

Although there are many kinds of viruses, most of them are composed of nothing but nucleic acid and a few proteins. Most viruses have relatively simple means of infecting their host cells. Some can infect a cell but postpone reproduction, lying low in the host chromosome until conditions are favorable. The simplest infective agents of all are viroids, which are made up only of genetic material.

Scientists studied viruses before they could see them

Most viruses are much smaller than most bacteria (Table 13.1). Viruses have become well understood only within the last half century, but the first step on this path of discovery was taken by the Russian botanist Dmitri Ivanovsky in 1892. He was trying to find the cause of *tobacco mosaic disease*, which results in the destruction of photosynthetic tissues and can devastate a tobacco crop. Ivanovsky passed an extract of diseased tobacco leaves through a fine porcelain filter, a technique that had been used previously by physicians and veterinarians to isolate disease-causing bacteria.

To Ivanovsky's surprise, the disease agent in this case did not stick to the filter: It passed through, and the liquid filtrate still caused tobacco mosaic disease. But instead of concluding that the agent was smaller than a bacterium, he assumed that his filter was faulty. Pasteur's recent demonstration that bacteria could cause disease was the dominant idea at the time, and Ivanovsky chose not to challenge it. But, as often happens in science, someone soon came along who did. In 1898, Martinus Beijerinck repeated Ivanovsky's experiment, and also showed that the tiny tobacco mosaic agent could diffuse through an agar gel. He called the agent *contagium vivum fluidum*, which later became shortened to **virus**.

Almost 40 years later, the infective agent was crystallized by Wendell Stanley (who won the Nobel Prize for his efforts). The crystalline viral preparation became infectious again when it was dissolved. It was soon shown that crys-

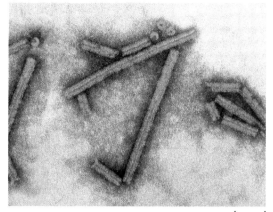

(a)

75 nm

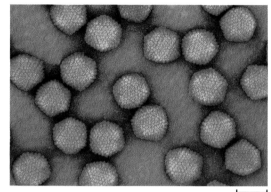

(b)

75 nm

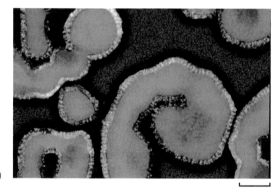

(c)

20 nm

13.1 Virions Come in Various Shapes
(a) The tobacco mosaic virus (a plant virus) consists of an inner helix of RNA covered with a helical array of protein molecules. (b) Many animal viruses, such as this adenovirus, have an icosahedral (20-sided) capsid as an outer shell. Inside the shell is a spherical mass of proteins and DNA. (c) Not all virions are regularly shaped. Wormlike virions of the influenza A virus infect humans, causing chills, fever and sometimes, death.

tallized viral preparations consist of protein and nucleic acid. Finally, direct observation of viruses with electron microscopes in the 1950s showed clearly how much they differ from bacteria and other organisms

Viruses reproduce only with the help of living cells

Unlike the organisms that make up the six taxonomic kingdoms of the living world, viruses are *acellular*; that is, they are not cells and do not consist of cells. Unlike cellular creatures, viruses do not metabolize energy—they neither produce ATP nor conduct fermentation, cellular respiration, or photosynthesis.

Whole viruses never arise directly from preexisting viruses. Viruses are *obligate intracellular parasites*; that is, they

13.1	**Common Sizes of Microorganisms**	
MICROORGANISM	**TYPE**	**TYPICAL SIZE RANGE (μm³)**
Protists	Eukaryote	5,000–50,000
Photosynthetic bacteria	Prokaryote	5–50
Spirochetes	Prokaryote	0.1–2.0
Mycoplasmas	Prokaryote	0.01–0.1
Poxviruses	Virus	0.01
Influenza virus	Virus	0.0005
Poliovirus	Virus	0.00001

13 The Genetics of Viruses and Prokaryotes

 JANET, A MEMBER OF HER UNIVERSITY'S CROSS-country team, entered the hospital just after final exams for some long-delayed surgery on a tendon in her knee. The tendon repair went well, but she left the hospital with something new: bacteria called *Pseudomonas aeruginosa* had infected the surgical wounds. The antibiotics typically used to kill these bacteria did not work. She ended up back in the hospital two weeks later, where she received intensive antibiotic therapy and ultimately recovered.

Janet developed what is called a *nosocomial infection*—an infection acquired as a result of a hospital stay. Why would a hospital, which we think of as a place to get better, sometimes—in fact, for about 10 percent of all patients—be a place where we get sick? Of course, the stresses of Janet's surgery could have reduced her immunity to the bacteria that are common everywhere in our environment. Increasingly, however, the heavy use of antibiotics in hospitals makes them breeding grounds for bacteria that have genes for resistance to those antibiotics.

How have bacteria acquired antibiotic resistance so rapidly, and how do they pass that acquired resistance along to other bacteria? The answer involves some DNA sequences called R factors. But before we can discuss these remarkable pieces of DNA, we must introduce the genetics of prokaryotes. Prokaryotes usually reproduce asexually by cell division, but can acquire new genes in several ways. These range from simple recombination in a sexual process to using infective viruses as carriers for prokaryotic genes. We also describe how the expression of prokaryotic genes is regulated, and what DNA sequencing has revealed about the prokaryotic genome.

Viruses are not prokaryotes. In fact, they are not even cells, but intracellular parasites that can reproduce only within living cells. We begin this chapter by examining the structures, classification, reproduction, and genetics of viruses.

Using Prokaryotes and Viruses to Probe the Nature of Genes

Prokaryotes such as *Escherichia coli* and the viruses that infect them have been important tools in discovering the structure, function, and transmission of genes, as we saw in Chapter 11. What are the advantages of working with prokaryotes and viruses?

First, it is easier to work with small amounts of DNA than with large amounts. A typical bacterium contains about a thousandth as much DNA as a single human cell, and a typical bacterial virus contains about a hundredth as much DNA as a bacterium. Second, data on large numbers of individuals can be obtained easily from prokaryotes. A single milliliter of medium can contain more than 10^9 *E. coli* cells or 10^{11} bacteriophages. In addition, most prokaryotes grow rapidly. A culture of *E. coli* can be grown under conditions that allow their numbers to double every 20 minutes. By contrast, 10^9 mice would cost more than 10^9 dollars, would require a cage that would cover about 3 square miles, and growing a generation of mice takes about 3 months, not 20 minutes. Third, prokaryotes and viruses are usually haploid, which makes genetic analyses easier.

The ease of growing and handling bacteria and their viruses permitted the explosion of genetics and molecular biology that began shortly after the mid-twentieth century. Their relative biological simplicity contributed immeasurably to discoveries about the genetic material, the replication of DNA, and the mechanisms of gene expression. Later, they were the first subjects of recombinant DNA technology (see Chapter 16).

Questions of interest to all biologists continue to be studied in prokaryotes, and prokaryotes continue to be important tools for biotechnology and for research on eukaryotes. Prokaryotes are important players in the environment, performing much of the cycling of elements in the atmosphere

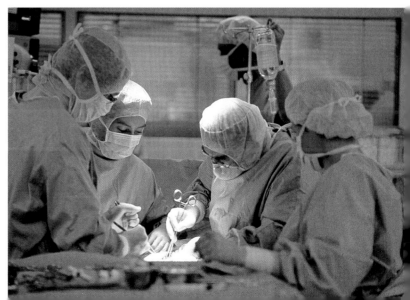

Are There Uninvited Guests Here?
A masked team performs surgery on a patient. But have harmful, drug-resistant bacteria invaded the surgical suite?

and water. And, as we saw at the opening of this chapter, infectious diseases caused by prokaryotes and viruses continue to challenge humankind.

Viruses: Reproduction and Recombination

Although there are many kinds of viruses, most of them are composed of nothing but nucleic acid and a few proteins. Most viruses have relatively simple means of infecting their host cells. Some can infect a cell but postpone reproduction, lying low in the host chromosome until conditions are favorable. The simplest infective agents of all are viroids, which are made up only of genetic material.

Scientists studied viruses before they could see them

Most viruses are much smaller than most bacteria (Table 13.1). Viruses have become well understood only within the last half century, but the first step on this path of discovery was taken by the Russian botanist Dmitri Ivanovsky in 1892. He was trying to find the cause of *tobacco mosaic disease*, which results in the destruction of photosynthetic tissues and can devastate a tobacco crop. Ivanovsky passed an extract of diseased tobacco leaves through a fine porcelain filter, a technique that had been used previously by physicians and veterinarians to isolate disease-causing bacteria.

To Ivanovsky's surprise, the disease agent in this case did not stick to the filter: It passed through, and the liquid filtrate still caused tobacco mosaic disease. But instead of concluding that the agent was smaller than a bacterium, he assumed that his filter was faulty. Pasteur's recent demonstration that bacteria could cause disease was the dominant idea at the time, and Ivanovsky chose not to challenge it. But, as often happens in science, someone soon came along who did. In 1898, Martinus Beijerinck repeated Ivanovsky's experiment, and also showed that the tiny tobacco mosaic agent could diffuse through an agar gel. He called the agent *contagium vivum fluidum*, which later became shortened to **virus**.

Almost 40 years later, the infective agent was crystallized by Wendell Stanley (who won the Nobel Prize for his efforts). The crystalline viral preparation became infectious again when it was dissolved. It was soon shown that crys-

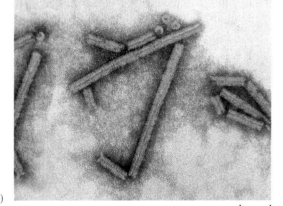

(a)

75 nm

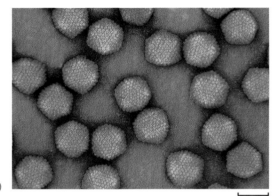

(b)

75 nm

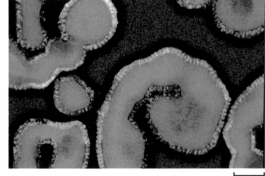

(c)

20 nm

13.1 Virions Come in Various Shapes
(a) The tobacco mosaic virus (a plant virus) consists of an inner helix of RNA covered with a helical array of protein molecules. (b) Many animal viruses, such as this adenovirus, have an icosahedral (20-sided) capsid as an outer shell. Inside the shell is a spherical mass of proteins and DNA. (c) Not all virions are regularly shaped. Wormlike virions of the influenza A virus infect humans, causing chills, fever and sometimes, death.

tallized viral preparations consist of protein and nucleic acid. Finally, direct observation of viruses with electron microscopes in the 1950s showed clearly how much they differ from bacteria and other organisms

Viruses reproduce only with the help of living cells

Unlike the organisms that make up the six taxonomic kingdoms of the living world, viruses are *acellular*; that is, they are not cells and do not consist of cells. Unlike cellular creatures, viruses do not metabolize energy—they neither produce ATP nor conduct fermentation, cellular respiration, or photosynthesis.

Whole viruses never arise directly from preexisting viruses. Viruses are *obligate intracellular parasites*; that is, they

13.1	*Common Sizes of Microorganisms*	
MICROORGANISM	TYPE	TYPICAL SIZE RANGE (μm^3)
Protists	Eukaryote	5,000–50,000
Photosynthetic bacteria	Prokaryote	5–50
Spirochetes	Prokaryote	0.1–2.0
Mycoplasmas	Prokaryote	0.01–0.1
Poxviruses	Virus	0.01
Influenza virus	Virus	0.0005
Poliovirus	Virus	0.00001

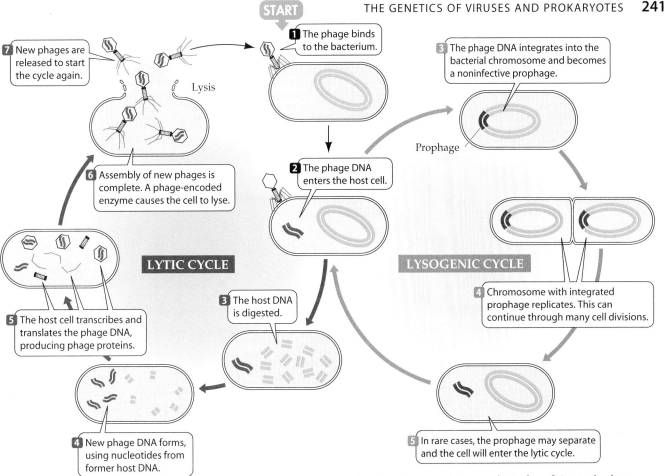

START

7 New phages are released to start the cycle again.

1 The phage binds to the bacterium.

3 The phage DNA integrates into the bacterial chromosome and becomes a noninfective prophage.

Lysis

Prophage

6 Assembly of new phages is complete. A phage-encoded enzyme causes the cell to lyse.

2 The phage DNA enters the host cell.

LYTIC CYCLE

LYSOGENIC CYCLE

3 The host DNA is digested.

4 Chromosome with integrated prophage replicates. This can continue through many cell divisions.

5 The host cell transcribes and translates the phage DNA, producing phage proteins.

4 New phage DNA forms, using nucleotides from former host DNA.

5 In rare cases, the prophage may separate and the cell will enter the lytic cycle.

13.2 The Lytic and Lysogenic Cycles of Bacteriophages
In the lytic cycle, infection by viral DNA leads directly to the multiplication of the virus and lysis of the host bacterial cell. In the lysogenic cycle, a prophage is replicated as part of the host's chromosome.

develop and reproduce only within the cells of specific hosts. The cells of animals, plants, fungi, protists, and prokaryotes (both bacteria and archaea) serve as hosts to viruses. When they reproduce, viruses usually destroy the host cell, releasing progeny viruses that then seek new hosts.

Many diseases of humans, animals, and plants are caused by viruses. Because they lack the distinctive cell wall and ribosomal biochemistry of bacteria, viruses are not affected by antibiotics.

Viruses outside of host cells exist as individual particles called **virions**. The virion, the basic unit of a virus, consists of a central core of either DNA or RNA (but not both) surrounded by a **capsid**, or coat, composed of one or more proteins. The way in which these proteins assemble gives each type of virion a characteristic shape (Figure 13.1). In addition, many animal viruses have a lipid and protein membrane acquired from host cell plasma membranes.

There are many kinds of viruses

A common way to classify viruses separates them by whether their genetic material is DNA or RNA, and then by whether their nucleic acid is single-stranded or double-stranded. Some RNA viruses have more than one molecule of RNA, and the DNA of one virus family is circular. Further levels of classification depend on factors such as the overall shape of the virus and the symmetry of the capsid (see Figure 13.1). Another level of classification is based on the presence or absence of a membranous envelope around the virion; still further subdivision is based on capsid size.

One way to classify viruses is based on the type of host. Let's see how reproductive cycles and other properties vary among the major groups of viruses: those that infect bacteria, animals, and plants.

Bacteriophages reproduce by a lytic cycle or a lysogenic cycle

Viruses that infect bacteria are known as **bacteriophages**. Bacteriophages recognize their hosts by means of specific binding between proteins in the capsid and receptor proteins in the host's cell wall. The virions, which must penetrate the cell wall, are often equipped with tail assemblies that insert their nucleic acid through the cell wall into the host bacterium. After the phage has injected its nucleic acid into the host, one of two things happens, depending on the kind of phage.

We saw one type of viral reproductive cycle when we studied the Hershey–Chase experiment (see Figure 11.3). That was the **lytic cycle**, so named because the infected bacterium *lyses* (bursts), releasing progeny phages. In the **lysogenic cycle**, the infected bacterium does not lyse, but instead harbors the viral nucleic acid for many generations. Some viruses reproduce only by the lytic cycle; others undergo both types of reproductive cycles (Figure 13.2).

THE LYTIC CYCLE. A phage that reproduces only by the lytic cycle is called a **virulent virus**. Once the phage has injected its nucleic acid into the host cell, the phage nucleic acid takes over the host's synthetic machinery. It does so in two stages (Figure 13.3):

▶ The early stage transcribes the virus's *early genes*. This part of the viral genome contains the promoter sequence that attracts host RNA polymerase. The early genes often include proteins that shut down host transcription, stimulate viral genome replication, and stimulate late gene transcription. Nuclease enzymes digest the host chromosome, providing nucleotides for the synthesis of viral genomes.

▶ The late stage transcribes the virus's *late genes*. These genes code for the proteins that package virions and lyse the host cell to release the new virions.

This sequence of transcriptional events is carefully controlled: Premature lysis of the host cell before virus particles are ready for release would stop the infection. The whole process—from binding and infection to lysis of the host—takes about half an hour.

Rarely, two viruses infect a cell at the same time. This is an unusual event, as once an infection cycle is under way, there is usually not enough time for an additional infection. In addition, an early gene product prevents further infections in some cases. The presence of two viral genomes in the same host cell affords the opportunity for genetic recombination by crossing over, as in prophase of meiosis. This enables genetically different viruses of the same kind to swap genes and create new strains.

THE LYSOGENIC CYCLE. Phage infection does not always result in lysis of the host cell. Some phages seem to disappear from a bacterial culture, leaving the bacteria immune to further attack by the same strain of phage. In such cultures, however, a few free phages are always present. Bacteria harboring phages that are not lytic are called *lysogenic*, and the phages are called **temperate viruses**.

Lysogenic bacteria contain a noninfective entity called a **prophage**: a molecule of phage DNA that has been integrated into the bacterial chromosome (see Figure 13.2). As part of the bacterial genome, the prophage can remain quiet within the bacteria through many cell divisions. However, an occasional lysogenic bacterium can be induced to activate its prophage. This results in a lytic cycle, releasing a large number of free phages, which can then infect other uninfected bacteria and renew the reproductive cycle.

This capacity to switch between the lysogenic and the lytic cycle is very useful to the phage, whose purpose is to reproduce as many offspring as possible. When its host cell is growing slowly and is low on energy, the phage becomes

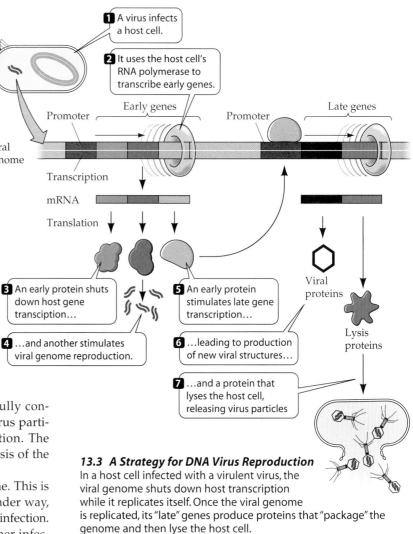

13.3 A Strategy for DNA Virus Reproduction In a host cell infected with a virulent virus, the viral genome shuts down host transcription while it replicates itself. Once the viral genome is replicated, its "late" genes produce proteins that "package" the genome and then lyse the host cell.

lysogenic. Then, when the host's health is restored to a level that provides maximal resources for phage reproduction, the prophage is released from its dormant state, and the lytic cycle proceeds. We will see how this switch works later in the chapter when we discuss control of gene expression.

Animal viruses have diverse reproductive cycles

Almost all vertebrates are susceptible to viral infections, but among invertebrates, such infections are common only in arthropods (the group that includes insects and crustaceans). One group of viruses, called *arboviruses* (short for "*arthropod-borne viruses*"), is transmitted to a mammalian host through an insect bite. Although carried within the arthropod host's cells, arboviruses apparently do not affect that host severely; they affect only the bitten and infected mammal. The arthropod acts as a **vector**—an intermediate carrier—transmitting the disease organism from one host to another.

Animal viruses are very diverse. Some are just particles of proteins surrounding a nucleic acid core. Others have a membrane derived from the host cell's plasma membrane. Some animal viruses have DNA as their genetic material; others have RNA. In all cases, the small viral genome has limited coding capacity, making only a few proteins.

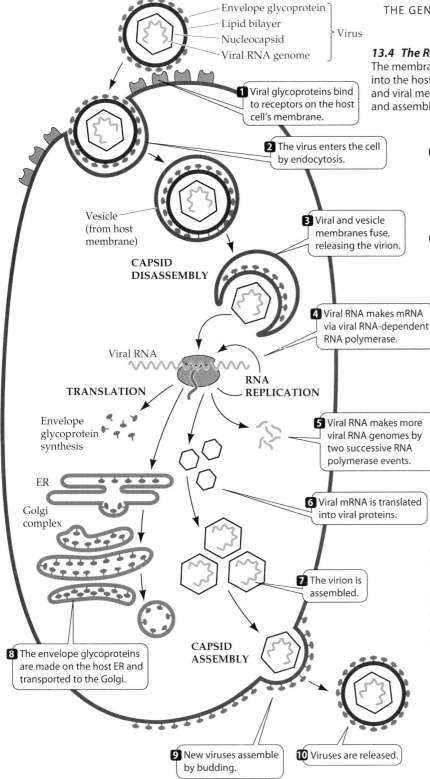

Envelope glycoprotein
Lipid bilayer
Nucleocapsid ⎫ Virus
Viral RNA genome ⎭

1 Viral glycoproteins bind to receptors on the host cell's membrane.

2 The virus enters the cell by endocytosis.

Vesicle (from host membrane)

3 Viral and vesicle membranes fuse, releasing the virion.

CAPSID DISASSEMBLY

4 Viral RNA makes mRNA via viral RNA-dependent RNA polymerase.

Viral RNA

TRANSLATION

RNA REPLICATION

Envelope glycoprotein synthesis

5 Viral RNA makes more viral RNA genomes by two successive RNA polymerase events.

ER

Golgi complex

6 Viral mRNA is translated into viral proteins.

7 The virion is assembled.

8 The envelope glycoproteins are made on the host ER and transported to the Golgi.

CAPSID ASSEMBLY

9 New viruses assemble by budding.

10 Viruses are released.

13.4 The Reproductive Cycle of the Influenza Virus
The membrane-enclosed, or enveloped, influenza virus is taken into the host cell by endocytosis. Once inside, fusion of the vesicle and viral membranes releases the RNA genome, which replicates and assembles new virions.

▶ Viruses with membranes (called *enveloped viruses*) may also be taken up by endocytosis (see Figure 13.4), and released from a vesicle. In these viruses, the viral membrane is studded with glycoproteins that bind to receptors on the host cell plasma membrane.

▶ More commonly, the host and viral membranes fuse, releasing the rest of the virion into the cell (see Figure 13.5).

Enveloped viruses usually escape from the host cell by budding through virus-modified areas of the host's plasma membrane. During this process, the completed virions acquire a membrane similar to that of the host cell.

The life cycles of influenza virus and HIV, two important RNA viruses, illustrate the two different styles of infection by enveloped viruses. Influenza virus is endocytosed into a membrane vesicle (Figure 13.4). Fusion of the viral and vesicle membranes releases the virion into the cell. The virus carries its own enzyme to replicate its RNA genome into a complementary strand. The latter is then used as mRNA to make, by complementary base pairing, more copies of the viral genome.

Retroviruses such as HIV have a more complex reproductive cycle (Figure 13.5). The virus enters a host cell by direct fusion of viral and cellular membranes. A major feature of the retroviral life cycle is the reverse transcription of retroviral RNA to produce a DNA *provirus* (cDNA), which is the form of the viral genome that gets integrated into the host DNA. The provirus may reside in the host chromosome permanently, occasionally being expressed to produce new virions. Almost every step in this complex cycle can, in principle, be attacked by therapeutic drugs; this fact is used by researchers in their quest to conquer AIDS, the deadly condition caused by HIV infection in humans. This medical battle will be discussed further in Chapter 19.

Like that of bacteriophages, the life cycle of animal viruses can be divided into early and late stages (see Figure 13.3). Animal viruses enter cells in one of three ways:

▶ A naked virion (without a membrane) is taken up by endocytosis, which traps it within a membranous vesicle inside the host cell. The membrane of the vesicle breaks down, releasing the virion into the cytoplasm, and the host cell digests the protein capsid, liberating the viral nucleic acid, which takes charge of the host cell.

Many plant viruses spread with the help of vectors

Viral diseases of flowering plants are very common. Plant viruses can be transmitted *horizontally*, from one plant to another, or *vertically*, from parent to offspring. To infect a plant cell, viruses must pass through a cell wall and through the host plasma membrane. Most plant viruses accomplish this penetration through their association with

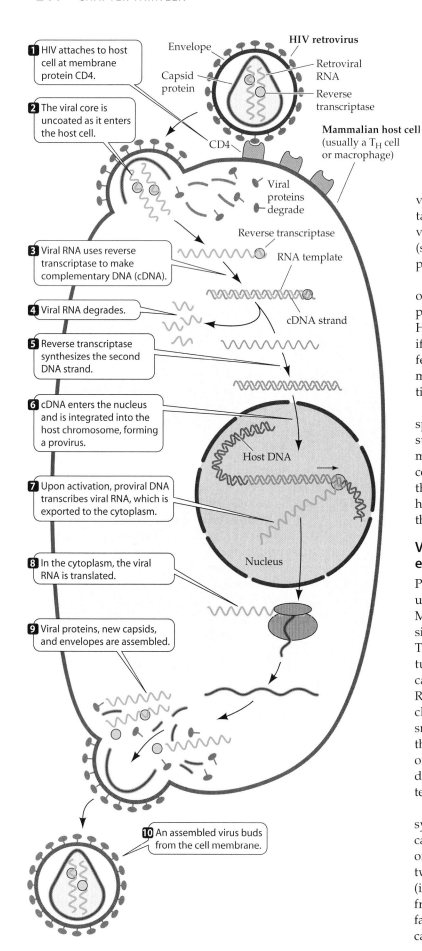

1 HIV attaches to host cell at membrane protein CD4.

2 The viral core is uncoated as it enters the host cell.

3 Viral RNA uses reverse transcriptase to make complementary DNA (cDNA).

4 Viral RNA degrades.

5 Reverse transcriptase synthesizes the second DNA strand.

6 cDNA enters the nucleus and is integrated into the host chromosome, forming a provirus.

7 Upon activation, proviral DNA transcribes viral RNA, which is exported to the cytoplasm.

8 In the cytoplasm, the viral RNA is translated.

9 Viral proteins, new capsids, and envelopes are assembled.

10 An assembled virus buds from the cell membrane.

Envelope

HIV retrovirus

Capsid protein

Retroviral RNA

Reverse transcriptase

CD4

Mammalian host cell (usually a T_H cell or macrophage)

Viral proteins degrade

Reverse transcriptase

RNA template

cDNA strand

Host DNA

Nucleus

13.5 The Reproductive Cycle of HIV

The retrovirus HIV enters a host cell via fusion of its membranes with the host's plasma membrane. Reverse transcription of retroviral RNA then produces a DNA provirus—a strand of complementary DNA that enters the host nucleus, where it transcribes viral RNA.

vectors. Infection of a plant usually results from attack by a virion-laden insect vector. When an insect vector penetrates a cell wall with its proboscis (snout), the virions can move from the insect into the plant.

Plant viruses can be introduced artificially, without insect vectors, by bruising a leaf or other plant part, then exposing it to a suspension of virions. Horizontal viral infections may also occur in nature if a bruised infected plant contacts an injured uninfected one. Vertical transmission of viral infections may occur through vegetative or sexual reproduction.

Once inside a plant cell, the virus reproduces and spreads to other cells in the plant. Within an organ such as a leaf, the virus spreads through the plasmodesmata, the cytoplasmic connections between cells. Because the viruses are too large to go through these channels, special proteins bind to them and help change their shape so that they can squeeze through the pores.

Viroids are infectious agents consisting entirely of RNA

Pure viral nucleic acids can produce viral infections under laboratory conditions, although inefficiently. Might there be infectious agents in nature that consist of nucleic acid without a protein capsid? In 1971, Theodore Diener of the U.S. Department of Agriculture reported the isolation of agents of this type, called viroids. **Viroids** are circular, single-stranded RNA molecules consisting of a few hundred nucleotides. They are one-thousandth the size of the smallest viruses. These RNA's are most abundant in the nuclei of infected cells. Viroids have been found only in plants, in which they produce a variety of diseases. Like plant viruses, viroids can be transmitted horizontally or vertically.

There is no evidence that viroids are translated to synthesize proteins, and it is not known how they cause disease. Viroids are replicated by the enzymes of their plant hosts. Similarities in base sequences between viroids and certain nontranslated sequences (introns) of plant genes suggest that viroids evolved from introns. This conclusion is supported by the fact that viroids, although made of RNA, are catalytically active in the way that some introns are.

Prokaryotes: Reproduction, Mutation, and Recombination

In contrast to viruses, bacteria and archaea are living cells. Prokaryotes carry out all the functions required for their own reproduction. They harvest and use energy, and they produce and use the molecular equipment that synthesizes their components and replicates their genes.

Prokaryotes usually reproduce asexually, but nonetheless have several ways of recombining their genes. Whereas in eukaryotes, genetic recombination occurs between the genomes of two parents, in prokaryotes it results from the interaction of the genome of one cell with a much smaller sample of genes from another cell.

The reproduction of prokaryotes gives rise to clones

Most prokaryotes reproduce by the division of single cells into two identical offspring (see Figure 9.3). In this way, a single cell gives rise to a clone—a population of genetically identical individuals. Prokaryotes reproduce very rapidly. A population of *E. coli*, for example, can double every 20 minutes, as long as conditions remain favorable.

Pure cultures of *E. coli* or other bacteria can be grown on the surface of a solid medium that contain a sugar, minerals, a nitrogen source such as ammonium chloride (NH_4Cl), and a solidifying agent such as agar (Figure 13.6). If the number of cells spread on the medium is small, each cell will give rise to a small, rapidly growing bacterial *colony*. If a large number of cells is poured onto the solid medium, their growth will produce one continuous colony—a bacterial "lawn." Bacteria can also be grown in a liquid nutrient medium. We'll see examples of all these techniques in this chapter.

Some bacteria conjugate, recombining their genes

The existence and heritability of mutations in bacteria attracted the attention of geneticists to these microbes. But if there were no form of exchange of genetic information between individuals, bacteria would not be useful for genetic analysis. Luckily, in 1946, Joshua Lederberg and Edward Tatum demonstrated that such exchanges do occur, although they are rare events.

Lederberg and Tatum grew two nutrient-requiring, or *auxotrophic*, strains of *E. coli*. Like the *Neurospora* in Figure 12.1, these strains will not grow on a minimal medium, but require supplementation with a nutrient that they cannot synthesize for themselves because of an enzyme defect. *E. coli* strain 1 requires the amino acid methionine and the vitamin biotin for growth, and its genotype is symbolized as *met⁻bio⁻*. Strain 2 requires neither of these substances, but cannot grow without the amino acids threonine and leucine. Considering all four factors, we say that strain 1 is *met⁻bio⁻thr⁺leu⁺* and strain 2 is *met⁺bio⁺thr⁻leu⁻*.

Lederberg and Tatum mixed these two mutant strains and cultured them together for several hours on a medium supplemented with methionine, biotin, threonine, and leucine, so that both strains could grow. The bacteria were then removed from the medium by centrifugation, washed, and transferred to minimal medium, which lacked all four supplements. Neither strain could grow on this medium because of their nutritional requirements. However, a few colonies *did* appear on the plates (Figure 13.7). Because they grew in the minimal medium, these colonies must have consisted of bacteria that were *met⁺bio⁺thr⁺leu⁺*; that is, they must have been *prototrophic*. These colonies appeared at a rate of approximately 1 for every 10 million cells originally put on the plates ($1/10^7$).

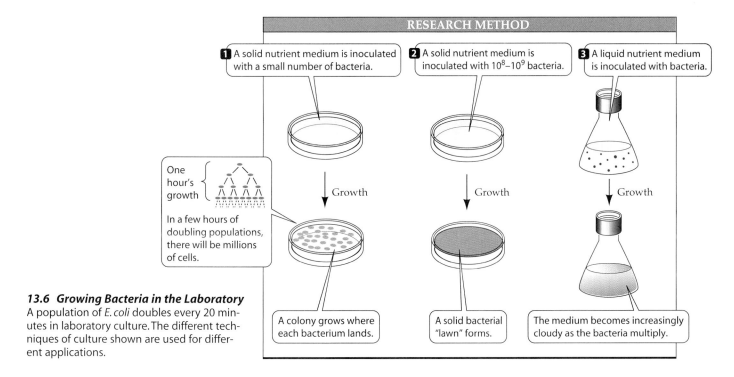

RESEARCH METHOD

1 A solid nutrient medium is inoculated with a small number of bacteria.

2 A solid nutrient medium is inoculated with 10^8–10^9 bacteria.

3 A liquid nutrient medium is inoculated with bacteria.

One hour's growth

In a few hours of doubling populations, there will be millions of cells.

Growth

Growth

Growth

A colony grows where each bacterium lands.

A solid bacterial "lawn" forms.

The medium becomes increasingly cloudy as the bacteria multiply.

13.6 Growing Bacteria in the Laboratory
A population of *E. coli* doubles every 20 minutes in laboratory culture. The different techniques of culture shown are used for different applications.

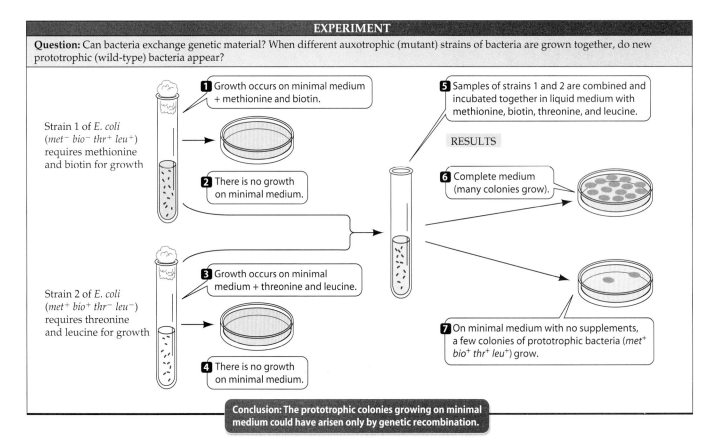

Question: Can bacteria exchange genetic material? When different auxotrophic (mutant) strains of bacteria are grown together, do new prototrophic (wild-type) bacteria appear?

Strain 1 of *E. coli* ($met^- bio^- thr^+ leu^+$) requires methionine and biotin for growth

1 Growth occurs on minimal medium + methionine and biotin.

2 There is no growth on minimal medium.

Strain 2 of *E. coli* ($met^+ bio^+ thr^- leu^-$) requires threonine and leucine for growth

3 Growth occurs on minimal medium + threonine and leucine.

4 There is no growth on minimal medium.

5 Samples of strains 1 and 2 are combined and incubated together in liquid medium with methionine, biotin, threonine, and leucine.

RESULTS

6 Complete medium (many colonies grow).

7 On minimal medium with no supplements, a few colonies of prototrophic bacteria ($met^+ bio^+ thr^+ leu^+$) grow.

Conclusion: The prototrophic colonies growing on minimal medium could have arisen only by genetic recombination.

13.7 Lederberg and Tatum's Experiment

After growing together, a mixture of complementary auxotrophic strains of *E. coli* contained a few cells that gave rise to new prototrophic colonies. This experiment proved that genetic recombination takes place in prokaryotes.

Where did these prototrophic colonies come from? Lederberg and Tatum were able to rule out mutation, and other investigators ruled out transformation. A third possibility is that the two strains of bacteria had exchanged genetic material, allowing it to mix and recombine to produce cells containing met^+ and bio^+ alleles from strain 2 and thr^+ and leu^+ alleles from strain 1 (see Figure 13.7). Later experiments showed that such an exchange, called **conjugation**, had indeed occurred. One bacterial cell—the *recipient*—had received DNA from the other cell—the *donor*—that included the two wild-type alleles that were missing in the recipient. Recombination then created a genotype with four wild-type alleles.

The physical contact required for conjugation can be observed under the electron microscope (Figure 13.8). It is initiated by a thin projection called a *pilus*. Then the actual transfer of DNA from one cell to another occurs by a thin conjugation tube. Since bacterial DNA is circular, it must be made linear (broken) so that it can pass through the tube. Contact between the cells is brief—certainly not long enough for all of the donor genome to enter the recipient cell. Therefore, the recipient cell usually receives only a portion of the donor DNA.

Once the donor fragment is inside the recipient cell, recombination can occur. In much the same way that chromosomes pair up, gene for gene, in prophase of meiosis, the donor DNA can line up beside its homologous gene in the recipient. Enzymes that can cut and rejoin DNA molecules are active in bacteria, and so gene(s) of the donor can end up integrated into the genome of the recipient, thus changing its genetic constitution (Figure 13.9).

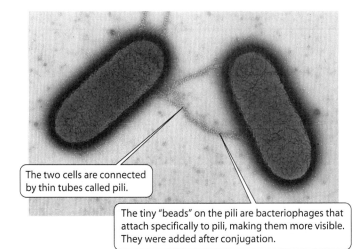

The two cells are connected by thin tubes called pili.

The tiny "beads" on the pili are bacteriophages that attach specifically to pili, making them more visible. They were added after conjugation.

13.8 Bacterial Conjugation

Pili draw two bacteria into close contact, and DNA is transferred from one cell to the other via a conjugation tube.

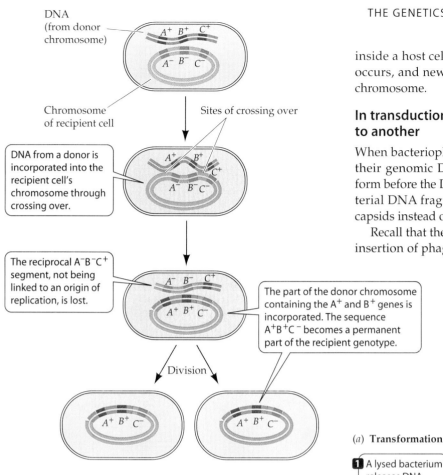

DNA (from donor chromosome)

Chromosome of recipient cell

Sites of crossing over

DNA from a donor is incorporated into the recipient cell's chromosome through crossing over.

The reciprocal A⁻B⁻C⁺ segment, not being linked to an origin of replication, is lost.

The part of the donor chromosome containing the A⁺ and B⁺ genes is incorporated. The sequence A⁺B⁺C⁻ becomes a permanent part of the recipient genotype.

Division

13.9 Recombination Following Conjugation
DNA from a donor cell may become incorporated into the recipient cell's chromosome through crossing over. Only about half the transferred genes become integrated in this way.

In transformation, cells pick up genes from their environment

Frederick Griffith obtained the first evidence for the transfer of prokaryotic genes more than 75 years ago when he discovered the transforming principle (see Figure 11.1). We now know the reason for his results: DNA had leaked from dead cells of virulent pneumococci and was taken up as free DNA by living nonvirulent pneumococci, which became virulent as a result. This phenomenon, called **transformation**, occurs in nature in some species of bacteria when cells die and their DNA leaks out (Figure 13.10a). Once transforming DNA is

inside a host cell, an event very similar to recombination occurs, and new genes can be incorporated into the host chromosome.

In transduction, viruses carry genes from one cell to another

When bacteriophages undergo a lytic cycle, they package their genomic DNA in capsids. These capsids generally form before the DNA is inserted into them. Sometimes, bacterial DNA fragments get inserted into the empty phage capsids instead of the phage DNA. (Figure 13.10b).

Recall that the binding of a phage to its host cell and the insertion of phage DNA are carried out by the capsid. So, when a phage capsid carries a piece of *bacterial* DNA, the latter is injected into the "infected" bacterium. This mechanism of DNA transfer is called **transduction**. Needless to say, it does not result in a productive viral infection. Instead, the incoming DNA fragment can recombine with the host chromosome.

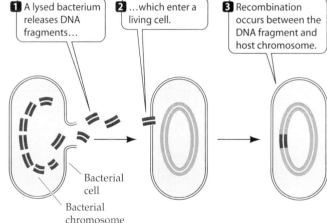

(a) **Transformation**

1 A lysed bacterium releases DNA fragments…

2 …which enter a living cell.

3 Recombination occurs between the DNA fragment and host chromosome.

Bacterial cell

Bacterial chromosome

13.10 Transformation and Transduction
After a new DNA fragment enters the host cell, recombination can occur. (a) Transforming DNA can leak from dead bacterial cells and be taken up by a living host bacterium, which may incorporate new genes into its chromosome. (b) In transduction, viruses carry DNA fragments from one cell to another.

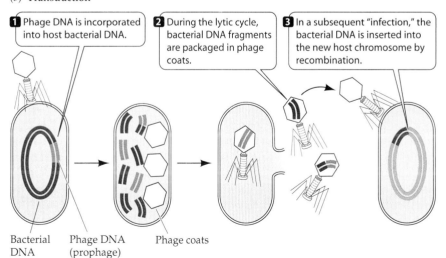

(b) **Transduction**

1 Phage DNA is incorporated into host bacterial DNA.

2 During the lytic cycle, bacterial DNA fragments are packaged in phage coats.

3 In a subsequent "infection," the bacterial DNA is inserted into the new host chromosome by recombination.

Bacterial DNA

Phage DNA (prophage)

Phage coats

Plasmids are extra chromosomes in bacteria

In addition to their main chromosome, many bacteria harbor additional smaller, circular chromosomes. These chromosomes, called **plasmids**, contain at most a few dozen genes, and, importantly, an origin sequence where DNA replication starts, which defines them as chromosomes. Usually, plasmids replicate at the same time as the host chromosome during the bacterial cell cycle, but this is not necessarily the case.

Plasmids are not viruses. They do not take over the cell's molecular machinery or make a protein coat to help them move from cell to cell. Instead, they can move between cells during conjugation, thereby adding some new genes to the recipient bacterium (Figure 13.11). Since plasmids exist independently of the main chromosome (the term *episomes* is sometimes used), they do not need to recombine with the main chromosome to add their genes to the cell's genome.

There are several types of plasmids, classified according to the kinds of genes they carry.

METABOLIC FACTORS CARRY GENES FOR UNUSUAL METABOLIC FUNCTIONS. Some plasmids, called *metabolic factors*, have genes that allow their recipients to carry out unusual metabolic functions. For example, there are many unusual hydrocarbons in oil spills. Some bacteria can actually thrive on these molecules, using them as a carbon source. The genes for the enzymes involved in these degradative pathways are carried on plasmids.

F FACTORS CARRY GENES FOR CONJUGATION. The "F" in F *factors* stands for fertility. Their approximately 25 genes include the ones that make both the pilus for attachment and the conjugation tube for DNA transfer to a recipient bacterium. A cell harboring the F factor is called F$^+$. It can transfer a copy of the F factor to an F$^-$ cell, making the recipient F$^+$. Sometimes the factor integrates into the main chromosome (at which point it is no longer a plasmid), and when it does, it can bring along some bacterial genes when it moves through the conjugation tube from one cell to another.

R FACTORS ARE RESISTANCE FACTORS. *R factors* may carry genes coding for proteins that destroy or modify antibiotics. Other R factors provide resistance to heavy metals that bacteria encounter in their environment.

R factors first came to the attention of biologists in 1957 during an epidemic in Japan, when it was discovered that some strains of the *Shigella* bacterium, which causes dysentery, were resistant to several antibiotics. Researchers found that resistance to the entire spectrum of antibiotics could be transferred by conjugation even when no genes on the main chromosome were transferred.

Eventually it was shown that the genes for antibiotic resistance are carried on plasmids. Each R factor carries one or more genes conferring resistance to particular antibiotics, as well as genes that code for proteins involved in the transfer of DNA to a recipient bacterium. As far as biologists can determine, R factors appeared long before antibi-

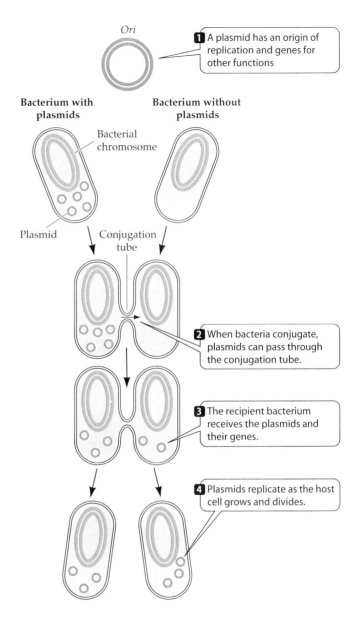

13.11 Gene Transfer by Plasmids
When plasmids enter a cell via conjugation, their genes can be expressed in the new cell.

otics were discovered, but they seem to have become more abundant in modern times, possibly because the heavy use of antibiotics in the hospital environment selects for bacterial strains bearing them.

R factors also pose a threat to people in the general clinical setting if antibiotics are used inappropriately. You probably have gone to see a physician because of a sore throat, which can have either a viral or a bacterial cause. The best way to know is for the doctor to take a small sample from your inflamed throat, culture it, and identify any bacteria that are present. But perhaps you cannot wait another day for the results. Impatient, you ask the doctor to give you something to make you feel better. She prescribes an antibiotic, which you take. The sore throat gradually gets better, and you think that the antibiotic did the job.

But suppose the infection was viral? In that case, the antibiotic did nothing to combat the disease, which just ran its normal course. However, it may have done something harmful: By killing many normal bacteria in your body, it may have exerted selection for bacteria harboring R factors. These bacteria may reproduce in the presence of the antibiotic, and may soon become quite numerous. The next time you got a *bacterial* infection, there would be a ready supply of R factors to be transferred to the invading bacteria, and antibiotics might be ineffective.

Transposable elements move genes among plasmids and chromosomes

As we have seen, plasmids, viruses, and even phage capsids (in the case of transduction) can transport genes from one bacterial cell to another. There is another type of "gene transport" that occurs *within* the individual cell. It relies on segments of chromosomal or plasmid DNA that can be inserted either at new locations on the same chromosome, or into another chromosome. These DNA sequences are called **transposable elements**. Their insertion often produces phenotypic effects by disrupting the genes into which they are inserted (Figure 13.12*a*).

The first transposable elements to be discovered in prokaryotes were large pieces of DNA, typically 1,000 to 2,000 base pairs long, found in many places in the *E. coli*

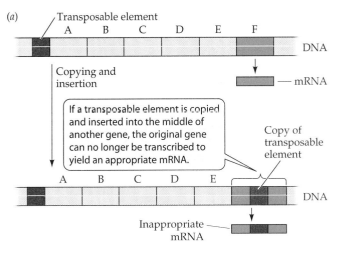

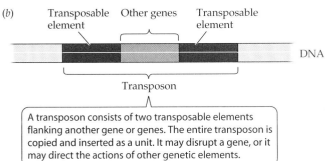

13.12 Transposable Elements and Transposons
(*a*) Transposable elements are segments of DNA that can be inserted at new locations, either on the same chromosome or on a different chromosome. (*b*) Transposons consist of transposable elements combined with other genes.

main chromosome. In one mechanism of transposition, the sequence of a transposable element replicates independently of the rest of the chromosome. The copy then inserts itself at other, seemingly random places in the chromosome. The genes encoding the enzymes necessary for this insertion are found within the transposable element itself. Some other transposable elements are cut from their original sites and inserted elsewhere without replication. Many transposable elements discovered later were longer (about 5,000 base pairs) and carried one or more additional genes with them. These longer elements with additional genes are called **transposons** (Figure 13.12*b*).

What do transposons and other transposable elements have to do with the genetics of prokaryotes—or with hospitals? Transposable elements have contributed to the evolution of plasmids. R plasmids probably originally gained their genes for antibiotic resistance through the activity of transposable elements. One piece of evidence for this conclusion is that each resistance gene in an R plasmid is part of a transposon.

Regulation of Gene Expression in Prokaryotes

Except for mutations, all cells of a bacterial species have the same DNA, and thus the capacity to make the same proteins. Yet the protein content of a bacterium can change rapidly when conditions warrant. For example, there are two ways for a bacterium to get the amino groups that it needs to make amino acids and proteins. One way involves taking N_2 from the air and "fixing" it into ammonia (NH_3), then using the ammonia as a source of amino groups. This reaction requires several enzymes and a lot of energy.

The other way to obtain amino groups is to take them from glutamine (see Table 3.2) and use them directly. This reaction requires only one enzyme and is not as endergonic. If there is a lot of glutamine around, the cell takes the easy way out, using the glutamine rather than the N_2 pathway. In fact, the enzymes that are involved in the N_2 pathway are not even made when glutamine is present.

There are several ways in which a prokaryotic cell could shut off the synthesis or activity of an unneeded protein:

▶ The cell could block the transcription of mRNA for that protein.
▶ The cell could hydrolyze the mRNA after it was made.
▶ The cell could prevent translation of the mRNA at the ribosome.
▶ The cell could hydrolyze the protein after it was made.

These methods would all have to be selective, responding to some biochemical signal. In the case of our two pathways for obtaining amino groups, the signal might be an increased concentration of glutamine.

Clearly, the earlier the cell intervenes in the process, the less energy it has to expend. Selective inhibition of transcription is far more efficient than transcribing the gene, translating the message, and then degrading the protein. While there are examples of all four methods of control of

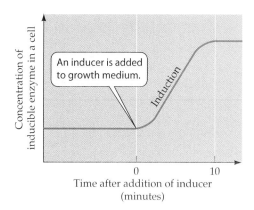

An inducer is added to growth medium.

Induction

Concentration of inducible enzyme in a cell

0 · 10

Time after addition of inducer (minutes)

13.13 An Inducer Stimulates the Synthesis of an Enzyme
It is most efficient for a cell to produce an enzyme only when it is needed. Some enzymes are induced by the presence of the substance they act upon (for example, β-galactosidase is induced by the presence of lactose).

protein levels in nature, prokaryotes generally use the most efficient one, transcriptional control.

Regulation of transcription conserves energy

As a normal inhabitant of the human intestine, *E. coli* has to adjust to sudden changes in its chemical environment. Its host may present it with one foodstuff one hour and another the next. This variability presents the bacterium with a metabolic challenge. Glucose is its preferred energy source, and is the easiest sugar to metabolize, but not all of its host's foods contain an abundant supply of glucose. For example, the bacteria may suddenly be deluged with milk, the main carbohydrate of which is the sugar lactose. Lactose is a β-galactoside—a disaccharide containing galactose β linked to glucose (see Chapter 3). Before lactose can be of any use to the bacteria, it must first be taken into the cells by a membrane transport carrier called β-galactoside permease. Then it must be hydrolyzed to glucose and galactose by the enzyme β-galactosidase. A third protein, the enzyme thiogalactoside transacetylase, is also required for lactose metabolism.

When *E. coli* is grown on a medium that does not contain lactose or other β-galactosides, the levels of all three of these enzymes within the bacterial cell are extremely low—the cell does not waste energy and materials making the unneeded proteins. If, however, the environment changes such that lactose is the predominant sugar and very little glucose is

present, the synthesis of all three of these enzymes begins promptly, and they increase rapidly in abundance. For example, there are only two molecules of β-galactosidase present in an *E. coli* cell when glucose is present in the medium. But when it is absent, lactose can induce the synthesis of 3,000 molecules of β-galactosidase per cell!

If lactose is removed from *E. coli*'s environment, synthesis of the three enzymes that process it stops almost immediately. The enzyme molecules that have already formed do not disappear; they are merely diluted during subsequent growth and reproduction until their concentration falls to the original low level within each bacterium.

Compounds that stimulate the synthesis of an enzyme (such as lactose in our example) are called **inducers** (Figure 13.13). The enzymes that are produced are called **inducible enzymes**, whereas enzymes that are made all the time at a constant rate are called **constitutive enzymes**.

We have now seen two basic ways to regulate the rates of metabolic pathways. Chapter 6 described allosteric regulation of enzyme *activity* (the rate of enzyme-catalyzed reactions); this mechanism allows rapid fine-tuning of metabolism. Regulation of protein synthesis—that is, regulation of the *concentration* of enzymes—is slower, but produces a greater savings of energy. Figure 13.14 compares these two modes of regulation.

A single promoter controls the transcription of adjacent genes

The genes that serve as blueprints for the synthesis of the three proteins that process lactose are called **structural genes**, indicating that they specify the primary structure (the amino acid sequence) of a protein molecule. In other words, structural genes are those that can be transcribed into mRNA. Three such genes are involved in the metabolism of lactose, and they lie adjacent to each other on the *E. coli* chromosome. This is no coincidence. Their DNA is transcribed into a single, continuous molecule of mRNA. Because this particular messenger governs the synthesis of all three lactose-metabolizing enzymes, either all or none of

13.14 Two Ways to Regulate a Metabolic Pathway
Feedback from the end product can block enzyme activity, or it can stop the transcription of genes that code for the enzyme.

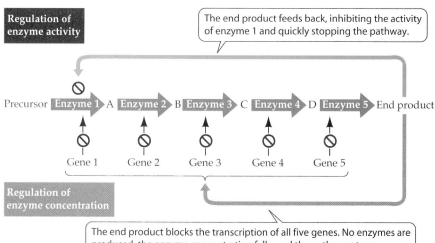

Regulation of enzyme activity

The end product feeds back, inhibiting the activity of enzyme 1 and quickly stopping the pathway.

Precursor ⊘ Enzyme 1 ▸ A ▸ Enzyme 2 ▸ B ▸ Enzyme 3 ▸ C ▸ Enzyme 4 ▸ D ▸ Enzyme 5 ▸ End product

⊘ Gene 1   ⊘ Gene 2   ⊘ Gene 3   ⊘ Gene 4   ⊘ Gene 5

Regulation of enzyme concentration

The end product blocks the transcription of all five genes. No enzymes are produced, the enzyme concentration falls, and the pathway stops.

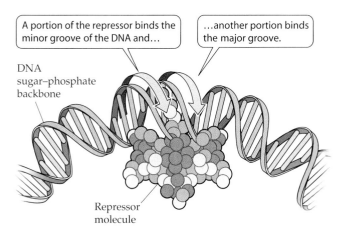

A portion of the repressor binds the minor groove of the DNA and...

...another portion binds the major groove.

DNA sugar–phosphate backbone

Repressor molecule

13.15 Repressor Bound to Operator Blocks Transcription
Portions of the repressor bind to the major and minor grooves in the DNA helix, preventing transcription.

the enzymes are made, depending on whether their common message—their mRNA—is present in the cell.

The three genes share a single promoter. Recall from Chapter 12 that a promoter is a site on DNA where RNA polymerase binds to initiate transcription. The promoter for these three structural genes is very efficient, so that the maximum rate of mRNA synthesis can be high, but there must also be a way to shut down mRNA synthesis when the enzymes are not needed.

Operons are units of transcription in prokaryotes

Prokaryotes shut down transcription by placing an obstacle between the promoter and its structural genes. A short stretch of DNA called the **operator** lies in this position. It can bind very tightly to a special type of protein molecule, called a **repressor**, to create such an obstacle. When the repressor protein is bound to the operator region of DNA, it blocks the transcription of mRNA (Figure 13.15). When the repressor is not attached to the operator, mRNA synthesis proceeds rapidly.

The whole unit, consisting of the closely linked structural genes and the stretches of DNA that control their transcription, is called an **operon** (Figure 13.16). An operon al-

ways consists of a promoter, an operator, and two or more structural genes. The promoter and operator are binding sites on DNA and are not transcribed.

E. coli has numerous ways to control the transcription of operons, and we will focus on three of them. Two ways depend on interactions of the repressor protein with the operator, and the third depends on interactions of other proteins with the promoter. Let's consider each of these three control systems in turn.

Operator–repressor control that induces transcription: The *lac* operon

The operon that controls and contains the structural genes for the three *lac*tose-metabolizing enzymes is called the *lac* operon (see Figure 13.16). As we have just learned, RNA polymerase can bind to the promoter, and a repressor can bind to the operator.

How is the operon controlled? The key lies in the repressor and its binding to the operator. The repressor protein has two binding sites: one for the operator and the other for inducers. The inducers of the *lac* operon, as we know, are molecules of lactose and certain other β-galactosides. Binding of an inducer changes the shape of the repressor (by allosteric modification; see Chapter 6). This change in shape prevents the repressor from binding to the operator (Figure 13.17). As a result, RNA polymerase can bind to the promoter and start transcribing the structural genes of the *lac* operon. The mRNA transcribed from these genes is translated on ribosomes, synthesizing the three proteins required for metabolizing lactose.

What happens if the concentration of lactose drops? As the lactose concentration decreases, the inducer (lactose) molecules separate from the repressor. Free of lactose molecules, the repressor returns to its original shape and binds to the operator, and transcription of the *lac* operon stops. Translation stops soon thereafter, because the mRNA that is already present breaks down quickly. Thus, it is the pres-

13.16 The lac Operon of E. coli and Its Regulator
The *lac* operon of *E. coli* is a segment of DNA that includes a promoter, an operator, and the three structural genes that code for the lactose-metabolizing enzymes.

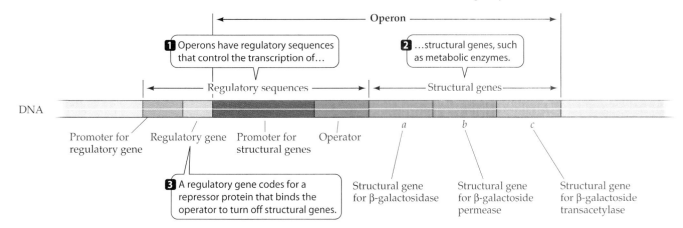

Operon

1 Operons have regulatory sequences that control the transcription of...

2 ...structural genes, such as metabolic enzymes.

Regulatory sequences

Structural genes

DNA

Promoter for regulatory gene

Regulatory gene

Promoter for structural genes

Operator

a

b

c

3 A regulatory gene codes for a repressor protein that binds the operator to turn off structural genes.

Structural gene for β-galactosidase

Structural gene for β-galactoside permease

Structural gene for β-galactoside transacetylase

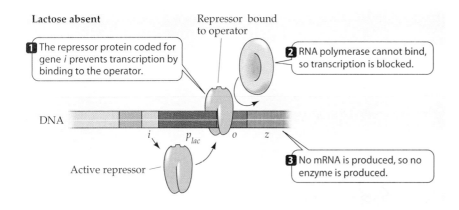

Lactose absent

Repressor bound to operator

1 The repressor protein coded for gene *i* prevents transcription by binding to the operator.

2 RNA polymerase cannot bind, so transcription is blocked.

DNA

i p_{lac} *o*   *z*

Active repressor

3 No mRNA is produced, so no enzyme is produced.

Lactose present

1 Lactose induces transcription by binding to the repressor, which cannot then bind to the operator.

Inducer (lactose)

RNA polymerase binds promoter

DNA

i p_{lac} *o*   *z*

Transcription proceeds

2 As long as the operator remains free of repressor, RNA polymerase can transcribe the genes for enzymes.

DNA

i p_{lac} *o*   *z*

3 mRNA transcript is produced.

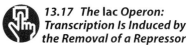

13.17 The lac Operon: Transcription Is Induced by the Removal of a Repressor
Lactose (the inducer) leads to enzyme synthesis by preventing the repressor protein (which would have stopped transcription) from binding to the operator.

▶ Control is exerted by a regulatory protein—the repressor—that turns the operon *off*.

▶ Some genes, such as *i*, produce proteins whose sole function is to regulate the expression of other genes.

▶ Certain other DNA sequences (operators and promoters) do not code for proteins but are binding sites for regulatory or other proteins.

▶ Adding inducer turns the operon *on*.

Operator–repressor control that represses transcription: The *trp* operon

We have seen that *E. coli* benefits from having an inducible system for lactose metabolism. Only when lactose is present does the system switch on. Equally valuable to a bacterium is the ability to switch *off* the synthesis of certain enzymes in response to the excessive accumulation of their end products. For example, if the amino acid tryptophan, an essential constituent of proteins, is present in ample concentration, it is advantageous to stop making the enzymes for tryptophan synthesis. When the synthesis of an enzyme can be turned off in response to such a biochemical cue, it is said to be **repressible**.

The French biochemist Jacques Monod, who had been part of the team that deciphered the *lac* operon, realized that repressible systems, such as the *trp* operon for *trp*tophan synthesis, could work by mechanisms similar to those of inducible systems. In repressible systems, the repressor protein cannot shut off its operon unless it first binds to a **corepressor**, which may be either the metabolite itself (tryptophan in this case) or an analog of it (Figure 13.18). If the metabolite is absent, the operon is transcribed at a maximum rate. If the metabolite is present, the operon is turned off.

The difference between inducible and repressible systems is small, but significant. In inducible systems, a substance in the environment (the inducer) interacts with the regulatory gene product (the repressor), rendering it *incapable* of binding to the operator and thus incapable of blocking transcription. In repressible systems, a substance in the cell (the corepressor) interacts with the regulatory gene product to make it *capable* of binding to the operator and

ence or absence of lactose—the inducer—that regulates the binding of the repressor to the operator, and therefore the synthesis of the proteins needed to metabolize it.

Repressor proteins are coded by **regulatory genes**. The regulatory gene that codes for the repressor of the *lac* operon is called the *i* (inducibility) gene. The *i* gene happens to lie close to the operon that it controls, but some other regulatory genes are distant from their operons. Like all other genes, the *i* gene itself has a promoter, which can be designated p_i. Because this promoter does not bind RNA polymerase very effectively, only enough mRNA to synthesize about ten molecules of repressor protein per cell per generation is produced. This quantity of the repressor is enough to regulate the operon effectively—to produce more would be a waste of energy. There is no operator between p_i and the *i* gene. Therefore, the repressor of the *lac* operon is constitutive; that is, it is made at a constant rate that is not subject to environmental control.

Let's review the important features of inducible systems such as the *lac* operon:

▶ In the absence of inducer, the *lac* operon is turned *off*.

Tryptophan absent

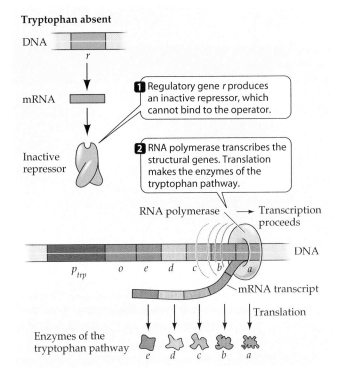

1 Regulatory gene *r* produces an inactive repressor, which cannot bind to the operator.

2 RNA polymerase transcribes the structural genes. Translation makes the enzymes of the tryptophan pathway.

13.18 The trp *Operon: Transcription Is Repressed by the Binding of a Repressor*
Because tryptophan activates an otherwise inactive repressor, it is called a corepressor.

Tryptophan present

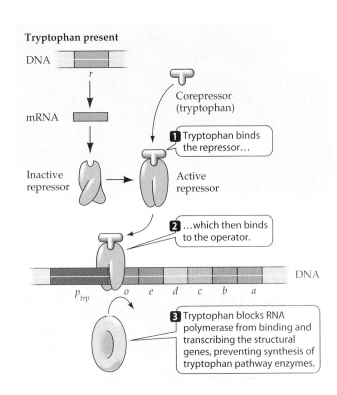

1 Tryptophan binds the repressor…

2 …which then binds to the operator.

3 Tryptophan blocks RNA polymerase from binding and transcribing the structural genes, preventing synthesis of tryptophan pathway enzymes.

blocking transcription. Although the effects of the substances are exactly opposite, the systems as a whole are strikingly similar.

In both the inducible lactose system and the repressible tryptophan system, the regulatory molecule functions by binding the operator. Next we'll consider an example of control by binding the *promoter*.

Protein synthesis can be controlled by increasing promoter efficiency

The mechanisms of transcriptional regulation that we have discussed thus far involve repressor–operator interactions that turn the operon on or off. Another way to regulate transcription is to make the promoter work more efficiently.

Suppose that a bacterial cell lacks a supply of glucose, its preferred energy source, but instead has access to another sugar (e.g., lactose or maltose) that can be broken down to enter an energy pathway. In operons such as the *lac* operon that have genes for enzymes that catabolize such alternative energy sources, the promoters bind RNA polymerase in a series of steps (Figure 13.19). First, a protein called **CRP** (short for

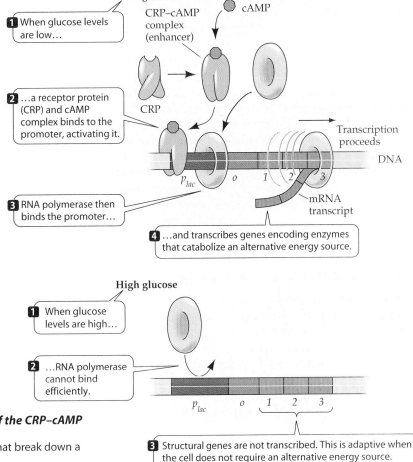

Low glucose

1 When glucose levels are low…

2 …a receptor protein (CRP) and cAMP complex binds to the promoter, activating it.

3 RNA polymerase then binds the promoter…

4 …and transcribes genes encoding enzymes that catabolize an alternative energy source.

High glucose

1 When glucose levels are high…

2 …RNA polymerase cannot bind efficiently.

3 Structural genes are not transcribed. This is adaptive when the cell does not require an alternative energy source.

13.19 Transcription Is Enhanced by the Binding of the CRP–cAMP Complex to the Promoter
The structural genes of this operon encode enzymes that break down a food source other than glucose.

13.2 The Relationships Between Positive and Negative Control in the lac Operon

GLUCOSE	cAMP LEVELS	RNA POLYMERASE BINDING TO PROMOTER	LACTOSE	LAC REPRESSOR	TRANSCRIPTION OF LAC GENES?	LACTOSE USED BY CELLS?
Present	Low	Absent	Absent	Active and bound to operator	No	No
Present	Low	Absent	Present	Inactive and not bound to operator	No	No
Absent	High	Present	Present	Inactive and not bound to operator	Yes	Yes
Absent	High	Absent	Absent	Active and bound to operator	No	No

cAMP receptor protein) binds the low-molecular-weight compound adenosine 3',5'-cyclic monophosphate, better known as **cyclic AMP**, or **cAMP**. Next, the CRP–cAMP complex binds to DNA just upstream of the promoter. This binding results in more efficient binding of RNA polymerase to the promoter, and an elevated level of transcription of the structural genes.

When glucose becomes abundant in the medium, the bacteria do not need to break down alternative food molecules, so the cell diminishes or ceases synthesizing the enzymes that catabolize these alternative sources. Glucose decreases the synthesis of these enzymes—a phenomenon called *catabolite repression*—by lowering the cellular concentration of cAMP.

As you will see in later chapters of this book, cAMP is a widely used signaling molecule in eukaryotes, as well as in prokaryotes. The use of this nucleotide in such widely diverse situations as a bacterium sensing glucose levels and humans sensing hunger demonstrates the prevalence of common themes in biochemistry and natural selection.

The *lac* and *trp* systems—the two operator–repressor systems—are examples of *negative* control of transcription because the regulatory molecule (the repressor) in each case *prevents* transcription. The promoter system is an example of *positive* control of transcription because the regulatory molecule (the CRP–cAMP complex) *enhances* transcription. The relationships between these positive and negative systems are summarized in Table 13.2.

Control of Transcription in Viruses

The mechanisms used by prokaryotes for the regulation of gene expression are also used by viruses. Even a "simple" biological agent such as a virus is faced with complicated molecular decisions when its genome enters a cell. For example, the viral genome must direct the shutdown of host transcription and translation, then redirect the host protein synthesis machinery to virus production. Genes must be activated in the right order; it makes little sense, for example, for the viral genome to transcribe and translate proteins that lyse the host cell membrane before the virus particles are assembled, ready for release. In temperate viruses,

which insert their genome (or a DNA copy) into the host chromosome, another issue arises: When should the provirus leave the host chromosome and undertake a lytic cycle?

Bacteriophage lambda is a temperate phage, which can undergo either a lytic or a lysogenic cycle (see Figure 13.2). When there is a rich medium and its host bacteria are growing, the phage takes advantage of its favorable environment (lots of resources for the phage in the host cell cytoplasm) and undergoes a lytic cycle. When the host bacteria are not as healthy, the phage senses this, and "lays low" as a lysogenic prophage. When things improve, the prophage leaves the host chromosome and becomes lytic.

The phage makes this decision by means of a "genetic switch": *Two* regulatory proteins compete for *two* operator/promoter sites on phage DNA. The two operators control the transcription of genes involved in the lytic and the lysogenic cycles, respectively, and the two regulatory proteins have *opposite* effects on the two operators (Figure 13.20):

PROTEIN	LYTIC OPERATOR/ PROMOTER	LYSOGENIC OPERATOR/PROMOTER
cI	Represses	Activates
Cro	Activates	Represses

So phage infection is a "race" between these two regulatory proteins. If cI "wins," which occurs when Cro synthesis is low in an unhealthy *E. coli* host cell, the phage enters a lysogenic cycle. If the host cell is healthy, a lot of Cro is made, lysogeny is blocked, and lysis ensues. These regulatory proteins, made very early in phage infection, both have binding domains for recognition of specific phage DNA sequences.

The life cycle of phage lambda, which has been greatly simplified here, is a paradigm for viral infections throughout the biological world. The lessons learned from transcriptional controls in this system have been applied again and again to other viruses, including HIV. The control of gene activity in eukaryotic cells is somewhat different, as we will see in the next chapter, but nevertheless usually involves protein–DNA interactions.

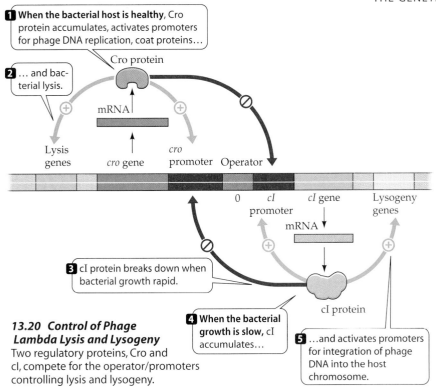

1 **When the bacterial host is healthy,** Cro protein accumulates, activates promoters for phage DNA replication, coat proteins…

2 … and bacterial lysis.

Cro protein

mRNA

Lysis genes

cro gene

cro promoter   Operator

0   *cI* promoter   *cI* gene   Lysogeny genes

mRNA

3 cI protein breaks down when bacterial growth rapid.

4 **When the bacterial growth is slow,** cI accumulates…

cI protein

5 …and activates promoters for integration of phage DNA into the host chromosome.

13.20 Control of Phage Lambda Lysis and Lysogeny
Two regulatory proteins, Cro and cI, compete for the operator/promoters controlling lysis and lysogeny.

Prokaryotic Genomes

When DNA sequencing first became possible in the late 1970s, the first biological agents to be sequenced were the simplest viruses.

Soon, over 150 viral genomes, including those of important animal and plant pathogens, were sequenced. Information on how they infected and reproduced came quickly as a result.

But manual sequencing was not up to the task of elucidating the genomes of prokaryotes and eukaryotes, the smallest of which are 100 times larger than those of a bacteriophage. In the past 6 years, however, the automation of sequencing has rapidly added many prokaryotic sequences to the biologists' store of knowledge.

In 1995, a team led by Craig Venter and Hamilton Smith determined the first sequence of a free-living organism, the bacterium *Haemophilus influenzae*. Many more sequences have followed, and they have revealed not only how these organisms apportion their genes to perform different cellular roles, but how their specialized functions are carried out. A beginning has even been made on the provocative question of what the minimal requirements for a living cell might be.

Three types of information can be obtained from a genomic sequence:

▶ *Open reading frames*, which are the coding regions of genes. For protein-coding genes, these regions can be recognized by the start and stop codons for translation.
▶ *Amino acid sequences*. For proteins, these can be deduced from the DNA sequence by looking up the genetic code.
▶ *Gene control sequences*, such as promoters and terminators for transcription.

Functional genomics relates gene sequences to functions

The only host for the bacterium *Haemophilus influenzae* is humans. It lives in the upper respiratory tract and can cause ear infections or, more seriously, meningitis in children. Its 1,830,137 base pairs are in a single circular chromosome (Figure 13.21). In addition to its origin of DNA replication and genes coding for rRNA's and tRNA's, this bacterial chromosome has 1,743 regions containing amino acid codons along with the transcriptional (promoter) and translational (start and stop codons) information needed for protein synthesis.

When this sequence was announced, only 1,007 (58%) of its genes had amino acid sequences that corresponded to proteins with known functions—in other words, were genes that researchers, based on their knowledge of the functions of bacteria, expected to find. Roles for most of the unknown proteins have been identified since then, a process known as *annotation*.

Functional genomics, the assignment of roles to genes and the description of how they work in the organism, has become the major occupation of many biologists.

Of the genes and proteins with known roles, most confirm a century of biochemical description of bacterial enzymatic pathways. For example, there are genes for the entire pathways of glycolysis, fermentation, and electron transport.

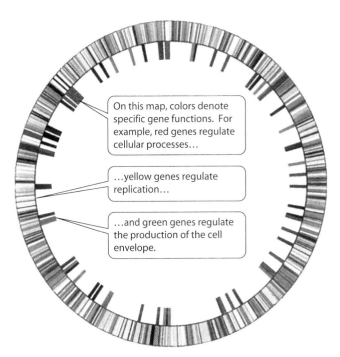

On this map, colors denote specific gene functions. For example, red genes regulate cellular processes…

…yellow genes regulate replication…

…and green genes regulate the production of the cell envelope.

13.21 Functional Organization of the Genome of H. influenzae
The entire DNA sequence has 1,830,137 base pairs.

Some of the gene sequences for unknown proteins may code for membrane proteins, possibly those involved in active transport. Another important finding is that highly infective strains of this bacterium have genes coding for surface proteins that attach them to the human respiratory tract, while noninfective strains lack those genes.

Soon after the sequence of *H. influenzae* was announced, smaller (*Mycoplasma genitalium*, 580,070 base pairs) and larger (*E. coli*, 4,639,221 base pairs) prokaryotic sequences were completed. Thus began a new era in biology, the era of **comparative genomics**, in which genome sequences are compared to see what genes one organism has or is missing, in order to relate the results to physiology.

M. genitalium, for example, lacks the enzymes needed to synthesize amino acids, which the other two organisms possess. This finding reveals that *M. genitalium* is a parasite, which must obtain all its amino acids from its environment, the human urogenital tract. *E. coli* has 55 genes coding for transcriptional activation and 58 repressors; *M. genitalium* has only 3 activators. Comparisons such as these have led to the formulation of specific questions about how an organism lives the way it does.

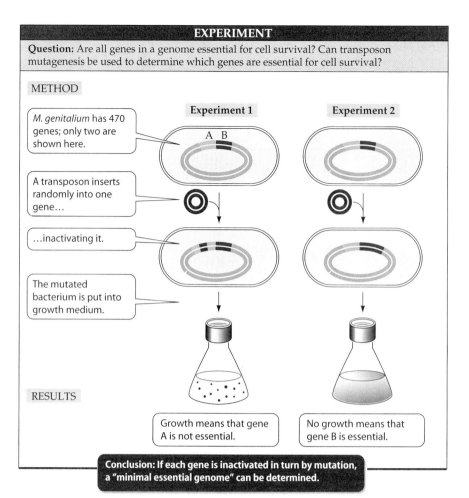

EXPERIMENT

Question: Are all genes in a genome essential for cell survival? Can transposon mutagenesis be used to determine which genes are essential for cell survival?

METHOD

Experiment 1   Experiment 2

M. genitalium has 470 genes; only two are shown here.

A transposon inserts randomly into one gene…

…inactivating it.

The mutated bacterium is put into growth medium.

RESULTS

Growth means that gene A is not essential.

No growth means that gene B is essential.

Conclusion: If each gene is inactivated in turn by mutation, a "minimal essential genome" can be determined.

13.22 Using Transposon Mutagenesis to Determine the Minimal Genome
By inactivating genes one by one, scientists can determine which ones are essential for the cell's survival.

The sequencing of prokaryotic genomes has medical applications

Sequencing has important ramifications for the study of prokaryotes that cause human diseases. Indeed, most of the early efforts in sequencing have focused on human pathogens.

▸ *Chlamydia trachomatis* causes the most common sexually transmitted disease in the United States. Because it is an intracellular parasite, it has been very hard to study. Among its 900 genes are ones for ATP synthesis—something scientists used to think this bacterium could not do:

▸ *Rickettsai prowazekii* causes typhus; it infects people bitten by lice vectors. Of its 634 genes, 6 code for proteins that are essential for its virulence. These genes are being used to develop vaccines.

▸ *Mycobacterium tuberculosis* causes tuberculosis. It has a large (for a prokaryote) genome, coding for 4,000 proteins. Over 250 of these are used to metabolize lipids, so this may be the main way that the bacterium gets its energy. Some genes coding for previously unidentified cell surface proteins are targets for potential vaccines. The sequence has enough similarities to those of mitochondria to lead to the proposal that this bacterium's ancestor was the one that colonized a cell to ultimately produce that organelle.

Sequencing has also provided the necessary information for the design of primers and hybridization probes used to detect these and other pathogens.

What genes are required for cellular life?

One striking conclusion arising from comparing the genomes of prokaryotes and eukaryotes is that there are some truly universal genes, present in all organisms. There are also some universal gene segments—coding for an ATP binding site, for example—that are present in many genes in many organisms. These findings suggest that there is some ancient, minimal set of DNA sequences that all cells must have. One way to identify these sequences is to look for them (or, more realistically, to have a computer look for them).

Another way to define a minimal genome is to take the simplest genome and deliberately mutate one gene at a time and see what happens. *M. genitalium*, with only 470 genes, has the smallest known genome of any organism. Some of its genes are dispensable under some circumstances: There are genes for using both glucose and fructose, and in the laboratory, the organism could survive on only one of those sugars, making the genes for the other unnecessary. But what about other genes? Experiments using transposons as mutagens have been performed to address this question. The transposons insert themselves into a gene at random, mutating and inactivating it (Figure 13.22). The resulting mutated cell is sequenced to determine which gene was mutated, and then examined for growth and survival.

The astonishing result of these studies is that *M. genitalium* can survive in the laboratory without the services of 133 of its genes, leaving a minimum of 337 genes! Putting it another way, these 337 genes could theoretically be spliced together (or even synthesized in the laboratory) to make totally artificial life. It is not surprising that the scientists involved in this research have convened a panel of theologians, philosophers and lawyers to advise them.

 Chapter Summary

Using Prokaryotes and Viruses to Probe the Nature of Genes

▶ Prokaryotes and viruses are useful for the study of genetics and molecular biology because they contain much less DNA than eukaryotes, they grow and reproduce rapidly, and they are haploid.

Viruses: Reproduction and Recombination

▶ Viruses were discovered as disease-causing agents small enough to pass through a filter that retains bacteria. They consist of a nucleic acid genome, which codes for a few proteins, and a protein coat. Some viruses also have a lipid membrane derived from host membranes.

▶ Viruses are obligate intracellular parasites, needing the biochemical machinery of living cells to reproduce.

▶ There are many types of viruses, classified by their size and shape, by their genetic material (RNA or DNA), or by their host organism. **Review Figure 13.1**

▶ Bacteriophages are viruses that infect bacteria. In the lytic cycle, the host cell breaks open, releasing many new phage particles. Some phages can also undergo a lysogenic cycle, in which their DNA is inserted into the host chromosome, where it replicates for generations. When conditions are appropriate, the lysogenic DNA exits the host chromosome and enters a lytic cycle. **Review Figure 13.2**

▶ Some viruses have promoters for host RNA polymerase, which they use to transcribe their own genes. **Review Figure 13.3**

▶ Most of the many types of RNA and DNA viruses that infect animals cause diseases. Some animal viruses are surrounded by membranes derived from the host's plasma membrane. Retroviruses, such as HIV, have RNA genomes that they reproduce through a DNA intermediate. Other RNA viruses use their RNA as mRNA to code for enzymes and replicate their genomes without using DNA. **Review Figures 13.4, 13.5, and Table 13.2**

▶ Many plant viruses are spread by other organisms, such as insects.

▶ Viroids are made only of RNA and infect plants, where they are replicated by the plant's enzymes.

Prokaryotes: Reproduction, Mutation, and Recombination

▶ When bacteria divide, they form clones of identical cells that can be observed as colonies when grown on solid media. **Review Figure 13.6**

▶ A bacterium can transfer its genes to another bacterium by conjugation, transformation, or transduction.

▶ In conjugation, a bacterium attaches to another bacterium and passes a partial copy of its DNA to the adjacent cell. **Review Figures 13.7, 13.8, 13.9**

▶ In transformation, genes are transferred between cells when fragments of bacterial DNA are taken up by a cell from the medium. These genetic fragments may recombine with the host chromosome, thereby permanently adding new genes. **Review Figure 13.10**

▶ In transduction, phage capsids carry bacterial DNA from one bacterium to another. **Review Figure 13.10**

▶ Plasmids are small bacterial chromosomes that are independent of the main chromosome. R factors, which are plasmids that carry genes for antibiotic resistance, are a serious public health threat. **Review Figure 13.11**

▶ Transposable elements are movable stretches of DNA that can jump from one place to another on the bacterial chromosome—either by actually moving or by making a new copy, which is inserted at a new location. **Review Figure 13.12**

Regulation of Gene Expression in Prokaryotes

▶ In prokaryotes, the expression of some genes is regulated; their products are made only when they are needed. Other genes, called constitutive genes, whose products are essential to the cell at all times, are constantly expressed. A compound that stimulates the synthesis of an enzyme needed to process it is called an inducer. **Review Figures 13.13, 13.14**

▶ An operon consists of a promoter, an operator, and a number of structural genes. Promoters and operators do not code for proteins, but serve as binding sites for regulatory proteins. When a repressor protein binds to the operator, transcription of the structural genes is inhibited. **Review Figures 13.15, 13.16**

▶ The expression of prokaryotic genes is regulated by three different mechanisms: inducible operator–repressor systems, repressible operator–repressor systems, and systems that increase the efficiency of a promoter.

▶ The *lac* operon is an example of an inducible system whose proteins allow bacteria to metabolize lactose. When lactose is absent, a repressor protein binds tightly to the operator. The repressor prevents RNA polymerase from binding to the promoter, turning transcription off. When glucose is absent and lactose is present, lactose acts as an inducer by binding to the repressor. This changes the repressor's shape so that it no longer recognizes the operator. With the operator unbound, RNA polymerase binds to the promoter, and transcription is turned on.

▶ Repressor proteins are coded by constitutive regulatory genes.

▶ The *trp* operon is a repressible system in which the presence of the end product of a biochemical pathway, tryptophan, represses the synthesis of enzymes involved in its own synthesis. Tryptophan acts as a corepressor by binding to an inactive repressor protein and making it active. When the activated repressor binds to the operator, transcription is turned off. **Review Figure 13.18**

▶ The efficiency of RNA polymerase can be increased by regulation of the level of cyclic AMP, which binds to a protein called CRP. The CRP–cAMP complex then binds to a site near the promoter of a target gene, enhancing the binding of RNA polymerase and hence transcription. **Review Figure 13.19**

Control of Transcription in Viruses

▶ In bacteriophages that can undergo a lytic or a lysogenic cycle, the decision as to which pathway to take is made by operator–regulatory protein interactions. **Review Figure 13.20**

Prokaryotic Genomes

▶ Functional genomics relates gene sequences to functions. **Review Figure 13.21**

▶ By mutating individual genes in a small genome, scientists can determine the minimal genome required for a prokaryote. **Review Figure 13.22**

For Discussion

1. Viruses sometimes carry DNA from one cell to another by transduction. Sometimes a segment of bacterial DNA is incorporated into a phage capsid without any phage DNA. These particles can infect a new host. Would the new host become lysogenic if the phage originally came from a lysogenic host? Why or why not?

2. Compare the life cycles of the viruses that cause influenza (Figure 13.4) and AIDS (Figure 13.5) with respect to how the virus enters the cell; how the virion is released into the cell; how the viral genome is replicated; and how new virus particles are produced.

3. Lederberg, Tatum, and colleagues were able to rule out new mutation and transformation as explanations for the prototrophic colonies that appeared when they mixed cultures of different auxotrophic *E. coli*. Propose experiments to rule out each of these alternatives.

4. Compare promoters adjacent to "early" and "late" genes in the viral life cycle.

5. The repressor protein that turns off the *lac* operon of *E. coli* is encoded by a regulatory gene. The repressor molecules are made in very small quantities and at a constant rate per cell. Would you surmise that the promoter for these repressor molecules is efficient or inefficient? Is synthesis of the repressor constitutive, or is it under environmental control?

6. A key characteristic of a repressible enzyme system is that the repressor molecule must react with a corepressor (typically, the end product of a metabolic pathway) before it can combine with the operator of an operon to shut the operon off. How is this different from an inducible enzyme system?

14 *The Eukaryotic Genome and Its Expression*

WHEN TOM WAS DIAGNOSED WITH LEUKEMIA—cancer of the blood cells—his initial treatment included chemotherapy. Combinations of powerful antimitotic drugs were administered to kill the rapidly dividing cancer cells that were spreading throughout his body. But the dosages his physicians prescribed were not up to the task, and the cells continued to spread. Higher dosages of these chemotherapeutic drugs would be lethal; they would kill not only the cancer cells, but the healthy and essential cells in the bone marrow that divide by the hundreds of millions to form blood cells. Without these cells, Tom's bone marrow would no longer produce red blood cells with their vital oxygen-carrying protein hemoglobin, nor would he be able to produce white blood cells, which make the proteins that combat infectious diseases as well as some tumors.

Tom's doctors tried a new approach. They extracted some of his bone marrow and removed the cancer cells from it, then stored the marrow in a refrigerator. Then they gave Tom extremely high doses of the chemotherapeutic drugs, which killed the cancer cells. Finally, the stored bone marrow was replaced in Tom's body. The healthy bone marrow cells began to divide, and after a few weeks they were forming populations of normal red and white blood cells. Tom's leukemia had disappeared.

The success of Tom's bone marrow transplant depended on many things, but the principle behind it is based on the specificity of gene expression during cell differentiation. What are the genetic mechanisms that ensure that healthy red blood cells will contain hemoglobin, and that white blood cells are able to create the vital antibody proteins of the immune system? What features of the DNA sequences of eukaryotic genes determine these mechanisms, and how do they differ from the genes that code for proteins in prokaryotes?

In this chapter, you will see that, although both prokaryotes and eukaryotes use DNA as genetic material, eukaryotic DNA differs from prokaryotic DNA in both its content and its organization. In addition to the genes for metabolism that prokaryotes have, eukaryotes have genes that mark them as complex cells: genes for addressing, or *targeting*, proteins to organelles, and genes for cell–cell interaction and cell differentiation.

Unlike prokaryotes, eukaryotes have repetitive sequences of DNA, many of which do not code for proteins. In addition, the transcription and later tailoring of mRNA is more complicated in eukaryotes than in prokaryotes. Elegant molecular machinery allows the precise regulation of gene expression needed for all the cells of these complex organisms to develop and function.

The Eukaryotic Genome

As biologists unraveled the intricacies of gene structure and expression in prokaryotes, they tried to generalize their findings by saying, "What's true for *E. coli* is also true for elephants." Although much of prokaryotic biochemistry does apply to eukaryotes, the old saying has its limitations. Table 14.1 lists some of the differences between prokaryotic and eukaryotic genomes.

The eukaryotic genome is larger and more complex than the prokaryotic genome

The fact that the genome of eukaryotes (in terms of haploid DNA content) is larger than that of prokaryotes might be

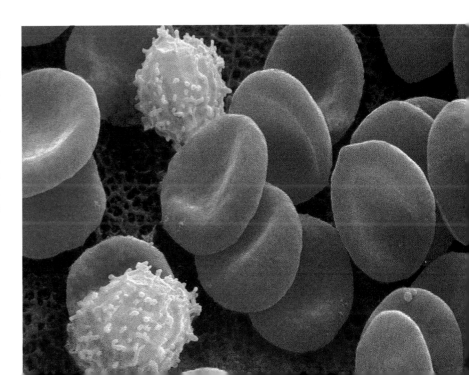

Two Cells, Two Different Protein Products
The red blood cells—erythrocytes—contain abundant hemoglobin, while the white blood cells synthesize proteins of the immune system.

14.1 A Comparison of Prokaryotic and Eukaryotic Genes and Genomes

CHARACTERISTIC	PROKARYOTES	EUKARYOTES
Genome size (base pairs)	10^4–10^7	10^8–10^{11}
Repeated sequences	Few	Many
Noncoding DNA within coding sequences	Rare	Common
Transcription and translation separated in cell	No	Yes
DNA segregated within a nucleus	No	Yes
DNA bound to proteins	Some	Extensive
Promoter	Yes	Yes
Enhancer/silencer	Rare	Common
Capping and tailing of mRNA	No	Yes
RNA splicing required	Rare	Common
Number of chromosomes in genome	One	Many

expected, given that in multicellular organisms there are many cell types, many jobs to do, and many proteins—all coded for by DNA—to do those jobs. A typical virus contains only enough DNA to code for a few proteins—about 10,000 base pairs (bp). The most thoroughly studied prokaryote, *E. coli*, has sufficient DNA (about 4.5 million bp) to make several thousand different proteins and regulate their synthesis. Humans have considerably more genes and regulators: Nearly 6 billion bp (2 meters of DNA) are crammed into each human cell. However, the idea of a more complex organism needing more DNA seems to break down with some plants. For example, the lily (which produces beautiful flowers each spring, but produces fewer proteins than a human does) has 18 times more DNA than humans have (Figure 14.1).

As we will see, the organization of the nuclear eukaryotic genome is fundamentally about regulation. The great complexity of eukaryotes requires a great deal of regulation, and this fact is evident in the many processes and points of control associated with the expression of the eukaryotic genome.

Unlike prokaryotic DNA, most eukaryotic DNA is noncoding. Interspersed throughout the eukaryotic genome are various kinds of repeated DNA sequences that are not transcribed into proteins. Even the coding regions of genes contain sequences that do not end up in mature mRNA.

Some of this noncoding DNA maintains structural integrity at the ends of chromosomes, and some regulates gene expression. But the presence of much of this noncoding DNA remains an enigma.

In contrast to the single main chromosome of most prokaryotes, the eukaryotic genome is partitioned into several separate chromosomes. In humans, each chromosome contains a double helix of DNA with 20 million to 100 million bp. This separation of genomic encyclopedia into multiple volumes requires that each chromosome have, at a minimum, three defining DNA sequences: *Recognition sequences* for the DNA replication machinery, a *centromere region* that holds the replicated sequences together before mitosis, and a *telomeric sequence* at each end of the chromosome. We have described the roles of the first two types of sequences in previous chapters, and will discuss telomeres later in this chapter.

In eukaryotes, the nuclear envelope separates DNA and its transcription (inside the nucleus) from the cytoplasmic sites where mRNA is translated into protein. This separation allows for many points of control in the synthesis, processing, and transport of mRNA to the cytoplasm.

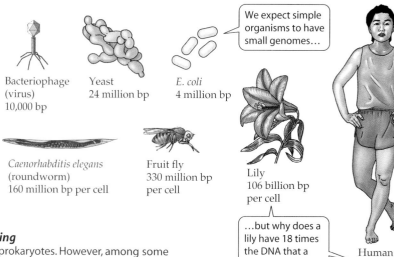

14.1 Amounts of Genomic DNA Can Be Deceiving
Eukaryotes have more DNA in their genomes than prokaryotes. However, among some eukaryotes—especially plants—there is no apparent relationship between diploid genome size and organism complexity.

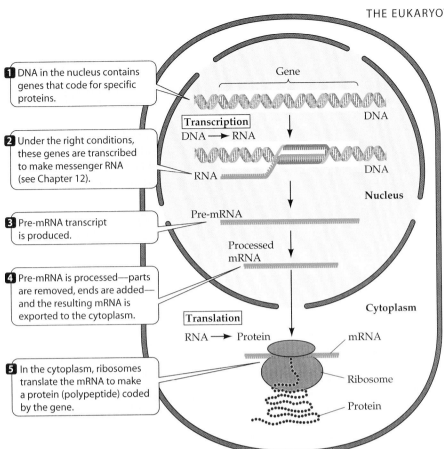

1 DNA in the nucleus contains genes that code for specific proteins.

2 Under the right conditions, these genes are transcribed to make messenger RNA (see Chapter 12).

3 Pre-mRNA transcript is produced.

4 Pre-mRNA is processed—parts are removed, ends are added—and the resulting mRNA is exported to the cytoplasm.

5 In the cytoplasm, ribosomes translate the mRNA to make a protein (polypeptide) coded by the gene.

Gene

Transcription
DNA → RNA

DNA

RNA

DNA

Nucleus

Pre-mRNA

Processed
mRNA

Cytoplasm

Translation
RNA → Protein

mRNA

Ribosome

Protein

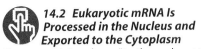

14.2 Eukaryotic mRNA Is Processed in the Nucleus and Exported to the Cytoplasm
Compare this "road map" to the prokaryotic one shown in Figure 12.3.

entire sequences of many prokaryotic genomes. A next step in size and complexity is the sequencing of simple eukaryotes. This has been achieved for a single-celled organism, yeast, as well as for the multicellular roundworm *Caenorhabditis elegans*. Further breakthroughs in the speed and sophistication of the equipment used to sequence DNA accelerated the work on complex eukaryotic genomes, and complete base sequencing of the fruit fly *Drosophila melanogaster* was completed in late 1999. And summer of the year 2000 saw the announcement, attended by frenzied media coverage, of the complete base sequencing of the human genome.

The yeast genome adds some eukaryotic functions onto a prokaryotic model

In comparison with *E. coli*, whose genome has about 4,500,000 bp on a single chromosome, the genome of budding yeast, *Saccharomyces cerevisiae*, has 16 chromosomes and a haploid content of more than 12,068,000 bp. More than 600 scientists around the world collaborated in mapping and sequencing the yeast genome. When they began, they knew of about 1,000 yeast genes coding for RNA or protein. The final sequence revealed 6,200 genes; sequence analyses have assigned probable roles to about 70 percent of them. The functions of the other 30 percent are being investigated by gene inactivation studies similar to those performed on prokaryotes (see Figure 13.22).

It is now possible to estimate what proportions of the yeast genome code for specific metabolic roles. Apparently, 11 percent of yeast proteins are for general metabolism, 3 percent for energy production and storage, 3 percent for DNA replication and repair, 12 percent for protein synthesis, and 6 percent for protein targeting and secretion. Many of the others are involved in cell structure, cell division, and the regulation of gene expression.

The most striking difference between the yeast genome and that of *E. coli* is in the genes for protein targeting (Table 14.2). Both of these single-celled organisms appear to use about the same number of genes to perform the basic functions of cell survival. It is the compartmentalization of the eukaryotic cell into organelles that requires it to have many more genes. This finding is direct, quantitative confirmation of something we have known for a century: The eukaryotic cell is structurally more complex than the prokaryotic cell.

Most eukaryotic DNA is not even fully exposed to the nuclear environment. Instead, it is extensively packaged by proteins into nucleosomes, chromatin fibers, and ultimately chromosomes (see Figure 9.7). This extensive compaction is a means of restricting access of the RNA synthesis machinery to the DNA, as well as a way of segregating replicated DNA's during mitosis and meiosis.

Like the genes of prokaryotes that code for proteins, eukaryotic genes are flanked by noncoding sequences that regulate their transcription. These include the promoter region, where RNA polymerase binds to begin transcription. Of equal importance in eukaryotes, but rare in prokaryotes, is a second set of regulatory DNA sequences, the enhancers and silencers. These sequences are often located quite far from the promoter, and they act by binding proteins that then stimulate or inhibit transcription.

The noncoding DNA sequences found within protein-coding genes present a special problem: How do cells ensure that transcripts of these noncoding regions do not end up in the mRNA that exits the nucleus? The answer lies in an elaborate cutting and splicing mechanism within the nucleus that modifies the initial transcript, called **pre-mRNA**, by cutting out the noncoding regions after transcription (Figure 14.2). Thus, in contrast to the "what is transcribed is what is translated" scheme of most prokaryotic genes, the mature mRNA that is translated at the eukaryotic ribosome is a modified and much smaller molecule than the one initially made in the nucleus.

As we described in Chapter 13, advances in the automation of DNA sequencing have made it possible to obtain the

14.2 Comparison of the Genomes of E. coli and Yeast

	E. COLI	YEAST
Genome length (base pairs)	4,640,000	12,068,000
Number of proteins	4,300	6,200
Proteins with roles in:		
Metabolism	650	650
Energy production/storage	240	175
Membrane transporters	280	250
DNA replication/repair/ recombination	120	175
Transcription	230	400
Translation	180	350
Protein targeting/secretion	35	430
Cell structure	180	250

The nematode genome adds developmental complexity

The presence of more than a single cell adds a new level of complexity to the genome. *Caenorhabditis elegans* is a 1–mm long nematode (roundworm) that normally lives in the soil. But it also lives in the laboratory, where it is a favorite organism of developmental biologists (see Chapter 16). It has a transparent body, through which scientists can watch over 3 days as its fertilized egg divides and forms an adult worm of 1,000 cells. In spite of its small number of cells, the adult worm has a nervous system, digests food, reproduces sexually, and ages. So it is not surprising that an intense effort was made to sequence the genome of this organism.

Just as with yeast and *E. coli*, the computer-based science of comparative genomics has given us much information on the *C. elegans* genome. It is eight times larger than that of yeast (97 million base pairs) and has four times more protein-coding genes (19,099). Once again, sequencing revealed far more than expected: When the sequencing effort began, researchers estimated that the worm would have about 6,000 proteins.

About 3,000 genes in the worm have direct homologs in yeast; these genes code for basic eukaryotic cell functions. What do the rest of the genes—the bulk of the worm genome—do? In addition to surviving, growing, and dividing, as single-celled organisms do, multicellular organisms must have genes for holding cells together to form tissues, for cell differentiation to divide up tasks in the organism, and for intercellular communication to coordinate its activities (Table 14.3). Many of the genes so far identified in *C. elegans* that are not present in yeast perform these roles, which will be described in the remainder of this chapter and the next one.

The fruit fly genome has surprisingly few genes

The fruit fly *Drosophila melanogaster* is a much larger organism than *C. elegans*, both in size (the fly has 10 times more cells) and complexity. Not surprisingly, the fly genome is also larger, about 180,000,000 base pairs. New technologies made it possible to sequence the entire *Drosophila* genome in about a year.

Even before the complete sequence was announced, decades of genetic studies had already identified some 2,500 different genes in the fly. These genes were all found in the complete DNA sequence, along with many other genes whose functions are as yet unidentified. Efforts are now under way to determine what these genes do in the life of the fly. (This process of discovering the protein product and function of a known gene sequence is called *annotation*.)

But the big surprise of the *Drosophila* genome sequence was the total number of protein-coding regions. Instead of being higher than the roundworm's (18,000 genes), the fly has only 13,600 genes. One reason for this is that the roundworm has some large *gene families*, which, as we will see later in this chapter, are groups of genes related in their sequence and function. For example, *C. elegans* has 1,100 genes involved in either nerve cell signaling or development; a fly has only 160 genes for these two functions. Another major expansion in the worm is in its genes coding for proteins that sense chemicals in its environment.

Many genes that are present in the worm genome have homologs with similar sequences in fly DNA, accounting for a third of the fly genes. And about half of the fly genes have mammalian homologs. An important medical contribution of comparative genomics has resulted from finding genes that are implicated in human diseases in other organisms. Often the roles of such genes can be elucidated in the simpler organism, providing a clue to how the gene might function in human disease. The fly genome contains 177 genes whose sequences are known to be directly involved in human diseases, including cancer and neurological conditions.

14.3 C. elegans Genes Essential to Multicellularity

FUNCTION	PROTEIN/DOMAIN	GENES
Transcription control	Zinc finger; homeobox	540
RNA processing	RNA binding domains	100
Nerve impulse transmission	Gated ion channels	80
Tissue formation	Collagens	170
Cell interactions	Extracellular domains; glycotransferases	330
Cell–cell signaling	G protein-linked receptors; protein kinases; protein phosphatases	1,290

Gene sequences for other organisms are rapidly becoming known

A "rough" human genome sequence is already available, with a more detailed one just a few years away. The human genome sequence and its myriad implications will be discussed in Chapter 18. Meanwhile, sequencing is proceeding rapidly for another model organism: the weedy plant *Arabidopsis thaliana* (130 million base pairs). These eukaryotic sequences will pose great challenges and opportunities for biologists in the next decades. "The sequence is not the end of the day," says Sydney Brenner, a leader of this effort. "It's the beginning of the day."

Repetitive Sequences in the Eukaryotic Genome

As we have mentioned, and as you have seen in the genome sequences we have examined, the eukaryotic genome has some base sequences that are repeated many times. Some of these sequences are present in millions of copies in a single genome. In this section, we will examine the organization and possible roles of these repetitive sequences.

Highly repetitive sequences are present in large numbers of copies

Three types of *highly repetitive sequences* are found in eukaryotes:

▸ *Satellites* are 5–50 base pairs long, repeated side by side up to a million times. For example, in guinea pigs, the satellite sequence is CCCTAA.* Satellites are usually present at the centromeres of chromosomes. Their role is not known.

▸ *Minisatellites* are 12–100 base pairs long and are repeated several thousand times. Because DNA polymerase tends to slip and make errors in copying these sequences, they are variable in the numbers of copies. For

*When a DNA sequence such as CCCTAA is written, the complementary bases on the other strand are assumed.

example, one person might have 300, and another, 500. This variation provides a set of molecular genetic markers for identifying an individual.

▸ *Microsatellites* are very short (1–5 base pairs) sequences, present in small clusters of 10–50 copies. They are scattered all over the genome, and have been used in human gene sequencing.

While laboratory scientists have made use of these sequences in genetic studies, their roles in eukaryotes are not clear.

Telomeres are repetitive sequences at the ends of chromosomes

There are several types of *moderately repetitive sequences* in the eukaryotic genome. One type is important in maintaining the ends of chromosomes when DNA is replicated. Recall from Chapter 11 that replication proceeds differently on the two strands of a DNA molecule. Both new strands form in the 5'-to-3' direction, but one strand (the leading strand) grows continuously from one end to the other, while the other (the lagging strand) grows as a series of short Okazaki fragments (see Figure 11.17).

In a eukaryotic chromosome, replication must begin with an RNA primer at the 5' end of the forming strand. The leading strand can grow without interruption to the very end, but on the lagging strand there is nothing beyond the primer in the 5' direction to replace the RNA. So the new chromosome formed after DNA replication lacks a bit of double-stranded DNA at each end. This situation signals DNA repair mechanisms in the cell, and the single-stranded regions, along with some of the intact double-stranded end, is cut off. In this way, the chromosome becomes shorter with each cell division.

In many eukaryotes, there are moderately repetitive sequences at the ends of chromosomes called **telomeres**. In humans, the sequence is TTAGGG, and it is repeated about 2,500 times (Figure 14.3a). These repeats bind special proteins that maintain the stability of chromosome ends. Otherwise, the DNA rapidly breaks down.

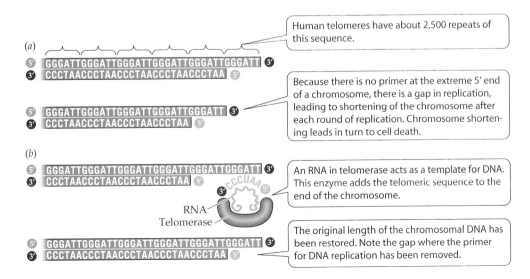

(a)

Human telomeres have about 2,500 repeats of this sequence.

5′ GGGATTGGGATTGGGATTGGGATTGGGATTGGGATT 3′
3′ CCCTAACCCTAACCCTAACCCTAA 5′

Because there is no primer at the extreme 5' end of a chromosome, there is a gap in replication, leading to shortening of the chromosome after each round of replication. Chromosome shortening leads in turn to cell death.

5′ GGGATTGGGATTGGGATTGGGATTGGGATT 3′
3′ CCCTAACCCTAACCCTAACCCTAA 5′

(b)

5′ GGGATTGGGATTGGGATTGGGATTGGGATTGGGATT 3′
3′ CCCTAACCCTAACCCTAACCCTAA 5′

RNA Telomerase

An RNA in telomerase acts as a template for DNA. This enzyme adds the telomeric sequence to the end of the chromosome.

5′ GGGATTGGGATTGGGATTGGGATTGGGATTGGGATT 3′
3′ CCCTAACCCTAACCCTAACCCTAACCCTAA 5′

The original length of the chromosomal DNA has been restored. Note the gap where the primer for DNA replication has been removed.

14.3 Telomeres and Telomerase
(a) The loss of moderately repetitive sequences from the telomere leads to cell death. (b) In cells that divide continuously (such as germ line cells), the enzyme telomerase prevents the loss of telomeric ends.

When human cells are removed from the body and put in a nutritious medium in the laboratory, they will grow and divide. But each chromosome can lose 50–200 bp of telomeric DNA after each round of DNA replication and cell division. This shortening compromises the stability of the chromosomes. After 20–30 divisions, chromosomes are unable to take part in cell division, and the cell dies. The same thing happens in the body, and explains in part why cells do not last the entire lifetime of the organism: Their telomeres shorten.

Yet constantly dividing cells, such as bone marrow cells and germ line cells, manage to maintain their moderately repetitive telomeric DNA. An enzyme, appropriately called *telomerase*, prevents the loss of this DNA by catalyzing the addition of any lost telomeric sequences (Figure 14.3b). Telomerase is made up not only of proteins, but also of an RNA sequence that acts as a template for the telomeric sequence addition.

Considerable interest has been generated by the finding that telomerase is expressed in more than 90 percent of human cancers. Telomerase may be an important factor in the ability of cancer cells to divide continuously. Since most normal cells do not have this ability, telomerase is an attractive target for drugs designed to attack tumors specifically.

There is also interest in telomerase and aging. When a gene expressing high levels of telomerase is added to human cells in culture, their telomeres do not shorten, and instead of dying after 20–30 cell generations, the cells become immortal. It remains to be seen how this finding relates to the aging of a large organism.

Some moderately repetitive sequences are transcribed

Some moderately repetitive DNA sequences code for tRNA's and rRNA's, which are used in protein synthesis (see Chapter 12). These RNA's are constantly being made, but even at the maximum rate of transcription, single copies of these sequences would be inadequate to supply the large amounts of these molecules needed by most cells; hence there are multiple copies of the DNA sequences coding for them. Since these moderately repetitive sequences are transcribed into RNA, they are properly termed "genes," and we can speak of rRNA genes and tRNA genes.

In mammals, there are four different rRNA molecules that make up the ribosome—the 18S, 5.8S, 28S, and 5S rRNA's.* The 18S, 5.8S, and 28S rRNA's are transcribed from a repeated sequence of DNA as a single precursor, which is twice the size of the three ultimate products (Figure 14.4). Several posttranscriptional steps cut this precursor into its final three rRNA's and discard the nonuseful, or "spacer," RNA. The DNA coding for these RNA's is moder-

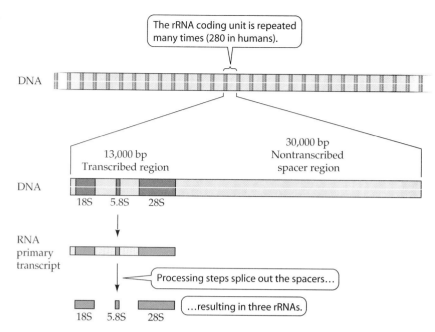

14.4 A Moderately Repetitive Sequence Codes for rRNA
This rRNA gene, along with its nontranscribed spacer, is repeated 280 times in the human genome.

ately repetitive in humans: A total of 280 copies of the sequence are located in clusters on five different chromosomes.

Other moderately repetitive sequences in mammals are not clustered, but instead are scattered throughout the genome. These DNA's usually are not transcribed and usually are short, about 300 bp long. In humans, half of these DNA's are of a single type, called the *Alu* family (because they contain a sequence that is recognized by a nuclease enzyme, Alu I). There are 300,000 copies of the *Alu* family in the genome, and they may act as multiple origins for DNA replication.

Transposable elements move about the genome

Most of the remaining scattered moderately repetitive DNA is not stably integrated into the genome. Instead, these DNA sequences can move from place to place in the genome. Such sequences are called **transposable elements**, or **transposons**.

There are four main types of transposable elements in eukaryotes:

▶ *SINEs* (short *in*terspersed *e*lements) are up to 500 bp long and are transcribed, but not translated.

▶ *LINEs* (long *in*terspersed *e*lements) are up to 7,000 bp long, and some are transcribed and translated into proteins. They constitute about 15 percent of the human genome.

Both of these elements are present in more than 100,000 copies. They move about the genome in a distinctive way: They make an RNA copy, which acts as a template for the

*The measure "S" refers to the movement of a molecule in a centrifuge: In general, larger molecules have a higher S value.

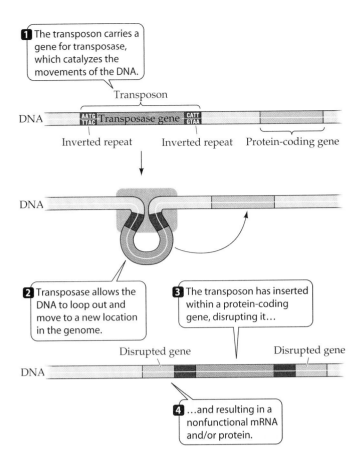

1 The transposon carries a gene for transposase, which catalyzes the movements of the DNA.

Transposon

DNA — AATG TTAC Transposase gene CATT GTAA

Inverted repeat   Inverted repeat   Protein-coding gene

DNA

2 Transposase allows the DNA to loop out and move to a new location in the genome.

3 The transposon has inserted within a protein-coding gene, disrupting it…

Disrupted gene   Disrupted gene

DNA

4 …and resulting in a nonfunctional mRNA and/or protein.

14.5 Transposons and Transposition
At the end of each transposable element is an inverted repeat sequence that helps in the transposition process.

new DNA, which then inserts itself at a new location in the genome.

▶ *Retrotransposons* also make an RNA copy when they move. They are rare in mammals, but are more common in other animals and yeasts. The genetic organization of *viral retrotransposons* resembles that of retroviruses such as HIV, but these segments lack the genes for protein coats and thus cannot produce viruses.

▶ *DNA transposons* are similar to their prokaryotic counterparts. They do not use an RNA intermediate, but actually move to a new spot in the genome without replicating (Figure 14.5).

What role do these moving sequences play in the cell? There are few answers to this question. The best answer so far seems to be that transposons are cellular parasites that simply replicate themselves. But these replications can lead to the insertion of a transposon at a new location, and this event has important consequences. For example, insertion of a transposon into the coding region of a gene causes a mutation because of the addition of new base pairs. This has been found in rare forms of several human genetic diseases, including hemophilia and muscular dystrophy.

If the insertion of a transposon takes place in the germ line, a gamete with a new mutation results. If the insertion takes place in a somatic cell, cancer may result. If a transposon replicates not just itself but also an adjacent gene, the result may be a gene duplication. A transposon can carry a

gene, or a part of it, to another location on a chromosome, shuffling genetic material and creating new genes. Clearly, transposition stirs the genetic pot in the eukaryotic genome and thus contributes to genetic variability.

In Chapter 4, we described the endosymbiosis theory of the origin of chloroplasts and mitochondria, which proposes that these organelles are the descendants of once free-living prokaryotes. Transposable elements may have played a role in this process. In living eukaryotes, although the organelles have some DNA, the nucleus contains most of the genes that encode the organelle proteins. If the organelles were once independent, they must originally have contained all of these genes. How did the genes move to the nucleus? The answer may lie in DNA transposition. Genes once in the organelle may have moved to the nucleus by well-known molecular events that still occur today. The DNA that remains in the organelles may be the remnants of more complete prokaryotic genomes.

The Structures of Protein-Coding Genes

Like their prokaryotic counterparts, many protein-coding genes in eukaryotes are single-copy DNA sequences. But eukaryotic genes have two distinctive characteristics that are uncommon among prokaryotes. First, they contain noncoding internal sequences, and second, they form gene families with structurally and functionally related cousins in the genome.

Protein-coding genes contain noncoding internal and flanking sequences

Preceding the coding region of a eukaryotic gene is a *promoter*, where RNA polymerase begins the transcription process. Unlike the prokaryotic enzyme, eukaryotic RNA polymerase does not recognize the promoter sequence by itself, but requires help from other molecules, as we'll see later. At the other end of the gene, after the coding region, is a DNA sequence appropriately called the *terminator*, which RNA polymerase recognizes as the end point for transcription (Figure 14.6). Neither the promoter nor the terminator sequence is transcribed into RNA.

Eukaryotic protein-coding genes also contain noncoding base sequences, called **introns**. One or more introns are interspersed with the coding regions—called **exons**—in most eukaryotic genes. Transcripts of the introns appear in the primary transcript of RNA—the pre-mRNA—within the nucleus, but by the time the mature mRNA exits the organelle, they have been removed. The transcripts of the introns are cut out of the pre-mRNA, and the transcripts of the exons are spliced together.

The locations of the introns can be determined by comparing the base sequences of a gene (DNA) with those of its final mRNA. Although direct sequencing of the DNA that codes for an mRNA is the easiest way to map the locations of introns within a gene, **nucleic acid hybridization** is the method that originally revealed the existence of introns in protein-coding genes. This method, outlined in

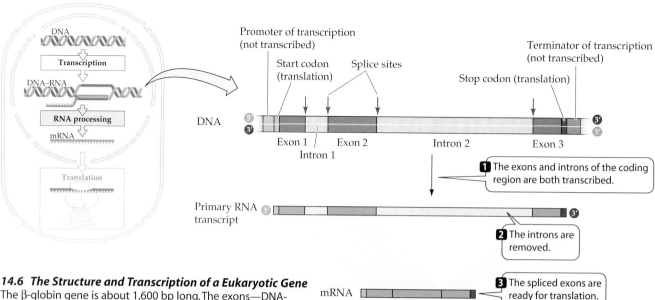

14.6 The Structure and Transcription of a Eukaryotic Gene
The β-globin gene is about 1,600 bp long. The exons—DNA-coding sequences— contain 441 base pairs (triplet codons for 146 amino acids plus a triplet stop codon). Noncoding sequences of DNA—introns—are initially transcribed between codons 30 and 31 (130 bp long) and 104 and 105 (850 bp long), but are spliced out of the final transcript.

Figure 14.7, has been crucial to genetic research; in later chapters we will see its use in localizing genes, testing for alleles, localizing mRNA's during development, and many other applications.

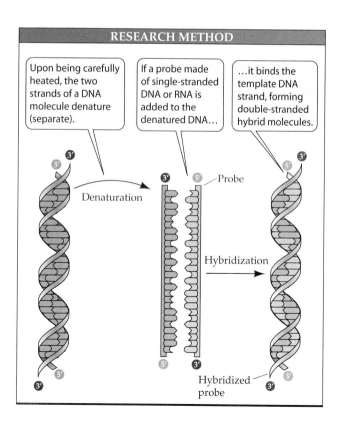

14.7 Nucleic Acid Hybridization
Base pairing permits the detection of a sequence complementary to the probe.

To examine the relationship between a gene and its transcript, biologists used nucleic hybridization to examine the gene for one of the globin proteins that make up hemoglobin (Figure 14.8). They first denatured the globin DNA by heating it, then added mature globin mRNA. As expected, the mRNA bound to the DNA by complementary base pairing. The researchers expected to obtain a linear matchup of the mRNA to the globin-coding DNA. They got their wish, in part: There were indeed stretches of RNA–DNA hybridization. But some looped structures were also visible. These loops were the introns, stretches of DNA that did not have complementary bases on the mRNA. Later studies showed that hybridization to the gene using pre-mRNA was complete, and that the introns were indeed transcribed. Somewhere on the path from transcript to mature mRNA, the introns had been removed, and the exons had been spliced together. We will examine this splicing process later in the chapter.

Most (but not all) vertebrate genes contain introns, as do many other eukaryotic genes (and even a few prokaryotic ones). Introns interrupt, but do not scramble, the DNA sequence that codes for a polypeptide chain. The base sequence of the exons, taken in order, is exactly complementary to that of the mature mRNA product. The introns, therefore, separate a gene's protein-coding region into distinct parts—the exons. In some cases, the separated exons code for different functional regions, or *domains*, of the protein. For example, the globin proteins that make up hemoglobin have two domains: one for binding to heme, and another for binding to the other globin chains. These two domains are coded for by different exons in the globin gene.

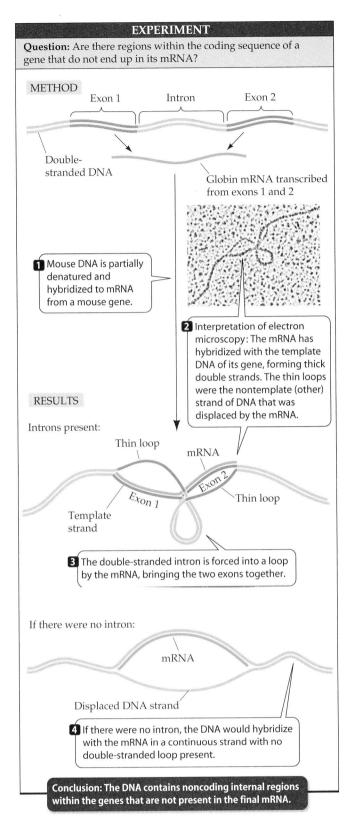

EXPERIMENT

Question: Are there regions within the coding sequence of a gene that do not end up in its mRNA?

METHOD

Exon 1 Intron Exon 2

Double-stranded DNA

Globin mRNA transcribed from exons 1 and 2

1 Mouse DNA is partially denatured and hybridized to mRNA from a mouse gene.

2 Interpretation of electron microscopy: The mRNA has hybridized with the template DNA of its gene, forming thick double strands. The thin loops were the nontemplate (other) strand of DNA that was displaced by the mRNA.

RESULTS

Introns present:

Thin loop

mRNA

Exon 1

Exon 2

Thin loop

Template strand

3 The double-stranded intron is forced into a loop by the mRNA, bringing the two exons together.

If there were no intron:

mRNA

Displaced DNA strand

4 If there were no intron, the DNA would hybridize with the mRNA in a continuous strand with no double-stranded loop present.

Conclusion: The DNA contains noncoding internal regions within the genes that are not present in the final mRNA.

Many eukaryotic genes are members of gene families

About half of all eukaryotic protein-coding genes are present in only one copy in the haploid genome. The rest have multiple copies. Often, inexact, nonfunctional copies of a particular gene, called **pseudogenes**, are located near it on a chromosome. These duplicates may have arisen by an ab-

14.8 Hybridization Revealed Noncoding DNA

When an mRNA transcript was experimentally hybridized to the double-stranded DNA of a gene, the introns from the DNA "looped out," demonstrating that the coding region of a eukaryotic gene can contain noncoding DNA that is not present in the mRNA transcript.

normal event in chromosomal crossing over during meiosis or by the action of retrotransposons.

In other cases, however, the genome contains slightly altered copies of a gene that are functional. A set of duplicated or related genes is called a **gene family**. Some families, such as the β-globins that are part of hemoglobin, contain only a few members; other families, such as the immunoglobulins that make up antibodies, have hundreds of members.

Like the members of any family, the DNA sequences in a gene family are usually different from one another to a certain extent. As long as one member retains the original DNA sequence and thus codes for the proper protein, the other members can mutate slightly, extensively, or not at all. The availability of such extra genes is important for "experiments" in evolution: If the mutated gene is useful, it may be selected for in succeeding generations. If the gene is a total loss (a pseudogene), the functional copy is still there to save the day.

The gene family for the *globins* is a good example of the gene families found in vertebrates. These proteins are found in hemoglobin, as well as in myoglobin (an oxygen-binding protein present in muscle). The globin genes probably all arose from a single common ancestor gene long ago. In humans, there are three functional members of the alpha-globin (α-globin) cluster and five in the beta-globin (β-globin) cluster (Figure 14.9). In a human adult, each hemoglobin molecule is a tetramer containing the heme pigments (each held inside a globin polypeptide), two identical α-globins, and two identical β-globins.

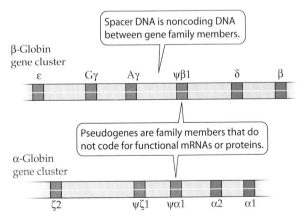

Spacer DNA is noncoding DNA between gene family members.

β-Globin gene cluster

ε Gγ Aγ ψβ1 δ β

Pseudogenes are family members that do not code for functional mRNAs or proteins.

α-Globin gene cluster

ζ2 ψζ1 ψα1 α2 α1

14.9 Gene Families

The human α-globin and β-globin gene clusters are located on different chromosomes. Each family is organized into a cluster of genes separated by noncoding "spacer" DNA. The nonfunctional pseudogenes are indicated by the Greek letter psi (ψ).

14.10 Differential Expression in the β-Globin Gene Family

During human development, different members of the β-globin gene family are expressed at different times and in different tissues.

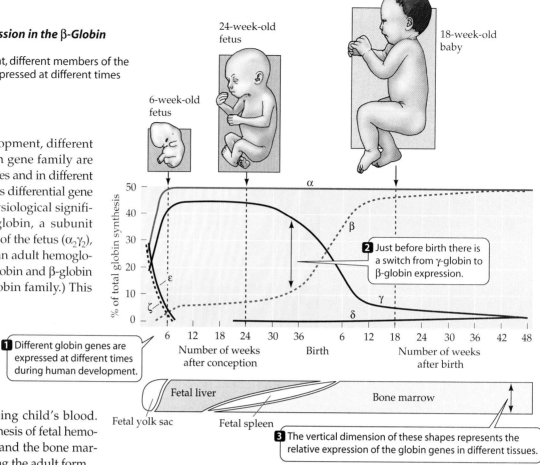

1 Different globin genes are expressed at different times during human development.

2 Just before birth there is a switch from γ-globin to β-globin expression.

3 The vertical dimension of these shapes represents the relative expression of the globin genes in different tissues.

6-week-old fetus

24-week-old fetus

18-week-old baby

During human development, different members of the β-globin gene family are expressed at different times and in different tissues (Figure 14.10). This differential gene expression has great physiological significance. For example, γ-globin, a subunit found in the hemoglobin of the fetus ($\alpha_2\gamma_2$), binds O_2 more tightly than adult hemoglobin ($\alpha_2\beta_2$) does. (Both γ-globin and β-globin are members of the β-globin family.) This specialized form of hemoglobin ensures that in the placenta, where the maternal and fetal circulation come near each other, O_2 will be transferred from the mother's to the developing child's blood. Just before birth, the synthesis of fetal hemoglobin in the liver stops, and the bone marrow cells take over, making the adult form.

In addition to genes that encode proteins, the globin family includes nonfunctional pseudogenes, designated with the Greek letter psi (φ). These pseudogenes are the "black sheep" of any gene family: They result from mutations that cause a *loss* of function rather than an enhanced or new function.

The DNA sequence of a pseudogene may not differ vastly from that of other family members. It may just lack a promoter, for example, and thus cannot be transcribed. Or it may lack the recognition sites for the removal of introns, and thus will be transcribed into pre-mRNA, but not correctly processed into a useful mRNA. In some gene families, pseudogenes outnumber functional genes. However,

since some members of the family are functional, there appears to be little selective pressure in evolution to eliminate pseudogenes.

RNA Processing

As we have seen, the primary RNA transcript (pre-mRNA) of a eukaryotic gene is not the same as the mature mRNA. To produce the mRNA, the primary transcript is processed by the addition of bases at both ends, and by the removal of introns and the joining of exons.

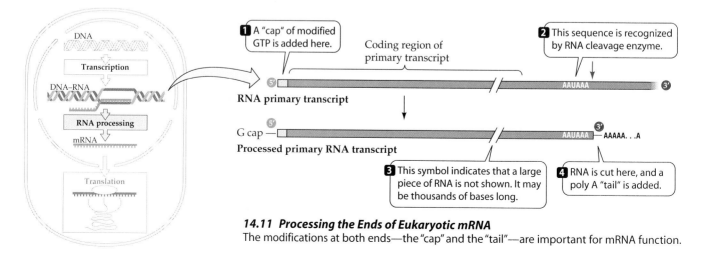

1 A "cap" of modified GTP is added here.

Coding region of primary transcript

2 This sequence is recognized by RNA cleavage enzyme.

RNA primary transcript

G cap — **Processed primary RNA transcript**

3 This symbol indicates that a large piece of RNA is not shown. It may be thousands of bases long.

4 RNA is cut here, and a poly A "tail" is added.

14.11 Processing the Ends of Eukaryotic mRNA

The modifications at both ends—the "cap" and the "tail"—are important for mRNA function.

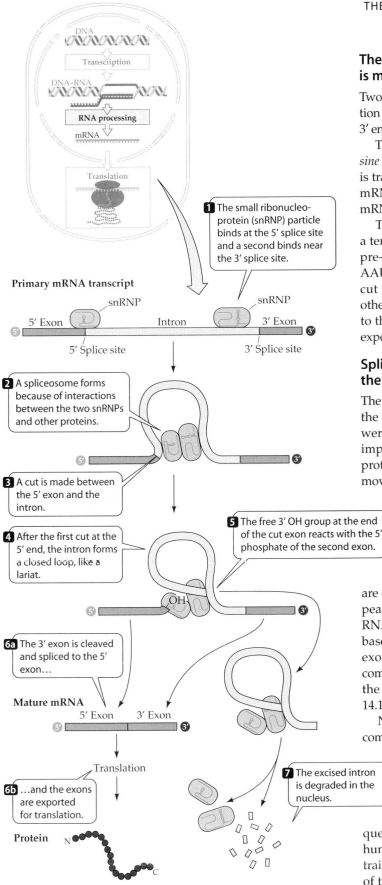

14.12 The Spliceosome, an RNA Splicing Machine
Binding of two snRNP's to consensus sequences on the pre-mRNA lines up the splicing machinery. After the snRNP's bind to pre-mRNA, other proteins join the complex to form a spliceosome.

The labels in the figure:

1 The small ribonucleo-protein (snRNP) particle binds at the 5' splice site and a second binds near the 3' splice site.

Primary mRNA transcript

snRNP — snRNP
5' Exon — Intron — 3' Exon
5' Splice site — 3' Splice site

2 A spliceosome forms because of interactions between the two snRNPs and other proteins.

3 A cut is made between the 5' exon and the intron.

4 After the first cut at the 5' end, the intron forms a closed loop, like a lariat.

5 The free 3' OH group at the end of the cut exon reacts with the 5' phosphate of the second exon.

6a The 3' exon is cleaved and spliced to the 5' exon…

Mature mRNA
5' Exon — 3' Exon

Translation

6b …and the exons are exported for translation.

Protein N ———— C

7 The excised intron is degraded in the nucleus.

The primary transcript of a protein-coding gene is modified at both ends

Two early steps in the processing of pre-mRNA are the addition of a "cap" at the 5' end and the addition of a "tail" at the 3' end (Figure 14.11).

The **G cap** is a chemically modified molecule of *guanosine triphosphate* (*GTP*). It is added to the 5' end as the RNA is transcribed. The cap apparently facilitates the binding of mRNA to the ribosome for translation and protects the mRNA from breaking down.

The **poly A tail** is added to the 3' end of pre-mRNA after a terminal sequence has been removed. Near the 3' end of pre-mRNA, and after the last codon, is the sequence AAUAAA. This sequence acts as a signal for an enzyme to cut the pre-mRNA. Immediately after this cleavage, another enzyme adds 100 to 300 residues of adenine (poly A) to the 3' end of the pre-mRNA. This tail may assist in the export of the mRNA from the nucleus.

Splicing removes introns from the primary transcript

The next step in processing of eukaryotic pre-mRNA within the nucleus is deleting the introns. If these RNA regions were not removed, a nonfunctional mRNA, producing an improper amino acid sequence and thus a nonfunctional protein, would result. The process called RNA **splicing** removes the introns and splices the exons together.

As soon as the pre-mRNA is transcribed, it is quickly bound to several **small nuclear ribonucleo-protein particles** (**snRNP's**, commonly pronounced "snurps"). There are several types of these RNA–protein particles in the nucleus.

At the boundaries between introns and exons are **consensus sequences**—short stretches of DNA that appear, with little variation, in many different genes. The RNA in one of the snRNP's (called U1) has a stretch of bases complementary to the consensus sequence at the 5' exon–intron boundary, and binds to the pre-mRNA by complementary base pairing. Another snRNP (U2) binds to the pre-mRNA near the 3' intron–exon boundary (Figure 14.12).

Next, other proteins bind, forming a large RNA–protein complex called a **spliceosome**. The spliceosome uses energy from ATP for its assembly. It cuts the RNA, releases the introns, and joins the ends of the exons together to produce mature mRNA.

Molecular studies of human diseases have been valuable tools in the investigation of consensus sequences and splicing machinery. Beta thalassemia is a human genetic disease inherited as an autosomal recessive trait. People with this disease make an inadequate amount of the β-globin subunit that is part of hemoglobin. These people suffer from severe anemia because they have an inadequate supply of red blood cells. In some cases, the genetic mutation that causes the disease occurs at a consensus sequence in the β-globin gene. Consequently, the pre-mRNA cannot be spliced correctly, and nonfunctional β-globin mRNA is made.

This finding is an excellent example of the use of mutations in determining a cause-and-effect relationship in biology. In the logic of science, merely linking two phenomena (for example, consensus sequences and splicing) does not prove that one is necessary for the other. In an experiment, the scientist alters one phenomenon (for example, the base sequence at the consensus region) to see whether the other event (for example, splicing) occurs. In beta thalassemia, nature has done the experiment for us.

After the processing events are completed in the nucleus, the mRNA exits the organelle, apparently through the nuclear pores (see Figure 4.9). A receptor at the nuclear pore recognizes the processed mRNA (or a protein bound to it). Unprocessed or incompletely processed pre-mRNA's remain in the nucleus.

Transcriptional Control

In a multicellular organism with specialized cells and tissues, every cell contains every gene in the organism's genome. For development to proceed normally, and for each cell to acquire and maintain its proper function, certain proteins must be synthesized at just the right times and in just the right cells. Thus, the expression of eukaryotic genes must precisely regulated.

Regulation of gene expression can occur at many points (Figure 14.13). This section describes the mechanisms that control the transcription of specific genes. These often involve nuclear proteins that alter chromosome function or structure. In some cases, the regulation of transcription involves changes in the DNA itself: Genes are selectively replicated to give more templates to transcribe, or even rearranged on the chromosome.

Posttranscriptional events can also regulate gene expression. As we have seen, the processing of pre-mRNA can be controlled after transcription. The transport of the mRNA into the cytoplasm, and how long it remains there, can also be controlled. The translation of mRNA into protein can also be regulated. Finally, once the protein itself is made, its structure can be modified, or it can be broken down and destroyed.

Specific genes can be selectively transcribed

The brain cells and the liver cells of a mouse have some proteins in common and some that are distinctive for each cell type. Yet both cells have the same DNA sequences and, therefore, the same genes. Are the differences in protein content due to *differential transcription* of genes? Or is it that all the genes are transcribed in both cell types, and a *posttranscriptional mechanism* is responsible for the differences in proteins?

These two alternatives—transcriptional or posttranscriptional control—can be distinguished by examination of the actual RNA sequences made within the nucleus of each cell type. Such analyses indicate that for some proteins, the

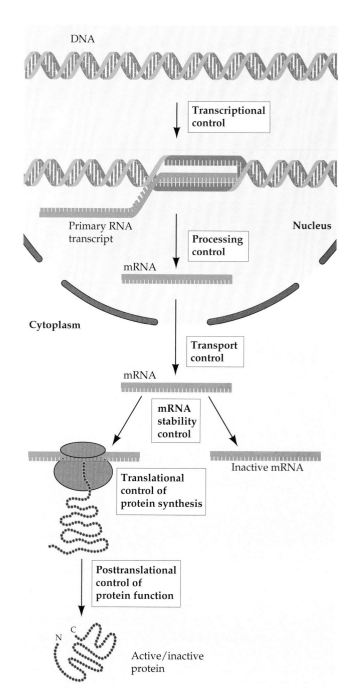

14.13 Potential Points for the Regulation of Gene Expression
Gene expression can be regulated at three levels: at transcription, at translation, or after translation.

mechanism of control is differential gene transcription. Both brain and liver cells, for example, transcribe "housekeeping" genes, such as those for glycolysis enzymes and ribosomal RNA's. But liver cells transcribe some genes for liver-specific proteins, and brain cells transcribe some genes for brain-specific proteins. And neither cell type transcribes the genes for proteins that are characteristic of muscle, blood, bone, and the other specialized cell types in the body.

CONTRASTING EUKARYOTES AND PROKARYOTES. Unlike prokaryotes, in which related genes are transcribed as a unit in operons, eukaryotes tend to have solitary genes. Thus, regulating several genes at once requires common control elements in each gene, which allow all of the genes to respond to the same signal.

In contrast to the single RNA polymerase in bacteria, eukaryotes have three different RNA polymerases. Each eukaryotic polymerase catalyzes the transcription of a specific type of gene. Only one (RNA polymerase II) transcribes protein-coding genes to mRNA. The other two transcribe the DNA that codes for rRNA (polymerase I) and for tRNA and small nuclear RNA's (polymerase III). The diversity of eukaryotic polymerases is reflected in the diversity of eukaryotic promoters, which tend to be much more variable than prokaryotic promoters. In addition, most eukaryotic genes have regulator, enhancer, and silencer elements (which we will discuss shortly) that can control the rate of transcription. Whether a eukaryotic gene is transcribed depends on the sum total of the effects of all of these DNA and protein elements; thus there are many points of possible control.

Finally, the transcription complex in eukaryotes is very different from that of prokaryotes, in which a single peptide subunit can cause RNA polymerase to recognize the promoter. In eukaryotes, many proteins are involved in initiating transcription. We will confine the following discussion to RNA polymerase II, which catalyzes the transcription of most protein-coding genes, but the mechanisms for the other two polymerases are similar.

TRANSCRIPTION FACTORS.   In prokaryotes, the promoter is a sequence of DNA near the 5′ end of the coding region of a gene where RNA polymerase begins transcription. A prokaryotic promoter has two essential regions. One is the *recognition sequence*—the sequence recognized by RNA polymerase. The second, closer to the initiation point, is the *TATA box* (so called because it is rich in AT base pairs), where DNA begins to denature so that its templates can be exposed. In eukaryotes, there is a TATA box about 25 bp away from the initiation site for transcription, and one or two recognition sequences of about 50 to 70 bp 5′ from the TATA box.

Eukaryotic RNA polymerase II cannot simply bind tightly to the promoter and initiate transcription. Rather, it binds and acts only after various regulatory proteins, called **transcription factors**, have assembled on the chromosome (Figure 14.14). First, the protein *TFIID* ("TF" stands for *transcription factor*) binds to the TATA box. Its binding changes both its own shape and that of the DNA, presenting a new surface that attracts the binding of other transcription factors. RNA polymerase II does not bind until several other proteins have already bound to this complex.

Some DNA sequences, such as the TATA box, are common to the promoters of many genes and are recognized by transcription factors that are found in all the cells of an organism. Other sequences in promoters are specific to only a few genes and are recognized by transcription factors found only in certain tissues. These specific transcription factors play an important role in *differentiation*, the specialization of cells during development.

REGULATORS, ENHANCERS, AND SILENCERS IN DNA.   In addition to the promoter, two other regions of DNA bind proteins that activate RNA polymerase. The recently discovered **regulator** regions are clustered just upstream of the promoter. Various regulator proteins (seven in the β-globin

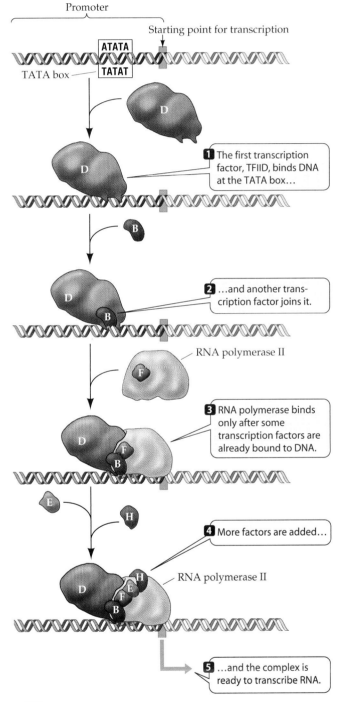

1 The first transcription factor, TFIID, binds DNA at the TATA box…

2 …and another transcription factor joins it.

RNA polymerase II

3 RNA polymerase binds only after some transcription factors are already bound to DNA.

4 More factors are added…

RNA polymerase II

5 …and the complex is ready to transcribe RNA.

14.14 The Initiation of Transcription in Eukaryotes
Except for TFIID, which also binds to the TATA box, each transcription factor has binding sites only for the other proteins and does not bind directly to DNA.

14.15 The Roles of Transcription Factors, Regulators, and Activators

The actions of many proteins determine whether and where RNA polymerase II will transcribe DNA.

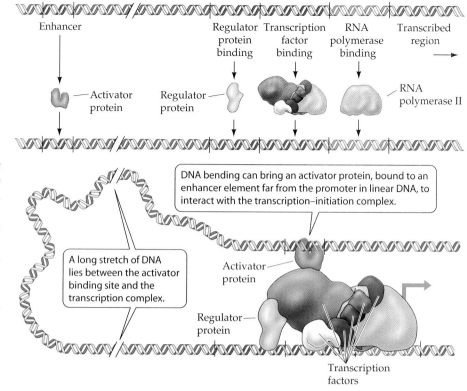

DNA bending can bring an activator protein, bound to an enhancer element far from the promoter in linear DNA, to interact with the transcription–initiation complex.

A long stretch of DNA lies between the activator binding site and the transcription complex.

gene) may bind to these regions (Figure 14.15). Their net effect is to bind to the adjacent transcription complex and activate it.

Much farther away—up to 20,000 bp away—are the **enhancer** regions. Enhancer regions bind *activator* proteins, and this binding strongly stimulates the transcription complex. How enhancers can exert this influence is not clear. In one proposed model, the DNA bends—it is known to do so—so that the activator is in contact with the transcription complex (see Figure 14.15).

Finally, there are negative regulatory regions on DNA called **silencers**, which have the reverse effect of enhancers. Silencers turn off transcription by binding proteins appropriately called *repressors*.

How do these proteins and DNA sequences—transcription factors, activators, repressors, regulators, enhancers, and silencers—regulate transcription? Apparently, all genes in most tissues can transcribe a small amount of RNA. But the right *combination* of the factors is what determines the maximum rate of transcription. In the immature red blood cells of bone marrow, for example, which make a large amount of β-globin, the transcription of globin genes is stimulated by the binding of 7 regulators and 6 activators. But in white blood cells in the same bone marrow, these 13 proteins are not made, and they do not bind to their sites adjacent to the β-globin genes; consequently, these genes are hardly transcribed at all.

COORDINATING THE EXPRESSION OF GENES. How do eukaryotic cells coordinate the regulation of several genes whose transcription must be turned on at the same time? In prokaryotes, in which related genes are linked together in an operon, a single regulatory system can regulate several adjacent genes. But in eukaryotes, the several genes whose regulation requires coordination may be on different chromosomes.

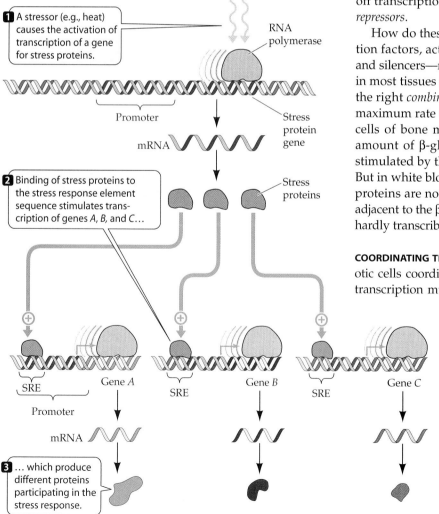

1 A stressor (e.g., heat) causes the activation of transcription of a gene for stress proteins.

2 Binding of stress proteins to the stress response element sequence stimulates transcription of genes *A*, *B*, and *C*...

3 ... which produce different proteins participating in the stress response.

14.16 Coordinating Gene Expression

A single signal, for example heat stress, causes the synthesis of a transcriptional regulator for many genes.

In such a case, regulation can be achieved if the various genes all have the same regulatory sequences near them, which bind to the same activators and regulators. One of the many examples of this phenomenon is provided by the response of organisms to a stressor—for example, drought in plants. Under conditions of drought stress, a plant must synthesize various proteins, but the genes for these proteins are scattered throughout the genome. However, each of these genes has a specific regulatory sequence near its promoter called the *stress response element* (SRE). The binding of a regulator protein to this element stimulates RNA synthesis (Figure 14.16). In the drought example, the proteins made are involved not only in water conservation, but also in protecting the plant against excess salt in the soil and against freezing. This finding has considerable importance for agriculture, in which crops are often grown under less than optimal conditions.

THE BINDING OF PROTEINS TO DNA.   A key to transcriptional control in eukaryotes is that transcription factors, regulators, activators, and repressors all bind to specific DNA sequences. In these proteins, there are four common structural themes in the domains that bind to DNA. These themes are called *motifs* and consist of combinations of structures and special components.

The *helix-turn-helix* motif involves several α-helices, one of which makes contact with DNA; the others stabilize the structure. This motif appears in the proteins that activate genes involved in embryonic development (homeobox proteins; see Chapter 16) and in the proteins that regulate the development of the immune and central nervous systems.

Helix-turn-helix motif

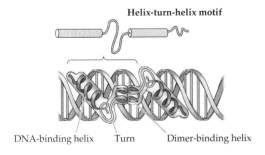

DNA-binding helix   Turn   Dimer-binding helix

The *zinc finger* motif has loops that form when a zinc ion is held by the amino acids cysteine and histidine. It occurs most notably in the receptors for steroid hormones (see Figure 15.9).

Zinc finger motif

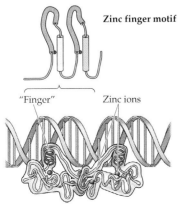

"Finger"   Zinc ions

The *leucine zipper* motif places hydrophobic leucine residues on one side of a polypeptide. Their presence allows two polypeptide chains to interact (zipper) hydrophobically, setting up the positively charged residues just past the zipper to bind to DNA. This motif occurs in many DNA-binding proteins—for example, the transcription factor AP-1, which binds near promoters of genes involved in mammalian cell growth and division. Overactivity of AP-1 has been linked to several types of cancer.

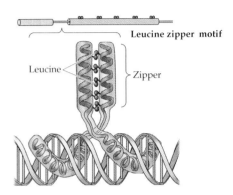

Leucine zipper motif

Leucine   Zipper

The *helix-loop-helix* motif is two helices separated by a loop. This region is adjacent to a stretch of amino acids that interact with DNA. This motif occurs in the activator proteins that bind to enhancers for the immunoglobulin genes that synthesize antibodies, as well as in the transcription factors involved in muscle protein synthesis.

Helix-loop-helix motif

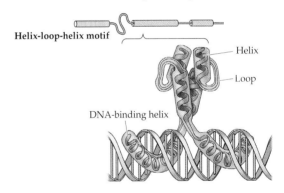

Helix

Loop

DNA-binding helix

Genes can be inactivated by chromatin structure

Chromatin contains nucleosomes and many other chromosomal proteins (see Chapter 9). The packaging of DNA by these nuclear proteins can make DNA physically inaccessible to RNA polymerase and the rest of the transcription apparatus, much as the binding of a repressor to the operator in the prokaryotic *lac* operon prevents transcription. Both local and global chromatin structure affect transcription.

LOCAL EFFECTS.   Nucleosomes inhibit both the initiation and elongation of transcription. To alleviate these blocks, cells recruit two protein complexes. One binds upstream of the initiation site, disaggregating the nucleosomes so that the large initiation complex can bind and begin transcription. The other binds once transcription is under way, allowing the transcription complex to move through these nucleosomes. These processes are called *chromatin remodeling* (Figure 14.17).

GLOBAL EFFECTS. Two kinds of chromatin can be distinguished by staining of the interphase nucleus: euchromatin and heterochromatin. *Euchromatin* is diffuse and stains lightly; it contains the DNA that is transcribed into mRNA. *Heterochromatin* stains densely and is generally not transcribed; any genes that it contains are thus inactivated. Perhaps the most dramatic example of heterochromatin is the inactive X chromosome of mammals.

A normal female mammal has two X chromosomes; a normal male, an X and a Y. The Y chromosome has only a few genes that are also present on the X, and is largely transcriptionally inactive in most cells. So there is a great difference between females and males in the "dosage" of X chromosome genes. In other words, each female cell has two copies of the genes on the X chromosome, and therefore has the potential to produce twice as much protein product of these genes as a male has. Yet X-linked gene expression is generally the same in males and females. How can this happen?

The answer was found in 1961 independently by Mary Lyon, Liane Russell, and Ernest Beutler. They suggested that one of the X chromosomes in each cell of an XX female is transcriptionally inactivated early in embryonic development. That copy of the X remains inactive in that cell, and in all the cells arising from it. In a given cell, the "choice" of which X in the pair of Xs to inactivate is usually random. Recall that one of the Xs in a female comes from her father and one from her mother. Thus, in one embryonic cell, the paternal X might be the one remaining active in mRNA synthesis, but in a neighboring cell, the maternal X might be active.

Interphase cells of XX females have a single, stainable nuclear body called a **Barr body**, after its discoverer, Murray Barr (Figure 14.18). This clump of heterochromatin, which is not present in males, is the inactivated X chromosome. The number of Barr bodies in each nucleus is equal to the number of X chromosomes minus one (the one represents the X chromosome that remains transcriptionally active). So a female with the normal two X chromosomes will have one Barr body, one with three X's will have two, an XXXX female will have three, and an XXY male will have one. We may infer that the interphase cells of each person, male or female, have a single *active* X chromosome, making the dosage of the expressed X chromosome genes constant across both sexes.

The mechanism of X inactivation involves chromosome condensation that makes the DNA sequences physically unavailable to the transcription machinery. One method may be the addition of a methyl group (—CH$_3$) to the 5' position of cytosine on DNA. Such *methylation* seems to be most prevalent in transcriptionally inactive genes. For example, most of the DNA of the inactive X chromosome has many cytosines methylated, while few of them on the ac-

INITIATION OF TRANSCRIPTION

1 DNA wraps around histone proteins, forming a nucleosome.

DNA

2 Nucleosomes block transcription.

3 Remodeling proteins bind, disaggregating the nucleosome.

Remodeling protein

4 Now the initiation complex can bind to begin transcription.

Histone protein

Initiation complex

mRNA

ELONGATION

5 A second remodeling complex can bind to the DNA–histone complex…

Remodeling protein

mRNA

6 …allowing transcription without disaggregation.

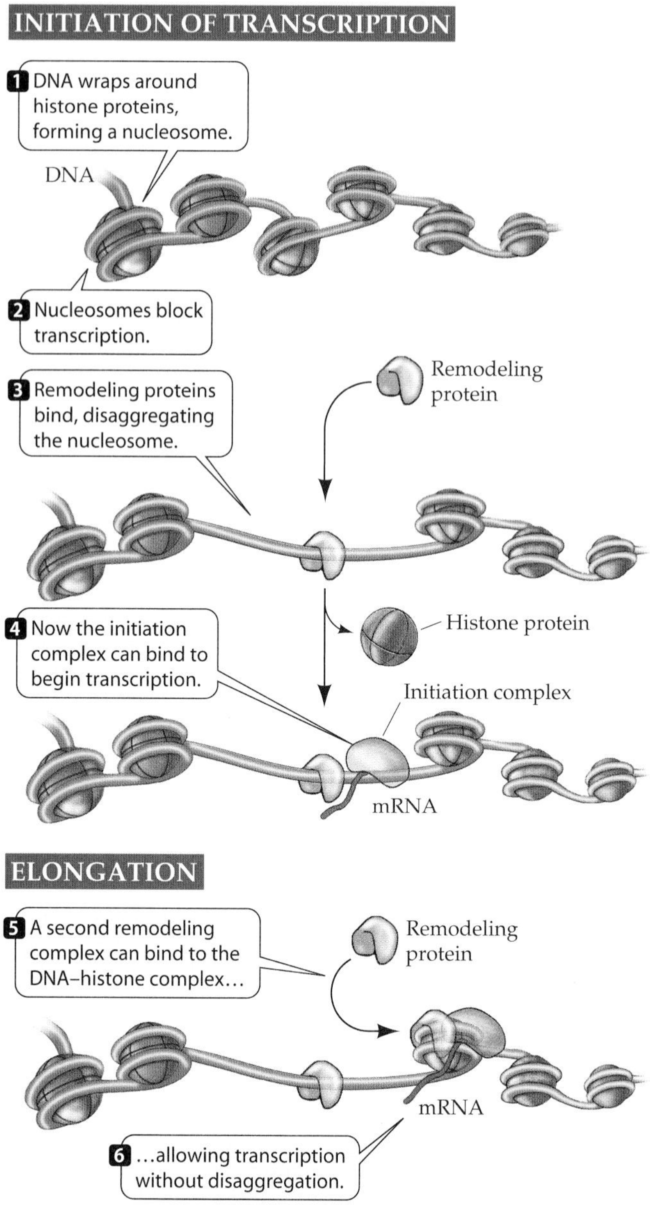

14.17 Local Remodeling of Chromatin for Transcription
Initiation of transcription requires that nucleosomes disaggregate. During elongation, however, they remain intact.

tive X are methylated. Methylated DNA appears to bind certain chromosomal proteins, which may be responsible for heterochromatin formation. But this seems to occur after the actual inactivation event, making methylation a way to keep genes turned off.

The otherwise inactive X chromosome has one gene that is only lightly methylated and *is* transcriptionally active. This gene is called *XIST* (for *X i*nactivation *s*pecific *t*ranscript), and it is heavily methylated on, and *not* transcribed from, the other, "active" X chromosome. The RNA transcribed from *XIST* does not leave the nucleus and is not an mRNA. Instead, it appears to bind to the X chromosome that transcribes it, and this binding somehow leads to a spreading of inactivation along the chromosome.

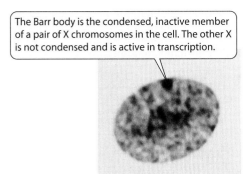

The Barr body is the condensed, inactive member of a pair of X chromosomes in the cell. The other X is not condensed and is active in transcription.

14.18 Barr Bodies in the Nuclei of Female Cells
The number of Barr bodies per nucleus is equal to the number of X chromosomes minus one. Thus males (XY) have no Barr body, whereas females (XX) have one.

A DNA sequence can move to a new location to activate transcription

In some instances, gene expression is regulated by the movement of a gene to a new location on the chromosome. An example of this mechanism is found in the yeast *Saccharomyces cerevisiae*. The haploid single cells of this fungus exist in two *mating types*, *a* and α, which fuse to form a diploid zygote. Although all yeast cells have an allele for each of these types, the allele that is expressed determines the mating type of the cell. In some yeasts, the mating type changes with almost every cell division cycle. How does it change so rapidly?

The yeast cell keeps the two different alleles (coding for type α and type *a*) at separate locations on its chromosome, away from a third site, the *MAT locus*. The two mating type alleles are usually transcriptionally silent because a repressor protein binds to them. However, when a copy of the α or *a* allele is inserted at the MAT region, the gene for proteins of the appropriate mating type is transcribed.

A change in mating type requires three steps:

▶ First, a new DNA copy of the nonexpressed allele is made (if the cell is now α, the new copy will be the *a* allele).

▶ Second, the current occupant of the MAT region (in this case, the α DNA) is removed by an enzyme.

▶ Third, the new allele (*a*) is inserted at the MAT region and transcribed. The *a* proteins are now made, and the mating type is changed.

DNA rearrangement is also important in producing the highly variable proteins that make up the human repertoire of antibodies, and in cancer, when inactive genes move to be adjacent to active promoters.

Selective gene amplification results in more templates for transcription

Another way for one cell to make more of a certain gene product than another cell does is to have more copies of the appropriate gene and to transcribe them all. The process of creating more copies of a specific gene in order to increase transcription is called **gene amplification**.

As described earlier, the genes that code for three of the four human ribosomal RNA's are linked together in a unit, and this unit is repeated several hundred times in the genome to provide multiple templates for rRNA synthesis (rRNA is the most abundant kind of RNA in the cell). In some circumstances, however, even this moderate repetition is not enough to satisfy the demands of the cell. For example, the mature eggs of frogs and fishes have up to a trillion ribosomes. These ribosomes are used for the massive protein synthesis that follows fertilization. The cell that differentiates into the egg contains fewer than 1,000 copies of the rRNA gene cluster, and would take 50 years to make a trillion ribosomes if it transcribed those rRNA genes at peak efficiency. How does the egg end up with so many ribosomes (and so much rRNA)?

The egg cell solves this problem by selectively amplifying its rRNA gene clusters until there are more than a million copies. In fact, this gene complex goes from being 0.2 percent of the total genome DNA to being 68 percent. These million copies transcribed at maximum rate (Figure 14.19) are just enough to make the necessary trillion ribosomes in a few days.

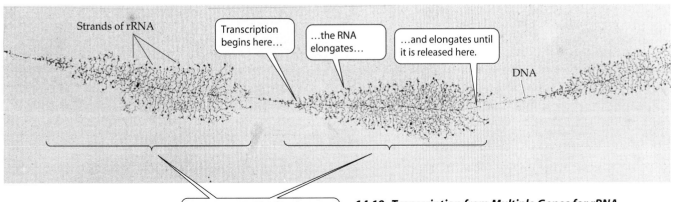

Strands of rRNA

Transcription begins here...

...the RNA elongates...

...and elongates until it is released here.

DNA

Multiple rRNA genes are actively transcribing rRNA precursors.

14.19 Transcription from Multiple Genes for rRNA
Elongating strands of rRNA transcripts form arrowhead-shaped regions, each centered on a strand of DNA that codes for rRNA.

The mechanism for selective overreplication of a single gene is not clearly understood, but it has important medical implications. As Chapter 18 will show, in some cancers, a cancer-causing gene called an oncogene becomes amplified. Also, in some tumors treated with a drug that targets a single protein, amplification of the gene for the target protein leads to an excess of that protein, and the cell becomes resistant to the prescribed dose of the drug.

Posttranscriptional Control

There are many ways to regulate the presence of mature mRNA in a cell even after a precursor has been transcribed. As we saw earlier, pre-mRNA can be processed by cutting out the introns and splicing the exons together. If the exons of the pre-mRNA are recombined in different ways by alternate splicing, different proteins can be synthesized.The longevity of mRNA in the cytoplasm can also be regulated. The longer an mRNA exists in the cytoplasm, the more of its coded protein can be made.

Different mRNA's can be made from the same gene by alternate splicing

Most primary transcripts contain several introns (see Figure 14.6). We have seen how the splicing mechanism recognizes the boundaries between exons and introns. What would happen if the β-globin pre-mRNA, which has two introns, was spliced from the start of the first intron to the end of the second? Not only the two introns but also the middle exon would be spliced out. An entirely new protein (certainly not a β-globin) would be made, and the functions of normal β-globin would be lost.

Alternate splicing can be a deliberate mechanism for generating a family of different proteins from a single gene. For example, a single pre-mRNA for the structural protein tropomyosin is alternatively spliced to give five different mRNA's and five different forms of tropomyosin found in five different tissues: skeletal muscle, smooth muscle, fibroblast, liver, and brain (Figure 14.20).

The stability of mRNA can be regulated

DNA, as the genetic material, must remain stable, and there are elaborate mechanisms for repairing it if it becomes damaged. RNA has no such repair system. After it arrives in the cytoplasm, mRNA is subject to breakdown catalyzed by ribonucleases, which exist both in the cytoplasm and in lysosomes. But not all eukaryotic mRNA's have the same life span. Differences in the stabilities of mRNA's provide another mechanism for posttranscriptional control of protein synthesis. The less time an mRNA spends in the cytoplasm, the less of its protein can be translated.

Tubulin is a protein that polymerizes to form microtubules, a component of the cytoskeleton (see Chapter 4). When a large pool of free tubulin is available in the cytoplasm, there is no particular need for the cell to make more of it. Under these conditions, some tubulin molecules bind to tubulin mRNA. This binding makes the mRNA especially susceptible to breakdown, and less tubulin is made.

Translational and Posttranslational Control

Is the amount of a protein in a cell determined by the amount of its mRNA? Recently, a survey was made of the relationships between mRNA's and proteins in yeast cells. Dozens of genes were surveyed. For about a third of them, the relationship between mRNA and protein held: More of one led to more of the other. But for two-thirds of the proteins, there was no apparent relationship. Their concentration in the cell must be determined by factors acting after the mRNA is made.

Just as proteins can control the synthesis of mRNA by binding to DNA, they can also control the translation of mRNA by binding to mRNA in the cytoplasm. This mode of control is especially important in long-lived mRNA's. A cell must not continue to make proteins that it does not

14.20 Alternate Splicing Results in Different mRNA's and Proteins
In mammals, the protein tropomyosin is coded for by a gene that has 11 exons. Different tissues splice tropomyosin pre-mRNA differently, resulting in five different forms of the protein.

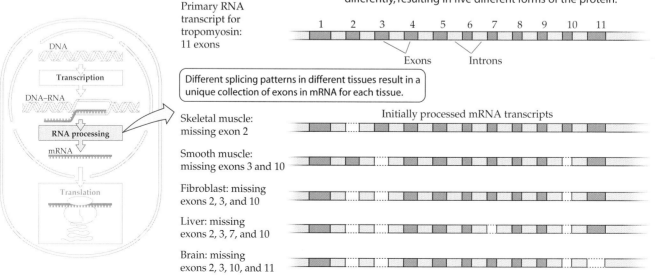

Primary RNA transcript for tropomyosin: 11 exons

Exons   Introns

Different splicing patterns in different tissues result in a unique collection of exons in mRNA for each tissue.

DNA
Transcription
DNA–RNA
RNA processing
mRNA
Translation

Initially processed mRNA transcripts

Skeletal muscle: missing exon 2

Smooth muscle: missing exons 3 and 10

Fibroblast: missing exons 2, 3, and 10

Liver: missing exons 2, 3, 7, and 10

Brain: missing exons 2, 3, 10, and 11

need. For example, mammalian cells respond to certain stimuli by making cyclins, which stimulate the events of the cell cycle. If the mRNA for a cyclin is still in the cytoplasm and available for translation long after the cyclin is needed, cyclin will be made and released inappropriately. Its presence might cause a target cell population to divide inappropriately, forming a tumor.

The translation of mRNA can be controlled

One way to control translation is by the capping mechanism on mRNA. As already noted, mRNA is capped at its 5′ end by a modified guanosine molecule (see Figure 14.11). Messenger RNA's that have unmodified caps are not translated. For example, stored mRNA in the oocyte of the tobacco hornworm moth has the guanosine added to its 5′ end, but the G is not modified. Hence, this stored mRNA is not translated. However, after fertilization, the cap is modified, allowing the mRNA to be translated to produce proteins needed for early embryogenesis.

Free iron ions (Fe^{2+}) within a mammalian cell are bound by a storage protein, *ferritin*. When iron is in excess, ferritin synthesis rises dramatically. Yet the amount of ferritin mRNA remains constant. The increase in ferritin synthesis is due to an increased rate of mRNA translation. When the iron level in the cell is low, a *translational repressor* protein binds to ferritin mRNA and prevents its translation by blocking its attachment to a ribosome. When iron levels rise, the excess iron binds to the repressor and alters its three-dimensional structure, causing it to detach from the mRNA, and translation of ferritin proceeds.

Translational control also acts in the synthesis of hemoglobin. As we described earlier, hemoglobin consists of four polypeptide chains and a nonprotein pigment, heme. If heme synthesis does not equal globin synthesis, some polypeptide chains stay free in the cell, waiting for a heme partner. Excess heme in the cell increases the rate of translation of globin mRNA by removing a block to the initiation of translation at the ribosome, helping to maintain the balance.

The proteasome controls the longevity of proteins after translation

We have considered how gene expression may be regulated by the control of transcription, RNA processing, and translation. However, the story does not end here, because most gene products—proteins—are modified after translation. Some of these changes are permanent, such as the addition of sugars (glycosylation), the addition of phosphate groups, or the removal of a signal sequence after a protein has crossed a membrane (see Figure 12.14).

An important way to regulate the action of a protein in a cell is to regulate its *lifetime* in the cell. Proteins involved in cell division (e.g., the cyclins) are hydrolyzed at just the right moment to time the sequence of events. Proteins identified for breakdown are often covalently linked to a 76-amino acid protein called *ubiquitin* (so called because it is ubiquitous, or widespread). The protein–ubiquitin complex then binds to a huge complex of several dozen polypeptide chains called a **proteasome** (Figure 14.21). The entryway to this "molecular chamber of doom" is a hollow cylinder, with ATPase activity, that cuts off the ubiquitin for recycling and unfolds its targeted protein victim. The protein then passes by three different proteases (thus the name of the complex) that digest it into small peptides and amino acids.

The cellular concentrations of many proteins are determined not by differential transcription of their genes, but by their degradation in proteasomes. For example, cyclins are degraded at just the right time during the cell cycle (see Figure 9.5). Transcription factors are broken down after they are used, lest the affected genes be always "on." Abnormal proteins are often targeted for destruction by a quality control mechanism. Human papillomavirus, which causes cervical cancer, targets the cell division inhibitory protein p53 for proteasomal degradation, so that unregulated cell division—and cancer—results.

14.21 The Proteasome Breaks Down Proteins
Proteins targeted for breakdown are bound to ubiquitin, which "leads" them to the proteasome, a complex composed of many polypeptides.

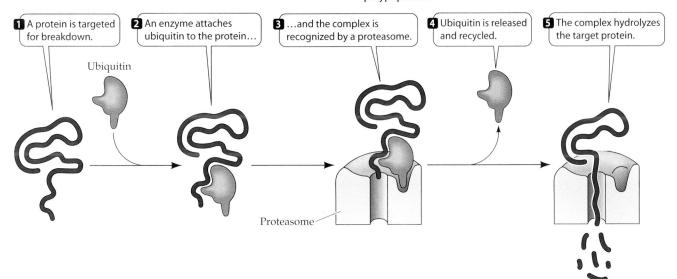

Chapter Summary

The Eukaryotic Genome

▶ Although eukaryotes have more DNA in their genomes than prokaryotes, in some cases there is no apparent relationship between genome size and organism complexity. **Review Figure 14.1**

▶ Unlike prokaryotic DNA, eukaryotic DNA is separated from the cytoplasm by being contained within a nucleus. The initial mRNA transcript of the DNA may be modified before it is exported from the cytoplasm. **Review Figure 14.2**

▶ The genome of the single-celled budding yeast contains genes for the same metabolic machinery as bacteria, with the addition of genes for protein targeting in the cell. **Review Table 14.2**

▶ The genome of the multicellular roundworm *Caenorhabditis elegans* contains genes required for intercellular interactions. **Review Table 14.3**

▶ The genome of the fruit fly has fewer genes than that of the roundworm. Many of its sequences are homologs of sequences on roundworm and mammalian genes.

Repetitive Sequences in the Eukaryotic Genome

▶ Highly repetitive DNA is present in up to millions of copies of short sequences. It is not transcribed. Its role is unknown.

▶ Telomeric DNA is found at the ends of chromosomes. Some telomeric DNA may be lost during each DNA replication, eventually leading to chromosome instability and cell death. The enzyme telomerase catalyzes the restoration of the lost telomeric DNA. Most somatic cells lack telomerase and thus have limited life spans. **Review Figure 14.3**

▶ Some moderately repetitive DNA sequences, such as those coding for rRNA's, are transcribed. **Review Figure 14.4**

▶ Some moderately repetitive DNA sequences are transposable, or able to move about the genome. **Review Figure 14.5**

The Structures of Protein-Coding Genes

▶ A typical protein-coding gene has noncoding internal sequences (introns) as well as flanking sequences that are involved in the machinery of transcription and translation. **Review Figures 14.6, 14.8**

▶ Nucleic acid hybridization is an important technique for analyzing eukaryotic genes. **Review Figure 14.7**

▶ Some eukaryotic genes form families of related genes that have similar sequences and code for similar proteins. These related proteins may be made at different times and in different tissues. Some sequences in gene families are pseudogenes, which code for nonfunctional mRNA's or proteins. **Review Figure 14.9**

▶ Differential expression of different genes in the β-globin family ensures important physiological changes during human development. **Review Figure 14.10**

RNA Processing

▶ After transcription, the pre-mRNA is altered by the addition of a G cap at the 5′ end and a poly A tail at the 3′ end. **Review Figure 14.11**

▶ The introns are removed from the mRNA precursor by the spliceosome, a complex of RNA's and proteins. **Review Figure 14.12**

Transcriptional Control

▶ Eukaryotic gene expression can be controlled at the transcriptional, posttranscriptional, translational, and posttranslational levels. **Review Figure 14.13**

▶ The major method of control of eukaryotic gene expression is selective transcription, which results from specific proteins binding to regulatory regions on DNA.

▶ A series of transcription factors must bind to the promoter before RNA polymerase can bind. Whether RNA polymerase will initiate transcription also depends on the binding of regulatory proteins, activator proteins (which are bound by enhancers and stimulate transcription), and repressor proteins (which are bound by silencers and inhibit transcription). **Review Figures 14.14, 14.15**

▶ The simultaneous control of widely separated genes is possible through proteins that bind to common sequences in their promoters. **Review Figure 14.16**

▶ The DNA-binding domains of most DNA-binding proteins have one of four structural motifs: helix-turn-helix, zinc finger, leucine zipper, or helix-loop-helix.

▶ Remodeling of chromatin occurs during transcription. **Review Figure 14.17**

▶ Heterochromatin is a condensed form of DNA that cannot be transcribed. It is found in the inactive X chromosome of female mammals. **Review Figure 14.18**

▶ The movement of a gene to a new location on a chromosome may alter its ability to be transcribed, as in the change from one mating type to another in yeast.

▶ Some genes become selectively amplified, and the extra copies result in increased transcription of their protein product. **Review Figure 14.19**

Posttranscriptional Control

▶ Because eukaryotic genes have several exons, alternate splicing can be used to produce different proteins. **Review Figure 14.20**

▶ The stability of mRNA in the cytoplasm can be regulated by the binding of proteins.

Translational and Posttranslational Control

▶ Translational repressors can inhibit the translation of mRNA.

▶ Proteasomes degrade proteins targeted for breakdown. **Review Figure 14.21**

For Discussion

1. In rats, a gene 1,440 bp long codes for an enzyme made up of 192 amino acid units. Discuss this apparent discrepancy. How long would the initial and final mRNA transcripts be?

2. The activity of the enzyme dihydrofolate reductase (DHFR) is high in some tumor cells. This activity makes the cells resistant to the anticancer drug methotrexate, which targets DHFR. Assuming that you had the complementary DNA for the gene that encodes DHFR, how would you show whether this increased activity was due to increased transcription of the single-copy DHFR gene or to amplification of the gene?

3. Describe the steps in the production of a mature, translatable mRNA from a eukaryotic gene that contains introns. Compare this to the situation in prokaryotes (see Chapter 13).

4. A certain protein-coding gene has three introns. How many different proteins can be made from alternate splicing of the pre-mRNA transcribed from this gene?

5. Most somatic cells in mammals do not express telomerase. Yet the germ line cells that produce gametes by meiosis do express this enzyme. Explain.

15 Cell Signaling and Communication

CAROL HAD INTENDED TO COMPLETE THE REport summarizing her semester's biology lab project long before it was due; it would, after all, count for significant points toward her grade. But between her other courses and a few interesting distractions, she kept putting off the report until "next week." Finally, at 9:30 the night before it was due, she sat down to create the 20-page report. By 11:00 she realized she still had hours of work to do. She made a frantic call to her lab partner, asking him to provide her with some data from one of their experiments. And Carol filled the coffee-brewing machine in her dorm room, knowing she would need a "caffeine jolt" to keep her awake so she could finish the job.

To understand how a "caffeine jolt" works, we must understand the pathways by which the body's cells respond to certain signals in their environment. The signals might be chemicals traveling between brain cells, or hormones produced in response to an outside event. There are three sequential processes involved in the cell's response to any signal. First, the signal binds to a receptor protein. Second, the binding of the signal causes a message to be conveyed to the cell's cytoplasm and amplified. Third, the cell changes its activity in response to the signal.

Caffeine acts in different ways in different tissues. First, a tired person's brain produces adenosine molecules that bind to specific receptor proteins, resulting in decreased brain activity and increased drowsiness. Caffeine's molecular structure is similar to that of adenosine, so it occupies the adenosine receptors without inhibiting brain cell function, and alertness is restored. Then, in the heart and liver, caffeine stimulates a pathway inside cells so that they do not need hormonal stimulation. In the heart, the result is an increased rate of beating; the liver is stimulated to release glucose into the bloodstream.

We begin this chapter with a discussion of signals that affect cells. As you will see, these range from chemicals such as hormones to physical entities such as light. Whatever the signal, it will affect a cell only if that cell has a receptor protein that binds to the signal. In addition to binding the signal, the receptor must somehow communicate to the rest of the cell that binding has occurred. This process of signal transduction often involves special small molecules called second messengers, which initiate a series of events that amplify the signal. As a result, the third phase—alteration of cell function—may involve many instances of the same event, such as the opening of an ion channel in the plasma membrane or increased transcription of a number of genes.

We close with a description of how cells communicate with one another directly through specialized channels in their adjacent plasma membranes.

Signals

Both prokaryotic and eukaryotic cells process information from their environment. This information can be in the form of a physical stimulus, such as the light reaching your eye as you read this book, or chemicals that bathe a cell, such as lactose in the medium surrounding *E. coli*. It may come from outside the organism, such as the ions dissolved in the water that bathes plant roots, or from a neighboring cell within the organism, as occurs in the heart, where thousands of cells contract in unison by transmitting signals to one another.

Multiple Signals at Work
The computer, the telephone, and the coffee all send signals to the brain. All may be necessary for the successful completion of a research project.

Of course, the mere presence of a signal does not mean that a cell will respond, just as you do not pay close attention to every sound in your environment as you study. To respond, the cell must have a specific **receptor protein** that can bind to the signal. In this section, we describe some of the signals different cells respond to and look at one model system of signal transduction.

Cells receive signals from the physical environment and from other cells

The physical environment is full of signals. Our sense organs allow us to respond to light, odors, touch, and sound. Bacteria and protists respond to even minute chemical changes in their environment. Plants respond to light as a signal. For example, at sunset, at night, or in the shade, not only the amount of sunlight but also the spectrum of light reaching the surface of Earth differs from that of full sunlight in the daytime. These variations are signals that affect plant growth and reproduction. Even magnetism can be a signal: Some bacteria and birds orient themselves to the Earth's magnetic poles, like a needle on a compass.

But a cell inside a large organism is far from the exterior environment. Instead, its environment is other cells and extracellular fluids. Cells receive their nutrients from, and

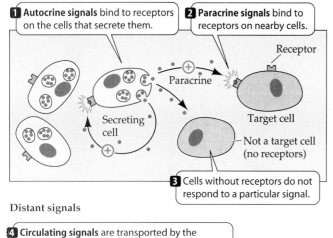
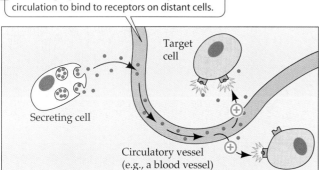

15.1 Chemical Signaling Systems
A signal molecule can act on the same cell that produces it, or on a nearby cell. Most signals act on distant cells, to which they are transported by the organism's circulatory system.

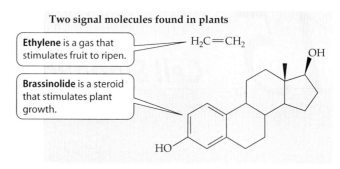

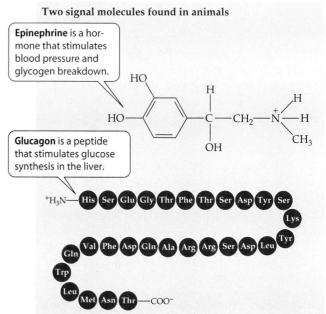

15.2 A Variety of Biological Signals
Many different kinds of molecules can serve as biological signals. The structure of glucagon is simplified so that a "bead" represents an amino acid, each about the size of the epinephrine molecule whose chemical formula is shown.

pass their wastes into, extracellular fluids. Cells also receive signals—mostly chemical signals—from their extracellular fluid environment. Most of these chemical signals come from other cells. In animal cells, they include hormones (Chapter 41), neurotransmitters (Chapter 44), and chemical messages from the immune system (Chapter 19). Cells also respond to chemical signals coming from the environment via the digestive and respiratory systems. And cells can respond to chemicals, such as CO_2 and H^+, whose presence in the extracellular fluids results from the metabolic activities of other cells.

Inside a large organism, chemical signals reach a target cell by local diffusion or by circulation within the blood. *Autocrine* signals affect the cells that make them. *Paracrine* signals diffuse to nearby cells. Signals to distant cells usually travel through the circulatory system (Figure 15.1).

The biological signals cells receive are diverse (Figure 15.2). In each case, the cell must be able to receive or sense the signal and respond to it. Depending on the cell and the signal, the responses range from entering the cell division cycle to heal a wound, to moving to a new location in the embryo to form a tissue, to releasing enzymes that digest food, to sending messages to the brain about the book you are reading. Clearly, signaling underlies a lot of biology. The whole process, from signal detection to final response, is called a **signal transduction pathway**.

Signaling involves a receptor, transduction, and effects

In Chapter 13, we saw that bacteria respond to changes in nutrients in their environment by altering their transcription of genes, as in the *lac* operon. In addition to responding to such changes, these same bacteria must be able to sense and respond to non-nutritive changes in their environment, such as changes in osmotic concentration. For example, if the solute concentration around *E. coli* rises far above that inside the cell, the law of diffusion tells us that water will diffuse out of the cell and solutes into the cell. Since the cell must maintain homeostasis in its cytoplasm, it must perceive and respond to this environmental change. The way in which this one-celled organism responds to such signals has much in common with signaling in more complex animals and plants (Figure 15.3).

The receptor protein in *E. coli* for osmotic changes is called EnvZ. It is a transmembrane protein that extends through the bacterium's plasma membrane into the space between the plasma membrane and a highly porous outer membrane that forms a complex with the cell wall. When the solute concentration of the extracellular environment rises, so does the concentration in the environment between the two membranes. This change in its aqueous medium causes the part of the receptor protein sticking into the intermembrane space to undergo a conformational change.

We saw in Chapter 6 that changing the tertiary structure of one part of a protein often leads to changes in distant parts of the protein. In the case of the bacterial EnvZ receptor, the conformational change in the intermembrane region of the protein is transmitted to the region that lies in the bacterium's cytoplasm. This change initiates the events of signal transduction. EnvZ becomes an active *protein kinase,* which catalyzes the addition of a phosphate group from ATP to one of EnvZ's own histidine residues. In other words, EnvZ phosphorylates itself.

The charged phosphate group added to the histidine causes the cytoplasmic tail of the EnvZ protein to change its shape again. It now binds to a second protein, OmpR,

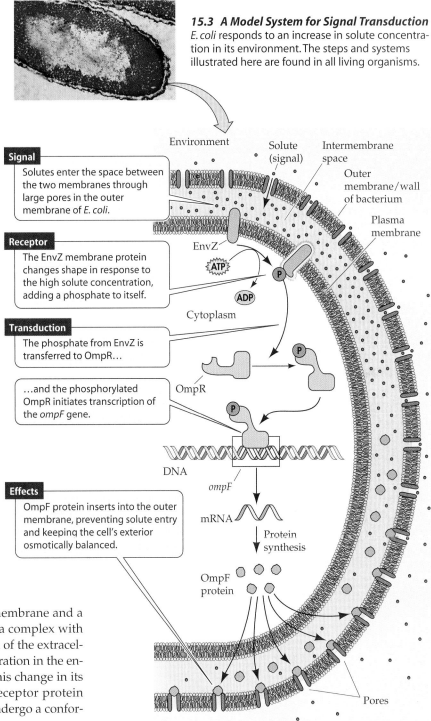

15.3 A Model System for Signal Transduction
E. coli responds to an increase in solute concentration in its environment. The steps and systems illustrated here are found in all living organisms.

Signal
Solutes enter the space between the two membranes through large pores in the outer membrane of *E. coli.*

Receptor
The EnvZ membrane protein changes shape in response to the high solute concentration, adding a phosphate to itself.

Transduction
The phosphate from EnvZ is transferred to OmpR…

…and the phosphorylated OmpR initiates transcription of the *ompF* gene.

Effects
OmpF protein inserts into the outer membrane, preventing solute entry and keeping the cell's exterior osmotically balanced.

which takes the phosphate group from EnvZ. OmpR also changes its structure due to this phosphorylation. This change is a key event in signaling for three reasons. First, the signal on the outside of the cell has now been *transduced* to a protein totally within the cell's cytoplasm. Second, OmpR can *do* something, and that is to bind to a promoter on *E. coli* DNA adjacent to the DNA coding for the protein OmpF. This binding begins the final phase of this signaling pathway: the *effect* of the signal, which is an alteration in cell function. Third, the signal has been *amplified.* EnvZ can alter the structure of many OmpR molecules.

Phosphorylated OmpR has the correct three-dimensional structure to bind to DNA at the *ompF* promoter, resulting in an increase in transcription of that gene. Translation of *ompF* mRNA results in the production of OmpF protein, which is inserted into the outer membrane and prevents solutes from entering the intermembrane space. Thus the *E. coli* cell can go on behaving just as if the environment has a normal osmotic concentration.

It is important to highlight the major features of this prokaryotic system, as they will reappear in many other signal transduction systems in animals and plants:

▶ A receptor changes its conformation upon interacting with the signal.
▶ A conformational change exposes a protein kinase activity.
▶ Phosphorylation alters the function of a protein.
▶ The signal is amplified.
▶ Transcription factors are activated.
▶ Altered synthesis of specific proteins occurs.
▶ Protein action alters cell activity.

Signal transduction pathways featuring these seven activities occur in all types of organisms. The emergence of these activities was an important event in the evolution of cellular life, as they allowed the organism to react to and survive in a rapidly changing environment.

Receptors

While a given cell is bombarded with many signals, it responds to only a few of them. The reason for this is that any given cell makes receptors for only some signals. Which cells make which receptors is genetically determined: If a cell transcribes the gene encoding a particular receptor and the resulting mRNA is translated, the cell will have that receptor. A receptor protein binds to a signal in much the same way as an enzyme binds to a substrate or a membrane transport protein binds to the molecule it is transporting.

Receptors have specific binding sites for their signals

A signaling molecule, usually called a **ligand**, fits into a site on its receptor much like a substrate fits into the active site of an enzyme (Figure 15.4). Whether the receptor protrudes from the plasma membrane surface or is located in the cytoplasm, the result of ligand binding is the same: The receptor protein changes its three-dimensional structure and initiates a cellular response. The ligand does not contribute further to this response. In fact, the ligand usually is not metabolized into useful products. Its role is purely to "knock on the door." This is in sharp contrast to enzyme–substrate interactions, in which the whole purpose is to change the substrate into a useful product.

Receptors bind to their ligands according to chemistry's law of mass action:

$$R + L \rightleftharpoons RL$$

This means that the binding is reversible, although for most ligand/receptor complexes, the equilibrium point is far to the right—that is, favoring binding. Release of the ligand is important because if it does not happen, the receptor will be continuously stimulated.

Just as with enzymes, *inhibitors* can bind to the ligand site on a receptor protein. Both natural and artificial inhibitors of receptor binding are important in medicine.

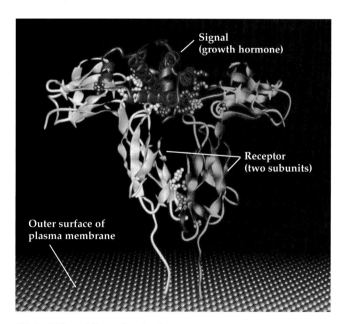

15.4 A Signal Bound to Its Receptor
Only the extracellular regions of the human growth hormone receptor are shown.

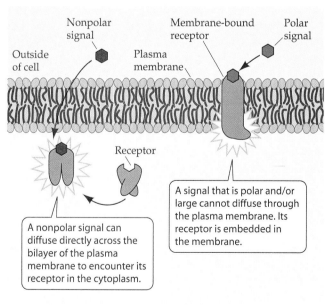

15.5 Two Locations for Receptors
Receptors can be located on the plasma membrane or in the cytoplasm of the cell.

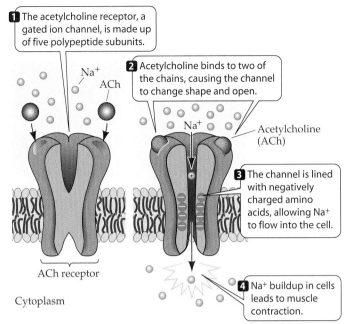

1 The acetylcholine receptor, a gated ion channel, is made up of five polypeptide subunits.

2 Acetylcholine binds to two of the chains, causing the channel to change shape and open.

Na⁺

ACh

Na⁺

Acetylcholine (ACh)

3 The channel is lined with negatively charged amino acids, allowing Na⁺ to flow into the cell.

ACh receptor

Cytoplasm

4 Na⁺ buildup in cells leads to muscle contraction.

15.6 An Ion Channel Receptor
The acetylcholine receptor is a channel for sodium ions that resembles a gate. The gate opens when its ligand, acetylcholine, binds to it, allowing Na⁺ to flow into the cell.

There are several types of receptors

A major division among receptors is in their cellular location, which largely depends on the nature of their ligands. The chemistry of signals is quite variable, but they can be divided into two classes: those that are nonpolar, and can cross the plasma membrane and enter the cell, and those that are large or polar, and cannot cross the membrane (Figure 15.5). Estrogen, for example, is a steroid and can easily diffuse across the plasma membrane and enter the cell; it binds to a receptor inside the cytoplasm. Insulin, on the other hand, is

a protein hormone that cannot diffuse through the plasma membrane; instead, it binds to a receptor that is a transmembrane protein with an extracellular binding region.

In more complex eukaryotes, there are three well-studied types of receptors on plasma membranes: ion channels, protein kinases, and G protein-linked receptors.

ION CHANNELS. In the plasma membranes of many types of cells, there are channel proteins that can be open or closed. These **ion channels** act as "gates," allowing ions such as Na⁺, K⁺, Ca²⁺, or Cl⁻ to enter or leave the cell. The gate-opening mechanism is an alteration in the three-dimensional structure of the channel protein upon ligand binding. Each ion channel has its own signal. These signals include sensory stimuli, such as light and sound, voltage differences across the plasma membrane, and chemical ligands such as small molecules and hormones.

An example of a gated ion channel is the acetylcholine receptor (Figure 15.6). This receptor is located at the plasma membranes of vertebrate skeletal muscle cells and binds the ligand acetylcholine, which is released from nerve cells. When two molecules of acetylcholine bind to the receptor, it opens for about a thousandth of a second. This is enough time for Na⁺, which is more concentrated outside the cell than inside, to rush into the cell. The change in Na⁺ concentration in the cell ultimately results in muscle contraction. Right after the channel opens, the ligand is released from the receptor and then degraded. This makes the receptor (and the cell) responsive to the next signal, so that the muscle can contract again.

PROTEIN KINASES. Like the activated EnvZ protein of *E. coli*, some eukaryotic receptor proteins become kinases when they are activated: That is, they catalyze the transfer of a phosphate group from ATP to a protein. The targets for the protein kinase activity are both the receptor itself and cytoplasmic molecules, which alter their shape and then act to change the cell's activities. While histidine is phosphorylated by EnvZ in *E. coli*, the amino acid targeted by the protein kinase receptors of animal cells is usually tyrosine. In plants, either serine or threonine is phosphorylated.

Insulin is a protein hormone made by the mammalian pancreas. The receptor for insulin is a protein consisting of two copies each of two different polypeptide subunits (Figure 15.7). As with acetylcholine, two molecules of insulin must bind to the receptor. After binding insulin on its extracellular surface, the receptor changes

1 The α subunit binds insulin (the signal).

2 The β subunit transmits a signal from bound insulin to the cytoplasm.

Insulin

β α   α β

β α   α β

Outside of cell

Membrane

Insulin receptor

Phosphate groups

3 The insulin signal activates the receptor's protein kinase domain in the cytoplasm.

Cytoplasm

Insulin response substrate (IRS)

4 Protein kinases from the receptor phosphorylate insulin-response substrates, triggering other chemical responses inside the cell.

Cellular responses

15.7 A Protein Kinase Receptor
The mammalian hormone insulin does not enter the cell, but is bound by a membrane receptor protein with four subunits (two α and two β). The β subunits transmit a signal that changes the cytoplasmic end of the receptor protein, activating a protein kinase domain and triggering further responses by the cell, eventually resulting in the transport of glucose across the membrane.

its shape to expose a cytoplasmic protein kinase active site. Like the EnvZ receptor described above, the insulin receptor self-phosphorylates. Then, as a protein kinase signal, it targets certain cytoplasmic proteins, appropriately called insulin response substrates. These proteins then initiate many cell responses, including the insertion of glucose transporters into the plasma membrane.

G PROTEIN-LINKED RECEPTORS.   A third category of eukaryotic plasma membrane receptor is the seven-spanning G protein-linked receptors. This long name identifies a fascinating group of receptors, all of which are composed of a single protein with seven regions that pass through the lipid bilayer, separated by short loops that extend either outside or inside the cell. Ligand binding on the extracellular side changes the shape of the receptor's cytoplasmic region, opening up a binding site for a mobile membrane protein.

This membrane protein, known as a **G protein**, has two important binding sites: one for the G protein-linked receptor, and the other for the nucleotide GDP/GTP (Figure 15.8). G proteins have several polypeptide subunits. When the G protein binds to the activated receptor, it also binds GTP to one of its subunits. At the same time, the ligand is released from the extracellular side of the receptor. The GTP-bound subunit of the G protein now separates from the parent G protein, diffusing in the plane of the lipid bilayer until it encounters an *effector protein* to which it can bind. Effector proteins are what their name implies: They cause an effect. The binding of the GTP-bearing G protein subunit activates the effector—which may be an ion channel or an enzyme—thereby causing changes in cell function.

After binding to the effector protein, the GTP on the G protein is hydrolyzed to GDP. The now inactive G protein subunit separates from the effector protein. The G protein subunit must form a complex with another subunit before binding to yet another activated receptor. When this activated receptor is bound, the G protein exchanges its GDP for GTP, and the cycle begins again.

By means of their diffusing subunits, G proteins can either activate or inhibit an effector. An example of an *activating* response involves the receptor for epinephrine (adrenaline), the famous "fight-or-flight" hormone made by the adrenal gland in response to stress or heavy exercise. In heart muscle, this hormone binds to its G protein-linked receptor, causing a G protein to become activated. The GTP-bound subunit then activates a membrane-bound enzyme to produce a small molecule, cyclic AMP (see below), which has many effects on the cell, including glucose mobilization for energy and muscle contraction.

An example of G protein-mediated *inhibition* occurs when the same hormone, epinephrine, binds to its receptor in the smooth muscle cells surrounding blood vessels lining the digestive tract. Again, the epinephrine-bound receptor changes its shape and binds a G protein, which then binds GTP, and the G protein subunit with its GTP binds to the target enzyme. But in this case, the enzyme is inhibited instead of being activated. As a result, the muscles relax, and the blood vessel diameter increases, allowing more nutrients to be carried away from the digestive system to the rest of the body. Thus the same signal and initial signaling mechanism can have different consequences in different cells, depending on the nature of the responding cell.

CYTOPLASMIC RECEPTORS.   Not all signals act at the plasma membrane. Some signals diffuse across the lipid bilayer of the plasma membrane and enter the cytoplasm. In these cases, the receptor protein lies inside the cytoplasm. Steroid hormones in animals, for example, enter the cytoplasm and bind to steroid hormone receptors. Binding to the ligand

15.8 A G Protein-Linked Receptor

Binding of an extracellular signal—in this case, a hormone—causes the activation of a seven-segmented G protein. The G protein then activates an enzyme that catalyzes a reaction in the cytoplasm, amplifying production of the product. The figure is a generalized diagram that could apply to any of the large family of G proteins and the signals they react with.

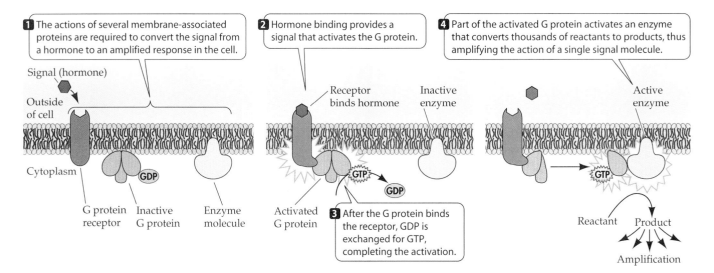

1 The actions of several membrane-associated proteins are required to convert the signal from a hormone to an amplified response in the cell.

2 Hormone binding provides a signal that activates the G protein.

4 Part of the activated G protein activates an enzyme that converts thousands of reactants to products, thus amplifying the action of a single signal molecule.

Signal (hormone)

Outside of cell

Cytoplasm

G protein receptor

Inactive G protein

GDP

Enzyme molecule

Receptor binds hormone

Inactive enzyme

Activated G protein

GTP

GDP

3 After the G protein binds the receptor, GDP is exchanged for GTP, completing the activation.

Active enzyme

GTP

Reactant   Product

Amplification

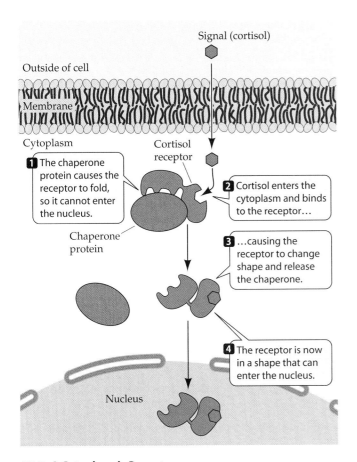

15.9 A Cytoplasmic Receptor
The receptor for cortisol is bound to a chaperone protein. Binding of the signal (which diffuses directly through the membrane) releases the chaperone and allows the receptor protein to enter the cell's nucleus, where it functions as a transcription factor.

causes the receptor to change its shape so that it can enter the cell nucleus, where it acts as a transcription factor (Figure 15.9). But this general view is somewhat simplified. The receptor for the hormone cortisol, for example, is bound to a chaperone protein, which blocks it from entering the nucleus. Binding of the hormone causes the receptor to change its shape so that the chaperone is released. This allows the receptor, which is a transcription factor, to fold into an appropriate configuration for entering the nucleus and initiating transcription.

Transducers

As previously mentioned, the same signal may produce different responses in different tissues. Acetylcholine, for example, can bind to receptors on skeletal muscle cells, where it stimulates muscle contraction, but on heart muscle cells, it slows contraction. These different responses to the same ligand/receptor complex are mediated by the events of signal transduction. These events, which are critical to the cell's response, may be either direct or indirect.

Direct transduction is a property of the receptor itself and occurs at the plasma membrane. In *indirect transduction*, which is more common, another molecule, termed a second messenger, mediates the interaction between receptor binding and cellular reaction. In neither case is transduction a single event. Rather, the signal initiates a cascade of events, in which proteins interact with other proteins until the final responses are achieved (Figure 15.10).

15.10 Direct and Indirect Signal Transduction
(*a*) All the events of direct transduction occur on the plasma membrane, at or near the receptor protein. (*b*) In indirect transduction, the binding of the signal to the receptor triggers formation of a "second messenger" molecule that works in the cytoplasm. It is the second messenger that sets off the necessary biochemical reactions.

(*a*) **Direct transduction**

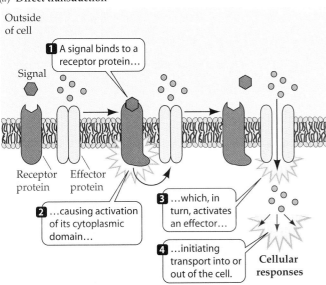

(*b*) **Indirect transduction**

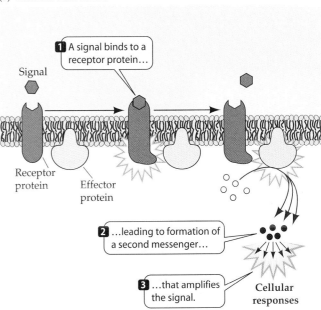

Protein kinase cascades amplify a response to receptor binding

We have seen that when a signal binds to a protein kinase receptor, the receptor changes its structure to expose a protein kinase active site, which catalyzes the phosphorylation of target proteins. This process is an example of direct signal transduction. Protein kinase receptors are important in binding ligands that stimulate cell division in both plants and animals. In Chapter 9, we described growth factors that were external inducers of the cell cycle. These factors stimulate cell division by binding to protein kinase receptors.

The complete signal transduction pathway that occurs after a protein kinase receptor is activated was worked out from studies on a cell that went wrong. Many human bladder cancers contain an abnormal protein called Ras (so named because it was first isolated from a *rat sarcoma* tumor). Investigations of these bladder cancers showed that the ras protein was a G protein, but was always active because it was permanently bound to GTP. So the abnormal ras protein caused continuous tumor cell division. If the cancer cells' Ras protein was inhibited, the tumor cells stopped dividing. This discovery has led to a major effort to develop specific Ras inhibitors for cancer treatment.

What does Ras do in normal, noncancerous cells? When scientists treated cells in a laboratory dish with both a Ras inhibitor and a growth factor, the expected cell division did not occur. Since growth factor binding is the first event in stimulating cell division, these results meant that Ras, like other G proteins, must be an intermediary between signal (growth factor) and response (cell division). After this discovery, the challenge was to work out what the activated growth factor receptor did to Ras, and what Ras did to stimulate further events in signal transduction.

This signaling pathway has been worked out, and it is an excellent example of a more general phenomenon, a **protein kinase cascade** (Figure 15.11). Such cascades are key to the external regulation of many cellular activities. Indeed, as we saw in Chapter 14, the eukaryotic genome codes for hundreds, even thousands, of such kinases. The unbound receptors for growth factors exist in the plasma membrane as separate polypeptide chains (subunits). When the growth factor signal binds to a subunit, it associates with another subunit to form a dimer, which changes shape to expose a protein kinase active site. The kinase activity sets off a series of events, activating several other protein ki-

15.11 A Protein Kinase Cascade
In a protein kinase cascade, a series of proteins becomes sequentially activated. In this example, the growth factor receptor protein stimulates the G protein Ras, which mediates a cascading series of reactions. The final product of the cascade, MAP kinase (MAPk), enters the nucleus and causes changes in transcription. Inactive forms of the proteins are on the left, activated forms are on the right.

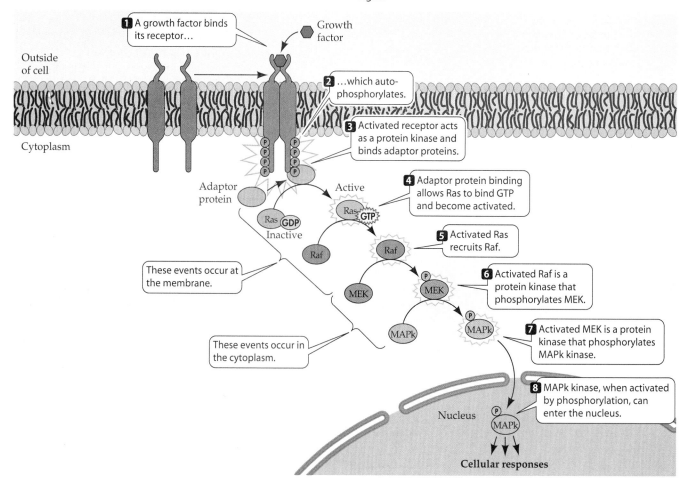

1. A growth factor binds its receptor…

Growth factor

Outside of cell

Cytoplasm

2. …which auto-phosphorylates.

3. Activated receptor acts as a protein kinase and binds adaptor proteins.

Adaptor protein

Active

4. Adaptor protein binding allows Ras to bind GTP and become activated.

Ras GDP
Inactive

Ras GTP

5. Activated Ras recruits Raf.

Raf

Raf

These events occur at the membrane.

MEK

MEK

6. Activated Raf is a protein kinase that phosphorylates MEK.

MAPk

MAPk

7. Activated MEK is a protein kinase that phosphorylates MAPk kinase.

These events occur in the cytoplasm.

8. MAPk kinase, when activated by phosphorylation, can enter the nucleus.

Nucleus

MAPk

Cellular responses

nases in turn. The final phosphorylated, activated protein—MAP kinase—moves into the nucleus and phosphorylates target proteins necessary for cell division.

Is the protein kinase cascade pathway universal in eukaryotes? Genome sequencing of the plant *Arabidopsis* has revealed proteins with strong homologies to many of the proteins in the mammalian pathway. A number of proteins that resemble tyrosine kinase receptors are also present. There is even a Ras-like protein. The functions of this pathway in *Arabidopsis* are under investigation.

Protein kinase cascades are very useful for signal transduction for three reasons:

▶ At each step in the cascade of events, the signal is amplified. Because each newly activated protein kinase is an enzyme, each can catalyze the phosphorylation of many target proteins.
▶ The information from the signal that was originally at the plasma membrane is communicated to the nucleus.
▶ The multitude of steps provides some specificity to the process. As we have seen with epinephrine, signal binding and receptor activation do not result in the same response in all cells. Different target proteins at every step in the cascade can provide variability of response.

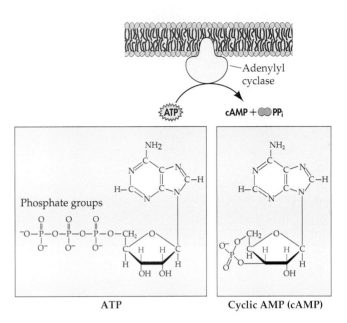

15.12 The Formation of Cyclic AMP
The formation of cAMP from ATP is catalyzed by adenylyl cyclase, an enzyme that is activated by G proteins.

Cyclic AMP is a common second messenger

As we have seen, protein kinase receptors stimulate the protein kinase cascade right at the plasma membrane. However, the stimulation of events in the cell is more often indirect. In a series of clever experiments, Earl Sutherland, Edwin Krebs, and Edmond Fischer showed that in many cases, there is a small, water-soluble chemical messenger between the membrane receptor and cytoplasmic events. These researchers were investigating the activation of the liver enzyme phosphorylase by the hormone epinephrine. Phosphorylase catalyzes the breakdown of glycogen stored in the liver so that carbohydrate can be released to the blood to fuel the fight-or-flight response.

The researchers found that phosphorylase could be activated in liver cells that had been broken open, but only if the entire cell contents, including the plasma membrane, were present. Epinephrine had bound to the plasma membrane, and phosphorylase was present in the cytoplasm. So they tried the steps of this experiment in sequence. First, they incubated membranes of broken liver cells with epinephrine. Then they removed the membranes, but kept the solution in which the membranes had been bathed. When they added this solution to the cytoplasm, the phosphorylase enzyme present became activated! Hormone binding to the membrane receptor had caused the production of a small, water-soluble molecule that then diffused to the cytoplasm, where it activated the enzyme.

This small molecule was identified as **cyclic AMP (cAMP)**, which we also encountered in the *lac* operon regulatory system in *E. coli*, where cAMP was working as a second messenger. **Second messengers** are substances re-

leased into the cytoplasm after the first messenger—the signal—binds its receptor.

In contrast to the uniqueness of receptor binding, *second messengers affect many processes in the cell, and allow a cell to respond to a single event at the plasma membrane with many events inside the cell.* Like the kinase cascade, second messengers amplify the signal—a single epinephrine molecule leads to the production of several dozen molecules of cAMP, which then activate many enzyme targets.

Adenylyl cyclase, the enzyme that catalyzes the formation of cAMP from ATP, is located on the cytoplasmic surface of the plasma membrane of target cells (Figure 15.12). Usually, it is activated by the binding of G proteins, themselves activated by receptors. Second messengers do not have enzymatic activity; rather, they act as cofactors or allosteric regulators of target proteins. In the case of cAMP, there are two major target types. In many kinds of sensory cells, cAMP binds to ion channels to cause them to open. A second major target type is cytoplasmic. Cyclic AMP binds to an enzyme such as a protein kinase, whose active site gets exposed as a result. The sequential activation of yet another kinase ensues, leading to the final effects in the cell.

Two second messengers are derived from lipids

Membrane phospholipids are involved in signal transduction in addition to their roles as structural components of the plasma membrane. When certain phospholipids are hydrolyzed into their component parts (see Figure 3.21) by enzymes called phospholipases, second messengers are formed. The best-studied of these come from hydrolysis of the lipid phosphatidyl inositol-bisphosphate (PTI), which

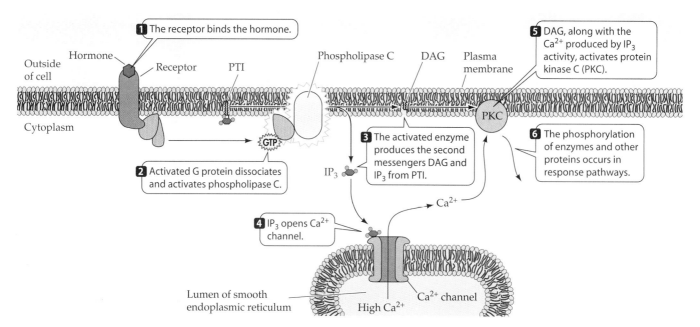

1 The receptor binds the hormone.

5 DAG, along with the Ca²⁺ produced by IP₃ activity, activates protein kinase C (PKC).

Outside of cell

Hormone

Receptor

PTI

Phospholipase C   DAG   Plasma membrane

Cytoplasm

GTP

2 Activated G protein dissociates and activates phospholipase C.

PKC

3 The activated enzyme produces the second messengers DAG and IP₃ from PTI.

IP₃

6 The phosphorylation of enzymes and other proteins occurs in response pathways.

Ca²⁺

4 IP₃ opens Ca²⁺ channel.

Ca²⁺ channel

Lumen of smooth endoplasmic reticulum

High Ca²⁺

15.13 The IP₃ and DAG Second Messenger System
Phospholipase C hydrolyzes the lipid phosphatidyl inositol-bis-phosphate (PTI) into its components IP₃ and DAG, both of which are second messengers. IP₃ and DAG act separately but in concert, ultimately producing a wide range of responses in the cell.

has two fatty acid chains (diacylglycerol, or DAG) embedded in the plasma membrane, and a hydrophilic inositol group (inositol triphosphate, or IP₃) projecting into the cytoplasm. There are over two dozen signals whose actions are mediated by the products of PTI hydrolysis. Once again, the receptors involved are often linked to G proteins. The activated G proteins diffuse through the plasma membrane and activate an enzyme, phospholipase C. This enzyme cleaves off the IP₃ from PTI, leaving the glycerol and the two attached fatty acids (DAG) embedded in the lipid bilayer:

$$\text{PTI} \xrightarrow{\text{Phospholipase C}} \text{IP}_3 + \text{DAG}$$

IP₃ and DAG are both second messengers and have different modes of action that build on each other (Figure 15.13). DAG activates a membrane-bound enzyme, protein kinase C (PKC), in much the same way that cAMP activates protein kinase A. PKC is dependent on Ca²⁺ (hence the "C"), and this is where IP₃ comes in. IP₃ is charged and diffuses through the cytoplasm to the endoplasmic reticulum, where it causes the release of Ca²⁺ into the cytoplasm. There, in combination with DAG, the Ca²⁺ causes PKC to become active. PKC then phosphorylates a wide variety of proteins, leading to the ultimate response of the cell (Figure 15.13).

Calcium ions are involved in many transduction pathways

Calcium ions can also act as a second messenger. They are scarce in most cells, with a cytoplasmic concentration of only about 0.1 μM, while the concentrations of Ca²⁺ outside

the cell and within the ER are usually much higher. This difference is maintained by active transport proteins at the plasma and ER membranes that pump the ion out of the cytoplasm. In contrast to cAMP and the lipid second messengers, the level of intracellular Ca²⁺ cannot be increased by making more of it. Instead, the opening and closing of channels and the action of membrane pumps regulate levels of the ion in a cellular compartment.

There are many signals that can cause Ca²⁺ channels to open, including IP₃ (see the previous section) and the entry of a sperm into an egg cell (Figure 15.14). Whatever the signal, the open channels result in a dramatic increase in cyto-

Sperm entry point

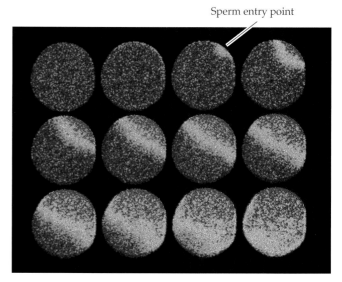

15.14 Calcium Ions as an Intracellular Messenger
The concentration of Ca²⁺ can be measured by a dye that fluoresces and turns red when it binds the ion. Here, photographed at 5-second intervals, fertilization causes a wave of Ca²⁺ to pass through the egg of a starfish. The message that fertilization is complete and development can begin is thus delivered.

plasmic Ca^{2+} concentration, up to a hundredfold within a fraction of a second. As we saw earlier, this ion activates protein kinase C. In addition, Ca^{2+} controls other ion channels and stimulates secretion by exocytosis.

A distinctive aspect of Ca^{2+} signaling is that the ion can stimulate its own release from intracellular stores. For example, in some plant leaf cells, the hormone abscisic acid binds to gated Ca^{2+} channels and opens them, causing the ion to rush into the cells. This influx is not enough to trigger the cell's response, however. The ion binds to Ca^{2+} channels in the endoplasmic reticulum and vacuolar membranes, causing those organelles to release their Ca^{2+} stores as well.

In some cases, Ca^{2+} ions act via a calcium-binding protein called *calmodulin*, and it is the Ca^{2+}–calmodulin complex that performs cellular functions by binding to target proteins. Calmodulin, which is present in many cells, has four binding sites for Ca^{2+}. When the cytoplasmic Ca^{2+} concentration is low, calmodulin does not bind enough of it to become activated. But when the cell is stimulated by a signal and the Ca^{2+} level rises, all four binding sites are filled. Then the calmodulin changes shape and binds to a number of cellular targets, activating them in turn. One such target is a protein kinase in smooth muscle cells that phosphorylates the muscle protein myosin, initiating muscle contraction.

Nitric oxide is a gas that can act as a second messenger

Pharmacologist Robert Furchgott, at the State University of New York in Brooklyn, was investigating how acetylcholine causes smooth muscles lining blood vessels to relax, thus allowing more blood to flow to certain organs. Acetylcholine appeared to stimulate the IP_3 signal transduction system to produce an influx of Ca^{2+}, which led to an increase in the level of an unusual second messenger, cyclic GMP (cGMP). This nucleotide bound to a protein kinase, which then stimulated a kinase cascade leading to muscle relaxation. So far, the pathway seemed straightforward.

But while this pathway seemed to work in intact animals, it did not work on isolated strips of artery tissue. When Furchgott switched to tubular sections of artery, however, signal transduction did occur. There turned out to be a crucial difference between these two tissue preparations: In the strips, the delicate inner layer of cells that lines the blood vessel had been lost. Furchgott hypothesized that this layer, the endothelium, was making something that diffused into the muscle cells and was needed for their response to acetylcholine. The substance was not easy to isolate. It seemed to break down quickly, with a half-life (the time in which half of it disappeared) of 5 seconds in living tissues. It turned out to be a gas, nitric oxide (NO), that had always been thought of as a toxic air pollutant!

In the body, NO is made via an enzyme, NO synthase. This enzyme is activated by Ca^{2+}, which enters the endothelial cell through a channel opened by IP_3, released after acetylcholine binds to its receptor. The NO formed is chem-

ically very unstable and although it diffuses readily, it does not get too far. Conveniently, the endothelial cells are close to the smooth muscle cells, where NO acts as a second messenger, stimulating the formation of cGMP (Figure 15.15).

The spectacular discovery of NO as a second messenger explained the action of nitroglycerin, a drug that has been used for over a century to treat angina, the chest pain caused by insufficient blood flow to the heart. Nitroglycerin releases NO, which results in relaxation of the blood vessels and increased blood flow.

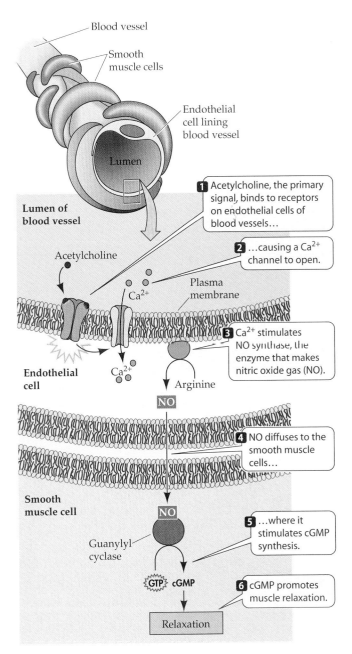

15.15 Nitric Oxide as a Second Messenger
Nitric oxide (NO) is an unstable gas, which nevertheless serves as a second messenger between a signal, acetylcholine, and its effect, the relaxation of smooth muscles. The endothelial tissue of blood vessels is a crucial intermediary in this communication between three types of tissue.

Signal transduction is highly regulated

There are several ways in which cells can regulate the activity of a signal transduction mechanism. The concentration of NO, which breaks down quickly, can be regulated only by how much of it is made. The level of Ca^{2+}, on the other hand, is determined by both membrane pumps and ion channels. For protein kinase cascades, G proteins, and cAMP, there are enzymes that convert the activated form back to its inactivated precursor:

▶ Protein phosphatases remove the phosphate groups from phosphorylated proteins.

▶ GTPases convert the GTP on an active G protein back to GDP, inactivating the protein.

▶ cAMP phosphodiesterase converts cAMP into its precursor, AMP, which has no second messenger activity.

These three inactivation systems are themselves under controls. For example, the major protein phosphatase can be inhibited by a protein whose activity is determined by phosphorylation by protein kinase A, which itself is under control of cAMP. So the cAMP pathway can intersect with the protein kinase cascade. On the other hand, some cAMP phosphodiesterases are stimulated by Ca^{2+}, thus showing an interaction between these two signaling pathways. The caffeine in coffee acts as a stimulant in part because it inhibits cAMP phosphodiesterase.

Effects

We have seen that the binding of a signal to its receptor initiates the response of a cell to an environmental signal, and how the direct or indirect transduction of this signal to the cytoplasm of the cell amplifies the stimulus. In this section, we consider the third and final step in the process, the actual effects of the signal on cell function. These effects are primarily the opening of membrane channels, changes in the activities of enzymes, and differential gene transcription.

Membrane channels are opened

The opening of ion channels is of great importance when the nervous system responds to a signal. Sensory nerve cells of the sense organs, for example, become stimulated through the opening of ion channels. We will focus here on one such signal transduction pathway, that for the sense of smell (Figure 15.16).

The sense of smell is well developed in mammals, some of which have an amazing *1,000 genes* for odorant receptors, the largest gene family known. Each of the thousands of nerve cells in the nose expresses one of these receptors. The identification of which chemical signal, or odorant, activates which receptor is just getting under way.

15.16 A Signal Transduction Pathway Leads to the Opening of Membrane Channels
In the signal transduction pathway for the sense of smell, the final effect is the opening of Na^+ channels. The resulting influx of Na^+ stimulates the transmission of a scent message to a specific region of the brain.

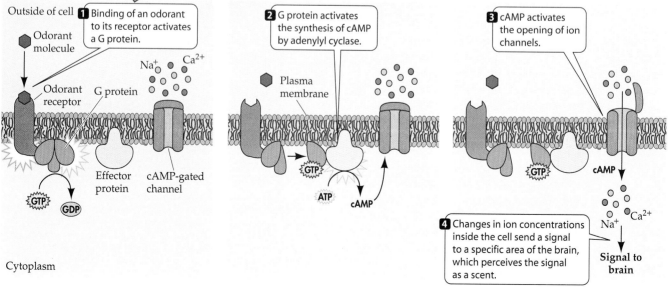

Outside of cell

1 Binding of an odorant to its receptor activates a G protein.

Odorant molecule

Na^+ Ca^{2+}

Odorant receptor   G protein

Effector protein   cAMP-gated channel

GTP   GDP

Cytoplasm

2 G protein activates the synthesis of cAMP by adenylyl cyclase.

Plasma membrane

GTP

ATP   cAMP

3 cAMP activates the opening of ion channels.

GTP   cAMP

Na^+ Ca^{2+}

4 Changes in ion concentrations inside the cell send a signal to a specific area of the brain, which perceives the signal as a scent.

Signal to brain

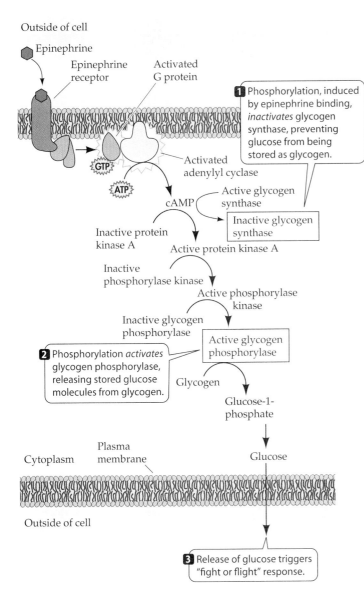

Outside of cell

Epinephrine

Epinephrine receptor

Activated G protein

1 Phosphorylation, induced by epinephrine binding, *inactivates* glycogen synthase, preventing glucose from being stored as glycogen.

GTP

Activated adenylyl cyclase

ATP

cAMP

Active glycogen synthase

Inactive glycogen synthase

Inactive protein kinase A

Active protein kinase A

Inactive phosphorylase kinase

Active phosphorylase kinase

Inactive glycogen phosphorylase

Active glycogen phosphorylase

2 Phosphorylation *activates* glycogen phosphorylase, releasing stored glucose molecules from glycogen.

Glycogen

Glucose-1-phosphate

Cytoplasm

Plasma membrane

Glucose

Outside of cell

Glucose

3 Release of glucose triggers "fight or flight" response.

15.17 A Cascade of Reactions Leads to Altered Enzyme Activity

Liver cells respond to epinephrine by activating G proteins, which in turn activate cAMP synthesis. The second messenger initiates a series of kinase reactions. The cascade both inhibits the continued storage of glucose molecules and stimulates the release of previously stored glucose.

The G protein-mediated protein kinase cascade stimulated by epinephrine in liver cells results in the phosphorylation of two key enzymes in glycogen metabolism (Figure 15.17). One of them, glycogen synthase, catalyzes the joining of glucose molecules to synthesize the energy-storing molecule glycogen, but it is inactivated by phosphorylation. Thus the epinephrine signal *prevents* glucose from being stored in glycogen. On the other hand, phosphorylase kinase becomes activated when a phosphate group is added to it, and goes on to stimulate a protein kinase cascade that ultimately leads to the activation by phosphorylation of glycogen phosphorylase, the other key enzyme in glucose metabolism. This enzyme *liberates* glucose molecules from glycogen. Thus the same signaling pathway inhibits the storage of glucose as glycogen (by inhibiting glycogen synthase) and promotes the release of glucose through glycogen breakdown (by activating glycogen phosphorylase). As we mentioned earlier, the released glucose fuels the ATP-requiring fight-or-flight response to epinephrine.

Phosphorylation by activated protein kinase A alters the activities of many other proteins, including enzymes involved in glycolysis, a ribosomal protein, and a receptor for a neurotransmitter. Likewise, Ca^{2+} binds to many proteins in the cell, changing their activities. In addition to protein kinase C, Ca^{2+}-activated targets include proteins that bind to and organize actin microfilaments and microtubules, as well as troponin, a modulator of muscle contraction.

Different genes are transcribed

Cell surface receptors are involved in activating a broad range of gene expression responses. For example, the Ras signaling pathway ends in the nucleus (see Figure 15.11). The final protein kinase enters the nucleus and phosphorylates a leucine zipper protein called AP-1. This activated protein stimulates the transcription of a number of genes involved in cell proliferation.

As we described earlier in this chapter, lipid-soluble hormones can diffuse directly through the plasma membrane and meet their receptors in the cytoplasm. Binding of the ligand allows the ligand/receptor complex to enter the nucleus, where it binds to hormone-responsive elements at the promoters of a number of genes. In some cases, transcription is stimulated, and in others it is inhibited.

In plants, light acts as a signal to initiate the formation of chloroplasts. Between this signal and response is a transcription-mediated signaling pathway. In bright sunlight, red wavelengths are absorbed by a receptor protein called phytochrome. We will say more about this important recep-

When an odorant molecule binds to its receptor, a G protein becomes activated, which in turns activates adenylyl cyclase to form the second messenger cAMP. This molecule then binds to an ion channel, causing it to open. The resulting influx of Na^+ causes the nerve cell to become stimulated so it sends a signal back to the brain that a particular odor is present.

Enzyme activities are changed

Proteins will change their shape, and their functioning, if they are modified either covalently or noncovalently. We have seen examples of both types of modificaton in signal transduction. Protein kinases add phosphate groups to a target protein, and the covalent change alters the protein's shape. Cyclic AMP binds to target proteins allosterically, and this noncovalent interaction changes the protein's shape. In both cases, previously inaccessible active sites are exposed, and the target protein goes on to perform a cellular role.

tor later in the book, but for now it is important to note that it is activated by red light. The activated phytochrome binds to cytoplasmic regulatory proteins, which enter the nucleus and bind to promoters for genes involved in the synthesis of important chloroplast proteins. Synthesis of these proteins is the key to the plant "greening".

Direct Intercellular Communication

Up to now, we have described how signals from the environment can influence a cell. But the environment of a cell in a multicellular organism is more than the extracellular medium. Most cells are in contact with their neighbors. In Chapter 5, we described how cells adhere to one another by the noncovalent interactions of recognition proteins protruding from the cell surface. There are also specialized cell junctions, such as tight junctions and desmosomes, that help "cement" the cells together (see Figure 5.7).

However, as we know from our own neighbors (and roommates), just being in proximity does not necessarily mean that there is functional communication. In this section, we look at the specialized junctions between cells that allow them to signal one another. In animals, these are gap junctions; in plants, they are plasmodesmata.

Animal cells communicate by gap junctions

Gap junctions are channels between adjacent cells that occur in many multicellular animals, occupying up to 25 percent of the area of the plasma membrane (Figure 15.18). Gap junctions traverse the 2-nm space between the plasma membranes of two cells (the "gap") by means of thin molecular channels called connexons. The walls of these tubes are composed of six subunits of an integral membrane protein appropriately named *connexin*. In two cells close to each another, two connexons come together, forming a channel that links the two cytoplasms. There may be hun-

dreds of these channels between a cell and its neighbors. The channels are quite narrow, about 1.5 nm in diameter. This is far too small for the passage of large molecules, such as proteins. But it is wide enough to allow small signal molecules and ions to pass between the cells. Experiments in which a labeled signal molecule or ion is injected into one cell show that it can readily pass into the adjacent cells if they are connected by gap junctions.

Gap junctions permit metabolic cooperation among linked cells. Such cooperation assures the sharing of important small molecules such as ATP, metabolic intermediates, amino acids, and coenzymes between cells. It may also assure that concentrations of ions and small molecules are similar in linked cells, thereby maintaining equivalent regulation of metabolism. It is not clear how important this is in many tissues, but it is known to be vital in some. For example, in the lens of the mammalian eye, only the cells at the periphery are close enough to the blood supply to allow diffusion of nutrients and wastes. But because lens cells are connected by large numbers of gap junctions, material can diffuse between them rapidly and efficiently.

There is evidence that signal molecules such as hormones and second messengers such as cAMP and IP_3 can move through gap junctions. If this is true, only a few cells would need to have receptors binding a signal in order for the stimulus to spread throughout the tissue. In this way, a tissue could have a coordinated response to the signal.

Plant cells communicate by plasmodesmata

Instead of gap junctions, plants have **plasmodesmata**, which are membrane-lined bridges spanning the thick cell walls that separate plant cells from one another. A typical plant cell has several thousand plasmodesmata.

Plasmodesmata differ from gap junctions in one fundamental way: Unlike gap junctions, in which the wall of the channel is made of integral proteins from the adjacent plasma membranes, plasmodesmata are lined by the fused plasma membranes themselves. Plant biologists are so familiar with the notion of a tissue as cells interconnected in this way that they refer to these continuous cytoplasms as a *symplast* (see Chapter 35).

The diameter of a plasmodesma is about 6 nm, far larger than the gap junction channel. But the actual space available for diffusion is about the same—1.5 nm—as with gap junctions. A look at the interior of the plasmodesma gives the reason for this reduction in pore size: A tubule called the *desmotubule*, apparently derived from the endoplasmic reticulum, fills up most of the opening of the plasmodesma (Figure 15.19). So, typically, only small metabolites and ions move between plant cells. This fact is important physiologically to plants, which lack the tiny circulatory vessels (capillaries) many animals use to bring gases and nutrients near enough to every cell.

Diffusion from cell to cell through plasma membranes is probably inadequate for hormonal responses in plants. Instead, they rely on more rapid diffusion through plasmo-

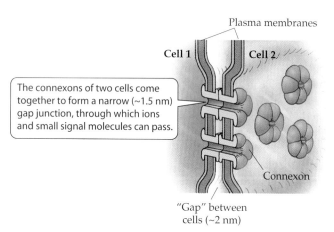

Plasma membranes

Cell 1   Cell 2

The connexons of two cells come together to form a narrow (~1.5 nm) gap junction, through which ions and small signal molecules can pass.

Connexon

"Gap" between cells (~2 nm)

15.18 Gap Junctions Connect Animal Cells
An animal cell may contain hundreds of gap junctions connecting it to neighboring cells. Gap junctions are too small for proteins to pass through, but small molecules such as ATP, metabolic intermediates, amino acids, and coenzymes can be shared in this way.

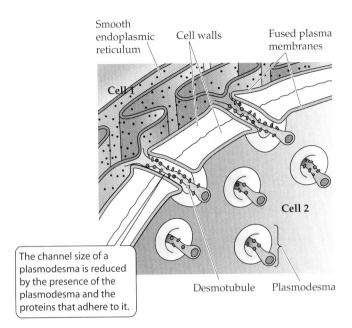

The channel size of a plasmodesma is reduced by the presence of the plasmodesma and the proteins that adhere to it.

15.19 Plasmodesmata Connect Plant Cells
The desmotubule, derived from the endoplasmic reticulum, fills up most of the space inside plasmodesmata, leaving a tiny gap through which small metabolites and ions can pass.

desmata to ensure that all cells of a tissue respond to a signal at the same time. In C_4 plants (see Chapter 8), there are abundant plasmodesmata between the mesophyll and bundle sheath cells, helping to rapidly move the fixed carbon in the former cell type to the latter. A similar transport system, found at the junctions of nonvascular tissues and phloem, conducts organic solutes throughout the plant.

Plasmodesmata are not merely passive channels, but can be regulated. Plant viruses may infect cells at one location, then spread rapidly through a plant organ by plasmodesmata until they reach the plant's vascular tissue (circulatory system). These viruses, and even their RNA, would appear to be many times too large to pass through the plasmodesmal channel. But they get through, apparently by making "movement proteins" that increase the pore size temporarily while attached to the viral genome. Similar movement proteins are involved in transporting mRNAs between plant cells. This finding opens up the possibility of long-distance regulation of translation.

Chapter Summary

Signals
▶ Cells receive many signals from both the physical environment and other cells. **Review Figures 15.1, 15.2**
▶ Signaling involves three steps: the binding of a signal by a receptor, the transduction of the signal within the cell, and the ultimate cellular response. **Review Figure 15.3**

Receptors
▶ Cells respond to signals only if they have specific receptor proteins that can bind to those signals. **Review Figure 15.4**
▶ Depending on the nature of the signal, the receptor may be at the plasma membrane or in the cytoplasm of the target cell. **Review Figure 15.5**
▶ Membrane receptors include ion channels, protein kinases. and G protein-linked receptors. **Review Figures 15.6, 15.7, 15.8**

Transducers
▶ The events of signal transduction may be direct, occurring at the plasma membrane, or indirect, involving the formation of a second messenger. **Review Figure 15.10**
▶ Protein kinase cascades amplify a response to receptor binding. **Review Figure 15.11**
▶ Second messengers include cyclic AMP, the lipid-derived substances phosphatidylinositol and diacylglycerol, calcium ions, and the gas nitric oxide. **Review Figures 15.12, 15.13, 15.14, 15.15**

Effects
▶ The ultimate cell response to a signal may be the opening of membrane channels, the alteration of enzyme activities, or changes in gene transcription. **Review Figures 15.16, 15.17**

Direct Intercellular Communication
▶ Animal cells can communicate directly, through small pores in their plasma membranes called gap junctions. Small molecules and ions can pass through these channels. **Review Figure 15.18**
▶ Plant cells are connected by somewhat larger pores called plasmodesmata, which traverse both membranes and cell walls. **Review Figure 15.19**

For Discussion

1. Like Ras itself, the various components of the Ras signaling pathway (see Figure 15.11) were discovered when tumors showed mutations in one or another of the components. What might be the biochemical consequences of mutations in the genes for (*a*) Raf and (*b*) MAP kinase that result in rapid cell division?

2. Cyclic AMP is a second messenger in many different responses, ranging from the sense of smell to the breakdown of glycogen. How can one messenger act in different ways in different cells?

3. Compare direct communication via plasmodesmata or gap junctions with ligand/receptor-mediated communication between cells. What are the advantages of one method over the other?

4. The tiny invertebrate *Hydra* has an apical region, which has tentacles, and a long, slender body. *Hydra* can reproduce asexually when cells on the body wall differentiate and form a bud, which breaks off as a new organism. Buds form only at a certain distance from the apex, leading to the idea that the apex releases a molecule that diffuses down the body and, at high concentrations (near the apex), inhibits bud formation. *Hydra* lacks a circulatory system, so the inhibitor must diffuse from cell to cell. If you had an antibody that binds to connexin and plugs up gap junctions, how would you show that the inhibitory factor passes through them?

16 Development: Differential Gene Expression

IT IS A DAY IN THE NOT-TOO-DISTANT FUTURE. Decades of eating fatty foods, combined with a genetically based tendency to deposit cholesterol in his arteries, finally catch up with 60-year-old Don: A blood clot closes off the blood flow to part of his heart, leading to a heart attack and irreversible tissue damage. Today, Don would be faced with a long period of rehabilitation, taking medications to manage his weakened heart. Instead, his physicians inject undifferentiated embryonic stem cells directly into his heart. The cells differentiate into cardiac muscle cells, replacing the ones that were lost to oxygen starvation, and full heart function is restored.

Embryonic stem cells, which are cells from a very young mammalian embryo, are able to form an entire organism if separated from one another. These cells can be removed from an embryo and maintained indefinitely in the laboratory. If they could be genetically altered to make them acceptable for transplants, these cells could be a source of tissue replacement not only for damaged hearts, but for the pancreas in people with diabetes and the brain in people with Alzheimer's disease.

While the application of stem cells to medicine is not yet possible, considerable new knowledge about the molecular biology of development has emerged. Much of this knowledge has come from studies on organisms such as the fruit fly *Drosophila*, the roundworm *Caenorhabditis elegans*, frogs, sea urchins, and a flowering plant, *Arabidopsis thaliana*. As we saw in Chapter 14, the genomes of eukaryotes are surprisingly similar, and the cellular and molecular principles underlying their development also turn out to be similar. Thus discoveries from one organism aid us in understanding other organisms, including ourselves.

Two major conclusions have emerged from studies of development. The first is that all types of somatic cells—all cells except gametes—in an organism retain all of the genes that were present in the fertilized egg. In other words, cell differentiation does not result from a loss of DNA. The second is that cellular changes during development and cell differentiation result from differential expression of genes. During development, the various mechanisms of transcriptional and translational control described in Chapter 14 and the signaling mechanisms described in Chapter 15 work together to produce a complex organism.

The Processes of Development

Development is a process in which an organism undergoes a series of progressive changes, taking on the successive forms that characterize its life cycle (Figure 16.1). In its earliest stages of development, a plant or animal is called an **embryo**. Sometimes the embryo is contained within a protective structure such as a seed coat, an eggshell, or a uterus. An embryo does not photosynthesize or feed actively; instead, it obtains its food from its mother directly or indirectly (by way of nutrients stored in the egg, for example). A series of embryonic stages may precede the birth of the new, independent organism. Most individual organisms continue to develop throughout their life cycle; development ceases only with death.

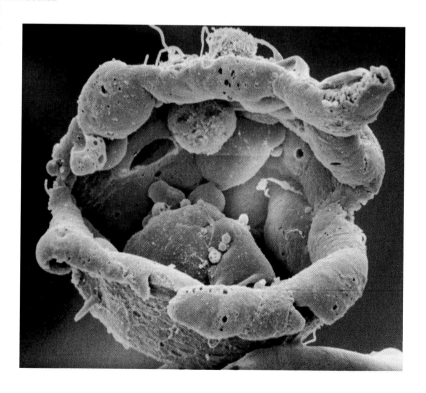

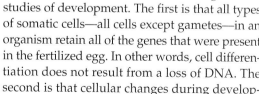

An Early Embryo
This mammalian embryo has been opened up to show the undifferentiated stem cells.

(*a*) **Flowering plant**
(*Arabidopsis*)

(*b*) **Insect**
(*Drosophila*)

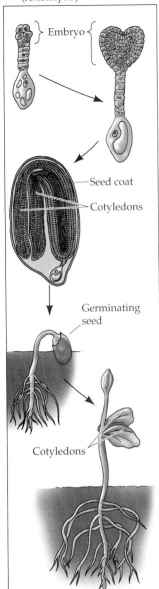

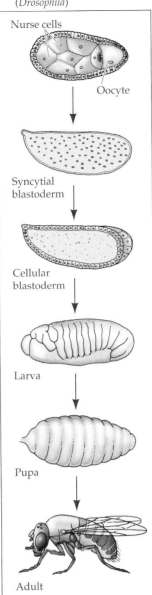

16.1 *Stages of Development*
Stages from embryo to adult are shown for a plant and an animal. Cell division and expansion, growth, cell differentiation, and the creation of the organs and tissues of the adult body are all part of the complex process of development.

Development consists of growth, differentiation, and morphogenesis

Growth (increase in size) occurs through cell division and cell expansion. It continues throughout the individual's life in some species, but reaches a more or less stable end point in others. Repeated mitotic divisions generate the multicellular body. In plants, unless the daughter cells become longer (expand) after they form, the embryo does not grow very much; thus in plant development, cell expansion begins shortly after the first divisions of the fertilized egg. In animal development, on the other hand, cell expansion is often slow to begin: The animal embryo may consist of

thousands of cells before it becomes larger than the fertilized egg.

Differentiation is the generation of cellular specializations; that is, *differentiation defines the specific structure and function of a cell*. Mitosis, as we have seen, produces daughter nuclei that are chromosomally and genetically identical to the nucleus that divides to produce them. However, the cells of an animal or plant body are obviously not all identical in structure or function. The human body, with its approximately 100 trillion (10^{14}) cells, consists of about 200 functionally distinct cell types—for example, muscle cells, blood cells, and nerve cells. This apparent contradiction results from regulation of the expression of various parts of the genome. When the embryo consists of only a few cells, each cell has the potential to develop in many different ways. As development proceeds, however, the possibilities available to individual cells gradually narrow, until each cell's *fate* is fully determined and the cell has differentiated.

Whereas differentiation gives rise to cells of different kinds, **morphogenesis** (literally, "creation of form") gives rise to the shape of the multicellular body and its organs. Morphogenesis results from *pattern formation*, the organization of differentiated tissues into specific structures. In plant development, cells are constrained by cell walls and do not move around, so organized division and expansion of cells are the major processes that build the plant body. In animals, cell movements are very important in morphogenesis. And in both plants and animals, programmed cell death is essential to orderly development.

In plants and animals alike, differentiation and morphogenesis result ultimately from the regulated activities of genes and their products, as well as the interplay of extracellular signals and their transduction in target cells.

As development proceeds, cells become more and more specialized

Marking specific cells of an early embryo with stains reveals which adult structures are derived from which part of the embryo. For instance, the shaded area of the frog embryo shown in Figure 16.2 normally becomes part of the skin of the tadpole larva. However, if we cut out a piece from this region and transplant it to another location on an early embryo, the type of tissue it becomes is determined by its new environment. The *developmental potential* of such cells—that is, their range of possible development—is thus greater than their actual *fate*, which is limited to what normally develops.

Does embryonic tissue retain its broad developmental potential? Generally speaking, the answer is no. The developmental potential of cells becomes restricted fairly early in normal development. Tissue from a later-stage frog embryo, for example, if taken from a region fated to develop into brain, becomes brain tissue even if transplanted to parts of an early-stage embryo destined to become other structures. The tissue of the later-stage embryo is thus said to be *determined*: Its fate has been sealed, regardless of its surroundings. By contrast, the younger transplant tissue in Figure 16.2 has not yet become determined.

Determination precedes differentiation

Determination, the commitment of a cell to a particular fate, is a process influenced by the extracellular environment and the contents of the cell acting on the cell's genome. *Determination* is not something that is visible under the microscope—cells do not change appearance when they become determined. Determination is followed by differentiation, the actual changes in biochemistry, structure, and function that result in cells of different types. *Differentiation* often involves a change in appearance as well as function. Determination precedes differentiation. Determination is a commitment; the final realization of this commitment is differentiation.

The Role of Differential Gene Expression in Cell Differentiation

Differentiated cells are recognizably different from one another, sometimes visually as well as in their protein products. Certain cells in our hair follicles continuously produce keratin, the protein that makes up hair, nails, feathers, and porcupine quills. Other cell types in the body do not produce keratin. In the hair follicle cells, the gene that encodes keratin is transcribed; in most other cells in our body, that gene is not transcribed. Activation of the keratin gene is a key step in the differentiation of the hair follicle cells.

Generalizing from examples like this one, we may say that *differentiation results from differential gene expression*—that is, from the differential regulation of transcription, posttranscriptional events such as mRNA splicing, and translation (see Chapter 14).

Because the fertilized egg, or zygote, has the ability to give rise to every type of cell in the adult body, we say it is **totipotent**. Its genome contains instructions for all of the structures and functions that will arise throughout the life cycle. Later in the development of animals (and probably to a lesser extent in plants), the cellular descendants of the zygote lose their totipotency and become determined—that is, committed to form only certain parts of the embryo. Determined cells differentiate into specific types of specialized cells, such as nerve cells or muscle cells. When a cell becomes specialized, it is said to have differentiated.

Differentiation usually does not include an irreversible change in the genome

Differentiation is irreversible in certain types of cells. Examples include the mammalian red blood cell, which loses its nucleus during development, and the tracheid, a water-conducting cell in vascular plants. Tracheid development culminates in the death of the cell, leaving only the pitted cell walls that formed while the cell was alive (see Chapter 34). In both of these extreme cases, the irreversibility of differentiation can be explained by the absence of a nucleus.

Generalizing about mature cells that retain functional nuclei is more difficult. We tend to think of plant differentiation as reversible and of animal differentiation as irreversible, but this is not a hard-and-fast rule. Why is differentiation apparently reversible in some cells but not in

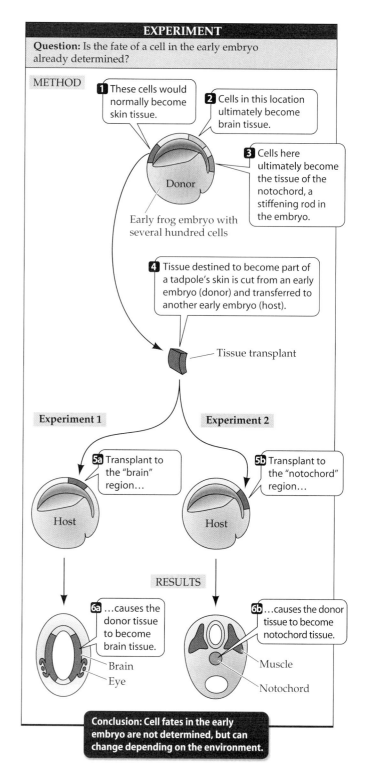

16.2 Developmental Potential in Early Frog Embryos
Cells that would be expected to form one kind of tissue can form completely different tissues when they are experimentally moved within the early embryo.

others? At some stage of development, do changes within the nucleus permanently commit a cell to specialization? For both higher plants and animals, the answer appears to be no. Under the right environmental circumstances, differentiation is reversible.

A food storage cell in a carrot root faces a dark future. It cannot photosynthesize or give rise to new carrot plants.

However, if we isolate that cell from the root, maintain it in a suitable nutrient medium, and provide it with appropriate chemical cues, we can "fool" the cell into acting as if it were a fertilized egg. It can divide and give rise to a normal carrot embryo and, eventually, a complete plant (Figure 16.3). Since the new plant is genetically identical to the somatic cell from which it came, we call the plant a *clone*.

The ability to clone an entire carrot plant from a differentiated root cell indicates that the cell contains the entire carrot genome and that it can express the appropriate genes in the right sequence. Many cells from other plant species show similar behavior in the laboratory, and this ability to generate a whole plant from a single cell has been invaluable in agricultural biotechnology (see Chapter 17).

These experiments with plants establish that a somatic *cell* is totipotent. A more direct demonstration that all the genetic material is present in differentiated cells has come from *nuclear* transplant experiments. Such experiments were first done on frogs by Robert Briggs and Thomas King, who asked whether the nuclei of early frog embryos had lost the ability to do what the totipotent zygote nucleus could do. They first removed the nucleus from an unfertilized egg, thus forming an *enucleated* egg. Then, with a very fine glass tube, they punctured a cell from an early embryo and drew up part of its contents, including the nucleus, which they injected into the enucleated egg.

More than 80 percent of these nuclear transplant operations resulted in the formation, from the egg and its new nucleus, of a normal early embryo. Of these embryos, more than half developed into normal tadpoles and, ultimately, normal adult frogs. These experiments showed that no information is lost from the nuclei of cells as they pass through the early stages of embryonic development, and that the cytoplasmic environment around a nucleus can modify its fate.

Similar experiments have been performed on rhesus monkeys, in which a single cell can be removed from an 8-celled embryo and fused with an enucleated egg, allowing the nucleus of the embryonic cell to enter the egg cytoplasm. The resulting cell acts like a zygote, forming an embryo, which can be implanted into a foster mother, who ultimately gives birth to a normal monkey. Each of the remaining 7 cells from the original embryo can similarly give rise to offspring by the same cell fusion technique.

In humans, the totipotency of early embryonic cells permits both genetic screening and in vitro fertilization. An 8-cell human embryo can be isolated in the laboratory and a single cell removed to determine whether a harmful genetic condition is present (Chapter 18). Each remaining cell, being totipotent, can be stimulated to divide and form an embryo, which can be implanted into the mother, where it develops into an infant (Chapter 42).

Later frog nuclear transplant experiments by John Gurdon showed that a donor nucleus obtained from later stages in the frog's life could occasionally give rise to a normal tadpole, again showing totipotency. But successful cloning of animals was very difficult until the late 1990s,

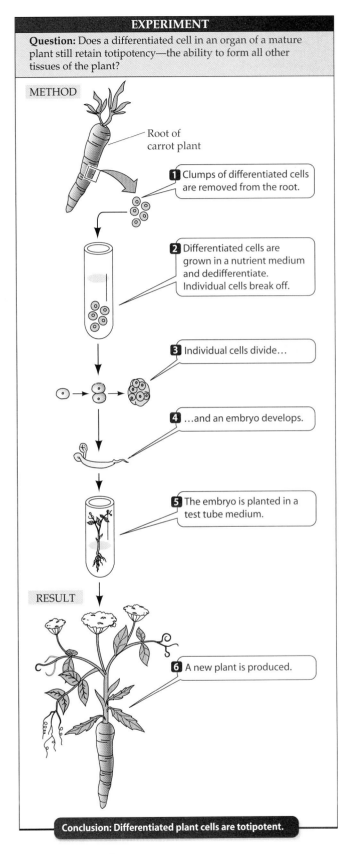

16.3 Cloning a Plant
Differentiated, specialized food storage cells from the root of a carrot can be induced to dedifferentiate, act like embryonic cells, and form a new plant.

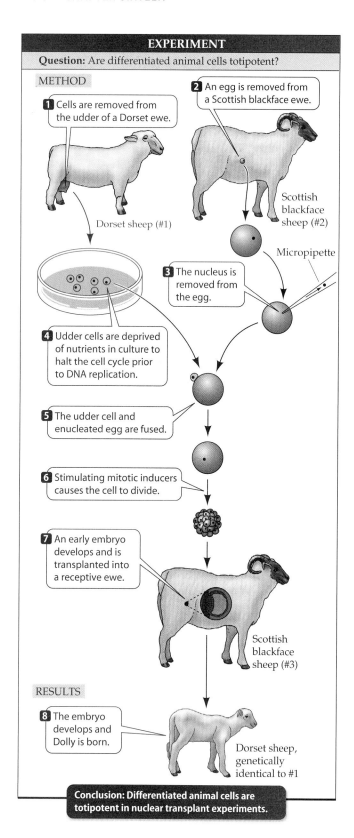

EXPERIMENT

Question: Are differentiated animal cells totipotent?

METHOD

1 Cells are removed from the udder of a Dorset ewe.

2 An egg is removed from a Scottish blackface ewe.

Dorset sheep (#1)

Scottish blackface sheep (#2)

Micropipette

3 The nucleus is removed from the egg.

4 Udder cells are deprived of nutrients in culture to halt the cell cycle prior to DNA replication.

5 The udder cell and enucleated egg are fused.

6 Stimulating mitotic inducers causes the cell to divide.

7 An early embryo develops and is transplanted into a receptive ewe.

Scottish blackface sheep (#3)

RESULTS

8 The embryo develops and Dolly is born.

Dorset sheep, genetically identical to #1

Conclusion: Differentiated animal cells are totipotent in nuclear transplant experiments.

16.4 A Clone and Her Offspring
Although Dolly herself (right) is a clone with only one parent, she has mated and given birth to "normal" offspring (the lamb on the left), proving the genetic viability of cloned mammals.

embryo. Apparently, when mammalian donor cells were in the G2 phase of the cell cycle and were fused with egg cytoplasm also in G2, some extra DNA replication took place that created havoc with the cell cycle in the egg when it attempted to divide.

Wilmut took differentiated cells from a ewe's udder and starved them of nutrients for a week, thus halting the cells in G1. After one of these cells was fused with an enucleated egg from a different breed of ewe, mitotic inducers in the egg cytoplasm (see Chapter 9) were stimulated and the donor nucleus entered S phase; the rest of the cell cycle proceeded normally. After several cell divisions, the resulting early embryo was transplanted into the womb of a surrogate mother. Out of 277 successful attempts to fuse adult cells with enucleated eggs, one lamb, named Dolly, survived to be born. DNA analyses confirmed that Dolly was genetically identical to the ewe from whose udder the donor nucleus had been obtained.

The purpose of Wilmut's experiment was to clone sheep that have been genetically programmed to produce products such as pharmaceuticals in their milk. The cloning procedure could make multiple, identical copies of sheep that are reliable producers of a drug such as α-1-antitrypsin, which is used to treat people with cystic fibrosis (see Figure 17.15).

The trick of starving donor cells for cloning has been applied to other mammals. Mice have been cloned using the cells lining the egg as a source of donor nuclei (Figure 16.5). Cows have been cloned to preserve a rare breed in New Zealand. Genetically engineered goats have been cloned to produce several useful proteins in their milk. This flurry of cloning has touched off a flurry of controversy, but cloning

when Ian Wilmut and his colleagues at a biotechnology company in Scotland used the cell fusion procedure to clone sheep (Figure 16.4). Previous attempts to produce mammals by this method had worked, as in the rhesus monkey case, only if the donor nucleus was from an early

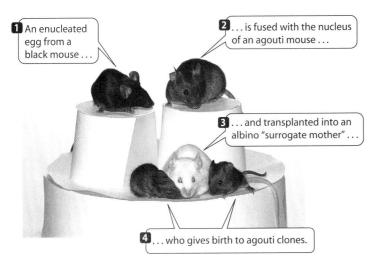

16.5 Cloned Mice
Because so much is known about mouse genetics and molecular biology, cloned mice may be useful in studies of basic biology.

is not a new scientific concept. The idea of totipotency was accepted long before Dolly was born, but achieving it is an impressive technical achievement.

An example of nuclear totipotency gone awry occurs in a human tumor called a *teratocarcinoma*. Here, a differentiated cell *dedifferentiates* to form an unspecialized cell. Then it divides, forming a tumor, as occurs in most cancers. But some cells in the tumor redifferentiate to form specialized tissue arrangements. So the tumor can be a single mass of cells inside the abdomen, with some of the cells forming kidney tubes, others hair, and still others teeth! How this occurs is not clear.

Stem cells can be induced to differentiate by environmental signals

Totipotency implies that a differentiated cell stays that way because of its environment, and that appropriate environmental changes could result in a new pattern of differentiation. In normal development, a complex series of signals and their transduction results in the patterns of differentiation we see in a newborn organism. If these signals could be described in enough detail, we should be able to understand how any cell type becomes any other.

Stem cells are undifferentiated, dividing cells that are found in adult animal tissues that need frequent cell replacement, such as skin, the inner lining of the intestine, and the blood system. As they divide, stem cells produce cells that differentiate to replace dead cells and maintain tissue homeostasis. In the body, stem cells have limited abilities to differentiate. The stem cells in bone marrow, for example, produce the various types of red and white blood cells, while stem cells in the nervous system produce the various differentiated nerve cells.

Can one kind of stem cell be manipulated by its environment to produce cells that differentiate into cells of another tissue? The answer appears to be yes. For example, when stem cells of the brain were transplanted into the bone marrow of mice whose bone marrow stem cells had been depleted, they proceeded to act like bone marrow stem cells, producing blood cells. In the reverse experiment, bone marrow stem cells were implanted into the brains of mice, where they formed brain cells. These experiments indicate that the environment—presumably intercellular signals—determines what a stem cell will do.

The stem cell populations closest to totipotency are not the ones in adults, but those of the early embryo. In mice, these embryonic stem cells can be removed from an early embryo (called a *blastocyst*) and then induced to differentiate in some particular way. Normally, these cells are formed a few days after fertilization, and soon become determined as to what their fate will be in the developing embryo. Before then, however, they are virtually totipotent. Such cells can be grown indefinitely in the laboratory and, when injected back into a mouse blastocyst, will mix with the resident cells and differentiate to form all the cells of the mouse. This kind of experiment shows that they have not lost any of their developmental potential while growing in the laboratory.

Embryonic stem cells growing in the laboratory can be induced to differentiate if the right signal is provided (Figure 16.6). For example, treatment of mouse embryonic stem cells with a derivative of vitamin A causes them to form nerve cells, while other growth factors induce them to form blood cells, again demonstrating their developmental potential and the roles of environmental signals. This finding raises the possibility of using stem cell cultures as sources of differentiated cells for clinical medicine. A key advance toward this use has been the ability to grow human embryonic stem cells in the laboratory. The age of custom-made cells to replace ones lost to disease or injury is rapidly approaching.

Genes are differentially expressed in cell differentiation

Experiments such as nuclear transplants in frogs and sheep—as well as plant cell cloning—point to the conclusion of genome constancy or equivalence in all somatic cells of an organism. Molecular experiments have provided even more convincing evidence. For example, the gene for β-globin, one of the protein components of hemoglobin, is present and expressed in red blood cells as they form in the bone marrow of mammals. Is the same gene also present—but unexpressed—in nerve cells in the brain, which do not make hemoglobin?

Nucleic acid hybridization (see Figure 14.7) can provide an answer. A probe for the β-globin gene can be applied to DNA from both brain cells and immature red blood cells (recall that mature red blood cells lose their nuclei and DNA). In both cases, the probe finds its complement, showing that the β-globin gene is present in both types of cells.

On the other hand, if the probe is applied to cellular mRNA rather than cellular DNA, it finds β-globin mRNA

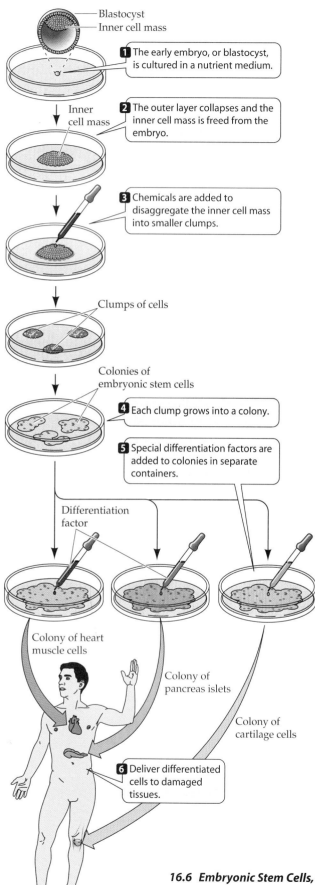

1 The early embryo, or blastocyst, is cultured in a nutrient medium.

Blastocyst
Inner cell mass

2 The outer layer collapses and the inner cell mass is freed from the embryo.

Inner cell mass

3 Chemicals are added to disaggregate the inner cell mass into smaller clumps.

Clumps of cells

Colonies of embryonic stem cells

4 Each clump grows into a colony.

5 Special differentiation factors are added to colonies in separate containers.

Differentiation factor

Colony of heart muscle cells

Colony of pancreas islets

Colony of cartilage cells

6 Deliver differentiated cells to damaged tissues.

16.6 Embryonic Stem Cells, Differentiation, and Medicine
While embryonic stem cells have been cultured from humans, their potential use in medicine was suggested by experiments in mice. This technique is under intensive investigation.

only in the red blood cells, and not in the brain cells. This result shows that the gene is expressed in only one of the two tissues. Many similar experiments have shown convincingly that differentiated cells lose none of the genes that were present in the fertilized egg.

What leads to this differential gene expression? One well-studied example is the conversion of undifferentiated muscle precursor cells, called *myoblasts*, into the large, multi-nucleated cells that make up mammalian skeletal muscle cells (called *muscle fibers*). The key event that starts this conversion is the expression of *MyoD1* (*My*oblast *D*etermining Gene *1*). The protein product of this gene is a transcription factor (MyoD1) with a helix-loop-helix domain (see Chapter 14) that not only binds to promoters of the muscle-determining genes to stimulate their transcription, but also acts on its own promoter to keep its levels high in the myoblasts and in their descendants.

Strong evidence for the controlling role of MyoD1 comes from experiments in which a gene containing an active promoter adjacent to *MyoD1* DNA is injected into the precursors of other cell types. For example, if *MyoD1* DNA is put into fat cell precursors, they are reprogrammed to become muscle cells. Genes such as *MyoD1*, which code for proteins that direct fundamental decisions in development, often by regulating genes on other chromosomes, usually code for transcription factors.

The Role of Polarity in Cell Determination

What initially stimulates the *MyoD1* promoter to begin transcription is not clear, but a chemical signal clearly is involved. In general, two overall mechanisms for producing such signals have been found. In **cytoplasmic segregation**, a factor within eggs, zygotes, or precursor cells is unequally distributed in the cytoplasm. After cell division, the factor ends up in some cells or regions of cells and not others. In **induction**, a factor is actively produced and secreted by certain cells to induce other cells to differentiate.

Polarity—the difference between one end of an embryo and the other—is obvious in development. Our heads are distinct from our feet, and the distal ends of our arms (wrists and fingers) differ from the proximal ends (shoulders). An animal's polarity develops early, even in the egg itself. Yolk may be distributed asymmetrically in the egg and embryo. In addition, other chemical substances may be confined to specific parts of the egg, or may be more concentrated at one pole than at the other.

In some animals, the original polar distribution of materials in the egg's cytoplasm changes as a result of fertilization. As cell division proceeds, the resulting cells contain unequal amounts of the materials that were not distributed uniformly in the zygote. As we learned from the work on

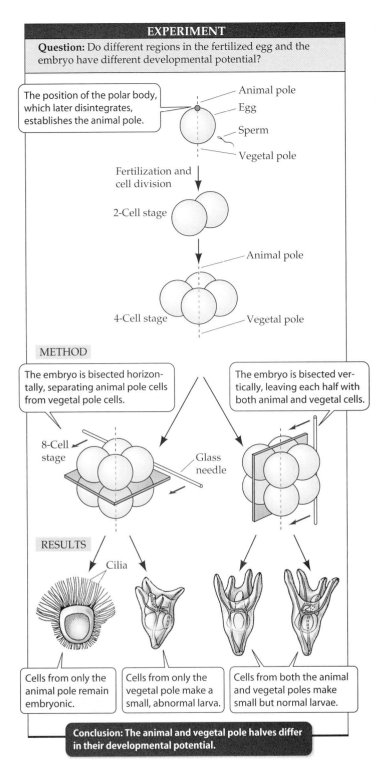

EXPERIMENT

Question: Do different regions in the fertilized egg and the embryo have different developmental potential?

The position of the polar body, which later disintegrates, establishes the animal pole.

Animal pole
Egg
Sperm
Vegetal pole

Fertilization and cell division

2-Cell stage

Animal pole

4-Cell stage — Vegetal pole

METHOD

The embryo is bisected horizontally, separating animal pole cells from vegetal pole cells.

The embryo is bisected vertically, leaving each half with both animal and vegetal cells.

8-Cell stage

Glass needle

RESULTS

Cilia

Cells from only the animal pole remain embryonic.

Cells from only the vegetal pole make a small, abnormal larva.

Cells from both the animal and vegetal poles make small but normal larvae.

Conclusion: The animal and vegetal pole halves differ in their developmental potential.

16.7 Early Asymmetry in the Embryo
The upper (animal pole) and lower (vegetal pole) halves of very early sea urchin embryos differ in their developmental potential. Cells from both halves are necessary to produce a normal larva.

cloning, cell nuclei do not always undergo irreversible changes during early development; thus we can explain some embryological events on the basis of the *cytoplasmic* differences in cells.

Even a structure as apparently uniform as a sea urchin egg has polarity. A striking difference between cells can be

demonstrated very early in embryonic sea urchin development. The development of embryos that have been divided in half at the 8-cell stage depends on how they are separated. If the embryo is split into "left" and "right" halves, with each half containing cells from both the upper and the lower halves, normal-shaped but dwarfed larvae develop. If, however, the cut separates the upper four cells from the lower four, the result is different. The upper four cells develop into an abnormal early embryo with large cilia at one end that cannot form a larva. The lower four cells develop into small, somewhat misshapen larvae with an oversized gut (Figure 16.7). These results shows that for fully normal development, factors from both the upper and lower parts of the embryo are necessary.

This and many other experiments established that certain materials, called **cytoplasmic determinants**, are distributed unequally in the egg cytoplasm, and that these materials play a role in directing the embryonic development of many organisms.

The Role of Embryonic Induction in Cell Determination

Experimental work on developing embryos has clearly established that in many cases, the fates of particular tissues are determined by interactions with other specific tissues in the embryo. In developing animal embryos there are many such instances of induction, in which one tissue causes an adjacent tissue to develop in a particular manner. These effects are mediated by intercellular biochemical communication—that is, signal transduction mechanisms. We will describe two examples of such induction: one in the developing vertebrate eye, and the other in a developing reproductive structure in the nematode *C. elegans*.

Tissues direct the development of their neighbors by secreting inducers

The development of the lens in the vertebrate eye is a classic example of induction. In a frog embryo, the developing forebrain bulges out at both sides to form the *optic vesicles*, which expand until they come into contact with the cells at the surface of the head (Figure 16.8). The surface tissue in the region of contact with the optic vesicles thickens, forming a *lens placode*. The lens placode bends inward, folds over on itself, and ultimately detaches from the surface to produce a structure that will develop into the lens.

If the growing optic vesicle is cut away before it contacts the surface cells, no lens forms. Placing an impermeable barrier between the optic vesicle and the surface cells also prevents the lens from forming. These observations suggest that the surface tissue begins to develop into a lens when it receives a signal—an embryonic **inducer**—from the optic vesicle.

The interaction of tissues in eye development is a two-way street: There is a "dialogue" between the developing optic vesicle and the surface tissue. The optic vesicle induces lens development, and the developing lens determines the

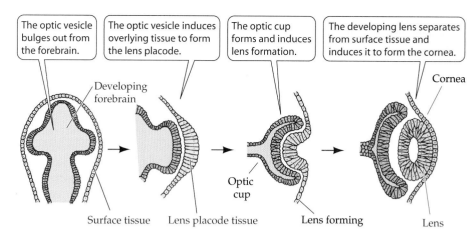

16.8 Inducers in the Vertebrate Eye
The eye of a frog develops as inducers take their turns.

The optic vesicle bulges out from the forebrain.

The optic vesicle induces overlying tissue to form the lens placode.

The optic cup forms and induces lens formation.

The developing lens separates from surface tissue and induces it to form the cornea.

Developing forebrain

Optic cup

Surface tissue   Lens placode tissue   Lens forming   Lens

Cornea

size of the optic cup that forms from the optic vesicle. If head surface tissue from a frog species with small eyes is grafted over the optic vesicle of one with large eyes, both lens and optic cup are of intermediate size.

The developing lens also induces the surface tissue over it to develop into a *cornea*, a specialized layer that allows light to pass through and enter the eye. Thus a chain of inductive interactions participates in the development of the parts required to make an eye. Induction triggers a sequence of gene expression in the responding cells. Tissues do not induce themselves; rather, different tissues interact and molecularly induce each other. We will return to embryonic induction in Chapter 43.

Single cells can induce changes in their neighbors

The tiny nematode roundworm *Caenorhabditis elegans* is used as a model organism in many biological studies, but it is especially useful for studying development. It normally lives in the soil, where it feeds on bacteria, but can also grow in the laboratory if supplied with its food source. The entire process of development from egg to larva takes about 8 hours. It is easily observed using a low-magnification dissecting microscope because the body covering is transparent (Figure 16.9a). Because *C. elegans* is easy to culture, develops rapidly, and is easily observed, it is a favorite experimental organism. The development of *C. elegans* does not vary, so it has been possible to identify the source of each of the 959 somatic cells of the adult form.

The adult nematode is hermaphroditic, containing both male and female reproductive organs. It lays eggs through a pore called the *vulva* on the ventral surface. During development, a single cell, called the *anchor cell*, induces the vulva to form. If the anchor cell is destroyed by laser surgery, no vulva forms. The eggs develop inside the parent, and a "bag of worms" that eventually consume the parent results.

The anchor cell controls the fates of six cells on the animal's ventral surface through two molecular switches. Each of these cells has three possible fates: It may become a primary vulval precursor, a secondary vulval precursor, or simply part of the worm's surface—an epidermal cell (Figure 16.9b).

The anchor cell produces an inducer that diffuses out of the cell and interacts with adjacent cells. Cells that receive enough of the inducer become vulval precursors; cells slightly farther from the anchor cell become epidermis. The first molecular switch, controlled by the inducer from the anchor cell, determines whether a cell takes the "track" toward becoming part of the vulva or the track toward becoming epidermis.

The cell closest to the anchor cell, having received the most inducer, differentiates into the primary vulval precursor and apparently produces its own inducer, which acts on the two neighboring cells and directs them to become secondary vulval precursors. Thus the primary vulval precursor cell controls a second molecular switch, determining whether a vulval precursor will take the primary track or the secondary track. The two inducers control the activation or inactivation of specific genes in the responding cells.

There is an important lesson to draw from this example: Much of development is controlled by molecular switches that allow a cell to proceed down one of two alternative tracks. One challenge for the developmental biologist is to find these molecular switches and determine how they work. The primary inducer for the *C. elegans* vulva appears to be a growth factor homologous to the mammalian epidermal growth factor (EGF). The nematode growth factor, called LIN-3, binds to a receptor on the surface of a vulval precursor cell. This binding sets in motion a signal transduction cascade involving the ras protein and MAP kinases (see Figure 15.11). The end result is increased transcription of genes involved in the differentiation of vulval cells.

The Role of Pattern Formation in Organ Development

Pattern formation, the spatial organization of a tissue or organism, is inextricably linked to *morphogenesis*, the appearance of body form. The differentiation of cells is beginning to be understood in terms of molecular events, but how do molecular events contribute to the organization of multitudes of cells into specific body parts, such as a leaf, a flower, a shoulder blade, or a tear duct?

Some cells are programmed to die

Apoptosis is programmed cell death, a series of events caused by the expression of certain genes (see Figure 9.20). Some of these "death genes" have been pinpointed, and related ones have been found in organisms as diverse as nematodes and humans.

Apoptosis is vital to the normal development of all animals. For example, the nematode *C. elegans* produces precisely 1,090 somatic cells as it develops from a fertilized egg to an adult (see Figure 16.9). But 131 of these cells die. The sequential expression of two genes called *ced-4* and *ced-3*

(a) **The nematode** *C. elegans*

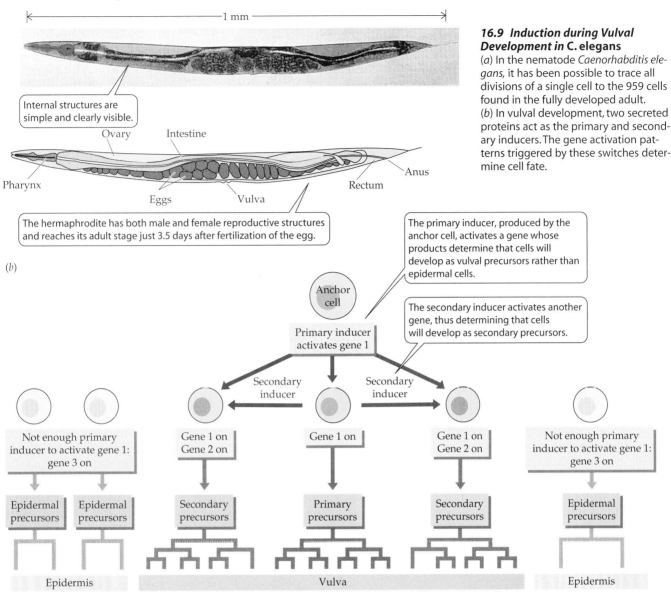

Internal structures are simple and clearly visible.

Ovary   Intestine

Pharynx

Eggs   Vulva   Rectum   Anus

The hermaphrodite has both male and female reproductive structures and reaches its adult stage just 3.5 days after fertilization of the egg.

***16.9 Induction during Vulval Development in* C. elegans**
(*a*) In the nematode *Caenorhabditis elegans*, it has been possible to trace all divisions of a single cell to the 959 cells found in the fully developed adult.
(*b*) In vulval development, two secreted proteins act as the primary and secondary inducers. The gene activation patterns triggered by these switches determine cell fate.

(*b*)

Anchor cell

The primary inducer, produced by the anchor cell, activates a gene whose products determine that cells will develop as vulval precursors rather than epidermal cells.

The secondary inducer activates another gene, thus determining that cells will develop as secondary precursors.

Primary inducer activates gene 1

Secondary inducer   Secondary inducer

Not enough primary inducer to activate gene 1: gene 3 on

Gene 1 on Gene 2 on

Gene 1 on

Gene 1 on Gene 2 on

Not enough primary inducer to activate gene 1: gene 3 on

Epidermal precursors   Epidermal precursors

Secondary precursors

Primary precursors

Secondary precursors

Epidermal precursors

Epidermis   Vulva   Epidermis

(for *cell death*) appears to control this process. In the nervous system, for example, there are 302 nerve cells that come from 405 precursors; thus 103 cells undergo apoptosis. If the protein coded for by either *ced-3* or *ced-4* is nonfunctional, all 405 cells form neurons, and disorganization results. A third gene, *ced-9*, codes for an inhibitor of apoptosis: that is, its protein blocks the function of the *ced-4* gene. So, where cell death is required, *ced-3* and *ced-4* are active and *ced-9* is inactive; where cell death does not occur, the reverse is true.

Remarkably, a similar system of cell death genes acts in humans. During early development, human hands and feet look like tiny paddles—the fingers and toes are linked by webbing. Between days 41 and 56 of development, the cells in the webbing die, freeing the individual fingers and toes (Figure 16.10). The protein—an enzyme called *caspase*—that stimulates this apoptosis is similar in amino acid sequence to the protein encoded by *ced-3*, and a human protein (*bcl-2*) that inhibits apoptosis is similar to *ced-9*. So humans and

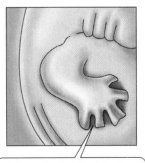

41 days after fertilization: Genes for programmed cell death are expressed in the tissue between the digits.

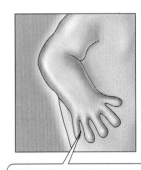

56 days after fertilization: Apoptosis is complete. Cells of the digits have absorbed the remains of the dead cells.

16.10 Apoptosis Removes the Webbing between Fingers
Early in the second month of human development, the webbing connecting the fingers is removed by apoptosis, freeing the individual fingers.

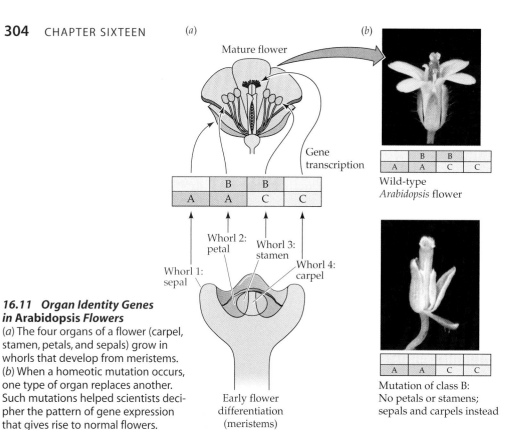

Wild-type
Arabidopsis flower

Mutation of class A:
No petals or sepals;
stamens and carpels instead

Mutation of class B:
No petals or stamens;
sepals and carpels instead

Mutation of class C:
No stamens or carpels;
petals and sepals instead

16.11 Organ Identity Genes in Arabidopsis Flowers
(a) The four organs of a flower (carpel, stamen, petals, and sepals) grow in whorls that develop from meristems. (b) When a homeotic mutation occurs, one type of organ replaces another. Such mutations helped scientists decipher the pattern of gene expression that gives rise to normal flowers.

nematodes, two creatures separated by more than 600 million years of evolutionary time, have similar genes controlling programmed cell death.

Apoptosis plays many other roles in your life. The dead cells that form the outermost layer of your skin and those from the uterine wall that are lost during menstruation have undergone apoptosis. White blood cells live only a few months in the circulation, then undergo apoptosis (see Figure 9.20). In a form of cancer called follicular large-cell lymphoma, these white blood cells do not die, but continue to divide. The reason is a mutation that causes the overexpression of *bcl-2*, the gene that inhibits cell death.

Plants have organ identity genes

Like animals, plants have organs—for example, leaves and roots. Many plants form flowers, and many flowers are composed of four types of organs: sepals, petals, stamens, and carpels. These floral organs occur in *whorls*, which are groups of each organ around a central axis. The whorls develop from groups of cells called *meristems* in the shape of domes, which develop at growing points on the stem (Figure 16.11a). How is the identity of a particular whorl determined? The answer appears to lie in the activities of a group of genes.

These genes have been best described in *Arabidopsis thaliana*, also called mouse-ear cress. This plant is very useful for studies of development because of its small size, abundant seed production (over 1,000 per plant), rapid development (from seed to plant to seed in 6 weeks), and small genome (10 chromosomes with 80 million base pairs of DNA). Finally, it is easy to produce mutations in this plant by treating the seeds with mutagens.

Normal *Arabidopsis* flowers have four whorls of organs, but there are mutant strains that have the wrong organs in particular whorls (Figure 16.11b). Such mutations, in which one organ is replaced with another, are called *homeotic mutations*. Studies on three mutant strains led to a model for the determination of floral organ identity. This model involves three organ identity genes, each of which is expressed in *two* of the whorls:

▶ A class A gene is expressed in whorls 1 and 2 (which normally form sepals and petals, respectively).
▶ A class B gene is expressed in whorls 2 and 3 (which normally form petals and stamens, respectively).
▶ A class C gene is expressed in whorls 3 and 4 (which normally form stamens and carpels, respectively).

Nucleic acid hybridization confirmed these locations for the mRNA transcripts of the three genes.

Not surprisingly, these three genes code for transcription factors, which are active as dimers. Possibly, gene regulation in these cases is *combinatorial*—that is, the composition of the dimer determines which other genes will be activated by the transcription factor. For example, a dimer made up only of transcription factor A would activate transcription of the genes that make sepals; a dimer made up of A and B would result in petals, and so forth.

A gene called *leafy* appears to control the transcription of the ABC genes. Plants with the *leafy* mutation are just that—they don't make flowers. The protein product of this gene acts as a transcription factor stimulating genes A, B, and C so that they produce the flower (Figure 16.12).

In addition to being fascinating to biologists, organ identity genes have caught the eye of horticultural and agricultural scientists. Flowers filled with petals instead of stamens and carpels often have mutations of the C genes. Many of the foods that make up the human diet come from fruits and seeds, such as the grains of wheat, rice, and corn. These fruits and seeds form from the carpels (the female reproductive organs) of the flower. Genetically modifying the number of these organs on a particular plant could increase the amount of grain a crop could produce.

Plants and animals use positional information

Certain cells in both plants and animals appear to "know" where they are with respect to the body as a whole. This spatial "sense" is called **positional information**. In plants, the pattern of development of two major types of conducting tissue suggested to scientists long ago that distance from the body surface may play a role in their formation. Cells destined to become water conductors are farther from the body surface than are those destined to become conductors of sucrose, a product of photosynthesis. Thus those cells destined to become water conductors are exposed to lower concentrations of O_2 and higher concentrations of CO_2. These differences may help determine which genes are expressed in which parts of the stem and root.

Recently it has been suggested that cells on the surface of the stem secrete a protein or other signal that is more concentrated close to the surface than deeper in the stem. Other signals may diffuse from the stem tip and root tip, establishing positional information along the plant's axis. These signals are called *morphogens*.

The wing of a chick develops from a round bud. The cells that become the bones and muscles of the wing must receive positional information. If they do not, the limbs will be totally disorganized (imagine fingers growing out of your shoulders). Three groups of cells—one at the junction of the bud and the body, a second at the tip of the bud, and a third on the surface of the bud—produce different morphogens that diffuse through the bud. Each cell in the bud receives unique concentrations of each morphogen. The

first morphogen determines the *proximal–distal* ("shoulder to fingertip") axis of the wing, the second determines the *anterior–posterior* ("thumb to little finger") axis, and the third determines the *dorsal–ventral* ("palm to knuckles") axis.

The signaling pathways involved in limb development have been conserved through animal evolution. Comparative genomic studies have revealed developmental signaling pathways using homologous morphogen proteins in organisms ranging from nematodes to humans.

The Role of Differential Gene Expression in Establishing Body Segmentation

Another experimental subject that developmental biologists have used to study pattern formation is the fruit fly, *Drosophila*. Insects (and many other animals) develop a highly modular body composed of different types of *segments*. Complex interactions of different sets of genes underlie the pattern formation of segmented bodies.

Unlike the body segments of segmented worms such as earthworms, the segments of the *Drosophila* body are clearly different from one another. The adult fly has a head (composed of several fused segments), 3 different thoracic segments, 8 abdominal segments, and a terminal segment at the posterior end. The 13 seemingly identical segments of the *Drosophila* larva correspond to these specialized adult segments. Several types of genes are expressed sequentially in the embryo to define these segments. The first step in this process is to establish the polarity of the embryo.

Maternal effect genes determine polarity

In *Drosophila* eggs and larvae, polarity is based on the distribution of morphogens, of which some are mRNA's and some are proteins. These molecules are products of specific **maternal effect genes** in the mother and are distributed to the eggs, often in a nonuniform manner. The maternal effect genes are transcribed in the *nurse cells*, which surround and nurture the developing egg and are localized at certain specific regions of the egg as it forms. Maternal effect genes produce their effects on the embryo regardless of the genotype of the father. Their products determine the dorsal–ventral (back–belly) and anterior–posterior (head–tail) axes of the embryo.

The fact that these morphogens specify these axes has been established by the results of experiments in which cytoplasm was transferred from one egg to another. Females homozygous for a particular mutation of the maternal effect gene *bicoid* produce larvae with no head and or no thorax. However, if eggs of homozygous *bicoid* mutant females are inoculated at the anterior end with cytoplasm from the anterior region of a wild-type egg, the treated eggs develop into normal larvae, with heads developing from the part of the egg that receives the wild-type cytoplasm. Conversely, removal of 5 percent or more of the cytoplasm from the anterior of a wild-type egg results in an abnormal larva that looks like a *bicoid* mutant larva.

Wild-type   *Leafy* mutant

16.12 A Nonflowering Mutant
Mutations in the *leafy* gene of *Arabidopsis* prevent the transcription of the ABC genes, and the resulting plant does not produce any flower.

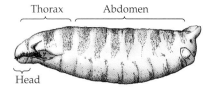

Thorax   Abdomen

Head

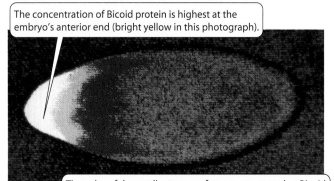

The concentration of Bicoid protein is highest at the embryo's anterior end (bright yellow in this photograph).

The color of the gradient moves from orange to red as Bicoid concentration decreases into the dark blue posterior end.

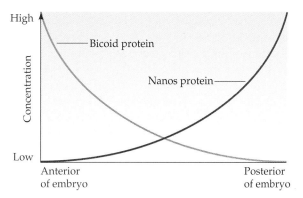

16.13 Bicoid and Nanos Protein Gradients Provide Positional Information
The anterior–posterior axis of *Drosophila* arises from morphogens produced by the maternal effect genes *bicoid* and *nanos*. The mRNA's are translated at the ends of the larva, and the resulting gradient controls the developing body's polarity.

Another maternal effect gene, *nanos*, plays a comparable role in the development of the posterior end of the larva. Eggs from homozygous *nanos* mutant females develop into larvae with missing abdominal segments. Injecting a *nanos* egg with cytoplasm from the posterior region of a wild-type egg allows normal development. These findings show that, in wild-type larvae, the overall framework of the anterior–posterior axis is laid down by the activity of these two maternal effect genes (Figure 16.13).

After the axes of the embryo are determined, the next step in pattern formation is to determine the larval segments.

Segmentation and homeotic genes act after the maternal effect genes

The number, boundaries, and polarity of the larval segments are determined by proteins encoded by the **segmentation genes**. These genes are expressed when there are

about 6,000 nuclei in the embryo. The nuclei all look the same, but in terms of gene expression, they are not.

The maternal effect genes set the segmentation genes in motion. Three classes of segmentation genes act, one after the other, to regulate finer and finer details of the segmentation pattern (Figure 16.14):

▶ First, **gap genes** organize large areas along the anterior–posterior axis. Mutations in gap genes result in gaps in the body plan—the omission of several larval segments.

▶ Second, **pair rule genes** divide the embryo into units of two segments each. Mutations in pair rule genes result in embryos missing every other segment.

▶ Third, **segment polarity genes** determine the boundaries and anterior–posterior organization of the segments. Mutations in segment polarity genes result in segments in which posterior structures are replaced by reversed (mirror-image) anterior structures.

▶ Finally, after the basic pattern of segmentation has been established by the segmentation genes, differences between the segments are mediated by the activities of **homeotic genes**. These genes are expressed in different combinations along the length of the body and tell each segment what to become. Homeotic genes are analogous to the organ identity genes of plants.

The maternal effect, segmentation, and homeotic genes interact to "build" a *Drosophila* larva step by step, beginning with the unfertilized egg.

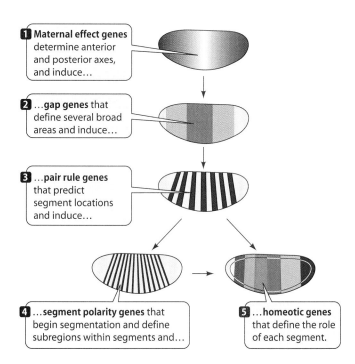

1 Maternal effect genes determine anterior and posterior axes, and induce…

2 …gap genes that define several broad areas and induce…

3 …pair rule genes that predict segment locations and induce…

4 …segment polarity genes that begin segmentation and define subregions within segments and…

5 …homeotic genes that define the role of each segment.

16.14 A Gene Cascade Controls Pattern Formation in the Drosophila Embryo
Gap, pair rule, and segment polarity genes are collectively referred to as the segmentation genes. The shading shows the locations of their gene products in the embryo.

Drosophila development results from a transcriptionally controlled cascade

One of the most striking and important observations about development in *Drosophila*—and in other animals—is that it results from a sequence of changes, with each change triggering the next. The sequence, or *cascade*, is largely controlled at the levels of transcription and translation.

In general, unfertilized eggs are storehouses of mRNA's, which are made prior to fertilization to support protein synthesis during the early stages of embryo development. Indeed, early embryos do not carry out transcription. After several cell divisions, mRNA production begins, forming the mRNA's needed for later development.

Some of the prefabricated mRNA's in the egg provide positional information. Before the egg is fertilized, mRNA for the Bicoid protein is localized at the end destined to become the anterior end of the fly. After the egg is fertilized and laid, nuclear divisions begin. (In *Drosophila*, cell divisions do not begin right away; until the thirteenth cell division, the embryo is a single, multinucleated cell called a *syncytium*.) At this early point, *bicoid* mRNA is translated, forming Bicoid protein, which diffuses away from the anterior end, establishing a gradient. At the posterior end, Nanos forms a gradient in the other direction. Thus each nucleus in the developing embryo is exposed to a different concentration ratio of Bicoid and Nanos proteins.

The two morphogens regulate the expression of the gap genes, although in different ways. Bicoid protein affects transcription, while Nanos affects translation. The high concentrations of Bicoid protein in the anterior portion of the egg turn on a gap gene called *hunchback*, while simultaneously turning off another gap gene, *Krüppel*. Nanos at the posterior end reduces the translation of *hunchback*, so a difference in concentration of these two gap gene products at the two ends is established.

The proteins encoded by the gap genes control the expression of the pair rule genes. Many pair rule genes in turn encode transcription factors that control the expression of the segment polarity genes, giving rise to a complex, striped pattern (see Figure 16.14) that foreshadows the segmented body plan of *Drosophila*.

By this point, each nucleus of the embryo is exposed to a distinct set of transcription factors. The segmented body pattern of the larva is established even before any sign of segmentation is visible. When the segments do appear, they are not all identical, because the homeotic genes specify the different structural and functional properties of each segment. Each homeotic gene is expressed over a characteristic portion of the embryo.

Let's turn now to the homeotic genes and see how their mutation can alter the course of development.

Homeotic mutations produce large-scale effects

Two bizarre homeotic mutations in *Drosophila* are the *Antennapedia* mutation, in which legs grow in place of antennae (Figure 16.15), and the *bithorax* mutation, in which an extra pair of wings grows in a thoracic segment (see Figure 24.9).

Edward Lewis at the California Institute of Technology found that *bithorax* was not a single gene, but a cluster of genes, each one determining the functional identity of a segment. Moreover, the genes were lined up along the chromosome in the order of the segments they determined: Genes at the beginning of the cluster determined thoracic segments, then the next genes determined the upper abdomen, and so on. A similar cluster of genes—the ones that are mutated in the *Antennapedia* flies—was found to determine the identities of the segments at the front of the fly. Lewis predicted that all of these genes might have come from the duplication of a single gene in an ancestral, unsegmented organism.

Molecular biologists confirmed Lewis's prediction. Using a nucleic acid hybridization probe for part of one of the genes of the cluster, several scientists found the probe binding not only to its own gene, but also to adjacent genes in its cluster and to genes in the other homeotic cluster. In

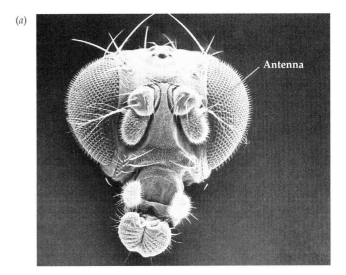

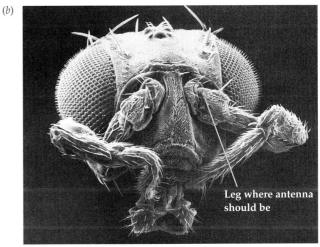

16.15 A Homeotic Mutation in Drosophila
Mutations of the homeotic genes cause body parts to form on inappropriate segments. (*a*) A wild-type fruit fly. (*b*) An *Antennapedia* mutant fruit fly.

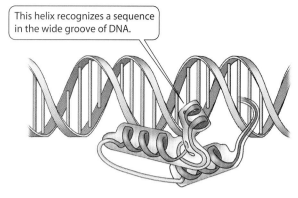

This helix recognizes a sequence in the wide groove of DNA.

16.16 A Homeodomain Binds to DNA
The homeodomain region has a helix-turn-helix motif. There are three α-helices, one of which is involved in recognition of a DNA sequence.

other words, there are DNA sequences that are common to all the homeotic genes in both clusters.

Homeobox-containing genes encode transcription factors

The 180-base-pair DNA sequence that is common to the *bithorax* and *Antennapedia* gene clusters is called the **homeobox**. It encodes a 60-amino-acid sequence, called the **homeodomain**, which binds to DNA (Figure 16.16). This sequence turns out to be present in other proteins involved in *Drosophila* pattern formation, such as *bicoid*. In all cases, the DNA-binding region of the protein has a helix-turn-helix motif (see Chapter 14). Each type of homeodomain recognizes a specific DNA sequence in target genes. The *bicoid* homeodomain, for example, recognizes TCCTAATCCC.

What do these proteins do when they recognize their target sequence in DNA? Not surprisingly, they are transcription factors, affecting genes involved in development. For example, Bicoid protein binds to promoters for the gap gene *hunchback*, activating its transcription. Hunchback protein is also a transcription factor, binding to enhancers and activating genes involved in head and thorax formation. In this way, the homeodomain proteins produce the cascade of events that controls *Drosophila* development.

Evolution and Development

Once DNA probes had identified the homeobox sequences involved in fruit

fly development, those probes were applied to other organisms. Soon, homeoboxes were found in over a hundred proteins from organisms as diverse as nematodes, frogs, mice, chickens, plants, and humans. As developmental biologists have described more and more intricate developmental systems at the molecular level, they have found that these systems are present in other organisms as well. These systems are an example of gene conservation. Comparative genomic studies have shown that, like the genes involved in biochemical pathways, the genes controlling development have much in common among different organisms. There may be only small differences that have turned one species into another, or a fin into a limb. This comparative approach has spawned a new discipline called *evolutionary developmental biology*, or "evo-devo."

HOMEOTIC GENE CLUSTERS.   As Lewis predicted, homeobox genes are involved in many developmental pathways. In the mouse, for example, there are 38 such genes in four clusters, each located on a different chromosome. As do the homeobox genes in *Drosophila*, these **Hox genes** control the development of specific regions of the mouse embryo, and are arranged in the same order on the chromosome as they are expressed, from anterior to posterior, in the developing animal (Figure 16.17). These four clusters are present in all vertebrate animals, and apparently arose from repeated duplications, followed by small mutations, of a single ancestral gene. This gene was recently found in the lancelet, a tiny marine organism thought to be the simplest living member of the chordates, a lineage that includes the vertebrates. As we will see in Chapter 24, gene duplication is a common mechanism of evolutionary change. Here, it allowed the formation of gene clusters that could determine increasingly complex organisms.

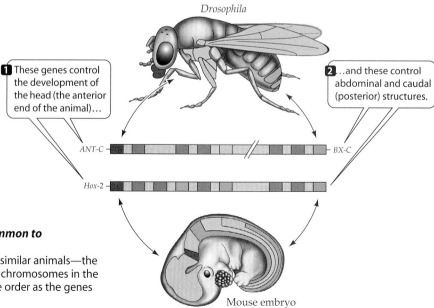

Drosophila

1 These genes control the development of the head (the anterior end of the animal)…

2 …and these control abdominal and caudal (posterior) structures.

ANT-C

BX-C

Hox-2

Mouse embryo

16.17 Homeobox-Containing Genes Are Common to Vastly Different Organisms
In both the fruit fly and the mouse—two very dissimilar animals—the homeobox-containing genes are lined up on the chromosomes in the same order, and produce their effects in the same order as the genes along the anterior–posterior axis.

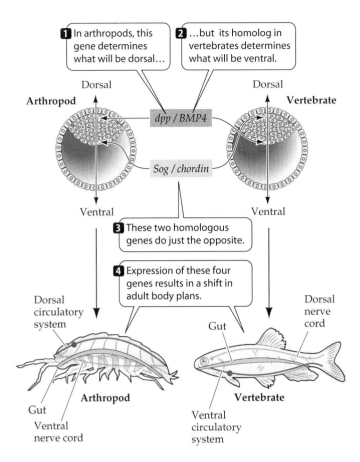

1 In arthropods, this gene determines what will be dorsal...

2 ...but its homolog in vertebrates determines what will be ventral.

Dorsal

Dorsal

Arthropod

Vertebrate

dpp / BMP4

Sog / chordin

Ventral

Ventral

3 These two homologous genes do just the opposite.

4 Expression of these four genes results in a shift in adult body plans.

Dorsal circulatory system

Gut

Dorsal nerve cord

Arthropod

Vertebrate

Gut

Ventral nerve cord

Ventral circulatory system

16.18 Shifts of Homologous Gene Expression and Shift in Body Plans of Arthropods and Vertebrates
Although arthropods and vertebrates have similar and analogous genes governing development, these genes determine opposite locations for the nervous and circulatory systems.

CONSERVATION OF A GENE FOR EYE FORMATION. The eye of an insect is very different in form and function from the eye of a mammal. Yet similar genes are involved in their formation. Mutations of a *Drosophila* gene called *eyeless* result in the reduction or absence of eyes in the adult fly. The protein encoded by the wild-type gene is a transcription factor that binds to promoters in over 1,500 different genes involved in eye formation. The homolog of *eyeless* in vertebrates, *Pax6*, is also involved in eye formation. In fact, mutations in the human version of *Pax6* result in a disease called aniridia—the partial or total absence of the iris of the eye. Remarkably, the *Pax6* and *eyeless* genes can be swapped for one another and yet remain functional in the other organism's development. This conservation of a global role for a gene for eye formation suggests that these "general instructions" evolved early and have been used repeatedly as different types of eyes have evolved.

DORSAL AND VENTRAL. In 1822, Geoffrey Saint-Hillaire noticed that in arthropods, such as the lobster, the nerve cord is on the same side as the mouth (ventral), with the circulatory system on the opposite side (dorsal) and the digestive system between the two. This pattern is the inverse of the one in vertebrates, in which the circulatory system is ventral and the nerve cord is dorsal (Figure 16.18). Molecular biologists have now described this "flip" in terms of genes that act during development. The signals for dorsal–ventral determination in arthropods and in vertebrates are the same; it's just that they have opposite mean-

ings! So the protein chordin determines dorsal identity in vertebrates, but its homolog, Sog, specifies the ventral region of arthropods. Likewise, BMP4 is a ventral determiner in vertebrates, but its relative, Dpp, is a dorsal determiner in arthropods. These findings suggest that the difference in body plan between arthropods and vertebrates involved an inversion of the axis of expression of these genes during evolution.

Chapter Summary

The Processes of Development

▶ A multicellular organism develops through a series of embryonic stages and eventually into an adult. Development continues until death. **Review Figure 16.1**

▶ Growth results from a combination of cell division and cell expansion.

▶ Differentiation produces specialized cell types.

▶ Morphogenesis—the creation of the overall form of the multicellular organism—is the result of pattern formation.

▶ In many organisms, the fates of the earliest embryonic cells have not yet been decided. These early embryonic cells may develop into different tissues if transplanted to other parts of an embryo. **Review Figure 16.2**

▶ As the embryo develops, its cells gradually become determined—committed to developing into particular parts of the embryo and particular adult structures. Following determination, cells eventually differentiate into their final, often specialized, forms.

The Role of Differential Gene Expression in Cell Differentiation

▶ The zygote is totipotent; it contains the entire genetic constitution of the organism and is capable of forming all adult tissues.

▶ Two lines of evidence show that differentiation does not involve permanent changes in the genome. First, nuclear transplant and cloning experiments show that the nucleus of a differentiated cell retains the ability to act like a zygote nucleus and stimulate the production of an entire organism. Second, molecular investigations have shown directly that all cells contain all genes for the organism, but that only certain genes are expressed in a given tissue. **Review Figures 16.3, 16.4, 16.5**

▶ Embryonic stem cells are totipotent, and can be cultured in the laboratory. With suitable environmental stimulation, these cells can be induced to form cells that differentiate. **Review Figure 16.6**

The Role of Polarity in Cell Determination

▶ Unequal distribution of cytoplasmic determinants in the egg, zygote, or embryo leads to cell determination in normal development. Experimentally altering this distribution can alter gene expression and produce abnormal or nonfunctional organisms. **Review Figure 16.7**

The Role of Embryonic Induction in Cell Determination

▶ Some embryonic animal tissues direct the development of their neighbors by secreting inducers.

▶ Induction is often reciprocal: One tissue induces a neighbor to change, and the neighbor, in turn, induces the first tissue to change, as in eye formation in vertebrate embryos. **Review Figure 16.8**

▶ The nematode *Caenorhabditis elegans* provides a striking example of induction. The adult consists of 959 cells that develop from the fertilized egg by a precise pattern of cell divisions and other events. **Review Figure 16.9**

▶ Induction in *C. elegans* can be very precise, with individual cells producing specific effects in just two or three neighboring cells. **Review Figure 16.9**

The Role of Pattern Formation in Organ Development

▶ In plants and animals, programmed cell death (apoptosis) is important in pattern formation. Some genes whose protein products regulate apoptosis have been identified. **Review Figure 16.10**

▶ Plants have organ identity genes that interact to cause the formation of sepals, petals, stamens, and carpels. Mutations of these genes may cause undifferentiated cells to form a different organ. **Review Figures 16.11, 16.12**

▶ Both plants and animals use positional information as a basis for pattern formation. Gradients of morphogens provide this information.

The Role of Differential Gene Expression in Establishing Body Segmentation

▶ The fruit fly *Drosophila melanogaster* has provided much information about the development of body segmentation. Some of this information applies to other animals.

▶ The first genes to act in determining *Drosophila* segmentation are maternal effect genes, such as *bicoid* and *nanos*, which encode morphogens that form gradients in the egg. These morphogens act on segmentation genes to define the anterior–posterior organization of the embryo. **Review Figures 16.13, 16.14, 16.15**

▶ Segmentation develops as the result of a transcriptionally controlled cascade, with the product of one gene promoting or repressing the expression of another gene. There are three kinds of segmentation genes, each responsible for a different step in segmentation. Gap genes organize large areas along the anterior–posterior axis, pair rule genes divide the axis into pairs of segments, and segment polarity genes see to it that each segment has an appropriate anterior–posterior axis. **Review Figure 16.14**

▶ The Bicoid and Nanos proteins act as a transcription factor and translation regulator, respectively, to control the level of expression of gap genes. Gap genes encode transcription fac-

tors that regulate the expression of pair rule genes. The products of the pair rule genes are transcription factors that regulate the segment polarity genes.

▶ Activation of the segmentation genes leads to the activation of the appropriate homeotic genes in different segments. The homeotic genes define the functional characteristics of the segments.

▶ Mutations in homeotic genes often have bizarre effects, causing structures to form in inappropriate parts of the body. **Review Figure 16.16**

▶ Homeotic genes contain the homeobox, which encodes an amino acid sequence that is part of many transcription factors. **Review Figure 16.17**

Evolution and Development

▶ The homeobox is found in key genes of distantly related species; thus numerous regulatory mechanisms may trace back to a single evolutionary precursor. **Review Figure 16.18**

▶ The activities of similar genes result in similar programs of development in different organisms. These genes include homeotic genes, genes involved in eye formation, and genes that determine dorsal and ventral surfaces. **Review Figure 16.19**

For Discussion

1. Molecular biologists can insert genes attached to high-level promoters into cells (see Chapter 16). What would happen if the following were inserted and overexpressed? Explain your answers.

   a. ced-9 in embryonic nerve cell precursors in *C. elegans*

   b. MyoD1 in undifferentiated myoblasts

   c. nanos at the anterior end of the *Drosophila* embryo

2. A powerful method to test for the function of a gene in development is to generate a "knockout" organism, in which the gene in question is inactivated. What do you think would happen in each of the following?

   a. C. elegans with *ced-9* knocked out

   b. Drosophila with *nanos* knocked out

   c. Drosophila with *bithorax* knocked out

3. During development, the potential of a tissue becomes ever more limited until, in the normal course of events, the potential is the same as the original fate. On the basis of what you have learned here and in Chapter 14, discuss possible mechanisms for the progressive limitation of developmental potential.

4. How were biologists able to obtain such a complete accounting of all the cells in *C. elegans*? Why can't we reason directly from studies of *C. elegans* to comparable problems in our own species?

17 Recombinant DNA and Biotechnology

AT THE BEGINNING OF CHAPTER 3, WE INTRO-duced you to spider silk, a family of proteins that show a fascinating combination of strength and elasticity. There is great interest in studying this biomaterial, not only to find out how it meets the structural challenges of the spider's web, but also because it has potential uses to humans, ranging from replacing expensive Kevlar in bulletproof vests to being used for surgical sutures and even to snag airplanes landing on aircraft carriers. The problem with spider silk is one of supply. "Milking" spiders, as has been done for centuries to get silk for fabrics from moth larvae, does not work on a commercial scale.

Because silk is made of a protein, one approach to the mass production of silk has been to use applied biology—biotechnology—to produce silk proteins in quantity and then develop ways to spin them into fibers, just as the spider does in its abdomen. This process has two parts. First, the gene for the silk protein is isolated from the spider genome. Then, the gene is put into a system that can express it in quantity. Silk genes turn out to be similar to the proteins they code for in that they are composed of repetitive domains. Through cutting and splicing of DNA, silk genes have been inserted into bacteria and into yeast, a relatively simple eukaryote. In both cases, an active promoter region was fused to the silk gene so that it would be expressed. Unfortunately, both of these types of host cells for this new DNA—called recombinant DNA—make insoluble silk that remains inside the cells.

Spider silk glands are similar in structure and function to animal mammary glands, in that both of them have epithelial cells that manufacture and secrete water-soluble, complex proteins in large amounts. Exploiting these similarities, scientists at a biotechnology company inserted the spider silk gene, with an accompanying mammary gland promoter for tissue-specific expression, into a goat, which produced abundant (10 g/L), soluble, easily purified silk in its milk. Creating such a transgenic goat is a tricky procedure, so a reliable silk producer goat has now been cloned, using the method described in the previous chapter. The next step, currently under way, is to develop a way to spin this silk protein into fibers.

This story—from problem to solution, from protein to expressed gene—has been repeated many times in the past two decades. The products of biotechnology range from life-saving drugs that there is no other way to make in adequate amounts to crop plants with improved agricultural characteristics. Although the basic techniques of DNA manipulation have been called revolutionary, most of them come from the knowledge of DNA transcription and translation that we described in earlier chapters.

We begin this chapter with a description of how DNA molecules can be cut into smaller fragments, and the fragments from different sources covalently linked to create recombinant DNA in the test tube. Recombinant (or any other) DNA can be introduced into a suitable prokaryotic or eukaryotic host cell. Sometimes, the purpose of adding new gene(s) to a host cell or organism is to ask an experimental question about the role of that gene, which can be answered by placing it in a new environment. In other instances, the purpose is to coax the host cell to make a new gene product.

Cleaving and Rejoining DNA

Scientists have long realized that the chemical reactions used in living cells for one purpose may be applied in the laboratory for other, novel purposes. Recombinant DNA technology—the manipulation and combination of DNA molecules from different sources—is based on this realization, and on an understanding of the properties of certain enzymes and of DNA itself.

A Factory for Spider Silk
These goats belong to a strain that is being genetically engineered to make spider silk in their milk.

As we saw in previous chapters, the nucleic acid base-pairing rules underlie many fundamental processes of molecular biology. The mechanisms of DNA replication, transcription, and translation rely on complementary base pairing. Similarly, all the key techniques in recombinant DNA technology—sequencing, rejoining, amplifying, and locating DNA fragments—make use of the complementary base pairing of A with T (or U) and of G with C.

In this section we will identify some of the numerous naturally occurring enzymes that cleave DNA, help it replicate, and repair it. Many of these enzymes have been isolated and purified, and are now used in the laboratory to manipulate and combine DNA. Then we will see how fragments of DNA can be separated and covalently linked to other fragments.

Restriction endonucleases cleave DNA at specific sequences

All organisms must have mechanisms to deal with their enemies. As we saw in Chapter 13, bacteria are attacked by viruses called bacteriophages that inject their genetic material into the host cell. Some bacteria defend themselves against such invasions by first altering their own DNA and then producing enzymes called **restriction endonucleases**, which catalyze the cleavage of double-stranded DNA molecules—such as those injected by phages—into smaller, noninfectious fragments (Figure 17.1). The bonds cut are between the 3' hydroxyl of one nucleotide and the 5' phosphate of the next one.

There are many such restriction enzymes, each of which cleaves DNA at a specific site defined by a sequence of bases called a **recognition site** or **restriction site**. The DNA of the host cell is not cleaved by its own restriction enzymes, because specific modifying enzymes called methylases add methyl (—CH$_3$) groups to certain bases at the restriction sites of the host's DNA when it is being replicated. The methylation of the host's bases makes the recognition sequence unrecognizable to the restriction endonuclease. But the unmethylated phage DNA is efficiently recognized and cleaved.

A specific sequence of bases defines each recognition site. For example, the enzyme *Eco*RI (named after its source, a strain of the bacterium *E. coli*) cuts DNA only where it encounters the following paired sequence in the DNA double helix:

5' … GAATTC … 3'
3' … CTTAAG … 5'

Notice that this sequence reads the same in the 5'-to-3' direction on both strands. It is palindromic, like the word "mom," in the sense that it is the same in both directions from the 5' end. *Eco*RI has two identical subunits that cleave the two strands between the G and the A.

This recognition sequence occurs on average about once in 4,000 base pairs in a typical prokaryotic genome—or about once per four prokaryotic genes. So *Eco*RI can chop a large piece of DNA into smaller pieces containing, on average, just a few genes. Using *Eco*RI in the laboratory to cut

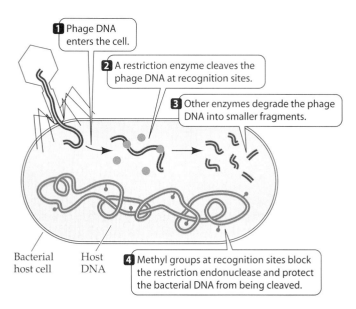

1 Phage DNA enters the cell.

2 A restriction enzyme cleaves the phage DNA at recognition sites.

3 Other enzymes degrade the phage DNA into smaller fragments.

Bacterial host cell   Host DNA

4 Methyl groups at recognition sites block the restriction endonuclease and protect the bacterial DNA from being cleaved.

17.1 Bacteria Fight Invading Viruses with Restriction Enzymes
Bacteria produce restriction enzymes that cleave and degrade phage DNA. Other enzymes protect the bacteria's own DNA from being cleaved.

small genomes, such as those of viruses that have only a few thousand base pairs, may result in a few fragments. For a huge eukaryotic chromosome with tens of millions of base pairs, the number of fragments will be very large.

Of course, "on average" does not mean that the enzyme cuts all stretches of DNA at regular intervals. The *Eco*RI recognition sequence does not occur even once in the 40,000 base pairs of the genome of a phage called T7—a fact that is crucial to the survival of this virus, since its host is *E. coli*. Fortunately for the *E. coli* that make *Eco*RI, the DNA of other phages does contain the recognition sequence.

Hundreds of restriction enzymes have been purified from various microorganisms. In the test tube, different restriction enzymes that recognize different recognition sequences can be used to cut the same sample of DNA. Thus cutting a sample of DNA in many different, specific places is an easy task, and restriction enzymes can be used as "knives" for genetic "surgery."

Gel electrophoresis identifies the sizes of DNA fragments

After a laboratory sample of DNA has been cut with a restriction enzyme, the DNA is in fragments, each of which is bounded at its ends by the recognition sequence. As we noted, these fragments are not all the same size, and this property provides a way to separate them from one another. Separating the fragments is necessary to determine the number and sizes (in base pairs) of fragments produced, or to identify and purify an individual fragment of particular interest.

The best way to separate DNA fragments is by **gel electrophoresis** (Figure 17.2). Because of its phosphate groups,

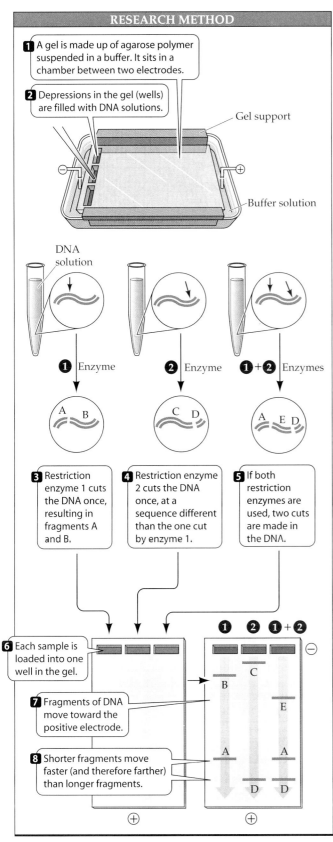

RESEARCH METHOD

1 A gel is made up of agarose polymer suspended in a buffer. It sits in a chamber between two electrodes.

2 Depressions in the gel (wells) are filled with DNA solutions.

Gel support

Buffer solution

DNA solution

1 Enzyme   **2** Enzyme   **1** + **2** Enzymes

A B   C D   A E D

3 Restriction enzyme 1 cuts the DNA once, resulting in fragments A and B.

4 Restriction enzyme 2 cuts the DNA once, at a sequence different than the one cut by enzyme 1.

5 If both restriction enzymes are used, two cuts are made in the DNA.

6 Each sample is loaded into one well in the gel.

1   **2**   **1** + **2**

7 Fragments of DNA move toward the positive electrode.

8 Shorter fragments move faster (and therefore farther) than longer fragments.

⊕   ⊕

17.2 Separating Fragments of DNA by Gel Electrophoresis
A mixture of DNA fragments is placed in a gel and an electric field is applied across the gel. The negatively charged DNA moves toward the positive end of the field, with smaller molecules moving faster than larger ones. When the electric power is shut off, the separate fragments can be analyzed.

DNA is negatively charged at neutral pH. A mixture of DNA fragments is placed in a porous gel, and an electric field (with positive and negative ends) is applied across the gel. Because opposite charges attract, the DNA moves toward the positive end of the field. Since the porous gel acts as a sieve, the smaller molecules move faster than the larger ones. After a fixed time, and while all fragments are still on the gel, the electric power is shut off. The separated fragments can then be examined or removed individually.

Different samples of fragmented DNA can be analyzed side by side on a gel. DNA fragments of known molecular size are often run in a lane on the gel next to the sample to provide a size reference. The separated DNA fragments can be visualized by staining them with a dye that fluoresces under ultraviolet light. Or a specific DNA sequence can be located by denaturing the DNA in the gel, affixing the denatured DNA to a nylon membrane to make a "blot" of the gel, and exposing the fragments to a single-stranded DNA *probe* with a sequence complementary to the one that is being sought (Figure 17.3). The probe can be labeled in some way—for example, with radioactive phosphorus (P^{32}). Therefore, after hybridization, the presence of radioactivity on the membrane indicates that the probe has hybridized to its target at that location. The gel region containing a desired fragment can be removed when the gel is sliced, and then the pure DNA fragment can be removed from the gel.

Recombinant DNA can be made in a test tube

Some restriction enzymes cut the DNA backbone cleanly, cutting both strands exactly opposite one another. Others make two staggered cuts, cutting one strand of the double helix several bases away from where they cut the other. Fragments cut in this manner are particularly useful in biotechnology.

*Eco*RI, for example, cuts DNA within its recognition sequence in a staggered manner, as shown at the top of Figure 17.4. After the two cuts in the opposing strands are made, the strands are held together only by the hydrogen bonding between four base pairs. The hydrogen bonds of these few base pairs are too weak to persist at warm temperatures (above room temperature), so the two strands of DNA separate, or denature. As a result, there are single-stranded

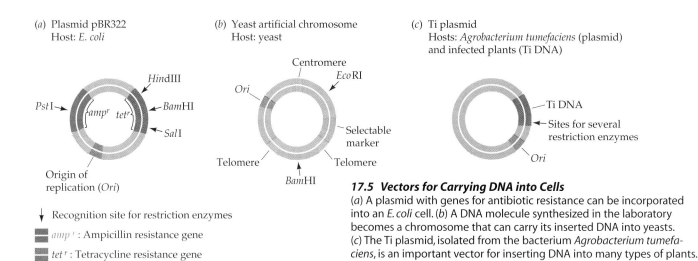

(*a*) Plasmid pBR322
Host: *E. coli*

(*b*) Yeast artificial chromosome
Host: yeast

(*c*) Ti plasmid
Hosts: *Agrobacterium tumefaciens* (plasmid)
and infected plants (Ti DNA)

*Hin*dIII

*Pst*I

amp^r *tet^r*

*Bam*HI

*Sal*I

Origin of
replication (*Ori*)

Centromere

*Eco*RI

Ori

Selectable
marker

Telomere   Telomere

*Bam*HI

Ti DNA

Sites for several
restriction enzymes

Ori

↓ Recognition site for restriction enzymes

amp ^{r} : Ampicillin resistance gene

tet ^{r} : Tetracycline resistance gene

17.5 *Vectors for Carrying DNA into Cells*
(*a*) A plasmid with genes for antibiotic resistance can be incorporated into an *E. coli* cell. (*b*) A DNA molecule synthesized in the laboratory becomes a chromosome that can carry its inserted DNA into yeasts. (*c*) The Ti plasmid, isolated from the bacterium *Agrobacterium tumefaciens*, is an important vector for inserting DNA into many types of plants.

VIRUSES AS VECTORS. Constraints on plasmid replication limit the size of the new DNA that can be spliced into a plasmid to about 5,000 base pairs. Although a prokaryotic gene may be this small, 5,000 base pairs is much smaller than most eukaryotic genes, with their introns and extensive flanking sequences. So, a vector that accommodates larger DNA inserts is needed.

Both prokaryotic and eukaryotic viruses are often used as vectors for eukaryotic DNA. Bacteriophage lambda, which infects *E. coli*, has a DNA genome of more than 45,000 base pairs. If the genes that cause the host cell to die and lyse—about 20,000 base pairs—are eliminated, the virus can still infect a host cell and inject its DNA. The deleted 20,000 base pairs can be replaced with DNA from another organism, thereby creating recombinant viral DNA.

Because viruses infect cells naturally, they offer a great advantage as vectors over plasmids, which often require artificial means to coax them to enter cells. As we will see in Chapter 18, viruses are important vectors for delivering new genes to people in gene therapy.

ARTIFICIAL CHROMOSOMES AS VECTORS. Bacterial plasmids are not good vectors for yeast hosts, because prokaryotic and eukaryotic DNA sequences use different origins of replication. Thus a recombinant bacterial plasmid will not replicate in yeast. To remedy this problem, scientists have created in the laboratory a "minimalist chromosome" called the *yeast artificial chromosome*, or *YAC* (Figure 17.5*b*). This DNA molecule contains not only the yeast origin of replication, but sequences for the yeast centromere and telomeres as well, making it a true eukaryotic chromosome. YACs also contain artificially synthesized single restriction sites and useful marker genes (for yeast nutritional requirements). YACs are only about 10,000 base pairs in size, but can accommodate 50,000 to 1.5 million base pairs of inserted DNA.

There has been considerable progress in creating a *human* artificial chromosome (HAC), which could someday be used as a gene therapy vector. Instead of yeast cen-

tromere and telomere sequences, their human counterparts have been used. The vector acts as a separate minichromosome in human cells, and can be maintained there for months.

PLASMID VECTORS FOR PLANTS. An important vector for carrying new DNA into many types of plants is a plasmid that is found in *Agrobacterium tumefaciens*. This bacterium lives in the soil and causes a plant disease called crown gall, which is characterized by the presence of growths, or tumors, in the plant. *A. tumefaciens* contains a plasmid called Ti (for *tumor-inducing*) (Figure 17.5*c*).

Part of the Ti plasmid is T DNA, a transposon that produces copies of itself in the chromosomes of infected plant cells. The T DNA has recognition sites for restriction enzymes, and new DNA can be spliced into the T DNA region of the plasmid. When the T DNA is thus replaced, the plasmid no longer produces tumors, but the transposon, with the new DNA, is inserted into the host cell's chromosomes. The plant cell containing this DNA can then be grown in culture or induced to form a new, transgenic, plant.

There are many ways to insert recombinant DNA into host cells

Although some vectors, such as viruses, can enter host cells on their own, most vectors require help to do so. A major barrier to DNA entry is that the exterior surface of the plasma membrane, with its phospholipid heads, is negatively charged, as is DNA. The resulting charge repulsion can be alleviated if the exterior of the cells and the DNA are both neutralized with Ca^{2+} (calcium) salts. The salts reduce the charge effect, and the plasma membrane becomes permeable to DNA. In this way, almost any cell, prokaryotic or eukaryotic, can take up a DNA molecule from its environment. In plants and fungi, the cell wall must first be removed by hydrolysis with fungal enzymes; the resulting wall-less plant cells are called *protoplasts*.

In addition to this "naked" DNA approach, DNA can be introduced into host cells by a variety of mechanical methods:

▸ In *electroporation*, host cells are exposed to rapid pulses of high-voltage current. This treatment temporarily renders the plasma membrane permeable to DNA in the surrounding medium.

▸ In *injection*, a very fine pipette is used to insert DNA into cells. This method is especially useful on large cells such as eggs.

▸ In *lipofection*, DNA is coated with lipid, which allows it to pass through the plasma membrane. For example, DNA can be encased in liposomes, bubbles of lipid that fuse with the membranes of the host cell.

▸ In *particle bombardment*, tiny high-velocity particles of tungsten or gold are coated with DNA and then shot into host cells. This "gene gun" approach must be undertaken with great care to prevent the cell contents from being damaged.

Genetic markers identify host cells that contain recombinant DNA

Even when a population of host cells is allowed to interact with an appropriate vector, only a small percentage of the cells actually take up the vector. Also, since the process of cutting the vector and inserting the new DNA to make recombinant DNA is far from perfect, only a few of the vectors that have moved into the host cells will actually contain the new DNA sequence. How can we select only the host cells that contain the recombinant DNA?

The experiment we are about to describe illustrates an elegant, commonly used approach to this problem. In this example, we use *E. coli* bacteria as hosts and a plasmid vector (see Figure 17.5*a*) that carries the genes for resistance to the antibiotics ampicillin and tetracycline.

When the plasmid is incubated with the restriction enzyme *Bam*HI, the enzyme encounters its recognition se-

quence, GGATCC, only once, at a site within the gene for tetracycline resistance. If foreign DNA is inserted into this restriction site, the presence of these "extra" base pairs within the tetracycline resistance gene inactivates it. So plasmids containing the inserted DNA will carry an intact gene for ampicillin resistance, but *not* an intact gene for tetracycline resistance. This is the key to the selection of host bacteria that contain the recombinant plasmid (Figure 17.6).

The cutting and splicing process results in three types of DNA, all of which can be taken up by the host bacteria:

▸ The recombinant plasmid—the one we want—turns out to be the rarest type of DNA. Its uptake confers on host *E. coli* resistance only to ampicillin.

▸ More common are bacteria that take up plasmids that have sealed their own ends back together. These plasmids retain intact genes for resistance to both ampicillin and tetracycline.

▸ Even more common are bacteria that take up the foreign DNA sequence alone, without the plasmid; since it is not part of a replicon, it does not survive as the bacteria divide. These host cells will remain susceptible to both antibiotics, as will the vast majority (more than 99.9 percent) of cells that take up no DNA at all.

So the unique drug resistance phenotype of the cells with recombinant DNA (tetracycline-sensitive and ampicillin-resistant) marks them in a way that can be detected by simply adding ampicillin and/or tetracycline to the medium surrounding the cells.

In addition to genes for antibiotic resistance, several other marker genes are used to detect recombinant DNA in host cells. Scientists have created several artificial vectors in the laboratory that include sites for restriction enzymes within the *lac* operon (see Chapter 13). When this gene is inactivated by the insertion of foreign DNA, the vector no longer carries this operon's function into the host cell. Other *reporter genes* that have been used in vectors include the gene for luciferase, the enzyme that causes fireflies to glow in the dark; this enzyme causes host cells to glow when supplied with its substrate. Green fluorescent protein, which normally occurs in the jellyfish *Aequopora victo-*

17.6 Marking Recombinant DNA by Inactivating a Gene
Scientists manipulate marker genes within plasmids so they will know which host cells have incorporated the recombinant genes. The host bacteria in this experiment could display any of the phenotypes indicated in the table. Assuming we wish to select only those that have taken up the recombinant plasmid, we can do so by adding antibiotics to the medium surrounding the cells.

1 A plasmid has genes for resistance to both ampicillin (*amp^r*) and tetracycline (*tet^r*).

2 Foreign DNA is inserted at the *Bam*HI recognition site, which is within the *tet^r* gene.

3 The resulting recombinant DNA has an intact functional gene for ampicillin resistance but not tetracycline resistance.

Detection of Recombinant DNA in *E. coli*

DNA TAKEN UP BY AMP^S AND TET^S *E. COLI*	PHENOTYPE FOR AMPICILLIN	PHENOTYPE FOR TETRACYCLINE
None	Sensitive	Sensitive
Foreign DNA only	Sensitive	Sensitive
pBR322 plasmid	Resistant	Resistant
pBR322 recombinant plasmid	Resistant	Sensitive

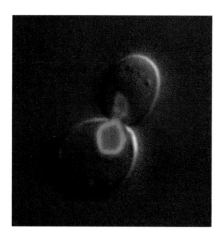

17.7 A Reporter Gene Announces the Presence of a Vector in Eukaryotic Cells
These cells have taken up a vector that expresses a gene producing green fluorescent protein.

riana, does not require a substrate to glow, and is now widely used (Figure 17.7).

Many vectors in common use contain only a *single* marker for antibiotic resistance, outside of the sites for foreign DNA insertion. In this case, the recombinant DNA will have the same antibiotic resistance gene that the nonrecombinant plasmid does. The formation of recombinant DNA in the ligase reaction is favored if there is a high concentration of foreign DNA fragments compared to the cut plasmid. So there will be a preponderance of colonies containing recombinant DNA among those that grow in the presence of the antibiotic.

After DNA uptake (or not), host cells are usually first grown on a solid medium. If the concentration of cells dispersed on the solid medium is low, each cell will divide and grow into a distinct colony (see Chapter 13). The colonies that contain recombinant DNA can be identified and removed from the medium, and then grown in large amounts in liquid culture. A quick examination of a plasmid can confirm whether the plasmids in the cells of the colony actually have the recombinant DNA. The power of bacterial transfection to amplify a gene is indicated by the fact that a 1-liter culture of bacteria harboring the human β-globin gene in the pBR322 plasmid has as many copies of the gene as the sum total of all the cells in a typical human being.

Sources of Genes for Cloning

The genes or DNA fragments used in recombinant DNA work are obtained from three principal sources. One source is random pieces of chromosomes maintained as gene libraries. The second source is complementary DNA, obtained by reverse transcription from specific mRNA's. The third source is DNA synthesized by organic chemists in the laboratory. Specific fragments can be deliberately modified to create mutations or to change a mutant sequence back to the wild type.

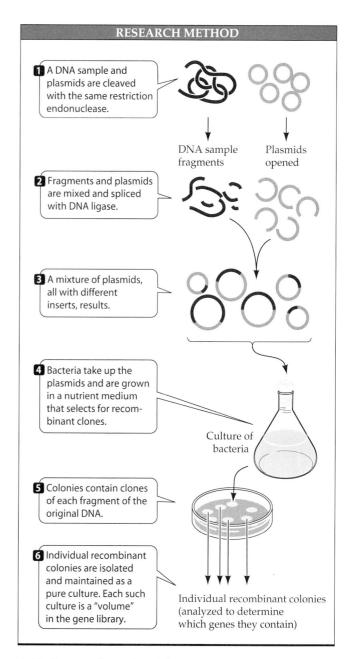

1 A DNA sample and plasmids are cleaved with the same restriction endonuclease.

DNA sample fragments   Plasmids opened

2 Fragments and plasmids are mixed and spliced with DNA ligase.

3 A mixture of plasmids, all with different inserts, results.

4 Bacteria take up the plasmids and are grown in a nutrient medium that selects for recombinant clones.

Culture of bacteria

5 Colonies contain clones of each fragment of the original DNA.

6 Individual recombinant colonies are isolated and maintained as a pure culture. Each such culture is a "volume" in the gene library.

Individual recombinant colonies (analyzed to determine which genes they contain)

17.8 Constructing a Gene Library
Human chromosomes are broken up into fragments of DNA that are inserted into vectors (plasmids are shown here) and taken up by host bacterial cells, each of which harbors a single fragment of the human DNA. The information in these bacterial colonies constitutes a gene library.

Gene libraries contain pieces of a genome

The 23 pairs of human chromosomes can be thought of as a library that contains the entire genome of our species. Each chromosome, or "volume" in the library, contains, on average, 80 million base pairs of DNA, encoding several thousand genes. Such a huge molecule is not very useful for studying genome organization or for isolating a specific gene. To address this problem, researchers can break each chromosome into smaller pieces using restriction enzymes, and then analyze each piece. These smaller fragments still represent a **gene library** (Figure 17.8); however, the information is now in many more volumes than 23. Each of the fragments can be inserted into a vector, which can then be taken up by a host bacterial cell. Each host cell colony, then, harbors a single fragment of human DNA.

Using plasmids, which are able to insert only a few thousand base pairs of foreign DNA into a bacterium, about a million separate fragments are required to make a library of the human genome. By using phage lambda, which can carry four times as much DNA as a plasmid, the number of volumes is reduced to about 250,000. Although this seems like a large number, a single growth plate can hold up to 80,000 phage colonies, or *plaques*, and is easily screened for the presence of a particular DNA sequence by denaturing the phage DNA and applying a particular probe for hybridization.

A DNA copy of mRNA can be made

A much smaller DNA library—one that includes only genes transcribed in a particular tissue—can be made from **complementary DNA**, or **cDNA** (Figure 17.9). Recall that most eukaryotic mRNA's have a poly A tail—a string of adenine residues at their 3′ end (see Figure 14.11). The first step in cDNA production is to extract mRNA from a tissue and allow it to hybridize with a molecule called oligo dT (the "d" indicates *d*eoxyribose), which consists of a string of thymine residues. After the oligo dT hybridizes with the poly A tail of the mRNA, it serves as a primer, and the mRNA as a template, for the enzyme reverse transcriptase, which synthesizes DNA from RNA. In this way, a cDNA strand complementary to the mRNA is formed.

A collection of cDNA's from a particular tissue at a particular time in the life cycle of an organism is called a **cDNA library**. mRNA's do not last long in the cytoplasm and are often present in small amounts, so a cDNA library is a "snapshot" that preserves the transcription pattern of the cell. cDNA libraries have been invaluable in comparisons of gene expression in different tissues at different stages of development. For example, their use has shown that up to one-third of all the genes of an animal are ex-

pressed only during prenatal development. Complementary DNA is also a good starting point for the cloning of eukaryotic genes. It is especially useful for genes expressed at low levels in only a few cell types.

DNA can be synthesized chemically in the laboratory

When we know the amino acid sequence of a protein, we can obtain the DNA that codes for it by simply making it in the laboratory, using organic chemistry techniques. DNA synthesis has even been automated, and at many institutions, a special service laboratory can make short-to-medium-length sequences overnight for any number of investigators.

How do we design a synthetic gene? Using the genetic code and the known amino acid sequence, we can figure out the most likely base sequence for the gene. With this sequence as a starting point, we can add other sequences, such as codons for translation initiation and termination and flanking sequences for transcription initiation, termination, and regulation. Of course, these noncoding DNA sequences must be the ones actually recognized by the host cell if the synthetic gene is to be transcribed. It does no good to have a prokaryotic promoter sequence near a gene if that gene is to be inserted into a yeast cell for expression. Codon usage is also important: Many amino acids are encoded by more than one codon, and different organisms stress the use of different synonymous codons.

DNA can be mutated in the laboratory

Mutations that occur in nature have been important in proving cause-and-effect relationships in biology. For ex-

17.9 Synthesizing Complementary DNA
Gene libraries that include only genes transcribed in a particular tissue at a particular time can be made from complementary DNA. cDNA synthesis is especially useful for identifying genes that are present only in a few copies, and is often a starting point for gene cloning.

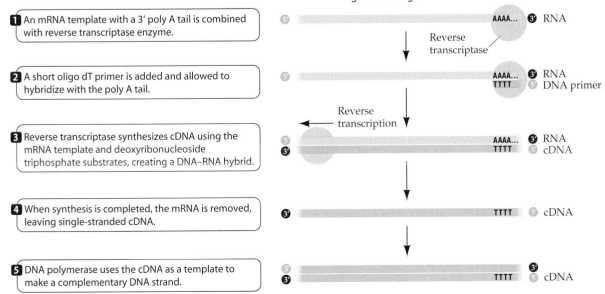

1 An mRNA template with a 3′ poly A tail is combined with reverse transcriptase enzyme.

2 A short oligo dT primer is added and allowed to hybridize with the poly A tail.

3 Reverse transcriptase synthesizes cDNA using the mRNA template and deoxyribonucleoside triphosphate substrates, creating a DNA–RNA hybrid.

4 When synthesis is completed, the mRNA is removed, leaving single-stranded cDNA.

5 DNA polymerase uses the cDNA as a template to make a complementary DNA strand.

ample, in Chapter 14, we learned that some people with the disease beta thalassemia have a mutation at the consensus sequence for intron removal and so cannot make proper β-globin mRNA. This example shows the importance of the consensus sequence.

Recombinant DNA technology has allowed us to ask "What if?" questions without having to look for mutations in nature. Because synthetic DNA can be made in any sequence desired, we can manipulate DNA to create specific mutations and then see what happens when the mutant DNA expresses itself in a host cell. Additions, deletions, and base-pair substitutions are all possible with isolated or synthetic DNA.

These mutagenesis techniques have allowed scientists to bypass the search for naturally occurring mutant strains, leading to many cause-and-effect proofs. For example, it was proposed that the signal sequence at the beginning of a secreted protein is essential to its passage through the endoplasmic reticulum membrane. So, a gene coding for such a protein, but with the codons for the signal sequence deleted, was made. Sure enough, when this gene was expressed in yeast cells, the protein did not cross the ER membrane. When the signal sequence codons were added to an unrelated gene encoding a soluble cytoplasmic protein, that protein crossed the ER membrane.

Mutagenesis has also begun to be useful in the design of specific drugs. The advent of a new branch of biology called *computational biology* has led to sophisticated studies of the three-dimensional shapes and chemical properties of enzymes, substrates, and their possible regulators. Attempts are being made to devise rules to predict the tertiary structure of a protein from its primary structure. For example, if we know the structure of an enzyme, the three-dimensional design of a polypeptide regulating that enzyme might be proposed. Mutant bacterial strains with genes coding for variants of this polypeptide could be made. Then, the variant polypeptides could be isolated and used to test the relationship between structure and activity.

Some Additional Tools for DNA Manipulation

Biological methods are not the only ways of manipulating DNA managed in the laboratory. In Chapter 11, we described DNA sequencing and the polymerase chain reaction, two applications of DNA replication techniques. Here, we examine three additional techniques. One is the use of genetic recombination to create an inactive, or "knocked-out," gene. The second is the use of "DNA chips" to detect the presence of many different sequences simultaneously. The third is the use of antisense RNA to block the translation of specific mRNA's.

Genes can be inactivated by homologous recombination

As we have seen, laboratory-created mutations are an excellent way to ask the "what if" questions about the role of a gene in cell function. *Homologous recombination* is used to ask these questions at the organism level (Figure 17.10). The

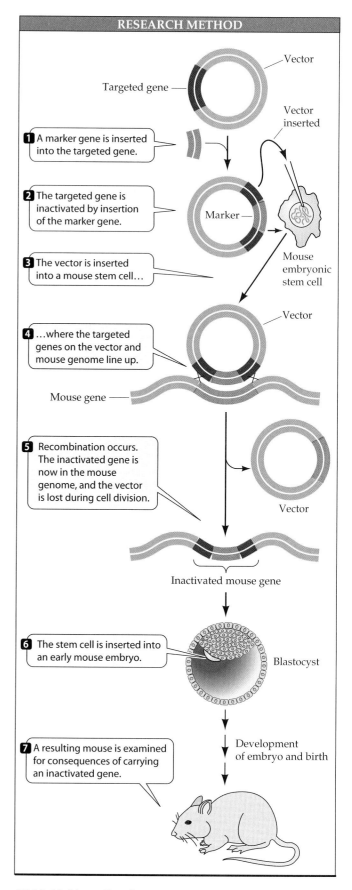

RESEARCH METHOD

Targeted gene — Vector

Vector inserted

1 A marker gene is inserted into the targeted gene.

2 The targeted gene is inactivated by insertion of the marker gene.

Marker — Mouse embryonic stem cell

3 The vector is inserted into a mouse stem cell...

Vector

4 ...where the targeted genes on the vector and mouse genome line up.

Mouse gene —

5 Recombination occurs. The inactivated gene is now in the mouse genome, and the vector is lost during cell division.

Vector

Inactivated mouse gene

6 The stem cell is inserted into an early mouse embryo.

Blastocyst

7 A resulting mouse is examined for consequences of carrying an inactivated gene.

Development of embryo and birth

17.10 Making a Knockout Mouse
Homologous recombination is used to replace a normal mouse gene with an inactivated copy, thus "knocking out" the active gene. Discovering what happens to a mouse with an inactive gene tells us much about the role of that gene.

aim of this technique is to replace a gene inside a cell with an inactivated form of that gene, and then see what happens when the inactive gene is part of an organism. Such a manipulation is called a *knockout* experiment.

Mice are frequently used in knockout experiments. The mouse gene to be tested is inserted into a plasmid. Restriction enzymes are then used to insert another fragment, containing a genetic marker, in the middle of this gene. Addition of extra DNA to the gene creates havoc with its transcription and translation; a functional mRNA is seldom made from such an interrupted gene. Next, the plasmid is transfected into a mouse embryonic stem cell (see Chapter 16). Because much of the targeted gene is still present in the plasmid (although in two separated regions), there is DNA sequence recognition between the gene on the plasmid and the homologous gene in the mouse genome. As in prophase I of meiosis, the plasmid and the mouse chromosomes line up, and, sometimes, a genetic exchange occurs in which the plasmid's inactive gene is swapped for the functional gene in the host cell. The genetic marker in the insert is used to identify those stem cells carrying the inactivated gene. The transfected stem cell is now inserted into an early mouse embryo, and through some clever tricks, a knockout mouse carrying the inactivated gene in homozygous form can be produced. The phenotype of the mutant mouse is an indication of the role of the gene in the normal, wild-type animal. The knockout technique has been very important in assessing the roles of genes during development.

DNA chips can reveal DNA mutations and RNA expression

The emerging science of genomics deals with two major quantitative circumstances: First, there are a large number of genes in eukaryotic genomes. Second, the pattern of gene expression in different tissues at different times is quite distinctive. For example, a skin cancer cell at its early stage may have a unique mRNA "fingerprint" that differs from that of both normal skin cells and more advanced skin cancer cells.

To find these patterns, scientists could isolate total cell mRNA and test it by hybridization with each gene in the genome, one gene at a time. But it would be far better to do these hybridizations all in one step. For this, one needs some way to arrange all the DNA sequences in a genome in an array on some solid support.

DNA chip technology has been developed to provide these large arrays of sequences for hybridization. The chips were developed by modifications of methods that have been used for several decades in the semiconductor industry. You may be familiar with the silicon microchip, in which an array of microscopic electric circuits is etched onto a tiny chip. In the same way, DNA chips are glass slides onto which are attached, in precise order, pre-established sequences of DNA (Figure 17.11). Typically, the slide is divided into 24×24 μM squares, each of which contains about 10 million copies of a particular sequence, up to 20 nucleotides long. A computer controls the addition of nucleotides in a predetermined pattern. Up to 60,000 different sequences can be put on a single chip.

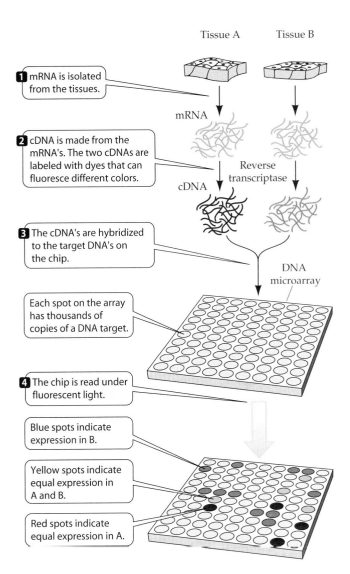

1 mRNA is isolated from the tissues.

2 cDNA is made from the mRNA's. The two cDNAs are labeled with dyes that can fluoresce different colors.

3 The cDNA's are hybridized to the target DNA's on the chip.

Each spot on the array has thousands of copies of a DNA target.

4 The chip is read under fluorescent light.

Blue spots indicate expression in B.

Yellow spots indicate equal expression in A and B.

Red spots indicate equal expression in A.

17.11 DNA on a Chip
Thousands of different DNA probes of known sequence can be attached to a glass slide.

If cellular mRNA is to be analyzed, it is usually incubated with reverse transcriptase (RT) to make cDNA (see Figure 17.9), and the cDNA is amplified by PCR prior to hybridization. This technique is called "RT-PCR," and it ensures that mRNA sequences present in only a few copies (or in a small sample such as a cancer biopsy) will be numerous enough to form a signal when hybridized. The amplified cDNA's are coupled to a fluorescent dye and then allowed to hybridize with DNA on a chip. Those DNA sequences that form a hybrid can be located by a sensitive scanner. With the number of genes on a chip approaching that of the largest genomes, these chips will result in an information explosion on mRNA transcription patterns in cells in different physiological states.

Another use for DNA chips is in detecting genetic variants. Suppose one wants to find out if a particular gene, which is 5,500 base pairs long, has any mutations in a particular individual. One way would be to sequence the entire gene, but this would be difficult to do, and would require a large tissue sample. On the other hand, chip technology can be used to make 20-nucleotide fragments along the gene in every possible mutant sequence. Then, probing with the individual's DNA might reveal a particular mutation if it hybridized to a mutant sequence on the chip. This method may provide a rapid way to detect mutations in people.

Antisense RNA and ribozymes can prevent the expression of specific genes

The base-pairing rules not only can be used to make genes; they can also be employed to stop the translation of mRNA. As is often the case, this technique is an example of scientists imitating nature. In normal cells, a rare method of controlling gene expression is the production of an RNA molecule that is complementary to mRNA. This complementary molecule is called **antisense RNA** because it binds by base pairing to the "sense" bases on the mRNA that code for a protein. The formation of a double-stranded RNA hybrid prevents tRNA from binding to the mRNA, and the hybrid tends to be broken down rapidly in the cytoplasm. So, although the gene continues to be transcribed, translation does not take place.

In the laboratory, after determining the sequence of a gene and its mRNA, scientists can add antisense RNA to a cell to prevent translation of the mRNA (Figure 17.12). The antisense RNA can be added as itself—RNA can be inserted into cells in the same way that DNA is—or it can be made in the cell by transcription from a DNA molecule introduced as a part of a vector.

Without this technique, repressing the synthesis of a specific protein would be very difficult. It is especially useful if a tissue-specific promoter is used to prime transcription of

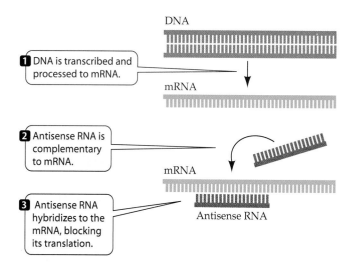

17.12 Using Antisense RNA to Block Translation of an mRNA
Once a gene's sequence is determined in the laboratory, the synthesis of its protein can be prevented using antisense RNA that is complementary to its mRNA.

the antisense RNA, so that its expression occurs only in a targeted tissue. An even more effective way to ensure that antisense RNA works is to couple the antisense sequence to a special RNA sequence—a *ribozyme*—that catalyzes the cleavage of its target RNA.

Antisense RNA (with or without a ribozyme) has been widely used to test cause-and-effect relationships. For example, when antisense RNA was used to block the synthesis of a protein essential for the growth of cancer cells, the cells reverted to a normal state.

Biotechnology: Applications of DNA Manipulation

Biotechnology is the use of microbial, plant, and animal cells to produce materials useful to people. These products include foods, medicines, and chemicals. We have been making some of them for a long time. For example, the use of yeasts to brew beer and wine dates back at least 8,000 years in human history, and the use of bacterial cultures to make cheese and yogurt is a technique many centuries old.

For a long time, people were not aware of the cellular bases of these biochemical transformations. About 100 years ago, thanks largely to Pasteur's work, it became clear that specific bacteria, yeasts, and other microbes could be used as biological converters to make certain products. Alexander Fleming's discovery that the mold *Penicillium* makes the antibiotic penicillin led to the large-scale commercial culture of microbes to produce antibiotics as well as other useful chemicals. Today, microbes are grown in vast quantities to make much of the industrial-grade alcohol, glycerol, butyric acid, and citric acid that are used by themselves or as starting materials in the manufacture of other products.

In the past, the list of such products was limited to those that were naturally made by microbes. The many products that eukaryotes make, such as hormones and certain enzymes, had to be extracted from those complex organisms. Yields were low, and purification was difficult and costly. All this has changed with the advent of gene cloning. The ability to insert almost any gene into bacteria or yeast, along with methods to induce the gene to make its product, has turned these microbes into versatile factories for important products.

Expression vectors can turn cells into protein factories

If a eukaryotic gene is inserted into a typical plasmid (see Figure 17.5a) and cloned into *E. coli*, little, if any, of the product of the gene will be made by the host cell. The reason is that the eukaryotic gene lacks the bacterial promoter for RNA polymerase binding, the terminator for transcription, and a special sequence on mRNA for ribosome binding. All of these are necessary for the gene to be expressed and its products synthesized in the bacterial cell.

Expression vectors can be made that have all the characteristics of typical vectors, as well as the extra sequences

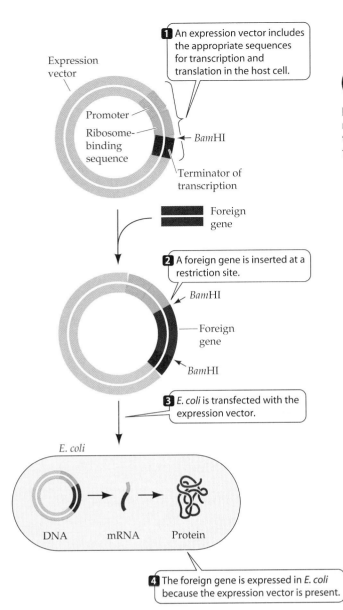

1 An expression vector includes the appropriate sequences for transcription and translation in the host cell.

2 A foreign gene is inserted at a restriction site.

3 *E. coli* is transfected with the expression vector.

4 The foreign gene is expressed in *E. coli* because the expression vector is present.

17.13 An Expression Vector Allows a Foreign Gene to Be Expressed in a Host Cell
An inserted eukaryotic gene may not be expressed in *E. coli* because it lacks the necessary bacterial sequences for promotion, termination, and ribosome binding. Expression vectors contain these additional sequences, enabling the eukaryotic protein to be synthesized in the prokaryotic cell.

hormone is added. An enhancer that responds to hormonal stimulation might also be added so that transcription and protein production will occur at very high rates—a goal of obvious importance in the manufacture of an industrial product.

A *tissue-specific promoter*, which is expressed only in a certain tissue at a certain time, can be used if localized expression in an organism is desired. For example, many seed proteins are expressed only in the plant embryo. So coupling a gene to a seed-specific promoter will allow the gene to be expressed only as a seed protein.

Targeting sequences can be added to the gene in the expression vector so that the protein product is directed to an appropriate destination. For example, in a large vessel containing yeast cells making a protein, it might be useful for the protein to be secreted into the extracellular medium for easier recovery.

Medically useful proteins can be made by DNA technology

Many medically useful products have been made by recombinant DNA technology (Table 17.1), and hundreds more are in various stages of development. We will describe three such products to illustrate the techniques that have been used in their development.

needed for the foreign gene to be expressed in the host cell. For bacteria, these additional sequences include the bacterial promoter, the transcription terminator, and the sequence for ribosome binding (Figure 17.13). For eukaryotes, expression vectors would include the poly A addition site, transcription factor binding sites, and enhancers. Once these sequences are placed at the appropriate location on the vector, the gene will be expressed in the host cell.

An expression vector can be refined in various ways. An *inducible promoter*, which responds to a specific signal, can be made part of an expression vector. For example, a specific promoter can be used that responds to hormonal stimulation so that the foreign gene can be induced to transcribe its mRNA when the

17.1	Some Medically Useful Products of Biotechnology	
PRODUCT	**USE**
Brain-derived neurotropic factor	Stimulates regrowth of brain tissue in patients with Lou Gehrig's disease
Colony-stimulating factor	Stimulates production of white blood cells in patients with cancer and AIDS
Erythropoietin	Prevents anemia in patients undergoing kidney dialysis
Factor VIII	Replaces clotting factor missing in patients with hemophilia A
Growth hormone	Replaces missing hormone in people of short stature
Insulin	Stimulates glucose uptake from blood in some people with diabetes
Platelet-derived growth factor	Stimulates wound healing
Tissue plasminogen activator	Dissolves blood clots after heart attacks and strokes
Vaccine proteins: Hepatitis B, herpes, influenza, Lyme disease, meningitis, pertussis, etc.	Prevent and treat infectious diseases

TISSUE PLASMINOGEN ACTIVATOR. In most people, when a wound begins bleeding, a blood clot soon forms to stop the flow. Later, as the wound heals, the clot dissolves. How does the blood perform these conflicting functions at the right times? Mammalian blood contains an enzyme called *plasmin* that catalyzes the dissolution of the clotting proteins. But plasmin is not always active; if it were, a blood clot would dissolve as soon as it formed! Instead, plasmin is "stored" in the blood in an inactive form called *plasminogen*. The conversion of plasminogen to plasmin is activated by an enzyme appropriately called *tissue plasminogen activator* (TPA), which is produced by cells lining the blood vessels. Thus, the reaction is

$$\text{plasminogen} \xrightarrow{\text{TPA}} \text{plasmin}$$
$$\text{(inactive)} \qquad\qquad \text{(active)}$$

Heart attacks and many strokes are caused by blood clots that form in important blood vessels leading to the heart or the brain, respectively. During the 1970s, a bacterial enzyme, streptokinase, was found to stimulate the quick dissolution of clots in some patients with these afflictions. Treating people with this enzyme saved lives, but there were side effects. The drug was a protein foreign to the body, so patients' immune systems reacted against it. More important, the drug sometimes prevented clotting throughout the circulatory system, leading to an almost hemophilia-like condition in some patients.

The discovery of TPA and its isolation from human tissues led to the hope that this enzyme could be used to bind specifically to clots, and that it would not provoke an immune reaction. But the amounts of TPA available from human tissues were tiny, certainly not enough to inject at the site of a clot in a patient in the emergency room.

Recombinant DNA technology solved the problem. TPA mRNA was isolated and used to make a cDNA copy, which was then inserted into an expression vector and introduced into *E. coli* (Figure 17.14). The transfected bacteria made the protein in quantity, and it soon became available commercially. This protein has had considerable success in dissolving blood clots in people undergoing heart attacks and, especially, strokes.

ERYTHROPOIETIN. Another protein made through recombinant DNA methods and widely used in medicine is *erythropoietin* (EPO). The kidneys produce this hormone, which travels through the blood to the bone marrow, where it stimulates the division of stem cells to produce red blood cells. People who have suffered kidney failure often require a procedure called kidney dialysis to remove toxins from the blood. However, because dialysis also removes EPO, these patients can become severely anemic (depleted of red blood cells).

As with TPA, the amounts of EPO that can be obtained from healthy people to give to people undergoing dialysis are extremely small, but once again, biotechnology has come to the rescue. The gene for EPO was isolated, inserted in an expression vector, and introduced into bacteria. Large

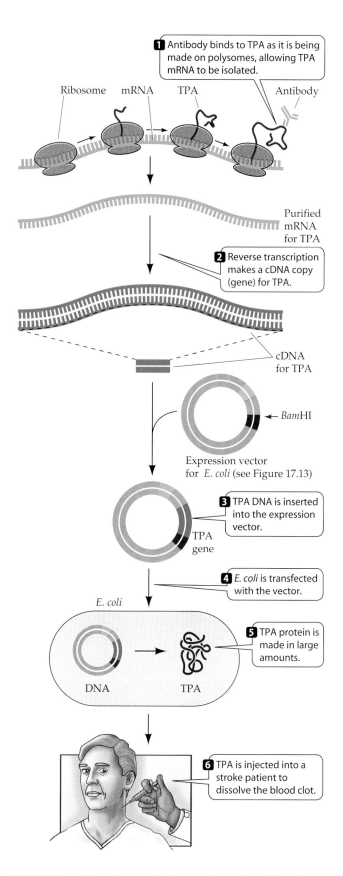

1 Antibody binds to TPA as it is being made on polysomes, allowing TPA mRNA to be isolated.

Ribosome   mRNA   TPA   Antibody

Purified mRNA for TPA

2 Reverse transcription makes a cDNA copy (gene) for TPA.

cDNA for TPA

*Bam*HI

Expression vector for *E. coli* (see Figure 17.13)

3 TPA DNA is inserted into the expression vector.

TPA gene

4 *E. coli* is transfected with the vector.

E. coli

5 TPA protein is made in large amounts.

DNA   TPA

6 TPA is injected into a stroke patient to dissolve the blood clot.

17.14 Tissue Plasminogen Activator: From Protein to Gene to Pharmaceutical
TPA is a naturally occurring human protein that prevents blood from clotting. Its isolation and use as a pharmaceutical agent for treating patients suffering from blood clotting in the brain or heart—in other words, strokes and heart attacks—was made possible by recombinant DNA technology.

amounts of the protein are now produced by bacteria and given to tens of thousands of people undergoing dialysis, with great success at reducing anemia.

HUMAN INSULIN.   One of the first important medications made by recombinant DNA methods was human insulin. This hormone, normally made by the pancreas, stimulates cells to take up glucose from the blood. People who have certain forms of diabetes mellitus have a deficiency of pancreatic insulin. Injections of the hormone can compensate for this deficiency.

In the past, the injected insulin was obtained from the pancreases of cows and pigs, which caused two problems. First, animal insulin is laborious to purify; second, it is slightly different in its amino acid sequence from human insulin. Some diabetics' immune systems detect these differences and react against the foreign protein.

The ideal solution is to use human insulin, but until the advent of recombinant DNA technology, it was available only in minuscule amounts. Since insulin is made up of only 51 amino acids, scientists were able to synthesize a gene for this protein in the laboratory. This gene (there were actually two of them, one for each polypeptide chain of the protein) was inserted into *E. coli* via an expression vector. Production of human insulin by the bacteria made widespread use of the human hormone by diabetics feasible.

DNA manipulation is changing agriculture

The cultivation of plants and husbanding of animals that constitute agriculture give us the world's oldest examples of biotechnology, dating back more than 8,000 years in human history. Over the centuries, people have adapted crops and farm animals to their needs. Through cultivation and breeding (artificial selection), desirable characteristics, such as ease of cooking the seeds or quality of the meat, have been imparted and improved. In addition, people have developed crops with desirable growth characteristics, such as a reliable ripening season and resistance to diseases.

Until recently, the most common way to improve crop plants and farm animals was to select and breed varieties with the desired phenotypes that existed in nature through mutational variation. The advent of genetics in the past century was followed by its application to plant and animal breeding. A crop plant or animal with a desirable gene could be identified, and through deliberate crosses, a single gene or, more usually, many genes could be introduced into a widely used variety of that crop.

Despite spectacular successes, such as the breeding of "supercrops" of wheat, rice, and corn, such deliberate crossing remains a hit-or-miss affair. Many desirable characteristics are complex in their genetics, and it is hard to predict accurately the results of a cross. Moreover, traditional crop plant breeding takes a long time: Many plants can reproduce only once or twice a year—a far cry from the rapid reproduction of bacteria or fruit flies.

Modern recombinant DNA technology has two advantages over traditional methods of breeding. First, the molecular approach allows a breeder to choose specific genes, making the process more precise and less likely to fail as a result of the incorporation of unforeseen genes. The ability to work with cells in the laboratory and then regenerate a whole plant by cloning makes the process much faster than the years needed for traditional breeding. The second advantage—and it is truly an amazing one—is that these molecular methods allow breeders to introduce any gene from any organism into a plant or animal species. This ability, combined with mutagenesis techniques, expands the range of possible new characteristics to an almost limitless horizon.

Biotechnology has found many applications in agriculture (Table 17.2), ranging from improving the nutritional properties of crops, to using animals as gene product factories, to using edible crops to make oral vaccines. We will describe a few examples here to demonstrate the approaches that have been used.

PLANTS THAT MAKE THEIR OWN INSECTICIDES.   Humans are not the only species that consumes crop plants. Plants are subject to infections by viruses, bacteria, and fungi, but probably the most important crop pests are herbivorous insects. From the locusts of biblical (and modern) times to the cotton boll weevil, insects have continually eaten the crops people grow.

The development of insecticides has improved the situation somewhat, but insecticides have their problems. Most, such as the organophosphates, are relatively nonspecific, killing not only the pests in the field but beneficial insects in the ecosystem as well. Some even have toxic effects on other organisms, including people. What's more, insecticides are applied to the surface of crop plants and tend to be blown away to adjacent areas, where they may have unforeseen effects.

17.2   ***Agricultural Applications of Biotechnology under Development***

PROBLEM	TECHNOLOGY/GENES
Improving the environmental adaptations of plants	Genes for drought tolerance, salt tolerance
Improving breeding	Male sterility for hybrid seeds
Improving nutritional traits	High-lysine seeds
Improving crops after harvest	Delay of fruit ripening; high-solids tomatoes; sweeter vegetables
Using plants as bioreactors	Plastics, oils, and drugs produced in plants
Controlling crop pests	Herbicide tolerance; resistance to viruses, bacteria, fungi, insects

Some bacteria have solved their own pest problem by producing proteins that kill insect larvae that eat them. For example, there are dozens of strains of *Bacillus thuringiensis*, each of which produces a protein toxic to the insect larvae that prey on it. The toxicity of this protein is 80,000 times that of the usual commercial insecticides. When a hapless larva eats the bacteria, the toxin becomes activated, binding specifically to the insect's gut to produce holes. The insect starves to death.

Dried preparations of *B. thuringiensis* have been sold for decades as a safe, biodegradable insecticide. But biodegradation is their limitation, because it means that the dried bacteria must be applied repeatedly during the growing season. A more permanent approach would be to have the crop plants make the toxin themselves.

The toxin genes from different strains of *B. thuringiensis* have been isolated and cloned. They have been extensively modified by the addition of plant promoters and terminators, plant poly A signals, plant codon usage, and plant regulatory elements on DNA. These modified genes have been introduced into plant cells in the laboratory using the Ti plasmid vector (see Figure 17.5c), and transgenic plants have been grown and tested for insect resistance in the field. So far, transgenic tomato, corn, potato, and cotton crops have been successfully shown to have considerable resistance to their insect predators.

CLONED ANIMALS THAT EXPRESS USEFUL GENES. As we described in Chapter 16, the cloning of Dolly the sheep was not done only out of scientific curiosity. One of the main objectives of the biotechnology company associated with this experiment is to make useful products in the milk of transgenic dairy animals.

This transgenic strategy is the one that was described in the opening of this chapter for making spider silk. It can also be used to make pharmaceutical products, such as human α-1-antitrypsin (α-1AT). This protein inhibits elastase, an enzyme that breaks down connective tissue. Elastase is found in excess on the surfaces of the lungs of people with cystic fibrosis, and is partly responsible for their severe breathing problems. Thus, using an inhibitor of elastase could alleviate these symptoms in these patients.

The problem is that it has been hard to get enough α-1AT from human serum. To overcome this problem, the gene for human α-1AT was introduced into the fertilized eggs of sheep, next to the promoter for lactoglobulin, a protein made in large amounts in milk (Figure 17.15). The resulting transgenic sheep made large amount of α-1AT in its milk. Since milk is produced in large amounts all year, this natural "bioreactor" produced a large supply of α-1AT, which was easily purified from the other components of the milk.

The production of animals with reliably integrated transgenes is difficult, however, so another approach is to make transgenic clones. In this case, the human gene (with its promoter) is inserted into sheep somatic cells. Those sheep cells that incorporate the transgene can then be used

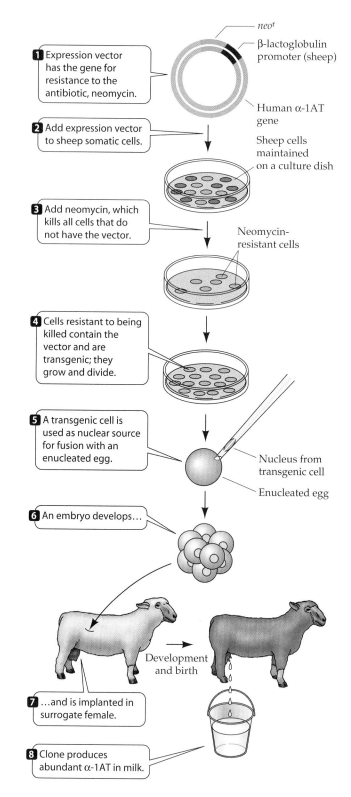

17.15 Production of Transgenic Clones for "Pharming"
The production of transgenic animals involves a combination of DNA technology and reproductive technology.

as the donor nucleus source for cloning (see Figure 16.4). This was the motivation behind the creation of Dolly.

Goats, sheep, and cows are all being used for what has come to be called "pharming," the production of medically useful products in milk. These products include blood clotting factors for treating hemophilia and antibodies for treating colon cancer.

CROPS THAT ARE RESISTANT TO HERBICIDES. Herbivorous insects are not the only threat to agriculture. Weeds may grow in fields and compete with crop plants for water and soil nutrients. Glyphosate (which is known by the trade name Roundup) is a widely used and effective weed killer, or herbicide. It works only on plants, by inhibiting an enzyme system in the chloroplast that is involved in the synthesis of amino acids. Glyphosate is truly a "miracle herbicide," killing 76 of the world's 78 most prevalent weeds. Unfortunately, it also kills crop plants, so great care must be taken with its use. In fact, it is best used to rid a field of weeds before the crop plant starts to grow. But as any gardener knows, when the crop begins to grow, the weeds reappear. So it would be advantageous if the crop were not affected by the herbicide. Then, the herbicide could be applied to the field at any time, and would kill only the weeds.

Fortunately, some soil bacteria have mutated to develop an enzyme that breaks down glyphosate. Scientists have isolated the gene for this enzyme, cloned it, and added plant sequences for transcription, translation, and targeting to the chloroplast. The gene has been inserted into corn, cotton, and soybean plants, making them resistant to glyphosate. In the late 1990s, this technology expanded so rapidly that half of the U.S. crops of these three plants are now transgenic in this way.

GRAINS WITH IMPROVED NUTRITIONAL CHARACTERISTICS. Humans must eat foods (or supplements) containing an adequate amount of β-carotene, which the body converts into vitamin A. About 400 million people worldwide suffer from vitamin A deficiency, which makes them susceptible to infections and blindness. One reason is that they eat rice grains, which do not contain β-carotene, but have only a precursor molecule for it. However, other organisms, such as the bacterium *Erwinia* and daffodil plants, have enzymes that can convert the precursor into β-carotene. The genes for this biochemical pathway are present in the bacterial and daffodil genomes, but not in the rice genome.

Scientists isolated two of the genes for the β-carotene pathway from the bacterium and the other two from daffodil plants. They added promoter signals for expression in the developing rice grain, and then added each gene to rice plants by using the vector *Agrobacterium tumefaciens* (see Figure 17.5c). The resulting rice plants produce grains that look yellow because of their high content of β-carotene (Figure 17.16). About 300 grams of this cooked rice a day can supply all the β-carotene a person needs. This new transgenic strain is now being crossed with more locally adapted strains, and it is hoped that the diets of millions of people will be improved soon.

There is public concern about biotechnology

When the initial experiments creating recombinant DNA in the laboratory were done in the 1970s, there was considerable concern, especially by the scientists involved, over the

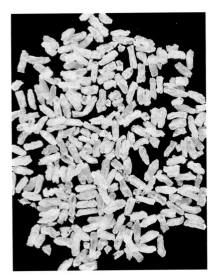

17.16 Grains From Transgenic Rice Rich in β-Carotene
The grains from this transgenic strain (right) are yellow because they make the pigment β-carotene, which is converted by humans into vitamin A. Normal rice (left) does not contain β-carotene.

safety of recombinant DNA. After all, the bacterium they used, *E. coli*, normally lives in the human intestine. What would happen if the laboratory strain shared its new genes with the bacteria living in humans? In response to this concern, the scientists involved initially stopped their research, took stock of the implications of what they were doing, and then took elaborate safety precautions to prevent accidental release of the recombinant organisms and their genes. For example, the strains of *E. coli* used in the lab have a number of mutations that make their survival in the human intestine impossible.

As biotechnology developed, it became apparent that these fears for safety were exaggerated. Accidental release of organisms and transfer of genes has not been a problem. Medical products made by DNA technology are widely used and accepted.

However, with the rapid expansion of genetically modified crops, new concerns have been raised. The issue now is a different one in that genetically modified organisms are being designed to be introduced into the natural environment. Indeed, some countries have banned foods that come from genetically modified crops. These concerns are centered on three claims:

▶ Genetic manipulation is an unnatural interference with nature.
▶ Genetically altered foods are unsafe to eat.
▶ Genetically altered plants are dangerous to the environment.

Advocates of biotechnology tend to agree with the first claim. However, they point out that all major crops are unnatural in the sense that they come from highly bred plants growing in a manipulated environment (a farmer's field). The new technology just adds another level of sophistication.

The concern about safety for humans is countered by the facts that only single genes are added, and that these genes are specific for plant function. For example, the *B. thuringiensis* toxin produced by transgenic plants does not have any effects on people. However, as plant biotechnology moves from adding genes to improve plant growth to adding genes that affect human nutrition, such concerns will become more pressing.

The third concern, about environmental effects, involves the possible "escape" of transgenes from crops to other species. If the gene for herbicide resistance, for example, was inadvertently transferred from a crop to a nearby weed, the latter could thrive in herbicide-treated areas. Or beneficial insects could eat plant materials containing *B. thuringiensis* toxin and die. Transgenic plants undergo extensive field testing before they are approved for use, but the complexity of the biological world makes it impossible to predict all potential environmental effects of transgenic organisms. Because of the potential benefits of agricultural biotechnology (see Table 17.1), scientists believe that it is wise to "proceed with caution."

DNA fingerprinting uses the polymerase chain reaction

"Everyone is unique." This old saying applies not only to human behavior, but also to the human genome. Mutations and recombination through sexual reproduction ensure that each member of a species (except identical twins) has a unique DNA sequence. An individual can be definitively characterized ("fingerprinted") by his or her DNA base sequence.

The ideal way to distinguish an individual from all the other people on Earth would be to describe his or her entire genomic DNA sequence. But since the human genome contains more than 3 billion nucleotides, this idea is clearly not practical. Instead, scientists have looked for genes that are highly polymorphic—that is, genes that have multiple alleles in the human population and are therefore different in different individuals.

One easily analyzed genetic system consists of short moderately repetitive DNA sequences that occur side by side in the chromosomes. These repeat patterns are inherited. For example, an individual might inherit a chromosome 15 with a short sequence repeated six times from her mother, and the same sequence repeated two times from her father. These repeats, called VNTRs (*variable number of tandem repeats*), are easily detectable if they lie between two recognition sites for a restriction enzyme. If the DNA from this individual is cut with the restriction enzyme, it will form two different-sized fragments: one larger (the one from the mother) and the other smaller (the one from the father). These patterns are easily seen by use of gel electrophoresis (Figure 17.17). With several different repeated sequences (as many as eight are used, each with numerous alleles), an individual's unique pattern becomes apparent.

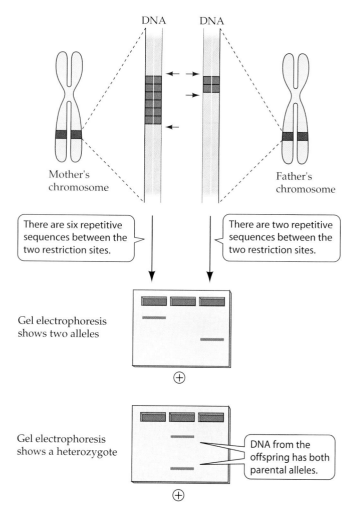

17.17 DNA Fingerprinting
The number of VNTRs inherited by an individual can be used to make a DNA fingerprint.

Typically, these methods require 1 μg of DNA, or the DNA content of about 100,000 human cells, but this amount is not always available. The power of the polymerase chain reaction (see Figure 11.21) permits the DNA from a single cell to be amplified, producing in a few hours the necessary 1 μg for restriction and gel analysis.

DNA fingerprints are used in forensics (crime investigation) to help prove the innocence or guilt of a suspect. For example, in a rape case, DNA can be extracted from dried semen or hair left by the attacker and compared with DNA from a suspect. So far, this method has been used to prove innocence (the DNA patterns are different) more often than guilt (the DNA patterns are the same). It is easy to exclude someone on the basis of these tests, but two people could theoretically have the same patterns, since what is being tested is just a small sample of the genome. Therefore, proving that a suspect is guilty cannot rest on DNA fingerprinting alone.

Two fascinating examples demonstrate the use of DNA fingerprinting in the analysis of historical events. Three hundred years of rule by the Romanov dynasty in Russia ended on July 16, 1918, when Tsar Nicholas II, his wife, and their five children were executed by a firing squad during the Communist revolution. A report that the bodies had been burned to ashes was never questioned until 1989, when a shallow grave with several skeletons was discovered several miles from the presumed execution site. Recent DNA fingerprinting of bone fragments found in this grave indicated that they came from an older man and woman and three female children, who were clearly related to each other (Figure 17.18) and were also related to several living descendants of the Tsar.

The other example involves Thomas Jefferson, the third president of the United States. In 1802, Jefferson was alleged to have fathered a son by his female slave, Sally Hemmings. Jefferson denied this, and his denial was accepted by many historians because of his vocal opposition to mixed-race relationships. But descendants of Hemmings' two oldest sons (the second was named Eston Jefferson) pressed their case. DNA fingerprinting was done using Y chromosome markers from descendants of these two sons as well as the president's paternal uncle (the president had no acknowledged sons). The results show that Thomas Jefferson may have been the father of the second son, but was not the father of the first son.

In addition to such highly publicized cases, there are many other applications of PCR-based DNA fingerprinting. In 1992, the California condor was extinct in the wild. There were only 52 California condors on Earth, all cared for by the San Diego and Los Angeles zoos. Scientists made DNA fingerprints of all these birds so that the geneticists at the zoos could select unrelated individuals for mating in order to increase genetic variation and increase the viability of the offspring. A number of these young birds have now been returned to the wild. A similar program is under way for the threatened Galápagos tortoises (see Chapter 58).

Plant scientists have found in nature or produced by artificial selection thousands of varieties of crops such as rice, wheat, corn, and grapes. The seeds of many of these varieties are kept in cold storage in "seed banks." Samples of these plants are being DNA-fingerprinted to determine which varieties are genetically the same and which are the most diverse, as a guide to future breeding programs.

A related use of PCR is in the diagnosis of infections. In this case, the test shows whether the DNA of an infectious agent is present in a blood or tissue sample. A primer strand matching the pathogen's DNA is added to the sample. If the pathogen is present, its DNA will serve as a template for the primer, and will be amplified. Because so little of the target sequence is needed, and because primers can be made to bind only to a specific viral or bacterial genome, the PCR-based test is extremely sensitive. If an organism is present in small amounts, PCR testing will detect it.

Finally, the isolation and characterization of genes for various human diseases, such as sickle-cell anemia and cystic fibrosis, has made PCR-based genetic testing a reality. We will discuss this subject in depth in the next chapter.

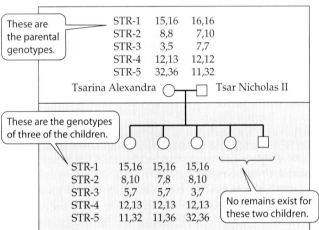

17.18 DNA Fingerprinting the Russian Royal Family
The skeletal remains of Tsar Nicholas II, his wife Alexandra, and three of their children were found in 1989 and subjected to DNA fingerprinting. Five VNTRs were tested. The results can be interpreted as follows. Using the VNTR STR-2 as an example, the parents had genotypes 8,8 (homozygous) and 7,10 (heterozygous). The three children all inherited type 8 from the Tsarina and either type 7 or type 10 from the Tsar.

Chapter Summary

Cleaving and Rejoining DNA

▶ Knowledge of DNA transcription, translation, and replication has been used to create recombinant DNA molecules, made up of sequences from different organisms.

▶ Restriction enzymes, which are made by microbes as a defense mechanism against viruses, bind to DNA at specific sequences and cut it. **Review Figure 17.1**

▶ DNA fragments generated from cleavage by restriction enzymes can be separated by size using gel electrophoresis. The sequences of these fragments can be further identified by hybridization with a probe. **Review Figures 17.2, 17.3**

▶ Many restriction enzymes make staggered cuts in the two strands of DNA, creating "sticky ends" with unpaired bases. The sticky ends can be used to create recombinant DNA if

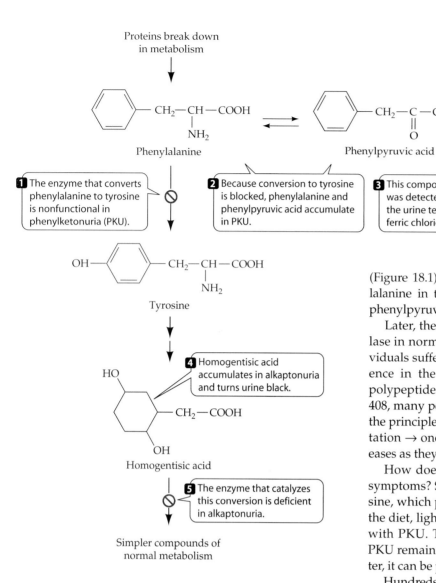

18.1 One Gene, One Enzyme in Humans
Phenylketonuria and alkaptonuria are both caused by abnormalities in a specific enzyme. Knowing the causes of such single-gene, single-enzyme metabolic diseases can aid in the development of screening tests and treatments.

Protein as Phenotype

As we saw in Chapters 11 and 12, genetic mutations are often expressed phenotypically as proteins that differ from the normal wild type. In the first section of this chapter, we identify and discuss the kinds of abnormal proteins that can result from inheritance of an abnormal allele or its origin by mutation. Then we will consider the role of the environment and of patterns of inheritance resulting from autosomal recessives, autosomal dominants, X linkages, and chromosomal abnormalities.

Many genetic diseases result from abnormal or missing proteins

Proteins have many roles in eukaryotic cells, and the genes that code for them can be mutated to cause genetic diseases. Enzymes, receptors, transport proteins, structural pro-

teins, and carriers such as hemoglobin have all been implicated in genetic diseases.

ENZYMES. Although Dr. Følling made his discovery in 1934, it was not until 1957 that the complex clinical phenotype of phenylketonuria (PKU) were traced back to its *primary phenotype*: a single abnormal protein. As Følling had predicted, phenylalanine hydroxylase, the enzyme that catalyzes the conversion of dietary phenylalanine to tyrosine, was not active in patients' livers (Figure 18.1). Lack of this conversion led to excess phenylalanine in the blood and explained the accumulation of phenylpyruvic acid.

Later, the protein sequences of phenylalanine hydroxylase in normal people were compared with those in individuals suffering from PKU. In many cases, the only difference in the 451 amino acids that constitute this long polypeptide chain was that instead of arginine at position 408, many people with PKU have tryptophan. Once again, the principles of one gene → one polypeptide and one mutation → one amino acid change hold true in human diseases as they do in studies of so many other organisms.

How does the abnormality in PKU lead to its clinical symptoms? Since the pigment melanin is made from tyrosine, which patients cannot synthesize but must obtain in the diet, lighter skin and hair color are observed in people with PKU. The exact cause of the mental retardation in PKU remains elusive, but as we will see later in this chapter, it can be prevented.

Hundreds of human genetic diseases that result from enzyme abnormalities have been discovered, many of which lead to mental retardation and premature death. Most of these diseases are rare; PKU, for example, shows up in one newborn infant out of every 12,000. But this is just the tip of the mutational iceberg. Undoubtedly, some mutations result in altered proteins that have no obvious clinical effects. For example, there could be many amino acid changes in phenylalanine hydroxylase that do not affect its catalytic activity.

Analysis of the same protein in different people often shows variations that have no functional significance. In fact, at least 30 percent of all proteins whose sequences are known show detectable amino acid differences among individuals. If one protein variant exists in less than 99 percent of a population (that is, if the protein has another variant at least 1 percent of the time), the protein is said to be *polymorphic*. The key point is that polymorphism does not necessarily mean disease.

HEMOGLOBIN. The first human genetic disease for which an amino acid abnormality was tracked down as the cause was not PKU. It was the blood disease *sickle-cell anemia*, which most often afflicts people whose ancestors came from

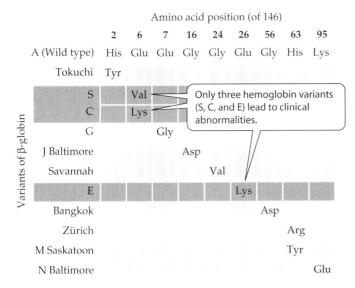

Amino acid position (of 146)	2	6	7	16	24	26	56	63	95
A (Wild type)	His	Glu	Glu	Gly	Gly	Glu	Gly	His	Lys
Tokuchi	Tyr								
S		Val							
C		Lys							
G			Gly						
J Baltimore				Asp					
Savannah					Val				
E						Lys			
Bangkok							Asp		
Zürich								Arg	
M Saskatoon								Tyr	
N Baltimore									Glu

Only three hemoglobin variants (S, C, and E) lead to clinical abnormalities.

18.2 Hemoglobin Polymorphism
Only three of the many variants of hemoglobin are known to lead to clinical abnormalities.

the Tropics or from the Mediterranean. Among African-Americans, about 1 in 655 are homozygous for the sickle allele and have the disease. The abnormal allele produces an abnormal protein that leads to sickled red blood cells (see Figure 12.17). These cells tend to block narrow blood capillaries, especially when the oxygen concentration of the blood is low, and the result is tissue damage.

Human hemoglobin is a protein with quaternary structure, containing four globin chains—two α chains and two β chains—as well as the pigment heme (see Figure 3.7). In sickle-cell anemia, one of the 146 amino acids in the β–globin chain is abnormal: At position 6, the normal glutamic acid has been replaced by valine. This replacement changes the charge of the protein (glutamic acid is negatively charged and valine is neutral; see Table 3.2), causing the protein to form long aggregates in the red blood cells. The result is anemia, a deficiency of normal red blood cells.

Because hemoglobin is easy to isolate and study, its variations in the human population have been extensively documented (Figure 18.2). Hundreds of single amino acid alterations in β-globin have been reported. Some of these polymorphisms are even at the same amino acid position. For example, at the same position that is mutated in sickle-cell anemia, the normal glutamic acid may be replaced by lysine, causing *hemoglobin C disease*. In this case, the anemia is usually not severe. Many alterations of hemoglobin have no effect on the protein's function, and thus no clinical phenotype. This is fortunate, because about 5 percent of all humans are carriers for one of these variants.

RECEPTORS AND TRANSPORT PROTEINS. Some of the most common human genetic diseases show their primary phenotype as altered membrane proteins. About one person in 500 is born with *familial hypercholesterolemia* (FH), in which levels of cholesterol in the blood are several times higher than normal. The excess cholesterol can accumulate on the inner walls of blood vessels, leading to complete blockage if a blood clot forms. If a blood clot forms in a major vessel serving the heart, the heart becomes starved of oxygen, and a heart attack results. If a blood clot forms in the brain, the result is a stroke. People with FH often die of heart attacks before the age of 45, and in severe cases, before they are 20 years old.

Unlike PKU, which is characterized by the inability to convert phenylalanine to tyrosine, the problem in FH is not an inability to convert cholesterol to other products. People with FH have all the machinery needed to metabolize cholesterol. The problem is that they are unable to transport the cholesterol into the liver cells that use it.

Cholesterol travels in the bloodstream in protein-containing particles called lipoproteins. One type of lipoprotein, *low-density lipoprotein*, carries cholesterol to the liver cells (Figure 18.3a). After binding to a specific receptor on the membrane of a liver cell, the lipoprotein is taken up by endocytosis and delivers its cholesterol to the interior of the cell. People with FH lack a functional version of

(a) Hypercholesterolemia

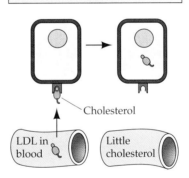

Normal liver cell: Cholesterol, as part of low-density lipoprotein (LDL), enters the cell after LDL binds to a receptor.

Familial hypercholesterolemia: Absence of an LDL receptor prevents cholesterol from entering the cells, and it accumulates in the blood.

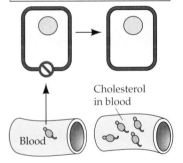

Cholesterol

LDL in blood

Little cholesterol

Cholesterol in blood

Blood

(b) Cystic fibrosis

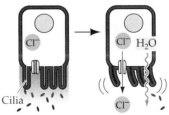

Normal cell lining the airway: Cl⁻ leaves the cell through a channel. Water follows by osmosis, and moist thin mucus allows cilia to beat and sweep away foreign particles, including bacteria.

Cystic fibrosis: Lack of a Cl⁻ channel causes a thick viscous mucus to form. Protective cilia cannot beat properly and remove bacteria; infections can easily take hold.

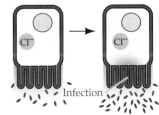

Cl^- Cl^- H_2O Cl^- Cl^-

Cilia Cl^- Infection

Thick mucus   Thin mucus   Thick mucus   Thick mucus

18.3 Genetic Diseases of Membrane Proteins
The left two panels illustrate normal cell function, while the two right panels show the abnormalities caused by (a) hypercholesterolemia and (b) cystic fibrosis.

the receptor protein. Of the 840 amino acids that make up the receptor, often only one is abnormal in FH, but this is enough to change its structure so that it cannot bind to the lipoprotein.

Among Caucasians, about one baby in 2,500 is born with *cystic fibrosis*. The clinical phenotype of this genetic disease is an unusually thick and dry mucus that lines organs such as the tubes that serve the respiratory system. Its dryness prevents cilia on the surfaces of the epithelial cells from working efficiently to clear out the bacteria and fungal spores that we take in with every breath. The results are recurrent and serious infections, as well as liver, pancreatic, and digestive failures. Patients often die in their twenties or thirties.

The reason for the thick mucus in patients with cystic fibrosis is a defective version of a membrane protein, the *chloride transporter* (Figure 18.3b). In normal cells, this membrane channel opens to release Cl⁻ to the outside of an epithelial cell. The imbalance of Cl⁻ ions (more are now on the outside of the cell than on the inside) causes water to leave the cell by osmosis, resulting in a moist mucus outside the cell. A single amino acid change in the transporter renders it nonfunctional, which leads to dry mucus and the consequent clinical problems.

STRUCTURAL PROTEINS. About one boy in 3,000 is born with *Duchenne's muscular dystrophy*. In this genetic disease, the problem is not an enzyme or receptor, but a protein involved in biological structure. People with this disease show progressively weaker muscles and are wheelchair-bound by their teenage years. Patients usually die in their twenties, when the muscles that serve their respiratory system fail. Most people have a protein in their skeletal muscles called dystrophin, which may bind the major muscle protein actin to the plasma membrane of the muscle cells. Patients with Duchenne's muscular dystrophy do not have a working copy of dystrophin, so their muscles do not work.

Coagulation proteins are involved in the clotting of blood at a wound. As we saw in Chapter 17, inactive clotting proteins are always present in the blood, and become active only at a wound. People with the genetic disease *hemophilia* lack one of the coagulation proteins. Some people with this disease risk death from even minor cuts, since they cannot stop bleeding.

Prion diseases are disorders of protein conformation

Transmissible spongiform encephalopathies (TSE's) are degenerative brain diseases that occur in many mammals, including humans. In these diseases, the brain gradually develops holes, leaving it looking like a sponge. Scrapie, a TSE that causes affected sheep and goats to rub the wool off their bodies, has been known for 250 years. In the 1980s, a TSE appeared in many cows in Britain and was traced to the cows eating products from sheep that had scrapie. Then, in the 1990s, some people who had eaten beef from cows with

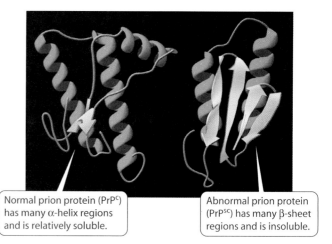

Normal prion protein (PrP^c) has many α-helix regions and is relatively soluble.

Abnormal prion protein (PrP^sc) has many β-sheet regions and is insoluble.

18.4 Prion Proteins
Normal prion proteins (PrP^c, left) can be converted to the disease-causing form (PrP^sc, right), which has a different three-dimensional structure.

TSE got a human version of the disease (dubbed "mad cow disease" by the media), again suggesting that the causative agent could cross species lines.

Another instance of humans consuming an infective agent and getting TSE was kuru, a disease resulting in dementia that occurred among the Fore tribe of New Guinea. In the 1950s, it was discovered that people with kuru had consumed the brains of people who had died of it. When this ritual cannibalism stopped, so did the epidemic of kuru.

Researchers found that TSE's could be transmitted from one animal to another via brain extracts from a diseased animal. But when Tikva Alper treated these extracts with high doses of ultraviolet light to inactivate nucleic acids, they still caused TSE's. She proposed that the causative agent for TSE's was a protein and not a virus, as had been suspected. Later, Stanley Prusiner purified the protein responsible and showed it to be free of DNA or RNA. He called it a **proteinaceous infective particle (prion)**.

Normal brain cells have a membrane protein called PrP^c. A protein with the same amino acid sequence is present in TSE-affected brain tissues, but this protein has an altered shape and is called PrP^sc (Figure 18.4). So TSE is not caused by a mutated gene (the primary structures of the two proteins are the same), but is somehow caused by altered protein conformation. The altered three-dimensional structure has profound effects on the protein's function in the cell. Insoluble PrP^sc piles up as fibers in brain tissue, causing cell death.

How can the exposure of a normal cell to material containing PrP^sc result in a TSE? The abnormal PrP^sc protein seems to induce a conformational change in the normal PrP^c protein so that it too becomes abnormal, just as one rotten apple results in a whole barrel full of rotten apples. Just how the conversion occurs and how it causes TSE are unclear.

Most diseases are caused by both heredity and environment

The human diseases for which clinical phenotypes can be traced to a single altered protein may number in the thousands, and in most cases they are dramatic evidence of the one-gene, one-polypeptide principle. Taken together, these diseases have a frequency of about 1 percent in the total population.

Far more common, however, are diseases that are *multifactorial*; that is, they are caused by many genes and proteins interacting with the environment. Although we tend to call individuals either normal (wild-type) or abnormal (mutant), the sum total of our genes is what determines which of us who eat a high-fat diet will die of a heart attack, or which of us exposed to infectious bacteria will come down with a disease. Estimates suggest that up to 60 percent of all people have diseases that are genetically influenced.

Human genetic diseases have several patterns of inheritance

As in any human genetic system, the alleles that cause diseases are inherited as dominants or recessives, and are carried on autosomes or sex chromosomes (see Chapter 10). In addition, some human diseases are caused by more extensive chromosomal abnormalities.

AUTOSOMAL RECESSIVES. PKU, sickle-cell anemia, and cystic fibrosis are all caused by autosomal recessive mutant alleles. Typically, both parents of an affected person are normal, heterozygous carriers of the abnormal allele. The parents have a 25 percent (one in four) chance of having an affected (homozygous) son or daughter. Because of this low probability and the fact that many families in Western societies now have fewer than four children, it is unusual for more than one child in a family to have an autosomal recessive disease.

In the cells of a person who is homozygous for an autosomal recessive mutant allele, only the nonfunctional, mutant version of the protein it encodes is made. Thus a biochemical pathway or important cell function is disrupted, and disease results. Not unexpectedly, heterozygotes, with one normal and one mutant allele, often have 50 percent of the normal level of functional protein. For example, people who are heterozygous for the allele for PKU have half the number of active molecules of phenylalanine hydroxylase in their liver cells as individuals who carry two normal alleles for this enzyme. But by one mechanism or another, this 50 percent suffices for relatively normal cellular function.

AUTOSOMAL DOMINANTS. An example of a disease caused by abnormal autosomal dominant alleles is familial hypercholesterolemia. In autosomal dominance, the presence of only one mutant allele is enough to produce the clinical phenotype. In people who are heterozygous for familial hypercholesterolemia having half the normal number of functional receptors for low-density lipoprotein on the surface of liver cells is simply not enough to clear cholesterol from the blood. In autosomal dominance, direct transmission from parent to offspring is the rule.

X-LINKED INHERITANCE. Both hemophilia and Duchenne's muscular dystrophy are inherited as X-linked recessives; that is, the mutant alleles responsible are located on the X chromosome. Thus a son who inherits a mutant allele on the X chromosome from his mother will have the disease, because the Y chromosome does not contain a normal allele. However, a daughter who inherits one mutant allele will be a heterozygous carrier, since she has two X chromosomes, and hence two alleles. Because, until recently, few males with these diseases lived to reproduce, the most common pattern of inheritance has been from carrier mother to son, and these diseases are much more common in males than in females.

CHROMOSOMAL ABNORMALITIES. Chromosomal abnormalities also cause human diseases. Such abnormalities include an excess or loss of one or two chromosomes (*aneuploidy*), loss of a piece of a chromosome (*deletion*), and transfer of a piece of one chromosome to another chromosome (*translocation*). About one newborn in 200 is born with a chromosomal abnormality. While some of them are inherited, many are the result of meiotic problems such as nondisjunction (see Chapter 9).

Many zygotes that have abnormal chromosomes do not survive development and are spontaneously aborted. Of the 20 percent of pregnancies that are spontaneously aborted during the first 3 months of human development, an estimated half of them have chromosomal aberrations. For example, more than 90 percent of human zygotes that have only one X chromosome and no Y (Turner syndrome) do not live beyond the fourth month of pregnancy.

A common cause of mental retardation is *fragile-X syndrome* (Figure 18.5). About one male in 1,500 and one female in 2,000 are affected. Near the tip of the abnormal X chromosome is a constriction that tends to break during preparation for microscopy, giving the name for this syndrome. Although the basic pattern of inheritance is that of an X-linked recessive trait, there are departures from this pat-

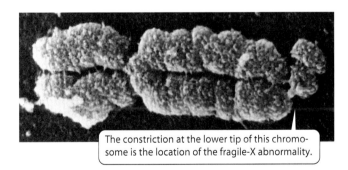

The constriction at the lower tip of this chromosome is the location of the fragile-X abnormality.

18.5 A Fragile-X Chromosome at Metaphase
The chromosomal abnormality that causes the mental retardation symptomatic of fragile-X syndrome shows up physically as a constriction.

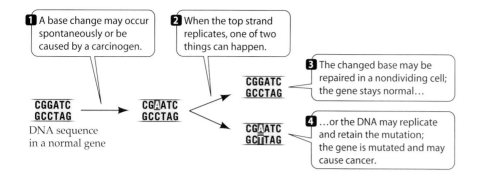

1 A base change may occur spontaneously or be caused by a carcinogen.

2 When the top strand replicates, one of two things can happen.

3 The changed base may be repaired in a nondividing cell; the gene stays normal…

4 …or the DNA may replicate and retain the mutation; the gene is mutated and may cause cancer.

CGGATC
GCCTAG
DNA sequence in a normal gene

CG**A**ATC
GCCTAG

CGGATC
GCCTAG

CG**A**ATC
GC**T**TAG

18.14 Dividing Cells Are Especially Susceptible to Genetic Damage and Cancer
A base change is more likely to be repaired in a nondividing cell.

changes in the nucleotides. In addition, certain substances called *carcinogens* cause mutations that lead to cancer. Familiar carcinogens include the chemicals that are present in tobacco smoke and meat preservatives, ultraviolet light from the sun, and ionizing radiation from sources of radioactivity. Less familiar, but just as harmful, are thousands of chemicals present naturally in the foods people eat. According to one estimate, these "natural" carcinogens account for well over 80 percent of the human exposure to agents that cause cancer.

The common theme in natural and human-made carcinogens is that almost all of them damage DNA, usually by causing changes from one base to another (Figure 18.14). In somatic cells that divide often, such as epithelial and bone marrow cells, there is less time for DNA repair mechanisms to work before replication occurs again. Therefore, such cells are especially susceptible to genetic damage.

Two kinds of genes are changed in many cancers

The changes in the control of cell division that lie at the heart of cancer can be likened to the controls of an automobile. To make a car move, two things must happen: The gas pedal must be pressed, and the brake must be released. In the human genome, some genes act as **oncogenes**, which act as the "gas pedal" to stimulate cell division, and some as **tumor suppressor genes**, which "put the brake on" to inhibit it.

ONCOGENES. The first hint that oncogenes (from the Greek *onco-*, "mass") were necessary for cells to become cancerous came with the identification of virally induced cancers in animals. In many cases, these viruses bring a new gene into their host cells, one that acts to stimulate cell division when it is expressed in the viral genome. It soon became apparent that these viral oncogenes had counterparts in the genomes of the host cells, called **proto-oncogenes**, that were not actively transcribed. So the search for genes that are damaged by carcinogens quickly zeroed in on the proto-oncogenes.

Proto-oncogenes are genes that have the capacity to stimulate cell division, but are normally turned "off" in differentiated, nondividing cells. Many are involved in growth factor stimulation (Figure 18.15). Some remarkable proto-oncogenes control apoptosis (programmed cell death). Acti-

vation of these genes by mutation causes them to prevent apoptosis, allowing cells that normally die to continue dividing.

Some proto-oncogenes can be activated by point mutations, others by chromosome changes such as translocations, and still others by gene amplification. Whatever the mechanism, the result is the same: The proto-oncogene becomes activated, and the "gas pedal" for cell division is pressed.

TUMOR SUPPRESSOR GENES. About 10 percent of all cancer is clearly inherited. Often the inherited form of the cancer is clinically similar to the noninherited form that occurs later in life, called the *sporadic form*. The major differences are that the inherited form strikes a person much earlier in life and usually shows up as multiple tumors.

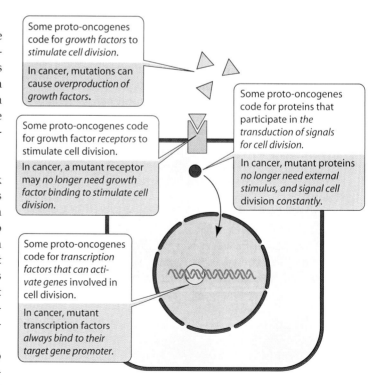

Some proto-oncogenes code for *growth factors* to *stimulate cell division*.

In cancer, mutations can cause *overproduction of growth factors*.

Some proto-oncogenes code for growth factor *receptors* to stimulate cell division.

In cancer, a mutant receptor may *no longer need growth factor binding to stimulate cell division*.

Some proto-oncogenes code for proteins that participate in *the transduction of signals for cell division*.

In cancer, mutant proteins *no longer need external stimulus, and signal cell division constantly*.

Some proto-oncogenes code for *transcription factors that can activate genes* involved in cell division.

In cancer, mutant transcription factors *always bind to their target gene promoter*.

18.15 Proto-Oncogene Products Stimulate Cell Division
Mutations can affect any of the several ways proto-oncogenes normally stimulate cell division, thus causing cancer.

(a)

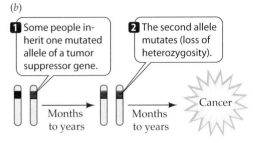

1 Most people are born with two normal alleles for a tumor suppressor gene.

2 One of the alleles mutates.

3 Later, the second allele mutates (loss of heterozygosity).

Months to years | Months to years | Months to years → Cancer

(b)

1 Some people inherit one mutated allele of a tumor suppressor gene.

2 The second allele mutates (loss of heterozygosity).

Months to years | Months to years → Cancer

18.16 The "Two-Hit" Hypothesis for Cancer
(a) Although a single mutation can activate a proto-oncogene, two mutations are needed to inactivate a tumor suppressor gene. (b) An inherited predisposition to cancer occurs in people born with one allele already mutated.

In 1971, Alfred Knudson used these observations to predict that for a tumor to occur, a tumor suppressor gene must be inactivated. But unlike oncogenes, in which one mutated allele is all that is needed for activation, the full inactivation of a tumor suppressor gene requires that both alleles be turned off, which requires two mutational events. It takes a long time for both alleles in a single cell to mutate and cause sporadic cancer. But in inherited cancer, people are born with one mutant allele for the tumor suppressor, and need just one more mutational event for full inactivation of the "brakes" (Figure 18.16).

Various tumor suppressor genes have been isolated and confirm Knudson's "two-hit" hypothesis. Some of these genes are involved in inherited forms of rare childhood cancers such as retinoblastoma (a tumor of the eye) and Wilms' tumor of the kidney, as well as in inherited breast and prostate cancers.

An inherited form of breast cancer demonstrates the effect of tumor suppressor genes. The 9 percent of women who inherit one mutated copy of the gene *BRCA1* have a 60 percent chance of having breast cancer by age 50 and an 82 percent chance of developing it by age 70. The comparable figures for women who inherit two normal copies of the gene are 2 percent and 7 percent, respectively.

How do tumor suppressor genes act in the cell? Like the proto-oncogenes, they are normally involved in vital cell functions (Figure 18.17). Some control progress through the cell cycle. The protein encoded by *Rb*, a gene that was first described for its contribution to retinoblastoma, is active during G1. In the active form, it encodes a protein that binds to and inactivates transcription factors that are necessary for progress to the S phase and the rest of the cell cycle. In nondividing cells, *Rb* remains active, preventing

cell division until the proper growth factor signals are present. When the Rb protein is inactivated by mutation, the cell cycle moves forward independent of growth factors.

The protein product of another widespread tumor suppressor gene, *p53*, also stops cells during G1. It does this by acting as a transcription factor, stimulating the production of (among other things) a protein that blocks the interaction of a cyclin and protein kinase needed for moving the cell cycle beyond G1. This gene is mutated in many types of cancers, including lung cancer and colon cancer.

The pathway from normal cell to cancerous cell is complex

The "gas pedal" and "brake" analogies for oncogenes and tumor suppressor genes, respectively, are elegant but simplified. There are many oncogenes and tumor suppressor genes, some of which act only in certain cells at certain times. Therefore, a sequence of events must occur before a normal cell becomes malignant.

Because colon cancers progress to full malignancy slowly, it is possible to describe the oncogenes and tumor suppressor genes at each stage in great molecular detail. Figure 18.18 outlines the progress of this form of cancer. At least three tumor suppressor genes and one oncogene must be mutated in sequence for an epithelial cell in the colon to become metastatic. Although the occurrence of all these events in a single cell might appear unlikely, remember that the colon has millions of cells, that the cells giving rise to epithelial cells are constantly dividing, and that these changes take

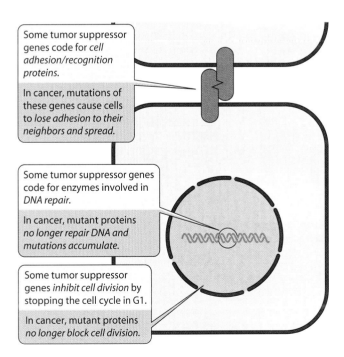

Some tumor suppressor genes code for *cell adhesion/recognition proteins.*

In cancer, mutations of these genes cause cells to *lose adhesion to their neighbors and spread.*

Some tumor suppressor genes code for enzymes involved in *DNA repair.*

In cancer, mutant proteins *no longer repair DNA and mutations accumulate.*

Some tumor suppressor genes *inhibit cell division* by stopping the cell cycle in G1.

In cancer, mutant proteins *no longer block cell division.*

18.17 Tumor Suppressor Gene Products Inhibit Cell Division and Cancer
Mutations can affect any of the several ways that tumor suppressor genes inhibit cell division, causing cells to divide and form a tumor.

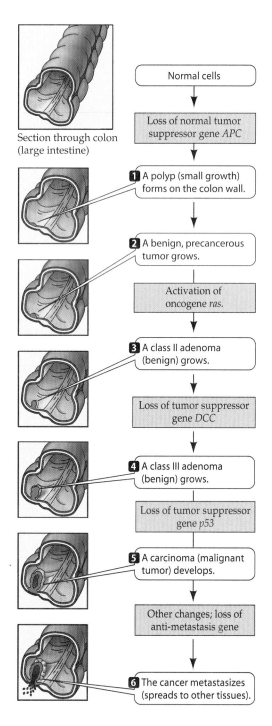

Section through colon (large intestine)

Normal cells

↓

Loss of normal tumor suppressor gene *APC*

↓

1 A polyp (small growth) forms on the colon wall.

↓

2 A benign, precancerous tumor grows.

↓

Activation of oncogene *ras*.

↓

3 A class II adenoma (benign) grows.

↓

Loss of tumor suppressor gene *DCC*

↓

4 A class III adenoma (benign) grows.

↓

Loss of tumor suppressor gene *p53*

↓

5 A carcinoma (malignant tumor) develops.

↓

Other changes; loss of anti-metastasis gene

↓

6 The cancer metastasizes (spreads to other tissues).

18.18 Multiple Mutations Transform a Normal Colon Epithelial Cell into a Cancerous Cell
In this form of cancer, at least five genes are mutated in a single cell.

place over many years of exposure to natural and human-made carcinogens and to spontaneous mutations.

The characterization of the molecular changes in tumor cells has opened up the possibility of genetic diagnosis and screening for cancer, as is done for genetic diseases. Many tumors are now commonly diagnosed in part by specific oligonucleotide probes for oncogene and/or tumor suppressor gene alterations. It is also possible to detect early in life whether an individual has inherited a mutated tumor sup-

pressor gene. For example, a person who inherits mutated copies of the tumor suppressor genes involved in colon cancer normally would have a high probability of developing this cancer by age 40. Surgical removal of the colon would prevent this metastatic tumor from arising.

Treating Genetic Diseases

Most treatments of genetic diseases try to alleviate the symptoms that affect the patient. But to effectively treat diseases caused by genes—whether they affect all cells, as in inherited disorders such as PKU, or only somatic cells, as in cancer—physicians must be able to diagnose the disease accurately, must know how the disease works at the biochemical level, and must be able to intervene early, before the disease ravages or kills the individual.

Basic research has provided the knowledge needed for accurate diagnostic tests, as well as a beginning at understanding *pathogenesis* (the cause of diseases) at the molecular level. Physicians are now applying this knowledge to treat genetic diseases. In this section, we will see that these treatments range from specifically modifying the mutant phenotype to supplying the normal version of a mutant gene.

One approach to treatment is to modify the phenotype

There are three ways to alter the phenotype of a genetic disease so that it no longer harms an individual.

RESTRICTING THE SUBSTRATE. Restricting the substrate of a deficient enzyme is the approach taken when a newborn is diagnosed with PKU. In this case, the deficient enzyme is phenylalanine hydroxylase, and the substrate is phenylalanine. The infant's inability to break down the phenylalanine in food leads to a buildup of the substrate, which causes the clinical symptoms. So the infant is immediately put on a special diet that contains only enough phenylalanine for immediate use.

Lofenelac, a milk-based product that is low in phenylalanine, is fed to these infants just like formula. Later, certain fruits, vegetables, cereals, and noodles low in phenylalanine can be added to the diet. Meat, fish, eggs, dairy products, and bread, which contain high amounts of phenylalanine, must be avoided, especially during childhood, when brain development is most rapid. The artificial sweetener aspartame must also be avoided, because it is made of two amino acids, one of which is phenylalanine.

People with PKU are generally advised to stay on a low-phenylalanine diet for life. Although maintaining these dietary restrictions may be difficult, it is effective. Numerous follow-up studies since newborn screening was initiated have shown that people with PKU who stay on the diet are no different from the rest of the population in terms of mental ability. This is an impressive achievement in public health, given the extent of mental retardation in untreated patients.

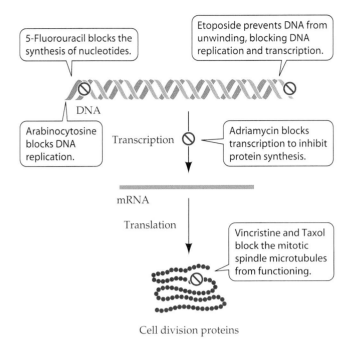

18.19 Drug Strategies for Killing Cancer Cells
The medications used against cancer attack rapidly dividing cancer cells in several ways. Unfortunately, most of them also affect noncancerous dividing cells.

METABOLIC INHIBITORS. As we described earlier, people with familial hypercholesterolemia accumulate dangerous levels of cholesterol in their blood. Not only are these people unable to metabolize dietary cholesterol, they also synthesize a lot of it. One effective treatment for people with this disease is the drug mevinolin, which blocks the patients' own cholesterol synthesis. Patients who receive this drug need only worry about cholesterol in their diet, and not about what their cells are making.

Metabolic inhibitors also form the basis of cancer therapy with drugs. The strategy is to kill rapidly dividing cells, since rapid cell division is the hallmark of malignancy. Many drugs kill dividing cells (Figure 18.19), but most of these drugs are given in the blood and thus also damage other, noncancerous, dividing cells in the body. Therefore, it is not surprising that people undergoing chemotherapy suffer side effects such as loss of hair (due to damage to the skin epithelium), digestive upsets (gut epithelial cells), and anemia (bone marrow stem cells). The effective dose of these highly toxic drugs for treating the cancer is often just below the dose that would kill the patient, so they must be used with utmost care. Often they can control the spread of cancer, but not cure it.

SUPPLYING THE MISSING PROTEIN. An obvious way to treat a disease phenotype in which a functional protein is missing is to supply the missing protein. This approach is the basis of treatment of hemophilia A, in which the missing blood clotting factor is supplied in pure form. The production of human clotting protein by biotechnology has made it possible for a pure protein to be given instead of crude blood products, which could be contaminated with the AIDS virus or other pathogens.

Unfortunately, the phenotypes of many diseases caused by genetic mutations are very complex. Simple interventions, such as some of those we have described, do not work for most such diseases. Indeed, a recent survey showed that current therapies for 351 diseases caused by single-gene mutations improved the patients' life span by only 15 percent.

Gene therapy offers the hope of specific treatments

Perhaps the most obvious thing to do when a cell lacks a functional allele is to provide one. Such **gene therapy** approaches are under intensive investigation for diseases ranging from the rare inherited disorders caused by single mutations, to cancer, AIDS, and atherosclerosis.

Gene therapy in humans seeks to insert a new gene that will be expressed in the host. Thus, the new DNA is often attached to a promoter that will be active in human cells. Presenting the DNA for cellular uptake follows the methods used in biotechnology: Calcium salts, liposomes, and viral vectors are used to get the "good gene" into the human cells. The physicians who are developing this "molecular medicine" are confronted by all the challenges of genetic engineering: effective vectors, efficient uptake, appropriate expression and processing of mRNA and protein, and selection within the body for the cells that contain the recombinant DNA.

Which human cells should be the targets of gene therapy? The best approach would be to replace the nonfunctional allele with a functional one in every cell of the body. But vectors to do this are simply not available, and delivery to every cell poses a formidable challenge. Until recently, the major attempts at gene therapy have been *ex vivo*. That is, physicians have taken cells from the patient's body, added the new gene to the cells in the laboratory, and then returned the cells to the patient in the hope that the correct gene product would be made (Figure 18.20).

A widely publicized example of this approach was the introduction of a functional gene for the enzyme adenosine deaminase into the white blood cells of a girl with a genetic deficiency of this enzyme. Unfortunately, these were mature white blood cells, and although they survived for a time in the girl, and provided some therapeutic benefit, they eventually died, as is the normal fate of such cells. It would be more effective to insert the functional gene into *stem cells*—the bone marrow cells that constantly divide to produce white blood cells. The use of stem cells is a major thrust of many current clinical experiments on gene therapy.

The other approach to gene therapy is to insert the gene directly into cells in the body of the patient. This *in vivo* approach is being attempted for various types of cancer. For example, lung cancer cells are accessible if the DNA or vector is given as an aerosol through the respiratory system. Vectors carrying functional alleles of the tumor suppressor genes that are mutated in the tumors, as well as vectors expressing antisense RNA's against oncogene mRNA's, have been successfully introduced in this way to patients with lung cancer, with some clinical improvement.

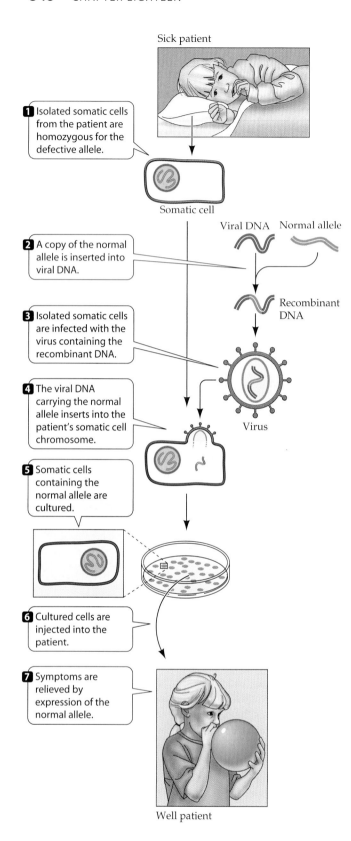

Sick patient

1 Isolated somatic cells from the patient are homozygous for the defective allele.

Somatic cell

Viral DNA   Normal allele

2 A copy of the normal allele is inserted into viral DNA.

Recombinant DNA

3 Isolated somatic cells are infected with the virus containing the recombinant DNA.

4 The viral DNA carrying the normal allele inserts into the patient's somatic cell chromosome.

Virus

5 Somatic cells containing the normal allele are cultured.

6 Cultured cells are injected into the patient.

7 Symptoms are relieved by expression of the normal allele.

Well patient

Several thousand patients, over half of them with cancer, have undergone gene therapy. Most of these clinical trials have been at a preliminary level, in which people are given the therapy to see whether it has any toxicity, and whether the new gene is actually incorporated into the genome of the patient. More ambitious trials are under way, in which a

18.20 Gene Therapy: The *Ex Vivo* Approach
New genes are added to somatic cells taken from a patient's body, then returned to the body to make the missing gene product.

larger number of patients receive the therapy with the hope that their disease will disappear, or at least improve.

Sequencing the Human Genome

In 1984, the United States government sponsored a conference to examine the problem of detecting DNA damage in people exposed to low levels of radiation, such as those who had survived the atomic bomb in Japan 39 years earlier. Scientists attending this conference quickly realized that the ability to detect such damage would also be useful in evaluating environmental mutagens. But in order to detect *changes* in the human genome, scientists first needed to know its *normal* sequence.

In 1986, Renato Dulbecco, who won the Nobel prize for his pioneering work on cancer viruses, suggested that determining the sequence of human DNA could also be a boon to cancer research. He proposed that the scientific community be mobilized for the task. The result was the publicly funded Human Genome Project, an international effort. In the 1990s, private industry launched their own sequencing effort. By the summer of 2000, "draft" sequences of the human genome were ready. The final sequence is expected to be completed by 2003.

There are two approaches to genome sequencing

Each human chromosome consists of one double-stranded molecule of DNA. Because of their differing sizes, the 46 human chromosomes can be separated from each other and identified (see Figure 9.14). So it is possible to carefully isolate the DNA of each chromosome for sequencing. The straightforward approach would be to start at one end and simply sequence the entire 50 million base pairs of the chromosome. Unfortunately, this approach does not work.

The DNA of a molecule that is 50 million base pairs long cannot be sequenced all at once; only about 700 base pairs at a time can be sequenced. (See Figure 11.21 to review the DNA sequencing technique.) To sequence the entire genome, chromosomal DNA is first cut into fragments about 500 base pairs long, then each fragment is sequenced. For the human genome, which has about 3 billion base pairs, there are more than 6 million such fragments.

The problem then becomes putting these millions of pieces of the jigsaw puzzle back together. This problem can be overcome by breaking up the DNA in several ways into "sub-jigsaws" that overlap, and aligning the overlapping fragments. There are two ways to do this.

HIERARCHICAL SEQUENCING.   The publicly funded effort first systematically identified short marker sequences along the chromosomes, so that every small sequenced section of DNA would contain a marker. Then sequences with the common

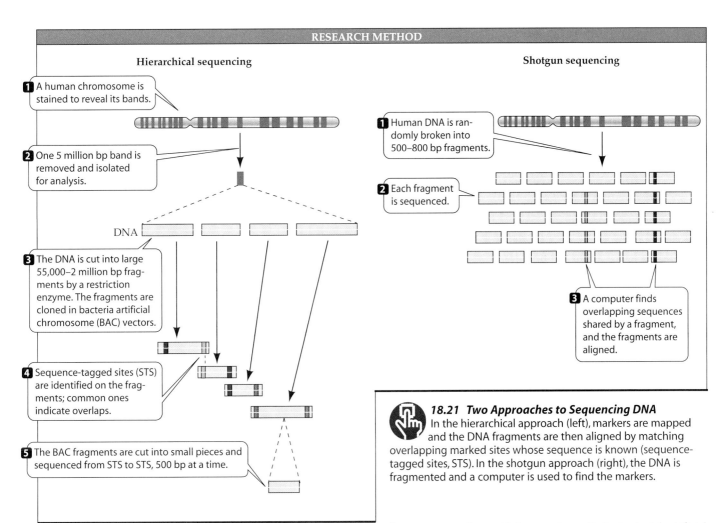

RESEARCH METHOD

Hierarchical sequencing

1 A human chromosome is stained to reveal its bands.

2 One 5 million bp band is removed and isolated for analysis.

DNA

3 The DNA is cut into large 55,000–2 million bp fragments by a restriction enzyme. The fragments are cloned in bacteria artificial chromosome (BAC) vectors.

4 Sequence-tagged sites (STS) are identified on the fragments; common ones indicate overlaps.

5 The BAC fragments are cut into small pieces and sequenced from STS to STS, 500 bp at a time.

Shotgun sequencing

1 Human DNA is randomly broken into 500–800 bp fragments.

2 Each fragment is sequenced.

3 A computer finds overlapping sequences shared by a fragment, and the fragments are aligned.

18.21 Two Approaches to Sequencing DNA
In the hierarchical approach (left), markers are mapped and the DNA fragments are then aligned by matching overlapping marked sites whose sequence is known (sequence-tagged sites, STS). In the shotgun approach (right), the DNA is fragmented and a computer is used to find the markers.

markers could then be aligned (Figure 18.21, left panel). There are two types of mapped markers along the chromosome:

▶ *Physical map markers* are chromosomal landmarks that can be ordered and the distances between them determined. The result can be compared to a road map showing towns with the mileage separating them. The "towns" are DNA markers, and the "mileage" is in base pairs.

▶ *Genetic map markers* are specific DNA sequences, such as RFLPs or short simple sequence repeats, whose locations are determined genetically.

The simplest of the physical map markers are the recognition sites for restriction enzymes. Since there are hundreds of these recognition site–restriction enzyme pairs, there are hundreds of ways to cut the DNA and then generate a restriction map. Restriction mapping has been useful in generating maps for relatively small chromosome regions of thousands of base pairs.

Some restriction enzymes recognize 8–12 base pairs in DNA, not just the usual 4–6 base pairs. A DNA molecule with several million base pairs will have relatively few of these larger sites, and thus the enzyme will generate a small number of relatively large fragments. These large

fragments can be put into a vector called a *bacterial artificial chromosome* (BAC), which can carry about 250,000 base pairs of inserted DNA in a single copy and cloned in bacteria to create a DNA library.

The books (fragments) in this library can be arranged in the proper order by using *sequence-tagged sites* (STS's). These are unique stretches of DNA, 60 to 100 base pairs long, whose sequences are known. About 41,000 STS's have been precisely mapped on human chromosomes, meaning that each is less than 100,000 base pairs (or just a few genes) away from the next STS.

To arrange DNA fragments on a map, libraries made from different restriction enzyme cuts are compared. If two large fragments of DNA cut with different enzymes have the same STS, they must overlap.

Genetic map markers are also useful tools in analyzing the genome. As we described earlier, linkage studies with markers have been very important in tracking down disease-causing genes by positional cloning. The genetic and physical map markers can complement each other—one providing new markers for the other.

SHOTGUN SEQUENCING. Instead of the "top-down" approach—getting map makers, then fragmenting the DNA and sequencing it—the "shotgun" approach cuts the DNA at random into small, sequence-ready fragments and lets

18.22 Is This the Future of Medicine?

The elucidation of the human genome sequence may result in an approach to medicine that is oriented to the genetic and functional individuality of each patient.

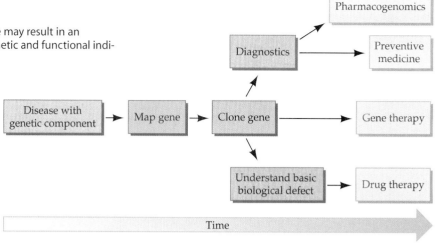

powerful computers determine markers that overlap (Figure 18.21, right panel). The fragments can then be aligned.

The shotgun method, which has been used by private industry, is much faster than the hierarchical approach because there is no need for physical or genetic maps. At first there was considerable skepticism about the method. There were concerns that without rigorous prior mapping of marker sites on the chromosomes, the computer might pick out repetitive sequences common to many DNA fragments and line the fragments up incorrectly. But the rapid rate of development of sophisticated computer and software power has allowed the technique to quickly be refined to a point where inaccurate alignment is not a major problem.

The entire 180 million-base-pair fruit fly genome (see Chapter 14) was sequenced by the shotgun method in little over a year. This proved that the rapid method might work for the much larger human genome, and in fact it did.

The human genome is more than just a sequence

Reading the human "book of life" is an achievement that ranks with other recent great events in exploration, such as landing on the moon. But gene sequencing and the tools developed to carry it out, are changing biology in many other ways as well:

▶ The sequences of other organisms have provided insights and practical information on both prokaryotic (Chapter 13) and eukaryotic (Chapter 14) genomes. Many genes sequenced and identified in "simpler" organisms have counterparts in humans, so these findings have facilitated the identification of human genes.

▶ Mapping technology has made the isolation of human genes by positional cloning much easier, because of the many chromosome markers available. Over 100 disease-related genes have been identified in this way.

▶ Genetic variability in drug metabolism has been a medical problem for a long time. The identification of the genes responsible is leading to tests that predict who will react best to which medications. This is the emerging science of *pharmacogenomics.*

▶ DNA chips (see Figure 17.12) are being used to analyze the specific expression of thousands of genes in cells in different biochemical states. For example, a Cancer Genome Anatomy Project seeks to make an mRNA "fingerprint" of a tumor at each stage of its development. Finding out which sequences are expressed at what stage will be important not only in diagnosis, but also to identify targets for gene therapy.

▶ The Human Genome Diversity Project is looking for important polymorphisms in specific human popula-tions. For example, the Pima Indians in Arizona have a high frequency of extreme obesity and diabetes. A search of their genomes might reveal genes predisposing them to these diseases.

The end result of all of this knowledge of the human genome may lead to a new approach to medical care, in which each person's genes will be used to prescribe lifestyles and treatments that can maximize that person's genetic potential (Figure 18.22).

How should genetic information be used?

After the primary genetic defect that causes cystic fibrosis was discovered, many people predicted a "tidal wave" of genetic testing for heterozygous carriers. Everyone would want the test, it was thought—especially the relatives of patients with the genetic disease. But the tidal wave has not developed. To find out why, a team of psychologists, ethicists, and geneticists interviewed 20,000 people in the United States. What the researchers found surprised them: Most people are simply not very interested in their genetic makeup, unless they have a close relative with a genetic disease and are involved in a decision about pregnancy.

There are other people, however, who might be very interested in the results of genetic testing. People who test positive for genetic abnormalities, from hypercholesterolemia to cancer, might be denied employment or health insurance. The linking of genetic abnormalities to behavioral characteristics, such as manic depression and schizophrenia, has led to the potential for screening and then social manipulations of those at risk. Consequently, many legislative bodies are considering and passing bills that prohibit genetic discrimination.

The Human Genome Diversity Project has raised many concerns about exploitation and commercialization of people's DNA sequences. Is a gene that confers resistance to cancer, for example, the property of an individual, an ethnic group in which it may be frequent, a pharmaceutical company that finds it, or humanity at large? This issue of ownership is being tested worldwide, perhaps no more acutely than in Iceland, most of whose 270,000 people trace their ancestry back to the first settlers that arrived on the island

1,000 years ago. Tissues from the entire population have been sampled and stored for several generations. This tissue bank is a potential gold mine for genetic prospectors, and a single company has been set up, with government support, to sell the knowledge that comes out of analyzing the DNA's of Iceland's people.

 *Chapter Summary*

Protein as Phenotype

▶ In many human genetic diseases, a single protein is missing or nonfunctional. Therefore, the one-gene, one-polypeptide relationship applies to human genetic diseases. **Review Figure 18.1**

▶ A mutation in a single gene causes alterations in its protein product that may lead to clinical abnormalities or have no effect. **Review Figure 18.2**

▶ Some diseases are caused by mutations that affect structural proteins; examples include Duchenne's muscular dystrophy and hemophilia.

▶ The genes that code for receptors and membrane transport proteins can also be mutated and cause diseases such as familial hypercholesterolemia and cystic fibrosis. **Review Figure 18.3**

▶ Prion diseases are caused by a protein with an altered shape that is transmitted from one person to another and alters the same protein in the second person. **Review Figure 18.4**

▶ Few human diseases are caused by a single-gene mutation. Most are caused by the interactions of many genes and proteins with the environment.

▶ Human genetic diseases show different patterns of inheritance. Mutant alleles may be inherited as autosomal recessives, autosomal dominants, X-linked conditions, or chromosomal abnormalities.

Mutations and Human Diseases

▶ Molecular biology techniques have made possible the isolation of many genes responsible for human diseases.

▶ One method of identifying the gene responsible for a disease is to isolate the mRNA for the protein in question and then use the mRNA to isolate the gene from a gene library. DNA from a patient that lacks a piece of a chromosome can be compared with DNA from a person who does not show this deletion to isolate a missing gene. **Review Figure 18.6**

▶ In positional cloning, DNA markers are used as guides to point the way to a gene. These markers may be restriction fragment length polymorphisms that are linked to a mutant gene. **Review Figure 18.7**

▶ Human mutations range from single point mutations to large deletions. Some of the most common mutations occur where the modified base 5-methylcytosine is converted to thymine. **Review Figure 18.8**

▶ The effects of the fragile-X chromosome worsen with each generation. This pattern is caused by a triplet repeat that tends to expand with each new generation. **Review Figure 18.9**

▶ Genomic imprinting results in a gene being differentially expressed depending on which parent it comes from.

Detecting Human Genetic Variations

▶ Genetic screening detects human gene mutations. Some protein abnormalities can be detected by simple tests, such as tests for the presence of excess substrate or lack of product. **Review Figure 18.10**

▶ The advantage of testing DNA for mutations directly is that any cell can be tested at any time in the life cycle.

▶ There are two methods of DNA testing: allele-specific cleavage and allele-specific oligonucleotide hybridization. **Review Figures 18. 11, 18.12**

Cancer: A Disease of Genetic Changes

▶ Tumors may be benign, growing only to a certain extent and then stopping, or malignant, spreading through organs and to other parts of the body.

▶ At least five types of human cancers are caused by viruses, which account for about 15 percent of all cancers. **Review Table 18.2**

▶ Eighty-five percent of human cancers are caused by genetic mutations of somatic cells. These mutations occur most commonly in dividing cells. **Review Figure 18.14**

▶ Normal cells contain proto-oncogenes, which, when mutated, can become activated and cause cancer by stimulating cell division or preventing cell death. **Review Figure 18.15**

▶ About 10 percent of all cancer is inherited as a result of the mutation of tumor suppressor genes, which normally act to slow down the cell cycle. For cancer to develop, both alleles of a tumor suppressor gene must be mutated.

▶ In inherited cancer, an individual inherits one mutant allele and then somatic mutation occurs in the second one. In sporadic cancer, two normal alleles are inherited, so two mutational events must occur in the same somatic cell. **Review Figures 18.16, 18.17**

▶ Mutations must activate several oncogenes and inactivate several tumor suppressor genes for a cell to produce a malignant tumor. **Review Figure 18.18**

Treating Genetic Diseases

▶ Most genetic diseases are treated symptomatically. However, as more knowledge is accumulated, specific treatments are being devised.

▶ One treatment approach is to modify the phenotype—for example, by manipulating the diet, providing specific metabolic inhibitors to prevent the accumulation of a harmful substrate, or supplying a missing protein. **Review Figure 18.19**

▶ In gene therapy, a mutant gene is replaced with a normal gene. Either the affected cells can be removed, the new gene added, and the cells returned to the body, or the new gene can be inserted via a vector directly into the patient. **Review Figure 18.20**

Sequencing the Human Genome

▶ Human genome sequencing is determining the entire human DNA sequence, which requires sequencing many 500-base-pair fragments and then fitting the sequences back together.

▶ In hierarchical gene sequencing, marker sequences are identified and mapped. The markers are then sought in sequenced fragments and are used to align the fragments. In the shotgun approach, the fragments are sequenced and then common markers are identified by computer. **Review Figure 18.21**

▶ The identification of more than 30,000 human genes may lead to a new molecular medicine. **Review Figure 18.22**

▶ As more genes relevant to human health are described, concerns about how such information is used are growing.

For Discussion

1. Compare the roles of proto-oncogenes and tumor suppressor genes in normal cells. How do these genes and their functions change in tumor cells? Propose targets for cancer therapy involving these gene products.

2. In the past, it was common for people with phenylketonuria (PKU) who were placed on a low-phenylalanine diet after birth to be allowed to return to a normal diet during their teenage years. Although the levels of phenylalanine in their blood were high, their brains were thought to be beyond the stage of being harmed. If a woman with PKU becomes pregnant, however, a problem arises. Typically, the fetus is heterozygous, but is unable at early stages of development to metabolize the high levels of phenylalanine that arrive from the mother's blood. Why is the fetus heterozygous? What do you think would happen to the fetus during this "maternal PKU" situa-

tion? What would be your advice to a woman with PKU who wants to have a child?

3. A "knockout" mouse has been made that has deletions in both copies of the gene coding for Pr^c. Would you expect this mouse to develop a prion disease if infected with Pr^{sc}? Explain your answer.

4. Cystic fibrosis is an autosomal recessive disease in which thick mucus is produced in the lungs and airway. The gene responsible for this disease codes for a protein composed of 1,480 amino acids. In most patients with cystic fibrosis, the protein has 1,479 amino acids: A phenylalanine is missing at position 508. How would you test the older brother of a baby with cystic fibrosis to determine whether he is a carrier for the disease? How would you design a gene therapy protocol to "cure" the cells in the lung and airway?

19 *Natural Defenses against Disease*

ON JANUARY 6, 1777, GEORGE WASHINGTON, commanding the Revolutionary army of the fledgling United States, made a fateful medical/military decision. As he wrote to his chief physician, "Finding smallpox to be spreading much, and fearing that no precaution can prevent it from running through the whole of our army, I have determined that the troops shall be inoculated. Should the disease rage with its usual virulence, we should have more to dread from it than the sword of the enemy."

Washington was speaking from experience. He himself had survived the disease in 1751. During 1776 his army lost 1,000 men in battle, but 10,000 to smallpox. This highly virulent disease, which killed up to 1 in 4 people who were exposed to it, had already figured prominently in American history. A century before, it had decimated the Native Americans, making colonization by Europeans easier. Two years previously at Quebec, it had laid waste to an American army that was trying to annex Canada by force.

The death rate due to smallpox in the Revolutionary army plummeted after Washington's order was carried out. How did inoculation, a practice that was learned from the people in the Near East and from African slaves, save the soldiers? And why was Washington himself immune to the disease as it ravaged his army?

The answers to these questions lie in the cells and molecules of the immune system. When Washington caught smallpox as a teenager, cells called macrophages engulfed some of the smallpox viruses by phagocytosis and partly digested them. The macrophages displayed fragments of the viruses on their cell surfaces. Specialized white blood cells called T cells recognized those fragments and were activated to divide and differentiate further. The descendants of these activated T cells then attacked Washington's virus-infected cells, preventing the lethal spread of the disease. Other descendants of the T cells persisted in his body as "memory cells" and rapidly defended him when he was exposed to the disease as an adult. Inoculation of his soldiers with powdered scabs containing dead viruses from smallpox patients stimulated the formation of these memory cells in their bodies, once again preventing the virus from spreading following infection. This practice had been

used for centuries, and was finally put on a more scientific basis by Edward Jenner two decades after Washington's army was inoculated.

These defensive events in the bodies of Washington and his soldiers required the participation of many kinds of proteins. Some cellular proteins functioned as specific receptors, such as the markers identifying Washington's cells, some as signals triggering events in the macrophages and T cells, and others as weapons leading to the breakdown of infected cells.

Animal defense systems are based on the distinction between *self* —the animal's own molecules—and *nonself*, or foreign, molecules. In this chapter we consider the mechanisms by which animals recognize nonself molecules and combat infection and disease. Many of these mechanisms are based on the principles of genetics and molecular biology that have been discussed in earlier chapters.

In general, there are two types of defenses. **Innate**, or nonspecific, defenses are general mechanisms that protect the body from many pathogens. An example is the skin, which acts as a barrier to stop potentially invading viruses

George Washington
Washington's decision to immunize his army against smallpox saved many lives and probably helped him win the Revolutionary War.

from entering the body. Most animals and plants have innate defenses. **Specific** defenses are aimed at a single target. For example, an antibody protein can be made that binds to a virus if that virus ever enters the bloodstream, and this binding results in the virus being destroyed. Specific defense mechanisms are present in vertebrate animals. In animals that have both, *innate and specific defenses operate together and offer a coordinated defense.*

Defensive Cells and Proteins

The components of the defense system are dispersed throughout the body and interact with almost all of its other tissues and organs. The *lymphoid tissues*, which include the thymus, bone marrow, spleen, and lymph nodes, are essential parts of our defense system (Figure 19.1), but central to their functioning are the blood and lymph.

Blood and lymph are fluid tissues that consist of water, dissolved solutes, and cells. *Blood plasma* is a yellowish solution containing ions, small molecular solutes, and soluble proteins. Suspended in the plasma are red blood cells, white blood cells, and platelets (cell fragments essential to

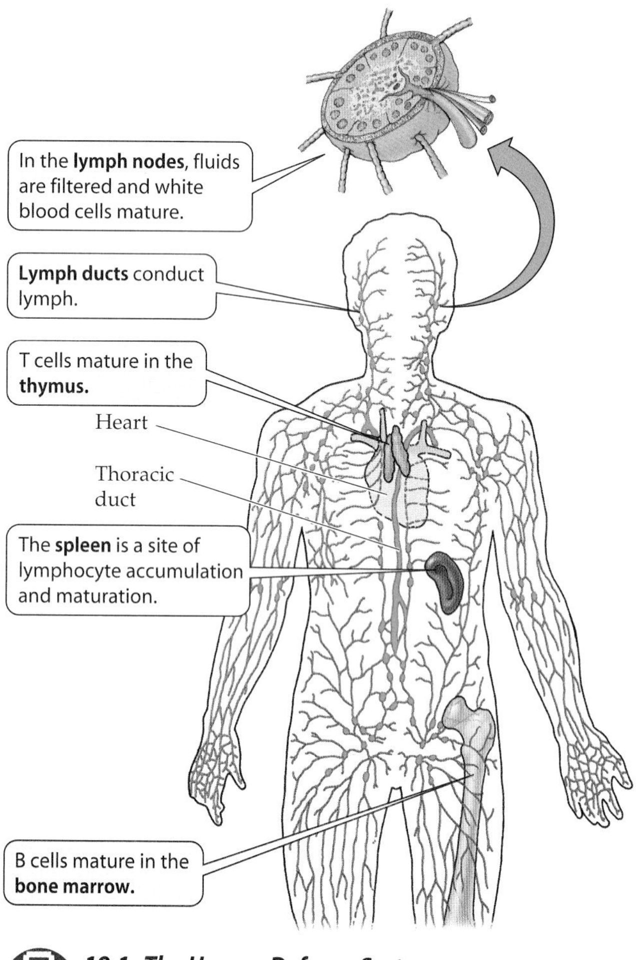

In the **lymph nodes**, fluids are filtered and white blood cells mature.

Lymph ducts conduct lymph.

T cells mature in the **thymus.**

Heart

Thoracic duct

The **spleen** is a site of lymphocyte accumulation and maturation.

B cells mature in the **bone marrow.**

19.1 The Human Defense System
A network of ducts collects lymph from the body's tissues and carries it toward the heart, where it mixes with blood to be pumped back to the tissues. The thymus, spleen, and bone marrow are essential to the body's defensive network.

clotting) (Figure 19.2). While red blood cells are normally confined to the *closed circulatory system* of (the heart, arteries, capillaries, and veins), white blood cells and platelets are also found in the lymph.

Lymph is a fluid derived from blood and other tissues. It accumulates in spaces outside the blood vessels and contains many of the components of blood, but not red blood cells. From the spaces around body cells, the lymph moves slowly into the lymphatic system. Tiny lymph capillaries conduct fluid to larger vessels that join together, forming one large lymph duct that joins a major vein (the left subclavian vein) near the heart. By this system of vessels, the lymphatic fluid is eventually returned to the blood and the circulatory system.

At many sites along the lymph vessels are small, roundish structures called **lymph nodes**, which contain a variety of white blood cells. As fluid passes through the node, it is filtered and "inspected" for foreign materials by these defensive cells.

White blood cells play many defensive roles

One milliliter of blood typically contains about 5 billion red blood cells and 7 million of the larger white blood cells. White blood cells have nuclei and are colorless, whereas mammalian red blood cells lose their nuclei during development. White blood cells can leave the circulatory system and enter extracellular spaces where foreign cells or substances are present. In response to invading pathogens, the number of white blood cells in the blood and lymph may rise sharply, providing medical professionals with a useful clue for detecting an infection.

Several types of white blood cells are important in the body's defenses. But they are all members of two broad groups, phagocytes and lymphocytes.

Phagocytes engulf and digest nonself materials. The most important phagocytes are the neutrophils and the macrophages. In addition to phagocytosis of nonself materials, **macrophages** have the important additional function of presenting partly digested nonself materials to the T cells.

Lymphocytes are the most abundant white cells. A healthy person has about a trillion lymphocytes, making them as numerous as brain cells. There are two types of lymphocytes, **B cells** and **T cells**. Both originate from cells in the bone marrow. Immature T cells migrate via the blood to the thymus, where they mature. They participate in specific defenses against foreign or altered cells, such as virus-infected cells and tumor cells. The B cells leave the bone marrow and circulate through the blood and lymph vessels. B cells make specialized proteins called **antibodies** that enter the blood and bind to nonself substances.

Immune system proteins bind pathogens or signal other cells

The cells that defend our bodies work together like cast members in a drama, interacting with one another and with the cells of invading pathogens. These cell–cell interactions are accomplished by a variety of key proteins, including re-

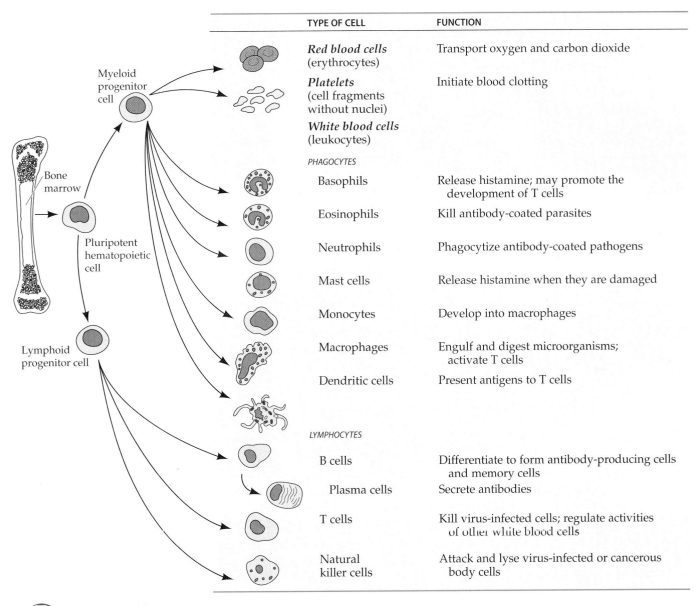

TYPE OF CELL	FUNCTION
Red blood cells (erythrocytes)	Transport oxygen and carbon dioxide
Platelets (cell fragments without nuclei)	Initiate blood clotting
White blood cells (leukocytes)	
PHAGOCYTES	
Basophils	Release histamine; may promote the development of T cells
Eosinophils	Kill antibody-coated parasites
Neutrophils	Phagocytize antibody-coated pathogens
Mast cells	Release histamine when they are damaged
Monocytes	Develop into macrophages
Macrophages	Engulf and digest microorganisms; activate T cells
Dendritic cells	Present antigens to T cells
LYMPHOCYTES	
B cells	Differentiate to form antibody-producing cells and memory cells
Plasma cells	Secrete antibodies
T cells	Kill virus-infected cells; regulate activities of other white blood cells
Natural killer cells	Attack and lyse virus-infected or cancerous body cells

19.2 Blood Cells
Pluripotent stem cells in the bone marrow can differentiate into red blood cells, platelets, and the various types of white blood cells.

ceptors, surface markers, signaling molecules, and toxins. They will be discussed later in the chapter, as they appear in the context of our story. However, let's take a brief look at four of the major players here.

▶ **Antibodies** are proteins that bind specifically to certain substances identified by the immune system as nonself or altered self. B cells secrete antibodies as defensive weapons.

▶ **T cell receptors** are proteins on the surfaces of T cells. They recognize and bind to nonself substances on the surfaces of other cells.

▶ **Major histocompatibility complex (MHC) proteins** protrude from the surfaces of most cells in the mammalian body. They are important "self"-identifying

labels and play major parts in coordinating interactions among lymphocytes and macrophages.

▶ **Cytokines** are soluble signal proteins released by T cells, macrophages, and other cells. They bind to and alter the behavior of their target cells. Different cytokines activate or inactivate B cells, macrophages, and T cells. Some cytokines limit tumor growth by killing tumor cells.

Innate Defenses

Innate defenses are general protection mechanisms to stop *pathogens*—harmful organisms that can cause disease—from invading the body. In humans, these defenses include physical barriers as well as cellular and chemical defenses. Table 19.1 provides a summary of the innate defense mechanisms.

19.1 Human Innate Defenses

DEFENSIVE AGENT	FUNCTION
Surface barriers	
Skin	Prevents entry of pathogens and foreign substances
Acid secretions	Inhibit bacterial growth on skin
Mucus membranes	Prevent entry of pathogens
Mucus secretions	Trap bacteria and other pathogens in digestive and respiratory tracts
Nasal hairs	Filter bacteria in nasal passages
Cilia	Move mucus and trapped materials away from respiratory passages
Gastric juice	Concentrated HCl and proteases destroy pathogens in stomach
Acid in vagina	Limits growth of fungi and bacteria in female reproductive tract
Tears, saliva	Lubricate and cleanse; contain lysozyme, which destroys bacteria
Nonspecific cellular, chemical, and coordinated defenses	
Normal flora	Compete with pathogens; may produce substances toxic to pathogens
Fever	Body-wide response inhibits microbial multiplication and speeds body repair processes
Coughing, sneezing	Expels pathogens from upper respiratory passages
Inflammatory response (involves leakage of blood plasma and phagocytes from capillaries)	Limits spread of pathogens to neighboring tissues; concentrates defenses; digests pathogens and dead tissue cells; released chemical mediators attract phagocytes and specific defense lymphocytes to site
Phagocytes (macrophages and neutrophils)	Engulf and destroy pathogens that enter body
Natural killer cells	Attack and lyse virus-infected or cancerous body cells
Antimicrobial proteins	
Interferon	Released by virus-infected cells to protect healthy tissue from viral infection; mobilizes specific defenses
Complement proteins	Lyse microorganisms, enhance phagocytosis, and assist in inflammatory response

Barriers and local agents defend the body against invaders

Skin is a primary innate defense against invasion. Fungi, bacteria, and viruses rarely penetrate healthy, unbroken skin. But damaged skin or internal surface tissue greatly increase the risk of infection by pathogenic agents.

The bacteria and fungi that normally live and reproduce in great numbers on our body surfaces without causing disease are referred to as *normal flora*. These natural occupants of our bodies compete with pathogens for space and nutrients, so normal flora are a form of innate defense.

The mucus-secreting tissues found in parts of the visual, respiratory, digestive, excretory, and reproductive systems have other defenses against pathogens. Tears, nasal mucus, and saliva possess an enzyme called *lysozyme* that attacks the cell walls of many bacteria. Mucus in the nose traps airborne microorganisms, and most of those that get past this filter end up trapped in mucus deeper in the respiratory tract. Mucus and trapped pathogens are removed by the beating of cilia in the respiratory passageway, which moves a sheet of mucus and the debris it contains up toward the nose and mouth. Sneezing is another way to remove microorganisms from the respiratory tract.

Pathogens that reach the digestive tract (stomach, small intestine, and large intestine) are met by other defenses.

The stomach is a deadly environment for many bacteria because of the hydrochloric acid and protein-digesting enzymes that are secreted into it. The intact lining of the small intestine is not normally penetrated by bacteria, and some pathogens are killed by bile salts secreted into this part of the tract. The large intestine harbors many bacteria, which multiply freely; however, these are usually removed quickly with the feces. Most of the bacteria in the large intestine are normal flora that provide benefits to their host.

All of these barriers and secretions are *nonspecific* defenses because they act on all invading pathogens in the same way. But there are more complicated nonspecific cellular chemical defenses.

Innate defenses include chemical and cellular processes

Pathogens that manage to penetrate the body's outer and inner surfaces encounter additional nonspecific defenses. These defenses include the secretion of various defensive proteins as well as cellular defenses involving phagocytosis.

COMPLEMENT PROTEINS. Vertebrate blood contains about 20 different antimicrobial proteins that make up the **complement system**. These proteins, in different combinations, provide three types of defenses. In each type, the comple-

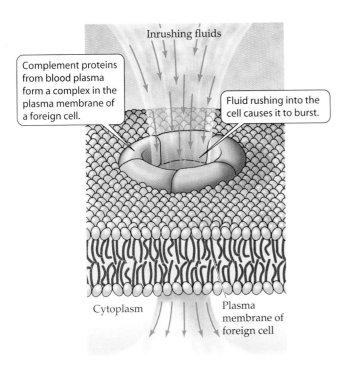

Inrushing fluids

Complement proteins from blood plasma form a complex in the plasma membrane of a foreign cell.

Fluid rushing into the cell causes it to burst.

Cytoplasm

Plasma membrane of foreign cell

19.3 Complement Proteins Destroy a Foreign Cell
Some complement proteins attach to foreign cells such as bacteria after antibodies have bound to them. The porelike structure of the protein allows fluids to pour into the foreign cell until it bursts.

ment proteins act in a characteristic sequence, or cascade, with each protein activating the next.

▶ Complement proteins attach to microbes, which helps phagocytes destroy them.
▶ Complement proteins activate the inflammatory response and attract phagocytes to sites of infection.
▶ Complement proteins, acting with antibodies, lyse (burst) invading cells such as bacteria (Figure 19.3).

INTERFERONS. When cells are infected by a virus, they produce small amounts of antimicrobial proteins called **interferons** that increase the resistance of neighboring cells to infection by the same *or other* viruses. Interferons have been found in many vertebrates and are one of the body's lines of nonspecific defense against viral infection.

Interferons differ from species to species, and each vertebrate species produces at least three different interferons. All interferons are glycoproteins (proteins with a carbohydrate component) consisting of about 160 amino acids. By binding to receptors in the plasma membranes of their target cells, interferons inhibit viral replication.

PHAGOCYTOSIS AND OTHER CELLULAR ASSAULTS. Phagocytes provide an important nonspecific defense against pathogens that penetrate the surface of the host. Some phagocytes travel freely in the circulatory system; others can move out of blood vessels and adhere to certain tissues. Pathogens such as large molecules, cells, and viruses become attached to the membrane of a phagocyte (Figure 19.4), which ingests them by endocytosis. After lysosomes fuse with the phagosome,

the pathogens are degraded by lysosomal enzymes (see Figure 4.15).

There are three types of phagocytes:

▶ **Neutrophils** are the most abundant phagocytes, but they are relatively short-lived. They recognize and attack pathogens in infected tissue.
▶ **Monocytes** mature into **macrophages**, which live longer than neutrophils and can consume large numbers of pathogens. Some macrophages roam through the body; others reside permanently in lymph nodes, the spleen, and certain other lymphoid tissues, "inspecting" the lymph for pathogens.
▶ **Eosinophils** are weakly phagocytic. Their primary function is to kill parasites, such as worms, that have been coated by antibodies.

A class of nonphagocytic white blood cells, known as **natural killer cells**, can distinguish virus-infected cells and some tumor cells from their normal counterparts and initiate the lysis of these target cells. In addition to this nonspecific action, natural killer cells are part of the specific defenses, as we will describe later in this chapter.

INFLAMMATION. The body employs the inflammation response in dealing with infection or with any other process that causes tissue injury either on the surface of the body or internally. You have experienced the symptoms of inflammation: redness and swelling, accompanied by heat and pain. The damaged body cells cause the inflammation by releasing various substances, such as the chemical signal **histamine**. Cells adhering to the skin and linings of organs, called **mast cells**, release histamine when they are damaged, as do white blood cells called **basophils**.

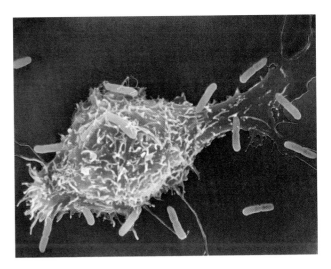

19.4 A Phagocyte and Its Bacterial Prey
Some bacteria (appearing yellow in this artificially colored scanning electron micrograph) have become attached to the surface of a phagocyte in the human bloodstream. Many of these bacteria will be engulfed by the phagocyte and destroyed before they can multiply and damage the human host. A single phagocyte can digest about 20 bacteria.

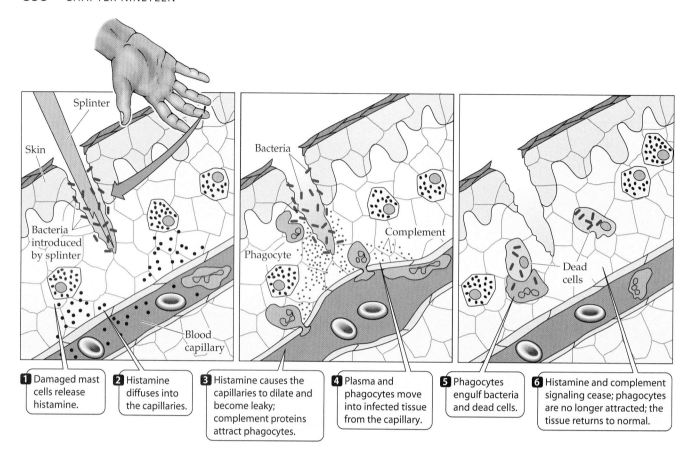

1 Damaged mast cells release histamine.

2 Histamine diffuses into the capillaries.

3 Histamine causes the capillaries to dilate and become leaky; complement proteins attract phagocytes.

4 Plasma and phagocytes move into infected tissue from the capillary.

5 Phagocytes engulf bacteria and dead cells.

6 Histamine and complement signaling cease; phagocytes are no longer attracted; the tissue returns to normal.

19.5 Interactions of Cells and Chemical Signals in Inflammation

The histamine-induced swelling of the inflammation reaction is accompanied by redness, heat, and pain. The chemical signals associated with the reaction attract the phagocytes that are largely responsible for healing the wound.

The redness and heat of inflammation result from histamine-induced dilation of blood vessels in the infected or injured area (Figure 19.5). Histamine also causes the capillaries (the smallest blood vessels) to become leaky, allowing blood plasma and phagocytes to escape into the tissue, causing the characteristic swelling. The pain of inflammation results from increased pressure (from the swelling) and from the action of leaked enzymes.

In damaged or infected tissue, complement proteins and other chemical signals attract phagocytes—neutrophils first, and then monocytes, which become macrophages. The macrophages engulf the invaders and debris and are responsible for most of the healing associated with inflammation. They produce several cytokines, which, among other functions, signal the brain to produce a fever, which inhibits the growth of the invading pathogen. Cytokines may also attract phagocytic cells to the site of injury and initiate a specific immune response to the pathogen.

Following inflammation, pus may accumulate. It is composed of dead cells (neutrophils and the damaged body cells) and leaked fluid. A normal result of inflammation, pus is gradually consumed and digested by macrophages.

Specific Defenses: The Immune Response

Nonspecific defenses are numerous and effective, but some invaders elude them and must be dealt with by defenses targeted against specific threats. The recognition and destruction of specific nonself substances is an important function of an animal's immune system.

Four features characterize the immune response

The characteristic features of the immune system are specificity, the ability to respond to an enormous diversity of foreign molecules and organisms, the ability to distinguish self from nonself, and immunological memory.

SPECIFICITY. **Antigens** are organisms or molecules that are recognized by and/or interact with the immune system to initiate an immune response. The specific sites on antigens that the immune system recognizes are called **antigenic determinants** (Figure 19.6). Chemically, an antigenic determinant is a specific portion of a large molecule, such as a sequence of amino acids that may be present in several proteins. A large antigen, such as a whole cell, may have many different antigenic determinants on its surface, each capable of being bound by a specific antibody or T cell. Even a single protein has multiple, different antigenic determinants. The host responds to the presence of an antigen by producing highly specific defenses—T cells or antibodies that correspond to the antigenic determinants of the antigen. Each T cell and each antibody is specific for a single antigenic determinant.

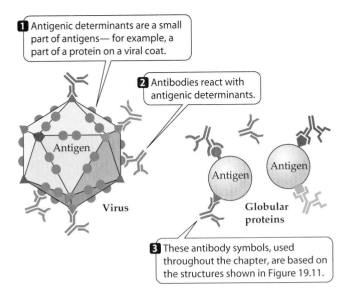

1. Antigenic determinants are a small part of antigens— for example, a part of a protein on a viral coat.

2. Antibodies react with antigenic determinants.

3. These antibody symbols, used throughout the chapter, are based on the structures shown in Figure 19.11.

19.6 Each Antibody Matches an Antigenic Determinant
Each antigen has many different antigenic determinants that are recognized by specific antibodies. An antibody recognizes and binds to its antigenic determinant to initiate defensive measures against the antigen.

DIVERSITY. Challenges to the immune system are legion: individual molecules, viruses, bacteria, protists, and multicellular parasites. Each of these types of potential pathogens includes many species; each species includes many subtly differing genetic strains; each strain possesses multiple surface features, each of which is presented to the immune system. Estimates vary, but a reasonable guess is that humans can respond *specifically* to 10 million different antigenic determinants. Upon recognition of an antigenic determinant, the immune system responds by activating lymphocytes of the appropriate specificity.

DISTINGUISHING SELF FROM NONSELF. The human body contains tens of thousands of different proteins, each with a specific three-dimensional structure capable of generating an immune response. Every cell in the body bears a tremendous number of antigenic determinants. A crucial attribute of an individual's immune system is that it recognizes the body's own antigenic determinants and does not attack them. Failure to make this distinction may lead to an *autoimmune disease*—an attack on one's own body. Such diseases include rheumatoid arthritis and lupus.

IMMUNOLOGICAL MEMORY. After responding to a particular type of pathogen once, the immune system "remembers" that pathogen and can usually respond more rapidly and powerfully to the same threat in the future. This immunological memory usually saves us from repeats of childhood diseases such as chicken pox. Vaccination and inoculation against disease work because the immune system "remembers" the antigenic determinants that are inoculated into the body.

There are two interactive immune responses

The immune system has two responses against invaders: the humoral immune response and the cellular immune response. The two responses operate in concert—simultaneously and cooperatively, sharing mechanisms.

In the **humoral immune response** (from the Latin *humor*, "fluid"), antibodies react with antigenic determinants on foreign invaders in blood, lymph, and tissue fluids. An animal produces a vast diversity of antibodies that, among them, can react with almost any conceivable antigen encountered. Some antibodies are soluble and travel free in the blood and lymph; others exist as integral membrane proteins on specialized lymphocytes called B cells.

The first time a specific antigen invades the body, it may be detected by and bind to a B cell whose membrane antibody recognizes one of its antegenic determinants. This activated B cell forms a **plasma cell** that makes multiple soluble copies of an antibody with the same specificity as the membrane antibody.

The **cellular immune response** is directed against an antigen that has become established within a cell of the host animal. It detects and destroys virus-infected or mutated cells.

Unlike the humoral response, the cellular immune response does not use antibodies. Instead, it is carried out by T cells within the lymph nodes, the bloodstream, and the intercellular spaces. The T cells have integral membrane proteins called T cell receptors—surface glycoproteins that recognize and bind to antigenic determinants while remaining part of the cell's plasma membrane. Like antibodies, T cell receptors have specific molecular configurations that bind to specific antigenic determinants. Once a T cell is bound to a determinant, it initiates an immune response.

Clonal selection accounts for the characteristic features of the immune response

Each person possesses an enormous number of different B cells and T cells, apparently capable of dealing with almost any antigen ever likely to be encountered. How does this diversity arise? How do B and T cells specific for certain antigens proliferate? And why don't our antibodies and T cells attack and destroy our own bodies? The versatility of immune responses, the proliferation of specific cells, the ability to distinguish between self and nonself, and immunological memory can all be explained by the theory of **clonal selection**.

According to clonal selection theory, each individual human contains an enormous variety of different B cells, and each type of B cell is able to produce *only one kind of antibody*. Thus there are millions of different B cells, each one producing a particular antibody and displaying it on the cell surface. When an antigen that fits this surface antibody binds to it, the B cell is activated. It divides to form a clone of cells, all of them producing that particular antibody. Thus the antigen "selects" a particular antibody-producing B cell for proliferation (Figure 19.7). In the same way, an antigenic cell "selects" a T cell expressing a particular T cell receptor on its surface for proliferation.

The clonal selection theory accounts nicely for the body's ability to respond rapidly to any of a vast number of different antigens. In the extreme case, even a single B cell might be sufficient for an immunological response by the body, provided that it encounters its antigen and then proliferates into a large clone rapidly enough to combat the invasion.

Immunological memory and immunity result from clonal selection

According to clonal selection theory, an activated lymphocyte (B cell or T cell) produces two types of daughter cells, effector cells and memory cells.

▶ **Effector cells** carry out the attack on the antigen. They are either plasma cells that produce antibodies, or T cells that, upon binding an antigenic determinant, release messenger molecules called *cytokines*. Effector cells live only a few days.

▶ **Memory cells** are long-lived cells that retain the ability to start dividing on short notice to produce more effector and more memory cells. Memory B and possibly T cells may survive for decades.

When the body first encounters a particular antigen, a *primary immune response* is activated, and lymphocytes produce clones of effector and memory cells. The effector cells destroy the invaders at hand and then die, but one or more clones of memory cells have now been added to the immune system and provide **immunological memory**.

After the body's first immune response to a particular antigen, subsequent encounters with the same antigen will result in a greater and more rapid production of antigen-specific antibody or T cells. This response is called the *secondary immune response*. The first time a vertebrate animal is exposed to a particular antigen, there is a time lag (usually several days) before the number of antibody molecules and T cells slowly increases (Figure 19.8). But for years afterward—sometimes for life—the immune system "remembers" that particular antigen. The secondary immune response has a shorter lag time, a greater rate of antibody production, and a larger production of total antibody or T cells than the primary response.

Thanks to immunological memory, recovery from many diseases, such as chicken pox, provides a *natural immunity* to those diseases. However, it is possible to protect against many life-threatening diseases, such as typhoid or tetanus, by *artificial immunity*. Artificial immunity can be acquired by the introduction of antigenic proteins or other molecular antigens into the body in a process called **immunization**, or by the introduction of whole pathogens, live or rendered harmless, which is called **vaccination**.

Immunization or vaccination initiates a primary immune response, generating memory cells without making the person ill. Later, if the same or very similar disease organisms attack, B memory cells already exist. They recognize the antigen and quickly overwhelm the invaders with a massive production of lymphocytes and antibodies.

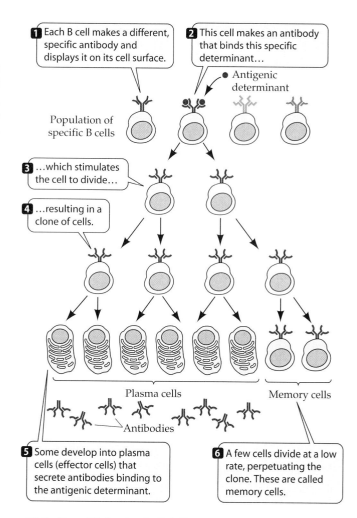

1 Each B cell makes a different, specific antibody and displays it on its cell surface.

2 This cell makes an antibody that binds this specific determinant…

Antigenic determinant

Population of specific B cells

3 …which stimulates the cell to divide…

4 …resulting in a clone of cells.

Plasma cells

Memory cells

Antibodies

5 Some develop into plasma cells (effector cells) that secrete antibodies binding to the antigenic determinant.

6 A few cells divide at a low rate, perpetuating the clone. These are called memory cells.

19.7 Clonal Selection in B Cells
The binding of an antigenic determinant to a specific antibody on the surface of a B cell stimulates the cell to divide, rapidly producing a clone of cells to fight the invader.

The ability of the human body to remember a specific antigen explains why immunization has almost completely wiped out deadly diseases such as diphtheria and polio in industrialized countries. In fact, smallpox has been eliminated worldwide, thanks to an international effort by the World Health Organization. As far as we know, the only remaining smallpox viruses on Earth are those kept in some laboratories.

Animals distinguish self from nonself and tolerate their own antigens

Given the presence of lymphocytes directed against so many antigens, why doesn't a healthy human produce self-destructive immune responses? The body is tolerant of its own molecules—the same molecules that would generate an immune response in another individual. Self-tolerance seems to be based on two mechanisms: clonal deletion and clonal anergy.

Clonal deletion physically removes B or T cells from the immune system at some point during their differentiation. For example, immature B cells in bone marrow may encounter self antigens. Any of these cells that shows the potential to mount an immune response against self antigens

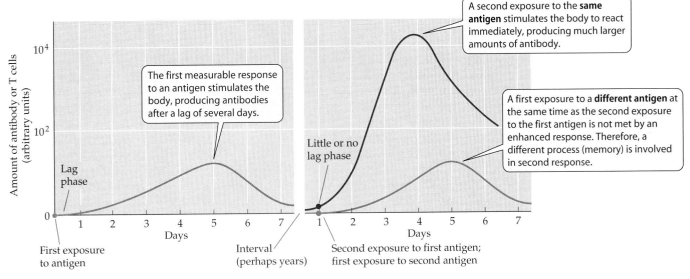

The first measurable response to an antigen stimulates the body, producing antibodies after a lag of several days.

A second exposure to the **same antigen** stimulates the body to react immediately, producing much larger amounts of antibody.

A first exposure to a **different antigen** at the same time as the second exposure to the first antigen is not met by an enhanced response. Therefore, a different process (memory) is involved in second response.

19.8 Immunological Memory
The ability of the body to remember an antigen to which it has been exposed is the basis for natural and artificial immunity against a disease.

undergoes programmed cell death (apoptosis) within a short time, and never differentiates enough to make antibodies. Thus, no clones of antiself lymphocytes normally appear in the bloodstream. Clonal deletion eliminates about 90 percent of all the B cells made in the bone marrow. A similar process occurs with T cells in the thymus.

Clonal anergy is the suppression of the immune response. For example, a mature T cell may encounter and recognize a self antigen on the surface of a body cell. But it does not send out the cytokines that signal the initiation of an immune response. Before it does so, the T cell must encounter not only an antigen, but also a second molecule called CD28 on the cell surface. This *co-stimulatory signal* is expressed only on certain cells called antigen-presenting

cells. So most body cells, lacking CD28, will not be attacked by the cellular immune system.

The phenomenon of **immunological tolerance** (Figure 19.9) was discovered through the observation that some *nonidentical* twin cattle with different blood types contained some of each other's red blood cells. Why did these "foreign" blood cells not cause immune responses resulting in their elimination? The hypothesis suggested was that the blood cells had passed between the fetal animals in the womb before the lymphocytes had matured. Thus each calf regarded the other's red blood cells as self. This hypothesis was confirmed when it was shown that injecting foreign

19.9 Making Nonself Seem Like Self
The ability of adult mice to recognize and reject grafts of foreign skin can be overcome by earlier exposure to nonself antigens. Both of the mouse strains used in this experiment are highly inbred, and so each member of a strain is genetically identical to the others in that strain. Strain B mice tolerate grafts from other strain B mice.

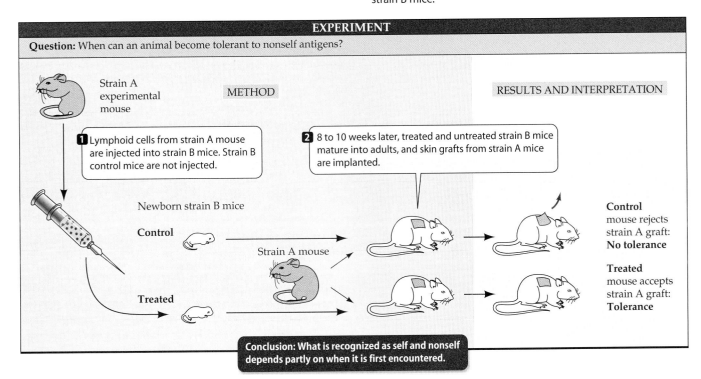

EXPERIMENT

Question: When can an animal become tolerant to nonself antigens?

METHOD

RESULTS AND INTERPRETATION

Strain A experimental mouse

1 Lymphoid cells from strain A mouse are injected into strain B mice. Strain B control mice are not injected.

2 8 to 10 weeks later, treated and untreated strain B mice mature into adults, and skin grafts from strain A mice are implanted.

Newborn strain B mice

Control

Strain A mouse

Treated

Control mouse rejects strain A graft: **No tolerance**

Treated mouse accepts strain A graft: **Tolerance**

Conclusion: What is recognized as self and nonself depends partly on when it is first encountered.

antigen into an animal early in fetal development caused that animal henceforth to recognize that antigen as self.

Tolerance must be established repeatedly throughout the life of the animal because lymphocytes are produced constantly. Continued exposure to self antigen helps maintain tolerance. For unknown reasons, tolerance to self antigens may be lost. When this happens, the body produces antibodies or T cells against its own proteins, resulting in an autoimmune disease.

B Cells: The Humoral Immune Response

Every day, billions of B cells survive the test of clonal deletion and are released from the bone marrow to enter the circulation. The B cells are the basis for the humoral immune response. Since each B cell expresses on its surface an antibody that is specific for a particular antigen, that antigen can bind to and activate the B cell, causing it to form a clone.

Some B cells develop into plasma cells

As described above, the activation of a B cell involves the binding of a particular antigenic determinant to the antibody protein on the surface of the B cell. Normally, for such

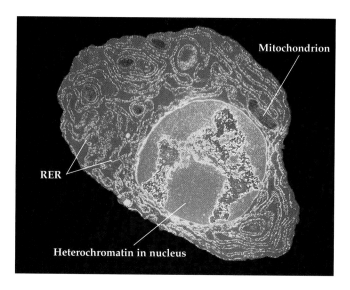

19.10 A Plasma Cell
The prominent nucleus with large amounts of heterochomatin (orange) and the cytoplasm (bright blue) crowded with rough endoplasmic reticulum are features of a cell that is actively synthesizing and exporting proteins—in this case, antibodies.

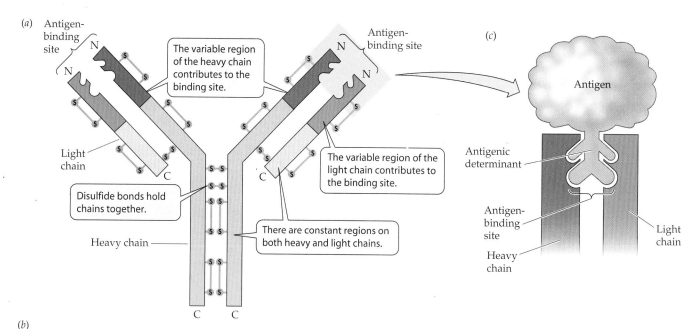

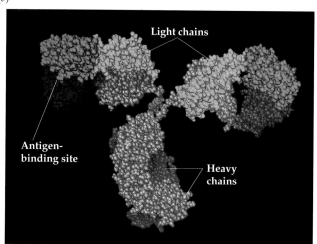

19.11 Structure of Immunoglobulins
In all these renderings, the light chains are shown in green and the heavy chains in blue. (a) The four polypeptide chains of an immunoglobulin molecule. (b) A more three-dimensional rendering of an immunoglobulin molecule in roughly the same orientation as (a). (c) The variable regions of both light and heavy chains participate in defining the antigen-binding site.

a B cell to develop into antibody-secreting plasma cells, a helper T cell (TH) must also bind to the same antigen on an antigen-presenting cell. The cellular division and differentiation of the B cells is stimulated by the receipt of chemical signals from the antigen-responsive T cell. These events lead to the formation of plasma cells (the effector cells) and memory cells (see Figure 19.7).

As plasma cells develop, the number of ribosomes and the amount of endoplasmic reticulum in their cytoplasm increase greatly (Figure 19.10). These increases prepare the cells for synthesizing large amounts of antibodies for secretion. All the plasma cells arising from a given B cell produce identical antibodies specific for the antigen that originally bound to the parent B cell. Thus antibody specificity is maintained as B cells proliferate.

Antibodies share a common structure, but may be of different classes

Antibodies are proteins called **immunoglobulins**. There are several types of immunoglobulins, but all contain a tetramer consisting of four polypeptides. Two of these polypeptides are identical *light chains,* and two are identical *heavy chains,* designated by their different sizes. Disulfide bonds (—S—S—) hold the chains together.

Each polypeptide chain consists of a constant region and a variable region (Figure 19.11). The **constant regions** of both light and heavy chains are similar in amino acid sequence in the immunoglobulins. The **variable regions** differ in their amino acid sequences. They contribute directly to the three-dimensional region where the antigen binds—

the *antigen-binding site*—and are responsible for the diversity of antibody specificity.

The amino acid sequence of the variable region is unique in each of the millions of antigen-specific immunoglobulins. Together the variable regions of a light and a heavy chain form a highly specific, three-dimensional structure. This part of a particular immunoglobulin molecule is what binds with a particular, unique antigenic determinant. The enormous range of antibody specificities is accomplished by a combination of rearrangements and mutations in the genes that encode the variable regions, as we will see later in the chapter.

Although the variable regions are responsible for the *specificity* of an immunoglobulin, the constant regions are also important. The constant regions determine whether the antibody remains part of the cell's plasma membrane or is secreted into the bloodstream. The constant regions also determine the type of action to be taken in eliminating the antigen, as we will see shortly.

The two antigen-binding sites on each immunoglobulin molecule are identical, permitting the formation of a large complex of antigen and antibody molecules. This complex is an easy target for ingestion and breakdown by phagocytic cells.

The five immunoglobulin classes are based on differences in the constant region of the heavy chain; they are described in Table 19.2.

IgG molecules make up about 85 percent of the total immunoglobulin content of the bloodstream. They are made in greatest quantity during a second immune response (see Figure 19.8). IgG defends the body in several ways. For ex-

19.2 Antibody Classes

CLASS	GENERAL STRUCTURE		LOCATION	FUNCTION
IgG	Monomer		Free in plasma; about 80 percent of circulating antibodies	Most abundant antibody in primary and secondary responses; crosses placenta and provides passive immunization to fetus
IgM	Pentamer	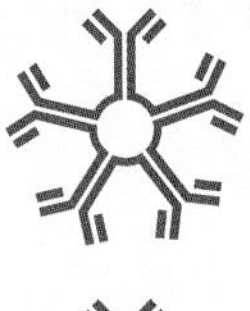	Surface of B cell; free in plasma	Antigen receptor on B cell membrane; first class of antibodies released by B cells during primary response
IgD	Monomer		Surface of B cell	Cell surface receptor of mature B cell; important in B cell activation
IgA	Dimer		Monomer found in plasma; polymers in saliva, tears, milk, and other body secretions	Protects mucosal surfaces; prevents attachment of pathogens to epithelial cells
IgE	Monomer		Secreted by plasma cells in skin and tissues lining gastrointestinal and respiratory tracts	Found on mast cells and basophils; when bound to antigens, triggers release of histamine from mast cell or basophil that contributes to inflammation and some allergic responses

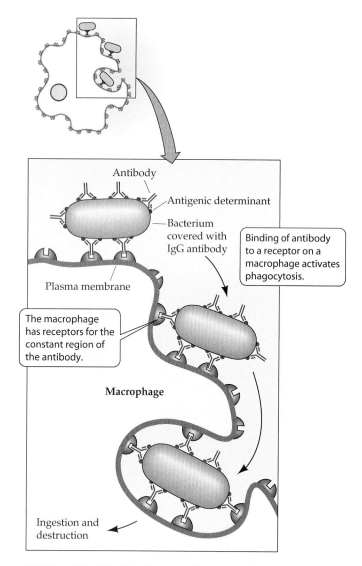

Antibody

Antigenic determinant

Bacterium covered with IgG antibody

Binding of antibody to a receptor on a macrophage activates phagocytosis.

Plasma membrane

The macrophage has receptors for the constant region of the antibody.

Macrophage

Ingestion and destruction

19.12 IgG Antibodies Promote Phagocytosis
When IgG antibodies cover a bacterium, receptors on a macrophage recognize them and engulf the cell they have coated.

ample, after some IgG molecules bind to antigens, they become attached by their heavy chains to macrophages. This attachment permits the macrophages to destroy the antigens by phagocytosis (Figure 19.12).

Hybridomas produce monoclonal antibodies

Because most antigens carry many different antigenic determinants, animals injected with a single antigen will produce a complex mixture of antibodies. Each of the antibodies is made by a clone of B cells. So the normal antibody response is said to be **polyclonal**.

We have learned in studies of biochemistry that many molecules share regions of similar structure. All human steroids, for example, have a similar multi-ring structure (see Figure 3.24). Some antibodies made by an animal against any steroid will bid to these common antigenic determinants. But a particular steroid has some parts that are

unique to that molecule, and some of the antibodies are directed against this region.

Suppose that a woman is infertile and her physician needs to measure the levels of the hormone estrogen in her blood. This could be done by using an antibody directed against estrogen as a reagent, and seeing how much antigen–antibody complex formed with a sample of the woman's blood. A polyclonal group of antibodies against estrogen would not be useful, however, because some of them, directed against common determinants, would bind to other steroid hormones as well as estrogen. Clearly, a clone of B cells making an antibody that binds only to a unique determinant—a **monoclonal antibody**—would be needed. But how can that clone be isolated and propagated?

Unfortunately, B cells cannot be cultured. On the other hand, cancerous tumors of plasma cells, called *myelomas*, grow rapidly in culture. Each tumor arises from a single plasma cell. Some myeloma cells cultured in laboratories have lost the ability to produce antibodies: These cells live for a long time, but they do not secrete immunoglobulins. Scientists use these myeloma cells and normal B lymphocytes to produce hybrid cells called *hybridomas*, which make specific normal antibodies in quantity and which, like the myeloma cells, proliferate rapidly and indefinitely in culture (Figure 19.13). These clones produce monoclonal antibodies in large quantities and can be preserved and stored by freezing.

Monoclonal antibodies have many practical applications. For example, they have been invaluable in the development of *immunoassays*, which use the great specificity of the antibodies to detect tiny amounts of molecules in tissues and fluids. This technique is used to quantify hormones such as estrogen. Most human pregnancy tests use a monoclonal antibody to a hormone made by the developing embryo.

Radioactively tagged monoclonal antibodies are used to target antigens on the surface of cancer cells, enabling precise imaging of the tumor so that the physician can monitor the progress of therapy. The cancer cell-targeted antibody, when attached to a poison, can be used to kill the tumor cells.

Monoclonal antibodies are also used for *passive immunization*—inoculation with specific antibody rather than with an antigen that causes the patient to develop his or her own antibody (as most vaccines are designed to do). Passive immunization is the approach used to treat the early symptoms of rabies infection or a rattlesnake bite, cases in which the toxic nature of the infection is so serious that there is not enough time to allow the person's immune system to mount its own defense.

T Cells: The Cellular Immune Response

Thus far we have been concerned primarily with the humoral immune response, whose effector molecules are the antibodies secreted by plasma cells that develop from acti-

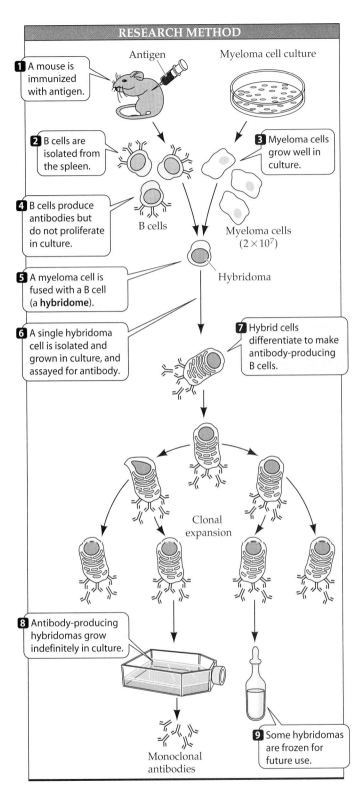

RESEARCH METHOD

Antigen

Myeloma cell culture

1 A mouse is immunized with antigen.

2 B cells are isolated from the spleen.

3 Myeloma cells grow well in culture.

B cells

Myeloma cells (2×10^7)

4 B cells produce antibodies but do not proliferate in culture.

5 A myeloma cell is fused with a B cell (a **hybridome**).

Hybridoma

6 A single hybridoma cell is isolated and grown in culture, and assayed for antibody.

7 Hybrid cells differentiate to make antibody-producing B cells.

Clonal expansion

8 Antibody-producing hybridomas grow indefinitely in culture.

9 Some hybridomas are frozen for future use.

Monoclonal antibodies

19.13 Creating Hybridomas for the Production of Monoclonal Antibodies
Cancerous myeloma cells and normal lymphocytes can be hybridized so that the proliferative properties of the myeloma cells are merged with the properties of of the antibody-producing lymphocytes.

quires special proteins encoded by the MHC (major histocompatibility complex) genes. These proteins underlie the immune system's tolerance for the cells of its own body and are responsible for the rejection of foreign tissues by the body.

T cell receptors are found on two types of T cells

Like B cells, T cells possess specific surface receptors. T cell receptors are not immunoglobulins, but glycoproteins with molecular weights about half that of an IgG. They are made up of two polypeptide chains, each encoded by a separate gene (Figure 19.14).

The genes that code for T cell receptors are similar to those for immunoglobulins, suggesting that both are derived from a single, evolutionarily more ancient group of genes. Like the immunoglobulins, T cell receptors include both variable and constant regions. The variable regions provide the specificity for reaction with a single antigenic determinant.

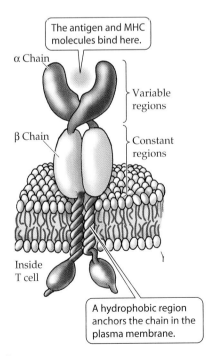

The antigen and MHC molecules bind here.

α Chain

Variable regions

β Chain

Constant regions

Inside T cell

A hydrophobic region anchors the chain in the plasma membrane.

19.14 A T Cell Surface Receptor
T cell receptors are glycoproteins, not immunoglobulins, although the structures of the two molecules are similar. The receptors are bound to the plasma membrane of the T cell that produces them.

vated B cells. T cells are the effectors of the cellular immune response, which is directed against any factor, such as a virus or mutation, that changes a normal cell into an abnormal cell.

In this section, we will describe two types of T cells (helper T cells and cytotoxic T cells). We will discover that the binding of a T cell receptor to an antigenic determinant re-

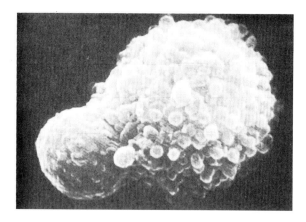

19.15 A Cytotoxic T Cell in Action
A cytotoxic T cell (the smaller sphere) has come into contact with a virus-infected cell, causing the infected cell to lyse. The blisters on the infected cell's surface indicate that it is beginning to break up.

There is a major difference between antibodies and T cell receptors: *While antibodies bind to an intact antigen, T cell receptors bind to a piece of the antigen displayed on the surface of an antigen-presenting cell.*

When T cells are activated by contact with a specific antigenic determinant, they proliferate and give rise to two types of effector cells:

▶ **Cytotoxic T cells**, or T_C **cells**, recognize virus-infected cells and kill them by causing them to lyse (Figure 19.15).
▶ **Helper T cells**, or T_H **cells**, assist both the cellular and humoral immune systems.

As mentioned already, a specific T_H cell must bind an antigen presented on a B cell before that B cell can become activated. The helper cell becomes the "conductor" of the "immunological orchestra," as it sends out chemical signals that not only result in its own clonal expansion, but also set in motion the actions of cytotoxic T cells as well as B cells.

Now that we are familiar with the major types of T cells, we can address the question of how T cells meet the antigenic determinants.

The major histocompatibility complex encodes proteins that present antigens to the immune system

We have seen that a body's defenses recognize its own cells as self—that proteins on our own cell surfaces are tolerated by our immune systems. There are several types of mammalian cell surface proteins, but we will focus on one very important group, the products of a cluster of genes called the major histocompatibility complex, or MHC.

The MHC gene products are plasma membrane glycoproteins—proteins with attached carbohydrate groups. In humans, the MHC molecules are called *human leukocyte antigens* (*HLA*), while in mice they are called *H-2 proteins*. Their major role is to "present" antigens on the cell surface to a T cell receptor.

There are three classes of MHC proteins:

▶ **Class I MHC proteins** are present on the surface of every nucleated cell in the animal. When cellular proteins are degraded to small peptide fragments in the proteasome (see Chapter 14), an MHC I protein may bind to a fragment and travel to the plasma membrane. There, the MHC I protein "presents" the cellular peptide to T_C cells. The T_C cells have a surface protein called CD8 that recognizes MHC I.
▶ **Class II MHC proteins** are found mostly on the surfaces of B cells, macrophages, and other antigen-presenting cells. When an antigen-presenting cell ingests an antigen, such as a virus, it is broken down in an endosome. An MHC II molecule may bind to one of the fragments and carry it to the cell surface, where it is presented to a T_H cell (Figure 19.16). T_H cells have a surface protein called CD4 that recognizes MHC II.
▶ **Class III MHC proteins** include some of the proteins of the complement system that interact with antigen–antibody complexes and result in the lysis of foreign cells (see Figure 19.4).

To accomplish their roles in antigen binding and presentation, both MHC I and MHC II have an *antigen-binding groove*, which can hold a peptide of about 10–20 amino acids (Figure 19.17). The T cell receptor recognizes not just the antigenic fragment, but the fragment *bound to an MHC I or MHC II molecule*. The table in Figure 19.17 summarizes the relationships of T cells and antigen-presenting cells.

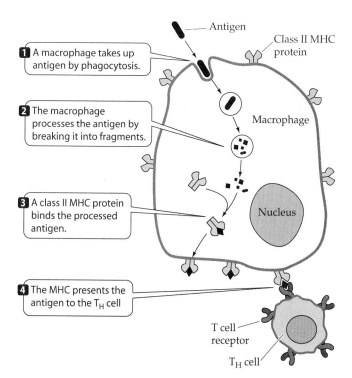

1 A macrophage takes up antigen by phagocytosis.

2 The macrophage processes the antigen by breaking it into fragments.

3 A class II MHC protein binds the processed antigen.

4 The MHC presents the antigen to the T_H cell

Antigen

Class II MHC protein

Macrophage

Nucleus

T cell receptor

T_H cell

19.16 Macrophages Are Antigen-Presenting Cells
Processed antigen is displayed by MHC II protein on the surface of a macrophage. Receptors on the helper T cell can then interact with the processed antigen/MHC II protein complex.

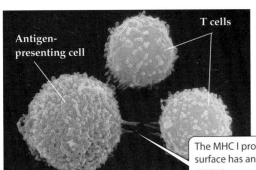

19.17 The Interaction between T Cells and Antigen-Presenting Cells
A groove in the MHC 1 protein holds an antigen, which it "presents" to a cytotoxic T cell. CD8 surface proteins on the T cells insure the specificity of interaction.

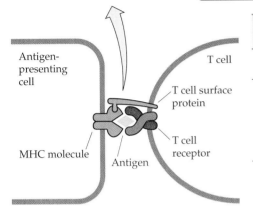

The MHC I protein on the cell's surface has an antigen-binding groove.

Antigen-Presenting and T Cell Types				
PRESENTING CELL TYPE	ANTIGEN PRESENTED	MHC CLASS	T CELL TYPE	T CELL SURFACE PROTEIN
Any cell	Intracellular protein fragment	Class I	Cytotoxic T cell (T_C)	CD8
Macrophages and B cells	Fragments from extracellular proteins	Class II	T helper cell (T_H)	CD4

There are three genetic loci each for MHC I and MHC II, and all six loci have as many as 100 different alleles. With so many possible allelic combinations, it is not surprising that different individuals are very likely to have different MHC genotypes. Similarities in base sequences between MHC genes and the genes coding for antibodies and T cell receptors suggest that all three may have descended from the same ancestral genes and are part of a "superfamily." Major aspects of the immune systems seem to be woven together by a common evolutionary thread.

Helper T cells and MHC II proteins contribute to the humoral immune response

When a T_H cell binds to an antigen-presenting macrophage, the T_H cell releases cytokines, which activate it to produce a clone of differentiated cells capable of interacting with B cells. The steps to this point constitute the **activation phase** of the response, and they occur in lymphatic tissue. Next comes the **effector phase**, in which B cells are activated to produce antibodies (Figure 19.18a).

B cells are also antigen-presenting cells. B cells take up by endocytosis antigen bound to their immunoglobulin receptors, process it, and display it on class II MHC proteins. When a T_H cell binds to the displayed antigen–MHC II complex, it releases cytokines that cause the B cell to produce a clone of plasma cells. Finally, the plasma cells secrete antibody, completing the effector phase of the humoral immune response.

Cytotoxic T cells and MHC I proteins contribute to the cellular immune response

Class I MHC molecules play a role in the cellular immune response that is similar to the role played by class II MHC molecules in the humoral immune response. In a virus-infected or mutated cell, "foreign" proteins or peptide frag-

ments combine with MHC class I molecules. The complex is displayed on the cell surface and presented to T_C cells. When a T_C cell binds to this complex, it is activated to proliferate and differentiate (Figure 19.18b).

In the effector phase of the cellular immune response, T_C cells produce molecules that lyse the target cell. In addition, the T_C cell can bind to a specific receptor (called Fas) on the target cell, that initiates apoptosis in the target cell. These two mechanisms, cell lysis and programmed cell death, work in concert to eliminate the altered host cell.

Because T cell receptors recognize self MHC molecules complexed with *nonself* antigens, they help rid the body of its own virus-infected cells. Because they also recognize MHC molecules complexed with *altered self* antigens (as a result of mutations), they help eliminate tumor cells, since most tumor cells have mutations.

In addition to the binding of an antigen–MHC complex to a cell surface receptor, T cells must receive a second signal for activation. This "co-stimulatory" signal occurs after the initial specific binding, and involves the interaction of additional proteins on the T cell and antigen-presenting cell. This second binding event leads to T cell activation, including cytokine production and cell division. It also sets in motion the production of an *inhibitor* of these events, so that the response is appropriately terminated. This inhibitor, a cell surface protein called CTLA4, is also important for the acquisition of tolerance, the capacity to avoid attacking one's own antigenic determinants.

MHC molecules underlie the tolerance of self

MHC molecules play a key role in establishing tolerance to self, without which an animal would be destroyed by its own immune system. Throughout the animal's life, developing T cells are tested in the thymus. One test question is, Can this cell recognize the body's MHC proteins? A T cell

(a)
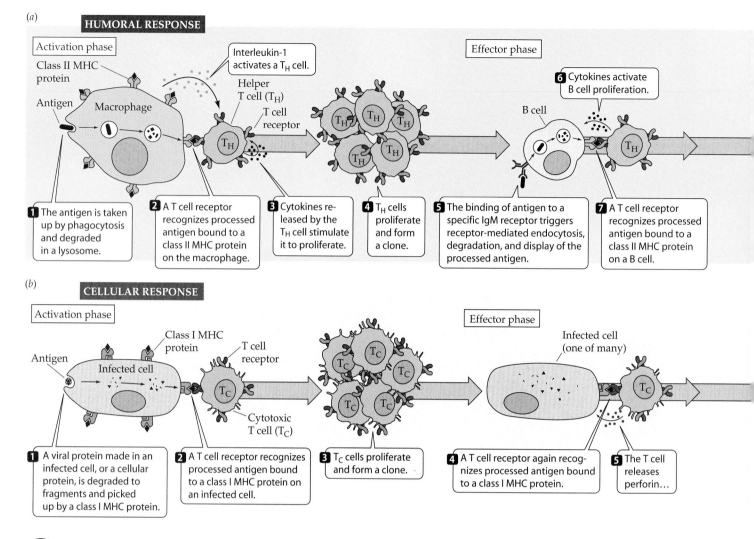

HUMORAL RESPONSE

Activation phase

Class II MHC protein

Antigen   Macrophage

Interleukin-1 activates a T_H cell.

Helper T cell (T_H)

T cell receptor

Effector phase

6 Cytokines activate B cell proliferation.

B cell

1 The antigen is taken up by phagocytosis and degraded in a lysosome.

2 A T cell receptor recognizes processed antigen bound to a class II MHC protein on the macrophage.

3 Cytokines released by the T_H cell stimulate it to proliferate.

4 T_H cells proliferate and form a clone.

5 The binding of antigen to a specific IgM receptor triggers receptor-mediated endocytosis, degradation, and display of the processed antigen.

7 A T cell receptor recognizes processed antigen bound to a class II MHC protein on a B cell.

(b)

CELLULAR RESPONSE

Activation phase

Class I MHC protein

T cell receptor

Antigen

Infected cell

Cytotoxic T cell (T_C)

Effector phase

Infected cell (one of many)

1 A viral protein made in an infected cell, or a cellular protein, is degraded to fragments and picked up by a class I MHC protein.

2 A T cell receptor recognizes processed antigen bound to a class I MHC protein on an infected cell.

3 T_C cells proliferate and form a clone.

4 A T cell receptor again recognizes processed antigen bound to a class I MHC protein.

5 The T cell releases perforin…

19.18 Phases of the Humoral and Cellular Immune Responses
Both immune responses have an activation phase and an effector phase.

unable to recognize self MHC proteins would be useless to the animal because it could not participate in any immune reactions. Such a T cell fails the test and dies within about 3 days.

The second question is, Does this cell bind to self MHC protein *and* to one of the body's own antigens? A T cell that satisfied both of these criteria would be harmful or lethal to the animal; it also fails the test and undergoes apoptosis. T cells that survive these tests mature into either T_C cells or T_H cells.

MHC molecules are responsible for transplant rejection

In humans, a major consequence of the MHC molecules became important with the development of organ transplant surgery. Because the proteins produced by the MHC are specific to each individual, they act as antigens if transplanted into another individual. An organ or a piece of skin transplanted from one person to another is recognized as nonself and soon provokes an immune response; the tissue then is killed, or "rejected," by the cellular immune system. But if the transplant is performed immediately after birth, or if it comes from a genetically identical person (an identical twin), the material is recognized as self and is not rejected.

The rejection problem can be overcome by treating a patient with drugs, such as cyclosporin, that suppress the immune system. However, this approach compromises the ability of patients to defend themselves against bacteria and viruses. Cyclosporin and some other immunosuppressants interfere with communication between cells of the immune system. Specifically, they inhibit the production of cytokines.

The Genetic Basis of Antibody Diversity

A newborn mammal possesses a full set of genetic information for immunoglobulin synthesis. At each of the loci coding for the heavy and light chains, it has an allele from the mother and one from the father. Throughout the animal's

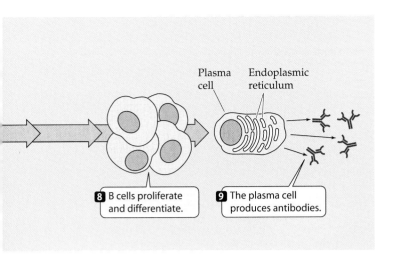

8 B cells proliferate and differentiate.

9 The plasma cell produces antibodies.

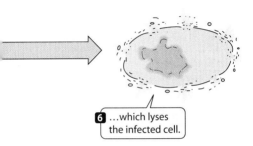

6 ...which lyses the infected cell.

life, each of its cells begins with the same full set of genes. However, as B cells develop, their genomes become modified in such a way that each cell eventually can produce one—and only one—specific type of antibody. In other words, different B cells develop slightly different genomes encoding different antibody specificities. How can a single organism produce millions of different immunoglobulins?

One hypothesis was that we simply have millions of antibody genes. However, a simple calculation (the number of base pairs needed per antibody gene multiplied by millions) shows that if this were true, our entire genome would be taken up by antibody genes. More than 30 years ago, an alternative hypothesis was proposed: A relatively small number of genes recombine at the DNA level to produce many unique combinations, and it is this shuffling of the genetic deck that produces antibody diversity. This is now the accepted theory.

In this section, we will describe the unusual events that generate the enormous antibody diversity normally characterizing each individual mammal. Then we will see how similar events produce the five classes of antibodies by producing slightly different "constant regions" that have special properties.

Antibody diversity results from DNA rearrangement and other mutations

In an unusual genetic process, functional immunoglobulin genes are assembled from DNA segments that initially are spatially separate. Every cell in the body has hundreds of DNA segments potentially capable of participating in the synthesis of the variable regions of the Ig molecule. In most body cells, these DNA sequences remain intact and separate from one another. During B cell development, however, these DNA segments are *rearranged and joined*. Pieces of the DNA are deleted, and DNA segments formerly distant from one another are joined together. In this fashion, an immunoglobulin gene is assembled from randomly selected pieces of DNA. Each B cell precursor in the animal assembles its own unique set of immunoglobulin genes. This remarkable process generates many diverse antibodies from the same starting genome. The same type of process also accounts for the diversity of T cell receptors.

In both humans and mice, the DNA segments coding for immunoglobulin heavy chains are on one chromosome and those for light chains are on others. The variable region of the light chain is encoded by two families of DNA segments, and the variable region of the heavy chain is encoded by three families.

Look at Figure 19.19 for an example of the gene families coding for the constant and variable regions of the heavy chain in mice. There are multiple genes coding for each of four kinds of segments in the polypeptide chain: 100 *V*, 30 *D*, 6 *J*, and 8 *C*. Each B cell that becomes committed to making an antibody randomly selects *one* gene for each of these segments to make the final heavy-chain coding sequence, *VDJC*. So the number of *different* heavy chains that can be made through this random recombination process is quite large.

Now consider that the light chains are similarly constructed, with a similar amount of diversity made possible by random recombination. If we assume that light-chain diversity is the same as heavy-chain diversity (144,000 possibilities), the number of possible combinations of light *and* heavy chains is 144,000 different light chains × 144,000 different heavy chains = 21 *billion* possibilities!

Even if this number is an overestimate by severalfold (and it is), the number of different immunoglobulin molecules that a B cell can make is huge. But there are still other mechanisms that generate even more diversity:

▶ When the DNA sequences for the *V*, *J*, and *C* regions are rearranged so that they are next to one another, the recombination event is not precise, and errors occur at the junctions. This *imprecise recombination* can create new codons at the junctions, with resulting amino acid changes.

▶ After the DNA fragments are cut out and before they are joined, an enzyme, *terminal transferase*, often adds some nucleotides to the free ends of the DNA's. These additional bases create insertion mutations.

▶ Finally, there is a relatively *high mutation rate* in immunoglobulin genes. Once again, this process creates many new alleles and antibody diversity.

19.19 Heavy-Chain Genes
Mouse immunoglobulin heavy chains have four segments, each of which is coded for by one of multiple genes.

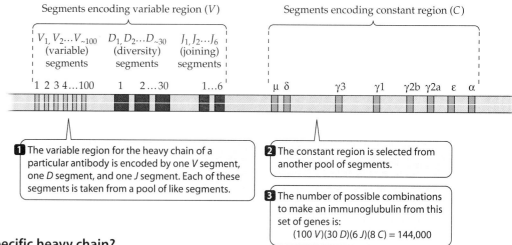

Segments encoding variable region (V)   Segments encoding constant region (C)

$V_1, V_2...V_{~100}$ $D_1, D_2...D_{~30}$ $J_1, J_2...J_6$
(variable)   (diversity)   (joining)
segments   segments   segments

1 2 3 4…100   1   2…30   1…6   μ δ   γ3   γ1   γ2b γ2a ε   α

1 The variable region for the heavy chain of a particular antibody is encoded by one V segment, one D segment, and one J segment. Each of these segments is taken from a pool of like segments.

2 The constant region is selected from another pool of segments.

3 The number of possible combinations to make an immunoglubulin from this set of genes is:
(100 V)(30 D)(6 J)(8 C) = 144,000

Adding these possibilities to the billions of combinations that can be made by random DNA rearrangements makes it not surprising that the immune system can mount a response to almost any natural or human-made substance.

How does a B cell produce a specific heavy chain?

As an example of how DNA rearrangement generates antibody diversity, let's consider how the heavy chain of IgM is produced in the mouse, a favorite subject for immunology studies.

The gene families governing all heavy-chain synthesis are on mouse chromosome 12, with the members arranged as shown in Figure 19.19. Light chains are produced from similar families, but they lack D segments.

How does order emerge from this seeming chaos of DNA segments? Two distinct types of nucleic acid rearrangements contribute to the formation of an antibody:

▶ DNA rearrangements, before transcription, join the V, D, and J segments.
▶ RNA splicing, after transcription, joins the variable region (VDJ) to the constant region.

First, substantial chunks of DNA are deleted from the chromosome during rearrangement of the segments. As a result of these deletions, a particular D segment ends up directly beside a particular J segment, and then the DJ segment ends up adjacent to one of the V segments. Thus, a single

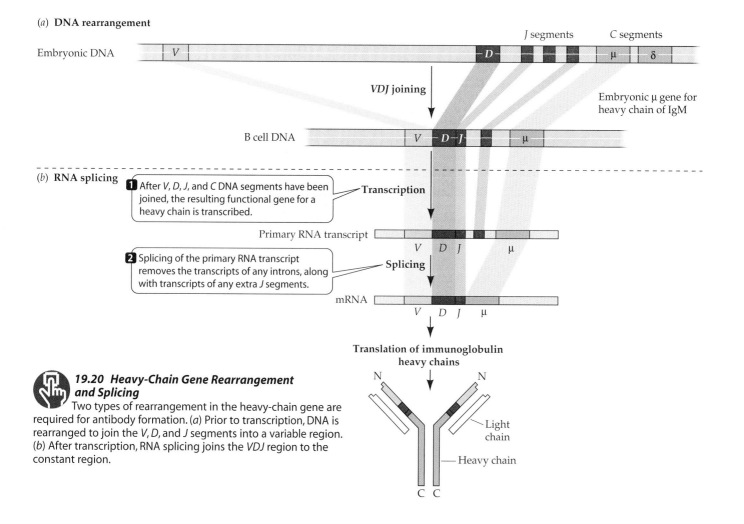

(a) **DNA rearrangement**

Embryonic DNA

J segments   *C* segments

V   D   μ   δ

VDJ joining

Embryonic μ gene for heavy chain of IgM

B cell DNA   V D J   μ

(b) **RNA splicing**

1 After V, D, J, and C DNA segments have been joined, the resulting functional gene for a heavy chain is transcribed.

Transcription

Primary RNA transcript   V   D J   μ

2 Splicing of the primary RNA transcript removes the transcripts of any introns, along with transcripts of any extra J segments.

Splicing

mRNA   V D J μ

Translation of immunoglobulin heavy chains

N   N

Light chain

Heavy chain

C C

19.20 Heavy-Chain Gene Rearrangement and Splicing
Two types of rearrangement in the heavy-chain gene are required for antibody formation. (a) Prior to transcription, DNA is rearranged to join the V, D, and J segments into a variable region. (b) After transcription, RNA splicing joins the VDJ region to the constant region.

"new" sequence, consisting of one *V*, one *D*, and one *J* segment, can now code for the variable region of the heavy chain. All the progeny of this cell constitute a clone having the same sequence for the variable region (Figure 19.20*a*).

The second step follows transcription. Splicing of the RNA transcript removes introns and any *J* segments lying between the selected *J* segment and the first constant region segment (Figure 19.20*b*). The result is an mRNA that can be translated, directly yielding the heavy chain of the cell's specific antibody.

The constant region is involved in class switching

In Table 19.2, we described the different classes of antibodies and their functions. Generally, a B cell makes only one class at a time. But **class switching** can occur, in which a B cell changes which antibody class it synthesizes.

Early in its life, a B cell produces IgM molecules, which are responsible for the specific recognition of a particular antigenic determinant. At this time, the constant region of the antibody's heavy chain is encoded by the first constant region segment, the μ segment (see Figure 19.19). If the B cell later becomes a plasma cell during an immunological response, another deletion commonly occurs in the cell's DNA, positioning the heavy-chain variable region gene (consisting of the same *V*, *D*, and *J* segments) next to a constant region segment farther down the original DNA, such as the γ, ε, or α segments (Figure 19.21). Such a DNA deletion results in the production of an antibody with a differ-

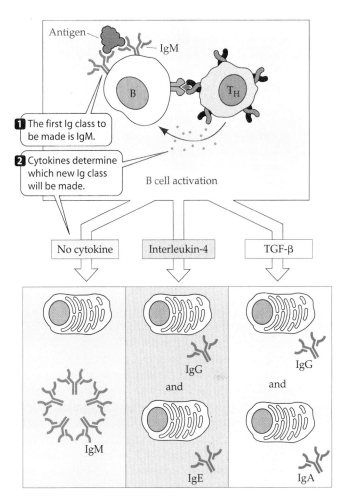

1 The first Ig class to be made is IgM.

2 Cytokines determine which new Ig class will be made.

B cell activation

No cytokine | Interleukin-4 | TGF-β

IgM

IgG and IgE

IgG and IgA

19.22 Cytokines Determine How the Antibody Switches Class
A T$_H$ cell initiates class switching in a B cell by secreting a cytokine. Different cytokines produce different switches.

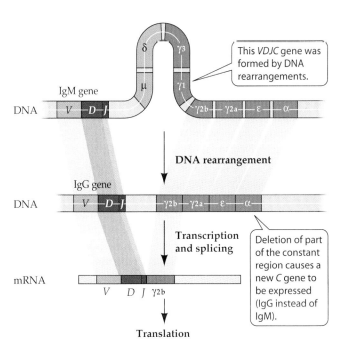

This *VDJC* gene was formed by DNA rearrangements.

IgM gene

DNA   *V — D — J*   δ   μ   γ3   γ1   γ2b   γ2a   ε   α

DNA rearrangement

IgG gene

DNA   *V — D — J*   γ2b   γ2a   ε   α

Transcription and splicing

Deletion of part of the constant region causes a new *C* gene to be expressed (IgG instead of IgM).

mRNA   *V   D   J   γ2b*

Translation

19.21 Class Switching
The gene produced by joining *V*, *D*, *J*, and *C* segments (see Figure 19.20) may later be modified, causing a different C segment to be transcribed. This modification, known as class switching, is accomplished by deletion of part of the constant region. Shown here is class switching from an IgM gene to an IgG gene.

ent constant region and *function*, but the same *antigen specificity*. The new antibody has *the same variable regions of the light and heavy chains, but a different constant region of the heavy chain*. This new antibody falls into one of the four other immunoglobulin classes (IgA, IgD, IgE, or IgG), depending on which of the constant region segments is placed adjacent to the variable region gene.

After switching classes, the plasma cell cannot go back to making the previous immunoglobulin class, because that part of the DNA has been lost. On the other hand, if additional constant region segments are still present, the cell may switch classes again.

What triggers class switching, and what determines the class to which a given B cell will switch? T$_H$ cells direct the course of an antibody response and determine the nature of the attack on the antigen. These T cells induce class switching by sending cytokine signals (Figure 19.22). These signals bind to receptors on the target B cells, generating a signal transduction cascade and resulting in altered transcription of immunoglobulin genes.

The Evolution of Animal Defense Systems

The strategy of rearranging DNA to make variable antibodies is used by all vertebrates with jaws, but nowhere else in the animal world. It is an "anticipatory" strategy, since the organism makes not only the defensive proteins that it needs, but also the antibodies and T cell receptors that it *might need*. Therefore, it must have provided an evolutionary advantage to the organisms that first had it.

This anticipatory strategy may first have arisen in an ancient creature resembling today's sharks, when a transposon inserted itself into a gene used for a defensive protein. Over time, this inserted element developed the ability to cut out adjacent DNA sequences and move them elsewhere in the genome.

The invertebrates, in all their diversity, have sturdy innate defense systems, and certain defense system elements are found even in unicellular protists. Many protists carry on phagocytosis, as do our own macrophages, and some protists use phagocytosis as a defense mechanism. Multicellular animals, both invertebrate and vertebrate, employ mobile phagocytic cells to patrol their bodies.

Like vertebrates, invertebrates (and probably some protists) distinguish between self and nonself. Making such distinctions enables invertebrates to reject tissue grafted from other individuals of the same species. Unlike vertebrates, however, invertebrates reject a second graft no more rapidly than a first graft—indicating that invertebrates lack immunological memory. This and other observations show that although immunological functions of invertebrates and vertebrates may be similar, their mechanisms often differ.

Invertebrates do not produce immunoglobulins, lymphocytes, or the complement system. However, they achieve similar defensive goals by analogous methods, and the analogs are probably evolutionary precursors of the systems found in vertebrates. Many invertebrates make proteins very similar to vertebrate cytokines, and those proteins play regulatory roles similar to those in humans.

Disorders of the Immune System

Immune deficiency diseases such as AIDS show us how much we depend on our immune system to protect us from pathogens. However, sometimes the immune system fails us in one way or another. It may overreact, as in an allergy; it may attack self antigens, as in an autoimmune disease; or it may function weakly or not at all, as in an immune deficiency disease. After a look at allergies and autoimmune conditions, we will examine the acquired immune deficiency that characterizes AIDS.

An inappropriately active immune system can cause problems

HYPERSENSITIVITY. A common type of condition can arise when the human immune system overreacts to a dose of antigen (**hypersensitivity**). Although the antigen itself may present no danger to the host, the inappropriate immune response may produce inflammation and other symptoms that can cause serious illness or death. **Allergies** are the most familiar examples of such a problem.

There are two types of allergic reactions. *Immediate hypersensitivity* occurs when an individual makes large amounts of IgE that bind to a molecule or structure in a food, pollen, or the venom of an insect. Mast cells in tissues and basophils in blood bind the IgE, which causes them to release amines such as histamine. The result is symptoms such as dilation of blood vessels, inflammation, and difficulty breathing. If not treated with antihistamines, a severe allergic reaction can lead to death.

Delayed hypersensitivity does not begin until hours after exposure to an antigen. In this case, the antigen is processed by antigen-presenting cells and a T cell response is initiated. This response can be so massive that the cytokines released cause macrophages to become activated and damage tissues. This is what happens when the bacteria that cause tuberculosis colonize the lung.

AUTOIMMUNITY. Sometimes clonal deletion fails, resulting in the appearance of one or more "forbidden clones" of B and T cells directed against self antigens (**autoimmunity**). This failure does not always result in disease, but in some instances it can.

People with *systemic lupus erythematosis* (*SLE*) have antibodies to many cellular components, including DNA and nuclear proteins. These antinuclear antibodies can cause serious damage when they link up with normal tissue antigens to form circulating immune complexes, which become stuck in tissues and provoke inflammation. A person with SLE has hyperactive B cells (thus the excess antibodies).

A person with *rheumatoid arthritis* has difficulty in shutting down a T cell response. We mentioned earlier that the inhibitor CTLA4 blocks T cells from reacting to self antigens. People with rheumatoid arthritis may have low CTLA4 activity, which results in inflammation of joints due to the infiltration of excess white blood cells.

Multiple sclerosis usually affects young adults, causing progressive damage to the nervous system. It involves both T cell- and B cell-mediated attack on two major proteins in myelin, the special membrane that coats some nervous tissues.

Insulin-dependent diabetes mellitus, or type I diabetes, occurs most often in children. It involves an immune reaction against several proteins in the cells of the pancreas that manufacture the protein hormone insulin. This reaction results in the cells being killed. These patients must take insulin daily in order to survive.

The causes of these autoimmune diseases are not known. They tend to "run in families," indicating a genetic component. Some alleles of MHC II are strongly linked to certain of these diseases. In some cases, the underlying reason may be molecular mimicry, in which T cells that recognize a nonself antigen also recognize something on the self that has a similar structure.

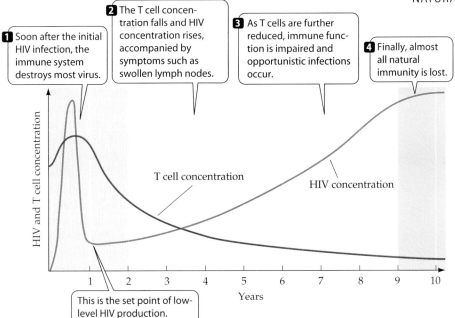

1 Soon after the initial HIV infection, the immune system destroys most virus.

2 The T cell concentration falls and HIV concentration rises, accompanied by symptoms such as swollen lymph nodes.

3 As T cells are further reduced, immune function is impaired and opportunistic infections occur.

4 Finally, almost all natural immunity is lost.

T cell concentration

HIV concentration

HIV and T cell concentration

This is the set point of low-level HIV production.

Years

19.23 The Course of an HIV Infection
HIV infection may be carried, unsuspected, for many years before the onset of symptoms. This long "dormant" period means the infection is often spread by people who are unaware that they are carrying the virus.

to the nodes and having cells in the nodes already receptive to virus infection—combine to ensure that HIV reproduces vigorously. Up to 10 billion viruses are made every day during this phase. The numbers of T_H cells quickly drop, and people show symptoms similar to mononucleosis, such as enlarged lymph nodes and fever.

These symptoms abate within 3 weeks, however, when T cells recognize infected lymphocytes, an immune response is mounted, and antibodies appear in the blood (Figure 19.23). By this time (several months after initial infection) the patient has a lot of circulating HIV complexed with antibodies that is gradually removed by the action of dendritic cells. But before they are filtered out, these antibody-complexed viruses can still infect T_H cells that come in contact with them. This secondary infection process reaches a low, steady-state level called the "set point." This point varies between people, and is a strong predictor of the rate of progression of the disease. For most people, it takes 8–10 years, even if untreated, for the more severe manifestations of AIDS to develop. In some, it can take as little as a year; in others, 20 years. During this "incubation period," infected people generally feel fine, and their T_H cell levels are adequate for them to mount immune responses.

However, in time, the virus destroys the T_H cells, and their numbers fall to dangerous levels. At this point, the infected patient is considered to have full-blown AIDS, and is susceptible to infections that the T_H cells would normally have been involved in eliminating (Figure 19.24). Most notable are the otherwise rare skin tumor called Kaposi's sarcoma caused by a herpesvirus, pneumonia caused by the fungus *Pneumocystis carinii*, and lymphoma tumors caused by Epstein-Barr virus. These are called **opportunistic infections**, because they take advantage of the crippled immune system of the host. They lead to death within a year or two.

AIDS is an immune deficiency disorder

There are various immune deficiency disorders, such as those in which T or B cells never form, and others in which B cells lose the ability to give rise to plasma cells. In either case, the affected individual is unable to mount an immune response and thus lacks a major line of defense against microbial pathogens.

Because of its essential roles in both the humoral and cellular immune responses, the T_H cell is perhaps the most central of all the components of the immune system—a significant cell to lose to an immune deficiency disorder. This cell is the target of HIV (*human immunodeficiency virus*), the virus that eventually results in AIDS (*acquired immune deficiency syndrome*).

HIV is transmitted from person to person several ways:

▶ Through blood, such as by an injection needle contaminated with the virus after being used by an infected individual

▶ Through exposure of broken skin, an open wound, or mucous membranes to body fluids, such as semen, containing HIV

▶ Through the blood of an infected mother to her baby during birth

HIV initially infects macrophages, T_H cells, and dendritic cells in blood and tissues. Dendritic cells are antigen-presenting cells with highly folded plasma membranes that can capture antigens (see Figure 19.2). These infected cells carry the virus to the lymphoid tissues (lymph nodes and spleen) where B cells mature and T cells reside.

Normally in the lymph node, the dendritic cells present their captured antigen to T_H cells, and this causes the T_H cells to divide and form a clone (see Figure 19.18). But HIV preferentially infects activated, and not resting, T_H cells. So the HIV arriving in the lymph nodes proceeds to infect the many activated T_H cells that are already responding to other antigens. These two processes—having cells take the virus

HIV infection and replication occur in T_H cells

As a retrovirus, HIV uses RNA as its genetic material. A central core, with a protein coat, contains two identical molecules of RNA as well as the enzymes reverse transcriptase, integrase, and a protease. An envelope, derived from the plasma membrane of the cell in which the virus was formed, surrounds the core. The envelope is studded with envelope proteins (gp120 and gp41, where "gp" stands for *glycoprotein*) that enable the virus to infect its target cells. The HIV replication cycle has several stages.

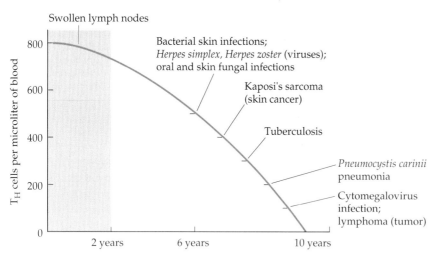

19.24 Relationship Between T_H Cell Count and Opportunistic Infections
As HIV kills more and more T_H cells, the immune system is less and less able to defend the body against various diseases.

VIRUS ENTRY INTO CELLS. HIV attaches to T_H cells and macrophages via CD4, which acts as a receptor for the viral envelope protein gp120. Following binding, the membrane that surrounds the HIV core particle fuses with the host cell plasma membrane, resulting in the entry of the core into the cytoplasm (Figure 19.25; see also Figure 13.7). These events require the participation of at least two other membrane proteins, one from the virus (gp41) and another from the host (appropriately called fusin in T_H cells).

REVERSE TRANSCRIPTION. HIV can insert a cDNA copy of its genome into the host cell's DNA. The process of making a cDNA copy from viral RNA occurs in the viral core particle in an infected cell. It requires the participation of three separate enzymes:

▶ Reverse transcriptase, to make single-stranded DNA from an RNA template
▶ RNAse H, to degrade the viral RNA
▶ DNA polymerase, to make the second strand of cDNA

HIV reverse transcriptase does not have the proofreading activity of many DNA polymerases, so the errors that inevitably creep into the process are not corrected. Up to 10 incorrect bases out of about 8,000 may end up in each cDNA produced. This is a great advantage to the virus, as genomic mutations allow its proteins to escape the host's immune response; however, the mutations present a challenge to scientists trying to design drugs and vaccines to bind to the constantly changing viral proteins.

INTEGRATION OF VIRAL cDNA INTO THE HOST GENOME. The viral core proteins have an amino acid sequence that is recognized by a receptor on the surface of the host cell's nucleus; thus the particle rapidly enters the nucleus. At this point, a viral enzyme called integrase catalyzes the breakage of host DNA and insertion of viral cDNA. This is similar to the way in which bacteriophage DNA becomes incorporated into a bacterial chromosome as a prophage (see Chapter 13). The cDNA thus becomes a permanent part of the T_H cell's DNA, replicating with it at each cell division, and may remain in the T_H cell genome for a decade or more.

VIRUS PRODUCTION. This latent period ends if the HIV-infected T_H cell becomes activated as it responds naturally to an antigen. The expression of viral genes requires the collaboration of host transcription factors that are made in activated T_H cells and a virally encoded protein called Tat. When the T_H cell is activated, the entire viral genome is transcribed into RNA, which can either remain as it is or be spliced. Unspliced RNA's become the genomes of new HIV particles; spliced RNA's make the viral structural proteins. An important activator of splicing is the viral protein called Rev.

The protease encoded by HIV is needed to complete the formation of individual viral proteins from larger products of translation. Packaging domains on viral proteins cause the RNA genomes to fold into them and form core particles. In the meantime, the viral membrane proteins are made on the endoplasmic reticulum of the host cell and transported to the plasma membrane via the Golgi complex. The cytoplasmic tails of the gp120 membrane proteins bind to the core particles, and the viruses bud from the infected cell, surrounding themselves with modified plasma membrane from the host.

Treatments for HIV infection rely on knowledge of its molecular biology

As the AIDS epidemic has grown, so has knowledge of HIV molecular biology. The general therapeutic strategy is to try to block a stage in the viral life cycle and hold HIV infection in check. Potential therapeutic agents that interfere with the major steps of the life cycle are being tested (see Figure 19.25). Of course, it is crucial to block only steps that are unique to the virus, so that drug therapies do not harm the patient by blocking a step in the patient's own metabolism.

Highly active antiretroviral therapy (HAART) was developed in the late 1990s and has had considerable success in delaying the onset of AIDS symptoms in people infected with HIV by 3 years or more, and in prolonging the lives of people with AIDS. The logic of HAART comes from cancer treatment: Employ a combination of drugs acting at different parts of the virus life cycle. Generally, the HAART regimen uses:

▶ A protease inhibitor. These drugs obstruct the active site of the HIV protease.
▶ Two reverse transcriptase inhibitors. These molecules are incorporated into the growing cDNA chain, but no nucleotides can be added to them, so reverse transcription stops.

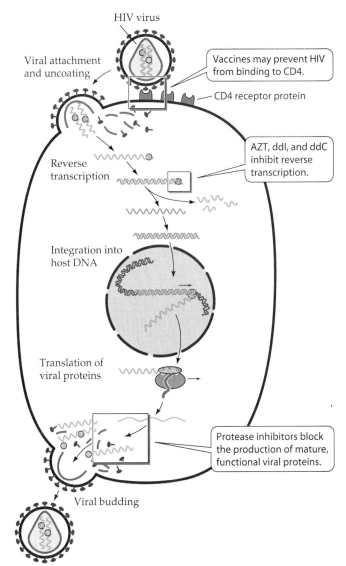

HIV virus

Viral attachment and uncoating

Vaccines may prevent HIV from binding to CD4.

CD4 receptor protein

Reverse transcription

AZT, ddI, and ddC inhibit reverse transcription.

Integration into host DNA

Translation of viral proteins

Protease inhibitors block the production of mature, functional viral proteins.

Viral budding

19.25 Strategies to Combat HIV Reproduction
Several widely used drugs block specific steps in the HIV life cycle.

These drugs have had such dramatic effects on patients that they may eliminate HIV entirely in some people, especially in those treated within the first few days after infection, before the virus has arrived in the lymph nodes. Most patients, however, face a lifetime of anti-HIV therapy.

The many treatments under development include:

▸ Vaccines to inhibit virus entry into cells and to form immune complexes with circulating viruses

▸ Integrase inhibitors, to block cDNA incorporation into the host genome

▸ Tat and Rev inhibitors, to block HIV transcription and splicing

▸ Antisense RNA and ribozymes directed against HIV RNA

What can be done until biomedical science provides the tools to bring the worldwide AIDS epidemic to an end? Above all, people must recognize that they are in danger whenever they have sex with a partner whose total sexual history is not known. The danger rises as the number of sex partners rises, and the danger is much greater if partners participating in sexual intercourse are not protected by a latex condom. The danger that heterosexual intercourse will transmit HIV rises tenfold to a hundredfold if either partner has another sexually transmitted disease.

Chapter Summary

▸ Animals defend themselves against pathogens by both innate (nonspecific) and specific means.

Defensive Cells and Proteins

▸ Many of our defenses are implemented by cells and proteins carried in the bloodstream and in the lymphatic system. **Review Figure 19.1**

▸ White blood cells, including lymphocytes (B and T cells) and phagocytes (such as neutrophils and macrophages), play many defensive roles. **Review Figure 19.2**

▸ Immune system cells produce several kinds of proteins. Antibodies and T cell receptors bind foreign substances, MHC proteins help coordinate the recognition of foreign substances and the activation of defensive cells, and cytokines alter the behavior of other cells.

Innate Defenses

▸ An animal's innate defenses include physical barriers, competing resident microorganisms, and local agents, such as secretions that contain an antibacterial enzyme. **Review Table 19.1**

▸ The complement system, composed of about 20 proteins, assembles itself in a cascade of reactions to cooperate with phagocytes or antibodies to lyse foreign cells. **Review Figure 19.3**

▸ Interferons produced by virus-infected cells inhibit the ability of viruses to replicate in neighboring cells

▸ Macrophages and neutrophils engulf invading bacteria. Natural killer cells attack tumor cells and virus-infected body cells.

▸ Macrophages play an important role in the inflammation response. Activated mast cells release histamine, which causes blood capillaries to leak. Complement proteins attract macrophages to the site, where they engulf bacteria and dead cells. **Review Figure 19.5**

Specific Defenses: The Immune Response

▸ Four features characterize the immune response: specificity, the ability to respond to an enormous diversity of antigens, the ability to distinguish self from nonself, and memory.

▸ The immune response is directed against antigens that evade the nonspecific defenses. Each antibody or T cell is directed against a particular antigenic determinant. **Review Figure 19.6**

▸ The immune response is highly diverse and can respond specifically to millions of different antigenic determinants.

▸ The immune response distinguishes its own cells from foreign cells, attacking only cells recognized as nonself.

▸ The immune system remembers; it can respond rapidly and effectively to a second exposure to an antigen.

▸ There are two interactive immune responses: the humoral immune response and the cellular immune response. The humoral immune response employs antibodies secreted by plasma cells to target antigens in body fluids. The cellular immune response employs T cells to attack body cells that have been altered by viral infection or mutation or to target antigens that have invaded the body's cells.

▶ Clonal selection theory accounts for the rapidity, specificity, and diversity of the immune response. It also accounts for immunological memory, which is based on the production of both effector and memory cells as T cell and B cell clones expand. **Review Figure 19.7**

▶ Immunological memory plays roles in both natural immunity and artificial immunity based on vaccination. **Review Figure 19.8**

▶ Clonal selection theory also accounts for the immune system's recognition of self. Tolerance of self results from clonal deletion of antiself lymphocytes. **Review Figure 19.9**

B Cells: The Humoral Immune Response

▶ Activated B cells form plasma cells, which synthesize and secrete specific antibodies.

▶ The basic unit of an antibody, or immunoglobulin, is a tetramer of four polypeptides: two identical light chains and two identical heavy chains, each consisting of a constant and a variable region. **Review Figure 19.11**

▶ The variable regions of the light and heavy chains collaborate to form the antigen-binding sites of an antibody. The variable regions determine an antibody's specificity; the constant region determines the destination and function of the antibody.

▶ There are five immunoglobulin classes. IgM, formed first, is a membrane receptor on B cells, as is IgD. IgG is the most abundant antibody class and performs several defensive functions. IgE takes part in inflammation and allergic reactions. IgA is present in various body secretions. **Review Table 19.2**

▶ Monoclonal antibodies consist of identical immunoglobulin molecules directed against a single antigenic determinant. Hybridomas are produced by fusing B cells with myeloma cells (from cancerous tumors of plasma cells). **Review Figure 19.13**

T Cells: The Cellular Immune Response

▶ The cellular immune response is directed against altered or antigen-infected cells of the body. T_C cells attack virus-infected or tumor cells, causing them to lyse. T_H cells activate B cells and influence the development of other T cells and macrophages. **Review Figure 19.14**

▶ T cell receptors in the cellular immune response are analogous to immunoglobulins in the humoral immune response.

▶ The major histocompatibility complex (MHC) encodes many membrane proteins. MHC molecules in macrophages, B cells, or body cells bind processed antigen and present it to T cells. **Review Figure 19.16, 19.17**

▶ The activation of the humoral immune response requires the collaboration of class II MHC molecules, the T cell surface protein CD4, and cytokines. The effector phase of the humoral immune response involves T cells, class II MHC molecules, B cells, and cytokines, and results in the formation of active plasma cells. **Review Figure 19.18**

▶ In the cellular immune response, class I MHC molecules, T_C cells, CD8, and cytokines collaborate to activate T_C cells with the appropriate specificity. **Review Figure 19.18**

▶ Developing T cells undergo two tests: They must be able to recognize self MHC molecules, and they must *not* bind to

both self MHC and any of the body's own antigens. T cells that fail either of these tests die.

▶ The rejection of organ transplants results from the genetic diversity of MHC molecules.

The Genetic Basis of Antibody Diversity

▶ Several gene families underlie the incredible diversity of antibody and T cell receptor specificities.

▶ Antibody heavy-chain genes are constructed from one each of numerous *V*, *D*, *J*, and *C* segments. The *V*, *D*, and *J* segments combine by DNA rearrangement, and transcription yields an RNA molecule that is spliced to form a translatable mRNA. Other gene families give rise to the light chains. **Review Figures 19.19, 19.20**

▶ There are millions of possible antibodies as a result of these DNA combinations. Imprecise DNA rearrangements, mutations, and random addition of bases to the ends of the DNA's before they are joined contribute even more diversity.

▶ A plasma cell produces IgM first, but later it may switch to the production of other classes of antibodies. This class switching, resulting in antibodies with the same antigen specificity but a different function, is accomplished by cutting and rejoining of the genes encoding the constant region. **Review Figure 19.21, 19.22**

The Evolution of Animal Defense Systems

▶ The DNA rearrangement mechanism for creating diversity among immune system molecules may have evolved from a DNA transposon.

▶ Invertebrate animals reject nonself tissues but lack immunological memory. They possess cells and molecules analogous, but not identical, to lymphocytes, immunoglobulins, and cytokines.

▶ Even the most evolutionarily ancient groups among today's vertebrates have immune systems more similar to those of humans than to those of invertebrates.

Disorders of the Immune System

▶ Allergies result from an overreaction of the immune system to an antigen.

▶ Autoimmune diseases result from a failure in the immune recognition of self, with the appearance of antiself B and T cells that attack the body's own cells.

▶ Immune deficiency disorders result from failures of one or another part of the immune system. AIDS is an immune deficiency disorder arising from depletion of the body's T_H cells as a result of infection with HIV. Depletion of the T_H cells weakens and eventually destroys the immune system, leaving the host defenseless against "opportunistic" infections. **Review Figures 19.23, 19.24**

▶ HIV inserts a copy of its genome into a chromosome of a macrophage or T_H cell, where it may lie dormant for years. When the viral genome is transcribed and translated, new viruses form. **Review Figure 19.25**

▶ All steps in the reproductive cycle of HIV are under investigation as possible targets for drugs. Currently the most effective drugs are those directed against reverse transcriptase and protease. **Review Figure 19.25**

▶ Some treatments may provide a dramatic reduction in HIV levels, but there is as yet no indication that we can prevent infection with HIV, as by vaccination. The only strategy currently available is for people to avoid behaviors that place them at risk.

For Discussion

1. Describe the part of an antibody molecule that interacts with an antigenic determinant. How is it similar to the active site of an enzyme? How does it differ from the active site of an enzyme?

2. Contrast immunoglobulins and T cell receptors with respect to structure and function.

3. Discuss the diversity of antibody specificities in an individual in relation to the diversity of enzymes. Does every cell in an animal contain genetic information for all the organism's enzymes? Does every cell contain genetic information for all the organism's immunoglobulins?

4. The gene family determining MHC on the cell surface in humans is on a single chromosome. A father's MHC type is A1, A3, B5, B7, D9, D11. His wife's phenotype is A2, A4, B6, B7, D11, D12. They have a child who is A1, A4, B6, B7, D11, D12. What are the parents' haplotypes—that is, which alleles are linked on the diploid chromosomes in each parent? Assuming that there is no recombination among the genes determining the MHC type, can these parents have a child who is A1, A2, B7, B8, D10, D11?

5. Is it true that any child can accept an organ transplant from either parent, but parents cannot accept a graft from a child? Explain your answer.

Part Three

EVOLUTIONARY PROCESSES

20 *The History of Life on Earth*

WHEN YOU WANT TO KNOW WHAT TIME IT IS, you probably look at your watch, or at the clock on the wall or on your computer. You could also listen to the radio or watch television to hear some announcement of time. But suppose the electric power system failed and you lost your watch. How could you tell time then? You would use the cues that people used during most of human history—the cycle of day and night. We are so accustomed to having time-measuring devices all around us that we forget these devices are recent inventions. When Galileo studied the motion of a ball rolling down an inclined plane 350 years ago, he used his pulse to mark off equal intervals of time.

The science of biology is intimately linked to concepts of time. Biology as we know it could not and did not develop very far until an appreciation of the age of Earth was provided by geologists more than 150 years ago. Until that time, most people believed that Earth was only a few thousand years old. Darwin could not have developed his theory of evolution by natural selection if he had not read the works of Charles Lyell, England's leading geologist, who believed that Earth was ancient. As we pointed out in Chapter 1, Darwin's theory was based on the assumption that Earth was very old and that life had existed for a very long time, during which it had steadily evolved.

The goals of Part Three are to document the history of life on Earth, to describe the processes of evolutionary change, and to discuss the agents that cause them. We begin in this chapter by asking the following questions: How do we know that Earth is ancient? What is the evidence that life evolved early during Earth's history and has continued to evolve since then? In the following chapter we discuss the processes by which life evolved. In subsequent chapters, we will see how biologists determine the evolutionary histories of organisms, and how the millions of species that live today (as well as those that became extinct) formed from a single common ancestor. Finally, in Chapter 25, we will examine how life probably arose from nonliving matter several billion years ago.

Understanding biological evolution is important because evolutionary changes are taking place all around us. These changes have powerful implications for human welfare. Our own attempts to control populations of undesirable species and increase populations of desirable ones make human beings powerful agents of evolutionary change. In addition to producing the results we desire, these efforts often cause undesirable outcomes, such as the evolution of resistance to medicines by pathogens and to pesticides by pests. Medicine and agriculture can respond creatively to the evolutionary changes they are causing only if their practitioners understand how and why those changes happen. But what exactly is biological evolution?

Biological evolution is a change over time in the genetic composition of a population. Changes that happen during the lifetimes of species constitute **microevolution**. Plant and animal breeding and changes occurring in response to environmental shifts over decades provide good examples of microevolution. Changes that involve the appearance of new species and evolutionary lineages are called **macro-**

The Hands of Time
London's Big Ben, perhaps the most recognizable timepiece in the world, epitomizes our worldview of the importance of hours and minutes. Geological time, however, is much more difficult to grasp but is essential to understanding biological evolution.

evolution. The fossil record provides the best evidence of macroevolutionary changes among organisms. Many of these changes are dramatic.

To understand the long-term patterns of evolutionary change that we will document in this chapter, we must think in time frames spanning many millions of years and imagine events and conditions very different from those we now observe. The Earth of the distant past is, to us, a foreign planet inhabited by strange organisms. The continents were not where they are today, and climates were different. One of the remarkable achievements of modern science has been the development of sophisticated techniques for inferring past conditions and dating them accurately.

In this chapter, we first examine how events in the distant past can be dated. Then we review the major changes in physical conditions on Earth during the past 4 billion years, look at how those changes affected life, and discuss the major patterns in the evolution of life.

How Do We Know Earth Is Ancient?

It is difficult to age rocks because a rock of a particular type could have been formed at any time during Earth's history. It is easier to determine the ages of rocks relative to one another. The first person to recognize that this could be done was the seventeenth-century Danish physician Nicolaus Steno. Steno realized that in an undisturbed sequence of sedimentary rocks, the oldest strata lie at the bottom and successively higher strata are progressively younger (Figure 20.1).

Geologists subsequently combined Steno's insights with their observations of the **fossils**—remains of ancient organisms—contained within rocks. They discovered that fossils of similar organisms were found in widely separated places on Earth, that certain organisms were always found in younger rocks than others, and that organisms in the most recent strata were more similar to modern organisms than those found in lower, more ancient strata. With this information, they were able to determine much about the relative ages of sedimentary rocks and about patterns in the evolution of life. But they still could not tell how old the rocks were. A method of dating rocks did not become available until the discovery of radioactivity at the turn of the twentieth century.

Radioactivity provides a way to date rocks

Radioactive isotopes decay in a regular pattern during successive, equal periods of time. During each successive time

20.1 Earth's Geological History

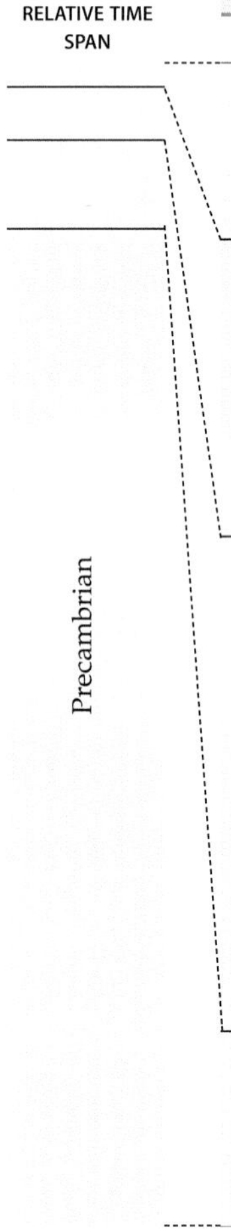

RELATIVE TIME SPAN	ERA	PERIOD	ONSET	MAJOR PHYSICAL CHANGES ON EARTH
	Cenozoic	Quaternary	1.8 mya	Cold/dry climate; repeated glaciations
		Tertiary	65 mya	Continents near current positions; climate cools
	Mesozoic	Cretaceous	144 mya	Northern continents attached; Gondwana begins to drift apart; meteorite strikes Yucatán Peninsula
		Jurassic	206 mya	Two large continents form: Laurasia (north) and Gondwana (south); climate warm
		Triassic	245 mya	Pangaea slowly begins to drift apart; hot/humid climate
Precambrian	Paleozoic	Permian	290 mya	Continents aggregate into Pangaea; large glaciers form; dry climates form in interior of Pangaea
		Carboniferous	354 mya	Climate cools; marked latitudinal climate gradients
		Devonian	409 mya	Continents collide at end of period; asteroid probably collides with Earth
		Silurian	440 mya	Sea levels rise; two large continents form; hot/humid climate
		Ordovician	510 mya	Gondwana moves over South Pole; massive glaciation, sea level drops 50 m
		Cambrian	543 mya	O_2 levels approach current levels
	Precambrian		600 mya	O_2 level at >5% of current level
			2.5 bya	O_2 level at >1% of current level
			3.8 bya	O_2 first appears in atmosphere
			4.5 bya	

20.1 Young Rocks Lie on Top of Old Rocks
The oldest rocks at the bottom of this photo of the North Rim of the Grand Canyon formed about 540 million years ago. The youngest rocks on top are about 500 million years old.

MAJOR EVENTS IN THE HISTORY OF LIFE

Humans evolve; large mammals become extinct

Radiation of birds, mammals, flowering plants, and insects

Dinosaurs continue to radiate; flowering plants and mammals diversify. **Mass Extinction** at end of period (≈76% of species disappear)

Diverse dinosaurs; first birds; two minor extinctions

Early dinosaurs; first mammals; marine invertebrates diversify. **Mass Extinction** at end of period (≈65% of species disappear)

Reptiles radiate; amphibians decline; **Mass Extinction** at end of period (≈96% of species disappear)

Extensive "fern" forests; first reptiles; insects radiate; earliest flowering plants

Fishes diversify; first insects and amphibians. **Mass Extinction** at end of period (≈75% of species disappear)

Jawless fishes diversify; first bony fishes; plants and animals colonize land

Mass Extinction at end of period (≈75% of species disappear)

Most animal phyla present; diverse algae

Ediacaran fauna
Eukaryotes evolve; several animal phyla appear
Origin of life; prokaryotes flourish

---

interval, an equal fraction of the remaining radioactive material of any radioisotope decays, either changing to another element or becoming the stable isotope of the same element. For example, in 14.3 days, one-half of any sample of phosphorus-32 (^{32}P) decays to its stable isotope, phosphorus-31 (^{31}P). During the next 14.3 days, one-half of the remaining half decays, leaving one-fourth of the original ^{32}P. The time it takes for half of an isotope to decay is that isotope's **half-life**. After 42.9 days, three half-lives have passed, so one-eighth (that is, $\frac{1}{2} \times \frac{1}{2} \times \frac{1}{2}$) of the original ^{32}P remains.

Each radioisotope has a characteristic half-life. Which isotope is used to estimate the age of an ancient material depends on how old the material is thought to be. Tritium (^{3}H) has a half-life of 12.3 years, and carbon-14 (^{14}C) has a half-life of about 5,700 years. The half-life of potassium-40 (^{40}K) is 1.3 billion years; the decay of potassium-40 to argon-40 has been used to date most of the ancient events in the evolution of life.

To use a radioisotope to date a past event, we must know or estimate the concentration of the isotope at the time of that event. In the case of carbon, we know that the production of new ^{14}C in the upper atmosphere (by the reaction of neutrons with ^{14}N) just balances the natural radioactive decay of ^{14}C. Therefore, the ratio of ^{14}C to its stable isotope, ^{12}C, exists in a more or less steady state in living organisms and their environment.

However, as soon as an organism dies, it ceases to exchange carbon compounds with the rest of the world. Its decaying ^{14}C is no longer replenished, and the ratio of ^{14}C to ^{12}C decreases. The ratio of ^{14}C to ^{12}C in fossil organisms can be used to date fossils (and thus the sedimentary rocks that contain those fossils) that are less than 50,000 years old with a fair degree of certainty.

Dating rocks more ancient than 50,000 years requires estimating isotope concentrations that exist in volcanic (but not in sedimentary) rocks. To date ancient sedimentary rocks, geologists search for places where volcanic ash or lava flows have intruded into beds of sedimentary rock. Radiometric dating, combined with observations of fossils, is the most powerful method of determining the ages of rocks.

But there are many places where sedimentary rocks do not contain suitable volcanic intrusions and few fossils are present. In these areas, dating methods other than radiometry must be used. One such method is based on the fact that Earth's magnetic poles move and occasionally reverse themselves. Because both sedimentary and igneous rocks preserve a record of Earth's magnetic field at the time they were formed, *paleomagnetism* helps determine the ages of those rocks. Other "time machines," which we will describe later, include continental drift, sea level changes, and molecular clocks.

Using these methods, geologists have divided Earth's history into eras, which in turn are subdivided into periods (Table 20.1). The boundaries between these divisions, which are based on major differences in the fossils con-

volcanic eruptions. The ash the volcanoes ejected into Earth's atmosphere reduced the penetration of sunlight to Earth's surface, lowering temperatures, reducing photosynthesis, and triggering massive glaciation. Massive volcanic eruptions also occurred as the continents drifted apart during the late Triassic period and again at the end of the Cretaceous.

External events have triggered other changes on Earth

At least 30 meteorites between the sizes of baseballs and soccer balls hit Earth each year, but collisions with large meteorites are rare. In 1980, Luis Alvarez and several of his colleagues at the University of California, Berkeley, proposed that the mass extinction at the end of the Cretaceous period, about 65 mya, might have been caused by the collision of Earth with a large meteorite. These scientists based their hypothesis on the finding of abnormally high concentrations of the element iridium in a thin layer separating the rocks deposited during the Cretaceous period from those of the Tertiary (Figure 20.5). Iridium is abundant in some meteorites, but is exceedingly rare on Earth's surface.

To account for the estimated amount of iridium in this layer, Alvarez postulated that a meteorite 10 km in diameter collided with Earth at a speed of 72,000 km per hour. The force of such an impact would have ignited massive fires, created great tidal waves, and sent up an immense dust cloud that blocked the sun, thus cooling the planet. As it settled, the dust would have formed the iridium-rich layer.

This hypothesis generated a great deal of controversy and stimulated much research. Some scientists searched for the site of impact of the supposed meteorite. Others worked to improve the precision with which events of that age could be dated. Still others tried to determine more exactly the speed with which extinctions occurred at the Cretaceous–Tertiary boundary. Progress on all three fronts has favored the meteorite theory.

The theory was supported by the discovery of a circular crater 180 km in diameter buried beneath the northern coast of the Yucatán Peninsula of Mexico, thought to have been formed 65 mya. Recent fossil evidence also suggests that there may have been a sudden extinction of organisms 65 mya, as required by the meteorite theory. Therefore, most scientists accept that the collision of Earth with a large meteorite contributed importantly to the mass extinctions at the boundary between the Cretaceous and Tertiary periods.

The Fossil Record

Geological evidence is a major source of information about changes on Earth during the remote past. But the fossils preserved in the rocks—not the rocks themselves—are what have enabled geologists to order those events in time. What are fossils, and what do they tell us about the influence of physical events on the evolution of life on Earth? After examining the conditions that preserve the remains of organ-

A thin band rich in iridium marks the boundary between rocks deposited in the Cretaceous and Tertiary periods.

20.5 *Evidence of a Meteorite Collision with Earth*
Iridium is a metal common in some meteorites, but rare on Earth. Its high concentrations in sediments deposited about 65 million years ago suggest the impact of a large meteorite.

isms, we will consider the completeness of the fossil record, and how that record reveals patterns in life's history.

An organism is most likely to become a fossil if its dead body is deposited in an environment that lacks oxygen. However, most organisms live in oxygenated environments and decompose when they die. Thus many fossil assemblages are collections of organisms that were transported by wind or water to sites that lacked oxygen. Occasionally, however, organisms, or imprints of them, are preserved where they lived. In such cases—especially if the environment in question was a cool, anaerobic swamp, where conditions for preservation were excellent—we can obtain a picture of communities of organisms that lived together.

How complete is the fossil record?

About 300,000 species of fossil organisms have been described, and the number is growing steadily. However, this number is only a tiny fraction of the species that have ever lived. We do not know how many species lived in the past, but we have ways of making reasonable estimates. Of the present-day biota—that is, all living species of all kinds—approximately 1.6 million species have been named. The actual number of living species is probably at least 10 million. It is possibly higher than 50 million, because most species of insects (the animal group with the largest number of species; see Chapter 32) have not yet been described. So the number of known fossil species is less than 2 percent of the probable minimum number of living species.

Because life has existed on Earth for about 3.8 billion years, and because most species exist, on average, for fewer than 10 million years, the species living on Earth must have turned over many times during geological history, and the total number of species that have lived over evolutionary time must vastly exceed the number living today.

The number of known fossils, although small in relation to the total number of extinct species, is higher for some

20.6 A Fossil Spider
Trapped in sap of a tree in what is now Arkansas about 50 mya, this spider is exquisitely preserved in the amber formed from the sap. The details of its external anatomy are clearly visible.

groups of organisms than for others. The record is especially good for marine animals that had hard skeletons. Among the nine major animal groups with hard-shelled members, approximately 200,000 species have been described from fossils, roughly twice the number of living marine species in these same groups. Paleontologists lean heavily on these groups in their interpretations of the evolution of life in the past. Insects and spiders are also relatively well represented in the fossil record (Figure 20.6).

The fossil record demonstrates several patterns

Despite its incompleteness, the fossil record reveals several patterns that are unlikely to be altered by future discoveries. First, great regularity exists. For example, organisms of particular types are found in rocks of specific ages, and new organisms appear sequentially in younger rocks. Second, as we move from ancient periods of geological time toward the present, fossil species increasingly resemble species living today. The fossil record also tells us that extinction is the eventual fate of all species.

The fossil record contains many series of fossils that demonstrate gradual change in lineages of organisms over time. A good example is the series of fossils showing the pathway by which whales evolved from hoofed terrestrial mammals, beginning about 50 mya. Fossils that are intermediate between whales and their terrestrial ancestors illustrate the major changes by which whales became adapted for aquatic existence and lost their hind limbs (Figure 20.7).

Interestingly, whales retain the genetic potential for developing legs; occasionally, living whales have been found with small hind legs that extend outside their bodies. The claim (made repeatedly by scientific creationists) that the fossil record does not contain examples of such intermediates is false. Intermediates abound, and more and more of them are being discovered.

But the incompleteness of the fossil record can mislead us when we try to interpret it. Organisms may have evolved in places where their fossils have not been discovered. Moreover, when a species that evolved in one place appears among the fossils at another site, it gives the false impression that it evolved very rapidly from one of the species that already lived there.

Horses, for example, evolved at varying rates over millions of years in North America. Many different lineages arose and died out (Figure 20.8). Ancestors of horses crossed the Bering land bridge into Asia at several different times, the most recent one only several million years ago. Evidence of each crossing appears suddenly in the Asian fossil record as a major new type of horse. If we lacked fossil evidence of horse evolution in North America, we might conclude that horses evolved very rapidly somewhere in Asia. On the other hand, an incomplete fossil record can also hide rapid changes.

By combining data about physical events during Earth's history with evidence from the fossil record, scientists can compose pictures of what Earth and its inhabitants looked like at different times. We know in general where the continents were and how life changed over time, but many of the details are poorly known, especially for events in the more remote past. In the next section we provide an overview of the major patterns in the history of life on Earth.

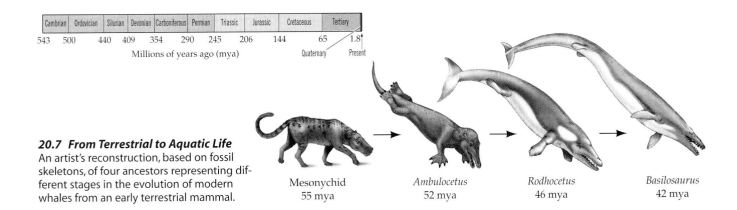

Cambrian	Ordovician	Silurian	Devonian	Carboniferous	Permian	Triassic	Jurassic	Cretaceous	Tertiary
543	500	440	409	354	290	245	206	144	65	1.8

Millions of years ago (mya)   Quaternary   Present

20.7 From Terrestrial to Aquatic Life
An artist's reconstruction, based on fossil skeletons, of four ancestors representing different stages in the evolution of modern whales from an early terrestrial mammal.

Mesonychid
55 mya

Ambulocetus
52 mya

Rodhocetus
46 mya

Basilosaurus
42 mya

Cambrian	Ordovician	Silurian	Devonian	Carboniferous	Permian	Triassic	Jurassic	Cretaceous	Tertiary

543   500   440   409   354   290   245   206   144   65   1.8

Millions of years ago (mya)

Quaternary   Present

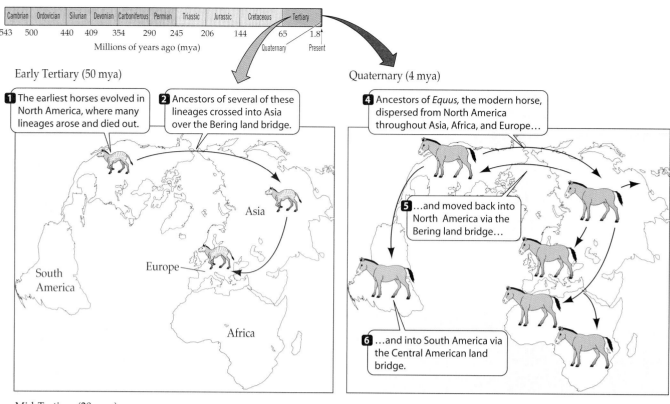

Early Tertiary (50 mya)

1 The earliest horses evolved in North America, where many lineages arose and died out.

2 Ancestors of several of these lineages crossed into Asia over the Bering land bridge.

Asia

South America

Europe

Africa

Quaternary (4 mya)

4 Ancestors of *Equus*, the modern horse, dispersed from North America throughout Asia, Africa, and Europe…

5 …and moved back into North America via the Bering land bridge…

6 …and into South America via the Central American land bridge.

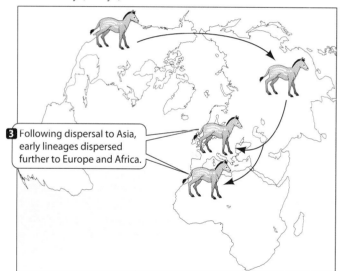

Mid-Tertiary (20 mya)

3 Following dispersal to Asia, early lineages dispersed further to Europe and Africa.

20.8 Horses Have a Complex Evolutionary History
Ancestors of horses crossed the Bering land bridge into Asia several times, the last one only a few million years ago. If we lacked the earlier fossil evidence of horse evolution in North America, we might reach the false conclusion that horses evolved rapidly somewhere in Asia.

Life in the Remote Past

Life first evolved about 3.8 billion years ago (bya). The major groups of eukaryotic organisms evolved during the Precambrian, about 2.5 bya. The fossil record of organisms that lived prior to the Cambrian period is fragmentary, but shows that the volume of organisms increased dramatically in late Precambrian times, about 650 mya (see Table 20.1). The shallow Precambrian seas teemed with life. Protists and small multicellular animals fed on floating algae. Living plankton and plankton remains were devoured by animals that filtered food from the water or ingested sediments and digested the organic material in them.

20.9 Ediacaran Animals
These fossils of soft-bodied invertebrates, excavated at Ediacara in southern Australia, formed 600 million years ago. They illustrate the diversity of life that evolved in Precambrian times.

The best fossil assemblage of Precambrian animals, all soft-bodied invertebrates, was discovered at Ediacara, in southern Australia (Figure 20.9). The Ediacaran fauna is very different from any assemblage of animals living today. Some of its members may represent animal lineages that have no living descendants.

Diversity exploded during the Cambrian

By the early Cambrian period (543–510 mya), oxygen levels in Earth's atmosphere approached their current concentrations, and the continental plates came together in several masses, the largest of which was Gondwana (Figure 20.10a). The three great evolutionary lineages of animals separated and began to radiate during this period. **Evolutionary radiation**—the proliferation of species within a single lineage—during this time resulted in the dramatic increase in diversity known as the Cambrian explosion. All of the major groups of animals that have species living today appeared during the Cambrian, as did animals belonging to many lineages that have left no surviving descendants.

The most extensive fossil evidence from the Cambrian period comes from an unusually well preserved fauna recently discovered in China (Figure 20.10b). Arthropods are the most diverse group in the Chinese fauna; some of them were large carnivores. A mass extinction occurred at the end of the Cambrian.

Major changes continued during the Paleozoic era

Because they have excellent fossil evidence and can date events relatively precisely, geologists have divided the remainder of the Paleozoic era into five periods: the Ordovician, Silurian, Devonian, Carboniferous, and Permian periods (see Table 20.1).

THE ORDOVICIAN (510–440 MYA). During the Ordovician period, the continents were located primarily in the Southern Hemisphere. Evolutionary radiation of marine organisms was spectacular during the early Ordovician, especially among animals (such as brachiopods and mollusks) that filter small prey from the water. All animals lived on the seafloor or burrowed in its sediments. Ancestors of club mosses and horsetails colonized wet terrestrial environments, but they were still relatively small. At the end of the Ordovician, sea levels dropped about 50 meters as massive glaciers formed over Gondwana, and ocean temperatures dropped. About 75 percent of the marine animal species became extinct, probably because of these major environmental changes.

THE SILURIAN (440–409 MYA). During the Silurian period, the northern continents coalesced, but the general positions of the continents did not change much. Marine life rebounded from the mass extinction at the end of the Ordovician. Animals able to swim and feed above the ocean bottom appeared for the first time, but no major new groups of marine organisms evolved. The tropical sea was uninterrupted by land barriers, and most marine genera were widely distributed. On land, the first terrestrial arthropods—scorpions and millipedes—appeared.

THE DEVONIAN (409–354 MYA). Rates of evolutionary change accelerated in many groups of organisms during the Devonian period. Northern and southern land masses slowly moved northward (Figure 20.11a). There was a great evolutionary radiation of corals and shelled squidlike cephalopods (Figure 20.11b). Fishes diversified as jawed forms replaced jawless ones, and heavy armor gave way to the less rigid outer coverings of modern fishes. All current major groups of fishes were present by the end of the period.

Terrestrial communities also changed dramatically during the Devonian. Club mosses, horsetails, and tree ferns became common, and some reached the size of trees. Their deep roots accelerated the weathering of rocks, resulting in the development of the first forest soils. Distinct floras

(a)

Cambrian	Ordovician	Silurian	Devonian	Carboniferous	Permian	Triassic	Jurassic	Cretaceous	Tertiary
543	500	440	409	354	290	245	206	144	65 / 1.8

Millions of years ago (mya)

Quaternary   Present

North Pole

The view of Earth has been distorted here so that you can see both poles.

South Pole

This group of land masses is gradually moving together to form Gondwana.

(b)

20.10 Cambrian Continents and Animals
(a) Positions of the continents during mid-Cambrian times (543–510 mya). (b) Fossil beds in China have yielded excellent remains of Cambrian animals including *Jianfangia*, a predatory arthropod.

Cambrian	Ordovician	Silurian	Devonian	Carboniferous	Permian	Triassic	Jurassic	Cretaceous	Tertiary
543	500	440	409	354	290	245	206	144	65	1.8

Millions of years ago (mya)   Quaternary   Present

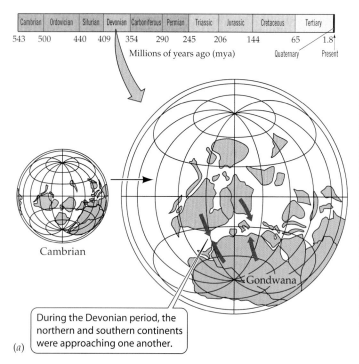

Cambrian

Gondwana

During the Devonian period, the northern and southern continents were approaching one another.

(a)

(b)

20.11 Devonian Continents and Marine Communities
(a) Positions of the continents during the Devonian period (409–354 mya). (b) This artist's reconstruction shows how a Devonian reef may have appeared.

evolved on the two land masses toward the end of the period, and the first gymnosperms appeared. The first known fossils of centipedes, spiders, pseudoscorpions, mites, and insects date to this period, and fishlike amphibians began to occupy the land.

An extinction of about 75 percent of all marine species marked the end of the Devonian. Paleontologists disagree on the cause of this mass extinction. Some believe that it was triggered by the collision of the two continents, which destroyed much of the existing shallow, warm-water marine environment. This hypothesis is supported by the fact that extinction rates were much higher among tropical than among cold-water species.

THE CARBONIFEROUS (354–290 MYA). Large glaciers formed over high-latitude Gondwana during the Carboniferous period, but extensive swamp forests grew on the tropical continents. These forests were not made up of the kinds of trees we know today, but were dominated by giant tree ferns and horsetails (see Figure 28.9). Fossilized remains of those "trees" formed the coal that we now mine for energy.

The diversity of terrestrial animals increased greatly. Snails, scorpions, centipedes,

20.12 A Carboniferous "Crinoid Meadow"
Crinoids, which were dominant marine animals during the Carboniferous (354–290 mya), may have formed communities that looked like this. Sharks and bony fishes were important members of these communities.

and insects were abundant and diverse. Insects evolved wings, which gave them access to tall plants; plant fossils from this period show evidence of insect damage. Amphibians became larger and better adapted to terrestrial existence. From one amphibian stock, the first reptiles evolved late in the period. In the seas, crinoids reached their greatest diversity, forming meadows on the seafloor (Figure 20.12).

THE PERMIAN (290–245 MYA). During the Permian period, the continents coalesced into a supercontinent—Pangaea. Massive volcanic eruptions resulted in outpourings of lava that covered large areas of Earth (Figure 20.13). The ash they produced blocked the sunlight, cooling the climate and resulting in the largest glaciers in Earth's history.

Permian deposits contain representatives of most modern groups of insects. By the end of the period, reptiles greatly outnumbered amphibians. Late in the period, the lineage leading to mammals diverged from one reptilian lineage. In fresh waters, the Permian period was a time of extensive radiation of bony fishes.

Cambrian	Ordovician	Silurian	Devonian	Carboniferous	Permian	Triassic	Jurassic	Cretaceous	Tertiary
543	500	440	409	354	290	245	206	144	65	1.8

Millions of years ago (mya)   Quaternary   Present

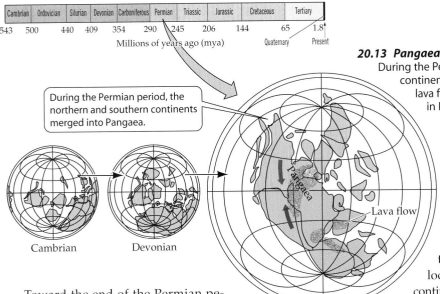

Cambrian	Ordovician	Silurian	Devonian	Carboniferous	Permian	Triassic	Jurassic	Cretaceous	Tertiary

543   500   440   409   354   290   245   206   144   65   1.8
Millions of years ago (mya)   Quaternary Present

During the Permian period, the northern and southern continents merged into Pangaea.

Cambrian   Devonian

Pangaea

Lava flow

20.13 Pangaea Formed in the Permian Period
During the Permian (290–245 mya), the interior of the "supercontinent" Pangaea experienced harsh climates. Massive lava flows spread over Earth, and the largest glaciers in Earth's history formed during this period.

Toward the end of the Permian period, two events may have caused separate mass extinctions. The first event was the massive outpouring of volcanic lava, which drastically reduced the oxygen content of deep ocean waters. The second was a rapid turnover of the oceans that brought oxygen-depleted deep waters to the surface. These waters released toxic concentrations of carbon dioxide and hydrogen sulfide into surface waters and the atmosphere, poisoning most species.

Geographic differentiation increased during the Mesozoic era

At the start of the Mesozoic era (250 mya), the few surviving organisms found themselves in a relatively empty world. As Pangaea slowly separated into individual continents, the glaciers melted, and the oceans rose and reflooded the continental shelves, forming huge, shallow inland seas. Life again proliferated and diversified, but different lineages came to dominate Earth. The large plants that dominated the great coal-forming forests, for example, were replaced by new plant lineages in which seeds had evolved.

During the Mesozoic, Earth's biota, which until that time had been relatively homogeneous, became increasingly provincialized. Distinct terrestrial floras and faunas evolved on each continent. The biotas of the shallow waters bordering the continents also diverged from one another. The localization that began during the Mesozoic continues to influence the geography of life today.

THE TRIASSIC (245–206 MYA).   During the Triassic period, many invertebrate lineages became more diverse, and many burrowing forms evolved from groups living on the surfaces of bottom sediments. On land, conifers and seed ferns became the dominant trees. The first frogs and turtles appeared. A great radiation of reptiles began, which eventually gave rise to dinosaurs, crocodilians, and birds. The end of the Triassic was marked by a mass extinction that eliminated about 65 percent of species on Earth. Why they went extinct is not known, but a meteor impact is suspected.

THE JURASSIC (206–144 MYA).   The mass extinction at the close of the Triassic was followed by another period of evolutionary diversification during the Jurassic period. Bony fishes began the great radiation that culminated in their dominance of the oceans. Salamanders and lizards first appeared. Flying reptiles evolved, and dinosaur lineages evolved into bipedal predators and large quadrupedal herbivores (Figure 20.14). Several groups of mammals first appeared during this time.

20.14 Mesozoic Dinosaurs
The dinosaurs of the Mesozoic era continue to capture our imagination. This painting illustrates some of the large species from the Jurassic period (206–144 mya).

Cambrian	Ordovician	Silurian	Devonian	Carboniferous	Permian	Triassic	Jurassic	Cretaceous	Tertiary

543   500   440   409   354   290   245   206   144   65   1.8
Millions of years ago (mya)   Quaternary Present

THE CRETACEOUS (144–65 MYA).   By the early Cretaceous period, the northern continents were completely separate from the southern ones, and a continuous sea encircled the Tropics (Figure 20.15). Sea levels were high, and Earth was warm and humid. Life proliferated both on land and in the oceans. Marine invertebrates increased in variety and number of species. On land, dinosaurs continued to diversify. The first snakes appeared during the Cretaceous, though their lineage did not radiate until much later.

Early in the Cretaceous, flowering plants—the angiosperms—evolved from gymnosperm ancestors and began the radiation that led to their current dominance on land. By the end of the period, many groups of mammals had evolved, but these mammals were generally small.

Another mass extinction took place at the end of the Cretaceous period. On land, all vertebrates larger than about 25 kg in body weight apparently became extinct. In the seas, many planktonic organisms and bottom-dwelling invertebrates became extinct. This mass extinction was probably caused by the large meteorite that collided with Earth off the Yucatán Peninsula, as described on page 384.

The modern biota evolved during the Cenozoic era

By the early Cenozoic era (65 mya), the positions of the continents resembled those of today, but Australia was still attached to Antarctica, the Atlantic Ocean was much narrower, and the northern continents were connected. The Cenozoic era was characterized by an extensive radiation of mammals, but other groups were also undergoing important changes. Flowering plants diversified extensively and dominated world forests, except in cool regions.

THE TERTIARY (65–1.8 MYA).   During the Tertiary period, Australia began its northward drift. By 20 mya it had nearly reached its current position. The map of the world during this period looks familiar to us. In the middle of the Tertiary, the climate became considerably drier and cooler. Many lineages of flowering plants evolved herbaceous

(nonwoody) forms, and grasslands spread over much of Earth.

By the beginning of the Cenozoic era, invertebrate faunas resembled those of today. It is among the vertebrates that evolutionary changes during the Tertiary were most rapid. Living groups of reptiles, including snakes and lizards, underwent extensive radiations during this period, as did birds and mammals.

THE QUATERNARY (1.8 MYA–PRESENT).   The current geological period, the Quaternary period, is subdivided into two epochs, the Pleistocene and the Holocene (also known as the Recent). The Pleistocene epoch, which began about 1.8 mya, was a time of drastic cooling and climatic fluctuations. During four major and about twenty minor episodes, massive glaciers spread across the continents. Earth became much cooler, and animal and plant populations shifted toward the equator. The last of these glaciers retreated from temperate latitudes less than 15,000 years ago. Organisms of the current Holocene epoch are still adjusting to these changes; many high-latitude ecological communities have occupied their current locations for no more than a few thousand years.

Interestingly, these climate fluctuations resulted in few extinctions. However, the Pleistocene was the scene of hominid evolution and radiation, resulting in the species *Homo sapiens*—modern humans (see Chapter 33). Many large birds and mammals became extinct in North and South America and in Australia when *H. sapiens* arrived on those continents. Human hunting may have caused these extinctions, although the existing evidence does not convince all paleontologists.

Rates of Evolutionary Change

Following each mass extinction, the diversity of life rebounded. How fast did evolution proceed during those times? Why did some lineages evolve rapidly while others

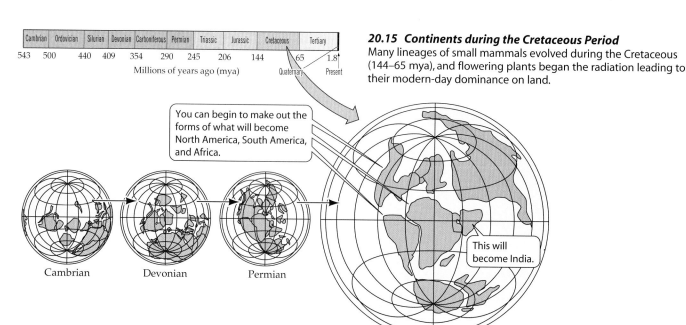

| Cambrian | Ordovician | Silurian | Devonian | Carboniferous | Permian | Triassic | Jurassic | Cretaceous | Tertiary |

543   500   440   409   354   290   245   206   144   65   1.8

Millions of years ago (mya)   Quaternary   Present

20.15 Continents during the Cretaceous Period
Many lineages of small mammals evolved during the Cretaceous (144–65 mya), and flowering plants began the radiation leading to their modern-day dominance on land.

You can begin to make out the forms of what will become North America, South America, and Africa.

This will become India.

Cambrian   Devonian   Permian

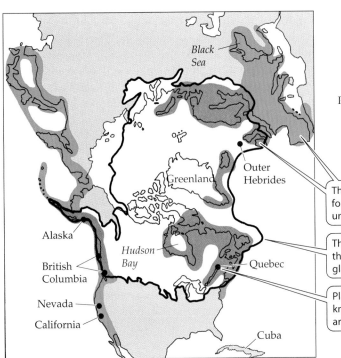

20.16 Natural Selection Acts on Stickleback Spines
Three-spined stickleback populations with reduced spines are found principally in young lakes that were covered by ice during the most recent glacial period. These lakes lack large predatory fishes, but contain predatory insects that capture the fish by grasping their spines.

The current range of sticklebacks includes formerly glaciated areas (lavender) and unglaciated areas (green).

The region of the Northern Hemisphere that was once covered by Pleistocene glaciers is outlined in black.

Places where sticklebacks are known to have reduced spines are indicated by circles.

did not? Scientists have made enough progress in studying evolution to be able to give at least tentative answers to these questions.

Evolutionary rates vary

The fossil record shows that rates of evolution have been uneven. Many species have experienced times of **stasis**, long periods during which they changed very little. For example, many marine lineages have evolved slowly. Horseshoe crabs that lived 300 mya are almost identical in appearance to those living today, and the chambered nautiluses of the late Cretaceous are indistinguishable from living species. Such "living fossils" are found today in harsh environments that have changed relatively little for millennia. The sandy coastlines where horseshoe crabs spawn have extremes in temperature and salt concentration that are lethal to many other organisms. Chambered nautiluses spend their days in deep, dark ocean waters, ascending to feed in food-rich surface waters only under the protective cover of darkness. Their intricate shells provide little protection against today's visually hunting fishes.

Periods of stasis may be broken by times during which changes, either in the physical or the biological environment, create conditions that favor new traits. How new conditions favor rapid evolutionary change is illustrated by the spines of the three-spined stickleback (*Gasterosteus aculeatus*). This widespread marine fish has repeatedly invaded fresh water throughout its evolutionary history (Figure 20.16).

Sticklebacks are tiny fish, usually less than 10 cm long. All marine and most freshwater populations have well-developed pelvic girdles with prominent spines that make it difficult for other fishes to swallow them. However, large predatory insects can readily grasp the stickleback's spines, and prey selectively on stickleback individuals with the largest spines. When stickleback populations invade freshwater habitats where predatory fish are absent but predatory insects are present, they rapidly evolve smaller spines. Populations with reduced spines are found primarily in young lakes that were covered by ice during the most recent glaciation, and hence do not have large predatory fishes.

The extensive fossil record of sticklebacks shows that spine reduction evolved many times in different populations that invaded fresh water. In addition, molecular data reveal that each freshwater population is most closely related to an adjacent marine population, not to other freshwater populations. Therefore, spine reduction has evolved rapidly many times in different places in response to the same ecological situation: the absence of predatory fish.

Extinction rates vary over time

More than 99 percent of the species that have ever lived are extinct. Species have become extinct throughout the history of life, but extinction rates have fluctuated dramatically over time; some groups had high extinction rates while others were proliferating.

Each mass extinction changed the flora and fauna of the next period by selectively eliminating some types of organisms, thereby increasing the relative abundance of others. For example, among the seashells of the Atlantic coastal plain of North America, species with broad geographic ranges were less likely to become extinct during normal periods (when no mass extinctions were taking place) than were species with small geographic ranges.

On the other hand, during the mass extinction of the late Cretaceous, groups of closely related species with large geographic ranges survived better than groups with small ranges, even if the individual species within the group had small ranges. Similar patterns are found in other molluscan groups elsewhere, suggesting that traits favoring long-term survival during normal times are often different from those that favor survival during times of mass extinctions.

At the end of the Cretaceous period, extinction rates on land were much higher among large vertebrates than among small ones. The same was true during the Pleistocene mass extinction, when extinction rates were high only among large mammals and large birds. During some mass extinctions, marine organisms were heavily hit while terrestrial organisms survived well. Other extinctions affected organisms living in both environments. These differences are not surprising, given that major changes on land and in the oceans did not always coincide.

Patterns of Evolutionary Change

Major new features, such as the feathers of birds or the legs of terrestrial vertebrates, that adapt organisms to a special way of life are called **evolutionary innovations**. How such novelties arise has been the subject of much debate from Darwin's time to the present. The variety of sizes and shapes among living organisms seems almost limitless, but the number of truly novel structures is remarkably small. As fiction writers often do, we can imagine unusual vertebrates with wings sprouting from their backs, but in reality the wings of vertebrates are always modified front legs. Modern mammals are highly varied in their shapes, but all

of their structures are modifications of structures found in ancestral mammals. As we saw earlier, even transforming a terrestrial mammal into a whale did not require a drastic reorganization of the mammalian body plan. Only a few evolutionary innovations, such as the notochord of chordates, do not appear to be modifications of a preexisting structure.

Three major faunas have dominated animal life on Earth

Only three events during the evolution of life have resulted in the evolution of major new faunas (Figure 20.17). The first one, the Cambrian explosion, took place about 540 mya. The second, about 60 million years later, resulted in the Paleozoic fauna. The great Permian extinctions 300 million years later were followed by the third event, the Triassic explosion, which led to our modern fauna.

During the Cambrian explosion, organisms representative of all major present-day lineages appeared, along with a number of lineages that subsequently became extinct. The Paleozoic and Triassic explosions greatly increased the number of families, genera, and species, but no new or dramatically different organismal body plans evolved. The later explosions resulted in many new organisms, but all of them were modifications of body plans that were already present when these great biological diversifications began.

Biologists have long puzzled over the striking differences between the Cambrian explosion and the two later explosions. A commonly accepted theory is that because the Cambrian explosion took place in a world that con-

20.17 Evolutionary Faunas
Representatives of the three major evolutionary faunas are shown together with a graphic illustration of the number of families in each fauna over time.

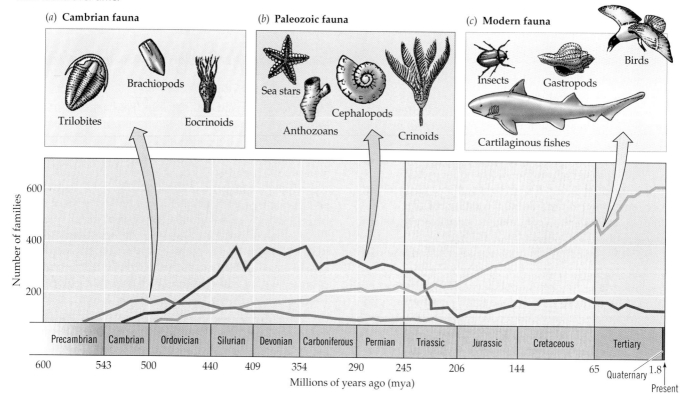

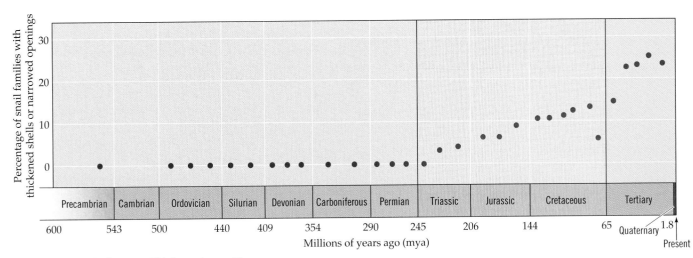

20.18 Snail Shells Have Thickened over Time
The percentage of families of snails that have internally thickened or narrowed openings to their shells has increased with evolutionary time—evidence that predation on shelled animals intensified.

tained only a few species of organisms, all of which were small, the ecological setting was favorable for the evolution of many new body plans and different ways of life. Many types of organisms were able to survive initially in this world, but as competition intensified and new types of predators evolved, many forms were unable to persist.

Although Earth was relatively poor in species at the time of the two later explosions as well, the species that were already present included a wide array of body plans and ways of life. As body plans became more specialized, major transformations of form became increasingly less likely. Therefore, major new innovations were less likely to evolve at these times than in the Cambrian.

The size and complexity of organisms have increased

The earliest organisms were small prokaryotes. A modest increase in size and a dramatic increase in structural complexity accompanied the evolution of the first eukaryotes 2.5 billion years ago. Since then, the maximum sizes of organisms in many lineages have increased, irregularly to be sure. The most striking exception to this trend is insects, which have remained relatively small throughout their evolutionary history.

The overall increase in body size is the result of two opposing forces. Within a species, selection often favors larger size because larger individuals can dominate smaller ones. But larger species on average survive for less time than small species do, which is one reason why Earth is not populated primarily by large organisms.

Predators have become more efficient

Over time, predators have evolved increasingly efficient methods of capturing prey, and prey, in turn, have evolved better defenses. During the Cretaceous, for example, many species of crabs with powerful claws evolved, and carnivorous marine snails able to drill holes in shells began to fill the seas. Skates, rays, and bony fishes with powerful teeth capable of crushing mollusk shells also evolved, and large, powerful marine reptiles—the placodonts—fed heavily on clams. The increasing thickness and narrowing openings of snail shells during the Cretaceous is evidence that predation rates intensified (Figure 20.18). Other evidence of heavy predation pressure is the increase in the percentage of fossil shells that show signs of having been repaired following an attack that did not kill the owner.

Although shell thickness provided some protection from predators, predators were so effective that clams disappeared from the surfaces of most marine sediments. The survivors were species that burrowed into the substratum, where they were more difficult to capture.

The Future of Evolution

The agents of evolution are operating today just as they have been since life first appeared on Earth. However, major changes are under way as a result of the dramatic increase of Earth's human population. Until recently, human-caused extinctions affected mostly large vertebrates, but these losses are now being compounded by increasing extinctions of small species, driven primarily by changes in Earth's vegetation. Deliberately or inadvertently, people are moving thousands of species around the globe, reversing the provincialization of Earth's biota that evolved during the Mesozoic era.

Humans have also taken charge of the evolution of certain valuable species by means of artificial selection and biotechnology. Our ability to modify species has been enhanced by modern molecular methods that enable us to move genes among species—even distantly related ones. In short, humans have become the dominant evolutionary agent on Earth today. How we handle our massive influence will powerfully affect the future of life on Earth.

Chapter Summary

▶ Changes that take effect during the lifetimes of species constitute microevolution. Changes that involve the appearance of new species and evolutionary lineages are called macroevolution.

How Do We Know that Earth is Ancient?

▶ The relative ages of rock layers in Earth's crust can be determined from their positions relative to one another and from their embedded fossils.

▶ Radioisotopes supplied the key for assigning absolute ages to rocks.

▶ Earth's geological history is divided into eras and periods. The boundaries between these units are based on differences between their fossil biotas. **Review Table 20.1**

How Has Earth Changed over Time?

▶ Unidirectional physical changes on Earth include gradual cooling and weakening of the forces that cause continental drift.

▶ Earth's early atmosphere lacked free oxygen. Oxygen accumulated after prokaryotes evolved the ability to use water as their source of hydrogen ions in photosynthesis. Increasing concentrations of atmospheric oxygen made possible the evolution of eukaryotes and multicellular organisms. **Review Figure 20.2**

▶ Throughout Earth's history the continents have moved about, sometimes separating from one another, at other times colliding. **Review Figures 20.10, 20.11, 20.13, 20.15**

▶ Earth has experienced periods of rapid climate change, massive volcanism, and major shifts in sea levels and ocean currents, all of which have had dramatic effects on the evolution of life. **Review Figures 20.3, 20.4**

▶ External events, such as collisions with meteorites, also have changed conditions on Earth. A meteorite may have caused the abrupt mass extinction at the end of the Cretaceous period.

The Fossil Record

▶ Much of what we know about the history of life on Earth comes from the study of fossils.

▶ The fossil record, although incomplete, reveals broad patterns in the evolution of life. About 300,000 fossil species have been described. The best record is that of hard-shelled marine animals.

▶ Fossils show that many evolutionary changes are gradual, but an incomplete record can falsely suggest or conceal times of rapid change. **Review Figures 20.7, 20.8**

Life in the Remote Past

▶ The fossil record for Precambrian times is fragmentary, but fossils from Australia show that many lineages that evolved then may not have left living descendants.

▶ Diversity exploded during the Cambrian period. **Review Figure 20.10**

▶ Geographic differentiation of biotas increased during the Mesozoic era.

▶ The modern biota evolved during the Cenozoic era.

Rates of Evolutionary Change

▶ Rates of evolutionary change have been very uneven.

▶ Rapid rates of evolution occur when changes to the physical or biological environment create conditions that favor new traits. **Review Figure 20.16**

Patterns of Evolutionary Change

▶ Truly novel features of organisms have evolved infrequently. Most evolutionary changes are the result of modifications of already existing structures.

▶ Three major faunas have dominated animal life on Earth. **Review Figure 20.17**

▶ Over evolutionary time, organisms have increased in size and complexity. Predation rates have also increased, resulting in the evolution of better defenses among prey species. **Review Figure 20.18**

The Future of Evolution

▶ The agents of evolution continue to operate today, but human intervention, both deliberate and inadvertent, now plays an unprecedented role in the history of life.

For Discussion

1. Some lineages of organisms have evolved to contain large numbers of species, whereas others have produced only a few species. Is it meaningful to consider the former more successful than the latter? What does the word "success" mean in evolution? How does your answer influence your thinking about *Homo sapiens*, the only surviving representative of the Hominidae—a family that never had many species in it?

2. Scientists date ancient events using a variety of methods, but nobody was present to witness or record those events. Accepting those dates requires us to believe in the accuracy and appropriateness of indirect measurement techniques. What other basic scientific concepts are based on the results of indirect measurement techniques?

3. Why is it useful to be able to date past events absolutely as well as relatively?

4. What factors favor increases in body size? Why might average body size among particular species in a lineage decrease even if natural selection favors larger body size in most species of that lineage?

5. The continents are still drifting today, but biologists ignore these movements when thinking about factors affecting current evolutionary changes. On what basis do they make that decision?

21

The Mechanisms of Evolution

 MOST SPECIES OF CUCKOOS AND COWBIRDS LAY their eggs in the nests of other species of birds. This behavior is known as brood parasitism. The host birds often incubate the eggs and raise the parasite nestlings. Host birds that accept parasite eggs and raise parasite chicks are likely to produce fewer offspring than hosts that recognize parasite eggs and push them out of the nest.

To investigate the evolution of such defensive behaviors, biologists studied cuckoos and their hosts in areas where brood parasitism had been occurring for different periods of time. In one valley in southern Spain, great spotted cuckoos and common magpies have lived together for many centuries. Here 78 percent of the magpies removed artificial cuckoo eggs experimenters placed in their nests. However, in another Spanish valley, where cuckoos did not arrive until the early 1960s, only 14 percent of magpies ejected the eggs.

Ejection of parasites' eggs evolved rapidly in Japan, where the ranges of the common cuckoo and the azure-winged magpie have only recently overlapped. In a region where cuckoos have parasitized magpies for 10 years, none of the magpies ejected cuckoo eggs, but in areas where they have been parasitized for 20 years, 42 percent of the magpies ejected cuckoo eggs.

What explains these differences in magpie behavior? Charles Darwin's main contribution to biology was to propose a plausible and testable hypothesis of a mechanism of evolutionary change that could result in the adaptation of organisms to their environments. Keep in mind that the environment of an organism includes the physical environment, individuals of other species, and individuals of the same species. All of these components influence the survival and reproductive success of individuals.

In this chapter we will review how Darwin developed his ideas, and then turn to the advances in understanding of evolutionary processes since Darwin's time. We will discuss the genetic basis of evolution and show how genetic variation within populations is measured. We will describe the agents of evolution and show how biologists design studies to investigate them. Finally, we will discuss constraints on the pathways evolution can take. When you understand these processes, you will understand the mechanisms of evolution.

Charles Darwin and Adaptation

The term **adaptation** has two meanings in evolutionary biology. The first refers to *traits* that enhance the survival and reproductive success of their bearers. For example, we believe that wings are adaptations for flight, a spider's web is an adaptation for capturing flying insects, and so forth. The second refers to the *process* by which these traits are acquired—that is, the evolutionary mechanisms that produced them.

Biologists regard an organism as being adapted to a particular environment when they can imagine—or better still, measure the performance of—a slightly different organism that reproduces and survives less well in that environment. That is, adaptation is a relative concept; to understand adaptation, biologists compare the performance of individuals that differ in traits within and among species. For example, to investigate the adaptive nature of spiders' webs, we would try to determine the effectiveness of slightly different webs spun by a given species in capturing insects. We would also measure changes in the webs of the species in different situations. With these data, we could understand how variations in web structure influence the survival and reproductive success of their makers.

A Magpie and a Cuckoo Chick
In parts of Japan, the azure-winged magpie has only recently experienced brood parasitism by the common cuckoo. This adult magpie will care for the cuckoo chick at the expense of its own offspring.

Darwin proposed a mechanism to explain adaptation

Charles Darwin was a keen naturalist who observed many examples of structures and behaviors that seemed to be designed to assist the survival and reproductive success of their bearers. He was given an unprecedented opportunity to study the adaptations of organisms in various parts of the world when in 1831, his Cambridge University botany professor, John Henslow, recommended him as a naturalist to Captain Robert Fitzroy, who was preparing to sail around the world on the survey ship *H.M.S. Beagle* (Figure 21.1). Whenever possible during the voyage, Darwin (who was often seasick) went ashore to observe and collect specimens of plants and animals.

Darwin spent most of his time ashore in South America, where the species he saw differed strikingly from those of Europe. He also noted that the species of the temperate regions of South America (Argentina and Chile) were more similar to those of tropical South America (Brazil) than they were to European species. When he explored the Galápagos archipelago, west of Ecuador, he noted that most of its animal species were found nowhere else, but were similar to those of the mainland of South America, 1,000 kilometers to the east. Darwin also observed that the animals of the archipelago differed from island to island. He postulated that some animals had dispersed from mainland South America and then evolved differently on different islands.

When he returned to England in 1836, Darwin continued to ponder his observations. Within a decade he had developed the main features of his theory of evolution, which had two major components:

▶ Species are not immutable, but change, or *adapt*, over time. (In other words, Darwin asserted that evolution is a historical fact.)
▶ The agent that produces the changes is *natural selection*.

Darwin wrote a long essay on natural selection and the origin of species in 1844, but, despite urging from his wife and colleagues, he was reluctant to publish it, preferring to assemble more evidence first.

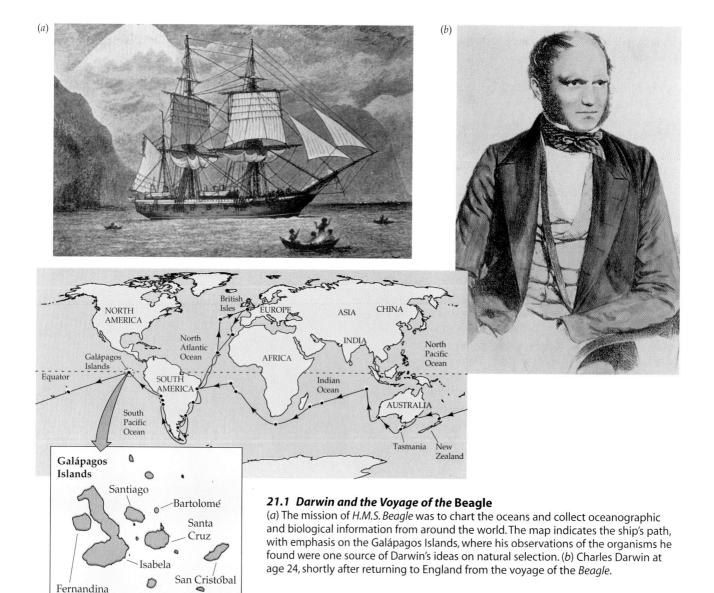

21.1 Darwin and the Voyage of the Beagle
(a) The mission of *H.M.S. Beagle* was to chart the oceans and collect oceanographic and biological information from around the world. The map indicates the ship's path, with emphasis on the Galápagos Islands, where his observations of the organisms he found were one source of Darwin's ideas on natural selection. (b) Charles Darwin at age 24, shortly after returning to England from the voyage of the *Beagle*.

21.2 Many Types of Pigeons Have Been Produced by Artificial Selection
Charles Darwin raised pigeons as a hobby, and he saw similar forces at work in artificial and natural selection. These are just some of over 300 varieties of pigeons that have been artificially selected by breeders to display different forms of traits such as color, size, and feather distribution.

Darwin's hand was forced in 1858 when he received a letter from Alfred Russel Wallace, who was studying plants and animals in the Malay Archipelago. Wallace asked Darwin to evaluate an enclosed manuscript, in which Wallace proposed a theory of natural selection almost identical to Darwin's. At first, Darwin was dismayed, believing that he had been preempted by Wallace. But extracts from Darwin's 1844 essay, together with Wallace's manuscript, were presented to the Linnaean Society of London on July 1, 1858, thereby giving credit for the idea to both men. Darwin then worked quickly to finish *The Origin of Species*, which was published the next year. Although both men conceived of natural selection independently, Darwin developed his ideas first, and his book provided a much more thorough justification of the concept—which is why natural selection is more closely associated with his name.

The facts that Darwin used to develop his theory of evolution by natural selection were familiar to most contemporary biologists. His insight was to perceive the significance of relationships among them. Darwin understood that populations of all species have the potential for exponential increases in numbers. To illustrate this point, he used the following example:

> Suppose … there are eight pairs of birds, and that only four pairs of them annually … rear only four young, and that these go on rearing their young at the same rate, then at the end of seven years (a short life, excluding violent deaths for any bird) there will be 2,048 birds instead of the original sixteen.

Yet such rates of increase are rarely seen in nature. Therefore, Darwin knew that death rates in nature must be high. Without high death rates, even the most slowly reproducing species would quickly reach enormous population sizes.

Darwin also observed that, although offspring tend to resemble their parents, the offspring of most organisms are not identical to one another or to their parents. He suggested that slight variations among individuals significantly affect the chance that a given individual will survive and the number of offspring it will produce. He called this differential reproductive success of individuals **natural selection**. Natural selection results from both differential survival and differential reproduction of individuals.

Darwin may have used the words "natural selection" because he was familiar with the artificial selection practices of animal and plant breeders. Many of Darwin's observations on the nature of variation came from domesticated plants and animals. Darwin was a pigeon breeder, and he knew firsthand the astonishing diversity in color, size, form, and behavior that could be achieved by humans selecting which pigeons to mate (Figure 21.2). He recognized close parallels between selection by breeders and selection in nature.

Darwin argued his case for natural selection in *The Origin of Species*:

> How can it be doubted, from the struggle each individual has to obtain subsistence, that any minute variation in structure, habits or instincts, adapting that individual better to the new conditions, would tell upon its vigour and health? In the struggle it would have a better chance of surviving; and those of its offspring which inherited the variation, be it ever so slight, would have a better chance.

That statement, written more than a hundred years ago, still stands as a good expression of the idea of evolution by natural selection.

Since Darwin wrote these words, biologists have developed a much deeper understanding of the genetic basis of evolutionary change and have assembled a rich array of examples of natural selection in action.

What have we learned about evolution since Darwin?

When Darwin proposed his theory of natural selection, he had no examples of selection operating in nature. He based his arguments on the results of selection on domesticated species. Since Darwin's time, many studies of the action of natural selection have been conducted; we will discuss some of them in this chapter.

We now know that biological evolution is a change over time in the genetic composition of a population. Darwin understood the importance of heredity for his theory, but he knew nothing of the mode of inheritance. He devoted considerable time to an attempt to develop a theory of heredity, but he failed to discover the laws of heredity, and he failed to understand the significance of Gregor Mendel's paper (see Chapter 10), which he apparently read.

Fortunately, the rediscovery of Mendel's publications in the 1900s paved the way for the development of **population genetics**, the field that provides a major underpinning for Darwin's theory. Population geneticists apply Mendel's laws to entire populations of organisms. They also study variation within and among species in order to understand the processes that result in evolutionary changes in species through time.

Genetic Variation within Populations

For a population to evolve, its members must possess variation, which is the raw material on which agents of evolution act. In everyday life, we do not directly observe the genetic composition of organisms or populations. What we do see in nature are *phenotypes*, the physical expressions of organisms' genes. The agents of evolution actually act on phenotypes, but for the moment we will concentrate on genetic variation within populations. We do so because genes are what is passed on to offspring via reproductive cells—eggs and sperm.

A *heritable trait* is a characteristic of an organism that is at least partly influenced by the organism's genes. The genetic constitution that governs a trait is called its *genotype*. *A population evolves when individuals with different genotypes survive or reproduce at different rates.*

Recall that different forms of a gene, called *alleles*, may exist at a particular locus. A single individual has only some of the alleles found in the population to which it belongs (Figure 21.3). The sum of all the alleles found in a population constitutes its **gene pool**. The gene pool contains the variation that produces the differing phenotypes on which agents of evolution act.

Fitness is the relative reproductive contribution of genotypes

The reproductive contribution of a genotype or phenotype to subsequent generations relative to the contribution of other genotypes or phenotypes in the same population is called **fitness**. The word "relative" is critical: The absolute number of offspring produced by an individual does not influence allele frequencies in the gene pool. Changes in absolute numbers of offspring are responsible for increases and decreases in the *size* of a population, but the relative success among genotypes within a population is what leads to changes in allele frequencies—that is, to evolution. When we discuss evolution, we talk about survival and reproductive success, because these rates determine how many genes different individuals contribute to subsequent generations.

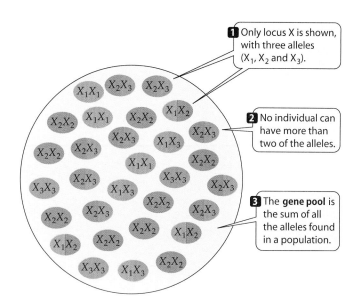

1 Only locus X is shown, with three alleles (X_1, X_2 and X_3).

2 No individual can have more than two of the alleles.

3 The **gene pool** is the sum of all the alleles found in a population.

21.3 A Gene Pool
The allele proportions in this gene pool are 0.20 for X_1, 0.50 for X_2, and 0.30 for X_3.

To contribute genes to subsequent generations, individuals must survive to reproductive age and produce offspring. The relative contribution of individuals of a particular genotype is determined by the probability that those individuals will survive times the average number of offspring they produce over their lifetimes. In other words, *the fitness of a genotype is determined by the average rates of survival and reproduction of individuals with that genotype.*

Most populations are genetically variable

Some level of genetic variation characterizes nearly all natural populations. Such variation has been demonstrated repeatedly for thousands of years by people attempting to develop desirable traits in plants and animals. For example, selection for different traits in a European wild mustard produced many important crop plants (Figure 21.4). Plant and animal breeders can achieve such results only if the original population has genetic variation for the traits of interest. Their success indicates that genetic variation is common, but it does not tell us how much variation there is.

Laboratory experiments also demonstrate that considerable genetic variation is present in most populations. In one such experiment, investigators chose as parents for subsequent generations of fruit flies (*Drosophila*) individuals with either high numbers or low numbers of bristles on their bodies. After 35 generations, flies in both lineages had bristle numbers that fell well outside the range found in the original population (Figure 21.5). These results show that there must have been considerable variation in the original fruit fly population for selection to act on.

To understand evolution, we need to know more precisely how much genetic variation populations contain, the sources of that genetic variation, and how genetic variation is maintained and expressed in populations in space and over time.

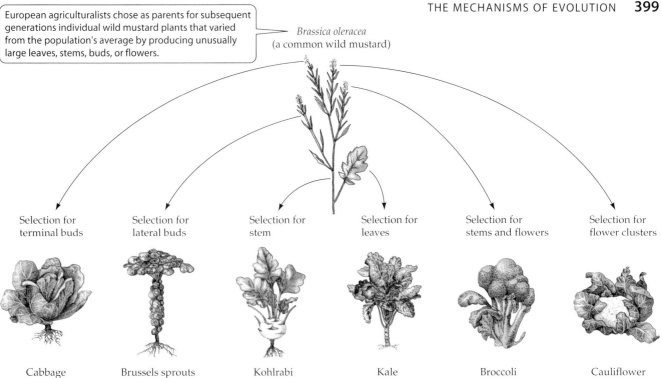

European agriculturalists chose as parents for subsequent generations individual wild mustard plants that varied from the population's average by producing unusually large leaves, stems, buds, or flowers.

Brassica oleracea
(a common wild mustard)

Selection for terminal buds — Cabbage

Selection for lateral buds — Brussels sprouts

Selection for stem — Kohlrabi

Selection for leaves — Kale

Selection for stems and flowers — Broccoli

Selection for flower clusters — Cauliflower

21.4 Many Vegetables from One Species
All of these crop plants have been derived from a single wild mustard species. They illustrate the vast amount of variation that can be present in a gene pool.

How do we measure genetic variation?

A locally interbreeding group within a geographic population is called a **Mendelian population**. Mendelian populations are often the subjects of evolutionary studies. To measure precisely the gene pool of a Mendelian population, we would need to count every allele at every locus in every organism in it. By measuring all the individuals, we could determine the relative proportions, or **frequencies**, of all alleles in the population.

Biologists can reliably estimate allele frequencies for a given locus by measuring numbers of alleles in a sample of individuals from a population. Measures of allele frequency range from 0 to 1; the sum of all allele frequencies at a locus is equal to 1. The frequencies of the different alleles at each locus and the frequencies of different genotypes in a Mendelian population describe its genetic structure.

An allele's frequency is calculated using the following formula:

$$p = \frac{\text{number of copies of the allele in the population}}{\text{sum of alleles in the population}}$$

If only two alleles (for example, *A* and *a*) for a given locus are found among the members of a diploid population, they may combine to form three different genotypes: *AA*,

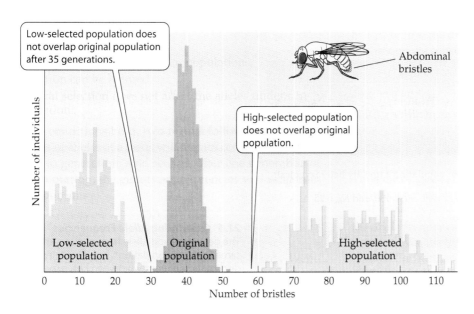

Low-selected population does not overlap original population after 35 generations.

High-selected population does not overlap original population.

Abdominal bristles

Number of individuals

Low-selected population

Original population

High-selected population

0   10   20   30   40   50   60   70   80   90   100   110
Number of bristles

21.5 Artificial Selection Reveals Genetic Variation
In laboratory experiments with *Drosophila*, changes in bristle number evolved rapidly when selected for artificially.

EXPERIMENT

Question: Have spider webs evolved to be efficient capturers of insects?

METHOD   **A computer experiment**

1. Write computer rules for web building based on observations of spider behavior.
2. Assign different rules to each of six "virtual" spiders to spin webs.
3. Bombard the webs with the same random distribution of "virtual" flies.

4. Count the number of flies caught.
5. Calculate quality of web as number of flies captured minus cost (length of silk).
6. Eliminate the least successful spiders and mate others to create new generations of spiders.
7. Repeat for many generations.

RESULTS

Virtual web   Actual web

After 40 generations the virtual webs are very similar to real webs.

Conclusion: This computer model suggests that the answer to the question is "yes": Spider webs have evolved to be efficient capturers of insects.

21.16 Computer "Models" Help Us Understand Natural Selection
An "experiment" using a computer program that modeled 40 generations of natural selection on spider webs resulted in webs remarkably similar to real ones.

real natural selection, they simulate the process well enough to lend support to the hypothesis that spider webs have evolved to be efficient at capturing insects.

Maintaining Genetic Variation

Random genetic drift, stabilizing selection, and directional selection all tend to reduce genetic variation within populations. Nevertheless, most populations have considerable

genetic variation. Why isn't the genetic variation of a species lost over time?

Sexual reproduction amplifies existing genetic variation

Recombination in sexually reproducing organisms amplifies existing genetic variation. In asexually reproducing organisms, the cells resulting from a mitotic division normally contain identical genotypes. Each new individual is genetically identical to its parent unless there has been a mutation. When organisms exchange genetic material during sexual reproduction, however, the offspring differ from their parents because chromosomes assort randomly during meiosis, crossing over occurs, and fertilization brings together material from two different cells (see Chapter 9).

Sexual reproduction generates an endless variety of genotypic combinations that increases the evolutionary potential of populations. Because it increases the variation among the offspring produced by an individual, sexual reproduction may improve the chance that at least some of the offspring will be successful in the varying and often unpredictable environments they will encounter. Sexual reproduction does not influence the frequencies of alleles; rather, it generates new combinations of alleles on which natural selection can act. It expands variation in a trait influenced by alleles at many loci by creating new genotypes. That is why selection for bristle number in *Drosophila* (see Figure 21.5) resulted in flies with more bristles than any flies in the initial population had.

Neutral genetic mutations accumulate within species

As we saw in Chapter 12, some mutations do not affect the functioning of the proteins the mutated genes encode. An allele that does not affect the fitness of an organism is called a **neutral allele**. Such alleles, untouched by natural selection, may be lost, or their frequencies may increase with time. Therefore, neutral alleles tend to accumulate in a population over time, providing it with considerable genetic variation.

Much of the variation in those traits we can observe with our unaided senses is not neutral. However, much variation at the molecular level apparently is neutral. Modern molecular techniques enable us to measure variation in neutral traits and provide a means by which we can distinguish adaptive from neutral variation. Chapter 24 will discuss how these techniques enable us to make such discriminations, and how variation in neutral traits can be used to estimate rates of evolution.

Much genetic variation is maintained in geographically distinct subpopulations

Much of the genetic variation in large populations is preserved as differences among subpopulations. Subpopulations often vary genetically because they are subjected to different selective pressures in different environments. Plant subpopulations, for example, may vary geographi-

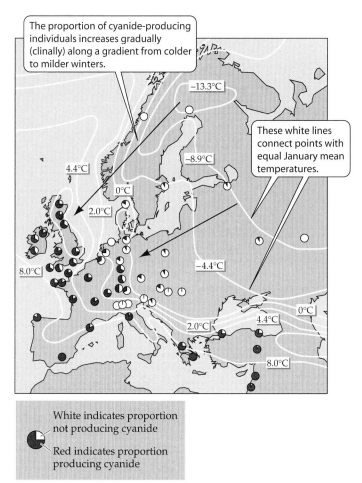

The proportion of cyanide-producing individuals increases gradually (clinally) along a gradient from colder to milder winters.

These white lines connect points with equal January mean temperatures.

-13.3°C

-8.9°C

4.4°C

0°C

2.0°C

8.0°C

-4.4°C

0°C

2.0°C

4.4°C

8.0°C

White indicates proportion not producing cyanide

Red indicates proportion producing cyanide

21.17 Geographic Variation in Poisonous Clovers
The frequency of cyanide-producing individuals in each population of white clover (*Trifolium repens*) is represented by the proportion of the circle that is red.

that of other genotypes (or phenotypes). This process is known as **frequency-dependent selection**.

A small fish that lives in Lake Tanganyika in east central Africa provides an example of frequency-dependent selection. The mouth of this scale-eating fish, *Perissodus microlepis*, opens either to the right or to the left as a result of an asymmetrical jaw joint (Figure 21.18). *Perissodus* approaches its prey (another fish) from behind and dashes in to bite off several scales from its flank. "Right-mouthed" individuals always attack from the victim's left; "left-mouthed" individuals always attack from the victim's right. The distorted mouth enlarges the area of teeth in contact with the prey's flank, but only if the scale eater attacks from the appropriate side.

Prey fish are alert to approaching scale eaters, so attacks are more likely to be successful if the prey must watch both flanks. Guarding by the prey favors equal numbers of right-mouthed and left-mouthed scale eaters, because if one form were more common than the other, prey fish would pay more attention to potential attacks from the corresponding flank. Therefore, success of individuals of the more common morph would be less than that of the less common morph. Over an 11-year period, the polymorphism was found to be stable: the two forms of *Perissodus* remained at about equal frequencies.

cally in the chemicals they synthesize to defend themselves against herbivores. Some individuals of the clover *Trifolium repens* produce the poisonous chemical cyanide. Poisonous individuals are less appealing to herbivores—particularly mice and slugs—than are nonpoisonous individuals. However, clover plants with cyanide are more likely to be killed by frost, because freezing damages cell membranes and releases the toxic cyanide into the plant's own tissues.

In populations of *Trifolium repens*, the frequency of cyanide-producing individuals increases gradually from north to south and from east to west across Europe (Figure 21.17). Poisonous individuals make up a large proportion of clover populations only in areas where the winters are mild. Cyanide-producing individuals are rare where winters are cold, even though herbivores graze clovers heavily in those areas.

Frequency-dependent selection maintains genetic variation within populations

Natural selection often preserves variation as **polymorphisms**—genetic differences within a population. A polymorphism may be maintained when the fitness of a genotype (or phenotype) varies with its frequency relative to

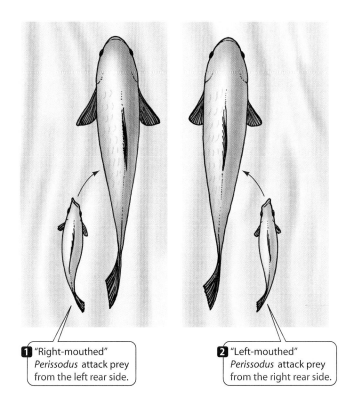

1 "Right-mouthed" *Perissodus* attack prey from the left rear side.

2 "Left-mouthed" *Perissodus* attack prey from the right rear side.

21.18 A Stable Polymorphism
Frequency-dependent selection maintains equal proportions of left-mouthed and right-mouthed individuals of the scale-eating fish *Perissodus microlepis*.

Leaves of a white oak (*Quercus alba*)

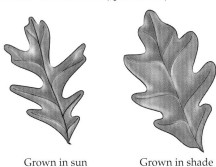

Grown in sun   Grown in shade

21.19 Environmentally Induced Variation
Traits may vary among genetically identical individuals or parts of individuals if they are exposed to different environments.

21.20 One Genotype: Two Seasonal Color Forms
The dry-season (left) and wet-season (right) form of the butterfly *Bicyclus anynana* have the same genotype. The environmental conditions experienced by a larva determine the form of the butterfly into which it develops.

How Do Genotypes Determine Phenotypes?

Genotypes do not uniquely determine phenotypes. If one allele is dominant to another, a particular phenotype can be produced by more than one genotype (for example, *AA* and *Aa* individuals may be phenotypically identical).

Similarly, different phenotypes can be produced by a given genotype, depending on the environment encountered during development. For example, the cells of the leaves on a tree or shrub are normally genetically identical. Yet leaves on the same tree often differ in shape and size. Leaves close to the top of an oak tree, where they are exposed to more wind and sunlight, may be more deeply lobed than leaves lower down on the same tree (Figure 21.19). The same differences can be seen between the leaves of individuals growing in sunny and shady sites.

Thus, the phenotype of an organism is the outcome of a complex series of developmental processes that are influenced by both environmental factors and genes. This nearly universal phenomenon is called **phenotypic plasticity**. Although variations in leaf shapes are not passed on to offspring, the ability to produce varied leaf shapes in response to environmental conditions *is* inherited. Phenotypic plasticity of leaf shapes benefits a tree because deeply lobed leaves offer less resistance to wind, absorb less sunlight, lose heat more rapidly by convection, and allow more sunlight to pass to lower leaves.

Because phenotypic plasticity is often adaptive, it may evolve under the influence of natural selection. A particularly thorough demonstration of the adaptive nature of phenotypic plasticity is provided by studies of the tropical African butterfly *Bicyclus anynana*. *B. anynana*, which lives in areas with distinct wet and dry seasons, has two distinct forms (Figure 21.20). The dry-season form, which rests on dried grasses and leaf litter and flies infrequently, has only one small wing spot. The wet-season form, which flies actively in lush, green vegetation, has many conspicuous spots on its wings.

The form of an adult butterfly is determined by the environmental conditions it encounters as a larva. Investigators compared the survival rates of the two forms in both seasons. The dry-season form, which closely matches the brown vegetation on which it rests, survives better during the dry season than does the wet-season form. On the other hand, the more active, conspicuous wet-season form does better during the wet season because its conspicuous spots, which resemble large eyes, deter some predators.

Constraints on Evolution

Thus far we have implicitly assumed that sufficient genetic variation always exists for the evolution of favored traits. A moment's reflection reveals that this assumption cannot be true. As we pointed out in the previous chapter, major evolutionary innovations are rare. Most changes are based on modifications of previously existing traits, even though those traits may come to serve new functions. In addition, evolutionary theory does not allow a population to temporarily become less well adapted. All intermediate forms must work; that is, all modifications must benefit their bearers in every generation.

A striking example that illustrates how natural selection operates by modifying existing states is provided by the evolution of fishes that spend most of their time resting on the sea bottom. One lineage, the bottom-dwelling skates and rays, is beautifully symmetrical (Figure 21.21*a*). These fishes are descended from sharks, which were already somewhat flattened and, therefore, able to lie on their bellies.

Plaice, sole, and flounders, on the other hand, are descendants of deep-bodied ancestors. Unlike sharks, these fishes cannot lie on their bellies; they must flop over on their sides. During development, the eyes of plaice and sole are grotesquely twisted around to bring both eyes to one side of the body (Figure 21.21*b*). No clever designer who was free of constraints would have designed plaice and sole as they are. But small shifts in the position of one eye probably helped ancestral flatfishes see better, resulting in the form found today.

Although constraints on evolution clearly exist, it is difficult to determine whether the absence of certain traits that

(a) *Taeniura lymma*

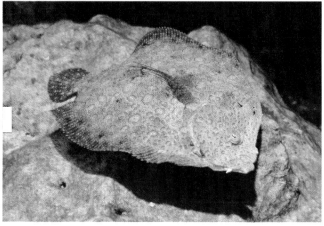

(b) *Bothus lunatus*

21.21 Two Solutions to a Single Problem
(a) Stingrays, whose ancestors were dorsally flattened, lie on their bellies. (b) Flounders, whose ancestors were laterally flattened, lie on their sides.

would seem to be desirable is due to some constraint or to our having wrongly guessed that the trait would be adaptive. A plausible answer to this question is now available for a puzzling pattern among amphibians. Many salamanders are *neotenic;* that is, individuals become sexually mature while still in their aquatic larval form. Why, then, are there no frogs and toads that reproduce when they are tadpoles? A universal constraint preventing frogs and toads from evolving neoteny may be their need for relatively high levels of thyroid hormones for sex differentiation and reproduction. Neotenic salamanders result only when levels of thyroid hormones are very low—too low to allow frogs and toads to become sexually mature. In many other cases, however, no plausible answer is yet available.

Short-Term versus Long-Term Evolution

Microevolutionary changes within populations are an important focus of study for evolutionary biologists. These changes can be observed directly, they can be manipulated experimentally, and they show the actual processes by which evolution occurs. Studies of these short-term changes identify the genetic bases of evolutionary changes

and demonstrate how natural selection acts. By themselves, however, they do not enable us to predict—or, more properly, "postdict" (because they have already happened)—the macroevolutionary changes we described in Chapter 20.

The reason for this is that patterns of macroevolutionary change can be strongly influenced by events that occur so infrequently or so slowly that they are unlikely to be observed during microevolutionary studies. In addition, the ways in which evolutionary agents act may change with time; even among the descendants of a single ancestral species, different lineages may evolve in different directions. Therefore, additional types of evidence, such as the occurrence of rare and unusual events and trends in the fossil record, must be gathered in order to understand the course of evolution over billions of years.

"Postdiction" problems in science are not unique to evolutionary studies. For example, volcanologists believe they understand the physical theory that explains why Mount St. Helens erupted in 1980, but they lack the detailed information necessary for them to "explain" why the mountain erupted on the exact day it did. Similarly, even though seismologists know the physical principles that govern earthquakes, they cannot predict exactly when or where an earthquake will happen.

In subsequent chapters we will discuss the kinds of information that biologists assemble to study long-term evolutionary changes and infer the processes that led to them.

Chapter Summary

Charles Darwin and Adaptation
▶ Darwin developed his theory of evolution by natural selection by carefully observing nature, especially during his voyage around the world on the *Beagle.*
▶ Darwin based his theory on well-known facts and some key inferences.
▶ Darwin had no examples of the action of natural selection, so he based his arguments on artificial selection by plant and animal breeders.
▶ Modern genetics has elucidated the mechanisms of heredity, which were unknown to Darwin but which have provided the solid base that supports and substantiates his theory.

Genetic Variation within Populations
▶ A single individual has only some of the alleles found in the population of which it is a member. **Review Figure 21.3**
▶ Genetic variation characterizes nearly all natural populations. **Review Figures 21.4, 21.5**
▶ Allele frequencies measure the amount of genetic variation in a population. Genotype frequencies show how a population's genetic variation is distributed among its members.
▶ Biologists estimate allele frequencies by measuring a sample of individuals from a population. The sum of all allele frequencies at a locus is equal to 1. **Review Figure 21.6**
▶ Populations that have the same allele frequencies may nonetheless have different genotype frequencies.

The Hardy–Weinberg Equilibrium
▶ A population that is not changing genetically is said to be at Hardy–Weinberg equilibrium.

▶ The assumptions that underlie the Hardy–Weinberg equilibrium are that the population is large, mating is random, there is no migration, mutation can be ignored, and natural selection is not acting on the population.

▶ In a population at Hardy–Weinberg equilibrium, allele frequencies remain the same from generation to generation. In addition, genotype frequencies remain in the proportions $p^2 + 2pq + q^2 = 1$. **Review Figure 21.7**

▶ Biologists can determine whether an agent of evolution is acting on a population by comparing the genotype frequencies of that population with Hardy–Weinberg equilibrium frequencies.

Microevolution: Changes in the Genetic Structure of Populations

▶ Changes in allele frequencies and genotype frequencies within populations are caused by the actions of several evolutionary agents: mutation, gene flow, random genetic drift, assortative mating, and natural selection.

▶ The origin of genetic variation is mutation. Most mutations are harmful or neutral to their bearers, but some are advantageous, particularly if the environment changes.

▶ Migration of individuals from one population to another, followed by breeding in the new location, produces gene flow. Immigrants may add new alleles to a population or may change the frequencies of alleles already present.

▶ Random genetic drift alters allele frequencies in all populations, but it overrides natural selection only in small populations. Organisms that normally have large populations may pass through occasional periods (population bottlenecks) when only a small number of individuals survive. **Review Figure 21.8**

▶ New populations established by a few founding individuals also have gene frequencies that differ from those in the parent population. **Review Figure 21.10**

▶ If individuals mate more often with individuals that have the same or different genotypes than would be expected on a random basis—that is, when mating is not random—frequencies of homozygous and heterozygous genotypes differ from Hardy–Weinberg expectations. **Review Figure 21.11**

▶ Self-fertilization, an extreme form of nonrandom mating, reduces the frequencies of heterozygous individuals below Hardy–Weinberg expectations without changing allele frequencies.

▶ Natural selection is the only agent of evolution that adapts populations to their environments. Natural selection may preserve allele frequencies or cause them to change with time.

▶ Stabilizing selection, directional selection, and disruptive selection change the distributions of phenotypes governed by more than one locus. **Review Figures 21.12, 21.13, 21.14**

Studying Microevolution

▶ Biologists study microevolution by measuring natural selection in the field, experimentally altering organisms, and building computer models. **Review Figures 21.15, 21.16**

Maintaining Genetic Variation

▶ Random genetic drift, stabilizing selection, and directional selection all tend to reduce genetic variation, but most populations are genetically highly variable.

▶ Sexual reproduction generates an endless variety of genotypic combinations that increases the evolutionary potential of populations, but it does not influence the frequencies of alleles. Rather, it generates new combinations of genetic material on which natural selection can act.

▶ Much genetic variation within many species is maintained in distinct subpopulations. **Review Figure 21.17**

▶ Genetic variation within a population may be maintained by frequency-dependent selection. **Review Figure 21.18**

How Do Genotypes Determine Phenotypes?

▶ Genotypes do not uniquely determine phenotypes. A given phenotype can be produced by more than one genotype.

▶ The phenotype of an organism is the result of a complex series of developmental processes that are influenced by both environmental factors and genes. **Review Figures 21.19, 21.20**

Constraints on Evolution

▶ Natural selection acts by modifying what already exists. A population cannot get temporarily worse in order to achieve some long-term advantage.

Short-Term versus Long-Term Evolution

▶ Patterns of macroevolutionary change can be strongly influenced by events that occur so infrequently or so slowly that they are unlikely to be observed during microevolutionary studies. Additional types of evidence must be gathered to understand why evolution in the long term took the particular course it did.

For Discussion

1. During the past 50 years, more than 200 species of insects that attack crop plants have become highly resistant to DDT and other pesticides. Using your recently acquired knowledge of evolutionary processes, explain the rapid and widespread evolution of resistance. Propose ways of using pesticides that would slow down the rate of evolution of resistance. Now that the use of DDT has been banned in the United States, what do you expect to happen to levels of resistance to DDT among insect populations? Justify your answer.

2. In what ways does artificial selection by humans differ from natural selection in nature? Was Darwin wise to base so much of his argument for natural selection on the results of artificial selection?

3. In nature, mating among individuals in a population is never truly random: Immigration and emigration are common, and natural selection is seldom totally absent. Why, then, does it make sense to use the Hardy–Weinberg model, which is based on assumptions known generally to be false? Can you think of other models in science that are based on false assumptions? How are such models used?

4. As far as we know, natural selection cannot adapt organisms to future events. Yet many organisms appear to respond to natural events before they happen. For example, many mammals go into hibernation while it is still quite warm. Similarly, many birds leave the temperate zone for their southern wintering grounds long before winter arrives. How can such "anticipatory" behaviors evolve?

5. Some people believe that species, like individual organisms, have life cycles. They believe that species are born by a process of speciation, grow and expand, and inevitably die out as a result of "species old age." Could any agents of evolution cause such a species life cycle? If not, how do you explain the high rates of extinction of species in nature?

22 Species and Their Formation

DURING THE 1940S, OFFICIALS IN TRINIDAD launched an intensive campaign to control malaria. Believing that malaria was being transmitted by *Anopheles albimanus*, a swamp-breeding mosquito that is the principal vector of malaria in Latin America, they spent a great deal of money spraying and draining marshes. The campaign failed, however, because the principal vector of malaria in Trinidad was *Anopheles bellator*, a mosquito species that breeds in water held within the leaves of bromeliad plants (relatives of pineapples) growing on tree branches.

Similarly, in Europe, people thought that malaria was transmitted only by mosquitoes of a single species: *Anopheles maculipennis*. European efforts to control malaria sometimes succeeded and sometimes failed, because *A. maculipennis* turned out to be not a single species, but a group of at least 18 species that can be distinguished only by examination of their chromosomes. Some of the species breed in fresh water, others in brackish water. Some enter houses, others do not. Furthermore, which mosquito species is the vector of malaria varies regionally. Control efforts are successful only when directed against the species that actually transmits malaria in that area.

Therefore, to control malaria, we need to know which species of mosquitoes are the vectors of the disease, as well as the details of their life cycles. But how did these many species of mosquitoes arise? What processes keep them cohesive and distinct?

All species, living and extinct, are believed to be descendants of a single ancestral species that lived more than 3 billion years ago. If speciation were a rare event, the biological world would be very different than it is today. Speciation is an essential ingredient of evolutionary diversification, and species are the fundamental units of the biological classification systems we will discuss in Chapter 23. But what are species? How did these millions of species form? How does one species become two? What factors stimulate such splitting? What conditions spur evolutionary radiations? These and related questions are the subject of this chapter.

Trinidad Rainforest
The mosquito that transmits malaria in Trinidad breeds in water held in the bases of leaves of bromeliad plants that grow on the trunks and branches of rainforest trees.

What Are Species?

The word **species** means, literally, "kinds." But what do we mean by "kinds"? Someone who is knowledgeable about a group of organisms, such as orchids or lizards, usually can distinguish the different species of that group found in a particular area simply by examining them superficially. The patterns of similarities and differences that unite groups of organisms and separate them from other groups are familiar to all of us. The standard field guides to birds, mammals, insects, and flowers are possible only because most species are cohesive units that change in appearance only gradually over large geographic distances. We can easily recognize red-winged blackbirds from New York and red-winged blackbirds from California as members of the same species (Figure 22.1).

But not all members of a species look that much alike. For example, males, females, and young individuals may not resemble one another. How do we decide whether similar but easily distinguished individuals should be assigned to different species or regarded as members of the same species? The concept that has guided these decisions for a long time is genetic integration. If individuals within a population mate with one another but not with individuals of

(a) *Agelaius phoeniceus*

(b) *Agelaius phoeniceus*

22.1 Redwings Are Redwings Everywhere
Both of these male birds are obviously red-winged blackbirds, even though (a) lives in the eastern United States and (b) lives in California. In parts of California, males have less yellow in their wings than males elsewhere in the broad range of the species.

other populations, they constitute a distinct group within which genes recombine; that is, they are independent evolutionary units. These independent evolutionary units are usually called species.

More than 200 years ago the Swedish biologist Carolus Linnaeus, who originated the system of naming organisms that we use today, described hundreds of species. Because he knew nothing about the mating patterns of the organisms he was naming, Linnaeus classified them on the basis of their appearances; in other words, he used a *morphological* concept of species. Many species that were classified by their appearances are actually independent evolutionary units. They look alike because they share many alleles that code for body structures. In many groups of organisms for which genetic data are unavailable, species are still recognized by their morphological traits.

A species definition that has been used by many biologists—the "biological" species definition—was proposed by Ernst Mayr in 1940. He stated, "Species are groups of actually or potentially interbreeding natural populations which are reproductively isolated from other such groups." The words *"actually or potentially"* assert that, even if some members of a species are not in the same place and hence are unable to mate, they should not be placed in separate species if they would be likely to mate if they were together. The word "natural" is an important part of the definition because only in nature does the exchange of genes affect evolutionary processes; the interbreeding of two different species in captivity does not. Gene exchange is the main reason why species are cohesive units.

Deciding whether two populations constitute different species can be difficult because speciation is often a gradual process (Figure 22.2). If a barrier divides one population into two populations, the daughter populations may evolve independently long before they become reproductively incompatible—or they may become reproductively incompatible before they evolve any noticeable morphological differences.

How Do New Species Arise?

Speciation is the process by which one species splits into two species, which thereafter evolve as distinct lineages. Not all evolutionary changes result in new species. A single lineage may change through time without giving rise to a new species. Although Charles Darwin entitled his book *The Origin of Species*, he did not discuss how a single species splits into two or more daughter species. Rather, he was concerned principally with demonstrating that species are altered by natural selection over time.

The critical process in the formation of new species is the separation of the gene pool of the ancestral species into two separate gene pools. Subsequently, within each isolated gene pool, allele and gene frequencies may change as a result of the action of evolutionary agents. If sufficient differences accumulate during this period of isolation, the two populations may not exchange genes if they come together again.

Gene flow among populations may be interrupted in several ways, each of which characterizes a mode of speciation. The next three sections focus on these modes of speciation: allopatric speciation, sympatric speciation, and parapatric speciation.

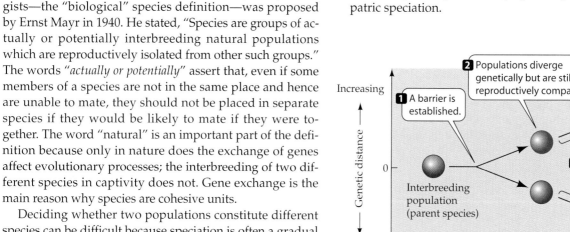
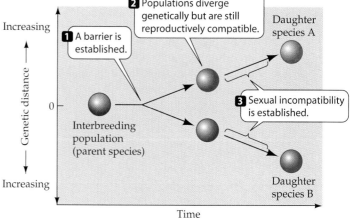

22.2 Speciation May Be a Gradual Process
In this hypothetical example, genetic divergence begins before reproductive incompatibility evolves.

Time

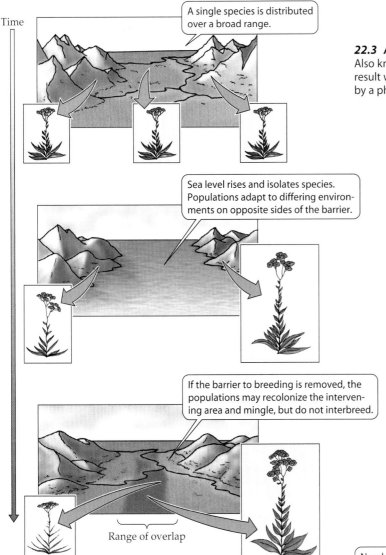

A single species is distributed over a broad range.

Sea level rises and isolates species. Populations adapt to differing environments on opposite sides of the barrier.

If the barrier to breeding is removed, the populations may recolonize the intervening area and mingle, but do not interbreed.

Range of overlap

Allopatric speciation requires total genetic isolation

Speciation that results when a population is divided by a geographic barrier is known as **allopatric speciation** (*allo-*, "different"; *patris*, "country"), or **geographic speciation** (Figure 22.3). Allopatric speciation is thought to be the dominant form of speciation among most groups of organisms. The range of a species may be divided by a barrier such as a water gap for terrestrial organisms, dry land for aquatic organisms, or a mountain range. Barriers can form when continents drift, sea levels rise, or climates change. Populations separated in this way are often large initially. They evolve differences because the places in which they live are, or become, different.

Alternatively, allopatric speciation may result when some members of a population cross an existing barrier and found a new population. Populations established in this way usually differ genetically from their parent populations because a small group of founding individuals has only an incomplete representation of the genes found in its parent population (see Chapter 21). Many of the hundreds of species of the fruit fly *Drosophila* in the Hawaiian Islands are restricted to a single island. They are almost certainly the re-

22.3 *Allopatric Speciation*
Also known as geographic speciation, allopatric speciation may result when a population is divided into two separate populations by a physical barrier such as rising seas.

sult of new populations founded by individuals dispersing among the islands, because the closest relative of a species on one island is often a species on a neighboring island, rather than a species on the same island. Biologists who have studied the chromosomes of picture-winged *Drosophila* believe that speciation among these flies has resulted from at least 45 such **founder events** (Figure 22.4).

The finches of the Galápagos archipelago, 1,000 km off the coast of Ecuador, demonstrate the importance of geographic isolation for speciation. Darwin's finches (as they are usually called, because Darwin was the first scientist to study them) arose on the Galápagos by speciation from a single South American species that colonized the islands. Today there are 14 species of Galápagos finches, all of which differ strikingly from the blue-black grassquit, their probable mainland ancestor (Figure 22.5).

The islands of the Galápagos archipelago are sufficiently isolated from one another that finches

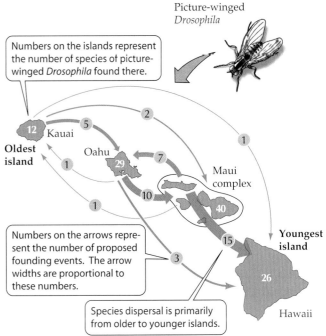

Picture-winged *Drosophila*

Numbers on the islands represent the number of species of picture-winged *Drosophila* found there.

Oldest island — Kauai 12, Oahu 29, Maui complex 40, Youngest island — Hawaii 26

Numbers on the arrows represent the number of proposed founding events. The arrow widths are proportional to these numbers.

Species dispersal is primarily from older to younger islands.

22.4 *Founder Events Lead to Allopatric Speciation*
The extremely high level of speciation found among picture-winged *Drosophila* in the Hawaiian Islands is almost certainly the result of founder events—new populations founded by individuals dispersing among the islands. The islands, which were formed in sequence as Earth's crust moved over a volcanic "hot spot," vary in age.

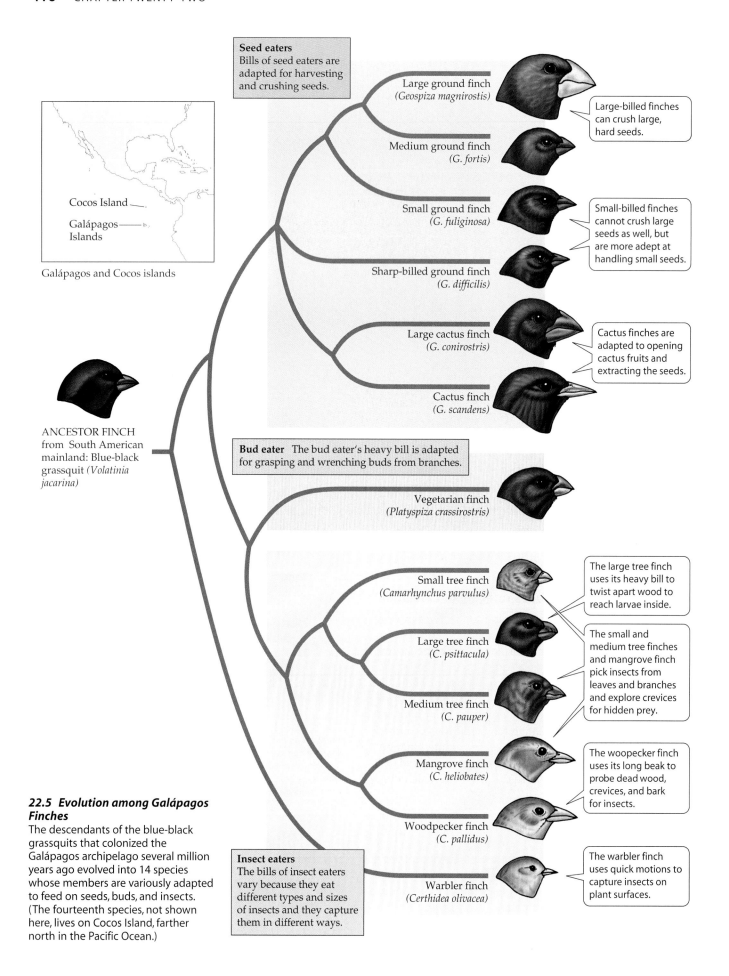

Galápagos and Cocos islands

ANCESTOR FINCH from South American mainland: Blue-black grassquit (*Volatinia jacarina*)

Seed eaters
Bills of seed eaters are adapted for harvesting and crushing seeds.

Large ground finch (*Geospiza magnirostis*)

Medium ground finch (*G. fortis*)

Large-billed finches can crush large, hard seeds.

Small ground finch (*G. fuliginosa*)

Sharp-billed ground finch (*G. difficilis*)

Small-billed finches cannot crush large seeds as well, but are more adept at handling small seeds.

Large cactus finch (*G. conirostris*)

Cactus finch (*G. scandens*)

Cactus finches are adapted to opening cactus fruits and extracting the seeds.

Bud eater The bud eater's heavy bill is adapted for grasping and wrenching buds from branches.

Vegetarian finch (*Platyspiza crassirostris*)

Small tree finch (*Camarhynchus parvulus*)

Large tree finch (*C. psittacula*)

The large tree finch uses its heavy bill to twist apart wood to reach larvae inside.

The small and medium tree finches and mangrove finch pick insects from leaves and branches and explore crevices for hidden prey.

Medium tree finch (*C. pauper*)

Mangrove finch (*C. heliobates*)

Woodpecker finch (*C. pallidus*)

The woopecker finch uses its long beak to probe dead wood, crevices, and bark for insects.

Insect eaters
The bills of insect eaters vary because they eat different types and sizes of insects and they capture them in different ways.

Warbler finch (*Certhidea olivacea*)

The warbler finch uses quick motions to capture insects on plant surfaces.

22.5 Evolution among Galápagos Finches
The descendants of the blue-black grassquits that colonized the Galápagos archipelago several million years ago evolved into 14 species whose members are variously adapted to feed on seeds, buds, and insects. (The fourteenth species, not shown here, lives on Cocos Island, farther north in the Pacific Ocean.)

seldom migrate between them. Also, environmental conditions differ among the islands. Some are relatively flat and arid; others have forested mountain slopes. Populations of finches on different islands have differentiated enough that when occasional immigrants arrive from other islands, they either do not breed with the residents, or if they do, the resulting offspring do not survive as well as those produced by pairs of island residents. The genetic distinctness and cohesiveness of different populations is thus maintained.

A barrier's effectiveness at preventing gene flow depends on the size and mobility of the species in question. What is an impenetrable barrier to a terrestrial snail may be no barrier at all to a butterfly or a bird. Populations of wind-pollinated plants are isolated at the maximum distance their pollen is blown by the wind, but individual plants are effectively isolated at much shorter distances. Among animal-pollinated plants, the width of the barrier is the distance that animals travel while carrying pollen or seeds. Even animals with great powers of dispersal are often reluctant to cross narrow strips of unsuitable habitat. For animals that cannot swim or fly, narrow water-filled gaps may be effective barriers.

Indirect evidence that most speciation among animals is allopatric is provided by patterns of species distributions. For example, 36 percent of Earth's 20,000 species of bony fishes live in fresh water, even though only 1 percent of Earth's surface is fresh water, and even though fish productivity and populations are higher in some marine environments than in most fresh waters. Because they are highly fragmented, fresh waters have provided abundant opportunities for fishes to form geographically isolated populations. Marine environments provide fewer such opportunities.

Sympatric speciation occurs without physical separation

The subdividing of a gene pool when members of the daughter species are not geographically separated is called **sympatric speciation** (*sym-*, "with"). The most common means of sympatric speciation is **polyploidy**, an increase in the number of chromosomes.

Polyploidy arises in two ways. One way is the accidental production during cell division of cells having four (tetraploid) instead of two (diploid) sets of chromosomes. This process produces an **autopolyploid** individual, one having more than two sets of chromosomes, all derived from a single species. This tetraploid individual cannot produce viable offspring by mating with diploids, but it can do so if it self-fertilizes or mates with other tetraploids.

A polyploid species can also be produced when individuals of two different species interbreed. The resulting offspring are usually sterile, because the chromosomes from one species do not pair properly with those from the other species during meiosis, but they may be able to reproduce asexually. After many generations, some of these individuals may become fertile as a result of further chromosome duplication. Species produced in this way are called **allopolyploids**.

Polyploidy can create new species among plants much more easily than among animals because plants of many species can reproduce by self-fertilization. If polyploidy arises in several offspring of a single parent, the siblings can fertilize one another. Speciation by polyploidy has been very important in the evolution of flowering plants. Botanists estimate that about 70 percent of flowering plant species and 95 percent of all fern species are polyploids. Most of these arose as a result of hybridization between two species, followed by self-fertilization.

The speed with which allopolyploidy can produce new species is illustrated by salsifies (*Tragopogon*), members of the sunflower family. Salsifies are weedy plants that thrive in disturbed areas around towns. People have inadvertently spread them around the world from their ancestral ranges in Eurasia. Three diploid species of salsify were introduced into North America early in the twentieth century: *T. porrifolius*, *T. pratensis*, and *T. dubius*. Two tetraploid hybrids—*T. mirus* and *T. miscellus*—between species of the original three were first reported in 1950. Both hybrids have spread since their discovery and today are more widespread than their diploid parents (Figure 22.6).

Studies of their genetic material have shown that both hybrids have been formed more than once. Some populations of *T. miscellus*—a hybrid of *T. pratensis* and *T. dubius*—have the chloroplast genome of *T. pratensis*, whereas other populations have the chloroplast genome of *T. dubius*. Such differences among local populations of *T. miscellus* show that this allopolyploid has evolved independently at least 21 times; *T. mirus* has formed 12 times! Scientists seldom know the dates and locations of species formation so well.

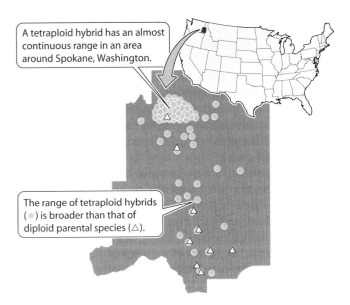

A tetraploid hybrid has an almost continuous range in an area around Spokane, Washington.

The range of tetraploid hybrids (●) is broader than that of diploid parental species (△).

22.6 Polyploids Can Outperform Their Parents
Tragopogon species (salsifies) are members of the sunflower family. The map shows the distribution of the diploid parent species and the tetraploid hybrid species of *Tragopogon* in eastern Washington and adjacent Idaho.

But the rate at which new species form also depends on how fast daughter populations diverge. The more rapidly they diverge, the sooner they are likely to evolve reproductive isolating mechanisms, and the less likely they are to hybridize if they again become sympatric. Shorter generation times result in more generations per unit of time and, as a result, generate the potential for more evolutionary changes per unit of time.

Evolutionary Radiations

As we learned in Chapter 20, the fossil record reveals that at certain times in some lineages, speciation rates have been much higher than extinction rates. The result is an *evolutionary radiation* that gives rise to a large number of daughter species. What conditions cause speciation rates to be much higher than extinction rates?

Evolutionary radiations are likely when a population colonizes an environment that has relatively few species. This condition typifies islands because many organisms disperse poorly across large water gaps. Because islands lack many plant and animal groups found on the mainland, ecological opportunities exist that may stimulate rapid evolutionary changes when a new species does reach them. Water barriers also restrict gene flow among islands in an archipelago, so populations on different islands can evolve adaptations to their local environments. Together these two factors make it likely that speciation rates on island archipelagos will exceed extinction rates.

Remarkable evolutionary radiations have occurred in the Hawaiian Islands, the most isolated islands in the world. The Hawaiian Islands lie 4,000 km from the nearest major land mass and 1,600 km from the nearest group of islands. The islands are arranged in a line of decreasing age—the youngest islands to the southeast, the oldest to the northwest. The archipelago is actually much older than the oldest existing islands because even older islands long ago eroded until they no longer rise above the sea surface.

The native biota of the Hawaiian Islands includes 1,000 species of flowering plants, 10,000 species of insects, 1,000 land snails, and more than 100 bird species. However, there were no amphibians, no terrestrial reptiles, and only one native mammal—a bat—until humans introduced additional species. The 10,000 known native species of insects on Hawaii are believed to have evolved from only about 400 immigrant species; only 7 immigrant species are believed to account for all the native Hawaiian land birds.

More than 90 percent of all plant species on the Hawaiian Islands are *endemic*—that is, they are found nowhere else. Several groups of flowering plants have more diverse forms and life histories on the islands and live in a wider variety of habitats than do their close relatives on the mainland. An outstanding example is the group of sunflowers called silverswords (the genera *Argyroxiphium*, *Dubautia*, and *Wilkesia*). Chloroplast DNA data show that these species share a relatively recent common ancestor, which is believed to be a species of tarweed from the Pacific coast of North America. Whereas all mainland tarweeds are small, upright, nonwoody plants (herbs), Hawaiian silversword species include prostrate and upright herbs, shrubs, trees, and vines (Figure 22.13). They occupy nearly all the habitats of the islands, from sea level to above timberline in the

22.13 Rapid Evolution among Hawaiian Plants

Three closely related genera of the sunflower family, two of which are illustrated here, are believed to have descended from a single ancestor, a tarweed (*Madia sativa*) that colonized Hawaii from the Pacific coast of North America. They appear more distantly related than they actually are.

Madia sativa (ancestral tarweed)

Argyoxiphium sandwichense

Dubautia menziesii

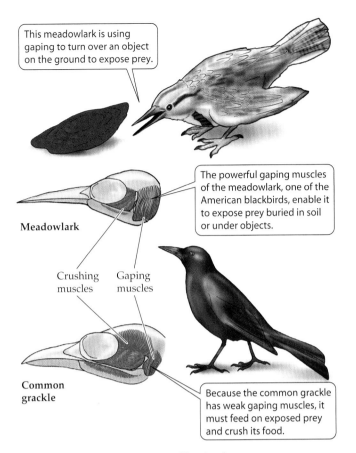

This meadowlark is using gaping to turn over an object on the ground to expose prey.

The powerful gaping muscles of the meadowlark, one of the American blackbirds, enable it to expose prey buried in soil or under objects.

Meadowlark

Crushing muscles   Gaping muscles

Common grackle

Because the common grackle has weak gaping muscles, it must feed on exposed prey and crush its food.

22.14 Blackbirds Expose Food by Gaping
The species of blackbirds that find food by gaping, here illustrated by a meadowlark, differ strikingly in the size and strength of their gaping muscles from blackbird species (such as grackles) that feed on exposed food.

mountains. Despite their extraordinary diversification, however, the silverswords have differentiated very little in their chloroplast genes.

The island silverswords are more diverse in size and shape than the mainland tarweeds because the original colonizers arrived on islands that had very few plant species. In particular, there were few trees and shrubs, because such large-seeded plants rarely disperse to oceanic islands. Many island trees and shrubs have evolved from nonwoody ancestors. On the mainland, however, tarweeds live in ecological communities that contain tree and shrub lineages older than their own—that is, where opportunities to exploit the tree way of life were already preempted.

Evolutionary lineages may also radiate when they acquire a new adaptation that enables them to use the environment in new and varied ways. For example, ancestors of the 95 species of American blackbirds evolved powerful muscles for opening their bills. These muscles enable the birds to obtain food by opening their bills forcibly against objects they wish to move, exposing otherwise hidden prey (Figure 22.14). This behavior is called *gaping*. Birds lacking these powerful muscles can find prey only on exposed surfaces of objects. Blackbirds gape into wood, fruits, leaf clusters, and stems of nonwoody plants; under sticks, stones, and animal droppings; and into the soil. With this feeding method, they have come to occupy nearly all habitat types

in North and South America, and they are among the most abundant birds throughout the region.

The Significance of Speciation

The result of speciation processes operating over billions of years is a world in which life is organized into millions of species, each adapted to live in a particular environment and to use environmental resources in a particular way. Earth would be very different if speciation had been a rare event in the history of life. How the millions of species are distributed over the surface of Earth and organized into ecological communities will be a major focus of Part Seven of this book, "Ecology and Biogeography." There we will also discuss how human activities are causing the extinction of species and what we can do to reduce the rate of species loss.

 Chapter Summary

What Are Species?

▶ Species are independent evolutionary units. A commonly accepted definition is that "species are groups of actually or potentially interbreeding natural populations which are reproductively isolated from other such groups."

▶ Because speciation is often a gradual process, it may be difficult to recognize boundaries between species. **Review Figure 22.2**

How Do New Species Arise?

▶ Not all evolutionary changes result in new species.

▶ Allopatric (geographic) speciation is the most important means of speciation among animals and is common in other groups of organisms. **Review Figures 22.3, 22.4, 22.5**

▶ Species may form sympatrically by a multiplication of chromosome numbers because the resulting polyploid organisms cannot interbreed with members of the parent species. Polyploidy has been a major factor in plant speciation, but is rare among animals. **Review Figure 22.6**

▶ Species may form parapatrically where marked environmental differences prevent gene flow among individuals living in adjacent environments.

Reproductive Isolating Mechanisms

▶ When previously allopatric species become sympatric, reproductive isolating mechanisms may prevent the exchange of genes.

▶ Barriers to gene exchange may operate before mating (prezygotic barriers) or after mating (postzygotic barriers).

▶ Hybrid zones may develop if barriers to gene exchange failed to develop during allopatry.

▶ Hybrids may form if separated populations come together again without sufficient genetic differences having accumulated. **Review Figure 22.9**

▶ The existence of hybrids tells us that the two hybridizing species are very similar genetically, but species that do not hybridize may also differ from one another very little genetically.

Variation in Speciation Rates

▶ Rates of speciation differ greatly among lineages of organisms. Speciation rates are influenced by the number of species in a lineage, their range sizes, their behavior, environmental changes, and generation times. **Review Figures 22.11, 22.12**

Evolutionary Radiations

▶ Evolutionary radiations happen when speciation rates exceed extinction rates.

▶ High speciation rates often coincide with low extinction rates when species invade islands that have few other species, or when a new way of exploiting the environment makes a different array of resources available to a species. **Review Figures 22.13, 22.14**

The Significance of Speciation

▶ As a result of speciation, Earth is populated with millions of species, each adapted to live in a particular environment and to use resources in a particular way.

For Discussion

1. Gene exchange between populations is prevented by geographical isolation, by behavioral responses before mating (for example, females may reject courting males of the other species), and by mechanisms that function after mating has occurred (for example, hybrid sterility). All of these are commonly called isolating mechanisms. In what ways are the three types very different? If you were to apply different names to them, which one would you call an isolating mechanism? Why? What names would you give the other types? Why?

2. The blue goose of North America has two distinct color forms, blue and white. Matings between the two color types are common. However, blue individuals pair with blue individuals and white individuals pair with white individuals much more frequently than would be expected by chance. Suppose that 75 percent of all mated pairs consisted of two individuals of the same color. What would you conclude about speciation processes in these geese? If 95 percent of pairs were the same color? If 100 percent of pairs were the same color?

3. Suppose pairs of blue geese of mixed colors were found only in a narrow zone within the broad Arctic breeding range of the geese, would you answer Question 2 the same way you did? Would your answer change if mixed-color pairs were widely distributed across the breeding range of the geese?

4. Although many species of butterflies are divided into local populations among which there is little gene flow, these butterflies often show relatively little geographic variation. Describe studies you would conduct to determine what maintains this morphological similarity?

5. Distinguish among allopatric, parapatric, and sympatric speciation. For each of the three statements below, indicate which type of speciation is implied:

    a. This process in nature is most commonly a result of polyploidy.

    b. The size of national parks and wildlife refuges may be too small to allow this type of speciation among organisms restricted to those areas.

    c. This process usually occurs in species that inhabit areas where sharp environmental contrasts exist.

6. Evolutionary radiations are common and easily studied on oceanic islands. In what types of mainland situations would you expect to find major evolutionary radiations? Why?

7. Fruit flies of the genus *Drosophila* are found worldwide, but most of the species in the genus are found on the Hawaiian Islands. Suggest a hypothesis that might account for this distribution pattern.

23

Reconstructing and Using Phylogenies

SCHISTOSOMIASIS IS A BLOOD INFECTION CAUSED BY a parasitic flatworm, *Schistosoma*. More than 200 million people in South America, Africa, China, Japan, and Southeast Asia have the disease. During part of its life cycle, *Schistosoma* inhabits a freshwater snail. People become infected when they come into contact with water where infested snails live. Larval *Schistosoma* swim from a snail and penetrate the skin. The worm matures and lives in the person's abdominal blood vessels. The disease is progressively debilitating, causing a slow death.

For most of the twentieth century, only one species, *Schistosoma japonicum*, was known to infect humans, and people believed that it was transmitted by a single species of snail in the genus *Oncomelania*. Then, in the 1970s, researchers discovered that a different snail was transmitting *Schistosoma* to humans in the Mekong River in Laos. This discovery stimulated extensive field surveys and anatomical, genetic, and geographic research on the worms and snails of Southeast Asia.

Investigators found that *S. japonicum* was actually a cluster of at least six species. They also discovered that evolutionary relationships among snails influenced which species could host *Schistosoma*. Evolutionary diversification from an ancestral stock of snails produced a group of species of modern snails. Of these, only three can host *Schistosoma*; ten have a genetic trait that allows them to resist invasion by the parasite.

This information is of great value in efforts to combat schistosomiasis. Few of the freshwater snail species in Southeast Asia have been described and named. By using information on evolutionary relationships among snails, scientists can quickly determine whether or not a newly discovered snail is likely to be a host for *Schistosoma*. Control efforts need to be directed toward only the snails that can transmit *Schistosoma* to humans, not all freshwater snails in the region.

How did investigators determine the evolutionary relationships among the snails that are hosts of *Schistosoma*? How could they determine the number of times that genes preventing snails from hosting *Schistosoma*

arose? How is knowledge of evolutionary relationships used to help answer other biological questions? How are evolutionary relationships expressed in systems of classification that help guide further studies of organisms?

In this chapter, we discuss systematics, the science that provides answers to these questions. We describe the methods systematists use to infer evolutionary relationships among organisms. Then we illustrate how knowledge of evolutionary relationships is used to solve other biological problems. Finally, we show how evolutionary relationships are incorporated into classification systems.

How Are Phylogenetic Trees Reconstructed?

Ever since its origin nearly 4 billion years ago, life has evolved under the influence of the evolutionary agents we described in Chapter 21. The incredible richness of today's biological world has resulted from millions of speciation events, determined by the processes we discussed in Chapter 22. Biologists have developed methods to trace the history of these processes and make sense of their results.

A **phylogeny** is a history of descent of a group of organisms from their common ancestor. Our understanding of the processes of speciation tells us that lineages of organisms can be represented as branching "trees." These **phylogenetic trees** show the order in which lineages split. A particular tree may portray the evolution of all life, of major

Asian Snails Can Transmit Schistosomiasis
Workers in the rice paddies of tropical Asia are at extreme risk of contracting schistosomiasis (known in some parts of the world as bilharzia). The disease is transmitted to humans via freshwater snails that thrive in the standing water of the paddies.

those characters. Recall from Chapter 10 that a *character* is a feature such as flower color; a *trait* is a particular form of a character, such as white flowers. A trait may be the presence or absence of a character, or the character may exist in more than one form. The next, and usually the most difficult, step is to determine the ancestral and derived traits. Finally, systematists must distinguish homologous from homoplastic traits.

Identifying ancestral traits

Distinguishing derived traits from ancestral traits may be difficult because traits often become so dissimilar that ancestral states are unrecognizable. For example, the leaves of plants have diverged to form many different structures. Several lines of evidence, especially details of their structure and development, indicate that protective spines, tendrils, and brightly colored structures that attract pollinators (Figure 23.4) are all modified leaves; they are *homologs* of one another even though they do not resemble one another closely.

One way to distinguish ancestral traits from derived traits is to assume that an ancestral trait should be found not only among the species of the focal group, but also in outgroups. An **outgroup** is a lineage that is closely related to the focal group , but which branched off from the lineage of the focal group below its base on the evolutionary tree. Traits found only within the focal group, on the other hand, are likely to be derived traits. Species that have a recent common ancestor should share very few homoplastic traits, because little time has been available for convergent evolution to produce them.

The spines of the barrel cactus and the bracts of *Heliconia* are both modified leaves.

Cheiridopsis tuberculata   *Heliconia* sp.

23.4 Homologous Structures Derived from Leaves
The leaves of plants have diverged during their evolution to form many different structures, some of which bear very little resemblance to each other. *Heliconia* bracts support flowers and attract pollinators.

The more traits that are measured, the more likely the data will support a single phylogenetic pattern, and the more readily biologists can distinguish between homologies and homoplasies. A few of the traits originally assumed to be homologies may turn out to be homoplasies, but the best way to determine the true status of shared traits is to assume that they are homologous until additional evidence suggests they are not.

 ## Reconstructing a simple phylogeny

To see how a phylogeny is constructed, consider eight vertebrate animals—hagfish, perch, pigeon, chimpanzee, salamander, lizard, mouse, and crocodile. We will assume initially that a given derived trait evolved only once during the evolution of these animals, and that no derived traits were lost from any of the descendant groups. For simplicity, we have selected traits that are either present (+) or absent (–). The traits we will consider are listed in Table 23.1.

As will become evident in Chapter 33, hagfishes are believed to be more distantly related to the other vertebrates than the other vertebrates are to each other. Therefore, we choose hagfishes as the outgroup for our analysis. Derived traits are those that have been acquired by other members of the lineage since they separated from hagfishes.

We begin by noting that the chimpanzee and the mouse share two unique traits, mammary glands and fur. Those traits are absent in both the outgroup and the other species whose relationships we are attempting to determine. Therefore, we infer that mammary glands and fur are derived traits that evolved in a common ancestor of chimpanzees and mice after that lineage separated from the ones leading to the other vertebrates. In other words, we provisionally assume that mammary glands and fur evolved only once among the animals we are classifying.

The pigeon has one unique trait: feathers. As before, we provisionally assume that feathers evolved only once, after the lineage leading to birds separated from that leading to the mouse, chimpanzee, and crocodile. By the same reasoning, we assume that four-chambered hearts evolved only once, after the lineage leading to crocodiles, birds, and mammals separated from the lineage leading to lizards. We assume that claws or nails evolved only once, after the lineage leading to salamanders separated from the lineage leading to those animals that have claws or nails. We make the same assumption for lungs and jaws, continuing to minimize the number of evolutionary events needed to produce the patterns of shared traits among these eight animals.

Using this information, we can reconstruct a provisional phylogeny. The group with no derived traits, the hagfish, is the outgroup, and we assume that the animals that share unique derived traits have a common ancestor not shared with animals lacking those traits. We assume, for exam-

23.1 Eight Vertebrates Ordered According to Unique Shared Derived Traits

TAXON	JAWS	LUNGS	CLAWS OR NAILS	FEATHERS	FUR	MAMMARY GLANDS	FOUR-CHAMBERED HEART
Hagfish (outgroup)	–	–	–	–	–	–	–
Perch	+	–	–	–	–	–	–
Salamander	+	+	–	–	–	–	–
Lizard	+	+	+	–	–	–	–
Crocodile	+	+	+	–	–	–	+
Pigeon	+	+	+	+	–	–	+
Mouse	+	+	+	–	+	+	+
Chimpanzee	+	+	+	–	+	+	+

[a] A plus sign indicates the trait is present, a minus sign that it is absent.

ple, that mice and chimpanzees, the only two animals that share fur and mammary glands, share a more recent common ancestor with each other than they do with birds and crocodiles. Otherwise we would need to assume that the ancestors of birds and crocodiles also had fur and mammary glands, but that those traits were subsequently lost—unnecessary additional assumptions.

A phylogeny for these eight vertebrates, based on the traits we used and the assumption that each derived trait evolved only once, is shown in Figure 23.5. Notice that the phylogeny does not describe the ancestors or date the splits between lineages. It shows only the sequential order of the splits: The oldest splits are to the left, and the more recent ones are to the right. Notice also that the y axis has no scale. In this and all other phylogenies in this book, vertical distances between groups do not correlate with degree of similarity or difference between them.

The phylogeny of these eight vertebrates was easy to construct because the traits we chose fulfilled the assumptions that derived traits appeared only once in the lineage and were never lost after they appeared. If we had included a snake in the group, however, our second assumption

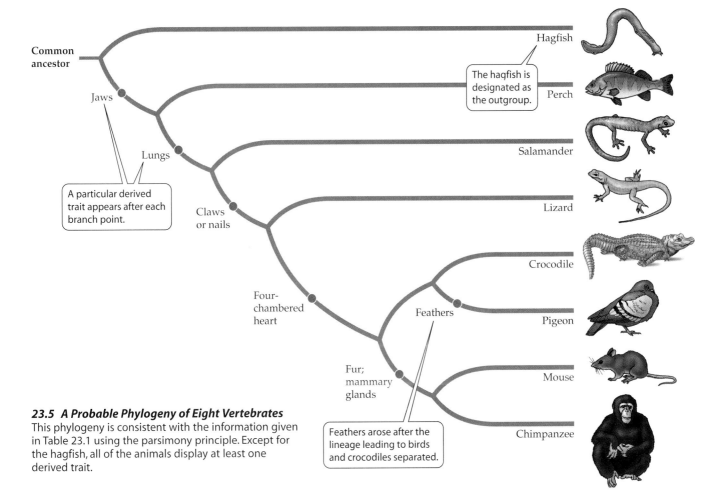

23.5 A Probable Phylogeny of Eight Vertebrates
This phylogeny is consistent with the information given in Table 23.1 using the parsimony principle. Except for the hagfish, all of the animals display at least one derived trait.

would have been violated, because the lizard ancestors of snakes had limbs, which were subsequently lost, along with their claws. We would need to examine additional traits to determine that the lineage leading to snakes separated from the one leading to lizards long after the lineage leading to lizards separated from the others. In fact, the analysis of many traits shows that snakes evolved from burrowing lizards that lost their limbs during a long period of subterranean existence.

Many traits must be analyzed to reconstruct a phylogeny, and systematists use various methods to combine information from the different traits. The simple method we used in our vertebrate example does not work in the vast majority of cases because we know from fossil and other evidence that traits can change more than once, or even undergo reversal. How do systematists deal with these complexities when they reconstruct phylogenies?

The most widely used methods of reconstructing phylogenetic trees employ the **parsimony principle**. In its most general form, the parsimony principle states that one should prefer the simplest hypothesis that is capable of explaining the known facts. Its application to the reconstruction of phylogenies means minimizing the number of evolutionary changes that need to be assumed over all characters in all groups in the tree—that is, the best hypothesis is the one that requires the fewest homoplasies.

Parsimony works best for morphological traits, whose evolutionary rates are generally slow enough that similarities due to homoplasies are uncommon relative to the number of traits retained because they were inherited from the common ancestor.

Another method, called the **maximum likelihood method**, is used primarily for the reconstruction of phylogenies based on molecular data. The computer programs employed in this method are complicated. They are designed to deal with the fact that mutations that result in substitutions of nucleotides are common, but that their frequencies can be estimated independently from other genetic information (see Chapter 24).

Using the parsimony principle is helpful not because evolutionary changes are necessarily parsimonious, but because it is generally wiser not to adopt complicated explanations when simpler ones explain the known facts. More complicated explanations are accepted only when evidence requires them. Phylogenetic trees are hypotheses about evolutionary relationships that are repeatedly tested and modified as additional traits are measured and as new fossil evidence becomes available.

Whatever method is employed, determining the most likely phylogeny for any group of organisms is difficult. For example, there are 34,459,425 possible phylogenetic trees for a lineage with only 11 species! Computer programs using the parsimony principle employ various search routines that calculate the shortest possible phylogenetic tree—that is, with the fewest homoplasies—for a given data set and then compare other possible phylogenies with the shortest one. If, as is usually the case, several

trees are of approximately equal length, they can be merged into a **consensus tree** that retains only those lineage splits that are found in all the most parsimonious trees. In a consensus tree, groups whose relationships differ among the trees form nodes with more than two branches. These nodes are considered "unresolved" because during speciation, a lineage typically splits into only two daughter species.

Traits Used in Reconstructing Phylogenies

Because organisms differ in many ways, systematists use many traits to reconstruct phylogenies. Some of these traits are readily preserved in fossils; others, such as behavior and molecular structure, rarely survive fossilization processes. Systematists take into consideration behavioral and molecular traits as well as structural traits in both living and fossil organisms. The more traits that are measured, the more inferred phylogenies should converge on one another and on the actual evolutionary pattern.

Morphology and development

An important source of information for systematists is **morphology**—that is, the sizes and shapes of body parts. Because living organisms have been studied for centuries, we have a wealth of morphological data, as well as extensive museum and herbarium collections of organisms whose traits can be measured. Sophisticated methods are now available for measuring and analyzing morphology and for estimating the amount of morphological variation among individuals, populations, and species.

The fossil record, which reveals when lineages diverged and began their independent evolutionary histories, can tell us the timing of evolutionary events. Fossils provide important evidence that helps us distinguish ancestral from derived traits. They provide the only available information about where and when organisms lived in the past and what they looked like. When available, this information is valuable, but sometimes few or no fossils have been found for a group whose phylogeny we wish to determine.

The early developmental stages of many organisms reveal similarities to other organisms that are lost by the time of adulthood. For example, the larvae of the marine creatures called sea squirts have a rod in the back—the *notochord*—that disappears as they develop into adults. Many other animals—all the animals called *vertebrates*—also have this structure at some time during their development. This shared structure is one of the reasons for believing that sea squirts are more closely related to vertebrates than would be suspected by examination of the adults only (Figure 23.6).

Molecular traits

Like the sizes and shapes of their body parts, the molecules that make up organisms are heritable characteristics that may diverge among lineages over evolutionary time. Molecular evolution will be discussed in detail in Chapter 24. The molecular traits most useful for constructing phyloge-

Sea squirt
(seen in section)

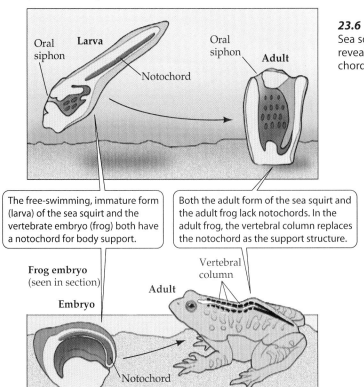

23.6 A Larva Reveals Evolutionary Relationships
Sea squirt larvae, but not adults, have a well-developed notochord that reveals their relationship with vertebrates, all of which have a notochord at some time during their life cycle.

The free-swimming, immature form (larva) of the sea squirt and the vertebrate embryo (frog) both have a notochord for body support.

Both the adult form of the sea squirt and the adult frog lack notochords. In the adult frog, the vertebral column replaces the notochord as the support structure.

moglobin pseudogene (a nonfunctional DNA sequence derived early in primate evolution by duplication of a hemoglobin gene). The outgroup in the analysis was the genus *Ateles*, the New World spider monkeys. The DNA data strongly indicate that chimpanzees and humans share a more recent ancestor with each other than they do with gorillas (Figure 23.7), a conclusion supported by other types of molecular data.

Phylogenetic Trees Have Many Uses

Phylogenetic trees contain information that is useful to scientists investigating a wide variety of biological questions. Here we illustrate how phylogenetic trees are being used to determine how many times a particular trait may have arisen during evolution, and to assess when lineages may have split.

nies are the structures of proteins and nucleic acids (DNA and RNA).

PROTEIN STRUCTURE. Relatively precise information about phylogenies can be obtained by comparison of the molecular structure of proteins. We can estimate genetic differences between two lineages by obtaining homologous proteins from both and determining the number of amino acids that have changed since the lineages diverged from a common ancestor.

DNA BASE SEQUENCES. The base sequences of DNA provide excellent evidence of evolutionary relationships among organisms. The cells of eukaryotes have genes in their mitochondria as well as in their nuclei; plant cells also have genes in their chloroplasts. The chloroplast genome (cpDNA), which is used extensively in phylogenetic studies of plants, has changed little over evolutionary time. Mitochondrial DNA (mtDNA), which evolved much more rapidly than cpDNA, has been used extensively for evolutionary studies of animals.

Relationships among apes and humans were investigated by sequencing more than 10,000 base pairs making up a segment of nuclear DNA that includes a he-

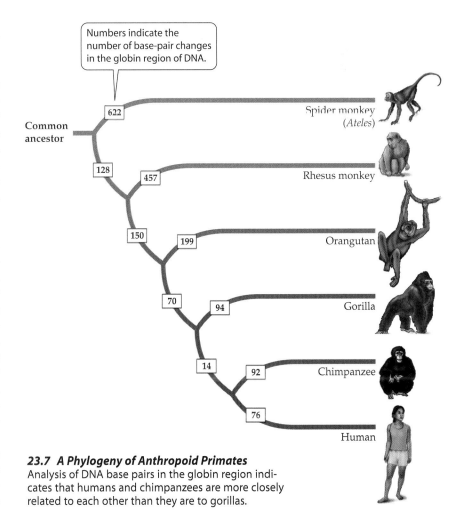

23.7 A Phylogeny of Anthropoid Primates
Analysis of DNA base pairs in the globin region indicates that humans and chimpanzees are more closely related to each other than they are to gorillas.

How Often Have Traits Evolved?

Most flowering plants reproduce by mating with another individual, or *outcrossing*, and have mechanisms to prevent self-fertilization. Many species, however, can fertilize themselves with their own pollen—they are *self-compatible*. How can we tell how often self-compatiblity has evolved in a lineage? We can do so by plotting on a phylogenetic tree which species are outcrossing and which are selfing.

The evolution of fertilization methods was examined in *Linanthus* (a genus in the phlox family), a lineage of plants with a diversity of breeding systems and pollination mechanisms. The outcrossing (self-incompatible) species of *Linanthus* have flowers with long tubes and are pollinated by long-tongued flies. The self-fertilizing (self-compatible) species all have short-tubed flowers.

Investigators reconstructed a phylogeny for 12 species in a section of the genus using the internal-transcriber-spacer (ITS) region of nuclear ribosomal DNA (Figure 23.8). This region was known to be useful for reconstructing species-level phylogenies in other plant groups and had already been used for constructing a phylogeny of the phlox family. The investigators determined whether each species was self-compatible by artificially pollinating flowers with their own or outcrossed pollen and observing the results.

Several lines of evidence suggested that self-incompatibility is the ancestral state in *Linanthus*. First, multiple origins of self-incompatibility are not known in any other flowering plant family. Second, self-incompatibility systems involve physiological mechanisms in both the pollen and the stigma and require the presence of at least three distinct alleles. Therefore, a change from self-incompatibility to self-compatibility is easier than the reverse change. Third, in all self-incompatible species of *Linanthus*, the site of pollen rejection is the stigma, even though sites of pollen rejection vary greatly among other plant groups.

Assuming that self-incompatibility is the ancestral state, the phylogeny suggests that self-compatibility has evolved three times in this *Linanthus* lineage (Figure 23.8). The change to self-compatibility has been accompanied by the evolution of reduced flower size. Interestingly, the striking similarity in flower form among the self-compatible groups had led to their classification as members of a single species. The phylogenetic analysis showed them to be members of three distinct lineages!

When Did Lineages Split?

How fossils can help us determine evolutionary pathways is illustrated by studies of lungfishes. A phylogenetic tree of the three extant genera of lungfishes, all of which are strictly limited to fresh water (Figure 23.9), indicates that the African *Protopterus* and the South American *Lepidosiren* (both in the family Lepidosirenidae) share a more recent common ancestor with each other than with the Australian *Neoceradotus* (family Ceratodontidae). Fossils of each genus are known only from the continent it now inhabits. By itself, this information suggests that the three genera were isolated by the breakup of Gondwana (see Chapter 20).

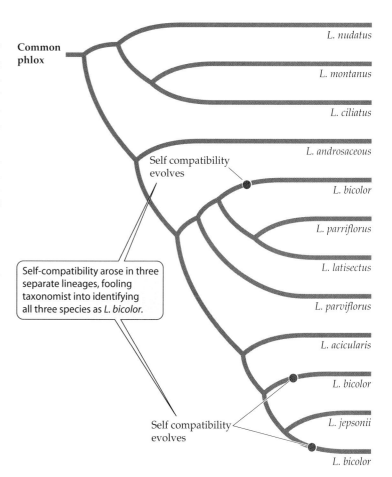

23.8 Phylogeny of a Section of the Phlox Genus Linanthus
Self-compatibility apparently evolved three times in this lineage. Because the form of flowers converged in the selfing lineages, taxonomists mistakenly thought that they were all members of a single species.

However, fossils of other members of the family Ceratodontidae have been found in all continents except South America and Antarctica, and fossil lepidosirenids have been found in both Europe and North America. Thus, the ancestors of both families probably ranged over much of Pangaea. The combination of the phylogenetic tree and fossil evidence informs us that their divergence happened long before the breakup of Gondwana.

Why Classify Organisms?

Classification systems are important for several reasons. They improve our ability to explain relationships among things. They are also an aid to memory. It is impossible to remember the characteristics of many different things unless we can group them into categories based on shared characteristics. They are also useful as predictors. For example, the discovery of biochemical precursors of the drug cortisone in certain species of yams (genus *Dioscorea*) stimulated a successful search for higher concentrations of the drug in other *Dioscorea* species. And, as we saw at the be-

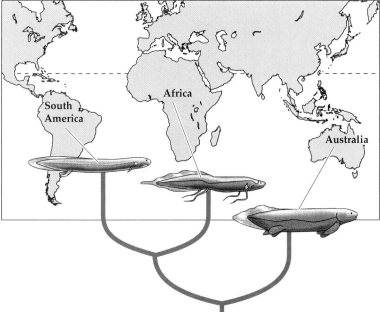

Linnaeus gave each species two names, one identifying the species itself and the other the genus to which it belongs. A **genus** (plural genera; adjectival form, generic) is a group of closely related species. In many cases the name of the taxonomist who first proposed the species name is added at the end. Thus, *Homo sapiens* Linnaeus is the name of the modern human species. *Homo* is the genus to which the species belongs, and *sapiens* identifies the species; Linnaeus proposed the species name *sapiens*. You can think of the generic name *Homo* as equivalent to your surname and the specific name *sapiens* as equivalent to your first name. The generic name is always capitalized; the species name is not. Both names are always italicized, whereas common names are not.

23.9 Evolutionary Pathways in Lungfish Species
In this phylogeny, the ancestor is at the bottom of the figure.

ginning of this chapter, a phylogeny of *Oncomelania* snails is helping to devise methods to control schistosomiasis.

Biological classification systems provide unique names for organisms. If the names are changed, the systems provide a means of tracing the changes. Common names, even if they exist (most organisms have none), are very unreliable and often confusing. For example, plants called bluebells are found in England, Scotland, Texas, and the Rocky Mountains—but none of the bluebells in any of those places is closely related evolutionarily to the bluebells in any of the other places (Figure 23.10).

Recognizing and interpreting similarities and differences among organisms is easier if the organisms are classified into groups that are ordered and ranked. Any group of organisms that is treated as a unit in a biological classification system is called a **taxon** (plural taxa). **Taxonomy** is the theory and practice of classifying organisms.

The Hierarchical Classification of Species

The biological classification system that is used today was developed by the Swedish biologist Carolus Linnaeus in 1758. His two-name system, referred to as *binomial nomenclature*, replaced the cumbersome descriptions biologists had previously used. For example, the honeybee, which had been named *Apis pubescens, thorace subgriseo, abdomine fusco, pedibus posticis glabris utrinque margine ciliatis*, became simply *Apis mellifera*. Binomial nomenclature is universally employed in biology today. Using this system, scientists throughout the world refer to the same organisms by the same names.

(a) *Campanula* sp.

(b) *Endymion nonscriptus*

(c) *Mertensia virginica*

23.10 Many Different Plants Are Called Bluebells
(a) These flowers from the plains of North Dakota are often called bluebells. (b) This English bluebell is a member of the lily family. (c) These are known as Virginia bluebells. None of these plants is closely related to the others.

24 *Molecular and Genomic Evolution*

NEANDERTHALS ARE A GROUP OF EXTINCT HU-man relatives that lived in Europe and eastern Asia from about 300,000 to 30,000 years ago. During part of that time they coexisted with *Homo sapiens*. Some researchers have identified Neanderthals as direct ancestors of modern humans. Others believe that Neanderthals contributed only a few genes to the human gene pool. Still others think that they contributed no genes.

In an attempt to discover which of these three hypotheses is correct, scientists extracted mitochondrial DNA from a section of a leg bone of a Neanderthal fossil between 30,000 and 100,000 years old. The base sequences of the Neanderthal mtDNA fell well outside the variation found in modern human mtDNA. From these results, investigators judged that Neanderthal mtDNA and modern human mtDNA have been evolving separately for at least 500,000 years. This finding provided evidence that Neanderthals contributed few or no genes to the human gene pool.

How can investigators compare the mtDNA of humans and Neanderthals? In this chapter we review how molecular biologists determine the structures of nucleic acids and proteins and use those structures to infer both the patterns and the causes of molecular evolution. With these insights, we explore how the functions of molecules change, where new genes come from, and the evolution of the genomes of organisms. Finally, we show how knowledge of the patterns of molecular evolution helps us solve other biological problems, including inferring phylogenetic relationships among organisms and determining how humans spread over Earth.

What Is Molecular Evolution?

The molecules of interest to molecular evolutionists are nucleotides, nucleic acids, amino acids, and proteins. Nucleic acids evolve by means of nucleotide base substitutions, which in turn result in changes in the amino acids they encode. Alterations in the structure and functioning of proteins result from changes in the ordering of the amino acids of which they are composed. Molecular evolutionists investigate the evolution of these macromolecules to determine how rapidly they have changed and why they have

changed. To do so, they must be able to characterize the precise structures of these macromolecules.

Molecular evolutionists also try to reconstruct the evolutionary histories of genes and organisms, a field known as **molecular phylogenetics**. These two components of the study of molecular evolution are intimately related because phylogenetic information is essential for determining the order of changes in molecular characters, and knowing the order of such changes is usually the first step in inferring their causes. Conversely, knowledge of the pattern and rate of change in a given molecule is crucial for attempts to reconstruct the evolutionary history of a group of organisms.

For most of its history, evolutionary biology depended on the study of the obvious morphological features of organisms. During his 5-year voyage aboard the *Beagle*, Charles Darwin observed morphological differences among species found in different geographic areas. He later synthesized these observations into descriptions of how species change over time. He was able to hypothesize *why*

Neanderthal Bones
DNA recovered from bones of Neanderthals can be used to infer whether Neanderthals contributed many or no genes to the modern human genome. This skeleton was unearthed in 1908 from a cave in France.

many of these morphological changes had happened, but he could not determine *how* they occurred. Understanding the mechanisms of morphological change had to await discoveries in biochemistry a century later.

Even though genetic differences underlie all components of the adaptive evolution of organisms, molecular evolution differs from phenotypic evolution in one important way. In addition to natural selection, random genetic drift and mutation exert important influences on the rates and directions of molecular evolution.

A *mutation*, as you know from Chapter 12, is a change in the sequence of a single copy of a gene (see pages 234–235). A **substitution** is the partial or complete replacement of a nucleotide base or longer sequence by another throughout an entire population or species. It is substitutions that are of interest to molecular evolutionists.

Many mutations in sequences of genes do not alter the proteins encoded by those genes. The reason is that most amino acids are specified by more than one codon. Leucine, for example, is specified by six different codons: UUA, UUG, CUU, CUC, CUA, and CUG (see Figure 12.5; in this and all other cases, most of the redundancy is in the third codon position).

When it occurs throughout a population, a nucleotide substitution that does not change the amino acid specified—UUA to UUG, for example—is known as a **synonymous** or **silent substitution**. Synonymous substitutions are unlikely to affect the functioning of the protein (and hence the organism) and are therefore unlikely to be influenced by natural selection.

Because they are unlikely to be influenced by natural selection, synonymous substitutions are free to accumulate in a population over evolutionary time at rates determined by rates of mutation and genetic drift. Because modern molecular techniques enable us to detect substitutions at the level of nucleotides, molecular evolutionists can measure even these nonfunctional changes.

The occurrence in a population of nucleotide substitutions that *do* change the amino acid that is specified—UUA to UCA, for example, which would result in serine rather than leucine—is known as **nonsynonymous substitution**. In general, nonsynonymous mutations are likely to be deleterious to the individual organism. But even an amino acid change does not necessarily change a protein's shape and, hence, its functional properties. Therefore, a nonsynonymous substitution may be selectively neutral, or nearly so.

Most natural populations of organisms harbor much more genetic variation than we would expect if genetic variation were influenced primarily by natural selection. This discovery, combined with the knowledge that many substitutions do not change molecular function, stimulated the development of the neutral theory of molecular evolution.

The neutral theory, first articulated by Motoo Kimura in 1968, postulates that, at the molecular level, the majority of mutations are selectively neutral: they confer neither an advantage nor a disadvantage on their bearers. If so, the majority of evolutionary changes in macromolecules, and much of the genetic variation within species, results from

neither positive selection of advantageous alleles nor stabilizing selection, but from random genetic drift.

To see why this is so, consider a population with a size of N and a rate of neutral mutation at a particular locus of μ per gamete per generation. The number of new mutations would on average be $\mu \times 2N$, because $2N$ gene copies are available to mutate. According to genetic drift theory (see Chapter 21), the probability that a mutation will be fixed by genetic drift is its frequency, p, which equals $1/(2N)$ for a newly arisen (and hence very rare) mutation. Therefore, the number of neutral mutations that arise per generation that are likely to become fixed is $2N\mu \times 1/(2N) = \mu$, which equals the mutation rate.

In other words, the rate of fixation of mutations is theoretically constant and is equal to the neutral mutation rate. This is the theoretical basis of the concept of the **molecular clock**, which states that macromolecules should diverge from one another over time at a constant rate. We will discuss molecular clocks later in this chapter and show how, with care, the concept can be used to study many features of molecular evolution.

According to the neutral theory of molecular evolution, most polymorphisms at specific genetic loci are transitory rather than stable, because the frequency of neutral alleles in a population should change slowly over time (Figure 24.1*a*). In contrast, advantageous mutations are rapidly fixed in a population, and deleterious mutations are quickly lost (Figure 24.1*b*). The neutral theory and the theory of natural selection agree that most mutations are deleterious, but the neutral theory asserts that the selective ad-

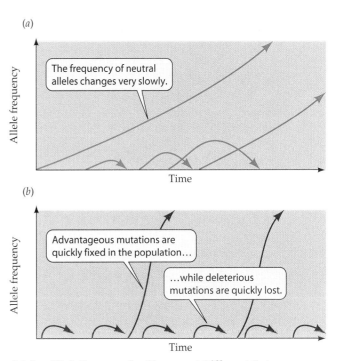

(a)

(b)

24.1 Allele Frequencies Change at Different Rates
(a) The frequencies of neutral alleles change slowly. Much polymorphism in these alleles is transitory. *(b)* Alleles carrying advantageous mutations become fixed in populations while disadvantageous mutations are eliminated; these allele shifts usually occur rapidly.

vantages or disadvantages of most molecular mutations are so small that selection on them is too weak to offset the influences of genetic drift.

Determining and Comparing the Structure of Macromolecules

To reveal patterns of molecular evolution, biologists may determine the precise structure of biological molecules. An investigator attempting to determine the structure of a nucleic acid or a protein begins by extracting and purifying it from a natural source. The molecule can then be analyzed by X ray crystallography. The molecule is crystallized, and the crystal is bombarded with a beam of X rays. The regularly spaced atoms in the crystal deflect the X rays into an orderly array of spots on a photographic film. With data from successive cross-sections through the crystal, a computer can generate a three-dimensional electron density map of the molecule. Graphics software enables the computer to create a picture showing the position of each atom in the molecule (Figure 24.2).

The base sequences of nucleic acids also provide important information about evolutionary histories. The invention of the polymerase chain reaction (PCR) technique (see Chapter 11) allowed biologists to determine the sequence of regions of DNA not only from living tissues, but also from fossilized remains, mummified tissues, dried skins in museums, and pressed plants in herbaria, even though these objects contain only tiny amounts of DNA. DNA has been extracted and amplified from human fossils more than

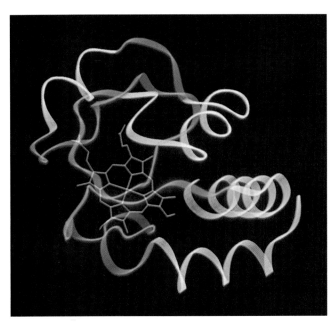

24.2 Computer Graphic Shows the Positions of Atoms in Molecules
The positions of atoms and the three-dimensional structure of tuna cytochrome *c* were computed from data generated by cross-sections through the crystallized molecule.

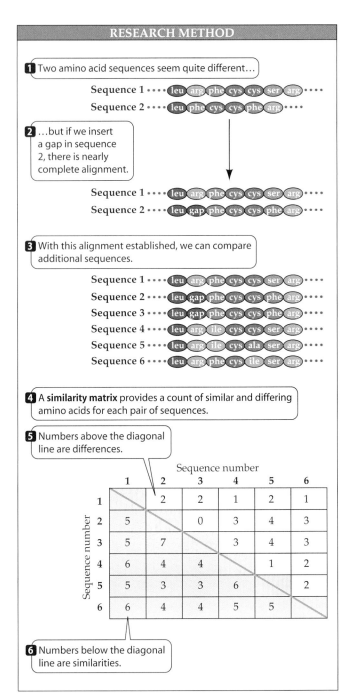

24.3 Amino Acid Sequence Alignment
Inserting a gap allows us to align two sequences so that we can compare homologous amino acids. Once the alignment is established, more sequences can be added and compared. The larger the number of similarities, the more recent the presumed common ancestor of the species.

30,000 years old, plant leaf fossils 40,000 years old, and insects fossilized in amber 135,000,000 years ago.

Once the sequences of amino acids in molecules from different organisms have been determined, they must be compared. A simple example illustrates how this is done. In Figure 24.3, two amino acid sequences (1 and 2) are compared. The two sequences come from homologous proteins in different organisms, and they differ in number and identity of amino acid residues. Our goal is to align these sequences so that we can compare homologous portions of the protein. To do so, we first observe that, although the sequences appear quite different, they would become similar if we were to insert a gap after the first amino acid in sequence 2 (after the leucine residue). In fact, these sequences then differ by only one amino acid at position 6 (serine or phenylalanine). A single insertion aligns the sequences in this case, but longer sequences and those that have diverged more extensively require more elaborate adjustments.

After we have aligned the sequences, we can compare them in several ways. First, we can simply count the number of nucleotides or amino acids that differ between the sequences. Let's add some more sequences to our previous example and compare them with our original two sequences. By adding up the number of similar and different amino acids in the sequences, we can construct a *similarity matrix* (see Figure 24.3). The assumption is that the longer the molecules have been evolving separately, the more differences they will have.

Enough analyses of mammalian genes have been performed to show that the rate of nonsynonymous nucleotide substitution in mammals varies from nearly zero to about 3×10^{-9} substitutions per site per year. Synonymous substitutions in the protein-coding regions of nuclear genes have occurred about 5 times more rapidly than nonsynonymous substitutions; in other words, substitution rates are highest at codon sites that do not change the amino acid being expressed (Figure 24.4). The rate of substitution is even higher in pseudogenes—duplicate copies of genes that have undergone one or more mutations that eliminate their ability to be expressed.

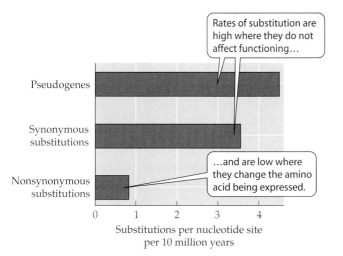

24.4 Rates of Base Substitution Differ
Rates of nonsynonymous substitutions in mammals are much slower than rates of synonymous substitutions and substitutions in pseudogenes.

Why do rates of nucleotide substitution vary so greatly?

The fact that rates of nucleotide substitution are highest at sites and in molecules where they have no functional significance is consistent with the hypothesis that substitution rates at these sites are driven primarily by a combination of mutation and genetic drift. The much slower rates of substitution at sites that *do* affect molecular function is consistent with the view that most nonsynonymous mutations are disadvantageous and are eliminated from the population by natural selection. An interesting consequence of these processes is that, in general, *the more essential a molecule is for cell functioning, the slower the rate of its evolution.*

A molecule that illustrates this principle is the enzyme cytochrome *c*, one component of the respiratory chain of mitochondria. Together with other proteins of the citric acid cycle and respiratory chain, cytochrome *c* is found in all eukaryotes. The amino acid sequences of cytochrome *c* are known for more than 100 species of organisms, including microbial eukaryotes, plants, fungi, and mammals. Within these cytochromes *c* are regions that accumulated changes relatively quickly; for example, positions 44, 89, and 100 differ among many of the organisms compared (Figure 24.5 on pages 442–443).

There are also invariant positions, such as 14, 17, 18, and 80. This particular set of invariant residues is known to interact with the iron-containing heme group that is essential for the functioning of the enzyme. Presumably, because any mutations that changed these amino acids diminished the functioning of the heme group, they were removed by natural selection when they arose.

Using biological molecules as molecular clocks

Earlier in this chapter, we stated the theoretical basis for expecting macromolecules to evolve at constant rates. But do they actually behave as the theory says they should? For example, if we plot the time since the divergence of certain organisms, as determined by the fossil record, against the number of amino acids by which their cytochromes *c* differ, we find that differences in cytochrome *c* sequences have evolved at a relatively constant rate (Figure 24.6).

Many other proteins show constancy in the rate at which they have accumulated changes over time. It would be convenient if the rates of change were the same for all protein molecules. Unfortunately, different molecular clocks tick at different rates. These differences exist because proteins differ in the nature of functional constraints on their evolution.

Despite these differences, the rates at which many molecular clocks tick appear to be relatively constant. This is especially true for nucleotide or amino acid substitutions that do not affect the functioning of the molecule and, hence, the fitness of the organism. Even if the rate of ticking of a molecular clock changes slightly over time, the variations may not be great enough to seriously affect our estimates of the dates of divergences of gene and organism

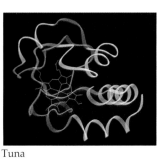

Tuna

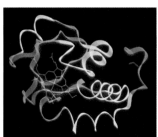

Rice

Acidic side chains:
- D Aspartic acid
- E Glutamic acid

Basic side chains:
- H Histidine
- K Lysine
- R Arginine

Hydrophobic side chains:
- F Phenylalanine
- I Isoleucine
- L Leucine
- M Methionine

- V Valine
- Y Tyrosine
- W Tryptophan
- A Alanine

Other:
- C Cysteine
- P Proline
- Q Glutamine
- N Asparagine
- S Serine
- T Threonine
- G Glycine

> The number 1 indicates an invariant position in the cytochrome *c* molecule (i.e., all the organisms have the same amino acid in this position) and that the position is probably functionally very significant.

> Side chains marked by red arrows interact with the heme group.

Position in sequence: 1 ... 5 ... 10 ... 15 ... 20 ... 25 ... 30

Number of amino acids in different organisms at the position shown: 1 3 5 5 5 1 3 3 4 1 4 3 2 1 3 3 1 1 2 4 3 4 2 3 4 2 1 4 1 1 2 1 5 1

Human, chimpanzee	G D V E K G K K I F I M K C S Q C H T V E K G G K H K T G P N L H G
Rhesus monkey	G D V E K G K K I F I M K C S Q C H T V E K G G K H K T G P N L H G
Horse	G D V E K G K K I F V Q K C A Q C H T V E K G G K H K T G P N L H G
Donkey	G D V E K G K K I F V Q K C A Q C H T V E K G G K H K T G P N L H G
Cow, pig, sheep	G D V E K G K K I F V Q K C A Q C H T V E K G G K H K T G P N L H G
Dog	G D V E K G K K I F V Q K C A Q C H T V E K G G K H K T G P N L H G
Rabbit	G D V E K G K K I F V Q K C A Q C H T V E K G G K H K T G P N L H G
Gray whale	G D V E K G K K I F V Q K C A Q C H T V E K G G K H K T G P N L H G
Gray kangaroo	G D V E K G K K I F V Q K C A Q C H T V E K G G K H K T G P N L N G
Chicken, turkey	G D I E K G K K I F V Q K C S Q C H T V E K G G K H K T G P N L H G
Pigeon	G D I E K G K K I F V Q K C S Q C H T V E K G G K H K T G P N L H G
Pekin duck	G D V E K G K K I F V Q K C S Q C H T V E K G G K H K T G P N L H G
Snapping turtle	G D V E K G K K I F V Q K C A Q C H T V E K G G K H K T G P N L N G
Rattlesnake	G D V E K G K K I F T M K C S Q C H T V E K G G K H K T G P N L H G
Bullfrog	G D V E K G K K I F V Q K C A Q C H T C E K G G K H K V G P N L Y G
Tuna	G D V A K G K K T F V Q K C A Q C H T V E N G G K H K V G P N L W G
Dogfish	G D V E K G K K V F V Q K C A Q C H T V E N G G K H K T G P N L S G
Samia cynthia (moth)	G N A E N G K K I F V Q R C A Q C H T V E A G G K H K V G P N L H G
Tobacco hornworm moth	G N A D N G K K I F V Q R C A Q C H T V E A G G K H K V G P N L H G
Screwworm fly	G D V E K G K K I F V Q R C A Q C H T V E A G G K H K V G P N L H G
Drosophila (fruit fly)	G D V E K G K K L F V Q R C A Q C H T V E A G G K H K V G P N L H G
Baker's yeast	G S A K K G A T L F K T R C E L C H T V E K G G P H K V G P N L H G
Candida krusei (yeast)	G S A K K G A T L F K T R C A E C H T I E A G G P H K V G P N L H G
Neurospora crassa (mold)	G D S K K G A N L F K T R C A E C H · E · · N L T Q K I G P A L H G
Wheat	G N P D A G A K I F K T K C A Q C H T V D A G A · · H K Q G P N L H G
Sunflower	G D P T T G A K I F K T K C A Q C H T V E K G A · · H K Q G P N L N G
Mung bean	G D S K S G E K I F K T K C A Q C H T V D K G A · · H K Q G P N L N G
Rice	G N P K A G E K I F K T K C A Q C H T V D K G A · · H K Q G P N L N G
Sesame	G D V K S G E K I F K T K C A Q C H T V D K G A · · H K Q G P N L N G

24.5 Amino Acid Sequences of Cytochrome c
The two computer graphics show how similar the three-dimensional structure of tuna and rice cytochrome *c* are. The amino acid sequences shown here were obtained from analyses of cytochromes *c* from 33 species of plants, fungi, and animals.

lineages. By comparing the rates of a variety of molecular clocks, further insights can be gained into why different protein molecules have evolved at such different rates.

Where Do New Genes Come From?

The earliest forms of life must have had very few organized nucleic acid sequences. Because we believe that life is monophyletic—that all living organisms arose from a single ancestor—the many thousands of different functional genes in modern organisms must have arisen from these few ancestral genes. How has this happened? By far the most important process appears to be gene duplication.

Gene duplication may involve part of a gene, a single gene, parts of a chromosome, an entire chromosome, or the whole genome (see Chapter 14). We saw in Chapter 22 that duplication of the entire genome (polyploidy) has been important in speciation. Polyploid individuals are usually vi-

able because all of their chromosomes are duplicated, so that they avoid imbalances in gene expression. As we have already discussed, polyploidy is widespread among plants. Genome duplication was probably widespread among animals before the sex chromosomes became differentiated. Among organisms with differentiated sex chromosomes, however, genome duplication disrupts the mechanisms of sex determination.

Duplications of part or all of a chromosome are probably unimportant as sources of new genes because they typically result in severe imbalances in gene products. *Drosophila* in which more than half of one arm of a chromosome is present in three doses (trisomy) do not survive. In humans, trisomies larger than one chromosome are lethal; even smaller ones result in sterility. For example, individuals having three copies of chromosome 21 have Down syndrome, and are usually sterile. Therefore, duplications of whole chromosomes or parts of chromosomes generally are not passed along to any offspring.

Duplication of genes can lead to new gene families

The two identical copies of a gene produced by gene duplication may retain their original function, with the result that the organism produces larger quantities of their RNA

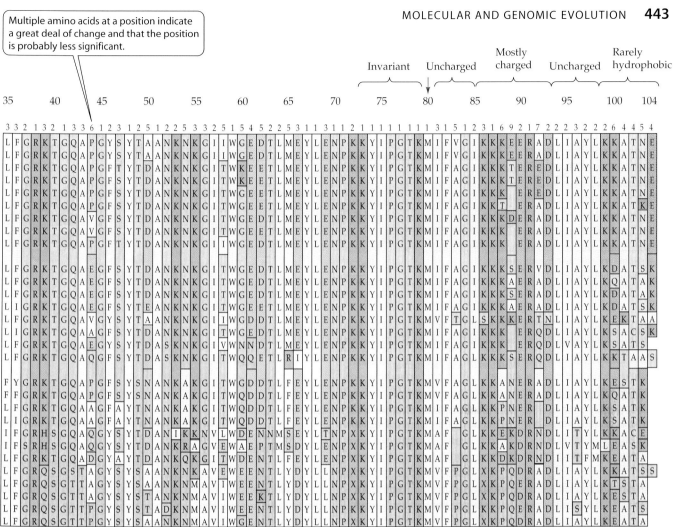

or protein product. Alternatively, one copy may be incapacitated by the accumulation of deleterious mutations and become a functionless pseudogene. More importantly for evolution, one copy may retain its original function while the other accumulates enough mutations that it can perform a different task. Several successive rounds of duplication may result in a **gene family**, a group of homologous genes with related functions. Members of a gene family are often arrayed in tandem along a chromosome.

Molecular evolution by gene duplication has been well studied in the globin gene family (see Chapter 14). Globins were among the first proteins to be sequenced and their amino acid sequences compared. Humans have three families of globin genes: the myoglobin family, whose single member is located on chromosome 22; the α-globin family, on chromosome 16; and the β-globin family, on chromosome 11 (see Figure 14.9).

Two types of proteins are produced by these three gene families: myoglobin and hemoglobin. Comparisons of their

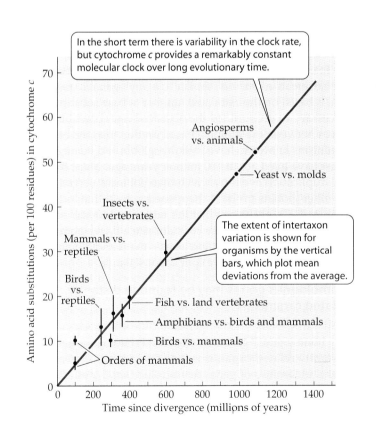

24.6 Cytochrome c Molecules Evolved at a Constant Rate
Rates of substitution in cytochrome *c* are constant enough that this molecule can be used as a molecular clock.

These data sound promising, but things aren't as easy as we might wish. When biologists examined other genes and RNA's, contradictions began to appear. Analyses of different nucleotide sequences suggested different phylogenetic patterns. How could such a situation have arisen?

 Lateral gene transfer muddied the phylogenetic waters

It is now clear that, from early in evolution to the present day, genes have been moving among prokaryotic species by **lateral gene transfer**. As we have seen, a gene from one species can become incorporated into the genome of another. Mechanisms of lateral gene transfer include transfer by plasmids and viruses and uptake of DNA by transformation. Such transfers are well documented, not just between bacterial species or archaean species, but also across the boundary between bacteria and archaea.

A gene that has been transferred will be inherited by the recipient's progeny and in time will be recognized as part of the normal genome of the descendants. Biologists are still assessing the extent of lateral gene transfer among prokaryotes and its implications for phylogeny, especially at the early stages of evolution.

There is great controversy now over prokaryotic phylogeny. Figure 26.11 is an overview of some major groups in the domains Bacteria and Archaea that we will discuss further in this chapter. Keep in mind that a new picture will likely emerge within the next decade, based on the addition of more nucleotide sequence data and on new information about the currently understudied archaea.

Mutations are the most important source of prokaryotic variation

Assuming that the prokaryote groups we are about to describe do indeed represent monophyletic groups, we discover that they are amazingly complex. A single group of bacteria or archaea may contain the most extraordinarily diverse species, and a species in one group may be phenotypically almost indistinguishable from one or many species in another group. What are the sources of this diversity?

Although prokaryotes can acquire new alleles by transformation, transduction, or conjugation, the most important source of genetic variation in populations of prokaryotes is probably mutation. Mutations, especially recessive mutations, are slow to make their presence felt in populations of humans and other diploid organisms. In contrast, a mutation in a prokaryote, which is haploid, has immediate consequences for that organism. If it is not lethal, it will be transmitted to and expressed in the organism's daughter cells—and in their daughter cells, and so forth. Thus, a beneficial mutant allele spreads rapidly.

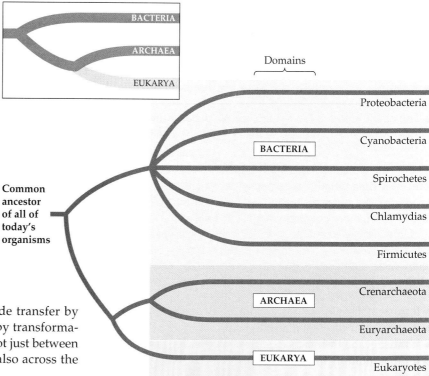

26.11 Two Domains: A Brief Overview
An abridged summary classification of the domains Bacteria and Archaea shows their relationships to each other and to the Eukarya. The relationships among the many lineages of bacteria, not all of which are listed here, are unresolved at this time.

The rapid multiplication of many prokaryotes, coupled with mutation, selection, and genetic drift, allows rapid phenotypic changes within their populations. Important changes, such as loss of sensitivity to an antibiotic, can occur over broad geographic areas in just a few years. Think how many significant metabolic changes could have occurred over even modest time spans in relation to the history of life on Earth. When we introduce the Proteobacteria, the largest group of bacteria, you will see that its different subgroups easily and rapidly adopted and abandoned metabolic pathways under selective pressure from their environments.

The Bacteria

The great majority of known prokaryotes are bacteria. Here we will describe bacterial diversity using a currently popular classification scheme that enjoys considerable support from nucleotide sequence data. More than a dozen monophyletic groups have been proposed under this scheme; we will describe just of a few of them here. The higher-order relationships among these groups of prokaryotes are not known. Some biologists describe them as kingdoms, some as subkingdoms, others as phyla. Here we call them groups. We'll pay the closest attention to the Proteobacteria, Cyanobacteria, Spirochetes, Chlamydias, and Firmicutes (see Figure 26.11), but first we mention one property that is shared by members of three other groups.

Some bacteria are heat lovers

Three of the bacterial groups that may have branched out earliest during bacterial evolution are all **thermophiles** (heat lovers), as are the most ancient of the archaea. This observation supports the hypothesis that the first living organisms were thermophiles that appeared in an environment much hotter than those that predominate today.

The Proteobacteria are a large and diverse group

By far the largest group of bacteria, in terms of number of described species, is the **Proteobacteria**, sometimes referred to as the *purple bacteria*. Among the proteobacteria are many species of Gram-negative, bacteriochlorophyll-containing, sulfur-using photoautotrophs. However, this group also includes a dramatically diverse group of bacteria that bear no resemblance to the purple bacteria in phenotype. The mitochondria of eukaryotes were derived from proteobacteria by endosymbiosis.

No characteristic demonstrates the diversity of the proteobacteria more clearly than their metabolic pathways (Figure 26.12). The common ancestor of all the proteobacteria was probably a photoautotroph. Early in evolution, two groups of proteobacteria lost their ability to photosynthesize and have been chemoheterotrophs ever since. The other three groups still have photoautotrophic members, but in *each* group, some evolutionary lines have abandoned photoautotrophy and taken up other modes of nutrition. There are chemoautotrophs and chemoheterotrophs in all three groups. Why? We can view each of the trends in Figure 26.12 as an evolutionary response to selective pressures encountered as these bacteria encountered new habitats that presented new challenges and opportunities.

Among the proteobacteria are some nitrogen-fixing genera such as *Rhizobium* (see Figure 34.10) and other bacteria that contribute to the global nitrogen and sulfur cycles. *E. coli*, one of the most studied organisms on Earth, is a proteobacterium. So, too, are many of the most famous human pathogens, such as *Yersinia pestis*, *Vibrio cholerae*, and *Salmonella typhimurium*, all mentioned in our discussion of pathogens.

Fungi cause most plant diseases, and viruses cause others, but about 200 plant diseases are of bacterial origin. *Crown gall*, with its characteristic tumors (Figure 26.13), is one of the most striking. The causal agent of crown gall is *Agrobacterium tumefaciens*, which harbors a plasmid used in recombinant DNA studies as a vehicle for inserting genes into new plant hosts (see Chapter 17).

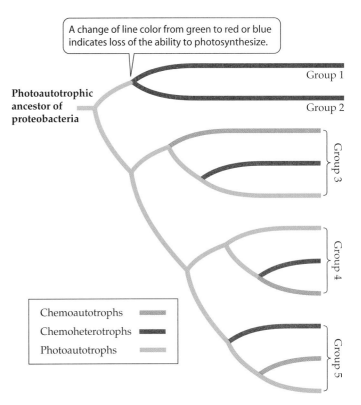

A change of line color from green to red or blue indicates loss of the ability to photosynthesize.

Photoautotrophic ancestor of proteobacteria

Group 1
Group 2
Group 3
Group 4
Group 5

Chemoautotrophs
Chemoheterotrophs
Photoautotrophs

26.12 The Evolution of Metabolism in the Proteobacteria
The common ancestor of all proteobacteria was probably a photoautotroph. As they encountered new environments, groups 1 and 2 lost the ability to photosynthesize; in the other three groups, some evolutionary lines became chemoautotrophs or chemoheterotrophs.

Proteobacteria
Cyanobacteria
Spirochetes
Chlamydias
Firmicutes
Crenarchaeota
Euryarchaeota
Eukaryotes

Crown gall is a plant disease caused by the Gram-negative rod Agrobacterium tumefaciens.

26.13 Crown Gall
This colorful tumor is a crown gall growing on the stem of a geranium plant.

The Cyanobacteria are important photoautotrophs

The **Cyanobacteria** (blue-green bacteria) require only water, nitrogen gas, oxygen, a few mineral elements, light, and carbon dioxide to survive. They use chlorophyll *a* for photosynthesis and liberate oxygen gas; many species also fix nitrogen. Their photosynthesis was the basis of the "oxygen revolution" that transformed Earth's atmosphere.

Cyanobacteria carry out the same type of photosynthesis that is characteristic of eukaryotic photosynthesizers. They contain elaborate and highly organized internal membrane systems called photosynthetic lamellae, or *thylakoids* (Figure 26.14). The chloroplasts of photosynthetic eukaryotes are derived from an endosymbiotic cyanobacterium.

Cyanobacteria may associate in colonies or live free as single cells. Depending on the species and on growth conditions, colonies of cyanobacteria may range from flat sheets one cell thick to spherical balls of cells.

Some filamentous colonies differentiate into three cell types: vegetative cells, spores, and heterocysts (Figure 26.15). Vegetative cells photosynthesize, **spores** are resting cells that can eventually develop into new filaments, and **heterocysts** are cells specialized for nitrogen fixation. All of the known cyanobacteria with heterocysts fix nitrogen. Heterocysts also have a role in reproduction: When filaments break apart to reproduce, the heterocyst may serve as a breaking point.

Proteobacteria
Cyanobacteria
Spirochetes
Chlamydias
Firmicutes
Crenarchaeota
Euryarchaeota
Eukaryotes

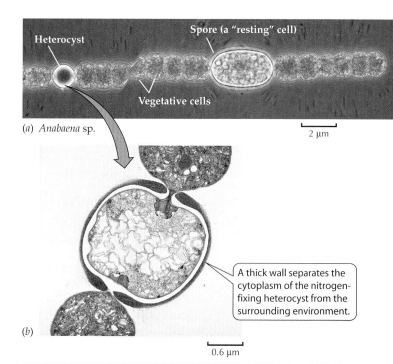

(a) *Anabaena* sp.

Heterocyst
Spore (a "resting" cell)
Vegetative cells

2 μm

(b)

A thick wall separates the cytoplasm of the nitrogen-fixing heterocyst from the surrounding environment.

0.6 μm

(c)

26.15 Cyanobacteria

(a) *Anabaena* is a genus of colonial, filamentous cyanobacteria. The vegetative cells are photosynthetic. (b) A thin neck attaches a heterocyst to each of two other cells in a colony. (c) Cyanobacteria appear in enormous numbers in some environments. This California pond has experienced eutrophication: Phosphorus and other nutrients generated by human activity have accumulated in the pond, feeding an immense green mat—commonly referred to as "pond scum"—made up of several species of unicellular cyanobacteria.

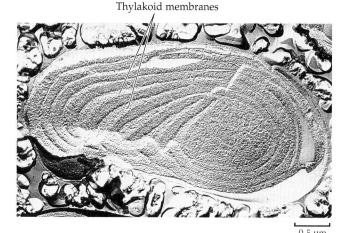

Thylakoid membranes

0.5 μm

26.14 Thylakoids in Cyanobacteria

This cyanobacterium was prepared by the freeze-etch technique to emphasize the extensive system of internal membranes. These photosynthetic thylakoid membranes are present through most of the cytoplasm and clearly identify the specimen as a cyanobacterium, even though the exact species is not identified here.

Spirochetes look like corkscrews

Spirochetes are Gram-negative bacteria characterized by unique structures called **axial filaments**, fibrils running through the periplasmic space (see Figure 26.5a). The cell body is a long cylinder coiled into a spiral (Figure 26.16). The axial filaments begin at either end of the cell and overlap in

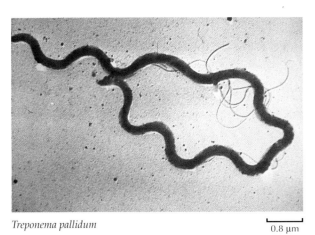

Treponema pallidum

0.8 μm

26.16 A Spirochete
This corkscrew-shaped spirochete causes syphilis in humans.

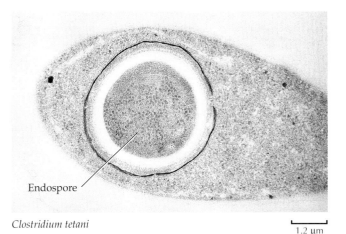

Endospore

Clostridium tetani

1.2 μm

26.18 The Endospore: A Structure for Waiting Out Bad Times
This firmicute, which causes tetanus, produces endospores as resistant resting structures.

the middle, and there are typical basal rings where they are attached to the cell wall. Many spirochetes live in humans as parasites. Others live free in mud or water.

Chlamydias are extremely small

Chlamydias are among the smallest bacteria (0.2–1.5 μm in diameter). They can live only as parasites within the cells of other organisms. These tiny spheres are unique prokaryotes because of their complex life cycle, which involves two different forms of cells (Figure 26.17). In humans, various strains of chlamydias cause eye infections (especially trachoma), sexually transmitted disease, and some forms of pneumonia.

Most Firmicutes are Gram-positive

The **Firmicutes** are sometimes referred to as the *Gram-positive bacteria*, but some firmicutes are Gram-negative, and some have no cell wall at all. Nonetheless, the firmicutes constitute a monophyletic group.

Some firmicutes produce **endospores** (Figure 26.18)—heat-resistant resting structures—when nutrients become scarce. The bacterium replicates its DNA and encapsulates one copy, along with some of its cytoplasm, in a tough cell wall heavily thickened with peptidoglycan and surrounded by a spore coat. The parent cell then breaks down, releasing the endospore. *Endospore production is not a reproductive process*; the endospore merely replaces the parent cell. The endospore can survive harsh environmental conditions, such as high or low temperatures or drought, because it is *dormant*—its normal activity is suspended. Later, if it encounters favorable conditions, the endospore becomes metabolically active and divides, forming new cells like the parent. Some endospores apparently can be reactivated even after more than a thousand years of dormancy.

Members of this endospore-forming group include the many species of *Bacillus* and *Clostridium*. The toxins produced by *C. botulinum* are among the most poisonous ever discovered; the lethal dose for humans is about one-millionth of a gram (1 μg).

The genus *Staphylococcus*—the staphylococci—includes firmicutes that are abundant on the human body surface; they are responsible for boils and

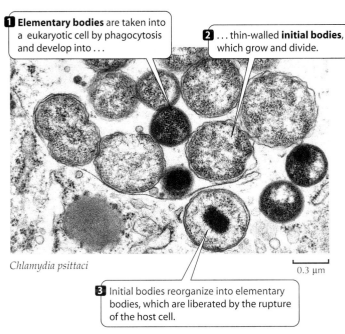

1 **Elementary bodies** are taken into a eukaryotic cell by phagocytosis and develop into . . .

2 . . . thin-walled **initial bodies**, which grow and divide.

3 Initial bodies reorganize into elementary bodies, which are liberated by the rupture of the host cell.

Chlamydia psittaci

0.3 μm

26.17 Chlamydias Change Form during Their Life Cycle
Elementary bodies and initial bodies are the two major phases of the life cycle of a chlamydia.

Proteobacteria

Cyanobacteria

Spirochetes

Chlamydias

Firmicutes

Crenarchaeota

Euryarchaeota

Eukaryotes

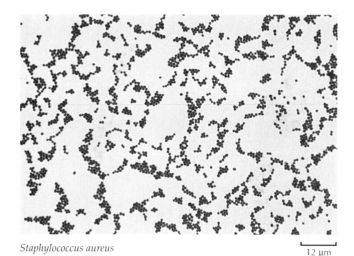

Staphylococcus aureus
12 μm

26.19 Gram-Positive Firmicutes
"Grape clusters" are the usual arrangement of Gram-positive staphylococci.

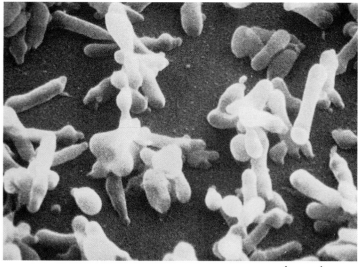

Mycoplasma gallisepticum
0.4 μm

26.21 The Tiniest Living Cells
Containing only about one-fifth as much DNA as *E. coli*, mycoplasmas are the smallest known bacteria.

many other skin problems (Figure 26.19). *S. aureus* is the best-known human pathogen; it is found in 20 to 40 percent of normal adults (and in 50 to 70 percent of hospitalized adults). It can cause respiratory, intestinal, and wound infections, in addition to skin diseases.

Actinomycetes are firmicutes that develop an elaborately branched system of filaments (Figure 26.20). These bacteria closely resemble the filamentous bodies of fungi. Some actinomycetes reproduce by forming chains of spores at the tips of the filaments. In the species that do not form spores, the branched, filamentous growth ceases and the structure breaks up into typical cocci or rods, which then reproduce by fission.

The actinomycetes include several medically important bacteria. *Mycobacterium tuberculosis* causes tuberculosis. *Streptomyces* produces streptomycin, as well as hundreds of

other antibiotics, including several dozen in general use. We derive most of our antibiotics from members of the actinomycetes.

Another interesting group of firmicutes, the **mycoplasmas**, lack cell walls, although some have a stiffening material outside the plasma membrane. Some of them are the smallest cellular creatures ever discovered—they are even smaller than chlamydias (Figure 26.21). The smallest mycoplasmas capable of growth have a diameter of about 0.2 μm, and they are small in another crucial sense: They have less than half as much DNA as do most other prokaryotes. It has been speculated that the amount of DNA in a mycoplasma may be the minimum amount required to code for the essential properties of a living cell.

The Archaea

The domain Archaea consists mainly of prokaryotic genera that live in habitats notable for characteristics such as extreme salinity (salt content), low oxygen concentration, high temperature, or high or low pH. On the face of it, the Archaea do not seem to belong together as a group; in fact, some evidence suggests that the domain Archaea is paraphyletic. One current classification scheme treats the domain as two kingdoms: **Euryarchaeota** and **Crenarchaeota**. In fact, we know very little about the phylogeny of archaea, in part because the study of archaea is still in its early stages. We do know that archaea share certain characteristics.

The Archaea share some unique characteristics

Two characteristics shared by all archaea are the absence of peptidoglycan in their cell walls and the presence of lipids of distinctive composition in their cell membranes (see Table 26.1). The base sequences of their ribosomal RNA's support a close evolutionary relationship among them. Their separation from the Bacteria and Eukarya was clari-

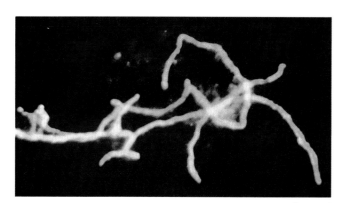

Actinomyces israelii
10 μm

26.20 Filaments of an Actinomycete
These branching filaments are visualized with a fluorescent stain. This species is part of the normal flora in the human tonsils, mouth, intestinal tract, and lungs, but will invade body tissues and cause severe abscesses when afforded the opportunity.

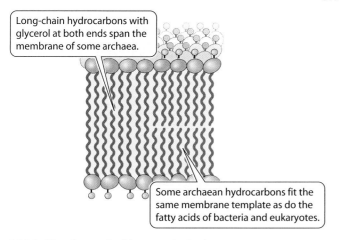

Long-chain hydrocarbons with glycerol at both ends span the membrane of some archaea.

Some archaean hydrocarbons fit the same membrane template as do the fatty acids of bacteria and eukaryotes.

26.22 Membrane Architecture in Archaea
The long-chain hydrocarbons of archaean membranes are branched, and may have glycerol at both ends. This structure still fits into a biological membrane, however; in fact, all three domains have similar membrane structures.

fied when biologists sequenced the first archaean genome: It consisted of 1,738 genes, *more than half of which* were unlike any genes ever found in the other two domains.

The unusual lipids in the membranes of archaea deserve some description. They are found in all archaea, and in no bacteria or eukaryotes. Most membrane lipids of bacteria and eukaryotes contain unbranched long-chain fatty acids connected to glycerol by **ester linkages**:

$$ \begin{array}{ccc} \text{O} & & \text{H} \\ \| & & | \\ -\text{C}-\text{O}-\text{C}- \\ & & | \\ & & \text{H} \end{array} $$

(see also Figure 3.19). In contrast, archaean membrane lipids contain long-chain hydrocarbons connected to glycerol by **ether linkages**:

$$ \begin{array}{ccc} \text{H} & & \text{H} \\ | & & | \\ -\text{C}-\text{O}-\text{C}- \\ | & & | \\ \text{H} & & \text{H} \end{array} $$

In addition, the long-chain hydrocarbons are branched in the archaea. One class of these lipids contains glycerol at *both* ends of the hydrocarbons. This structure still fits in a biological membrane, as shown in Figure 26.22. In spite of the striking difference in membrane lipids, all three domains have lipid bilayer membranes with similar overall structures, dimensions, and functions.

Proteobacteria

Cyanobacteria

Spirochetes

Chlamydias

Firmicutes

Crenarchaeota

Euryarchaeota

Eukaryotes

Most Crenarchaeota live in hot, acidic places

Most known Crenarchaeota are both thermophilic (heatloving) and *acidophilic* (acid-loving). Members of the genus *Sulfolobus* live in hot sulfur springs at temperatures of 70–75°C. They die of "cold" at 55°C (131°F). Hot sulfur springs are also extremely acidic. *Sulfolobus* grows best in the pH range from 2 to 3, but it readily tolerates pH values as low as 0.9. Some acidophilic hyperthermophiles maintain an internal pH near 7 (neutral) in spite of the acidity of their environment. These and other hyperthermophiles thus thrive where very few other organisms can even survive (Figure 26.23).

The Euryarchaeota live in many amazing places

Some species of Euryarchaeota, once assigned to unrelated bacterial groups, share the property of producing methane (CH_4) by reducing carbon dioxide. All of these *methanogens* are obligate anaerobes, and methane production is the key step in their energy metabolism. Comparison of rRNA nucleotide sequences revealed a close evolutionary relationship among all these methanogens.

Methanogens release approximately 2 billion tons of methane gas into Earth's atmosphere each year, accounting for all the methane in our air, including that associated with mammalian belching. Approximately a third of this methane comes from methanogens in the guts of grazing herbivores such as cows.

26.23 Some Would Call It Hell; Archaea Call It Home
Masses of heat- and acid-loving archaea form an orange mat inside a volcanic vent on the island of Kyushu, Japan. Sulfurous residue is visible at the edges of the archaean mat.

26.24 Extreme Halophiles
Commercial seawater evaporating ponds, such as these in San Francisco Bay, are attractive homes for salt-loving archaea.

One methanogen, *Methanopyrus*, lives on the ocean bottom near blazing volcanic vents. *Methanopyrus* can survive and grow at 110°C. It grows best at 98°C and not at all at temperatures below 84°C.

Another group of Euryarchaeota, the *extreme halophiles* (salt lovers), lives exclusively in very salty environments. Because they contain pink carotenoids, they can be seen easily under some circumstances (Figure 26.24). Halophiles grow in the Dead Sea and in brines of all types: Pickled fish may sometimes show reddish pink spots that are colonies of halophilic archaea. Few other organisms can live in the saltiest of the homes that the strict halophiles occupy; most would "dry" to death, losing too much water to the hypertonic environment. Strict halophiles have been found in lakes with pH values as high as 11.5—the most alkaline environment inhabited by living organisms, almost as alkaline as household ammonia.

Some of the extreme halophiles have a unique system for trapping light energy and using it to form ATP—without using any form of chlorophyll—when oxygen is in short supply. They use the pigment *retinal* (also found in the vertebrate eye) combined with a protein to form **bacteriorhodopsin**, and form ATP by a chemiosmotic mechanism of the sort described in Figure 7.12.

Another member of the Euryarchaeota, *Thermoplasma*, has no cell wall. It is thermophilic and acidophilic, its metabolism is aerobic, and it lives in coal deposits. It has the smallest genome among the archaea, and perhaps the smallest (along with the mycoplasmas) of any free-living organisms—1,100,000 base pairs.

This chapter has provided a brief summary of two of the three domains of the living world. The world of the eukaryotes, both unicellular and multicellular, will be the subject of the next seven chapters.

Chapter Summary

Why Three Domains?

▶ Living organisms can be divided into three domains: Bacteria, Archaea, and Eukarya. Both the Archaea and the Bacteria are prokaryotic, but they differ from each other more radically than do the Archaea from the Eukarya, which constitute the rest of the living world.

▶ The evolutionary relationships of the three domains were first revealed by their rRNA sequences. The common ancestor of all three domains lived more than 3 billion years ago, and the common ancestor of the Archaea and Eukarya at least 2 billion years ago. **Review Figure 26.2 and Table 26.1**

General Biology of the Prokaryotes

▶ The prokaryotes are the most numerous organisms on Earth, and they occupy an enormous variety of habitats.

▶ Most prokaryotes are cocci, bacilli, or spiral forms. Some link together to form associations, but very few are truly multicellular. **Review Figure 26.3**

▶ Prokaryotes lack nuclei, membrane-enclosed organelles, and cytoskeletons. Their chromosomes are circular. They often contain plasmids. Some prokaryotes contain internal membrane systems. **Review Figure 26.4**

▶ Many prokaryotes move by means of flagella, gas vesicles, or gliding mechanisms. Prokaryotic flagella rotate rather than beat.

▶ Prokaryotic cell walls differ from those of eukaryotes. Bacterial cell walls generally contain peptidoglycan. Differences in peptidoglycan content result in different reactions to the Gram stain. **Review Figure 26.7**

▶ Prokaryotes reproduce asexually by fission, but also exchange genetic information.

▶ Prokaryotes have diverse metabolic pathways and nutritional modes. They include obligate anaerobes, facultative anaerobes, and obligate aerobes. The major nutritional types are photoautotrophs, photoheterotrophs, chemoautotrophs, and chemoheterotrophs. Some prokaryotes base their energy metabolism on nitrogen- or sulfur-containing ions. **Review Figure 26.8 and Table 26.2**

Prokaryotes in Their Environments

▶ Some prokaryotes play key roles in global nitrogen and sulfur cycles. Important players in the nitrogen cycle are the nitrogen fixers, nitrifiers, and denitrifiers. **Review Figure 26.10**

▶ Photosynthesis by cyanobacteria generated the oxygen gas that permitted the evolution of aerobic respiration and the appearance of present-day eukaryotes.

▶ Many prokaryotes live in or on other organisms, with neutral, beneficial, or harmful effects.

▶ A small minority of bacteria are pathogens. Pathogens vary with respect to their invasiveness and toxigenicity. Some produce endotoxins, which are rarely fatal; others produce exotoxins, which tend to be highly toxic.

Prokaryote Phylogeny and Diversity

▶ Phylogenetic classification of prokaryotes is now based on rRNA sequences and other molecular evidence.

▶ Lateral gene transfer among prokaryotes, which has occurred throughout evolutionary history, makes it difficult to infer prokaryote phylogeny.

▶ Evolution, powered by mutation, natural selection, and genetic drift, can proceed rapidly in prokaryotes because they are haploid and can multiply rapidly.

The Bacteria

▶ There are far more known bacteria than known archaea. One phylogenetic classification of the domain Bacteria groups them into more than a dozen groups.

▶ The most ancient bacteria, like the most ancient archaea, may be thermophiles, suggesting that life originated in a hot environment.

▶ All four nutritional types occur in the largest bacterial group, the Proteobacteria. Metabolism in different groups of proteobacteria has evolved along different lines. **Review Figure 26.12**

▶ Cyanobacteria, unlike other bacteria, photosynthesize using the same pathways plants use. Many cyanobacteria fix nitrogen.

▶ Spirochetes move by means of axial filaments.

▶ Chlamydias are tiny parasites that live within the cells of other organisms.

▶ Firmicutes are diverse; some of them produce endospores as resting structures that resist harsh conditions. Actinomycetes, some of which produce important antibiotics, grow as branching filaments.

▶ Mycoplasmas, the tiniest living things, lack conventional cell walls. They have very small genomes.

The Archaea

▶ Archaea have cell walls lacking peptidoglycan, and their membrane lipids differ from those of bacteria and eukaryotes, containing branched long-chain hydrocarbons connected to glycerol by ether linkages. **Review Figure 26.22**

▶ The domain Archaea can be divided into two kingdoms: Crenarchaeota and Euryarchaeota.

▶ Crenarchaeota are heat-loving and often acid-loving archaea.

▶ Methanogens produce methane by reducing carbon dioxide. Some methanogens live in the guts of herbivorous animals; others occupy high-temperature environments on the ocean floor.

▶ Extreme halophiles are salt lovers that often lend a pinkish color to salty environments; some halophiles also grow in extremely alkaline environments.

▶ Archaea of the genus *Thermoplasma* lack cell walls, are thermophilic and acidophilic, and have a tiny genome (1,100,000 base pairs).

For Discussion

1. Why do systematic biologists find rRNA sequence data more useful than data on metabolism or cell structure for classifying prokaryotes?

2. Why does lateral gene transfer make it so difficult to arrive at agreement on phylogeny?

3. Differentiate among the members of the following sets of related terms:

   a. prokaryotic/eukaryotic

   b. obligate anaerobe/facultative anaerobe/obligate aerobe

   c. photoautotroph/photoheterotroph/chemoautotroph/chemoheterotroph

   d. Gram-positive/Gram-negative

4. Why are the endospores of firmicutes not considered to be reproductive structures?

5. Until fairly recently, the cyanobacteria were called blue-green algae and were not grouped with the bacteria. Suggest several reasons for this (abandoned) tendency to separate the bacteria and cyanobacteria. Why are the cyanobacteria now grouped with the other bacteria?

6. The actinomycetes are of great commercial interest. Why?

7. Hyperthermophiles are of great interest to molecular biologists and biochemists. Why? What practical concerns might motivate that interest?

27 *Protists and the Dawn of the Eukarya*

THE BACTERIA AND THE ARCHAEA HAD the living world to themselves for more than a billion years. As we saw in Chapter 26, members of these two domains differ sharply in several important ways—but neither of these prokaryotes is like the single-celled organism shown here. What strikes you the most about this amoeba? Probably the most obvious visible difference between it and the prokaryotes is that the amoeba has numerous compartments—membrane-enclosed organelles.

What are compartments useful for? For one thing, they keep items separate—like keeping greasy tools away from clean socks by storing them in different cabinets. Compartments also keep things together when that makes sense, such as keeping all your socks in one drawer, or keeping all the files for your term paper in a single directory on your computer. Rooms are another example of compartments, and they can be specialized for different activities: One room in the Biological Sciences Building might be set up as a laboratory, while another room serves as a lecture hall, and still another contains special protective seals that prevent radiation or pathogens from escaping.

The single-celled amoeba is an example of an organism that compartmentalizes: It has a cytoskeleton, a nucleus enclosed by a nuclear envelope, and several kinds of organelles. Amoebas are members of the domain Eukarya, and they differ from members of the two prokaryotic domains in other important ways as well.

The flexibility and options that arose once the eukaryotic cell had evolved resulted in a profusion of body forms and myriad specialized functions. Eukaryotic evolution has produced great diversity, especially among the multicellular lineages, but even among the unicellular protists. In both multicellular and unicellular forms, however, there are also many cases of convergent evolution; for example, organisms with an amoeba-like body form arose several times.

Protists Defined

Many modern members of the Eukarya are familiar to us—trees, dogs, and mushrooms, not to mention ourselves. These members of the kingdoms Plantae, Animalia, and Fungi are not strange to us. However, amoebas and a dazzling assortment of other eukaryotes, mostly microscopic organisms, don't fit into these three kingdoms. We call all those eukaryotes that are neither plants, animals, nor fungi **protists** (Figure 27.1; Table 27.1). *The protists are not a monophyletic group.* Some protists are more closely related to the animals than they are to other protists. Some protists are motile, while others are stationary; some are photosynthetic, while others are heterotrophic; most are unicellular, while some giant kelps are not only multicellular but also huge, sometimes achieving lengths greater than that of a football field.

The origin of the eukaryotic cell was one of the pivotal events in evolutionary history. In this chapter on the protists we describe and celebrate the origin and early diversification of the eukaryotes and the complexity achieved by some single cells. We'll explore some of the diversity of protist body forms, and we'll try to give a sense of developing current views of the evolutionary relationships of some of the protists.

The Origin of the Eukaryotic Cell

The eukaryotic cell differs in many ways from prokaryotic cells. How did it originate? Given the nature of evolutionary processes, the differences cannot all have arisen simultaneously. We think we can make some reasonable guesses

An Amoeba
Amoebas have several kinds of organelles (seen here as bubble-like "compartments"). Their flowing pseudopods are constantly changing shape as the amoeba moves and feeds.

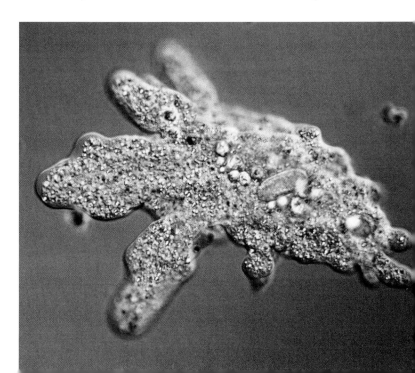

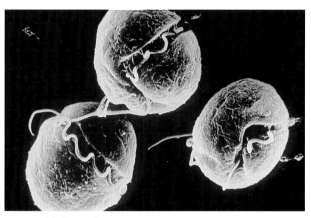

(a) *Gonyaulax* sp.

27.1 Three Eukaryote Protists
(a) Dinoflagellates are photosynthetic unicellular protists. (b) *Giardia* is a unicellular parasite of humans. (c) Giant kelps are some of the world's longest organisms.

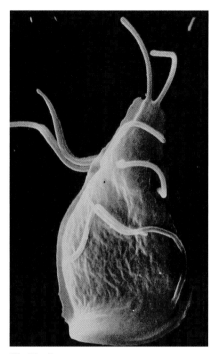

(b) *Giardia* sp.

(c) *Macrocystis* sp.

about the important events, bearing in mind that the global environment underwent an enormous change—from anaerobic to aerobic—during the course of these events. As you read this chapter, keep in mind that the steps we suggest are just that: guesses. This version of the story is one of a few under current consideration. We present it as a framework for thinking about this challenging problem, not as a set of facts.

27.1 Major Monophyletic Protist Groups

GROUP	COMMON NAME	ATTRIBUTES	EXAMPLES
Euglenozoa		Unicellular, with flagella	
Euglenoids		Mostly photoautotrophic	*Euglena*
Kinetoplastids		Have a single, large mitochondrion	*Trypanosoma*
Alveolata		Unicellular; cavities (alveoli) below cell surface	
Pyrrophyta	Dinoflagellates	Pigments give golden-brown color	*Gonyaulax*
Apicomplexa		Apical complex for penetration of host	*Plasmodium*
Ciliophora	Ciliates	Cilia; two types of nuclei	*Paramecium*
Stramenopila		Two unequal flagella, one with hairs	
Bacillariophyta	Diatoms	Unicellular; photoautotrophic; two-part walls	
Phaeophyta	Brown algae	Multicellular; marine; photoautotrophic	*Fucus, Macrocystis*
Oomycota	Water molds, powdery mildews	Mostly coenocytic; heterotrophic	*Saprolegnia*
Rhodophyta	Red algae	No flagella; photoautrophic; phycocyanin	*Chondrus*
Chlorophyta	Green algae[a]	Photoautotrophic	*Chlamydomonas, Ulothrix*
Choanoflagellida		Resemble sponge cells; heterotrophic	

[a] The green algae do not constitute a monophyletic group. The Chlorophyta are one lineage of green algae that qualifies as a monophyletic group; another lineage gave rise to the plant kingdom.

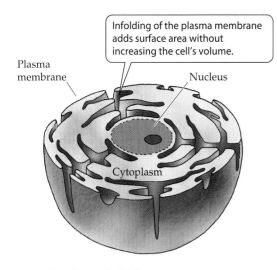

27.2 Membrane Infolding
The loss of the rigid prokaryotic cell wall allowed the plasma membrane to elaborate inward and create more surface area.

The modern eukaryotic cell arose in several steps

The essential steps in the origin of the eukaryotic cell include

▶ The origin of a flexible cell surface
▶ The origin of a cytoskeleton
▶ The origin of a nuclear envelope
▶ The appearance of digestive vesicles
▶ The endosymbiotic acquisition of certain organelles

WHAT A FLEXIBLE CELL SURFACE ALLOWS. Many ancient fossil prokaryotes look like rods, and we presume that they, like most present-day prokaryotic cells, had firm cell walls. The first step toward the eukaryotic condition may have been the loss of the cell wall by an ancestral prokaryotic cell. This may not seem like an obvious first step, but consider the possibilities open to a flexible cell without a wall.

First, think of cell size. As a cell grows, its surface area-to-volume ratio decreases (see Figure 4.2). Unless the surface is flexible and can fold inward and elaborate itself, creating more surface area for gas and nutrient exchange (Figure 27.2), the cell volume will reach an upper limit. With a surface flexible enough to allow infolding, the cell can exchange materials with its environment rapidly enough to sustain a larger volume and more rapid metabolism. Further, a flexible surface can pinch off bits of the environment, bringing them into the cell by endocytosis (Figure 27.3).

Also recall that the chromosome of a prokaryotic cell is attached to a site on its plasma membrane (see Figure 26.4). If that region of the plasma membrane were to fold into the cell, the first step would be taken

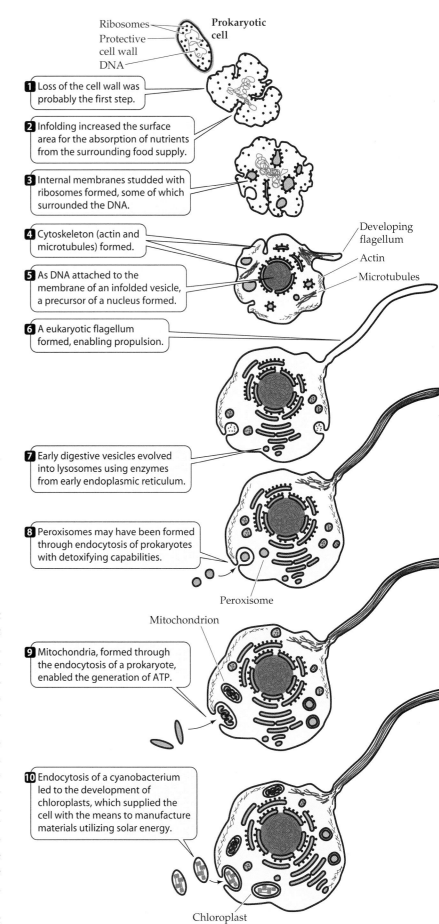

27.3 From Prokaryotic Cell to Eukaryotic Cell
One possible evolutionary sequence is shown here. The exact sequence, of course, is not known.

toward the evolution of a nucleus, the key feature of the eukaryotic cell.

CHANGES IN CELL STRUCTURE AND FUNCTION.   Early steps in the evolution of the eukaryotic cell are likely to have included three advances: the formation of ribosome-studded internal membranes, some of which surrounded the DNA (see Figure 27.3); the appearance of a cytoskeleton and the evolution of digestive vesicles.

A cytoskeleton made up of actin fibers and microtubules would allow the cell to manage changes in shape, to distribute daughter chromosomes, and to move materials from one part of the now much larger cell to other parts. The origin of the cytoskeleton remains a mystery, heightened by the fact that the genes that encode the cytoskeleton are present in neither bacteria nor archaea. An intriguing and controversial suggestion is that a fourth domain of life, now long extinct, originated these genes and transferred them laterally to an ancestor of the early eukaryotes.

From an intermediate kind of cell, the next advance was probably to a cell that we could call a *phagocyte*—a motile cell that could prey on other cells by engulfing and digesting them. The first true eukaryote possessed a cytoskeleton and a nuclear envelope. It may have had an associated endoplasmic reticulum and Golgi apparatus, and perhaps one or more flagella of the eukaryotic type. Notice how much of the progress to this point was made possible by the loss of the cell wall and the elaboration of what was originally the plasma membrane.

ENDOSYMBIOSIS AND ORGANELLES.   While the processes already outlined were taking place, the cyanobacteria were very busy, generating oxygen gas as a product of photosynthesis. The increasing O_2 levels in the atmosphere had disastrous consequences for most other living things, because most living things of the time (archaea and bacteria) were unable to tolerate the newly aerobic, oxidizing environment. But some prokaryotes managed to cope, and—fortunately for us—so did some of the ancient phagocytes.

According to one hypothesis, the key to the survival of early phagocytes was the ingestion and incorporation of a prokaryote that became symbiotic within the phagocyte and evolved into the peroxisomes of today (see Figure 27.3). These organelles were able to disarm the toxic products of oxygen action, such as hydrogen peroxide. This association may have been the first important endosymbiosis in the evolution of the eukaryotic cell.

In Chapter 4 we introduced the concept of endosymbiosis (organisms living together, one inside the other). A crucial endosymbiotic event in the history of the Eukarya was the incorporation of a proteobacterium that evolved into the mitochondrion. Upon completion of this step, the basic modern eukaryotic cell was complete. Some very important eukaryotes are the result of yet another endosymbiotic step, the incorporation of prokaryotes related to today's cyanobacteria, which became chloroplasts. We'll see how this happened later in the chapter.

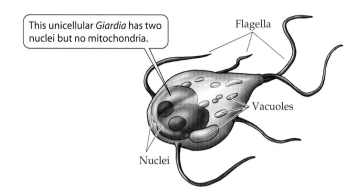

This unicellular *Giardia* has two nuclei but no mitochondria.

Flagella

Vacuoles

Nuclei

27.4 *Giardia*: A Protist without Mitochondria
Current evidence indicates that *Giardia* is descended from an ancestor that possessed mitochondria.

"Archaezoa": The little kingdom that was

The hypothesis that the eukaryotic nucleus evolved before the mitochondrion gained early support from the existence of a few unicellular eukaryotes, such as *Giardia*, that lack mitochondria. *Giardia lamblia* is a familiar parasite that contaminates water supplies and causes the intestinal disease giardiasis (Figure 27.4). This tiny organism has no mitochondria, chloroplasts, or other membrane-enclosed organelles, but it contains two nuclei bounded by nuclear envelopes, and it has a cytoskeleton. Some biologists treated such eukaryotes without mitochondria as the modern descendants of a hypothetical ancient group, which they called "archaezoan." It was later learned that at least some archaezoans may have descended from eukaryotes that lost their mitochondria. Research sometimes takes surprising twists and turns! We no longer speak of a *kingdom* of archaezoans. However, the existence of such organisms today shows that eukaryotic life is feasible without mitochondria, and the eukaryotes that lack mitochondria are the focus of much attention.

Many uncertainties remain

Several uncertainties cloud our current understanding of the origins of eukaryotic cells. Lateral gene transfer complicates the study of eukaryotic origins just as it complicates the study of relationships among the prokaryote lineages. At the same time, it may not have been extensive enough to account for the fact that, as genetic studies advance, more and more genes of bacterial origin are being found in eukaryotes.

An endosymbiotic origin of mitochondria and chloroplasts accounts for the presence of bacterial genes encoding enzymes for energy metabolism (respiration and photosynthesis), but it does not explain the presence of many other bacterial genes. The eukaryotic genome clearly is a mixture of genes with two distinct origins. A recent suggestion is that the Eukarya might have arisen from the mutualistic fusion (not endosymbiosis) of a Gram-negative bacterium and an archaean. There are many interesting ideas about eukaryotic origins awaiting additional data and analysis.

We can expect that these and other questions will yield to additional research. Let's leave our speculations about the origin of the protists for the moment and examine what we do know about them.

General Biology of the Protists

Most protists are aquatic. Some live in marine environments, others in fresh water, and still others in the body fluids of other organisms. The slime molds inhabit damp soil and the moist, decaying bark of rotting trees. Many other protists also live in soil water, some of them contributing to the global nitrogen cycle by preying on soil bacteria and recycling their nitrogen compounds to nitrates.

Protists are strikingly diverse in their structure, but not so diverse in their metabolism as the prokaryotes—in fact, some of the eukaryotes' most important metabolic pathways were "borrowed" from bacteria through endosymbiosis. However, protists do display a number of nutritional modes. Some are autotrophs, some are heterotrophs, and some switch with ease between the autotrophic and heterotrophic modes of nutrition.

Some protists, formerly classified as animals, are sometimes referred to as **protozoans**, although biologists increasingly regard this term as inappropriate because it lumps together protist groups that are phylogenetically distant from one another. Most protozoans are ingestive heterotrophs. There are several kinds of photosynthetic protists that some biologists still refer to as **algae** (singular alga). Although these two terms—protozoans and algae—are useful in some contexts, they do not correspond with natural phy-

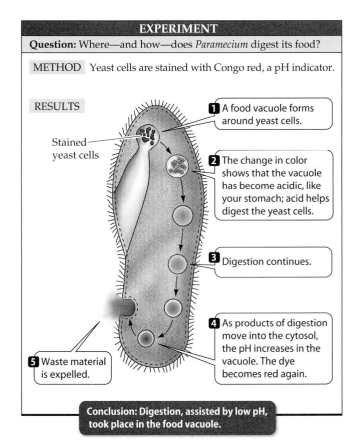

Question: Where—and how—does *Paramecium* digest its food?

METHOD   Yeast cells are stained with Congo red, a pH indicator.

RESULTS

Stained yeast cells

1 A food vacuole forms around yeast cells.

2 The change in color shows that the vacuole has become acidic, like your stomach; acid helps digest the yeast cells.

3 Digestion continues.

4 As products of digestion move into the cytosol, the pH increases in the vacuole. The dye becomes red again.

5 Waste material is expelled.

Conclusion: Digestion, assisted by low pH, took place in the food vacuole.

27.6 Food Vacuoles Handle Digestion and Excretion
An experiment with *Paramecium* demonstrates the function of food vacuoles. *Paramecium* ingests food by way of the oral groove at the left. The dye Congo red turns green at acidic pH and red at neutral or basic pH.

logeny, and we generally avoid them in this book except as parts of descriptive names such as "brown algae."

Protists have diverse means of locomotion

Although a few protist groups consist entirely of nonmotile organisms, most groups include cells that move, either by amoeboid motion, by ciliary action, or by means of flagella.

In amoeboid motion, the cell forms **pseudopods** ("false feet") that are extensions of its constantly changing body mass. Cells such as the amoeba on page 476 simply extend a pseudopod and then flow into it. Cilia are tiny, hairlike organelles that beat in a coordinated fashion to move the cell forward or backward (see Figure 4.24). A eukaryotic flagellum moves like a whip; some flagella *push* the cell forward, others *pull* the cell forward.

Vesicles perform a variety of functions

Unicellular organisms tend to be of microscopic size. As we noted above, an important reason that cells are small is that they need enough membrane surface area in relation to their volume to support the exchange of materials required for their existence. Many relatively large unicellular protists minimize this problem by having membrane-enclosed **vesicles** of various types that increase their effective surface area.

As we saw in Chapter 5, organisms living in fresh water are hypertonic to their environments. Many freshwater pro-

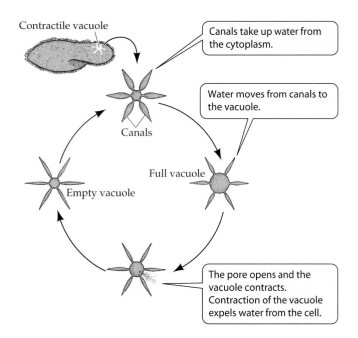

Contractile vacuole

Canals take up water from the cytoplasm.

Water moves from canals to the vacuole.

Canals

Full vacuole

Empty vacuole

The pore opens and the vacuole contracts. Contraction of the vacuole expels water from the cell.

27.5 Contractile Vacuoles Bail Out Excess Water
Water constantly enters freshwater protists by osmosis. A pore in the cell surface allows the contractile vacuole to expel the water it accumulates.

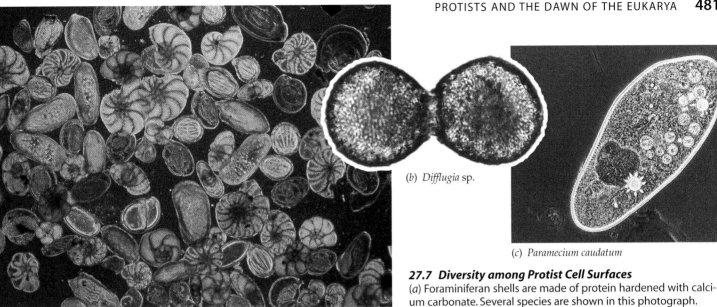

(a)

(b) *Difflugia* sp.

(c) *Paramecium caudatum*

27.7 Diversity among Protist Cell Surfaces
(*a*) Foraminiferan shells are made of protein hardened with calcium carbonate. Several species are shown in this photograph.
(*b*) This amoeba constructed its shell by cementing sand grains together. (*c*) Spirals of protein make this *Paramecium*'s surface—known as its pellicle—flexible but resilient.

tists address this problem by means of vesicles that contract to excrete excess water. Members of several protist groups have such **contractile vacuoles**. Because these organisms have a higher concentration of solutes than their freshwater environment does, they constantly take in water by osmosis. The excess water collects in the contractile vacuole and is then pushed out (Figure 27.5).

It is easy to confirm that bailing out water is the principal function of the contractile vacuole. First, we can observe some protists under a light microscope and note the rate at which the vacuoles are contracting—they look like little eyes winking. Then we can place other protists of the same species in solutions of differing osmotic potential. The less negative the osmotic potential of the surrounding solution, the more hypertonic the cells are, and the faster water rushes into them, causing the contractile vacuoles to pump more rapidly. Conversely, the contractile vacuoles will stop pumping if the solute concentration of the medium is increased so that it is equal to that of the cells.

A second important type of vesicle found in many protists is the **food vacuole**. Protists such as *Paramecium* engulf solid food by endocytosis, forming a food vacuole within which the food is digested (Figure 27.6). Smaller vesicles containing digested food pinch away from the food vesicle and enter the cytoplasm. These tiny vesicles provide a large surface area across which the products of digestion may be absorbed by the rest of the cell.

The cell surfaces of protists are diverse

A few protists, such as some amoebas, are surrounded by only a plasma membrane, but most have stiffer surfaces that maintain the structural integrity of the cell. Many protists have cell walls, which are often complex in structure. Other protists that lack cell walls have a variety of ways of strengthening their surfaces. Some have internal "shells," which the organism either produces itself, as foraminiferans do, or makes from bits of sand and thickenings imme-

diately beneath the plasma membrane, as some amoebas do (Figure 27.7).

Many protists contain endosymbionts

Endosymbiosis is very common among the protists, and in some instances both the host and the endosymbiont are protists. Many radiolarians, for example, harbor photosynthetic protists (Figure 27.8). As a result, these radiolarians appear greenish or yellowish, depending on the type of endosymbiont they contain. This arrangement is beneficial to the radiolarian, for it can make use of the food produced by its photosynthetic guest. The guest, in turn, may make use of metabolites made by the host, or it may simply receive physical protection. In other cases, the guest may be a victim, exploited for its photosynthetic products while receiving no benefit itself.

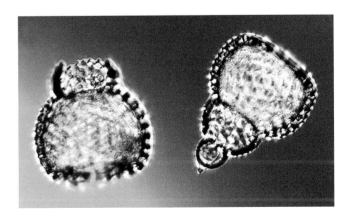

27.8 Protists within Protists
Photosynthetic organisms living as endosymbionts within these radiolarians provide food for the radiolarians, as well as part of the pigmentation seen through their glassy skeletons. Both the endosymbionts and the radiolarians are protists.

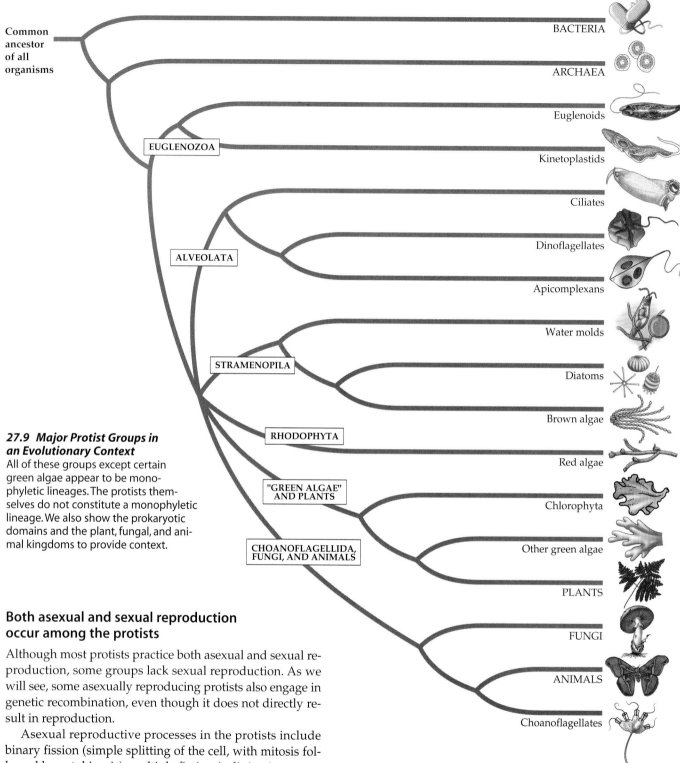

27.9 Major Protist Groups in an Evolutionary Context
All of these groups except certain green algae appear to be monophyletic lineages. The protists themselves do not constitute a monophyletic lineage. We also show the prokaryotic domains and the plant, fungal, and animal kingdoms to provide context.

Both asexual and sexual reproduction occur among the protists

Although most protists practice both asexual and sexual reproduction, some groups lack sexual reproduction. As we will see, some asexually reproducing protists also engage in genetic recombination, even though it does not directly result in reproduction.

Asexual reproductive processes in the protists include binary fission (simple splitting of the cell, with mitosis followed by cytokinesis), multiple fission (splitting into more than two cells), budding (the outgrowth of a new cell from the surface of an old one), and the formation of spores (cells that are capable of developing into new organisms). Sexual reproduction also takes various forms. In some protists, as in animals, the gametes are the only haploid cells. In some other protists, by contrast, both diploid and haploid cells undergo mitosis, giving rise to alternation of generations, which will be described later in the chapter.

The diversity of form, habitat, metabolism, locomotion, reproduction, and life cycles found among the protists reflects the diversity of avenues pursued during the early evolution of eukaryotes. Many of these avenues led to great success, judging from the abundance and diversity of today's protists and other eukaryotes.

Protist Diversity

As we have seen, the phylogeny of protists is an area of exciting, challenging research. The marvelous diversity of protist body forms and metabolic lifestyles seems reason enough for a fascination with these organisms, but questions about how the multicellular eukaryotic kingdoms originated from the protists stimulate further interest. Fortunately, the tools of molecular biology, such as rRNA sequencing, make it possible to explore evolutionary relationships among the protists in greater detail and with somewhat greater confidence than previously (see Chapters 24 and 26).

We will discuss several apparently monophyletic groups of protists, as well as a few other groups of more uncertain phylogenetic status. Some biologists refer to many of these monophyletic groups as kingdoms; others refer to them as subkingdoms; still others refer to them as phyla. This choice of words is not of immediate concern to us here, so we'll just call them "groups." We'll describe the Euglenozoa, Alveolata, Stramenopila, Rhodophyta, Chlorophyta, and Choanoflagellida (Figure 27.9).

As we shall see, some of the monophyletic protist groups consist of organisms with very diverse body plans. On the other hand, certain body plans, such as those of amoebas and those of slime molds, have arisen again and again during evolution, in groups only distantly related to one another.

Euglenozoa

The **Euglenozoa** are a monophyletic group of *flagellates:* unicellular organisms with flagella. They reproduce asexually by binary fission. There are two subgroups of Euglenozoa: euglenoids and kinetoplastids.

Euglenoids have anterior flagella

The **euglenoids** possess flagella arising from a pocket at the anterior end of the cell. Euglenoids used to be claimed by the zoologists as animals and by the botanists as plants. They are unicellular flagellates, but many members of the group are photosynthetic.

Figure 27.10 depicts a cell of the genus *Euglena*. Like most other euglenoids, this common freshwater organism has a complex cell structure. It propels itself through the water with one of its two flagella, which may also serve as an anchor to hold the organism in place. The flagellum provides power by means of a wavy motion that spreads from base to tip. The second flagellum is often rudimentary.

Euglena has very flexible nutritional requirements. Many species are always heterotrophic. Other species are fully autotrophic in sunlight, using chloroplasts to synthe-

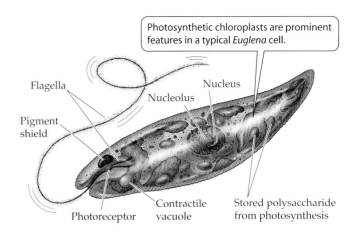

Photosynthetic chloroplasts are prominent features in a typical *Euglena* cell.

Flagella
Pigment shield
Photoreceptor
Nucleolus
Nucleus
Contractile vacuole
Stored polysaccharide from photosynthesis

27.10 A Photosynthetic Euglenoid
Several *Euglena* species are among the best-known flagellates. In this species, the second flagellum is rudimentary.

size organic compounds through photosynthesis. The chloroplasts of euglenas are surrounded by three membranes (unlike plant chloroplasts, which have only two). When kept in the dark, these euglenas lose their photosynthetic pigment and begin to feed exclusively on dead organic material floating in the water around them. Such a "bleached" *Euglena* resynthesizes its photosynthetic pigment when it is returned to the light and becomes autotrophic again. But *Euglena* cells treated with certain antibiotics or mutagens lose their photosynthetic pigment completely; neither they nor their descendants are ever autotrophs again. However, those descendants function well as heterotrophs.

Kinetoplastids have mitochondria that edit their own RNA

The **kinetoplastids** are unicellular, parasitic flagellates with a single, large mitochondrion. That mitochondrion contains a *kinetoplast*—a unique structure housing DNA and associated proteins. The kinetoplast DNA is of two types, called minicircles and maxicircles. The maxicircles encode enzymes associated with oxidative metabolism, and the minicircles encode "guides" that accomplish a remarkable type of RNA editing within the mitochondrion.

Some kinetoplastids are human pathogens. Sleeping sickness, one of the most dreaded diseases of Africa, is caused by the parasitic kinetoplastid *Trypanosoma* (Figure 27.11). The vector (intermediate host) for *Trypanosoma* is an insect, the tsetse fly. Carrying its deadly cargo, the tsetse fly bites livestock, wild animals, and even humans, infecting them with the parasite. *Trypanosoma* then multiplies in the mammalian bloodstream and produces toxins. When these parasites invade the nervous system, the neurological symptoms of sleeping sickness appear and are followed by death. Other trypanosomes cause leishmaniasis, Chagas' disease, and East Coast fever; all are major diseases in the tropics.

Euglenozoa
Alveolata
Stramenopila
Rhodophyta
Chlorophyta

Undulating membrane of trypanosome

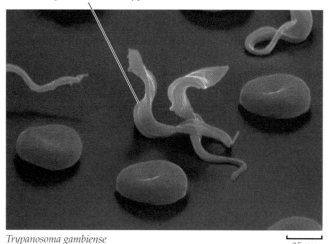

Trypanosoma gambiense

25 μm

27.11 A Parasitic Kinetoplastid
Trypanosomes, shown here among human red blood cells, cause sleeping sickness in mammals. A flagellum runs along one edge of the cell as part of a structure called the undulating membrane.

Alveolata

The **Alveolata** are a monophyletic group of unicellular organisms characterized by the possession of cavities called *alveoli* just below their plasma membranes. They are diverse in body form. The alveolate groups we'll consider here are the dinoflagellates, apicomplexans, and ciliates.

Dinoflagellates are unicellular marine organisms with two flagella

The **dinoflagellates** are all unicellular, and most are marine organisms. A distinctive mixture of photosynthetic and accessory pigments gives their chloroplasts a golden-brown color. The dinoflagellates are of great ecological, evolutionary, and morphological interest. They are among the most important primary photosynthetic producers of organic matter in the oceans.

Euglenozoa
Alveolata
Stramenopila
Rhodophyta
Chlorophyta

Many dinoflagellates are endosymbionts, living within the cells of other organisms, including various invertebrates and even other marine protists. Dinoflagellates are particularly common endosymbionts in corals, to whose growth they contribute by photosynthesis. Some dinoflagellates are nonphotosynthetic and live as parasites within other marine organisms.

Dinoflagellates have a distinctive appearance (see Figure 27.1*a*). They have two flagella, one in an equatorial groove around the cell, the other starting at the same point as the first and passing down a longitudinal groove before extending into the surrounding medium.

Some dinoflagellates reproduce in enormous numbers in warm and somewhat stagnant waters. The result can be a "red tide," so called because of the reddish color of the sea that results from pigments in the dinoflagellates (Figure 27.12). During a red tide, the concentration of dinoflagellates may reach 60 million cells per liter of ocean water. Certain red tide species produce a potent nerve toxin that can kill tons of fish. The genus *Gonyaulax* produces a toxin that can accumulate in shellfish in amounts that, although not fatal to the shellfish, may kill a person who eats the shellfish.

Many dinoflagellates are bioluminescent. In complete darkness, cultures of these organisms emit a faint glow. If you suddenly stir or bubble air through the culture, the organisms each emit numerous bright flashes. A ship passing through a tropical ocean that contains a rich growth of these species produces a bow wave and wake that glow eerily as billions of these dinoflagellates discharge their light systems.

Apicomplexans are parasites with unusual spores

Exclusively parasitic organisms, the **apicomplexans** derive their name from the *apical complex*, a mass of organelles contained within the apical end of their spores. These organelles help the apicomplexan spore invade its host tissue. Unlike many other protists, apicomplexans lack contractile vacuoles.

Apicomplexans generally have an amorphous body form like that of an amoeba. This body form has evolved over and over again in parasitic protists. It appears even among parasitic dinoflagellates, a group of organisms whose nonparasitic relatives have highly distinctive, complex body forms.

Like many obligate parasites, apicomplexans have elaborate life cycles featuring asexual and sexual reproduction

A red tide along the coast of Baja California

27.12 A Red Tide of Dinoflagellates
By reproducing in astronomical numbers, the dinoflagellate *Gonyaulax tamarensis* can cause a toxic red tide.

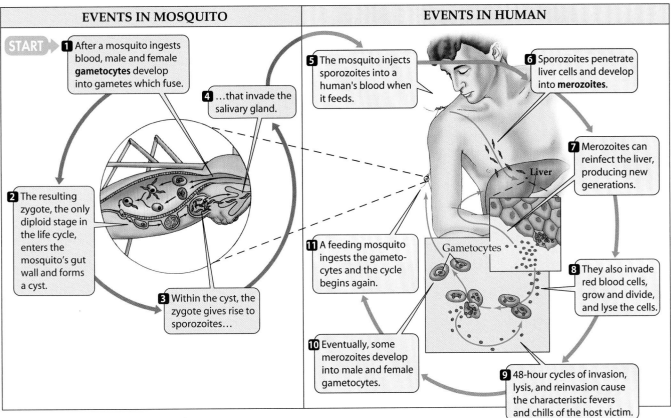

EVENTS IN MOSQUITO	EVENTS IN HUMAN

START ▶ **1** After a mosquito ingests blood, male and female **gametocytes** develop into gametes which fuse.

4 ...that invade the salivary gland.

5 The mosquito injects sporozoites into a human's blood when it feeds.

6 Sporozoites penetrate liver cells and develop into **merozoites**.

2 The resulting zygote, the only diploid stage in the life cycle, enters the mosquito's gut wall and forms a cyst.

7 Merozoites can reinfect the liver, producing new generations.

Liver

3 Within the cyst, the zygote gives rise to sporozoites...

11 A feeding mosquito ingests the gameto-cytes and the cycle begins again.

Gametocytes

8 They also invade red blood cells, grow and divide, and lyse the cells.

10 Eventually, some merozoites develop into male and female gametocytes.

9 48-hour cycles of invasion, lysis, and reinvasion cause the characteristic fevers and chills of the host victim.

27.13 The Life Cycle of an Apicomplexan
Malaria-causing *Plasmodium* species spend part of their life cycle in humans and part in mosquitoes. Sporozoites and merozoites are spores with apical complexes.

by a series of very dissimilar life stages. Often these stages are associated with two different types of host organism.

The best-known apicomplexans are the malarial parasites of the genus *Plasmodium*, a highly specialized group of organisms that spend part of the life cycle within human red blood cells (Figure 27.13). Although it has been almost eliminated from the United States, malaria continues to be a major problem in many tropical countries. In terms of the number of people infected, malaria is one of the world's most serious diseases.

Female mosquitoes of the genus *Anopheles* transmit *Plasmodium* to humans. The parasite enters the human circulatory system when an infected *Anopheles* mosquito penetrates the human skin in search of blood. The parasites find their way to cells in the liver and the lymphatic system, change their form, multiply, and reenter the bloodstream, attacking red blood cells. The apical complex enables *Plasmodium* to enter human liver cells and red blood cells.

The parasites multiply inside red blood cells, which then burst, releasing new swarms of parasites. If another *Anopheles* bites the victim, the mosquito takes in some of the parasitic *Plasmodium* cells along with blood. The infecting cells develop into gametes, which unite to form zygotes that lodge in the mosquito's gut, divide several times, and

move into its salivary glands, from which they can be passed on to another human host. Thus, *Plasmodium* is an extracellular parasite in the mosquito vector and an intracellular parasite in the human host.

Malaria kills more than a million people each year, and *Plasmodium* has proved to be a singularly difficult pathogen to attack. The *Plasmodium* life cycle is best broken by the removal of stagnant water, in which mosquitoes breed. The use of insecticides to reduce the *Anopheles* population can be effective, but their benefits must be weighed against the possible ecological, economic, and health risks posed by the insecticides themselves. However, there is now new hope in the form of a genome-sequencing project that targets the common form of *Plasmodium*. Scheduled to be completed by the year 2002, this project may provide the information needed to end the epidemic.

Ciliates have two types of nuclei

The **ciliates** are so named because they characteristically have hairlike cilia. This group is noteworthy for its diversity and ecological importance (Figure 27.14). Almost all ciliates are heterotrophic (a few contain photosynthetic endosymbionts), and they are much more specialized in body form than are most flagellates and other protists.

The definitive characteristic of ciliates is the possession of two types of nuclei, from one to as many as a thousand large **macronuclei** and, within the same cell, from one to eighty **micronuclei**. The micronuclei, which are typical eukaryotic nuclei, are essential for genetic recombination. The

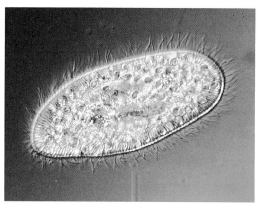

(a) *Paramecium bursaria*

10 μm

Cilia

(b) *Epistylis* sp.

27.14 Diversity among the Ciliates
(a) A free-swimming organism, this para-mecium belongs to a ciliate group whose members have many cilia of uniform length. (b) Members of this subgroup have cilia on their mouthparts. (c) In this group, tentacles replace cilia as develop-ment proceeds. (d) This ciliate "walks" on fused cilia called cirri that project from its body. Other cilia are fused into flat sheets that sweep food particles into the oral cavity.

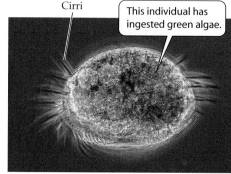

Cirri

This individual has ingested green algae.

(d) *Euplotes* sp.

25 μm

Tentacles

(c) *Paracineta* sp.

20 μm

paramecium can direct the beat of its cilia to propel itself either forward or backward in a spiraling manner (Fig-ure 27.16). It can also back off swiftly when it encounters a barrier or a neg-ative stimulus. The coordination of ciliary beating is probably the result of a differential distri-bution of ion channels in the plasma membrane near the two ends of the cell.

REPRODUCTION WITHOUT SEX, AND SEX WITHOUT REPRODUCTION. Paramecia reproduce asexually by binary fission. The micro-

macronuclei are derived from micronuclei. Each macronu-cleus contains many copies of the genetic information, packaged in units containing very few genes each; the macronuclear DNA is transcribed and translated to regu-late the life of the cell. Although we do not know how this system of macro- and micronuclei came into being, we do know something about the behavior of these nuclei, which we will discuss after describing the body plan of one im-portant ciliate, *Paramecium*.

A CLOSER LOOK AT ONE CILIATE. *Paramecium*, a frequently studied ciliate genus, exemplifies the complex structure and behavior of ciliates (Figure 27.15 *a*). The slipper-shaped cell is covered by an elaborate **pellicle**, a structure composed prin-cipally of an outer membrane and an inner layer of closely packed, membrane-enclosed sacs (the alveoli) that surround the bases of the cilia. Defensive organelles called *trichocysts* are also present in the pellicle. In response to a threat, a microscopic explosion expels the trichocysts in a few mil-liseconds, and they emerge as sharp darts, driven forward at the tip of a long, expanding filament (Figure 27.15*b*).

The cilia provide a form of locomotion that is generally more precise than locomotion by flagella or pseudopods. A

27.15 Anatomy of Paramecium
(a) The major structures of a typical paramecium. (b) A tri-chocyst discharged from beneath the pellicle of a parame-cium has a sharp point and a straight filament.

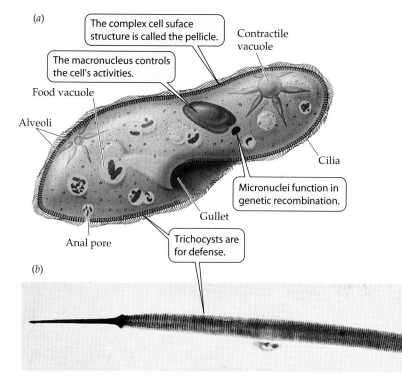

(a)

The complex cell suface structure is called the pellicle.

The macronucleus controls the cell's activities.

Contractile vacuole

Food vacuole

Alveoli

Cilia

Micronuclei function in genetic recombination.

Gullet

Anal pore

Trichocysts are for defense.

(b)

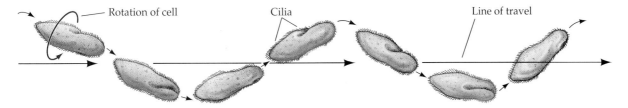

27.16 "Swimming" with Cilia

Beating its cilia in coordinated waves that progress from one end of the cell to the other, a paramecium can move in either direction with respect to the long axis of the cell. The cell rotates in a spiral as it travels.

nuclei divide mitotically. The macronuclei divide by a still unknown mechanism following a round of DNA replication.

Paramecia also have an elaborate sexual behavior called **conjugation**. Two paramecia line up tightly against each other and fuse in the oral region of the body. Nuclear material is extensively reorganized and exchanged over the next several hours (Figure 27.17). As a result of this process, each cell ends up with two haploid micronuclei, one from itself and one from the other cell, which fuse to form a new diploid micronucleus. New macronuclei develop from the micronuclei through a series of dramatic chromosomal rearrangements. The exchange of nuclei is fully reciprocal—each of the two paramecia gives and receives an equal amount of DNA. The two organisms then separate and go their own ways, each equipped with new combinations of alleles.

Conjugation in *Paramecium* is a *sexual* process of genetic recombination, but it is not a *reproductive* process. The same two cells that begin the process are there at the end, and no new cells are created. As a rule, each clone of paramecia must periodically conjugate. Experiments have shown that if some species are not permitted to conjugate, the asexual

27.17 Paramecia Achieve Genetic Recombination by Conjugating

Conjugating *Paramecium* individuals exchange micronuclei, thereby permitting genetic recombination. After conjugation, the cells separate and continue their lives as two individuals.

clones can live through no more than approximately 350 cell divisions before they die out.

Stramenopila

The **stramenopiles** include three prominent groups, two of which are photosynthetic. The two flagella of a stramenopile cell are typically unequal in length. The longer of the two bears rows of tubular hairs. Some stramenopiles lack flagella, but they are presumed to be descended from ancestors that possessed typical stramenopile flagella. The stramenopiles include the diatoms and the brown algae, which are photosynthetic, and the oomycetes, which are not. Other, smaller stramenopile groups include some that are nonphotosynthetic. Some botanists prefer to call the stramenopiles the "brown plant kingdom."

Diatoms are everywhere in the marine environment

Diatoms (Bacillariophyta) are single-celled organisms, although some species associate in filaments. Many have sufficient carotenoids in their chloroplasts to give them a yellow or brownish color. All make chrysolaminarin (a carbohydrate) and oils as photosynthetic storage products. They lack flagella.

Euglenozoa

Alveolata

Stramenopila

Rhodophyta

Chlorophyta

Architectural magnificence on a microscopic scale is the hallmark of the diatoms (Figure 27.18*a*). Many diatoms deposit silicon in their cell walls. The cell wall of some species

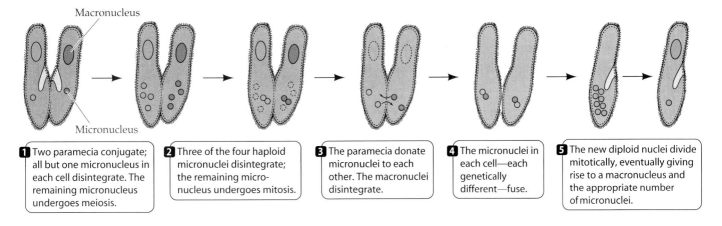

Macronucleus

Micronucleus

1 Two paramecia conjugate; all but one micronucleus in each cell disintegrate. The remaining micronucleus undergoes meiosis.

2 Three of the four haploid micronuclei disintegrate; the remaining micronucleus undergoes mitosis.

3 The paramecia donate micronuclei to each other. The macronuclei disintegrate.

4 The micronuclei in each cell—each genetically different—fuse.

5 The new diploid nuclei divide mitotically, eventually giving rise to a macronucleus and the appropriate number of micronuclei.

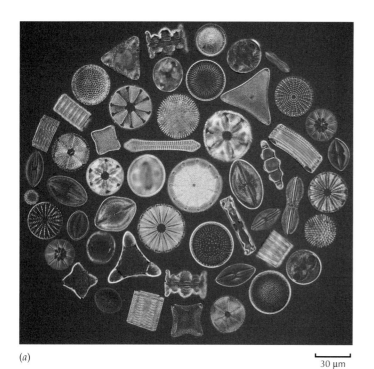

(a)

30 μm

27.18 Diatom Diversity
(a) Diatoms exhibit a splendid variety of species-specific forms.
(b) This artificially colored scanning electron micrograph shows the intricate patterning of diatom cell walls.

(b)   7 μm

is constructed in two pieces, with the wall of the top overlapping the wall of the bottom like the top and bottom of a petri plate. The silicon-impregnated walls have intricate, unique patterns (Figure 27.18b). Despite their remarkable morphological diversity, however, all diatoms are symmetrical—either bilaterally (with "right" and "left" halves) or radially (with the type of symmetry possessed by a circle).

Diatoms reproduce both sexually and asexually. Asexual reproduction is by cell division and is somewhat constrained by the stiff, silica-containing cell wall. Both the top and the bottom of the "petri plate" become tops of new "plates" without changing appreciably in size; as a result, the new cells made from former bottoms are smaller than the parent cells (Figure 27.19). If this process continued indefinitely, one cell line would simply vanish, but sexual reproduction largely solves this potential problem. Gametes are formed, shed their cell walls, and fuse. The resulting zygote then increases substantially in size before a new cell wall is laid down.

Diatoms are everywhere in the marine environment and are frequently present in great numbers, making them major photosynthetic producers in coastal waters. Diatoms are also common in fresh water. Because the silicon-containing walls of dead diatom cells resist decomposition, certain sedimentary rocks are composed almost entirely of diatom skeletons that sank to the seafloor over time. Diatomaceous earth, which is obtained from

such rocks, has many industrial uses, such as insulation, filtration, and metal polishing. It has also been used as an "Earth-friendly" insecticide that clogs the tracheae (breathing structures) of insects.

The brown algae include the largest protists

All the **brown algae** (Phaeophyta) are multicellular and composed either of branched filaments (Figure 27.20) or of leaflike growths called **thalli** (singular thallus) (Figure 27.21a). The brown algae obtain their namesake color from the carotenoid *fucoxanthin*, which is abundant in their chloro-

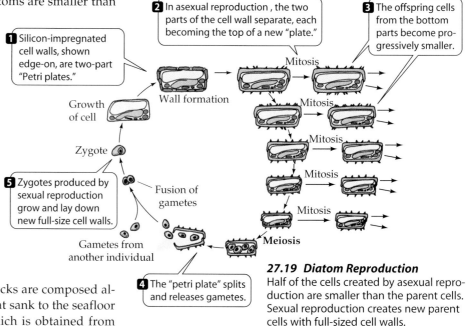

1 Silicon-impregnated cell walls, shown edge-on, are two-part "Petri plates."

2 In asexual reproduction, the two parts of the cell wall separate, each becoming the top of a new "plate."

3 The offspring cells from the bottom parts become progressively smaller.

Growth of cell

Wall formation

Mitosis

Mitosis

Mitosis

Mitosis

Mitosis

Meiosis

Zygote

5 Zygotes produced by sexual reproduction grow and lay down new full-size cell walls.

Fusion of gametes

Gametes from another individual

4 The "petri plate" splits and releases gametes.

27.19 Diatom Reproduction
Half of the cells created by asexual reproduction are smaller than the parent cells. Sexual reproduction creates new parent cells with full-sized cell walls.

(a) *Hormosira banksii*

(b) *Ectocarpus* sp.

27.20 Brown Algae
(a) A filamentous brown alga growing in Australia. This species is sometimes called "Neptune's necklace." (b) A filamentous brown alga seen through a light microscope.

plasts. The combination of this yellow-orange pigment with the green of chlorophylls *a* and *c* yields a brownish tinge.

The brown algae include the largest of the protists. Giant kelps, such as those of the genus *Macrocystis*, may be up to 60 meters long (see Figure 27.1*c*). The brown algae are almost exclusively marine. Some float in the open ocean; the most famous example is the genus *Sargassum*, which forms dense mats of vegetation in the Sargasso Sea in the mid-Atlantic. Most brown algae, however, are attached to rocks near the shore. A few thrive only where they are regularly exposed to heavy surf; a notable example is the sea palm *Postelsia palmaeformis* of the Pacific coast (Figure 27.21*a*). All

of the attached forms develop a specialized structure, called a **holdfast**, that literally glues them to the rocks (Figure 27.21*b*).

Some brown algae differentiate extensively into stemlike stalks and leaflike blades, and some develop gas-filled cavities or bladders. For biochemical reasons that are only poorly understood, these gas cavities often contain as much as 5 percent carbon monoxide—a concentration high enough to kill a human. In addition to organ differentiation, the larger brown algae also exhibit considerable tissue differentiation. Most of the giant kelps have photosynthetic filaments only in the outermost regions of their stalks and blades. Within the photosynthetic region lie filaments of long cells that closely resemble the food-conducting tissue of plants. Called *trumpet cells* because they have flaring ends, these tubes rapidly conduct the products of photosynthesis through the body of the organism.

The cell walls of brown algae may contain as much as 25 percent *alginic acid*, a gummy polymer of sugar acids. Alginic acid cements cells and filaments together and provides good holdfast glue. It is used commercially as an emulsifier in ice cream, cosmetics, and other products.

Many protist and all plant life cycles feature alternation of generations

Brown algae, like many photosynthetic protists and all plants, exhibit a type of life cycle known as **alternation of generations**, in which a multicellular, diploid, spore-producing organism gives rise to a multicellular, haploid, gamete-producing organism. When two gametes fuse (a process called *syngamy*), a diploid organism is formed (Figure 27.22). The haploid organism, the diploid organism, or both may also reproduce asexually.

(a) *Postelsia palmaeformis*

The leaflike structures are the thalli of sea palm.

(b)

27.21 Brown Algae in a Turbulent Environment
Brown algae growing in the intertidal zone on an exposed rocky shore take a tremendous pounding by the surf. (a) Sea palm growing along the California coast. (b) The tough, branched holdfast that anchors the sea palm.

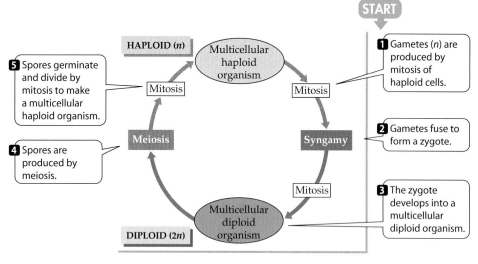

27.22 Alternation of Generations
In many multicellular photosynthetic protists and all plants, a diploid generation that produces spores alternates with a haploid generation that produces gametes.

The two generations (spore-producing and gamete-producing) differ genetically (one has haploid cells and the other has diploid cells), but they may or may not differ morphologically. In **heteromorphic** alternation of generations, the two generations differ morphologically; in **isomorphic** alternation of generations they do not, despite their genetic difference. We will see examples of both heteromorphic and isomorphic alternation of generations in some representative brown and green algae. In discussing the life cycles of plants and multicellular photosynthetic protists, we will use the terms **sporophyte** ("spore plant") and **gametophyte** ("gamete plant") to refer to the multicellular diploid and haploid generations, respectively.

Gametes are not produced by meiosis because the gamete-producing generation is already haploid. Instead, specialized cells of the diploid sporophyte, called **sporocytes**, divide meiotically to produce four haploid spores. The spores may eventually germinate and divide mitotically to produce multicellular haploid gametophytes, which produce gametes by mitosis and cytokinesis.

Gametes, unlike spores, can produce new organisms only by fusing with other gametes. The fusion of two gametes produces a diploid zygote, which then undergoes mitotic divisions to produce a diploid organism: the sporophyte generation. The sporocytes of the sporophyte generation then undergo meiosis and produce haploid spores, starting the cycle anew.

The brown algae exemplify the extraordinary diversity found among the photosynthetic protists. One genus of simple brown algae is *Ectocarpus* (see Figure 27.20*b*). Its branched filaments, a few centimeters long, commonly grow on shells and stones. The gametophyte and sporophyte generations of *Ectocarpus* can be distinguished only by chromosome number or reproductive products (spores or gametes). Thus the generations are isomorphic.

By contrast, some kelps of the genus *Laminaria* and some other brown algae show a more complex heteromorphic alternation of generations. The larger and more obvious generation of these species is the sporophyte. Meiosis in special fertile regions of the leaf-like fronds produces haploid **zoospores**—motile spores that are propelled by flagella. These germinate to form a tiny, filamentous gametophyte that produces either eggs or sperm. The eggs and sperm of brown algae typically have flagella.

The oomycetes include water molds and their relatives

A nonphotosynthetic stramenopile group called **oomycetes** consists in large part of the water molds and their terrestrial relatives, such as the downy mildews. Water molds are filamentous and stationary, and they feed by absorption. If you have seen a whitish, cottony mold growing on dead fish or dead insects in water, it was probably a water mold of the common genus *Saprolegnia* (Figure 27.23).

The oomycetes are **coenocytes**: They have many nuclei enclosed in a single plasma membrane. Their filaments have no cross-walls to separate the many nuclei into discrete cells. Their cytoplasm is continuous throughout the body of the mold, and there is no single structural unit with a single nucleus, except in certain reproductive stages. A distinguishing feature of the oomycetes is their flagellated reproductive cells. Oomycetes are diploid throughout most of their life cycle and have cellulose in their cell walls.

The water molds, such as *Saprolegnia*, are all aquatic and **saprobic** (they feed on dead organic matter). Some other

Saprolegnia sp.

27.23 An Oomycete
The filaments of a water mold radiate from the carcass of an insect.

oomycetes are terrestrial. Although most terrestrial oomycetes are harmless or helpful decomposers of dead matter, a few are serious plant parasites that attack crops such as avocados, grapes, and potatoes. The mold *Phytophthora infestans*, for example, is the causal agent of late blight of potatoes, which brought about the great Irish potato famine of 1845–1847. *P. infestans* destroyed the entire Irish potato crop in a matter of days in 1846. Among the consequences of the famine were a million deaths from starvation and the emigration of about 2 million people, mostly to the United States.

Rhodophyta

Almost all **red algae** (Rhodophyta) are multicellular (Figure 27.24). Some botanists now refer to the red algae as the "red plant kingdom." Their characteristic color is a result of the photosynthetic pigment phycoerythrin, which is found in relatively large amounts in the chloroplasts of many species. In addition to phycoerythrin, red algae contain phycocyanin, carotenoids, and chlorophyll.

The red algae include species that grow in the shallowest tide pools, as well as the algae found deepest in the ocean (as deep as 260 meters if nutrient conditions are right and the water is clear enough to permit the penetration of light). Very few red algae inhabit fresh water. Most grow attached to a substrate by a holdfast.

Euglenozoa

Alveolata

Stramenopila

Rhodophyta

Chlorophyta

In a sense the red algae, like several other groups of algae, are misnamed. They have the capacity to change the relative amounts of their various photosynthetic pigments depending on the light conditions where they are growing. Thus the leaflike *Chondrus crispus*, a common North Atlantic red alga, may appear bright green when it is growing at or near the surface of the water and deep red when growing at greater depths. The ratio of pigments present depends to a remarkable degree on the intensity of the light that reaches the alga. In deep water, where the light is dimmest, the alga accumulates large amounts of phycoerythrin, an accessory photosynthetic pigment (see Figure 8.7). Algae in deep water have as much chlorophyll as the green ones near the surface, but the accumulated phycoerythrin makes them look red.

In addition to being the only photosynthetic protists with phycoerythrin and phycocyanin among their pigments, the red algae have two other unique characteristics: They store the products of photosynthesis in the form of *floridean starch*, which is composed of very small, branched chains of approximately 15 glucose units. And they produce no motile, flagellated cells at any stage in their life cycle. The male gametes lack cell walls and are slightly amoeboid; the female gametes are completely immobile.

Some red algal species enhance the formation of coral reefs. Like the coral animals, they possess the biochemical machinery for depositing calcium carbonate both in and around their cell walls. After the death of the corals and algae, the calcium carbonate persists, sometimes forming substantial rocky masses.

Some red algae produce large amounts of mucilaginous polysaccharide substances, which contain the sugar galactose with a sulfate group attached. This material readily forms solid gels and is the source of agar, a substance widely used in the laboratory for making a solid aqueous medium on which tissue cultures and many microorganisms can be grown.

Certain red algae became endosymbionts, long ago, within the cells of other, nonphotosynthetic protists, eventually giving rise to chloroplasts. This was the evolutionary origin of the distinctive chloroplasts of the photosynthetic stramenopiles (the brown algae and the diatoms).

(a) *Palmaria palmata*

(b) *Polysiphonia* sp.

27.24 Red Algae
(a) Dulse, a large, edible red alga, is growing here on rocks in New Brunswick, Canada. (b) Both vegetative and reproductive structures of this alga can be seen under the light microscope.

Chlorophyta

The "green algae" do not form a monophyletic group, but include at least two lineages. One major lineage constitutes the **Chlorophyta**, a monophyletic group. A sister lineage to the Chlorophyta consists of other green algal lineages and the plant kingdom. The green algal lineages share characters that distinguish them from other protists: Like the plants, they contain chlorophylls *a* and *b*, and their reserve of photosynthetic products is stored as starch in plastids. There are more than 17,000 species of chlorophytes. Most chlorophytes are aquatic—some are marine, but more are freshwater forms—but others are terrestrial, living in moist environments. The chlorophytes range in size from microscopic unicellular forms to multicellular forms many centimeters in length.

Chlorophytes vary in shape and cellular organization

We find in the Chlorophyta an incredible variety in shape and construction of the algal body. *Chlorella* is an example of the simplest type: unicellular and flagellated.

Surprisingly large and well-formed colonies of cells are found in such freshwater groups as the genus *Volvox*. These cells are not differentiated into tissues and organs, as in plants and animals, but the colonies show vividly how the preliminary step of this great evolutionary development might have been taken. In *Volvox*, the origins of cell specialization can be seen as certain cells within the colony (Figure 27.25*a*) are specialized for reproduction.

While *Volvox* is colonial and spherical, *Oedogonium* is multicellular and filamentous, and each of its cells has only one nucleus. *Cladophora* is multicellular, but each cell is multinucleate. *Bryopsis* is tubular and coenocytic, forming cross-walls only when reproductive structures form. *Acetabularia* is a single, giant uninucleate cell a few centimeters long that becomes multinucleate only at the end of the reproductive stage. *Ulva lactuca* is a membranous sheet two cells thick; its unusual appearance justifies its common name: sea lettuce (Figure 27.25*b*).

Chlorophyte life cycles are diverse

The life cycles of chlorophytes show great diversity. Let's examine two chlorophyte life cycles in detail, beginning with that of the sea lettuce *Ulva lactuca* (Figure 27.26). The diploid sporophyte of this common seashore organism is a thin cellular sheet a few centimeters in diameter. Some of its cells (sporocytes) differentiate and undergo meiosis and cytokinesis, producing motile haploid spores (zoospores). These swim away, each propelled by four flagella, and some eventually find a suitable place to settle. The spores then lose their fla-

(a) *Volvox* sp.   Parent colony   Somatic cells

Specialized reproductive cells produce and release daughter colonies.

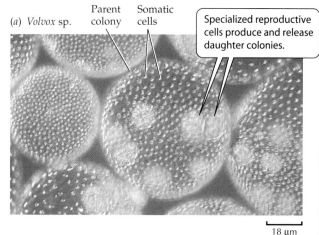

18 μm

(b) *Ulva lactuca*

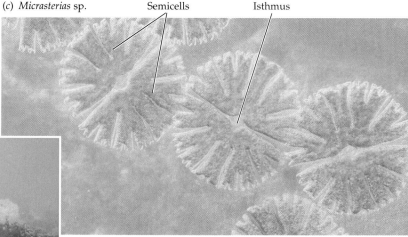

(c) *Micrasterias* sp.   Semicells   Isthmus

4 μm

27.25 Chlorophytes
(a) *Volvox* colonies are precisely spaced arrangements of cells. Several daughter colonies can be seen within the parent colonies. (b) Submerged in a tidal pool, this large sea lettuce appears "leafy" but is actually made up of membranous sheets of cells. (c) Each of these "constricted" desmids is a microscopic, unicellular alga comprising two semicells. The central isthmus contains the cell's nucleus. A single large, ornate chloroplast fills much of the volume of each semicell.

Euglenozoa

Alveolata

Stramenopila

Rhodophyta

Chlorophyta

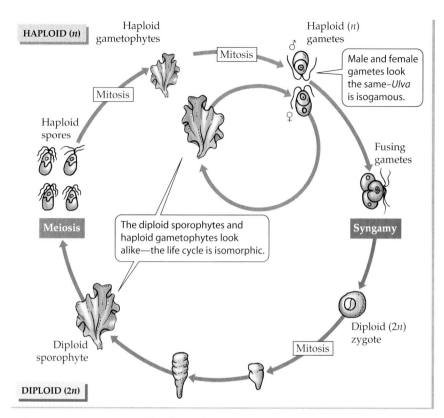

27.26 An Isomorphic Life Cycle
The life cycle of *Ulva lactuca* is an example of isomorphic alternation of generations.

flagella, undergo mitosis, and produce a new gametophyte directly; in other words, the gametes can also function as zoospores. Few chlorophytes other than *Ulva* have motile gametes that can also function as zoospores.

In contrast to the isomorphic life cycle of *Ulva*, many other chlorophytes have a heteromorphic life cycle: Sporophyte and gametophyte generations differ in structure. In one variation of the heteromorphic life cycle—the **haplontic** life cycle (Figure 27.27)—a multicellular haploid individual produces gametes that fuse to form a zygote. The zygote functions directly as a sporocyte, undergoing meiosis to produce spores, which in turn produce a new haploid individual. In the entire haplontic life cycle, only one cell—the zygote—is diploid. The filamentous organisms of the genus *Ulothrix* are examples of haplontic chlorophytes.

Other chlorophytes have a **diplontic** life cycle like that of many animals. In a diplontic life cycle, meiosis of sporocytes

gella and begin to divide mitotically, producing a thin filament that develops into a broad sheet only two cells thick. The gametophyte thus produced looks just like the sporophyte—in other words, *Ulva* has an isomorphic life cycle.

An individual gametophyte can produce only male or female gametes—never both. The gametes arise mitotically within single cells (called *gametangia*), rather than within a specialized multicellular structure, as in plants. Both types of gametes bear two flagella (in contrast to the four flagella of a haploid spore) and hence are motile.

In most species of *Ulva* the female and male gametes are indistinguishable structurally, making those species **isogamous**—having gametes of identical appearance. Other chlorophytes, including some other species of *Ulva*, are **anisogamous**—having female gametes that are distinctly larger than the male gametes.

Female and male gametes come together and unite, losing their flagella as the zygote forms and settles. After resting briefly, the zygote begins mitotic division, producing a multicellular sporophyte. Any gametes that fail to find partners can settle down on a favorable substrate, lose their

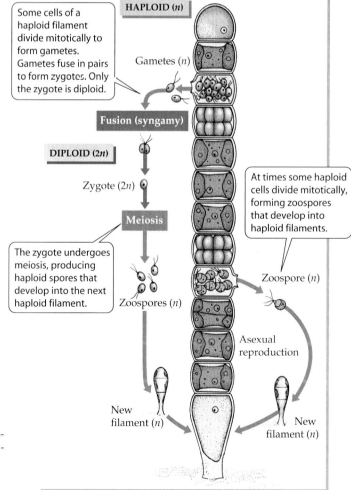

27.27 A Haplontic Life Cycle
In the life cycle of *Ulothrix*, a filamentous, multicellular gametophyte generation alternates with a sporophyte generation consisting of a single cell.

by bu
fuels
fuels
grew

In
their
other

Pe
form:
the g
press
proxi
1 per

Sp
yet it
adap
"plur
their
one c
duce,
ent o
becor
is ava

Th
fectiv
ronm
tion.
moss

Ea
abou
ance:
some
20.1)
their
tive
need
plant

Fuel i
Peat, '
moss
of fue
coast

(a) *Chara* sp. (stonewort)

28.1 The Closest Relatives of Land Plants
The plant kingdom probably evolved from a common ancestor shared with the charophytes, a green algal group. (a) Molecular evidence seems to favor stoneworts of the genus *Chara*. (b) Evidence from morphology indicates that the group including this coleochaete alga may be the ancestor of land plants.

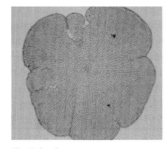

(b) *Coleochaete* sp.

sister group of the plants is a group of charophytes that includes the genus *Coleochaete* (Figure 28.1b). *Coleochaete*-like algae have features found in plants, such as plasmodesmata and a tendency to protect the young sporophyte.

Whether more similar to stoneworts or to *Coleochaete*, the ancestral green algae lived at the margins of ponds or marshes, ringing them with a green mat. From these margin habitats, which were sometimes wet and sometimes dry, early plants made the move onto land.

The Conquest of the Land

Plants or their immediate ancestors in the green mat pioneered and modified the terrestrial environment. That environment differs dramatically from the aquatic environment. The most obvious difference is the availability of the water that is essential for life: It is everywhere in the aquatic environment, but hard to find and to retain in the terrestrial environment. Water also provides aquatic organisms with support against gravity; a plant on land, however, must either have some other support system or sprawl unsupported on the ground. A land plant must also use different mechanisms for dispersing its gametes and progeny than its aquatic relatives use. How did terrestrial organisms arise from aquatic ancestors to thrive in such a challenging environment?

Adaptations to life on land distinguish plants from green algae

Most of the characteristics that distinguish plants from green algae are evolutionary adaptations to life on land (Figure 28.2). Many of the characteristics that proved adaptive to land plants probably evolved before the appearance of any of the plant groups we will discuss in this chapter. These characteristics include:

▶ The *cuticle*, a waxy covering that retards desiccation (drying).
▶ *Gametangia*, cases that enclose plant gametes and prevent them from drying. Eggs are housed in archegonia, sperm in antheridia.
▶ *Embryos*, which are young sporophytes contained within in a protective structure.
▶ Certain *pigments* that afford protection against the mutagenic ultraviolet radiation that bathes the terrestrial environment.
▶ *Thick spore walls* that prevent desiccation and resist decay.

All these characteristics were probably shared by a plant ancestral to today's plants. Further adaptations to the terrestrial environment appeared as plant evolution continued. We will identify the most important ones in this and the next chapter. For now, let's look at one of the key later adaptations: the appearance of vascular tissues.

Most present-day plants have vascular tissue

The first plants were truly nonvascular, lacking both water-conducting and food-conducting cells. Although the term "nonvascular plants" is a time-honored name, it is misleading to apply it to the entire nontracheophyte lineage, because some mosses (unlike liverworts and hornworts) do have a limited amount of vascular-like tissue. Thus the more unwieldy name nontracheophyte is more descriptive. The first true tracheophytes—possessing specialized conducting cells called tracheids—arose later.

The nontracheophytes (the liverworts, hornworts, and mosses) have never been large plants. Except for some of the mosses, they have no water-transporting tissue, yet some are found in dry environments. Many grow in dense masses (see Figure 28.7a), through which water can move by capillary action. Nontracheophytes have leaflike structures that readily catch and hold any water that splashes onto them. These plants are small enough that minerals can be distributed internally by diffusion. They lack the leaves, stems, and roots that characterize tracheophytes, although they have structures analogous to each.

Familiar tracheophytes include the ferns, conifers, and flowering plants. Tracheophytes differ from liverworts, hornworts, and mosses in crucial ways, one of which is the possession of a well-developed **vascular system** consisting of specialized tissues for the transport of materials from one part of the plant to another. One such tissue, the *phloem*, conducts the products of photosynthesis from sites where they are produced or released to sites where they are used

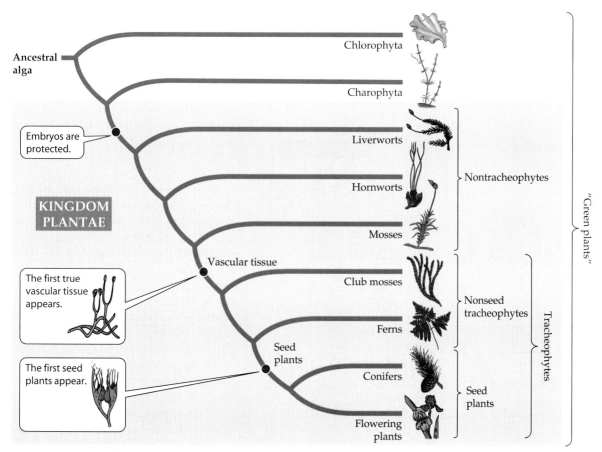

28.2 From Green Algae to Plants
Green algae called charophytes are sister to the plants; green algae called chlorophytes are sister to the lineage that includes the charophytes and plants.

or stored. The other vascular tissue, the *xylem*, conducts water and minerals from the soil to aerial parts of the plant; because some of its cell walls are stiffened by a substance called *lignin*, xylem also provides support in the terrestrial environment.

Nontracheophyte plants evolved tens of millions of years before the earliest tracheophytes, even though tracheophytes appear earlier in the fossil record. The oldest tracheophyte fossils date back more than 410 million years, whereas the oldest nontracheophyte fossils are only about 350 million years old, dating from a time when tracheophytes were already widely distributed. This simply means that, given their different structures and the chemical makeup of their cell walls, tracheophytes are more likely to form fossils than nontracheophytes are.

We will examine the adaptations of the tracheophytes later in the chapter, concentrating first on the nontracheophytes.

The Nontracheophytes: Liverworts, Hornworts, and Mosses

Most liverworts, hornworts, and mosses grow in dense mats, usually in moist habitats (see Figure 28.7a). The largest of these plants are only about 1 meter tall, and most are only a few centimeters tall or long. Why have no large nontracheophytes ever evolved? The probable answer is that they lack an efficient system for conducting water and minerals from the soil to distant parts of the plant body. However, to limit water loss, layers of maternal tissue protect the embryos of all nontracheophytes. All nontracheophyte lineages also have a cuticle, although it is often very thin (or even absent in some species) and thus not highly effective in retarding water loss.

Most nontracheophytes live on the soil or on other plants, but some grow on bare rock, dead and fallen tree trunks, and even on buildings. Nontracheophytes are widely distributed over six continents and exist very locally on the coast of the seventh (Antarctica). They are very successful plants, well-adapted to their environments. Most are terrestrial. Some live in wetlands. Although a few nontracheophyte species live in fresh water, these aquatic forms are descended from terrestrial ones. There are no marine nontracheophytes.

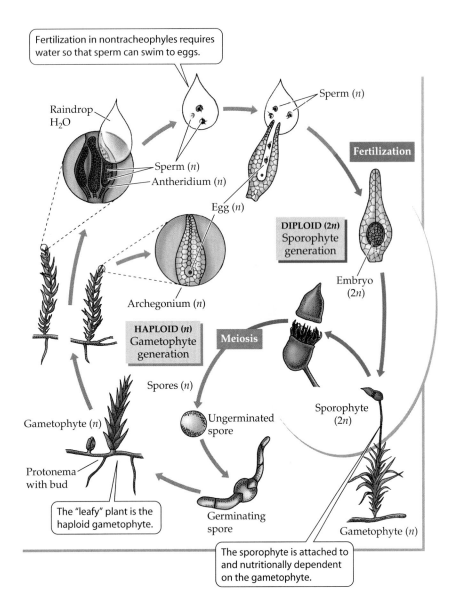

Fertilization in nontracheophyles requires water so that sperm can swim to eggs.

Raindrop H₂O

Sperm (n)

Sperm (n)
Antheridium (n)

Fertilization

Egg (n)

DIPLOID (2n) Sporophyte generation

Archegonium (n)

Embryo (2n)

HAPLOID (n) Gametophyte generation

Meiosis

Spores (n)

Gametophyte (n)

Ungerminated spore

Sporophyte (2n)

Protonema with bud

The "leafy" plant is the haploid gametophyte.

Germinating spore

Gametophyte (n)

The sporophyte is attached to and nutritionally dependent on the gametophyte.

28.3 A Nontracheophyte Life Cycle
The life cycle of nontracheophytes, illustrated here by a moss, is dependent on an external source of liquid water. The visible green structure of nontracheophytes is the gametophyte; in nontracheophyte plants, the "leafy" structures are sporophytes.

Once released, the sperm must swim or be splashed by raindrops to a nearby archegonium on the same or a neighboring plant. The sperm are aided in this task by chemical attractants released by the egg or the archegonium. Before sperm can enter the archegonium, certain cells in the neck of the archegonium must break down, leaving a water-filled canal through which the sperm swim to complete their journey. Note that all of these events require liquid water.

On arrival at the egg, one of the sperm nuclei fuses with the egg nucleus to form the zygote. Mitotic divisions of the zygote produce a multicellular, diploid sporophyte embryo. The base of the archegonium grows to protect the embryo during its early growth. Eventually the developing sporophyte elongates sufficiently to break out of the archegonium, but it remains connected to the gametophyte by a "foot" that is embedded in the parent tissue and absorbs water and nutrients from it (see Figure 28.3). The sporophyte remains attached to the gametophyte throughout its life. The sporophyte produces a sporangium, or **capsule**, within which meiotic divisions produce spores and thus the next gametophyte generation.

The structure and pattern of elongation of the sporophyte differ among the three phyla of nontracheophytes—the liverworts (Hepatophyta), hornworts (Anthocerophyta), and mosses (Bryophyta). The evolutionary relationships of the three phyla and the tracheophytes can be seen in Figure 28.2.

Nontracheophyte sporophytes are dependent on gametophytes

In nontracheophytes, the conspicuous green structure visible to the naked eye is the gametophyte (Figure 28.3). In contrast, the familiar forms of tracheophytes such as ferns and seed plants are sporophytes. The gametophyte of nontracheophytes is photosynthetic and therefore nutritionally independent, whereas the sporophyte may or may not be photosynthetic but is *always* dependent on the gametophyte and remains permanently attached to it.

A sporophyte produces unicellular, haploid spores as products of meiosis. A spore germinates, giving rise to a multicellular, haploid gametophyte whose cells contain chloroplasts and are thus photosynthetic. Eventually gametes form within specialized sex organs, the **gametangia**. The **archegonium** is a multicellular, flask-shaped female sex organ with a long neck and a swollen base (Figure 28.4a). The base contains a single egg. The **antheridium** is a male sex organ in which sperm, each bearing two flagella, are produced in large numbers (Figure 28.4b).

Liverworts are the most ancient surviving plant lineage

The gametophytes of some liverworts (phylum Hepatophyta) are green, leaflike layers that lie flat on the ground (Figure 28.5a). The simplest liverwort gametophytes, however, are flat plates of cells, a centimeter or so long, that produce antheridia or archegonia on their upper surfaces and water-absorbing filaments called **rhizoids** on the lower. Liverwort sporophytes are shorter than those of mosses and hornworts, rarely exceeding a few millimeters.

The sporophyte has a stalk that connects capsule and foot. The stalk elongates and thus raises the capsule above

(a)

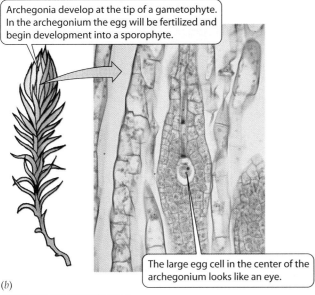

Archegonia develop at the tip of a gametophyte. In the archegonium the egg will be fertilized and begin development into a sporophyte.

The large egg cell in the center of the archegonium looks like an eye.

(b)

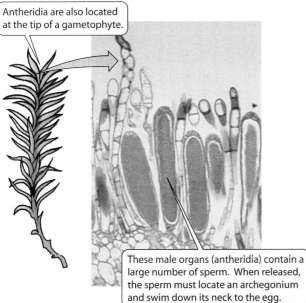

Antheridia are also located at the tip of a gametophyte.

These male organs (antheridia) contain a large number of sperm. When released, the sperm must locate an archegonium and swim down its neck to the egg.

28.4 Sex Organs in Plants
Archegonia (a) and antheridia (b) of the moss *Mnium* (phylum Bryophyta). Gametophytes of all plants have archegonia and antheridia, but they are much reduced in seed plants.

ground level, favoring dispersal of spores when they are released. The capsules of liverworts are simple: a globular capsule wall surrounding a mass of spores. In some species of liverworts, spores are not released by the sporophyte until the surrounding capsule wall rots.

In other liverworts, however, the spores are disseminated by structures called **elaters** located within the capsule. Elaters are long cells that have a helical thickening of the cell wall. As an elater loses water, the whole cell shrinks to a fraction of its former length, thus compressing the helical thickening like a spring. When the stress becomes sufficient, the compressed "spring" snaps back to its resting position, throwing spores in all directions.

Among the most familiar liverworts are species of the genus *Marchantia* (Figure 28.5a). *Marchantia* is easily recognized by the characteristic structures on which its male and female gametophytes bear their antheridia and archegonia (Figure 28.5b). Like most liverworts, *Marchantia* also reproduces vegetatively by simple fragmentation of the gametophyte. Along with sexual reproduction, *Marchantia* and some other liverworts and mosses also reproduce vegetatively by means of **gemmae** (singular gemma), which are lens-shaped clumps of cells. In a few liverworts the gemmae are loosely held in structures called gemma cups, which promote dispersal by raindrops (Figure 28.5c).

Hornworts evolved stomata as an adaptation to terrestrial life

The phylum Anthocerophyta comprises the hornworts, so named because their sporophytes look like little horns (Figure 28.6). Hornworts appear at first glance to be liverworts with very simple gametophytes. These gametophytes consist of flat plates of cells, a few cells thick.

However, the hornworts, along with the mosses and tracheophytes, share an advance over the liverwort lineage in their adaptation to life on land. They have *stomata*—pores

28.5 Liverwort Structures
Members of the phylum Hepatophyta display various characteristic structures. (a) Gametophytes. (b) Structures bearing antheridia and archegonia. (c) Gemmae.

The finger-headed structures bear archegonia.

The disc-headed structures bear antheridia.

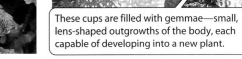

These cups are filled with gemmae—small, lens-shaped outgrowths of the body, each capable of developing into a new plant.

(a) *Marchantia* sp.   (b) *Marchantia* sp.   (c) *Lunularia* sp.

The sporophytes of hornworts can reach 20 cm in height.

Gametophytes are flat plates a few cells thick.

Anthoceros dieteret

28.6 A Hornwort
The sporophytes of hornworts can resemble little horns.

that, when open, allow the uptake of CO_2 for photosynthesis and the release of O_2, but that can close to prevent excessive water loss.

Hornworts have two characteristics that distinguish them from both liverworts and mosses. First, the cells of hornworts each contain a single large, platelike chloroplast, whereas the other nontracheophytes contain numerous small, lens-shaped chloroplasts. Second, of all the nontracheophyte sporophytes, those of the hornworts come closest to being capable of indefinite growth (without a set limit).

The stalk of either the liverwort or moss sporophyte stops growing as the capsule matures, so elongation of the sporophyte is strictly limited. In a hornwort such as *Anthoceros*, however, there is no stalk, but a basal region of the capsule remains capable of indefinite cell division, continuously producing new spore-bearing tissue above.

Sporophytes of some hornworts growing in mild and continuously moist conditions can become as tall as 20 centimeters. Eventually the sporophyte's growth is limited by the lack of a transport system. To support their growth, the hornworts need access to nitrogen. Hornworts have internal cavities filled with a mucilage; these cavities are often populated by cyanobacteria that fix atmospheric nitrogen gas into a nutrient form usable by the host plant.

We have presented the hornworts as sister to the lineage consisting of mosses and tracheophytes, but this is only one possible interpretation of the current data. The exact evolutionary status of the hornworts is still in doubt.

Water- and sugar-transport mechanisms emerged in the mosses

The most familiar nontracheophytes are the mosses (phylum Bryophyta). There are more species of mosses than of liverworts and hornworts combined, and these hardy little plants are found in almost every terrestrial environment. They often are found on damp, cool ground, where they form thick mats (Figure 28.7a). The mosses are sister to the tracheophytes (see Figure 28.2).

Many mosses contain a type of cell called a *hydroid*, which dies and leaves a tiny channel through which water may travel. The hydroid likely is a progenitor of the tracheid, the characteristic water-conducting cell of the tracheophytes, but it lacks lignin (a waterproofing substance) and the wall structure found in tracheids. The possession of hydroids and of a limited system for transport of sucrose by some mosses (via cells called *leptoids*) shows that the old term "nonvascular plant" is somewhat misleading when applied to mosses.

In contrast to liverworts and hornworts, the sporophytes of mosses and tracheophytes grow by **apical cell division**. A region at the growing tip provides an organized pattern of cell division, elongation, and differentiation. This allows extensive and sturdy vertical growth of sporophytes.

The moss gametophyte that develops following spore germination is a branched, filamentous structure, or **protonema** (see Figure 28.3). Although the protonema looks much like a filamentous green alga, it is unique to the mosses. Some of the filaments contain chloroplasts and are photosynthetic; others, called rhizoids, are nonphotosynthetic and anchor the protonema to the substrate. After a period of linear growth, cells close to the tips of the photosynthetic filaments divide rapidly in three dimensions to form buds. The buds eventually differentiate a distinct tip, or apex, and produce the familiar leafy moss shoot with leaflike structures arranged spirally.

These leafy shoots produce antheridia or archegonia (see Figure 28.4). The antheridia release sperm that travel through liquid water to the archegonia, where they fertilize the eggs. Sporophyte development in most mosses follows a precise pattern, resulting ultimately in the formation of an absorptive foot, a stalk, and, at the tip, a swollen capsule. In contrast to hornworts, which grow from the base, the moss sporophyte stalk grows at its apical end, as tracheophytes do. Cells at the tip of the stalk divide, supporting elongation of the structure and giving rise to the capsule. For a while, archegonial tissue grows rapidly as the stalk elongates, but eventually the archegonium is outgrown and is torn apart by the expanding sporophyte.

The top of the capsule is shed after completing meiosis and spore development. Groups of cells just below the lid form a series of toothlike structures surrounding the opening. Highly responsive to humidity, these structures dig into the mass of spores when the atmosphere is dry; then, when the atmosphere becomes moist, they fling out, scooping out the spores as they go (Figure 28.7b). The spores are thus dispersed when the surrounding air is

(a)

(b)

> The teeth of this moss capsule help expel the thousands of spores.

28.7 The Mosses
(a) Dense moss forms hummocks in a valley on New Zealand's South Island. (b) The moss capsule, from which spores are dispersed, grows at the tip of the plant.

moist—that is, when conditions favor their subsequent germination.

Only a few mosses depart from this pattern of capsule development. A familiar exception is the genus *Sphagnum*, which we discussed at the beginning of this chapter. Species in this genus have a simple capsule with an air chamber in it. Air pressure builds up in this chamber, eventually causing the capsule lid to pop open, dispersing the spores with an audible explosion.

With their simple system of internal transport, the mosses are in a sense vascular plants. However, they are not tracheophytes, because they lack true xylem and phloem.

Introducing the Tracheophytes

Although an extraordinarily large and diverse group, the tracheophytes can be said to have been launched by a single evolutionary event. Sometime during the Paleozoic era, probably well before the Silurian period (440 mya), the sporophyte generation of a now long-extinct organism produced a new cell type, the **tracheid**. The tracheid is the principal water-conducting element of the xylem in all tracheophytes except the angiosperms (flowering plants); and even in angiosperms the tracheid persists alongside a more specialized and efficient system of vessels and fibers derived from tracheids.

The evolutionary appearance of a tissue composed of tracheids had two important consequences. First, it provided a pathway for long-distance transport of water and mineral nutrients from a source of supply to regions of need. Second, it provided something almost completely lacking—and unnecessary—in the largely aquatic green algae: rigid structural support. Support is important in a terrestrial environment because plants tend to grow upward as they compete for sunlight to power photosynthesis. Thus the tracheid set the stage for the complete and permanent invasion of land by plants.

The tracheophytes feature a further evolutionary novelty: *a branching, independent sporophyte*. A branching sporophyte can produce more spores than an unbranched body, and it can develop in complex ways. The sporophyte of a tracheophyte is nutritionally independent of the gametophyte.

The present-day evolutionary descendants of the early tracheophytes belong to nine distinct phyla (Figure 28.8). We can sort these phyla into two groups: those that produce seeds and those that do not. The nonseed tracheophytes include ferns, horsetails, club mosses, and whisk ferns. In the nonseed tracheophytes, the haploid and diploid generations are independent at maturity. The sporophyte is the large and obvious plant that one normally notices in nature (in contrast to the nontracheophyte sporophyte, which is attached to, dependent on, and usually much smaller than the gametophyte). Gametophytes of the nonseed tracheophytes are rarely more than 1 or 2 centimeters long and are short-lived, whereas their sporophytes are often highly visible; the sporophyte of a tree fern, for example, may be 15 or 20 meters tall and may live for many years.

The most prominent resting stage in the life cycle of a nonseed tracheophyte is the single-celled spore. This feature makes this life cycle similar to those of the fungi, the green algae, and the nontracheophytes but not, as we will see in the next chapter, to that of the seed plants. Nonseed tracheophytes must have an aqueous environment for at least one stage of their life cycle because fertilization is accomplished by a motile, flagellated sperm.

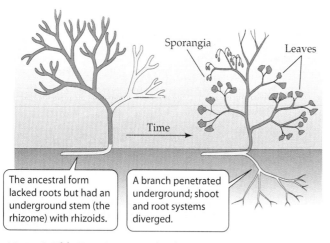

28.11 Is This How Roots Evolved?
According to Lignier's hypothesis, branches from ancestral root-less plants could have penetrated the soil, where they gradually evolved into a root system.

Early tracheophytes added new features

Within a few tens of millions of years, during the Devonian period, three new phyla of tracheophytes—Lycophyta, Sphenophyta, and Pterophyta—appeared on the scene, arising from rhyniophyte-like ancestors. These new groups featured specializations not found in the rhyniophytes, including one or more of the following: true roots, true leaves, and a differentiation between two types of spores.

THE ORIGIN OF ROOTS.   *Rhynia* and its close relatives lacked true roots. They had only rhizoids arising from a rhizome (Figure 28.11, left) with which to gather water and minerals. How, then, did subsequent groups of tracheophytes come to have the complex roots we see today?

In 1903, a French botanist, E. A. O. Lignier, proposed an attractive hypothesis that is still widely accepted today. Lignier argued that the ancestors of the first tracheophytes grew by branching dichotomously. This explanation is supported by the dichotomous branching observed in the rhyniophytes. Lignier suggested that such a branch could bend, penetrate the soil, and branch there (Figure 28.11,

right). The underground portion could anchor the plant firmly, and even in this primitive condition it could absorb water and minerals. The subsequent discovery of fossil plants from the Devonian period, all having horizontal stems (rhizomes) with both underground and aerial branches, supported Lignier's hypothesis.

Underground and aboveground branches, growing in sharply different environments, were subjected to very different selection during the succeeding millions of years. Thus the two parts of the plant axis (the shoot and root systems) diverged in structure and evolved distinct internal and external anatomies. In spite of these differences, scientists believe that the root and shoot systems of tracheophytes are homologous—that they were once part of the same organ.

THE ORIGIN OF TRUE LEAVES.   Thus far we have used the term "leaf" rather loosely. We spoke of "leafy" mosses and commented on the absence of "true leaves" in rhyniophytes. In the strictest sense, a **leaf** is a flattened photosynthetic structure emerging laterally from a main axis or stem and possessing true vascular tissue. Using this precise definition as we take a closer look at true leaves in the tracheophytes, we see that there are two different types of leaves, very likely of different evolutionary origins.

The first type, the *simple leaf,* is usually small and only rarely has more than a single vascular strand, at least in plants alive today. Plants in the phylum Lycophyta (club mosses), of which only a few genera survive, have such leaves. The evolutionary origin of simple leaves is thought by some biologists to be sterile sporangia (Figure 28.12a). The principal characteristic of this type of leaf is that its vascular strand departs from the vascular system of the stem in such a way that the structure of the stem's vascular system is scarcely disturbed. This was true even in the fossil lycopod trees of the Carboniferous period, many of which had leaves many centimeters long.

28.12 The Evolution of Leaves
(a) Simple leaves are thought to have evolved from sterile sporangia. (b) The complex leaves of ferns and seed plants may have arisen as photosynthetic tissue developed between complex branching patterns.

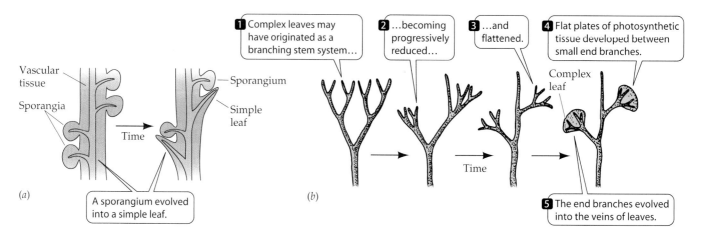

The other type of leaf is encountered in ferns and seed plants. This larger, more *complex leaf* is thought to have arisen from the flattening of a dichotomously branching stem system, with the development of extensive photosynthetic tissue between the branch members (Figure 28.12*b*). The complex leaf may have evolved several times, in different phyla of tracheophytes.

HOMOSPORY AND HETEROSPORY.   In the most ancient of the present-day tracheophytes, both the gametophyte and the sporophyte are independent and usually photosynthetic. Spores produced by the sporophytes are of a single type, and they develop into a single type of gametophyte that bears both female and male reproductive organs. Such plants, which bear a single type of spore, are said to be **homosporous** (Figure 28.13*a*). The sex organs on the gametophytes of homosporous plants are of two types. The female organ is a multicellular archegonium, typically containing a single egg. The male organ is an antheridium, containing many sperm.

A different system, with two distinct types of spores, evolved somewhat later. Plants of this type are said to be **heterosporous** (Figure 28.13*b*). One type of spore, the **megaspore**, develops into a larger, specifically female gametophyte (megagametophyte) that produces only eggs. The other type, the **microspore**, develops into a smaller, male gametophyte (microgametophyte) that produces only sperm. The sporophyte produces megaspores in small numbers in megasporangia on the sporophyte, and microspores in large numbers in microsporangia.

The most ancient tracheophytes were all homosporous. Heterospory evidently evolved independently several times in the early evolution of the tracheophytes descended from the rhyniophytes. The fact that heterospory evolved repeatedly suggests that it affords selective advantages. Subsequent evolution in the plant kingdom featured ever greater specialization of the heterosporous condition.

The Surviving Nonseed Tracheophytes

Today ferns are the most abundant and diverse phylum of nonseed tracheophytes, but club mosses and horsetails were once dominant elements of Earth's vegetation. A fourth phylum, the whisk ferns, contains only two genera. In this section we'll look at the characteristics of these four phyla and at some of the evolutionary advances that appeared in them.

The club mosses are sister to the other tracheophytes

The club mosses (lycopods, phylum Lycophyta) diverged earlier than all other living tracheophytes—that is, the remaining tracheophytes share an ancestor that was not ancestral to the Lycophyta. There are relatively few surviving species of club mosses. They have roots that branch dichoto-

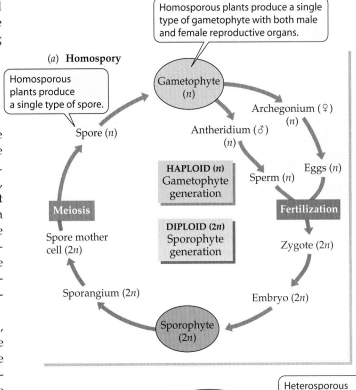

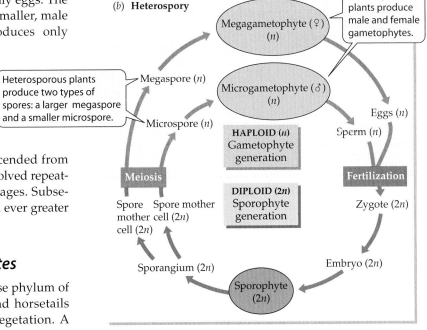

28.13 Homospory and Heterospory
(*a*) Homosporous plants bear a single type of spore. Each gametophyte has two types of sex organs, antheridia (male) and archegonia (female). (*b*) Heterospory, with two types of spores that develop into distinctly male and female gametophytes, evolved later.

mously. They bear only simple leaves, and the leaves are arranged spirally on the stem. Growth in club mosses comes entirely from groups of dividing cells at the tips of the stems and thus is apical, as it is in many flowering plants.

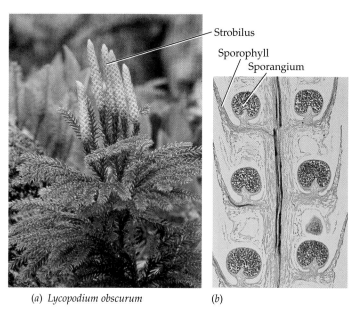

(a) *Lycopodium obscurum*   (b)

28.14 Club Mosses
(a) Strobili are visible at the tips of this club moss. Club mosses have simple leaves arranged spirally on their stems. (b) Thin section through a strobilus. Specialized leaves called sporophylls enclose each sporangium.

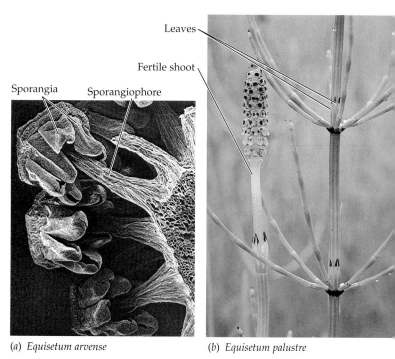

(a) *Equisetum arvense*   (b) *Equisetum palustre*

28.15 Horsetails
(a) Sporangia and sporangiophores of a horsetail. (b) Vegetative and fertile shoots of the marsh horsetail. Leaves form in whorls at nodes on the stems of the vegetative shoot on the right; the fertile shoot on the left is ready to disperse its spores.

The sporangia in most club mosses are contained within conelike structures called **strobili** (singular strobilus; Figure 28.14) and are tucked in the upper angle between a specialized leaf and the stem. This placement contrasts with the terminal sporangia of the rhyniophytes (see Figure 28.10). There are both homosporous species and heterosporous species of club mosses. Like all the nonseed tracheophytes, they have a large, independent sporophyte and a small, independent gametophyte.

Although only a minor element of present-day vegetation, the Lycophyta are one of two phyla that appear to have been the dominant vegetation during the Carboniferous period. One abundant type of coal (Cannel coal) is formed almost entirely from fossilized spores of a tree lycopod named *Lepidodendron*—which gives us an idea of the abundance of this genus in the forests of that time. The other major element of the Carboniferous vegetation was the phylum Sphenophyta, the horsetails.

Liverworts
Hornworts
Mosses
Club mosses
Horsetails
Whisk ferns
Ferns

Horsetails grow at the bases of their segments

Like the club mosses, the horsetails (phylum Sphenophyta) are represented by only a few present-day species. They are sometimes called scouring rushes because silica deposits found in the cell walls made them useful for cleaning. They have true roots that branch irregularly, as do the roots of all tracheophytes except the club mosses. Their sporangia curve back toward the stem on the ends of short stalks (sporangiophores) (Figure 28.15a). Horsetails have a large sporophyte and a small gametophyte, both independent.

The leaves of horsetails are simple and form distinct whorls (circles) around the stem (Figure 28.15b). Growth in horsetails originates to a large extent from discs of dividing cells just above each whorl of leaves, so each segment of the stem grows from its base. Such basal growth is uncommon in plants, although it is found in the grasses, a major group of flowering plants.

Present-day whisk ferns resemble the most ancient tracheophytes

There once was some disagreement about whether rhyniophytes are entirely extinct. The confusion arose because of the existence today of two genera of rootless, spore-bearing plants, *Psilotum* and *Tmesipteris*. *Psilotum nudum* (Figure 28.16) has only minute scales instead of true leaves, but plants of the genus *Tmesipteris* have flattened photosynthetic organs with well-developed vascular tissue. Are these two genera the living relics of the rhyniophytes, or do they have more recent origins?

Psilotum and *Tmesipteris* once were thought to be evolutionarily ancient descendants of anatomically simple ancestors. That hypothesis was weakened by an enormous hole in the geologic record between the rhyniophytes, which apparently became extinct more than 300 million years ago, and *Psilotum* and *Tmesipteris*, which are modern plants. DNA sequence data finally settled the question in favor

Psilotum nudum

28.16 A Whisk Fern
Aerial branches of a whisk fern, a plant once considered by some to be a surviving rhyniophyte and by others to be a fern. It is now included in the phylum Psilotophyta, and is widespread in the tropics and subtropics.

of a more modern origin from fernlike ancestors. Most botanists now treat these two genera as their own phylum, the Psilotophyta (whisk ferns) rather than as relatives of the rhyniophtes.

We now consider the whisk ferns to be highly specialized plants that evolved fairly recently from anatomically more complex ancestors. Whisk fern gametophytes live below the surface of the ground and lack chlorophyll. They depend upon fungal partners for their nutrition.

Ferns evolved large, complex leaves

The sporophytes of the ferns and seed plants have roots, stems, and leaves. Their leaves are typically large and have branching vascular strands. Some species have small leaves as a result of evolutionary reduction, but even the small leaves have more than one vascular strand.

The true ferns constitute the phylum Pterophyta, which first appeared during the Devonian period and today consists of about 12,000 species. The Pterophyta are probably not a monophyletic group. Ferns are characterized by

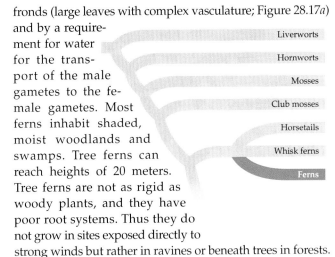

fronds (large leaves with complex vasculature; Figure 28.17*a*) and by a requirement for water for the transport of the male gametes to the female gametes. Most ferns inhabit shaded, moist woodlands and swamps. Tree ferns can reach heights of 20 meters. Tree ferns are not as rigid as woody plants, and they have poor root systems. Thus they do not grow in sites exposed directly to strong winds but rather in ravines or beneath trees in forests.

During its development, the fern frond unfurls from a tightly coiled "fiddlehead" (Figure 28.17*b*). Some fern leaves become climbing organs and may grow to be as much as 30 meters long. The sporangia are found on the undersurfaces of the leaves, sometimes covering the whole undersurface and sometimes only at the edges; in most species the sporangia are clustered in groups called **sori** (singular, sorus) (Figure 28.18).

The sporophyte generation dominates the fern life cycle

Inside the sporangia, fern cells undergo meiosis to form haploid spores. Once shed, spores travel great distances and eventually germinate to form independent gametophytes. Old World climbing fern, *Lygodium microphyllum,* is currently spreading disastrously through the Florida Everglades, choking off the growth of other plants. This rapid spread is testimony to the effectiveness of wind-borne spores.

28.17 Fern Fronds Take Many Forms
(*a*) Fronds of maidenhair fern form a pattern in this photograph. (*b*) The "fiddlehead" (developing frond) of a common forest fern; this structure will unfurl and expand to give rise to a complex adult frond such as those in (*a*). (*c*) The tiny fronds of a water fern.

(*a*) *Adiantum* sp.

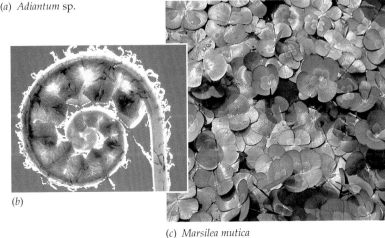

(*b*)

(*c*) *Marsilea mutica*

Sori

Dryopteris intermedia

28.18 Fern Sori Contain Sporangia
Sori, each with many spore-producing sporangia, form on the underside of a frond of the Midwestern fancy fern.

Fern gametophytes produce antheridia and archegonia, although not necessarily at the same time or on the same gametophyte. Sperm swim through water to archegonia, often on other gametophytes, where they unite with an egg.

The resulting zygote develops into a new sporophyte embryo. The young sporophyte sprouts a root and can thus grow independently of the gametophyte. In the alternating generations of a fern, the gametophyte is small, delicate, and short-lived, but the sporophytes can be very large and can sometimes survive for hundreds of years (Figure 28.19).

Most ferns are homosporous. However, two groups of aquatic ferns, the Marsileales and Salviniales, are derived from a common ancestor that evolved heterospory. Megaspores and microspores of these plants (which germinate to produce female and male gametophytes, respectively) are produced in different sporangia, and the microspores are always much smaller and greater in number than the megaspores.

A few genera of ferns produce a tuberous, fleshy gametophyte instead of the characteristic flattened, photosynthetic structure described earlier. Like the gametophytes of whisk ferns, these tuberous gametophytes depend on a mutualistic fungus for nutrition;* in some genera, even the sporophyte embryo must become associated with the fun-

*In a mutualistic association, both partners—here, the gametophyte and the fungus—profit.

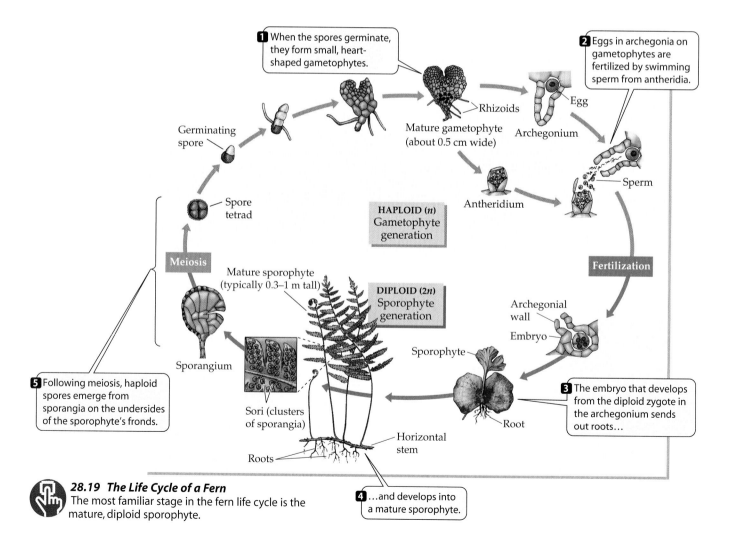

1 When the spores germinate, they form small, heart-shaped gametophytes.

2 Eggs in archegonia on gametophytes are fertilized by swimming sperm from antheridia.

Germinating spore

Rhizoids

Egg

Mature gametophyte (about 0.5 cm wide)

Archegonium

Spore tetrad

Sperm

HAPLOID (*n*)
Gametophyte generation

Antheridium

Meiosis

Fertilization

Mature sporophyte (typically 0.3–1 m tall)

DIPLOID (2*n*)
Sporophyte generation

Archegonial wall

Embryo

Sporophyte

Sporangium

5 Following meiosis, haploid spores emerge from sporangia on the undersides of the sporophyte's fronds.

Sori (clusters of sporangia)

3 The embryo that develops from the diploid zygote in the archegonium sends out roots…

Root

Horizontal stem

Roots

28.19 The Life Cycle of a Fern
The most familiar stage in the fern life cycle is the mature, diploid sporophyte.

4 …and develops into a mature sporophyte.

gus before extensive development can proceed. In Chapter 30 we will see that there are many important plant–fungus mutualisms.

All the tracheophytes we have discussed thus far disperse themselves by spores. In the next chapter we discuss the plants that dominate most of Earth's vegetation today, the seed plants, whose seeds afford new sporophytes protection unavailable to the nonseed tracheophytes.

Chapter Summary

▶ Plants are photosynthetic eukaryotes that use chlorophylls *a* and *b*, store carbohydrates as starch, and develop from embryos protected by parental tissue.

▶ Plant life cycles feature alternation of gametophyte (haploid) and sporophyte (diploid) generations.

▶ There are twelve surviving phyla of plants grouped into two main categories, nontracheophytes and tracheophytes. **Review Table 28.1**

▶ Plants arose from a common green algal ancestor, either of the stoneworts or of *Coleochaete*. Descendants of this ancestral charophyte colonized the land.

The Conquest of the Land

▶ Moving toward today's plants, early steps in plant evolution included the acquisition of a cuticle, gametangia, a protected embryo, protective pigments, and thick spore walls.

▶ Tracheophytes are characterized by possession of a vascular system, consisting of water- and mineral-conducting xylem and nutrient-conducting phloem. Nontracheophytes lack a vascular system. **Review Figure 28.2**

Nontracheophytes: Liverworts, Hornworts, and Mosses

▶ The nontracheophytes include the liverworts (phylum Hepatophyta), hornworts (phylum Anthocerophyta), and mosses (phylum Bryophyta). **Review Table 28.1**

▶ Nontracheophytes either lack vascular tissues completely or, in the case of certain mosses, have only a rudimentary system of water- and food-conducting cells.

▶ The nontracheophyte sporophyte generation is smaller than the gametophyte generation and depends on the gametophyte for water and nutrition. **Review Figures 28.3, 28.4**

▶ Liverwort sporophytes have no specific growing zone. Hornwort sporophytes grow at their basal end, and moss sporophytes grow at their apical end. **Review Figure 28.6**

▶ Beginning with hornworts, all plants have surface pores (stomata) that allow gas exchange and minimize water loss.

▶ Beginning with mosses, the sporophytes of all plants grow by apical cell division.

▶ The hydroids of mosses, through which water may travel, may have arisen from cells also ancestral to the water-conducting cells of the tracheophytes.

Introducing the Tracheophytes

▶ The tracheophytes have vascular tissue with tracheids and other specialized cells designed to conduct water, minerals, and foods.

▶ Present-day tracheophytes are grouped into nine phyla that form two major groups: nonseed tracheophytes and seed plants. **Review Figure 28.8**

▶ In tracheophytes the sporophyte generation is larger than the gametophyte and independent of the gametophyte generation.

▶ The earliest tracheophytes, known to us only in fossil form, lacked roots and leaves. Roots may have evolved from branches that penetrated the ground. Simple leaves are thought to have evolved from sporangia, and complex leaves may have resulted from the flattening and reduction of a branching stem system. **Review Figures 28.10, 28.11, 28.12**

▶ Heterospory, the production of distinct female megaspores and male microspores, evolved on several occasions from homosporous ancestors. **Review Figure 28.13**

The Surviving Nonseed Tracheophytes

▶ Club mosses (phylum Lycophyta) have simple leaves arranged spirally. Horsetails (phylum Sphenophyta) have simple leaves in whorls. Whisk ferns (phylum Psilotophyta) lack roots; one genus has minute scales rather than leaves, and the other has leaves with vascular tissue. Leaves with more complex vasculature are characteristic of all other phyla of tracheophytes. **Review Table 28.1**

▶ Ferns (phylum Pterophyta) are probably not a monophyletic group. They have complex leaves with branching vascular strands. **Review Figure 28.19**

For Discussion

1. Mosses and ferns share a common trait that makes water droplets a necessity for sexual reproduction. What is this trait?

2. Are the mosses well adapted to terrestrial life? Justify your answer.

3. Ferns display a dominant sporophyte stage (with large fronds). Describe the major advance in anatomy that enables most ferns to grow much larger than mosses.

4. What features distinguish club mosses from horsetails? What features distinguish these groups from rhyniophytes and psilotophytes? From ferns?

5. Why did some botanists once believe that psilotophytes should be classified together with the rhyniophytes?

6. Contrast simple leaves with complex leaves in terms of structure, evolutionary origin, and occurrence among plants.

29

The Evolution of Seed Plants

A VIOLENT THUNDERSTORM MOVES through forested hills and valleys where summer rain has been scarce. A jagged fork of lightning strikes a tree and it bursts into flame. Soon the flames reach dead and dry underbrush and fire spreads to surrounding trees. The fire rages rapidly through the forest, leaving a blackened and smoking landscape behind.

Though devastating, such fires are a natural part of the forest ecosystem. Life returns quickly following a fire in a natural grassland or forest, in part because some plants have adaptations that enable them to live with fire. One example, obvious from its common name, is fireweed. The seeds of fireweed not only survive fires, but are encouraged by high temperatures to break their dormancy and sprout. Another example is the lodgepole pine tree, which covers vast fire-prone areas in the Rocky Mountains and elsewhere. Its cones will not release their seeds unless the heat of a fire causes them to open.

Seeds are remarkable structures. They protect the plant embryo within them from environmental extremes through what may be a very long resting period. This and other properties contribute to making seed plants the predominant plants on Earth. All of today's forests are dominated by seed plants.

In this chapter we describe the defining characteristics of the seed plants as a group. We survey the diversity of seed plants and describe the flowers and fruits that are characteristic of the flowering plants. Finally, we consider some of the unsolved problems in seed plant evolution.

General Characteristics of the Seed Plants

The most recent group to appear in the evolution of the tracheophytes is the seed plants: the **gymnosperms** (such as pines and cycads) and the **angiosperms** (flowering plants). There are four living phyla of gymnosperms and one of angiosperms (Figure 29.1). The phylogenetic relationships among these five lineages have not yet been resolved.

In seed plants, the gametophyte generation is reduced even further than it is in the ferns (Figure 29.2). The haploid gametophyte develops partly or entirely while attached to and nutritionally dependent on the diploid sporophyte. Among the seed plants, only the earliest types of gymnosperms and their few survivors had swimming sperm. All other seed plants have evolved other means of bringing female and male gametes together. The culmination of this striking evolutionary trend in plants was independence from the liquid water that earlier plants needed for sexual reproduction.

Seed plants are heterosporous, forming separate megasporangia and microsporangia on structures that are grouped on short axes, such as the cones of conifers and the flowers of angiosperms.

As in other plants, the spores of seed plants are produced by meiosis within the sporangia, but in seed plants, the megaspores are not shed. Instead, the female gametophytes develop within the megasporangia and depend on them for food and water. In most species only one of the meiotic products in a megasporangium survives. The surviving haploid nucleus divides mitotically, and the resulting cells divide again to produce a multicellular female gametophyte. In the angiosperms, female gametophytes normally contain eight nuclei. The female gametophyte is retained within the megasporangium, where it matures and

A Forest Ablaze
Fires like this one in a northern Arizona forest can pose dangers to human life and property. But they play an essential role in the life cycles of many fire-adapted seed plants.

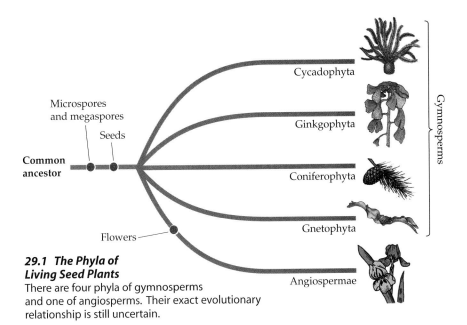

29.1 The Phyla of Living Seed Plants
There are four phyla of gymnosperms and one of angiosperms. Their exact evolutionary relationship is still uncertain.

When the tip of the pollen tube reaches the female gametophyte, two sperm are released from the tube and fertilization occurs. The resulting diploid zygote divides repeatedly, forming a young sporophyte that develops to an embryonic stage at which growth becomes temporarily suspended (often referred to as a dormant stage). The end product at this stage is a **seed**.

A seed may contain tissues from three generations. The seed coat and megasporangium develop from tissues of the diploid sporophyte parent (the integument). Within the megasporangium is the haploid female gametophytic tissue from the next generation. (This tissue is fairly extensive in most gymnosperm seeds. In angiosperm seeds its place is taken by a tissue called endosperm, which we will discuss shortly.) In the center of the seed package is the third generation, in the form of the embryo of the new diploid sporophyte.

The multicellular seed of a gymnosperm or an angiosperm is a well-protected resting stage. The seeds of some species may remain viable (capable of growth and development) for many years, germinating when conditions are favorable for the growth of the sporophyte. In contrast, the embryos of nonseed plants develop directly into sporophytes, which either survive or die, depending on environmental conditions; there is no resting stage in the life cycle.

houses the early development of the next sporophyte generation following fertilization of the egg. The megasporangium itself is surrounded by sterile sporophyte structures that form a protective **integument**.

Within the microsporangium, the meiotic products are microspores, which divide within the microspore wall one or a few times to form male gametophytes called **pollen grains** (Figure 29.3). Distributed by wind, an insect, a bird, or a plant breeder, a pollen grain that reaches the appropriate surface of a sporophyte develops further. It produces a slender **pollen tube** that elongates and digests its way through the sporophytic tissue toward the female gametophyte.

29.2 The Relationship between Sporophyte and Gametophyte Has Evolved
In seed plants, the gametophyte (shown in yellow) is nutritionally dependent on the sporophyte (shown in blue).

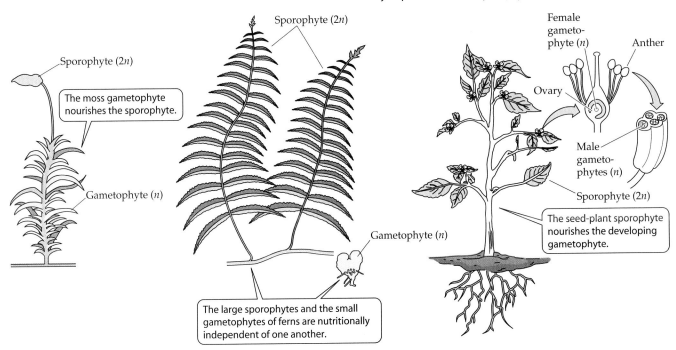

The moss gametophyte nourishes the sporophyte.

The large sporophytes and the small gametophytes of ferns are nutritionally independent of one another.

The seed-plant sporophyte nourishes the developing gametophyte.

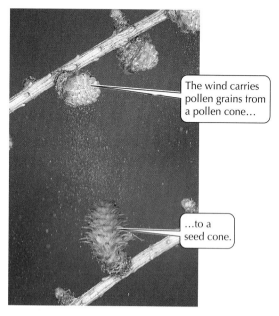

The wind carries pollen grains from a pollen cone...

...to a seed cone.

29.3 Pollen Grains
Pollen grains are the male gametophytes of seed plants. Conifers have separate seed cones (which contain the female gametophyte) and pollen cones; their pollen is dispersed by the wind.

During the dormant stage, the seed coat protects the embryo from excessive drying and may also protect against potential predators that would otherwise eat the embryo and its food reserves. Many seeds have structural adaptations that promote dispersal by wind or, more often, by animals. When the young sporophyte resumes growth, it draws on food reserves in the seed. The possession of seeds is a major reason for the enormous evolutionary success of seed plants, which are the dominant life forms of Earth's modern land flora in most areas.

The Gymnosperms: Naked Seeds

The gymnosperms are a group of seed plants that do not form flowers. Although there are probably fewer than 750 species of living gymnosperms, these plants are second only to the angiosperms (flowering plants) in their dominance of the terrestrial environment.

There are four phyla of living gymnosperms today. The cycads (phylum Cycadophyta) are palmlike plants of the tropics, growing as tall as 20 meters (Figure 29.4a). Ginkgos (phylum Ginkgophyta), which were common during the Mesozoic era, are represented today by a single genus and species, *Ginkgo biloba*, the maidenhair tree (Figure 29.4b). There are both microsporangiate and megasporangiate maidenhair trees. The difference is determined by X and Y sex chromo-

Cycadophyta
Ginkgophyta
Coniferophyta
Gnetophyta
Angiospermae

somes, as in humans; few other plants have sex chromosomes. The phylum **Gnetophyta** consists of three very different genera that share certain characteristics with the angiosperms. One of the gnetophytes is *Welwitschia* (Figure 29.4c), a long-lived desert plant with just two straplike leaves that sprawl on the sand and can become as long as 3 meters. Far and away the most abundant of the gymnosperms are the **conifers** (phylum **Coniferophyta**), cone-bearing plants such as pines and redwoods (Figure 29.4d).

All living gymnosperms have stems and roots that grow larger in diameter (called *secondary growth*), and all but the Gnetophyta have only tracheids as water-conducting and support cells in their xylem. Although the gymnosperm water transport and support system may seem less effective than that of the angiosperms, it serves some of the tallest trees known. The coastal redwoods of California are the tallest gymnosperms; the largest are well over 100 m tall. Secondary xylem—wood—produced by gymnosperms is the principal resource of the timber industry.

Before examining the conifer life cycle, we'll take a brief look at the fossil history of gymnosperms.

We know the early gymnosperms only as fossils

The earliest fossil evidence of gymnosperms is found in Devonian rocks. The early gymnosperms combined characteristics of rhyniophytes and heterosporous ferns, but they had tracheids of the same type found in modern gymnosperms. They also differed from the plants around them by their extensively thickened woody stems, which resulted from proliferation of xylem.

By the Carboniferous period, several new lines of gymnosperms had evolved, including various seed ferns that possessed fernlike foliage but had characteristic gymnosperm seeds attached to their leaves. The first true conifers appeared somewhat later. Either they were not dominant trees or they did not grow where conditions were right for fossilization, so we have few preserved examples. During the Permian period, however, the conifers and cycads flourished. Gymnosperm forests changed with time as the gymnosperm groups evolved, and they dominated the Mesozoic era, in which the continents drifted apart and dinosaurs strode the Earth. Gymnosperms dominated all forests until less than 100 million years ago, and they still dominate some present-day forests.

Conifers have cones but no motile cells

The great Douglas fir and cedar forests of the northwestern United States and the massive boreal forests of pine, fir, and spruce that clothe the northern continental regions and upper slopes of mountain ranges rank among the great vegetation formations of the world. All these trees belong to one phylum of gymnosperms, Coniferophyta—the conifers, or cone-bearers. A **cone** is an axis bearing a tight cluster of scales or leaves specialized for reproduction. Megaspores and microspores are produced in separate seed and pollen cones. Seed cones are much larger than pollen cones (see Figure 29.3).

(a) *Cycas* sp.

(b) *Ginkgo biloba*

(d) *Sequoiadendron giganteum*

(c) *Welwitschia mirabilis*

29.4 Diversity among the Gymnosperms

(a) This palm belongs to the cycads, the least changed group of present-day gymnosperms. Many cycads have growth forms that resemble both ferns and palms. (b) The characteristic fleshy seed coat and broad leaves of the maidenhair tree. (c) A gnetophyte growing in the Namib Desert of Africa. Two huge, straplike leaves grow throughout the life of the plant, breaking and splitting as they grow. (d) A dramatic conifer, this giant sequoia grows in Yosemite National Park, California.

We will use the life cycle of a pine to illustrate reproduction in gymnosperms (Figure 29.5). The production of male gametophytes in the form of pollen grains frees the plant completely from its dependence on liquid water for fertilization. Instead of water, wind assists conifer pollen grains in their first stage of travel to the female gametophyte inside the seed cone. The pollen tube provides the means for the last stage of travel by elongating and digesting its way through maternal sporophytic tissue. When it reaches the female gametophyte, it releases two sperm, one of which degenerates after the other unites with the egg.

The megasporangium, which will form the female gametophyte containing eggs within archegonia, is enclosed in a layer of sporophytic tissue—the integument—that will eventually develop into the seed coat. The integument, the megasporangium inside it, and the tissue attaching it to the maternal sporophyte constitute the **ovule**. The pollen grain enters through a small opening in the integument at the tip of the ovule, the **micropyle**.

Gymnosperms derive their name (which means "naked-seeded") from the fact that their ovules and seeds are not protected by flower or fruit tissue. Most conifer ovules (which upon fertilization develop into seeds) are borne ex-

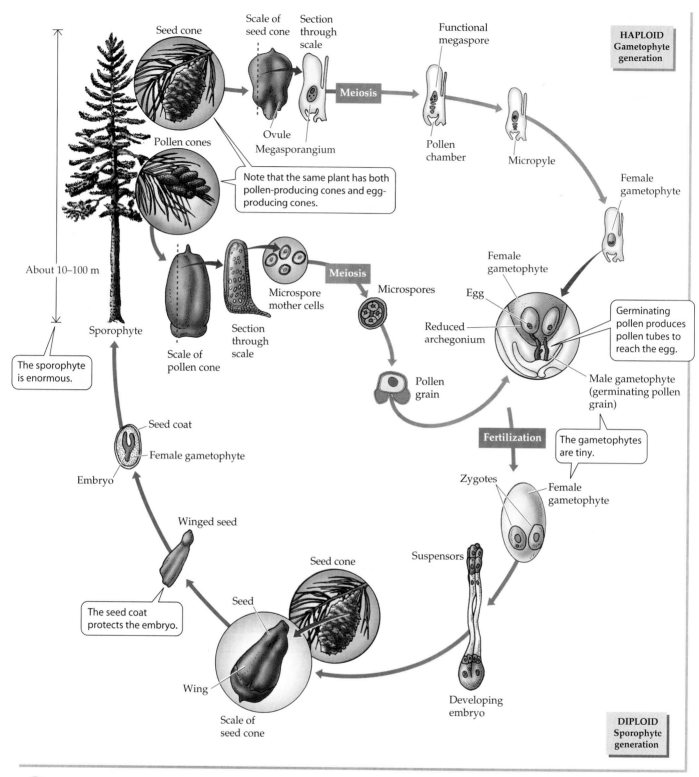

Seed cone

Scale of seed cone

Section through scale

Functional megaspore

HAPLOID Gametophyte generation

Ovule

Megasporangium

Meiosis

Pollen chamber

Micropyle

Female gametophyte

Pollen cones

Note that the same plant has both pollen-producing cones and egg-producing cones.

About 10–100 m

Sporophyte

Microspore mother cells

Scale of pollen cone

Section through scale

Meiosis

Microspores

Female gametophyte

Egg

Female gametophyte

Reduced archegonium

Germinating pollen produces pollen tubes to reach the egg.

The sporophyte is enormous.

Pollen grain

Male gametophyte (germinating pollen grain)

Seed coat

Female gametophyte

Fertilization

The gametophytes are tiny.

Embryo

Zygotes

Female gametophyte

Winged seed

Seed cone

Suspensors

The seed coat protects the embryo.

Seed

Wing

Scale of seed cone

Developing embryo

DIPLOID Sporophyte generation

29.5 The Life Cycle of a Pine Tree
The gametophytes are microscopically small and nutritionally dependent on the sporophyte generation.

posed on the upper surfaces of modified branches called *cone scales*. Their only protection from the environment lies in the fact that the scales are tightly pressed against each other within the cone. As we have seen, some pines, such as

the lodgepole pine, have such tightly closed seed cones that only fire suffices to split them open and release the seeds.

About half of the conifer species have soft, fleshy fruitlike tissues associated with their seeds; examples are the "berries" of juniper and yew. Animals may eat these tissues and then disperse the seeds in their feces, often carrying them considerable distances from the parent plant. These tissues, however, are not true fruits, which are characteristic of the plant phylum that is dominant today: the angiosperms.

The Angiosperms: Flowering Plants

The phylum **Angiospermae** consists of the flowering plants, also commonly known as the **angiosperms**. This highly diverse phylum includes more than 230,000 species. The oldest evidence of angiosperms dates to the late Jurassic period, more than 140 mya. The angiosperms radiated explosively and, over a period of only about 60 million years, became the dominant plant life of the planet. In later chapters, when we mention "plants," we are generally referring to the angiosperms.

The angiosperms represent the current extreme of an evolutionary trend that runs throughout the tracheophytes: *The sporophyte generation becomes larger and more independent of the gametophyte, while the gametophyte generation becomes smaller and more dependent on the sporophyte.*

Angiosperms differ from other plants in several ways:

▶ They have double fertilization.
▶ They produce a triploid endosperm.
▶ Their ovules and seeds are enclosed in a carpel.
▶ They have flowers.
▶ They produce fruit.
▶ Their xylem contains vessel elements and fibers.
▶ Their phloem contains companion cells.

Double fertilization was long considered the single most reliable distinguishing characteristic of the angiosperms. Two male gametes, contained within a single microgametophyte (pollen grain), participate in fertilization events within the megagametophyte of an angiosperm. One sperm combines with the egg to produce a diploid zygote, the first cell of the sporophyte generation. In most angiosperms, the other sperm nucleus combines with two other haploid nuclei of the female gametophyte to form a triploid (3n) nucleus. This nucleus, in turn, divides to form a triploid tissue, the **endosperm**, that nourishes the embryonic sporophyte during its early development.

Double fertilization occurs in all present-day angiosperms. We are not sure when and how it evolved because there is no fossil evidence on this point. It probably first resulted in two embryos, as it does in the three existing genera of Gnetophyta: *Ephedra, Gnetum,* and *Welwitschia*. Both of the fertilizations in gnetophytes produce diploid products.

The formation of an extensive triploid endosperm is one of the

most definitive angiosperm traits, although it is not universal.

The name angiosperm ("enclosed seed") is drawn from another diagnostic character: The ovules and seeds of these plants are enclosed in a modified leaf called a **carpel**. Besides protecting the ovules and seeds, the carpel often interacts with incoming pollen to prevent self-pollination, thus favoring cross-pollination and increasing genetic diversity. Of course, the most evident diagnostic feature of angiosperms is that they have flowers. Production of a fruit is another unique characteristic of the angiosperms.

Angiosperms are also distinguished by the possession of specialized water-transporting cells called **vessel elements** in their xylem, but these cells are also found, in anatomically different form, in gnetophytes and a few ferns. A second distinctive cell type in angiosperm xylem is the **fiber**, which plays an important role in supporting the plant body. Angiosperm phloem possesses another unique cell type, called a **companion cell**.

In the following sections we'll examine the structure and function of flowers, evolutionary trends in flower structure, the functions of pollen and fruits, the angiosperm life cycle, the two major groups of angiosperms, and the origin and evolution of flowering plants.

The sexual structures of angiosperms are flowers

If you examine any familiar flower, you will notice that the outer parts look somewhat like leaves. In fact, all the parts of a flower *are* modified leaves.

A generalized flower (for which there is no exact counterpart in nature) is shown in Figure 29.6 for the purpose of identifying its parts. The structures bearing microsporangia are called **stamens**. Each stamen is composed of a **filament** bearing an **anther** that contains pollen-producing microsporangia. The structures bearing megasporangia are the **carpels**. A structure composed of one carpel or two or more fused carpels is called a **pistil**. The swollen base of the pistil, containing one or more ovules (each containing a

29.6 A Generalized Flower
Not all flowers possess all the structures shown here, but they must possess a stamen (male), pistil (female), or both in order to play their role in reproduction. Flowers that have both, as this one does, are referred to as perfect.

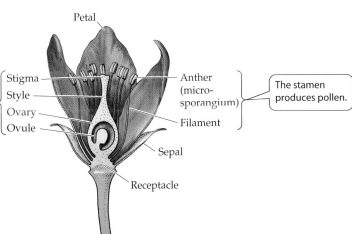

Cycadophyta
Ginkgophyta
Coniferophyta
Gnetophyta
Angiospermae

Petal

Stigma — The pistil receives pollen.
Style
Ovary
Ovule

Anther (microsporangium) — The stamen produces pollen.
Filament

Sepal

Receptacle

Umbels

Disk flowers (many)

Ray flowers

Compound umbel

(a) *Daucus carota*

(b) *Echinacea purpurea*

Spikes

(c) *Penniselum setaceum*

29.7 Inflorescences

(a) The inflorescence of Queen Anne's lace is an umbel. Each umbel bears flowers on stalks that arise from a common center. (b) Cornflowers are members of the aster family; their inflorescence is a head. In a head, each of the long, petal-like structures is a ray flower; the central portion of the head consists of dozens to hundreds of disc flowers. (c) Grasses such as this fountain grass have inflorescences called spikes.

megasporangium), is called the **ovary**. The apical stalk of the pistil is the **style**; and the terminal surface that receives pollen grains is called the **stigma**.

In addition, a flower often has several specialized sterile (non-spore-bearing) leaves: The inner ones are called **petals** (collectively, the **corolla**), and the outer ones **sepals** (collectively, the **calyx**). The corolla and calyx, which can be quite showy, often play roles in attracting animal pollinators to the flower. The calyx more commonly protects the immature flower in bud. From base to apex, the sepals, petals, stamens, and carpels (which are referred to as the floral organs; see Figure 15.11) are usually in circular arrangements called whorls and attached to a central stalk called the **receptacle**.

The generalized flower shown in Figure 29.6 has both megasporangia and microsporangia; such flowers are referred to as **perfect**. Many angiosperms produce two types of flowers, one with only megasporangia and the other with only microsporangia. Consequently, either the stamens or the carpels are nonfunctional or absent in a given flower, and the flower is referred to as **imperfect**.

Species such as corn or birch, in which both megasporangiate and microsporangiate flowers occur on the same plant, are said to be **monoecious** (meaning "one-housed"—but, it must be added, one house with separate rooms). Complete separation is the rule in some other angiosperm species, such as willows and date palms; in these species, a given plant produces either flowers with stamens or flowers with pistils, but never both. Such species are said to be **dioecious** ("two-housed").

Flowers come in an astonishing variety of forms, as you will realize if you think of some of the flowers you recognize. The generalized flower shown in Figure 29.6 has distinct petals and sepals arranged in distinct whorls. In

nature, however, petals and sepals sometimes are indistinguishable. Such appendages are called **tepals**. In other flowers, petals, sepals, or tepals are completely absent.

Flowers may be single, or grouped together to form an **inflorescence**. Different families of flowering plants have their own, characteristic types of inflorescences, such as the umbels of the carrot family, the heads of the aster family, and the spikes of many grasses (Figure 29.7).

Flower structure has evolved over time

The flowers that are evolutionarily the most ancient have a large and variable number of tepals (or sepals and petals), carpels, and stamens (Figure 29.8a). Evolutionary change within the angiosperms has included some striking modifications from this early condition: reduction in the number of each type of organ to a fixed number; differentiation of petals from sepals; and change in symmetry from radial (as in a lily or magnolia) to bilateral (as in a sweet pea or orchid), often accompanied by an extensive fusion of parts (Figure 29.8b).

According to one theory, the first carpels to evolve were modified simple leaves, folded but incompletely closed, and thus differing from the scales of the gymnosperms. In the groups of angiosperms that evolved later, the carpels fused and became progressively more buried in receptacle tissue (Figure 29.9a). In the flowers of the latest groups to evolve, the other flower parts are attached at the very top of the ovary, rather than at the bottom as in Figure 29.6. The stamens of the most ancient flowers may have appeared leaflike (Figure 29.9b), little resembling those of the generalized flower in Figure 29.6.

Why do so many flowers have pistils with long styles and anthers with long filaments? Natural selection has favored length in both of these structures, probably because length increases the likelihood of successful pollination. Long filaments may bring the anthers into contact with insect bodies, or they may place the anthers in a better position to catch the wind. Similar arguments apply to long styles.

(a)

(b)

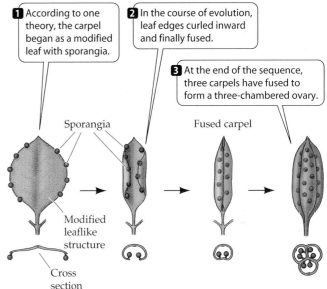

29.8 Flower Form and Evolution
(a) A magnolia flower shows the major features of early flowers: It is radially symmetrical, and the individual tepals, carpels, and stamens are separate, numerous, and attached at their bases. (b) Orchids have a bilaterally symmetrical structure that evolved much later than the form of the magnolia flower in (a). One of the three petals evolved into the complex lower "lip." Inside, the stamen and pistil are fused, and there is a single anther in this species.

A long style may serve another purpose as well. If several pollen grains land on one stigma, a pollen tube will start growing from each grain toward the ovary. If there are more pollen grains than ovules, there is a "race" to fertilize the ovules. The race down the style can be viewed as "mate selection" by the plant bearing that style.

Angiosperms have coevolved with animals

Pollen has played another crucial role in the evolution of the angiosperms. Whereas many gymnosperms are wind-pollinated, most angiosperms are animal-pollinated. Animals visit flowers to obtain nectar or pollen, and in the process often carry pollen from one flower to another, or from one plant to another. Thus, in its quest for food, the animal contributes to the genetic diversity of the plant population. Insects, especially bees, are among the most important pollinators; birds and some species of bats also play major roles.

For more than 130 million years, angiosperms and their animal pollinators have coevolved in the terrestrial environment. The animals have affected the evolution of the plants, and the plants have affected the evolution of the animals. Flower structure has become incredibly diverse under these selection pressures.

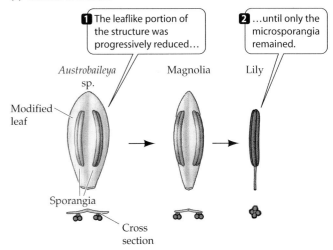

29.9 Carpels and Stamens Evolved from Leaflike Structures
(a) Possible stages in the evolution of a carpel from a more leaflike structure. (b) The stamens of three modern plants show the various stages in the evolution of that organ.

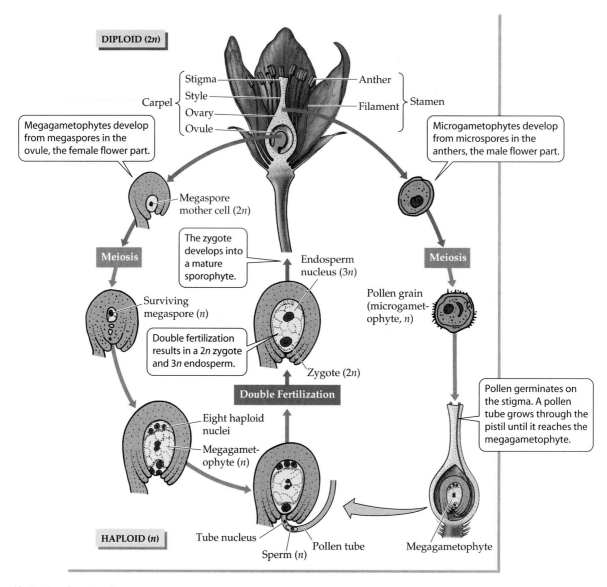

29.10 The Life Cycle of an Angiosperm
The formation of a triploid endosperm distinguishes the angiosperms from the gymnosperms.

Some of the products of coevolution are highly specific; for example, some yucca species are pollinated by only one species of moth. Pollination by just one or a very few animal species provides a plant species with a reliable mechanism for transferring pollen from one to another of its members.

Most plant–pollinator interactions are much less specific; that is, many different animal species pollinate the same plant species, and the same animal species pollinate many plant species. However, even these less specific interactions have developed some specialization. Bird-pollinated flowers are often red and odorless. Insect-pollinated flowers often have characteristic odors, and bee-pollinated flowers may have conspicuous markings, or *nectar guides*, that are evident only in the ultraviolet region of the spectrum, where bees have better vision than in the red region. Co-

evolution and other aspects of plant–animal interactions are covered in more detail in Chapter 55.

The angiosperm life cycle features double fertilization

The life cycle of the angiosperms is summarized in Figure 29.10. The angiosperm life cycle will be considered in detail in Chapter 38, but let's look at it briefly here and compare it with the conifer life cycle in Figure 29.5.

Like all seed plants, angiosperms are heterosporous. The female gametophyte is even more reduced than that of the gymnosperms. The ovules are contained within carpels, rather than being exposed on the surfaces of scales, as in most gymnosperms. The male gametophytes are, again, pollen grains.

The ovule develops into a seed containing the products of the double fertilization that characterizes angiosperms. The triploid endosperm serves as storage tissue for starch or lipids, proteins, and other substances that will be needed by the developing embryo.

(a)

(b)

29.11 Fleshy Fruits Come in Many Forms and Flavors
(a) A simple fruit (sour cherries). (b) An aggregate fruit (raspberries). (c) A multiple fruit (pineapple). (d) An accessory fruit (pear).

The diploid zygote develops into an embryo, consisting of an embryonic axis and one or two **cotyledons**. Also called seed leaves, the cotyledons have different fates in different plants. In many, they serve as absorptive organs that take up and digest the endosperm. In others, they enlarge and become photosynthetic when the seed germinates. Often they play both roles (see Chapter 37).

Angiosperms produce fruits

The ovary of a flowering plant (together with the seeds it contains) develops into a fruit after fertilization. A **fruit** may consist only of the mature ovary and its seeds, or it may include other parts of the flower or structures associated with it. A *simple fruit*, such as a cherry (Figure 29.11a), is one that develops from a single carpel or several united carpels. A raspberry is an example of an *aggregate fruit* (Figure 29.11b)—one that develops from several separate carpels of a single flower. Pineapples and figs are examples of *multiple fruits* (Figure 29.11c), formed from a cluster of flowers (an inflorescence). Fruits derived from parts in addition to the carpel and seeds are called *accessory fruits* (Figure 29.11d); examples are apples, pears, and strawberries. The development, ripening, and dispersal of fruits will be considered in Chapters 37 and 38.

(c)

Determining the oldest living angiosperm lineage

Which angiosperms were the first flowering plants was long a matter of great controversy. Two leading candidates were the magnolia family (see Figure 29.8a) and another family, the Chloranthaceae, whose flowers are much simpler than those of the magnolias. At the close of the twentieth century, an impressive convergence of evidence led to the conclusion that the base of the angiosperm phylogenetic tree belongs to neither of those families, but rather to a lineage that today consists of just a single species of the

(d)

genus *Amborella* (Figure 29.12). This woody shrub, with cream-colored flowers, lives only on New Caledonia, an island in the South Pacific. Its 5 to 8 carpels are in a single whorl, and it has 30 to 100 stamens. The xylem of *Amborella* lacks vessel elements, which appeared later in angiosperm evolution. The characteristics of *Amborella* give us a good sense of what the first angiosperms might have been like.

(a)

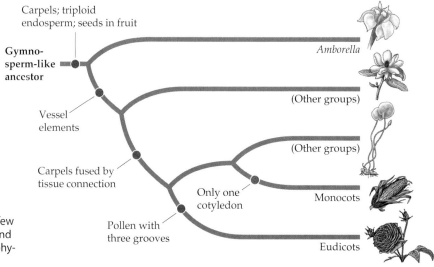

29.12 The First Angiosperm
(*a*) *Amborella*, a shrub, is the closest living relative of the first angiosperms. (*b*) A flower of *Amborella*.

(b)

There are two large monophyletic groups of angiosperms

There are two large lineages that include the great majority of angiosperm species: the **monocots** and the **eudicots**. Both are monophyletic groups (Figure 29.13). The monocots are so called because they have a single embryonic cotyledon; the eudicots have two. The cotyledons of some, but not all, eudicots store the reserves originally present in the endosperm. There are several other differences between the two lineages, which we will describe in Chapter 34. Some familiar plants, including magnolias and water lilies, belong to lineages more ancient than either the monocots or the eudicots.

The monocots (Figure 29.14) include grasses, cattails, lilies, orchids, and palm trees. The eudicots (Figure 29.15) include the vast number of familiar seed plants, including most of the herbs, vines, trees,

and shrubs. Among them are oaks, willows, violets, snapdragons, and sunflowers.

The origin of the angiosperms remains a mystery

We have learned a lot about evolution within the angiosperm lineage. The most important unanswered question about the evolution of seed plants is this: How did the angiosperms first arise—to which gymnosperm phylum are they sister? You might think that, given the advances in techniques of molecular genetics and computer technology, as well as new fossil finds, we would have opened this century with the answer to this question well in hand. A very few years ago, it seemed that we were on the verge of answering it. Although an answer may be agreed upon before the present decade ends, the puzzle is as vexing today as it was before.

Why should this be? Different phylogenetic methods, applied by different investigators, have produced apparently contradictory results. It might seem a simple matter to rectify this situation, but several questions complicate such efforts: What morphological characters should be selected as important, or should they all be treated as equally important? What algorithms should be applied to computerized analysis of data? Are all molecular differences and similarities significant, or are some of them incidental? Which fossils should be chosen for comparisons? What is the likelihood that we can find evidence of double fertilization in ancient fossils?

We are left with the question: Where did the first angiosperm come from? The angiosperms may be most closely related to the gnetophytes, or they may be more closely related to the conifers, or to another gymnosperm phylum. Current progress in methodology gives us reason to hope that our picture of seed plant evolution will be much more complete before this decade ends. We will see in Chapters 31–33 whether our understanding of animal evolution is more complete.

29.13 Evolutionary Relationships among the Angiosperms
The monocots and the eudicots are the largest monophyletic groups among the angiosperms. A few lineages that differ from the monocots, eudicots, and *Amborella* remain to be placed accurately on the phylogenetic tree.

(a) *Phoenix dactylifera*

29.14 Monocots

(a) Palms are among the few monocot trees. Date palms are a major food source in some areas of the world. (b) Grasses such as this cultivated wheat and the fountain grass in Figure 29.7c are monocots. (c) Monocots include popular garden flowers such as these daylilies. Orchids (Figure 29.8b) are another highly prized monocot flower.

(b) *Triticum* sp.

(c) *Hemerocallis* sp.

(a)

(b) *Cornus florida*

(c) *Rosa rugosa*

29.15 Eudicots

(a) The cactus family is a large group of eudicots, with about 1,500 species in the Americas. This cactus bears scarlet flowers for a brief period of the year. (b) The flowering dogwood is a small eudicot tree. (c) Climbing Cape Cod roses are members of the eudicot family Rosaceae, as are the familiar roses from your local florist.

Chapter Summary

General Characteristics of the Seed Plants

▶ The seed plants (gymnosperms and angiosperms) are heterosporous and have greatly reduced gametophytes. **Review Figures 29.1, 29.2**

▶ Most modern seed plants have no swimming gametes and do not require liquid water for fertilization. The male gametophyte—the pollen grain—is dispersed by wind or by animals. **Review Figure 29.3**

▶ The seed is a well-protected resting stage that often contains food that supports the growth of the embryo.

The Gymnosperms: Naked Seeds

▶ The gymnosperms, once the dominant vegetation on Earth, still dominate forests in the northern parts of the Northern Hemisphere and at high elevations.

▶ The four surviving gymnosperm phyla are the Cycadophyta (the most ancient), Ginkgophyta (consisting of a single species, the maidenhair tree), Gnetophyta (which has some characters in common with the angiosperms), and Coniferophyta (the familiar cone-bearing trees).

▶ Modern gymnosperms all have abundant xylem and extensive secondary growth.

▶ Conifers have a life cycle in which naked seeds are produced on the scales of female cones. Pollen cones are smaller than seed cones. Pollen is transferred from pollen cones to seed cones by wind. **Review Figure 29.5**

The Angiosperms: Flowering Plants

▶ Angiosperms (phylum Angiospermae) are distinguished by double fertilization, which results in a triploid nutritive tissue, the endosperm. Double fertilization is also characteristic of the Gnetophyta. **Review Figure 29.10**

▶ The ovules and seeds of angiosperms are enclosed by a carpel. Angiosperms are also characterized by the production of flowers and fruits.

▶ The vascular tissues of angiosperms contain three characteristic cell types: vessel elements, fibers, and companion cells.

▶ Flowers are made up of various combinations of carpels, stamens, petals, and sepals. Perfect flowers have both carpels (female parts) and stamens (male parts). **Review Figure 29.6**

▶ Monoecious plant species have both female and male flowers on the same plant. Dioecious species have separate female and male plants.

▶ Carpels and stamens may have evolved from leaflike structures. **Review Figure 29.9**

▶ Angiosperms and the animals that pollinate them have coevolved.

▶ *Amborella*, a tropical shrub, is the sole living representative of the first angiosperm lineage.

▶ There are two major lineages of flowering plants: monocots and eudicots. **Review Figure 29.13**

▶ The evolutionary origin of the angiosperms remains a mystery.

For Discussion

1. In most seed plant species, only one of the products of meiosis in the megasporangium survives. How might this be advantageous?

2. Suggest an explanation for the great success of the angiosperms in occupying terrestrial habitats.

3. In many locales, large gymnosperms predominate over large angiosperms. Under what conditions might gymnosperms have the advantage, and why?

4. Not all flowers possess all of the following parts: sepals, petals, stamens, and carpels. What kind or kinds of flower parts do you think might be found in the flowers that have the smallest number of kinds? Discuss the possibilities, both for a single flower and for a species.

5. The problem of the origin of the angiosperms has long been "an abominable mystery," as Charles Darwin once put it. Scientists still do not know the nearest relatives of the angiosperms. It has often been suggested (correctly or incorrectly) that the gnetophytes are sister to the angiosperms. What pieces of evidence suggested this connection?

30 Fungi: Recyclers, Killers, and Plant Partners

WHAT ARE THE LARGEST ORGANISMS you can think of? Whales? Trees? Some of the largest organisms on Earth are fungi. One such fungus, growing in Michigan, covers an area of 37 acres. Its effect on green plants is evident from the air, but from ground level, it is difficult to realize how large the fungus is. At the surface, you see only seemingly isolated clumps of mushrooms. But the vast body of the fungus *Armillariella*, which weighs approximately the same as a blue whale, grows underground and consists almost entirely of microscopic filaments.

Molecular studies indicate that this giant fungus is or was a single individual that arose from a single spore. It is possible that fragmentation over time may have broken it into a few separate—but still gigantic—individuals. Another, larger fungus of the same genus, growing in the state of Washington, occupies parts of three counties. But not all fungi are huge. Molds and mushrooms are fungi, as are the microscopic, unicellular yeasts.

Every breath we take contains large numbers of fungal spores. Some of those spores can be dangerous, and fungal diseases of humans, some of which are as yet uncurable, have become a major global threat. However, other fungi are of immense commercial importance to us. Fungi are essential to plants as well. Fungi interact with roots, greatly enhancing the roots' ability to take up water and mineral nutrients.

Earth would be a messy place without the fungi. They are at work in forests, fields, and garbage dumps, breaking down the remains of dead organisms (and even manufactured substances such as some plastics). For almost a billion years, the ability of fungi to decompose substances has been important for life on Earth, chiefly because by breaking down carbon compounds, they return carbon and other elements to the environment, where they can be used again by other organisms.

In this chapter we will examine the general biology of the kingdom Fungi, which differs in interesting ways from the other kingdoms. We will also explore the diversity of body forms, reproductive structures, and life cycles of the four phyla of fungi, as well as the mutually beneficial associations of certain fungi with other organisms. As we begin our study, recall that the fungi and the animals are descended from a common ancestor—we are more closely related to molds and mushrooms than we are to the flowers we admired in the last chapter.

General Biology of the Fungi

The fungi are superbly adapted for absorptive nutrition: They secrete digestive enzymes that break down large food molecules in the environment, then absorb the breakdown products. The kingdom Fungi encompasses *heterotrophic organisms with absorptive nutrition*. Many fungi are saprobes that absorb nutrients from dead matter, others are parasites that absorb nutrients from living hosts (Figure 30.1), and still others live in mutually beneficial symbioses with other organisms.

All fungi form spores, but only in one phylum (Chytridiomycota) do spores or gametes possess flagella. Fungi reproduce sexually in a variety of ways. Their cell walls contain at least some **chitin**, a polysaccharide that is also found in the skeletons of arthropods and in some protists. Most fungi have complex body forms.

These criteria enable us to distinguish between the fungi and some protists that resemble them. The slime molds consist of two protist groups whose members take up food

The Tip of the Iceberg
These fungal fruiting bodies of *Armillariella* are only a hint of the presence of a vast underground network of microscopic filaments extending over many acres.

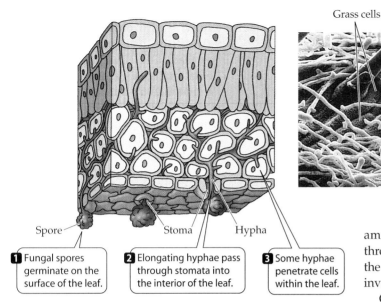

Grass cells

1 Fungal spores germinate on the surface of the leaf.

2 Elongating hyphae pass through stomata into the interior of the leaf.

3 Some hyphae penetrate cells within the leaf.

Spore   Stoma   Hypha

30.4 A Fungus Attacks a Leaf
The white structures in the micrograph are hyphae of the fungus *Blumeria graminis*, which is growing on the dark surface of the leaf of a grass.

Fungal hyphae

the main components of plant cell walls (most bacteria cannot break down these materials).

Because many saprobic fungi are able to grow on artificial media, we can perform experiments to determine their exact nutritional requirements. Sugars are their favored source of carbon. Most fungi obtain nitrogen from proteins or the products of protein breakdown. Many fungi can use nitrate (NO_3^-) or ammonium (NH_4^+) ions as their sole source of nitrogen. No known fungus can get its nitrogen directly from nitrogen gas, as can some bacteria and plant–bacteria associations (see Chapter 36). Nutritional studies also reveal that most fungi are unable to synthesize their own thiamin (vitamin B_1) or biotin (another B vitamin), and must absorb these vitamins from their environment. On the other hand, fungi can synthesize some vitamins that animals cannot. Like all organisms, fungi also require some mineral elements.

Nutrition in the parasitic fungi is particularly interesting to biologists. **Facultative** parasites can be grown by themselves on defined artificial media. **Obligate** parasites cannot be grown on any available medium; they can grow only on their specific living hosts, usually plants. Because their growth is limited to living hosts, they must have unusual nutritional requirements.

Some fungi have adaptations that enable them to function as active predators, trapping nearby microscopic protists or animals, from which they obtain nitrogen and energy. The most common strategy is to secrete sticky substances from the hyphae so that passing organisms stick tightly to them. The hyphae then quickly invade the prey, growing and branching within it, spreading through its body, absorbing nutrients, and eventually killing it.

A more dramatic adaptation for predation is the constricting ring formed by some species of *Arthrobotrys*, *Dactylaria*, and *Dactylella* (Figure 30.5). All of these fungi grow in soil. When nematodes (tiny roundworms) are present in the soil, these fungi form three-celled rings with a di-

ameter that just fits a nematode. A nematode crawling through one of these rings stimulates it, causing the cells of the ring to swell and trap the worm. Fungal hyphae quickly invade and digest the unlucky victim.

Certain highly specific associations between fungi and other organisms have nutritional consequences for the fungal partner. **Lichens** are associations of a fungus with a cyanobacterium, a unicellular photosynthetic eukaryote, or both. **Mycorrhizae** (singular mycorrhiza) are associations between specific fungi and the roots of plants. In such associations the fungus obtains organic compounds from its photosynthetic partner, but provides it with minerals and water so that the partner's nutrition is also promoted. We will discuss lichens and mycorrhizae more thoroughly later in this chapter.

Most fungi reproduce both asexually and sexually

Both asexual and sexual reproduction are common among the fungi. Asexual reproduction takes several forms:

▸ The production of (usually) haploid spores within structures called sporangia.

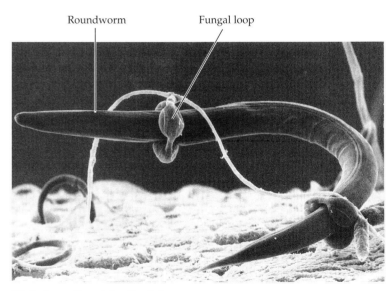

Roundworm   Fungal loop

30.5 Some Fungi Are Predators
A nematode (roundworm) is trapped in sticky loops of the soil-dwelling fungus *Arthrobotrys anchonia*.

► The production of naked spores (not enclosed in sporangia) at the tips of hyphae; such spores are called **conidia** (from the Greek *konis*, "dust").
► Cell division by unicellular fungi—either a relatively equal division or an asymmetrical division in which a tiny bud is produced.
► Simple breakage of the mycelium.

Sexual reproduction in many fungi features an interesting twist. There is often no morphological distinction between female and male structures, or between female and male individuals. Rather, there is a genetically determined distinction between two or more **mating types**. Individuals of the same mating type cannot mate with one another, but they can mate with individuals of another mating type. This distinction prevents self-fertilization. Individuals of different mating types differ genetically from one another, but are often visually and behaviorally indistinguishable. Many protists also have mating type systems.

In many fungi, the zygote nuclei formed by sexual reproduction are the only diploid nuclei in the life cycle. These nuclei undergo meiosis, producing haploid nuclei that become incorporated into spores. Haploid fungal spores, whether produced sexually in this manner or asexually, germinate, and their nuclei divide mitotically to produce hyphae.

Many fungal life cycles include a dikaryon stage

The hyphae of some Zygomycota, Ascomycota, and Basidiomycota have a nuclear configuration other than the familiar haploid or diploid. In these fungi, sexual reproduction begins in an unusual way: The cytoplasms of two individuals of opposite mating types fuse (*plasmogamy*) long before their nuclei fuse (*karyogamy*), so that *two genetically different haploid nuclei exist within the same hypha*. This hypha is called a **dikaryon** (having *two* nuclei). Because the two nuclei differ genetically, the hypha is also called a **heterokaryon** (having *different* nuclei).

Eventually, specialized fruiting structures form, within which the pairs of dissimilar nuclei—one from each parent—fuse, giving rise to zygotes long after the original "mating." The zygote nucleus undergoes meiosis, producing four haploid nuclei. The mitotic descendants of those nuclei become the nuclei of the next generation of hyphae.

The reproduction of such fungi displays several unusual features. First, there are no gamete *cells*, only gamete *nuclei*. Second, there is never any true diploid tissue, although for a long period the genes of both parents are present in the dikaryon and can be expressed. In effect, these hyphae are neither diploid (2*n*) nor haploid (*n*); rather, they are *dikaryotic* (*n* + *n*). A harmful recessive mutation in one nucleus may be compensated for by a normal allele on the same chromosome in the other nucleus. Dikaryosis is perhaps the most significant of the genetic peculiarities of the fungi.

Finally, although Zygomycota, Ascomycota, and Basidiomycota grow in moist places, their gamete nuclei are not motile and are not released into the environment. Therefore, liquid water is not required for fertilization.

Some fungi are pathogens

Fungal pathogens are a major cause of death among people with compromised immune systems. Most patients with AIDS die of fungal diseases, such as the pneumonia caused by *Pneumocystis carinii* or the incurable diarrhea caused by some other fungi. *Candida albicans* and certain other yeasts also cause severe diseases in individuals with AIDS and in individuals taking immunosuppressive drugs. Such fungal diseases are a growing international health problem. Our limited understanding of the basic biology of these fungi still hampers our ability to treat the diseases they cause.

Various fungi cause other, less threatening human diseases, such as ringworm and athlete's foot. Still others are responsible for plant diseases that affect human food supplies. These diseases include black stem rust of wheat and other diseases of wheat, corn, and oats. Fungal diseases of plants have cost billions of dollars in crop losses.

Diversity in the Kingdom Fungi

Each of the four phyla of the kingdom Fungi appears to be monophyletic (Figure 30.6). Because the imperfect fungi (deuteromycetes) are polyphyletic, we will not give them phylum status. In this section on fungal diversity, we'll consider the four phyla—Chytridiomycota, Zygomycota, Ascomycota, and Basidiomycota—and we'll discuss the status of the deuteromycetes.

Chytrids probably resemble the ancestral fungi

The earliest-diverging fungal lineage is the **chytrids** (phylum **Chytridiomycota**). These aquatic microorganisms have sometimes been classified as protists. We place chytrids among the fungi because their cell walls consist primarily of chitin and because molecular evidence indicates that they and the other fungi form a monophyletic group.

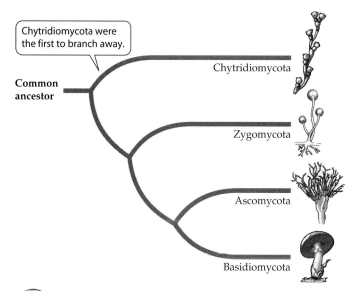

Chytridiomycota were the first to branch away.

Common ancestor

Chytridiomycota
Zygomycota
Ascomycota
Basidiomycota

30.6 Phylogeny of the Fungi
Four phyla are recognized among the fungi. In addition, the imperfect fungi, or Deuteromycetes, functions as a "holding group" for fungal species whose status is yet to be determined.

Chytrids are either parasitic (on organisms such as algae, mosquito larvae, and nematodes) or saprobic, obtaining nutrients by breaking down dead organic matter. (Chytrids in the compound stomachs of foregut-fermenting animals may be an exception, living in a mutualistic association with their hosts.) Most chytrids live in freshwater habitats or in moist soil, but some are marine. Some chytrids are unicellular; others have mycelia made up of branching chains of cells. Chytrids reproduce both sexually and asexually, but they do not have a dikaryon stage.

Allomyces, a well-studied genus of chytrids, displays alternation generations. A haploid **zoospore** (spore with flagella) comes to rest on dead plant or animal material in water and germinates to form a small haploid organism. That produces female and male **gametangia** (gamete cases) (Figure 30.7). The male gametangia are smaller than the female gametangia and possess a light orange pigment. Mitosis in the gametangia results in the formation of haploid gametes, each with a single nucleus.

Both female and male gametes have flagella. The motile female gamete produces a *pheromone*, a chemical that attracts the swimming male gamete. The gametes fuse in pairs, and then their nuclei fuse to form a diploid zygote. Mitosis and cytokinesis in the zygote gives rise to a small diploid organism, which produces numerous diploid flagellate zoospores. These diploid zoospores disperse and germinate to form more diploid organisms. Eventually, the diploid organism produces thick-walled resting sporangia that can survive unfavorable conditions such as dry weather or freezing. Nuclei in the resting sporangia eventually undergo meiosis, giving rise to haploid zoospores that are released into the water and begin the cycle anew.

Chytrids are the only fungi that have flagella at any life cycle stage. We speculate that the protist ancestor of the fungi possessed flagella, because the phylum Chytridiomycota was the first fungal group to diverge from the others (see Figure 30.6). The same protist ancestor gave rise to the protist group Choanoflagellida and to the animal kingdom. A key event in the evolution of the fungi after the chytrids diverged from the others was the loss of flagella.

Zygomycetes reproduce sexually by fusion of two gametangia

Most **zygomycetes** (phylum **Zygomycota**) have coenocytic hyphae (hyphae without regularly occurring septa). They produce no motile cells, and only one diploid cell—the zygote—appears in the entire life cycle. The mycelium of a zygomycete spreads over its substrate, growing forward by means of specialized hyphae. Most zygomycetes do not form a fleshy fruiting body; rather, the hyphae spread in an apparently random fashion, with occasional stalked **sporangiophores** reaching up into the air (Figure 30.8).

Almost 900 species of zygomycetes have been described. A very important group of zygomycetes serves as the fungal partners in the most common type of mycorrhizal association with plant roots. A zygomycete that you may be more familiar with is *Rhizopus stolonifer*, the black bread mold. *Rhizopus* reproduces asexually by producing many stalked sporangiophores, each bearing a single sporangium containing hundreds of minute spores (Figure 30.9*a*). Other zygomycetes have sporangiophores with many sporangia. As in other filamentous fungi, the spore-forming structure is separated from the rest of the hypha by a wall.

Zygomycetes reproduce sexually when adjacent hyphae of two different mating types release pheromones, which cause them to grow together. These hyphae produce ga-

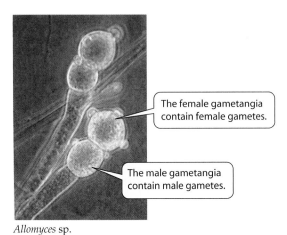

Allomyces sp.

30.7 Reproductive Structures of a Chytrid
The haploid gametes produced in these gametangia will fuse with another gamete to form a diploid organism.

The female gametangia contain female gametes.

The male gametangia contain male gametes.

Phycomyces sp.

Sporangia

Sporangiophores

30.8 A Zygomycete
This small forest of filamentous structures is made up of sporangiophores. The stalks end in tiny, rounded sporangia.

(a)

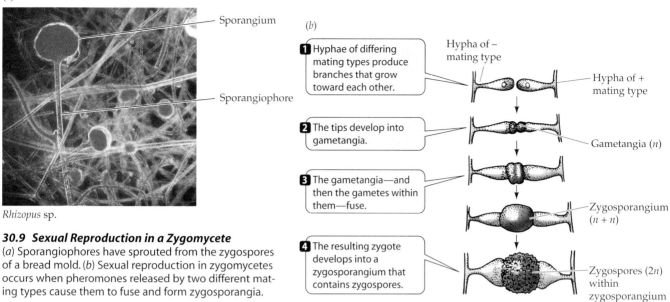

Rhizopus sp.

30.9 Sexual Reproduction in a Zygomycete
(a) Sporangiophores have sprouted from the zygospores of a bread mold. (b) Sexual reproduction in zygomycetes occurs when pheromones released by two different mating types cause them to fuse and form zygosporangia.

(b)

1 Hyphae of differing mating types produce branches that grow toward each other.

Hypha of – mating type

Hypha of + mating type

2 The tips develop into gametangia.

Gametangia (n)

3 The gametangia—and then the gametes within them—fuse.

4 The resulting zygote develops into a zygosporangium that contains zygospores.

Zygosporangium (n + n)

Zygospores (2n) within zygosporangium

metangia that fuse to form zygosporangia containing zygospores (Figure 30.9b). The zygosporangia develop thick, multilayered walls that protect the zygospores. The highly resistant zygospores may remain dormant for months before their nuclei undergo meiosis and a sporangium sprouts. The sporangium contains the products of meiosis: haploid nuclei that are incorporated into spores. These spores disperse and germinate to form a new generation of haploid hyphae.

The sexual reproductive structure of ascomycetes is an ascus

The **ascomycetes** (phylum **Ascomycota**) are a large and diverse group of fungi distinguished by the production of sacs called **asci** (singular ascus) (Figure 30.10). The ascus is the characteristic sexual reproductive structure of the ascomycetes. Ascomycete hyphae are segmented by more or

less regularly spaced septa. A pore in each septum permits extensive movement of cytoplasm and organelles (including the nuclei) from one segment to the next.

The approximately 30,000 known species of ascomycetes can be divided into two broad groups, depending on whether the asci are contained within a specialized fruiting structure. Species that have this fruiting structure, the **ascocarp**, are collectively called **euascomycetes** ("true ascomycetes"); those without ascocarps are called **hemiascomycetes** ("half ascomycetes").

Chytridiomycota

Zygomycota

Ascomycota

Basidiomycota

Ascus

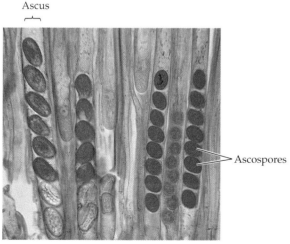

Ascospores

30.10 Asci and Ascospores
The ascomycetes are characterized by the production of ascospores within sacs called asci. Ascospores are the products of meiosis followed by a single mitotic division. Ascospores and asci do not mature all at once, and they may abort, so not every ascus in this micrograph contains eight mature ascospores.

HEMIASCOMYCETES. Most hemiascomycetes are microscopic, and many species are unicellular. Perhaps the best known are the ascomycete yeasts, especially baker's or brewer's yeast (*Saccharomyces cerevisiae*; see Figure 30.2). These yeasts are among the most important domesticated fungi. *S. cerevisiae* metabolizes glucose obtained from its environment to ethanol and carbon dioxide. It forms carbon dioxide bubbles in bread dough and gives baked bread its light texture. Although baked away in bread making, the ethanol and carbon dioxide are both retained in beer. Other yeasts live on fruits such as figs and grapes and play an important role in the making of wine.

Hemiascomycete yeasts reproduce asexually either by fission or by **budding** (the outgrowth of a new cell from the surface of an old one; see Figure 30.2). Sexual reproduction takes place when two adjacent haploid cells of opposite mating types fuse. (We discussed the genetics of yeast mating types in Chapter 14.) In some species, the resulting zygote buds to form a diploid cell population; in others, the zygote nucleus undergoes meiosis immediately. When these diploid nuclei undergo meiosis, the entire cell becomes an ascus. Depending on whether the products of meiosis then undergo mitosis, a yeast ascus usually has either eight or four **ascospores** (see Figure

(a) *Morchella esculenta*

(b) *Sarcoscypha coccinea*

30.11 Two Ascomycetes
(a) Morels, which have spongelike caps and a subtle flavor, are considered a delicacy by humans. The brilliant red cups in (b) are cup fungi, as are the three yellow morels in (a).

30.10). The ascospores germinate to become haploid cells. Hemiascomycetes have no dikaryon stage.

Yeasts, especially *Saccharomyces cerevisiae*, are frequently used in molecular biological research. Just as *E. coli* is the best-studied prokaryote, *S. cerevisiae* is the most completely studied eukaryote.

EUASCOMYCETES. The euascomycetes include the filamentous fungi known as molds. Among them are several common molds, including *Neurospora*, the pink molds, one of which Beadle and Tatum used in their pioneering work on biochemical genetics (see Figure 12.1). Many euascomycetes are parasites on higher plants. Chestnut blight and Dutch elm disease are caused by euascomycetes. The powdery mildews are euascomycetes that infect cereal grains, lilacs, and roses, among many other plants. They can be a serious problem to grape growers, and a great deal of research has focused on ways to control these agricultural pests.

The euascomycetes also include the cup fungi (Figure 30.11a and b). In most of these organisms the fruiting structures are cup-shaped and can be as large as several centimeters across. The inner surfaces of the cups are covered with a mixture of both sterile filaments and asci, and they produce huge numbers of spores. Although these fleshy structures appear to be composed of distinct tissue layers, microscopic examination shows that their basic organization is still filamentous—a tightly woven mycelium.

Two particularly delicious cup fungus fruiting structures are morels (Figure 30.11a) and truffles. Truffles grow underground, in a mutualistic association with the roots of some species of oaks. Europeans traditionally used pigs to find truffles because some truffles secrete a substance that has

an odor similar to a pig's sex attractant. Unfortunately, pigs also eat truffles, so dogs are now the usual truffle hunters.

Penicillium is a genus of green molds, of which some species produce the antibiotic penicillin, presumably for defense against competing bacteria. Two species, *P. camembertii* and *P. roquefortii*, are the organisms responsible for the characteristic flavors of Camembert and Roquefort cheeses, respectively.

Brown molds of the genus *Aspergillus* are important in some human diets. *A. tamarii* acts on soybeans in the production of soy sauce, and *A. oryzae* is used in brewing the Japanese alcoholic beverage sake. Some species of *Aspergillus* that grow on nuts such as peanuts and pecans produce extremely carcinogenic (cancer-inducing) compounds called aflatoxins.

The euascomycetes reproduce asexually by means of mating structures called conidia that form at the tips of specialized hyphae (Figure 30.12). Small chains of conidia are

30.12 Conidia
Chains of conidia are developing on stalks called hyphae arising from this powdery mildew growing on a leaf.

Erysiphe sp.

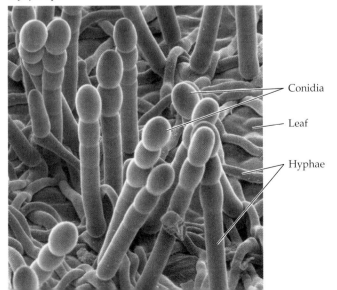

Conidia

Leaf

Hyphae

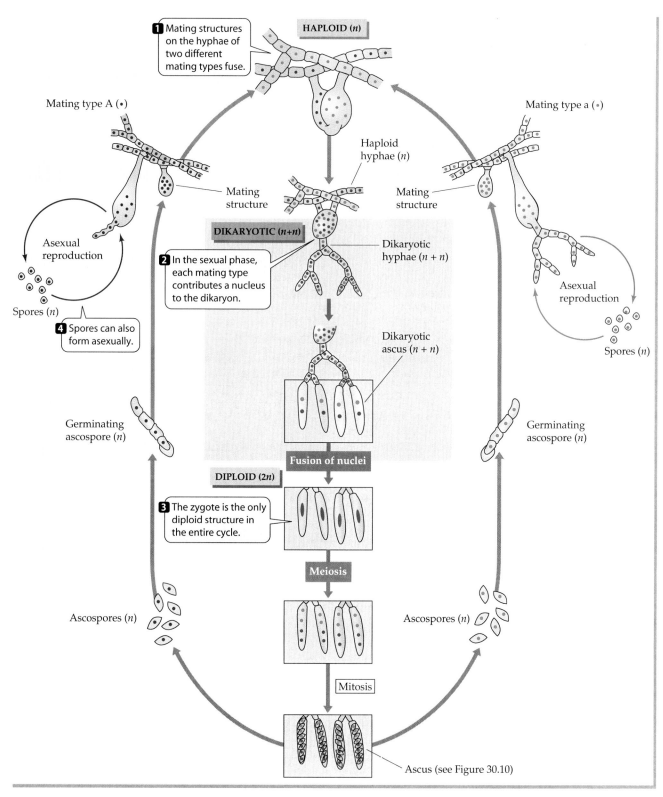

HAPLOID (n)

1 Mating structures on the hyphae of two different mating types fuse.

Mating type A (•)

Mating structure

Haploid hyphae (n)

Mating structure

Mating type a (•)

Asexual reproduction

Spores (n)

DIKARYOTIC (n+n)

Dikaryotic hyphae (n + n)

Asexual reproduction

Spores (n)

2 In the sexual phase, each mating type contributes a nucleus to the dikaryon.

4 Spores can also form asexually.

Dikaryotic ascus (n + n)

Germinating ascospore (n)

Germinating ascospore (n)

Fusion of nuclei

DIPLOID (2n)

3 The zygote is the only diploid structure in the entire cycle.

Meiosis

Ascospores (n)

Ascospores (n)

Mitosis

Ascus (see Figure 30.10)

30.13 The Life Cycle of a Euascomycete
Neurospora crassa, the species represented by this life cycle, is a bread mold that is often used in genetics experiments.

produced by the millions and can survive for weeks in nature. The conidia are what give molds their characteristic colors.

The sexual reproductive cycle of euascomycetes includes the formation of a dikaryon (Figure 30.13). Most euascomycetes form mating structures, some "female" and some "male."

Nuclei from a male structure on one hypha enter a female mating structure on a hypha of a compatible mating type. Dikaryotic *ascogenous* (ascus-forming) hyphae de-

velop from the now dikaryotic female mating structure. The introduced nuclei divide simultaneously with the host nuclei. Eventually asci form at the tips of the ascogenous hyphae. Only with the formation of asci do the nuclei finally fuse. Both nuclear fusion and the subsequent meiosis of the resulting diploid nucleus take place within individual asci. The meiotic products are incorporated into as-

(a) *Lycoperdon perlatum*

30.14 Basidiomycete Fruiting Structures

The fruiting structures of the basidiomycetes are probably the most familiar structures produced by fungi.
(a) When raindrops hit them, these puffballs will release
clouds of spores for dispersal. (b) A member of a highly
poisonous mushroom genus, *Amanita*. (c) This edible
bracket fungus is parasitizing a tree.

(b) *Amanita muscaria*

(c) *Laetiporus sulphureus*

cospores that are ultimately shed by the ascus to begin the
new haploid generation.

The sexual reproductive structure of basidiomycetes is a basidium

About 25,000 species of **basidiomycetes** (phylum **Basid-iomycota**) have been described. Basidiomycetes produce
some of the most spectacular fruiting structures found any-
where among the fungi. These fruiting structures include
puffballs (which may be more than half a meter in diameter), mushrooms of all kinds, and the giant bracket fungi
often encountered on trees and fallen logs in a damp forest
(Figure 30.14). There are more than 3,250 species of mushrooms, including the familiar *Agaricus bisporus* you may
enjoy on your pizza, as well as poisonous species, such as
members of the genus *Amanita*. Bracket fungi do great
damage to cut lumber and stands of timber. Some of the
most damaging plant pathogens are basidiomycetes, including the smut fungi (see Figure 30.1a) that parasitize cereal grains. In contrast, other basidiomycetes contribute to
the well-being of plants as fungal partners in mycorrhizae.

Basidiomycete hyphae characteristically have septa with
small, distinctive pores. The **basidium** (plural basidia), a
swollen cell at the tip of a hypha, is the characteristic sexual
reproductive structure of
the basidiomycetes.
It is the site of nu-
clear fusion and meiosis.
Thus, the basidium plays
the same role in the basidiomycetes as the ascus does in the
ascomycetes and the zygosporangium
does in zygomycetes.

Chytridiomycota
Zygomycota
Ascomycota
Basidiomycota

The life cycle of the basidiomycetes is shown in Figure
30.15. After nuclei fuse in the basidium, the resulting

diploid nucleus undergoes meiosis, and the four resulting
haploid nuclei are incorporated into haploid **basidiospores**,
which form on tiny stalks. These basidiospores typically are
forcibly discharged from their basidia and then germinate,
giving rise to haploid hyphae. As these hyphae grow, haploid hyphae of different mating types meet and fuse, forming dikaryotic hyphae, each cell of which contains two nuclei, one from each parent hypha. The dikaryotic mycelium
grows and eventually produces fruiting structures. The
dikaryotic phase may persist for years—some basidiomycetes live for decades or even centuries.

The elaborate fruiting structure of some fleshy basidiomycetes, such as the gill mushroom in Figure 30.15, is
topped by a cap, or *pileus*, which has structures called *gills*
on its underside. Enormous numbers of basidia develop on
the surfaces of the gills. The basidia discharge their spores
into the air spaces between adjacent gills, and the spores
sift down into air currents for dispersal and germination as
new haploid mycelia.

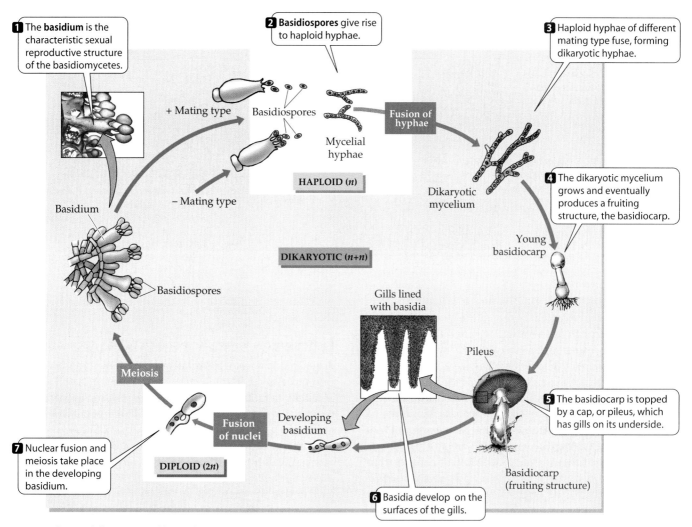

1 The **basidium** is the characteristic sexual reproductive structure of the basidiomycetes.

2 **Basidiospores** give rise to haploid hyphae.

3 Haploid hyphae of different mating type fuse, forming dikaryotic hyphae.

+ Mating type

Basidiospores

Mycelial hyphae

Fusion of hyphae

– Mating type

HAPLOID (*n*)

Basidium

DIKARYOTIC (*n+n*)

Basidiospores

Dikaryotic mycelium

4 The dikaryotic mycelium grows and eventually produces a fruiting structure, the basidiocarp.

Young basidiocarp

Meiosis

Gills lined with basidia

Pileus

Fusion of nuclei

Developing basidium

7 Nuclear fusion and meiosis take place in the developing basidium.

DIPLOID (2*n*)

5 The basidiocarp is topped by a cap, or pileus, which has gills on its underside.

Basidiocarp (fruiting structure)

6 Basidia develop on the surfaces of the gills.

30.15 The Basidiomycete Life Cycle
Basidiospores form on tiny stalks and are then forcibly dispersed to germinate into haploid hyphae, from which the familiar fruiting structure eventually grows.

Imperfect fungi lack a sexual stage

Mechanisms of sexual reproduction readily distinguish members of the four phyla of fungi from one another. But many fungi, including both saprobes and parasites, lack sexual stages entirely; presumably these stages have been lost during evolution or have not yet been found. Classifying these fungi as belonging to any of the four major phyla was at one time difficult, but biologists now can classify most such fungi on the basis of DNA sequences.

Fungi that have not yet been placed in any of the existing phyla are grouped together as the imperfect fungi, or deuteromycetes. Thus, the deuteromycete group is a holding area for species whose status is yet to be resolved. At present, about 25,000 species are classified as imperfect fungi. Some taxonomists, preferring to emphasize convenience of identification over strict phylogenetic considerations, treat the imperfect fungi as a paraphyletic phylum, Deuteromycota.

If sexual structures are found on a fungus classified as a deuteromycete, the fungus is reassigned to the appropriate phylum. That happened, for example, with a fungus that produces plant growth hormones called gibberellins. Originally classified as the deuteromycete *Fusarium moniliforme*, this fungus was later found to produce asci, whereupon it was renamed *Gibberella fujikuroi* and transferred to the phylum Ascomycota.

Fungal Associations

Earlier in this chapter we mentioned mycorrhizae and lichens, in which fungi live in intimate association with other organisms. Now that we have learned a bit about fungal diversity, let's consider mycorrhizae and lichens in greater detail.

Mycorrhizae are essential to many plants

Almost all tracheophytes enjoy a mutually beneficial symbiotic association with fungi. Unassisted, the root hairs of such plants do not absorb enough water or minerals to sustain maximum growth. However, the roots usually become infected with fungi, forming an association called a mycorrhiza.

30.16 Mycorrhizal Associations

(a) Ectomycorrhizal fungi wrap themselves around the plant root, increasing the area available for absorption of water and nutrients. (b) Endomycorrhizae infect the root internally.

(a)

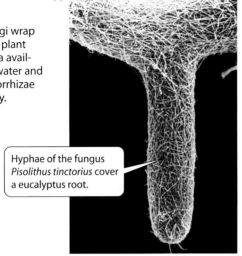

Hyphae of the fungus *Pisolithus tinctorius* cover a eucalyptus root.

(b)

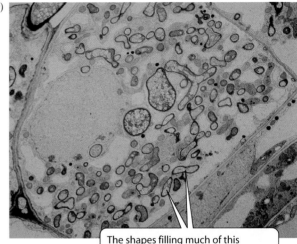

The shapes filling much of this soybean root cell are sections through the hyphae of the endomycorrhizal fungus *Glomus caledonius*.

In *ectomycorrhizae*, the fungus wraps around the root, and its mass is often as great as that of the root itself (Figure 30.16*a*). The hyphae of the fungi attached to the root increase the surface area for the absorption of water and minerals, and the mass of the mycorrhiza, like a sponge, holds water efficiently in the neighborhood of the root. Infected roots characteristically branch extensively and become swollen and club-shaped. In *endomycorrhizae*, the infection is internal to the root, with no hyphae visible on the root surface (Figure 30.16*b*).

The symbiotic fungus–plant association of a mycorrhiza is important to both partners. The fungus obtains important organic compounds, such as sugars and amino acids, from the plant. In return, the fungus greatly increases the absorption of water and minerals (especially phosphorus) by the plant. The fungus may also provide certain growth hormones, and may protect the plant against attack by microorganisms. Plants that have active mycorrhizae typically are a deeper green and may resist drought and temperature extremes better than plants of the same species that have little mycorrhizal development. Attempts to introduce some plant species to new areas have failed until a bit of soil from the native area (presumably containing the fungus necessary to establish mycorrhizae) was provided.

The partnership between plant and fungus results in a plant better adapted for life on land. It has been suggested that the evolution of this symbiotic association was the single most important step leading to the colonization of the terrestrial environment by living things. Fossils of mycorrhizal structures more than 300 million years old have been found, and some rocks dating back 460 million years contain structures that appear to be fossilized fungal spores.

Some liverworts, which are among the most ancient terrestrial plants (see Chapter 28), form mycorrhizae. Certain plants that live in nitrogen-poor habitats, such as cranberry bushes and orchids, invariably have mycorrhizae. Orchid seeds will not germinate in nature unless they are already infected by the fungus that will form their mycorrhizae. Plants that lack chlorophyll always have mycorrhizae, which they often share with the roots of green, photosynthetic plants.

Lichens grow where no eukaryote has succeeded

A lichen is not a single organism, but rather a meshwork of two radically different organisms: a fungus and a photosynthetic microorganism. Together the organisms constituting a lichen can survive some of the harshest environments on Earth. The flora of Antarctica, for example, features more than 100 times as many species of lichens as of plants.

In spite of this hardiness, lichens are very sensitive to air pollution because they are unable to excrete toxic substances that they absorb. Hence they are not common in industrialized cities. Because of their sensitivity, lichens are good biological indicators of air pollution.

The fungal components of most lichens are ascomycetes, but some are basidiomycetes or imperfect fungi (only one zygomycete serving as the fungal component of a lichen has been reported). The photosynthetic component may be either a cyanobacterium or a unicellular green alga. Relatively little experimental work has focused on lichens, perhaps because they grow so slowly—typically less than 1 centimeter per year.

There are about 13,500 "species" of lichens; their fungal components may constitute as many as 20 percent of all fungal species. Lichens are found in all sorts of exposed habitats: on tree bark, open soil, or bare rock. Reindeer "moss" (actually not a moss at all, but the lichen *Cladonia subtenuis*) covers vast areas in arctic, subarctic, and boreal regions, where it is an important part of the diets of reindeer and other large mammals. Lichens come in various forms and colors. *Crustose* (crustlike) lichens look like colored powder dusted over their substrate (Figure 30.17*a*); *foliose* (leafy) and *fruticose* (shrubby) lichens may have complex forms (Figure 30.17*b*).

The most widely held interpretation of the lichen relationship is that it is a type of mutually beneficial symbiosis. The hyphae of the fungal mycelium are tightly pressed against the photosynthetic cells of the alga or cyanobacterium and sometimes even invade them. The bacterial or

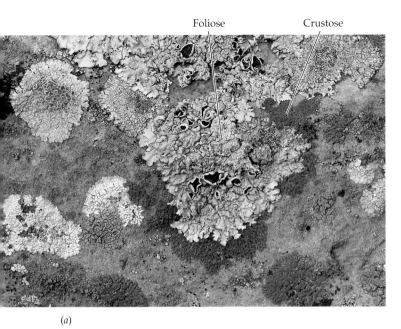

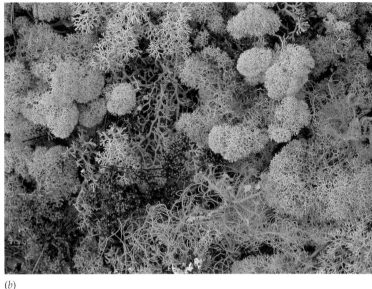

(b)

(a)

30.17 Lichen Body Forms
Lichens fall into three principal classes based on their body form. (a) Foliose and crustose lichens grow on otherwise bare rock. (b) A miniature jungle of fruticose lichens.

algal cells not only survive these indignities, but continue their growth and photosynthesis. In fact, algal cells in a lichen "leak" photosynthetic products at a greater rate than do similar cells growing on their own. On the other hand, photosynthetic cells from lichens grow more rapidly on their own than when combined with a fungus. On this basis, we could consider lichen fungi as parasitic on their photosynthetic partners.

Lichens can reproduce simply by fragmentation of the vegetative body, which is called the *thallus*, or by means of

specialized structures called **soredia** (singular soredium). Soredia consist of one or a few photosynthetic cells surrounded by fungal hyphae (Figure 30.18*a*). The soredia become detached, are dispersed by air currents, and upon arriving at a favorable location, develop into a new lichen. Alternatively, if the fungal partner is an ascomycete or a basidiomycete, it may go through its sexual cycle, producing either ascospores or basidiospores. When these spores are discharged, however, they disperse alone, unaccompanied by the photosynthetic partner, and thus may not be capable of reestablishing the lichen association. Nevertheless, many lichens produce characteristic fruiting structures containing asci or basidia.

Visible in a cross section of a typical foliose lichen are a tight upper region of fungal hyphae, a layer of cyanobacte-

30.18 Lichen Anatomy
(a) Soredia of a fruticose lichen. (b) Cross section showing the layers of a foliose lichen.

(a)

(b)

Each soredium consists of one or a few photosynthetic cells surrounded by fungal hyphae.

Soredia detach readily from the parent lichen and travel in air currents, founding new lichens when they settle in a suitable environment.

Hyphae

Soredium

Lichens are arranged in distinct layers.

Upper layer of fungal hyphae

Photosynthetic cell layer

Loose layer of fungal hyphae

Lower layer of fungal hyphae

Substrate

ria or algae, a looser hyphal layer, and finally hyphal rhizoids that attach the whole structure to its substrate (Figure 30.18*b*). The meshwork of fungal hyphae takes up some nutrients needed by the photosynthetic cells and provides a suitably moist environment for them by holding water tenaciously. The fungi derive fixed carbon from the photosynthesis of the algal or cyanobacterial cells.

Lichens are often the first colonists on new areas of bare rock. They satisfy most of their needs from the air and from rainwater, augmented by minerals absorbed from dust. A lichen begins to grow shortly after a rain, as it begins to dry. As it grows, the lichen acidifies its environment slightly, and this acid contributes to the slow breakdown of rocks, an early step in soil formation. After further drying, the lichen's photosynthesis ceases. The water content of the lichen may drop to less than 10 percent of its dry weight, at which point it becomes highly insensitive to extremes of temperature.

Whether living on their own or in symbiotic associations, fungi have spread successfully over much of Earth since their origin from a protist ancestor. That ancestor also gave rise to the choanoflagellates and the animal kingdom, the group we'll consider in the next three chapters.

Chapter Summary

▶ Fungi are the principal degraders of dead organic matter in the biosphere. Fungi are nutritional partners of almost all vascular plants. Some fungi are serious pathogens of plants and animals, including humans.

General Biology of the Fungi

▶ Fungi are heterotrophic eukaryotes with absorptive nutrition. They may be saprobes, parasites, or mutualists. **Review Figure 30.4**

▶ The yeasts are unicellular.

▶ The bodies of other fungi are composed of chitinous-walled, multinucleate hyphae, often massed to form a mycelium. The filamentous hyphae give fungi a large surface area-to-volume ratio, enhancing their ability to absorb nutrients. The hyphae usually have incomplete partitions (septa) that do not divide them into separate cells. **Review Figure 30.3**

▶ Fungi reproduce asexually by means of spores formed within sporangia, by conidia formed at the tips of hyphae, by budding, or by fragmentation.

▶ Fungi reproduce sexually when hyphae or motile cells of different mating types meet and fuse.

▶ In addition to the haploid and diploid states, many fungi demonstrate a third nuclear condition: the dikaryotic, or *n* + *n*, state. **Review Figure 30.13**

Diversity in the Kingdom Fungi

▶ The kingdom Fungi consists of four phyla: Chytridiomycota, Zygomycota, Ascomycota, and Basidiomycota. These phlya differ in their reproductive structures, mechanisms of spore formation, and less importantly, the presence and form of septa in their hyphae. **Review Figure 30.6, Table 30.1**

▶ Chytrids, with their flagellated zoospores and gametes, probably resemble the ancestral fungi.

▶ Zygomycetes reproduce sexually by fusion of gametangia. **Review Figure 30.9**

▶ The sexual reproductive structure of ascomycetes is an ascus containing ascospores. The ascomycetes are divided into two groups, euascomycetes and hemiascomycetes, on the basis of whether they have an ascocarp, or fruiting structure. **Review Figure 30.13**

▶ The sexual reproductive structure of basidiomycetes is a basidium, a swollen cell bearing basidiospores. **Review Figure 30.15**

▶ Imperfect fungi (deuteromycetes) lack sexual structures, but DNA sequencing can sometimes identify the phylum to which they belong.

Fungal Associations

▶ Mycorrhizae, associations of fungi with plant roots, enhance the ability of the roots to absorb water and nutrients.

▶ Lichens, mutualistic combinations of a fungus with a cyanobacterium or a green alga, are found in some of the most inhospitable environments on the planet. **Review Figure 30.18**

For Discussion

1. You are shown an object that looks superficially like a pale green mushroom. Describe at least three criteria (including anatomical and chemical traits) that would enable you to tell whether the object is a piece of a plant or a piece of a fungus.

2. Differentiate among the members of the following pairs of related terms:

   a. hypha/mycelium

   b. euascomycete/hemiascomycete

   c. ascus/basidium

   d. ectomycorrhiza/endomycorrhiza

3. For each type of organism listed below, give a single characteristic that may be used to differentiate it from the other, related organism(s) in parentheses.

   a. Zygomycota (Ascomycota)

   b. Basidiomycota (deuteromycetes)

   c. Ascomycota (Basidiomycota)

   d. baker's yeast (*Neurospora crassa*)

4. Many fungi are dikaryotic during part of their life cycle. Why are dikaryons described as *n* + *n* instead of 2*n*?

5. If all the fungi on Earth were suddenly to die, how would the surviving organisms be affected? Be thorough and specific in your answer.

6. How might the first mycorrhizae have arisen?

7. What might account for the ability of lichens to withstand the intensely cold environment of Antarctica? Be specific in your answer.

31 Animal Origins and Lophotrochozoans

IN 1822, NEARLY 40 YEARS BEFORE DARWIN wrote *The Origin of Species*, French naturalist E. Geoffroy Saint-Hilaire was examining a lobster. He noticed that when he viewed the lobster with its ventral surface up, its central nervous system was located above its digestive tract, which in turn was located above its heart—the same relative positions these systems have in mammals viewed *dorsally*. His observation led Saint-Hilaire to conclude that the differences between arthropods and vertebrates could be explained if the embryos of one of those groups had been inverted during development.

Saint-Hilaire's suggestion was regarded as totally preposterous at the time and was largely dismissed until recently. However, the discovery of two genes that influence a system of extracellular signals involved in development has lent new support to Saint-Hilaire's seemingly outrageous hypothesis.

A vertebrate gene called *chordin* helps establish cells on one side of the embryo as dorsal, the other as ventral. A probably homologous gene in fruit flies,* called *sog*, acts in a similar manner, but has the opposite effect. Fly cells where *sog* is active become ventral, whereas vertebrate cells expressing *chordin* become dorsal (see Figure 16.19). However, when *sog* mRNA is injected into the frog *Xenopus*, a vertebrate, it causes dorsal development. *Chordin* mRNA injected into flies promotes ventral development. In both cases, injection of the mRNA promotes the development of the portion of the embryo that contains the central nervous system!

Chordin and *sog* are among the many genes that appear to regulate similar functions in very different organisms. There are several almost universal animal genes that help transform a single-celled egg into a multicellular adult. Such genes are providing evolutionary biologists with information that can help them understand relationships among animal lineages that separated from one another in ancient times. As we saw in Chapter 23, new knowledge about gene functions and gene sequences provides some of the most powerful data being used in modern phylogenetic investigations to infer evolutionary relationships among organisms.

In this chapter we will first discuss how biologists infer evolutionary relationships among animals and review the defining characteristics of the animal way of life. Then we'll describe several lineages of simple animals. Finally, we'll describe the lophotrochozoans, one of the three great evolutionary lineages of animals. The next two chapters will discuss the other two great animal lineages, the ecdysozoans and the deuterostomes.

Descendants of a Common Ancestor

Biologists have long debated whether animals arose once or several times from protist ancestors, but enough molecular and morphological evidence has now been assembled to indicate that, with the possible exception of sponges (Porifera), the kingdom Animalia is a monophyletic group—that is, all animals are descendants of a single ancestral lineage.

---

*Insects (such as fruit flies) are arthropods and belong to the same evolutionary lineage as crustaceans (such as the lobster), as we will discuss in Chapter 32.

Genes that Control Development
The human and the lobster carry similar genes that control the development of the body axis. A lobster's nervous system runs up its ventral (belly) surface, while its circulatory system is dorsal (down its back). In vertebrates such as humans, similar genes position these two systems inversely to those of the lobster.

This conclusion is supported by the fact that all animals share a set of derived traits:

▶ Similarities in their 5S and 18S ribosomal RNAs

▶ Special types of cell–cell junctions: tight junctions, desmosomes, and gap junctions (see Figure 5.6)

▶ A common set of extracellular matrix molecules, including collagen (see Figure 4.28)

Animals evolved from ancestral colonial flagellated protists as a result of division of labor among their aggregated cells. Within these ancestral colonies of cells—perhaps analogous to those still existing in the chlorophyte *Volvox* or some colonial choanoflagellates (see Figures 27.25*a* and 27.28)—some cells became specialized for movement, others for nutrition, and still others differentiated into gametes. Once the division of labor had begun, these units continued to differentiate while improving their coordination with other working groups of cells. Such coordinated groups of cells evolved into the larger and more complex organisms that we now call animals.

The Animal Way of Life

What traits characterize the organisms we call animals? Animals are multicellular organisms that must take in preformed organic molecules because they cannot synthesize them from inorganic chemicals. They acquire these organic molecules by ingesting other organisms, either living or dead, and digesting them inside their bodies. To acquire these organic molecules, animals must expend energy to move themselves through the environment to find food, to position themselves where food will pass by them, or to move the environment and the food it contains to them.

The foods animals eat include most other members of the animal kingdom, as well as members of all other evolutionary lineages. Much of the diversity of animal sizes and shapes evolved as animals acquired the ability to capture and eat many different kinds of foods, and to avoid becoming food for other animals.

The need to move in search for food has favored sensory structures that provide animals with detailed information about their environment, and nervous systems able to receive and coordinate this information. Consequently, most animals are behaviorally much more complex than plants. Because animals ingest chemically complex foods, they expend considerable energy to maintain relatively constant internal conditions while taking in foods that vary chemically.

A real appreciation of animal structure and functioning is best achieved through firsthand experience in the field and laboratory. The accounts in this chapter and the following two serve as an orientation to the major groups of animals, their similarities and differences, and the evolutionary pathways that resulted in the current richness of animal evolutionary lineages and species. But how do biologists infer evolutionary relationships among animals?

Clues to Evolutionary Relationships among Animals

Biologists use a variety of traits to infer animal phylogenies. As we discussed in Chapters 23 and 24, clues to these relationships are found in the fossil record, in patterns of embryonic development, in the comparative morphology and physiology of living and fossil animals, and in the structure of their molecules.

Patterns of early development evolved very slowly in some animal lineages. For this reason, biologists have traditionally based their classifications of the major lineages of animals on developmental patterns. More recently, comparative molecular data from small subunit rRNA and mitochondrial genes also have been used. These two types of evidence suggest similar animal phylogenies.

31.1 Animal Body Cavities
The three major types of animal body cavities. (*a*) Acoelomates do not have enclosed body cavities. (*b*) Pseudocoelomates have only one layer of muscle, lying outside the body cavity. (*c*) Coelomates have a peritoneum surrounding the internal organs; the body cavities of some, such as this earthworm, are segmented.

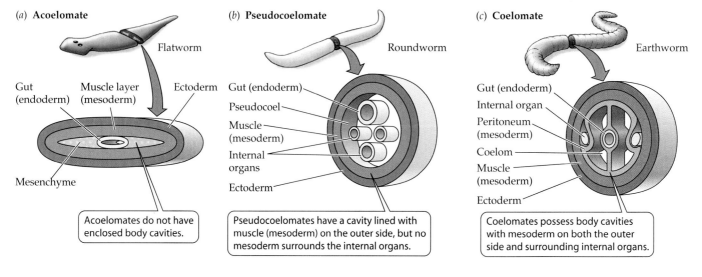

(*a*) **Acoelomate**

Flatworm

Gut (endoderm)   Muscle layer (mesoderm)   Ectoderm

Mesenchyme

Acoelomates do not have enclosed body cavities.

(*b*) **Pseudocoelomate**

Roundworm

Gut (endoderm)
Pseudocoel
Muscle (mesoderm)
Internal organs
Ectoderm

Pseudocoelomates have a cavity lined with muscle (mesoderm) on the outer side, but no mesoderm surrounds the internal organs.

(*c*) **Coelomate**

Earthworm

Gut (endoderm)
Internal organ
Peritoneum (mesoderm)
Coelom
Muscle (mesoderm)
Ectoderm

Coelomates possess body cavities with mesoderm on both the outer side and surrounding internal organs.

Using this wide variety of comparative data, zoologists have concluded that the sponges, cnidarians, and ctenophores separated from the remaining animal lineages early in evolutionary history. They have divided the remaining animals into two major lineages: the **protostomes** and the **deuterostomes**.

In the common ancestor of the protostomes and the deuterostomes, the pattern of early cell division in the fertilized egg—called *cleavage*—was radial. During *radial cleavage*, cells divide along a plane either parallel to or at right angles to the long axis of the fertilized egg. This pattern persisted during the evolution of deuterostomes and in many protostome lineages, but *spiral cleavage* evolved in one major protostome lineage. In spiral cleavage, the plane of cell division is oblique to the long axis of the egg, causing the cells to be arranged in a spiral pattern.

Other developmental patterns typically differ between protostomes and deuterostomes. Cleavage of the fertilized egg in protostomes is *determinate*; that is, if the egg is allowed to divide a few times and the cells are then separated, each cell develops into only a partial embryo. In contrast, cleavage in deuterostomes typically is *indeterminate*; cells separated after several cell divisions can still develop into complete embryos. We see this phenomenon in humans in identical twins. Among deuterostomes, the mouth of the embryo originates some distance away from the embryonic structure called the blastopore, which becomes the anus. Among protostomes, the mouth arises from or near the blastopore.

During development from a single-celled zygote to a multicellular adult, animals form layers of cells. The embryos of **diploblastic** animals have only two cell layers: an outer *ectoderm* and an inner *endoderm*. The embryos of **triploblastic** animals have a third layer, the *mesoderm*, which lies between the ectoderm and the endoderm.

Fluid-filled spaces, called **body cavities**, lie between the cell layers of the bodies of many kinds of animals. The type of body cavity an animal has strongly influences how it can move.

▶ Animals that lack an enclosed body cavity are called **acoelomates** In these animals, the space between the gut and the body wall is filled with masses of cells called *mesenchyme* (Figure 31.1*a*).

▶ Another group of animals, the **pseudocoelomates**, have a body cavity called the *pseudocoel*. The pseudocoel is a liquid-filled space in which many of the body organs

are suspended, but control over body shape is crude because a pseudocoel has muscles only on the outside (Figure 31.1*b*).

▶ **Coelomate** animals have a *coelom*, a body cavity that develops within the embryonic mesoderm. It is lined with a special structure called the *peritoneum*, and has muscles both inside and outside. The internal organs of coelomates are slung in pouches of peritoneum rather than being suspended within the body cavity (Figure 31.1*c*).

An animal with a coelom has better control over the movement of the fluids it contains, but control is limited if the animal has only a single, large body cavity. Control is improved if the coelom is separated into compartments or segments so that circular and longitudinal muscles in each individual segment can change its shape independently of the other segments. Segmentation of the coelom evolved several different times among both protostomes and deuterostomes.

The phylogeny of animals we adopt in this book is based on analyses of many developmental, structural (from both living and fossil animals), and molecular traits. Figure 31.2 shows the postulated order of splitting of the major lineages in animal evolution. New information continues to modify and refine our understanding of the details of phylogenetic relationships among animals. Nonetheless, the division of the animals into the lineages shown here is supported by many types of data.

Body Plans Are Basic Structural Designs

The entire structure of an animal, its organ systems, and the integrated functioning of its parts are known as its **body plan**. Animals in many (but not all) lineages have evolved greater body complexity over time.

A fundamental aspect of an animal's body plan is its overall shape, described as its **symmetry**. A symmetrical animal can be divided along at least one plane into similar halves. Animals that have no plane of symmetry are said to

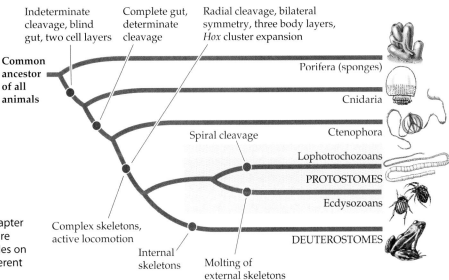

31.2 A Probable Phylogeny of Animals
The evolutionary tree that we will use in this chapter and the following two postulates that animals are monophyletic. The traits highlighted by red circles on the tree will be explained as we discuss the different phyla.

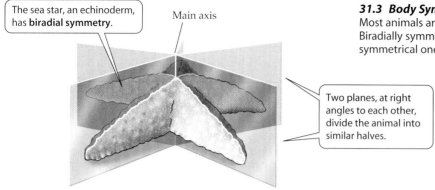

The sea star, an echinoderm, has **biradial symmetry**.

Main axis

Two planes, at right angles to each other, divide the animal into similar halves.

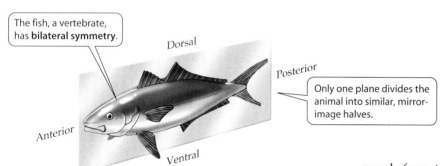

The fish, a vertebrate, has **bilateral symmetry**.

Dorsal

Posterior

Only one plane divides the animal into similar, mirror-image halves.

Anterior

Ventral

31.3 Body Symmetry
Most animals are either biradially or bilaterally symmetrical. Biradially symmetrical animals appear similar to radially symmetrical ones.

be **asymmetrical**. Many sponges are asymmetrical, but most animals have some kind of symmetry.

The simplest form of symmetry is **spherical symmetry**, in which body parts radiate out from a central point. An infinite number of planes passing through the central point can divide a spherically symmetrical organism into similar halves. Spherical symmetry is widespread among protists, but most animals possess other forms of symmetry.

An organism with **radial symmetry** has one main axis around which its body parts are arranged. A perfectly radially symmetrical animal can be divided into similar halves by any plane that contains the main axis. Some simple sponges and a few other animals, such as some sea anemones, have true radial symmetry.

Most radially symmetrical animals are modified such that only two planes, at right angles to each other, can divide them into similar halves. These animals are said to have **biradial symmetry** (Figure 31.3a). Three animal phyla—Cnidaria, Ctenophora, and Echinodermata—are composed primarily of radially or biradially symmetrical animals. These animals move slowly or not at all.

Bilateral symmetry is a common characteristic of animals that move freely through their environments. A bilaterally symmetrical animal can be divided into mirror images (left and right sides) by only a single plane that passes through the midline of its body from the front (anterior) to the back (posterior) end (Figure 31.3b). A plane at right angles to the first divides the body into two dissimilar sides; the side of a bilaterally symmetrical animal without a mouth is its dorsal (back) surface; the side with a mouth is its ventral (belly) surface.

Bilateral symmetry is strongly correlated with **cephalization**: the presence of a head, bearing sensory organs and central nervous tissues, at the anterior end of the animal. Cephalization may have been evolutionarily advantageous because the anterior end of a freely moving animal typically encounters new environments first.

Speed is often advantageous for both prey and the predators that pursue them. Fast-moving prey and predators had evolved by the early Cambrian period. To move rapidly, an animal needs some type of skeleton that supports its body and allows body parts to be moved relative to one another. A skeleton may be internal or external, rigid or flexible, and be composed of one, two, or more elements.

The fluid-filled body cavities of early animals functioned as **hydrostatic skeletons**. Because fluids are relatively incompressible, they move to another part of the cavity when muscles surrounding them contract. If the body tissues around the cavity are flexible, fluids moving from one region cause some other region to expand. Moving fluids can thus move specific body parts, or even the whole animal, provided that temporary attachments can be made to the substrate.

Other forms of skeletons developed in many animal lineages, either as substitutes for, or in combination with, hydrostatic skeletons. Some of these skeletons consist of a single element (snail shells); some have two elements (clam shells); others have many elements (centipedes). Some are internal (vertebrate bones); others are external (crab shells).

The form of the body cavity also changed in many animal lineages. Many became divided into compartments. The form of its skeleton and body cavities strongly influences the degree to which an animal can control and change its shape, and thus the complexity of the movements it can perform. What type of body plan and symmetry did the common ancestors of all animals possess? We are not certain, but because evidence suggests that animals evolved from colonies of flagellated cells, they may have been similar in structure to living flagellates.

Sponges: Loosely Organized Animals

The difference between protist colonies and simple multicellular animals is that the cells of animals are differentiated and their activities are coordinated. The lineage leading to modern sponges separated from the lineage leading to all other animals early during animal evolution. Some living

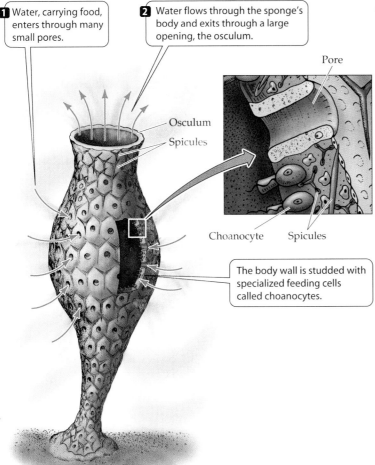

1 Water, carrying food, enters through many small pores.

2 Water flows through the sponge's body and exits through a large opening, the osculum.

Pore

Osculum

Spicules

Choanocyte   Spicules

The body wall is studded with specialized feeding cells called choanocytes.

31.4 The Sponge Body Plan
The flow of water through the sponge is shown by blue arrows. The body wall is studded with choanocytes, a type of specialized feeding cell that may be a link between animals and protists (see Figure 27.28).

sponges are still very similar to the probable ancestral colonial protists. Sponges are loosely organized. Even if a sponge is completely disassociated by being strained through a filter, its cells can reassociate into a new sponge.

The **sponges** (phylum **Porifera**, Latin for "pore bearers") are **sessile**. They live attached to the substrate and do not move about. The body plan of all sponges—even large ones, which may reach more than a meter in length—is an aggregation of cells built around a water canal system. A sponge feeds by drawing water into itself and filtering out the small organisms and nutrient particles that flow past the walls of its inner cavity.

Porifera
Cnidaria
Ctenophora
Lophotrochozoans
PROTOSTOMES
Ecdysozoans
DEUTEROSTOMES

Feeding cells with a collar and a flagellum, called *choanocytes*, line the inside of the water canals. By beating their flagella, the choanocytes cause water to flow into the animal, either by way of small pores that perforate special epidermal cells (in simple sponges) or through intercellular pores (in complex sponges). Water passes into small chambers within the body where food particles are captured by the choanocytes. Water then exits through one or more larger openings called *oscula* (Figure 31.4).

Between the thin epidermis and the choanocytes is a layer of cells, some of which are similar to amoebas and move about within the body. A supporting skeleton is also present, either in the form of simple or branching spines called *spicules* or as an elastic, often complex, network of fibers.

Most of the 10,000 species of sponges are marine animals; only about 50 species live in fresh water. Sponges come in a wide variety of sizes and shapes that are adapted to different patterns of water movement (Figure 31.5). Sponges living in intertidal or shallow subtidal environ-

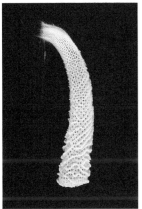

(a) *Euplectella aspergillum*

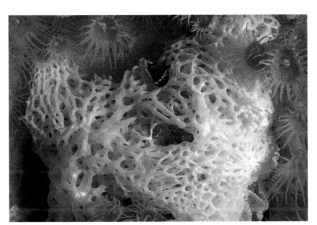

(b) *Clathrina coriacea*

31.5 Sponges Differ in Size and Shape
(a) Glass sponges are named after their glasslike spicules, which are formed of silicon. (b) The spicules of this marine sponge are made of calcium carbonate. (c) The brown volcano sponge is typical of many simple marine sponges.

(c) *Aplysina fistularia*

(a) *Gonothyraea loveni*

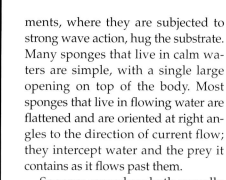

(c) *Urticina lofotensis*

ments, where they are subjected to strong wave action, hug the substrate. Many sponges that live in calm waters are simple, with a single large opening on top of the body. Most sponges that live in flowing water are flattened and are oriented at right angles to the direction of current flow; they intercept water and the prey it contains as it flows past them.

Sponges reproduce both sexually and asexually. In most species, a single individual produces both eggs

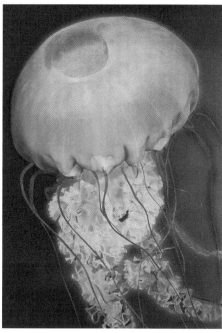

(b) *Chrysaora fuscescens*

31.6 Diversity among Cnidarians
(a) The structure of the polyps on a North Atlantic coastal hydrozoan is visible here. (b) This sea nettle jellyfish illustrates the complexity of some scyphozoan medusae. (c) The nematocyst-studded tentacles of this white-spotted anemone from British Columbia are poised to capture large prey carried to the animal by water movement.

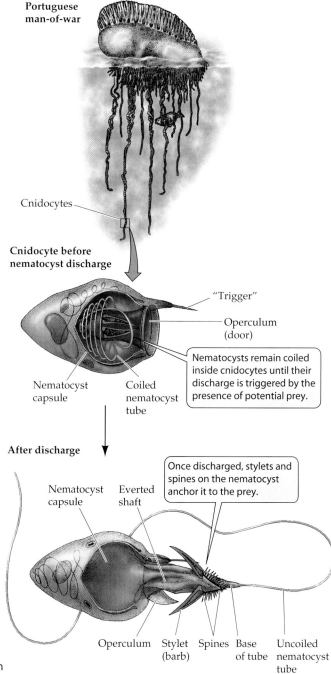

Portuguese man-of-war

Cnidocytes

Cnidocyte before nematocyst discharge

"Trigger"

Operculum (door)

Nematocyst capsule

Coiled nematocyst tube

Nematocysts remain coiled inside cnidocytes until their discharge is triggered by the presence of potential prey.

After discharge

Nematocyst capsule

Everted shaft

Once discharged, stylets and spines on the nematocyst anchor it to the prey.

Operculum Stylet (barb) Spines Base of tube Uncoiled nematocyst tube

31.7 Nematocysts Are Potent Weapons
Possessing a large number of nematocysts, cnidarians such as jellyfish can subdue and eat very large prey.

and sperm. Water currents carry sperm from one individual to another. Asexual reproduction is by budding and fragmentation.

Cnidarians: Cell Layers and Blind Guts

Animals in all phyla other than the Porifera have distinct cell layers and symmetrical bodies. The next lineage to split off from the main line of animal evolution after the sponges resulted in a phylum of animals—the **cnidarians** (phylum **Cnidaria**)—having only two cell layers (diploblastic) and a blind gut (with only one entrance). Within the constraints of this simple body organization, cnidarians evolved a wide variety of ways of making a living.

Cnidarians are simple but specialized carnivores

Cnidarians (phylum **Cnidaria**) appeared early in evolutionary history and radiated in the late Precambrian. About 10,000 cnidarian species—jellyfishes, sea anemones, corals, and hydrozoans—are living today (Figure 31.6). All but a few are marine. The smallest cnidarians can hardly be seen without a microscope; the largest known jellyfish is 2.5 meters in diameter. These animals are simple but specialized carnivores. The cnidarian body plan combines a low metabolic rate with the ability to capture large prey. These traits allow cnidarians to survive in environments where prey are scarce.

A key feature of cnidarians is tentacles that bear *cnidocytes*, specialized cells that contain stinging structures called *nematocysts* that can discharge toxins into their prey (Figure

31.7). Cnidocytes allow cnidarians to capture prey larger and more complex than themselves. Nematocysts are responsible for the sting that some jellyfishes and other cnidarians can inflict on human swimmers.

The mouth of a cnidarian is connected to a blind sac called the *gastrovascular cavity*, which functions in digestion, circulation, and gas exchange. The single opening serves as both mouth and anus. Cnidarians also have epithelial cells with muscle fibers whose contractions enable the animals to move, as well as nerve nets that integrate their body activities.

 Cnidarian life cycles

The generalized cnidarian life cycle has two distinct stages, the polyp and the medusa (Figure 31.8), although many species lack one of the stages.

▶ The sessile **polyp** stage has a cylindrical stalk attached to the substrate, with tentacles surrounding a mouth located at the opposite end from the site of attachment. This stage is usually asexual, but individual polyps may reproduce by budding, thereby forming a colony.

▶ The **medusa** is a free-swimming, sexual stage shaped like a bell or an umbrella. It typically floats with its mouth and tentacles facing downward. Medusae produce eggs and sperm and release them into the water. When an egg is fertilized, it develops into a free-swimming, ciliated larva called a **planula** that eventually settles to the bottom and transforms into a polyp.

Although the polyp and medusa stages appear very different, they share a similar body plan. A medusa is essentially a polyp without a stalk. Most of the outward differences between polyps and medusae are due to the *mesoglea*, a mass of jellylike material that lies between the two cell layers. The mesoglea contains few cells and has a low metabolic rate. In polyps, the mesoglea is usually thin; in medusae it is very thick, constituting the bulk of the animal.

HYDROZOANS. Life cycles are diverse among the **hydrozoans** (class **Hydrozoa**), a group containing the only freshwater cnidarians. The polyp commonly dominates the life cycle, but some species have only medusae and others only polyps. A few species have solitary polyps, but most hydrozoans are colonial. A single planula eventually gives rise to a colony of many polyps, all interconnected and sharing a continuous gastrovascular cavity (Figure 31.9). Within such a colony, some polyps have tentacles with many nematocysts; they capture prey for the colony. Others lack ten-

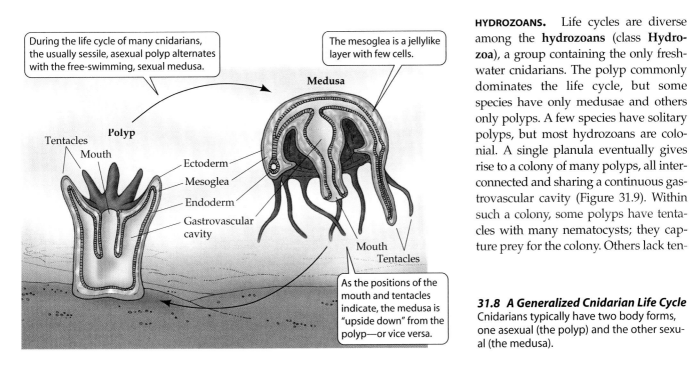

During the life cycle of many cnidarians, the usually sessile, asexual polyp alternates with the free-swimming, sexual medusa.

The mesoglea is a jellylike layer with few cells.

Medusa

Polyp

Tentacles
Mouth

Ectoderm
Mesoglea
Endoderm
Gastrovascular cavity

Mouth
Tentacles

As the positions of the mouth and tentacles indicate, the medusa is "upside down" from the polyp—or vice versa.

31.8 A Generalized Cnidarian Life Cycle
Cnidarians typically have two body forms, one asexual (the polyp) and the other sexual (the medusa).

31.9 Hydrozoans Often Have Colonial Polyps

The polyps with a hydrozoan colony may differentiate to perform specialized tasks.

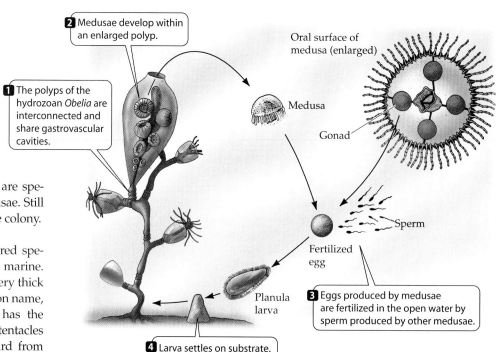

2 Medusae develop within an enlarged polyp.

1 The polyps of the hydrozoan *Obelia* are interconnected and share gastrovascular cavities.

Oral surface of medusa (enlarged)

Medusa

Gonad

Sperm

Fertilized egg

Planula larva

3 Eggs produced by medusae are fertilized in the open water by sperm produced by other medusae.

4 Larva settles on substrate.

tacles and are unable to feed, but are specialized for the production of medusae. Still others are fingerlike and defend the colony.

SCYPHOZOANS. The several hundred species of the class **Scyphozoa** are all marine. The mesoglea of their medusae is very thick and firm, giving rise to their common name, jellyfishes. The medusa typically has the form of an inverted cup, and the tentacles with nematocysts extend downward from the margin of the cup.

The medusa, rather than the polyp, dominates the life cycle of scyphozoans. An individual medusa is male or female, releasing eggs or sperm into the open sea. The fertilized egg develops into a small planula that quickly settles on a substrate and changes into a small polyp. This polyp feeds and grows and may produce additional polyps by budding. After a period of growth, the polyp begins to bud off small medusae (Figure 31.10). These small medusae feed, grow, and transform themselves into adult medusae, which are commonly seen during summer in harbors and bays. Thus a polyp that grows from a single fertilized egg is capable of producing many genetically identical medusae that will eventually reproduce sexually.

ANTHOZOANS. The roughly 6,000 species of sea anemones and corals that constitute the **anthozoans** (class **Anthozoa**) are all marine. Unlike other cnidarians, anthozoans entirely lack the medusa stage of the life cycle. The polyp produces eggs and sperm, and the fertilized egg develops into a planula that develops directly into another polyp. Many species can also reproduce asexually by budding or fission.

Sea anemones (see Figure 31.6c) are solitary. They are widespread in both warm and cold ocean waters. Many sea anemones are able to crawl slowly on the discs with which they attach themselves to the substrate. A few species can swim; some can burrow.

Corals, by contrast, are usually sessile and colonial. The polyps of most corals secrete a matrix of organic molecules upon which calcium carbonate—the eventual skeleton of the coral colony—is deposited. The forms of

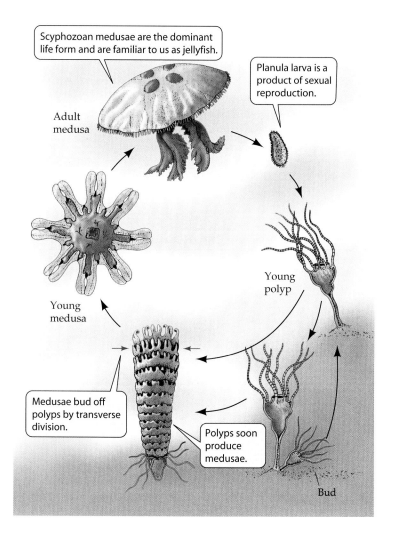

Scyphozoan medusae are the dominant life form and are familiar to us as jellyfish.

Planula larva is a product of sexual reproduction.

Adult medusa

Young medusa

Young polyp

Medusae bud off polyps by transverse division.

Polyps soon produce medusae.

Bud

31.10 Medusae Dominate Scyphozoan Life Cycles

Scyphozoan medusae are the familiar jellyfish of coastal waters. The small, sessile polyps quickly produce medusae.

(a)

(b)

31.11 Corals
(a) Many different species of corals and sponges grow together on this reef in the Bahama Islands. (b) The green plates of cabbage coral (*Turbinaria* sp.) and the branching staghorn coral (*Acrophora* sp.) are oriented to intercept sunlight in this Papua New Guinea reef.

coral skeletons are species-specific and highly diverse (Figure 31.11a). The common names of coral groups—horn corals, brain corals, staghorn corals, organ pipe corals, sea fans, and sea whips, among others—describe their appearance.

As a coral colony grows, old polyps die, but their calcareous skeletons remain. The living members form a layer on top of a growing reef of skeletal remains, eventually forming chains of islands and reefs. Corals are especially abundant in the Indo-Pacific region. The Great Barrier Reef along the northeastern coast of Australia is a system of coral formations more than 2,000 km long and as wide as 150 km. A continuous coral reef hundreds of kilometers long in the Red Sea has been calculated to contain more material than all the buildings in the major cities of North America combined.

Corals flourish in nutrient-poor, clear, tropical waters. For a long time scientists wondered how corals obtain enough nutrients to grow rapidly. The answer is that photosynthetic dinoflagellates live symbiotically within a coral's cells. They provide the corals with products of photosynthesis and contribute to calcium deposition. In turn, the corals protect the dinoflagellates from predators. This symbiotic relationship explains why reef-forming corals are restricted to clear surface waters, where light levels are high enough to allow photosynthesis (Figure 31.11b).

Coral reefs throughout the world are being threatened by both global warming, which is raising the temperatures of tropical shallow ocean waters, and nutrient runoff from developments on adjacent shorelines. An overabundance of

nitrogen gives an advantage to algae, which overgrow the corals and smother them.

Ctenophores: Complete Guts and Tentacles

Ctenophores (phylum **Ctenophora**) were the next lineage to separate from the lineage leading to all other animals. Ctenophores, also known as comb jellies, have body plans that are superficially similar to those of cnidarians. Both have two cell layers separated by a thick, gelatinous mesoglea, and both have radial symmetry and feeding tentacles. Like cnidarians, ctenophores have low metabolic rates because they are composed primarily of inert mesoglea. Unlike cnidarians, however, ctenophores have a complete gut. Food enters through a mouth and wastes are voided through two anal pores.

Ctenophores have eight comblike rows of fused plates of cilia, called *ctenes*. They move by beating these cilia rather than by use of muscular contractions. Ctenophoran tentacles do not have nematocysts; rather, they are covered with sticky filaments to which prey adhere (Figure 31.12). After capturing its prey, a ctenophore retracts its tentacles to bring the food to its mouth. In some species, the entire surface of the body is coated with a sticky mucus that captures prey. All of the 100 known species of ctenophores are marine carnivores. They are common in open seas, where prey are often scarce. Most ctenophores cannot capture large prey.

Ctenophore life cycles are simple. Gametes from gonads located on the walls of the gastrovascular cavity are released into the cavity and then discharged through the

Porifera

Cnidaria

Ctenophora

Lophotrochozoans
PROTOSTOMES
Ecdysozoans

DEUTEROSTOMES

(a)

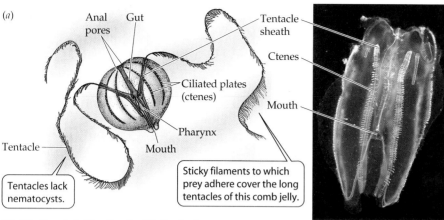

(b) Leucothea sp.

31.12 Comb Jellies Feed with Tentacles
(a) The body plan of a typical ctenophore. (b) This comb jelly has much shorter tentacles than many other ctenophores.

mouth or through pores. Fertilization takes place in the open seawater. In nearly all species, the fertilized egg develops directly into a miniature ctenophore that gradually grows into an adult.

The Evolution of Bilaterally Symmetrical Animals

The phylogenetic tree in Figure 31.2 postulates a common ancestor of all bilaterally symmetrical animals, but it does not tell us what that common ancestor looked like. Evolutionary biologists have attempted to infer the nature of those ancestral animals, which they call **urbilateria**, using evidence from the genes, development, and structure of existing animals.

One clue is provided by the fact that all living bilaterally symmetrical animals have an array of intercellular signaling systems and many homeobox gene families (see Chapter 16). The simplest bilaterally symmetrical animals have only a few homeobox genes, but some of them are shared with more complex animals. The mechanisms that regulate embryonic development in the protostomes and deuterostomes are governed by homologous homeobox genes. Such regulatory genes with similar functions are unlikely to have evolved independently in several different animal lineages.

Some evidence that urbilaterians may have been relatively complex is provided by fossilized traces of their movements. Fossilized trails from late Precambrian times exhibit complex search patterns, transverse furrows, and longitudinal ridges (Figure 31.13). They were made by organisms that were at least several centimeters long. The complexity of the movements recorded by the tracks suggests that urbilaterians had circulatory systems, systems of antagonistic muscles, and a tissue or fluid-filled body cavity. Some of their descendants subsequently lost some of those traits, but they retain signatures of their past in their genes.

Protostomes and Deuterostomes: An Early Lineage Split

The next major lineage split in the evolution of animals separated two groups that have been evolving separately ever since the Cambrian period. These two major lineages—the protostomes and deuterostomes—dominate today's biota. Members of both lineages are bilaterally symmetrical and have definite heads (cephalization). Because their skeletons and body cavities are more complex than those of the animals we have discussed so far, they are capable of more elaborate movements.

The most important shared, derived traits that unite the protostomes are a central nervous system consisting of an anterior brain that surrounds the entrance to the digestive tract; a ventral nervous system consisting of paired or fused longitudinal nerve cords; and a free-floating larva with a food-collecting system consisting of compound cilia on multiciliate cells. The major shared, derived traits that unite the deuterostomes are a dorsal nervous system and larvae with a food-collecting system consisting of cells with a single cilium.

Porifera

Cnidaria

Ctenophora

Lophotrochozoans
PROTOSTOMES
Ecdysozoans

DEUTEROSTOMES

Hiemalora   1 cm

31.13 Fossilized Trail of an Urbilaterian
These tracks indicate that their maker was able to crawl.

31.14 A Phylogeny of Lophotrochozoans
Three major lineages, including the lopho-phorate and spiralian phyla, dominate the tree. Some small phyla are not included in this diagram.

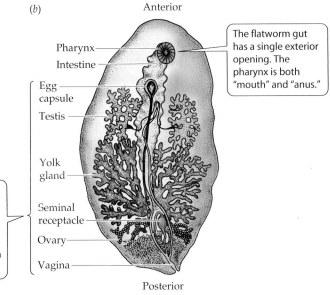

Most of the world's living animal species are protostomes. The diversity of protostome body plans and lifestyles has posed many challenges to zoologists attempting to infer the evolutionary relationships among these animals. Developmental, structural, and molecular data all suggest that protostomes split into two major lineages that have been evolving independently since ancient times: the lophotrochozoans and the ecdysozoans.

Lophotrochozoans, the animals we will discuss in the remainder of this chapter, grow by adding to the size of their skeletal elements. They use cilia for loco-motion, and many lineages have a type of free-living larva known as a **trochophore**. The phylogeny of lophotrochozoans we will use in this chapter is shown in Figure 31.14. In contrast, **ecdysozoans**, the animals we will discuss in the next chapter, increase in size by molting their external skeletons. They move by mechanisms other than ciliary action, and they share a common set of homeobox genes.

Simple Lophotrochozoans

Flatworms move by beating cilia

Members of the phylum **Platyhelminthes**, or **flatworms**, are the simplest lopho-trochozoans (Figure 31.15). They are bilat-erally symmetri-cal animals that have no enclosed body cavity. They lack organs for transporting oxygen to internal tissues, and they have only simple organs for excreting metabolic wastes. This body plan dictates that each cell must be near a body surface, a re-quirement met by the flattened body form.

The digestive tract of a flatworm consists of a mouth opening into a blind sac. However, the sac is often highly branched, forming intricate patterns that increase the sur-face area available for absorption of nutrients. All living flatworms feed on animal tissues—living or dead. Motile flatworms glide over surfaces, powered by broad bands of cilia. This form of movement is very slow, but it is sufficient

31.15 Flatworms Live Freely and Parasitically
(a) Some flatworm species are free-living, like this marine flatworm of the South Pacific. (b) The flatworm diagrammed here lives para-sitically in the gut of sea urchins. It is representative of parasitic flukes. Because their hosts provide all the nutrition they need, intestinal parasites do not require elaborate feeding or digestive organs.

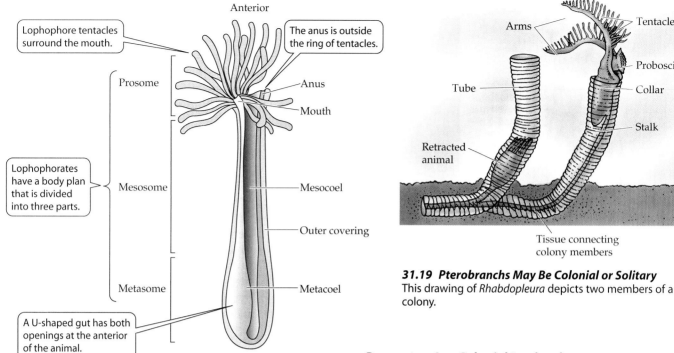

Anterior

Lophophore tentacles surround the mouth.

The anus is outside the ring of tentacles.

Prosome

Lophophorates have a body plan that is divided into three parts.

Mesosome

Metasome

A U-shaped gut has both openings at the anterior of the animal.

Anus

Mouth

Mesocoel

Outer covering

Metacoel

Posterior

31.18 Lophophore Artistry
The lophophore dominates the anatomy of a phoronid. The phoronid gut is U-shaped.

phoronids (see Figure 31.18). Cilia drive water into the top of the lophophore. Water exits through the narrow spaces between the tentacles. Suspended food particles are caught and transported by ciliary action to the food groove and into the mouth.

There are only 10 living species of **pterobranchs** (phylum **Pterobranchia**). Pterobranchs are sedentary animals up to 12 mm in length that live in tubes secreted by a proboscis, which is homologous to the prosome of phoronids. Some species are solitary; others form colonies of individuals joined together (Figure 31.19). Behind the proboscis is a collar with 1–9 pairs of arms bearing long tentacles that capture prey and permit gas exchange.

Arms

Tentacles

Proboscis

Tube

Collar

Stalk

Retracted animal

Tissue connecting colony members

31.19 Pterobranchs May Be Colonial or Solitary
This drawing of *Rhabdopleura* depicts two members of a colony.

Bryozoans Are Colonial Lophophorates

Bryozoans (phylum **Ectoprocta**) are colonial lophophorates that live in a "house" secreted by the body wall. A colony consists of many small individuals connected by strands of tissue along which materials can be moved (Figure 31.20*a*). Most bryozoans are marine, but a few live in fresh water. They are able to completely retract the lophophore, which they can also rock and rotate to increase contact with prey (Figure 31.20*b*).

A colony of bryozoans is created by the asexual reproduction of its founding members. One colony may contain as many as 2 million individuals. In some species, individual colony members are specialized for feeding, reproduction, defense, or support. Bryozoans reproduce sexually by releasing sperm into the water, where they are collected by other individuals. Eggs are fertilized internally, and devel-

31.20 Bryozoans
(*a*) Branching colonies of bryozoans may appear plantlike.
(*b*) Bryozoans have greater control over the movement of their lophophores than members of other lophophorate phyla.

(a) *Iodyticium* sp.

(b)

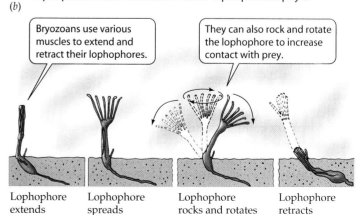

Bryozoans use various muscles to extend and retract their lophophores.

They can also rock and rotate the lophophore to increase contact with prey.

Lophophore extends

Lophophore spreads

Lophophore rocks and rotates

Lophophore retracts

oping embryos are brooded before they exit as larvae to seek suitable sites for attachment.

Brachiopods Superficially Resemble Bivalve Mollusks

Brachiopods (phylum **Brachiopoda**) are solitary, marine lophophorate animals that superficially resemble bivalve mollusks (Figure 31.21). Most brachiopods are between 4 and 6 cm long, but some are as long as 9 cm. Brachiopods have a shell divided into two parts connected by a ligament. The two halves can be pulled shut to protect the soft body. The shell differs from that of mollusks in that the two halves are dorsal and ventral rather than lateral. The two-armed lophophore of a brachiopod is located within the shell. The beating of cilia on the lophophore draws water into the slightly opened shell. Food is trapped in the lophophore and directed to a ridge along which it is transferred to the mouth.

Brachiopods are either attached to a solid substrate or embedded in soft sediments. Most species are attached by means of a short, flexible stalk that holds the animal above the substrate. Gases are exchanged across body surfaces, especially the tentacles of the lophophore. Most brachiopods release their gametes into the water, where they are fertilized. The larvae remain in the plankton for only a few days before they settle and change into adults.

Brachiopods reached their peak abundance and diversity in Paleozoic and Mesozoic times. More than 26,000 fossil species have been described. Only about 350 species survive, but they are common in some marine environments.

Spiralians: Wormlike Body Plans

The spiralian lineage gave rise to many phyla. Members of more than a dozen of these phyla are wormlike; that is, they are bilaterally symmetrical, legless, soft-bodied, and at least several times longer than they are wide. This body form enables animals to move efficiently through muddy and sandy marine sediments. Most of these phyla have no more than several hundred species, even though the lineages have been evolving independently since early animal evolution.

The carnivorous **ribbon worms** (phylum **Nemertea**) are dorsoventrally flattened and have nervous and excretory systems similar to those of flatworms but, unlike flatworms, they have a complete digestive tract with a mouth at one end and an anus at the other. Food moves in one direction through the digestive tract of a ribbon worm and is acted on by a series of digestive enzymes. Small ribbon worms move by beating their cilia. Larger ones employ waves of contraction of body mus-

Platyhelminthes

Rotifera

Bryozoa

Brachiopoda

Phoronida

Pterobranchia

Nemertea

Annelida

Mollusca

Laqueus sp.

31.21 Brachiopods
You can see the lophophore of this North Pacific brachiopod between the valves of its shell.

cles to move on the surface of sediments or to burrow. Movement by both of these methods is slow.

Within the body of almost all 900 species of ribbon worms is a fluid-filled cavity called the *rhynchocoel*, within which floats a hollow, muscular *proboscis*. The proboscis, which is the feeding organ, may extend much of the length of the worm. Contraction of the muscles surrounding the rhynchocoel causes the proboscis to be everted explosively through an anterior opening (Figure 31.22) without moving

(a)

Floating in a cavity called the rhynchocoel, the proboscis can be moved rapidly. The worm, however, moves slowly.

Rhynchocoel Proboscis Proboscis retractor muscle

Proboscis pore Mouth Intestine Anus

Retractor muscle

Mouth Intestine

Everted proboscis

(b)

31.22 Ribbon Worms
(a) The proboscis is the ribbon worm's feeding organ. (b) The full length of this ribbon worm from Oregon is impressive.

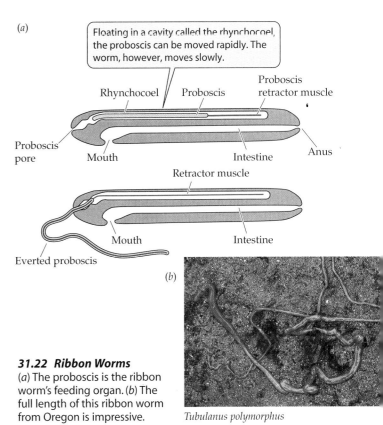

Tubulanus polymorphus

the rest of the animal. The proboscis of most ribbon worms is armed with a sharp stylet that pierces the prey. Paralysis-causing toxins produced by the proboscis are discharged into the wound made by the stylet.

Segmented Bodies: Improved Locomotion

A body cavity divided into segments allows an animal to alter the shape of its body in complex ways and to control its movements precisely. Fossils of segmented worms are known from the middle Cambrian; the earliest forms are thought to have been burrowing marine animals. Segmentation evolved several times among spiralians.

Annelids have many-segmented bodies

The **annelids** (phylum **Annelida**) are a diverse group of segmented worms (Figure 31.23). The approximately 15,000 known annelid species live in marine, freshwater, and terrestrial environments. A separate nerve center called a *ganglion* controls each segment, but the ganglia are connected by nerve cords that coordinate their functioning. The coelom in each segment is isolated from those in other segments. Most annelids lack a rigid, external protective covering. The thin body wall serves as a surface for gas exchange in most species, but this thin, permeable body surface restricts annelids to moist environments; they lose body water rapidly in dry air.

POLYCHAETES. More than half of all annelid species are members of the class **Polychaeta**. Nearly all polychaetes are marine animals. Most have one or more pairs of eyes and one or more pairs of tentacles at the anterior end of the body. The body wall in most segments extends laterally as a series of thin outgrowths, called *parapodia*, that contain many blood vessels. The parapodia function in gas exchange, and some species use them to move. Stiff bristles called *setae* protrude from each parapodium, forming temporary attachments to the substrate and preventing the animal from slipping backward when its muscles contract.

Many polychaete species live in burrows in soft sediments and filter prey from the surrounding water with elaborate feathery tentacles (Figure 31.24a). Typically, males and females release gametes into the water, where the eggs are fertilized and develop into a trochophore larva. As the larva develops, it forms body segments at its posterior end, eventually changing into a small adult worm.

OLIGOCHAETES. More than 90 percent of the approximately 3,000 described species of **oligochaetes** (class **Oligochaeta**) live in freshwater or terrestrial habitats. Oligochaetes have no parapodia, eyes, or anterior tentacles, and they have relatively few setae. Earthworms—the most familiar oligochaetes—are scavengers and ingesters of soil, from which they extract food particles.

Unlike polychaetes, all oligochaetes are *hermaphroditic*: Each individual is both male and female. Sperm are exchanged simultaneously between two copulating individuals (Figure 31.24b). Eggs are laid in a cocoon outside the adult's body. The cocoon is shed, and when development is complete, miniature worms emerge and begin independent life.

LEECHES. Leeches (class **Hirudinea**) probably evolved from oligochaete ancestors. Most species live in freshwater or terrestrial habitats and, like oligochaetes, lack parapodia and tentacles. Like oligochaetes, leeches are hermaphroditic. The coelom of leeches is not divided into compartments, and the coelomic space is largely filled with mesenchyme tissue. Groups of segments at each end of a leech are modified to form suckers, which serve as temporary anchors that aid in movement (Figure 31.24c). With its posterior sucker attached to a substrate, the leech extends its body by contracting its circular muscles. The anterior sucker is

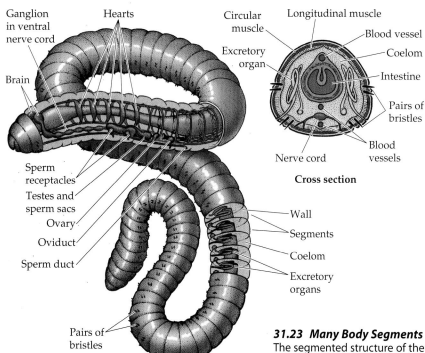

31.23 Many Body Segments
The segmented structure of the annelids is apparent both externally and internally. Most organs of this earthworm are repeated serially.

(a) *Spirobranchus* sp.

(b) *Lumbricus* sp.

(c) Australian tiger leech

31.24 Diversity among the Annelids
(a) The "feather duster" worm is a marine annelid with striking feeding tentacles. (b) Individual earthworms are hermaphroditic (simultaneously both male and female). When they copulate, each individual both donates and receives sperm. (c) This Australian tiger leech is attached to a leaf by its posterior sucker as it waits for a mammalian "victim." (d) Vestimentiferans live around thermal vents deep in the ocean. Their skin secretes chitin and other substances, forming tubes from which they extend feeding tentacles.

(d) *Riftia* sp.

om of a vestimentiferan consists of an anterior compartment, into which the tentacles can be withdrawn, and a long, subdivided cavity that extends much of the length of its body. Experiments using radioactively labeled molecules have shown that vestimentiferans take up dissolved organic matter at high rates from either the sediments in which they live or the surrounding water.

Vestimentiferans were not discovered until the twentieth century, when deep-ocean exploration revealed them living many thousands of meters below the surface. In these deep oceanic sediments they are abundant, reaching densities of many thousands per square meter. About 145 species have been described. The largest and most remarkable vestimentiferans, which grow to 2 meters in length, live near deep-ocean hydrothermal vents—openings in the seafloor through which hot, sulfide-rich water pours. The tissues of these species harbor endosymbiotic prokaryotes that fix carbon using energy obtained from the oxidation of hydrogen sulfide (H_2S).

then attached, the posterior one detached, and the leech shortens itself by contracting its longitudinal muscles.

Many leeches are external parasites of other animals, although some species eat snails and other invertebrates. A parasitic leech makes an incision in its host to expose its blood. It can ingest so much blood in a single feeding that its body may enlarge several times. A substance called hirudin secreted by the leech into the wound keeps the host's blood flowing (and gives this class its name; see the opening page of Chapter 6). For hundreds of years leeches were widely employed in medicine for bloodletting. Even today leeches are used to reduce fluid pressure and prevent blood clotting in damaged tissues and to eliminate pools of coagulated blood.

VESTIMENTIFERANS. Members of one lineage of annelids, the **vestimentiferans**, evolved into burrowing forms with a crown of tentacles through which gases are exchanged, and entirely lost their digestive systems (Figure 31.24d). The coel-

Mollusks lost segmentation but evolved shells

Mollusks (phylum **Mollusca**) range in size from snails only a millimeter high to giant squids more than 18 meters long—the largest known protostomes. Beginning with a segmented common ancestor, mollusks underwent one of the most

Platyhelminthes
Rotifera
Bryozoa
Brachiopoda
Phoronida
Pterobranchia
Nemertea
Annelida
Mollusca

Generalized molluscan body plan

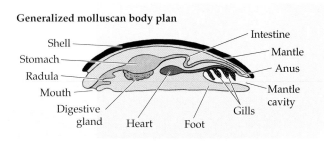

Chitons

In all mollusk lineages, a **mantle** covers the internal organs of the visceral mass.

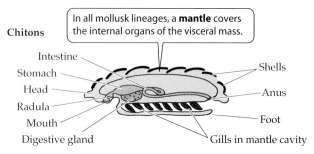

Gastropods

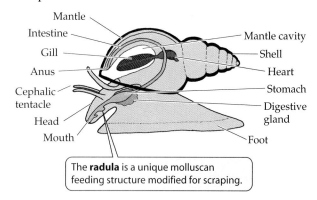

The **radula** is a unique molluscan feeding structure modified for scraping.

Bivalves

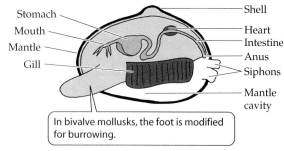

In bivalve mollusks, the foot is modified for burrowing.

Cephalopods

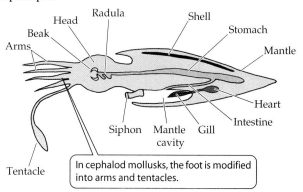

In cephalod mollusks, the foot is modified into arms and tentacles.

dramatic of animal evolutionary radiations, based on a body plan with three major structural components: a foot, a mantle, and a visceral mass. Animals that appear very different, such as snails, clams, and squids, are all built from these three components (Figure 31.25).

▶ The molluscan **foot** is a large, muscular structure that originally was both an organ of locomotion and a support for the internal organs.

In the lineage leading to squids and octopuses, the foot was modified to form arms and tentacles borne on a head with complex sensory organs. In other groups, such as clams, the foot was transformed into a burrowing organ. In some lineages the foot is greatly reduced.

▶ The **mantle** is a fold of tissue that covers the **visceral mass** of internal organs. In many mollusks, the mantle extends beyond the visceral mass to form a *mantle cavity.*

The gills, which are used for gas exchange and, in some species, for feeding, lie in the mantle cavity. When the cilia on the gills beat, they create a flow of oxygenated water over the gills.

The coelom of mollusks is much reduced, but the open circulatory system has large fluid-filled cavities that are major components of a hydraulic skeleton.

The mollusks also developed a rasping feeding structure known as the **radula**. The radula was originally an organ for scraping algae from rocks, a function it retains in many living mollusks. However, in some mollusks, it has been modified into a drill or a poison dart. In others, such as clams, it is absent.

Although individual components have been lost in some lineages, these three unique shared derived characteristics are why zoologists believe that all 100,000 species of mollusks share a common ancestor. A small sample of these species is shown in Figure 31.26.

MONOPLACOPHORANS. Monoplacophorans (class **Monoplacophora**) were the most abundant mollusks during the Cambrian period, but today there only a few surviving species. Unlike all other living mollusks, the surviving monoplacophorans have multiple gills, muscles, and excretory structures that are repeated over the length of the body. The gills are located in a large cavity under the shell, through which oxygen-bearing water circulates.

CHITONS. Chitons (class **Polyplacophora**) have multiple gills and segmented shells, but their other body parts are not segmented (Figure 31.26*a*). The chiton body is bilaterally symmetrical, and its internal organs, particularly the digestive and nervous systems, are relatively simple. The

31.25 Molluscan Body Plans
The diverse modern mollusks are all variations on a general body plan that includes a foot, a mantle, and a visceral mass of internal organs.

(a) *Tonicella lineata*

(c) *Hypsclodoris* sp.

(b) *Tridacna gigas*

(d) *Monadenia fidelis*

(e) *Octopus cyanea*

(f) *Nautilus belavensis*

31.26 Diversity among the Mollusks

(a) Chitons are common in the intertidal zones of the North American coast. (b) The giant clam of Indonesia is among the largest of the bivalve mollusks. (c) Slugs are terrestrial and marine gastropods that have lost their shells; this shell-less sea slug is very conspicuously colored. (d) Land snails are shelled, terrestrial gastropods. (e) Cephalopods such as the octopus are active predators. (f) The boundaries of its chambers are clearly visible on the outer surface of this shelled *Nautilus*, another cephalopod.

trochophore larvae of chitons are almost indistinguishable from those of annelids. Most chitons are marine herbivores that scrape algae from rocks with their sharp radulae. An adult chiton spends most of its life glued tightly to rock surfaces by its large, muscular, mucus-covered foot. It moves slowly by means of rippling waves of muscular contraction in the foot.

BIVALVES. One lineage of early mollusks developed a hinged, two-part shell that extended over the sides of the body as well as the top, giving rise to the **bivalves** (class **Bivalvia**), which include the familiar clams, oysters, scallops, and mussels (Figure 31.26*b*). Bivalves are largely sedentary and have greatly reduced heads. The foot is compressed and, in many clams, is used for burrowing into mud and sand. Bivalves feed by bringing water in through an opening called a *siphon* and extracting food from the water using their large gills, which are also the main sites of gas exchange. Water exits through another siphon.

GASTROPODS. Another lineage of early mollusks gave rise to the **gastropods** (class **Gastropoda**), which includes the snails. Most gastropods are motile, using the large foot to move slowly across the substrate or to burrow through it. Gastropods are the most species-rich and widely distributed of the molluscan classes (Figure 31.26*c,d*). Some species, such as snails, whelks, limpets, slugs, abalones, and the often brilliantly ornamented nudibranchs, can crawl. Others—the sea butterflies and heteropods—have a modified foot that functions as a swimming organ with which they move through open ocean waters. The only mollusks that live in terrestrial environments—land snails and slugs—are gastropods. In these terrestrial species the mantle cavity is modified into a highly vascularized lung.

CEPHALOPODS. In one lineage of mollusks, the **cephalopods** (class **Cephalopoda**), the exit siphon, which initially may have simply improved the flow of water over the gills, became modified to allow the early cephalopods to control the water content of the mantle cavity. The modification of the mantle into a device for forcibly ejecting water from the cavity enabled cephalopods to move rapidly through the water. Furthermore, as fluid moves out of a chamber, gases diffuse into it, changing the buoyancy of the animal. Thus, by pumping out water, the animals could also control their buoyancy. Together, these adaptations allowed cephalopods to live in open water.

With their greatly enhanced mobility, some cephalopods, such as squids and octopuses, became the major predators in open ocean waters (Figure 31.26*e*). They are still important marine predators today. Cephalopods capture and subdue their prey with their tentacles; octopuses use theirs to move over the substrate. As is typical of active predators, cephalopods have complex sensory organs, most notably eyes that are comparable to those of vertebrates in their ability to resolve images. The cephalopod head is closely associated with a large, branched foot that bears tentacles and a siphon. The large, muscular mantle is a solid external supporting structure. The gills hang within the mantle cavity.

Cephalopods appeared about 600 million years ago, near the beginning of the Cambrian period, and by the Ordovician period a wide variety of types were present. They were the first large, shelled animals able to move vertically in the ocean. The earliest cephalopod shells were divided by partitions penetrated by tubes through which liquids could be moved. Nautiloids (genus *Nautilus*) are the only cephalopods with external chambered shells that survive today (Figure 31.26*f*). Increases in size and reductions in external hard parts characterize the subsequent evolution of many lineages.

 *Chapter Summary*

Descendants of a Common Ancestor

▸ All members of the kingdom Animalia are believed to have a common flagellated protist ancestor.

▸ The specialization of cells by function made possible the complex, multicellular body plan of animals.

The Animal Way of Life

▸ Animals obtain their food—complex organic molecules—by active expenditure of energy.

Clues to Evolutionary Relationships among Animals

▸ Morphological, developmental, and molecular data support similar animal phylogenies.

▸ The body cavity of an animal is strongly correlated with its ability to move. On the basis of their body cavities, animals are classified as acoelomates, pseudocoelomates, or coelomates. **Review Figure 31.1**

▸ The two major animal lineages—protostomes and deuterostomes—are believed to have separated early in animal evolution; they differ in several components of their early embryological development. **Review Figure 31.2**

Body Plans Are Basic Structural Designs

▸ Most animals have either radial or bilateral symmetry. Radially symmetrical animals move slowly or not at all. Bilateral symmetry is strongly correlated with more rapid movement and the development of sensory organs at the anterior end of the animal. **Review Figure 31.3**

Sponges: Loosely Organized Animals

▸ Sponges (phylum Porifera) are simple animals that lack cell layers and body symmetry, but have several different cell types.

▸ Sponges feed via choanocytes, feeding cells that draw water through the sponge body and filter out small organisms and nutrient particles. **Review Figure 31.4**

Cnidarians: Cell Layers and Blind Guts

▸ Cnidarians (phylum Cnidaria) are radially symmetrical and have only two cell layers, but with their nematocyst-studded tentacles they can capture prey larger and more complex than themselves. **Review Figure 31.7**

► Most cnidarian life cycles have a sessile polyp and a free-swimming, sexual medusa stage, but some species lack one of the stages. **Review Figures 31.8, 31.9, 31.10**

Ctenophores: Complete Guts and Tentacles

► Ctenophores (phylum Ctenophora), descendants of the first split in the lineage of bilaterally symmetrical animals, are marine carnivores that have simple life cycles. **Review Figure 31.12**

The Evolution of Bilaterally Symmetrical Animals

► The common ancestors of bilateral animals, called urbilaterians, were probably simple, bilaterally symmetrical animals composed of flattened masses of cells.

Protostomes and Deuterostomes: An Early Lineage Split

► Protostomes and deuterostomes are monophyletic lineages that have been evolving separately since the Cambrian period. Their members are structurally more complex than cnidarians and ctenophores. Protostomes have a ventral nervous system, paired nerve cords, and larvae with compound cilia. Deuterostomes have a dorsal nervous system and larvae with single cilia.

► Protostomes split into two major clades—lophotrochozoans and ecdysozoans. **Review Figure 31.14**

Simple Lophotrochozoans

► Flatworms (phylum Platyhelminthes) have no body cavity, lack organs for oxygen transport, have only one entrance to the gut, and move by beating their cilia. Many species are parasitic. **Review Figures 31.15, 31.16**

► Although no larger than many ciliated protists, rotifers (phylum Rotifera) have highly developed internal organs. **Review Figure 31.17**

Lophophorates: An Ancient Body Plan

► The lophotrochozoan lineage split into two branches whose descendants became the lophophorates and the spiralians.

► The lophophore dominates the anatomy of many lophophorate animals. **Review Figure 31.18**

► Bryozoans are colonial lophophorates that can move their lophophores. **Review Figure 31.20**

► Brachiopods, which superficially resemble bivalve mollusks, were much more abundant in the past than they are today.

Spiralians: Wormlike Body Plans

► The spiralian lineage gave rise to many phyla, most of whose members have wormlike body forms.

► Ribbon worms (phylum Nemertea) have a complete digestive tract and capture prey with an eversible proboscis. **Review Figure 31.22**

Segmented Bodies: Improved Locomotion

► Annelids (phylum Annelida) are a diverse group of segmented worms that live in marine, freshwater, and terrestrial environments. **Review Figure 31.23**

► Mollusks (phylum Mollusca) evolved from segmented ancestors but subsequently became unsegmented. The molluscan body plan has three basic components: foot, mantle, and visceral mass. **Review Figure 31.25**

► The molluscan body plan has been modified to yield a diverse array of animals that superficially appear very different from one another.

For Discussion

1. Differentiate among the members of each of the following sets of related terms:
   a. radial symmetry/bilateral symmetry
   b. protostome/deuterostome
   c. indeterminate cleavage/determinate cleavage
   d. spiral cleavage/radial cleavage
   e. coelomate/pseudocoelomate/acoelomate

2. For each of the types of organisms listed below, give a single trait that may be used to distinguish them from the organisms in parentheses:
   a. cnidarians (sponges)
   b. gastropods (all other mollusks)
   c. polychaetes (other annelids)

3. In this chapter we listed some of the traits shared by all animals that convince most biologists that all animals are descendants of a single common ancestral lineage. In your opinion, which of these traits provides the most compelling evidence that animals are monophyletic?

4. Describe some features that allow animals to capture prey that are larger and more complex than they themselves are.

5. Animals in many phyla have wormlike, or vermiform, shapes. Why has this body form met the needs of species in so many different lineages of animals? In what types of environments does the worm shape function well? Why?

6. Having a complete digestive tract in which materials enter at an anterior mouth and move in one direction until they exit from a posterior opening would appear to be a very efficient way to digest food and rid the body of the indigestible residues. Nonetheless, several successful phyla of animals with a blind gut must void their digestive wastes via the same opening through which food entered. Why has this type of digestive system persisted? What limitations does it impose on the types of food animals can eat and the way in which the food is treated?

32 *Ecdysozoans: The Molting Animals*

A FIRM, NONLIVING COVERING THAT IS DIFFICULT to penetrate—an **exoskeleton**—provides an animal with both protection and support. Its very attributes, however, pose a huge problem: An exoskeleton cannot grow as the animal body inside it grows. Ancestors of today's ecdysozoan animals evolved a solution. They shed, or **molt**, the outgrown exoskeleton and expand and harden a new, larger one.

The new exoskeleton is already in place, growing underneath the old one. Directly after molting, the animal is very vulnerable. With its soft, new armor it can move only very slowly for a while. Despite this constraint, the lineages of Ecdysozoa—the molting animals—have more species than all other animal lineages combined.

An increasingly rich array of molecular and genetic evidence, including a common set of homeobox genes, suggests that molting may have evolved only once during animal evolution. The exoskeletons of ecdysozoan animals range from thin and flexible to thick, hard, and rigid.

The presence of an exoskeleton presented new problems and opportunities in other areas of the body plan besides growth. Unlike the lophotrochozoans, ecdysozoans cannot use cilia for locomotion; new forms of locomotion evolved in these lineages. And because hard exoskeletons impede the passage of oxygen into the animal, new mechanisms for respiration evolved.

In this chapter we will review the characteristics of animals in the various ecdysozoan phyla and show how developing an exoskeleton has influenced the evolution of these animals. The phylogeny we follow here is presented in Figure 32.1. The latter half of the chapter details the characteristics of several ecdysozoan lineages that have traditionally been classified in the phylum Arthropoda—the arthropods.

Collectively, arthropods (which include the terrestrial insects and the marine crustaceans) are the dominant animals on Earth, both in number of species (some 1.5 million) and number of individuals (estimated at some 10^{18} individuals, or a billion billion). The highly successful arthropod body plan is based on three elements: the rigid exoskeleton that marks them as ecdysozoans; segmenta-

tion; and jointed appendages, which immensely enhance their powers of locomotion, and which we will encounter again in Chapter 33 when we cover another major lineage, the vertebrates.

We close the chapter with an overview of evolutionary themes found in the evolution of the protostomate phyla, including both the lophotrochozoan and ecdysozoan lineages.

Cuticles: Flexible, Unsegmented Exoskeletons

Some ecdysozoans have wormlike bodies covered by exoskeletons that are relatively thin and flexible. These exoskeletons, called **cuticles**, protect the animal, but do not provide support for the bodies. The action of circular and longitudinal muscles on fluid in the body cavity provides a hydrostatic skeleton for many of these animals, which can

Molting the Exoskeleton
This green darner dragonfly (genus *Anax*) has just emerged from its larval exoskeleton and is pumping fluids into its expanding wings. At this stage the insect can move only very slowly.

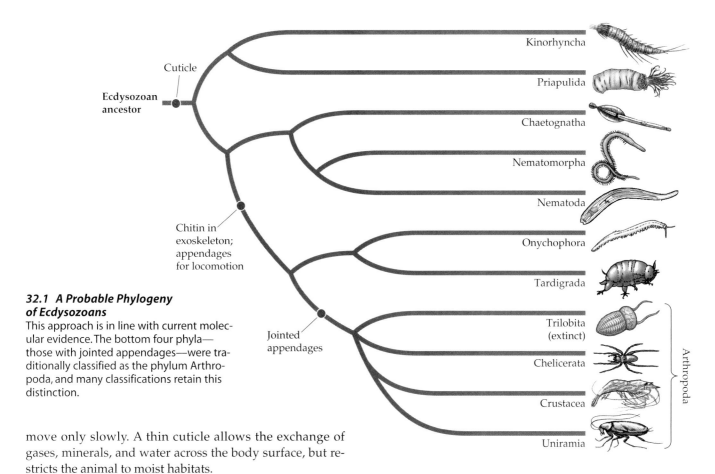

Ecdysozoan ancestor

Cuticle

Chitin in exoskeleton; appendages for locomotion

Jointed appendages

Kinorhyncha

Priapulida

Chaetognatha

Nematomorpha

Nematoda

Onychophora

Tardigrada

Trilobita (extinct)

Chelicerata

Crustacea

Uniramia

Arthropoda

32.1 A Probable Phylogeny of Ecdysozoans

This approach is in line with current molecular evidence. The bottom four phyla—those with jointed appendages—were traditionally classified as the phylum Arthropoda, and many classifications retain this distinction.

move only slowly. A thin cuticle allows the exchange of gases, minerals, and water across the body surface, but restricts the animal to moist habitats.

Some marine phyla have few species

Several phyla of marine wormlike animals (that is, they are long and slender, without appendages) branched off early within the ecdysozoan lineage. These phyla contain only a few species. They have relatively thin cuticles that are molted periodically as they grow to full size. Their bodies are supported primarily by their hydrostatic skeletons, not by the cuticle.

PRIAPULIDS AND KINORHYNCHS. The 16 species of **priapulids** (phylum **Priapulida**) are cylindrical, unsegmented, wormlike animals that range in size from half a millimeter to 20 centimeters in length. They burrow in fine marine sediments.

About 150 species of **kinorhynchs** (phylum **Kinorhyncha**) have been described. They are all less than 1 millimeter in length and live in marine sands or muds. Their bodies are divided into 13 segments by a series of cuticular plates that are periodically molted during growth (Figure 32.2). Kinorhynchs feed by ingesting the substratum and digesting the organic material found within it, which may include living algae as well as dead matter.

Kinorhyncha
Priapulida
Chaetognatha
Nematomorpha
Nematoda
Onychophora
Tardigrada
Trilobita
Chelicerata
Crustacea
Uniramia

ARROW WORMS. Arrow worms (phylum **Chaetognatha**) have three-part, streamlined bodies. Their body plan is based on a coelom that is divided into head, trunk, and tail compartments. Most of them swim in the open sea, but a few live on the seafloor. Their abundance as fossils indicates that they were already common more than 500 million years ago. The 100 or so living species of arrow worms are so small—less than 12 centimeters long—that their gas exchange and excretion requirements can be met by diffusion through the body surface. Arrow worms lack a circulatory system. Wastes and nutrients are moved around the body in the coelomic fluid, which is propelled by cilia that line the coelom. Arrow

32.2 A Kinorhynch
Kinorhynchs are tiny (less than a millimeter long) marine worms. Their segmented bodies are covered with plates of cuticle that are periodically molted.

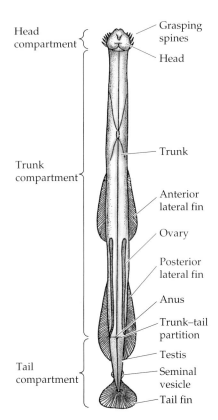

32.3 An Arrow Worm
Arrow worms have a three-part body plan. The fins and grasping spines are adaptations for a predatory life.

worms are stabilized in the water by means of one or two pairs of lateral fins and a "tail" fin (Figure 32.3). There is no distinct larval stage; miniature adults hatch directly from eggs that are released into the water.

Arrow worms are major predators of small organisms in the open oceans. Their prey range from small protists to young fish as large as an arrow worm. An arrow worm typically lies motionless in the water until movement of the water signals the approach of prey. The arrow worm then darts forward and grasps the prey with the stiff spines adjacent to its mouth.

Tough cuticles evolved in some unsegmented worms

Tough external cuticles evolved in some members of an ecdysozoan lineage whose descendants have colonized freshwater and terrestrial environments as well as marine ones. Two extant phyla represent this lineage.

Kinorhyncha
Priapulida
Chaetognatha
Nematomorpha
Nematoda
Onychophora
Tardigrada
Trilobita
Chelicerata
Crustacea
Uniramia

HORSEHAIR WORMS.   About 230 species of horsehair worms (phylum **Nematomorpha**) have been described. As their name implies, they are extremely thin and range in length from a

few millimeters to up to a meter (Figure 32.4). Most adult horsehair worms live in fresh water among litter and algal mats near the edges of streams and ponds. The larvae of horsehair worms are all internal parasites of terrestrial and aquatic insects and crabs. The much reduced gut has no mouth opening and is probably nonfunctional. Horsehair worms may feed only as larvae, absorbing nutrients from their hosts across their body wall, but many continue to grow after they have left their hosts, suggesting that adults may also absorb nutrients from their environment.

ROUNDWORMS.   Roundworms (phylum **Nematoda**) have a thick, multilayered cuticle secreted by the underlying epidermis that gives their body its shape (Figure 32.5). As a roundworm grows, it sheds and re-secretes its cuticle four times. The largest known roundworm, which reaches a length of 9 meters, is a parasite in the placentas of female sperm whales. About 20,000 species of roundworms have been described, but the actual number of living species may be more than a million.

Roundworms exchange oxygen and nutrients with their environment through both the cuticle and the intestine, which is only one cell layer thick. Materials are moved through the gut by rhythmic contraction of a highly muscular organ, the *pharynx*, at the worm's anterior end. Roundworms move by contracting their longitudinal muscles.

Roundworms are one of the most abundant and universally distributed of all animal groups. Countless roundworms live as scavengers in the upper layers of the soil, on the bottoms of lakes and streams, and as parasites in the bodies of most kinds of plants and animals. The flesh of a single rotting apple found on the ground in an orchard contained 90,000 roundworms, and 1 square meter of mud off the coast of the Netherlands yielded 4,420,000 individuals. The topsoil of rich farmland has up to 3 billion nematodes per acre.

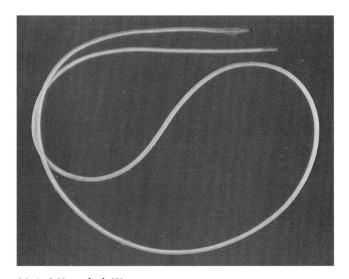

32.4 A Horsehair Worm
How these worms got their name is evident from this photograph.

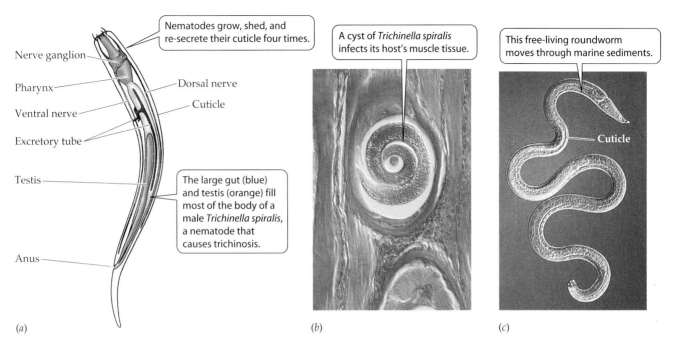

Nematodes grow, shed, and re-secrete their cuticle four times.

Nerve ganglion

Pharynx

Ventral nerve

Excretory tube

Testis

Anus

Dorsal nerve

Cuticle

The large gut (blue) and testis (orange) fill most of the body of a male *Trichinella spiralis*, a nematode that causes trichinosis.

A cyst of *Trichinella spiralis* infects its host's muscle tissue.

This free-living roundworm moves through marine sediments.

Cuticle

(*a*)
(*b*)
(*c*)

32.5 Nematodes
(*a*,*b*) *Trichinella* is an example of a parasitic roundworm that infects mammals, including humans. (*c*) Free-living roundworms have a body plan similar to the adult parasite's.

The diets of roundworms are as varied as their habitats. Many are predators, preying on protists and other small animals (including other roundworms). Many roundworms live parasitically within their hosts. The roundworms that are parasites of humans (causing diseases such as trichinosis, filariasis, and elephantiasis), domestic animals, and economically important plants have been studied intensively in an effort to find ways of controlling them. One soil-inhabiting nematode species, *Caenorhabditis elegans*, has been intensely studied in the laboratory by geneticists and developmental biologists.

The structure of parasitic roundworms is similar to that of free-living species, but the life cycles of many parasitic species have special stages that facilitate their transfer among hosts. *Trichinella spiralis*, the species that causes the human disease trichinosis, has a relatively simple life cycle. A person may become infected by eating the flesh of an animal (usually a pig) containing larvae of *Trichinella* encysted in its muscles.

The larvae are activated in the mammalian digestive tract, leave their cysts, and attach to the person's intestinal wall, where they feed. Later they bore through the intestinal wall and are carried in the bloodstream to the muscles, where they form cysts (Figure 32.5*b*). If present in great numbers, these cysts can cause severe pain or even death.

Trichinella can infect a number of mammal species; there is no special stage in the life cycle that lives in a particular alternate host. Other roundworm life cycles are more complex, involving one or more alternate hosts.

Arthropods and Their Relatives: Segmented External Skeletons

In Precambrian times, the body coverings of some wormlike ecdysozoan lineages became thickened by the incorporation of layers of protein and a strong, flexible, waterproof polysaccharide called **chitin**. After this change, which initially probably had a protective function, the rigid body covering acquired both support and locomotory functions.

A rigid body covering precludes wormlike movement. To move, these animals require **appendages** that can be manipulated by muscles. Such appendages evolved several times in late Precambrian times, leading to the phyla collectively called **arthropods**. The divisions among arthropod lineages are so ancient that we divide them into a number of phyla. However, as indicated at the opening of this chapter, many zoologists treat these groups as members of a single phylum, **Arthropoda**.

The bodies of arthropods are divided into segments. Their muscles attach to the inside of the skeleton, and each segment has muscles that operate that particular segment and the appendages attached to it (Figure 32.6). The appendages of most present-day arthropods have joints, although those of some lineages do not. Arthropod appendages serve many functions, including walking and swimming, food capture and manipulation, copulation, and sensory perception.

The sturdy exoskeleton had a profound influence on arthropod evolution. Encasement within armor provides support for walking on dry land, and, with special waterproofing, it keeps the animal from dehydrating in dry air. Aquatic arthropods were, in short, excellent candidates to invade the terrestrial environment, and as we will see, they did so several times.

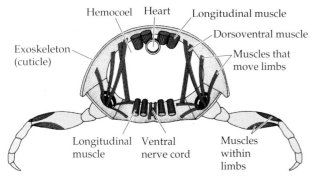

32.6 Arthropods Have Rigid, Segmented Exoskeletons
This cross section through a segment of a generalized arthropod shows the typical structure of an arthropod body, which is characterized by a rigid exoskeleton and jointed appendages.

(a) *Peripatodes novazealaniae*

Related lineages had unjointed legs

Although they were once thought to be closely related to segmented annelid worms, the molecular evidence links the 80 species of **onychophorans** (phylum **Onychophora**) to the arthropod lineages. Onychophorans have soft bodies that are covered by a thin, flexible cuticle that contains chitin. Onychophorans use their body cavities as hydrostatic skeletons. Their soft, fleshy, unjointed, claw-bearing legs are formed by outgrowths of the body (Figure 32.7a). They are probably similar in appearance to ancestral arthropods.

Kinorhyncha
Priapulida
Chaetognatha
Nematomorpha
Nematoda
Onychophora
Tardigrada
Trilobita
Chelicerata
Crustacea
Uniramia

(b) *Echinisucus springer*   50 µm

32.7 Arthropod Relatives with Unjointed Appendages
(a) Onychophorans have unjointed legs and use the body cavity as a hydrostatic skeleton. (b) The appendages and general anatomy of a water bear (phylum Tardigrada) superficially resemble those of onychophorans.

Like the onychophorans, **water bears** (phylum **Tardigrada**) have fleshy, unjointed legs and use their fluid-filled body cavities as hydrostatic skeletons (Figure 32.7b). Unlike onychophorans, water bears are all extremely small (0.1–0.5 mm in length), and they lack circulatory systems and gas exchange organs. The 600 extant species of water bears live in marine sands and on temporary water films on plants. When these films dry out, the water bears also lose water and shrink to small, barrel-shaped objects that can survive for at least a decade in a dehydrated resting state. They may occur at densities as high as 2,000,000 per square meter of moss.

Jointed legs appeared in the trilobites

Once the dominant line of arthropods, the **trilobites** (phylum **Trilobita**) flourished in Cambrian and Ordovician seas but were extinct by the close of the Paleozoic era. Trilobites were heavily armored, and their body segmentation and appendages followed a relatively simple, repetitive plan. But their appendages were jointed, giving them added flexibility, and the beginnings of specialization—using different appendages for different functions—can be discerned.

Why trilobites declined in abundance and eventually became extinct is unknown. However, because their heavy external skeletons provided ideal material for fossilization, they left behind a vivid record of their presence (Figure 32.8).

Odontochile rugosa

32.8 A Trilobite
The relatively simple, repetitive segments of the now-extinct trilobites are illustrated here by a fossil trilobite from the shallow seas of the Devonian period.

(a) *Decalopoda* sp.

32.9 Minor Chelicerate Phyla
(a) Although they are not true spiders, it is easy to see why sea spiders were given their common name. (b) This spawning aggregation of horseshoe crabs was photographed on the New Jersey coast.

(b) *Limulus polyphemus*

Chelicerates Invaded the Land

The bodies of all **chelicerates** (phylum **Chelicerata**) are divided into two major regions. The anterior region bears two pairs of appendages, modified to form mouthparts, and four pairs of walking legs. The 63,000 described species are usually placed in three classes: Pycnogonida, Arachnida, and Merostomata. Only the class Arachnida contains many species.

The **pycnogonids** (class **Pycnogonida**), or sea spiders, are a small group of marine species that are seldom seen except by marine biologists (Figure 32.9a). The class **Merostomata** contains a single order, the Xiphosura, or horseshoe crabs. These marine animals, which have changed very little during their long fossil history, have a large horseshoe-shaped covering over most of the body. They are common in shallow waters along the eastern coasts of North America and Southeast Asia, where they scavenge and prey on bottom-dwelling invertebrates. Periodically they crawl into the intertidal zone to mate and lay eggs (Figure 32.9b).

Arachnids (class **Arachnida**) are abundant in terrestrial environments. Most arachnids have a simple life cycle in which miniature adults hatch from eggs and begin independent lives almost immediately. Some arachnids retain their eggs during development and give birth to live young. The most species-rich and abundant arachnids are the scorpions, harvestmen, spiders, mites, and ticks (Figure 32.10).

Spiders are important terrestrial predators. Some have excellent vision that enables them to chase and seize their prey. Others spin elaborate webs made of protein threads to snare prey. The webs of different groups of spiders are strikingly varied and enable spiders to position their snares in many different environments. Spiders also use protein threads to construct safety lines during climbing and as homes, mating structures, protection for developing young, and means of dispersal. The threads are produced by modified abdominal appendages that are connected to internal glands that secrete the proteins of which the threads are constructed.

Kinorhyncha
Priapulida
Chaetognatha
Nematomorpha
Nematoda
Onychophora
Tardigrada
Trilobita
Chelicerata
Crustacea
Uniramia

Crustaceans: Diverse and Abundant

Crustaceans (phylum **Crustacea**) are the dominant marine arthropods. The most familiar crustaceans are decapods (shrimps, lobsters, crayfishes, and crabs; Figure 32.11a); isopods (sow bugs; Figure 32.11b); and amphipods (sand fleas; see Figure 1.11). Also included among the crustaceans are a wide variety of other small species, many of which superficially resemble shrimps (Figure 32.11c). The individuals of one group alone, the copepods (class Copepoda), are so numerous that they may be the most abundant of all animals.

Kinorhyncha
Priapulida
Chaetognatha
Nematomorpha
Nematoda
Onychophora
Tardigrada
Trilobita
Chelicerata
Crustacea
Uniramia

Barnacles (class Cirripedia) are unusual crustaceans that are sessile as adults (Figure 32.11d). With their calcareous shells, they superficially resemble mollusks, but, as the zoologist Louis Agassiz remarked more than a century ago, a barnacle is "nothing more than a little shrimp-like animal, standing on its head in a limestone house and kicking food into its mouth."

(a) *Uroctonus mondax*

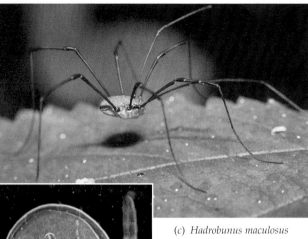

(c) *Hadrobunus maculosus*

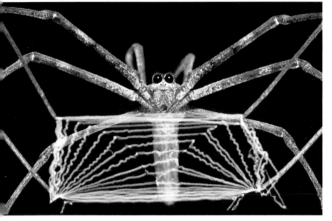

(b) *Deinopis* sp.

(d) *Ixodes ricinus*

32.10 Diversity among the Arachnids
(a) Scorpions are nocturnal predators. (b) Many spiders use webs to snare and envelop their prey. This ogre-faced spider uses its specialized web like a net. (c) Harvestmen, often called daddy longlegs, are scavengers. (d) Ticks are blood-sucking, external parasites on vertebrates. This wood tick is piercing the skin of its human host.

32.11 Diversity among the Crustaceans
(a) This crayfish is a decapod crustacean. (b) This sow bug is a common isopod found in grasslands. (c) A typical planktonic copepod from the deep ocean. (d) The appendages of these gooseneck barnacles protrude from their shells to capture prey.

(a) *Orconectes palmeri*

(b) *Armadillidium vulgare*

(c) *Megacalanus princeps*

(d) *Lepas pectinata*

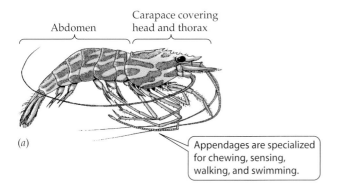

(a)

Abdomen | Carapace covering head and thorax

> Appendages are specialized for chewing, sensing, walking, and swimming.

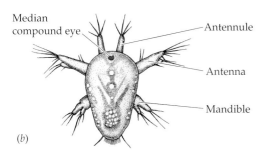

Median compound eye — Antennule

Antenna

Mandible

(b)

32.12 Crustacean Structure
(a) The bodies of most crustaceans are divided into three regions, each segment of which bears appendages. (b) A nauplius larva has one simple eye and three pairs of appendages.

Most of the 40,000 described species of crustaceans have a body that is divided into three regions: *head, thorax,* and *abdomen.* The segments of the head are fused together, and the head bears five pairs of appendages. Each of the multiple thoracic and abdominal segments usually bears one pair of appendages. In many species, a fold of the exoskeleton, the *carapace,* extends dorsally and laterally back from the head to cover and protect some of the other segments (Figure 32.12a).

The fertilized eggs of most crustacean species are attached to the outside of the female's body, where they remain during their early development. At hatching, the young of some species are released as larvae; those of other species are released as juveniles that are similar in form to the adults. Still other species release fertilized eggs into the water or attach them to an object in the environment. The typical crustacean larva, called a **nauplius**, has three pairs of appendages and one simple eye (Figure 32.12b). In many crustaceans, the nauplius larva develops within the egg before it hatches.

Uniramians are Primarily Terrestrial

The body of a **uniramian** (phylum **Uniramia**) is divided into either two or three regions (in myriapods and insects, respectively). The anterior regions have few segments, but the posterior region—the abdomen—has many segments. Uniramians are primarily terrestrial animals; most have elaborate systems of channels that bring oxygen to the cells of their internal organs.

Myriapods have many legs

Centipedes, millipedes, and the two other groups of animals in the subphylum **Myriapoda** have two body regions: a head and a trunk. Centipedes and millipedes have a well-formed head and a long, flexible, segmented trunk that bears many pairs of legs (Figure 32.13). Centipedes prey on insects and other small animals. Millipedes scavenge and eat plants. More than 3,000 species of centipedes and 10,000 species of millipedes have been described; many more species probably remain unknown. Although most myriapods are less than a few centimeters long, some tropical species are ten times that size.

Kinorhyncha
Priapulida
Chaetognatha
Nematomorpha
Nematoda
Onychophora
Tardigrada
Trilobita
Chelicerata
Crustacea
Uniramia

Insects are the dominant uniramians

The 1.5 million species of **insects** (subphylum **Insecta**) that have been described are believed to be only a small fraction of the total number living on Earth today. Insects are found in nearly all terrestrial and freshwater habitats, and they utilize as food nearly all species of plants and many species of animals. Some are internal parasites of plants and animals; others suck their host's blood or consume its surface

(a) *Scolopendra heros*

(b) *Harapaphe haydeniana*

32.13 Myriapods
(a) Centipedes have powerful jaws for capturing active prey.
(b) Millipedes, which are scavengers and plant eaters, have smaller jaws and legs.

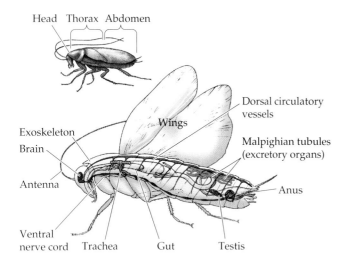

32.14 Structure of an Insect
The body plan of an insect differs in many details from that of other arthropods, but the basic theme of a segmented body with modified jointed appendages is shared with most arthropod lineages.

body tissues. Insects transmit many viral, bacterial, and protist diseases among plants and animals. Very few insect species are oceanic. In freshwater environments, on the other hand, they are sometimes the dominant animals, burrowing through the muddy substrate, extracting suspended prey from the water, or actively pursuing other animals.

Insects, like the crustaceans, have three basic body parts: head, thorax, and abdomen. They have a single pair of antennae on the head, and three pairs of legs attached to the thorax (Figure 32.14). An insect exchanges gases by means of air sacs and tubular channels called *tracheae* (singular trachea) that extend from external openings inward to tissues throughout the body. The adults of most flying insects have two pairs of stiff, membranous wings attached to the thorax. The exceptions are flies, which have only one pair of wings, and beetles, in which the forewings form heavy, hardened wing covers.

Wingless insects (class **Apterygota**) include firebrats and collembolans (Figure 32.15*a*). Of the modern insects, they are probably the most similar in form to insect ancestors. Apterygote insects have a simple life cycle, hatching from their eggs looking like small adults.

Development in the winged insects (class **Pterygota**) is more complex. The hatchlings are not similar to adults, and they undergo substantial changes at each molt in the process of growing larger. The immature stages of insects between molts are called **instars**. A substantial change that occurs between one developmental stage and another is called **metamorphosis**. When the change from one instar to the next is gradual, an insect is said to undergo **incomplete metamorphosis**.

In some insect genera, the larvae and adult forms can appear to be completely different animals. The most familiar example of such **complete metamorphosis** occurs in members of the order Lepidoptera, when the larval caterpillar trans-

forms itself into the adult butterfly (see Figure 1.6). During complete metamorphosis, the wormlike larva transforms itself during a specialized phase, called the **pupa**, in which many larval tissues are broken down and the adult form develops.

Entomologists divide the winged insects into about 28 different orders. We can make sense out of this bewildering variety by recognizing three major lineages:

▸ Winged insects that cannot fold their wings back against the body.
▸ Winged insects that can fold their wings and that undergo incomplete metamorphosis.
▸ Winged insects that can fold their wings and that undergo complete metamorphosis.

Because they can fold their wings over their backs, flying insects belonging to the second and third lineages are able to tuck their wings out of the way upon landing and crawl into crevices and other tight places.

The only surviving groups of the first lineage are the orders Odonata (dragonflies and damselflies; Figure 32.15*b*) and Ephemeroptera (mayflies). All members of these two orders have aquatic larvae that metamorphose into flying adults after they crawl out of the water. Although many of these insects are excellent flyers, they require a great deal of open space in which to maneuver. Dragonflies and damselflies are active predators as adults, but adult mayflies lack functional digestive tracts and do not eat, living only long enough to mate and lay eggs.

The second lineage includes the orders Orthoptera (grasshoppers, crickets, roaches, mantids, and walking sticks; Figure 32.15*c*), Isoptera (termites), Plecoptera (stone flies), Dermaptera (earwigs), Thysanoptera (thrips), Hemiptera (true bugs; Figure 32.15*d*), and Homoptera (aphids, cicadas, and leafhoppers). Hatchlings are sufficiently similar in form to adults to be recognizable. They acquire adult organ systems, such as wings and compound eyes, gradually through several juvenile instars.

Insects belonging to the third lineage have different life stages specialized for living in different environments and using different food sources. In many species the larvae are adapted for feeding and growing, and the adults are specialized for reproduction and dispersal. The adults of some species do not feed at all, living only long enough to mate, disperse, and lay eggs. In many species whose adults do feed, adults and larvae use different food resources. About 85 percent of all species of winged insects belong to this lineage. Familiar examples are the orders Neuroptera (lacewings

32.15 Diversity among the Insects   ▸
(*a*) This silverfish is a typical member of the apterygote order Thysanura. (*b*) Unlike most insects, this adult dragonfly (order Odonata) cannot fold its wings over its back. Representatives of some of the largest insect orders are (*c*) a broad-winged katydid (Orthoptera), (*d*) harlequin bugs (Hemiptera), (*e*) a predaceous diving beetle (order Coleoptera), (*f*) a Great Mormon butterfly (Lepidoptera), (*g*) a hoverfly (Diptera), and (*h*) a honeybee (Hymenoptera).

(a) *Lepisma saccharina*

(e) *Dysticus marginalis*

(b) *Sympetrum vulgatum*

(f) *Papilio memnon*

(c) *Microcentrum rhombifolium*

(g) Family Syrphidae

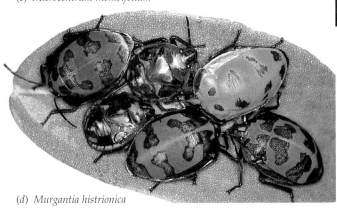

(d) *Murgantia histrionica*

(h) *Apis mellifera*

32.1 General Characteristics of the Major Protostomate Phyla[a]

PHYLUM	BODY CAVITY	DIGESTIVE TRACT	CIRCULATORY SYSTEM
Lophotrochozoans			
Platyhelminthes	None	Dead-end sac	None
Rotifera	Pseudocoelom	Complete	None
Bryozoa	Coelom	Complete	None
Brachiopoda	Coelom	Complete	None
Phoronida	Coelom	Complete	None
Pterobranchia	Coelom	Complete	None
Nemertea	Coelom	Complete	Closed
Annelida	Coelom	Complete	Closed or open
Mollusca	Reduced coelom	Complete	Open except in cephalopods
Ecdysozoans			
Chaetognatha	Coelom	Complete	None
Nematomorpha	Pseudocoelom	Greatly reduced	None
Nematoda	Pseudoceolom	Greatly reduced	None
Chelicerata	Hemocoel	Complete	Open
Crustacea	Hemocoel	Complete	Open
Uniramia	Hemocoel	Complete	Open

[a]All have bilateral symmetry.

and their relatives), Coleoptera (beetles; Figure 32.15e), Trichoptera (caddisflies), Lepidoptera (butterflies and moths; Figure 32.15f), Diptera (flies; Figure 32.15g), and Hymenoptera (sawflies, bees, wasps, and ants; Figure 32.15h).

There are also several orders of pterygote insects, including the Phthiraptera (lice) and Siphonaptera (fleas) that are parasitic. Although descended from flying ancestors, these insects have lost the ability to fly.

Why have the insects undergone such incredible evolutionary diversification? Insects may have originated from a centipede-like ancestor as far back as the Devonian period. The terrestrial environments penetrated by the arthropods were like a new planet, an ecological world with more complexity than the seas they came from, but one containing relatively few species other of animals. The evolution of the ability to fly allowed the insects to escape from potential predators and to traverse boundaries that might otherwise have been insurmountable—both very highly adaptive features. The numbers and diversity of insect species attest to the supreme success of this highly visible and dominant animal group.

Themes in Protostome Evolution

Most of protostome evolution took place in the oceans. As we have seen, early animals used fluids within their body cavities as the basis for support and movement. Subdivisions of the body cavity allowed better control of movement and permitted different parts of the body to be moved independently of one another. Thus some protostome lineages gradually evolved the ability to change their shape in complex ways and to move with greater speed on and through sediments or in the water.

During much of animal evolution, the only food available in the water consisted of dissolved organic matter and very small organisms. Consequently, many different lineages of animals evolved feeding structures designed to extract small prey from water, as well as structures for moving water through or over their prey-collecting devices. Animals that feed in this manner are abundant and widespread in marine waters today.

Because water flows readily, bringing food with it, sessile lifestyles evolved repeatedly during lophotrochozoan and ecdysozoan evolution. Most protostome phyla today have at least some sessile members. Sessile lifestyles have both advantages and disadvantages. A sessile animal gains access to local resources, but forfeits access to more distant resources. Sessile animals cannot come together to mate; instead, they must rely on the fertilization of gametes that they have ejected into the water. Some species eject both eggs and sperm into the water; others retain their eggs within their bodies and extrude only their sperm, which are carried by the water to other individuals. Species whose adults are sessile often have motile larvae, many of which have complicated mechanisms for locating suitable sites on which to settle. Many colonial sessile protostomes are able to grow in the direction of better resources or into sites offering better protection.

A frequent consequence of a sessile existence is competition for space. Such competition is intense among plants in most terrestrial environments. In the sea, especially in shallow waters, animals also compete directly for space. They have evolved mechanisms for overgrowing one another and for engaging in toxic warfare where they come into contact.

Individual members of sessile colonies, if they are directly connected, can share resources. The ability to share resources enables some individuals to specialize for particular functions, such as reproduction, defense, or feeding. The nonfeeding individuals derive their nutrition from their feeding associates.

Predation may have been the major selective pressure behind the development of external body coverings. Such coverings evolved independently in many lophotrochozoan and ecdysozoan lineages. In addition to providing protection, they became key elements in the development of new systems of locomotion. Locomotory abilities permitted prey to escape more readily from predators, but also allowed predators to pursue their prey more effectively. Thus, the evolution of animals has been, and continues to be, a complex arms race among predators and prey.

Although we have concentrated on the evolution of greater complexity in animal lineages, many lineages that remained simple have been very successful. Cnidarians are common in the oceans; roundworms abound in most aquatic and terrestrial environments. Parasites lost complex body plans but evolved complex life cycles.

The characteristics of the major existing phyla of protostomate animals are summarized in Table 32.1. All the phyla had evolved by the Cambrian period, but extinction and diversification within these lineages continue.

Many of the evolutionary trends demonstrated by protostomes also dominated the evolution of deuterostomes, the lineage that includes the chordates, the group to which humans belong. Hard external body coverings evolved and were later abandoned by many lineages. We will consider the evolution of the deuterostomes in the next chapter.

 *Chapter Summary*

▶ A major innovation during animal evolution was the development of a sturdy, nonliving external cover—an exoskeleton. An animal with an exoskeleton grows by periodically molting its exoskeleton and replacing it with a larger one.

▶ The presence of an exoskeleton opened avenues for the evolution of new body plans in the ecdysozoan lineage. **Review Figure 32.1**

Animals with Flexible Exoskeletons

▶ Tough cuticles are found in members of two phyla that live in freshwater, marine, and terrestrial environments.

▶ Roundworms (phylum Nematoda) are one of the most abundant and universally distributed of all animal groups. Many are parasites. **Review Figure 32.5**

Arthropods and Their Relatives: Segmented External Skeletons

▶ The body coverings of one ecdysozoan lineage, the arthropods, became thickened and made rigid by the incorporation of layers of protein and the polysaccharide chitin.

▶ Animals with rigid exoskeletons cannot move in a worm-like fashion. To move, they have appendages that can be manipulated by muscles. **Review Figure 32.6**

▶ Onychophorans have soft, fleshy, unjointed legs. They are probably similar to ancestral arthropods.

▶ The tiny and abundant water bears (phylum Tardigrada) also have unjointed legs.

▶ Jointed legs with specialized functions appeared among the trilobites (phylum Trilobita). Trilobites flourished in Cambrian and Ordovician seas, but became extinct by the close of the Paleozoic era.

Chelicerates: Invasion of the Land

▶ The bodies of all chelicerates (phylum Chelicerata) are divided into two major regions, the anterior of which bears four pairs of jointed legs.

▶ Arachnids—scorpions, harvestmen, spiders, mites, and ticks—are abundant in terrestrial environments.

Crustaceans: Diverse and Abundant

▶ Most of the 40,000 described species of crustaceans (phylum Crustacea) have a body that is divided into three regions: head, thorax, and abdomen. **Review Figure 32.12**

▶ The most familiar crustaceans are shrimps, lobsters, crayfishes, crabs, sow bugs, and sand fleas.

Uniramians are Primarily Terrestrial

▶ The body of a uniramian (phylum Uniramia) is divided into two or three regions; the posterior region has paired legs.

▶ Myriapods (centipedes and millipedes) have many segments and many pairs of legs.

▶ About 1.5 million species of insects (subphylum Insecta) have been described, but that is probably only a small fraction of the total number of species living on Earth today.

▶ Insects have three body regions (head, thorax, abdomen), a single pair of antennae on the head, and three pairs of legs attached to the thorax. **Review Figure 32.14**

▶ Wingless insects (class Apterygota) look like little adults when they hatch from their eggs. Hatchlings of many winged insects (class Pterygota) do not resemble adults and undergo substantial changes at each molt.

▶ Entomologists divide the winged insects into three major subgroups and about 28 different orders. Members of one subgroup cannot fold their wings back against the body; members of the other two groups can.

Themes in Protostome Evolution

▶ Most evolution of protostomes took place in the oceans.

▶ Early animals used fluid-filled spaces as hydrostatic skeletons. Subdivision of the body cavity allowed better control of movement and permitted different parts of the body to be moved independently of one another.

▶ Predation may have been the major selective pressure for the development of hard, external body coverings.

▶ During much of animal evolution, the only food in the water consisted of dissolved organic matter and very small organisms.

▶ Flowing water brings food with it, so many animals are sessile.

▶ All the phyla of protostomate animals had evolved by the Cambrian period.

For Discussion

1. Segmentation has arisen several times during animal evolution. What advantages does segmentation provide? Given these advantages, why do so many unsegmented animals survive?

2. Many animals extract food from the surrounding medium. What phyla contain animals that extract suspended food from the water column? What structures do these animals use to capture prey?

3. An animal that sheds its external skeleton in order to grow in size is virtually helpless during the time that its new, larger exoskeleton is hardening. Give some examples of how predators take advantage of this vulnerable stage of their prey. Include at least one example of predation by humans.

4. The British biologist J. B. S. Haldane is reputed to have quipped that "God was unusually fond of beetles." Beetles are, indeed, the most species-rich lineage of organisms. What features of beetles have contributed to the generation and survival of so many species?

5. In Part Three we pointed out that major structural novelties have arisen infrequently during the course of evolution. Which of the features of protostomes do you think are major evolutionary novelties? What criteria do you use to judge whether a feature is a major as opposed to a minor novelty?

6. A frequent consequence of sessile existence is competition for space. How do plants and animals differ in the ways in which they compete for space?

7. There are more described and named species of insects than of all other animals lineages combined. However, only a very few species of insects live in marine environments, and those species are restricted to the intertidal zone or the ocean surface. What factors may have contributed to the inability of insects to be successful in the oceans?

33 Deuterostomate Animals

THERE ARE ABOUT 25,000 SPECIES OF ray-finned fishes—more species than exist in all other vertebrate groups combined. Ray-finned fishes include almost all fish species with bony skeletons (as opposed to the sharks, whose skeletons are made of cartilage). Most ray-finned fishes have excellent color vision, and they use their brightly colored bodies to advertise their presence, species identity, and sex. Some go through dramatic color changes at different stages of their lives, and some are even able to change colors quickly when they are ready to mate, fight, or flee.

Part of the reason for the richness of ray-finned species may be that fishes are an ancient lineage that has had many millions of years in which to radiate in Earth's oceans and fresh waters. But part of the reason may be genetic. Most vertebrates have only four clusters of homeobox genes, but some ray-finned fishes have seven. The entire genome of these fishes was apparently duplicated about 300 million years ago, providing new opportunities for genetic variability that may have helped drive their explosive evolutionary radiation.

There are fewer major lineages and many fewer species among deuterostomes than among protostomes (Table 33.1), but we have a special interest in deuterostomes because we are members of that lineage. In this chapter, we first discuss some evolutionary themes shared by protostomes and deuterostomes, then describe and discuss the deuterostome phyla Echinodermata, Hemichordata, and Chordata, with special attention to the primate lineage of Chordata that gave rise to our own species.

Deuterostomes and Protostomes: Shared Evolutionary Themes

Deuterostome evolution paralleled protostome evolution in several important ways. Both lineages exploited the abundant food supplies buried in soft marine sediments, attached to rocks, or suspended in water. Because of the ease with which water can be moved, many groups in both lineages developed elaborate structures for moving water and extracting prey from it.

In lineages of both groups, the body became divided into compartments that allowed better control of shape and movement. Some members of both groups evolved mechanisms for controlling their buoyancy in water, using gas-filled internal spaces. Planktonic larval stages evolved in marine members of many protostome and deuterostome phyla; these all fed on tiny planktonic organisms while floating freely in the open water.

The ancestral traits shared by all members of the deuterostome lineage include indeterminate cleavage in the early embryo, a blastopore that becomes the anus, three body layers (they are triploblastic), formation of the mesoderm from an outpocketing of the embryonic gut, and a well-developed coelom (see Chapter 31). No fossils of ancestral deuterostomes that lived before the lineage split into two major lineages (echinoderms and chordates; Figure 33.1) have been found, so we can only deduce what they must have been like from these shared traits.

Both protostomes and deuterostomes colonized the land—the former via beaches, the latter via fresh water—but the consequences of these colonizations were very different. The jointed external skeletons of arthropods, although they provide excellent support and protection in air, cannot support large animals. The internal skeletons developed by deuterostomes are capable of supporting large bodies. The largest terrestrial deuterostomes to ever live were some of the dinosaurs; elephants are the largest living terrestrial animals.

Terrestrial deuterostomes recolonized aquatic environments a number of times. Suspension feeding re-evolved in several of these lineages. The largest living animals, the baleen (toothless) whales, feed on relatively small prey that they extract from the water with large straining structures in their mouths.

Two Colors, One Fish
The spotted puffer fish, *Arothron meleagris*, changes color during the course of its life cycle. The individuals shown here are in two different color phases of the cycle.

33.1 Summary of Living Members of the Kingdom Animalia[a]

PHYLUM	NUMBER OF LIVING SPECIES DESCRIBED	SUBGROUPS
Porifera: Sponges	10,000	
Cnidaria: Cnidarians	10,000	Hydrozoa: Hydras and hydroids Scyphozoa: Jellyfishes Anthozoa: Corals, sea anemones
Ctenophora: Comb jellies	100	
PROTOSTOMES		
Lophotrochozoans		
Platyhelminthes: Flatworms	20,000	Turbellaria: Free-living flatworms Trematoda: Flukes (all parasitic) Cestoda: Tapeworms (all parasitic) Monogenea (ectoparasites of fishes)
Rotifera: Rotifers	1,800	
Ectoprocta: Bryozoans	4,500	
Brachiopoda: Lamp shells	340	More than 26,000 fossil species described
Phoronida: Phoronids	20	
Pterobranchia: Pterobranchs	10	
Nemertea: Ribbon worms	900	
Annelida: Segmented worms	15,000	Polychaeta: Polychaetes (all marine) Oligochaeta: Earthworms, freshwater worms Hirudinea: Leeches
Mollusca: Mollusks	50,000	Monoplacophora: Monoplacophorans Polyplacophora: Chitons Bivalvia: Clams, oysters, mussels Gastropoda: Snails, slugs, limpets Cephalopoda: Squids, octopuses, nautiloids
Ecdysozoans		
Kinorhyncha: Kinorhynchs	150	
Chaetognatha: Arrow worms	100	
Nematoda: Roundworms	20,000	
Nematomorpha: Horsehair worms	230	
Onychophora: Onychophorans	80	
Tardigrada: Water bears	600	
Chelicerata: Chelicerates	70,000	Merostomata: Horseshoe crabs Arachnida: Scorpions, harvestmen, spiders, mites, ticks
Crustacea	50,000	Crabs, shrimps, lobsters, barnacles, copepods
Uniramia	1,500,000	Myriapoda: Millipedes, centipedes Insecta: Insects
DEUTEROSTOMES		
Echinodermata: Echinoderms	7,000	Crinoidea: Sea lilies, feather stars Ophiuroidea: Brittle stars Asteroidea: Sea stars Concentricycloidea: Sea daisies Echinoidea: Sea urchins Holothuroidea: Sea cucumbers
Hemichordata: Hemichordates	85	Acorn worms
Chordata: Chordates	50,000	Urochordata: Sea squirts Cephalochordata: Lancelets Agnatha: Lampreys, hagfishes Chondrichthyes: Cartilaginous fishes Osteichthyes: Bony fishes Amphibia: Amphibians Reptilia: Reptiles Aves: Birds Mammalia: Mammals

[a]Some small phyla are not included.

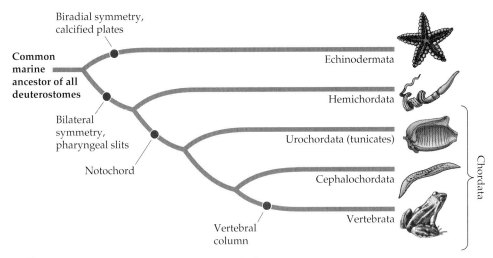

Biradial symmetry,
calcified plates

Common
marine
ancestor of all
deuterostomes

Bilateral
symmetry,
pharyngeal slits

Notochord

Echinodermata

Hemichordata

Urochordata (tunicates)

Cephalochordata

Vertebrata

Chordata

Vertebral
column

33.1 A Probable Deuterostomate Phylogeny
There are fewer major lineages and many fewer species
of deuterostomes than of protostomes.

Echinoderms: Complex Biradial Symmetry

The ancestors of one deuterostome lineage, the **echinoderms**
(phylum **Echinodermata**), were probably sluggish animals.
They evolved into more aggressive and active forms as a result of two major
structural features.
One is a system
of calcified internal
plates covered by thin
layers of skin and some
muscles. The calcified plates of
early echinoderm ancestors became
enlarged and thickened until they fused
inside the entire body, giving rise to an internal skeleton.

Echinodermata
Hemichordata
Urochordata (tunicates)
Cephalochordata
Vertebrata

The other major innovation
was the evolution of a **water
vascular system**, a network of
calcified hydraulic canals leading to extensions called **tube
feet**. The water vascular system
functions in gas exchange, locomotion, and feeding (Figure
33.2*a*). Seawater enters the water
vascular system through a perforated *sieve plate*. A calcified
canal leads from the sieve plate
to another canal that rings the
esophagus. Other canals radiate
from this *ring canal* extending
through the arms (in species that
have arms) and connecting with
the tube feet. The development of these two structural innovations—calcified internal skeleton and water vascular
system—resulted in one of the most striking of evolutionary radiations.

Echinoderms have an extensive fossil record. About 23
classes have been described, of which only 6 survive today.
About 7,000 species of echinoderms exist today, but 13,000
species—probably only a small fraction of those that actually lived—have been described from their fossil remains.
Nearly all living species have a bilaterally symmetrical, ciliated larva that feeds for some time as a planktonic organism before settling and transforming into a biradially symmetrical adult (Figure 33.2*b*).

The living echinoderms are divided into two lineages:
Pelmatozoa and **Eleutherozoa**. The two lineages differ in
the number of arms they have and the form of their water
vascular systems. Pelmatozoa consists only of the crinoids,
whereas several groups are included in the Eleutherozoa.

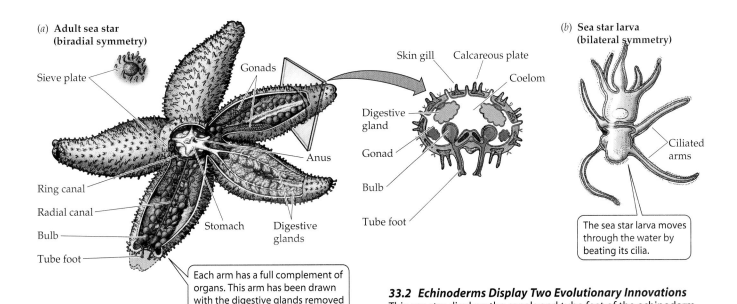

(*a*) **Adult sea star
(biradial symmetry)**

Sieve plate

Gonads

Skin gill   Calcareous plate

Coelom

Digestive
gland

Gonad

Bulb

Tube foot

Anus

Ring canal

Radial canal

Bulb

Tube foot

Stomach   Digestive
glands

Each arm has a full complement of
organs. This arm has been drawn
with the digestive glands removed
to show the organs lying below.

(*b*) **Sea star larva
(bilateral symmetry)**

Ciliated
arms

The sea star larva moves
through the water by
beating its cilia.

33.2 Echinoderms Display Two Evolutionary Innovations
This sea star displays the canals and tube feet of the echinoderm
water vascular system, as well as a calcified internal skeleton.

(a) *Oxycomanthus bennetti*

(b) *Strongylocentrotus purpuratus*

(c) *Bohadschia argus*

(d) *Henricia leviuscula*

(e) *Opiothrix suemsonii*

33.3 Diversity among the Echinoderms
(a) The flexible arms of the golden feather star are clearly visible. (b) Purple sea urchins are important grazers of algae in the intertidal zone of the Pacific Coast of North America. (c) This sea cucumber lives on rocky substrates in seas around Papua New Guinea. (d) The blood sea star is typical of many sea stars; some species, however, have more than five arms. (e) This brittle star is resting on a sponge.

Pelmatozoans have jointed arms

Sea lilies and feather stars (class **Crinoidea**) are the only surviving pelmatozoans. Sea lilies were abundant 300–500 mya, but only about 80 species survive today. Most sea lilies attach to a substratum by means of a flexible stalk consisting of a stack of calcareous discs. The main body of the animal is a cup-shaped structure that contains a tubular digestive system. Five to several hundred arms, usually in multiples of five, extend outward from the cup. The jointed calcareous plates of the arms enable them to bend. A groove runs down the center of each arm to the mouth. On both sides of the groove are tube feet covered with mucus-secreting glands.

A sea lily feeds by orienting its arms in passing water currents. Food particles strike and stick to the tube feet, which transfer the particles to the grooves in the arms, where the action of cilia carries the food to the mouth. The tube feet of sea lilies are also used for gas exchange and elimination of nitrogenous wastes.

Feather stars are similar to sea lilies, but they have flexible appendages with which they grasp the substratum while they are feeding and resting (Figure 33.3a). Feather stars feed in much the same manner as sea lilies. They can walk on the tips of their arms or swim by rhythmically beating their arms. About 600 living species of feather stars have been described.

Eleutherozoans are the dominant echinoderms

Most surviving echinoderms are members of the eleutherozoan lineage. Biochemical data suggest that the ancestors of sea urchins and sand dollars (class **Echinoidea**) were the first to split off from the lineage leading to the other eleutherozoans. Sea urchins and sand dollars lack arms, but they share a five-part body plan with all other echinoderms. Sea urchins are hemispherical animals that are covered with spines attached to the underlying skeleton via ball-and-socket joints (Figure 33.3*b*). The spines of sea urchins come in varied sizes and shapes; a few produce highly toxic substances. Many sea urchins consume algae, which they scrape from the rocks with a complex rasping structure. Others feed on small organic debris that they collect with their tube feet or spines. Sand dollars, which are flattened and disc-shaped, feed on algae and fragments of organic matter on the seafloor.

The tube feet of sea cucumbers (class **Holothuroidea**; Figure 33.3*c*) are used primarily for attaching to the substratum rather than for moving. The anterior tube feet are modified into large, feathery, sticky tentacles that can be protruded around the mouth. Periodically, a sea cucumber withdraws the tentacles into its mouth, wipes off the material that has adhered to them, and digests it.

Sea daisies (class **Concentricycloidea**) were not discovered until 1986. Little is known about them. They have tiny disc-shaped bodies with a ring of marginal spines, and two ring canals, but no arms. Sea daisies are found on rotting wood in ocean waters. They apparently feed on prokaryotes, which they digest outside their bodies and absorb either through a membrane that covers the oral surface or via a shallow saclike stomach.

The most familiar echinoderms are the sea stars (class **Asteroidea**; Figure 33.3*d*; see also Figure 33.2). Their tube feet serve as organs of locomotion and, because their walls are thin, they are important sites for gas exchange. Each tube foot of a sea star is also an adhesive organ, consisting of an internal bulb connected by a muscular tube to an external sucker. A tube foot is moved by expansion and contraction of the circular and longitudinal muscles of the tube. It can adhere to a surface by secreting a sticky substance around the sucker.

Many sea stars prey on polychaetes, gastropods, bivalves, and fishes. They are important predators in many marine environments, such as coral reefs and rocky intertidal zones. With hundreds of tube feet acting simultaneously, a sea star can exert an enormous and continuous force. It can grasp a clam in its arms, anchor the arms with its tube feet, and, by steady contraction of the muscles in the arms, gradually exhaust the muscles with which the clam keeps its shell closed. Sea stars that feed on bivalves are able to push the stomach out through the mouth and

then through the narrow space between the two halves of the shell. The stomach secretes digestive enzymes into the soft parts of the bivalve, digesting it.

Brittle stars (class **Ophiuroidea**) are similar in structure to sea stars, but their flexible arms are composed of jointed hard plates (Figure 33.3*e*). Brittle stars generally have five arms, but each arm may divide a number of times. Most of the 2,000 species of brittle stars ingest particles from the surfaces of sediments and assimilate the organic material from them, but some species remove suspended food particles from the water; others capture small animals. They eject the indigestible particles through their mouths because, unlike most other echinoderms, brittle stars have only one opening to their digestive tract.

Chordates: New Ways of Feeding

The second major lineage of deuterostomes, the phylum Chordata, evolved several different modifications of the coelomic cavity that provided new ways of capturing and handling food. Some living representatives of one early lineage—acorn worms—live buried in marine sand or mud, under rocks, or attached to algae. They may be similar to the ancestors of the chordate lineage, but are currently classed in their own lineage as hemichordates ("half-chordates"). Animals in the chordate lineage evolved a strikingly different body plan from the acorn worms, characterized by an internal dorsal supporting structure, which in the vertebrates evolved into the spinal column.

Acorn worms capture prey with a proboscis

The **acorn worms** (phylum **Hemichordata**) have a three-part body consisting of a proboscis, collar, and trunk (Figure 33.4). The 70 species of acorn worms live in burrows in muddy and sandy sediments.

Echinodermata
Hemichordata
Urochordata (tunicates)
Cephalochordata
Vertebrata

The large proboscis of acorn worms is a digging organ. It is coated with a sticky mucus that traps prey items in the sediment. The mucus and its attached prey are conveyed by cilia to the mouth. In the esophagus, the food-laden mucus is compacted into a ropelike mass that is moved through the digestive tract by ciliary action. Behind the mouth is a **pharynx** that opens to the outside through a number of **pharyngeal slits** through which water can exit. Highly vascularized tissue surrounding the pharyngeal slits serves as a gas exchange apparatus. An acorn worm breathes by pumping water into its mouth and out through its pharyngeal slits.

The pharynx becomes a feeding device

The same property required for effective gas exchange—a large surface area—also serves well for capturing prey. The pharyngeal slits, which originally functioned as sites for

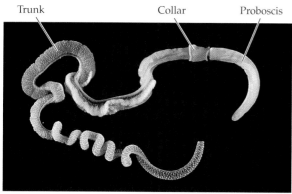

Trunk   Collar   Proboscis

Saccoglossus kowaleskii

33.4 A Hemichordate
The proboscis (right) of this acorn worm is modified for digging. This individual has been extracted from its burrow.

33.5 Tunicates
Pharyngeal baskets occupy most of the body cavities of these transparent sea squirts. The blue color is a reflection of the environment in this photograph.

gas exchange and eliminating water, as they do in modern acorn worms, were enlarged in a sister lineage. This enlargement of the pharyngeal slits eventually led to the remarkable evolutionary developments that gave rise to the chordates.

Chordates (phylum **Chordata**) are bilaterally symmetrical animals whose body plans are characterized by several shared features:

▶ Pharyngeal slits (at some stage of their development).

▶ A dorsal, hollow *nerve cord.*

▶ A ventral heart.

▶ A tail that extends beyond the anus.

▶ A dorsal supporting rod, the **notochord**.

The notochord is the most important derived trait of the Chordata and is unique to that phylum. In some species, such as tunicates, the notochord is lost during metamorphosis to the adult stage. In the vertebrates, it is replaced by skeletal structures that provide support for the body.

A notochord appears in tunicates and lancelets

The **tunicates** (subphylum **Urochordata**) may be similar to the ancestors of all chordates. All 2,500 species of tunicates are marine animals, most of which are sessile as adults. It is their swimming, tadpole-like larvae that reveal the close evolutionary relationships between tunicates and other chordates.

A tunicate larva has pharyngeal slits, a dorsal, hollow nerve cord, and a notochord. Muscles are attached to the notochord, providing the body with relatively rigid support. After a short time floating in the water, the larva settles on the seafloor and becomes a sessile adult. The nerve cord and notochord disappear in the adult animal, which

feeds by extracting plankton from the water. An adult's pharynx is enlarged into a *pharyngeal basket* lined with cilia, whose beating moves water through the animal.

More than 90 percent of known species of tunicates are sea squirts (class Ascidiacea). Some sea squirts are solitary,

Echinodermata
Hemichordata
Urochordata (tunicates)
Cephalochordata
Vertebrata

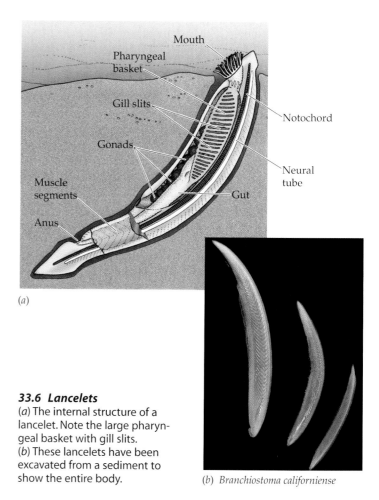

Mouth
Pharyngeal basket
Gill slits
Notochord
Gonads
Neural tube
Muscle segments
Gut
Anus

(a)

33.6 Lancelets
(*a*) The internal structure of a lancelet. Note the large pharyngeal basket with gill slits.
(*b*) These lancelets have been excavated from a sediment to show the entire body.

(b) Branchiostoma californiense

but others produce colonies by asexual budding from a single founder. Individual sea squirts range in size from less than 1 mm to 60 cm in length, but colonies may measure several meters across. The baglike body of an adult is surrounded by a tough tunic, composed of protein and a complex polysaccharide, which is secreted by the epidermal cells. Much of the body is occupied by the large pharyngeal basket (Figure 33.5).

The 25 species of **lancelets** (subphylum **Cephalochordata**) are small animals that rarely exceed 5 cm in length. Their notochord extends the entire length of the body throughout their lives, and they resemble small fishes. Lancelets live partly buried in soft marine sediments. They extract small prey from the water with their pharyngeal baskets (Figure 33.6).

Origin of the Vertebrates

In one chordate lineage, the pharyngeal basket became enlarged. With its many exit openings, an enlarged basket was effective in extracting prey from mud, which contains many inedible particles along with food. This lineage gave rise to the **vertebrates** (subphylum **Vertebrata**). In the late Cambrian period, these early vertebrates evolved improved structures for extracting food from mud and sand and for moving over the surface of the substratum.

Vertebrates take their name from a jointed, dorsal **vertebral column**, which replaced the notochord as their primary support. The vertebrate body plan (Figure 33.7) can be characterized as follows:

▶ With the vertebral column as its anchor, a rigid *internal skeleton* provides support and mobility.

▶ Two pairs of *appendages* are attached to the vertebral column.

▶ The faster locomotion made possible by appendages favored the evolution of an *anterior skull with a large brain* and highly developed sensory receptors.

▶ The internal organs are suspended in a large coelom.

▶ A well-developed *circulatory system*, driven by contractions of a ventral heart, delivers oxygen to internal organs.

The filter-feeding ancestral vertebrates lacked jaws. They probably swam over the sediments, sucking up mud and extracting microscopic food from it. These animals gave rise to the **fishes**. The lineage leading to modern hagfishes probably separated first from

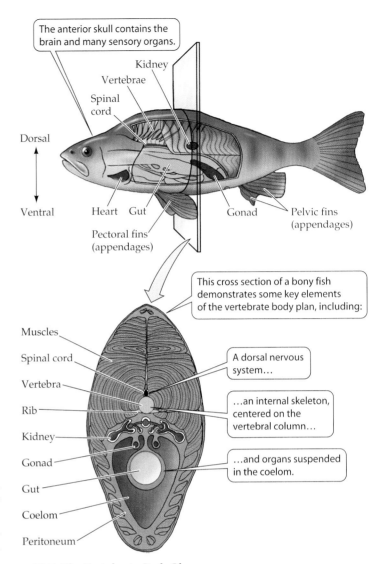

33.7 The Vertebrate Body Plan
A bony fish is used here to illustrate the structural elements common to all vertebrates.

the other groups (Figure 33.8). One early group of jawless fishes, called ostracoderms, meaning "shell-skinned," evolved a bony external armor that protected them from predators. With their heavy armor, these small fishes could swim only slowly, but they could safely swim above the substratum, which was easier than having to burrow through it, as all previous sediment feeders had done.

The new mobility of jawless fishes enabled them to exploit their environments in new ways. They could attach to dead organisms and use the pharynx to create suction to pull fluids and partly decomposed tissues into the mouth. Hagfishes and lampreys, the only jawless fishes to survive beyond the Devonian period, feed on both dead and living organisms in this way (Figure 33.9). These fishes have tough, scaly skins instead of external armor. The round mouth is a sucking organ with which the animals attach to their prey and rasp at the flesh. Lampreys live in both fresh and salt water; many species move between the two environments, laying their eggs in rivers and maturing in the sea.

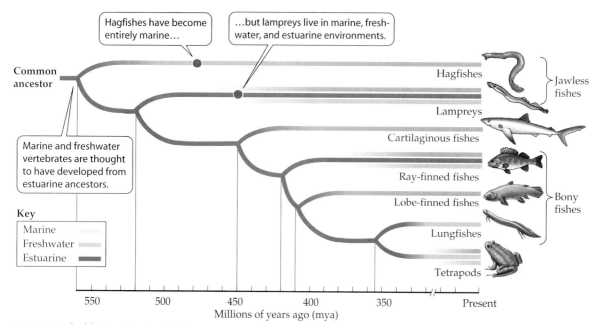

33.8 A Probable Vertebrate Phylogeny
This phylogeny incorporates the view that vertebrates evolved in estuaries, where their ability to handle varying salinities allowed them to exploit habitats not available to marine animals.

Jaws improve nutrition

During the Devonian period, many new kinds of fishes evolved in the seas, estuaries, and fresh waters. Although most of these were jawless, members of one lineage evolved jaws from some of the skeletal arches that supported the gill region (Figure 33.10). A jaw allows a fish to grasp and subdue relatively large, living prey. Further development of jaws and teeth among fishes led to the ability to chew both soft and hard body parts of prey. Chewing aided chemical digestion and improved the ability of fishes to obtain nutrients from prey.

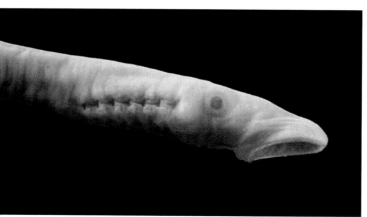

Petromyzon marinus

33.9 A Modern Jawless Fish
This sea lamprey uses its large, jawless mouth to suck blood and flesh from other fishes.

The dominant early jawed fishes were the heavily armored **placoderms** (class **Placodermi**). Some of these fishes evolved elaborate appendages and relatively sleek body forms that improved their ability to maneuver in open water. A few became huge (10 meters long) and, together with squids (cephalopod mollusks), were probably the major predators in the Devonian oceans. Despite their early abundance, however, most placoderms disappeared by the end of the Devonian period; none survived to the end of the Paleozoic era.

Fins improve mobility

Two other groups of fishes—the bony fishes and the cartilaginous fishes, both of which survive today—became abundant during the Devonian period.

   Cartilaginous fishes (class **Chondrichthyes**)—the sharks, skates and rays, and chimaeras (Figure 33.11)—have a skeleton composed entirely of a firm but pliable material called *cartilage*. Their skin is flexible and leathery, sometimes bearing bristly projections that give it the consistency of sandpaper. The loss of external armor increased their mobility and ability to escape from predators.

   In the cartilaginous fishes and their descendants, swimming is controlled by pairs of unjointed appendages called *fins*: a pair of pectoral fins just behind the gill slits and a pair of pelvic fins just in front of the anal region. A dorsal median fin stabilizes the fish as it moves. Sharks move forward by means of their tail and pelvic fins. Skates and rays propel themselves by means of the undulating movements of their greatly enlarged pectoral fins.

Jawless fishes (agnathans)
Extinct and living forms

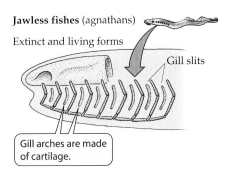

Gill slits

Gill arches are made of cartilage.

Early jawed fishes (placoderms)

Extinct

Gill slits

Some anterior gill arches became modified to form jaws.

Modern jawed fishes (cartilaginous and bony fishes)

Living forms

Additional gill arches were incorporated to form heavier, more efficient jaws.

33.10 Jaws from Gill Arches
This illustrates one probable scenario for the evolution of jaws from the anterior gill arches of fishes.

Most sharks are predators, but some feed by filtering plankton from the water. The world's largest fish, the whale shark (*Rhincodon typhus*), is a filter feeder. It may grow to more than 15 meters in length and weigh more than 9,000 kilograms. Most skates and rays live on the ocean floor, where they feed on mollusks and other invertebrates buried in the sediments. Nearly all cartilaginous fishes live in the oceans.

Swim bladders allow control of buoyancy

The **bony fishes** (class **Osteichthyes**) have internal skeletons of bone rather than cartilage, giving them their common name. Their bony skeleton is lighter than that of the cartilaginous fishes. In most species, the outer surface is covered with flat, smooth, thin, lightweight scales that provide some protection. The gills of bony fishes open into a single chamber covered by a hard flap. Movement of the flap improves the flow of water over the gills, where gas exchange takes place.

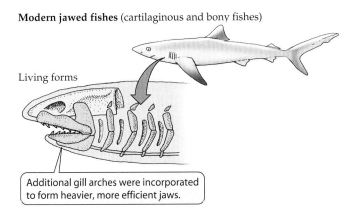

Hagfishes

Lampreys

Cartilaginous fishes

Ray-finned fishes

Lobe-finned fishes

Lungfishes

Tetrapods

Early bony fishes also evolved gas-filled sacs that supplemented the action of the gills in respiration. These features enabled early bony fishes to live where oxygen was periodically in short supply, as it often is in estuarine and freshwater environments. They still serve this function in lungfishes and a few other modern fishes, but in the ray-finned fishes—a group that includes most of the many species of bony fish—these lunglike sacs evolved into **swim bladders,**

33.11 Cartilaginous Fishes
(*a*) Most sharks, such as this whitetip reef shark, are active marine predators. (*b*) Skates and rays, represented here by a stingray, feed on the ocean bottom. Their modified pectoral fins are used for propulsion.

(*a*) *Triaenodon obesus*

(*b*) *Trygon pastinaca*

(a) *Ocyurus chrysurus*

(b) *Plectorhinchus chaetodonoides*

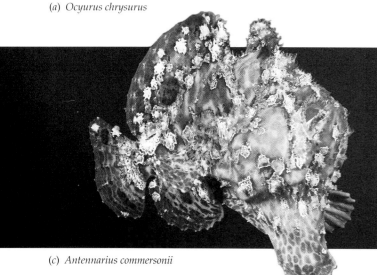

(c) *Antennarius commersonii*

(d) *Phyllopteryx taeniolatus*

33.12 Diversity among Bony Fishes
(a) The yellowtail snapper has a "typical" fish shape. (b) The coral grouper lives on tropical coral reefs. (c) Commerson's frogfish can change its color in a range from pale yellow to orange-brown to deep red, thus enhancing its camouflage abilities. (d) This weedy sea dragon is difficult to see when it hides in vegetation. It is a larger "cousin" of the more familiar seahorse.

which serve as organs of buoyancy. By adjusting the amount of gas in its swim bladder, a fish can control the depth at which it is suspended in the water without expending energy.

With their lighter skeletons and protective coverings and their swim bladders, ray-finned bony fishes evolved a remarkable diversity of sizes, shapes, and lifestyles (Figure 33.12). The smallest are less than 1 cm long as adults; the largest are ocean sunfishes that weigh up to 900 kilograms. Ray-finned fishes exploit nearly all types of aquatic food sources. In the oceans they filter plankton from the water, rasp algae from rocks, eat corals and other colonial invertebrates, dig invertebrates from soft sediments, and prey upon virtually all other vertebrates except large whales and dolphins. In fresh water they eat plankton, devour insects

of all aquatic orders, eat fruits that fall into the water in flooded forests, and prey on other aquatic vertebrates.

Some fishes live buried in soft sediments, capturing passing prey or emerging at night to feed. Many are solitary, but in open water others form large aggregations called *schools*. Many fishes perform complicated behaviors by means of which they maintain schools, build nests, court and choose mates, and care for their young.

With their fins and swim bladders, fishes can readily control their positions in open water, but their eggs tend to sink. Therefore, most fishes attach their eggs to the substratum, although a few species discharge their small eggs directly into surface waters where they are buoyant enough to complete their development before they sink very far. Most marine fishes move to food-rich shallow waters to lay their eggs, which is why coastal waters and estuaries are so important in the life cycles of many species. Some, such as salmon, abandon salt water when they breed, ascending rivers to spawn in freshwater streams and lakes.

Colonizing the Land: Obtaining Oxygen from the Air

Although the evolution of lunglike sacs was a response to the inadequacy of gills for respiration in oxygen-poor waters, it also set the stage for the invasion of land. Some early bony fishes probably used their lung sacs to supplement their gills when oxygen levels in the water were low. This ability would also have allowed them to breathe air, and to leave the water temporarily when pursued by predators unable to do so. But with their unjointed fins, bony fishes could only flop around on land, as most fishes do today if placed out of water. Changes in the structure of fins would help such fishes move about on land.

Two lineages of bony fishes evolved jointed fins: the **lobe-finned fishes** (subclass **Crossopterygii**) and the lung-fishes. The lobe-fins flourished from the Devonian period until about 25 mya, when they were thought to have become extinct. However, in 1938, a lobe-fin was caught by commercial fishermen off the Comoro Islands in the Indian Ocean. Since that time, several dozen specimens of this extraordinary fish, *Latimeria chalumnae*, have been collected. *Latimeria*, a predator on other fishes, reaches a length of about 1.5 meters and weighs up to 82 kilograms (Figure 33.13). Although it belongs to the bony fish lineage, the skeleton of *Latimeria* is composed mostly of cartilage, not bone.

Some descendants of early fishes with jointed fins began to use terrestrial food sources and over time became more fully adapted to life on land. This lineage became the **tetrapods**: the four-legged amphibians, reptiles, birds, and mammals that are common today.

Amphibians invade the land

During the Devonian period, **amphibians** (class **Amphibia**) arose from ancestors they shared with the lungfishes. In this lineage, the stubby, jointed fins of their ancestors evolved into walking legs. The design of those legs has

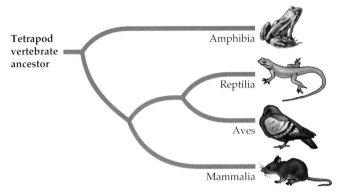

33.14 A Probable Phylogeny of Tetrapods
In the birds (Class Aves), the paired forelimbs evolved into wings.

remained largely unchanged throughout the evolution of terrestrial vertebrates (Figure 33.14).

The Devonian predecessors of amphibians were probably able to crawl from one pond or stream to another by pulling themselves along on their finlike legs, as do some modern species of catfishes. They gradually evolved to be able to live on swampy land and, eventually, on dry land. Living amphibians have relatively small lungs, and most species exchange gases through their skins. Most terrestrial species are confined to moist environments because they lose water rapidly through their skins when exposed to dry air.

About 4,500 species of amphibians live on Earth today, many fewer than the number known only from fossils. Living amphibians belong to three orders (Figure 33.15): the wormlike, tropical, burrowing caecilians (order Gymnophiona); frogs and toads (order Anura, which means "tailless"); and salamanders (order Urodela, which means "tailed"). Most species of frogs and toads live in tropical and warm temperate regions, although a few are found at very high latitudes and altitudes. Salamanders are more diverse in temperate regions, but many species are found in cool, moist environments in the mountains of Central America.

Most species of amphibians live in water at some time in their lives. In the typical amphibian life cycle, part or all of the adult stage is spent on land, usually in a moist habitat, but adults return to fresh water to lay their eggs (Figure 33.16). An amphibian egg can survive only in a moist environment because it is enclosed by a delicate envelope that cannot prevent water loss in dry conditions. The fertilized eggs of most species give rise to larvae that live in water until they change into terrestrial adults.

Amphibians are the focus of much attention today because populations of many species are declining rapidly. The golden toad, for example, has disappeared from the Monteverde Cloud Forest Reserve in Costa Rica, which was established primarily to protect this rare species. Several possible reasons for the declines, including drought, in-

33.13 A Modern Lobe-Fin
Latimeria chalumnae, found in deep waters of the Indian Ocean, is the sole surviving species of its lineage, which had been thought to be extinct.

(a) *Dermophus mexicanus*

(b) *Scaphiophryne gottlebei*

33.15 Diversity among the Amphibians
(a) Burrowing caecilians superficially look more like worms than amphibians.
(b) A rare frog species discovered in a national park on the island of Madagascar.
(c) A European fire salamander.

(c) *Salamandra salamandra*

Amniotes colonize dry environments

Most amphibians, as we have just seen, are limited to moist environments. Two morphological changes contributed to the ability of one lineage of vertebrates to control water loss and, therefore, to exploit a wider range of terrestrial habitats. One was the evolution of an egg with a shell that is relatively impermeable to water. The other was a combination of traits that reduced water loss, including a tough skin impermeable to water and kidneys that could excrete concentrated urine. The vertebrates that evolved both of these traits are called **amniotes**. They were the first vertebrates to become widely distributed over the terrestrial surface of Earth.

creased ultraviolet radiation, and diseases, have been identified. Biologists are monitoring amphibian populations closely to learn more about the causes of their difficulties and to determine the implications of their declines for other organisms, including humans.

The amniote egg has a leathery or brittle calcium-impregnated shell that retards evaporation of the fluids inside, but permits O_2 and CO_2 to pass through. Such an egg does not require a moist environment, but can be laid anywhere. Within the shell and surrounding the embryo are membranes that protect the embryo from desiccation and assist its respiration and excretion of waste nitrogen. The egg also stores large quantities of food as yolk, permitting the embryo to attain a relatively advanced state of development before it hatches and must feed itself (Figure 33.17).

An early amniote lineage, the **reptiles** (class **Reptilia**) arose from the tetrapods during the Carboniferous period. As we discussed in Chapter 23, Rep-

START

Adult

9 The frog respires with lungs; its tail is resorbed.

1 Adults spawn in water.

Sperm

♂   ♀

Eggs

2 The fertilized egg develops in water.

3 The embryo develops.

4 A tailbud forms.

5 The egg hatches.

6 The larva (tadpole) respires with external gills.

7 The tadpole respires with internal gills; hind legs appear.

8 Front legs appear.

33.16 In and Out of the Water
Most stages in the life cycle of temperate-zone frogs take place in water. The aquatic tadpole is transformed into a terrestrial adult through metamorphosis.

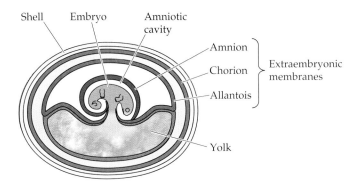

33.17 An Egg for Dry Places
The evolution of the amniote egg, with its shell, three extraembryonic membranes and embryo-nourishing yolk, was a major step in the colonization of the terrestrial environment.

tilia, as we use the term here, is a paraphyletic group because it does not include the birds, a major lineage that split off relatively recently during reptilian evolution. (Figure 33.18). However, because all reptiles are structurally similar, they serve as a convenient example for discussing the characteristics of amniotes. Therefore, we use this traditional classification as a basis for our discussion while recognizing that the birds should technically be included within it.

About 6,000 species of reptiles live today. Most reptiles do not care for their eggs after laying them. In some species the eggs do not develop shells, but are retained inside the female's body until they hatch. Still other species evolved structures called **placentas** that nourish the developing embryos.

The skin of a reptile is covered with horny scales that greatly reduce loss of water from the body surface. These scales, however, make the skin unavailable as an organ of gas exchange. In reptiles, gases are exchanged almost entirely by the lungs, which are proportionally much larger in surface area than those of amphibians. A reptile forces air into and out of its lungs by bellows-like movements of its ribs. Unlike the amphibian heart, the reptilian heart is divided into chambers that partially separate oxygenated from unoxygenated blood. With this type of heart, reptiles can generate higher blood pressures than amphibians and can sustain higher levels of muscular activity.

Reptilian lineages diverge

The lineages leading to modern reptiles began to diverge about 250 mya when the ancestors of the subclass **Squamata** (lizards, snakes, and amphisbaenians—a group of legless, wormlike, burrowing animals with greatly reduced eyes) diverged from the lineage leading to all other reptiles (Figure 33.19a, b). Most lizards are insectivores, but some are herbivores, and a few prey upon other vertebrates. The largest lizards, which may grow as long as 3 meters, are some species of monitors that live in the East Indies. Most lizards walk on four limbs, but some are limbless, as are all snakes, which are descendants of burrowing lizards.

All snakes are carnivores, and many can swallow objects much larger in diameter than themselves. The combination of poison glands and the ability to inject venom rapidly into their prey evolved several times. The largest snakes are pythons more than 10 meters long.

The tuataras (subclass **Sphenodontida**), which today are represented by only two species restricted to a few islands off the coast of New Zealand (Figure 33.19c), superficially resemble lizards, but differ from them in several internal anatomical features. Their phylogenetic relationships are uncertain.

Considerable uncertainty surrounds the next lineage split. Traditionally turtles were thought to have separated from other reptiles early in the history of the group, but new molecular analyses suggest that turtles are closely related to crocodilians, which diverged later. Both turtles and ancestral crocodiles have dorsal and ventral armored plates, and such plates have characterized those groups for many millions of years.

The dorsal and ventral bony plates of modern turtles and tortoises (subclass **Chelonia**) form a shell into which the head and limbs can be withdrawn (Fig-

33.18 The Reptiles Form a Paraphyletic Group
The traditional classification of the reptiles creates the paraphyletic group Reptilia. As used here, Reptilia does not include the birds (Aves), even though this major lineage split off from the crocodilian reptiles relatively recently (in evolutionary terms).

(a) *Trimeresurus sumatranus*

(d) *Chelonia mydas*

(b) *Ocyurus chrysurus*

(e) *Alligator mississippiensis*

(c) *Sphenodon punctatus*

Figure 33.19 Reptilian Diversity
(a) This Sumatran pit viper is prepared to strike. (b) This African chameleon, a lizard, has a long tail with which it can grasp branches and large eyes that move independently in their sockets. (c) The tuatara looks like a typical lizard, but it is one of only two survivors of a lineage that separated from lizards long ago. (d) The green sea turtle is widely distributed in tropical oceans. (e) Most crocodilians are tropical; alligators live in warm temperate environments in China and, like this one, in the southeastern United States.

ure 33.19*d*). Most turtles live in lakes and ponds, but tortoises are terrestrial; some live in deserts. Sea turtles spend their entire lives at sea except when they come ashore to lay eggs; all seven species are endangered. A few species of turtles and tortoises are carnivores, but most species are omnivores that eat a variety of aquatic and terrestrial plants and animals.

The crocodilians (subclass **Crocodylia**)—crocodiles, caimans, gharials, and alligators—are confined to tropical and warm temperate environments (Figure 33.19*e*). Crocodilians spend much of their time in water, but they build nests on land or on floating piles of vegetation. Their eggs are warmed by heat generated by the decay of organic matter that they place in the nest. Typically the eggs are guarded by the female until they hatch. All crocodilians are carnivorous; they prey on vertebrates of all classes, including large mammals.

Another lineage led to the *dinosaurs*, reptiles that rose to dominance about 215 mya and dominated terrestrial environments for about 150 million years. During this time, virtually all terrestrial animals more than 1 meter in length were dinosaurs. Some of the largest dinosaurs weighed up to 100 tons. Many were agile and could run rapidly. Some small predatory dinosaurs evolved feathers.

The ability to breathe and run simultaneously, which we take for granted, was a major innovation in the evolution of terrestrial vertebrates. Not until the evolution of the lineages leading to the mammals, dinosaurs, and birds did the legs assume vertical positions, which reduced the lateral forces on the body during locomotion. Special ventilatory muscles that enabled the lungs to be filled and emptied while the limbs moved also evolved. These muscles are visible in living birds and mammals; we can infer their existence in dinosaurs from the structure of the vertebral column in their fossils and the capacity of many dinosaurs for bounding bipedal (using two legs) locomotion.

Birds: Feathers and Flight

During the Mesozoic era, a dinosaur lineage gave rise to the birds (subclass **Aves**). The oldest known avian fossil, *Archaeopteryx*, which lived about 150 mya, was covered with feathers that are virtually identical to those of modern birds. It also had well-developed wings, a long tail (Figure 33.20*a*), and a wishbone, which in modern birds serves as an anchoring site for flight muscles. *Archaeopteryx* had typical perching bird claws, suggesting that it lived in trees and shrubs and used the clawed fingers on its forearms to assist it in clambering over branches.

Another early bird, *Confuciusornis sanctus*, which is known from hundreds of complete fossils from China, lived only slightly more recently than *Archaeopteryx*. Well-preserved fossils show that males had greatly elongated tail feathers (Figure 33.20*b*), which they probably used in communal courtship displays. Large numbers of these fossils have been found together, as would be expected if a number of males displayed together on communal display grounds.

Because the avian lineage separated from the other reptiles long before *Archaeopteryx* lived, existing data are insufficient to identify the ancestors of birds with certainty. Most paleontologists (scientists who study fossils) believe that birds evolved from terrestrial bipedal dinosaurs that used their forelimbs for capturing prey. According to this view, these small dinosaurs evolved feathers for insulation or display, and eventually were able to become airborne for short distances.

During the Cretaceous period, birds underwent an extensive evolutionary radiation. The dominant Cretaceous lineage was the "opposite birds," so named because the tarsal bones of their legs fused in the opposite direction from the way fusion happens in all modern birds. All lineages of opposite birds died out at the end of the Cretaceous,

(a)

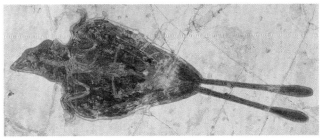

(b)

33.20 Mesozoic Birds
(*a*) An artist's recreation of *Archaeopteryx* shows its modern feathers, arboreal habits, and flight. (*b*) The elongated tail feathers of a male *Confuciusornis sanctus* ("sacred bird of Confucius") fossil suggest that the males used them in courtship displays.

but scientists disagree over how many other avian lineages survived the mass extinction. Paleontologists believe that members of only one lineage, collectively known as the transitional shorebirds, survived the end of the Cretaceous, because no later fossils of other lineages have been found. Other scientists, who base their conclusions on molecular clocks, believe that at least some representatives of many Cretaceous avian lineages must have survived, because modern birds are too different to have come from a single lineage.

(a) *Pygoscelis papua*

33.21 Diversity among the Birds
(a) Penguins such as these gentoos are widespread in the cold waters of the southern hemisphere. They are expert swimmers, although they have lost the ability to fly. (b) Parrots are a diverse group of birds, especially in the tropics of Asia and the Pacific islands. This Australian king parrot is one member of Australia's rich parrot fauna. (c) Perching birds, represented here by a male northern cardinal, are the most species-rich of all the bird lineages.

(c) *Cardinalis cardinalis*

(b) *Alisterus scapularis*

As a group, birds eat almost all types of animal and plant material. A few aquatic species have bills modified for filtering small food particles from the water. Insects are the most important dietary item for terrestrial birds. In addition, they eat fruits and seeds, nectar and pollen, leaves and buds, carrion, and other vertebrates. Birds are major predators of flying insects during the day, and some species exploit that food source at night. By eating the fruits and seeds of plants, birds serve as major agents of seed dispersal.

The feathers developed by some dinosaurs may originally have had thermoregulatory or display functions. Birds use them for these purposes as well as for flying. The flying surface of the wings is created by large quills that arise from the forelimbs. Other strong feathers sprout like a fan from the shortened tail and serve as stabilizers during flight. The contour feathers and down feathers provide insulation to control loss of body heat.

The bones of birds are also modified for flight. They are hollow and have internal struts for strength. The sternum (breastbone) forms a large, vertical keel to which the breast muscles are attached. These muscles pull the wings downward during the main propulsive movement in flight.

Flight is metabolically expensive, and a flying bird consumes energy at a very high rate—about eight times the amount of energy per day as a lizard of the same weight! Because birds have such high metabolic rates, they generate large amounts of heat. They control the rate of heat loss using their feathers, which may be held close to the body or elevated to alter the amount of air trapped as insulation. The brain of a bird is larger in proportion to its body size than lizard or crocodile brains, primarily because the cerebellum, the center of sight and muscular coordination, is enlarged. The beaks of modern birds lack teeth.

Most birds lay their eggs in a nest, where they are warmed by the body heat of an adult that sits on them. Because birds have high body temperatures, the eggs of most species develop rapidly, hatching in less than 2 weeks. The offspring of many species hatch at a relatively helpless stage and are fed for some time by their parents. The young of other bird species, such as chickens, sandpipers, and ducks, can feed themselves shortly after hatching. Adults of all species attend their offspring for some time, warning them of and protecting them from predators, leading them to good foraging places, sheltering them from bad weather, and feeding them.

As adults, birds range in size from the 2-gram bee hummingbird of the West Indies to the 150-kg ostrich. Some flightless birds of Madagascar and New Zealand known from fossils were even larger, but they were exterminated by the humans that first colonized those islands. There are about 9,600 species of living birds, more than in any other major vertebrate group except fishes (Figure 33.21).

The Origin and Diversity of Mammals

Mammals (class **Mammalia**) appeared in the early part of the Mesozoic era, branching from a lineage of mammal-like reptiles. Small mammals coexisted with reptiles and dinosaurs for at least 150 million years. After the large reptiles and dinosaurs disappeared during the mass extinction at the close of the Mesozoic, mammals increased dramatically in numbers, diversity, and size.

Amphibia

Reptilia

Aves

Mammalia

Skeletal simplification accompanied the evolution of small mammals from their larger reptilian ancestors. During mammalian evolution, bones from the lower jaw were incorporated into the middle ear, leaving a single bone in the lower jaw, and the number of bones in the skull decreased. The bulk of both the limbs and the bony girdles from which they are suspended was reduced. Mammals have far fewer, but more highly differentiated teeth than reptiles. Differences in the number, type, and arrangement of teeth in mammals reflect their varied diets.

These skeletal features are readily preserved as fossils, but the soft parts of mammals are seldom fossilized. Therefore we do not know when mammalian features such as mammary glands, sweat glands, hair, and a four-chambered heart evolved. As is the case with birds, the mammalian fossil record suggests that most of the modern mammalian orders evolved rapidly after the end of the Cretaceous, whereas molecular data suggest that they had already evolved during the Cretaceous.

Mammals are unique among animals in providing their young with a nutritive fluid (milk) secreted by mammary glands. Mammalian eggs are fertilized within the female's body, and the embryos undergo some development within a uterus prior to being born. In addition, mammals have a protective and insulating covering of hair, which is luxuriant in some species but has been almost entirely lost in whales, dolphins, and humans. In whales and dolphins thick layers of insulating fat (blubber) replace hair. Clothing assumes the same role for humans.

Mammals range in size from tiny shrews weighing only about 2 grams to the endangered blue whale, which measures up to 31 meters long and weighs up to 160,000 kilograms—the largest animal ever to live on Earth. The approximately 4,000 species of living mammals are divided into two major subclasses: Prototheria and Theria.

The subclass **Prototheria** contains a single order, the Monotremata, with only three species, which are found only in Australia and New Guinea. These mammals, the duck-billed platypus and spiny anteaters, or echidnas, differ from other mammals in that they lack a placenta, lay eggs, and have legs that poke out to the side (Figure 33.22). Monotremes nurse their young on milk, but they have no nipples on their mammary glands; rather, the milk simply oozes out and is lapped off the fur by the offspring.

(a) *Tachyglossus aculeata*

(b) *Ornithorhyncus anatinus*

33.22 Monotremes
(a) The short-beaked echidna is one of two surviving species of echidnas. (b) The duck-billed platypus is the third surviving monotreme species.

Two major groups of mammals are members of the subclass **Theria**. Females of one group, the **Marsupialia**, have a ventral pouch in which they carry and feed their offspring (Figure 33.23a). Gestation (pregnancy) in marsupials is short; the young are born tiny but with well-developed forelimbs, with which they climb to the pouch. They attach to a nipple but cannot suck. The mother ejects milk into the tiny offspring until they grow large enough to suckle. Once her offspring have left the uterus, a female marsupial may become sexually receptive again. She can then carry fertilized eggs capable of initiating development and replacing the offspring in the pouch should something happen to them.

There are about 240 living species of marsupials. At one time marsupials were widely distributed on Earth, but today the majority of species are restricted to the Australian region, with a modest representation in South America

(c) *Sarcophilus harrisii*

33.23 Marsupials
(a) Australia's kangaroos are thought of as the typical marsupial, but the marsupial radiation also produced (b) arboreal species such as this South American opossum and (c) carnivores such as the Tasmanian devil.

(a) *Macropus rufus*   (b) *Calumorys phicander*

(Figure 33.23b). Only one species, the Virginia opossum, is widely distributed in the United States. Marsupials radiated to become terrestrial herbivores, insectivores, and carnivores, but no species live in the oceans or can fly, although some are gliders. The largest living marsupial is the red kangaroo of Australia, which weighs up to 90 kilo-

33.24 Diversity among the Eutherians
(a) The Arctic ground squirrel is one of many species of small, diurnal rodents of western North America. (b) Temperate-zone bats are all insectivores, but many tropical bats such as this leaf-nosed bat eat fruit. (c) Dolphins represent a eutherian lineage that returned to the marine environment. (d) Large hoofed mammals are important herbivores over much of Earth. This caribou bull is grazing by himself, although caribou are often seen in huge herds.

(a) *Citellus parryi*   (b) *Carollia perspicillapa*

(c) *Tursiops truncatus*

(d) *Rangifer tardanus*

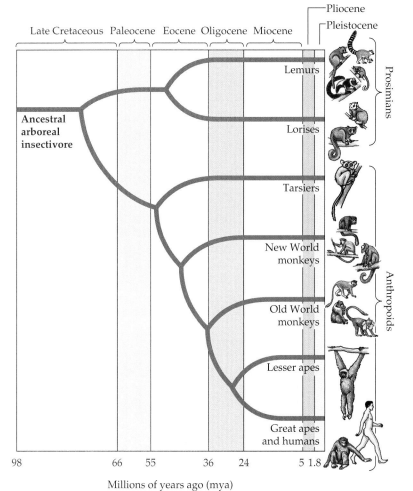

33.25 A Probable Primate Phylogeny
Too few fossil primates have been discovered to reveal with certainty their evolutionary relationships, but this phylogenetic tree is consistent with existing evidence.

Primates and the Origins of Humans

Another eutherian lineage that has had dramatic effects on ecosystems worldwide is the **primate** lineage, to which humans belong. Primates have undergone extensive recent evolutionary radiation. They probably descended from small arboreal (tree-living) insectivores sometime during the Cretaceous period. The major traits that distinguish primates from other mammals are all adaptations to arboreal life. They include:

▶ Dexterous hands with opposable thumbs that can grasp branches and manipulate food.
▶ Nails rather than claws.
▶ Eyes on the front of the face that provide good depth perception.
▶ Very small litters of offspring (usually just one) that receive extended parental care.

Early in its evolutionary history, the primate lineage split into two main branches, prosimians and anthropoids (Figure 33.25). **Prosimians**—lemurs, bush babies, and lorises—once lived on all continents, but today they are restricted to Africa, Madagascar, and tropical Asia. All of the mainland species are arboreal and nocturnal (Figure 33.26). However, on Madagascar, the site of a remarkable prosimian radiation, there are also diurnal and terrestrial species.

The **anthropoids**—tarsiers, monkeys, apes, and humans—evolved from an early primate lineage about

grams, but much larger marsupials existed in Australia until they were exterminated by humans soon after they reached Australia about 50,000 years ago.

Most living mammals are **eutherians**. (Eutherians are sometimes called *placental mammals,* but this name is not accurate because some marsupials also have placentas.) Eutherians are more highly developed at birth than are marsupials, and no external pouch houses them after birth. The nearly 4,000 species of eutherians are divided into 16 major groups, the largest of which is the rodents, with about 1,700 species (Figure 33.24*a*). The next largest group, the bats, has about 850 species (Figure 33.24*b*), followed by the insectivores (moles and shrews) with slightly more than 400 species.

Eutherians are extremely varied in form and ecology. Several lineages of terrestrial mammals subsequently colonized marine environments, to become whales, dolphins, seals, and sea lions (Figure 33.24*c*). Eutherian mammals are—or were, until they were greatly reduced in numbers by humans—the most important grazers and browsers in most terrestrial ecosystems (Figure 33.24*d*). Grazing and browsing have been an evolutionary force intense enough to select for the spines, tough leaves, and difficult-to-eat growth forms found in many plants, a striking example of coevolution.

(a) *Propithecus verreauxi*   (b) *Galago senegalensis*

33.26 Prosimians
(*a*) The sifaka lemur is one of the many lemur species of Madagascar, where they are part of a unique asemblage of plants and animals. (*b*) The lesser bush baby is common in savannas over much of Africa. Its large eyes tell us that it is nocturnal.

(a) *Leontopithecus rosalia*

(b) *Macaca sylvanus*

33.27 Monkeys

(a) Golden lion tamarins, are endangered New World monkeys living in coastal Brazilian rainforests. (b) Many Old World species, such as these Barbary macaques, live in social groups. Here two members of a group groom each other.

55 million years ago in Africa or Asia. New World monkeys have been evolving separately from Old World monkeys long enough that they could have reached South America from Africa when those two continents were still close to each other. Perhaps because tropical America has been heavily forested for a long time, all New World monkeys are arboreal (Figure 33.27*a*). Many of them have long, prehensile tails with which they can grasp branches. Many Old World primates are arboreal as well, but a number of species are terrestrial. Some of these species, such as baboons and macaques, live and travel in large groups (Figure 33.27*b*). No Old World primates have prehensile tails.

About 20 mya, the lineage that leads to modern

(a) *Hylobates lar*

(b) *Pan paniscus*

(c) *Pongo pygmaeus*

(d) *Gorilla gorilla*

33.28 Apes

(a) Gibbons are the smallest of the apes. This common gibbon is found in Asia, from India to Borneo. (b) Chimpanzees, our closest relatives, are found in forested regions of Africa. (c) Orangutans live in the forests of Indonesia. (d) Gorillas, the largest apes, are restricted to humid African forests. This male is a lowland gorilla.

apes separated from the other Old World primates. The first apes were arboreal, but some species came to live in drier habitats with scattered trees, where they obtained most of their food on the ground. Apes are known to have lived in Africa, the Near East, and Asia 15–20 mya. Africa was especially rich in ape species, but the DNA sequences of living primates and some fossil evidence suggests that a European ape that dispersed into Africa about 10 mya may be the ancestor of modern apes. Four of the living genera of apes—gorillas (*Gorilla*), chimpanzees (*Pan*), orangutans (*Pongo*), and gibbons (*Hylobates*)—are restricted to tropical Africa and Asia (Figure 33.28). The fifth (*Homo*) has a worldwide distribution.

Human ancestors descended to the ground

The primate lineage that led to humans began with the **ardipithecines**. These apes had distinct morphological adaptations for **bipedalism**—locomotion in which the body is held erect and moved exclusively by movements of the hind legs. Bipedal locomotion frees the hands to manipulate objects and to carry them while walking. It also elevates the eyes, enabling the animal to see over tall vegetation to spot predators and prey. Both advantages were probably important for early ardipithecines and their descendants, the **australopithecines**.

The first australopithecine skull was found in South Africa in 1924; since then other fragments have been found in a number of sites in Africa (Figure 33.29). The most complete fossil skeleton of an australopithecine, approximately 3.5 million years old, was discovered in Ethiopia in 1974. That individual, a young female known to the world as Lucy, attracted a great deal of attention because her remains were so complete and well preserved. Lucy has been assigned to the species *Australopithecus afarensis*, the most likely ancestor of humans. All the evidence from different parts of her skeleton suggests that Lucy was only about 1 meter tall and walked upright.

From *Australopithecus afarensis* ancestors, a number of species of australopithecines evolved. Several million years ago, two distinct types of australopithecines lived together over much of eastern Africa. The larger type (about 40 kg) is represented by at least two species, both of which died out suddenly about 1.5 million years ago. The 25–30-kg *A. africanus* is much rarer as a fossil, suggesting that it was less common than the other species.

Humans arose from australopithecine ancestors

Many experts believe that the recently discovered *Australopithecus garhi* or a similar species gave rise to the genus *Homo*. *A. garhi*, a small-brained, big-toothed hominid with

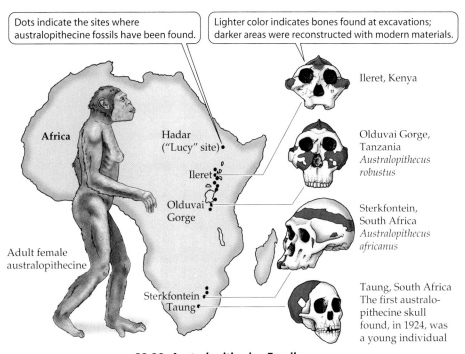

33.29 Australopithecine Fossils
Few fossilized remains are complete, but skull shapes can be reconstructed accurately.

humanlike leg proportions, began using tools to obtain food about 2.5 mya. They butchered animals to get at bone marrow, which is rich in fat.

Early **hominids**—members of the genus *Homo*—lived contemporaneously with australopithecines for perhaps half a million years. Two major changes accompanied the evolution of *Homo* from *Australopithecus*: an increase in body size and a doubling of brain size.

The oldest fossil remains of a member of the genus *Homo*, named *H. habilis*, were discovered in the Olduvai Gorge, Tanzania, and are estimated to be 2 million years old. Other fossils of *H. habilis* have been found in Kenya and Ethiopia. Tools used by these early hominids were found with the fossils. *H. habilis* lived in relatively dry areas where, for much of the year, the main food reserves are subterranean roots, bulbs, and tubers. To exploit these food resources, an animal must dig into hard, dry soils, something that cannot be done with an unaided primate hand. However, roots can be dug in large quantities in a relatively short time by an individual with a simple digging tool. *H. habilis* females carrying infants could have done so, freeing males to hunt animal prey to provide the proteins that roots lack.

The only other known extinct species of our genus, *Homo erectus*, evolved in Africa about 1.6 million years ago. Soon thereafter it had spread as far as eastern Asia. As it expanded its range and increased in abundance, *H. erectus* may have exterminated *H. habilis*. Members of *H. erectus* were as large as modern people, but their bones were considerably heavier. *H. erectus* used fire for cooking and for hunting large animals, and made characteristic stone tools

33.30 A Probable Human Phylogeny The evolution of *Homo* from *Australopithecus* was marked by an increase in body size and a doubling of brain size.

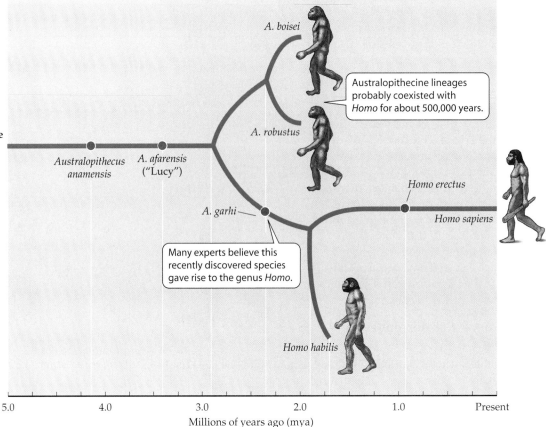

Ardipithecine ancestor (bipedalism)

Australopithecus anamensis

A. afarensis ("Lucy")

A. boisei

A. robustus

Australopithecine lineages probably coexisted with *Homo* for about 500,000 years.

A. garhi

Many experts believe this recently discovered species gave rise to the genus *Homo*.

Homo erectus

Homo sapiens

Homo habilis

Millions of years ago (mya)
5.0   4.0   3.0   2.0   1.0   Present

that have been found in many parts of the Old World. These tools were probably used for digging, capturing animals, cleaning and cutting meat, scraping hides, and cutting wood. Although *H. erectus* survived in Eurasia until about 250,000 years ago, it was replaced in tropical regions by our species, *Homo sapiens*, about 200,000 years ago.

Brains steadily became larger

The trends that accompanied the transition from *Australopithecus* to *H. erectus* continued during the evolution of our own species (Figure 33.30). The earliest members of *Homo sapiens* had larger brains than members of the earlier species of *Homo*, a change that was probably favored by an increasingly complex social life. The ability of group members to communicate with one another was valuable for cooperative hunting and gathering and for improving one's status in the complex social interactions that must have characterized those societies, just as they do ours today.

Several types of *H. sapiens* existed during the mid-Pleistocene epoch, from about 1.5 million to about 300,000 years ago. All were skilled hunters of large mammals, but plants continued to be important components of their diets. During this period another distinctly human trait emerged: rituals and a concept of life after death. Deceased individuals were buried with tools and clothing, presumably for their existence in the next world.

One type of *H. sapiens*, generally known as Neanderthals because they were first discovered in the Neander Valley in Germany, was widespread in Europe and Asia between about 75,000 and 30,000 years ago. Neanderthals were short, stocky, and powerfully built humans whose massive skulls housed brains somewhat larger than our own. They manufactured a variety of tools and hunted large mammals, which they probably ambushed and subdued in close combat. For a short time, their range overlapped that of a more modern form of *H. sapiens* known as Cro-Magnons, but then the Neanderthals abruptly disappeared. Many scientists believe that they were exterminated by the Cro-Magnons, just as *H. habilis* may have been exterminated by *H. erectus*.

Cro-Magnon people made and used a variety of sophisticated tools. They created the remarkable paintings of large mammals, many of them showing scenes of hunting, that have been discovered in caves in various parts of Europe (Figure 33.31a). The animals depicted were characteristic of the cold steppes and grasslands that occupied much of Europe during periods of glacial expansion. Cro-Magnon people spread across Asia, reaching North America perhaps as early as 20,000 years ago, although the date of their arrival in the New World is still uncertain. Within a few thousand years they had spread southward through North America to the southern tip of South America.

Humans evolved language and culture

As our ancestors evolved larger brains, their behavioral capabilities increased, especially the capacity for language. Most animal communication consists of a limited number of signals, which pertain mostly to immediate circumstances

(a)

(b)

33.31 Hunting, Pastoralism, and Agriculture

(a) Cro-Magnon cave drawings such as those found in Lascaux Cave in France typically depict the large mammals that they hunted. (b) The Masai are a pastoral people living on East African savannas, where they and their cattle typically coexist with native grazing and browsing mammals. (c) Intense agricultural development totally transforms the landscape. These rice terraces are on the island of Bali in Indonesia.

and are associated with changed emotional states induced by those circumstances. Human language is far richer in its symbolic character than any other animal vocalizations. Our words can refer to past and future times and to distant places. We are capable of learning thousands of words, many of them referring to abstract concepts. We can rearrange words to form sentences with complex meanings.

The expanded mental abilities of humans are largely responsible for the development of **culture**, the process by which knowledge and traditions are passed along from one generation to another by teaching and observation. Culture can change rapidly because genetic changes are not necessary for a cultural trait to spread through a population. The primary disadvantage of culture is that its norms must be taught to each generation. The tools and other implements associated with human fossils, as well as the cave paintings early humans created, reveal cultural traditions.

Cultural learning greatly facilitated the spread of domesticated plants and animals and the resultant conversion of most human societies from ones in which food was obtained by hunting and gathering to ones in which *pastoralism* (herding large animals) and *agriculture* dominated (Figure 33.31b,c). The development of agriculture led to an increasingly sedentary life, the growth of cities, greatly expanded food supplies, a rapid increase in the human population, and the appearance of occupational specializations, such as artisans and healers.

Agriculture developed in the Middle East approximately 11,000 years ago. From there it spread rapidly northwest-

(c)

ward across Europe, finally reaching the British Isles about 4,000 years ago. The first plants and animals to be domesticated were cereal grains such as wheat and barley; legumes (beans, lentils, and peas); and woody plant crops such as grapes and olives. Others, such as rye, cabbage, celery, and carrots, were domesticated later. Cattle, sheep, goats, horses, dogs, and cats were the most important domesticated animals.

Agriculture developed independently in eastern Asia, contributing to our modern diet soybeans, rice, citrus fruits,

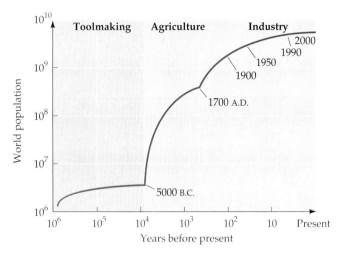

33.32 Human Population Surges
The human population surged following (1) the invention of tools, (2) the domestication of plants and animals, and (3) the Industrial Revolution.

mangoes, pigs, and chickens. There was some exchange, even at early times, among agricultural centers in the Old World, but when people spread across the cold and barren Bering land bridge into the New World, they apparently brought no domesticated plants with them. These people subsequently developed rich and varied agricultural systems based on corn, tomatoes, kidney and lima beans, peanuts, potatoes, chili peppers, and squashes. The largest animals domesticated by humans in the New World were llamas and alpacas in the Andes of South America—animals not large enough to carry a person. The Amerindians of Mexico and Central America had no domesticated animals larger than turkeys.

The human population has grown rapidly

The human population has experienced three major phases of increase (Figure 33.32). The first, stimulated by tool use, lasted about a million years. At the end of that period, the human population is estimated to have been approximately 5 million. During the second surge, which followed the domestication of plants and animals and the invention of agriculture, the human population may have increased to about 500 million people within 8,000 years.

We are currently in the middle of the third great population surge, triggered in the eighteenth century by the Industrial Revolution. In the industrialized countries of the world, death rates fell and life spans increased. By the end of the nineteenth century, human numbers had passed 1 billion. Despite the devastation of two World Wars and countless minor conflicts, the twentieth century saw the human population reach its current level of more than 6 billion. It is projected to increase to more than 11 billion by 2050.

The first two human population surges were followed by periods of relative stability. Whether the current surge will follow the same pattern, at what size it might level off, what hardships might ensue if it does not, and what the

consequences will be for other species are questions that are fiercely debated. We will discuss these issues further in Chapter 58.

 ## Chapter Summary

Deuterostomes and Protostomes

▶ The deuterostome lineage separated from the protostome lineage early in the history of animal life on Earth.

▶ There are fewer major lineages and fewer species of deuterostomes than protostomes, but as members of the deuterostome lineage we have a special interest in its members. **Review Figure 33.1**

Echinoderms: Complex Biradial Symmetry

▶ Echinoderms have a radially symmetrical body plan, a unique water vascular system, and a calcified internal skeleton. **Review Figure 33.2a**

▶ Nearly all living species of echinoderms have a bilaterally symmetrical, ciliated larva that feeds as a planktonic organism. **Review Figure 33.2b**

▶ Six major groups of echinoderms survive today, but 23 other lineages existed in the past. Some groups of echinoderms have arms, but others do not.

Chordates: New Ways of Feeding

▶ Evolution among the hemichordates and chordates led to new ways of capturing and handling food.

▶ The large proboscis of acorn worms is both a digging and a food-capturing organ.

▶ Members of the chordates evolved enlarged pharyngeal slits as feeding devices and a dorsal supporting rod, the notochord.

▶ Tunicates are sessile as adults and filter prey from seawater with large pharyngeal baskets. Their larvae have notochords and dorsal, hollow nerve cords.

Origin of the Vertebrates

▶ Vertebrates evolved jointed internal skeletons centered around a vertebral column, a body plan that enabled them to swim rapidly. Early vertebrates fed by filtering small animals from mud. **Review Figure 33.7**

▶ Jaws evolved from anterior gill arches and enabled their possessors to grasp and chew their prey, expanding food sources and improving nutrition. Jawed fishes rapidly became the dominant animals in both marine and fresh waters. **Review Figures 33.8, 33.10**

▶ Fishes evolved unjointed fins with which they could control their swimming movements and stabilize themselves in the water, and lunglike sacs that helped them stay suspended in open water.

▶ Bony fishes come in a wide variety of sizes and shapes, and many species have complex behaviors.

Colonizing the Land: Obtaining Oxygen from the Air

▶ Two fish lineages—lobe-fins and lungfishes—evolved jointed fins. Amphibians, the first terrestrial vertebrates, arose from one of these lineages. **Review Figure 33.14**

▶ Most amphibians live in water at some time in their lives, and their eggs must remain moist. **Review Figure 33.16**

▶ About 4,500 species of amphibians live today. They belong to three orders: caecilians, frogs and toads, and salamanders.

▶ Amniotes evolved eggs with shells impermeable to water and thus became the first vertebrates to be independent of water for breeding. **Review Figure 33.17**

▶ Modern reptiles are members of four lineages—turtles and tortoises; tuataras; snakes and lizards; and crocodilians. **Review Figure 33.18**

▶ Dinosaurs rose to dominance about 215 mya and dominated terrestrial environments for about 150 million years until their extinction at the end of the Mesozoic era.

Birds: Feathers and Flight

▶ Birds arose about 175 mya, but much controversy surrounds their origins.

▶ The 9,600 species of living birds are characterized by feathers, high metabolic rates, and parental care.

The Origin and Diversity of Mammals

▶ Mammals evolved during the Mesozoic era, about 225 mya.

▶ Eggs of mammals are fertilized within the bodies of females, and embryos develop there for some time before being born. Mammals are unique in suckling their young with milk secreted by mammary glands.

▶ The three species of mammals in subclass Prototheria lay eggs, but all other mammals give birth to developed young.

▶ Therian mammals are divided into two major groups: the marsupials, which give birth to tiny young that are raised in a pouch on the female's belly, and eutherians, which give birth to relatively well-developed offspring.

Primates and the Origins of Humans

▶ The primates split into two major lineages, one leading to the prosimians—lemurs, lorises, and pottos—and the other leading to the anthropoids—tarsiers, monkeys, apes, and humans. **Review Figure 33.25**

▶ Hominids evolved in Africa from terrestrial, bipedal ancestors. **Review Figures 33.29, 33.30**

▶ Early humans evolved large brains, language, and culture. They made and used tools, developed rituals, and domesticated plants and animals. In combination, these traits enabled humans to increase greatly in numbers.

▶ The human population has increased greatly three times. We are currently in the middle of the third population surge. When and how it will end is hotly debated. **Review Figure 33.32**

For Discussion

1. In what animal phyla has the ability to fly evolved? How do structures used for flying differ among these animals?

2. Extracting suspended food from the water column is a common mode of foraging among animals. Which groups contain species that extract prey from the air? Why is this mode of obtaining food so much less common than extracting prey from water?

3. Large size both confers benefits and poses certain risks. What are these risks and benefits?

4. Amphibians have survived and prospered for many millions of years, but today many species are disappearing and populations of others are declining seriously. What features of amphibian life histories might make them especially vulnerable to the kinds of environmental changes now happening on Earth?

5. The evolution of jaws allowed vertebrates to utilize a remarkably wide array of food types; yet jawless animals are able to eat most of those kinds of foods. Compare the ways that jawed and jawless animals would eat the following kinds of food:
   a. a snail
   b. an insect
   c. a fish
   d. a bird
   e. a plant leaf

6. The body plan of all vertebrates is based on four appendages. Describe the varied forms that these appendages take and how they are used. How do the vertebrates that have lost their four appendages move?

7. Compare the ways that different animal lineages colonized the land. How were those ways influenced by the body plans of animals in the different lineages?

Part Five

THE BIOLOGY OF FLOWERING PLANTS

34 *The Plant Body*

THE OLDEST KNOWN PLANT IS A BRISTLECONE pine that has been living for more than 4,900 years—almost 50 centuries. In contrast, it is doubtful that any animal has ever lived as long as 2 centuries. The extreme ages achieved by some trees prove that plants can cope very successfully with their environments.

Plants cannot move, but they have mechanisms for coping with environmental changes that they can't escape. They create and maintain an internal environment that differs from the external environment. They also regulate their own metabolism, which enables them to perform their necessary functions.

Motion is not a characteristic of plants; instead, we may think of plants as "growing machines." By growing, plants accomplish some of the same things that animals achieve through motion. Growing roots, for example, can reach into new supplies of water and nutrients.

Although they have simpler nutritional needs than animals do, plants must nevertheless obtain nutrients—not only the raw materials of photosynthesis (carbon dioxide and water), but also mineral elements such as nitrogen, potassium, and calcium. Seed plants—even the tallest trees—transport water from the soil to their tops, and they transport the products of photosynthesis from the leaves to their roots and other parts.

Plants also interact with their living and nonliving environments. They respond to environmental cues as they grow and develop. Their responses are mediated by chemical signals that move within cells and throughout the plant body. Among the resulting changes are ones that lead to growth, development, and reproduction.

Because we can understand the function of these growing machines only in terms of their underlying structure, this chapter focuses on the structure of the plant body, with primary emphasis on flowering plants. We'll examine plant structure at the level of

the organs, cells, tissues, and tissue systems. Then we'll see how meristems—organized groups of dividing cells—serve the growth of the plant body, both in length and, in woody plants, in width. The chapter concludes with a consideration of how leaf structure supports photosynthesis.

Vegetative Organs of the Flowering Plant Body

You will recall from Chapter 29 that flowering plants (angiosperms) are tracheophytes that are characterized by double fertilization, a triploid endosperm, and seeds en-

An Ancient Individual
Bristlecone pines (*Pinus aristata*) can live for centuries. The oldest known living organism is a bristlecone pine that has been alive for almost 5,000 years—long enough to have witnessed all of recorded human history.

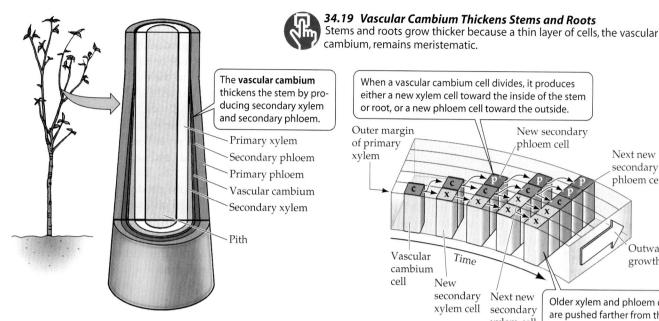

34.19 Vascular Cambium Thickens Stems and Roots
Stems and roots grow thicker because a thin layer of cells, the vascular cambium, remains meristematic.

The **vascular cambium** thickens the stem by producing secondary xylem and secondary phloem.

Primary xylem
Secondary phloem
Primary phloem
Vascular cambium
Secondary xylem

Pith

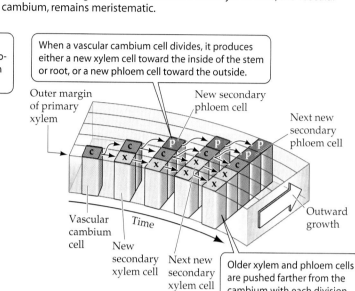

When a vascular cambium cell divides, it produces either a new xylem cell toward the inside of the stem or root, or a new phloem cell toward the outside.

Outer margin of primary xylem

New secondary phloem cell

Next new secondary phloem cell

Vascular cambium cell

Time

Outward growth

New secondary xylem cell

Next new secondary xylem cell

Older xylem and phloem cells are pushed farther from the cambium with each division of the cambium.

bium divides, are rows of living parenchyma cells that run perpendicular to the xylem vessels and phloem sieve tubes (Figure 34.20). As the root or stem continues to increase in diameter, new vascular rays are initiated so that this storage and transport tissue continues to meet the needs both of the bark and of the living cells in the xylem.

The vascular cambium itself increases in circumference with the growth of the root or stem. To do this, some of its cells divide in a plane at right angles to the plane that gives rise to secondary xylem and phloem. The products of each of these divisions lie within the vascular cambium itself and increase its circumference.

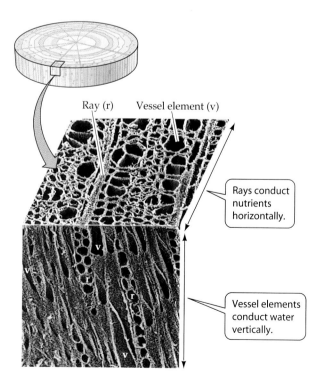

Ray (r)   Vessel element (v)

Rays conduct nutrients horizontally.

Vessel elements conduct water vertically.

34.20 Vascular Rays and Vessel Elements
Wood of the tulip poplar, showing that the orientation of xylem vessels is perpendicular to that of vascular rays. The rays transport sieve tube sap horizontally from the phloem to storage parenchyma cells.

Only eudicots have a vascular cambium and a cork cambium and thus undergo secondary growth. In the rare cases in which monocots form thickened stems—palm trees, for example—they do so without using vascular cambium or cork cambium. Palm trees have a very wide apical meristem that produces a wide stem, and dead leaf bases also add to the diameter of the stem. Basically, monocots grow in the same way as do other angiosperms that lack secondary growth.

Wood and bark are unique to plants showing secondary growth. These tissues have their own patterns of organization and development.

WOOD.   Cross sections of most tree trunks (stems) in temperate-zone forests show *annual rings* (Figure 34.21), which result from seasonal environmental conditions. In spring, when water is relatively plentiful, the tracheids or vessel elements produced by the vascular cambium tend to be large in diameter and thin-walled. As water becomes less available during the summer, narrower cells with thicker walls are produced, making this summer wood darker and perhaps more dense than the wood formed in spring. Thus each year is usually recorded in a tree trunk by a clearly visible annual ring consisting of one light and one dark layer. Trees in the wet tropics do not lay down such obvious regular rings.

The difference between old and new regions also contributes to the appearance of wood. As a tree grows in di-

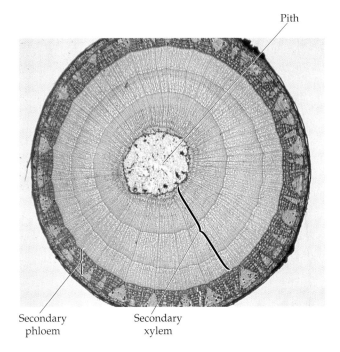

Pith

Secondary phloem

Secondary xylem

34.21 Annual Rings
Rings of secondary xylem are the most noticeable feature of this cross section from a 3-year-old basswood stem.

BARK. As secondary growth of stems or roots continues, the expanding vascular tissue stretches and breaks the epidermis and cortex, which ultimately flake away and are lost. Tissue derived from the secondary phloem then becomes the outermost part of the stem. Before the dermal tissues are broken away, cells lying near the surface of the secondary phloem begin to divide and produce layers of **cork**, a tissue composed of cells with thick walls, waterproofed with suberin. The cork soon becomes the outermost tissue of the stem or root. The dividing cells, derived from the secondary phloem, form a cork cambium. Sometimes the cork cambium produces cells to the inside as well as to the outside; these cells constitute what is known as the **phelloderm**.

Cork, cork cambium, and phelloderm make up the periderm of the secondary plant body. As the vascular cambium continues to produce secondary vascular tissue, the corky layers are in turn lost, but the continuous formation of cork cambium in the underlying phloem gives rise to new corky layers.

When bark forms on stems and roots, the underlying tissues still need to release carbon dioxide and take up oxygen. **Lenticels** are spongy regions in the cork of stems and roots that allow such gas exchange (Figure 34.22).

Leaf Anatomy Supports Photosynthesis

We can think of roots and stems as important supporting actors that sustain the activities of the real stars of the plant body, the leaves—the organs of photosynthesis. Leaf anatomy is beautifully adapted to carry out photosynthesis and to support photosynthesis by exchanging the gases O_2 and CO_2 with the environment, limiting evaporative water loss, and transporting the products of photosynthesis to the rest of the plant. Figure 34.23a shows a typical eudicot leaf in cross section.

ameter, the xylem toward the center becomes clogged with water-insoluble substances and ceases to conduct water and minerals; this is *heartwood* and appears darker in color. The portion of the xylem that is actively conducting all water and minerals in the tree is called *sapwood* and is lighter in color and more porous than heartwood.

The knots that we find attractive in knotty pine but regard as a defect in structural timbers are cross sections of branches: As a trunk grows, the bases of branches become buried in the trunk's new wood and appear as knots when the trunk is cut lengthwise.

(a)

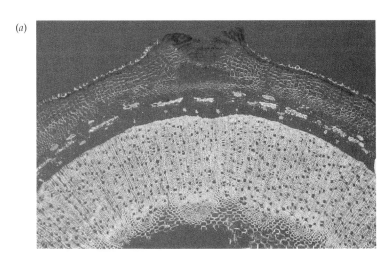

(b)

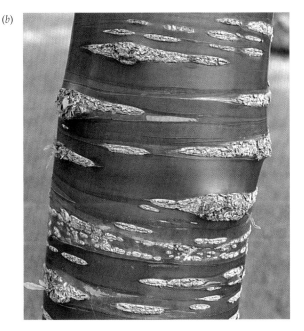

34.22 Lenticels Allow Gas Exchange through the Periderm
(a) The region of periderm that appears broken open is a lenticel in a year-old elder twig; note the spongy tissue that constitutes the lenticel. (b) The rough areas on the trunk of this Chinese plum tree are lenticels. Most tree species have lenticels much smaller than these.

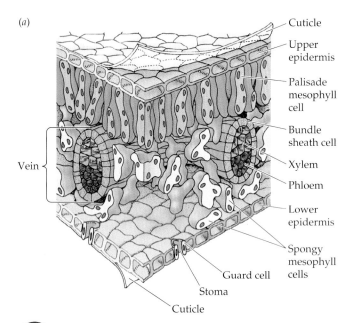

(a)

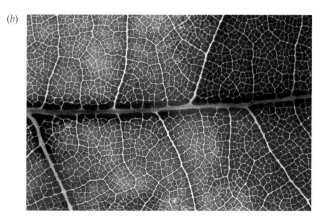

(b)

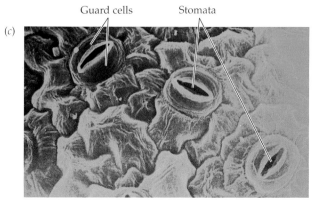

(c)

Guard cells   Stomata

34.23 The Eudicot Leaf

(a) Cross section of a eudicot leaf. *(b)* The network of fine veins in this maple leaf carries water to the mesophyll cells and carries photosynthetic products away from them. *(c)* These paired cells on the lower epidermis of a eudicot leaf are guard cells; the gaps between them are stomata, through which carbon dioxide enters the leaf.

Most eudicot leaves have two zones of photosynthesizing parenchyma tissue referred to as **mesophyll**, which means "middle of the leaf." The upper layer or layers of mesophyll consist of roughly cylindrical cells. This zone is referred to as *palisade mesophyll*. The lower layer or layers consist of irregularly shaped cells; this zone is called *spongy mesophyll*. Within the mesophyll is a great deal of air space through which carbon dioxide can diffuse to surround all photosynthesizing cells.

Vascular tissue branches extensively in the leaf, forming a network of **veins** (Figure 34.23*b*). Veins extend to within a few cell diameters of all the cells of the leaf, ensuring that the mesophyll cells are well supplied with water and minerals. The products of photosynthesis are loaded into the phloem of the veins for export to the rest of the plant.

Covering the entire leaf is a layer of nonphotosynthetic cells constituting the epidermis. The epidermal cells have an overlying waxy cuticle that is highly impermeable to water. But this impermeability poses a problem: While keeping water in the leaf, the epidermis also keeps carbon dioxide, the other raw material of photosynthesis, out.

The problem of balancing water retention and carbon dioxide availability is solved by an elegant regulatory system that will be discussed in more detail in the next chapter. *Guard cells* are modified epidermal cells that change their shape, thereby opening or closing pores called *stomata*, which serve as passageways between the environment and the leaf's interior (Figure 34.23*c*). When the stomata are open, carbon dioxide can enter and oxygen can leave, but some water can also be lost.

In Chapter 8 we described C_4 plants, which can fix carbon dioxide efficiently even when the carbon dioxide supply in the leaf decreases to a level at which the photosynthesis of C_3 plants is inefficient. One adaptation that helps C_4 plants do this is their modified leaf anatomy, as shown in Figure 8.19. Notice that the photosynthetic cells in the C_4 leaf are grouped around the veins in concentric layers, forming an outer mesophyll layer and an inner *bundle sheath*. These layers each contain different types of chloroplasts, leading to the biochemical division of labor described in Chapter 8.

Leaves receive water and mineral nutrients from the roots by way of the stems. In return, the leaves export products of photosynthesis, providing a supply of chemical energy to the rest of the plant body. And, as we have just seen, leaves exchange gases, including water vapor, with the environment by way of the stomata. All three of these processes will be considered in detail in the next chapter.

Chapter Summary

Vegetative Organs of the Flowering Plant Body

▶ Monocots typically have a single cotyledon, narrow leaves with parallel veins, flower parts in threes or multiples of three, and stems with scattered vascular bundles.

▶ Eudicots typically have two cotyledons, broad leaves with netlike veins, flower parts in fours or fives, and vascular bundles in a ring. Flowering plants that are neither monocots nor eudicots are generally similar in structure to eudicots. **Review Figure 34.1**

▶ The vegetative organs of flowering plants are roots, which form a root system, and stems and leaves, which form a shoot system. **Review Figure 34.2**

▶ Roots anchor the plant and take up water and minerals.

▶ Stems bear leaves and buds. Lateral buds form branches. Apical buds produce cells that contribute to the elongation of the stem.

▶ Leaves are responsible for most photosynthesis, for which their flat blades, oriented perpendicular to the sun's rays, are well adapted. **Review Figure 34.5**

Plant Cells

▶ The walls of plant cells have a structure that often corresponds to the special functions of the cell. The walls of individual cells are separated by a middle lamella common to two neighboring cells; each cell also has its own primary wall. **Review Figure 34.6**

▶ Some cells produce a thick secondary wall. Adjacent cells are connected by plasmodesmata that extend through both cell walls. **Review Figures 34.7, 34.8**

▶ Parenchyma cells have thin walls. Many parenchyma cells store starch or lipids; some others carry out photosynthesis. **Review Figure 34.9a**

▶ Collenchyma cells provide flexible support. **Review Figure 34.9b**

▶ Sclerenchyma cells provide strength and function when dead. **Review Figure 34.9c, d**

▶ Tracheids and vessel elements are xylem cells that conduct water and minerals after the cells die. **Review Figures 34.9e, f, 34.10**

▶ Sieve tube members are the conducting cells of the phloem. Their activities are often controlled by companion cells. **Review Figure 34.11**

Plant Tissues and Tissue Systems

▶ Three tissue systems extend throughout the plant body. The vascular tissue system, consisting of xylem and phloem, conducts water, minerals, and the products of photosynthesis. The dermal tissue system protects the body surface. The ground tissue system produces and stores food materials and performs other functions. **Review Figure 34.12**

Forming the Plant Body

▶ The apical–basal pattern and the radial pattern are parts of the plant body plan; they arise through orderly development. **Review Figure 34.13**

▶ The plant body is modular, and the growth of stems and roots is indeterminate. Leaves, flowers, and fruits show determinate growth.

▶ Meristems are localized regions of cell division. A hierarchy of meristems generates the plant body.

▶ Apical meristems at the tips of stems and roots produce the primary growth of those organs. **Review Figure 34.14**

▶ Shoot apical meristems and root apical meristems give rise to primary meristems: the protoderm, the ground meristem, and the procambium. **Review Figure 34.15**

▶ In some plants, the products of primary growth constitute the entire plant body. Many other plants show secondary growth. Two lateral meristems, the vascular cambium and cork cambium, are responsible for secondary growth. **Review Figure 34.14**

▶ The young root has an apical meristem that gives rise to the root cap and to the three primary meristems, which in turn produce the three tissue systems. The protoderm produces the dermal tissue system, the ground meristem produces the ground tissue system, and the procambium produces the vascular tissue system. **Review Figure 34.15**

▶ Root tips have three overlapping zones: the zone of cell division, the zone of cell elongation, and the zone of cell differentiation. **Review Figure 34.15**

▶ The dermal tissue system consists of the epidermis, part of which forms the root hairs that are responsible for absorbing water and minerals. **Review Figure 34.16**

▶ The ground tissue system of a young root is the cortex, whose innermost cell layer, the endodermis, controls access to the stele.

▶ The stele, consisting of the pericycle, xylem, and phloem, is the root's vascular tissue system. Lateral roots arise in the pericycle. Monocot roots have a central pith region. **Review Figure 34.17**

▶ The shoot apical meristem also gives rise to three primary meristems, with roles similar to their counterparts in the root. Leaf primordia on the sides of the apical meristem develop into leaves.

▶ The vascular tissue in young stems is divided into vascular bundles, each containing both xylem and phloem. Pith occupies the center of the eudicot stem. Cortex lies to the outside of the vascular bundles in eudicots, with pith rays lying between the vascular bundles. **Review Figure 34.18**

▶ Many eudicot stems and roots show secondary growth, in which vascular and cork cambia give rise to secondary xylem and secondary phloem. As secondary growth continues, the products are wood and bark. **Review Figure 34.19**

▶ The vascular cambium lays down layers of secondary xylem and phloem. Living cells within these tissues are nourished by vascular rays. **Review Figure 34.20**

▶ The periderm consists of cork, cork cambium, and phelloderm, all pierced at intervals by lenticels that allow gas exchange.

Leaf Anatomy Supports Photosynthesis

▶ The photosynthetic tissue of a leaf is called mesophyll. Veins bring water and minerals to the mesophyll and carry the products of photosynthesis to other parts of the plant body.

▶ A waxy cuticle prevents water loss from the leaf, but is impermeable to carbon dioxide. Guard cells control the opening of stomata, openings in the leaf that allow CO_2 to enter but also allow some water to escape. **Review Figure 34.23**

For Discussion

1. When a young oak was 5 m tall, a thoughtless person carved his initials in its trunk at a height of 1.5 m above the ground. Today that tree is 10 m tall. How high above the ground are those initials? Explain your answer in terms of the manner of plant growth.

2. Consider a newly formed sieve tube member in the secondary phloem of an oak tree. What kind of cell divided to produce the sieve tube member? What kind of cell divided to produce that parent cell? Keep tracing back until you arrive at a cell in the apical meristem.

3. Distinguish between sclerenchyma cells and collenchyma cells in terms of structure and function.

4. Distinguish between primary and secondary growth. Do all angiosperms undergo secondary growth? Explain.

5. What anatomical features make it possible for a plant to retain water as it grows? Describe the plant tissues and how and when they form.

35 *Transport in Plants*

 About 40 years ago the biologist Per Scholander was studying water movement to the top of an 80-meter Douglas fir. To collect samples rapidly from the treetop, he hired a sharpshooter, who aimed a high-powered rifle at a twig high in the tree and fired. From high above, a twig fluttered to the ground, and Scholander quickly inserted it into an instrument for measuring tension in the xylem sap. As we will soon see, Scholander's measurements increased our understanding of how water and minerals reach the tops of tall trees.

The water and minerals in a plant's xylem must be transported to the entire shoot system, all the way to the highest leaves and apical buds. Carbohydrates produced in all the leaves, including the highest, must be translocated to all the living nonphotosynthetic parts of the plant. Before we consider the mechanisms underlying these processes, we should consider two questions: How much water is transported? And how high can water be transported?

In answer to the first question, consider the following example: A single maple tree 15 meters tall was estimated to have some 177,000 leaves, with a total leaf surface area of 675 square meters—half again the area of a basketball court. During a summer day, that tree lost 220 liters of water *per hour* to the atmosphere by evaporation from the leaves. To prevent wilting, the xylem needed to transport 220 liters of water from the roots to the leaves every hour. (By comparison, a 50-gallon drum holds 189 liters.)

The second question can be rephrased: How tall are the tallest trees? The tallest gymnosperms, the coast redwoods—*Sequoia sempervirens*—exceed 110 meters in height, as do the tallest angiosperms, the Australian *Eucalyptus regnans*. Any successful explanation of water transport in the xylem must account for transport to these great heights.

In this chapter we consider the uptake and transport of water and minerals by plants, the control of evaporative water loss through the stomata, and the translocation of substances in the phloem.

A Long Way to the Top
Water and minerals must defy gravity and climb over 80 meters to reach the top branches of these Douglas firs (*Pseudotsuga menziesii*).

Uptake and Transport of Water and Minerals

Terrestrial plants obtain both water and mineral nutrients from the soil, usually by way of their roots. You already know that water is one of the ingredients required for carbohydrate production by photosynthesis in the leaves. Water is also essential for transporting solutes, for cooling the plant, and for developing the internal pressure that supports the plant body.

How do leaves high in a tree obtain water from the soil? What are the mechanisms by which water and mineral ions enter the plant body through the roots and ascend as sap in the xylem? Because neither water nor minerals can move through the plant into the xylem without crossing at least

one plasma membrane, we focus first on osmosis. Then we examine the uptake of mineral ions, and follow the pathway by which both water and minerals move through the root to gain entry to the xylem.

Water moves through a membrane by osmosis

Osmosis, the movement of water through a membrane in accordance with the laws of diffusion, was described in Chapter 5. The **solute potential** (osmotic potential) of a solution is a measure of the effect of dissolved solutes on the osmotic behavior of the solution. The greater the solute concentration of a solution, the more negative its solute potential, and the greater the tendency of water to move into it from another solution of lower solute concentration (and less negative solute potential). For osmosis to occur, the two solutions must be separated by a membrane permeable to water but relatively impermeable to the solute. Recall, too, that osmosis is a passive process—energy is not required.

Unlike animal cells, plant cells are surrounded by a relatively rigid cell wall. As water enters a plant cell, the entry of more water is increasingly resisted by an opposing **pressure potential** (turgor pressure), owing to the rigidity of the wall. As more and more water enters, the pressure potential becomes greater and greater.

Pressure potential is a hydraulic pressure analogous to the air pressure in an automobile tire; it is a real pressure that can be measured with a pressure gauge. Cells with walls do not burst when placed in pure water; instead, water enters by osmosis until the pressure potential exactly balances the solute potential. At this point, the cell is *turgid*; that is, it has a high pressure potential.

The overall tendency of a solution to take up water from pure water, across a membrane, is called its **water potential**, represented as ψ, the Greek letter psi (Figure 35.1). The water potential

is simply the sum of the (negative) solute potential (ψ_s) and the (usually positive) pressure potential (ψ_p):

$$\psi = \psi_s + \psi_p$$

For pure water under no applied pressure, all three of these parameters are zero.

We can measure solute potential, pressure potential, and water potential in *megapascals* (MPa), a unit of pressure. (Atmospheric pressure is about 0.1 MPa, or 14.7 pounds per square inch; typical pressure in an automobile tire is about 0.2 MPa.)

In all cases in which water moves between two solutions separated by a membrane, the following rule of osmosis applies: *Water always moves across a differentially permeable membrane toward the region of more negative water potential.*

Osmotic phenomena are of great importance to plants. The structure of many plants is maintained by the pressure potential of their cells; if the pressure potential is lost, a plant *wilts*. Within living tissues, the movement of water from cell to cell by osmosis follows a gradient of *water potential*. Over longer distances, in open tubes such as xylem vessels and phloem sieve tubes, the flow of water and dissolved solutes is driven by a gradient in *pressure potential*. The movement of a solution due to a difference in pressure potential between two parts of a plant is called **bulk flow**.

Uptake of mineral ions requires transport proteins

Mineral ions, which carry electric charges, cannot move across a plasma membrane unless they are aided by trans-

2 Because of the difference in ψ between the solution and the distilled water, water moves from the beaker to the tube.

3 Water entering the tube dilutes the solution, making its ψ_s less negative. As the solution rises in the tube, **pressure potential** (ψ_p) builds up until it balances the ψ_s. This pressure corresponds to turgor pressure in plants. At equilibrium, ψ in the solution is equal to ψ in the beaker.

$$\psi_p = 0.15$$
$$\psi_s = -0.15$$
$$\psi = 0$$

$$\psi = 0$$

1 The solution in the tube has a negative **solute potential** (ψ_s) due to the presence of dissolved solutes; its $\psi_p = 0$; thus its ψ is negative. The beaker contains distilled water ($\psi = 0$). The two liquids are not at equilibrium.

$$\psi_p = 0$$
$$\psi_s = -0.4$$
$$\psi = -0.4$$

$$\psi = 0$$

Membrane

4 A piston resists the entry of water, as does the wall of a plant cell. The solution in the tube is not diluted, so its ψ_s does not change. However, the system is not initially at equilibrium. Enough water squeezes in to raise ψ_p until equilibrium is reached, with equal water potentials.

Piston

$$\psi_p = 0.4$$
$$\psi_s = -0.4$$
$$\psi = 0$$

$$\psi = 0$$

35.1 Water Potential, Solute Potential, and Pressure Potential
Water potential (ψ) is the tendency of a solution to take up water from pure water. The water potential is the sum of the solute potential (ψ_s) and the pressure potential (ψ_p). For pure water under no applied pressure, all three of these parameters are equal to zero.

port proteins. (You may wish to review the description of transport proteins in Chapter 5.) When the concentration of these charged ions in the soil is greater than that in the plant, ion channels and carrier proteins can move them into the plant by facilitated diffusion.

The concentrations of some ions in the soil solution, however, are lower than those required inside the plant. Thus the plant must take up these ions against a concentration gradient. Electric potential also plays a role in this process: To move a negatively charged ion into a negatively charged region is to move it against an electrical gradient. The combination of concentration and electrical gradients is called an *electrochemical gradient*. Uptake against an electrochemical gradient is active transport, an energy-requiring process, depending on cellular respiration for ATP. Active transport, of course, requires specific carrier proteins.

Unlike animals, plants do not have a sodium–potassium pump for active transport. Rather, plants have a **proton pump**, which uses energy obtained from ATP to move protons out of the cell against a proton concentration gradient (Figure 35.2a). Because protons (H^+) are positively charged, their accumulation on one side of a membrane has two results:

▶ The region outside the membrane becomes positively charged with respect to the region inside.

▶ A proton concentration gradient develops.

Each of these results has consequences for the movement of other ions. Because of the charge difference across the membrane, the movement of positively charged ions, such as potassium (K^+), into the cell through their membrane channels is enhanced. These positive ions move into the now more negatively charged interior of the cell by facilitated diffusion (Figure 35.2b). In addition, the proton concentration gradient can be harnessed to drive secondary active transport, in which negatively charged ions such as

chloride (Cl^-) are moved into the cell against an electrochemical gradient by a symport protein that couples their movement with that of H^+ (Figure 35.2c). In sum, there is vigorous traffic of ions across plant membranes.

The proton pump and the coordinated activities of other membrane transport proteins cause the interior of a plant cell to be strongly negative with respect to the exterior. Such a difference in charge across a membrane is called a *membrane potential*. Biologists can measure the membrane potential of a plant cell with microelectrodes, just as they can measure similar charge differences in neurons (nerve cells) and other animal cells. Most plant cells have a membrane potential of at least –120 millivolts, and they maintain it at this level. The membrane potential difference affects the movements of mineral ions into and out of cells.

Water and ions pass to the xylem by way of the apoplast and symplast

Mineral ions enter and move through plants in various ways. Where bulk flow of water is occurring, dissolved minerals are carried along in the stream. Where water is moving more slowly, minerals move by diffusion. At certain points, where plasma membranes are being crossed, some mineral ions are moved by active transport. One such point is the surface of a root hair, where mineral ions first enter the cells of the plant. Later, within the stele, the ions must cross a plasma membrane before entering the lifeless cells of the xylem.

The movement of ions across membranes can also result in the movement of water. Water moves into a root because the root has a more negative water potential than does the

35.2 The Proton Pump and Its Effects
The buildup of hydrogen ions transported across the plasma membrane by the proton pump (*a*) triggers the movement of both cations (*b*) and anions (*c*) into the cell.

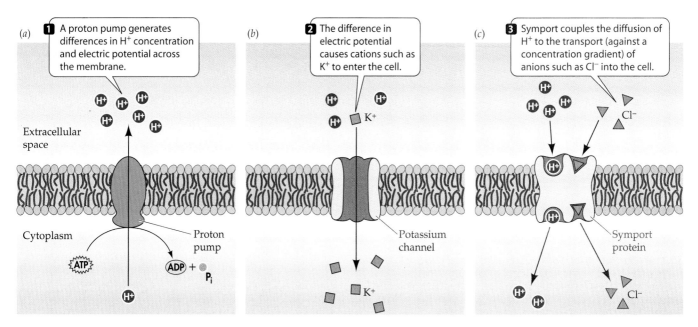

(*a*) **1** A proton pump generates differences in H^+ concentration and electric potential across the membrane.

Extracellular space

Cytoplasm

Proton pump

ATP

ADP + P_i

(*b*) **2** The difference in electric potential causes cations such as K^+ to enter the cell.

Potassium channel

(*c*) **3** Symport couples the diffusion of H^+ to the transport (against a concentration gradient) of anions such as Cl^- into the cell.

Symport protein

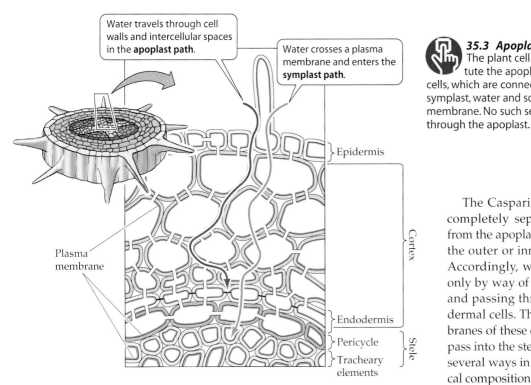

Water travels through cell walls and intercellular spaces in the **apoplast path**.

Water crosses a plasma membrane and enters the **symplast path**.

Plasma membrane

Epidermis

Cortex

Endodermis

Pericycle

Tracheary elements

Stele

35.3 *Apoplast and Symplast*
The plant cell walls and intercellular spaces constitute the apoplast. The symplast comprises the living cells, which are connected by plasmodesmata. To enter the symplast, water and solutes must pass through a plasma membrane. No such selective barrier limits movement through the apoplast.

soil solution. Water moves from the cortex of the root into the stele (which is where the vascular tissues are located) because the stele has a more negative water potential than does the cortex.

Water and minerals from the soil may pass through the dermal and ground tissues to the stele via two pathways: the apoplast and the symplast. Plant cells are surrounded by cell walls that lie outside the plasma membrane, and intercellular spaces (spaces between cells) are common in many tissues. The walls and intercellular spaces together constitute the **apoplast** (from the Greek for "away from living material"). The apoplast is a continuous meshwork through which water and dissolved substances can flow or diffuse without ever having to cross a membrane (Figure 35.3). Movement of materials through the apoplast is thus unregulated.

The remainder of the plant body is the **symplast** (from the Greek for "together with living material"). The symplast is the portion of the plant body enclosed by membranes—the continuous cytoplasm of the living cells, connected by plasmodesmata (see Figure 35.3). The selectively permeable plasma membranes of the cells control access to the symplast, so movement of water and dissolved substances into the symplast is tightly regulated.

Water and minerals can pass from the soil solution through the apoplast as far as the endodermis, the innermost layer of the cortex. The endodermis is distinguished from the rest of the ground tissue by the presence of **Casparian strips**. These waxy, suberin-containing structures impregnate the endodermal cell wall and form a belt surrounding the endodermal cells. The Casparian strips act as a gasket that prevents water and ions from moving between the cells (Figure 35.4).

The Casparian strips of the endodermis thus completely separate the apoplast of the cortex from the apoplast of the stele. They do not obstruct the outer or inner faces of the endodermal cells. Accordingly, water and ions can enter the stele only by way of the symplast—that is, by entering and passing through the cytoplasm of the endodermal cells. Thus transport proteins in the membranes of these cells determine which mineral ions pass into the stele, and at what rates. This is one of several ways in which plants regulate their chemical composition and ensure an appropriate balance of their constituents. This balance is essential to plant life.

Once they have passed the endodermal barrier, water and minerals leave the symplast. Parenchyma cells in the pericycle or xylem help mineral ions move back into the apoplast. Some of these parenchyma cells, called **transfer cells**, are structurally modified for transporting mineral ions from their cytoplasm (part of the symplast) into their cell walls (part of the apoplast). The wall that receives the transported ions has many knobby extensions projecting into the transfer cell, increasing the surface area of the plasma membrane, the

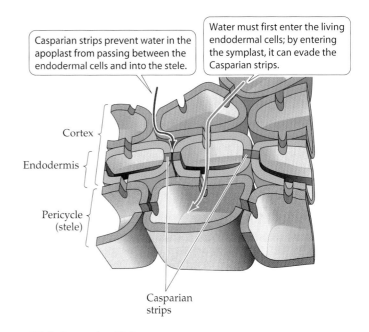

Casparian strips prevent water in the apoplast from passing between the endodermal cells and into the stele.

Water must first enter the living endodermal cells; by entering the symplast, it can evade the Casparian strips.

Cortex

Endodermis

Pericycle (stele)

Casparian strips

35.4 *Casparian Strips*
Suberin-impregnated Casparian strips prevent water and ions from moving between the endodermal cells.

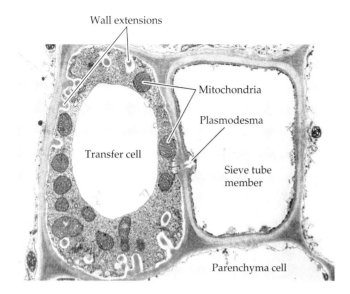

35.5 A Transfer Cell
Three walls of this transfer cell in a pea leaf have knobby extensions that face the cells from which the transfer cell imports
solutes. This transfer cell exports the solutes to the neighboring
sieve tube member.

number of transport proteins, and thus the rate of transport
(Figure 35.5).

Transfer cells also have many mitochondria that produce
the ATP needed to power the active transport of mineral
ions. As mineral ions move into the solution in the walls,
the water potential of the wall solution (apoplast) becomes
more negative; thus water moves out of the cells and into
the apoplast by osmosis. Active transport of ions moves the
ions directly, and water follows passively. The end result is
that water and minerals end up in the xylem, where they
constitute the *xylem sap.*

We have just seen that proteins regulate the movement
of ions across membranes. We shall now see that even
water movement itself is regulated by proteins.

Aquaporins control the rate, but not the direction, of water movement

Aquaporins are membrane channel proteins through which
water can traverse a membrane without interacting with the
hydrophobic environment of its phospholipid bilayer. These
proteins, important in both plants and animals, allow water
to move rapidly from environment to cell and from cell to
cell. The permeability of some aquaporins is subject to regulation, changing the *rate* of osmosis across the membrane.
However, water movement through aquaporins is always
passive, so the *direction* of water movement is unchanged by
alterations in aquaporin permeability.

Transport of Water and Minerals in the Xylem

So far in this chapter we've described the movement of
water and minerals into plant roots and their entry into the
root xylem. Now we will consider how xylem sap moves

throughout the remainder of the plant. Let's first consider
some early ideas about the ascent of sap and then turn to
our current understanding of how it works. We'll describe
the experiments that ruled out some early models as well
as some evidence in support of the current model—and
we'll find out what Per Scholander's sharpshooter was up
to in the story that opened this chapter.

Experiments ruled out some early models of transport in the xylem

Some of the earliest attempts to explain the rise of sap in
the xylem were based on a hypothetical pumping action by
living cells in the stem, which pushed the sap upward.
However, experiments conducted and published in 1893 by
the German botanist Eduard Strasburger definitively ruled
out such models.

Strasburger worked with trees about 20 meters tall. He
sawed them through at their bases and plunged the cut
ends into buckets containing solutions of poisons such as
picric acid. The solutions rose through the trunks, as was
readily evident from the progressive death of the bark
higher and higher up. When the solutions reached the
leaves, the leaves died, too, at which point the solutions
stopped being transported (as shown by the liquid levels in
the buckets, which stopped dropping).

This simple experiment established three important
points:

▶ Living, "pumping" cells were not responsible for the
  upward movement of the solutions, because the solutions themselves killed all living cells with which they
  came in contact.

▶ The leaves play a crucial role in transport. As long as
  they were alive, solutions continued to be transported
  upward; when the leaves died, transport ceased.

▶ Transport was not caused by the roots, because the
  trunks had been completely separated from the roots.

Root pressure does not account for xylem transport

In spite of Strasburger's observations, some plant physiologists turned to a model of transport based on **root pressure**—pressure exerted by the root tissues that would force
liquid up the xylem. The basis for root pressure is a higher
solute concentration, and accordingly a more negative
water potential, in the xylem sap than in the soil solution.
This negative potential draws water into the stele; once
there, the water has nowhere to go but up.

There is good evidence for root pressure—for example,
the phenomenon of *guttation*, in which liquid water is
forced out through openings in the leaves (Figure 35.6).
Guttation occurs only under conditions of high atmospheric humidity and plentiful water in the soil, which occur
most commonly at night. Root pressure is also the source of
the sap that oozes from the cut stumps of some plants, such
as *Coleus*, when their tops are removed. Root pressure,
however, cannot account for the ascent of sap in trees.

Root pressure seldom exceeds 0.1–0.2 MPa (one or two
times atmospheric pressure). If root pressure were driving

35.6 Guttation
Root pressure is responsible for forcing water through openings in the tips of this strawberry leaf.

sap up the xylem, we would observe a positive pressure potential in the xylem at all times. In fact, as we are about to see, the xylem sap is under tension—a negative pressure potential—when it is ascending. Furthermore, as Stras-

burger had already shown, materials can be transported upward in the xylem even when the roots have been removed. If the roots aren't pushing the xylem sap upward, what does cause it to rise?

The transpiration–cohesion–tension mechanism accounts for xylem transport

The obvious alternative to pushing is pulling: The leaves pull the xylem sap upward. **Transpiration**, the evaporative loss of water from the leaves, generates a pulling force (tension) on the water in the apoplast of the leaves. Hydrogen bonding between water molecules makes the sap in the xylem cohesive enough to withstand the tension and rise by bulk flow. Let's see how this **transpiration–cohesion–tension** mechanism works.

We'll start with transpiration. Water vapor diffuses from the intercellular spaces of the leaf, by way of the stomata, to the outside air because the water vapor concentration is greater inside the leaf than outside. Where did this water vapor come from?

As water vapor diffuses out of the leaf, more water evaporates from the moist walls of the mesophyll cells (Figure 35.7). Evaporation of water from the thin film surrounding the cell wall causes the film to shrink into the cellulose meshwork of the wall. The surface of the film curves where the water retreats into microscopic pores. The surface tension of the curved surfaces generates a **tension**—a negative pressure potential, a pull—in the film. The tension increases as more water leaves the film. This tension is what causes the bulk flow of water all the way from the roots.

The tension in the mesophyll draws water from the vessels or tracheids in the xylem of the nearest vein. The water, with its dissolved solutes, moves by bulk flow through the apoplast. The removal of water from the xylem of the veins establishes tension on the entire column of water contained within the xylem, so the column is drawn upward all the way from the roots.

The ability of water to be pulled upward through tiny tubes results from the remark-

Leaf Stoma START
1 Water vapor diffuses out of the stomata.

Mesophyll cell
2 Water evaporates from mesophyll cell walls.

Vein
3 Tension pulls water from the veins into the apoplast about the mesophyll cells.

4 Tension pulls the water column upward and outward in the xylem of veins in the leaves.

Stem
Xylem
5 Tension pulls the water column upward in the xylem of the stem.

Root
6 Tension pulls the water column upward in the xylem of the root.

7 Water molecules form a cohesive column.

8 Water moves into the stele by osmosis.

H_2O
Root hair   Xylem

35.7 Water Transport in Plants
Evaporation from surface cells, tension generated by the curvature of the shrinking surface film, and the cohesive nature of water molecules all account for the bulk flow of water from the soil to the atmosphere.

able cohesiveness of water—the tendency of water molecules to cohere to one another through hydrogen bonding. The narrower the tube, the greater the tension the water column can withstand without breaking. The integrity of the column is also maintained by the adhesion of water to the cell walls. In the tallest trees, such as a 100-meter redwood, the difference in pressure potential between the top and the bottom of the column may be as great as 3 MPa. The cohesiveness of water in the xylem is great enough to withstand even that great a tension.

In summary, the key elements of water transport in the xylem are:

▶ *Transpiration*, followed by evaporation from the moist cell walls in the leaves, resulting in…

▶ *tension* in the remainder of the xylem's water owing to the…

▶ *cohesion* of water, which pulls up more water to replace water that has been lost.

These elements require no work—no expenditure of energy—on the part of the plant. At each step between soil and atmosphere, water moves passively to a region with a more strongly negative water potential. Dry air has the most negative water potential (–95 MPa at 50% relative humidity), and the soil solution has the least negative water potential (between –0.01 and –3 MPa). Xylem sap has a water potential more negative than that of cells in the root, but less negative than that of mesophyll cells in the leaf.

Mineral ions contained in the xylem sap rise passively with the solution as it ascends from root to leaf. In this way the nutritional needs of the shoot are met. Some of the mineral elements brought to the leaves are subsequently redistributed to other parts of the plant by way of the phloem, but the initial delivery from the roots is through the xylem.

In addition to promoting the transport of minerals, transpiration contributes to temperature regulation. As water evaporates from mesophyll cells, heat is taken up from the cells, and the leaf temperature drops. This cooling effect is important in enabling plants to live in hot environments. A farmer can hold a leaf between thumb and forefinger to estimate its temperature; if the leaf doesn't feel cool, that means that transpiration is not occurring, so it must be time to water.

A pressure bomb measures tension in the xylem sap

The transpiration–cohesion–tension model can be true only if the column of sap in the xylem is under tension (negative pressure potential). The most elegant demonstrations of this tension, and of its adequacy to account for the ascent of sap in tall trees, were performed by Per Scholander. He measured tension in stems with an instrument called a **pressure bomb**.

The principle of the pressure bomb is as follows: Consider a stem in which the xylem sap is under tension. If the stem is cut, the sap pulls away from the cut, into the stem. Now the stem is placed in a device called a pressure bomb,

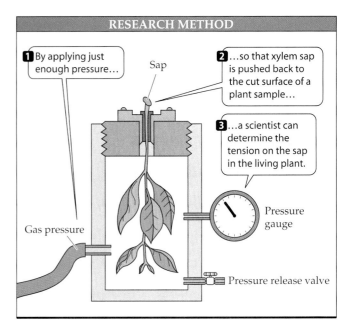

1 By applying just enough pressure…

Sap

2 …so that xylem sap is pushed back to the cut surface of a plant sample…

3 …a scientist can determine the tension on the sap in the living plant.

Pressure gauge

Gas pressure

Pressure release valve

35.8 A Pressure Bomb
The amount of tension on the sap in different types of plants can be measured with this laboratory device.

in which the pressure may be raised. The cut surface remains outside the bomb. As pressure is applied to the plant parts within the bomb, the xylem sap is forced back to the cut surface. When the sap first becomes visible again at the cut surface, the pressure in the bomb is recorded. This pressure is equal in magnitude but opposite in sign to the tension (negative pressure potential) originally present in the xylem (Figure 35.8).

Scholander used the pressure bomb to study dozens of plant species, from diverse habitats, growing under a variety of conditions. In all cases in which xylem sap was ascending, it was found to be under tension. The tension disappeared in some of the plants at night, when transpiration ceased. In developing vines, the xylem sap was under no tension until leaves formed. Once leaves developed, transport in the xylem began, and tensions were recorded.

Suppose you wanted to measure tensions in the xylem at various heights in a large tree, to confirm that the tensions are sufficient to account for the rate at which sap is moving up the trunk. How would you obtain stem samples for measurement? Scholander used surveying instruments to determine the heights of particular twigs and then had a sharpshooter shoot the twigs from the tree with a high-powered rifle. As quickly as the twigs fell to the ground, Scholander inserted them in the pressure bomb and recorded their xylem tension. In every case, the differences in tensions at different heights were great enough to keep the xylem sap ascending.

Although transpiration provides the impetus for transport of water and minerals in the xylem, it also results in the loss of tremendous quantities of water from the plant. How do plants control this loss?

Transpiration and the Stomata

The epidermis of leaves and stems minimizes transpirational water loss by secreting a waxy cuticle, which is impermeable to water. However, the cuticle is also impermeable to carbon dioxide. This poses a problem: How can the leaf balance its need to retain water with its need to obtain carbon dioxide for photosynthesis?

Plants have evolved an elegant compromise in the form of **stomata** (singular stoma), or gaps, in the epidermis. A pair of specialized epidermal cells called **guard cells** controls the opening and closing of each stoma (Figure 35.9*a*). When the stomata are open, carbon dioxide can enter the leaf by diffusion, but water vapor is also lost in the same way. Closed stomata prevent water loss, but also exclude carbon dioxide from the leaf.

Most plants open their stomata only when the light intensity is sufficient to maintain a moderate rate of photosynthesis. At night, when darkness precludes photosynthesis, the stomata remain closed; no carbon dioxide is needed at this time, and water is conserved. Even during the day, the stomata close if water is being lost at too great a rate.

The stoma and guard cells in Figure 35.9*a* are typical of eudicots. Monocots typically have specialized epidermal cells associated with their guard cells. The principle of operation, however, is the same for both monocot and eudicot stomata. In what follows, we describe the regulation and mechanism of stomatal opening, the normal cycle of opening and closing, and the modified cycle used by some plants that live in dry or saline environments.

The guard cells control the size of the stomatal opening

Light causes the stomata of most plants to open, admitting carbon dioxide for photosynthesis. Another cue for stomatal opening is the level of carbon dioxide in the spaces inside the leaf. A low level favors opening of the stomata, thus allowing the uptake of more carbon dioxide.

Water stress is a common problem for plants, especially on hot, sunny, windy days. Plants have a protective response to these conditions, using the water potential of the mesophyll cells as a cue. Even when the carbon dioxide level is low and the sun is shining, if the mesophyll is too dehydrated—that is, if the water potential of the mesophyll is too negative—the mesophyll cells release a plant hormone called *abscisic acid*. Abscisic acid acts on the guard cells, causing them to close the stomata and prevent further drying of the leaf. This response reduces the rate of photosynthesis, but it protects the plant.

The increasing internal concentration of potassium ions makes the water potential of the guard cells more negative. Water enters the guard cells by osmosis, increasing their pressure potential. Their cell walls contain cellulose microfibrils that cause the cells to respond to this increase by changing their shapes so that a gap—the stoma—appears between them.

The stoma closes by the reverse process when the proton pump is no longer active. Potassium ions diffuse passively

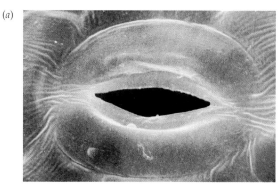

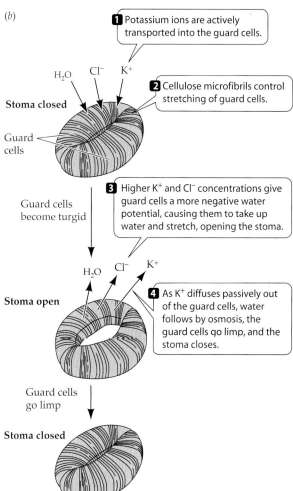

35.9 Stomata
(*a*) A scanning electron micrograph of a gaping stoma between two sausage-shaped guard cells. (*b*) Potassium ion (K⁺) concentrations and water potential control the opening and closing of stomata. Negatively charged ions traveling with K⁺ maintain electrical balance and contribute to the changes in osmotic potential that open and close the stomata.

out of the guard cells, water follows by osmosis, the pressure potential decreases, and the guard cells sag together and seal off the stoma. Negatively charged chloride ions and organic ions also move out of the guard cells with the potassium ions, maintaining electrical balance and contributing to the change in the solute potential of the guard cells.

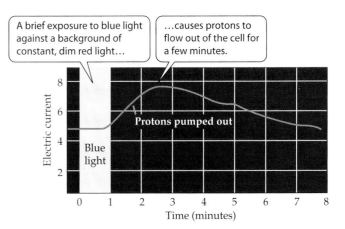

35.10 Light-Induced Proton Pumping in a Guard Cell Membrane
This graph shows a trace of the tiny electric current that results from the flow of protons across the plasma membrane of a guard cell when it is exposed briefly to blue light.

What drives the opening and closing of the stomata? Certain wavelengths of blue light, absorbed by a pigment in the guard cell plasma membrane, activate a proton pump, which actively transports protons (H^+) out of the guard cells and into the surrounding epidermis (Figure 35.10). The resulting proton gradient drives the accumulation of potassium ions (K^+) (Figure 35.9b) in the guard cell.

Antitranspirants decrease water loss

Stomata are the referees of a compromise between the admission of CO_2 for photosynthesis and the loss of water by transpiration. Farmers would like their crops to transpire less, thus reducing the need for irrigation. Similarly, nurseries and gardeners would like to be able to reduce the amount of water lost by plants that are to be transplanted, because transplanting often damages the roots, causing the plant to wilt or die. What we need is a good **antitranspirant**: a compound that can be applied to plants, reducing water loss from the stomata without producing disastrous side effects by excessively limiting carbon dioxide uptake.

Abscisic acid and its commercial chemical analogs have been found to work as antitranspirants in small-scale tests, but their high cost has precluded commercial use. What about making plants more sensitive to their own abscisic acid? The guard cells of plants with a genetic mutation called *era* are highly sensitive to abscisic acid. These plants are resistant to wilting during drought stress.

A totally different type of antitranspirant seals off the leaves from the atmosphere for a time. Growers use a variety of compounds, most of which form polymeric films around leaves, to form a barrier to evaporation. These compounds cause undesirable side effects, however, and can be used only for relatively short periods of time. Their most common use is in the transplanting of nursery stock.

Crassulacean acid metabolism correlates with an inverted stomatal cycle

Most plants open and close their stomata on a schedule like that shown by the blue curve in Figure 35.11. The stomata are typically open for much of the day and closed at night. (They may also close during very hot days to reduce water loss.) But not all plants follow this pattern.

Many plants that live in dry areas or near the ocean have some unusual biochemical and behavioral features. One particularly surprising feature is their "backward" stomatal cycle: Their stomata are open at night and closed by day (as shown by the red curve in Figure 35.11). This behavior is part of the phenomenon of **crassulacean acid metabolism (CAM)**, which was described in Chapter 8 (see Figure 8.21).

At night, while the stomata are open, carbon dioxide diffuses freely into the leaves of CAM plants and reacts in the mesophyll cells with phosphoenolpyruvic acid to produce organic acids. These acids accumulate to high concentrations. At daybreak the stomata close. Throughout the day, the organic acids are broken down to release the carbon dioxide they contain—behind closed stomata. Because the carbon dioxide cannot diffuse out of the plant, it is available for photosynthesis.

CAM is well adapted to environments where water is scarce: A leaf with its stomata open only at night—when the environment is cooler—loses much less water than does a leaf with its stomata open by day.

In both CAM and non-CAM plants, carbon dioxide is fixed and converted to the products of photosynthesis. How are these products delivered to other parts of the plant?

Translocation of Substances in the Phloem

Substances in the phloem move from sources to sinks. A **source** is an organ (such as a mature leaf or a storage root) that produces (by photosynthesis or by digestion of stored

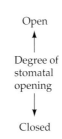

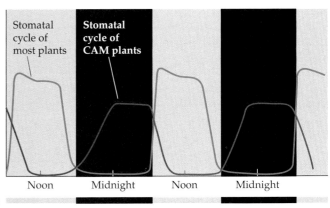

35.11 Stomatal Cycles
Most plants open their stomata during the day. CAM plants reverse this stomatal cycle: Their stomata open during the night.

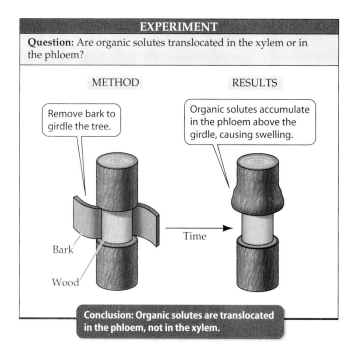

EXPERIMENT

Question: Are organic solutes translocated in the xylem or in the phloem?

METHOD

Remove bark to girdle the tree.

Bark

Wood

Time

RESULTS

Organic solutes accumulate in the phloem above the girdle, causing swelling.

Conclusion: Organic solutes are translocated in the phloem, not in the xylem.

35.12 Girdling Blocks Translocation in the Phloem
By removing a ring of bark (containing the phloem), Malpighi blocked the translocation of organic solutes in a tree.

the bark below the girdle died because it no longer received sugars from the leaves.

Any model to explain translocation of organic solutes must account for a few important facts:

▶ Translocation stops if the phloem tissue is killed by heating or other methods; thus the mechanism must be different from that of transport in the xylem.
▶ Translocation often proceeds in both directions—up and down the stem—simultaneously.
▶ Translocation is inhibited by compounds that inhibit respiration and thus limit the ATP supply in the source.

To investigate translocation, plant physiologists needed to obtain samples of pure sieve tube sap from individual sieve tube members. This difficult task was simplified when scientists discovered that a common garden pest, the aphid, feeds by drilling into a sieve tube. An aphid inserts its stylet, or feeding organ, into a stem until the stylet enters a sieve tube (Figure 35.13a). Within the sieve tube, the pressure is much greater than in the surrounding plant tissues, so nutritious sieve tube sap is forced up the stylet and into the aphid's digestive tract. So great is the pressure that sugary liquid is forced through the insect's body and out the anus (Figure 35.13b).

Plant physiologists use aphids to collect sieve tube sap. When liquid appears on the aphid's abdomen, indicating that the insect has connected with a sieve tube, the physiologist quickly freezes the aphid and cuts its body away from the stylet, which remains in the sieve tube member. For hours, sieve tube sap continues to exude from the cut stylet, where it may be collected for analysis. Chemical analysis of sieve tube sap collected in this manner reveals the contents of a single sieve tube member over time. We can also infer the rates at which different substances are translocated by measuring how long it takes for radioactive tracers administered to a leaf to appear at stylets at different distances from the leaf.

These methods have allowed us to understand how, at times, different substances might move in opposite directions in the phloem of a stem. Experiments with aphid stylets have shown that all the contents of any given sieve tube member move in the same direction. Thus, bidirectional translocation can be understood in terms of different sieve tubes conducting sap in opposite directions. Data obtained by these and other means led to the general adoption of the pressure flow model as an explanation for translocation in the phloem.

reserves) more sugars than it requires. A **sink** is an organ (such as a root, a flower, a developing tuber, or an immature leaf) that does not make enough sugar for its own growth and storage needs. Sugars (primarily sucrose), amino acids, some minerals, and a variety of other substances are translocated between sources and sinks in the phloem.

How do we know that such organic solutes are translocated in the phloem, rather than in the xylem? Just over 300 years ago, the Italian scientist Marcello Malpighi performed a classic experiment in which he removed a ring of bark (containing the phloem) from the trunk of a tree—that is, he *girdled* the tree (Figure 35.12). The bark in the region above the girdle swelled over time. We now know that the swelling resulted from the accumulation of organic solutes that came from higher up the tree and could no longer continue downward because of the disruption of the phloem. Later,

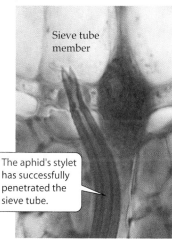

(a)

Sieve tube member

The aphid's stylet has successfully penetrated the sieve tube.

(b) *Longistigma caryae*

Sap droplet

35.13 Aphids Collect Sieve Tube Sap
(a) Aphids feed on phloem sap drawn from the sieve tube, which they penetrate with a modified feeding organ, the stylet. (b) Pressure inside the sieve tube forces sap through the aphid's digestive tract, from which it can be harvested.

The pressure flow model appears to account for phloem translocation

The tonoplast breaks down during sieve tube member development, allowing the contents of the central vacuole to combine with much of the cytosol to form the sieve tube sap (see Chapter 34). The sap flows under pressure through the sieve tubes. It moves from one sieve tube member to the next by bulk flow through the sieve plates, without crossing a membrane.

Two steps in sieve tube sap flow require metabolic energy:

▶ Transport of sucrose and other solutes into the sieve tubes (**loading**) at sources

▶ Removal (**unloading**) of the solutes where the sieve tubes enter sinks

According to the **pressure flow model** of translocation in the phloem, sucrose is actively transported into sieve tube members at sources, giving these cells a much greater sucrose concentration than surrounding cells. Water therefore enters the sieve tube members by osmosis. The entry of this water causes a greater pressure potential at the source end, so the entire fluid content of the sieve tube is pushed to-

ward the sink end of the tube—that is, the sap moves by bulk flow (Figure 35.14).

The pressure flow model of translocation in the phloem is contrasted with the transpiration–cohesion–tension model of xylem transport in Table 35.1.

Testing the pressure flow model

The pressure flow model was first proposed more than half a century ago, but some of its features are still debated. Other mechanisms have been proposed to account for translocation in sieve tubes. Some have been disproved, and none of the rest have been supported by a weight of evidence comparable to that for the pressure flow model, which must meet two requirements:

▶ The sieve plates must be open, so that bulk flow from one sieve tube member to the next is possible.

▶ There must be an effective method for loading sucrose and other solutes into the phloem in source tissues and removing them in sink tissues.

Let us see whether these requirements are met.

ARE THE SIEVE PLATES CLOGGED OR OPEN? Early electron microscopic studies of phloem samples cut from plants produced results that seemed to contradict the pressure flow

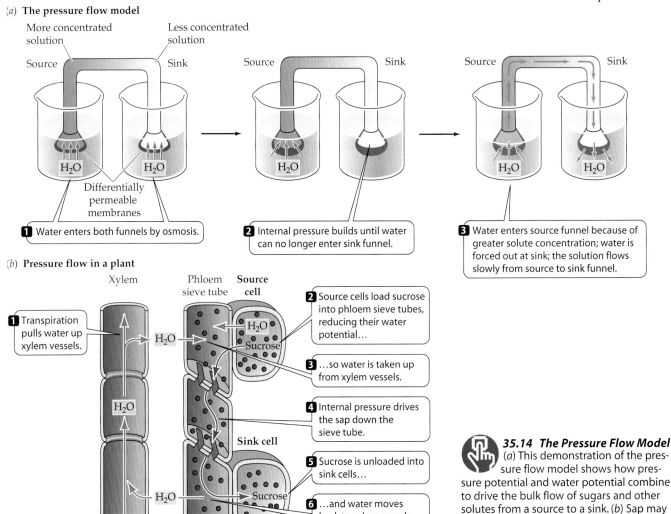

(a) **The pressure flow model**

More concentrated solution   Less concentrated solution

Source   Sink

Source   Sink

Source   Sink

Differentially permeable membranes

1 Water enters both funnels by osmosis.

2 Internal pressure builds until water can no longer enter sink funnel.

3 Water enters source funnel because of greater solute concentration; water is forced out at sink; the solution flows slowly from source to sink funnel.

(b) **Pressure flow in a plant**

Xylem   Phloem sieve tube   **Source cell**

1 Transpiration pulls water up xylem vessels.

2 Source cells load sucrose into phloem sieve tubes, reducing their water potential…

3 …so water is taken up from xylem vessels.

4 Internal pressure drives the sap down the sieve tube.

Sink cell

5 Sucrose is unloaded into sink cells…

6 …and water moves back to xylem vessels.

35.14 The Pressure Flow Model (a) This demonstration of the pressure flow model shows how pressure potential and water potential combine to drive the bulk flow of sugars and other solutes from a source to a sink. (b) Sap may flow through sieve tubes in this manner.

35.1 *Mechanisms of Bulk Flow in Plant Vascular Tissues*

	XYLEM	PHLOEM
Source of bulk flow	Transpiration from leaves	Active transport of sucrose at source
Site of bulk flow	Dead vessel elements and tracheids	Living sieve tube members
Pressure potential in sap	Negative (pull from top)	Positive (push from source)

model. The pores in the sieve plates always appeared to be plugged with masses of a fibrous protein, suggesting that sieve tube sap could not flow freely. But what is the function of that fibrous protein?

One possibility is that this protein is usually distributed more or less at random throughout the sieve tube members until the sieve tube is damaged; then the sudden surge of sap toward the cut surface carries the protein into the pores, blocking them and preventing the loss of valuable nutrients. In other words, perhaps the protein does *not* block the pores unless the phloem is damaged. How might this possibility be tested? Could we obtain phloem for microscopic observation without causing the sap to surge to the cut surface?

One way to prevent the surge of the sap is to freeze plant tissue before cutting it. Another way is to let the tissue wilt so that there is no pressure in the phloem before cutting. When these methods were used, the sieve plates were not clogged by the protein. Thus, the first condition of the pressure flow model is met.

NEIGHBORING CELLS LOAD AND UNLOAD THE SIEVE TUBE MEMBERS.   If the pressure flow model is correct, there must be mechanisms for loading sugars and other solutes into the phloem in source regions and for unloading them in sink regions. One pathway of phloem loading has been demonstrated in some plant species.

Sugars and other solutes pass from cell to cell through the symplast in the mesophyll. When these substances reach cells adjacent to the ends of leaf veins, they leave the mesophyll cells and enter the apoplast, sometimes with the help of transfer cells. Then specific sugars and amino acids are actively transported into cells of the phloem, thus reentering the symplast (Figure 35.15).

Passage through the apoplast and back into the symplast selects substances to be accumulated for translocation because substances can enter the phloem only after passing through a differentially permeable membrane. In many plants, solutes reenter the symplast at the companion cells (see Chapter 34), which then transfer the solutes to the adjacent sieve tube members. As Figure 35.15 shows, in other plant species, sucrose or other sugars move from the mesophyll to the sieve tube members entirely within the symplast; that is, transfer of solutes from symplast to apoplast and back again is not a universal feature of phloem loading.

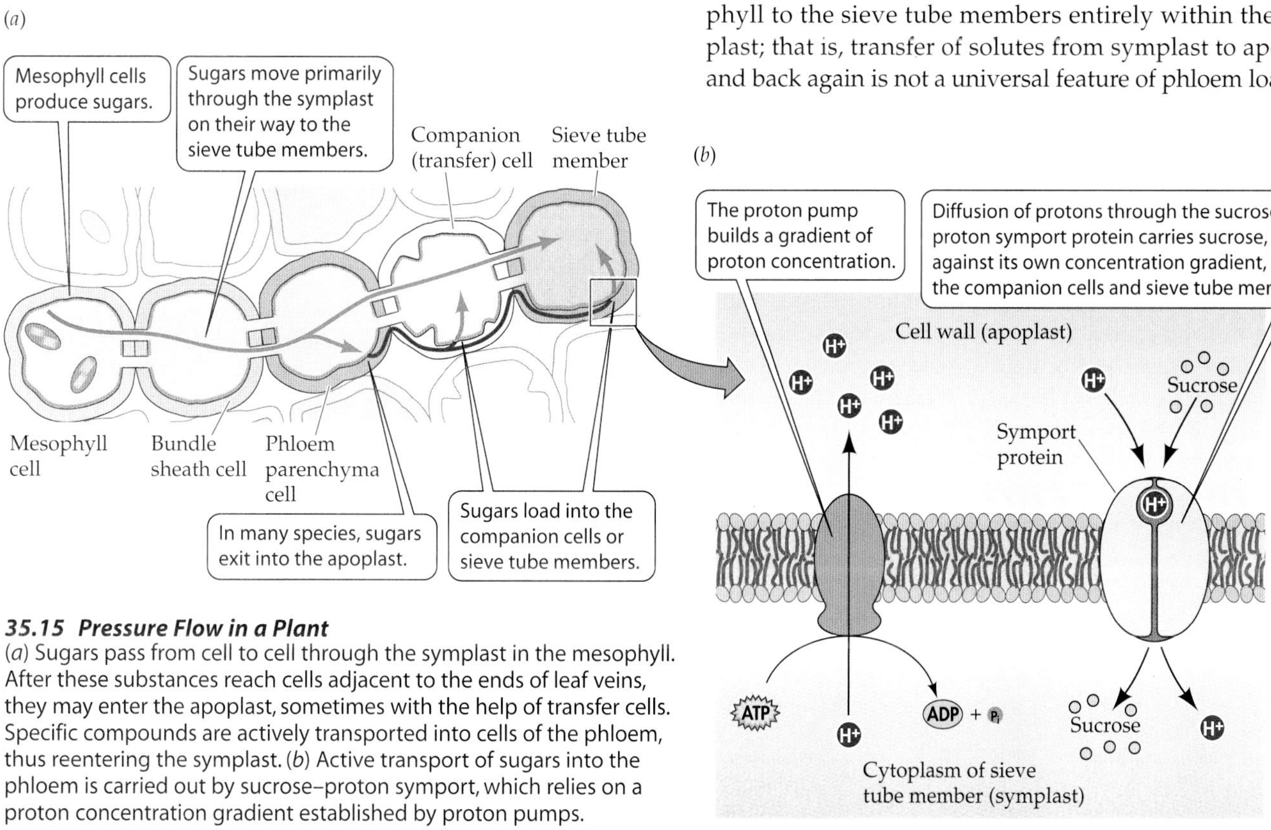

(a)

(b)

35.15 Pressure Flow in a Plant
(a) Sugars pass from cell to cell through the symplast in the mesophyll. After these substances reach cells adjacent to the ends of leaf veins, they may enter the apoplast, sometimes with the help of transfer cells. Specific compounds are actively transported into cells of the phloem, thus reentering the symplast. (b) Active transport of sugars into the phloem is carried out by sucrose–proton symport, which relies on a proton concentration gradient established by proton pumps.

A form of secondary active transport (see Chapter 5, pages 88–90) loads sucrose into the companion cells and sieve tube members. Sucrose is carried through the plasma membrane from apoplast to symplast by sucrose–proton symport; thus the entry of sucrose and of protons is strictly coupled. For this symport to work, the apoplast must have a high concentration of protons; the protons are supplied by a primary active transport system, the proton pump. The protons then diffuse back into the cell through the symport protein, bringing sucrose with them.

In sink regions, the solutes are actively transported out of the sieve tube members and into the surrounding tissues. This unloading serves two purposes: It helps maintain the gradient of solute potential and hence of pressure potential in the sieve tubes, and it promotes the buildup of sugars and starch to high concentrations in storage regions, such as developing fruits and seeds.

Plasmodesmata and material transfer between cells

Many substances move from cell to cell within the symplast by way of plasmodesmata (see Figure 34.7). Among their other roles, plasmodesmata participate in the loading and unloading of sieve tube members. Mechanisms vary among plant species, but the story in tobacco plants is a common one. In tobacco, sugars and other compounds in source tissues enter companion cells by active transport from the apoplast and move on to the sieve tube members through plasmodesmata. In sink tissues, plasmodesmata connect sieve tube members, companion cells, and the cells that will receive and use the transported compounds.

Plasmodesmata undergo developmental changes as an immature sink leaf matures into a mature source leaf. Plasmodesmata in sink tissues favor rapid unloading: They are more abundant, and they allow the passage of larger molecules. Plasmodesmata in source tissues are few in number.

It was long thought that only substances with molecular weights less than 1,000 could fit through a plasmodesma. Then biologists discovered that cells infected with tobacco mosaic virus (TMV) could allow molecules with molecular weights of as much as 20,000 to exit. We now know that TMV encodes a "movement protein" that produces this change in the permeability of the plasmodesmata—and that plants themselves normally produce at least one such movement protein. Even large molecules such as proteins and RNAs, with molecular weights up to at least 50,000, can thus move between living plant cells. We will see some consequences of this movement of macromolecules through plasmodesmata in later chapters. Biologists are exploring possible ways to regulate the permeability, number, and form of plasmodesmata as a means of modifying traffic in the plant. Such modifications might, for example, allow the diversion of more of a grain crop's photosynthetic products into the grain, increasing the crop yield.

Chapter Summary

Uptake and Transport of Water and Minerals
▶ Plant roots take up water and minerals from the soil.
▶ Water moves through biological membranes by osmosis, always moving toward cells with a more negative water potential. The water potential of a cell or solution is the sum of the solute potential and the pressure potential. All three parameters are expressed in megapascals (MPa). **Review Figure 35.1**
▶ Mineral uptake requires transport proteins. Some minerals enter the plant by facilitated diffusion; others enter by active transport. A proton pump facilitates the active transport of many solutes across membranes in plants. **Review Figure 35.2**
▶ Water and minerals pass from the soil to the xylem by way of the apoplast and symplast. In the root, water and minerals may pass from the cortex into the stele only by way of the symplast because Casparian strips in the endodermis block water and solute movement in the apoplast. **Review Figures 35.3, 35.4**

Transport of Water and Minerals in the Xylem
▶ Early experiments established that sap does not move via the pumping action of living cells.
▶ Root pressure is responsible for guttation and for the oozing of sap from cut stumps, but it cannot account for the ascent of sap in trees.
▶ Xylem transport is the result of the combined effects of transpiration, cohesion, and tension. Evaporation in the leaf produces tension in the surface film of water on the moist-walled mesophyll cells, and thus pulls water—held together by its cohesiveness—up through the xylem from the root. Dissolved minerals go along for the ride. **Review Figure 35.7**
▶ Support for the transpiration–cohesion–tension model of xylem transport came from studies using a pressure bomb. **Review Figure 35.8**

Transpiration and the Stomata
▶ Evaporation of water cools the leaves, but a plant cannot afford to lose too much water. Transpirational water loss is minimized by the waxy cuticle of the leaves.
▶ Stomata allow a compromise between water retention and carbon dioxide uptake. A pair of guard cells controls the size of the stomatal opening. A proton pump, activated by blue light, pumps protons from the guard cells to surrounding epidermal cells. As a result, the guard cells take up potassium ions, causing water to follow osmotically, swelling the cells and opening the stomata. Carbon dioxide level and water availability also affect stomatal opening. **Review Figures 35.9, 35.10**
▶ In most plants the stomata are open during the day and closed at night. CAM plants have an inverted stomatal cycle, enabling them to conserve water. **Review Figure 35.11**

Translocation of Substances in the Phloem
▶ Products of photosynthesis, and some minerals, are translocated through sieve tubes in the phloem by way of living sieve tube members. Translocation proceeds in both directions in the stem, although in a single sieve tube it goes only one way. Translocation requires a supply of ATP.

▶ Translocation in the phloem proceeds in accordance with the pressure flow model: The difference in solute concentration between sources and sinks allows a difference in pressure potential along the sieve tubes, resulting in bulk flow. **Review Figure 35.14, Table 35.1**

▶ The pressure flow model succeeds because the sieve plates are normally open, allowing bulk flow, and because neighboring cells load organic solutes into the sieve tube members in source regions and unload them in sink regions. **Review Figure 35.15**

▶ The distribution and properties of plasmodesmata differ between source and sink tissues. It may become possible to regulate plasmodesmata in crop plants.

For Discussion

1. Epidermal cells protect against excess water loss. How do they perform this function?

2. Phloem transports material from sources to sinks. What is meant by "source" and "sink"? Give examples of each.

3. What is the minimum number of plasma membranes a water molecule would have to cross in order to get from the soil solution to the atmosphere by way of the stele? To get from the soil solution to a mesophyll cell in a leaf.

4. Transpiration exerts a powerful pulling force on the water column in the xylem. When would you expect transpiration to proceed most rapidly? Why? Describe the source of the pulling force.

5. Plants that perform crassulacean acid metabolism (CAM plants) are adapted to environments in which water supply is limited; these plants open their stomata only at night. Could a non-CAM plant, such as a pea plant, enjoy an advantage if it opened its stomata only at night? Explain.

36 *Plant Nutrition*

AN INSECT HAS STEPPED ON A TRIGGER HAIR on the leaf of a Venus flytrap—a big mistake for the insect. The trigger hair sends an electrical signal that springs a mechanical trap. The two halves of the leaves close, and spiny outgrowths at the margins of the leaves interlock to imprison the insect. The leaf secretes enzymes that will digest its prey. The leaf then absorbs the products of digestion, especially amino acids, and uses them as a nutritional supplement.

Why does the Venus flytrap go to all this trouble? Few other plants are carnivorous—your petunia plant is not stalking you, after all. But the Venus flytrap (*Dionaea muscipula*) lives on soils in which nitrogen is scarce. Its carnivorous adaptation gives it another way to obtain needed nitrogen.

Why do plants need nitrogen? The answer is simple if we recall the chemical structures of proteins and nucleic acids that we looked at in Chapter 3. These vital components of all living things contain nitrogen, as do chlorophyll and many other important biochemical compounds. If a plant cannot get enough nitrogen, it cannot synthesize these compounds at a rate adequate to keep itself healthy.

In addition to nitrogen, plants need other materials from their environment. In this chapter we explore the differences between the basic strategies of plants and animals for obtaining nutrition. Then we look at what nutrients plants require, and how they acquire them. Because most nutrients come from the soil, we discuss the formation of soils and the effects of plants on soils. As any farmer can tell you, nitrogen is the nutrient that most often limits plant growth, so we devote a section specifically to nitrogen metabolism in plants. The chapter concludes with a look at plants that use means other than photosynthesis to supplement their nutrition.

The Acquisition of Nutrients

Every living thing must obtain raw materials from its environment. These **nutrients** include the major ingredients of macromolecules: carbon, hydrogen, oxygen, and nitrogen. Carbon and oxygen enter the living world through the carbon-fixing reactions of photosynthesis, in which photosynthetic organisms obtain them from atmospheric carbon dioxide. Hydrogen enters living systems through the light reactions of photosynthesis, which split water. For carbon, oxygen, and hydrogen, photosynthesis is the gateway to the living world.

The movement of nitrogen into organisms begins with processing by some highly specialized bacteria living in the soil. Some of these bacteria act on nitrogen gas, converting it into a form usable by plants. The plants in turn provide organic nitrogen and carbon to animals, fungi, and many microorganisms.

In addition to carbon, oxygen, hydrogen, and nitrogen, other **mineral nutrients** are essential to living systems. The proteins of organisms contain sulfur (S), and their nucleic acids contain phosphorus (P). There is magnesium (Mg) in chlorophyll, and iron (Fe) in many important compounds, such as the cytochromes. Within the soil, these and other minerals dissolve in water, forming a solution—called the **soil solution**—that contacts the roots of plants. Plants take up most of these mineral nutrients from the soil solution in ionic form.

A Meat-Eating Plant
Dionaea muscipula, the Venus flytrap, has adapted to a nitrogen-poor environment by becoming carnivorous. It obtains this necessary mineral from the bodies of insects trapped inside the plant when its hinges snap shut.

Autotrophs make their own organic compounds

The plants provide carbon, oxygen, hydrogen, nitrogen, and sulfur to most of the rest of the living world. Plants, some protists, and some bacteria are *autotrophs*; that is, they make their own organic compounds from simple inorganic nutrients—carbon dioxide, water, nitrate or ammonium ions containing nitrogen, and a few other soluble mineral nutrients (Figure 36.1). *Heterotrophs* are organisms that require preformed *organic* compounds (compounds that contain carbon) as food. Herbivores and carnivores are heterotrophs that depend directly or indirectly on autotrophs as their source of nutrition.

Most autotrophs are *photosynthesizers*—that is, they use light as the source of energy for synthesizing organic compounds from inorganic raw materials. Some autotrophs, however, are *chemosynthesizers*, deriving their energy not from light, but from reduced inorganic substances, such as hydrogen sulfide (H_2S), in their environment. All chemosynthesizers are bacteria. Some chemosynthetic bacteria in the soil contribute to the nutrition of plants by increasing the availability of nitrogen and sulfur. But how does a plant obtain its nutrients, whether they come with or without bacterial action?

How does a stationary organism find nutrients?

An organism that cannot move must exploit energy that is somehow brought to it. Most sessile animals depend primarily on the movement of water to bring energy, in the form of food, to them, but a plant's supply of energy arrives at the speed of light from the sun. A plant's supply of nutrients, however, is strictly local, and the plant may use up the water and mineral nutrients in its local environment as it develops. How does a plant cope with such a problem?

One answer is to extend itself by growing into new resources. *Growth is a plant's version of locomotion*. Root systems mine the soil; by growing, they reach new sources of mineral nutrients and water. Growth of leaves helps a plant secure light and carbon dioxide. A plant may compete with other plants for light by outgrowing them, both capturing more light for itself and preventing the growth of its neighbors by shading them.

As it grows, a plant—or even a single root—must deal with environmental heterogeneity. Animal droppings create high local concentrations of nitrogen. A particle of calcium carbonate in the soil may make a tiny area alkaline, while dead organic matter may make a nearby area acidic.

Mineral Nutrients Essential to Plants

What important mineral nutrients do plants take up from their environment, and what are their roles? Table 36.1 lists the mineral nutrients that have been proved to be essential for plants. Except for nitrogen, they all come from the soil solution and derive ultimately from rock.

There are three criteria for calling something an **essential element**:

▸ The element must be *necessary* for normal growth and reproduction.
▸ The element cannot be *replaceable* by another element.
▸ The requirement must be *direct*—that is, not the result of an indirect effect, such as the need to relieve toxicity caused by another substance.

In this section, we'll consider the symptoms of particular mineral deficiencies, the roles of some of the mineral nutrients, and the technique by which the essential elements for plants were identified.

The essential elements in Table 36.1 are divided into two categories: the macronutrients and the micronutrients. Plant tissues need **macronutrients** in concentrations of at least 1 gram per kilogram of their dry matter, and they need **micronutrients** in concentrations of less than 100 milligrams per kilogram of their dry matter. (Dry matter, or dry weight, is what remains after all the water has been removed from a tissue sample.) Both the macronutrients and the micronutrients are essential for the plant to complete its life cycle from seed to seed.

Deficiency symptoms reveal inadequate nutrition

Before a plant that is deficient in an essential element dies, it usually displays characteristic deficiency symptoms. Table 36.2 describes the symptoms of some common mineral deficiencies. Such symptoms help horticulturists diagnose mineral nutrient deficiencies in plants.

36.1 What Do Plants Need?
To survive, plants require only light plus carbon dioxide, water, and several essential mineral elements. These plants are growing on nothing more than a solution that contains water and mineral elements. This technique is known as hydroponic culture.

36.1   Mineral Elements Required by Plants

ELEMENT	ABSORBED FORM	MAJOR FUNCTIONS
Macronutrients		
Nitrogen (N)	NO_3^- and NH_4^+	In proteins, nucleic acids, etc.
Phosphorus (P)	$H_2PO_4^-$ and HPO_4^{2-}	In nucleic acids, ATP, phospholipids, etc.
Potassium (K)	K^+	Enzyme activation; water balance; ion balance; stomatal opening
Sulfur (S)	SO_4^{2-}	In proteins and coenzymes
Calcium (Ca)	Ca^{2+}	Affects the cytoskeleton, membranes, and many enzymes; second messenger
Magnesium (Mg)	Mg^{2+}	In chlorophyll; required by many enzymes; stabilizes ribosomes
Micronutrients		
Iron (Fe)	Fe^{2+}	In active site of many redox enzymes and electron carriers; chlorophyll synthesis
Chlorine (Cl)	Cl^-	Photosynthesis; ion balance
Manganese (Mn)	Mn^{2+}	Activation of many enzymes
Boron (B)	$B(OH)_3$	Possibly carbohydrate transport (poorly understood)
Zinc (Zn)	Zn^{2+}	Enzyme activation; auxin synthesis
Copper (Cu)	Cu^{2+}	In active site of many redox enzymes and electron carriers
Nickel (Ni)	Ni^{2+}	Activation of one enzyme
Molybdenum (Mo)	MoO_4^{2-}	Nitrogen fixation; nitrate reduction

Nitrogen deficiency is the most common mineral deficiency in plants. Plants in natural environments are almost always deficient in nitrogen, but they seldom display deficiency symptoms. Instead, their growth slows to match the available supply of nitrogen. Crop plants, on the other hand, show deficiency symptoms if a formerly abundant supply of nitrogen runs out. The visible symptoms of nitrogen deficiency include uniform yellowing, or *chlorosis*, of older leaves. Chlorophyll, which is responsible for the green color of leaves, contains nitrogen. Without nitrogen there is no chlorophyll, and without chlorophyll, the yellow pigments become visible.

36.2   Some Mineral Deficiencies in Plants

DEFICIENCY	SYMPTOMS
Calcium	Growing points die back; young leaves are yellow and crinkly
Iron	Young leaves are white or yellow with green veins
Magnesium	Older leaves have yellow in stripes between veins
Manganese	Younger leaves are pale with stripes of dead patches
Nitrogen	Oldest leaves turn yellow and die prematurely; plant is stunted
Phosphorus	Plant is dark green with purple veins and is stunted
Potassium	Older leaves have dead edges
Sulfur	Young leaves are yellow to white with yellow veins
Zinc	Young leaves are abnormally small; older leaves have many dead spots

Inadequate available iron in the soil can also cause chlorosis because, although it is not contained in the chlorophyll molecule, iron is required for chlorophyll synthesis. However, iron deficiency commonly causes chlorosis of the *youngest* leaves, with their veins sometimes remaining green. The reason for this difference is that nitrogen is readily translocated in the plant and can be redistributed from older tissues to younger tissues to favor their growth. Iron, on the other hand, cannot be readily redistributed. Younger tissues that are actively growing and synthesizing compounds needed for their growth show iron deficiency before older leaves, which have already completed their growth.

Several essential elements fulfill multiple roles

Essential elements can play several different roles—some structural, others catalytic. Magnesium, as we have mentioned, is a constituent of the chlorophyll molecule and hence is essential to photosynthesis. It is also required as a cofactor by numerous enzymes in cellular respiration and other metabolic pathways.

Phosphorus, usually in phosphate groups, is found in many organic compounds, particularly in nucleic acids and in the intermediates of the energy pathways of photosynthesis and glycolysis. The transfer of phosphate groups occurs in many energy-storing and energy-releasing reactions, notably those that use or produce ATP. Other roles of phosphate groups include the activation and inactivation of enzymes.

Calcium plays many roles in plants. Its function in the processing of hormonal and environmental cues is the subject of great biological interest, as we'll see in the next chapter. Calcium also affects membranes and cytoskeleton activity, participates in spindle formation for mitosis and meiosis, and is a constituent of the middle lamella of cell

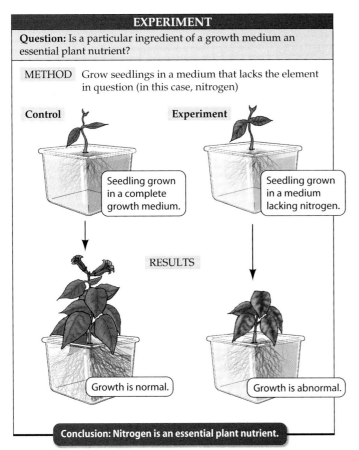

EXPERIMENT

Question: Is a particular ingredient of a growth medium an essential plant nutrient?

METHOD   Grow seedlings in a medium that lacks the element in question (in this case, nitrogen)

Control

Experiment

Seedling grown in a complete growth medium.

Seedling grown in a medium lacking nitrogen.

RESULTS

Growth is normal.

Growth is abnormal.

Conclusion: Nitrogen is an essential plant nutrient.

36.2 Identifying Essential Elements for Plants
The diagram shows the procedure for identifying nutrients essential to plants, using nitrogen as an example. The environment in such experiments must be rigorously controlled because some essential elements are needed in only tiny amounts, and may be present in sufficient quantities as contaminants.

walls. Other elements, such as iron and potassium, also play multiple roles.

All of these elements are essential to the life of all plants. How did biologists discover which elements are essential?

The identification of essential elements

An element is considered essential if a plant fails to complete its life cycle, or grows abnormally, when that element is not available, or is not available in sufficient quantities. Plant physiologists identified most of the essential elements for plants by the technique outlined in Figure 36.2. This technique is limited, however, by the possibility that some elements thought to be absent from the test solutions are actually present. Impurities and contamination are always possible.

In early experiments on plant nutrition, some of the chemicals used were so

impure that they provided micronutrients that the investigators thought they had excluded. Some mineral nutrients are required in such tiny amounts that there may be enough in a seed to supply the embryo and the resultant plant throughout its lifetime and leave enough in the next seed to get the next generation well started. Simply touching a plant may give it a significant dose of chlorine in the form of chloride ions from sweat. Only rarely are new essential elements reported now. Either the list is nearly complete, or more likely, we will need more sophisticated techniques to add to it.

Where does the plant find its essential mineral nutrients? How does it absorb them?

Soils and Plants

Soils are very important to plants, and plant interactions with the soil are complex. Plants obtain their mineral nutrients from the soil solution or the water in which they grow. Water for terrestrial plants also comes from the soil, as does the supply of oxygen for the roots. Soil also provides mechanical support for plants on land, and it harbors bacteria that perform chemical reactions leading to products required for plant growth. On the other hand, soil may also contain organisms harmful to plants.

In the pages that follow, we'll examine the composition and structure of soils, their formation, their role in plant nutrition, their care and supplementation in agriculture, and their modification by the plants that grow in them.

Soils are complex in structure

Soils are complex systems of living and nonliving components. The living components include plant roots, as well as populations of bacteria, fungi, and animals such as earthworms and insects (Figure 36.3). The nonliving portion of the soil includes rock fragments ranging in size from large

Soil consists of more than inorganic mineral particles such as clay and quartz.

Quartz

Living organisms such as bacteria are present.

Air

Organic matter

Organic matter (from plants, animals, and fungi) can also be found.

Air

H_2O

Air

Air and water are present in pores in soil crumbs like this one.

Aggregates of clay particles

25 μm

36.3 The Complexity of Soil
Soil has both organic and inorganic components.

36.3 Two Systems for Classifying Soil Particles

UNITED STATES DEPARTMENT OF AGRICULTURE		INTERNATIONAL SOCIETY FOR SOIL SCIENCE	
SOIL TYPE	PARTICLE SIZE (MM)	SOIL TYPE	PARTICLE SIZE (MM)
Sand	0.05–2.0	Coarse sand	0.2–2.0
		Fine sand	0.02–0.2
Silt	0.002–0.05	Silt	0.002–0.02
Clay	<0.002	Clay	<0.002

boulders to tiny particles called **clay** that are 2 μm or less in diameter (Table 36.3). Soils also contain water and dissolved mineral nutrients, air spaces, and dead organic matter. The air spaces are crucial sources of oxygen for plant roots. Soils change constantly through natural causes—such as rain, temperature extremes, and the activities of plants and animals—as well as human activities—farming in particular.

The structure of many soils changes with depth, revealing a **soil profile**. Although soils differ greatly, almost all soils consist of two or more **horizons**—recognizable horizontal layers—lying on top of one another. Mineral nutrients tend to be **leached**—dissolved in rain or irrigation water and carried to deeper horizons.

Soil scientists recognize three major zones (A, B, and C) in the profile of a typical soil (Figure 36.4). **Topsoil** is the A horizon, from which mineral nutrients may be depleted by leaching. Most of the organic matter in the soil is in the A horizon, as are most roots, earthworms, insects, nematodes, and microorganisms. Successful agriculture depends on the presence of a suitable A horizon. Pure sand contains plenty of air spaces, but is low in water and mineral nutrients. Clay contains lots of nutrients and more water than sand does, but it is low in air. A little bit of clay goes a long way in affecting soil properties. A **loam** has significant amounts of sand, silt, and clay, and thus has good levels of air, water, and nutrients for plants. Most of the best topsoils for agriculture are loams.

Below the A horizon is the B horizon, or subsoil, which is the zone of infiltration and accumulation of materials leached from above. Farther down, the C horizon is the original parent rock from which the soil is derived. Some deep-growing roots extend into the B horizon, but roots rarely enter the C horizon.

Soils form through the weathering of rock

The type of soil in a given area depends on the type of rock from which it formed, the climate, the landscape features, the organisms living there, and the length of time that soil-forming processes have been acting (sometimes millions of years). Rocks are broken down in part by **mechanical weathering**, which is the physical breakdown—without any accompanying chemical changes—of materials by wetting, drying, and freezing. The most important parts of soil formation, however, include **chemical weathering**, the chemical alteration of at least some of the materials in the rocks.

The key process is the formation of clay. Both the physical and the chemical properties of soils depend on the amount and kind of clay particles they contain. Just grinding up rocks does not produce a clay that binds mineral nutrients and aggregates into particles. Such a clay results only from chemical weathering.

Soils are the source of plant nutrition

The supply of mineral nutrients for plants depends on the presence of clay particles in the soil. Many of the minerals that are important for plant nutrition, such as potassium (K^+), magnesium (Mg^{2+}), and calcium (Ca^{2+}), exist in soil as positively charged ions, or *cations*. Clay particles have a net negative charge, which they get from negatively charged ions that are permanently attached to them. Cations in solution are attracted to these negative ions. To become available to plants, the cations must be detached from the clay particles.

A horizon
Topsoil

B horizon
Subsoil

C horizon
Weathering
parent rock
(bedrock)

36.4 A Soil's Profile
The A, B, and C horizons can sometimes be seen in road cuts such as this one in Australia. The dark upper layer (A horizon) is home to most of the living organisms in the soil.

This task is accomplished by reactions with protons (hydrogen ions, H^+). Roots release protons into the soil, and they are also released by the ionization of carbonic acid (H_2CO_3), which is formed whenever CO_2 from respiring roots or from the atmosphere dissolves in water. Protons bond more strongly to the clay particles than do the mineral cations, so they trade places with the cations, thus putting the nutrients back into the soil solution. This trading of places is called **ion exchange** (Figure 36.5). The fertility of a soil is determined in part by its ability to provide nutrients in this manner.

Clay particles effectively hold and exchange positively charged ions, but there is no comparable mechanism for exchanging negatively charged ions. As a result, important negative ions such as nitrate (NO_3^-) and sulfate (SO_4^{2-})—the primary and direct sources of nitrogen and sulfur—leach rapidly from soil, whereas positive ions tend to be retained in the A horizon. The reservoir of soil nitrogen is the organic matter in the soil, which slowly decomposes to release nitrogen in a form available to plants.

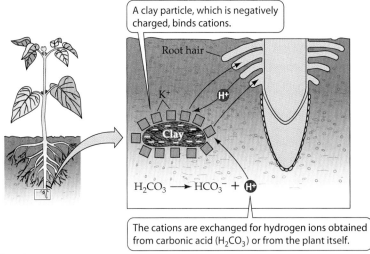

A clay particle, which is negatively charged, binds cations.

Root hair

$H_2CO_3 \longrightarrow HCO_3^- + H^+$

The cations are exchanged for hydrogen ions obtained from carbonic acid (H_2CO_3) or from the plant itself.

36.5 Ion Exchange
Plants obtain mineral nutrients from the soil primarily in the form of positive ions; potassium is the example shown here.

Fertilizers and lime are used in agriculture

Agricultural soils often require fertilizers because irrigation and rainwater leach mineral nutrients from the soil, and the harvesting of crops removes the nutrients that the plants took up from the soil during their growth. Crop yields decrease if any essential element is depleted. Mineral nutrients may be replaced by organic fertilizers, such as rotted manure, or inorganic fertilizers of various types. The three elements most commonly added to agricultural soils are nitrogen (N), phosphorus (P), and potassium (K). Commercial fertilizers are characterized by their "N-P-K" percentages. A 5-10-10 fertilizer, for example, contains 5 percent nitrogen, 10 percent phosphate (P_2O_5), and 10 percent potash (K_2O) by weight.* Sulfur, in the form of a sulfate, is also occasionally added to soils.

Either organic or inorganic fertilizers can provide the necessary mineral nutrients for plants. Organic fertilizers release nutrients slowly, which results in less leaching than a one-time application of inorganic fertilizer. Organic fertilizers also contain materials that improve the physical properties of the soil, providing spaces for gas movement, root growth, and drainage. Inorganic fertilizers, on the other hand, provide an almost instantaneous supply of soil nutrients and can be formulated to meet the requirements of a particular soil and a particular crop.

The availability of nutrient ions, whether they are naturally present in the soil or added as fertilizer, is altered by changes in soil pH. The optimal soil pH for most crops is about 6.5, but so-called acid-loving crops such as blueberries prefer a pH closer to 4. Rainfall and the decomposition of

*The analysis is by weight and is not reported as elemental N, P, and K. A 5-10-10 fertilizer actually *does* contain 5 percent nitrogen, but only 4.3 percent phosphorus and 8.3 percent potassium on an elemental basis.

organic substances in the soil lower its pH. Such acidification of the soil can be reversed by **liming**—the application of compounds commonly known as lime, such as calcium carbonate, calcium hydroxide, or magnesium carbonate. The addition of these compounds leads to the removal of H^+ ions from the soil. Liming also increases the availability of calcium to plants, which require it as a macronutrient.

It is easy to guess how humans learned to use fertilizer: It didn't take much insight to notice improved plant growth around animal feces. Perhaps a similar observation of limestone, or chalk, or oyster shells—all sources of calcium carbonate—led to the practice of liming. Sometimes, on the other hand, a soil is not acidic enough. In this case, a farmer can add sulfur, and soil bacteria will convert it to sulfuric acid. Iron and some other elements are more available to plants at a slightly acidic pH. Soil pH testing is useful for home gardens and lawns as well as for agriculture.

Spraying leaves with a nutrient solution is another effective way to deliver some essential elements to growing plants. Plants take up more copper, iron, and manganese when these elements are applied as *foliar* (leaf) sprays than when they are added to the soil as fertilizer. Adjusting the concentrations of nutrient ions and pH in order to optimize uptake and to minimize toxicity can yield excellent results. Such foliar application of mineral nutrients is increasingly used in wheat production, but fertilizer is still delivered most commonly by way of the soil.

Plants affect soils

The soil that forms in a particular place also depends on the types of plants growing there. Plant litter, such as dead fallen leaves, is the major source of carbon-rich materials that break down to form **humus**—dark-colored organic material, each particle of which is too small to be recognizable with the naked eye. Soil bacteria and fungi produce humus by breaking down plant litter, animal feces, and other organic material. Humus is rich in mineral nutrients, especially nitrogen. Humus in combination with clay promotes a soil structure favorable to plant growth, promoting adequate supplies of both water and oxygen to the roots.

Plants affect the pH of the soil in which they grow. Roots maintain a balance of electric charges. If they absorb more cations than anions, they excrete H^+ ions, thus lowering the soil pH. If they absorb more anions than cations, they excrete OH^- ions or HCO_3^- ions, raising the soil pH.

The mineral nutrient most commonly limiting, in both natural and agricultural situations, is nitrogen. Let's consider how nitrogen is made available to plants.

Nitrogen Fixation

Earth's atmosphere is a vast reservoir of nitrogen in the form of nitrogen gas (N_2). N_2 constitutes almost four-fifths of the atmosphere. However, plants cannot use N_2 directly as a nutrient. It is a highly unreactive substance—the triple bond linking the two nitrogen atoms is extremely stable, and a great deal of energy is required to break it.

A few species of bacteria have an enzyme that enables them to convert N_2 into a more reactive form by a process called **nitrogen fixation**. These prokaryotic organisms—*nitrogen fixers*—convert N_2 to ammonia (NH_3). There are relatively few species of nitrogen fixers, and their biomass is small relative to the mass of other organisms that depend on them for survival on Earth. This talented group of prokaryotes is just as essential to the biosphere as are the photosynthetic autotrophs.

Nitrogen fixers make all other life possible

By far the greatest share of total world nitrogen fixation is performed biologically by nitrogen-fixing organisms, which fix approximately 170 million Mg (megagrams, metric tons) of nitrogen per year. About 80 million Mg is fixed industrially by humans. A smaller amount of nitrogen is fixed in the atmosphere by nonbiological means such as lightning, volcanic eruption, and forest fires. Rain brings these atmospherically formed products to the ground.

Several groups of bacteria fix nitrogen. In the oceans, various photosynthetic bacteria, including cyanobacteria, fix nitrogen. In fresh water, cyanobacteria are the principal nitrogen fixers. On land, free-living soil bacteria make some contribution to nitrogen fixation, but they fix only what they need for their own use and release the fixed nitrogen only when they die. Other nitrogen-fixing bacteria live in close association with plant roots. They release up to 90 percent of the nitrogen they fix to the plant and excrete some amino acids into the soil, making nitrogen immediately available to other organisms.

Bacteria of the genus *Rhizobium* fix nitrogen only in close association with the roots of plants in the legume family (Figure 36.6). The legumes include peas, soybeans, clover, alfalfa, and many tropical shrubs and trees. The bacteria infect the plant's roots, and the roots develop nodules in re-

36.6 Root Nodules
Large, round, tumorlike nodules are visible in the root system of a broad bean. These nodules house nitrogen-fixing bacteria.

sponse to their presence. The various species of *Rhizobium* show a fairly high specificity for the species of legume they infect. Farmers and gardeners coat legume seeds with *Rhizobium* to make sure the bacteria are present. Some farmers alternate their crops, planting clover or alfalfa occasionally to increase the available nitrogen content of the soil.

The legume–*Rhizobium* association is not the only bacterial association that fixes nitrogen. Some cyanobacteria fix nitrogen in association with fungi in lichens or with ferns, cycads, or nontracheophytes. Rice farmers can increase crop yields by growing the water fern *Azolla*, with its symbiotic nitrogen-fixing cyanobacterium, in the flooded fields where rice is grown. Another group of bacteria, the filamentous actinomycetes, fix nitrogen in association with root nodules on woody species such as alder and mountain lilacs.

How does biological nitrogen fixation work? In the four sections that follow, we'll consider the role of the enzyme nitrogenase, the mutualistic collaboration of plant and bacterial cells in root nodules, the need to supplement biological nitrogen fixation in agriculture, and the contributions of plants and bacteria to the global nitrogen cycle.

Nitrogenase catalyzes nitrogen fixation

Nitrogen fixation is the reduction of nitrogen gas. It proceeds by the stepwise addition of three pairs of hydrogen atoms (Figure 36.7). In addition to N_2, these reactions require a strong reducing agent to transfer hydrogen atoms to nitrogen and the intermediate products, as well as a great deal of energy, which is supplied by ATP. Depending on the species of nitrogen fixer, either respiration or photosynthe-

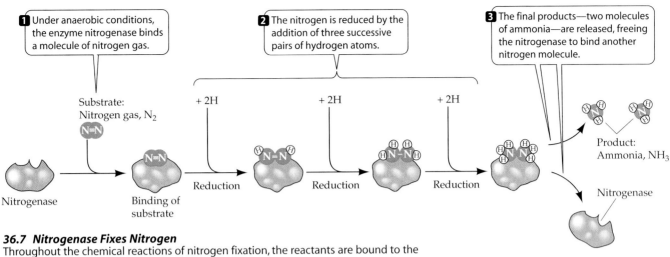

1 Under anaerobic conditions, the enzyme nitrogenase binds a molecule of nitrogen gas.

2 The nitrogen is reduced by the addition of three successive pairs of hydrogen atoms.

3 The final products—two molecules of ammonia—are released, freeing the nitrogenase to bind another nitrogen molecule.

Substrate:
Nitrogen gas, N_2

$+ 2H$ $+ 2H$ $+ 2H$

Nitrogenase   Binding of substrate   Reduction   Reduction   Reduction   Product: Ammonia, NH_3

Nitrogenase

36.7 Nitrogenase Fixes Nitrogen
Throughout the chemical reactions of nitrogen fixation, the reactants are bound to the enzyme nitrogenase. A reducing agent transfers hydrogen atoms to nitrogen, and eventually the final product—ammonia—is released.

sis may provide both the necessary reducing agent and ATP. The reactants are firmly bound to the surface of a single enzyme, called **nitrogenase**.

Nitrogenase is so strongly inhibited by oxygen (O_2) that its discovery was delayed because investigators had not thought to seek it under anaerobic conditions. Because nitrogenase cannot function in the presence of oxygen, it is not surprising that many nitrogen fixers are anaerobes. Legumes respire aerobically, as do *Rhizobium*. Within a root nodule, oxygen is maintained at a level sufficient to support respiration but not so high as to inactivate nitrogenase.

Some plants and bacteria work together to fix nitrogen

The legume nodule provides an excellent example of *symbiosis*, in which two different organisms live in physical contact. In the form of symbiosis called *mutualism*, both organisms benefit from their relationship. The legume obtains fixed nitrogen from the bacterium, and the bacterium obtains the products of photosynthesis from the plant. Neither free-living *Rhizobium* species nor uninfected legumes can fix nitrogen. Only when the two are closely associated in root nodules does the reaction take place.

The establishment of this symbiosis between *Rhizobium* and a legume requires a complex series of steps, with active contributions by both the bacteria and the plant root (Figure 36.8). First the root releases flavonoids and other chemical signals that attract the *Rhizobium* to the vicinity of the root. Flavonoids trigger the transcription of bacterial *nod* genes, which encode Nod (nodulation) factors. These factors, secreted by the bacteria, cause cell divisions in the root cortex, leading to the formation of a primary nodule meristem. Within the nodules, the bacteria take the form of **bacteroids** within membranous vesicles. Bacteroids are swollen, deformed bacteria that can fix nitrogen.

Before the bacteroids can begin to fix nitrogen, the plant must produce the protein **leghemoglobin**, which sur-

rounds the bacteroids. Leghemoglobin is a close relative of hemoglobin, the oxygen-carrying pigment of animals. Some plant nodules contain enough of it to be bright pink when viewed in cross section. Leghemoglobin, with its iron-containing heme, transports oxygen to the bacteroids to support their respiration.

The partnership between bacterium and plant in nitrogen-fixing nodules is not the only case in which plants depend on other organisms for assistance with their nutrition. Another example that we considered earlier is that of *mycorrhizae*, root–fungus associations in which the fungus greatly increases the absorption of water and minerals (especially phosphorus) by the plant (see Figure 30.16). A growing body of evidence suggests that nodule formation depends on some of the same genes and mechanisms that allow mycorrhizae to develop.

Biological nitrogen fixation does not always meet agricultural needs

Bacterial nitrogen fixation is not sufficient to support the needs of agriculture. Traditional farmers used to plant dead fish along with corn so that the decaying fish would release fixed nitrogen that the developing corn could use. Industrial nitrogen fixation is becoming ever more important to world agriculture because of the degradation of soils and the need to feed a rapidly expanding population. Research on biological nitrogen fixation is being vigorously pursued, with commercial applications very much in mind.

Most industrial nitrogen fixation is done by a chemical process called the Haber process, which requires a great deal of energy. An alternative is urgently needed because of the cost of energy. At present, the manufacture of nitrogen-containing fertilizer takes more energy than does any other aspect of crop production in the United States.

One line of investigation centers on recombinant DNA technology as a means of engineering new plants that produce their own nitrogenase. Workers in many industrial

36.11 A Carnivorous Sundew
Sundews trap insects on their sticky hairs. Secreted enzymes will digest the carcasss externally.

now known to get its nutrients, with the help of fungi, from nearby actively photosynthesizing plants. Hence it too is a parasite.

Some other plants that do not live by photosynthesis alone are the 450 or so carnivorous species—those that augment their nitrogen and phosphorus supply by capturing and digesting flies and other insects (Figure 36.11; also shown at the start of this chapter). The best-known carnivorous plants are Venus flytraps (genus *Dionaea*), sundews (genus *Drosera*), and pitcher plants (genus *Sarracenia*).

Carnivorous plants are normally found in boggy regions where the soil is acidic. Most decay-causing organisms require a less acidic pH to break down the bodies of dead organisms, so relatively little nitrogen is recycled into these acidic soils. Accordingly, the carnivorous plants have adaptations that allow them to augment their supply of nitrogen by capturing animals and digesting their proteins.

Sarracenia produces pitcher-shaped leaves that collect small amounts of rainwater. Insects are attracted into the pitchers either by bright colors or by scent and are prevented from getting out again by stiff, downward-pointing hairs. The insects eventually die and are digested by a combination of enzymes and bacteria in the water. Even rats have been found in large pitcher plants.

Sundews have leaves covered with hairs that secrete a clear, sticky, sugary liquid. An insect touching one of these hairs becomes stuck, and more hairs curve over the insect and stick to it as well. The plant secretes enzymes to digest the insect and later absorbs the carbon- and nitrogen-containing products of digestion.

None of the carnivorous plants must feed on insects. They grow adequately without insects, but in their natural habitats they grow faster and are a darker green when they succeed in catching insects. The additional nitrogen from the insects is used to make more proteins, chlorophyll, and other nitrogen-containing compounds.

Chapter Summary

The Acquisition of Nutrients

▶ Plants are photosynthetic autotrophs that can produce all the compounds they need from carbon dioxide, water, and minerals, including a nitrogen source. They obtain energy from sunlight, carbon dioxide from the atmosphere, and nitrogen-containing ions and mineral nutrients from the soil.

▶ Plants explore their surroundings by growing rather than by locomotion.

Mineral Nutrients Essential to Plants

▶ Plants require 14 essential mineral elements, all of which come from the soil. Several essential elements fulfill multiple roles. **Review Table 36.1**

▶ The six mineral nutrients required in substantial amounts are called macronutrients; the eight required in much smaller amounts are called micronutrients. **Review Table 36.1**

▶ Deficiency symptoms suggest what essential element a plant lacks. **Review Table 36.2**

▶ Biologists discovered the essentiality of each mineral nutrient by growing plants on nutrient solutions lacking the test element. **Review Figure 36.2**

Soils and Plants

▶ Soils are complex in structure, with living and nonliving components. They contain water, gases, and inorganic and organic substances. They typically consist of two or three horizontal zones called horizons. **Review Figures 36.3, 36.4, and Table 36.3**

▶ Soils form by mechanical and chemical weathering of rock.

▶ Plants obtain some mineral nutrients by ion exchange between the soil solution and the surface of clay particles. **Review Figure 36.5**

▶ Farmers use fertilizer to make up for deficiencies in soil mineral nutrient content, and they apply lime to raise low soil pH.

▶ Plants affect soils in various ways, helping them form, adding material such as humus, and removing nutrients (especially in agriculture).

Nitrogen Fixation

▶ A few species of soil bacteria are responsible for almost all nitrogen fixation. Some nitrogen-fixing bacteria live free in the soil; others live symbiotically as bacteroids within the roots of plants.

▶ In nitrogen fixation, nitrogen gas (N_2) is reduced to ammonia (NH_3) or ammonium ions (NH_4^+) in a reaction catalyzed by nitrogenase. **Review Figure 36.7**

▶ Nitrogenase requires anaerobic conditions, but the bacteroids in root nodules require oxygen for their respiration. Leghemoglobin helps maintain the oxygen supply to the bacteroids.

▶ The formation of a nodule requires an interaction between the root system of a legume and *Rhizobium* bacteria. **Review Figure 36.8**

▶ Nitrogen-fixing bacteria reduce atmospheric N_2 to ammonia, but most plants take up both ammonium ions and nitrate ions. Nitrifying bacteria oxidize ammonia to nitrate. Plants take up nitrate and reduce it back to ammonia, a feat of which animals are incapable. **Review Figure 36.9**

▶ Denitrifying bacteria return N_2 to the atmosphere, completing the biological nitrogen cycle. **Review Figure 36.9**

Sulfur Metabolism

▶ Plants take up sulfate ions and reduce them, forming the amino acids cysteine and methionine. Cysteine is the major precursor for other sulfur-containing compounds in plants and in animals, which must obtain their organic sulfur from plants.

Heterotrophic and Carnivorous Seed Plants

▶ A few heterotrophic plants are parasitic on other plants.

▶ Carnivorous plant species are autotrophs that supplement their nitrogen supply by feeding on insects.

For Discussion

1. Methods for determining whether a particular element is essential have been known for more than a century. Since these methods are so well established, why was the essentiality of some elements discovered only recently?

2. If a Venus flytrap were deprived of soil sulfates and hence made unable to synthesize the amino acids cysteine and methionine, would it die from lack of protein?

3. Soils are dynamic systems. What changes might result when land is subjected to heavy irrigation for agriculture after being relatively dry for many years? What changes in the soil might result when a virgin deciduous forest is cut down and replaced by crops that are harvested each year?

4. We mentioned that important positively charged ions are held in the soil by clay particles, but other, equally important, negatively charged ions are leached deeper into the soil's B horizon. Why doesn't leaching cause an electrical imbalance in the soil? (*Hint*: Think of the ionization of water.)

5. The biosphere of Earth as we know it depends on the existence of a few species of nitrogen-fixing prokaryotes. What do you think might happen if one of these species were to become extinct? If all of them were to disappear?

37 *Plant Growth Regulation*

 MORE THAN A CENTURY AGO, CHARLES Darwin and his son studied the growth of plant shoots toward the light. Their findings, which we will detail in this chapter, pointed the way to the eventual discovery of the photoreceptor molecules that capture light signals and the hormones that transmit those signals to other parts of the plant. Light and hormones affect processes in plants as diverse as stem growth, flowering, bud dormancy, and the dropping of leaves in autumn. Several of the hormones now find important commercial applications, including the regulation of fruit ripening and enhanced germination of barley for the brewing industry.

Recent advances in understanding plant development have come largely from work with *Arabidopsis thaliana*, a little mustard-like weed. This plant is useful to researchers because its body and seeds are tiny, and its genome is unusually small for a flowering plant. It also flowers and forms seeds in a relatively short time after growth begins. *Arabidopsis* mutants with altered developmental patterns provide evidence for the existence of hormones and for the mechanisms of hormone and photoreceptor action.

In this chapter we first give a brief overview of the life of a flowering plant and its developmental stages. We explore the environmental cues, photoreceptors, and hormones that regulate plant development, and consider the multiple roles that each plays in normal development.

Interacting Factors in Plant Development

The *development* of a plant—the series of progressive changes that take place throughout its life—is regulated in complex ways. Four factors take part in this regulation:

▶ The plant senses and responds to *environmental cues*.
▶ The plant's *genome* encodes enzymes that catalyze the biochemical reactions of development, including the ones that make hormones and receptors, produce chemical building blocks, and participate in protein synthesis and energy metabolism.
▶ In order to sense environmental cues, the plant uses *receptors*, such as photoreceptors that absorb light.
▶ Chemical messages, or *hormones*, mediate the effects of the environmental cues sensed by the receptors.

Several hormones and photoreceptors regulate plant growth

Hormones are regulatory compounds that act at very low concentrations at sites distant from where they are produced. They mediate many developmental phenomena in plants, such as stem growth and autumn leaf fall. Unlike

Catching Some Rays
Most of us have observed the manner in which plants turn toward sunlight. Light signals caught by photoreceptor proteins are transmitted by hormones to other parts of the plant in a finely tuned developmental dance.

37.1 *Plant Hormones*

HORMONE	TYPICAL ACTIVITIES
Abscisic acid	Maintains seed dormancy and winter dormancy; closes stomata
Auxin	Promotes stem elongation, adventitious root initiation, and fruit growth; inhibits lateral bud outgrowth and leaf abscission
Brassinosteroids	Promote elongation of stems and pollen tubes; promote vascular tissue differentiation
Cytokinins	Inhibit leaf senescence; promote cell division and lateral bud outgrowth; affect root growth
Ethylene	Promotes fruit ripening and leaf abscission; inhibits stem elongation and gravitropism
Gibberellins	Promote seed germination, stem growth, and fruit development; break winter dormancy; mobilize nutrient reserves in grass seeds
Jasmonates	Trigger defenses against pathogens and herbivores
Oligosaccharins	Trigger defenses against pathogens; limit effects of high auxin concentrations; regulate cell differentiation
Salicylic acid	Triggers resistance to pathogens
Systemin	Causes jasmonate production in response to tissue damage

animals, which produce each hormone in a specific part of the body, plants produce hormones in many of their cells. Each plant hormone plays multiple regulatory roles, affecting several different aspects of development (Table 37.1). Interactions among the hormones are often complex.

Like hormones, **photoreceptors** regulate many developmental processes in plants. Unlike the hormones, which are small molecules, plant photoreceptors are proteins. Light (an environmental cue) acts directly on photoreceptors, which in turn regulate processes such as the many changes accompanying the growth of a young plant out of the soil and into the light.

No matter what cues direct development, ultimately the plant's genome determines the limits within which the plant and its parts will develop. The genome encodes the master plan, but its interpretation depends on conditions in the environment. It is also the target for some hormone actions. For several decades hormones and photoreceptors were the focus of most work on plant development, but recent advances in molecular genetics allow us to focus on underlying processes such as signal transduction pathways.

Signal transduction pathways mediate hormone and photoreceptor action

We introduced the topic of signal transduction pathways in Chapter 15. Plants, like other organisms, make extensive use of these pathways. Cell signaling in plant development involves three steps: a receptor (for a hormone or for light), a signal transduction pathway, and the ultimate cellular response (see Figure 15.3). Protein kinase cascades amplify responses to receptor binding in plants, as they do in other organisms (see Figure 15.11). The signal transduction pathways of plants differ from those of animals only in the details; for example, their protein kinases

phosphorylate the amino acid residues serine or threonine but not tyrosine.

Before concerning ourselves with molecular details, let's set a broader context. What is the general pattern of plant development?

From Seed to Death: An Overview of Plant Development

Let's review the life history of a flowering plant, from seed to death, focusing on how the developmental events are regulated. As plants develop, environmental cues, photoreceptors, and hormones affect three fundamental processes: cell division, cell expansion, and cell differentiation.

The seed germinates and forms a growing seedling

All developmental activity may be suspended in a seed, even when conditions appear to be suitable for its growth. In other words, a seed may be **dormant**. Typically, only 5 to 20 percent of a seed's weight is water, whereas most plant parts contain far more water.

Cells in dormant seeds do not divide, expand, or differentiate. For the embryo to begin developing, seed dormancy must be broken by one of several physical mechanisms, such as exposure to light, mechanical abrasion, fire, or leaching of inhibitors by water.

As the seed **germinates** (begins to develop), it first imbibes (takes up) water. The growing embryo must then obtain building blocks—monomers—for its development by digesting the polysaccharides, fats, and proteins stored in the cotyledons or in the endosperm. The embryos of some plant species secrete hormones that direct the mobilization of these reserves.

If the seed germinates underground, the new seedling must elongate rapidly and cope with life in darkness or dim

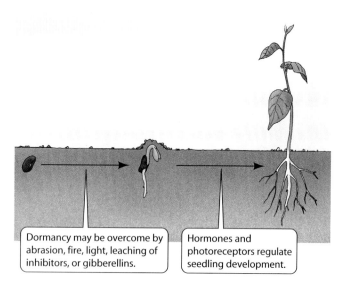

Dormancy may be overcome by abrasion, fire, light, leaching of inhibitors, or gibberellins.

Hormones and photoreceptors regulate seedling development.

37.1 From Seed to Seedling
Environmental factors, hormones, and photoreceptors regulate the first stages of plant growth.

light. A photoreceptor controls this stage, and ends it when the shoot is exposed to sufficient light to begin photosynthesis (Figure 37.1).

Early shoot development varies among the flowering plants. Figure 37.2 presents the distinctive shoot development patterns of monocots and eudicots. Plant growth from seedling to adult, both in darkness and in light, also involves several hormones.

The plant flowers and sets fruit

Flowering—the formation of reproductive organs—may be initiated when the plant reaches an appropriate age or size.

Some plant species, however, flower at particular times of the year, meaning that the plant must sense the appropriate time. In these plants, the leaves measure the length of the night (shorter in summer, longer in winter) with great precision. Light absorbed by photoreceptors affects this time-measuring process.

Once a leaf has determined that it is time for the plant to flower, that information must be transported as a signal to the places where flowers will form. The means by which this signal is transmitted remains a mystery, but it is likely that a "flowering hormone" travels from the leaf to the point of flower formation.

After flowers form, hormones play further roles. Hormones and other substances control the growth of a pollen tube down the style of a pistil. Following fertilization, a fruit develops and ripens under hormonal control (Figure 37.3).

The plant senesces and dies

Some plants, known as **perennials**, continue to grow year after year. Many perennials have buds that enter a state of winter dormancy during the cold season. A hormone called abscisic acid helps maintain this dormancy.

In many species, leaves **senesce** (deteriorate because of aging) and fall at the end of the growing season, shortly before the onset of the severe conditions of winter. Leaf fall (abscission) is regulated by an interplay of the hormones ethylene and auxin. Finally, the entire plant senesces and dies.

37.2 Patterns of Early Shoot Development
(a) In grasses and some other monocots, growing shoots are protected by a coleoptile until they reach the surface. (b) In most eudicots, the growing point of the shoot is protected by the cotyledons. (c) In some other eudicots, the cotyledons remain in the soil, and the growing point is protected by the first true leaves.

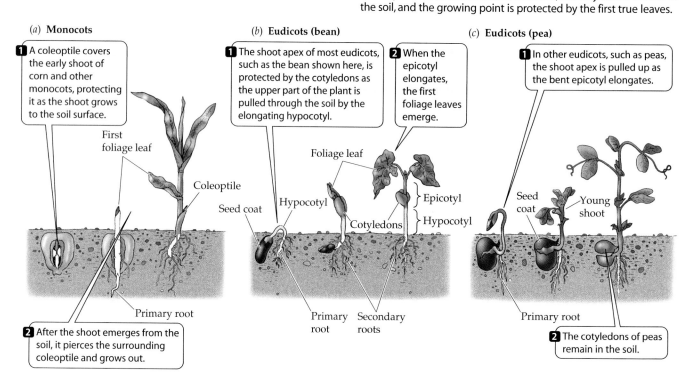

(a) **Monocots**

1 A coleoptile covers the early shoot of corn and other monocots, protecting it as the shoot grows to the soil surface.

First foliage leaf

Coleoptile

Primary root

2 After the shoot emerges from the soil, it pierces the surrounding coleoptile and grows out.

(b) **Eudicots (bean)**

1 The shoot apex of most eudicots, such as the bean shown here, is protected by the cotyledons as the upper part of the plant is pulled through the soil by the elongating hypocotyl.

2 When the epicotyl elongates, the first foliage leaves emerge.

Foliage leaf

Seed coat

Hypocotyl

Cotyledons

Epicotyl

Hypocotyl

Primary root

Secondary roots

(c) **Eudicots (pea)**

1 In other eudicots, such as peas, the shoot apex is pulled up as the bent epicotyl elongates.

Seed coat

Young shoot

Primary root

2 The cotyledons of peas remain in the soil.

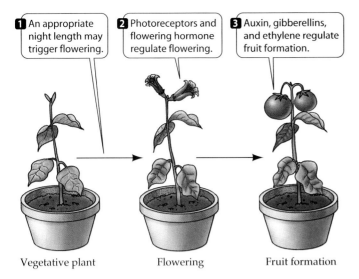

1 An appropriate night length may trigger flowering.

2 Photoreceptors and flowering hormone regulate flowering.

3 Auxin, gibberellins, and ethylene regulate fruit formation.

Vegetative plant   Flowering   Fruit formation

37.3 Flowering and Fruit Formation
Environmental cues, photoreceptors, and hormones regulate plant reproduction.

Death, which may be initiated by signals from the environment, follows senescent changes that are controlled by hormones such as ethylene. This life history pattern appears to be an adaptation for producing more offspring by pumping energy (food) and nutrients into the seeds; in so doing, the parent plant essentially starves itself to death.

We have reached the end of the plant's life history. Now let's examine how the various steps are regulated. We'll begin with regulation at the start of the life history—the seed and its germination.

Ending Seed Dormancy and Beginning Germination

The seeds of some species are, in effect, instant plants: All they need for germination is water. But many other species have seeds whose germination is regulated in more complex ways.

Seed dormancy may last for weeks, months, years, or even centuries. The mechanisms of dormancy are numerous and diverse, but three principal strategies dominate:

▶ Exclusion of water or oxygen from the embryo by means of an impermeable seed coat

▶ Mechanical restraint of the embryo by means of a tough seed coat

▶ Chemical inhibition of embryo development

The dormancy of seeds with impermeable coats can be broken if the seed coat is abraded as the seed tumbles across the ground or through creek beds, or passes through the digestive tract of an animal. Soil microorganisms probably play a major role in softening seed coats. Fire can release mechanical restraint. It can also melt wax in seed coats, removing the waterproofing and allowing water to reach the embryo (Figure 37.4). *Leaching*—prolonged exposure to water—is one way to reduce the level of a water-soluble chemical inhibitor and end dormancy. Scorching of seeds by fire can also break down some inhibitors.

Seed dormancy affords adaptive advantages

What are the potential advantages of seed dormancy? For many species, dormancy assures survival through unfavorable conditions and results in germination when conditions are more favorable. To avoid germination in the dry days of late summer, for example, some seeds must be exposed to a long cold period before they will germinate. Other seeds will not germinate until a certain amount of time has passed, regardless of how they are treated. This strategy prevents germination while the seed of a cereal grain, for example, is still attached to the parent plant.

Seeds that must be scorched by fire in order to germinate avoid competition by germinating only when an area has been cleared by fire. Light-requiring seeds, which germinate only at or near the surface of the soil, are generally tiny seeds with few food reserves. Conversely, germination of some seeds is inhibited by light; these seeds germinate only when buried and thus kept in darkness. Light-inhibited seeds are usually large and well stocked with nutrients.

Seed dormancy helps annual plants counter the effects of year-to-year variation in the environment. The seeds of some annuals remain dormant throughout an unfavorable year. The seeds of other plants germinate at different times

37.4 Fire and Seed Germination
This fireweed germinated and flourished after a great fire along the Alaska Highway.

37.5 Leaching of Germination Inhibitors
The seeds of the cypress, a swamp-adapted tree, germinate only after being leached by water, which increases the chances that they will germinate in a situation suitable for their growth.

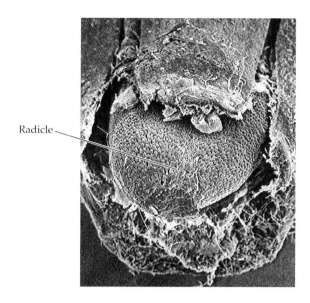

Radicle

37.6 The Radicle Emerges
The tip of this barley seed's radicle has just broken through its protective sheath. The appearance of the radicle—the embryonic root—is one of the first externally visible events in seed germination.

during the year, increasing the likelihood that at least some of the seedlings will encounter favorable conditions.

Dormancy may also increase the likelihood of a seed's germinating in the right place. Some cypress trees, for example, grow in standing water, and their seeds germinate only if inhibitors are leached by water (Figure 37.5).

Seed germination begins with the uptake of water

The first step in seed germination is the uptake of water, called **imbibition**. A seed's water potential (see Chapter 35) is very negative, and water can be taken up readily if the seed coat allows it. The magnitude of this water potential is demonstrated by the force exerted by seeds expanding in water. Cocklebur seeds that are imbibing can exert a pressure of up to 1,000 atmospheres (about 100 megapascals) against a restraining force.

As a seed takes up water, it undergoes metabolic changes: Certain existing enzymes become activated, RNA and then proteins are synthesized, the rate of cellular respiration increases, and other metabolic pathways become activated. In many seeds there is no DNA synthesis and no cell division during these early stages of germination. Initially, growth results solely from the expansion of small, preformed cells. DNA is synthesized only after the embryonic root, called the **radicle**, begins to grow and poke out beyond the seed coat (Figure 37.6).

The embryo must mobilize its reserves

Until the young plant (the **seedling**) becomes able to photosynthesize, it depends on reserves stored in the endosperm or cotyledons. The principal reserve of energy and carbon

in many seeds is starch. Other seeds store fats or oils. Usually, the endosperm of the seed holds amino acid reserves in the form of proteins, rather than as free amino acids.

The giant molecules of starch, lipids, and proteins must be digested by enzymes into monomers that can enter the cells of the embryo. The polymer starch yields glucose for energy metabolism. The digestion of reserve proteins provides the amino acids the embryo needs to synthesize its own proteins. The digestion of lipids releases glycerol and fatty acids, both of which can be metabolized for energy. Glycerol and fatty acids can also be converted to glucose, which permits fat-storing plants to make all the building blocks they need for growth.

In germinating barley and other cereal seeds, the embryo secretes **gibberellins**, one of several classes of plant growth hormones. Gibberellins diffuse through the endosperm to a surrounding tissue called the **aleurone layer**, which lies inside the seed coat. The gibberellins trigger a crucial series of events in the aleurone layer, culminating in the release of enzymes that digest proteins and starch stored in the endosperm (Figure 37.7). Commercially, gibberellins are used in the brewing industry to enhance the "malting" (germination) of barley and the breakdown of its endosperm, producing sugar that is fermented to alcohol.

Gibberellins: Regulators from Germination to Fruit Growth

Gibberellins produce a wide variety of effects on plant development in addition to triggering digestive enzyme synthesis. We begin our discussion of the different plant growth hormones by discussing the discovery of the gibberellins, as well as their many effects.

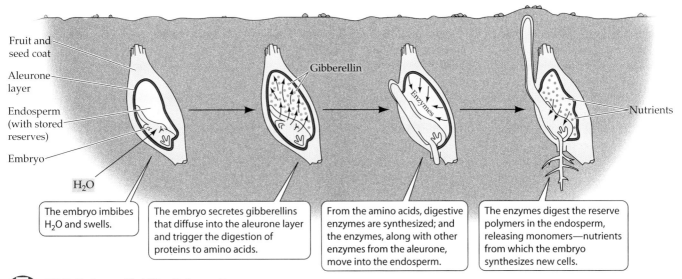

Fruit and seed coat
Aleurone layer
Endosperm (with stored reserves)
Embryo
H_2O
Gibberellin
Enzymes
Nutrients

The embryo imbibes H_2O and swells.

The embryo secretes gibberellins that diffuse into the aleurone layer and trigger the digestion of proteins to amino acids.

From the amino acids, digestive enzymes are synthesized; and the enzymes, along with other enzymes from the aleurone, move into the endosperm.

The enzymes digest the reserve polymers in the endosperm, releasing monomers—nutrients from which the embryo synthesizes new cells.

37.7 Embryos Mobilize Polymer Reserves
Seed germination in cereal grasses consists of a cascade of processes. Gibberellins trigger the conversion of reserve polymers into monomers that can be used by the developing embryo.

Foolish seedlings led to the discovery of the gibberellins

The gibberellins are a large family of closely related compounds. Some are found in plants and others in a pathogenic (disease-causing) fungus, where they were first discovered.

Gibberellin A_1
(important in stem growth)

Gibberellin A_3
(commercially available)

In 1809, the study of the gibberellins began indirectly with observations of the *bakanae*, or "foolish seedling," disease of rice. Seedlings affected by this disease grow tall more rapidly than their healthy neighbors, but this rapid growth gives rise to spindly plants that die before producing seed (the rice grain used for food). The disease has had considerable economic impacts in several parts of the world. It is caused by the ascomycete fungus *Gibberella fujikuroi*.

In 1925, the Japanese biologist Eiichi Kurosawa grew *G. fujikuroi* on a liquid medium, then separated the fungus from the medium by filtering. He heated the filtered medium to kill any remaining fungus, but the resulting heat-treated filtrate still caused rapid growth in rice seedlings. Medium that had never contained the fungus did not stimulate seedling growth. This experiment established that *G. fujikuroi* produces a growth-promoting chemical substance, which Kurosawa called a gibberellin.

Were the gibberellins simply exotic products of an obscure fungus, or did they play a more general role in the growth of plants? Bernard O. Phinney of the University of California, Los Angeles, answered this question in part in 1956, when he reported the spectacular growth-promoting effect of gibberellins on dwarf corn seedlings. He used plants that were known to be genetic dwarfs; each phenotype was produced when a particular recessive allele (say, *d1*) was present in the homozygous condition (*d1/d1*). Gibberellins applied to nondwarf—normal—corn seedlings had almost no effect, but gibberellins applied to the dwarfs caused them to grow as tall as their normal relatives. (A comparable effect of gibberellins applied to a dwarf tomato plant is shown in Figure 37.8.)

22 days after being sprayed with a tiny amount of gibberellin, this plant reached the size of a nondwarf plant.

This untreated control plant remained dwarf.

37.8 The Effect of Gibberellins on Dwarf Plants
In this experiment, the effect of gibberellins was tested on two dwarf tomato plants. Both plants were the same size when the one on the right was treated with gibberellins.

This result suggested to Phinney (1) that gibberellins are normal constituents of corn, and perhaps of all plants, and (2) that dwarf plants are short because they cannot produce their own gibberellins. According to Phinney's hypothesis, nondwarf plants manufacture enough gibberellins to promote their full growth, but dwarf plants do not. Extracts from numerous plant species were found to promote growth in dwarf corn. These findings provided direct evidence that plants that are not genetic dwarfs contain gibberellin-like substances. Phinney's work set the stage for today's use of mutant plants to investigate the control of plant development.

The roots, leaves, and flowers of a dwarf corn plant appear normal, but the stems are much shorter than those of wild-type plants. All parts of the dwarf plant contain a much lower concentration of gibberellins than do the organs of a wild-type plant. We may infer, then, that stem elongation *requires* gibberellins or the products of gibberellin action. We can further conclude that gibberellins play a less essential role in the development of roots, leaves, and flowers.

Although more than 80 gibberellins have been identified, only one, *gibberellin A_1*, actually controls stem elongation in most plants. The other gibberellins found in stems are simply intermediates in the production of gibberellin A_1. As we will see in the next section, gibberellins affect processes other than stem elongation, but we do not yet know which gibberellin has any other particular effect.

The gibberellins have many effects

Gibberellins and other hormones regulate the growth of fruits. It has long been known that seedless grapes (an inbred strain) form smaller fruit than their seeded relatives. Experimental removal of seeds from very young seeded grapes prevented normal fruit growth, suggesting that the seeds are sources of a fruit growth regulator. It was then shown that spraying young seedless grapes with a gibberellin solution caused them to grow as large as seeded ones. It is now a standard commercial procedure to spray seedless grapes with gibberellins. Subsequent biochemical studies showed that the developing seeds produce gibberellins, which diffuse out into the immature fruit tissue.

Some biennial plants respond dramatically to an increased level of gibberellins. **Biennials** grow vegetatively in their first year and flower and die in their second year. In the second year, the apical meristems of biennials respond to environmental cues by producing elongated shoots that eventually bear flowers. This elongation is called **bolting**. When the plant senses the appropriate environmental cue—longer days or a sufficient winter chilling—it produces more gibberellins, raising the gibberellin concentration to a level that causes the shoot to bolt. Plants of some biennial species will bolt when sprayed with a gibberellin solution without the environmental cue (Figure 37.9).

Gibberellins also cause fruit to grow from unfertilized flowers, promote seed germination in lettuce and some other species, and help bring spring buds out of winter dor-

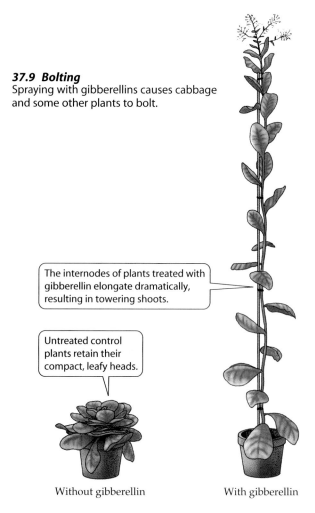

37.9 Bolting
Spraying with gibberellins causes cabbage and some other plants to bolt.

The internodes of plants treated with gibberellin elongate dramatically, resulting in towering shoots.

Untreated control plants retain their compact, leafy heads.

Without gibberellin   With gibberellin

mancy. Most hormones have multiple effects within the plant, and they often interact with one another in regulating developmental processes. In controlling stem elongation, for example, gibberellins interact with another hormone, auxin.

Auxin Affects Plant Growth and Form

If you pinch off the apical bud at the top of a bean plant, inactive lateral buds become active, developing into branches. Similarly, pruning a shrub causes an increase in branching. If you cut off the blade of a leaf but leave its petiole (stalk) attached to the plant, the petiole drops off sooner than it would have if the leaf were intact. If a plant is kept indoors, its shoot system grows toward a window. These diverse responses of shoot systems are all mediated by a plant hormone called **auxin**, or *indoleacetic acid* (IAA).

Auxin (indoleacetic acid)

H_2C—COOH

In the discussions that follow, we will look at the discovery of auxin, its transport within the plant, and its role as mediator of the effects of light and gravity on plant growth. We'll discover its many effects on vegetative growth and on fruit development. Then we'll examine its mechanism of action.

Plant movements led to the discovery of auxin

The discovery of auxin and its numerous physiological effects can be traced back to work done in the 1880s by Charles Darwin and his son Francis. The Darwins were interested in plant movements. One type of movement they studied was **phototropism**, the growth of plant structures toward light (as in most shoots) or away from it (as in roots). They asked, What part of the plant senses the light?

To answer this question, the Darwins worked with canary grass (*Phalaris canariensis*) seedlings grown in the dark. A young grass seedling has a **coleoptile**—a cylindrical

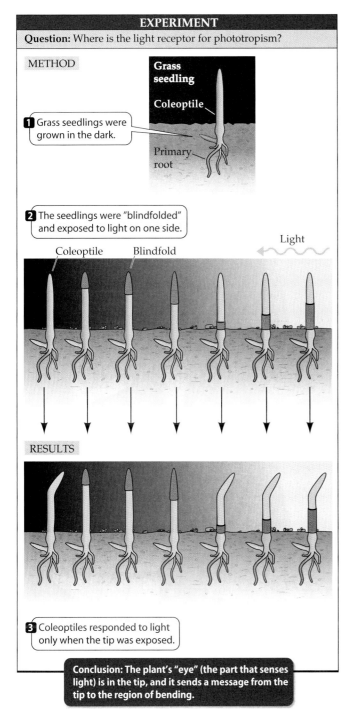

EXPERIMENT

Question: Where is the light receptor for phototropism?

METHOD

Grass seedling

Coleoptile

1 Grass seedlings were grown in the dark.

Primary root

2 The seedlings were "blindfolded" and exposed to light on one side.

Coleoptile   Blindfold   Light

RESULTS

3 Coleoptiles responded to light only when the tip was exposed.

Conclusion: The plant's "eye" (the part that senses light) is in the tip, and it sends a message from the tip to the region of bending.

sheath a few cells thick that protects the delicate shoot as it pushes through the soil (see Figure 37.2a). When the coleoptile breaks through the surface of the soil, it soon stops growing, and the first leaves emerge unharmed. The coleoptiles of grasses are *phototropic*—they grow toward the light.

To find the light-receptive region of the coleoptile, the Darwins tried "blindfolding" the coleoptiles of dark-grown canary grass seedlings in various places, then illuminating them from one side (Figure 37.10). The coleoptile grew toward the light whenever its tip was exposed. If the top millimeter or more of the coleoptile was covered, however, there was no phototropic response. Thus the tip contains the photoreceptor that responds to light. The actual bending toward the light, however, takes place in a growing region a few millimeters below the tip. Therefore, the Darwins reasoned, some type of message must travel within the coleoptile from the tip to the growing region. Others later demonstrated that the message is a chemical substance by showing that it can move through certain nonliving materials, such as gelatin, but not through others, such as a metal barrier.

Further experiments showed that the tip of the coleoptile produces a hormone that moves down the coleoptile to the growing region. If the tip is removed, the growth of the coleoptile is sharply inhibited. If the tip is carefully replaced, growth resumes, even if the tip and base are separated by a thin layer of gelatin. The hormone moves down from the tip, but it does not move from one side of the coleoptile to the other. If the tip is cut off and replaced so that it covers only one side of the cut end of the coleoptile, the coleoptile curves as the cells on the side below the replaced tip grow more rapidly than those on the other side.

The Dutch botanist Frits W. Went removed coleoptile tips and placed their cut surfaces on a block of gelatin. Then he placed pieces of the gelatin block on decapitated coleoptiles—positioned to cover only one side, just as coleoptile tips had been placed in earlier experiments (Figure 37.11). As they grew, the coleoptiles curved toward the side away from the gelatin. This curvature demonstrated that a hormone had indeed diffused into the gelatin block from the isolated coleoptile tips. Went had at last isolated a hormone from a plant. Later chemical analysis showed that this hormone, named auxin, was indoleacetic acid.

Auxin transport is polar

Since being isolated, auxin has been intensively studied. Early experiments showed that its movement through certain plant tissues is strictly *polar*—that is, unidirectional along a line from apex to base. By inverting some plants or plant parts, scientists determined that the apex-to-base di-

37.10 The Darwins' Phototropism Experiment
The top drawings show some of the ways in which seedlings grown in the dark were "blindfolded"; the lower drawings show what the Darwins observed in each case. Their observations led them to hypothesize the existence of a growth-promoting "messenger" substance produced by the coleoptile.

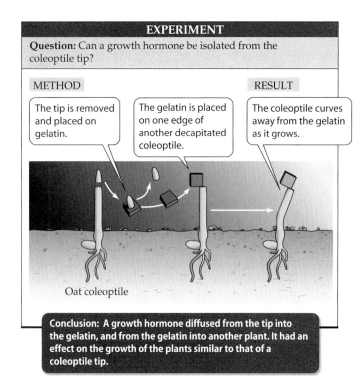

EXPERIMENT

Question: Can a growth hormone be isolated from the coleoptile tip?

METHOD

The tip is removed and placed on gelatin.

The gelatin is placed on one edge of another decapitated coleoptile.

RESULT

The coleoptile curves away from the gelatin as it grows.

Oat coleoptile

Conclusion: A growth hormone diffused from the tip into the gelatin, and from the gelatin into another plant. It had an effect on the growth of the plants similar to that of a coleoptile tip.

37.11 Went's Experiment
Went succeeded in isolating the growth-promoting chemical substance whose existence the Darwins had hypothesized by placing coleoptile tips on a block of gelatin.

rection of auxin movement has nothing to do with gravity; the polarity of this movement is a totally biological matter. Many plant parts show complete or partial polarity of auxin transport. For example, in most leaf petioles auxin moves only from the blade end toward the stem end.

Auxin carrier proteins move auxin into and out of cells

In one of the most intense areas of current research on auxin, biologists are using *Arabidopsis* plants that have mutations affecting the transport of auxin. By cloning genes from these plants and characterizing their products, they are finding a growing number of auxin carrier proteins. In polar transport in the stem, a carrier protein imports auxin at the end of the cell toward the shoot apex, and it or another carrier exports auxin at the other end of the cell. Auxin carrier proteins contribute to the establishment of auxin gradients in the plant. As a result, auxin acts as a morphogen, telling cells where they lie within the plant and determining how they differentiate. There are probably auxin carrier proteins specific to different tissues and different cells, participating in different auxin responses.

Light and gravity affect the direction of plant growth

While polar auxin transport establishes the orientation of growth, *lateral* (side-to-side) redistribution of auxin appears to be the mechanism that explains both phototropism and

another response depending on differential growth, gravitropism. This redistribution may be carried out by other auxin carrier proteins.

When light strikes a coleoptile from one side, auxin at the tip moves laterally toward the shaded side. The imbalance thus established is maintained down the coleoptile, so that in the growing region below, there is more auxin on the shaded side, causing the unequal growth that results in a coleoptile bent toward the light. This bending toward light is phototropism (Figure 37.12a). If you have noticed a house plant bending and pointing toward a window, you have seen phototropism.

Even in the dark, auxin moves to the lower side of a shoot that has been tipped over, causing more rapid growth in the lower side and, hence, an upward bending of the shoot. Such growth in a direction determined by gravity is called **gravitropism** (Figure 37.12b). The upward gravitropic response of shoots is defined as negative; the gravitropism of roots, which bend downward, is positive.

(a) **Phototropism**

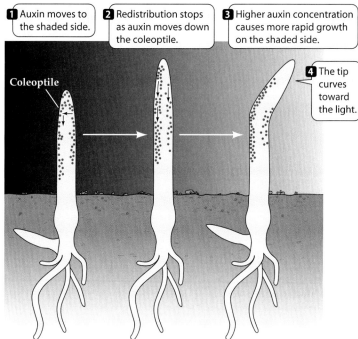

1 Auxin moves to the shaded side.

2 Redistribution stops as auxin moves down the coleoptile.

3 Higher auxin concentration causes more rapid growth on the shaded side.

Coleoptile

4 The tip curves toward the light.

(b) **Gravitropism**

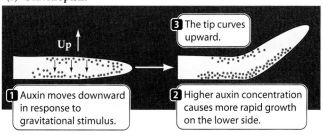

Up

3 The tip curves upward.

1 Auxin moves downward in response to gravitational stimulus.

2 Higher auxin concentration causes more rapid growth on the lower side.

37.12 Plants Respond to Light and Gravity
Phototropism and gravitropism occur in response to a redistribution of auxin.

Auxin affects vegetative growth in several ways

Like the gibberellins, auxin has many roles in plant development. It affects the vegetative growth of plants in several ways, including:

▶ Initiating root growth in cuttings
▶ Stimulating the detachment of old leaves from their stems (abscission)
▶ Maintaining apical dominance
▶ Promoting stem elongation and inhibiting root elongation

Let's examine each of these aspects in turn.

Cuttings from the shoots of some plants can produce roots and grow into entire new plants. For this to happen, certain undifferentiated cells in the *interior* of the shoot, originally destined to function only in food storage, must set off on an new mission: They must differentiate and become organized into the apical meristem of a new root.

These changes are similar to those in the pericycle of a root when a lateral root forms (see Chapter 34). Shoot cuttings of many species can be stimulated to grow profuse roots by dipping the cut surfaces into an auxin solution; this observation suggests that the plant's own auxin plays a role in the initiation of lateral roots. Commercial preparations that enhance the rooting of plant cuttings typically contain mostly synthetic auxins.

The effect of auxin on the detachment of old leaves from stems is quite different from root initiation. This process, called **abscission**, is the cause of autumn leaf fall. Leaves consist of a blade and a petiole that attaches the blade to the stem. Abscission results from the breakdown of a specific part of the petiole, the *abscission zone* (Figure 37.13). If the blade of a leaf is cut off, the petiole falls from the plant more rapidly than if the leaf had remained intact. If the cut surface is treated with an auxin solution, however, the petiole remains attached to the plant, often longer than an intact leaf would have (Figure 37.14). The time of abscission of leaves in nature appears to be determined in part by a decrease in the movement of auxin, produced in the blade, through the petiole.

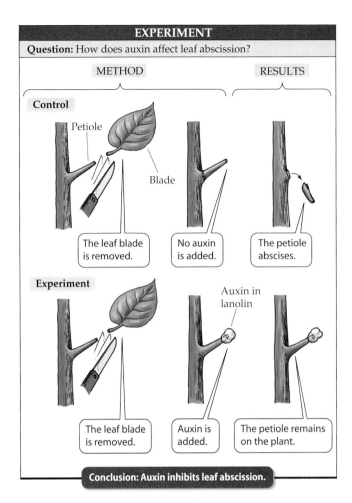

37.14 Auxin and Leaf Abscission
The leaf blade is a source of auxin throughout the growing season; without auxin, the petiole falls from the plant.

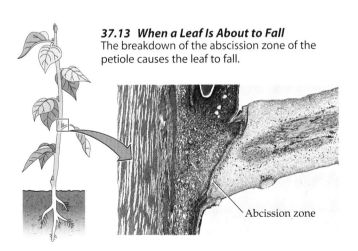

37.13 When a Leaf Is About to Fall
The breakdown of the abscission zone of the petiole causes the leaf to fall.

Auxin maintains **apical dominance**, a phenomenon in which apical buds inhibit the growth of lateral buds. This phenomenon can be demonstrated by an experiment with young seedlings. If the plant remains intact, the stem elongates, and the lateral buds remain inactive. Removal of the apical bud—the major site of auxin production—permits the lateral buds to grow out vigorously. If the cut surface of the stem is treated with an auxin solution, however, the lateral buds do not grow (Figure 37.15). Apical buds of branches also exert apical dominance: The lateral buds on the branch are inactive unless the apex of the branch is removed.

In the two experiments on leaves and stems that we have just discussed, removal of a particular part of the plant produces an effect—abscission or loss of apical dominance—and that effect is prevented by treatment with auxin. These results are consistent with other data showing that the excised part of the leaf or stem is an auxin source and that auxin in the intact plant helps maintain apical dominance and delays the abscission of leaves. As we will discover later, other hormones can modify the effects of auxin. *Plant*

37.15 Auxin and Apical Dominance
Auxin produced by the apical bud maintains apical dominance—the growth of a single main stem with minimal branching.

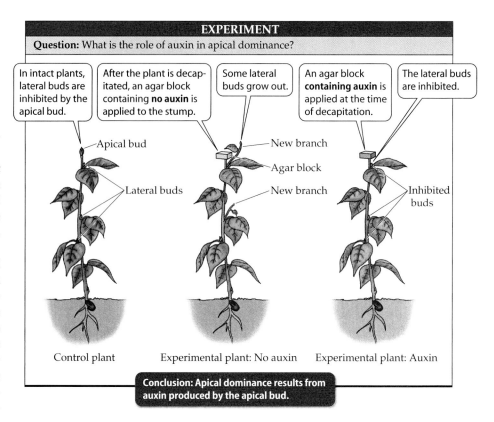

EXPERIMENT

Question: What is the role of auxin in apical dominance?

In intact plants, lateral buds are inhibited by the apical bud.

After the plant is decapitated, an agar block containing **no auxin** is applied to the stump.

Some lateral buds grow out.

An agar block **containing auxin** is applied at the time of decapitation.

The lateral buds are inhibited.

Apical bud

Lateral buds

New branch

Agar block

New branch

Inhibited buds

Control plant

Experimental plant: No auxin

Experimental plant: Auxin

Conclusion: Apical dominance results from auxin produced by the apical bud.

growth is regulated more by hormone interactions than by a single hormone.

Auxin promotes stem elongation, but it inhibits the elongation of roots. The question of why different organs should respond in opposite ways to the same chemical signal remains unanswered, but is a subject of current research.

Many synthetic auxins—chemical analogs of indoleacetic acid—have been produced and studied. One of them, 2,4-dichlorophenoxyacetic acid (2,4-D), has the striking property of being lethal to eudicots at concentrations that are harmless to monocots. This property made 2,4-D an effective *selective herbicide* that could be sprayed on a lawn or a cereal crop to kill those weeds that are eudicots. However, because 2,4-D takes a long time to break down, it pollutes the environment, so scientists are seeking new approaches to selective weed killing.

Auxin controls the development of some fruits

Although fruit development normally depends on prior fertilization of the egg, in many species treatment of an unfertilized ovary with auxin or gibberellins causes **parthenocarpy**—fruit formation without fertilization of the egg. Parthenocarpic fruits form spontaneously in some plants, including dandelions, seedless grapes, and cultivated bananas.

All of these activities illustrate the great diversity of important roles that auxin plays. Now let's see *how* auxin plays one of its roles—promoting stem elongation through effects on the cell wall.

Auxin promotes growth by acting on cell walls

CELL WALLS ARE A KEY TO PLANT GROWTH. The principal strengthening component of the plant cell wall is *cellulose*, a large polymer of glucose. In the wall, cellulose molecules tend to associate in parallel with one another. Bundles of approximately 250 cellulose molecules make up *microfibrils*

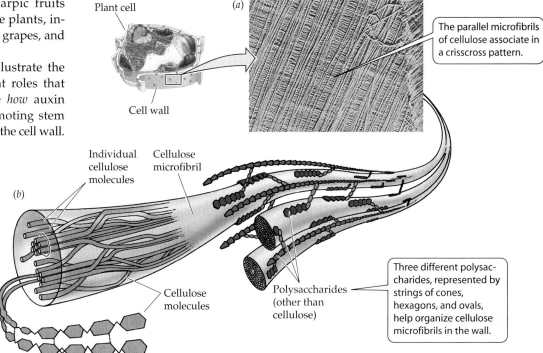

(*a*)

Plant cell

Cell wall

The parallel microfibrils of cellulose associate in a crisscross pattern.

(*b*)

Individual cellulose molecules

Cellulose microfibril

Cellulose molecules

Polysaccharides (other than cellulose)

Three different polysaccharides, represented by strings of cones, hexagons, and ovals, help organize cellulose microfibrils in the wall.

37.16 Cellulose in the Cell Wall
The plant cell wall is a network of cellulose microfibrils linked by other polysaccharides.

that are visible with an electron microscope (Figure 37.16). What makes the cell wall rigid is a network of cellulose microfibrils connected by bridges of other, smaller polysaccharides (Figure 37.16*b*). The orientation of the cellulose microfibrils determines the direction of cell expansion (Figure 37.17).

The growth of a plant cell is driven primarily by the uptake of water, which enters the cytoplasm of the cell and accumulates in its central vacuole. As the vacuole expands, the cell grows rapidly, with the vacuole often making up more than 90 percent of the volume of a mature cell. As the vacuole expands, it presses the cytoplasm against the cell wall, and the wall resists this force.

For the cell to grow, its wall must loosen and be stretched. If the wall simply stretched, it would become thinner. However, new polysaccharides are deposited throughout the wall and new cellulose microfibrils are deposited at the inner surface of the wall, maintaining its thickness. Thus the cellulose microfibrils in the outermost part of the wall are the oldest, and those in the innermost part the youngest.

The cell wall plays key roles in controlling the rate and direction of growth of a plant cell. How does the plant determine the behavior of its cell walls?

AUXIN LOOSENS THE CELL WALL. Experiments with segments of oat coleoptiles showed that plant cell walls recover incompletely from being stretched (Figure 37.18). Reversible stretching is called *elasticity*, and irreversible stretching is called *plasticity*. Pretreating the coleoptile segments with auxin significantly increased their plasticity; in other words, it loosened the cell walls. This result suggested that auxin-induced cell expansion might result from just such a loosening effect.

Auxin acts by causing the release of a "wall-loosening factor" from the cytoplasm. Studies in the 1970s indicated that the wall-loosening factor was sometimes simply hydrogen ions (protons, H^+). Acidifying the growth medium (that is, adding H^+) causes segments of stems or coleoptiles to grow as rapidly as segments treated with auxin. Furthermore, treating coleoptile segments with auxin causes acidification of the growth medium. Treatments that block acidification by auxin also block auxin-induced growth. It was suggested that hydrogen ions secreted into the cell wall as a result of auxin action might activate one or more proteins in the wall.

Proteins called *expansins*, isolated from plant cell walls in the 1990s, were found to cause the extension of isolated cell walls of several species. Expansins are widespread among land plants. Expansin action is pH-dependent, and the expansins appear to be activated by hydrogen ions. These proteins apparently modify hydrogen bonding between polysaccharides in the plant cell wall. The changed hydrogen bonding pattern may allow the polysaccharide macromolecules to slip past each other, so that the wall stretches and the cell expands.

37.17 Plant Cells Expand
The orientation of cellulose microfibrils in the plant's cell walls determines the direction of cell expansion.

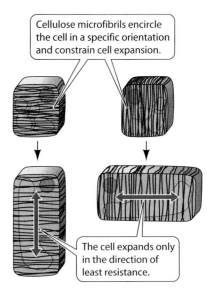

Cellulose microfibrils encircle the cell in a specific orientation and constrain cell expansion.

The cell expands only in the direction of least resistance.

Plants contain specific auxin receptor proteins

The initial step in the action of any plant hormone is the binding of the hormone to specific receptor proteins. Several proteins can bind various plant hormones, but some of this binding may be nonspecific. It must be shown that auxin-binding proteins actually mediate the effects of auxin.

Plant molecular biologists showed that the protein ABP1 (*Auxin-Binding Protein 1*) functions as an auxin receptor. They inserted the *ABP1* gene of *Arabidopsis* into other species and then induced the expression of the gene in cells that normally show a limited response to auxin. Upon expression of the inserted *ABP1* gene, the cells showed

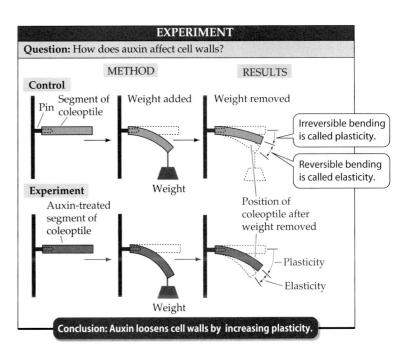

EXPERIMENT

Question: How does auxin affect cell walls?

METHOD · RESULTS

Control

Pin · Segment of coleoptile · Weight added · Weight removed

Irreversible bending is called plasticity.

Reversible bending is called elasticity.

Weight

Experiment

Auxin-treated segment of coleoptile

Position of coleoptile after weight removed

Plasticity

Elasticity

Weight

Conclusion: Auxin loosens cell walls by increasing plasticity.

37.18 Auxin Affects Cell Walls
Auxin increases the plasticity, but not the elasticity, of cell walls.

greater responses to both endogenous and applied auxin. Subsequent work has conclusively shown the existence and importance of other auxin receptor proteins. Given the number of processes regulated by auxin, it is hardly surprising that there appear to be multiple receptors and signal transduction pathways for this hormone.

Auxin and other hormones evoke differentiation and organ formation

What plant substance signals the different types of cells and organs to form? Much of the research on such questions has been done with plant tissues grown in culture outside the plant body. One easily grown tissue is pith—the spongy, innermost tissue of a stem. Pith tissue cultures proliferate rapidly, but show no differentiation. All the cells are similar and unspecialized; they grow into a lump on the surface of a culture medium.

Cutting a notch in the cultured pith tissue and inserting a stem tip into the notch causes the pith cells below the inserted tip to differentiate. Some of them differentiate to form water-conducting xylem cells. Differentiation of pith cells can also be initiated by adding to the notch a mixture of auxin and coconut milk (a rich source of plant hormones).

A similar effect can be observed in intact plants. If notches are cut in the stems of *Coleus blumei* plants, interrupting some of the strands of vascular tissue, the strands gradually regenerate from the upper side of the cut to the lower (recall that auxin moves from the tip to the base of a stem). If the leaves above the cut are removed, regeneration is slowed. However, when the missing leaves are replaced with an auxin solution, vascular tissue regenerates. Auxin and other plant hormones signal the formation of specific cell types.

Experiments with cultured tissues have helped clarify which hormones control organ formation. Undifferentiated cultures of tobacco pith form roots when treated with an appropriate concentration of auxin. Another group of hormones—the **cytokinins**—causes buds and then shoots to form in such cultures. The pattern of organ formation depends on the ratio of auxin to cytokinin in the medium. A high proportion of auxin favors roots, and a high proportion of cytokinins favors buds, but both processes are most active when both hormones are present.

Cytokinins Are Active from Seed to Senescence

Besides stimulating bud formation, the cytokinins promote cell division in cultured plant tissues, an activity that led to their discovery. In addition, cytokinins aid germination, inhibit stem elongation, stimulate lateral bud growth, and delay leaf senescence.

Cytokinins are derivatives of adenine. In studies of plant cell division, botanists discovered a substance that powerfully stimulated cell division in tissue cultures. This compound, *kinetin*, consists of adenine with an attached group. We now know that kinetin is just one of a family of compounds, which are now called cytokinins. Kinetin may be considered a synthetic cytokinin, because it has never been isolated from plant tissue. However, two closely related compounds, called *zeatin* and *isopentenyl adenine*, occur naturally in plants.

Kinetin
(a cytokinin discovered in aged DNA)

Zeatin
(a naturally occurring cytokinin in plants)

Cytokinins form primarily in the roots and move to other parts of the plant. They have several effects:

▶ Adding an appropriate combination of auxin and cytokinins to a growth medium yields rapid growth of plant tissues.

▶ Cytokinins can cause certain light-requiring seeds to germinate when the seeds are kept in constant darkness.

▶ Cytokinins usually inhibit the elongation of stems, but they cause lateral swelling of stems and roots (the fleshy roots of radishes are an extreme example).

▶ Cytokinins stimulate lateral buds to grow into branches; thus the balance between auxin and cytokinin levels controls the bushiness of a plant.

▶ Cytokinins increase the expansion of cut pieces of leaf tissue, and may regulate normal leaf expansion.

▶ Cytokinins delay the senescence of leaves. If leaf blades are detached from a plant and floated on water or a nutrient solution, they quickly turn yellow and show other signs of senescence. If instead they are floated on a solution containing a cytokinin, they remain green and senesce much more slowly.

Ethylene: A Gaseous Hormone That Promotes Senescence

Whereas the cytokinins oppose or delay senescence, another plant hormone promotes it. This hormone is the gas **ethylene**, which is sometimes called the senescence hormone. Ethylene can be produced by all parts of the plant, and like all plant hormones, it has several effects.

Ethylene
(the "senescence hormone")

Back when streets were lit by gas rather than by electricity, leaves on trees near street lamps abscised earlier than those on trees farther from the lamps. We now know that ethylene, a combustion product of the illuminating gas, is what caused the abscission. Auxin delays leaf abscission, but ethylene strongly promotes it; thus a balance of auxin and ethylene controls abscission.

Ethylene hastens the ripening of fruit

By promoting senescence, ethylene speeds the ripening of fruit. The old saying "one rotten apple spoils the barrel" is true. That rotten apple is a rich source of ethylene, which speeds the ripening and subsequent rotting of the others in the barrel. As the fruit ripens, it loses chlorophyll and its cell walls break down. Ethylene produced in the fruit tissue promotes both processes. Ethylene also causes an increase in its own production. Thus, once ripening begins, more and more ethylene forms, and because it is a gas, it diffuses readily throughout the fruit and even to neighboring fruits on the same or other plants.

Farmers in ancient times used to slash developing figs to hasten their ripening. We now know that wounding causes an increase in ethylene production by the fruit, and that the raised ethylene level promotes ripening. Today commercial shippers and storers of fruit hasten ripening by adding ethylene to storage chambers. This use of ethylene is the single most important use of a plant hormone in agriculture and commerce. Ripening can also be delayed by the use of "scrubbers" and adsorbents to remove ethylene from the atmosphere in fruit storage chambers.

As flowers senesce, their petals may abscise, to the detriment of the cut-flower industry. Florists or their suppliers often spray their flowers with dilute solutions of silver thiosulfate. Silver salts inhibit ethylene action, probably by interacting directly with the ethylene receptor, and thus delay senescence—enabling florists to keep their wares salable longer.

Ethylene affects stems in several ways

Although associated primarily with senescence, ethylene is active at other stages of plant development as well. The stems of many eudicot seedlings form an **apical hook** that protects the delicate shoot apex while the stem grows through the soil (Figure 37.19). The apical hook is maintained through an asymmetrical production of ethylene gas, which inhibits the elongation of cells on the inner surface of the hook. Once the seedling breaks through the soil surface and is exposed to light, ethylene synthesis stops, and the cells of the inner surface are no longer inhibited. These cells now elongate, and the hook opens, raising the shoot apex and expanding leaves into the sun.

37.19 The Apical Hook of a Eudicot
Asymmetrical production of ethylene is responsible for the apical hook of this seedling, which was grown in the dark.

Ethylene also inhibits stem elongation in general, promotes lateral swelling of stems (as do the cytokinins), and causes stems to lose their sensitivity to gravitropic stimulation.

The ethylene signal transduction pathway is well understood

Analysis of *Arabidopsis* mutants has revealed the steps in the mechanism of ethylene action. Some of these mutants do not respond to applied ethylene, and others act as if they have been exposed to ethylene even though they haven't. Studies of genes from these mutants and their protein products, coupled with comparisons of their amino acid sequences with those of other known proteins, have revealed some of the details of the signal transduction pathway through which ethylene produces its effects (Figure 37.20).

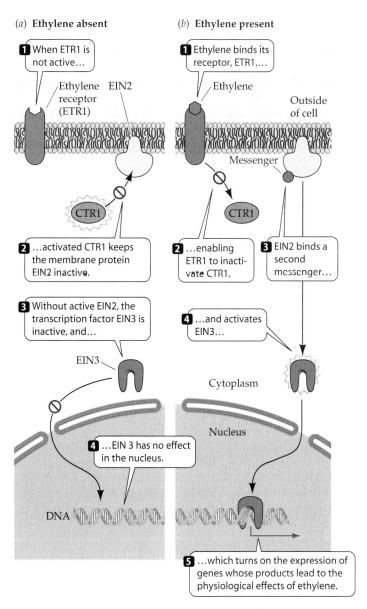

37.20 A Signal Transduction Pathway for Ethylene
This slightly simplified scheme shows the roles of four gene products (ETR1, CTR1, EIN2, and EIN3) in the signal transduction pathway through which ethylene exerts its effects.

The pathway includes two membrane proteins: The first is an ethylene receptor (ETR1), and the second is a channel (EIN2) that acts through a second messenger to activate a transcription factor that turns on genes. The resolution of this pathway has been one of the high points of plant biological research over the last 30 years.

Abscisic Acid: The Stress Hormone

Abscisic acid is another hormone that has multiple effects in the living plant. During embryo formation, abscisic acid promotes the accumulation of storage proteins in seeds by allowing the expression of the genes that encode those proteins. It is generally present in high concentrations in dormant buds and some dormant seeds, and it is the most common inhibitor of seed germination. Abscisic acid also inhibits stem elongation. It is sometimes referred to as the stress hormone of plants because it accumulates when plants are deprived of water and because of its possible role in maintaining the winter dormancy of buds.

Abscisic acid (the "stress hormone")

Some mutant corn plants, called *vp* mutants, have seeds that germinate while still attached to the cob, on the intact plant—a condition called *vivipary*. Several *vp* mutants are naturally deficient in abscisic acid, and a different kind of *vp* mutant fails to respond in any way to applied abscisic acid. Applying abscisic acid to the abscisic acid-deficient mutants reduces their tendency to show vivipary. The results of applying abscisic acid to both kinds of mutants indicate that it is the inhibitor that normally prevents seeds from germinating on the plant rather than in the soil.

Abscisic acid also regulates gas and water vapor exchange between leaves and the atmosphere through its effects on the guard cells of the leaf stomata (see Chapter 35). Abscisic acid causes stomata to close, and it also prevents stomatal opening normally caused by light. Both of these processes involve ion channels in the plasma membrane of the guard cells. The first response of a guard cell to abscisic acid is the opening of calcium channels and the entry of calcium into the cell. This calcium causes the cell's vacuole to release calcium, too. The increased concentration of calcium leads to a chain of events that result in the opening of potassium channels and the release of K^+ and of water, and the closing of the stoma as the guard cells sag together.

Hormones in Plant Defenses

When bacteria, viruses, or fungi attack a plant, the plant responds in several ways, as we will see in Chapter 39. One of its first responses is to release hormones called **oligosaccharins**. These hormones, as their name implies, are oligosaccharides—compounds consisting of a few sugar or derivative sugar units. They are actually fragments of the cell wall, which are released when enzymes from an attacker degrade it. They act as signals that trigger the plant's defenses.

Because auxin modifies the cell wall, it is not surprising that auxin, too, causes the release of oligosaccharins. The interactions between auxin and oligosaccharins may be complex. One oligosaccharin, at an extremely low concentration, has been found to inhibit auxin-induced growth promotion. Other oligosaccharins may regulate aspects of cell differentiation.

Three other hormones—*jasmonates, salicylic acid*, and *systemin*—serve as important signals in plant defenses. Their activities will be discussed in Chapter 39.

Brassinosteroids: "New" Hormones with Multiple Effects

More than 20 years ago, biologists isolated an interesting steroid from the pollen of rape, a member of the Brassicaceae, or mustard family. When applied to various plant tissues, this **brassinosteroid** stimulated cell elongation, pollen tube elongation, and vascular tissue differentiation, and it inhibited root elongation. Since then, dozens of chemically related and growth-affecting brassinosteroids have been found in plants. Treatment with as little as a few nanograms of brassinosteroid per plant is enough to promote growth. However, the brassinosteroids were not at first regarded as plant hormones, in part because of similarities between their effects and those of auxin.

Brassinolide (a brassinosteroid)

The properties of an *Arabidopsis* mutant called *det2* made it clear that brassinosteroids are naturally occurring plant hormones. When grown in darkness, seedlings homozygous for the *det2* allele differ dramatically from wild-type seedlings: In many respects, they look like wild-type seedlings grown in the light. Treatment of dark-grown *det2* mutant seedlings with brassinosteroids causes them to grow normally—that is, like wild-type plants grown in the dark. The *det2* plants are unable to synthesize their own brassinosteroids, and the lack of the hormone results in abnormal growth.

Brassinosteroids will probably be important in agriculture. They have increased the yields of some crops in field tests. Could they also be useful for keeping some plants small? What about limiting the growth rate of lawns and the height of trees and hedges? Joanne Chory and her colleagues at the University of California, San Diego and the

Salk Institute found a way to do this. They showed that a mutation of a gene called *bas-1* in *Arabidopsis* results in a dwarfed plant because the gene's product inactivates brassinosteroids in the stem. By introducing the *bas-1* mutation into selected plants, agriculturists could produce slow-growing plants and then adjust their growth rate by treatment with brassinosteroids.

Chory and others have shown that some of the effects of light on plant development result from effects on the signal transduction pathway for brassinosteroids. Let's now look more closely at the effects of environmental cues such as light.

Light and Photoreceptors

The length of the night determines the onset of winter dormancy. As summer wears on, the days become shorter (that is, the nights become longer). Leaves have a mechanism for measuring the length of the night, as we will see in the next chapter. Measuring night length is an accurate way to determine the season of the year. If a plant determined the season by the temperature, it could be fooled by a winter warm spell or by unseasonably cold weather in the summer. The length of the night, on the other hand, is determined by Earth's rotation around the sun and does not vary. Plants use the environmental cue of night length to time several aspects of their growth and development.

Length of the night is one of several environmental cues detected by plants, or by individual parts such as leaves. Light—its presence or absence, its intensity, its color, and its duration—provides cues to various conditions. Temperature, too, provides important environmental cues, both by its value at any particular time and by the distribution of warmer and colder stretches over a period of time. The plant "reads" an environmental cue and then "interprets" it, often by stepping up or decreasing its production of hormones.

We'll discuss an example of a temperature cue in the next chapter. Here, we'll see how certain photoreceptors interpret light, its duration, and its wavelength distribution.

Light regulates many aspects of plant development in addition to phototropism. The affected processes range from seed germination to shoot elongation to the initiation of flowering. Several photoreceptors take part in these and other processes. Five **phytochromes** mediate the effects of red and dim blue light. Three or more **blue-light receptors**, discovered more recently, mediate the effects of higher-intensity blue light.

37.21 Sensitivity of Seeds to Light
In each case, the final exposure reverses the preceding exposure; seeds respond only to the wavelength of the final light exposure.

Phytochromes mediate the effects of red and far-red light

Some seeds will not germinate in darkness, but do so readily after even a brief exposure to light. Blue and red light are highly effective in promoting germination, whereas green light is not.

Of particular importance to plants is the fact that far-red light *reverses* the effect of a prior exposure to red light. Far-red light is a very deep red, bordering on the limit of human vision and centered on a wavelength of 730 nm; red wavelengths are around 660 nm. If exposed to brief, alternating periods of red and far-red light in close succession, lettuce seeds respond only to the final exposure: If it is red, they germinate; if it is far-red, they remain dormant (Figure 37.21). This reversibility of the effects of red and far-red light regulates many other aspects of plant development, including flowering and seedling growth.

The basis for the red and far-red effects resides in certain bluish photoreceptor proteins called **phytochromes**. They are blue because they absorb red and far-red light and transmit other light. In the cytosol of plants are two interconvertible forms of phytochromes. Light drives the interconversion of the two forms. The form that absorbs principally red light is called P_r. Upon absorption of a photon of red light, a molecule of P_r is converted into P_{fr}. The P_{fr} form absorbs far-red light; when it does so, it is converted to P_r.

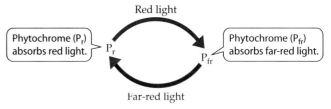

Red light

Phytochrome (P_r) absorbs red light. P_r P_{fr} Phytochrome (P_{fr}) absorbs far-red light.

Far-red light

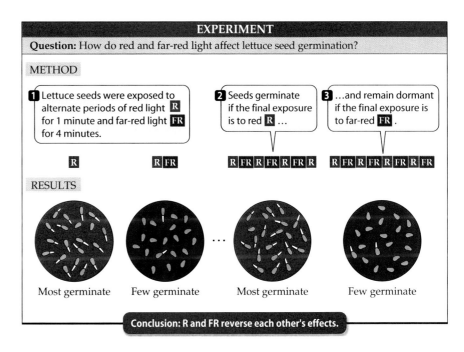

EXPERIMENT

Question: How do red and far-red light affect lettuce seed germination?

METHOD

1 Lettuce seeds were exposed to alternate periods of red light **R** for 1 minute and far-red light **FR** for 4 minutes.

2 Seeds germinate if the final exposure is to red **R** …

3 …and remain dormant if the final exposure is to far-red **FR** .

R   **R** **FR**   **R** **FR** **R** **FR** **R** **FR** **R**   **R** **FR** **R** **FR** **R** **FR** **R** **FR**

RESULTS

Most germinate   Few germinate   …   Most germinate   Few germinate

Conclusion: R and FR reverse each other's effects.

P_{fr} has some important biological effects. As we have just seen, one of them is to initiate germination in certain seeds, such as lettuce.

Phytochromes have many effects

Phytochromes help to regulate a seedling's early growth. The radicle, or embryonic root, is the first portion of the seedling to escape the seed coat (see Figure 37.6); the shoot emerges later. When seeds germinate in the dark below the soil surface, a pale and spindly seedling forms, with undeveloped leaves. Such an **etiolated** plant cannot carry out photosynthesis. The seedling shoot must reach the soil surface and begin photosynthesis before its nutrient reserves are expended and it starves.

Plants have evolved a variety of ways to cope with the problem of germinating underground. Etiolated flowering plants, for example, do not form chlorophyll. They turn green only when exposed to light, thereby conserving the resources needed to make chlorophyll, which would be useless in the dark. An etiolated shoot uses stored resources to elongate rapidly and hasten its arrival at the soil surface, where photosynthesis quickly begins. To break through soil yet protect delicate, underdeveloped leaves, the shoot of an etiolated eudicot seedling forms an apical hook (see Figure 37.19).

All of these etiolation phenomena (lack of chlorophyll, rapid shoot elongation, production of an apical hook, delayed leaf expansion) are regulated by the phytochromes. In a seedling that has never been exposed to light, all the phytochrome is in the red-absorbing (P_r) form. Exposure to light converts P_r to P_{fr} (the far-red-absorbing form), and the P_{fr} initiates reversal of the etiolation phenomena: Chlorophyll synthesis begins, shoot elongation slows, the apical hook straightens out, and the leaves start to expand.

There are multiple phytochromes

For years, plant biologists had difficulty accounting for some aspects of phytochrome action. A solution to these problems may lie in the discovery of multiple forms of phytochromes and other photoreceptors. *Arabidopsis* has five genes that encode different phytochromes, and this diversity has been found throughout the plant kingdom and in algae as well.

The several phytochromes may play differing roles in various phytochrome-controlled responses. Some of them may even play off each other to fine-tune plant growth during the day. Consider, for example, the light spectrum available to a seedling that is growing in the shade of other plants. Because chlorophyll in the leaves above it absorbs the light first, the shaded seedling "sees" a spectrum relatively rich in far-red (and poor in red); the ratio of far-red to red is increased as much as 10-fold to 20-fold in the shade. The interplay among signal transduction pathways initiated by the different phytochromes may lead to an increased rate of stem elongation that tends to bring the leaf into full sunlight.

We do not yet know how the various phytochromes produce their many effects, although it is evident that phy-

tochromes act through the plant's genome. Phytochromes appear to activate one or more G proteins. G proteins are membrane proteins that must bind to guanosine triphosphate (GTP) to exert their effects (see Chapter 15). The phytochrome-activated G proteins may convert GTP into the second messenger cGMP (cyclic guanosine monophosphate) and open channels that admit calcium ions into the cell, where they bind to the protein calmodulin. Both cGMP and the calcium–calmodulin complex can trigger changes leading eventually to the activation of specific genes.

Cryptochromes and phototropin are blue-light receptors

Cryptochromes are yellow photoreceptor pigments that absorb blue and ultraviolet light. They affect some of the same developmental processes, including seedling development and flowering, as do phytochromes. Unlike phytochromes, cryptochromes are present and play important roles in animals as well as plants.

In contrast to phytochromes, cryptochromes are located primarily in the plant nucleus. The exact mechanism of cryptochrome action is not yet known. It may be significant that phytochromes behave like protein kinases, and that cryptochromes can be substrates of such enzymes. It is likely that both classes of photoreceptors participate in protein kinase-based signaling pathways (see Chapter 15).

We began this chapter with a photo of a plant's phototropic response. Later we saw that the study of phototro-

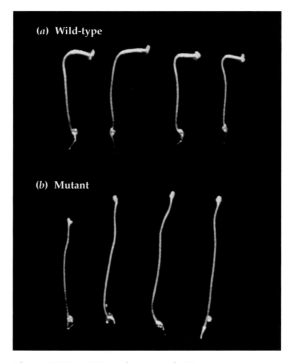

Figure 37.22 A Nonphototropic Mutant
(*a*) The four etiolated wild-type *Arabidopsis* seedlings in the top row are demonstrating normal phototropism. (*b*) These mutant seedlings cannot produce phototropin, the photoreceptor that signals the plant to curve toward light.

pism led to the discovery of auxin. But a question remained: What is the photoreceptor for phototropism? Plant scientists working with phototropic mutants of *Arabidopsis* have recently showed convincing evidence that it is a yellow protein, which they named **phototropin**. Upon absorbing blue light, phototropin initiates a signal transduction pathway leading to phototropic curvature (Figure 37.22). Still another type of blue-light receptor may be responsible for the light-induced closure of stomata.

Plants respond to light in many ways, and their responses are mediated by interactions of several photoreceptors, as we have seen.

Chapter Summary

Interacting Factors in Plant Development

▶ The environment, photoreceptors, hormones, and the plant's genome all play roles in the regulation of plant development.

▶ Hormones mediate many developmental phenomena in plants. Each plant hormone plays multiple regulatory roles, affecting several different aspects of development. Interactions among the hormones are often complex. **Review Table 37.1**

▶ Hormones and photoreceptors act through signal transduction pathways.

From Seed to Death: An Overview of Plant Development

▶ Cell division, cell expansion, and cell differentiation all contribute to plant development.

▶ The dormant seed eventually germinates and forms a growing seedling. Photoreceptors and hormones regulate seedling development, including growth. **Review Figures 37.1, 37.2**

▶ Eventually the plant flowers and forms fruit. Flowering in some plants is controlled by the length of the night. Hormones, probably including a flowering hormone, play roles in plant reproduction. **Review Figure 37.3**

▶ Some plant buds demonstrate winter dormancy. Eventually, all plants senesce and die. Dormancy and senescence are triggered by environmental cues, mediated by photoreceptors and hormones.

Ending Seed Dormancy and Beginning Germination

▶ Seed dormancy may be caused by exclusion of water or oxygen from the embryo, mechanical restraint of the embryo, or chemical inhibition of embryo development. In nature, seed dormancy is broken in various ways, including scarification, fire, leaching, and low temperatures.

▶ Seed dormancy offers adaptive advantages, such as an increased likelihood of germination in a place and at a time favorable for seedling growth.

▶ Seed germination begins with the imbibition of water. Then the embryo mobilizes its reserves to obtain building blocks and energy. The embryos of cereal seeds secrete gibberellins, which cause the aleurone layer to synthesize and secrete digestive enzymes that break down large molecules stored in the endosperm. **Review Figure 37.7**

Gibberellins: Regulators from Germination to Fruit Growth

▶ There are dozens of gibberellins. One, gibberellin A_1, regulates stem growth in most plants.

▶ Mutant plants that cannot produce normal amounts of gibberellins are dwarfs: Their stems are shorter than wild-type stems.

▶ Gibberellins regulate the growth of some fruits and cause bolting in some biennial plants. **Review Figure 37.9**

Auxin Affects Plant Growth and Form

▶ Studies of phototropism led to the discovery and isolation of auxin (indoleacetic acid). In grass seedlings, the photoreceptor for phototropism is in the tip of the coleoptile, and auxin is a messenger from the photoreceptor to the growing region of the coleoptile. **Review Figures 37.10, 37.11**

▶ Auxin transport is polar. Lateral movement of auxin establishes shoot and root responses to light and gravity: phototropism and gravitropism, respectively. Auxin carrier proteins move auxin into and out of cells. **Review Figure 37.12**

▶ Auxin plays roles in root formation, leaf abscission, apical dominance, and parthenocarpic fruit development. Certain synthetic auxins are used as selective herbicides. **Review Figures 37.13, 37.14**

▶ The arrangement of cellulose microfibrils in the plant cell wall limits the rate and direction of cell growth. Auxin increases the plasticity of the cell wall, promoting cell expansion. Part of the auxin response results from the pumping of protons from the cytoplasm into the cell wall, where the lowered pH activates proteins called expansins. **Review Figures 37.15, 37. 16, 37.17**

▶ Like all plant hormones, auxin is bound by receptor proteins.

▶ Auxin and other plant hormones signal cell differentiation and organ formation.

Cytokinins Are Active from Seed to Senescence

▶ Cytokinins are adenine derivatives. Zeatin and isopentenyl adenine are naturally occurring cytokinins, and kinetin is a synthetic cytokinin.

▶ First studied as promoters of plant cell division, cytokinins also promote seed germination in some species, inhibit stem elongation, promote lateral swelling of stems and roots, stimulate the growth of lateral buds, promote the expansion of leaf tissue, and delay leaf senescence.

Ethylene: A Gaseous Hormone That Promotes Senescence

▶ A balance between auxin and ethylene controls leaf abscission.

▶ Ethylene promotes senescence and fruit ripening.

▶ Ethylene causes the formation of a protective apical hook in eudicot seedlings that have not been exposed to light. In stems, it inhibits elongation, promotes lateral swelling, and causes a loss of gravitropic sensitivity.

▶ Ethylene acts through a signal transduction pathway that includes two proteins in the plasma membrane and that leads to the expression of genes. **Review Figure 37.20**

Abscisic Acid: The Stress Hormone

▶ Abscisic acid appears to maintain winter dormancy in buds. It prevents seeds from germinating while still attached to the parent plant, and it inhibits stem elongation. Through its effects on stomatal opening, it also regulates gas and water exchange between leaves and the atmosphere.

Hormones in Plant Defenses

▶ Oligosaccharins are hormones released by the cell wall in response to an attack by a pathogen. They participate in plant defenses against pathogens, and they interact in complex ways with auxin.

Brassinosteroids: "New" Hormones with Multiple Effects

▶ There are dozens of brassinosteroids. They affect cell elongation, pollen tube elongation, vascular tissue differentiation, and root elongation.

Light and Photoreceptors

▶ Phytochromes are bluish proteins found in the cytosol. Each phytochrome exists in two forms, P_r and P_{fr}, that are interconvertible by light. P_r absorbs red light (with a maximum at 660 nm), and P_{fr} absorbs far-red light (730 nm). **Review Figure 37.21**

▶ Phytochromes have many effects, including the various manifestations of etiolation.

▶ There are five phytochromes. They may play different roles in development, and their signal transduction pathways may interact to mediate the effects of light environments of differing spectral distribution. They mediate the effects of red and low-energy blue light.

▶ Cryptochromes, yellow photoreceptor proteins that absorb blue and ultraviolet light, interact with phytochromes in controlling seedling development and floral initiation. Cryptochromes mediate high-energy blue light effects.

▶ The signaling pathways for phytochromes and cryptochromes are based on protein kinases.

▶ Phototropin, another yellow protein, is the photoreceptor for phototropism.

For Discussion

1. How may it be advantageous for some species to have seeds whose dormancy is broken by fire?

2. Cocklebur fruits contain two seeds each, and the two seeds are kept dormant by two different mechanisms. How may this use of two mechanisms of dormancy be advantageous to cockleburs?

3. Corn stunt virus causes a great reduction in the growth rate of infected corn plants, so the diseased plants take on a dwarfed form. Since their appearance is reminiscent of the genetically dwarfed corn studied by Phinney, you suspect that the virus may inhibit the synthesis of gibberellins by the corn plants. Describe two experiments you might conduct to test this hypothesis, only one of which should require chemical measurement.

4. Whereas relatively low concentrations of auxin promote the elongation of segments cut from young plant stems, higher concentrations generally inhibit growth, as shown in the figure.

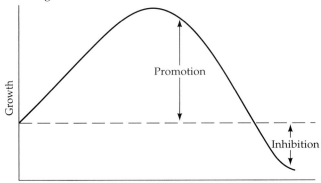

In some plants, the inhibitory effects of high auxin concentrations appear to be secondary: High auxin concentrations cause the synthesis of ethylene, which is what causes the growth inhibition. Silver thiosulfate inhibits ethylene action. How do you think the addition of silver thiosulfate to the solutions in which the stem segments grew would affect the appearance of the above graph?

5. Some etiolated seedlings develop hairs on their epidermis when exposed to dim light. Describe an experiment to test the hypothesis that a phytochrome is the photoreceptor for this effect.

38 *Reproduction in Flowering Plants*

BIOLOGISTS HAVE KNOWN FOR MORE THAN 60 years that the leaves of some plants, such as the cocklebur, contain built-in timers that measure the length of every night. When the night is of the appropriate length, the leaves—even a single leaf on a plant stripped of all other leaves—send a signal to other parts of the plant, telling them to form flowers. The evidence for this signal is substantial and convincing, yet nobody has been able to isolate and identify it.

After years of frustration, we may soon solve this mystery. The probable key lies in recent discoveries, described earlier in this book, about the functioning of plasmodesmata, the minute passageways connecting adjacent plant cells. Studies using mutant plants may allow scientists to identify the signal and learn how it triggers flowering. This knowledge will be a major advance in our understanding of reproduction in plants.

Why do plants expend energy and resources to produce flowers? The answer is simple: Flowers are sexual reproductive structures, and reproduction is one of the most important events in a plant's—or any organism's—life.

In this chapter we look at several aspects of plant reproduction, including some that are still not well understood. We contrast sexual and asexual reproduction, and we consider sexual reproduction in detail. In doing so, we look at angiosperm gametophytes, pollination, double fertilization, embryonic development, and the roles of fruits in seed dispersal. The transition to the flowering state is a key event in plant development, and we'll see how changing seasons trigger flowering in some plants—and speculate on the existence of a flowering hormone. We conclude the chapter with an examination of asexual reproduction in nature and in agriculture.

Many Ways to Reproduce

Plants have many ways of reproducing themselves—and with humans helping, there are even more ways. Flowers contain the sex organs of plants; it is thus no surprise that almost all flowering plants reproduce sexually. But many reproduce asexually as well; some even reproduce asexually most of the time. What are the advantages and disadvantages of these two kinds of reproduction? The answers to this question involve genetic recombination. As we have seen, sexual reproduction produces new genetic combinations and diverse phenotypes. Asexual reproduction, in contrast, produces a clone of genetically identical individuals.

Both sexual and asexual reproduction are important in agriculture. Many important annual crops are grown from seeds, which are the products of sexual reproduction. Seed-grown crops include wheat, rice, millet, and corn—the great grain crops, all of which are grasses—as well as plants in other families, such as soybeans and safflower. Other crops are produced asexually from grafts, or by other asexual means.

Orange trees, which have been under cultivation for centuries, can be grown from seed—except for one type, the navel orange. This plant apparently arose only once in history. Early in the nineteenth century, on a plantation on the Brazilian coast, one seed gave rise to one tree that had aberrant flowers. Parts of the flowers aborted, and seedless

Where It All Began
Some of the best early evidence for a flowering hormone came from studies of cockleburs (*Xanthium* sp.).

fruits formed. Every navel orange in the world comes from a navel orange tree derived asexually from that original Brazilian tree. Asexual reproduction is the only way of propagating this plant.

Unlike navel oranges, strawberries need not be propagated asexually, because they are capable of forming seeds. Nonetheless, asexual propagation of strawberries is common because vast numbers of plants that are genetically identical to a particularly desirable plant can be produced in this way.

We will treat asexual reproduction in greater detail at the end of this chapter. We begin, however, by considering sexual reproduction.

Sexual Reproduction

Sexual reproduction provides genetic diversity through recombination (see Chapter 9). Meiosis and mating shuffle genes into new combinations, giving a population a variety of genotypes in each generation. This genetic diversity may serve the population well as the environment changes or as the population expands into new environments. The adaptability resulting from genetic diversity is the major advantage of sexual reproduction over asexual reproduction.

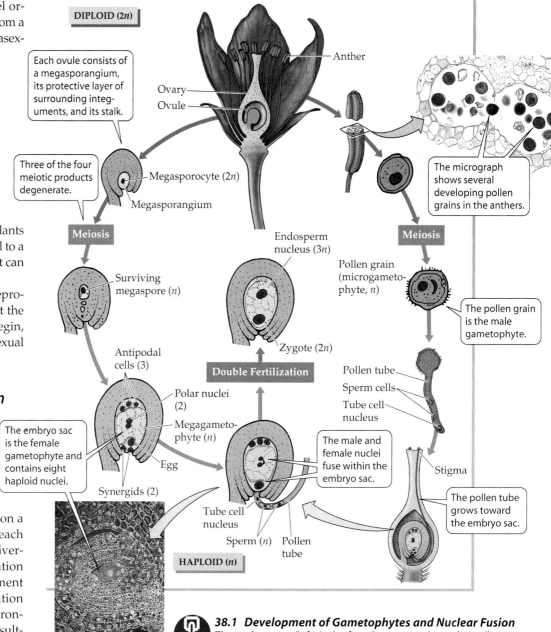

38.1 Development of Gametophytes and Nuclear Fusion
The embryo sac (left) is the female gametophyte; the pollen grain (right) is the male gametophyte. The male and female nuclei meet and fuse within the embryo sac.

The flower is an angiosperm's device for sexual reproduction

A complete flower consists of four groups of organs that are modified leaves: the carpels, stamens, petals, and sepals (see Figure 29.6 for review). The *carpels* and *stamens* are, respectively, the female and male sex organs. A *pistil* is a structure composed of one or more carpels. The base of the pistil, called the *ovary*, contains one or more *ovules*, each of which contains a megasporangium. The stalk of the pistil is the *style*, and the end of that stalk is the *stigma*. Each stamen is composed of a *filament* bearing a two-lobed *anther*, which consists of four microsporangia fused together.

The *petals* and *sepals* of many flowers are arranged in whorls (circles) around the carpels and stamens. Together, the petals constitute the *corolla*. Below them, the sepals constitute the *calyx*. The petals are often colored; the sepals are often green and photosynthetic. All the parts of the flower are borne on a stem tip, the *receptacle*.

Flowering plants have microscopic gametophytes

Before reading this section, you may wish to review the section in Chapter 28 entitled "Life cycles of plants feature alternation of generations." The concept of alternation of generations is central to an understanding of plant reproduction.

In plants, the sporophyte generation produces flowers. The flowers produce spores, which develop into tiny gametophytes. The flower is more than just a place where the egg and sperm are eventually found—it is also the place where the alternate generation resides.

The gametophytes—the gamete-producing generation—of flowering plants develop from haploid spores in sporangia within the flower (Figure 38.1).

▶ Female gametophytes (megagametophytes), which are called **embryo sacs**, develop in megasporangia.
▶ Male gametophytes (microgametophytes), which are called **pollen grains**, develop in microsporangia.

Within the ovule, a megasporocyte—a cell within the megasporangium—divides meiotically to produce four haploid megaspores. All but one of these megaspores then degenerate. The surviving megaspore undergoes mitotic divisions, usually producing eight haploid nuclei, all initially contained within a single cell—three nuclei at one end, three at the other, and two in the middle. Subsequent cell wall formation leads to an elliptical, seven-celled megagametophyte with a total of eight nuclei.

▶ At one end of the elliptical megagametophyte are three tiny cells: the egg and two cells called **synergids**. The egg is the female gamete, and the synergids participate indirectly in fertilization.
▶ At the opposite end of the megagametophyte are three **antipodal cells**, which eventually degenerate.
▶ In the large central cell are two **polar nuclei**.

The embryo sac is the entire seven-celled, eight-nucleus structure. (Follow the arrows down the left-hand side of Figure 38.1 to review the development of the embryo sac.)

The male gametophyte, or pollen grain, consists of fewer cells than the female gametophyte. The development of a pollen grain begins when a microsporocyte within the anther divides meiotically. Each resulting haploid microspore normally undergoes one mitotic division within the spore wall before the anthers open and release these two-celled pollen grains. Further development of the pollen grain, which we will describe shortly, is delayed until the pollen arrives at a stigma. In angiosperms, the transfer of pollen from the anther to the stigma is referred to as **pollination**.

Pollination enables fertilization in the absence of liquid water

Gymnosperms and angiosperms evolved independence from liquid water as a medium for gamete travel and fertilization—a freedom not shared by other plant groups. The male gametes of gymnosperms and angiosperms travel within pollen grains (Figure 38.2). But how do angiosperm pollen grains travel from an anther to a stigma?

Many different mechanisms have evolved for pollen transport. In some plants, such as peas and their relatives, pollination is accomplished before the flower bud opens.

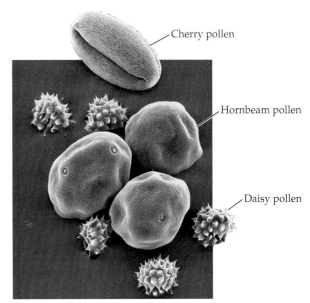

38.2 A Pollen Grain Sampler Each species' pollen has a characteristic size, shape, and cell wall formation.

100 μm

Pollen is transferred by the direct contact of anther and stigma within the same flower, resulting in *self-fertilization*.

Wind is the vehicle for pollen transport in many species. Wind-pollinated flowers have sticky or featherlike stigmas, and they produce pollen grains in great numbers (Figure 38.3). Some aquatic angiosperms are pollinated by water action, with water carrying pollen grains from plant to

Corylus cornuta

38.3 Wind Pollination The numerous anthers on these inflorescences (groups of flowers) of a hazelnut tree all point away from the stalk and stand free of the plant, promoting dispersal of the pollen by wind.

plant. Animals, including insects, birds, and bats, carry pollen among the flowers of many plants.

Some plants practice "mate selection"

In our discussion of Mendel's work (see Chapter 10), we saw that some plants can reproduce sexually either by cross-pollinating or by self-pollinating. Many plants demonstrate **self-incompatibility**; that is, they reject pollen from their own flowers. This rejection promotes genetic variation and limits inbreeding. A single gene, the *S* gene, is responsible for self-incompatibility. The *S* gene has dozens of alleles. A pollen grain is haploid and possesses a single *S* allele; the recipient stigma is diploid. In self-incompatible plants, pollen fails to germinate, or develops abnormally, on a stigma that possesses the same *S* allele (Figure 38.4).

The stigma plays an important role in "mate selection" by flowering plants. The stigmas of wind-pollinated plants are exposed to the pollen of many other species as well as their own, and even the flowers of plants with coevolved, specific animal pollinators may receive pollen from other plant species. Pollen from the same species binds strongly to the stigma due to cell–cell signaling by the cell wall of pollen grains of the same species. In contrast, foreign pollen falls off readily, without germinating.

A pollen tube delivers male cells to the embryo sac

When a pollen grain lands on the stigma of a compatible pistil, a **pollen tube** develops from the pollen grain (Figure 38.5). The pollen tube either digests its way through the spongy tissue of the style or, if the style is hollow, grows downward on the inner surface of this female organ. The pollen tube grows millimeters or even centimeters in the process .

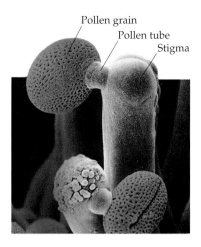

38.5 Pollen Tubes Begin to Grow
Pollen grains have landed on hairlike structures on the stigma of an *Arabidopsis* flower. Pollen tubes have started to form.

The rapid growth of the pollen tube requires calcium ions, taken up at the growing tip of the tube, as well as cell adhesion proteins. The downward growth of the pollen tube is guided by a long-distance signal from the ovule.

Angiosperms perform double fertilization

The pollen grain consists of two cells. The larger **tube cell** encloses the much smaller **generative cell** (Figure 38.6). Guided by the tube cell nucleus, the pollen tube eventually grows through megasporangial tissue and reaches the embryo sac. The generative cell meanwhile has undergone one mitotic division and cytokinesis to produce two **sperm cells**.

Both of the sperm cells enter the embryo sac, where they are released into the cytoplasm of one of the synergids. This synergid degenerates, releasing the sperm cells. Each sperm cell then fuses with a different cell of the embryo sac. One sperm cell fuses with the egg cell, producing the diploid zygote. The other fuses with the central cell, and that sperm cell nucleus and the two polar nuclei unite to form a triploid ($3n$) nucleus. While the zygote nucleus begins division to form the new sporophyte embryo, the triploid nucleus undergoes rapid mitosis to form a specialized nutritive tissue, the **endosperm**. The antipodal cells and the remaining synergid eventually degenerate, as does the pollen tube nucleus.

The fusion of a sperm cell nucleus with polar nuclei to form endosperm takes place only in angiosperms. This and the possession of flowers are the two most definitive characteristics shared by all angiosperms.

Embryos develop within seeds

Shortly after fertilization, highly coordinated growth and development of embryo, endosperm, integuments, and carpel ensues. The integuments develop into a double-layered seed coat, and the carpel ultimately becomes the wall of the fruit that encloses the seed.

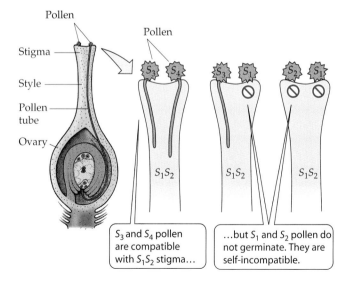

38.4 Self-Incompatibility
Pollen grains do not germinate normally if their *S* allele matches one of the *S* alleles of the stigma. Thus, the egg cannot be fertilized by a sperm from the same plant.

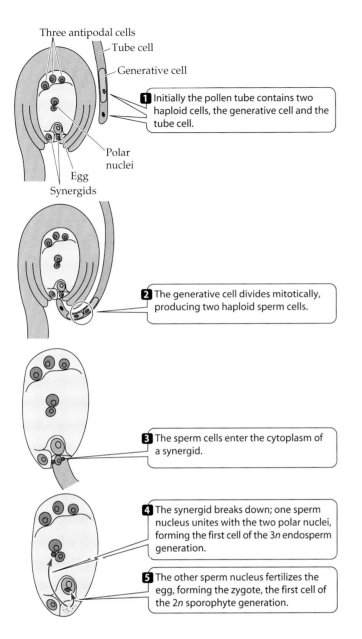

Three antipodal cells

Tube cell

Generative cell

1 Initially the pollen tube contains two haploid cells, the generative cell and the tube cell.

Polar nuclei

Egg

Synergids

2 The generative cell divides mitotically, producing two haploid sperm cells.

3 The sperm cells enter the cytoplasm of a synergid.

4 The synergid breaks down; one sperm nucleus unites with the two polar nuclei, forming the first cell of the 3*n* endosperm generation.

5 The other sperm nucleus fertilizes the egg, forming the zygote, the first cell of the 2*n* sporophyte generation.

38.6 Pollen Nuclei and Double Fertilization
The sperm nuclei contribute to the formation of the diploid zygote and the triploid endosperm. Double fertilization is a characteristic feature of angiosperm reproduction.

The first step in the normal formation of the embryo is a mitotic division of the zygote—the fertilized egg—giving rise to two daughter cells. Even at this stage, the two cells face different fates. An asymmetrical (uneven) distribution of cytoplasm within the zygote causes one end to produce the embryo proper and the other end to produce a supporting structure, the **suspensor** (Figure 38.7). The suspensor pushes the embryo against or into the endosperm, and provides one route by which nutrients enter the embryo.

With the asymmetrical division of the zygote, polarity has been established, as has the longitudinal axis of the new plant. A filamentous suspensor and a globular embryo are distinguishable after just four mitotic divisions. The suspensor soon ceases to elongate. In the embryo, the primary meristems form. As development continues, the first organs take form within the embryo.

In eudicots (monocots are somewhat different), the initially globular embryo takes on a characteristic *heart-stage* form as the cotyledons start to grow. Further elongation of the cotyledons and of the main axis of the embryo gives rise to what is called the *torpedo* stage (see Figure 38.7), during which some of the internal tissues begin to differentiate. The elongating region below the cotyledons is the *hypocotyl*. At the top of the hypocotyl, between the cotyledons, is the shoot apex; at the other end is the root apex. Each of these apical regions contains an apical meristem whose dividing cells will give rise to the organs of the mature plant.

Large amounts of nutrients are moved in from other parts of the plant, and the endosperm accumulates starch,

38.7 Early Development of a Eudicot
The embryo develops through intermediate stages, including a characteristic heart-shaped form, to reach the torpedo stage.

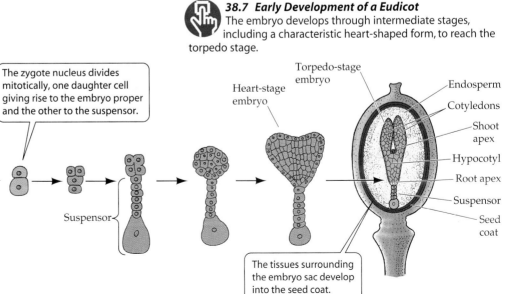

The zygote nucleus divides mitotically, one daughter cell giving rise to the embryo proper and the other to the suspensor.

Torpedo-stage embryo

Heart-stage embryo

Endosperm

Cotyledons

Shoot apex

Hypocotyl

Root apex

Suspensor

Seed coat

Endosperm nucleus

Embryo sac

Zygote

Suspensor

The tissues surrounding the embryo sac develop into the seed coat.

lipids, and proteins. In many species, the cotyledons absorb the nutrient reserves from the surrounding endosperm and grow very large in relation to the rest of the embryo (Figure 38.8*a*). In others, the cotyledons remain thin (Figure 38.8*b*); they draw on the reserves in the endosperm as needed when the seed germinates.

In the late stages of embryonic development, the seed loses water—sometimes as much as 95 percent of its original water content. In its dried state, the embryo is incapable of further development. It remains in this quiescent state until the conditions are right for germination. (Recall from Chapter 37 that a necessary first step in seed germination is the massive imbibition of water.)

Some fruits assist in seed dispersal

After fertilization, the ovary wall of a flowering plant—together with its seeds—develops into a fruit. A **fruit** may consist of only the mature ovary and its seeds, or it may include other parts of the flower or structures that are closely related to it. Some major variations on this theme are illustrated in Figure 29.11, which shows only fleshy, edible fruits. Many other fruits are dry or inedible.

Some fruits help disperse seeds over substantial distances. Various trees, including ash, elm, maple, and tree of heaven, produce a dry, winged fruit that may be blown some distance from the parent tree by the wind (Figure 38.9*a*). Water disperses some fruits; coconuts have been spread in this way from island to island in the Pacific (Figure 38.9*b*). Still other fruits travel by hitching rides with animals—either inside or outside them. Fleshy fruits such as berries provide food for mammals or birds; their seeds travel safely through the animal's digestive tract and are deposited some distance from the parent plant.

We have traced the sexual life cycle from the flower to the fruit to the dispersal of seeds. We discussed seed germination and vegetative development of the seedling in

(a)

(b)

38.9 Dispersal of Fruits
(*a*) A samara is a winged fruit characteristic of the maple family. (*b*) A coconut seed germinates where it washed ashore on a beach in the South Pacific.

Chapter 37. Now let's complete the sexual life cycle by considering the transition from the vegetative to the flowering state, and how this transition is regulated.

The Transition to the Flowering State

Flowering may terminate, interrupt, or accompany vegetative growth. The transition to the flowering state marks the end of vegetative growth for some plants. If we view a plant as something produced by a seed for the purpose of bearing more seeds, then the act of flowering is one of the supreme events in a plant's life.

Apical meristems can become inflorescence meristems

The first visible sign of the transition to the flowering state may be a change in one or more apical meristems in the shoot system. During vegetative growth, an apical meristem continually produces leaves, lateral buds, and internodes (regions of stem between the nodes where leaves and buds form: Figure 38.10*a*). This unrestricted growth is *indeterminate* (see Chapter 34).

Flowers may appear singly or in an orderly cluster that constitutes a structure called an **inflorescence**. If a vegetative meristem becomes an **inflorescence meristem**, it generally produces several other structures: smaller leafy structures called **bracts**, as well as new meristems in the angles between the bracts and the internodes (Figure 38.10*b*).

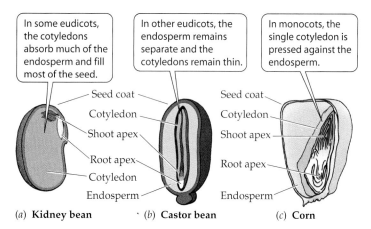

In some eudicots, the cotyledons absorb much of the endosperm and fill most of the seed.

In other eudicots, the endosperm remains separate and the cotyledons remain thin.

In monocots, the single cotyledon is pressed against the endosperm.

Seed coat

Cotyledon

Shoot apex

Root apex

Cotyledon

Endosperm

Seed coat

Cotyledon

Shoot apex

Root apex

Endosperm

Seed coat

Cotyledon

Shoot apex

Root apex

Endosperm

(*a*) **Kidney bean**   (*b*) **Castor bean**   (*c*) **Corn**

38.8 Variety in Angiosperm Seeds
In some seeds, such as kidney beans (*a*), the nutrient reserves of the endosperm are absorbed by the cotyledons at the seed stage. In others, such as castor beans (*b*) and corn (*c*), the reserves in the endosperm will be drawn on throughout the course of development.

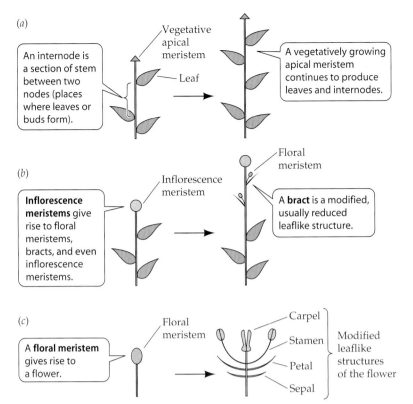

38.10 Flowering and the Apical Meristem
A vegetative apical meristem (*a*) grows without producing flowers. Once the transition to the flowering state is made, inflorescence meristems (*b*) give rise to bracts and to floral meristems (*c*), which become the flowers.

These new meristems may also be inflorescence meristems, or they may be **floral meristems**, which give rise to the flowers themselves.

Each floral meristem typically produces four consecutive whorls of organs—the sepals, petals, stamens, and carpels—separated by very short internodes, keeping the flower compact (Figure 38.10*c*). In contrast to vegetative meristems and some inflorescence meristems, floral meristems are responsible for *determinate* growth—the limited growth of the flower to a particular size and form.

A cascade of gene expression leads to flowering

How do apical meristems become inflorescence meristems, and how do inflorescence meristems give rise to floral meristems? How does a floral meristem give rise, in short order, to four different organs? How does each flower come to have the correct number of each of the floral organs? Numerous genes collaborate to produce these results. We'll refer here to some of the genes whose actions have been most thoroughly understood in *Arabidopsis* and snapdragons.

In order for an inflorescence meristem to give rise to a floral meristem, a group of *floral meristem identity genes* must be expressed. Expression of these genes initiates a cascade of further gene expression. Another set of genes, part of this cascade, participates in *pattern formation*—the spatial organization of the whorls of organs, which are still to be

determined. These genes, in turn, trigger the expression of a group of *organ identity genes* that work in concert to specify the successive whorls (see Figure 16.11). They are homeotic genes, and their products are transcription factors that mediate the expression of still further genes.

Now that we have seen how flowering occurs, we will consider how the transition from the vegetative to the flowering state is initiated.

Photoperiodic Control of Flowering

The life cycles of flowering plants fall into three categories: annual, biennial, and perennial. **Annuals**, such as many food crops, complete their life cycle (seed to flower) in less than a year. **Biennials**, such as carrots and cabbage, grow for all or part of one year and live on into a second year, during which they flower, form seeds, and die. **Perennials**, such as oak trees, live for a few to many years, during which both growth and flowering occur. What control systems give rise to these and other differences in flowering behavior?

In 1920, W. W. Garner and H. A. Allard of the U.S. Department of Agriculture studied the behavior of a newly discovered mutant tobacco plant. The mutant, named 'Maryland Mammoth,' had large leaves and exceptional height. When the other plants in the field flowered, the 'Maryland Mammoth' continued to grow. Garner and Allard took cuttings of the 'Maryland Mammoth' into their greenhouse, and the plants that grew from the cuttings finally flowered in December.

Garner and Allard guessed that this pattern had something to do with the seasons. They tested several likely seasonal variables, such as temperature, but the key variable proved to be the length of the day (as they saw it). By moving plants between light and dark rooms at different times to vary the day length artificially, they were able to establish a direct link between flowering and day length. (We now know that the key variable is the length of the *night*, rather than the day, but Garner and Allard did not make that distinction.)

The 'Maryland Mammoth' plants did not flower if the light period was longer than 14 hours each day, but flowering commenced after the days became shorter than 14 hours. Thus, the **critical day length** for 'Maryland Mammoth' tobacco is 14 hours (Figure 38.11). This phenomenon of control by the length of day or night is called **photoperiodism**.

There are short-day, long-day, and day-neutral plants

Poinsettias, chrysanthemums, and 'Maryland Mammoth' tobacco are **short-day plants** (SDP's), which flower only when the day is *shorter* than a critical *maximum*. Spinach and clover are examples of **long-day plants** (LDP's), which flower only when the day is *longer* than a critical *minimum*.

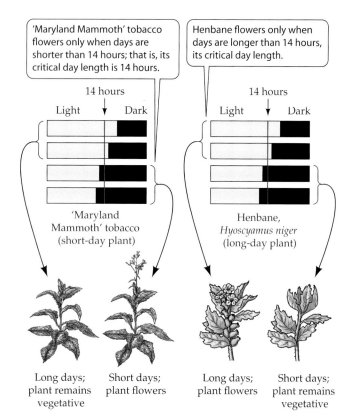

'Maryland Mammoth' tobacco flowers only when days are shorter than 14 hours; that is, its critical day length is 14 hours.

Henbane flowers only when days are longer than 14 hours, its critical day length.

14 hours

Light   Dark

14 hours

Light   Dark

'Maryland Mammoth' tobacco (short-day plant)

Henbane, *Hyoscyamus niger* (long-day plant)

Long days; plant remains vegetative

Short days; plant flowers

Long days; plant flowers

Short days; plant remains vegetative

38.11 Day Length and Flowering
By artificially varying the length of the day, Garner and Allard showed that the flowering of 'Maryland Mammoth' tobacco is initiated when the days become shorter than a critical length. 'Maryland Mammoth' tobacco is thus called a short-day plant. Henbane, a long-day plant, shows an inverse pattern of flowering.

Generally, LDP's are triggered to flower in midsummer and SDP's in late summer, or sometimes in the spring.

Some plants require photoperiodic signals that are more complex than just short or long days in order to flower. One group, the *short–long-day plants*, must first experience short days and then long ones. Accordingly, white clover and other short–long-day plants flower during the long days before midsummer. Another group, the *long–short-day plants*, cannot flower until the long days of summer have been followed by shorter ones, so they bloom only in the fall. *Kalanchoe*, seen in Figure 38.17b, is a long–short-day plant.

Other effects besides flowering are also under photoperiodic control. We have learned, for example, that short days trigger the onset of winter dormancy in plants. (Animals, too, show a variety of photoperiodic behaviors.)

The flowering of some angiosperms, such as corn and tomatoes, is not photoperiodic. In fact, there are more of these **day-neutral plants** than there are short-day and long-day plants. Some plants are photoperiodically sensitive only when young and become day-neutral as they grow older. Others require specific combinations of day length and other factors—especially temperature—to flower.

The length of the night determines whether a plant will flower

The terms "short-day plant" and "long-day plant" became entrenched before scientists learned that plants actually measure the length of the *night*, or of a period of darkness, rather than the length the of day. This fact was demonstrated by Karl Hamner of the University of California at Los Angeles and James Bonner of the California Institute of Technology (Figure 38.12).

Working with cocklebur, an SDP, Hamner and Bonner ran a series of experiments using two sets of conditions:

▶ The light period was kept constant—either shorter or longer than the critical day length—and the dark period was varied.

▶ The dark period was kept constant and the light period was varied.

The plants flowered under all treatments in which the dark period exceeded 9 hours, regardless of the length of the light period. Thus it is the length of the *night* that matters; for cocklebur, the *critical night length* is about 9 hours. Thus, it would be more accurate to call cocklebur a "long-night plant" than a short-day plant.

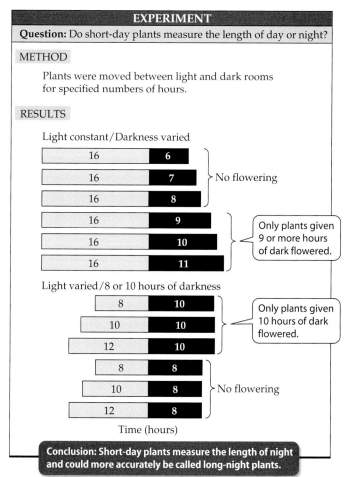

EXPERIMENT

Question: Do short-day plants measure the length of day or night?

METHOD

Plants were moved between light and dark rooms for specified numbers of hours.

RESULTS

Light constant/Darkness varied

16	6
16	7
16	8

} No flowering

16	9
16	10
16	11

Only plants given 9 or more hours of dark flowered.

Light varied/8 or 10 hours of darkness

8	10
10	10
12	10

Only plants given 10 hours of dark flowered.

8	8
10	8
12	8

} No flowering

Time (hours)

Conclusion: Short-day plants measure the length of night and could more accurately be called long-night plants.

38.12 Night Length and Flowering
The length of the dark period, not the length of the light period, determines flowering.

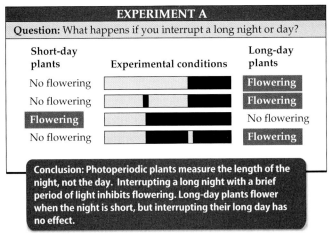

EXPERIMENT A

Question: What happens if you interrupt a long night or day?

Short-day plants	Experimental conditions	Long-day plants
No flowering		Flowering
No flowering		Flowering
Flowering		No flowering
No flowering		Flowering

Conclusion: Photoperiodic plants measure the length of the night, not the day. Interrupting a long night with a brief period of light inhibits flowering. Long-day plants flower when the night is short, but interrupting their long day has no effect.

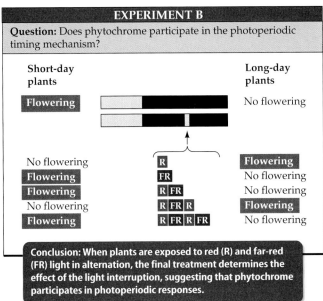

EXPERIMENT B

Question: Does phytochrome participate in the photoperiodic timing mechanism?

Short-day plants		Long-day plants
Flowering		No flowering
No flowering	R	Flowering
Flowering	FR	No flowering
Flowering	R FR	No flowering
No flowering	R FR R	Flowering
Flowering	R FR R FR	No flowering

Conclusion: When plants are exposed to red (R) and far-red (FR) light in alternation, the final treatment determines the effect of the light interruption, suggesting that phytochrome participates in photoperiodic responses.

38.13 The Effect of Interrupted Days and Nights
(a) Experiments suggest that plants are able to measure the length of a continuous dark period and use this information to trigger flowering. (b) Phytochromes seem to be involved in the photoperiodic timing mechanism.

In cocklebur, a single long night is enough of a photoperiodic stimulus to trigger full flowering some days later, even if the intervening nights are short ones. Most plants are less sensitive than cocklebur, requiring from two to many nights of appropriate length to induce flowering. For some plants, a single shorter night in a series of long ones, even one day before flowering would have commenced, inhibits flowering.

Hamner and Bonner showed that plants measure the length of the night using another method as well. They grew SDP's and LDP's under a variety of light conditions. Under some conditions, the dark period was interrupted by a brief exposure to light; in others, the light period was interrupted briefly by darkness. Interruptions of the light period by darkness had no effect on the flowering of either short-day or long-day plants. Even a brief interruption of

the dark period by light, however, completely nullified the effect of a long night (Figure 38.13a). An SDP flowered only if the long nights were uninterrupted. An LDP experiencing long nights flowered if those nights were broken by exposure to light. Thus a plant must have a timing mechanism that measures the length of a continuous dark period. Despite much study, the nature of this timing mechanism is still unknown.

Phytochromes and blue-light receptors, which affect several aspects of plant development (see Chapter 37), also participate in the photoperiodic timing mechanism. In the interrupted-night experiments, the most effective wavelengths of light were in the red range (Figure 38.13b), and the effect of a red-light interruption of the night could be fully reversed by a subsequent exposure to far-red light. It was once thought that the timing mechanism might simply be the slow conversion of phytochrome during the night from the P_{fr} form—produced during the light hours—to the P_r form. But this suggestion is inconsistent with most of the experimental observations and must be wrong. Phytochrome must be only a photoreceptor. The timekeeping role must be played by a biological clock.

Circadian rhythms are maintained by a biological clock

It is abundantly clear that organisms have some way of measuring time, and that they are well adapted to the 24-hour day–night cycle of our planet. Some sort of biological clock resides within the cells of all eukaryotes. The major outward manifestations of this clock are known as **circadian rhythms** (from the Latin *circa*, "about," and *dies*, "day").

We can characterize circadian rhythms, as well as other regular biological cycles, in two ways: The **period** is the length of one cycle, and the **amplitude** is the magnitude of the change over the course of a cycle (Figure 38.14).

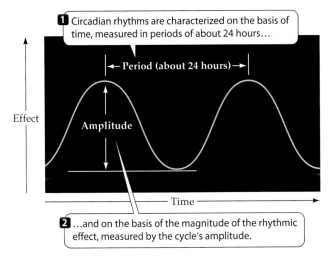

1 Circadian rhythms are characterized on the basis of time, measured in periods of about 24 hours...

← Period (about 24 hours) →

Effect

Amplitude

Time

2 ...and on the basis of the magnitude of the rhythmic effect, measured by the cycle's amplitude.

38.14 Features of Circadian Rhythms
The circadian rhythms of plants, like those of other organisms, can be characterized in two ways.

Circadian rhythms of protists, animals, fungi, and plants have been found to share some important characteristics:

► The period is remarkably *insensitive to temperature*, although lowering the temperature may drastically reduce the amplitude of the fluctuation.

► Circadian rhythms are *highly persistent*; they continue even in an environment in which there is no alternation of light and dark.

► Circadian rhythms can be *entrained*, within limits, by light–dark cycles that differ from 24 hours. That is, the period an organism expresses can be made to coincide with that of the light–dark regime.

► A brief exposure to light can shift the rhythm—it can cause a *phase shift*.

Plants provide innumerable examples of approximately 24-hour cycles. The leaflets of a plant such as clover or the tropical tree *Albizia* normally hang down and fold at night and rise and expand during the day. Flowers of many plants show similar "sleep movements," closing at night and opening during the day. They continue to open and close on an approximately 24-hour cycle even when the light and dark periods are experimentally modified (Figure 38.15).

The period of circadian rhythms in nature is approximately 24 hours. If an *Albizia* tree, for example, were to be placed under electric light on a day–night cycle totaling exactly 24 hours, the rhythm expressed would show a period of exactly 24 hours. However, if an experimenter used a day–night cycle of, say, 22 hours, then over time the rhythm would change—it would be **entrained** to a 22-hour period.

If an organism is maintained under constant darkness, with its circadian rhythm being expressed on the approximately 24-hour period, a brief exposure to light can cause a **phase shift**—that is, it can make the next peak of activity appear either later or earlier than expected, depending on when the exposure is given. Moreover, the organism does not then return to its old schedule if it remains in darkness. If the first peak is delayed by 6 hours, the subsequent peaks are all 6 hours late. Such phase shifts are permanent—until the organism receives more exposures to light.

Phytochromes and blue-light receptors are known to affect the period of the biological clock, with the different pigments reporting on different wavelengths and intensities of light. Perhaps this diversity of photoreceptors is an adaptation to the changes in the light environment that a plant experiences.

There is now ample evidence that the photoperiodic behavior of plants is based on the interaction of night length with the biological clock. But how the clock is coupled with flowering remains unclear.

Is there a flowering hormone?

Is the timing device for flowering located in a particular part of an angiosperm, or are all parts able to sense the length of the night? This question was resolved by "blind-folding" different parts of the plant.

It quickly became apparent that each leaf is capable of timing the night. If a short-day plant is kept under a regime of short nights and long days, but a leaf is covered so as to give it the needed long nights, the plant will flower (Experiment A in Figure 38.16). This type of experiment works best if only one leaf is left on the plant. If one leaf is given a photoperiodic treatment conducive to flowering—an *inductive* treatment—other leaves kept under noninductive conditions will tend to inhibit flowering.

Although it is the leaves that sense an inductive dark period, the flowers form elsewhere on the plant. Thus a message must be sent from the leaf to the site of flower formation. Three lines of evidence suggest that this message is a chemical substance—a flowering hormone.

► If a photoperiodically induced leaf is removed from the plant shortly after the inductive dark period, the plant does not flower. If, however, the induced leaf remains attached to the plant for several hours, the plant flowers. This result suggests that something must be synthesized in the leaf in response to the inductive dark period, then move out of the leaf to induce flowering.

► If two cocklebur plants are grafted together, and if one plant is given inductive long nights and its graft partner is given noninductive short nights, both plants flower (Experiment B in Figure 38.16).

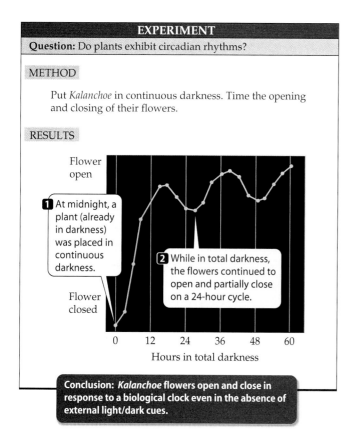

EXPERIMENT

Question: Do plants exhibit circadian rhythms?

METHOD

Put *Kalanchoe* in continuous darkness. Time the opening and closing of their flowers.

RESULTS

Flower open

1 At midnight, a plant (already in darkness) was placed in continuous darkness.

2 While in total darkness, the flowers continued to open and partially close on a 24-hour cycle.

Flower closed

0 12 24 36 48 60

Hours in total darkness

Conclusion: *Kalanchoe* flowers open and close in response to a biological clock even in the absence of external light/dark cues.

38.15 *Plants Can Measure Time*
Even when *Kalanchoe* is placed in continuous darkness, the opening and closing of its flowers continues to exhibit a circadian rhythm.

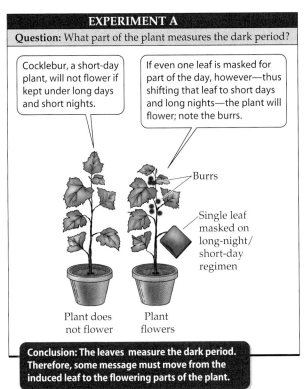

EXPERIMENT A

Question: What part of the plant measures the dark period?

Cocklebur, a short-day plant, will not flower if kept under long days and short nights.

If even one leaf is masked for part of the day, however—thus shifting that leaf to short days and long nights—the plant will flower; note the burrs.

Burrs

Single leaf masked on long-night/short-day regimen

Plant does not flower

Plant flowers

Conclusion: The leaves measure the dark period. Therefore, some message must move from the induced leaf to the flowering parts of the plant.

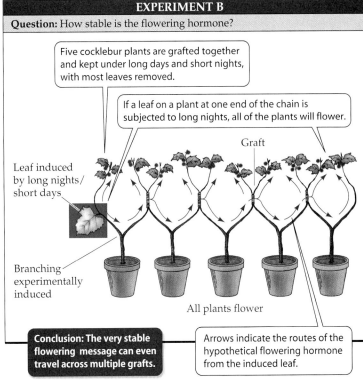

EXPERIMENT B

Question: How stable is the flowering hormone?

Five cocklebur plants are grafted together and kept under long days and short nights, with most leaves removed.

If a leaf on a plant at one end of the chain is subjected to long nights, all of the plants will flower.

Graft

Leaf induced by long nights/short days

Branching experimentally induced

All plants flower

Conclusion: The very stable flowering message can even travel across multiple grafts.

Arrows indicate the routes of the hypothetical flowering hormone from the induced leaf.

38.16 Evidence for a Flowering Hormone
If even a single leaf is exposed to inductive conditions, a "message" travels to the entire plant (and even to other plants, in grafting experiments), inducing it to flower.

▶ In at least one species, if an induced leaf from one plant is grafted onto another, noninduced plant, the host plant flowers.

Jan A. D. Zeevaart, a plant physiologist now at Michigan State University, performed this last experiment. He exposed a single leaf of the SDP *Perilla* to a short-day/long-night regime, inducing the plant to flower. Then he detached this leaf and grafted it onto another, noninduced, *Perilla* plant—which responded by flowering. The same leaf grafted onto successive hosts caused each of them to flower in turn. As long as 3 months after the leaf was exposed to the short-day/long-night regime, it could still cause plants to flower.

Experiments such as Zeevaart's suggest that the photoperiodic induction of a leaf causes a more or less permanent change in it, inducing it to start and continue producing a flowering hormone that is transported to other parts of the plant, switching those target parts to the reproductive state. So reasonable is this idea that biologists have named this hormone **florigen**, even though, after decades of active searching, it has not been isolated and characterized.

An elegant experiment suggested that the florigen of short-day plants is identical to that of long-day plants, even though SDP's produce it only under long nights and LDP's only under short nights. An SDP and an LDP were grafted together, and both flowered, as long as the photoperiodic conditions were inductive for one of the partners. Either the SDP or the LDP could be the one induced, but both would always flower. These results suggest that a flowering hormone—the elusive florigen—was being transferred from one plant to the other.

The direct demonstration of florigen activity remains a cherished goal of plant physiologists. For a long time it was thought that florigen could be neither a protein nor an RNA because these molecules were too large to pass from one living plant cell to another. However, we now know that such macromolecules can be transferred by way of plasmodesmata, and biologists are reexamining the possibility that an RNA or a protein is the long-sought florigen.

Vernalization and Flowering

In both wheat and rye, we distinguish two categories of flowering behavior. Spring wheat, for example, is sown in the spring and flowers in the same year. It is an annual plant. Winter wheat is biennial and must be sown in the fall; it flowers in the following summer. If winter wheat is not exposed to cold after its first year, it will not flower normally the next year. The implications of this finding were of great agricultural interest in Russia because winter wheat is a better producer than spring wheat, but it cannot be grown in some parts of Russia because the winters there are too cold for its survival.

Several studies performed in Russia during the early 1900s demonstrated that if seeds of winter wheat were premoistened and prechilled, they could be sown in the spring, and would develop and flower normally the same year. Thus, high-yielding winter wheat could be grown even in previously hostile regions. This induction of flowering by low temperatures is called **vernalization**.

Vernalization may require as many as 50 days of low temperatures (in the range from about –2° to +12°C). Some plant species require both vernalization and long days to flower. There is a long wait from the cold days of winter to the long days of summer, but because the vernalized state easily lasts at least 200 days, these plants do flower when they experience the appropriate night length.

Asexual Reproduction

Although sexual reproduction takes up most of the space in this chapter, asexual reproduction is responsible for many of the new plant individuals appearing on Earth. This fact suggests that in some circumstances, asexual reproduction must be advantageous.

Consider genetic recombination. When a plant self-fertilizes, there are fewer opportunities for genetic recombination than there are with cross-fertilization. A self-fertilizing plant that is heterozygous for a certain locus can produce among its progeny both kinds of homozygotes for that locus plus the heterozygote, but it cannot produce any progeny that carry alleles that it does not itself possess. Yet many plants continue to be self-compatible.

Asexual reproduction goes farther than self-fertilization: It eliminates genetic recombination altogether. When a plant reproduces asexually, it produces a *clone* of progeny with genotypes identical to its own. If a plant is well adapted to its environment, asexual reproduction may spread its genotype throughout that environment. This ability to exploit a particular environment is an advantage of asexual reproduction.

There are many forms of asexual reproduction

We call stems, leaves, and roots *vegetative organs*, distinguishing them from flowers, the reproductive parts of the plant. The modification of a vegetative organ is what makes **vegetative reproduction** possible. The stem is the organ that is modified in many cases. Strawberries and some grasses produce *stolons* (runners), horizontal stems that form roots at intervals and establish potentially independent plants (see Figure 34.4b). *Tip layers* are upright branches whose tips sag to the ground and put out roots, as in blackberry and forsythia.

Some plants, such as potatoes, form *tubers*, enlarged fleshy tips of underground stems (see Figure 34.4a). *Rhizomes* are horizontal underground stems that can give rise to new shoots. Bamboo is a striking example of a plant that reproduces vegetatively by means of rhizomes. A single bamboo plant can give rise to a stand—even a forest—of plants constituting a single, physically connected entity.

Whereas stolons and rhizomes are horizontal stems, bulbs and corms are short, vertical, underground stems. Lilies and onions form *bulbs* (Figure 38.17a), short stems with many fleshy, modified leaves. The leaves make up most of the bulb. Bulbs are thus large buds that store nutrients. They can give rise to new plants by dividing or by producing new bulbs from lateral buds. Crocuses, gladioli, and many other plants produce *corms*, underground stems that function very much as bulbs do. Corms are disclike and consist primarily of stem tissue; they lack the fleshy modified leaves that are characteristic of bulbs.

Not all vegetative organs modified for reproduction are stems. Leaves may also be the source of new plantlets, as in the succulent plants of the genus *Kalanchoe* (Figure 38.17b). Many kinds of angiosperms, ranging from grasses to trees such as aspens and poplars, form interconnected, genetically homogeneous populations by means of *suckers*—shoots produced by roots. What appears to be a whole stand of aspen trees, for example, may be a clone derived from a single tree by suckers (see Figure 54.1b).

Plants that reproduce vegetatively often grow in physically unstable environments, such as eroding hillsides. Plants with stolons or rhizomes, such as beach grasses,

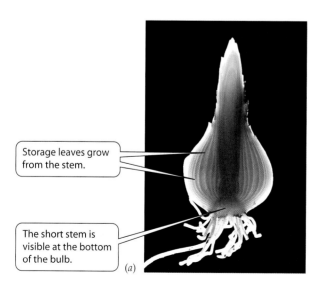

Storage leaves grow from the stem.

The short stem is visible at the bottom of the bulb.

(a)

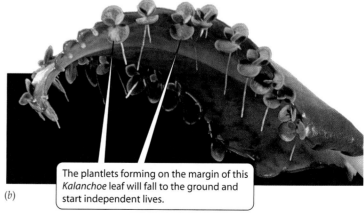

(b)

The plantlets forming on the margin of this *Kalanchoe* leaf will fall to the ground and start independent lives.

38.17 Vegetative Organs Modified for Reproduction
(a) Bulbs are short stems with large buds that store food and can give rise to new plants. (b) In *Kalanchoe*, new plantlets can form on leaves.

rushes, and sand verbena, are common pioneers on coastal sand dunes. Rapid vegetative reproduction enables these plants, once introduced, not only to multiply but also to survive burial by the shifting sand; in addition, the dunes are stabilized by the extensive network of rhizomes or stolons.

Dandelions, citrus trees, and some other plants reproduce by **apomixis**, the asexual production of seeds. As we have seen, meiosis reduces the number of chromosomes in gametes, and fertilization restores the sporophytic number of chromosomes in the zygote. Some plants can skip over *both* meiosis and fertilization and still produce seeds. Apomixis produces seeds within the female gametophyte without the mingling and segregation of chromosomes and without the union of gametes. The ovule simply develops into a seed, and the ovary wall develops into a fruit. An apomictic embryo has the sporophytic number ($2n$) of chromosomes. The result of apomixis is a fruit with seeds that are genetically identical to the parent plant.

Interestingly, apomixis sometimes requires pollination. In some apomictic species, a sperm nucleus must combine with the polar nuclei in order for the endosperm to form. In other apomictic species, the pollen provides the signals for embryo and endosperm formation, although neither sperm nucleus participates in fertilization. Pollination and fertilization are not the same thing!

Asexual reproduction is important in agriculture

Farmers take advantage of some natural forms of vegetative reproduction. Farmers and scientists have also developed new types of asexual reproduction by manipulating plants. One of the oldest methods of vegetative reproduction used in agriculture consists simply of making cuttings of stems, inserting them in soil, and waiting for them to form roots and thus become autonomous plants. The cuttings are usually encouraged to root by treatment with a plant hormone, auxin, as described in Chapter 37.

Horticulturists reproduce many woody plants by **grafting**—attaching a bud or a piece of stem from one plant to the root-bearing stem of another plant. The part of the resulting plant that comes from the root-bearing "host" is called the *stock*; the part grafted on is called the *scion* (Figure 38.18). In order for a graft to succeed, the cambium of the scion must become associated with the cambium of the stock. By cell division, both cambia form masses of wound tissue. If the two masses meet and fuse, the resulting continuous cambium can produce xylem and phloem, allowing transport of water and minerals to the scion and of photosynthate to the stock. Grafts are most often successful when the stock and scion belong to the same or closely related species.

Most fruit grown for market in the United States is produced on trees grown from grafts. There are many reasons for grafting plants for fruit production. The most common is the desire to combine a hardy root system with a shoot system that produces the best-tasting fruit. This motive is

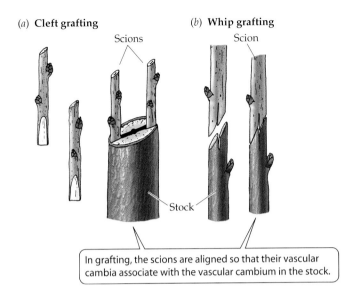

(a) **Cleft grafting**   *(b)* **Whip grafting**

In grafting, the scions are aligned so that their vascular cambia associate with the vascular cambium in the stock.

38.18 Grafting
Grafting—attaching a piece of a plant to the stem or root of another plant—is common in agriculture. The "host" stem or root is the stock; the grafted piece is the scion.

illustrated by the story of the wine grape *Vitis vinifera*. In 1863, plant lice of the genus *Phylloxera* inflicted great damage on the root systems of grapevines in French vineyards. More than 2.5 million acres of vines were destroyed. The problem was overcome by importing *V. vinifera* plants, which have *Phylloxera*-resistant root systems, from California. These plants were used as stocks to which French vines were grafted as scions. Thus the fine French grapes could be grown using roots resistant to the lice. (But the battle continues; in recent years, a new strain of *Phylloxera* has been damaging grapevines in California.)

Scientists in universities and industrial laboratories have been developing new ways to produce valuable plant materials via **tissue culture**. Because many plant cells are totipotent (see Figure 16.3), cultures of undifferentiated tissue can give rise to entire plants, as can small pieces of tissue cut directly from a parent plant. Tissue cultures are used commercially to produce orchids, rhododendrons, and many crops without resorting to seeds.

Culturing tiny bits of apical meristem can produce plants free of viruses. Because apical meristems lack developed vascular tissues, viruses tend not to enter them. Such meristem cultures have been used to increase the yields of potatoes and other crops.

Recombinant DNA techniques applied to tissue cultures can provide plants with capabilities they previously lacked such as resistance to pests, or increased nutritive value to humans. There is also interest in making certain valuable, sexually reproducing plants capable of apomikis. By causing cells of different types to fuse, one can obtain plants with exciting new combinations of properties.

Chapter Summary

Many Ways to Reproduce

▶ Almost all flowering plants reproduce sexually, and many also reproduce asexually.

▶ Both sexual and asexual reproduction are important in agriculture.

Sexual Reproduction

▶ Sexual reproduction promotes genetic diversity in a population, which may give the population an advantage under changing environmental conditions.

▶ The flower is an angiosperm's device for sexual reproduction.

▶ Flowering plants have microscopic gametophytes that develop in the flowers of the sporophytes. The megagametophyte is the embryo sac, which typically contains eight nuclei in a total of seven cells. The microgametophyte is the pollen grain, which delivers two sperm cells to the megagametophyte by means of a long pollen tube. **Review Figure 38.1**

▶ Pollination enables fertilization in the absence of liquid water.

▶ In self-incompatible species, the stigma rejects pollen from the same plant. **Review Figure 38.4**

▶ Angiosperms perform double fertilization: One sperm nucleus fertilizes the egg, forming a zygote, and the other sperm nucleus unites with the two polar nuclei to form a triploid endosperm nucleus. **Review Figure 38.6**

▶ The zygote develops into an embryo (with an attached suspensor), which remains quiescent in the seed until conditions are right for germination. The endosperm is the nutritive reserve upon which the embryo depends at germination. **Review Figures 38.7, 38.8**

▶ Flowers develop into seed-containing fruits, which often play important roles in the dispersal of the species.

The Transition to the Flowering State

▶ For a vegetatively growing plant to flower, an apical meristem in the shoot system must become an inflorescence meristem, which gives rise to bracts and more meristems. The meristems it produces may become floral meristems or additional inflorescence meristems. **Review Figure 38.10**

▶ Flowering results from a cascade of gene expression. Organ identity genes are expressed in floral meristems that give rise to sepals, petals, stamens, and carpels.

Photoperiodic Control of Flowering

▶ Photoperiodic plants regulate their flowering by measuring the length of light and dark periods.

▶ Short-day plants flower when the days are shorter than a species-specific critical day length; long-day plants flower when the days are longer than a critical day length. **Review Figure 38.11**

▶ Some angiosperms have more complex photoperiodic requirements than short-day or long-day plants have, but most are day-neutral.

▶ The length of the *night* is what actually determines whether a photoperiodic plant will flower. **Review Figure 38.12**

▶ Interruption of the nightly dark period by a brief exposure to light undoes the effect of a long night. **Review Figure 38.13**

▶ The mechanism of photoperiodic control involves a biological clock and phytochromes. **Review Figures 38.14, 38.15**

▶ Evidence suggests that there is a flowering hormone, called florigen, but the substance has yet to be isolated from any plant. **Review Figure 38.16**

Vernalization and Flowering

▶ In some plant species, exposure to low temperatures—vernalization—is required for flowering.

Asexual Reproduction

▶ Asexual reproduction allows rapid multiplication of organisms well suited to their environment.

▶ Vegetative reproduction involves the modification of a vegetative organ—usually the stem—for reproduction. Stolons, tip layers, tubers, rhizomes, bulbs, corms, and suckers are means by which plants may reproduce vegetatively.

▶ Some plant species produce seeds asexually by apomixis.

▶ Agriculturalists use natural and artificial techniques of asexual reproduction to reproduce particularly desirable plants.

▶ Horticulturists often graft different plants together to take advantage of favorable properties of both stock and scion. **Review Figure 38.18**

▶ Tissue culture techniques, based on the totipotency of many plant cells, are used to propagate plants asexually, to produce virus-free clones of crop plants, and to manipulate plants by recombinant DNA technology.

For Discussion

1. For a crop plant that reproduces both sexually and asexually, which method of reproduction might the farmer prefer?

2. Thompson seedless grapes are produced by vines that are triploid. Think about the consequences of this chromosomal condition for meiosis in the flowers. Why are these grapes seedless? Describe the role played by the flower in fruit formation when no seeds are being formed. How do you suppose Thompson seedless grapes are propagated?

3. Poinsettias are popular ornamental plants that typically bloom just before Christmas. Their flowering is photoperiodically controlled. Are they long-day or short-day plants? Explain.

4. You plan to induce the flowering of a crop of long-day plants in the field by using artificial light. Is it necessary to keep the lights on continuously from sundown until the point at which the critical day length is reached?

39 Plant Responses to Environmental Challenges

IF YOU ARE ATTACKED, IT MAKES SENSE TO call for help. Plants do this, too. When caterpillars begin to chew on the leaves of corn, cotton, or some other plant species, the plants synthesize and release chemical signals into the atmosphere. These substances attract other insects that feed on the caterpillars.

Herbivores aren't the only challenges plants face, however. The environment teems with plant pathogens. We know of more than a hundred diseases that can kill a tomato plant, each of them caused by a different pathogen (including various bacteria, fungi, protists, and viruses). Like animals, plants have a variety of defenses against pathogens. And, like the defenses of our own bodies, these mechanisms are not perfect, but they keep the plant world in competitive balance with its pathogens.

Environmental challenges to plants aren't limited to herbivores and pathogens. Some physical conditions pose substantial problems for plants and thus limit the places where different kinds of plants can live. The most challenging physical environments include ones that are very dry (deserts), that are water-saturated, that are dangerously salty, that contain high concentrations of toxic substances such as heavy metals, and that are very hot or very cold.

This chapter focuses on how plants meet the myriad challenges presented by their biological and physical environments. We begin by examining interactions between plants and pathogens and go on to consider interactions between plants and herbivores. Then we discuss the adaptations of some types of plants to their physical environments.

Plant–Pathogen Interactions

Plants and pathogens have evolved together in a continuing "arms race." Pathogens have evolved mechanisms by which to attack plants, and plants have evolved defenses against them. Each set of mechanisms uses information from the other. The pathogen's enzymes break down the plant's cell walls, for example, and the breakdown products signal to the plant that it is under attack. In turn, the plant's defenses alert the pathogen that it is under attack.

Calling In an Air Strike
As this caterpillar of a corn earworm moth (*Helicoverpa zea*) munches on a cotton boll, it is triggering a series of reactions in the plant that may end in the attraction of other insects that will attack the caterpillar.

What determines the outcome of a battle between a plant and a pathogen? The key to success for the plant is to respond to the information about the pathogen quickly and massively. Plants use both mechanical and chemical defenses in this effort.

Plants seal off infected parts to limit damage

Tissues such as epidermis or cork protect the outer surfaces of plants, and these tissues are generally covered by cutin, suberin, or waxes. This protection is comparable to the nonspecific defenses of animals. When pathogens pass these barriers, other nonspecific plant defenses are activated.

The defense systems of plants and animals differ. Animals generally repair tissues that have been damaged by pathogens, but plants do not. Instead, they seal off and sacrifice the damaged tissue so that the rest of the plant does not become infected. This approach works because most plants, unlike most animals, are modular and can replace damaged parts by growing new ones.

One of a plant cell's first defensive responses is the rapid deposition of additional polysaccharides to the cell wall, reinforcing this barrier to invasion by the pathogen (Figure 39.1). These polysaccharides block the plasmodesmata, limiting the ability of viral pathogens to move from cell to cell. They also serve as a base upon which lignin may be laid down. Lignin enhances the mechanical barrier, and the toxicity of lignin building blocks makes the cell inhospitable to some pathogens.

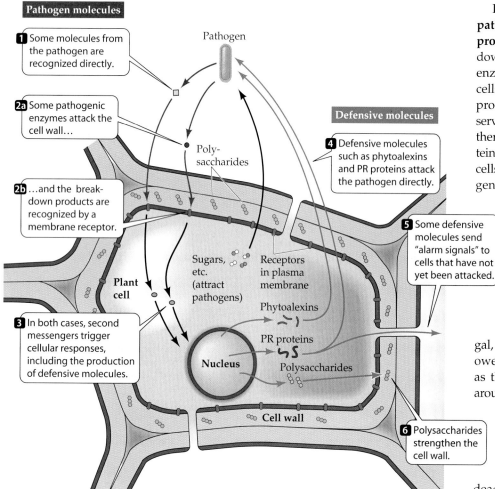

Pathogen molecules

1 Some molecules from the pathogen are recognized directly.

Pathogen

2a Some pathogenic enzymes attack the cell wall...

2b ...and the break-down products are recognized by a membrane receptor.

Poly-saccharides

Defensive molecules

4 Defensive molecules such as phytoalexins and PR proteins attack the pathogen directly.

Plant cell

Sugars, etc. (attract pathogens)

Receptors in plasma membrane

Phytoalexins

5 Some defensive molecules send "alarm signals" to cells that have not yet been attacked.

3 In both cases, second messengers trigger cellular responses, including the production of defensive molecules.

PR proteins

Nucleus

Polysaccharides

Cell wall

6 Polysaccharides strengthen the cell wall.

39.1 Signaling between Plants and Pathogens
Chemical interactions between plants and pathogens are highly coevolved. Plants produce molecules such as sugars that attract pathogens. But the presence of a pathogen stimulates the plant to produce defensive molecules that can work in many different ways.

Plants have potent chemical defenses against pathogens

When infected by certain fungi and bacteria, plants produce a variety of defensive compounds, among which are small molecules called phytoalexins and larger proteins called pathogenesis-related proteins (see Figure 39.1).

Phytoalexins are toxic to many fungi and bacteria. (Most are phenolics or terpenes, compounds that are also used to protect plants against herbivores; see Table 39.1.) They are produced by infected cells and their immediate neighbors within hours of the onset of infection. Enzymes from a pathogenic fungus can cause plant cell walls to release hormones called oligosaccharins (see Chapter 37), which trigger phytoalexin production. Because their antimicrobial activity is nonspecific, phytoalexins can destroy many species of fungi and bacteria in addition to the one that originally triggered their production. Physical injuries, viral infections, and chemical compounds produced in response to damage by herbivores can also induce the production of phytoalexins.

Plants also produce several types of **pathogenesis-related proteins**, or **PR proteins**. Some are enzymes that break down the cell walls of pathogens. These enzymes destroy some of the invading cells, and in some cases the breakdown products of the pathogen's cell walls serve as chemical signals that trigger further defensive responses. Other PR proteins may serve as alarm signals to plant cells that have not yet been attacked. In general, PR proteins appear not to be rapid-response weapons; rather, they act more slowly, perhaps after other mechanisms have blunted the pathogen's attack.

The hypersensitive response is a localized containment strategy

Plants that are resistant to fungal, bacterial, or viral diseases generally owe this resistance to what is known as the **hypersensitive response**. Cells around the site of microbial infection die, preventing the spread of the pathogen by depriving it of nutrients. Some of the cells produce phytoalexins and other chemicals before they die. The dead tissue, called a *necrotic lesion*, contains and isolates what is left of the microbial invasion (Figure 39.2). The rest of the plant remains free of the infecting microbe.

One of the chemicals produced during the hypersensitive response is a close relative of aspirin. Since ancient times, people in Asia, Europe, and the Americas have used willow (*Salix*) leaves and bark to relieve pain and fever. The active ingredient in willow is **salicylic acid**, the same substance from which aspirin is derived:

Salicylic acid

It now appears that all plants contain at least some salicylic acid. This compound plays a hormonal role in the plants' own defenses, often leading to a long-lasting effect that makes them resistant to later attacks by pathogens.

Systemic acquired resistance is a form of long-term "immunity"

Systemic acquired resistance is a general increase in the resistance of the entire plant to a wide range of pathogen species. It is not limited to the pathogen that originally triggered it or to the site of the original infection.

39.2 The Aftermath of a Hypersensitive Response
The necrotic spots on these leaves are a response to the fungus that causes strawberry blight.

The systemic acquired resistance that sometimes follows the hypersensitive response is accompanied by the synthesis of PR proteins. Treatment of plants with salicylic acid or aspirin leads to the production of PR proteins and to a resistance to pathogens. Salicylic acid treatment provides substantial protection against tobacco mosaic virus (a well-studied plant pathogen) and some other viruses.

Salicylic acid also serves as a hormone for disease resistance. In some cases, microbial infection in one part of a plant leads to the export of salicylic acid to other parts of the plant, where it causes the production of PR proteins before the infection can spread. The PR proteins then limit the extent of the infection. Infected plant parts also produce the closely related *methyl salicylate* (also known as oil of wintergreen). This volatile substance travels to other plant parts through the air. It may be that methyl salicylate can also trigger the production of PR proteins in neighboring plants that have not yet been infected.

Some plant genes match up with pathogen genes

Many plants use the hypersensitive response and systemic acquired resistance as nonspecific defenses against various pathogens. However, the triggering of these responses resides in a highly specific mechanism, called **gene-for-gene resistance**. In gene-for-gene resistance, the ability of a plant to defend itself against a specific strain of

a pathogen depends on the plant's having a particular allele of a gene that corresponds to a particular allele of a gene in the pathogen (Figure 39.3). Let's see how this matching works.

Plants have a large number of **R genes** (resistance genes), and many pathogens have sets of **Avr genes** (avirulence genes). Dominant R alleles favor resistance, and dominant Avr alleles make a pathogen less effective. If a particular plant has the dominant allele of an R gene and a pathogen strain infecting it has the dominant allele of the corresponding Avr gene, the plant will be resistant to that strain. This is true even when none of the other R–Avr pairs features corresponding dominant alleles. (This effect, one R–Avr pair overruling the others, is an example of epistasis, which was discussed in Chapter 10.)

The mechanism of gene-for-gene resistance is not completely understood. There are thousands of specific R genes among the plants, and their products have different functions. The Avr genes in pathogens are simply genes that cause the pathogen to produce a substance that elicits a defensive response in the plant. Most gene-for-gene interactions trigger the hypersensitive response.

Not all biological threats to plants come from microorganisms and viruses that cause diseases. Many animals, from inchworms to elephants, *eat* plants.

Plants and Herbivores: Benefits and Losses

Herbivores—animals that eat plants—depend on plants for energy and nutrients. Plants have many defense mechanisms that protect them against herbivores, as we will see. First, let's consider how herbivores can have a *positive* effect on the plants they eat.

Grazing increases the productivity of some plants

In **grazing**, a predator eats part of a plant, such as the leaves, without killing its prey, which then has the potential

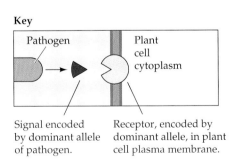

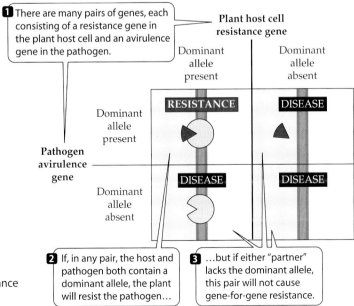

39.3 Gene-for-Gene Resistance
A single pair of corresponding dominant alleles promotes resistance even if all the other pairs are mismatches.

to grow back (Figure 39.4). What are the consequences of grazing? Is it always detrimental to plants, or are they somehow adapted to their place in the food chain? Certain plants and their predators evolved together, each acting as the agent of natural selection on the other (see Chapter 29). Because of this coevolution, grazing actually increases photosynthetic production in some plant species.

Removing some leaves from a plant may increase the rate of photosynthesis of the remaining leaves. This phenomenon probably is the result of several factors. First, nitrogen obtained from the soil by the roots no longer needs to be divided among so many leaves. Second, the export of sugars and other photosynthetic products from the leaves may be enhanced because the demand for those products in the roots is undiminished, while the sources for such products—leaves—have been decreased. The remaining leaves may compensate by photosynthesizing more rapidly.

A third and particularly significant factor increasing photosynthesis, especially in grasses, is an increase in the availability of light to the younger, more active leaves or leaf parts. The removal of older or dead leaves by a grazer decreases the shading of younger leaves. Unlike most other plants, which grow from their shoot and leaf tips, grasses grow from the base of the shoot and leaf, so their growth is not cut short by grazing.

Mule deer and elk graze many plants, including one called scarlet gilia. Their grazing removes about 95 percent of the aboveground part of each plant (Figure 39.5), but each plant quickly regrows not one, but four, replacement stems. Grazed plants produce three times as many fruits by the end of the growing season as do ungrazed plants.

39.4 Is Grazing Helpful or Harmful to Plants?
Grazing mammals such as this North American elk exist in virtually all of Earth's biomes, and the plants they feed on have evolved along with them.

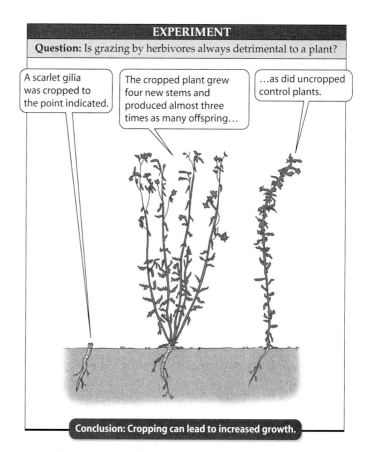

EXPERIMENT

Question: Is grazing by herbivores always detrimental to a plant?

A scarlet gilia was cropped to the point indicated.

The cropped plant grew four new stems and produced almost three times as many offspring...

...as did uncropped control plants.

Conclusion: Cropping can lead to increased growth.

39.5 Overcompensation for Being Eaten
Experiments confirm that some plants benefit from the effects of grazing.

Some grazed trees and shrubs continue to grow until much later in the season than do ungrazed but otherwise similar plants. This longer growing season results in part because the removal of apical buds by the grazers stimulates lateral buds to become active, producing a more heavily branched plant. Leaves on ungrazed plants may also die earlier in the growing season than leaves on grazed plants.

A plant may benefit from moderate herbivory by attracting animals that spread its pollen or that eat its fruit and thus disperse its seeds. Nevertheless, resisting attack by herbivores is often to the advantage of a plant.

Some plants produce chemical defenses

Although a plant cannot flee its herbivorous enemies, it may be able to defend itself chemically. Many plants attract, resist, and inhibit other organisms by producing special chemicals known as **secondary products**. *Primary products* are substances, such as proteins, nucleic acids, carbohydrates, and lipids, that are produced and used by all living things. Although all organisms use the same kinds of primary products, plants can differ as radically in their secondary products as they do in their external appearance.

The more than 10,000 known secondary plant products range in molecular weight from about 70 to more than 400,000, but most are of low molecular weight. Some are produced by only a single species, while others are characteristic of an entire genus or even family. These compounds help plants compensate for being unable to move.

39.1 *Secondary Plant Products in Defense*

CLASS	TYPE	ROLE	EXAMPLE
Nitrogen-containing	Alkaloids	Affect herbivore nervous system	Nicotine in tobacco
	Glycosides	Release cyanide or sulfur compounds	Dhurrin in sorghum
	Nonprotein amino acids	Disrupt herbivore protein structure	Canavanine in jack bean
Phenolics	Flavonoids	Phytoalexins	Capsidol in peppers
	Quinones	Inhibit competing plants	Juglone in walnut
	Tannins	Herbivore and microbe deterrents	Many woods, such as oak
Terpenes	Monoterpenes	Insecticides	Pyrethroids in chrysanthemum
	Sesquiterpenes	Antiherbivores	Gossypol in cotton
	Steroids	Mimic insect hormones and disrupt insect life cycle	
	Polyterpenes	Feeding deterrent?	Rubber in rubber tree

The effects of defensive secondary products on animals are diverse. Some secondary products act on the nervous systems of herbivorous insects, mollusks, or mammals. Others mimic the natural hormones of insects, causing some larvae to fail to develop into adults. Still others damage the digestive tracts of herbivores. Some secondary products are toxic to fungal pests. Humans make commercial use of many secondary plant products as fungicides, insecticides, rodenticides, and pharmaceuticals.

While many secondary products have protective functions, others are essential as attractants for pollinators and seed dispersers. Table 39.1 lists the major classes of defensive secondary plant products and their biological roles.

Let's look at a specific example of an insecticidal secondary product, canavanine.

Some secondary products play multiple roles

Canavanine is an amino acid that is not found in proteins, but is closely similar to the amino acid arginine, which is found in almost all proteins. Canavanine has two important roles in those plants that produce it in significant quantities. The first role is as a nitrogen-storing compound in seeds. The second, defensive role is based on the similarity of canavanine to arginine:

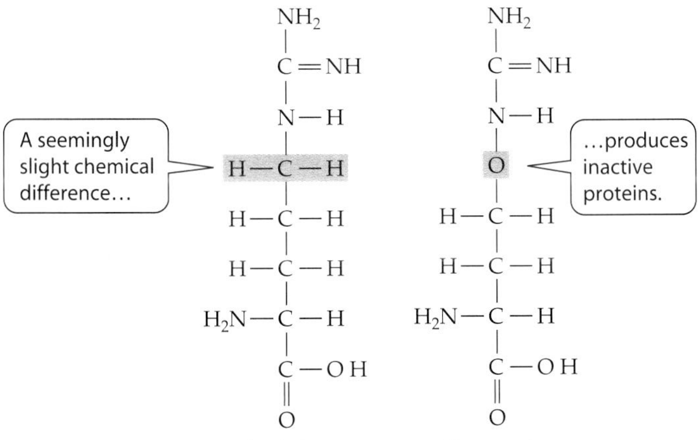

Many insect larvae that consume canavanine-containing plant tissue are poisoned. The canavanine is incorporated into the insect's proteins in some of the places where the DNA has coded for arginine because the enzyme that charges the tRNA specific for arginine fails to discriminate accurately between the two amino acids. The structure of canavanine is different enough from that of arginine that some of the resulting proteins end up with a modified tertiary structure and hence reduced biological activity. These defects in protein structure and function lead to developmental abnormalities that kill the insect.

A few insect larvae are able to eat canavanine-containing plant tissue and still develop normally. How can this be? In these larvae, the enzyme that charges the arginine tRNA discriminates correctly between arginine and canavanine. The canavanine they ingest is thus not incorporated into the proteins they form, and the larvae are not harmed.

Many defenses depend on extensive signaling

Plant defenses result from a series of signals. Insects feeding on tomato leaves damage the cells, leading to a chain of events including the formation of hormones and ending with the production of an insecticide. The signaling steps in the production of one defensive compound, shown in Figure 39.6, involve two hormones. **Systemin** is a polypeptide hormone—the first polypeptide hormone to be discovered in plants. **Jasmonates** are formed from the unsaturated fatty acid linolenic acid. The final step in this series is the production of a protease inhibitor. The inhibitor, once in an insect's gut, interferes with the digestion of proteins and thus stunts the insect's growth.

Jasmonic acid (a jasmonate)

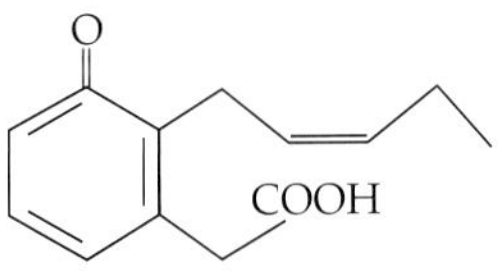

Jasmonates also take part in the "call for help" described at the beginning of this chapter. In that case, a substance re-

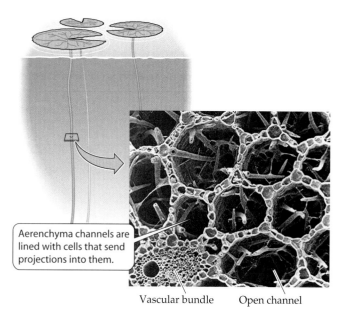

Aerenchyma channels are lined with cells that send projections into them.

Vascular bundle   Open channel

39.14 Aerenchyma Lets Oxygen Reach Submerged Tissues
The scanning electron micrograph, a cross section of a petiole of the yellow water lily, shows a vascular bundle and aerenchyma.

Too Much Salt: Saline Environments

Worldwide, no toxic substance restricts flowering plant growth more than salt (sodium chloride) does. *Saline*—salty—habitats support, at best, sparse vegetation. Saline habitats themselves are diverse, ranging from hot, dry, salty deserts to moist, cool, salty marshes. Along the seashore are saline environments created by ocean spray. The ocean itself is a saline environment, as are river estuaries, where fresh and salt water meet and mingle. The salinization of agricultural land is an increasing global problem. Even where crops are irrigated with fresh water, sodium ions from the water accumulate in the soil to ever greater concentrations as the water evaporates.

Saline environments pose an osmotic problem for plants. Because of its high salt concentration, a saline environment has an unusually negative water potential. To obtain water from such an environment, a plant must have an even more negative water potential than that of a plant in a nonsaline environment; otherwise, it will lose water, wilt, and die. A second problem is the potential toxicity of high concentrations of certain ions, notably sodium and chloride.

The **halophytes**—plants adapted to saline habitats—belong to a wide variety of flowering plant groups. How can these plants cope with a highly saline environment?

Most halophytes accumulate salt

Most halophytes share one adaptation: They accumulate sodium and, usually, chloride ions, and transport these ions to the leaves. The accumulated ions are stored in the central vacuoles of leaf cells, away from more sensitive parts of the cells. Nonhalophytes accumulate relatively little sodium, even when placed in a saline environment; of the sodium

that is absorbed by their roots, very little is transported to the shoot. The increased salt concentration in the tissues of halophytes makes their water potential more negative, so they can take up water more easily from the saline environment.

In 1999, scientists reported the first success in causing overexpression of a gene in *Arabidopsis* that enables sodium uptake. This gene encodes a Na^+/H^+ antiport protein in the tonoplast (the membrane surrounding the central vacuole). By making the gene produce a greater than normal number of these antiport proteins, the scientists increased sodium transport in *Arabidopsis*, converting this nonhalophyte into a halophyte. Further research along this line may result in a great boost to agriculture in saline environments. Biologists in Israel and elsewhere have had some success in breeding crops that can be watered with seawater or diluted seawater.

Some halophytes have other adaptations to life in saline environments. Some, for example, have **salt glands** in their leaves. These glands excrete salt, which collects on the leaf surface until it is removed by rain or wind (Figure 39.15). This adaptation, which reduces the danger of poisoning by accumulated salt, is found both in some desert plants, such as tamarisk, and in some mangroves growing in seawater in the Tropics.

Salt glands can play multiple roles, as in the desert shrub *Atriplex halimus*. This shrub has glands that secrete salt into small bladders on the leaves, where, by increasing the gradient in water potential, the salt helps the leaves obtain water from the roots. At the same time, by making the water potential of the leaves more negative, the salt reduces the transpirational loss of water to the atmosphere.

The adaptations we have just discussed are specific to halophytes. Several other adaptations are shared by halophytes and xerophytes.

Halophytes and xerophytes have some similar adaptations

Many halophytes, like some xerophytes, accumulate the amino acid proline in their cell vacuoles, making the water

39.15 Secreting Salt
This salty mangrove has special salt glands that secrete salt, which appears here as crystals on the leaves.

potential of their tissues more negative. Unlike sodium, proline is relatively nontoxic.

Succulence is another adaptation that halophytes and xerophytes have in common, as might be expected, since saline environments, like dry ones, make water uptake difficult. Succulence characterizes many halophytes that occupy salt marshes. There the salt concentration in the soil solution may change throughout the day; while the tide is out, for instance, evaporation increases the salt concentration. Succulence may offer a reserve of water for the plant during the period of maximum salinity; when the salinity drops as the tide comes in, the leaf's store of water is replenished. Many succulents—both xerophytes and halophytes—use crassulacean acid metabolism (CAM) and have reversed stomatal cycles that enable them to conserve water by closing their stomata in the daytime (see Figure 35.11). Other general adaptations to a saline environment include high root-to-shoot ratios, sunken stomata, reduced leaf areas, and thick cuticles.

Salt is not the only toxic solute found in soils. Some heavy metal ions are more toxic than sodium at equivalent concentrations.

Habitats Laden with Heavy Metals

High concentrations of some heavy metal ions, such as aluminum, mercury, lead, and cadmium, poison most plants. Some geographic sites are naturally rich in heavy metals as a result of normal geological processes. Acid rain leads to the release of toxic aluminum ions in the soil. Other human activities, notably the mining of metallic ores, leave localized areas—known as *tailings*—with substantial concentrations of heavy metals and low concentrations of nutrients. Such sites are hostile to most plants, and seeds falling on them generally do not produce adult plants.

Mine tailings rich in heavy metals, however, generally are not completely barren (Figure 39.16). They may support healthy plant populations that differ genetically from populations of the same species on the surrounding normal soils. How can these plants survive?

Initially, some plants were thought to tolerate heavy metals by excluding them: By not taking up the metal ions, it was believed, the plant avoided being poisoned. However, measurements have shown that tolerant plants growing on mine tailings do take up heavy metals, accumulating them to concentrations that would kill most plants. Thus the tolerant plants must have a mechanism for dealing with the heavy metals they take up. Such tolerant plants may be found to be useful agents for *bioremediation*, a decontamination process by which the heavy metal content of some contaminated soils is decreased by living organisms.

We know the mechanism of at least one case of tolerance to a heavy metal. The roots of a buckwheat grown in China secrete oxalic acid soon after they are exposed to aluminum concentrations high enough to inhibit root growth in other plants. Oxalic acid combines with aluminum ions, forming a complex that does not inhibit growth.

39.16 Life after Strip Mining
Although high concentrations of heavy metals kill most plants, grass is colonizing this eroded strip mine in North Park, Colorado.

From mine to mine, the heavy metals in the soil differ. In Wales and Scotland, bent grass (*Agrostis*) grows near many mines. Samples of bent grass from several such sites were tested for their ability to grow in various solutions, each containing only one heavy metal. In general, the plants tolerated a particular heavy metal—the one most abundant in their habitat—but were sensitive to others. That is, they tolerated only one or two heavy metals, rather than heavy metals as a group.

Tolerant plant populations can evolve and colonize an area surprisingly rapidly. The bent grass population around a particular copper mine in Wales is resistant to copper and is relatively abundant, even though the copper-rich soil dates from mining done only a century ago.

Hot and Cold Environments

Temperatures that are too high or too low can stress plants and even kill them. Plants differ in their sensitivity to heat and cold, but all plants have their limits.

Any temperature extreme can damage cellular membranes:

- High temperatures destabilize membranes and denature many proteins, especially some of the enzymes of photosynthesis.
- Low temperatures cause membranes to lose their fluidity and alter their permeabilities to solutes.
- Freezing temperatures may cause ice crystals to form, damaging cellular membranes.

Plants have ways of coping with high temperatures

Transpiration, the evaporative loss of water, can cool a plant, but it also increases the plant's need for water. Therefore, it is not surprising that many plants living in hot environments have adaptations similar to those of xerophytes. These adaptations include epidermal hairs and spines that radiate heat, modified leaf displays that intercept less direct sunlight, and others.

Plants respond within minutes to high temperatures by producing several kinds of **heat shock proteins**. Among these are chaperonins (see Chapter 3), which help other proteins maintain their structures and avoid denaturation. Threshold temperatures for the production of heat shock proteins vary, but 40°C is sufficient to induce them in most plants. We have much to learn about the dozens of heat shock proteins, but we do know that some other types of stress also induce their formation. Among these are chilling and freezing.

Some plants are adapted to survival at low temperatures

Low temperatures above freezing injure many plants, including important crops such as rice, corn, and cotton. Many plant species can be modified to resist the effects of cold spells by a process called **cold-hardening**, which involves repeated exposure to cool, but not injurious, temperatures. The hardening process is a slow one, requiring many days. A key change that occurs during the hardening process is an increase in the relative fraction of unsaturated fatty acids in membranes. Unsaturated fatty acids solidify at lower temperatures than do saturated ones. Thus, the membranes retain their fluidity and function normally at cooler temperatures.

Low temperatures induce the formation of certain heat shock proteins that protect against chilling damage. There are also cases of "cross-protection" by heat shock proteins that are induced by one type of stress and that protect against other stresses. Tomatoes shocked by 2 days of high temperatures, for example, formed heat shock proteins and became resistant to chilling damage for the next 3 weeks.

If ice crystals form within cells, they can kill the cells by puncturing organelles and plasma membranes. Even outside cells, the growth of ice crystals can draw water from the cells and dehydrate them. Freezing-tolerant plants have a variety of adaptations to cope with these problems. A common one is the production of *antifreeze proteins* that inhibit the growth of ice crystals.

Plants have many effective mechanisms for coping with environmental challenges of many kinds. Their success is obvious—just look around you.

 *Chapter Summary*

Plant–Pathogen Interactions

▶ Plants and pathogens evolve together. **Review Figure 39.1**

▶ Plants can strengthen their cell walls when attacked.

▶ Plant chemical defenses include PR proteins and phytoalexins.

▶ In the hypersensitive response, cells produce phytoalexins and then die, trapping the pathogens in dead tissue.

▶ The hypersensitive response is often followed by systemic acquired resistance, in which the hormone salicylic acid activates further synthesis of PR proteins and triggers responses in other parts of the plant.

▶ The hypersensitive response is nonspecific. A more specific response, called gene-for-gene resistance, matches up alleles in a plant's resistance genes and a pathogen's avirulence genes. **Review Figure 39.3**

Plants and Herbivores: Benefits and Losses

▶ Grazing by herbivores increases the productivity of some plants. **Review Figure 39.5**

▶ Some plants produce secondary products that function as chemical defenses against herbivores. **Review Table 39.1**

▶ Various hormones, including systemin and jasmonates, participate in the pathways leading to the production of defensive chemicals. **Review Figure 39.6**

▶ To avoid poisoning themselves, plants may confine the toxic substances they produce to special compartments, or they may produce the substances only after cells have been damaged, or they may form enzymes and receptors that are not affected by the substances.

Water Extremes: Dry Soils and Saturated Soils

▶ Desert annuals evade drought by living only long enough to take advantage of the brief period during which the soil has enough moisture to support them.

▶ Some leaves have special adaptations to dry environments: a thickened cuticle, epidermal hairs, sunken stomata, fleshy leaves and stems, spines, and altered leaf display angles.

▶ Other adaptations to dry environments include long taproots and root systems that die back seasonally.

▶ The submerged roots of some plants form pneumatophores to allow oxygen uptake from the air. Aerenchyma in submerged plant parts stores and permits the diffusion of oxygen. **Review Figure 39.14**

Too Much Salt: Saline Environments

▶ A saline environment restricts the availability of water to plants. Halophytes are plants that are adapted to such environments.

▶ Most halophytes accumulate salt, and some have salt glands that excrete the salt to the leaf surface.

▶ Halophytes and xerophytes have some adaptations in common.

Habitats Laden with Heavy Metals

▶ Aluminum, mercury, lead, and cadmium are among the heavy metals that are toxic to plants at high concentrations.

▶ Rather than excluding heavy metals, tolerant plants deal with them after taking them up. A given plant's tolerance is limited to only one or two heavy metals.

Hot and Cold Environments

▶ High temperatures destabilize cell membranes and some proteins.

▶ Adaptations to elevated temperatures include the production of heat shock proteins.

▶ Low temperatures cause membranes to lose their fluidity.

▶ Ice crystals can puncture organelles and plasma membranes.

▶ Adaptations to low temperatures and freezing include a change in membrane fatty acid composition and the production of antifreeze proteins.

For Discussion

1. We mentioned the possibility of designing crop plants that produce their own pesticides. Now chemical companies are designing crop plants capable of detoxifying weed killers, so that crops grow after farmers have destroyed competing vegetation. Discuss the likely usefulness and possible drawbacks of such applications of recombinant DNA technology.

2. How might plant adaptations affect the evolution of herbivores? How might adaptions of herbivores affect plant evolution?

3. The stomata of the common oleander, *Nerium oleander*, are sunk in crypts in its leaves. Whether or not you know what an oleander is, you should be able to descibe an important feature of its natural habitat; what is this feature?

4. Explain why halophytes often use the same mechanisms for coping with their challenging environments as xerophytes do for coping with theirs.

5. In ancient times, people used less sophisticated methods for mining than we use today. Thus ancient mines often yield substantial profits to modern-day miners who find and work them. On the basis of material in this chapter, how might you try to locate the site of an ancient mine?

(a)

Cuboidal cells
in simple epithelium

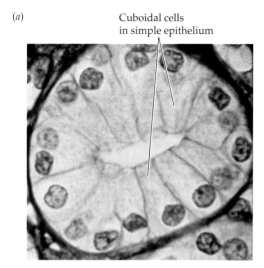

(b)

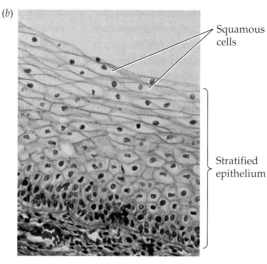

Squamous
cells

Stratified
epithelium

40.3 Epithelial Tissue
(a) A single layer of
cuboidal cells forms a simple epithelium lining the
collecting ducts of a
human kidney. (b) Multiple
layers of squamous cells
form a stratified epithelium. (c) The columnar cells
of this pseudostratified
epithelium lining the respiratory tract give the
appearance of multiple
layers.

(c)

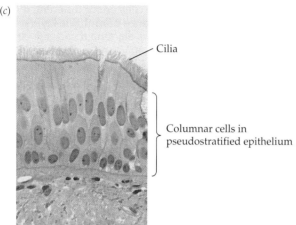

Cilia

Columnar cells in
pseudostratified epithelium

Connective tissues support and reinforce other tissues

In contrast to the densely packed epithelial tissues, **connective tissues** consist of dispersed populations of cells embedded in an extracellular matrix that they secrete. The composition and properties of the matrix differ among types of connective tissues.

An important component of the extracellular matrix is protein fibers secreted by the connective tissue cells. The dominant protein in the extracellular matrix is *collagen*, which is, in fact, the most abundant protein in the body (representing 25 percent of total body protein). Collagen fibers have high tensile strength. They give the *dense connective tissue* of skin, tendons, and ligaments resistance to stretch. Similarly, the collagen fibers of *reticular connective tissue* provide a netlike framework for organs, giving them shape and structural strength. *Loose connective tissue* fills spaces between organs and has a low density of collagen fibers.

Another type of protein fiber in the extracellular matrix of connective tissues is the stretchable protein *elastin*. It can be stretched to several times its resting length and then recoil. Fibers composed of elastin are most abundant in *elastic*

connective tissue, such as that in the walls of the lungs and the large arteries. Elastin fibers in the skin are responsible for its ability to snap back when stretched, and gradual loss of these fibers with age causes gradual loss of the resiliency of the skin.

Proteoglycans are extracellular proteins that give connective tissues resistance to compression. Proteoglycans are abundant in the extracellular matrix of the connective tissues lining joints.

Cartilage and bone are connective tissues that provide rigid structural support. In **cartilage**, a network of collagen fibers is embedded in a rather flexible matrix consisting of a protein–carbohydrate complex called chondroitin sulfate. The cells that form cartilage are called *chondrocytes*, and they exist in small cavities in the cartilage (Figure 40.4a). Cartilage forms the entire skeletal system of sharks and rays, which are therefore called cartilaginous fishes.

Cartilage forms the skeletons of the early developmental stages of more complex vertebrates, but it is gradually replaced by **bone**, a harder connective tissue (Figure 40.4b). Adult vertebrates retain cartilage as the support for flexible structures such as external ears, noses, and the windpipe. The extracellular matrix in bone also contains many collagen fibers, but it is hardened by the deposition of the mineral calcium phosphate. We will discuss bone in greater detail in Chapter 44.

Adipose tissue is a form of loose connective tissue that includes adipose cells, which form and store droplets of lipids (Figure 40.4c). Adipose tissue is a major source of stored energy, but it also serves to cushion organs, and layers of adipose tissue under the skin can provide a barrier to heat loss.

Blood is a connective tissue consisting of cells dispersed in an extensive extracellular matrix: the blood plasma (Figure 40.4d). The blood plasma is much more liquid than the extracellular matrices of the other connective tissues, but it too contains an abundance of proteins. One of those proteins, *fibrinogen*, serves a structural function when it is stimulated to polymerize and form a blood clot. Many of

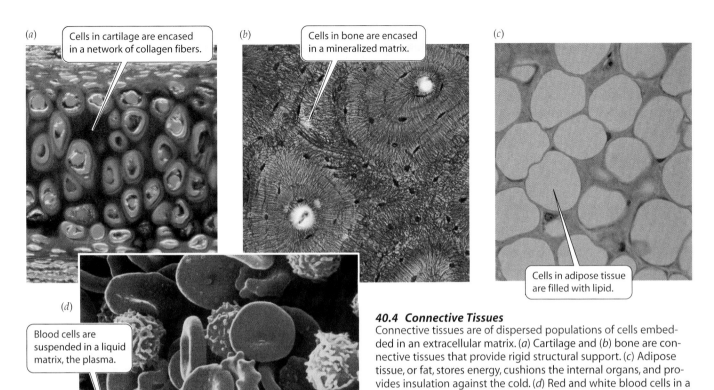

(a) Cells in cartilage are encased in a network of collagen fibers.

(b) Cells in bone are encased in a mineralized matrix.

(c) Cells in adipose tissue are filled with lipid.

(d) Blood cells are suspended in a liquid matrix, the plasma.

40.4 Connective Tissues
Connective tissues are of dispersed populations of cells embedded in an extracellular matrix. (a) Cartilage and (b) bone are connective tissues that provide rigid structural support. (c) Adipose tissue, or fat, stores energy, cushions the internal organs, and provides insulation against the cold. (d) Red and white blood cells in a liquid plasma matrix make up the connective tissue blood.

the proteins and cellular elements of the blood were presented in Chapter 19, and blood will be discussed again in Chapter 49.

Muscle tissues contract

Muscle tissues consist of elongated cells that can contract and cause movement. Muscle tissues are the most abundant tissues in the body, and they use most of the energy produced in the body. The contraction of muscle cells depends on intracellular protein filaments that can slide past each other. In Chapter 47, we will encounter three types of muscle tissues:

▸ **Skeletal muscle** connects bones to bones and is responsible for the body movements that constitute behavior.
▸ **Smooth muscle** is found in internal organs and is not under voluntary control; it performs functions such as moving food through the gut and constricting blood vessels.
▸ **Cardiac muscle** makes up the mass of the heart and pumps the blood.

Nervous tissues process information

There are two basic cell types in nervous tissues: neurons and glial cells. **Neurons**, which are extremely diverse in size and form, generate electrochemical signals. Responding to specific types of stimuli, such as light, sound, pressure, or certain molecules, neurons generate sudden voltage changes across their plasma membranes. These nerve impulses can be conducted via long extensions of the neurons to other parts of the body, where they are communicated to other neurons, muscle cells, or secretory cells. Neurons are involved in controlling the activities of most organ systems to achieve homeostasis.

Glial cells do not generate or conduct electric signals, but they provide a variety of supporting functions for neurons. There are more glial cells than neurons in our nervous systems. We will detail and illustrate the properties of nervous tissues in Chapters 44, 45, and 46.

Organs consist of multiple tissues

A discrete structure that carries out a specific function in the body is called an **organ**. Examples are the stomach, the heart, the liver, or the kidney. Most organs include all four tissue types. The wall of the stomach is a good example (see Figure 40.2). The inner surface of the stomach that contacts food is lined with a simple epithelium. Some of the epithelial cells secrete mucus, enzymes, or stomach acid.

Beneath the epithelial lining is connective tissue. Within this connective tissue are nerves, glands (secretory epithelial cells), and blood vessels. Concentric layers of muscle tissue enable the stomach to contract to mix food with the digestive juices. A network of neurons between the muscle layers controls these movements and also partially controls the secretions of the stomach. Surrounding the stomach is a layer of connective tissue called the serosa.

40.1 The Major Organ Systems of Mammals

SYSTEM	ORGANS	FUNCTIONS
Nervous system	Brain, spinal cord, sensory organs, peripheral nerves	Receives, integrates, stores information and controls muscles and glands. Chapters 44, 45, 46
Endocrine system	Glands: pituitary, thyroid, parathyroid, pineal, adrenal, testes, ovaries, pancreas	A system of glands releases chemical messages (hormones) that control and regulate other tissues and organs. Chapter 41
Muscle system	Skeletal muscle, smooth muscle, cardiac muscle	Produces forces and motion. Chapter 47
Skeletal system	Bones	Provides structural support for the body. Chapter 47
Reproductive system	Female: ovaries, oviducts, uterus, vagina, mammary glands Male: testes, sperm ducts, accessory glands, penis	Produces sex cells and hormones necessary to procreate and nurture offspring. Chapter 42
Digestive system	Mouth, esophagus, stomach, intestines, liver, pancreas, rectum, anus	Acquires and digests food, absorbs and stores nutrients, then makes them available to the cells of the body. Chapter 50
Gas exchange system	Airways, lungs, diaphragm	Exchanges respiratory gases with the environment. Chapter 48
Circulatory system	Heart and blood vessels	Transports respiratory gases, nutrients, hormones, and heat around the body. Chapter 49
Lymphatic system	Lymph and lymph vessels, lymph nodes, spleen	Brings extracellular fluids back into the circulatory system; helps the immune system fight invading organisms. Chapters 40 and 19
Immune system	Many types of white blood cells	Fights invading organisms and infections. Chapter 19
Skin system	Skin, sweat glands, hair	Protects the body from invading organisms and harsh physical conditions, helps regulate body temperature. Chapter 40
Excretory system	Kidneys, bladder, ureter, urethra	Regulates the composition of the extracellular fluids; excretes waste products. Chapter 51

An individual organ is usually part of an **organ system**—a group of organs that function together. The stomach is part of the digestive system, which also includes the food tube (esophagus), the small and large intestines, the pancreas, which secretes digestive enzymes, and the liver, which secretes bile. The major organ systems of mammals are outlined in Table 40.1.

Physiological Regulation and Homeostasis

Homeostasis depends on the ability to regulate the functions of the organs and organ systems to counteract influences that would change the physical or chemical composition of the internal environment. In this section we discuss the general properties of physiological regulatory systems, and then consider temperature regulation as a specific example.

Set points and feedback information are required for regulation

In addition to control mechanisms, regulation requires information. You can regulate the speed of a car only if you know the speed at which you are traveling and the speed you wish to maintain. The desired speed is a **set point**, and the reading on your speedometer is **feedback information**. When the set point and the feedback are compared, any difference between them is an *error signal*. Error signals suggest corrective actions, which you make by using the accelerator or brake (Figure 40.5).

Physiological regulation requires actions of cells, tissues, and organs, which are called *effectors* because they effect changes. Effectors are also referred to as **controlled systems** because their activities are controlled by commands coming from **regulatory systems**. Regulatory systems obtain, process, and integrate information and issue commands to the controlled systems. A fundamental way to analyze a regulatory system is to identify its source of feedback information.

Negative feedback is the most common type of feedback information in regulatory systems. The word "negative" indicates that this feedback information causes the effectors to reduce or reverse the process or influence that created the signal. In our car analogy, the recognition that you are going too fast is negative feedback if it causes you to slow down.

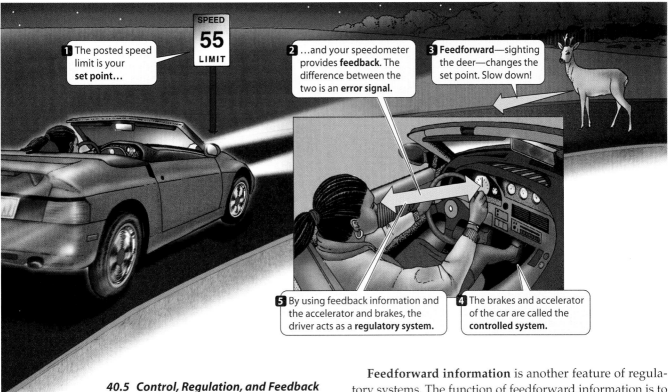

1 The posted speed limit is your **set point**...

2 ...and your speedometer provides **feedback**. The difference between the two is an **error signal**.

3 **Feedforward**—sighting the deer—changes the set point. Slow down!

5 By using feedback information and the accelerator and brakes, the driver acts as a **regulatory system**.

4 The brakes and accelerator of the car are called the **controlled system**.

40.5 Control, Regulation, and Feedback
A driver uses information and control mechanisms to regulate the speed of the car.

Thermostats regulate temperature

The thermostat that is part of the heating–cooling system of a house is a regulatory system. It has upper and lower set points that you can adjust, and it receives feedback information from a sensor that measures room temperature. The circuitry of the thermostat converts differences between the set points and feedback information into signals that activate the controlled systems—the furnace and the air conditioner.

When room temperature rises above the upper set point, the thermostat activates the air conditioner to reduce the temperature; when room temperature falls below that upper set point, the air conditioner is turned off. If temperature falls below the lower set point, the furnace is activated, raising room temperature. The mechanism that senses room temperature provides negative feedback that is used to regulate both the air conditioner and the furnace. **Negative feedback** is a stabilizing influence in physiological regulatory systems. It contributes to homeostasis by stimulating actions that return a variable to its set point.

Is there any such thing as positive feedback in physiology? Although not as common as negative feedback, it does exist. Rather than returning a system to a set point, **positive feedback** amplifies a response. Examples of regulatory systems that use positive feedback are the responses that empty body cavities, such as urination, defecation, sneezing, and vomiting. Another example is sexual behavior, in which a little stimulation causes more behavior, which causes more stimulation, and so on.

Feedforward information is another feature of regulatory systems. The function of feedforward information is to change the set point. Seeing a deer ahead on the road when you are driving is an example of feedforward information (see Figure 40.5); this information takes precedence over the posted speed limit, and you change your set point to a slower speed. If you want the temperature of your house to be lower at night than during the day, you can add a clock to the thermostat to provide feedforward information about time of day.

These principles of control and regulation help organize our thinking about physiological systems. Once we understand how an organ or an organ system works, we can then ask how is it regulated. As an example, we will discuss in detail the system that regulates body temperature. But first, why is it necessary to regulate body temperature?

Temperature and Life

Over the face of Earth, temperatures vary enormously, from the boiling hot springs of Yellowstone National Park to the interior of Antarctica, where the temperature can fall below −80°C. Because heat always moves from a warmer to a cooler object, any change in the temperature of the environment causes a change in the temperature of an organism in that environment—unless the organism does something to regulate its temperature.

Living cells function over only a narrow range of temperatures. If cells cool to below 0°C, ice crystals damage their structures, possibly fatally. Some animals have adaptations such as antifreeze molecules in their blood that help them resist freezing; others have adaptations that enable them to survive freezing. Generally, however, cells must remain above 0°C to stay alive.

The upper temperature limit is less than 45°C for most cells. Some specialized algae can grow in hot springs at 70°C, and some archaea can live at near 100°C, but in general, proteins begin to denature and lose their function as temperatures approach 45°C. Most cellular functions are limited to the range between 0°C and 45°C, which are considered the thermal limits for life. A particular species, however, generally has much narrower limits.

Q_{10} is a measure of temperature sensitivity

Even within the range 0° to 45°C, temperature changes create problems for animals. Most physiological processes, like the biochemical reactions that constitute them, are temperature-sensitive, going faster at higher temperatures (see Figure 6.26). The temperature sensitivity of a reaction or process can be described in terms of Q_{10}, a quotient calculated by dividing the rate of a process or reaction at a certain temperature, R_T, by the rate of that process or reaction at a temperature 10°C lower, R_{T-10}:

$$Q_{10} = \frac{R_T}{R_{T-10}}$$

Q_{10} can be measured for a simple enzymatic reaction or for a complex physiological process, such as the rate of oxygen consumption. If a reaction or process is not temperature-sensitive, it has a Q_{10} of 1. Most biological Q_{10} values are between 2 and 3, which means that reaction rates double or triple as temperature increases by 10°C (Figure 40.6).

Changes in temperature can be particularly disruptive to an animal's functioning because all the component reactions in the animal do not have the same Q_{10}. Individual reactions with different Q_{10}'s are linked together in complex networks that carry out physiological processes. Changes in temperature shift the rates of some reactions

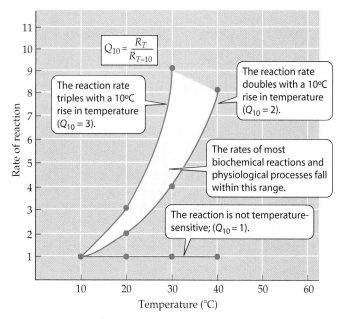

40.6 Q_{10} and Reaction Rate
The larger the Q_{10}, the faster the reaction rate rises in response to an increase in temperature.

more than those of others, thus disrupting the balance and integration that the processes require. For homeostasis, organisms must be able to compensate for or prevent changes in temperature.

An animal's sensitivity to temperature can change

The body temperature of some animals is tightly coupled to the environmental temperature. Think of a fish in a temperate-zone pond. As the temperature of the pond changes from 4°C in midwinter to 24°C in midsummer, the body temperature of the fish does the same (Figure 40.7). We can bring the fish into the laboratory in the summer and measure its **metabolic rate** (the sum total of the energy turnover of its cells, often measured by O_2 consumption). If we measure the metabolic rate at different water temperatures, we might plot our data as shown by the red line in Figure 40.7 and calculate a Q_{10} of 2. We predict from our graph that in winter, when the temperature is 4°C, the fish's metabolic rate will be only one-fourth of what it was in the summer. We then return the fish to its pond.

When we bring the fish back to the laboratory in the winter and repeat the measurements, we find, as the blue line shows, that its metabolic rate at 4°C is not as low as we predicted; rather, it is almost the same as it was at 24°C in the summer. If we repeat the measurement over a range of temperatures, we find that the fish's metabolic rate is always higher than the rate we predicted from the measurement we took at the same temperature in the summer. This difference is due to **acclimatization**, the process of physiological and biochemical change that an animal undergoes in response to seasonal changes in climate.

Seasonal acclimatization in the fish has produced **metabolic compensation**, which readjusts the biochemical machinery to counter the effects of temperature. What might account for such a change? Look again at Figure 6.26, which shows the different optimal temperatures of enzymes. If the fish can express similar enzymes that operate at different optimal temperatures, it can compensate metabolically by catalyzing reactions with one set of enzymes in summer and another set in winter. The end result is that metabolic functions are much less sensitive to long-term changes in temperature than they are to short-term thermal fluctuations.

Maintaining Optimal Body Temperature

Animals can be classified by how they respond to environmental temperatures.

▶ A **homeotherm** is an animal that maintains a constant body temperature.

▶ A **poikilotherm** is an animal whose body temperature changes when the temperature of its environment (the *ambient temperature*) changes.

This system of classification says something about the biology of the animals, but it presents problems. Should a fish in the deep ocean, where the temperature changes very little, be called a homeotherm? Should a hibernating mam-

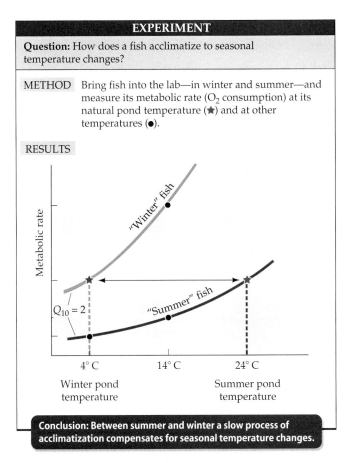

EXPERIMENT

Question: How does a fish acclimatize to seasonal temperature changes?

METHOD   Bring fish into the lab—in winter and summer—and measure its metabolic rate (O_2 consumption) at its natural pond temperature (★) and at other temperatures (●).

RESULTS

Conclusion: Between summer and winter a slow process of acclimatization compensates for seasonal temperature changes.

40.7 Metabolic Compensation
In its natural environment, a fish's metabolism readjusts, or acclimatizes, to compensate for seasonal changes in temperature.

mal that allows its body temperature to drop to nearly the temperature of its environment be called a poikilotherm? The problem posed by the hibernator has been solved by creating a third category: the **heterotherm**, an animal that regulates its body temperature at a constant level *some* of the time.

Another set of terms classifies animals on the basis of the sources of heat that determine their body temperatures.

▶ **Ectotherms** depend largely on external sources of heat, such as solar radiation, to maintain their body temperatures above the environmental temperature.

▶ **Endotherms** can regulate their body temperatures by producing heat metabolically or by mobilizing active mechanisms of heat loss.

Mammals and birds are endotherms; animals of all other species behave as ectotherms most of the time.

Ectotherms and endotherms respond differently in metabolic chambers

A small lizard is an example of an ectotherm. We can compare it with a mouse, which is an endotherm of the same body size. We can put each animal in a metabolic chamber and measure body temperatures and metabolic rates as we change the temperature of the chamber from 0°C to 35°C.

The results obtained from the two species differ. The body temperature of the lizard equilibrates with that of the chamber, whereas the body temperature of the mouse remains at 37°C (Figure 40.8*a*). The metabolic rate of the lizard decreases as the temperature decreases (Figure 40.8*b*). In contrast, the mouse's metabolic rate increases as chamber temperature falls below about 27°C (notice that you must read the graph right to left to see this). The lizard apparently cannot regulate its body temperature or metabolism independently of environmental temperature. The mouse, however, regulates its body temperature by in-

40.8 Ectotherms and Endotherms
The body temperatures of a lizard and a mouse of the same body size respond differently to changes in environmental temperature.

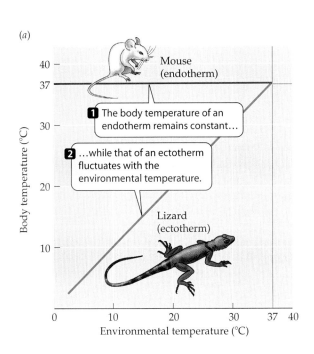

(*a*)

Mouse (endotherm)

1 The body temperature of an endotherm remains constant…

2 …while that of an ectotherm fluctuates with the environmental temperature.

Lizard (ectotherm)

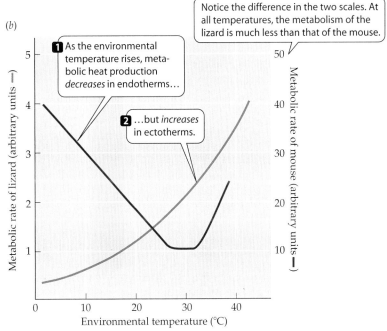

(*b*)

Notice the difference in the two scales. At all temperatures, the metabolism of the lizard is much less than that of the mouse.

1 As the environmental temperature rises, metabolic heat production *decreases* in endotherms…

2 …but *increases* in ectotherms.

creasing its metabolic rate, which increases its production of body heat.

Ectotherms and endotherms use behavior to regulate body temperature

We can test our laboratory conclusion that the lizard cannot regulate its body temperature. To do this, we release the lizard in its desert habitat and measure its temperature as it goes about its normal behavior in this environment, where temperature can change 40°C in a few hours (Figure 40.9).

Unlike what we observed in the metabolic chamber, the body temperature of the lizard is at times considerably different from the environmental temperature. At night, the temperature in the desert may drop close to freezing, but the temperature of the lizard remains stable at 16°C. No mystery here: The lizard spends the night in a burrow, where the soil temperature is a constant 16°C.

Early in the morning, soon after sunrise, the lizard emerges from its burrow. The air temperature is still cool, but the lizard's body temperature rises above 30°C in less than 30 minutes. The lizard achieves this by basking on a rock with maximum exposure to the sun. As its skin absorbs solar radiation, its body temperature rises considerably above the air temperature. By altering its exposure to the sun, the lizard maintains its body temperature around 35°C all morning.

By noon, the air temperature near the surface of the desert has risen to 50°C, but the lizard maintains its body temperature around 35°C by staying mostly in the shade and frequently in the branches of bushes, where there is a cooling breeze. As afternoon progresses, the air cools, and the lizard again spends more of its time in the sun and on hot rocks to maintain its body temperature around 35°C. The lizard returns to its burrow just before sunset, and its body temperature rapidly drops to 16°C.

Our conclusion must be that the lizard can regulate its body temperature quite well by behavioral mechanisms rather than by internal metabolic mechanisms. In our laboratory experiment, the lizard in the chamber could not use its thermoregulatory behavior, but in its natural environment it could move to different places to alter the heat exchange between its internal and external environments.

But behavioral thermoregulation is not the exclusive domain of ectotherms. It is also the first line of

defense for endotherms. When the option is available, most animals select the thermal microenvironments that are best for them. They may change their posture, orient to the sun, move between sun and shade, and move between still air and moving air, as demonstrated by the lizard in our field experiment. Examples of more complex thermoregulatory behavior are nest construction and social behavior such as huddling. Humans select appropriate clothing and heat or cool their buildings. Behavioral thermoregulation is widespread in the animal kingdom (Figure 40.10).

Both ectotherms and endotherms control blood flow to the skin

Just as behavioral thermoregulation is not the exclusive domain of ectotherms, physiological thermoregulation is not the exclusive domain of endotherms. Both ectotherms and endotherms can alter the rate of heat exchange between their bodies and their environments by controlling the flow of blood to the skin.

The skin is the interface between the internal and the external environment, and heat exchanges that alter body temperature occur across this interface. Heat exchange between the skin and the external environment occurs through four mechanisms: radiation, conduction, convection, and evaporation (Figure 40.11).

Heat exchange between the internal environment and the skin occurs largely through blood flow. For example, when a person's body temperature rises as a result of exercise, blood flow to the skin increases, and the skin surface becomes quite warm. The heat brought to the skin by the blood is lost to the environment, and this loss helps to bring the body temperature back to normal. In contrast, when a person is exposed to cold, the blood vessels supplying the skin constrict, decreasing blood flow and heat transport to the skin and reducing heat loss to the environment.

40.9 An Ectotherm Uses Behavior to Regulate Its Body Temperature
The lizard's body temperature is dependent on environmental heat, but it can regulate its temperature by moving between different environments.

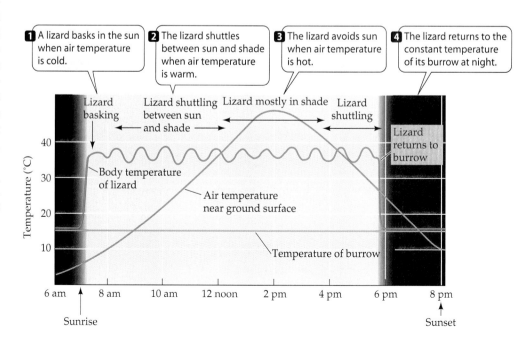

1 A lizard basks in the sun when air temperature is cold.

2 The lizard shuttles between sun and shade when air temperature is warm.

3 The lizard avoids sun when air temperature is hot.

4 The lizard returns to the constant temperature of its burrow at night.

(a)

(b)

40.10 Endotherms Can Use Behavior to Thermoregulate
(*a*) Humans must put on many layers of insulating clothing to help their thermoregulatory mechanisms keep pace with the extreme cold of western Siberia. (*b*) When air temperatures on the African savanna soar, an elephant may use a cool shower to thermoregulate.

The control of blood flow to the skin can be an important adaptation for an ectotherm like the marine iguana of the Galápagos archipelago. The Galápagos are volcanic islands that lie on the equator, but they are bathed by very cold oceanic currents. Marine iguanas are reptiles that bask on black lava rocks on shore and swim in the cold ocean, where they feed on algae. When the iguanas are feeding,

they cool to the temperature of the sea, which makes them slower and more vulnerable to predators, and probably incapable of efficient digestion. They therefore alternate between feeding in the cold sea and basking in the sun on the hot rocks. It is advantageous for iguanas to retain body heat as long as possible while swimming and to warm up as fast as possible when basking. They adjust by changing their heart rate, and therefore their blood flow (Figure 40.12).

Some ectotherms produce heat

Some ectotherms raise their body temperatures by producing heat. For example, the powerful flight muscles of many insects must reach 35°–40°C before the insects can fly, and they must maintain these high temperatures during flight, even at air temperatures around 0°C. Such insects produce the required heat by contracting their flight muscles in a manner analogous to shivering in mammals (Figure 40.13). The heat-producing ability of these insects can be quite remarkable. Probably the most impressive case is a species of scarab beetle that lives mostly underground in mountains north of Los Angeles, California. To mate, these beetles come above ground, and males fly in search of females. They un-

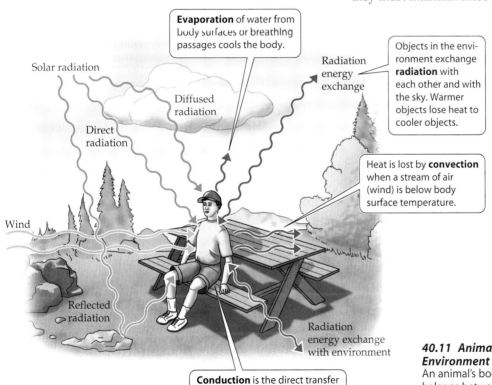

Evaporation of water from body surfaces or breathing passages cools the body.

Objects in the environment exchange **radiation** with each other and with the sky. Warmer objects lose heat to cooler objects.

Heat is lost by **convection** when a stream of air (wind) is below body surface temperature.

Solar radiation

Diffused radiation

Direct radiation

Wind

Reflected radiation

Radiation energy exchange

Radiation energy exchange with environment

Conduction is the direct transfer of heat when objects of different temperatures come into contact.

40.11 Animals Exchange Heat with the Environment
An animal's body temperature is determined by the balance between internal heat production and the avenues of heat exchange with the environment: radiation, conduction, convection, and evaporation.

39

°C   37

35

Body temperature

0   10   20   30   40 (°C)

Metabolic rate

1 Within the **thermoneutral zone**, body temperature is regulated by altering heat loss through the skin.

Lower critical temperature

Upper critical temperature

Metabolic rate

Basal metabolic rate

RIP

RIP

0   10   20   30   40

Environmental temperature (°C)

2 Below the lower critical temperature, the animal produces metabolic heat to compensate for increased heat loss to the environment.

3 Above the upper critical temperature, the animal must expend energy to lose heat by panting or sweating, which makes its metabolic rate increase.

40.15 Environmental Temperature and Mammalian Metabolic Rates

Outside the thermoneutral zone, maintaining a constant body temperature requires the expenditure of energy.

mammals, but almost no reptiles or amphibians, live in very cold habitats. What adaptations besides endothermy characterize species that live in the cold?

The most important adaptations of endotherms to cold environments are those that reduce heat loss to the environment. Since most heat is lost from the body surface, many cold-climate species have a smaller surface area than their warm-climate cousins, even when their body masses are the same. Rounder body shapes and shorter appendages reduce the surface area–to–volume ratios of some cold-climate species; compare, for example, the San Joaquin kit fox and the arctic fox (Figure 40.17).

Another means of decreasing heat loss is to increase thermal insulation. Animals adapted to cold have much thicker layers of fur, feathers, or fat than do their warm-climate relatives. The fur of an arctic fox or a northern sled dog provides such good thermal insulation that those animals don't even begin to shiver until the air temperature drops as low as –20°C to –30°C.

Fur and feathers are good insulators because they trap a layer of still, warm air close to the skin surface. If that air is displaced by water, insulation is drastically reduced. In many species, oil secretions spread through fur or feathers by grooming are critical for resisting wetting and maintain-

expenditure due to increased muscle tone and increased body movements also contributes to increased heat production in the cold.

Most nonshivering heat production occurs in specialized adipose tissue called **brown fat** (Figure 40.16). This tissue looks brown because of its abundant mitochondria and rich blood supply. In brown fat cells, a protein called *thermogenin* uncouples proton movement from ATP production, allowing protons to leak across the inner mitochondrial membrane rather than having to pass through the ATP synthase protein and generate ATP (review the discussion of respiration in Chapter 7). As a result, metabolic fuels are consumed without producing ATP, but heat is still released. Brown fat is especially abundant in newborn infants of many mammalian species, in some adult mammals that are small and acclimatized to cold, and in mammals that hibernate.

Decreasing heat loss is important for life in the cold

The coldest habitats on Earth are in the Arctic, the Antarctic, and at the peaks of high mountains. Many birds and

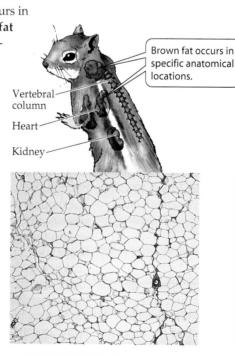

Brown fat occurs in specific anatomical locations.

Vertebral column

Heart

Kidney

40.16 Brown Fat
In many mammals, specialized brown fat tissue produces heat.

White fat viewed through a light microscope. Each cell is filled with a globule of lipid and has few organelles. The tissue has few blood vessels.

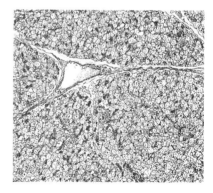

Brown fat viewed through a light microscope at the same magnification reveals cells with many intracellular structures and multiple droplets of lipid.

(a) *Vulpes macrotis*

(b) *Alopex lagopus*

40.17 Adaptations to Hot and Cold Climates
(*a*) The San Joaquin kit fox, a desert dweller, has a large surface area for its body. The large ears serve as heat exchangers, passing heat from the fox's blood to the surrounding air. (*b*) The thick fur of the arctic fox provides insulation in the frigid winter. Its ears and extremities are relatively smaller than those of the kit fox.

ing a high level of insulation.

Humans change their thermal insulation by putting on or taking off clothes. How do other animals do it? We have already discussed one example, the ectothermic marine iguana, which controls blood flow to its skin. Increasing or decreasing blood flow to the skin is an important thermoregulatory adaptation for endotherms as well. In a hot environment, your skin feels hot because there is a high rate of blood flow through it, but when you sit in an overly airconditioned theater, your hands, feet, and other body surfaces feel cold as blood flow to those areas decreases.

Evaporation of water is an effective way to lose heat

For highly insulated arctic animals, and for many large mammals in all climates, getting rid of excess heat can be a serious problem, especially during exercise. Arctic species usually have an area on the body surface, such as the abdomen, that has only a thin layer of fur and can act as a window for heat loss. Large mammals, such as elephants, rhinoceroses, and water buffalo, have little or no fur and seek places where they can wallow in water when the air

temperature is too high. Having water in contact with the skin greatly increases heat loss because water has a much greater capacity for absorbing heat than air does.

A gram of water absorbs about 580 calories of heat when it evaporates. Water is heavy, however, so animals do not carry an excess supply of it. Furthermore, hot environments tend to be arid places where water is a scarce resource. Therefore, evaporation of water by sweating or panting is usually a last resort for animals adapted to hot environments (recall the camels at the beginning of this chapter).

Sweating and panting are active processes that require the expenditure of metabolic energy. That's why the metabolic rate increases when the upper critical temperature is exceeded (see Figure 40.15). A sweating or panting animal is producing heat in the process of dissipating heat, which can be a losing battle. Animals can survive in environments that are below their lower critical temperature much better than they can in environments above their upper critical temperature.

The Vertebrate Thermostat

The thermoregulatory mechanisms and adaptations we have discussed are the controlled systems for the regulation of body temperature. These controlled systems must receive commands from a regulatory system that integrates information relevant to the regulation of body temperature. Such a regulatory system can be thought of as a *thermostat*. All animals that thermoregulate, both vertebrate and invertebrate, must have regulatory systems, but here we will focus on the vertebrate thermostat.

Where is the vertebrate thermostat? Its major integrative center is at the bottom of the brain in a structure called the **hypothalamus**. If you slide your tongue back as far as possible along the roof of your mouth, it will be just a few centimeters below your hypothalamus. The hypothalamus is a part of many regulatory systems, so we will refer to it again in the chapters to come. If the hypothalamus of a mammal's brain is damaged, the animal loses its ability to regulate its body temperature, which then rises in warm environments and falls in cold ones.

The vertebrate thermostat uses feedback information

What information does the vertebrate thermostat use? In many species, the temperature of the hypothalamus itself is the major source of feedback information to the thermostat. Cooling the hypothalamus causes fish and reptiles to seek a warmer environment, and heating the hypothalamus causes them to seek a cooler environment. In mammals, cooling the hypothalamus can stimulate constriction of the blood vessels supplying the skin and increase metabolic heat production. Because it activates these thermoregulatory responses, cooling the hypothalamus causes the body temperature to rise. Conversely, warming the hypothalamus stimulates dilation of blood vessels supplying the skin and sweating or panting, and the overall body temperature falls (Figure 40.18).

EXPERIMENT

Question: Does the hypothalamus act as a thermostat?

METHOD Heat and cool hypothalamus and measure metabolic heat production.

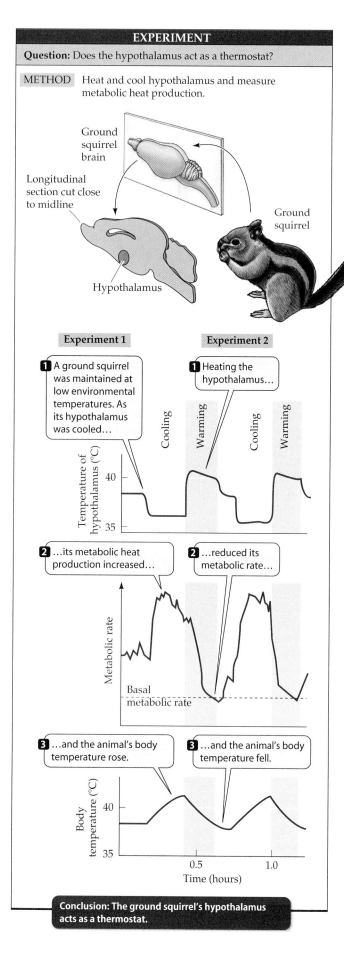

Conclusion: The ground squirrel's hypothalamus acts as a thermostat.

The hypothalamus appears to generate a set point like a setting on the thermostat of a house. When the temperature of the hypothalamus exceeds or drops below that set point, thermoregulatory responses (the controlled system) are activated to reverse the direction of temperature change. Hence, hypothalamic temperature is a negative feedback signal.

Heating and cooling the hypothalamus show that an animal has separate set points for activating different thermoregulatory responses. If the hypothalamus of a mammal is cooled, the vessels supplying blood to the skin constrict at a specific hypothalamic temperature. A slightly lower hypothalamic temperature initiates shivering. If the hypothalamic temperature is then raised, shivering ceases; then blood vessels supplying the skin dilate; and at still higher hypothalamic temperatures, panting starts.

We can describe the characteristics of hypothalamic control of each thermoregulatory response. For example, if we measure metabolic heat production while heating and cooling the hypothalamus (see Figure 40.18), we can describe the results graphically (Figure 40.19). Within a certain range of hypothalamic temperatures, metabolic heat production remains low and constant, but cooling the hypothalamus below a certain level—a set point—stimulates increased metabolic heat production. The increase in heat production is proportional to how much the hypothalamus is cooled below the set point. This regulatory system is much more sophisticated than a simple on–off thermostat like the one in a house.

The vertebrate thermoregulatory system integrates other sources of information in addition to hypothalamic temperature. It uses information about the temperature of the environment as registered by temperature sensors in the skin. Changes in environmental temperature shift the hypothalamic set points for thermoregulatory responses. As Figure 40.19 shows, in a warm environment you might have to cool the hypothalamus of a mammal to stimulate it to shiver, but in a cold environment you would have to warm the hypothalamus of the same animal to stop it from shivering. The set point for the metabolic heat production response is higher when the skin is cold and lower when the skin is warm.

The temperature of the skin can be considered feedforward information that adjusts the hypothalamic set point. Many other factors also shift hypothalamic set points for responses. Set points are higher during wakefulness than during sleep, and they are higher during the active part of the daily cycle than during the inactive part, even if the animal is awake at both times.

40.18 The Hypothalamus Regulates Body Temperature
The observation that damage to the hypothalamus disrupts thermoregulation led to the finding that hypothalamus acts as a thermostat in the vertebrate body.

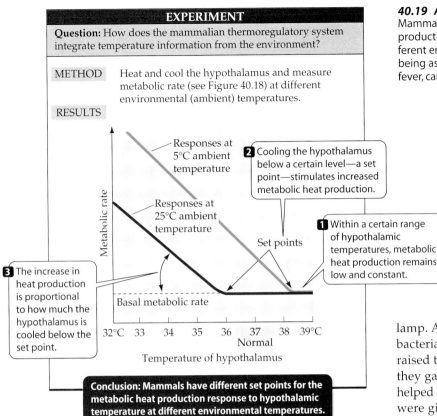

EXPERIMENT

Question: How does the mammalian thermoregulatory system integrate temperature information from the environment?

METHOD   Heat and cool the hypothalamus and measure metabolic rate (see Figure 40.18) at different environmental (ambient) temperatures.

RESULTS

Responses at 5°C ambient temperature

Responses at 25°C ambient temperature

Set points

Basal metabolic rate

2 Cooling the hypothalamus below a certain level—a set point—stimulates increased metabolic heat production.

1 Within a certain range of hypothalamic temperatures, metabolic heat production remains low and constant.

3 The increase in heat production is proportional to how much the hypothalamus is cooled below the set point.

Metabolic rate

Temperature of hypothalamus

32°C 33 34 35 36 37 38 39°C
Normal

Conclusion: Mammals have different set points for the metabolic heat production response to hypothalamic temperature at different environmental temperatures.

40.19 *Adjustable Set Points*
Mammals have different set points for the metabolic heat production response to hypothalamic temperature at different environmental temperatures. Other factors, such as being asleep or awake, the time of day, or the presence of a fever, can also affect the set point.

Fevers help the body fight infections

A **fever** is a rise in body temperature in response to substances called *pyrogens* that are derived from bacteria or viruses that invade the body. We respond to many infectious illnesses by getting a fever. Growing evidence suggests that fevers are adaptive responses that help the body fight disease-causing organisms.

The presence of a pyrogen in the body causes a rise in the hypothalamic set point for the heat production response. As a result, you shiver, put on a sweater, or crawl under a blanket, and your body temperature rises until it matches the new set point. At the higher body temperature you no longer feel cold, and you may not feel hot, but someone touching your forehead will say that you are "burning up." If you take an aspirin, it lowers your set point to normal. Now you feel hot, take off clothes, and even sweat until your elevated body temperature returns to normal.

Why do we take aspirin for fevers and "feeling crummy"? The pyrogens entering the body are attacked by cells of the immune system called macrophages (see Chapter 19). One of the things the macrophages do is to release chemicals called *interleukins*, which sound the alarm to other cells of the immune system throughout the body and trigger responses that contribute to feeling crummy. The interleukins also raise the hypothalamic set point for metabolic heat production. Among the intracellular signals trig-

gered by interleukins are *prostaglandins*. Aspirin is a potent inhibitor of prostaglandin synthesis, thus explaining how this miracle drug reduces fever and makes us feel better.

Evidence suggests that moderate fevers help the body fight an infection. Some interesting studies were done on lizards that were given access to a heat lamp. These animals kept their body temperatures at about 38°C by adjusting their position with respect to the lamp. After they were injected with disease-causing bacteria, they spent more time close to the lamp and raised their body temperatures to 40°C and higher—they gave themselves fevers. To find out if the fever helped the lizards fight the bacteria, groups of lizards were given equal inoculations of bacteria, but were then placed in different incubators at 34°, 36°, 38°, 40°, and 42°C, respectively. All of the lizards at 34°C and 36°C died, about 25 percent at 38°C survived, and about 75 percent at 40°C and 42°C survived. Apparently fever helped the lizards fight the disease organisms.

However, extreme fevers (for example, 40°C) can be dangerous to humans and must be reduced. Even more modest fevers can be dangerous to people who have weakened hearts or those who are seriously ill. A fetus can be endangered when a pregnant woman has a fever. Fever-reducing drugs may be important in such cases.

Animals can save energy by turning down the thermostat

Hypothermia is the condition in which body temperature is below normal. It can result from a natural turning down of the thermostat, or from traumatic events such as starvation (lack of metabolic fuel), exposure, serious illness, or treatment by anesthesia. Many species of birds and mammals use regulated hypothermia as a means of surviving periods of cold and food scarcity. Some become hypothermic on a daily basis. Hummingbirds, for example, are very small endotherms and have a high metabolic rate. They could exhaust their metabolic reserves just getting through a single day without food. Hummingbirds and other small endotherms can extend the period over which they can survive without food by dropping their body temperature during the portion of day when they would normally be inactive. This adaptive hypothermia is called **daily torpor**. Body temperature can drop 10°–20°C during daily torpor, resulting in an enormous saving of metabolic energy.

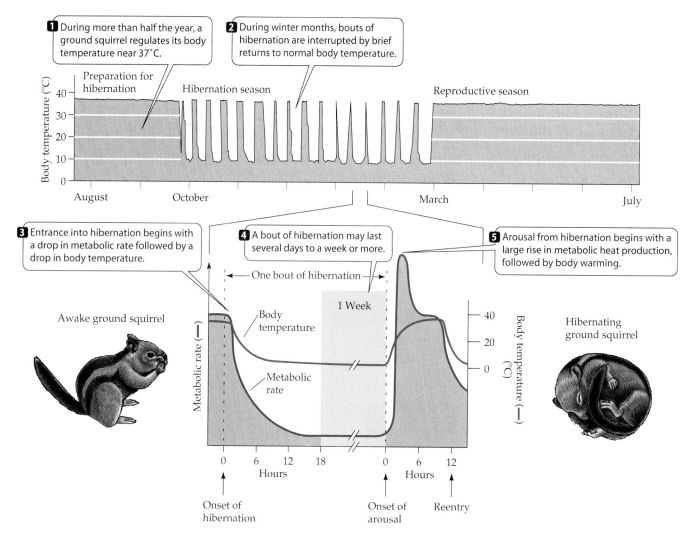

1 During more than half the year, a ground squirrel regulates its body temperature near 37°C.

2 During winter months, bouts of hibernation are interrupted by brief returns to normal body temperature.

3 Entrance into hibernation begins with a drop in metabolic rate followed by a drop in body temperature.

4 A bout of hibernation may last several days to a week or more.

5 Arousal from hibernation begins with a large rise in metabolic heat production, followed by body warming.

Awake ground squirrel

Hibernating ground squirrel

One bout of hibernation

1 Week

Body temperature

Metabolic rate

Onset of hibernation

Onset of arousal

Reentry

40.20 A Ground Squirrel Enters Repeated Bouts of Hibernation during Winter
At the beginning of each bout of hibernation, the ground squirrel's metabolic rate and body temperature fall. Its body temperature may come into equilibrium with the temperature of its nest and stay at that level for days. The bout is ended by a rise in metabolic heat production that returns body temperature to a normal level.

Regulated hypothermia can also last for days or even weeks, with drops to very low temperatures; this phenomenon is called **hibernation** (Figure 40.20). During the deep sleep of hibernation, the body's thermostat is turned down to an extremely low level to maximize energy conservation. Arousal from hibernation occurs when the hypothalamic set point returns to a normal level.

Many hibernators maintain body temperatures close to the freezing point during hibernation. The metabolic rate needed to sustain an animal in hibernation may be only one-fiftieth its basal metabolic rate, an enormous saving of metabolic energy. Many species of mammals, such as bats, bears, and ground squirrels, hibernate, but only one species of bird, the poorwill, has been shown to hibernate. The ability of hibernators to reduce their thermoregulatory set points so dramatically probably evolved as an extension of

the set point decrease that accompanies sleep even in non-hibernating species of mammals and birds.

Chapter Summary

Homeostasis: Maintaining the Internal Environment

▶ The internal environment consists of the extracellular fluids. Organs and organ systems have specialized functions to keep certain aspects of the internal environment in a constant state. **Review Figure 40.1**

▶ Homeostasis is the maintenance of constancy in the internal environment. Homeostasis depends on the ability to control and regulate the functions of organs and organ systems.

Tissues, Organs, and Organ Systems

▶ Cells that have a similar structure and function make up a tissue. There are four general types of tissues: epithelial, connective, muscle, and nervous. **Review Figure 40.2**

▶ Epithelial tissues are sheets of tightly connected cells that cover the body surfaces and line hollow organs. **Review Figure 40.3**

▶ Connective tissues support and reinforce other tissues. They generally consist of dispersed cells in an extracellular matrix. Examples are cartilage, bone, blood, and adipose tissue. **Review Figure 40.4**

▶ Muscle tissues contract. There are three types: skeletal, cardiac, and smooth.

▶ There are two types of cells in nervous tissues: Neurons generate and transmit electrochemical signals, and glial cells provide supporting functions for neurons.

▶ Organs consist of multiple tissue types, and organs make up organ systems. **Review Table 40.1**

Physiological Regulation and Homeostasis

▶ Regulatory systems have set points and respond to feedback information. Negative feedback corrects deviations from the set point, positive feedback amplifies responses, and feedforward information changes the set point. **Review Figure 40.5**

Temperature and Life

▶ Living systems require a range of temperatures between the freezing point of water and the temperatures that denature proteins.

▶ Most biological processes and reactions are temperature-sensitive. Q_{10} is a measure of temperature sensitivity. **Review Figure 40.6**

▶ Animals that cannot avoid seasonal changes in body temperature have biochemical adaptations that compensate for those changes. These adaptations enable animals to acclimatize to seasonal changes. **Review Figure 40.7**

Maintaining Optimal Body Temperature

▶ Homeotherms maintain a fairly constant body temperature most of the time; poikilotherms do not. Endotherms produce metabolic heat to elevate body temperature; ectotherms depend mostly on environmental sources of heat. **Review Figure 40.8**

▶ Ectotherms and endotherms can regulate body temperature through behavior. **Review Figure 40.9**

▶ Heat exchange between a body and the environment is via radiation, conduction, convection, and evaporation. **Review Figure 40.11**

▶ Ectotherms and endotherms can control heat exchange with the environment by altering blood flow to the skin. **Review Figure 40.12**

▶ Some ectotherms, such as bees, nocturnal moths, and beetles, can produce metabolic heat to raise their body temperatures. **Review Figure 40.13**

▶ Some fish have circulatory systems that function as countercurrent heat exchangers to conserve heat produced by muscle metabolism. **Review Figure 40.14**

Thermoregulation in Endotherms

▶ Endotherms have high basal metabolic rates. Over a range of environmental temperatures called the thermoneutral zone, the metabolic rate of resting endotherms remains at basal levels. **Review Figure 40.15**

▶ When the environmental temperature falls below a lower critical temperature, endotherms maintain their body temperatures through shivering and nonshivering metabolic heat production. **Review Figure 40.16**

▶ When the environmental temperature rises above an upper critical temperature, metabolic rate increases as a consequence of active evaporative water loss through sweating or panting.

▶ Endotherms that live in cold climates have adaptations that minimize heat loss, including a reduced surface area-to-volume ratio and increased insulation.

▶ Endotherms may dissipate excess heat generated by exercise or the environment via evaporation. The water loss involved in this process can be dangerous to endotherms in dry environments.

The Vertebrate Thermostat

▶ The vertebrate thermostat is located in the hypothalamus. It has set points for activating thermoregulatory responses. Hypothalamic temperature provides negative feedback information.

▶ Cooling the hypothalamus induces the constriction of blood vessels and increased metabolic heat production. Heating the hypothalamus induces the dilation of blood vessels and active evaporative water loss. Thermoregulatory behaviors are also induced by changes in hypothalamic temperature. **Review Figure 40.18**

▶ Changes in set point reflect the integration of information, such as environmental temperature and time of day, that is relevant to the regulation of body temperature. **Review Figure 40.19**

▶ Fever, which results from a rise in set point, helps the body fight infections.

▶ Adaptations in which set points are reduced to conserve energy include daily torpor and hibernation. **Review Figure 40.20**

For Discussion

1. In some sheets of epithelial tissue, the cells are joined together with dense extracellular proteins that form "tight junctions," which are extremely impermeable (see Chapter 5). In other epithelial sheets the cells are joined by filamentous extracellular proteins that are strong, but not as impermeable. What do you think are the functions of tight junctions, and where would you expect to find them? Where might you expect to find epithelial sheets with the leakier connections?

2. If the major adaptation of endotherms to cold climates is their insulation, how would you compare the cold adaptations of a polar bear and a seal?

3. Why is an environment above its upper critical temperature more dangerous to an endotherm than an environment below its lower critical temperature?

4. We discussed the vertebrate thermostat by describing experiments done on mammals. Lizards also have a temperature-sensitive hypothalamus. How would you design an experiment on a lizard to see if the temperature of its hypothalamus was important feedback information for its thermoregulation? How would you modify your experiment to see if the lizard also used information from temperature sensors in its skin?

5. If the hypothalamic temperature of a mammal is the feedback information for its thermostat, why does the hypothalamic temperature scarcely change when that animal moves between environments hot enough and cold enough to stimulate the animal to pant and to shiver, respectively?

41 Animal Hormones

IN SHALLOW POOLS AROUND THE EDGE OF Lake Tanganyika in east central Africa, brightly colored male cichlid fish stake out territories and vigorously defend them against neighboring males. These dominant males constantly patrol their territories and display their colorful sexual adornments for the benefit of females, who assemble in groups at the edge of the cichlid colony. The females are hard to see because they are inactive and protectively colored. When a female is ready to spawn, and is impressed by a male's territory and display, she enters his territory and lays her eggs in a spawning pit that the male has prepared. The male then fertilizes her eggs.

At any one time, only about 10 percent of the males in the colony are displaying and holding territories. All the other males are small, nondescript, and nonaggressive like the females. If a dominant male is removed by a predator, however, the nondescript males fight over the vacated territory. The winner rapidly assumes the appearance and behavior of a dominant male: brightly colored, big, aggressive, and attractive to females.

What accounts for this dramatic change? Russell Fernald and his students at Stanford University have shown that soon after the nondescript male's victory, certain cells in his brain enlarge and secrete a chemical message. This message triggers cells in the pituitary gland, which is outside of the brain, to secrete chemical messages in turn. Although secreted in tiny quantities, these molecules enter the blood and are transported around the body. The responses of cells to these chemical messages produce the characteristics of a dominant male.

This change in the male cichlid is one example of how chemical messages, or hormones, can produce and coordinate anatomical, physiological, and behavioral changes in an animal. We explore many other examples in this chapter.

We look first at the nature of hormones and their evolution, and then examine some of their roles in the control of invertebrate life cycles. Most of this chapter is devoted to vertebrate hormones: their functions, control, and molecular mechanisms of action. We pay particular attention to the extensive interactions between the systems of neural and hormonal information. In the process, we discuss several human diseases involving hormonal dysfunction.

Dominant and Non-Dominant Male Cichlids
A dominant male cichlid (*Haplochromis burtoni*) displays bright colors that attract females to his spawning pit.

Hormones and Their Actions

In Chapter 40, we learned that control and regulation require information. This information is transmitted mostly as electric signals and as chemical signals. *Electric signals* are nerve impulses, a major focus of later chapters on the nervous system. Nerve impulses can be rapidly conducted over long distances to specific targets. *Chemical signals* are **hormones**, which are secreted by cells, diffuse locally in the extracellular fluid until they are picked up by the blood, and are distributed by the circulatory system.

Because the secretion, diffusion, and circulation of hormones is much slower than the transmission of nerve impulses, hormones are not useful for controlling rapid actions such as those involved in cichlid fighting. Hormone action *is* good for coordinating longer-term developmental processes, such as the transition of a nondescript cichlid into a dominant, territorial, breeding male. Hormones control many long-term physiological responses, such as the secretion of digestive enzymes by our guts and the reproductive cycles of many species.

A hormone is a chemical signal produced by certain cells of a multicellular organism and received by cells of the same organism. Cells that secrete hormones are called **endocrine cells**. To receive the hormonal message, a **target cell** must have appropriate *receptors* to which the hormone can bind. The binding of a hormone to its receptor activates mechanisms within the target cell that eventually lead to a response, which may be developmental, physiological, or behavioral. (In the case of the male cichlid, hormone release stimulates all three types of responses.)

Hormonal signaling systems can be distinguished according to the distance over which their messages operate. Some hormones only act on target cells close to their sites of release; others act on target cells at distant locations in the body. Some chemical messages, called *pheromones*, even exert their effects on other individuals.

Most hormones are distributed in the blood

The classic hormone is a chemical message secreted by cells in minute amounts and distributed throughout the body by the circulatory system (Figure 41.1*a*). Wherever such a hormone encounters a cell with a receptor to which it can bind, it triggers a response. The nature of the response depends on the responding cell. The same hormone can cause different responses in different types of cells.

Consider the hormone epinephrine. If you step off a curb without looking and a car screeches to a halt right next to you, you jump, your heart starts to thump, and a whole set of protective actions are set in motion. The jump and the initial heart thumping are driven by your nervous system, which can react very quickly. Simultaneously, the nervous system stimulates endocrine cells just above your kidneys (adrenal cells) to secrete epinephrine. Within seconds, epinephrine is diffusing into your blood and circulating around your body to activate the many components of the *fight-or-flight response*.

Epinephrine acts on the heart, blood vessels, liver, and fat cells. When it binds to its receptors in the heart, it causes the heart to beat faster and more strongly. It binds to receptors in the vessels that supply blood to your digestive tract, causing those vessels to constrict (digestion can wait!). Your heart is pumping more blood, and a greater percentage of that blood is going to the muscles needed for your escape. In the liver, epinephrine stimulates the breakdown of glycogen into glucose for a quick energy supply. In fatty tissue, it stimulates the breakdown of fats as another source of energy. These are just some of the many actions triggered by one hormone. They all contribute to increasing your chances of escaping a dangerous situation.

Whether or not a cell responds to a hormone depends on whether it has receptors for that hormone, and how it responds depends on what kind of a cell it is. A single hormone can stimulate many different responses.

Some hormones act locally

Some hormones are released into the extracellular fluids in such tiny quantities, or they are so rapidly inactivated by degradative enzymes, or they are taken up so efficiently by local cells, that the circulation never has the chance to distribute them to distant target cells. Thus, these hormones act only locally. When a hormone affects cells near the secreting cell, it is said to have **paracrine** function (Figure 41.1*b*).

An example of a paracrine hormone is **histamine**, one of the mediators of inflammation. Histamine is released in damaged tissues by specialized cells called *mast cells* (see Chapter 19). When the skin is cut, the area around the cut becomes inflamed—red, hot, and swollen. Histamine causes this response by dilating the local blood vessels and making them more permeable ("leaky"), which allows

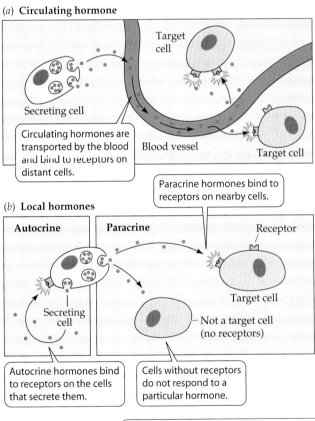

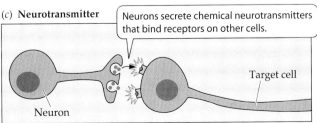

41.1 Chemical Signaling Systems
(*a*) The classic hormone is a secreted chemical message that is distributed throughout the body by the circulatory system. (*b*) An autocrine hormone influences the cell that releases it; paracrine hormones influence nearby cells. (*c*) Neurotransmitters (see Chapter 44) can be considered paracrine hormones.

blood plasma, including protective blood proteins and white blood cells, to move into the damaged tissue.

A major class of paracrine hormones consists of the various **growth factors**, which stimulate the growth and differentiation of cells. Growth factors were first discovered when scientists attempted to culture cells outside of the body. Even when given all sorts of nutrients and optimal conditions, the cells did not grow well unless blood plasma or a tissue extract was added to the medium. The components necessary for growth were found to be specific molecules present in very small quantities. At present, about 50 specific growth factors are known, along with a complex group of receptors. Some examples are:

▶ *Nerve growth factor* (*NGF*), which promotes the survival and growth of nerve cells.
▶ *Epidermal growth factor* (*EGF*), which stimulates many kinds of cells to divide
▶ *Vascular endothelial growth factor*, which stimulates the growth and branching of blood vessels

The nerve cells called *neurons* can also be considered paracrine cells. As we will see in Chapter 44, a neuron communicates with another cell by means of a chemical message called a *neurotransmitter*, which travels over a very small distance to the target cell (Figure 41.1*c*).

In cases in which receptors for the hormone are on the secreting cell itself, the hormone acts as an **autocrine** message (see Figure 41.1*b*). Growth factors are examples of local chemical messages that can also act as autocrine messages for the purpose of negative feedback. The autocrine response prevents the secretory cell from secreting too much of the hormone.

Hormones do not evolve as rapidly as their functions

Chemical signaling between cells exists even in single-celled organisms. Recall the life cycle of slime molds as described in Chapter 27. These protists lead solitary lives and reproduce by mitosis and fission as long as conditions are good. But when food and moisture become scarce, they secrete 3′,5′-cyclic adenosine monophosphate (cAMP), which acts as a chemical signal for the individual cells to aggregate into a slug, form a fruiting structure, and release spores. Thus, in this protist, a chemical message passed between cells influences and coordinates behavior, development, and physiology.

The molecule responsible for this very primitive form of chemical communication between slime mold cells—cAMP—is involved in many hormonal signaling systems in multicellular animals. As you learned in Chapter 15, many molecular signals cause the production of cAMP within cells. This "second messenger" mediates a variety of responses within the cell via the phosphorylation of enzymes.

With the evolution of increasingly complex multicellular animals, more and more molecules acquired signaling functions. Also, as physiological systems changed through evolution, many existing molecular signals acquired new functions. As a result, the same chemical substances are used as

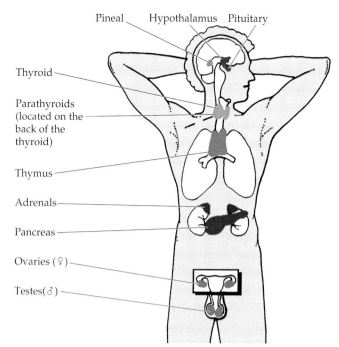

41.2 The Endocrine System of Humans
The endocrine system broadcasts chemical signals (hormones) that are received by cells with appropriate receptors. There are nine glands in the human endocrine system, but hormones are also secreted by tissues that are not part of discrete glands.

hormones in widely divergent species, but they may have completely different actions in different species.

The hormone thyroxine, for example, is found in animal species ranging from mollusks and tunicates (sea squirts) to humans. Its function is unknown in invertebrates, but it is produced in an organ that is involved in feeding. In frogs, thyroxine is essential for the metamorphosis from tadpole to adult. In mammals, thyroxine elevates cellular metabolism.

Prolactin is another example of a hormone that evolved different functions in different species. Prolactin stimulates milk production in female mammals after they give birth. In pigeons and doves, prolactin stimulates the production of crop milk, a substance secreted from the crop that is fed to the young. Prolactin causes amphibians to prepare for reproduction by seeking water. In fishes, such as salmon, that migrate between salt water and fresh water to breed, prolactin regulates the mechanisms that maintain osmotic balance with the changing environment. In all of these cases prolactin is involved in reproductive processes, but as those processes have changed through evolution, so has the information signaled by the hormone.

In summary, the structures of the molecules involved in chemical signaling are highly conserved—they have changed little throughout evolution—but their functions have changed dramatically.

Endocrine glands secrete hormones

Some endocrine cells are distributed as single cells within a tissue. Many hormones of the digestive tract, for example, are produced and secreted by isolated cells in the lining of the tract. As the contents of the digestive tract come into contact with these cells, they release their hormones, which enter the blood and, like epinephrine, circulate throughout the body and activate cells that have appropriate receptors. Many hormones, however, are secreted by aggregations of endocrine cells that form secretory organs called endocrine glands.

Animals have two types of glands. **Exocrine glands**, such as sweat glands and salivary glands, release secretions that are not hormones through ducts that lead outside the body. Sweat gland ducts, for example, open onto the surface of the skin, and salivary gland ducts open into the mouth. Glands that secrete hormones and do not have ducts are called **endocrine glands**; they secrete their products directly into the extracellular fluid. Vertebrates have nine discrete endocrine glands, which collectively make up the **endocrine system** (Figure 41.2).

Hormonal Control of Molting and Development in Insects

Many hormones of invertebrate animals have multiple functions. In this chapter we cannot do justice to the diversity of hormones in the invertebrates, but we'll discuss two important aspects of the lives of many invertebrates that are controlled by hormonal mechanisms: molting and metamorphosis.

Hormones from the head control molting in insects

Because insects have rigid exoskeletons, their growth is episodic, punctuated with *molts* (shedding) of the exoskeleton (see Chapter 32). Each growth stage between two molts is called an *instar*. The British physiologist Sir Vincent Wigglesworth was a pioneer in the study of the hormonal control of growth and development in insects.

Wigglesworth conducted experiments on the blood-sucking bug *Rhodnius*, which undergoes *incomplete metamorphosis*. Upon hatching, *Rhodnius* is nearly a miniature version of an adult, but it lacks some adult features. *Rhodnius* molts five times before developing into a mature adult; a blood meal triggers each episode of molting and growth.

Rhodnius is a hardy experimental animal; it can live a long time even after it is decapitated. If decapitated about an hour after it has a blood meal, *Rhodnius* may live for up to a year, but it does not molt. If decapitated a week after its blood meal, it does molt (Figure 41.3, Experiment 1). These observations led Wigglesworth to the hypothesis that something diffusing slowly from the head controls molting.

The proof of this hypothesis came from a clever experiment in which Wigglesworth decapitated two *Rhodnius*: one that had just had its blood meal and another that had had its blood meal a week earlier. The two decapitated bodies

41.3 A Diffusible Substance Triggers Molting
The effect of time since the last blood meal on *Rhodnius* molting led Sir Vincent Wigglesworth to hypothesize that some substance was diffusing slowly through the insect's body. Further experiments showed that molting is indeed controlled by a substance—a hormone—diffusing from the head.

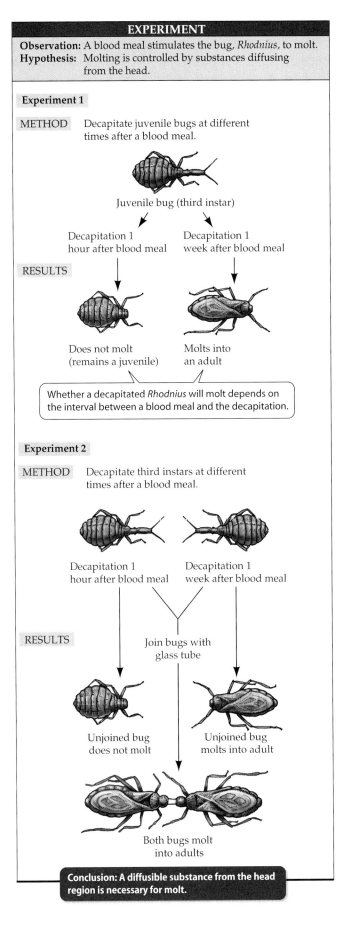

EXPERIMENT

Observation: A blood meal stimulates the bug, *Rhodnius*, to molt.
Hypothesis: Molting is controlled by substances diffusing from the head.

Experiment 1

METHOD   Decapitate juvenile bugs at different times after a blood meal.

Juvenile bug (third instar)

Decapitation 1 hour after blood meal

Decapitation 1 week after blood meal

RESULTS

Does not molt (remains a juvenile)

Molts into an adult

Whether a decapitated *Rhodnius* will molt depends on the interval between a blood meal and the decapitation.

Experiment 2

METHOD   Decapitate third instars at different times after a blood meal.

Decapitation 1 hour after blood meal

Decapitation 1 week after blood meal

RESULTS

Join bugs with glass tube

Unjoined bug does not molt

Unjoined bug molts into adult

Both bugs molt into adults

Conclusion: A diffusible substance from the head region is necessary for molt.

were connected with a short piece of glass tubing—and they both molted (Figure 41.3, Experiment 2). Thus one or more substances from the bug fed a week earlier crossed through the glass tube and stimulated molting in the other bug.

We now know that two hormones working in sequence regulate molting:

▶ Cells in the brain produce **brain hormone**.
▶ Brain hormone is transported to and stored in a pair of structures attached to the brain, the *corpora cardiaca* (singular corpus cardiacum).
▶ After appropriate stimulation (which for *Rhodnius* is a blood meal) the corpora cardiaca release brain hormone, which diffuses to an endocrine gland, the *prothoracic gland.*
▶ Brain hormone stimulates the prothoracic gland to release the hormone **ecdysone**.
▶ Ecdysone diffuses to target tissues and stimulates molting.

The control of molting by brain hormone and ecdysone is a general mechanism in insects. The nervous system receives various types of information relevant in determining the optimal timing for growth and development. It makes sense, therefore, that the nervous system should control the endocrine gland that produces the hormone that orchestrates all the physiological processes involved in development and molting. Later in this chapter we will see similar links between the nervous system and endocrine glands in vertebrates.

Juvenile hormone controls development in insects

The *Rhodnius* decapitation experiments yielded a curious result: Regardless of the instar used, the decapitated bug always molted directly into an adult form. Additional experiments by Wigglesworth demonstrated that a hormone other than those responsible for molting determines whether a bug molts into another juvenile instar or into an adult.

Because the head of *Rhodnius* is long, it was possible to remove just the front part of the head, which contains the cells that secrete and release brain hormone, while leaving intact the rear part. That rear part contains two other endocrine structures called the *corpora allata* (singular corpus allatum). When fourth-instar bugs that had been fed one week earlier were partly decapitated, leaving the corpora allata intact, they molted into fifth instars, not into adults.

This experiment, was followed up by more experiments using glass tubes to connect individual bugs, allowing body fluid transfer between them. When an unfed, completely decapitated, fifth-instar bug was connected to a fourth-instar bug that had been fed and had only the front part of its head removed, both bugs molted into juvenile forms. A substance coming from the rear part of the head of the fourth-instar bug prevented the expected result that both bugs would molt into adult forms.

We now know that the substance is **juvenile hormone** and that it comes from the corpora allata. As long as juve-

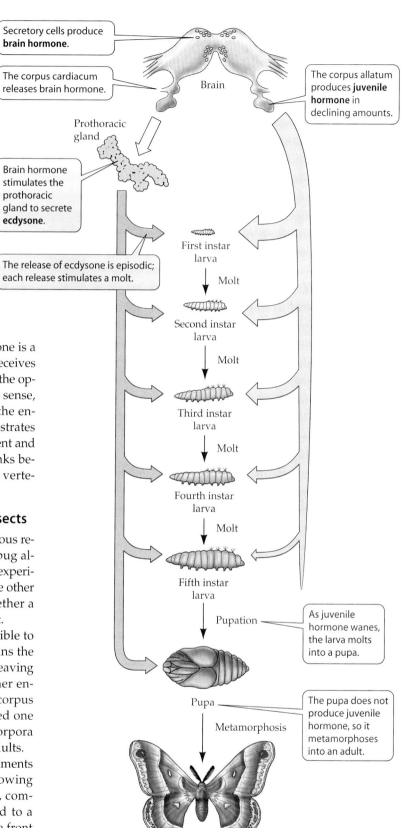

Secretory cells produce **brain hormone**.

The corpus cardiacum releases brain hormone.

Brain

Prothoracic gland

The corpus allatum produces **juvenile hormone** in declining amounts.

Brain hormone stimulates the prothoracic gland to secrete **ecdysone**.

The release of ecdysone is episodic; each release stimulates a molt.

First instar larva

Molt

Second instar larva

Molt

Third instar larva

Molt

Fourth instar larva

Molt

Fifth instar larva

Pupation

As juvenile hormone wanes, the larva molts into a pupa.

Pupa

Metamorphosis

The pupa does not produce juvenile hormone, so it metamorphoses into an adult.

Adult

41.4 Complete Metamorphosis
Butterflies and moths undergo complete metamorphosis, in which the feeding larvae (caterpillars) bear no resemblance to the reproductive adult. Three hormones control molting and metamorphosis in the silkworm moth *Hyalophora cecropia*.

nile hormone is present, *Rhodnius* molts into another juvenile instar. The corpora allata normally stop producing juvenile hormone during the fifth instar. If juvenile hormone is absent, the bug molts into the adult form.

The control of development by juvenile hormone is more complex in insects that, like butterflies, undergo *complete metamorphosis*. These animals undergo dramatic developmental changes between instars. The fertilized egg hatches into a *larva*, which feeds and molts several times, becoming bigger and bigger. Then it enters an inactive stage called a *pupa*. It undergoes major body reorganization as a pupa, and finally emerges as an *adult*.

An excellent example of complete metamorphosis is provided by the silkworm moth, *Hyalophora cecropia* (Figure 41.4). As long as juvenile hormone is present in high concentrations, larvae molt into larvae. When the level of juvenile hormone falls, larvae molt into pupae. Because no juvenile hormone is produced in pupae, they molt into adults.

In our perpetual war against insects, juvenile hormone is a new weapon. Synthetic forms of juvenile hormone can be distributed in the environment to prevent the development of juvenile insects into adults capable of reproduction. However, as you might expect from the fact that hormone structures are highly conserved, such a weapon is not without potentially serious side effects. First, it is not selective in its effects on insects and can affect species that are beneficial as well as those that are pests. Second, the synthetic juvenile hormone could have actions in vertebrates that are not yet known.

The existence and function of insect hormones was experimentally demonstrated many years before the hormones were identified chemically. That is not surprising when you consider the tiny amounts of certain hormones that exist in an organism. In one of the earliest studies of ecdysone, biochemists produced only 250 mg of pure ecdysone (about one-fourth the weight of an apple seed) from 4 tons of silkworms!

Vertebrate Endocrine Systems

The list of chemical messages in the bodies of vertebrates is long and growing longer. To make the subject manageable, we focus mostly on the hormones of humans—how they function and how they are controlled. Table 41.1 presents an overview of the hormones of humans (most of which are found in all other mammals as well). Notice that the column listing the target tissues of these hormones includes every organ system of the body.

We begin this survey with the **pituitary gland** because it plays a central role in the endocrine system. The pituitary is a link between the nervous system and many endocrine glands. It secretes some hormones that are actually produced by neurons in the brain, and under the influence of still other brain hormones, it produces a number of its own hormones, which control the activities of various endocrine glands throughout the body. For these reasons, the pituitary has been called "the master gland."

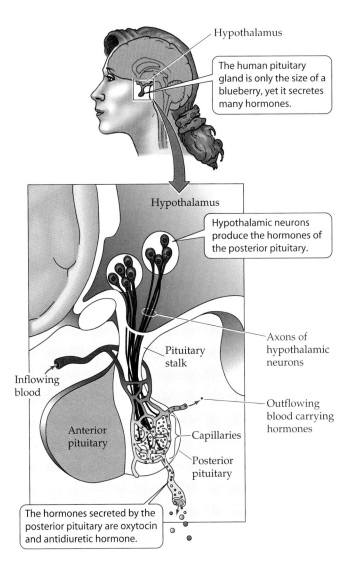

41.5 The Posterior Pituitary Releases Neurohormones
The two hormones stored and released by the posterior pituitary are neurohormones produced in the hypothalamus.

The pituitary develops from outpocketings of the mouth and brain

The pituitary gland sits in a depression at the bottom of the skull just over the back of the roof of the mouth (Figure 41.5). It is attached to the part of the brain called the *hypothalamus*, which is involved in many homeostatic regulatory systems (see Chapter 40).

The pituitary has two distinct parts that have different functions and separate origins during development. The **anterior pituitary** originates as an outpocketing of the embryonic mouth cavity, and the **posterior pituitary** originates as an outpocketing of the developing brain in the region that becomes the hypothalamus.

THE POSTERIOR PITUITARY. The posterior pituitary releases two hormones, antidiuretic hormone and oxytocin. Both are small peptides synthesized in neurons in the hypothalamus.

41.1 Principal Hormones of Humans

SECRETING TISSUE OR GLAND	HORMONE	CHEMICAL NATURE	TARGET(S)	IMPORTANT PROPERTIES OR ACTIONS
Hypothalamus	Releasing and release-inhibiting hormones (see Table 41.2)	Peptides	Anterior pituitary	Control secretion of hormones of anterior pituitary
	Oxytocin, antidiuretic hormone	Peptides	(See Posterior pituitary)	Stored and released by posterior pituitary
Anterior pituitary: Tropic hormones	Thyrotropin	Glycoprotein	Thyroid gland	Stimulates synthesis and secretion of thyroxine
	Adrenocorticotropin (ACTH)	Polypeptide	Adrenal cortex	Stimulates release of hormones from adrenal cortex
	Luteinizing hormone (LH)	Glycoprotein	Gonads	Stimulates secretion of sex hormones from ovaries and testes
	Follicle-stimulating hormone (FSH)	Glycoprotein	Gonads	Stimulates growth and maturation of eggs in females; stimulates sperm production in males
Anterior pituitary: Other hormones	Growth hormone (GH)	Protein	Bones, liver, muscles	Stimulates protein synthesis and growth
	Prolactin	Protein	Mammary glands	Stimulates milk production
	Melanocyte-stimulating hormone	Peptide	Melanocytes	Controls skin pigmentation
	Endorphins and enkephalins	Peptides	Spinal cord neurons	Decrease painful sensations
Posterior pituitary	Oxytocin	Peptide	Uterus, breasts	Induces birth by stimulating labor contractions; causes milk flow
	Antidiuretic hormone (ADH) (vasopressin)	Peptide	Kidneys	Stimulates water reabsorption and raises blood pressure
Thyroid	Thyroxine	Iodinated amino acid derivative	Many tissues	Stimulates and maintains metabolism necessary for normal development and growth
	Calcitonin	Peptide	Bones	Stimulates bone formation; lowers blood calcium
Parathyroids	Parathormone	Protein	Bones	Absorbs bone; raises blood calcium
Thymus	Thymosins	Peptides	Immune system	Activate immune responses of T cells in the lymphatic system
Pancreas	Insulin	Protein	Muscles, liver, fat, other tissues	Stimulates uptake and metabolism of glucose; increases conversion of glucose to glycogen and fat
	Glucagon	Protein	Liver	Stimulates breakdown of glycogen and raises blood sugar
	Somatostatin	Peptide	Digestive tract; other cells of the pancreas	Inhibits insulin and glucagon release; decreases secretion, motility, and absorption in the digestive tract

Hormones that are produced and released by neurons are called **neurohormones**. Antidiuretic hormone and oxytocin move down long extensions (axons) of the neurons that produce them, through the pituitary stalk into the posterior pituitary, where they are stored in the nerve endings (see Figure 41.5). How do they move down the axons? In the bodies of the neurons, these neurohormones are packaged into vesicles. Proteins called *kinesins* grab onto the vesicles and, powered by ATP, "walk" step by step down microtubules in the axons (see Figure 4.25).

The main action of **antidiuretic hormone (ADH)** is to increase the amount of water conserved by the kidneys. When ADH secretion is high, the kidneys resorb more water and produce only a small volume of highly concentrated urine. When ADH secretion is low, the kidneys produce a large volume of dilute urine.

The posterior pituitary increases its release of ADH whenever blood pressure falls or the blood becomes too salty. We will discuss the mechanism of ADH action in Chapter 51. ADH is also known as *vasopressin* because it

41.1 Principal Hormones of Humans (continued)

SECRETING TISSUE OR GLAND	HORMONE	CHEMICAL NATURE	TARGET(S)	IMPORTANT PROPERTIES OR ACTIONS
Adrenal medulla	Epinephrine, norepinephrine	Modified amino acids	Heart, blood vessels, liver, fat cells	Stimulate fight-or-flight reactions: increase heart rate, redistribute blood to muscles, raise blood sugar
Adrenal cortex	Glucocorticoids (cortisol)	Steroids	Muscles, immune system, other tissues	Mediate response to stress; reduce metabolism of glucose, increase metabolism of proteins and fats; reduce inflammation and immune responses
	Mineralocorticoids (aldosterone)	Steroids	Kidneys	Stimulate excretion of potassium ions and reabsorption of sodium ions
Stomach lining	Gastrin	Peptide	Stomach	Promotes digestion of food by stimulating release of digestive juices; stimulates stomach movements that mix food and digestive juices
Lining of small intestine	Secretin	Peptide	Pancreas	Stimulate secretion of bicarbonate solution by ducts of pancreas
	Cholecystokinin	Peptide	Pancreas, liver, gallbladder	Stimulates secretion of digestive enzymes by pancreas and other digestive juices from liver; stimulates contractions of gallbladder and ducts
	Enterogastrone	Polypeptide	Stomach	Inhibits digestive activities in the stomach
Pineal	Melatonin	Modified amino acid	Hypothalamus	Involved in biological rhythms
Ovaries	Estrogens	Steroids	Breasts, uterus, other tissues	Stimulate development and maintenance of female characteristics and sexual behavior
	Progesterone	Steroid	Uterus	Sustains pregnancy; helps maintain secondary female sexual characteristics
Testes	Androgens	Steroids	Various tissues	Stimulate development and maintenance of male sexual behavior and secondary male sexual characteristics; stimulate sperm production
Many cell types	Prostaglandins	Modified fatty acids	Various tissues	Have many diverse actions
Heart	Atrial natriuretic hormone	Peptide	Kidneys	Increases sodium ion excretion

also causes the constriction of peripheral blood vessels as a means of elevating blood pressure.

When a woman is about to give birth, her posterior pituitary releases **oxytocin**, which stimulates the contractions of the muscles that push the baby out of her body. Oxytocin also brings about the flow of milk from the mother's breasts. The baby's suckling stimulates nerve cells in the mother, causing the secretion of oxytocin. Even the sight and sounds of her baby can cause a nursing mother to secrete oxytocin and release milk from her breasts.

THE ANTERIOR PITUITARY. Four hormones released by the anterior pituitary (*thyrotropin, adrenocorticotropin, luteinizing hormone,* and *follicle-stimulating hormone*) control the activities of other endocrine glands and thus are called **tropic hormones** (see Figure 41.7). Each tropic hormone is produced by a different type of pituitary cell. We will say more about these tropic hormones when we describe their target

glands (thyroid, adrenal cortex, testes, and ovaries) later in this chapter and in the next.

The other hormones produced by the anterior pituitary influence tissues that are not endocrine glands. These hormones are growth hormone, prolactin, melanocyte-stimulating hormone, endorphins, and enkephalins.

Growth hormone (GH) consists of about 200 amino acids and acts on a wide variety of tissues to promote growth directly and indirectly. One of its important direct effects is to stimulate cells to take up amino acids. Growth hormone promotes growth indirectly by stimulating the liver to produce chemical messages that stimulate the growth of bone and cartilage. Thus, in some of its actions, growth hormone can also be considered a tropic hormone.

Overproduction of growth hormone in children causes *gigantism,* and underproduction causes *dwarfism* (Figure 41.6). Beginning in the late 1950s, children diagnosed as having a serious deficiency of growth hormone were treated

41.6 Effects of Abnormal Amounts of Growth Hormone
(a) Overproduction of growth hormone in childhood causes gigantism. This photo from 1939 shows a young man who is more than 8 feet tall standing next to his father, who is just under 6 feet tall. (b) Underproduction of growth hormone during childhood results in pituitary dwarfism. The man on the left is P. T. Barnum, the circus entrepreneur. The man on the right is Charles Stratton, a dwarf, who appeared in Barnum's circus under the name General Tom Thumb.

(a)

(b)

with human growth hormone extracted from human pituitaries in cadavers. The treatment was successful in stimulating substantial growth, but it could be made available to only small numbers of patients. A year's supply of human growth hormone for one individual required up to 50 pituitaries. In the mid-1980s, scientists using genetic engineering technology isolated the gene for human growth hormone and introduced it into bacteria, which produced enough of the hormone to make it widely available.

Preventing pituitary dwarfism is now feasible and affordable, but the availability of growth hormone raises new questions. Should every child at the lower end of the height charts be treated? Should a normal child whose parents think basketball stardom is assured if she is tall be given growth hormone? These types of questions are impossible to answer with scientific data alone.

Prolactin, another hormone produced by the anterior pituitary, stimulates the production and secretion of milk in female mammals. In some mammals, prolactin also functions as an important hormone during pregnancy. In human

males, prolactin plays a role along with other pituitary hormones in controlling the endocrine function of the testes.

Endorphins and **enkephalins** are the body's "natural opiates." In the brain, these molecules act as neurotransmitters in pathways that control pain. The significance of their release from the anterior pituitary is unknown. Interestingly, the production of endorphins and enkephalins in the pituitary is encoded by the same gene that encodes at least two other pituitary hormones. The gene actually encodes a large parent molecule called *pro-opiomelanocortin*. This large protein molecule is cleaved to produce several peptides, some of which have hormonal functions. Adrenocorticotropin, melanocyte-stimulating hormone, endorphins, and enkephalins all result from the cleavage of pro-opiomelanocortin.

THE ANTERIOR PITUITARY IS CONTROLLED BY HYPOTHALAMIC NEUROHORMONES   Because the anterior pituitary produces tropic hormones that control other endocrine tissues, it acquired the designation "master gland." But we now

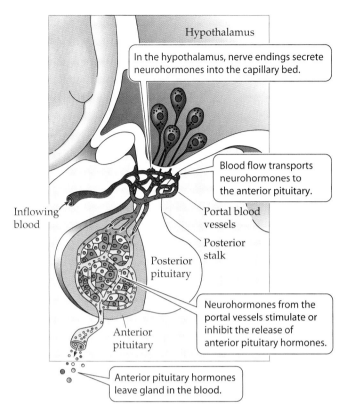

In the hypothalamus, nerve endings secrete neurohormones into the capillary bed.

Blood flow transports neurohormones to the anterior pituitary.

Inflowing blood

Portal blood vessels

Posterior stalk

Posterior pituitary

Neurohormones from the portal vessels stimulate or inhibit the release of anterior pituitary hormones.

Anterior pituitary

Anterior pituitary hormones leave gland in the blood.

41.7 Hormones from the Hypothalamus Control the Anterior Pituitary
Neurohormones produced in tiny quantities by cells in the hypothalamus are transported to the anterior pituitary through a system of portal blood vessels. These releasing and release-inhibiting hormones control the activities of anterior pituitary endocrine cells.

know that this "master" is under still higher control by the hypothalamus, and that their interaction integrates nervous system and endocrine system functions.

The hypothalamus receives information about conditions in the body and in the external environment through the nervous system. If the connection between the hypothalamus and the pituitary is experimentally cut, pituitary hormones are no longer released in response to changes in the environment or in the body. If pituitary cells are maintained in culture, extracts of hypothalamic tissue stimulate

some of those cells to release their hormones into the culture medium. Therefore, scientists hypothesized that secretions of the hypothalamic cells control the activities of anterior pituitary cells.

Although hypothalamic neurons do not extend into the anterior pituitary as they do into the posterior pituitary, a special set of **portal blood vessels** connects the hypothalamus and the anterior pituitary (Figure 41.7). It was thus proposed that secretions from nerve endings in the hypothalamus enter the blood and are conducted down the portal vessels to the anterior pituitary, where they cause the release of anterior pituitary hormones.

In the 1960s, two large teams of scientists, led by Roger Guillemin and Andrew Schally, initiated the search for the hypothalamic neurohormones. Because the amounts of such hormones in any individual mammal would be tiny, massive numbers of hypothalami from pigs and sheep were collected from slaughterhouses and shipped to laboratories in refrigerated trucks. One extraction effort began with the hypothalami from 270,000 sheep and yielded only 1 mg of purified **thyrotropin-releasing hormone**, or **TRH**, which was the first hypothalamic releasing (that is, release-stimulating) hormone isolated and characterized. Biochemical analysis of this pure sample revealed that TRH is a simple tripeptide consisting of glutamine, histidine, and proline. TRH causes certain anterior pituitary cells to release the tropic hormone *thyrotropin*, which in turn stimulates the activity of the thyroid gland.

Soon after discovering thyrotropin-releasing hormone, Guillemin's and Schally's teams identified **gonadotropin-releasing hormone**, which stimulates certain anterior pituitary cells to release the tropic hormones that control the activity of the gonads (the ovaries and the testes). For these discoveries, Guillemin and Schally received the 1972 Nobel prize in medicine. Many more hypothalamic neurohormones, including both releasing hormones and release-inhibiting hormones, are now known (Table 41.2).

Negative feedback loops control hormone secretion

As well as being controlled by hypothalamic releasing and release-inhibiting hormones, the endocrine cells of the ante-

41.2 *Releasing and Release-Inhibiting Neurohormones of the Hypothalamus*

NEUROHORMONE	ACTION
Thyrotropin-releasing hormone (TRH)	Stimulates thyrotropin release
Gonadotropin-releasing hormone (GnRH)	Stimulates release of follicle-stimulating hormone and luteinizing hormone
Prolactin release-inhibiting hormone	Inhibits prolactin release
Prolactin-releasing hormone	Stimulates prolactin release
Somatostatin (growth hormone release-inhibiting hormone)	Inhibits growth hormone release; interferes with thyrotropin release
Growth hormone–releasing hormone	Stimulates growth hormone release
Adrenocorticotropin-releasing hormone	Stimulates adrenocorticotropin release
Melanocyte-stimulating hormone release-inhibiting hormone	Inhibits release of melanocyte-stimulating hormone

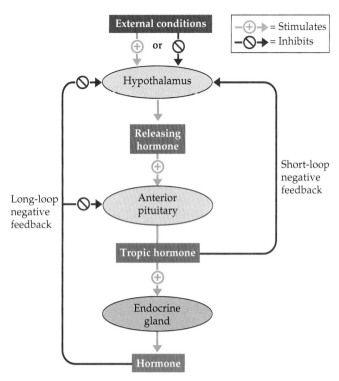

41.8 Multiple Feedback Loops Control Hormone Secretion
Multiple feedback loops regulate the chain of command from hypothalamus to anterior pituitary to endocrine glands.

rior pituitary are also under negative feedback control by the hormones of the target glands they stimulate (Figure 41.8). For example, the hormone cortisol, produced by the adrenal gland in response to adrenocorticotropin, returns to the pituitary in the circulating blood and inhibits further adrenocorticotropin release. Cortisol also acts as a negative feedback signal at the level of the hypothalamus, inhibiting the release of adrenocorticotropin-releasing hormone. In some cases a tropic hormone of the anterior pituitary also exerts negative feedback control on the hypothalamic cells producing the corresponding releasing hormone.

Thyroxine controls cell metabolism

The **thyroid gland** wraps around the front of the windpipe (*trachea*) and expands into a lobe on either side (see Figure 41.2). The thyroid gland produces the hormones thyroxine and calcitonin.

Thyroxine is synthesized in thyroid cells from two molecules of diiodotyrosine, which is the amino acid tyrosine with two atoms of iodine chemically bonded to it. Thus, a thyroxine molecule has four atoms of iodine, and is called T_4:

HO — ⬡ — O — ⬡ — CH_2 — CH — C — OH
                                    | ||
                                   NH_2 O

Thyroxine (T_4)

Thyroid cells also produce triiodothyronine, a version of thyroxine that has only three atoms of iodine and is called T_3:

HO — ⬡ — O — ⬡ — CH_2 — CH — C — OH
                                    | ||
                                   NH_2 O

Triiodothyronine (T_3)

The thyroid usually makes and releases about four times as much T_4 as T_3. T_3 is the more active hormone in the cells of the body, but when T_4 is in circulation, it can be converted to T_3 by an enzyme. Therefore, when you read about thyroxine, keep in mind that the actions discussed are primarily due to T_3.

Thyroxine in mammals plays many roles in regulating cell metabolism. It elevates the metabolic rates of most cells and tissues and promotes the use of carbohydrates rather than fats for fuel. Exposure to cold for several days leads to an increased release of thyroxine, an increased conversion of T_4 to T_3, and an increase in basal metabolic rate. Thyroxine is especially crucial during development and growth, as it promotes amino acid uptake and protein synthesis by cells. Insufficient thyroxine in a human fetus or growing child greatly retards physical and mental growth, resulting in a condition known as *cretinism*.

The tropic hormone **thyrotropin** from the anterior pituitary activates the thyroid cells that produce thyroxine (Figure 41.9). TRH (thyrotropin-releasing hormone) produced in the hypothalamus and transported to the anterior pituitary through the portal blood vessels activates the thyrotropin-producing pituitary cells. The brain uses environmental information such as temperature or day length to determine whether to increase or decrease the secretion of TRH. There is a very important negative feedback loop in this sequence of steps: Circulating thyroxine inhibits the response of the pituitary cells to TRH. Less thyrotropin is released when thyroxine levels are high, and more thyrotropin is released when thyroxine levels are low.

Thyroid dysfunction causes goiter

A *goiter* is an enlarged thyroid gland, which causes a pronounced bulge on the front and sides of the neck. Goiter can be associated with either **hyperthyroidism** (very high levels of thyroxine) or **hypothyroidism** (very low levels of thyroxine). The control diagram in Figure 41.9 helps explain how two very different conditions can result in the same symptom.

Hyperthyroid goiter results when the negative feedback mechanism fails to turn off the thyroid cells even though blood levels of thyroxine are high. The most common cause of hyperthyroidism is an autoimmune disease in which an antibody to the thyrotropin receptor is produced. This antibody can bind to the receptor and cause the thyroid cells to produce and release thyroxine. Even though blood levels of thyrotropin may be quite low because of the negative feed-

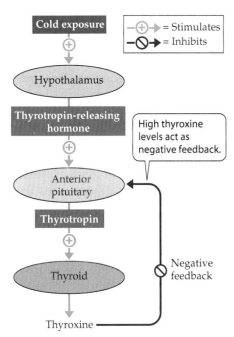

41.9 Regulation of Thyroid Function in Response to Cold
Exposure to cold temperatures stimulates the hypothalamus to produce thyrotropin-releasing hormone, which stimulates anterior pituitary cells to secrete thyrotropin, which in turn stimulates the thyroid to release thyroxine.

back from thyroxine, the thyroid remains maximally stimulated, and it grows bigger. Hyperthyroid patients have high metabolic rates, are jumpy and nervous, usually feel hot, and may have a buildup of fat behind the eyeballs, causing their eyes to bulge.

Hypothyroid goiter results when there is not enough circulating thyroxine to turn off thyrotropin production. Its most common cause is a deficiency of dietary iodide, without which the thyroid gland cannot make thyroxine. Without sufficient thyroxine, thyrotropin levels remain high, so the thyroid continues to produce large amounts of nonfunctional thyroxine and becomes very large. The symptoms of hypothyroidism are low metabolism, intolerance of cold, and general physical and mental sluggishness.

Worldwide, goiter affects about 5 percent of the population. The addition of iodide to table salt has greatly reduced the incidence of the condition in industrialized nations, but goiter is still common in the less industrialized countries of the world.

Calcitonin reduces blood calcium

Another hormone released by the thyroid gland is **calcitonin**, although it is not produced by the same cells that produce thyroxine. Calcitonin helps reduce the levels of calcium circulating in the blood (Figure 41.10).

Bone is a huge repository of calcium in the body and is continually being remodeled. Cells called **osteoclasts** break down bone and release calcium; **osteoblasts**, on the other hand, use circulating calcium to deposit new bone. Calcitonin decreases the activity of osteclasts and stimulates the activity of osteoblasts, thus shifting the

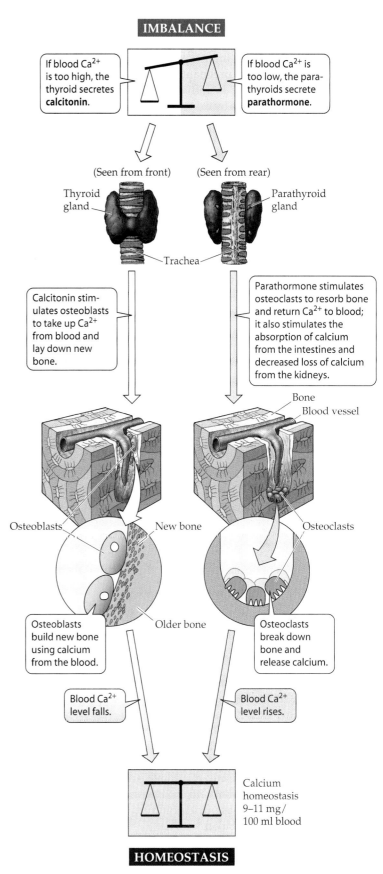

41.10 Hormonal Regulation of Calcium
Calcitonin and parathormone help regulate blood calcium levels. Bone can be a source (site of production) or a sink (site of utilization or storage) for calcium.

Aldosterone, the main mineralocorticoid, stimulates the kidney to conserve sodium and to excrete potassium. If the adrenal glands are removed from an animal, it must have sodium added to its diet, or its sodium will be depleted and it will die. One human patient with a nonfunctional adrenal gland compensated by salting her food heavily and, in addition, ate a 60-pound block of salt in the course of a year.

The main glucocorticoid, **cortisol**, is critical for mediating the body's response to stress. As we have seen, your immediate reaction to a frightening situation is stimulated by your nervous system and by the release of epinephrine. This fight-or-flight response ensures that your muscles will have enough oxygen and glucose to fuel your escape. You have a limited amount of blood glucose, however, and you need to conserve it for your muscles and your brain. Within minutes of the frightening stimulus, blood cortisol level rises. Cortisol stimulates cells not critical for your escape to decrease their use of blood glucose and shift instead to utilizing fats and proteins for energy. This is not a time to feel sick, have allergic reactions, or heal wounds, so cortisol also blocks immune system reactions. This is why cortisol is useful for reducing inflammations and allergies.

Cortisol release is controlled by ACTH from the anterior pituitary, which in turn is controlled by the hypothalamic **adrenocorticotropin-releasing hormone**. Because the cortisol response to a stressor has this chain of steps, each involving secretion, diffusion, circulation, and cell activation, it is much slower than the epinephrine response.

Turning off the cortisol response is as important as turning it on. A study of stress in rats showed that old rats could turn on their stress responses as effectively as young rats, but that they had lost the ability to turn them off as rapidly. As a result, they suffered from the well-known consequences of stress: ulcers, cardiovascular problems, strokes, impaired immune system function, and increased susceptibility to cancers and other diseases. Further research showed that turning off stress responses involves the long-loop negative feedback action of cortisol on cells in the brain, which causes a decrease in the release of adrenocorticotropin-releasing hormone (see Figure 41.8). Repeated activation of this negative feedback mechanism leads to a gradual loss of cortisol-sensitive cells in the brain, and therefore a decreased ability to terminate stress responses.

The sex steroids are produced by the gonads

The **gonads**—the testes of the male and the ovaries of the female—produce hormones as well as gametes. Most of the gonadal hormones are steroids synthesized from cholesterol (see Figure 41.12). The male steroids are collectively called **androgens**, and the dominant one is **testosterone**. The female steroids are **estrogens** and **progesterone**. The dominant estrogen is **estradiol**.

The sex steroids have important developmental effects: They determine whether a fetus develops into a female or a male. (A *fetus* is the latter stage of an embryo; a human embryo is called a fetus from the eighth week of pregnancy to the moment of birth.) After birth, the sex steroids control the maturation of the reproductive organs and the development and maintenance of secondary sexual characteristics, such as breasts and facial hair.

The sex steroids begin to exert effects in the human embryo in the seventh week of development. Until that time, the embryo has the potential to develop into either sex. In mammals and birds, the ultimate instructions for sex determination reside in the genes. In mammals, individuals that receive two X chromosomes normally become females, and individuals that receive an X and a Y chromosome normally become males (Figure 41.13). These genetic instructions are carried out through the production and action of the sex steroids, and the potential for error exists.

The presence of a Y chromosome normally causes the embryonic, undifferentiated gonads to begin producing androgens in the seventh week. In response to the androgens, the reproductive system develops into that of a male. If androgens are not produced at that time, female reproductive structures develop. In other words, androgens are required to trigger male development in humans. The opposite situation exists in birds: Male characteristics develop unless estrogens are present to trigger female development.

Occasionally the hormonal control of sexual development does not work perfectly, resulting in *intersex* individuals. The most extreme (but rare) case is a true **hermaphrodite**, who has both testes and ovaries. **Pseudohermaphrodites** have the gonads of one sex and the external sex organs of the other. For example, an XY fetus will develop testes, but if his tissues are insensitive to the androgens they produce because his androgen receptors do not function, the testes will remain within the abdomen, and the external sex organs and the secondary sexual characteristics of a female will develop.

Changes in control of sex steroid production initiate puberty

Sex steroids have dramatic effects at **puberty**—the time of sexual maturation in humans. Sex steroids are produced at low levels by the juvenile gonads, but their production increases rapidly at the beginning of puberty—around the age of 12 to 13 years. Why does this sudden increase occur?

In the juvenile, as in the adult, the production of sex steroids by the ovaries and testes is controlled by the anterior pituitary tropic hormones **luteinizing hormone (LH)** and **follicle-stimulating hormone (FSH)**, which together are called the **gonadotropins**. The production of these tropic hormones is under the control of the hypothalamic gonadotropin-releasing hormone (GnRH). Prior to puberty, the gonads are capable of responding to gonadotropins, and the pituitary is capable of responding to GnRH. But prior to puberty the hypothalamus produces only very low levels of GnRH. Puberty is initiated by a reduction in the sensitivity of hypothalamic GnRH-producing cells to negative feedback from sex steroids and from gonadotropins. As a result, GnRH release increases, stimulating increased pro-

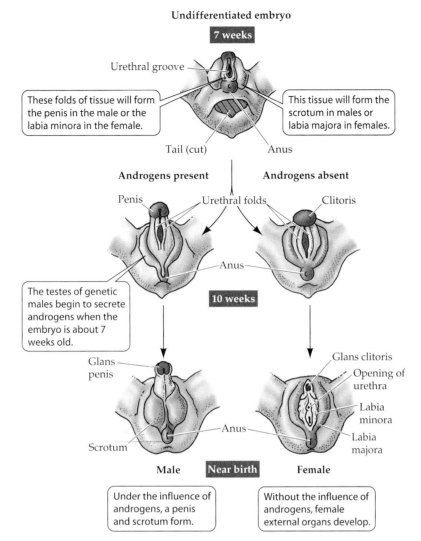

Undifferentiated embryo

7 weeks

Urethral groove

These folds of tissue will form the penis in the male or the labia minora in the female.

This tissue will form the scrotum in males or labia majora in females.

Tail (cut)   Anus

Androgens present   **Androgens absent**

Penis   Urethral folds   Clitoris

Anus

10 weeks

The testes of genetic males begin to secrete androgens when the embryo is about 7 weeks old.

Glans penis   Glans clitoris

Opening of urethra

Labia minora

Anus

Labia majora

Scrotum

Male   Near birth   **Female**

Under the influence of androgens, a penis and scrotum form.

Without the influence of androgens, female external organs develop.

41.13 The Development of Human Sex Organs
The sex organs of early human embryos are similar. Male sex steroids (androgens) promote the development of male sex organs. Without androgen action, female sex organs form, even in genetic males.

duction of gonadotropins and hence increased production of sex steroids.

In females, increasing levels of LH and FSH at puberty stimulate the ovaries to begin producing the female sex hormones. The increased circulating levels of these hormones initiate the development of the traits of a sexually mature woman: enlarged breasts, vagina, and uterus; broad hips; increased subcutaneous fat; pubic hair; and the initiation of the menstrual cycle.

In the male, an increasing level of LH stimulates groups of cells in the testes to synthesize androgens, which in turn initiate the profound physiological, anatomical, and psychological changes associated with adolescence. The voice deepens, hair begins to grow on the face and body, and the testes and penis grow. Androgens also help skeletal muscles grow, especially when they are exercised regularly.

Natural muscle development can be exaggerated by both men and women who want to increase their maximum strength in athletic competition if they take synthetic androgens—called *anabolic steroids*. However, anabolic steroids have serious negative side effects. In women, their use causes the breasts and uterus to shrink, the clitoris to enlarge, menstruation to become irregular, facial and body hair to grow, and the voice to deepen. In men, the testes shrink, hair loss increases, the breasts enlarge, and sterility can result. You can understand the causes of some of these side effects by considering the negative feedback effects of sex steroids on the production of LH and FSH. Other side effects are even more serious. Continued use of anabolic steroids greatly increases the risk of heart disease, certain cancers, kidney damage, and personality disorders such as depression, mania, psychoses, and extreme aggression. Most official athletic organizations, including the International Olympic Committee, ban the use of anabolic steroids.

Melatonin is involved in biological rhythms and photoperiodicity

The **pineal gland** is situated between the cerebral hemispheres of the brain on a little stalk. It produces the hormone **melatonin** from the amino acid tryptophan. In various vertebrates, melatonin is involved in biological rhythms and photoperiodicity. **Photoperiodicity** is the phenomenon whereby seasonal changes in day length cause physiological changes in animals. Many species, for example, come into reproductive condition when the days begin to get longer (Figure 41.14). Humans are not photoperiodic, but melatonin in humans may play a role in entraining the daily rhythm of physiological and behavioral activities to the daily cycle of light and dark.

The release of melatonin by the pineal occurs in the dark and therefore marks the length of the night. Exposure to light inhibits the release of melatonin. The pineal of birds is directly sensitive to light, but in mammals, the light response is mediated through the eyes via a group of cells at the base of the brain, which generates a daily rhythm for many physiological functions of the body. We will learn more about this brain structure in Chapter 52.

The list of other hormones is long

We have discussed the major endocrine glands and the "classic" hormones in this chapter, but there are many hormones we have not mentioned. Examples include the hormones produced in the digestive tract that help organize the way the gut processes food (see Table 41.1 and Chapter 50). Even the heart has endocrine functions. When blood pressure rises and causes the walls of the heart to stretch, certain cells in the walls of the heart release *atrial natriuretic hormone*. This hormone increases the excretion of sodium ions and water by the kidneys, thereby lowering blood volume and blood pressure. As we discuss the organ systems

42 Animal Reproduction

(a)

Plasma melatonin

(b)

41.

Se

(a)
sur
day
phy
Sib
the

of
m
th

M

Tr
qu
su
sp
th
ti
3(
h

NATURAL SELECTION HAS CREATED SOME AMAZ-ing and bizarre adaptations, but among the most unusual and diverse are the methods some animals use to reproduce. Just as "unmanned" submersibles are used in deep ocean exploration, some species of polychaete worms use "unwormed" submersibles to reproduce. The adults of these marine worms live in burrows on the ocean floor or in reefs. Predators make it dangerous for them to leave their burrows to seek a mate, and if they simply released their eggs and sperm at the mouth of the burrow, they would have a poor chance of successful fertilization. So both males and females develop specialized body segments that form at the worm's posterior end and become stuffed with sperm or eggs. These segments develop sensory organs but no mouth or gut, since they will not need to feed.

When the time is right—full moon for some species, new moon for others—these "sex-cell transporters" break loose from the main body of the worm, leave the burrow, swim up into the water column, swarm with more of their kind, and release their sperm or eggs. The sex-cell transporters die soon after they release their cargo. Union of sperm and eggs takes place in the water column, and fertilized eggs may drift a long way before they descend to the ocean floor and develop into adult worms. But in many places, the native people know when and where the sex-cell transporters will swarm, and people harvest them for food.

In this chapter you will learn how animals reproduce. We first examine *asexual* mechanisms of reproduction, in which only a single parent is involved, and then turn to *sexual* reproduction, which requires two parents. Sexually reproducing organisms produce haploid sex cells—sperm and eggs—through the process of meiosis. An egg and a sperm must unite through the process of fertilization to create a new diploid individual.

As we will see, much of the diversity in reproductive systems is in mechanisms for getting sperm and eggs together. This chapter, however, focuses the most attention on the anatomy, function, and endocrine control of the human re-

productive system. This information will allow us to understand the technologies we use both to limit and to overcome infertility. We end the chapter with a discussion of sexual health and sexually transmitted diseases.

Asexual Reproduction

Sexual reproduction is a nearly universal trait of animals, although many species can reproduce asexually as well. Offspring produced asexually are genetically identical to one another and to their parents. Asexual reproduction is highly efficient because there is no mating. Mating requires energy, involves risks, and requires that resources be devoted to a large population of males, who do not produce offspring. Asexual populations can use resources efficiently because all individuals in the population can convert resources to offspring. However, asexual reproduction does not generate genetic diversity, and this can be a disadvantage in changing environments. As we learned in Chapter

Feasting on Sex Cells
During the final quarter of November's moon, the people of Samoa and Fiji harvest the reproductive segments of the palolo worm, *Eunice viridis*. The adult worms release specialized reproductive vehicles into the ocean according to a precise cycle that native people have understood for centuries. The protein-rich worm segments are prepared by roasting or frying and are eaten as a delicacy.

42.1 Asexual Reproduction in Animals
(a) Budding: A new individual forms as an outgrowth from an adult hydra. (b) Regeneration: This five-armed sea star is generating three new arms to replace amputated limbs and form an entire animal. (c) Parthenogenesis: When reproductive conditions are favorable, female aphids produce genetically identically offspring.

(a) *Hydra* sp.

(b) *Linickia* sp.

(c) *Macrosiphum rosae*

21, genetic diversity enables natural selection to shape adaptations in response to environmental change.

A variety of species, mostly invertebrates, reproduce asexually. They tend to be species that are sessile and cannot search for mates, or species that live in sparse populations and rarely encounter potential mates. Furthermore, asexually reproducing species are likely to be found in relatively constant environments, in which the potential for rapid evolutionary change is not as important as in more variable environments.

There are three common modes of asexual reproduction:

▸ *Budding*, in which new individuals form by mitotic cell division.
▸ *Regeneration*, in which a piece or section of an organism can generate an entire new individual.
▸ *Parthenogenesis*, in which individuals develop from unfertilized eggs.

Budding and regeneration produce new individuals by mitosis

Many simple multicellular animals produce offspring by **budding**; new individuals form as outgrowths of the bodies of older animals. These buds grow by mitotic cell division, and the cells differentiate before the buds break away from the parent (Figure 42.1a). The bud is genetically identical to the parent, and it may grow as large as the parent before it becomes independent.

Regeneration is usually thought of as the replacement of damaged tissues or lost limbs, but in some cases pieces of an organism can regenerate complete individuals. In a classic experiment demonstrating regeneration, a sponge is pushed through a cloth mesh, producing many little clusters of cells. Each cluster grows into a small but complete sponge. The ability of sponges to regenerate was used off the coast of Florida to restore the commercial bath-sponge

fishery, which was endangered by overfishing. Echinoderms also have remarkable abilities to regenerate. If sea stars are cut into pieces, each piece that includes a portion of the central disc grows into a new animal (Figure 42.1b).

Regeneration frequently results when an animal is broken by an outside force. A storm, for example, can cause a heavy surf that breaks colonial cnidarians such as corals. Pieces broken off the colony can regenerate into new colonies. In some species, the breakage occurs in the absence of external forces. Some species of segmented marine worms related to the ones we discussed at the beginning of this chapter develop segments with rudimentary heads bearing sensory organs, then break apart. Each fragmented segment forms a new worm.

Parthenogenesis is the development of unfertilized eggs

Not all eggs have to be fertilized to develop. A common mode of asexual reproduction in arthropods is the development of offspring from unfertilized eggs. This phenomenon, called **parthenogenesis**, also occurs in some species of fish, amphibians, and reptiles. Most species that reproduce parthenogenetically also engage in sexual reproduction or sexual behavior.

The aphids that can rapidly populate your rosebushes in the spring and summer reproduce parthenogenetically while conditions are favorable (Figure 42.1c). Some of the unfertilized eggs laid in spring and summer develop into male aphids, others into females. As conditions become less favorable, the aphids mate, and the females lay fertilized eggs. These eggs do not hatch until the following spring, and they yield only females.

In some species, parthenogenesis is part of the mechanism that determines sex. For example, in many hymenopterans (ants, and most species of bees and wasps), males develop from unfertilized eggs and are haploid.

(a)

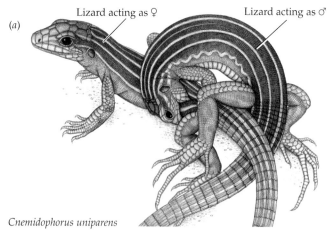

Lizard acting as ♀   Lizard acting as ♂

Cnemidophorus uniparens

(b)

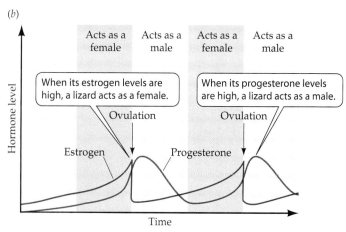

42.2 Sexual Behavior May Be Required for Asexual Reproduction
(*a*) Parthenogenetic whiptail lizards are all female, but take turns acting the male role in reproductive behavior. (*b*) The stage of the ovarian cycle determines the role an individual plays.

Females develop from fertilized eggs and are diploid. Most females are sterile workers, but a select few become fertile queens. After a queen mates, she has a supply of sperm that she controls, enabling her to produce either fertilized or unfertilized eggs. Thus the queen determines when and how much of the colony resources are expended on males.

Parthenogenetic reproduction in some species requires sexual activity even though this activity does not fertilize eggs. The eggs of parthenogenetically reproducing ticks and mites, for example, develop only after the animals have mated, even though the eggs remain unfertilized. One case that has been investigated by David Crews and his students at the University of Texas is parthenogenetic reproduction in a species of whiptail lizard. There are no males in this species, but females act as males, engaging in all aspects of courtship display and mating, even though no sperm are produced or transferred (Figure 42.2). Whether a specific female acts as a female or as a male depends on her hormonal state at the time, but sexual activity is required to stimulate ovulation.

Sexual Reproduction

A large portion of the time and energy budgets of sexually reproducing animals goes into sexual behavior, which exposes them to predation, can result in physical damage, and detracts from other useful activities such as feeding, building secure living places, and caring for existing offspring. In spite of all of these disadvantages, there is an overwhelming evolutionary advantage to sexual reproduction: It produces genetic diversity.

Sexual reproduction requires the joining of two haploid sex cells to form a diploid individual. These haploid cells, or *gametes*, are produced through **gametogenesis**, a process that involves meiotic cell divisions. Two events in meiosis contribute to genetic diversity: *crossing over* of homologous chromosomes, and the *independent assortment* of chromo-

somes. Both of these genetic phenomena were described in Chapter 10.

Mating behavior also contributes to genetic diversity in sexually reproducing species. The genetic variation in the gametes of a single individual and the genetic variation between any two parents produce an enormous potential for genetic variation between any two offspring of a sexually reproducing pair of individuals. This genetic diversity is the raw material for natural selection; thus evolutionary change in sexually reproducing animals can be quite rapid.

There are three fundamental phenomena of sexual reproduction in animals:

▶ *Gametogenesis* (making sex cells)
▶ *Mating* (getting sex cells together)
▶ *Fertilization* (getting sex cells to fuse)

There is not a great deal of diversity in gametogenesis when we compare different groups of animals. Processes of fertilization are also rather similar in widely different species. Therefore, although the discussion of gametogenesis that follows is primarily derived from information from mammals, the facts would not be terribly different if we focused on a different group of animals.

Mating, on the other hand, shows incredible evolutionary diversity. Our discussion of mating in this chapter will focus on a few specific examples as representative of the fascinating diversity that exists.

Eggs and sperm form through gametogenesis

Gametogenesis occurs in the primary sex organs, the **gonads**, which are **testes** (singular testis) in males and **ovaries** in females. The tiny gametes of males, called **sperm**, are motile and move by beating their flagella. The much larger female gametes are **eggs**, or **ova** (singular ovum), and are nonmotile (Figure 42.3).

Gametes are produced from **germ cells**, which have their origin in the earliest cell divisions of the embryo and remain distinct from the rest of the body. All the rest of the cells of the embryo are called **somatic cells**. Germ cells are sequestered in the body of the embryo until its gonads begin to form. The germ cells then migrate to the gonads, where they take up residence and proliferate by mitosis,

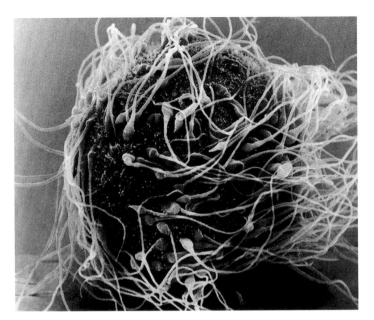

42.3 Gametes Differ in Size
Mammalian sperm (white) are much smaller than the mammalian egg (blue), as illustrated by this artificially colored micrograph of human fertilization.

producing **oogonia** (singular oogonium) in females and **spermatogonia** (singular spermatogonium) in males. Oogonia and spermatogonia, which are diploid, multiply by mitosis in turn, eventually producing **primary oocytes** and **primary spermatocytes**, which are still diploid cells.

Meiosis, the next step in gametogenesis, reduces the chromosomes to the haploid number, and these haploid cells mature into sperm and ova. (You may want to review the discussion of meiosis in Chapter 9 before reading further.) Although the steps of meiosis are very similar in males and females, there are some significant differences in gametogenesis.

SPERMATOGENESIS PRODUCES SPERM. Primary spermatocytes undergo the first meiotic division to form **secondary spermatocytes**, which are haploid. The second meiotic division produces four haploid **spermatids** for each primary spermatocyte that entered meiosis. In mammals, these cells remain connected by cross-bridges of cytoplasm after each division (Figure 42.4a).

The reason that mammalian spermatocytes remain in cytoplasmic contact throughout their development probably is the asymmetry of sex chromosomes in the males of

42.4 Gametogenesis
(a) Diploid spermatogonia develop into haploid spermatids. Spermatids differentiate into sperm. (b) Diploid oogonia develop into haploid secondary oocytes, which mature into ova.

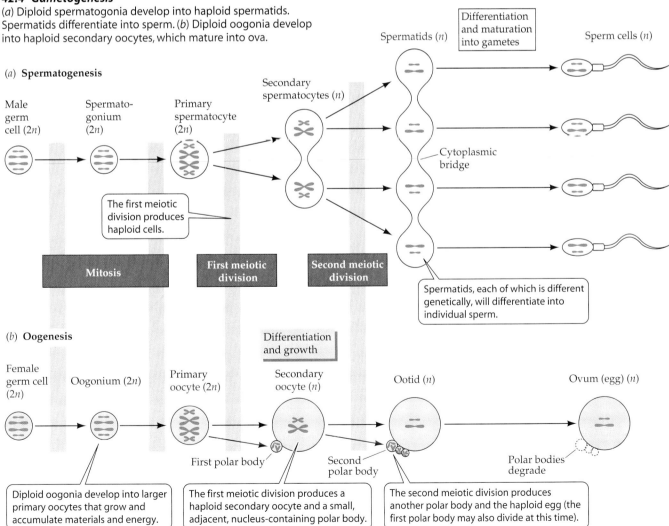

(a) **Spermatogenesis**

Male germ cell (2n) → Spermato-gonium (2n) → Primary spermatocyte (2n) → Secondary spermatocytes (n) → Spermatids (n) → Sperm cells (n)

The first meiotic division produces haploid cells.

Mitosis | **First meiotic division** | **Second meiotic division**

Differentiation and maturation into gametes

Cytoplasmic bridge

Spermatids, each of which is different genetically, will differentiate into individual sperm.

(b) **Oogenesis**

Female germ cell (2n) → Oogonium (2n) → Primary oocyte (2n) → Secondary oocyte (n) → Ootid (n) → Ovum (egg) (n)

Differentiation and growth

First polar body

Second polar body

Polar bodies degrade

Diploid oogonia develop into larger primary oocytes that grow and accumulate materials and energy.

The first meiotic division produces a haploid secondary oocyte and a small, adjacent, nucleus-containing polar body.

The second meiotic division produces another polar body and the haploid egg (the first polar body may also divide at this time).

most species. Half of the secondary spermatocytes receive an X chromosome, the other half a Y chromosome. The Y chromosome contains fewer genes than the X chromosome, and apparently some of the products of genes not included in the Y chromosome are essential for spermatocyte development. By remaining in cytoplasmic contact, all spermatocytes can share the gene products of the X chromosomes, even though only half of them have an X chromosome.

Just after being produced by meiosis, a spermatid bears little resemblance to a sperm. Through further differentiation, it will become compact, streamlined, and motile. We will look at the differentiation of human sperm in more detail below.

OOGENESIS PRODUCES EGGS.   Oogonia, like spermatogonia, proliferate through mitosis. The resulting egg precursor cells differentiate into primary oocytes, which immediately enter prophase of the first meiotic division. In many species, including humans, the development of the oocyte is arrested at this point, and may remain so for days, months, or years. In contrast, there is no arrest in the development of male gametes; the process goes steadily to completion once the primary spermatocyte has differentiated. In the human female, as we will see, some primary oocytes may remain in arrested prophase I for 50 years!

During this prolonged prophase I, or shortly before it ends, the primary oocyte undergoes its major growth phase. It grows larger due to increased production of ribosomes, RNA, cytoplasmic organelles, and energy stores. At this time, the primary oocyte acquires all of the energy, raw materials, and RNA that the egg will need to survive its first cell divisions after fertilization. In fact, the nutrients in the egg will have to nourish the embryo until it is either nourished by the maternal system or can feed on its own.

When a primary oocyte resumes meiosis, its nucleus completes the first meiotic division near the surface of the cell. The daughter cells of this division receive grossly unequal shares of cytoplasm. This asymmetry represents another major difference from spermatogenesis, in which cell divisions apportion cytoplasm equally. The daughter cell that receives almost all of the cytoplasm becomes the **secondary oocyte**, and the one that receives almost none forms the *first polar body* (see Figure 42.4b).

The second meiotic division of the large secondary oocyte is also accompanied by an asymmetrical division of the cytoplasm. One daughter cell forms the large, haploid **ootid**, which eventually differentiates into a mature ovum, and the other forms the *second polar body*. Polar bodies degenerate, so the end result of oogenesis is that each primary oocyte produces only one mature egg. However, that egg is a very large, well-provisioned cell.

A second period of arrested development occurs after the first meiotic division forms the secondary oocyte. The egg may be expelled from the ovary in this condition. In many species, including humans, the second meiotic division is not completed until the egg is fertilized by a sperm.

A single body can function as both male and female

Sexual reproduction requires both male and female haploid gametes. In most species, these gametes are produced by individuals that are either male or female. Species that have male and female members are called **dioecious** (from the Greek for "two houses"). In some species, a single individual may possess both female and male reproductive systems. Such species are called **monoecious** ("one house") or **hermaphroditic**.

Almost all invertebrate groups have hermaphroditic species. An earthworm is an example of a **simultaneous hermaphrodite**, meaning that it is both male and female at the same time. When two earthworms mate, they exchange sperm, and as a result, the eggs of each are fertilized (see Figure 31.24b). Some animals are **sequential hermaphrodites**, meaning that individuals function as a male or as a female at different times in their lives.

What is the selective advantage of hermaphroditism? Some simultaneous hermaphrodites have a low probability of meeting a potential mate. An example is a parasitic tapeworm. Even though it may be large and cause lots of trouble for its host, it may be the only tapeworm in the host. Tapeworms can fertilize their own eggs. Most simultaneous hermaphrodites must mate with another individual, but since each member of the population is both male and female, the probability of encountering a possible mate is double what it would be in monoecious species.

Sequential hermaphroditism can reduce the possibility of inbreeding among siblings by making them all the same sex at the same time and therefore incapable of mating with one another. In a species in which only a few males fertilize all females, sequential hermaphroditism can maximize reproductive success by making it possible for an individual to reproduce as a female until the opportunity arises for it to function successfully as a male.

Anatomical and behavioral adaptations bring eggs and sperm together

Sexual reproduction requires that two haploid gametes join together to form a diploid **zygote**. The purpose of mating behavior is to get eggs and sperm close enough together that this process—called **fertilization**—can occur. Many anatomical and behavioral adaptations have evolved to support mating. The simplest distinction in mating systems is whether fertilization occurs externally or internally.

EXTERNAL FERTILIZATION REQUIRES AN AQUATIC HABITAT.   In an aquatic environment, animals can simply release their gametes into the water. *External fertilization* is common among simple aquatic animals that are not very mobile (Figure 42.5). These animals produce huge numbers of gametes. A female oyster, for example, may produce 100 million eggs in a year, and the number of sperm produced by a male oyster is astronomical.

But numbers alone do not guarantee that gametes will meet. Timing is also important. The reproductive activities

Acropora sp.

42.5 External Fertilization Is Common in Aquatic Species
External fertilization requires an aqueous environment. These staghorn corals are all releasing sperm–egg bundles into the oceans of Ningaloo Reef, Australia.

of the males and females of a population must be synchronized. Seasonal breeders may use day length, changes in temperature, or changes in weather to time their production and release of gametes. Social stimulation is also important. Sexual activity on the part of one member of a population can stimulate others to engage in mating.

Behavior can play an important role in bringing gametes together even when fertilization is external. Many species travel great distances to congregate with potential mates and release their gametes at the same time in a suitable environment. Salmon are an extreme example, traveling hundreds of miles to spawn in the stream where they hatched.

INTERNAL FERTILIZATION ENABLES TERRESTRIAL LIFE. Sperm can move only through liquid, and delicate gametes released into air would dry out and die. Terrestrial animals avoid these problems by engaging in *internal fertilization.*

Animals have evolved an incredible diversity of behavioral and anatomical adaptations toget male gametes into the female reproductive tract. As we saw above, gametogenesis occurs in the primary sex organs, the gonads. All of the additional anatomical components of an animal's reproductive system are called **accessory sex organs**. An obvious accessory sex organ in the male is a tubular structure called the **penis**, which enables the male to deposit sperm in the female's accessory sex organ, the **vagina** (or, in some species, the **cloaca**, a cavity common to the digestive, urinary, and reproductive systems). Accessory sex organs include a variety of glands, tubules, ducts, and other structures.

Copulation is the physical joining of the male and female accessory sex organs. Transfer of sperm in internal fertilization can also be indirect. Males of some species of mites and scorpions (among the arthropods) and salaman-

ders (among the vertebrates) deposit **spermatophores**— containers filled with sperm—in the environment. When a female mite finds a spermatophore, she straddles it and opens a pair of plates in her abdomen so that the tip of the spermatophore enters her reproductive tract and allows the sperm to enter. Some female salamanders use the lips of their cloacae to scoop up the spermatophore.

Male squid and spiders play a more active role in spermatophore transfer. The male spider secretes a drop containing sperm onto a bit of web; then, with a special structure on his foreleg, he picks up the sperm-containing web and inserts it through the female's genital opening. Male squid use one special tentacle to pick up a spermatophore and insert it into the female's genital opening.

Most male insects copulate and transfer sperm to the female's vagina through a tubular penis. The **genitalia**— external sex organs—of insects often have species-specific shapes that match in a lock-and-key fashion. This mechanism ensures a tight, secure fit between the mating pair during the prolonged period of sperm transfer. The males of some insect species have elaborate structures on their penises that can scoop sperm deposited by other males out of the female's reproductive tract.

The evolution of vertebrate reproductive systems parallels the move to land

The earliest vertebrates evolved in aquatic environments. The closest living relatives of those earliest vertebrates are modern-day fishes. They remain exclusively aquatic animals, and most practice external fertilization. The most primitive of the fishes, the lampreys and hagfishes, broadcast their gametes into the environment, as do many aquatic invertebrates. In most fishes, however, fertilization is more selective: Mating behaviors bring females and males into close proximity at the time of gamete release.

In some sharks and rays, certain fins have evolved into structures that hold the male and female together and enable sperm to be transferred directly into the female reproductive tract. This internal fertilization in sharks and rays has made it possible for the females of some species to enclose fertilized eggs in protective egg cases before depositing them in the environment.

Amphibians were the first vertebrates to live in terrestrial environments. They dealt with the challenge of a dry environment by returning to water to reproduce, as most amphibians still do today. Exceptions are the terrestrial salamanders that use spermatophores to transfer sperm, as mentioned earlier. The spermatophore provides a protective, non-desiccating environment for the sperm. Other amphibians, like most fishes, rely on sexual behavior to bring eggs and sperm together. Frog mating behavior is characterized by *amplexus*, a behavior in which a male grasps a female around the middle with his forelegs and holds on until she releases her egg mass, at which time he releases his sperm (Figure 42.6).

Reptiles were the first vertebrate group to solve the problem of reproduction in the terrestrial environment.

Agalynchnis calcarifer

42.6 Getting Sperm and Eggs Together
Fertilization in frogs is external, but amplexus—a behavior in which the male holds the female with his forelegs until she releases her egg mass—helps guarantee that sperm and eggs will get together.

Their solution, the shelled egg, is shared by birds (Figure 42.7). But the shelled egg created a new problem for fertilization: Sperm cannot penetrate the shell, so they have to reach the egg before the shell forms. Hence the need for internal fertilization and the evolution of the necessary accessory sex organs.

Male snakes and lizards have paired *hemipenes*, which can be filled with blood and thereby extruded from the male's cloaca to form intromittent organs. Only one hemi-

pene is inserted in the female's cloaca at a time. It is usually rough or spiny at the end to achieve a secure hold while sperm are transferred down a groove on its surface. Retractor muscles pull the hemipene back into the male's body when mating is completed. Birds have erectile penises that channel sperm along a groove into the female cloaca.

All mammals use internal fertilization, but except for the monotremes, they have done away with the shelled egg. They keep the developing embryo in the female reproductive tract, at least through the early stages of development. Mammalian species differ enormously as to the developmental stage of the offspring at the time of birth.

Reproductive systems are distinguished by where the embryo develops

Two patterns of care and nurture of the embryo have evolved in animals: oviparity (egg bearing) and viviparity (live bearing). **Oviparous** animals lay eggs in the environment, and their embryos develop outside the mother's body. Oviparity is possible because eggs are stocked with abundant nutrients to supply the needs of the embryo.

Oviparous terrestrial animals such as insects, reptiles, and birds protect their eggs from desiccation with tough, waterproof membranes or shells. However, these egg coverings must be permeable to oxygen and carbon dioxide.

(a) *Cheloria mydas*

42.7 The Shelled Egg
The shelled egg was a major evolutionary step that allowed reptiles and birds to reproduce in the terrestrial environment. (*a*) A female green sea turtle deposits her eggs in the sand. (*b*) Because the terrestrial environment offers no water to bring sperm and egg together, fertilization must take place internally, as with these penguins.

(b) *Aptenodytes patagonicus*

Some oviparous animals engage in various forms of parental behavior to protect their eggs, but until the eggs hatch, the embryos depend entirely on the nutrients stored in the egg. The only oviparous mammalian species are the monotremes: the echidnas and the duck-billed platypus (see Figure 33.22).

Viviparous animals retain the embryo within the mother's body during its early developmental stages. Most mammals are viviparous. There are examples of viviparity in all other vertebrate groups except the crocodiles, turtles, and birds. Even some sharks retain fertilized eggs in their bodies and give birth to free-living offspring. But there is a big difference between viviparity in mammals and in other species. Mammals (except monotremes) have a specialized portion of the female reproductive tract, the **uterus**, that holds the embryo and enables it to derive nutrients from and deliver wastes to the maternal blood. In contrast, non-mammalian viviparous animals simply retain the fertilized eggs in the mother's body until they hatch. The embryos still receive their nutrition from the stores in the egg, so this reproductive adaptation is called **ovoviviparity**.

Among mammals there are various degrees of uterine adaptation. In *marsupials*, such as kangaroos and koalas, the uterus simply holds the embryo and has a limited capability for exchanging nutrients and wastes. Marsupials are born at a very early developmental stage, crawl into a pouch called a *marsupium* on the mother's belly, attach to a nipple, and complete development outside of the mother's uterus (see Figure 33.23). Mammals other than monotremes and marsupials are called *eutherians*. They are characterized by an intimate association of the blood supplies of mother and embryo in the walls of the uterus. We will now look at the reproductive system of eutherians in greater depth, using *Homo sapiens* as our model.

The Human Reproductive System

So far we have seen a small sampling of the fascinating diversity of animal reproductive systems. In this section we describe the structures and functions of the male and female sex organs in eutherian mammals, specifically in human beings, and discuss hormonal regulation of both male and female systems.

Male sex organs produce and deliver semen

Semen is the product of the male reproductive system. Besides sperm, semen contains a complex mixture of fluids and molecules that support the sperm and facilitate fertilization. Sperm make up less than five percent of the volume of the semen.

Sperm are produced in the testes, the paired male gonads. In all mammals except bats, elephants, and marine mammals, the testes are located outside the body cavity in a pouch of skin, the **scrotum** (Figure 42.8). The optimal temperature for spermatogenesis in most mammals is slightly lower than the normal body temperature. The scrotum keeps the testes at this optimal temperature. Muscles in the scrotum contract in a cold environment, bringing the testes closer to the warmth of the body; in a hot environment they relax, suspending the testes farther from the body.

A testis consists of tightly coiled **seminiferous tubules** within which spermatogenesis takes place. Each tubule is lined with a stratified epithelium. Spermatogonia reside in the outer layers of this epithelium, and moving from these outer layers toward the lumen of the tubule, we find germ cells in successive stages of spermatogenesis (Figure 42.9). These germ cells are intimately associated with **Sertoli cells**, which protect them by providing a barrier between them and any noxious substances that might be circulating in the blood. Sertoli cells also provide nutrients for the developing sperm and are involved in the hormonal control

42.8 The Reproductive Tract of the Human Male Front and side views of the male reproductive organs.

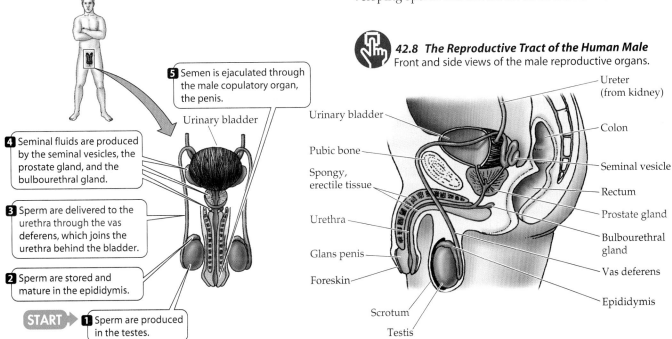

5 Semen is ejaculated through the male copulatory organ, the penis.

Urinary bladder

4 Seminal fluids are produced by the seminal vesicles, the prostate gland, and the bulbourethral gland.

3 Sperm are delivered to the urethra through the vas deferens, which joins the urethra behind the bladder.

2 Sperm are stored and mature in the epididymis.

START ▶ 1 Sperm are produced in the testes.

Urinary bladder
Pubic bone
Spongy, erectile tissue
Urethra
Glans penis
Foreskin
Scrotum
Testis

Ureter (from kidney)
Colon
Seminal vesicle
Rectum
Prostate gland
Bulbourethral gland
Vas deferens
Epididymis

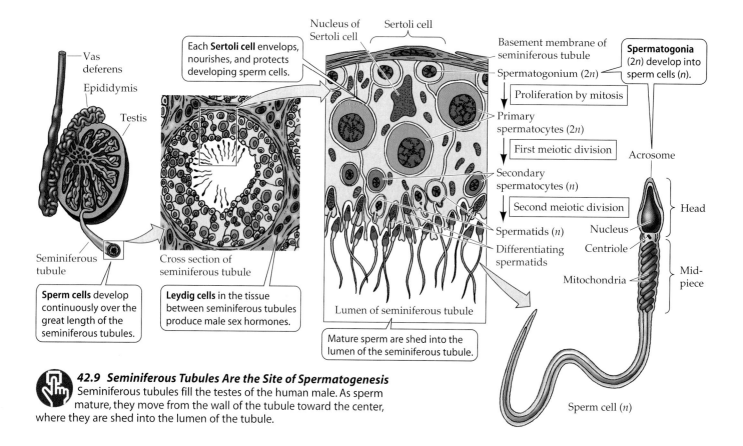

42.9 Seminiferous Tubules Are the Site of Spermatogenesis
Seminiferous tubules fill the testes of the human male. As sperm mature, they move from the wall of the tubule toward the center, where they are shed into the lumen of the tubule.

of spermatogenesis. Between the seminiferous tubules are clusters of **Leydig cells**, which produce male sex hormones.

With completion of the second meiotic division, each primary spermatocyte has given rise to four spermatids (see Figure 42.4a), which develop into sperm as they continue to migrate toward the lumen of the seminiferous tubule. The nucleus in what will become the head of the sperm becomes compact, and the surrounding cytoplasm is lost (see Figure 42.9). A flagellum, or tail, develops. The mitochondria, which will provide energy for tail motility, become condensed into a midpiece between the head and the tail. A cap, called an *acrosome*, forms over the nucleus in the head of the sperm. The acrosome contains enzymes that enable the sperm to digest a path through protective layers surrounding the egg. Fully differentiated sperm are shed into the lumen of the seminiferous tubule.

From the tubules, sperm move into a storage structure called the **epididymis**, where they mature and become motile. The epididymis connects to the **urethra** by a tube called the **vas deferens** (plural vasa deferentia). The urethra originates in the bladder, runs through the penis, and opens to the outside of the body at the tip of the penis. It serves as the common duct for the urinary and reproductive systems (see Figure 42.8).

The penis and the scrotum are the male genitalia. The shaft of the penis is covered with normal skin, but the tip, or **glans penis**, is covered with thinner, more sensitive skin that is especially responsive to sexual stimulation. A fold of skin called the *foreskin* covers the glans of the human penis.

The cultural practice of circumcision removes a portion of the foreskin.

Sexual arousal triggers responses in the the autonomic nervous system that result in the **erection** of the penis. The vessels carrying blood into the penis dilate, and this increased blood flow fills and swells shafts of spongy, erectile tissue located along the length of the penis. The enlargement of these blood-filled cavities compresses the vessels that normally carry blood out of the penis. As a result, the erectile tissue becomes more and more engorged with blood. The penis becomes hard and erect, facilitating its insertion into the female's vagina.*

The culmination of the male sex act propels semen through the vasa deferentia and the urethra in two steps, emission and ejaculation. During **emission**, rhythmic contractions of the smooth muscles of the ducts containing sperm and of the accessory glands move sperm and the various secretions into the urethra at the base of the penis. **Ejaculation**, which follows emission, is caused by contractions of other muscles at the base of the penis surrounding the urethra. The rigidity of the erect penis allows these contractions to force the gelatinous mass of semen through the urethra and out of the body.

Once a climax has been achieved, the autonomic nervous system switches signaling and causes the vessels leading

---

*In some species of mammals—but not humans—the penis contains a bone called the *baculum* or the *os penis*; however, even those species depend on erectile tissue for copulation.

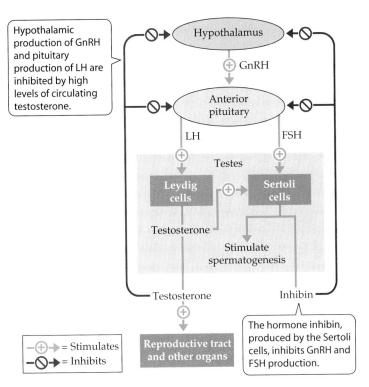

Hypothalamic production of GnRH and pituitary production of LH are inhibited by high levels of circulating testosterone.

The hormone inhibin, produced by the Sertoli cells, inhibits GnRH and FSH production.

= Stimulates
= Inhibits

42.10 Hormones Control the Male Reproductive System
The male reproductive system is under hormonal control by the hypothalamus and the anterior pituitary.

Male sexual function is controlled by hormones

Spermatogenesis and maintenance of male secondary sexual characteristics depend on testosterone, which is produced by Leydig cells in the testes. In Chapter 41 we learned that increased production of testosterone at puberty is due to an increased release of gonadotropin-releasing hormone (GnRH) by the hypothalamus, which in turn stimulates cells in the anterior pituitary to increase their secretion of luteinizing hormone (LH) and follicle-stimulating hormone (FSH) (Figure 42.10). Negative feedback loops help to regulate testis functions.

The Leydig cells are stimulated by LH to produce testosterone. The rise in the level of testosterone in the prepubertal male causes the development of secondary sexual characteristics and the pubertal growth spurt, promotes increased muscle mass, and stimulates growth and maturation of the testes. If a male is castrated (has his testes removed) before puberty, he will not develop a deep voice, typical patterns of body hair, or a muscular build, and his external genitalia will remain childlike. Continued production of testosterone after puberty is essential to maintain secondary sexual characteristics and to produce sperm. Spermatogenesis is controlled by the influence of FSH and testosterone on Sertoli cells in the seminiferous tubules.

Female sex organs produce eggs, receive sperm, and nurture the embryo

When an egg matures, it is released from the ovary directly into the body cavity. But the egg can't go far. Each ovary is enveloped by the undulating, fringed opening of an **oviduct** (also known as a *fallopian tube*), which sweeps the egg into the tube (Figure 42.11). Cilia lining the oviduct propel the egg slowly toward the uterus, or *womb*, which is a muscular, thick-walled cavity shaped like an upside-down pear. The uterus is where the embryo develops if the egg is fertilized. At the bottom of the uterus is an opening called the **cervix**, which leads into the vagina. Sperm are ejaculated into the vagina during copulation, and the fetus passes through the vagina during birth.

Two sets of skin folds surround the opening of the vagina and the opening of the urethra, through which urine passes. The inner, more delicate folds are the **labia minora** (singular labium minus); the outer, thicker folds are the **labia majora** (singular labium majus). At the anterior tip of the labia minora is the **clitoris**, a small bulb of erectile tissue that is the anatomical homolog of the penis. The clitoris is highly sensitive and plays an important role in sexual response. The labia minora and the clitoris become engorged with blood in response to sexual stimulation.

The opening of an infant female's vagina is partly covered by a thin membrane, the *hymen*. Eventually the hymen becomes ruptured by vigorous physical activity or first sexual intercourse; it can sometimes make first intercourse difficult or painful for the female.

To fertilize an egg, sperm swim and are propelled by contractions of the female reproductive tract up from the vagina, through the cervix, the uterus, and most of the

into the penis to constrict. The resulting decrease in blood pressure in the erectile tissue relieves the compression of the blood vessels leaving the penis, and the erection declines.

The components of the semen other than sperm come from several accessory glands that contribute secretions to the urethra. A relatively small volume of fluid comes from the **bulbourethral glands**. This alkaline and mucoid secretion precedes other secretions; it neutralizes acidity in the urethra and lubricates the tip of the penis. About two-thirds of the volume of semen is seminal fluid, which comes from the **seminal vesicles**. Seminal fluid is thick because it contains mucus and protein. It also contains fructose, an energy source for the sperm, which are too small to carry much of their own fuel. Semen also carries a message for the female reproductive tract in the form of chemicals called **prostaglandins**. Prostaglandins stimulate rhythmic contractions in the female reproductive tract that help move the sperm up into the regions where fertilization can take place.

One-fourth to one third of the volume of semen is a thin, milky fluid that comes from the **prostate gland**. Prostate fluid makes the uterine environment more hospitable to sperm. The prostate also secretes a clotting enzyme that works on the protein in seminal fluid to convert semen into a gelatinous mass. The prostate gland completely surrounds the urethra as it leaves the bladder. This gland tends to enlarge in men over 40 years of age, creating a condition known as *benign prostate hyperplasia* (*BPH*). A seriously enlarged prostate can block the urethra and make urination difficult. Unrelated to BPH, prostate cancer is the second most common cancer in men. It is relatively easy to diagnose, however, and is highly curable if detected early.

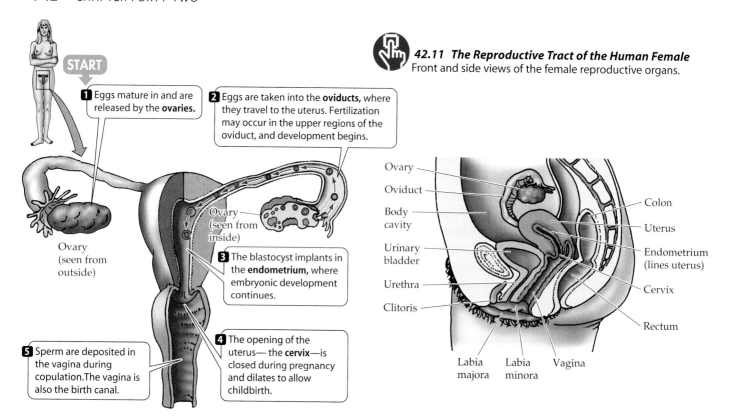

START

1 Eggs mature in and are released by the **ovaries.**

2 Eggs are taken into the **oviducts,** where they travel to the uterus. Fertilization may occur in the upper regions of the oviduct, and development begins.

Ovary (seen from inside)

3 The blastocyst implants in the **endometrium,** where embryonic development continues.

Ovary (seen from outside)

5 Sperm are deposited in the vagina during copulation. The vagina is also the birth canal.

4 The opening of the uterus— the **cervix**—is closed during pregnancy and dilates to allow childbirth.

42.11 The Reproductive Tract of the Human Female
Front and side views of the female reproductive organs.

Ovary · Oviduct · Body cavity · Urinary bladder · Urethra · Clitoris · Labia majora · Labia minora · Vagina · Colon · Uterus · Endometrium (lines uterus) · Cervix · Rectum

---

oviduct. The egg (actually a secondary oocyte in humans; see Figure 42.4*b*) is fertilized in the upper region of the oviduct. Fertilization stimulates the completion of the second meiotic division, after which the haploid nuclei of the sperm and the egg can fuse to produce a diploid zygote nucleus. Still in the oviduct, the zygote undergoes its first few cell divisions to become a **blastocyst.** The blastocyst moves down the oviduct to the uterus, where it attaches itself to the epithelial lining, the **endometrium.** The endometrium and the cells of the uterine wall are stimulated by estrogen to proliferate and grow many new blood vessels in anticipation of receiving a blastocyst.

Once attached to the endometrium, the blastocyst burrows into it, a process called **implantation,** and forms a structure called the **placenta.** The placenta exchanges nutrients and waste products between the mother's blood and the baby's blood. If a blastocyst does not arrive in the uterus, the endometrium regresses or is sloughed off. Thus the female reproductive cycle actually consists of two linked cycles: an ovarian cycle that produces eggs and hormones, and a uterine cycle that creates an appropriate environment for the embryo should fertilization occur.

The ovarian cycle produces a mature egg

An **ovarian cycle** is about 28 days long in the human female,* but there is considerable variation among individu-

---

*Some mammals have ovarian cycles shorter than 28 days, others have longer ones. Rats and mice have ovarian cycles of about 4 days; many seasonally breeding mammals have only one ovarian cycle per year.

als. During the first half of the cycle, at least one primary oocyte matures into a secondary oocyte (egg) and is expelled from the ovary. During the second half of the cycle, cells in the ovary that were associated with the maturing oocyte develop endocrine functions and then regress if the egg is not fertilized. The progression of these events is shown diagrammatically in Figure 42.12.

At birth, a human female has about a million primary oocytes in each ovary. By the time she reaches sexual maturity, she has only about 200,000; the rest have degenerated. During a woman's fertile years, her ovaries will go through about 450 ovarian cycles, and during each of these cycles one oocyte will mature and be released. At about 50 years of age, she reaches **menopause,** the end of fertility, and may have only a few oocytes left in each ovary. Throughout a woman's life, oocytes are degenerating, and no new ones are produced.

Each primary oocyte in the ovary is surrounded by a layer of follicle cells. An oocyte and its follicle cells constitute the functional unit of the ovary, the **follicle.** Between puberty and menopause, six to twelve follicles begin to mature each month. In each of these follicles, the oocyte enlarges and the surrounding cells proliferate. After about a week, one of these follicles is larger than the rest and continues to grow, while the others cease to develop and shrink. In the enlarged follicle, the follicle cells nurture the growing egg, supplying it with nutrients and with macromolecules and proteins that it will use in early stages of development if it is fertilized.

After 2 weeks of follicular growth, **ovulation** occurs— the follicle ruptures, and the egg is released. Following ovulation, the follicle cells continue to proliferate and form a mass of endocrine tissue about the size of a marble. This

42.12 The Ovarian Cycle
The ovarian cycle progresses from the development of a follicle to ovulation and finally to growth and degeneration of the corpus luteum. The micrograph shows a mature mammalian follicle; the oocyte is in the center.

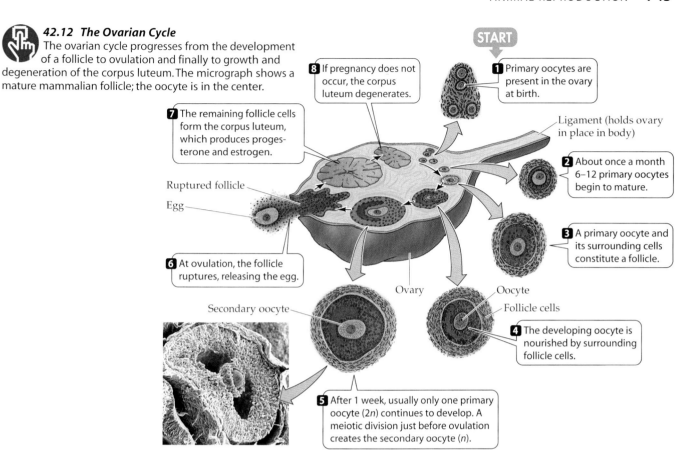

START

1 Primary oocytes are present in the ovary at birth.

8 If pregnancy does not occur, the corpus luteum degenerates.

7 The remaining follicle cells form the corpus luteum, which produces progesterone and estrogen.

Ligament (holds ovary in place in body)

2 About once a month 6–12 primary oocytes begin to mature.

Ruptured follicle

Egg

3 A primary oocyte and its surrounding cells constitute a follicle.

6 At ovulation, the follicle ruptures, releasing the egg.

Ovary

Oocyte

Follicle cells

Secondary oocyte

4 The developing oocyte is nourished by surrounding follicle cells.

5 After 1 week, usually only one primary oocyte (2*n*) continues to develop. A meiotic division just before ovulation creates the secondary oocyte (*n*).

structure, which remains in the ovary, is the **corpus luteum** (plural corpora lutea). It functions as an endocrine gland, producing estrogen and progesterone for about 2 weeks. It then degenerates unless the egg is fertilized.

The uterine cycle prepares an environment for the fertilized egg

The **uterine cycle** of human females parallels the ovarian cycle, and consists first of a buildup and then of a breakdown of the endometrium, or uterine lining (Figure 43.13). About 5 days into the ovarian cycle, the endometrium starts to grow in preparation for receiving a blastocyst. The uterus attains its maximum state of preparedness about 5 days after ovulation and remains in that state for another 9 days. If a blastocyst has not arrived by that time, the endometrium begins to break down, slough off, and flow from the body through the vagina—the process of **menstruation** (from *menses*, the Latin word for "months").

The uterine cycles of mammals other than humans do not include menstruation; instead, the uterine lining is resorbed. In these species the most obvious correlate of the ovarian cycle is a state of sexual receptivity called **estrus** around the time of ovulation. When the female comes into estrus, or "heat," she actively solicits male attention and may be aggressive to other females. The human female is unusual among mammals in that she is potentially sexually receptive throughout her ovarian cycle and at all seasons of the year.

Hormones control and coordinate the ovarian and uterine cycles

The ovarian and uterine cycles of human females are coordinated and timed by the same hormones that initiate sexual maturation. Gonadotropins secreted by the anterior pituitary are the central elements of this control. Before puberty (that is, before about 11 years of age), the secretion of gonadotropins is low, and the ovaries are inactive. At puberty, the hypothalamus increases its release of gonadotropin-releasing hormone (GnRH), thus stimulating the anterior pituitary to secrete follicle-stimulating hormone (FSH) and luteinizing hormone (LH).

In response to FSH and LH, ovarian tissue grows and produces estrogen. The rise in estrogen causes the development of female secondary sexual characteristics, including growth of the uterus. Between puberty and menopause, interactions of gonadotropin-releasing hormone, gonadotropins, and sex steroids control the ovarian and uterine cycles.

Menstruation marks the beginning of the uterine and ovarian cycles (see Figure 42.13). A few days before menstruation begins, the anterior pituitary begins to increase its secretion of FSH and LH. In response, some follicles begin to mature in the ovaries, and follicle cells gradually increase production of estrogen. After about a week of growth, usually all but one of these follicles wither away. Occasionally more than one follicle continues to develop, making it possible for the woman to bear fraternal (nonidentical) twins.

1 The posterior epiblast thickens.

2 Cells move toward the primitive streak, down through it, and forward.

3 The primitive streak narrows and lengthens…

4 …forming the primitive groove—the chick's blastopore.

5 Cells passing over Hensen's node form head structures and notochord.

Anterior

Embryo

Yolk

Posterior

Chick embryo viewed from above

Primitive streak

Hensen's node

Hensen's node

43.16 Gastrulation in Birds
Because of the large yolk mass in bird and reptile eggs, these embryos display a pattern of gastrulation very different from that of sea urchins and amphibians.

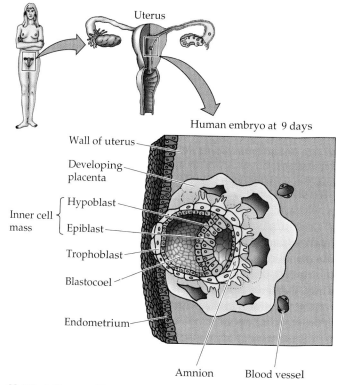

Primitive streak

Epiblast

Blastocoel

Yolk

Hypoblast

These cells will form mesoderm.

These will become endoderm.

Cross section through chick embryo

Gastrulation begins with a thickening of a posterior region of the epiblast caused by the movement of cells toward the midline and then forward along the midline (Figure 43.16). The result is a midline ridge called the **primitive streak**. A depression called the **primitive groove** forms along the length of the primitive streak. The primitive groove becomes the blastopore as cells migrate through it into the blastocoel to become endoderm and mesoderm.

In the avian embryo, no archenteron forms, but the prospective endoderm and mesoderm migrate forward to form gut and other structures. At the extreme forward end of the primitive groove is a thickening called **Hensen's node**, which is the equivalent of the dorsal lip of the amphibian blastopore. In fact, many signaling molecules that have been identified in the frog organizer are also expressed in Hensen's node. Cells passing over Hensen's node become determined to differentiate into the tissues and structures that make up the head and the dorsal midline of the embryo.

Mammals have no yolk, but retain the avian–reptilian gastrulation pattern

Mammals and birds both evolved from reptilian ancestors, so it is not surprising that they share patterns of early development, even though the eggs of placental mammals have no yolk. Earlier we described the development of the mammalian trophoblast and the inner cell mass, which is the equivalent of the avian epiblast. Keeping avian gastrulation in mind, think of the mammalian inner cell mass as sitting on top of an imaginary body of yolk (Figure 43.17).

As in avian development, the inner cell mass splits into an upper layer called the epiblast and a lower layer called the hypoblast with a fluid-filled cavity, or blastocoel, between them. The embryo will form from the epiblast, and the hypoblast will contribute to the extraembryonic membranes. The epiblast also contributes to the extraembryonic

Uterus

Human embryo at 9 days

Wall of uterus

Developing placenta

Inner cell mass {
Hypoblast
Epiblast
}

Trophoblast

Blastocoel

Endometrium

Amnion

Blood vessel

43.17 A Human Blastocyst before Gastrulation
The mammalian inner cell mass becomes the epiblast; it can be compared to the avian embryo by picturing it sitting on top of a mass of imaginary yolk.

membranes; specifically, it splits off an upper layer of cells that will form the *amnion*. The amnion will grow to surround the developing embryo as a sac filled with amniotic fluid.

Gastrulation occurs in the mammalian epiblast just as it does in the avian epiblast. A primitive groove forms, and epiblast cells migrate through the groove to become layers of endoderm and mesoderm. The cells migrating over Hensen's node become the cells that form dorsal structures such as the brain and spinal cord.

Neurulation: Initiating the Nervous System

Gastrulation produces an embryo with three germ layers that are positioned to influence one another through inductive tissue interactions. During the next phase of development, called **organogenesis**, many organs and organ systems develop simultaneously and in coordination with one another. An early process of organogenesis that is directly related to gastrulation is **neurulation**, the initiation of the nervous system. We will examine this event in the amphibian embryo, but it occurs in a similar fashion in reptiles, birds, and mammals. Many of the genes involved are highly conserved all the way from worms to humans.

The stage is set by the dorsal lip of the blastopore

The cells that pass through the dorsal lip of the blastopore and move anteriorly in the blastocoel during gastrulation are determined to become mesoderm. The dorsal mesoderm closest to the midline is further determined to become **chordomesoderm**, which forms a rod along the dorsal midline. This rod, called the **notochord**, gives structural support to the developing embryo. The notochord eventually will be replaced by the vertebral column, but after gastrulation it induces the overlying ectoderm to begin forming the nervous system. Neurulation involves the formation of an internal tube from an external sheet of cells.

The first signs of neurulation are flattening and thickening of the ectoderm overlying the notochord; this thickened area forms the **neural plate** (Figure 43.18). The edges of the neural plate that run in an anterior–posterior direction continue to thicken to form ridges or folds. Between the folds a groove forms and deepens as the folds roll over it to converge on the midline. The folds fuse, forming a cylinder, the **neural tube**, and a continuous overlying layer of epidermal

43.18 Neurulation in the Frog
Continuing the sequence from Figures 43.9 and 43.13, these drawings outline the development of the frog's neural tube. The midsagittal section (*b*) shows the development of the notochord and its position relative to the neural tube.

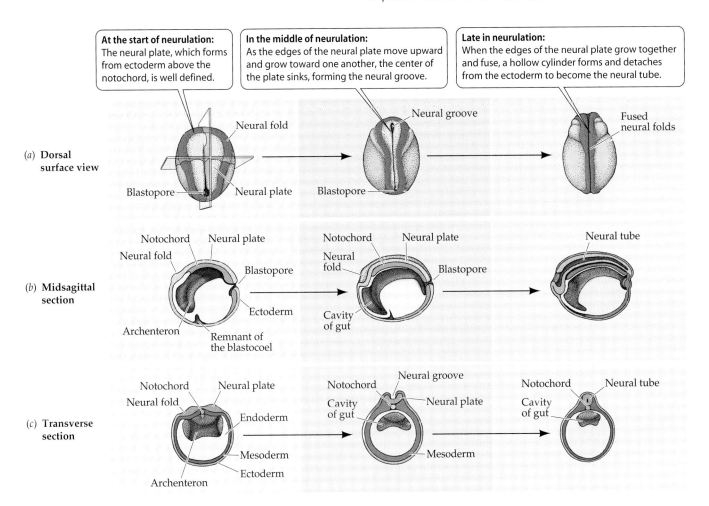

At the start of neurulation:
The neural plate, which forms from ectoderm above the notochord, is well defined.

In the middle of neurulation:
As the edges of the neural plate move upward and grow toward one another, the center of the plate sinks, forming the neural groove.

Late in neurulation:
When the edges of the neural plate grow together and fuse, a hollow cylinder forms and detaches from the ectoderm to become the neural tube.

(a) Dorsal surface view

Neural fold
Blastopore
Neural plate
Neural groove
Blastopore
Fused neural folds

(b) Midsagittal section

Notochord
Neural plate
Neural fold
Blastopore
Ectoderm
Archenteron
Remnant of the blastocoel
Notochord
Neural plate
Neural fold
Blastopore
Cavity of gut
Neural tube

(c) Transverse section

Notochord
Neural plate
Neural fold
Endoderm
Mesoderm
Ectoderm
Archenteron
Notochord
Neural groove
Neural plate
Cavity of gut
Mesoderm
Notochord
Neural tube
Cavity of gut

ectoderm. The neural tube develops bulges at the anterior end, which become the major divisions of the brain; the rest of the tube becomes the spinal cord.

Failure of the neural tube to develop normally can result in serious birth defects. If the neural tube fails to close in a posterior region, the result is a condition known as *spina bifida*. If it fails to close at the anterior end, an infant can develop without a forebrain—a condition called *anencephaly*. Whereas several genetic factors have been identified that can cause neural tube defects, there are also environmental factors, including dietary ones. The incidence of neural tube defects used to be about 1 in 300 live births, but we now know that this incidence can be cut in half if pregnant women have an adequate amount of folic acid (a B vitamin) in their diets.

Body segmentation develops during neurulation

Like the fruit flies whose development we traced in Chapter 16, vertebrates have a body plan consisting of repeating segments that are modified during development. These segments are most evident as the repeating patterns of vertebrae, ribs, nerves, and muscles along the anterior–posterior axis.

As the neural tube forms, mesodermal tissues gather along the sides of the notochord to form separate blocks of cells called **somites** (Figure 43.19). The somites produce cells that will become vertebrae, ribs, and muscles of the trunk and limbs.

The nerves that connect the brain and spinal cord with tissues and organs throughout the body are also arranged segmentally. The somites help guide the organization of these peripheral nerves, but the nerves are not of mesodermal origin. When the neural tube closes, cells adjacent to the line of closure break loose and migrate inward between the epidermis and the somites and under the somites. These cells, called **neural crest cells**, give rise to a number of structures, including the peripheral nerves, which grow out to the body tissues and back into the spinal cord.

As development progresses, the segments of the body become different. Regions of the spinal cord differ, regions of the vertebral column differ in that some vertebrae grow ribs of various sizes and others do not, forelegs arise in the anterior part of the embryo, and hind legs arise in the posterior region. How does a somite in the anterior part of a mouse embryo "know" to produce forelegs rather than hind legs?

Central to the process of anterior–posterior determination and differentiation are homeotic genes (see Chapter 16). We have seen how these genes control body segmentation in *Drosophila*. In the mouse, four families of similar genes, called homeobox or Hox genes, control differentiation along the anterior–posterior body axis.

Each family of mammalian Hox genes resides on a different chromosome and consists of about 10 genes. What is remarkable is that the temporal and spatial expression of these genes follows the same pattern as their linear order on their chromosomes. As a result, different segments of

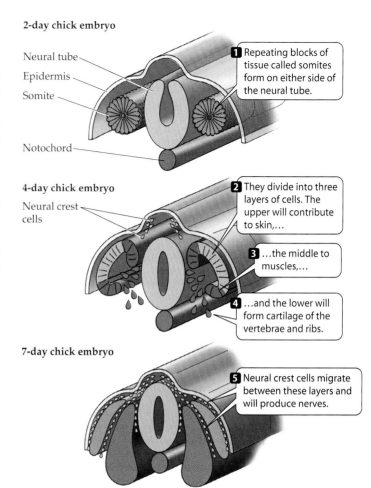

2-day chick embryo

Neural tube
Epidermis
Somite
Notochord

1 Repeating blocks of tissue called somites form on either side of the neural tube.

4-day chick embryo

Neural crest cells

2 They divide into three layers of cells. The upper will contribute to skin,…

3 …the middle to muscles,…

4 …and the lower will form cartilage of the vertebrae and ribs.

7-day chick embryo

5 Neural crest cells migrate between these layers and will produce nerves.

43.19 The Development of Body Segmentation
Repeating blocks of tissue called somites form on either side of the neural tube. Skin, muscle, and bone form from the somites.

the embryo receive different combinations of Hox gene products, which serve as transcription factors (Figure 43.20). What causes the linear, sequential expression of Hox genes is unclear.

Whereas Hox genes give cells information about their positional location on the anterior–posterior body axis, other genes give information about dorsal–ventral position. Tissues in each segment of the body differentiate according to their dorsal–ventral location. In the spinal cord, for example, sensory connections develop in the dorsal region and motor connections develop in the ventral region. In the somites, dorsal cells develop into skin and muscle and ventral cells develop into cartilage and bone.

An example of a gene that provides dorsal–ventral information in vertebrates is *sonic hedgehog*, which is expressed in the mammalian notochord and induces cells in the overlying neural tube to follow fates characteristic of ventral spinal cord cells. (As with the Hox genes, *sonic hedgehog* is homologous to a *Drosophila* gene, which is known simply as *hedgehog*.)

A family of homeobox genes, the Pax genes, play many roles in nervous system and somite development. One of these genes, *Pax3*, is expressed in neural tube cells that develop into dorsal spinal cord structures. *Sonic hedgehog* represses the expression of *Pax3*, and their interaction is one

(a)

1 Hox genes are clustered on four chromosomes.

	a1	a2	a3	a4	a5	a6	a7	a9	a10	a11	a13

Hoxa genes

	b1	b2	b3	b4	b5	b6	b7	b8	b9

Hoxb genes

	c4	c5	c6	c8	c9	c10	c11	c12	c13

Hoxc genes

	d1	d3	d4	d8	d9	d10	d11	d12	d13

Hoxd genes

3′ —— Hindbrain ————————— Trunk —— 5′

2 The genes closest to the 3′ end are expressed in the anteriormost positions…

3 …and those closest to the 5′ end are expressed more posteriorly.

43.20 Hox Genes Control Body Segmentation
The expression of Hox genes is patterned along the anterior–posterior axis of the embryo and from the 3′ to the 5′ ends of the chromosomes.

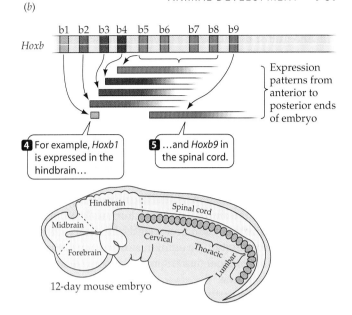

(b)

Hoxb b1 b2 b3 b4 b5 b6 b7 b8 b9

Expression patterns from anterior to posterior ends of embryo

4 For example, *Hoxb1* is expressed in the hindbrain…

5 …and *Hoxb9* in the spinal cord.

12-day mouse embryo

source of dorsal–ventral information for differentiation of the spinal cord.

With the development of body segmentation, the formation of organs and organ systems progresses rapidly. The development of an organ involves extensive inductive tissue interactions, which are a current focus of study for developmental biologists. In Chapter 16, you encountered the organogenesis of the vertebrate limb. In the next chapter you will learn about the development of the brain—another example of organogenesis.

Extraembryonic Membranes

The embryos of reptiles, birds, and mammals are surrounded by several **extraembryonic membranes**, which originate from the embryo but are not part of it. The ex-

traembryonic membranes play important roles in development, especially in mammals, in which they constitute the placenta that nourishes the embryo.

Four extraembryonic membranes form with contributions from all germ layers

We will use the chick to demonstrate how the extraembryonic membranes form from the germ layers created during gastrulation. The **yolk sac** is the first extraembryonic membrane to form, and it does so through extensions of the endodermal tissue of the hypoblast layer, which enclose the entire body of yolk in the egg (Figure 43.21, left). This yolk sac constricts at the top to create a tube that is continuous

43.21 The Extraembryonic Membranes
In birds, reptiles, and mammals, the embryo constructs four extraembryonic membranes. The yolk sac encloses the yolk, and the amnion and chorion enclose the embryo. Fluids secreted by the amnion fill the amniotic cavity, providing an aqueous environment for the embryo. The chorion, along with the allantois, mediates gas exchange between the embryo and its environment. The allantois stores the embryo's waste products.

5-Day embryo

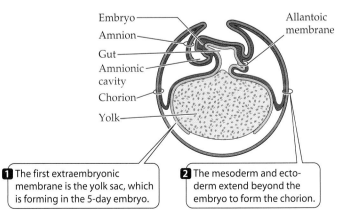

Embryo
Amnion
Gut
Amnionic cavity
Chorion
Yolk
Allantoic membrane

1 The first extraembryonic membrane is the yolk sac, which is forming in the 5-day embryo.

2 The mesoderm and ectoderm extend beyond the embryo to form the chorion.

9-Day embryo

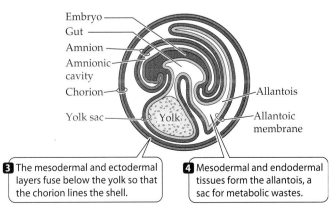

Embryo
Gut
Amnion
Amnionic cavity
Chorion
Yolk sac
Yolk
Allantois
Allantoic membrane

3 The mesodermal and ectodermal layers fuse below the yolk so that the chorion lines the shell.

4 Mesodermal and endodermal tissues form the allantois, a sac for metabolic wastes.

Thousands of synapses impinge on most individual neurons. A neuron generally receives information (synaptic inputs) from many sources before producing nerve impulses that travel down its single axon to target cells. We will discuss synaptic transmission in more detail later in the chapter.

Glial cells are also important components of nervous systems

Neurons are not the only type of cell in the nervous system. In fact, there are more **glial cells** than neurons in the human brain. Like neurons, glial cells come in several forms and have a diversity of functions. Some glial cells physically support and orient the neurons and help them make the right contacts during their embryonic development. Other glial cells insulate axons.

In the peripheral nervous system, **Schwann cells** wrap around the axons of neurons, covering them with concentric layers of insulating plasma membrane (Figure 44.4). Other glial cells called **oligodendrocytes** perform a similar function in the central nervous system. The covering produced by Schwann cells and oligodendrocytes, called **myelin**, gives many parts of the nervous system a glistening white appearance. Later in the chapter we will see how the electrical insulation provided by myelin increases the speed with which axons can conduct nerve impulses.

Glial cells are well known for the many housekeeping functions they perform. Some glial cells supply neurons with nutrients; others consume foreign particles and cell debris. Glial cells also help maintain the proper ionic environment around neurons. Although they have no axons and do not generate or conduct nerve impulses, some glial cells communicate with one another electrically through a special type of contact called a gap junction, a connection that enables ions to flow between cells (see Chapter 5).

Glial cells called **astrocytes** (because they look like stars) contribute to the **blood–brain barrier**, which protects the brain from toxic chemicals in the blood. Blood vessels throughout the body are very permeable to many chemicals, including toxic ones, which would reach the brain if this special barrier did not exist. Astrocytes help form this barrier by surrounding the smallest, most permeable blood vessels in the brain.

Protection of the brain is crucial because, unlike other tissues of the body, the brain has a limited capacity to recover from damage by generating new neurons and new neuronal connections. Throughout life, neurons are progressively lost. Without the blood–brain barrier, the rate of neuron loss could be much greater. However, the barrier is not perfect. Since it consists of plasma membranes, it is permeable to fat-soluble substances. Anesthetics and alcohol, both of which have well-known effects on the brain, are fat-soluble chemicals.

Neurons function in networks

As we learn more about the properties of neurons, it is important to keep in mind that nervous systems depend on

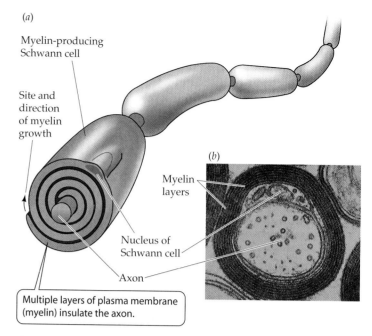

44.4 Wrapping Up an Axon
(a) Schwann cells wrap axons with layers of myelin, a type of plasma membrane that provides electrical insulation. (b) A group of myelinated axons, seen in cross section through an electron microscope.

neurons working together. The simplest *neuronal network* consists of three cells: a sensory neuron connected to a motor neuron connected to a muscle cell. Most of the neuronal networks that carry out the functions of the human nervous system are much more complex and consist of many more neurons. The human brain contains between 10^9 and 10^{11} neurons. Most of these neurons receive information from a thousand or more synapses. Thus there may be as many as 10^{14} synapses in the human brain. Therein lies the incredible ability of the brain to process information.

This astronomical number of neurons and synapses is divided into thousands of distinct but interacting networks that function in parallel and accomplish the many different tasks of the nervous system. But before we can understand how even one of these circuits works, we must understand the properties of individual neurons.

Neurons: Generating and Conducting Nerve Impulses

In this section, we explore the electrical properties of cells. After reviewing some basic electrical concepts, we examine in detail the roles of ions, ion pumps, and ion channels in establishing and altering the electrical properties of neurons. These electrical changes generate action potentials, the language by which the nervous system processes and communicates information.

The insides of cells are electrically negative in comparison to the outsides. The difference in voltage across the plasma membrane of a neuron is called its **membrane potential**. In an *unstimulated* neuron, this voltage difference is called a **resting potential**.

Membrane potentials can be measured with *electrodes*. An electrode can be made from a glass pipette pulled to a very sharp tip and filled with a solution containing ions that conduct electric charge. Using such electrodes, we can record very tiny local electrical events that occur across plasma membranes. If a pair of electrodes is placed one on each side of the plasma membrane of a resting axon, they measure a voltage difference of about –60 millivolts (mV)—the resting potential (Figure 44.5).

The resting potential provides a means for neurons to respond to a stimulus. A neuron is sensitive to any chemical or physical factor that causes a change in the resting potential across a portion of its plasma membrane. The most extreme change in membrane potential is a nerve impulse, which is a sudden and rapid reversal in the voltage across a portion of the plasma membrane. For a brief moment—1 or 2 milliseconds—the inside of a part of the cell becomes *more positive* than the outside. Nerve impulses are also called **action potentials**, a name that conveys their contrast with the resting potential. An action potential can travel along the plasma membrane from one part of a neuron to its farthest extensions.

Simple electrical concepts underlie neuronal functions

To understand how resting potentials are created, how they are perturbed, and how action potentials are generated and conducted along plasma membranes, it is necessary to know a little about electricity, ions, and the special ion channel proteins in the plasma membranes of neurons.

Voltage (potential) is the tendency for electrically charged particles like electrons or ions to move between two points. Voltage is to the flow of electrically charged particles what pressure is to the flow of water. If the negative and the positive poles of a battery are connected by a copper wire, electrons flow from negative to positive because there is a voltage difference between them. This flow of electrons is an electric current, and it can be used to do work, just as a current of water can be used to do work such as turning a turbine.

Electric charges move across cell membranes not as electrons, but as charged ions. The major ions that carry electric charges across the plasma membranes of neurons are sodium (Na^+), chloride (Cl^-), potassium (K^+), and calcium (Ca^{2+}). It is also important to remember that ions with opposite charges attract each other. With these basics of bioelectricity in mind, we can ask how the resting potential of the neuronal plasma membrane is created, and how the flow of ions through membrane channels is turned on and off to generate action potentials.

Ion pumps and channels generate resting and action potentials

The plasma membranes of neurons, like those of all other cells, are lipid bilayers that are impermeable to ions. However, these impermeable lipid bilayers contain many protein molecules that serve as ion channels and ion pumps (see Chapter 5). Ion pumps and channels are responsible for resting and action potentials.

44.5 Measuring the Resting Potential
The difference in electric charge across the plasma membrane of a neuron can be measured using two electrodes, one inside and one outside the cell. In an unstimulated neuron, this difference is constant (about –60 mV), and is known as the resting potential.

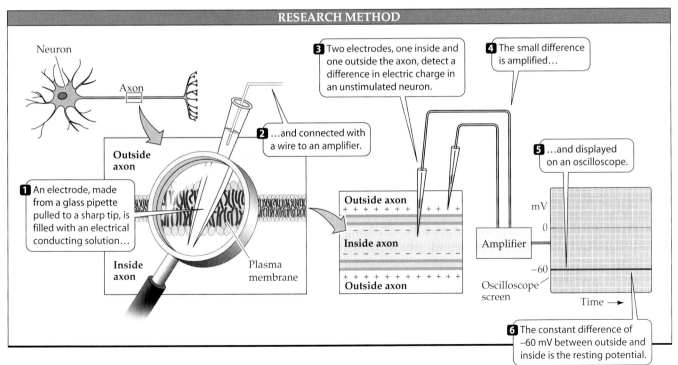

RESEARCH METHOD

Neuron

Axon

Outside axon

Inside axon

Plasma membrane

1 An electrode, made from a glass pipette pulled to a sharp tip, is filled with an electrical conducting solution…

2 …and connected with a wire to an amplifier.

3 Two electrodes, one inside and one outside the axon, detect a difference in electric charge in an unstimulated neuron.

4 The small difference is amplified…

5 …and displayed on an oscilloscope.

Outside axon
+ + + + + + + + +
Inside axon
– – – – – – – – –
+ + + + + + + + + + +
Outside axon

Amplifier

mV
0

–60

Oscilloscope screen

Time →

6 The constant difference of –60 mV between outside and inside is the resting potential.

Glutamate is an excitatory neurotransmitter, so activation of all glutamate receptors results in sodium entry into the neuron and depolarization. But the *timing* of the response to activation of these different types of receptors differs significantly. The AMPA receptors, for example, allow a rapid influx of Na^+ into the postsynaptic cell. The NMDA receptors allow a slower and longer-lasting influx of Na^+. The NMDA receptors also require the cell to be somewhat depolarized through the action of other receptors before they will open channels and permit Na^+ influx. When they do respond, the channels they open also allow Ca^{2+} to enter the cell. Ca^{2+} ions act as second messengers in the cell and can trigger a variety of long-term cellular changes.

Figure 44.19 shows how the AMPA and NMDA receptors can work in concert. The critical difference is that at normal resting potential, the NMDA channel is blocked by a magnesium ion (Mg^{2+}). Slight depolarization of the neuron due to other inputs displaces Mg^{2+} from the NMDA channels and allows Na^+ and Ca^{2+} to pass through the channels when they are activated by glutamate. These special properties of the NMDA receptor are probably involved in learning and memory.

Most of the synaptic events we have studied so far happen very quickly. It is therefore a special challenge to understand how the messages carried by action potentials can result in long-term events such as learning and memory. Our understanding of these processes has been greatly af-

44.1 Some Well-Known Neurotransmitters

NEUROTRANSMITTER	ACTIONS	COMMENTS
Acetylcholine	The neurotransmitter of vertebrate motor neurons and of some neural pathways in the brain	Broken down in the synapse by acetylcholinesterase; blockers of this enzyme are powerful poisons
Monoamines		
Norepinephrine	Used in certain neural pathways in the brain. Also found in the peripheral nervous system, where it causes gut muscles to relax and the heart to beat faster	Related to epinephrine and acts at some of the same receptors
Dopamine	A neurotransmitter of the central nervous system	Involved in schizophrenia. Loss of dopamine neurons is the cause of Parkinson's disease
Histamine	A minor neurotransmitter in the brain	Thought to be involved in maintaining wakefulness
Serotonin	A neurotransmitter of the central nervous system that is involved in many systems, including pain control, sleep/wake control, and mood	Certain medications that elevate mood and counter anxiety act by increasing serotonin levels
Purines		
ATP	Co-released with many neurotransmitters	Large family of receptors may shape postsynaptic responses to classical neurotransmitters
Adenosine	Transported across cell membranes; not synaptically released	Largely inhibitory effects on postsynaptic cells
Amino acids		
Glutamate	The most common excitatory neurotransmitter in the central nervous system	Some people have reactions to the food additive monosodium glutamate because it can affect the nervous system
Glycine Gamma-aminobutyric acid (GABA)	Common inhibitory neurotransmitters	Drugs called benzodiazepines, used to reduce anxiety and produce sedation, mimic the actions of GABA
Peptides		
Endorphins Enkephalins Substance P	Used by certain sensory nerves, especially in pain pathways	Receptors are activated by narcotic drugs: opium, morphine, heroin, codeine
Gas		
Nitric oxide	Widely distributed in the nervous system	Not a classic neurotransmitter, it diffuses across membranes rather than being released synaptically. A means whereby a postsynaptic cell can influence a presynaptic cell

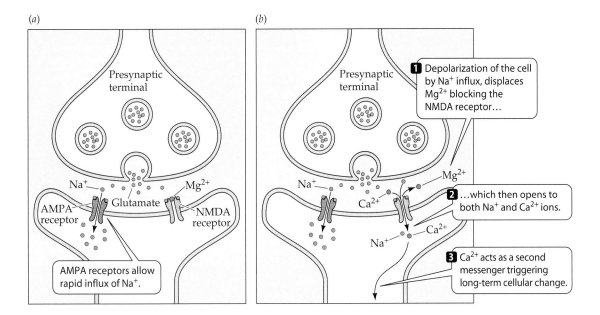

44.19 Two Ionotropic Glutamate Receptors
(*a*) AMPA receptors allow rapid influx of Na$^+$ into the postsynaptic cell. (*b*) NMDA receptors allow both Na$^+$ and Ca^{2+} to enter the cell.

fected by a phenomenon called **long-term potentiation**, or **LTP**, that was discovered by neurobiologists working with slices of brain that they kept alive in dishes of culture medium. Using these brain slice preparations, it is possible to stimulate and record from specific brain regions, or even specific cells.

In the studies leading to the discovery of LTP, experimenters repeatedly stimulated the synaptic inputs to a particular neuron and observed the usual action potential response. When the neuron was stimulated repeatedly in rapid succession, however, they found that the properties of the responding neuron changed. The size of the postsynaptic response was enhanced, or *potentiated*, and this change lasted for days or weeks (Figure 44.20).

How does this potentiation occur? The answer in some areas of the brain now seems quite clear. With low levels of stimulation, the glutamate released by presynaptic cells activates only the AMPA receptors, and the postsynaptic cell simply responds with action potentials. With higher levels of stimulation, however, the NMDA receptor is activated, allowing both Na$^+$ and Ca^{2+} ions to enter the postsynaptic neuron. The Ca^{2+} ions induce long-term changes in the postsynaptic neuron that make it more sensitive to synaptic input.

Exploiting the LTP system, Dr. Joe Tsien and his students and collaborators at Princeton University have made mice smarter by increasing the ability of their NMDA receptors to induce long-term changes in synaptic transmission. The experimenters genetically engineered mice so that their NMDA receptors had a slightly altered structure and were activated for a longer time whenever they bound a molecule of glutamate. The mice with these altered NMDA receptors learned tasks better, ran mazes faster, and remembered the mazes longer than normal mice. Maybe the world does not need smarter mice, but these exciting experiments confirm that we are on the right track to understanding how the brain achieves learning and memory.

To turn off responses, synapses must be cleared of neurotransmitter

Turning off the action of neurotransmitters is as important as turning it on. If released neurotransmitter molecules simply remained in the synaptic cleft, the postsynaptic membrane would become saturated with neurotransmitter, and receptors would be constantly bound. As a result, the postsynaptic cell would remain hyperpolarized or depolarized and would be unresponsive to short-term changes in the presynaptic cell. The more discrete each separate neural signal is, the more information can be processed in a given time. Thus neurotransmitter must be cleared from the synaptic cleft shortly after it is released by the axon terminal.

Neurotransmitter action may be terminated in several ways. First, enzymes may destroy the neurotransmitter. As we have seen, acetylcholine is rapidly destroyed by the enzyme acetylcholinesterase, which is present in the synaptic cleft in close association with the acetylcholine receptors on the postsynaptic membrane. Some of the most deadly nerve gases that were developed for chemical warfare work by inhibiting acetylcholinesterase. As a result, acetylcholine lingers in the synaptic clefts, causing the victim to die of spastic (contracted) muscle paralysis. Some agricultural insecticides, such as malathion, also inhibit acetylcholinesterase and can poison farm workers if used without safety precautions.

Second, neurotransmitter may simply diffuse away from the cleft. Third, neurotransmitter may be taken up via ac-

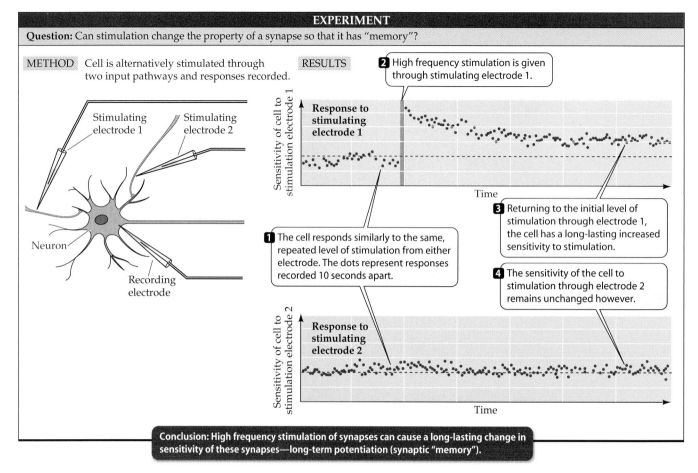

EXPERIMENT

Question: Can stimulation change the property of a synapse so that it has "memory"?

METHOD   Cell is alternatively stimulated through two input pathways and responses recorded.

RESULTS

2 High frequency stimulation is given through stimulating electrode 1.

Sensitivity of cell to stimulation electrode 1

Response to stimulating electrode 1

Time

3 Returning to the initial level of stimulation through electrode 1, the cell has a long-lasting increased sensitivity to stimulation.

1 The cell responds similarly to the same, repeated level of stimulation from either electrode. The dots represent responses recorded 10 seconds apart.

4 The sensitivity of the cell to stimulation through electrode 2 remains unchanged however.

Stimulating electrode 1

Stimulating electrode 2

Neuron

Recording electrode

Sensitivity of cell to stimulation electrode 2

Response to stimulating electrode 2

Time

Conclusion: High frequency stimulation of synapses can cause a long-lasting change in sensitivity of these synapses—long-term potentiation (synaptic "memory").

44.20 Repeated Stimulation Can Cause Long-Term Potentiation
When a cell receives regular synaptic input, the resulting postsynaptic potential remains constant. If, however, that same synaptic pathway is stimulated briefly at a high frequency, the subsequent sensitivity of the postsynaptic cell to the original level of synaptic input is potentiated for a long time.

tive transport by nearby cell membranes. The mode of action of Prozac, a commonly prescribed drug for depression, is to *slow* the reuptake of the neurotransmitter serotonin, thus enhancing its activity at the synapse.

Neurons in Networks

Because neurons can interact in the complex ways we have just discussed, networks of neurons can process and integrate information. Multiple neuronal networks constitute the nervous systems of animals. In subsequent chapters, we will see many examples of how neurons work together in networks to accomplish specific tasks. These networks use all of the mechanisms we have discussed: excitatory and inhibitory synapses, presynaptic excitation and inhibition, and mechanisms of long-term potentiation (and depression). Through these operations, our brains solve puzzles, create inventions, remember experiences, fall in love, and learn about biology. The challenge for the future is to understand how these networks work.

Chapter Summary

Nervous Systems: Cells and Functions

▶ Nervous systems consist of cells that process and transmit information.

▶ Sensory cells transduce information from the environment and the body and communicate commands to effectors such as muscles or glands.

▶ The nervous systems of different species vary, but all are composed of cells called neurons. **Review Figures 44.1, 44.2**

▶ In vertebrates, the brain and spinal cord form the central nervous system, which communicates with other tissues of the body via the peripheral nervous system.

▶ Neurons receive information mostly via their dendrites and transmit information over their axons. Neurons function in networks. **Review Figure 44.3**

▶ The information that neurons process is in the form of electrical events in their plasma membranes. Where neurons and other cells meet, information is transmitted between them, mostly by the release of chemical signals called neurotransmitters.

▶ Glial cells physically support neurons and perform many housekeeping functions. Schwann cells and oligodendrocytes produce myelin, which insulates neurons. Astrocytes create the blood–brain barrier. **Review Figure 44.4**

Neurons: Generating and Conducting Nerve Impulses

▶ Neurons have an electric charge difference across their plasma membranes. This resting potential is created by ion pumps and ion channels. **Review Figure 44.5**

▶ The sodium–potassium pump concentrates K^+ on the inside of neurons and Na^+ on the outsides. Ion channels allow K^+ to leak out, leaving behind unbalanced negative charges and leading to the resting potential. **Review Figures 44.6, 44.7**

▶ A potassium equilibrium potential exists when the electric charge that develops across the membrane is sufficient to prevent a net diffusion of K^+ down its concentration gradient. This potential can be calculated with the Nernst equation. **Review Figure 44.8**

▶ The resting potential is perturbed when ion channels open or close, thus changing the permeability of the plasma membrane to charged ions. Through this mechanism, neurons become depolarized or hyperpolarized in response to stimuli. **Review Figure 44.9**

▶ Rapid reversals in charge across portions of the plasma membrane resulting from the opening and closing of voltage-gated sodium and potassium channels produce action potentials. These changes in voltage-gated channels occur when the plasma membrane depolarizes to a threshold level. **Review Figure 44.10**

▶ Action potentials are conducted down axons because of local current flow, which depolarizes adjacent regions of membrane and brings them to threshold for the opening of voltage-gated sodium channels. **Review Figure 44.11**

▶ Patch clamping allows us to study single ion channels. **Review Figure 44.12**

▶ In myelinated axons, the action potentials appear to jump between nodes of Ranvier, patches of plasma membrane that are not covered by myelin. **Review Figure 44.13**

Neurons, Synapses, and Communication

▶ Neurons communicate with each other and with other cells at specialized junctions called synapses, where the plasma membranes of two cells come close together.

▶ The classic chemical synapse is the neuromuscular junction, a synapse between a motor neuron and a muscle cell. Its neurotransmitter is acetylcholine, which causes a depolarization of the postsynaptic membrane when it binds to its receptor. **Review Figure 44.14**

▶ When an action potential reaches an axon terminal of the presynaptic cell, it causes the release of neurotransmitters, chemical signals that diffuse across the synaptic cleft and bind to receptors on the postsynaptic membrane. **Review Figures 44.15, 44.16**

▶ Synapses between neurons can be either excitatory or inhibitory. Synapses can also form on presynaptic membranes and thereby influence the release of neurotransmitter by the presynaptic cell.

▶ A postsynaptic neuron integrates information by summing its synaptic inputs in both space and time. **Review Figure 44.17**

▶ Ionotropic neurotransmitter receptors are ion channels. Metabotropic receptors influence the postsynaptic cell through various signal transduction pathways that involve G proteins. These pathways can result in changes in ion channels, alterations of enzyme activity, and even gene expression. The actions of ionotropic synapses are generally faster than those of metabotropic synapses. **Review Figures 44.18, 44.19**

▶ Electrical synapses pass electric signals between cells without the use of neurotransmitters.

▶ There are many different neurotransmitters and even more receptors. The action of a neurotransmitter depends on the receptor to which it binds. **Review Table 44.1**

▶ Glutamate binds to both ionotropic and metabotropic receptors, and may be involved in learning and memory. **Review Figure 44.19**

▶ With repeated stimulation, a neuron can become more sensitive to its inputs. Since this increased sensitivity can last a long time, it is called long-term potentiation, or LTP. The properties of the NMDA glutamate receptor appear to explain LTP. **Review Figure 44.20**

▶ In chemical synapses, the transmitter must be cleared rapidly from the synapse. Some poisons and some drugs act by blocking or slowing the clearance of transmitter from the synapse.

Neurons in Networks

▶ Neurons work together in networks to accomplish specific tasks. These networks use all of the mechanisms we have discussed in this chapter.

For Discussion

1. The language of the nervous system consists of one "word," the action potential. How can this single message convey a diversity of information, how can that information be quantitative, and how can it be integrated?

2. If you stimulate an axon in the middle, action potentials are conducted in both directions. Yet when an action potential is generated at the axon hillock, it goes only *toward* the axon terminals and does not backtrack. Explain why action potentials are bidirectional in the first example and unidirectional in the second.

3. The nature of synapses presents various opportunities for plasticity in the nervous system. Discuss at least four synaptic mechanisms that could be altered to change the response of a neuron to a specific input.

4. If Dr. Tsien had genetically engineered the AMPA receptor to remain open longer when activated, would it have made his mice smarter? Why or why not?

45 *Sensory Systems*

ANIMALS PERCEIVE THE WORLD THROUGH THEIR senses. Different species look through different sensory windows, so their views of the world are not the same. Dogs, for example, do not see color well, but they have keener senses of hearing and smell than humans do. While you enjoy a beautiful sunset, your dog might be enjoying sniffing around in the bushes and listening for the sounds of small animals in the underbrush.

Human hunters have exploited the remarkable sensory abilities of dogs for thousands of years. Most recently that hunt has extended to illicit drugs, smuggled contraband, bombs, and firearms. Dogs can be trained to detect the signature odors of such items, so they are used by police, customs agents, and other investigators to identify those odors wherever suspicious activities are occurring.

A black Labrador named Charlie (badge K9-001) was the first dog trained by the U.S. Treasury Department's Bureau of Alcohol, Tobacco, and Firearms to sniff out firearms and explosives. Charlie has sniffed out more than 200 illegal guns and 500 pounds of hidden explosives. With a nose that outperforms electronic sensors, Charlie helped solve a terrorist bombing case by discovering a tiny fragment of the bomb hundreds of yards from the site of the explosion. Charlie's nose is never off duty; on a recreational visit to a U.S. Civil War battlefield, it smelled out cannonball fragments that had been buried for 130 years. ATF dogs receive a lot of expert training, but their careers are based on their remarkable sense of smell.

In this chapter, we look at the general properties of sensory cells and see how they convert environmental stimuli to neural information. Sensory cells are generally called *receptors*, which creates some confusion with the *receptor proteins* that bind signaling molecules. To avoid this confusion, we use the terms *sensory cells* or *receptor cells* in this chapter. We examine in detail the cells responsible for chemoreception, mechanoreception, and photoreception and see how they are incorporated into sensory systems that provide the CNS with information about the world around and within us. In the course of our study of sensory systems, we will learn about the unusual sensory abilities of many animals.

Sensory Cells, Sensory Organs, and Transduction

Sensory cells *transduce* (convert) physical or chemical stimuli into signals that are transmitted to other parts of the nervous system for processing and interpretation. Most sensory cells are modified neurons, but some are other types of cells closely associated with neurons. Sensory cells are specialized for detecting specific kinds of stimuli, such as pressure, heat, or light.

Most sensory cells possess a membrane receptor protein that detects a stimulus and responds by altering the flow of ions across the plasma membrane (Figure 45.1). The resulting change in membrane potential causes the sensory cell either to fire action potentials itself or to change its secretion of neurotransmitter onto an associated cell that fires action potentials. The intensity of the stimulus is encoded in the frequency of action potentials.

Sensation depends on which neurons in the CNS receive action potentials from sensory cells

If the messages derived from all sensory cells are the same—action potentials—how can we perceive different sensations? Sensations such as heat, itch, pressure, pain, light, smell, and sound differ because the messages from sensory cells arrive at different places in the central nervous

Special Agent K9-001
Charlie's remarkable sense of smell enables him and his partner to discover illicit firearms and explosives.

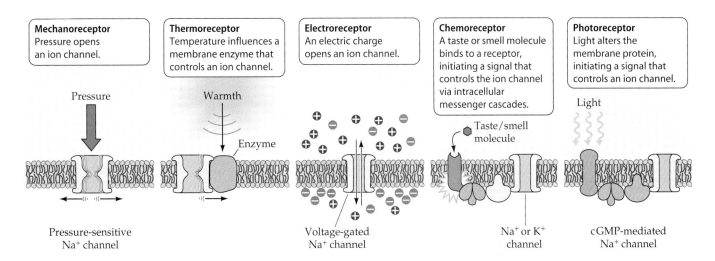

45.1 Sensory Cell Membrane Receptor Proteins Respond to Stimuli
The receptors in mechanoreceptors, thermoreceptors, and electroreceptors are themselves ion channels. The activated receptor proteins of chemoreceptors and photoreceptors initiate biochemical cascades that eventually open or close ion channels.

system (CNS). Action potentials arriving in the visual cortex are interpreted as light, in the auditory cortex as sound, in the olfactory bulb as smell, and so forth.

A small patch of skin on your arm contains sensory cells that increase their firing rates when the skin is warmed. Others increase their activity when the skin is cooled. Other types of sensory cells in the same patch of skin respond to touch, movement of hairs, irritants such as mosquito bites, and pain from cuts or burns. These sensory cells in your arm transmit their messages through axons that enter the CNS through the spinal cord. The synapses made by those axons in the spinal cord and the subsequent pathways of transmission determine whether the stimulation of the patch of skin on your arm is perceived as warmth, cold, pain, touch, itch, or tickle.

The specificity of these sensory circuits is dramatically illustrated in persons who have had a limb or part of a limb amputated. Although the sensory cells from that region are gone, the axons that communicated information from those sensory cells to the CNS may remain. If those axons are stimulated, the person feels specific sensations as if they were coming from the limb that is no longer there—a phantom limb.

The messages from some sensory cells communicate information about internal conditions in the body, but we may not be consciously aware of that information. The brain receives continuous information about body temperature, blood sugar, blood carbon dioxide and oxygen concentration, arterial pressure, muscle tension, and the position of the limbs. All this information is important for the maintenance of homeostasis, but we don't have to think about it. If we did, there would be no time to think about anything else. All sensory cells produce information that

the nervous system can use, but that information does not always result in conscious sensation.

Sensory organs are specialized for detecting specific stimuli

Some sensory cells are assembled with other types of cells into *sensory organs*, such as eyes, ears, and noses, that enhance the ability of the sensory cells to collect, filter, and amplify stimuli. A photoreceptor cell, for example, detects electromagnetic radiation of only a particular range of wavelengths. This selectivity is the basis for color vision and explains why some insects can see ultraviolet light. In some simple organisms photoreceptors sense only the presence of light, but in more complex animals, photoreceptors are combined with other cell types into eyes. We'll learn how eyes collect light and focus it onto sheets of photoreceptors so that patterns of light can be detected.

Sensory transduction involves changes in membrane potentials

In this chapter we examine several sensory cell types and the sensory organs with which they are associated. In each case we can ask the same general question: How does the sensory cell transduce energy from a stimulus into action potentials? The details differ for different sensory cells, but those details all fit into a general pattern.

We have already seen the first step of sensory transduction in Figure 45.1: the activation of a receptor protein. A receptor protein in the plasma membrane of the sensory cell is activated by a specific stimulus. The activated receptor protein opens or closes ion channels in the membrane by one of several mechanisms. The receptor protein may be part of an ion channel, much like an ionotropic synaptic receptor, and by changing its conformation it may open or close the channel directly. Alternatively, the activated receptor protein may function like a metabotropic synaptic receptor coupled to a G protein, setting in motion intracellular events that eventually affect ion channels (see Figures 15.8 and 15.10). Figure 45.2 reviews these first steps of sensory transduction and outlines the subsequent steps.

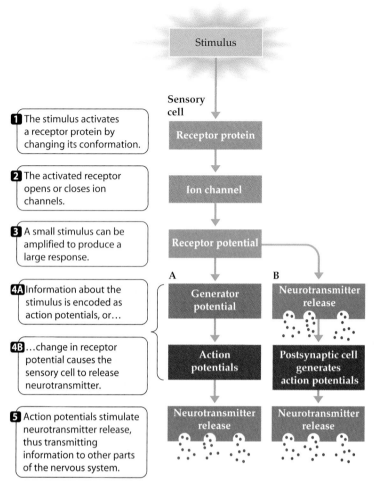

1 The stimulus activates a receptor protein by changing its conformation.

2 The activated receptor opens or closes ion channels.

3 A small stimulus can be amplified to produce a large response.

4A Information about the stimulus is encoded as action potentials, or…

4B …change in receptor potential causes the sensory cell to release neurotransmitter.

5 Action potentials stimulate neurotransmitter release, thus transmitting information to other parts of the nervous system.

45.2 Sensory Transduction Is a Series of Steps
The sensory cell itself may generate action potentials as a result of stimulation, or it may vary its release of neurotransmitter in response to changes in its membrane potential, and the neurotransmitter may induce another cell to generate action potentials.

A good example of generator potentials is found in the stretch receptors of crayfishes (Figure 45.3). By placing an electrode in the cell body of a crayfish stretch receptor cell, we can record the changes in the receptor potential that result from stretching the muscle to which the dendrites of the cell are attached. These changes in receptor potential become a generator potential at the base of the sensory cell's axon, where there are voltage-gated sodium channels. Action potentials generated here travel down the axon to the CNS. The rate at which action potentials are fired by the axon depends on the magnitude of the generator potential; that, in turn, depends on how much the muscle is stretched.

In sensory cells that do not fire action potentials, the spreading receptor potential reaches a presynaptic patch of plasma membrane and induces the release of a neurotransmitter.

Whether or not the sensory cell itself fires action potentials, ultimately the stimulus is transduced into action potentials and the intensity of the stimulus is encoded by the frequency of action potentials.

The opening or closing of ion channels in response to a stimulus changes the membrane potential of the sensory cell, which is called the **receptor potential**. Such changes in membrane potential can spread by current flow over short distances, but to travel long distances in the nervous system, receptor potentials must be converted into action potentials. The intracellular events involved in the conversion of the original stimulus-induced alteration of ion channels to action potentials can *amplify* the signal. In other words, the energy in the output of the sensory cell can be greater than the energy in the stimulus.

Receptor potentials produce action potentials in two ways: by generating action potentials within the sensory cell itself, or by releasing a neurotransmitter that induces an associated neuron to generate action potentials (see Figure 45.2). In the first case, the sensory cell has a region of plasma membrane with voltage-gated sodium channels. A receptor potential that spreads to this region is called a **generator potential** because it generates action potentials by causing the voltage-gated sodium channels to open.

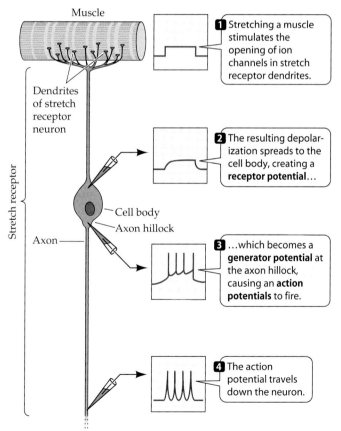

1 Stretching a muscle stimulates the opening of ion channels in stretch receptor dendrites.

2 The resulting depolarization spreads to the cell body, creating a **receptor potential**…

3 …which becomes a **generator potential** at the axon hillock, causing an **action potentials** to fire.

4 The action potential travels down the neuron.

45.3 Stimulating a Sensory Cell Produces a Generator Potential
The stretch receptor of a crayfish produces a generator potential when the muscle is stretched. At the axon hillock, the receptor potential becomes a generator potential, firing action potentials that travel down the axon.

Many receptors adapt to repeated stimulation

Some sensory cells give gradually diminishing responses to maintained or repeated stimulation. This phenomenon is known as **adaptation**, and it enables an animal to ignore background or unchanging conditions while remaining sensitive to changes or to new information. (Note that this use of the term "adaptation" is different from its application in an evolutionary context.) When you dress, you feel each item of clothing touch your skin, but the sensation of clothes touching your skin is not constantly on your mind throughout the day. You are immediately aware, however, when a seam rips, your shoe comes untied, or someone lightly touches your back.

The ability of animals to discriminate between continuous and changing stimuli is due partly to the fact that some sensory cells adapt; it is also due to information processing by the CNS. Some sensory cells adapt very little or very slowly; examples are pain receptors and mechanoreceptors for balance.

In the rest of this chapter we will learn how sensory systems gather and filter stimuli, transduce specific stimuli into action potentials, and transmit action potentials to the CNS.

Chemoreceptors: Responding to Specific Molecules

Animals receive information about chemical stimuli through **chemoreceptors**. Chemoreceptors are responsible for smell, taste, and the monitoring of aspects of the internal environment such as the level of carbon dioxide in the blood. Chemosensitivity is universal among animals.

A colony of corals responds to a small amount of meat extract in the seawater around it by extending bodies and tentacles and searching for food. A solution of a single amino acid can stimulate this response. Humans have similar reactions to chemical stimuli. When we smell freshly baked bread, we salivate and feel hungry, but we gag and retch when we smell diamines from rotting meat. Information from chemoreceptors can cause powerful behavioral and physiological responses.

Arthropods provide good examples for studying chemosensation

Arthropods use chemical signals to attract mates. These signals, called *pheromones*, demonstrate the sensitivity of chemosensory systems. One of the best-studied examples of this phenomenon is the silkworm moth.

To attract a mate, the female silkworm moth releases a pheromone called bombykol from a gland at the tip of her abdomen. The male silkworm moth has receptors for this molecule on his antennae (Figure 45.4). Each feathery antenna carries about 10,000 bombykol-sensitive hairs. A single molecule of bombykol is sufficient to generate action potentials in the antennal nerve that transmits the signal to the CNS. Because of the male's high degree of sensitivity, the sexual message of a female moth is likely to reach any male that happens to be within a downwind area stretching over several kilometers. When approximately 200 hairs per second are activated, the male flies upwind in search of the female. Because the rate of firing in the male's sensory nerves is proportional to bombykol concentrations in the air, he can follow a concentration gradient and home in on the signaling female.

Many arthropods have chemoreceptor hairs, each containing one or more specific types of receptors. Crabs and flies, for example, have chemoreceptor hairs on their feet; these hairs have receptors for sugars, amino acids, salts, and other molecules. A fly tastes a potential food by stepping in it.

Olfaction is the sense of smell

The sense of smell, known as **olfaction**, also depends on chemoreceptors. In vertebrates, the olfactory sensors are neurons embedded in a layer of epithelial cells at the top of the nasal cavity. The axons of these neurons project to the olfactory bulb of the brain, and their dendrites end in olfac-

45.4 Some Scents Travel Great Distances
Mating in silkworm moths of the genus *Bombyx* is coordinated by a pheromone called bombykol.

The female moth releases a pheromone from a gland at the tip of the abdomen.

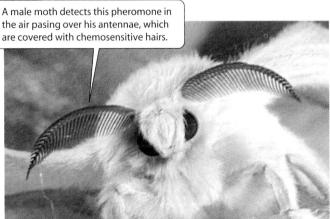

A male moth detects this pheromone in the air pasing over his antennae, which are covered with chemosensitive hairs.

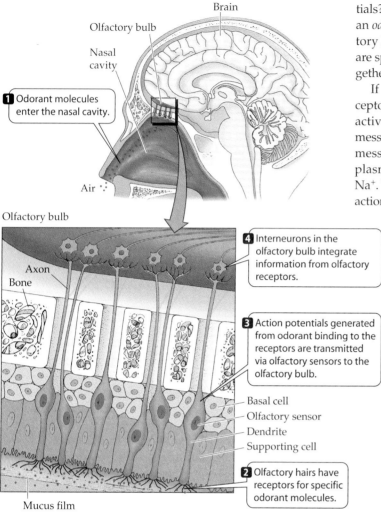

1 Odorant molecules enter the nasal cavity.

Brain

Olfactory bulb

Nasal cavity

Air

Olfactory bulb

Axon

Bone

4 Interneurons in the olfactory bulb integrate information from olfactory receptors.

3 Action potentials generated from odorant binding to the receptors are transmitted via olfactory sensors to the olfactory bulb.

Basal cell

Olfactory sensor

Dendrite

Supporting cell

2 Olfactory hairs have receptors for specific odorant molecules.

Mucus film

45.5 Olfactory Receptors Communicate Directly with the Brain
The receptors of the human olfactory system are embedded in tissues lining the nasal cavity and send their axons to the olfactory bulb of the brain.

tory hairs at the surface of the nasal epithelium. A protective layer of mucus covers the epithelium. Molecules from the environment must diffuse through this mucus to reach the receptor proteins on the olfactory hairs. When you have a cold, the amount of mucus increases and the epithelium swells. With this in mind, study Figure 45.5 and you will easily understand why respiratory infections can cause you to lose your sense of smell.

A dog has up to 40 million nerve endings per square centimeter of nasal epithelium, many more than we do. Humans have a fairly sensitive olfactory system, but we are unusual among mammals in that we depend more on vision than on olfaction (we tend to join bird-watching societies more often than mammal-smelling societies). Whales and porpoises have no olfactory receptors and hence no sense of smell.

How does an olfactory sensory cell transduce the structure of a molecule from the environment into action poten-

tials? A molecule that triggers an olfactory receptor is called an *odorant*. Odorants bind to receptor proteins on the olfactory hairs of the sensory cells. Olfactory receptor proteins are specific for particular odorant molecules—the two fit together like a lock and key.

If a "key" (an odorant molecule) fits the "lock" (the receptor protein), then a G protein is activated, which in turn activates an enzyme that causes an increase of a second messenger in the cytoplasm of the sensory cell. The second messenger binds to sodium channels in the sensory cell's plasma membrane and opens them, causing an influx of Na^+. The sensory cell thus depolarizes to threshold and fires action potentials (see Figure 15.16).

The olfactory world has an enormous number of "keys"—molecules that produce distinct smells. The number of "locks"—receptor proteins—is large, but not nearly as large as the number of possible odorants. A family of about a thousand genes codes for olfactory receptor proteins. Each receptor protein is expressed in a limited number of sensory cells in the olfactory epithelium, and those cells all project to the same regions in the olfactory bulb. A given odorant molecule may bind to one or to more than one receptor protein. Therefore, each odorant molecule can excite a unique selection of cells in the olfactory bulb, so an olfactory system with a thousand different receptor proteins can discriminate a large number of smells.

How does the sensory cell signal the intensity of a smell? It responds in a graded fashion to the concentration of odorant molecules: The more odorant molecules that bind to receptors, the more action potentials are generated and the greater the intensity of the perceived smell.

Gustation is the sense of taste

The sense of taste, or **gustation**, in humans and other vertebrates depends on clusters of sensory cells called **taste buds**. The taste buds of terrestrial vertebrates are confined to the mouth cavity, but some fishes have taste buds in the skin that enhance their ability to sense their environment. Some fishes living in murky water are very sensitive to small amounts of amino acids in the water around them and can find food without the use of vision. The duck-billed platypus, a monotreme mammal (see Figure 33.22*b*), has similar talents as a result of taste buds on the sensitive skin of its bill. What is a taste bud and how does it work?

A taste bud is a cluster of sensory cells (Figure 45.6). A human tongue has approximately 10,000 taste buds. The taste buds are embedded in the epithelium of the tongue, and many are found on the raised papillae of the tongue. (Look at your tongue in a mirror—the papillae make it look fuzzy.) Each papilla has many taste buds. The outer surface of a taste bud has a pore that exposes the tips of the sensory cells. Microvilli (tiny hairlike projections) increase the surface area of the sensory cells where their tips converge at the pore. These sensory cells, unlike olfactory receptors, are not

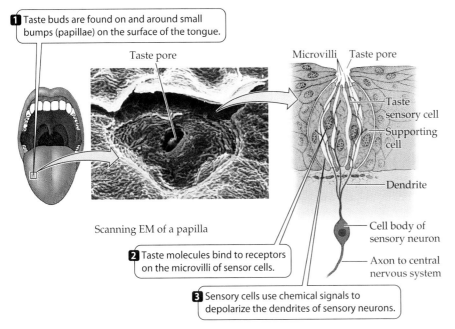

1 Taste buds are found on and around small bumps (papillae) on the surface of the tongue.

Taste pore

Microvilli Taste pore

Taste sensory cell

Supporting cell

Dendrite

Cell body of sensory neuron

Axon to central nervous system

Scanning EM of a papilla

2 Taste molecules bind to receptors on the microvilli of sensor cells.

3 Sensory cells use chemical signals to depolarize the dendrites of sensory neurons.

45.6 Taste Buds Are Clusters of Sensory Cells
Each taste bud contains a number of sensory cells that are not neurons.

neurons. At their bases, they form synapses with dendrites of sensory neurons.

Gustation begins with receptor proteins in the membranes of the microvilli. As with olfactory transduction, receptor proteins on the sensory cells bind specific molecules (such as sugar). This binding causes changes in the membrane potential of the sensory cells, which release neurotransmitters onto the dendrites of the sensory neurons. The sensory neurons fire action potentials that are conducted to the CNS.

The tongue does a lot of hard work, so its epithelium is shed and replaced at a rapid rate. Individual taste buds last only a few days before they are replaced, but the sensory neurons associated with them live on, always forming new synapses as new taste buds form.

You have probably heard that humans can perceive only four tastes: sweet, salty, sour, and bitter. In actuality, taste buds can distinguish among a variety of sweet-tasting molecules and a variety of bitter-tasting molecules. The full complexity of the chemosensitivity that enables us to enjoy the subtle flavors of food comes from the combined activation of gustatory and olfactory receptors; that is the reason you lose some of your sense of taste when you have a cold.

Why does a snake continually sample the air by darting its forked tongue in and out? The forks of the snake's tongue fit into cavities in the roof of its mouth that are richly endowed with olfactory receptors. The tongue samples the air and presents the sample directly to the olfactory receptors. Thus the snake is really using its tongue to smell its environment, not to taste it. Why doesn't the snake simply use the flow of air to and from its lungs, as we do, to smell the environment? In reptiles, air flows to and from the lungs slowly (and can even stop entirely for long periods of time), but the tongue can dart in and out many times in a second. It is a quick source of olfactory information.

Mechanoreceptors: Detecting Stimuli that Distort Membranes

Mechanoreceptors are cells that are sensitive to mechanical forces. In the skin, different kinds of mechanoreceptors are responsible for the perception of touch, pressure, and tickle. Stretch receptors in muscles, tendons, and joints provide information about the position of the parts of the body in space and the forces acting on them. Stretch receptors in the walls of blood vessels signal changes in blood pressure. "Hair" cells with extensions that are sensitive to bending are incorporated into mechanisms for hearing and for signaling the body's position in space.

Physical distortion of a mechanoreceptor's plasma membrane causes ion channels to open, altering the membrane potential of the cell, which in turn leads to the generation of action potentials. The rate of action potentials tells the CNS the strength of the stimulus exciting the mechanoreceptor.

Many different sensory cells respond to touch and pressure

Objects touching the skin generate varied sensations because skin is packed with diverse mechanoreceptors (Figure 45.7). The outer layers of skin contain whorls of nerve endings enclosed in connective tissue capsules. These very sensitive mechanoreceptors, called **Meissner's corpuscles**, respond to objects that touch the skin even lightly. Meissner's corpuscles adapt very rapidly. That is why you roll a small object between your fingers, rather than holding it still, to discern its shape and texture. As you roll it, you continue to stimulate these receptors anew.

Also in the outer regions of the skin are **expanded-tip tactile receptors** of various kinds. They differ from Meissner's corpuscles in that they adapt only partly and slowly. They are useful for providing steady-state information about objects that continue to touch the skin.

The density of these tactile mechanoreceptors varies across the surface of the body. A two-point discrimination test demonstrates this fact. If you lightly touch someone's back with two toothpicks, you can determine how far apart the two stimuli have to be before the person can distinguish whether he or she was touched with one or two points. On the back, the stimuli have to be rather far apart. The same test applied to the person's lips or fingertips reveals finer spatial discrimination; that is, the person can identify as separate two stimuli that are close together.

Deep in the skin, the dendrites of neurons wrap around hair follicles. When the hairs are displaced, those neurons are stimulated. Also deep within the skin is another type of mechanoreceptor, the **Pacinian corpuscle**. Pacinian corpuscles look like onions because they are made up of concentric layers of connective tissue, which encapsulate an extension of a sensory neuron. Pacinian corpuscles respond especially well to vibrations applied to the skin, but they adapt rapidly to steady pressure. The connective tissue capsule is important in this adaptation. An initial pressure distorts the corpuscle and the plasma membrane of the neuron at its core, but the layers of the capsule rapidly rearrange to redistribute the force, thus eliminating the distortion of the neuronal plasma membrane.

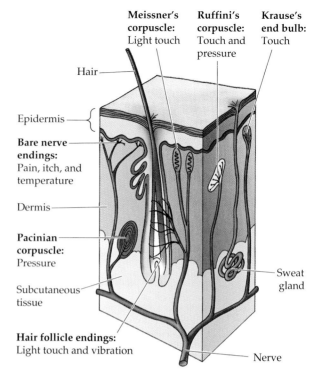

45.7 The Skin Feels Many Sensations
Even a very small patch of skin contains a diversity of sensory cells.

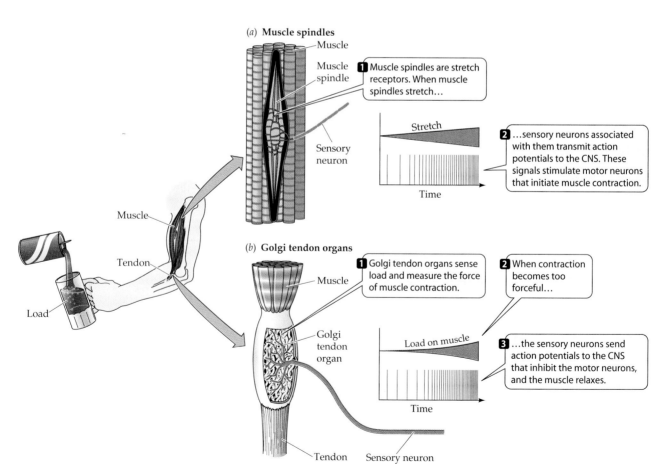

45.8 Stretch Receptors Are Activated When Limbs Are Stretched
Stretch receptors provide information about the stresses on muscles and joints in an animal's limbs. (a) Signals from muscle spindles to the CNS initiate muscle contraction. (b) Golgi tendon organs in tendons and ligaments inhibit a contraction that becomes too forceful, triggering relaxation and protecting the muscle from tearing.

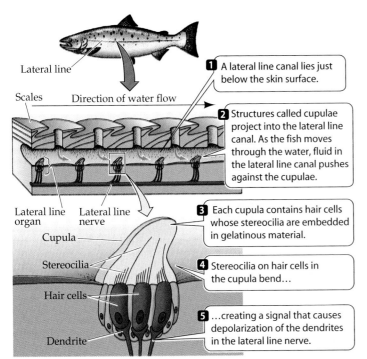

Lateral line

Scales — Direction of water flow

1 A lateral line canal lies just below the skin surface.

2 Structures called cupulae project into the lateral line canal. As the fish moves through the water, fluid in the lateral line canal pushes against the cupulae.

Lateral line organ — Lateral line nerve

Cupula

Stereocilia

Hair cells

Dendrite

3 Each cupula contains hair cells whose stereocilia are embedded in gelatinous material.

4 Stereocilia on hair cells in the cupula bend…

5 …creating a signal that causes depolarization of the dendrites in the lateral line nerve.

45.9 The Lateral Line System Contains Mechanoreceptors
Hair cells in the lateral line of a fish detect movement of the water around the animal, giving the fish information about its own movements and the movements of objects nearby.

Stretch receptors are found in muscles, tendons, and ligaments

An animal receives information from **stretch receptors** about the position of its limbs and the stresses on its muscles and joints. These mechanoreceptors are activated by being stretched. They continuously feed information to the CNS, and that information is essential for the coordination of movements.

The stretch receptors in skeletal muscle are called **muscle spindles**. They are embedded in connective tissue within skeletal muscle. They consist of modified muscle fibers that are innervated in the center with extensions of sensory neurons. Whenever the muscle stretches, muscle spindles are also stretched and the neurons transmit action potentials to the CNS (Figure 45.8a). Earlier in this chapter, we saw how crayfish stretch receptors transduce physical force into action potentials (see Figure 45.3). The actions of muscle spindles are similar.

Another stretch receptor, the **Golgi tendon organ**, is found in tendons and ligaments. It provides information about the force generated by a contracting muscle. When a contraction becomes too forceful, the information from the Golgi tendon organ feeds into the spinal cord, inhibits the motor neuron, and causes the contracting muscle to relax, thus protecting the muscle from tearing (Figure 45.8b).

Hair cells provide information about balance, orientation in space, and motion

Hair cells are also mechanoreceptors. Projecting from the surface of each hair cell is a set of **stereocilia**, which looks like a set of organ pipes. When these stereocilia (which are

really microvilli) are bent, they alter receptor proteins in the hair cell's plasma membrane. When the stereocilia of some hair cells are bent in one direction, the receptor potential becomes more negative; when they are bent in the opposite direction, it becomes more positive. When the receptor potential becomes more positive, the hair cell releases neurotransmitter to the sensory neuron associated with it and the sensory neuron sends action potentials to the CNS.

Hair cells are found in the **lateral line** sensory system of fishes. The lateral line consists of a canal just under the surface of the skin that runs down each side of the fish (Figure 45.9). The lateral line provides information about movements of the fish through the water, as well as about moving objects, such as predators or prey, that cause pressure waves in the surrounding water.

Many invertebrates have equilibrium organs called **statocysts** that use hair cells to signal the position of the animal with respect to gravity (Figure 45.10). A statocyst is a chamber lined with hair cells that contains a dense object called a *statolith*. As the animal changes its position, the statolith moves in response to gravity, stimulating the hair cells below it. Replacing the statoliths of a lobster with iron filings and holding a magnet over the animal causes it to swim upside down. When a magnet is held to the lobster's side, it swims on its side.

Vertebrates also have equilibrium organs. The mammalian inner ear has two equilibrium organs that use hair cells to detect the position of the body with respect to gravity: semicircular canals and the vestibular apparatus. The

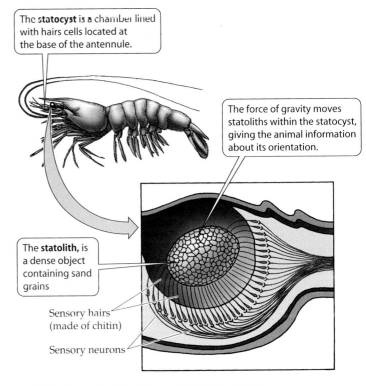

The **statocyst** is a chamber lined with hairs cells located at the base of the antennule.

The force of gravity moves statoliths within the statocyst, giving the animal information about its orientation.

The **statolith**, is a dense object containing sand grains

Sensory hairs (made of chitin)

Sensory neurons

45.10 How a Lobster Knows Which Way Is Up
The statocyst is an equilibrium-sensing organ found in many invertebrates.

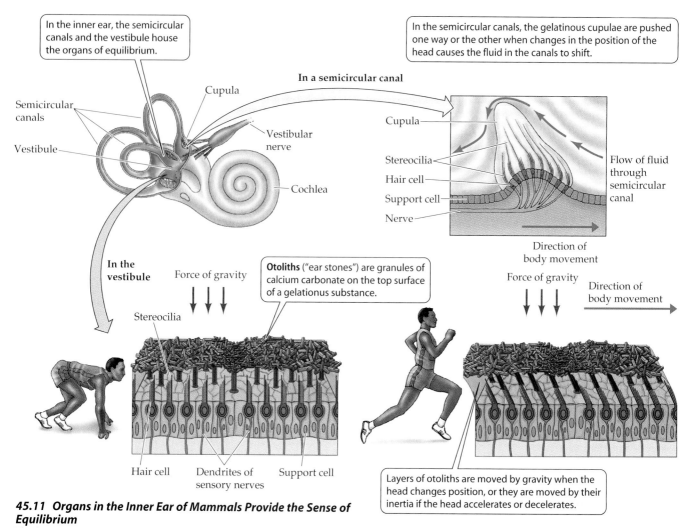

In the inner ear, the semicircular canals and the vestibule house the organs of equilibrium.

In the semicircular canals, the gelatinous cupulae are pushed one way or the other when changes in the position of the head causes the fluid in the canals to shift.

In a semicircular canal

Semicircular canals

Cupula

Vestibular nerve

Vestibule

Cochlea

Cupula

Stereocilia

Hair cell

Support cell

Nerve

Flow of fluid through semicircular canal

Direction of body movement

Force of gravity

Direction of body movement

In the vestibule

Force of gravity

Otoliths ("ear stones") are granules of calcium carbonate on the top surface of a gelationus substance.

Stereocilia

Hair cell   Dendrites of sensory nerves   Support cell

Layers of otoliths are moved by gravity when the head changes position, or they are moved by their inertia if the head accelerates or decelerates.

45.11 Organs in the Inner Ear of Mammals Provide the Sense of Equilibrium
The bony inner ear has three parts: the snail-shaped cochlea, the semicircular canals, and the vestibule. The cochlea is for hearing; the semicircular canals and the vestibular apparatus provide the sense of equilibrium.

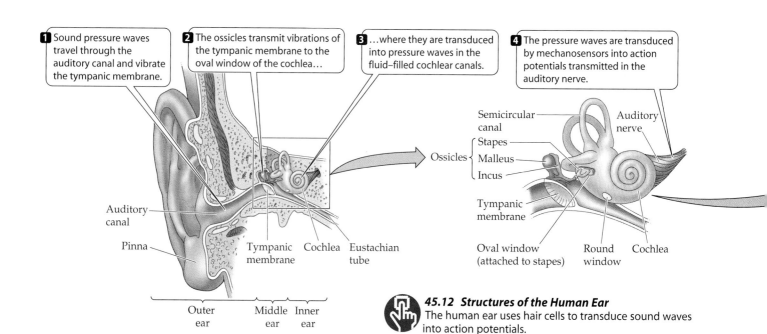

1 Sound pressure waves travel through the auditory canal and vibrate the tympanic membrane.

2 The ossicles transmit vibrations of the tympanic membrane to the oval window of the cochlea…

3 …where they are transduced into pressure waves in the fluid–filled cochlear canals.

4 The pressure waves are transduced by mechanosensors into action potentials transmitted in the auditory nerve.

Semicircular canal

Auditory nerve

Stapes

Ossicles { Malleus

Incus

Tympanic membrane

Oval window (attached to stapes)

Round window

Cochlea

Auditory canal

Pinna

Tympanic membrane

Cochlea

Eustachian tube

Outer ear   Middle ear   Inner ear

45.12 Structures of the Human Ear
The human ear uses hair cells to transduce sound waves into action potentials.

inner ear contains three **semicircular canals** at right angles to one another. The structure and function of these organs are described in Figure 45.11. The **vestibular apparatus** has two chambers that perform a function similar to that of the statocysts of invertebrates.

Auditory systems use hair cells to sense sound waves

The stimuli that animals perceive as sounds are pressure waves. **Auditory systems** use mechanoreceptors to transduce pressure waves into action potentials. Auditory systems include special structures to gather sound waves, direct them to the sensory organ, and amplify their effect on the mechanoreceptors.

Human hearing provides a good example of an auditory system. The organs of hearing are the ears. The two prominent structures on the sides of our heads usually thought of as ears are the **ear pinnae**. The pinna of an ear collects sound waves and directs them into the auditory canal, which leads to the actual hearing apparatus in the middle ear and the inner ear (Figure 45.12). If you have ever watched a rabbit, a horse, or a dog change the orientation of its ear pinnae to focus on a particular sound, then you have witnessed the role of ear pinnae in hearing.

The eardrum, or **tympanic membrane**, covers the end of the auditory canal. The tympanic membrane vibrates in response to pressure waves traveling down the auditory canal. The middle ear, an air-filled cavity, lies on the other side of the tympanic membrane.

The middle ear is open to the throat at the back of the mouth through the *eustachian tube*. Because air flows through the eustachian tube, pressure equilibrates between the middle ear and the outside world. When you have a cold or allergy, the tube can become blocked by mucus or by tissue swelling, so you have difficulty "clearing your

ears," or equilibrating the pressure in the middle ear with the outside air pressure. As a result, the flexible tympanic membrane bulges in or out, dampening your hearing and sometimes causing earaches.

The middle ear contains three delicate bones called the **ear ossicles**, individually named the *malleus* (hammer), *incus* (anvil), and *stapes* (stirrup). The ossicles transmit the vibrations of the tympanic membrane to another flexible membrane called the **oval window**. The leverlike action of the ossicles amplifies the vibrations of the tympanic membrane about 20-fold in transmitting them to the oval window membrane. Behind the oval window lies the fluid-filled inner ear. Movements of the oval window result in pressure changes in the inner ear. These pressure waves are transduced into action potentials.

The inner ear is a long, tapered, coiled chamber called the **cochlea** (from Latin and Greek words for "snail" or "shell"). A cross section of this chamber reveals that it is composed of three parallel canals separated by two membranes: **Reissner's membrane** and the **basilar membrane** (see Figure 45.12). Sitting on the basilar membrane is the **organ of Corti**, the apparatus that transduces pressure waves into action potentials in the auditory nerve, which in turn conveys information from the ear to the brain. The organ of Corti contains hair cells whose stereocilia are in contact with an overhanging, rigid shelf called the **tectorial membrane**. Whenever the basilar membrane flexes, the tectorial membrane bends the hair cell stereocilia. As a consequence, the hair cells depolarize or hyperpolarize, altering the rate of action potentials transmitted to the brain by their associated sensory neurons.

Stereocilia of a hair cell

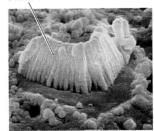

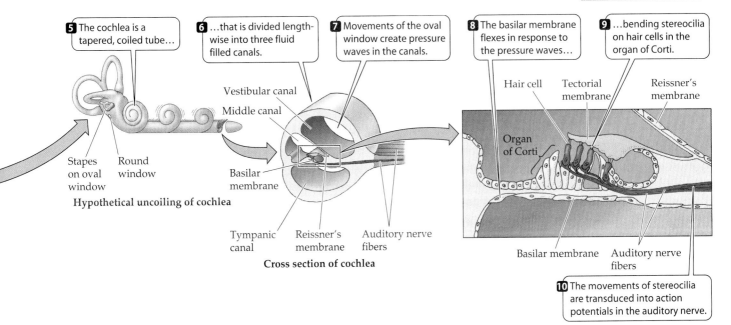

5 The cochlea is a tapered, coiled tube…

6 …that is divided lengthwise into three fluid filled canals.

7 Movements of the oval window create pressure waves in the canals.

8 The basilar membrane flexes in response to the pressure waves…

9 …bending stereocilia on hair cells in the organ of Corti.

Vestibular canal

Middle canal

Stapes on oval window

Round window

Basilar membrane

Hypothetical uncoiling of cochlea

Tympanic canal

Reissner's membrane

Auditory nerve fibers

Cross section of cochlea

Hair cell

Tectorial membrane

Reissner's membrane

Organ of Corti

Basilar membrane

Auditory nerve fibers

10 The movements of stereocilia are transduced into action potentials in the auditory nerve.

What causes the basilar membrane to flex, and how does this mechanism distinguish sounds of different frequencies? In Figure 45.13, the cochlea is shown uncoiled to make it easier to understand its structure and function. To simplify matters, we have left out Reissner's membrane, thus combining the upper and the middle canals into one upper canal. The purpose of Reissner's membrane is to contain a specific aqueous environment for the organ of Corti separate from the aqueous environment in the rest of the cochlea.

The simplified diagram of the cochlea shown in Figure 45.13 reveals two additional features that are important to its function. First, the upper and lower chambers separated by the basilar membrane are joined at the distal end of the cochlea (the end farthest from the oval window), making one continuous canal that folds back on itself. Second, just as the oval window is a flexible membrane at the beginning of the cochlea, the **round window** is a flexible membrane at the end of the long cochlear canal.

Air is highly compressible, but fluids are not. Therefore, a sound pressure wave can travel through air without much displacement of the air, but a sound pressure wave in fluid causes displacement of the fluid. When the stapes pushes the oval window in, the fluid in the upper canal of the cochlea is displaced. The cochlear fluid displacement travels down the upper canal, around the bend, and back through the lower canal. At the end of the lower canal, the displacement is absorbed by the outward bulging of the round window.

If the oval window vibrates in and out rapidly, the waves of fluid displacement do not have enough time to travel all the way to the end of the upper canal and back through the lower canal. Instead, they take a shortcut by crossing the basilar membrane, causing it to flex. The more rapid the vibration, the closer to the oval and round windows the wave of displacement will flex the basilar membrane. Thus different pitches of sound flex the basilar membrane at different locations and activate different sets of hair cells (see Figure 45.13).

The ability of the basilar membrane to respond to vibrations of different frequencies is enhanced by its structure. Near the oval and round windows, at the proximal end, the basilar membrane is narrow and stiff, but it gradually becomes wider and more flexible toward the opposite (distal) end. So it is easier for the proximal basilar membrane to resonate with high frequencies and for the distal basilar membrane to resonate with lower frequencies. A complex sound made up of many frequencies distorts the basilar membrane at many places simultaneously and activates a unique subset of hair cells. Action potentials stimulated by the mechanoreceptors at different positions along the organ of Corti travel to the brain stem along the auditory nerve.

Deafness, the loss of the sense of hearing, has two general causes. *Conduction deafness* is caused by the loss of function of the tympanic membrane and the ossicles of the middle ear. Repeated infections of the middle ear can cause scarring of the tympanic membrane and stiffening of the

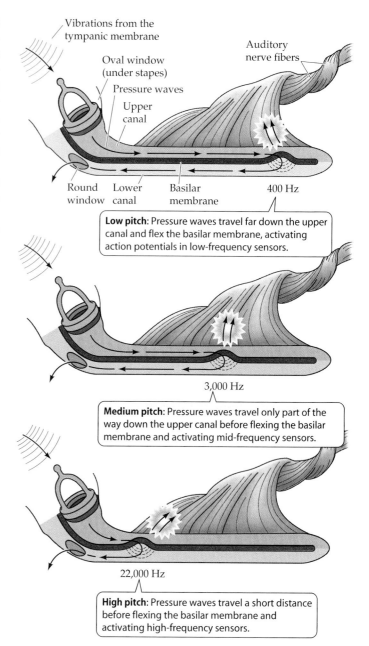

Low pitch: Pressure waves travel far down the upper canal and flex the basilar membrane, activating action potentials in low-frequency sensors.

Medium pitch: Pressure waves travel only part of the way down the upper canal before flexing the basilar membrane and activating mid-frequency sensors.

High pitch: Pressure waves travel a short distance before flexing the basilar membrane and activating high-frequency sensors.

45.13 Sensing Sound Pressure Waves in the Inner Ear
For simplicity, this diagram illustrates the cochlea as uncoiled, and leaves out Reissner's membrane. Pressure waves of different frequencies flex the basilar membrane at different locations. Information about sound frequency is specified by which hair cells are activated.

connections between the ossicles. The consequence is less efficient conduction of sound waves from the tympanic membrane to the oval window. With increasing age, the ossicles progressively stiffen, resulting in a gradual loss of the ability to hear high-frequency sounds. *Nerve deafness* is caused by damage to the inner ear or the auditory pathways. A common cause of nerve deafness is damage to the hair cells of the delicate organ of Corti by exposure to loud sounds such as jet engines, pneumatic drills, or highly amplified rock music. This damage is cumulative and permanent.

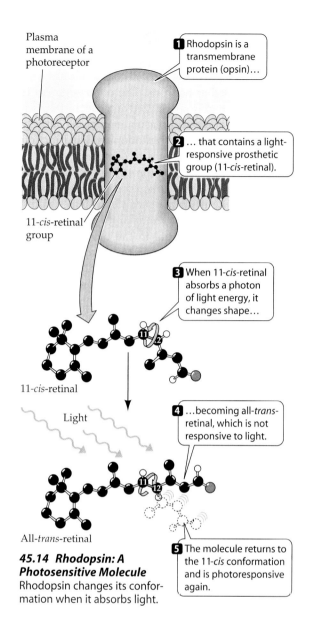

1 Rhodopsin is a transmembrane protein (opsin)…

Plasma membrane of a photoreceptor

2 … that contains a light-responsive prosthetic group (11-*cis*-retinal).

11-*cis*-retinal group

3 When 11-*cis*-retinal absorbs a photon of light energy, it changes shape…

11-*cis*-retinal

Light

4 …becoming all-*trans*-retinal, which is not responsive to light.

All-*trans*-retinal

45.14 Rhodopsin: A Photosensitive Molecule
Rhodopsin changes its conformation when it absorbs light.

5 The molecule returns to the 11-*cis* conformation and is photoresponsive again.

Photoreceptors and Visual Systems: Responding to Light

Sensitivity to light—**photosensitivity**—confers on the simplest animals the ability to orient to the sun and sky and gives more complex animals rapid and extremely detailed information about objects in their environment. It is not surprising that both simple and complex animals can sense and respond to light. What is remarkable is that across the entire range of animal species, evolution has conserved the same basis for photosensitivity: the family of pigments called **rhodopsins**.

In this section we will learn how rhodopsin molecules respond when stimulated by light energy and how that response is transduced into neural signals. We will also examine the structures of eyes, the organs that gather and focus light energy onto photoreceptor cells.

Rhodopsin is responsible for photosensitivity

Photosensitivity depends on the ability of rhodopsins to absorb photons of light and to undergo a change in conformation. A rhodopsin molecule consists of a protein, **opsin** (which by itself does not absorb light), and a light-absorbing prosthetic group, **11-*cis*-retinal**. The light-absorbing group is cradled in the center of the opsin and the entire rhodopsin molecule sits within the plasma membrane of a photoreceptor cell (Figure 45.14).

When the 11-*cis*-retinal absorbs a photon of light energy, its shape changes into a different isomer of retinal—all-*trans*-retinal. This change puts a strain on the bonds between retinal and opsin, changing the conformation of opsin. This change in conformation signals the detection of light. In vertebrate eyes, the retinal and the opsin eventually separate from each other—a process called *bleaching*, which causes the molecule to lose its photosensitivity. When the retinal spontaneously returns to its 11-*cis* isomer and recombines with opsin, it once again becomes photosensitive rhodopsin.

How does the light-induced conformational change of rhodopsin transduce light into a cellular response? After retinal converts from the 11-*cis* into the all-*trans* form, its interactions with opsin pass through several unstable intermediate stages. One of these stages is known as

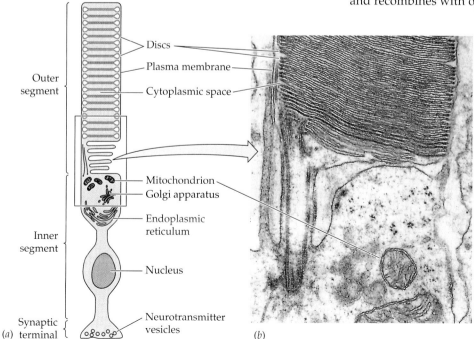

Discs
Plasma membrane
Cytoplasmic space

Outer segment

Mitochondrion
Golgi apparatus
Endoplasmic reticulum

Inner segment

Nucleus

Synaptic terminal
(*a*)
Neurotransmitter vesicles

(*b*)

45.15 The Rod Cell: A Vertebrate Photoreceptor
(*a*) The rod cell of the vertebrate retina is a neuron modified for photosensitivity. The membranes of a rod cell's discs are densely packed with rhodopsin. (*b*) A transmission electron micrograph of a section through a photoreceptor.

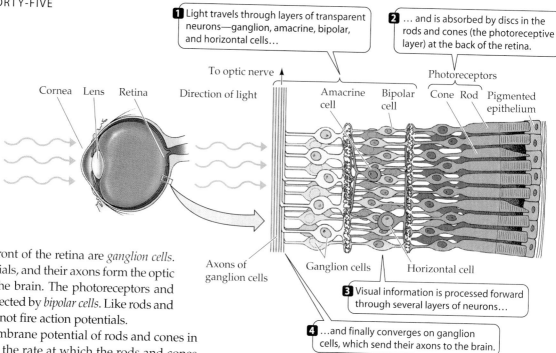

1 Light travels through layers of transparent neurons—ganglion, amacrine, bipolar, and horizontal cells...

2 ... and is absorbed by discs in the rods and cones (the photoreceptive layer) at the back of the retina.

To optic nerve

Direction of light

Cornea Lens Retina

Photoreceptors

Amacrine cell Bipolar cell Cone Rod Pigmented epithelium

Axons of ganglion cells

Ganglion cells Horizontal cell

3 Visual information is processed forward through several layers of neurons...

4 ...and finally converges on ganglion cells, which send their axons to the brain.

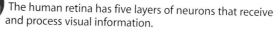

45.24 The Retina
The human retina has five layers of neurons that receive and process visual information.

tials. The cells at the front of the retina are *ganglion cells*. They fire action potentials, and their axons form the optic nerves that travel to the brain. The photoreceptors and ganglion cells are connected by *bipolar cells*. Like rods and cones, bipolar cells do not fire action potentials.

Changes in the membrane potential of rods and cones in response to light alter the rate at which the rods and cones release neurotransmitter at their synapses with the bipolar cells. In response to neurotransmitter from the photoreceptors, the membrane potentials of the bipolar cells change, altering the rate at which they release neurotransmitter onto ganglion cells. The ganglion cells generate action potentials, and the rate of neurotransmitter release from the bipolar cells determines the rate at which they do so. Thus the direct flow of information in the retina is from photoreceptor to bipolar cell to ganglion cell. Ganglion cells send the information to the brain.

The other two cell layers, the horizontal cells and the amacrine cells, communicate laterally across the retina. *Horizontal cells* connect neighboring pairs of photoreceptors and bipolar cells. Thus the communication between a photoreceptor and its bipolar cell can be influenced by the amount of light absorbed by neighboring photoreceptors. This lateral flow of information enables the retina to sharpen the perception of contrast between light and dark patterns.

Amacrine cells connect neighboring pairs of bipolar cells and ganglion cells. The role of amacrine cells is still poorly understood. Some amacrine cell types are highly sensitive to changing illumination or to motion. Others assist in adjusting the sensitivity of the eyes according to the overall level of light falling on the retina. When background light levels change, amacrine cell connections to the ganglion cells help the ganglion cells remain sensitive to temporal changes in stimulation. Thus even with large changes in background illumination, the eyes are sensitive to smaller, more rapid changes in the pattern of light falling on the retina.

INFORMATION PROCESSING IN THE RETINA. Knowing the paths of information flow through the retina still doesn't tell us how that information is processed. What does the eye tell the brain in response to a pattern of light falling on the retina? One aspect of information processing in the retina is *convergence of information*. There are more than 100 million photoreceptors in each retina, but only about 1 million ganglion cells sending messages to the brain. How is the information from all those photoreceptors integrated by the ganglion cells?

This question was addressed in some elegant, classic experiments in which electrodes were used to record the activity of single ganglion cells in living animals while their retinas were stimulated with spots of light. These studies revealed that each ganglion cell has a well-defined **receptive field** that consists of a specific group of photoreceptor cells. Stimulating these photoreceptors with light activates the ganglion cell (Figure 45.25). Information from many photoreceptor cells is integrated in this way to produce a single message.

The receptive fields of many ganglion cells are circular, but the way a spot of light influences the activity of the ganglion cell depends on where in the receptive field it falls. The receptive field of a ganglion cell can be divided into two concentric areas, called the *center* and the *surround*. There are two kinds of receptive fields, *on-center* and *off-center*. Stimulating the center of an on-center receptive field excites the ganglion cell, and stimulating the surround inhibits it. Stimulating the center of an off-center receptive field inhibits the ganglion cell, and stimulating the surround excites it. Center effects are always stronger than surround effects.

The response of a ganglion cell to stimulation of the center of its receptive field depends on how much of the surround area is also stimulated. A small dot of light directly on the center has the maximal effect. A bar of light hitting the center and pieces of the surround has less of an effect, and a large, uniform patch of light falling equally on center and surround has very little effect. Ganglion cells communicate information about contrasts between light and dark that fall on different regions of their receptive fields.

How are receptive fields related to the connections among the neurons of the retina? The photoreceptors in the center of the receptive field of a ganglion cell are connected to that ganglion cell by bipolar cells. The photoreceptors in the surround send information to the center photoreceptors, and thus to the ganglion cell, through the lateral connections of horizontal cells. Thus the receptive field of a ganglion cell consists of a pattern of synapses among photoreceptors, horizontal cells, bipolar cells, and ganglion cells.

The receptive fields of neighboring ganglion cells can overlap greatly; a given photoreceptor can be effectively connected to several ganglion cells. Thus the ganglion cells send simple messages to the brain about the pattern of light

45.25 What Does the Eye Tell the Brain?
When the retina is stimulated with dots and rings of light, individual ganglion cells show different responses.

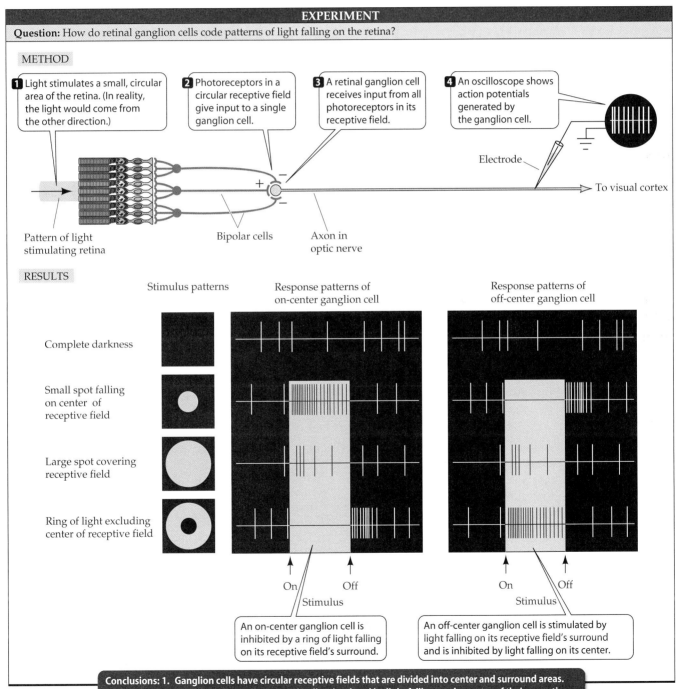

intensities falling on small, circular patches of retina. In Chapter 47 we will see how the brain reassembles that information into our view of the world.

Sensory Worlds Beyond Our Experience

Humans make use of only a subset of the information available to us in the environment. Other animals have sensory systems that enable them to use different subsets and different types of information.

Some species can see infrared and ultraviolet light

When discussing vision, we use the term "visible light," but what we really mean is light visible to humans. Our visible spectrum is a very narrow region of the entire, continuous range of electromagnetic radiation in the environment (see Figure 8.5). We cannot see ultraviolet radiation, for example, but many other animals can.

One of the seven photoreceptors in each ommatidium of a fruit fly is sensitive to ultraviolet light. The visual sensitivity of many pollinating insects includes the ultraviolet part of the spectrum. Some flowers have patterns that are invisible to us but show up if we photograph them with film that is sensitive to ultraviolet light. Those patterns provide information to prospective pollinators, but humans are not equipped to receive that information.

At the other end of the spectrum is infrared radiation, which we sense as heat. Other animals extract much more information from infrared radiation—especially that emitted by potential prey. Pit vipers such as rattlesnakes have *pit organs*, one just in front of each eye, that use highly sensitive heat detectors and a simple pinhole camera arrangement to sense and locate infrared radiation (Figure 45.26). In total darkness, these snakes can locate a prey animal such as a mouse, orient to it, and strike it with great accuracy.

Echolocation is sensing the world through reflected sound

Some species emit intense sounds and create images of their environments from the echoes of those sounds. Bats, porpoises, dolphins, and (to a lesser extent) whales are accomplished echolocators. Some species of bats have elaborate modifications of their noses to direct the sounds they emit, as well as impressive ear pinnae to collect the returning echoes. The high-frequency sounds they emit as pulses (about 20 to 80 per second) are above the range of human hearing, but they are extremely loud in contrast to the resulting faint echoes bouncing off small insects. An echolocating bat is similar to a construction worker who is trying to overhear a whispered conversation on a street corner while using a pneumatic drill. To avoid deafening themselves, bats use muscles in their middle ears to dampen their sensitivity while they emit sounds, then relax them quickly enough to hear the echoes. The ability of bats to use echolocation to sense their environment is so good that in a totally dark room strung with fine wires, they can capture tiny flying insects while navigating around the wires.

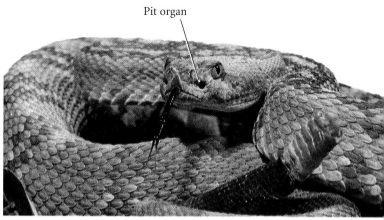

Pit organ

Crotalus molossus

45.26 Stalking in the Dark
The black-tailed rattlesnake of the southwestern United States is a pit viper. Pit vipers can locate prey in total darkness on the basis of infrared radiation they sense through their pit organs.

Some fish can sense electric fields

We discussed the mechanoreceptors in the lateral lines of fishes (see Figure 45.9). The lateral lines of some species, especially those such as catfish that live in murky waters, also contain electroreceptors. These sensory cells enable the fish to detect weak electric fields, which can help them locate prey.

The use of electroreceptors is even more sophisticated in species called electric fishes. These fishes have evolved electric organs in their tails that generate a continuous series of electric pulses, creating a weak electric field around their bodies. Any objects in the environment, such as rocks, plants, or other fish, disrupt the electric fish's electric field, and the electroreceptors of the lateral line detect those disruptions. In some electric fish species, each individual in a group emits its electric pulses at a different frequency. If a new fish is added to the group, they all readjust their frequencies.

Chapter Summary

Sensory Cells, Sensory Organs, and Transduction

▶ Sensory cells transduce information about an animal's external and internal environment into action potentials. **Review Figures 45.1, 45.2**

▶ The interpretation of action potentials as particular sensations depends on which neurons in the CNS receive them.

▶ Sensory cells have membrane receptor proteins that cause ion channels to open or close, generating receptor potentials. Receptor potentials can spread to regions of the sensory cell plasma membrane that generate action potentials, or they can influence the release of neurotransmitter from the sensory cell. **Review Figure 45.3**

▶ Adaptation enables the nervous system to ignore irrelevant stimuli while remaining responsive to relevant or to new stimuli.

Chemoreceptors: Responding to Specific Molecules

▶ Smell, taste, and the sensing of pheromones are examples of chemosensation. Chemoreceptor cells have receptor proteins that can bind to specific molecules that come into contact with the sensory cell membrane. **Review Figure 45.5**

▶ The binding of an odorant molecule to a receptor protein causes the production of a second messenger in the chemoreceptor cell. The second messenger alters ion channels and creates a receptor potential. **Review Figure 15.17**

▶ Chemoreceptors in the mouth cavities of vertebrates are responsible for the sense of taste. **Review Figure 45.6**

Mechanoreceptors: Detecting Stimuli that Distort Membranes

▶ In the skin there are a diversity of mechanoreceptors that respond to touch and pressure. The density of mechanoreceptors in any skin area determines the sensitivity of that area to touch and pressure. **Review Figure 45.7**

▶ Stretch receptors in muscles, tendons, and ligaments inform the CNS of the positions of and the loads on parts of the body. **Review Figure 45.8**

▶ Hair cells are mechanoreceptors that are not neurons. The bending of their stereocilia alters their membrane proteins and therefore their receptor potentials. Hair cells are found in organs of equilibrium and orientation such as the lateral line system of fishes, the statocysts of invertebrates, and the semicircular canals and vestibular apparatus of mammals. **Review Figures 45.9, 45.10, 45.11**

▶ Hair cells are responsible for mammalian auditory sensitivity. Ear pinnae collect and direct sound waves to the tympanic membrane, which vibrates in response to sound waves. The movements of the tympanic membrane are amplified through a chain of ossicles that conduct the vibrations to the oval window. Movements of the oval window create pressure waves in the fluid-filled cochlea. **Review Figure 45.12**

▶ The basilar membrane running down the center of the cochlea is distorted at specific locations that depend on the frequency of the pressure wave. These distortions cause the bending of hair cells in the organ of Corti, which rests on the basilar membrane. Changes in hair cell receptor potentials create action potentials in the auditory nerve, which conducts the information to the CNS. **Review Figure 45.13**

Photoreceptors and Visual Systems: Responding to Light

▶ Photosensitivity depends on the capture of photons of light by rhodopsin, a photoreceptor molecule that consists of a protein called opsin and a light-absorbing prosthetic group called retinal. Absorption of light by retinal is the first step in a cascade of intracellular events leading to a change in the receptor potential of the photoreceptor cell. **Review Figure 45.14**

▶ When excited by light, vertebrate photoreceptor cells hyperpolarize and release less neurotransmitter onto the neurons with which they form synapses. They do not fire action potentials. **Review Figures 45.15, 45.16, 45.17**

▶ Vision results when eyes focus patterns of light onto layers of photoreceptors. Eyes vary from the simple eye cups of flatworms, which enable the animal to sense the direction of a light source, to the compound eyes of arthropods, which enable the animal to detect shapes and patterns, to the lensed eyes of cephalopods and vertebrates. **Review Figures 45.18, 45.19**

▶ The eyes of vertebrates and cephalopods focus detailed images of the visual field onto dense arrays of photoreceptors that transduce the visual image into neural signals. **Review Figures 45.20, 45.21**

▶ The vertebrate photoreceptors are rod cells, responsible for dim light and black-and-white vision, and cone cells, responsible for color vision by virtue of their spectral sensitivities. **Review Figure 45.23**

▶ The vertebrate retina is a dense array of neurons lining the back of the eyeball. It consists of five layers of cells. The outermost layer consists of the rods and cones. The innermost layer consists of the ganglion cells, which send their axons in the optic nerve to the brain. Between the photoreceptors and the ganglion cells are neurons that process the information from the photoreceptors. **Review Figure 45.24**

▶ The area of the retina that receives light from the center of the visual field, the fovea, has the greatest density of photoreceptors. In humans the fovea contains almost exclusively cone cells, which are responsible for color vision but are not very sensitive in dim light.

▶ Each ganglion cell is stimulated by light falling on a small circular patch of photoreceptors called a receptive field. Receptive fields have a center and a surround, which have opposing effects on the ganglion cell. If the center is excitatory, the surround is inhibitory, and vice versa. **Review Figure 45.25**

Sensory Worlds Beyond Our Experience

▶ Many animals have sensory abilities that we do not share. Bats echolocate, insects see ultraviolet radiation, pit vipers "see" infrared radiation, and fish sense electric fields.

For Discussion

1. Compare and contrast the functioning of olfactory receptors and photoreceptors. How do these sensory cells enable the CNS to discriminate between an apple and an orange?

2. Amplification of signal is an important feature of sensory systems. Compare mechanisms of amplification in olfactory, visual, and auditory systems.

3. If you were blindfolded and placed in a wheelchair, how would you know if you were being pushed forward or backward?

4. Describe and contrast two sensory systems that enable animals to "see" in the dark. What problems or limitations are inherent in these systems in comparison with vision?

5. Communication is the transfer of information from one animal to another. Animals can use visual, olfactory, tactile, and auditory signals to communicate. From what you know about these sensory systems, discuss the relative advantages and disadvantages of these systems for communication.

46 The Mammalian Nervous System: Structure and Higher Functions

 PHINEAS GAGE WAS AN INDUSTRIOUS, RE-sponsible, considerate young man. He was 25 years old in 1848, working as a railroad construction foreman. He had the respect of his men, and he looked out for them to the extent that he took on himself the most dangerous tasks associated with blasting the rocks in the path of the railroad.

Late one afternoon the last hole had been drilled for the day. Gage poured blasting powder into the hole and tamped it with a meter-long, 3-cm wide iron rod. The tamping iron hit the side of the hole, struck a spark, and ignited the powder. The explosion shot the rod out of the hole like a bullet. It struck Phineas below his left eye, penetrated his skull, passed through the part of his brain behind his forehead, and exited out the top of his head. Was this the end of Phineas Gage?

Gage regained consciousness within minutes and could speak. He was taken to a hotel, where a physician dressed his wounds, but the doctor could do little else. Infections were a problem, but Gage's senses and memory were intact. In 3 weeks he left his bed, but he did not return to his work at the railroad. The recovered Phineas Gage was quarrelsome, bad-tempered, lazy, and irresponsible. He was impatient and obstinate, and he used profane language, which he had never done before.

The body of Phineas Gage survived the accident, but he was an entirely different person. He spent the rest of his days as a drifter, earning money by telling his story, exhibiting his scars and his tamping iron. If you are in Cambridge, Massachusetts, you can pay him a visit. His skull, death mask, and the tamping iron are on display in the Museum of the Medical College of Harvard University.

The sad story of Phineas Gage reveals that the essence of individuality and personality resides in the brain. What is this miraculous organ, and what does it do? The human brain weighs about 1.5 kg, is mostly water, and has the consistency of custard. Yet the complexity of this small mass of tissue exceeds that of any other known matter. The work of the brain is to process and store information and to control the physiology and behavior of the body. The brain is con-stantly receiving, integrating, and interpreting information from all the senses. To respond to that information, it generates commands to the muscles and organs of the body.

The unit of function of the brain is the neuron. The human brain consists of about 10 billion neurons, which account for its ability to handle vast amounts of information. In the previous two chapters we learned about the cellular properties of neurons: how they generate and conduct action potentials, how they transduce sensory information, how they communicate with each other at synapses, and how information is integrated at synapses. In this chapter we take on the challenge of understanding the functions of the human nervous system in terms of these cellular mechanisms.

The Nervous System: Structure, Function, and Information Flow

The human nervous system is more than the brain. The brain and spinal cord together constitute the **central nervous system** (**CNS**). Information is brought to and from the

An Unintentional Experiment
In a nineteenth-century railroad construction accident, an explosion blew a tamping iron through the brain of Phineas Gage. Unbelievably, Gage survived, but his personality was radically changed. This drawing of Gage's skull was made at the time of his death.

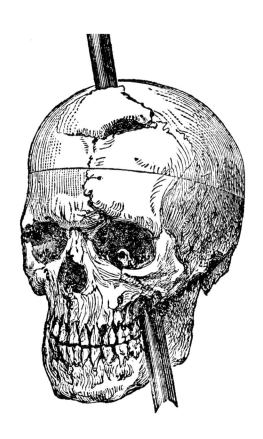

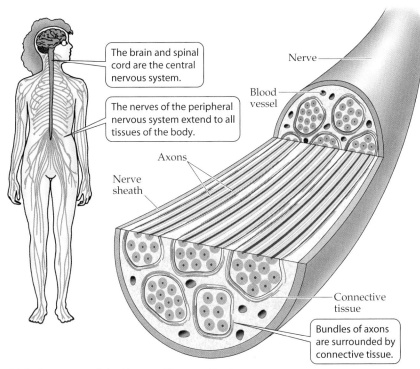

46.1 Anatomy of the Human Nervous System
(a) Information is communicated between the central nervous system and the other tissues of the body through the peripheral nervous system. (b) A nerve contains the axons of many neurons. Some of these neurons conduct information to and others from the central nervous system.

cal regulation (for example, blood pressure, deep body temperature, blood oxygen supply).

The **efferent** portion of the peripheral nervous system carries information *from the CNS* to the muscles and glands of the body. Efferent pathways can be divided into a **voluntary** division, which executes our conscious movements, and an **involuntary**, or **autonomic**, division, which controls physiological functions.

In addition to neural information, the CNS receives chemical information in the form of hormones circulating in the blood. Neurohormones released by neurons into the extracellular fluids of the brain can send chemical information to other neurons in the brain or can leave the brain and enter the circulation. In Chapter 41 we learned of the important role of neurohormones in the control of the anterior pituitary and saw that other neurohormones are released from the posterior pituitary into the circulation.

Now we can begin to translate our conceptual scheme of information flow into an anatomical view of the nervous system. It can be difficult to learn the relationships between the different structures of the adult nervous system, but the task is much easier if we begin with the development of the nervous system from a simple tubular structure that forms in the embryo.

CNS by means of an enormous network of nerves that make up the **peripheral nervous system** (Figure 46.1*a*). The peripheral nervous system reaches every tissue of the body. It connects to the CNS via spinal nerves and cranial nerves.

A **nerve** is a bundle of axons (Figure 46.1*b*) that carries information about many things simultaneously. It is important to distinguish between the axon of a single neuron and a nerve. Some axons in a nerve may be carrying information to the CNS while other axons in the same nerve are carrying information from the CNS to the organs of the body.

A conceptual diagram of the nervous system traces information flow

The nervous system is an information processing system—a very complex one that handles many tasks simultaneously. It will help to organize our thinking about the nervous system by beginning with a conceptual diagram of information flow (Figure 46.2). We can then plug anatomical and functional details into this general model.

The **afferent** portion of the peripheral nervous system carries information *to the CNS*. We are consciously aware of much of the information that moves through these afferent pathways (for example, vision, hearing, temperature, pain, the position of limbs), but we are not consciously aware of other afferent information that is important for physiologi-

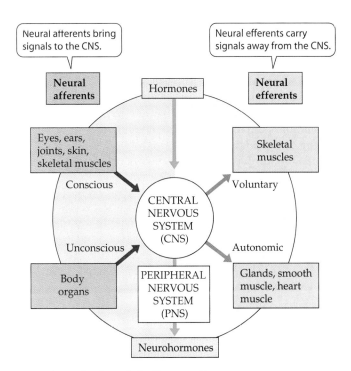

46.2 Organization of the Nervous System
The peripheral nervous system carries information both to and from the central nervous system. The CNS also receives hormonal inputs and produces hormonal outputs.

The vertebrate CNS develops from the embryonic neural tube

Early in the development of all vertebrate embryos, a hollow tube of neural tissue forms (see Chapter 43). This neural tube runs the length of the embryo on its dorsal side. At the anterior end of the embryo, the neural tube forms three swellings that become the basic divisions of the brain: the **hindbrain**, the **midbrain**, and the **forebrain**. The rest of the neural tube becomes the spinal cord (Figure 46.3). The cranial and spinal nerves, which make up the peripheral nervous system, sprout from the neural tube and grow throughout the embryo.

Each of the three regions of the embryonic brain develops into several structures in the adult brain. From the hindbrain come the **medulla**, the **pons**, and the **cerebellum**. The medulla is continuous with the spinal cord. The pons is in front of the medulla, and the cerebellum is a dorsal outgrowth of the pons. The medulla and pons contain distinct groups of neurons that are involved in the control of physiological functions such as breathing and circulation or basic motor patterns such as swallowing and vomiting. All neural information traveling between the spinal cord and higher brain areas must pass through the pons and the medulla.

The cerebellum is like the conductor of an orchestra; it receives "copies" of the commands going to the muscles from higher brain areas, and it receives information coming up the spinal cord from the joints and muscles. Thus it can compare the motor "score" with the actual behavior of the muscles and refine the motor commands.

From the embryonic midbrain come structures that process aspects of visual and auditory information. In addition, all information traveling between higher brain areas and the spinal cord must pass through the midbrain.

The embryonic forebrain develops a central region called the **diencephalon** and a surrounding structure called the **telencephalon**. The diencephalon is the core of the forebrain and consists of an upper structure called the **thalamus** and a lower structure called the **hypothalamus**. The thalamus is the final relay station for sensory information going to the telencephalon, and the hypothalamus is responsible for the regulation of many physiological functions and biological drives.

The telencephalon consists of two **cerebral hemispheres**, left and right (also referred to as the **cerebrum**). In humans, the telencephalon is by far the largest part of the brain and plays major roles in sensory perception, learning, memory, and conscious behavior.

Understanding the relationships among the many structures of the complex adult brain is a little easier if you keep this linear organization of the neural axis in mind: Communication between the spinal cord and the telencephalon travels through the medulla, pons, midbrain, and diencephalon. The medulla, pons, and midbrain are referred to collectively as the **brain stem**. In general, more primitive and autonomic functions are localized farther down this neural axis, while more complex and evolutionarily advanced functions are found higher on the axis.

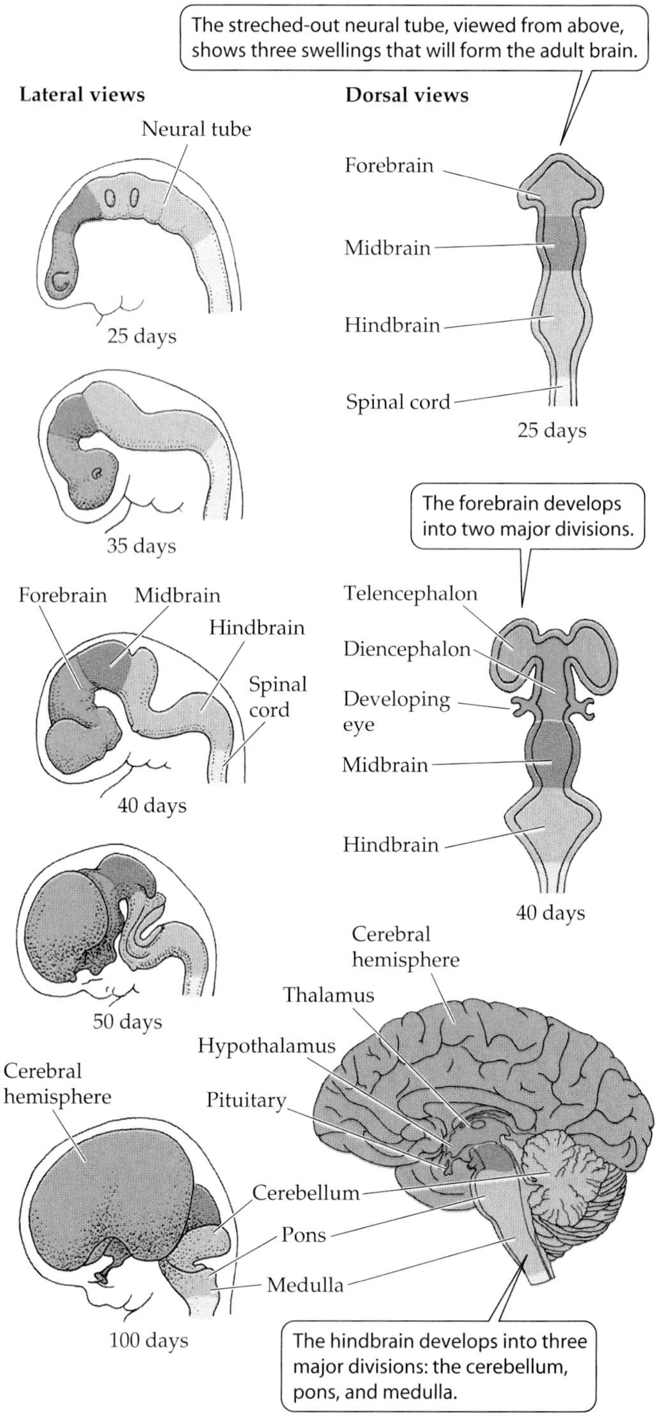

The streched-out neural tube, viewed from above, shows three swellings that will form the adult brain.

Lateral views

Neural tube
25 days

35 days

Forebrain Midbrain
Hindbrain
Spinal cord
40 days

50 days

Cerebral hemisphere
100 days

Dorsal views

Forebrain
Midbrain
Hindbrain
Spinal cord
25 days

The forebrain develops into two major divisions.

Telencephalon
Diencephalon
Developing eye
Midbrain
Hindbrain
40 days

Cerebral hemisphere
Thalamus
Hypothalamus
Pituitary
Cerebellum
Pons
Medulla

The hindbrain develops into three major divisions: the cerebellum, pons, and medulla.

46.3 Development of the Human Nervous System
Three swellings at the anterior end of the hollow neural tube in early vertebrate embryos develop into the parts of the adult brain. The final view is an adult human brain section cut in half through the midline.

As we go up the vertebrate phylogenetic scale from fish to mammals, the telencephalon increases in size, complexity, and importance. The forebrain dominates the nervous systems of mammals, and damage to this region results in severe impairment of sensory, motor, or cognitive functions, and even coma. In contrast, a shark with its telencephalon removed can swim almost normally.

46.4 The Spinal Cord Processes Information
Sensory information (afferent) enters through the dorsal horns (blue pathway), and motor output (efferent) leaves via the ventral horns (orange and red pathways). The extensor component of the knee-jerk response is a monosynaptic reflex circuit, but the flexor inhibition component involves a spinal interneuron (black).

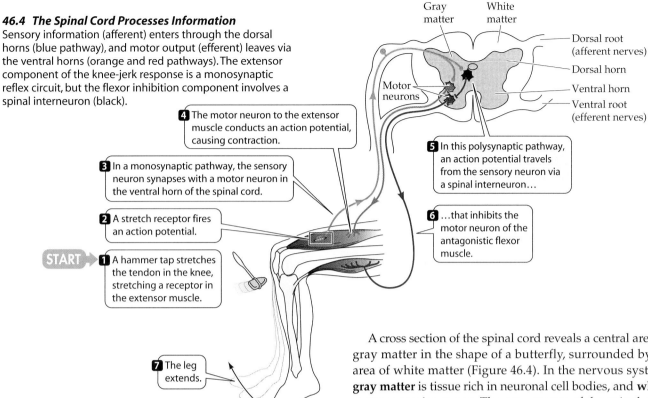

Gray matter   White matter

Dorsal root (afferent nerves)
Dorsal horn
Ventral horn
Ventral root (efferent nerves)

Motor neurons

4 The motor neuron to the extensor muscle conducts an action potential, causing contraction.

3 In a monosynaptic pathway, the sensory neuron synapses with a motor neuron in the ventral horn of the spinal cord.

2 A stretch receptor fires an action potential.

5 In this polysynaptic pathway, an action potential travels from the sensory neuron via a spinal interneuron…

6 …that inhibits the motor neuron of the antagonistic flexor muscle.

START ▶ 1 A hammer tap stretches the tendon in the knee, stretching a receptor in the extensor muscle.

7 The leg extends.

Functional Subsystems of the Nervous System

When we talk about the development of the nervous system, we describe it in terms of anatomically distinct structures. However, the nervous system is always engaged in many tasks at the same time—a phenomenon called *parallel processing of information*. Any one task usually involves many different anatomical structures. Understanding the nervous system is made simpler if we recognize its functional subsystems, such as the spinal cord, reticular system, limbic system, and cerebrum. Any one anatomical structure may play roles in several functional subsystems.

The spinal cord receives and processes information from the body

The spinal cord conducts information in both directions between the brain and the organs of the body. It also integrates a great deal of the information coming from the peripheral nervous system and responds to that information by issuing motor commands.

The conversion of afferent to efferent information in the spinal cord without participation of the brain is called a **spinal reflex**. The simplest type of spinal reflex involves only two neurons and one synapse and is therefore called a **monosynaptic reflex**. An example is the knee-jerk reflex, which your physician checks by tapping just below your knee with a small mallet. We can diagram the wiring of a monosynaptic reflex by following the flow of information through the spinal cord.

A cross section of the spinal cord reveals a central area of gray matter in the shape of a butterfly, surrounded by an area of white matter (Figure 46.4). In the nervous system, **gray matter** is tissue rich in neuronal cell bodies, and **white matter** contains axons. The gray matter of the spinal cord contains the cell bodies of the spinal neurons; the white matter contains the axons that conduct information up and down the spinal cord. Spinal nerves leave the spinal cord at regular intervals on each side. Each spinal nerve has two roots connecting to the gray matter—one connecting with the *dorsal horn*, the other with the *ventral horn*. Each spinal nerve carries both afferent and efferent information. The afferent axons enter the spinal cord through the dorsal roots and the efferent axons leave the spinal cord through the ventral roots.

In the case of the knee-jerk reflex, sensory information comes from stretch receptors in the leg muscle that is suddenly stretched when the mallet strikes the tendon that runs over the knee. Each stretch receptor initiates action potentials that are conducted by the axon of a sensory neuron in through the dorsal horn of the spinal cord and all the way to the ventral horn. In the ventral horn, the sensory neuron synapses with motor neurons, causing them to fire action potentials that are then conducted back to the leg extensor muscle, causing it to contract. The function of this simple circuit is to sense an increased load on the limb, and to cause the muscle to increase its strength of contraction to compensate for the added load.

Most spinal circuits are more complex than this monosynaptic reflex. We can demonstrate that by building on the circuit we have just traced. Limb movement is controlled by *antagonistic* sets of muscles—muscles that work against each other. When one member of an antagonistic set of muscles contracts, it bends or flexes the limb; it is therefore called a *flexor*. The antagonist to this muscle straightens or extends the limb and is called an *extensor*. For a limb to move, one muscle of the pair must relax while the other

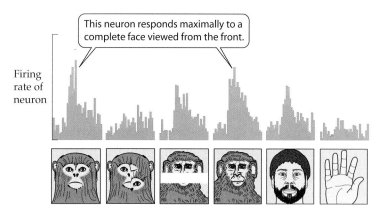

46.8 Neurons in One Region of the Temporal Lobe Respond to Faces
The traces represent the firing rate of a neuron in the temporal lobe of a monkey in response to the pictures shown below them.

THE FRONTAL LOBE.   A strip of the frontal lobe cortex just in front of the central sulcus is called the *primary motor cortex* (see Figure 46.7b). The neurons in this region have axons that project to muscles in specific parts of the body. The parts of the body can be mapped onto the primary motor cortex, from the head region on the lower side to the lower part of the body at the top. Areas with fine motor control, such as the face and hands, have the greatest representation (Figure 46.9). If a neuron in the primary motor cortex is electrically stimulated, the response is the twitch of a muscle, but not a coordinated, complex behavior.

The association functions of the frontal lobe are diverse and are best described as having to do with planning. The story of Phineas Gage at the beginning of this chapter demonstrates the effects of damage to these association areas. People with such deficits have drastic alterations of personality because they cannot create an accurate view of themselves in the context of the world around them and cannot plan for future events.

THE PARIETAL LOBE.   The frontal and parietal lobes are separated by a deep valley called the *central sulcus*. The strip of parietal lobe cortex just behind the central sulcus is the *primary somatosensory cortex* (see Figure 46.7b). This area receives information through the thalamus about touch and pressure sensations.

The whole body surface is represented in the primary somatosensory cortex—the head at the bottom and the legs at the top (see Figure 46.9). Areas of the body that have lots of sensory neurons and are capable of making fine distinctions in touch (such as the lips and the fingers) have disproportionately large representation. If a

very small area of the primary somatosensory cortex is stimulated electrically, the subject reports feeling specific sensations, such as touch, from a very localized part of the body.

A major association function of the parietal lobe is attending to complex stimuli. Damage to the right parietal lobe causes a condition called *contralateral neglect syndrome*, in which the individual tends to ignore stimuli from the left side of the body or the left visual field. Such individuals have difficulty performing complex tasks such as dressing the left side of the body; an afflicted man may not be able to shave the left side of his face. When asked to copy simple drawings, a person who exhibits this syndrome can do well with the right side of the drawing, but not the left (Figure 46.10). The parietal cortex is not symmetrical with respect to its role in attention. Damage to the left parietal cortex does not cause neglect of the right side of the body. We will see similar asymmetries in cortical function when we discuss language.

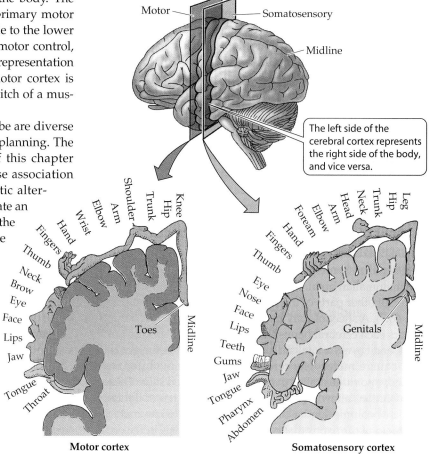

46.9 The Body Is Represented in the Primary Motor Cortex and the Primary Somatosensory Cortex
Cross sections through the primary motor and primary somatosensory cortexes can be represented as maps of the human body. Body parts are shown in relation to the brain area devoted to them.

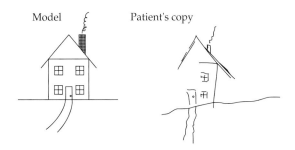

46.10 Contralateral Neglect Syndrome
A person with damage to the right parietal association cortex will neglect the left side of a drawing when asked to copy a model.

THE OCCIPITAL LOBE. The occipital lobes receive and process visual information. We'll learn more about the details of that process later in this chapter. The association areas of the occipital cortex are essential for making sense out of the visual world and translating visual experience into language. Some deficits resulting from damage to association areas of the occipital cortex are quite specific. In one case, a woman with limited damage was unable to see motion. Her vision was intact, but she could see a waterfall only as a still image, and a car approaching only as a series of scenes of a stationary object at different distances.

The cerebrum has increased in size and complexity

As mentioned earlier, the size of the telencephalon relative to the rest of the brain increases substantially as we go from fishes to amphibians, to reptiles, to birds and mammals. Even when we consider only mammals, the cerebral cortex increases in size and complexity when we compare animals such as rodents, whose behavioral repertoires are relatively simple, with animals such as primates that have much more complex behavior.

The most dramatic increase in the size of the cerebral cortex took place during the last several million years of human evolution. The incredible intellectual capacities of *Homo sapiens* are associated with enlargement of the cerebral cortex. Humans do not have the largest brains in the animal kingdom; elephants, whales, and porpoises have larger brains in terms of mass. If we compare brain size to body size, however, humans and dolphins top the list. Humans have the largest ratio of brain size to body size, and they have the most highly developed cerebral cortex. Another feature of the cerebral cortex that reflects increasing behavioral and intellectual capabilities is the ratio of association cortex to primary somatosensory and motor cortexes. Humans have the largest relative amount of association cortex.

Information Processing by Neuronal Networks

In Chapter 44 we learned how neurons interact to process information. A goal of neurobiology is to understand the complex functions of the nervous system in terms of the properties of neurons and synapses between them. We will use two subsystems as examples to show how the functions of the nervous system can be understood in terms of neuronal networks. The first example, the autonomic nervous system, consists of efferent pathways; the second, the visual system, consists of afferent and integrative pathways. Techniques that have allowed neurobiologists to trace neuronal connections, chemically characterize synapses, and record the activities of single cells and groups of cells have advanced our understanding of how certain subsystems of the nervous system work.

The autonomic nervous system controls organs and organ systems

The autonomic nervous system is divided into two parts: the **sympathetic** and **parasympathetic** divisions. These two divisions work in opposition to each other in their effects on most organs, one causing an increase in activity and the other causing a decrease. The best-known functions of the autonomic nervous system are those of the sympathetic division called the "fight-or-flight" mechanisms, which increase heart rate, blood pressure, and cardiac output and prepare the body for emergencies (see Chapter 41). In contrast, the parasympathetic division slows the heart and lowers blood pressure.

It is tempting to think of the sympathetic division as the one that speeds things up and the parasympathetic division as the one that slows things down, but that is not always a correct distinction. The sympathetic division slows the digestive system and the parasympathetic division accelerates it. The two divisions of the autonomic nervous system are easily distinguished from each other by their anatomy, their neurotransmitters, and their actions (Figure 46.11).

Both divisions of the autonomic nervous system are efferent pathways. Each autonomic efferent pathway begins with a neuron that has its cell body in the brain stem or spinal cord and uses acetylcholine as its neurotransmitter. These cells are called **preganglionic neurons** because the second neuron in the pathway with which they synapse resides in a **ganglion** (a collection of neuronal cell bodies that is outside of the CNS). The second neuron is called a **postganglionic neuron** because its axon extends out from the ganglion. The axons of the postganglionic neurons end on the cells of the target organs.

The postganglionic neurons of the sympathetic division use norepinephrine as their neurotransmitter; those of the parasympathetic division use acetylcholine. In organs that receive both sympathetic and parasympathetic input, the target cells respond in opposite ways to norepinephrine and to acetylcholine. A region of the heart called the *pacemaker*, which generates the heartbeat, is an example. Stimulating the sympathetic nerve to the heart or dripping norepinephrine onto the pacemaker region depolarizes the pacemaker cells, increases their firing rate, and causes the heart to beat faster. Stimulating the parasympathetic nerve to the heart or dripping acetylcholine onto the pacemaker region hyperpolarizes the pacemaker cells, decreases their firing rate, and causes the heart to beat slower. In contrast,

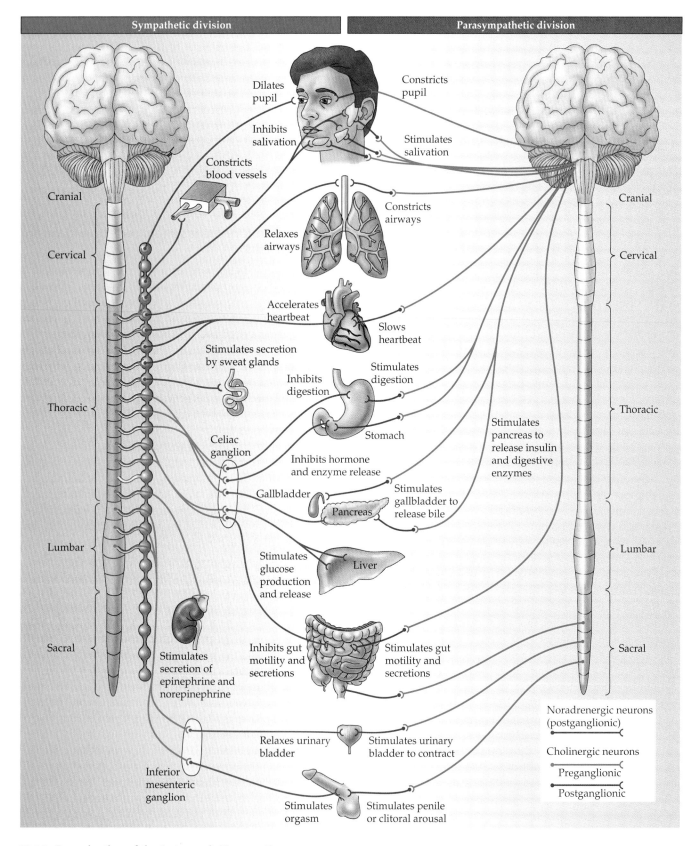

Sympathetic division | **Parasympathetic division**

Cranial

Cervical

Thoracic

Lumbar

Sacral

Dilates pupil

Inhibits salivation

Constricts blood vessels

Relaxes airways

Accelerates heartbeat

Stimulates secretion by sweat glands

Inhibits digestion

Celiac ganglion

Inhibits hormone and enzyme release

Gallbladder

Stimulates glucose production and release

Liver

Pancreas

Stimulates secretion of epinephrine and norepinephrine

Inhibits gut motility and secretions

Inferior mesenteric ganglion

Relaxes urinary bladder

Stimulates orgasm

Constricts pupil

Stimulates salivation

Constricts airways

Slows heartbeat

Stimulates digestion

Stomach

Stimulates pancreas to release insulin and digestive enzymes

Stimulates gallbladder to release bile

Stimulates gut motility and secretions

Stimulates urinary bladder to contract

Stimulates penile or clitoral arousal

Cranial

Cervical

Thoracic

Lumbar

Sacral

Noradrenergic neurons (postganglionic)

Cholinergic neurons
Preganglionic
Postganglionic

46.11 Organization of the Autonomic Nervous System
The autonomic nervous system is divided into the sympathetic and parasympathetic divisions, which work in opposition to each other in their effects on most organs (one causing an increase and the other a decrease in activity).

in the digestive tract, norepinephrine hyperpolarizes muscle cells, which slows digestion, and acetylcholine depolarizes muscle cells, which accelerates digestion.

The sympathetic and parasympathetic divisions of the autonomic nervous system can also be distinguished by anatomy (see Figure 46.11). The preganglionic neurons of the parasympathetic division come from the brain stem and the last segment of the spinal cord. The preganglionic neurons of the sympathetic division come from the upper regions of the spinal cord below the neck—the thoracic and lumbar regions. The ganglia of the sympathetic division are mostly lined up in two chains, one on either side of the spinal cord. The parasympathetic ganglia are close to—sometimes sitting on—the target organs.

The autonomic nervous system is an important link between the CNS and many physiological functions of the body. Its control of diverse organs and tissues is crucial to homeostasis. In spite of its complexity, work by neurobiologists and physiologists over many decades has made it possible to understand its functions in terms of neuronal properties and circuits. In Chapter 49, for example, we will see how information from pressure receptors in the blood vessels is transmitted to the CNS, where it produces autonomic signals that control the rate of the heartbeat.

Neurons and circuits in the occipital cortex integrate visual information

In Chapter 45 we learned that the information conveyed to the brain in the optic nerve consists of action potentials that are stimulated by light falling on small circular areas of the retina called receptive fields. A receptive field contains many photoreceptor cells connected together in a circuit in such a way that the signals they produce are integrated and transmitted to the brain by a single retinal ganglion cell. The axon of each ganglion cell travels to the brain in the optic nerve. How does the brain construct visual images from information about circular patches of light falling on the retina?

Information from the retina is transmitted through the optic nerve to a relay station in the thalamus, and then to the brain's visual processing area, in the occipital cortex at the back of the cerebral hemispheres (see Figure 46.7*b*). David Hubel and Torsten Wiesel of Harvard University studied the activity of neurons in this *visual cortex*. They recorded the activities of single cells in the brains of living animals while they stimulated the animals' retinas with spots and bars of light. They found that cells in the visual cortex, like retinal ganglion cells, have receptive fields—specific areas of the retina that, when stimulated by light, influence the rate at which the cells fire action potentials.

Cells in the visual cortex, however, have receptive fields that differ from the simple circular receptive fields of retinal ganglion cells. Cortical cells called **simple cells** are maximally stimulated by bars of light that have specific orientations. Simple cells probably receive input from several ganglion cells whose circular receptive fields are lined up in a row.

Complex cells in the visual cortex are also maximally stimulated by a bar of light with a particular orientation, but

the bar may fall anywhere on a large area of retina described as that cell's receptive field. Complex cells receive input from several simple cells that share a certain stimulus orientation, but have receptive fields in different places on the retina (Figure 46.12). Some complex cells respond most strongly when the bar of light moves in a particular direction.

The concept that emerges from these experiments is that the brain assembles a mental image of the visual world by analyzing edges of patterns of light falling on the retina.

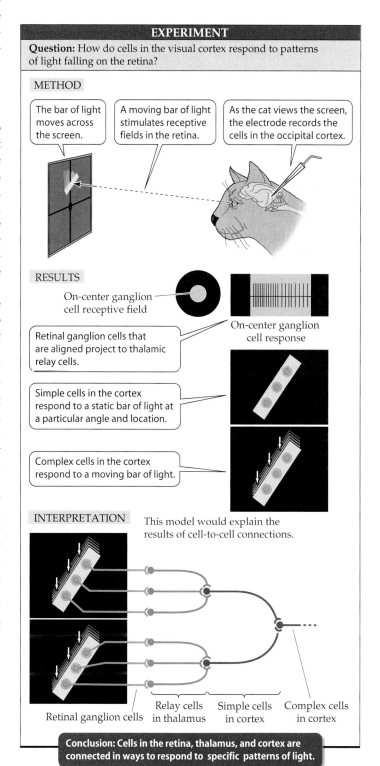

EXPERIMENT

Question: How do cells in the visual cortex respond to patterns of light falling on the retina?

METHOD

The bar of light moves across the screen.

A moving bar of light stimulates receptive fields in the retina.

As the cat views the screen, the electrode records the cells in the occipital cortex.

RESULTS

On-center ganglion cell receptive field

On-center ganglion cell response

Retinal ganglion cells that are aligned project to thalamic relay cells.

Simple cells in the cortex respond to a static bar of light at a particular angle and location.

Complex cells in the cortex respond to a moving bar of light.

INTERPRETATION This model would explain the results of cell-to-cell connections.

Retinal ganglion cells | Relay cells in thalamus | Simple cells in cortex | Complex cells in cortex

Conclusion: Cells in the retina, thalamus, and cortex are connected in ways to respond to specific patterns of light.

46.12 Receptive Fields of Cells in the Visual Cortex
Cells in the visual cortex respond to specific patterns of light falling on the retina. Ganglion cells that project information about circular receptive fields converge on simple cells in the cortex in such a way that the simple cells have linear receptive fields. Simple cells project to complex cells in such a way that the complex cells can respond to linear stimuli falling on different areas of the retina.

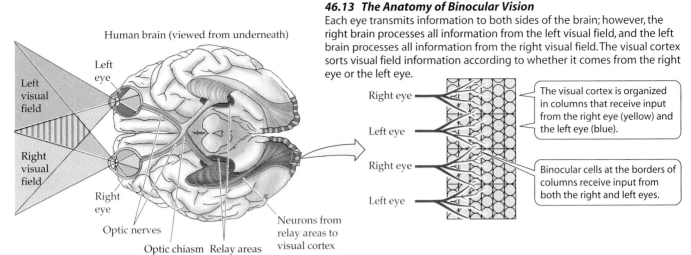

46.13 The Anatomy of Binocular Vision
Each eye transmits information to both sides of the brain; however, the right brain processes all information from the left visual field, and the left brain processes all information from the right visual field. The visual cortex sorts visual field information according to whether it comes from the right eye or the left eye.

Human brain (viewed from underneath)

Left eye

Left visual field

Right visual field

Right eye

Optic nerves

Optic chiasm   Relay areas in thalamus

Neurons from relay areas to visual cortex

Right eye
Left eye
Right eye
Left eye

The visual cortex is organized in columns that receive input from the right eye (yellow) and the left eye (blue).

Binocular cells at the borders of columns receive input from both the right and left eyes.

This analysis is conducted in a massively parallel fashion. Each retina sends a million axons to the brain, but there are hundreds of millions of neurons in the visual cortex. Each bit of information from a retinal ganglion cell is received by hundreds of cortical cells, each responsive to a different combination of orientation, position, and even movement of contrasting lines in the pattern of light falling on the retina.

Cortical cells receive input from both eyes

How do we see objects in three dimensions? The quick answer is that our two eyes see overlapping, yet slightly different, visual fields; that is, we have *binocular vision*. Turn a typical conical flowerpot upside down and look down at it so that the bottom of the pot is exactly in the center of your overall field of vision. You see the bottom of the pot, and you see equal amounts of the sides and rim of the pot as concentric circles. Now, if you close your left eye, you see more of the right side and right rim of the pot. With your right eye closed, you see more of the left side and left rim of the pot. The discrepancies in the information coming from your two eyes are interpreted by the brain to provide information about the depth and the three-dimensional shape of the flowerpot. If you are blind in one eye, you have great difficulty discriminating distances. Animals whose eyes are on the sides of their heads have nonoverlapping fields of vision and, as a result, poor depth vision, but they can see predators creeping up from behind.

The story of how the brain integrates information from two eyes begins with the paths of the optic nerves. If you look at the underside of the brain, the optic nerves from the two eyes appear to join together just under the hypothalamus and then separate again. The place where they join is called the **optic chiasm** (Figure 46.13). Axons from the half of each retina closest to your nose cross in the optic chiasm and go to the opposite side of your brain. The axons from the other half of each retina go to the same side of the brain.

The result of this division of axons in the optic chiasm is that all visual information from your left visual field (everything left of straight ahead) goes to the right side of your brain, as shown in red in Figure 46.13. All visual information from your right visual field goes to the left side of your brain, as indicated in green in the figure. Both eyes transmit information about a specific spot in your right visual field, for example, to the same place in the left visual cortex. How are the two sources of information integrated?

Cells in the visual cortex are organized in columns. These columns alternate: left eye, right eye, left eye, right eye, and so on. Cells closest to the border between two columns receive input from both eyes and are therefore called **binocular cells**. Binocular cells interpret distance by measuring the disparity between where the same stimulus falls on the two retinas.

What is disparity? Hold your finger out in front of you and look at it, closing one eye and then the other. Your finger appears to jump back and forth because its image falls on a different position on each retina. Repeat the exercise with an object at a distance. It doesn't appear to jump back and forth as much because there is less disparity in the positions of the image on the two retinas. Certain binocular cells respond optimally to a stimulus falling on both retinas with a particular disparity. Which set of binocular cells is stimulated depends on how far away the stimulus is.

When we look at something, we can detect its shape, color, depth, and movement. Where does all this information come together? Is there a single cell that fires only when a red sports car drives by? Probably not. A specific visual experience comes from simultaneous activity in a large collection of cells. In addition, most visual experiences are enhanced by information from the other senses and from memory as well. This realization helps explain why about 75 percent of the cerebral cortex is association cortex.

Understanding Higher Brain Functions in Cellular Terms

Very few functions of the nervous system have been worked out to the point of identifying the underlying neuronal networks. The processes responsible for the higher brain functions discussed in the remaining pages are undeniably complex. Nevertheless, neurobiologists, using a wide range of techniques, are making considerable progress in understanding some of the mechanisms involved. The following discussion presents several complex aspects of brain and behavior that present challenges to neurobiologists: sleep and dreaming, learning and memory, language use, and consciousness.

Sleeping and dreaming involve electrical patterns in the cerebrum

A dominant feature of our behavior is the daily cycle of sleep and waking. All birds and mammals, and probably all other vertebrates, sleep. We spend one-third of our lives sleeping, yet we do not know why or how.

We do know, however, that we need to sleep. Loss of sleep impairs alertness and performance. Most people in our society—certainly most college students—are chronically sleep-deprived. Large numbers of accidents and serious mistakes that endanger lives can be attributed to impaired alertness due to sleep loss. Yet insomnia (difficulty in falling or staying asleep) is one of the most common medical complaints. To discover ways of dealing with these problems, it is important to learn more about the neural control of sleep.

THE ELECTROENCEPHALOGRAM. A common tool of sleep researchers is the *electroencephalogram* (*EEG*). To record an EEG, electrodes are placed at different locations on the scalp, and changes in the electric potential differences between electrodes are recorded through time. These electric potential differences reflect the electrical activity of the neurons in the brain regions under the electrodes, primarily regions of the cerebral cortex. Pens writing on a moving chart are used to record the patterns of electric potential differences between electrodes (Figure 46.14*a,b*). Usually, the

electrical activity of one or more skeletal muscles is also recorded on the chart; this record is called an *electromyogram* (*EMG*).

EEG and EMG patterns reveal the transition from being awake to being asleep. They also reveal that there are different states of sleep. In mammals other than humans, two major sleep states are easily distinguished. They are called **slow-wave sleep** and **rapid-eye-movement** (**REM**) **sleep**. In humans, we characterize sleep states as **non-REM sleep** and **REM sleep**. Human non-REM sleep is divided into four stages. Only the two deepest stages are considered true slow-wave sleep.

When a person falls asleep at night, the first sleep state entered is non-REM sleep, which progresses from stage 1 to stage 4. Stages 3 and 4 are deep, restorative, slow-wave sleep. After this first episode of non-REM sleep follows an episode of REM sleep. Throughout the night, we have four or five cycles of non-REM and REM sleep (Figure 46.14*c*). About 80 percent of our sleep is non-REM sleep and 20 percent is REM sleep.

We have vivid dreams and nightmares during REM sleep, which gets its name from the jerky movements of the eyeballs that occur during this state. The most remarkable feature of REM sleep is that inhibitory commands from the brain almost completely paralyze the skeletal muscles. Occasional muscle twitches break through the paralysis, as can be seen in a dog that appears to be trying to run in its sleep. If you look closely at a sleeping dog when its legs and paws are twitching, you will be able to see the rapid eye movements as well. Probably the function of muscle paralysis during REM sleep is to prevent the acting out of dreams. Sleepwalking, however, occurs during non-REM sleep.

The EEG characterization of sleep raises many questions. Why do we have such very different states of sleep?

(*a*)

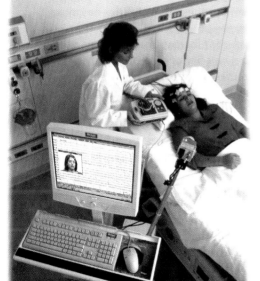

46.14 Patterns of Electrical Activity in the Cerebral Cortex Characterize Stages of Sleep
(*a*) Electrical activity in the cerebral cortex is detected by electrodes placed on the scalp and recorded on moving chart paper by a polygraph. (*b*) The resulting record is an electroencephalogram (EEG). (*c*) During a night, we cycle through the different stages of sleep.

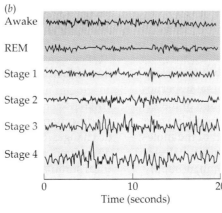

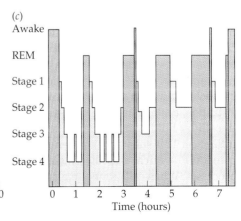

memories. Such observations were the first evidence that memories have anatomical locations in the brain and exist as properties of neurons and networks of neurons. Yet the destruction of a small area does not completely erase a memory, so it is postulated that memory is a function distributed over many brain regions and that a memory may be stimulated via many different routes.

You can recognize several forms of memory from your own experience. There is *immediate memory* for events that are happening now. Immediate memory is almost perfectly photographic, but it lasts only seconds. *Short-term memory* contains less information, but it lasts longer—on the order of 10 to 15 minutes. If you are introduced to a group of new people, you may remember most of their names for 5 or 10 minutes, but you will have forgotten them in an hour or so if you have not repeated them, written them down, or used them in a conversation. Repetition, use, or reinforcement by something that gets your attention (such as the title President) facilitates the transfer of short-term memory to *long-term memory*, which can last for days, months, or years.

Knowledge about neural mechanisms for the transfer of short-term memory to long-term memory has come from observations of persons who have lost parts of the limbic system, notably the hippocampus. A famous case is that of a man identified as H.M., whose hippocampus on both sides of the brain was removed in an effort to control severe epilepsy. Since that surgery, H.M. has not been able to transfer information to long-term memory. If someone is introduced to him, has a conversation with him, and then leaves the room, when that person returns he or she is unknown to H.M.—it is as if the previous conversation had never taken place. H.M. retains memories of events that happened before his surgery, but he remembers post-surgery events for only 10 or 15 minutes.

Memory of people, places, events, and things is called *declarative memory* because you can consciously recall and describe them. Another type of memory, called *procedural memory*, cannot be consciously recalled and described: It is the memory of how to perform a motor task. When you learn to ride a bicycle, ski, or use a computer keyboard, you form procedural memories. Although H.M. is incapable of forming declarative memories, he is capable of forming procedural memories. When taught a motor task day after day, he cannot recall the lessons of the previous day, yet his performance steadily improves. Thus procedural learning and memory must involve mechanisms different from those used in declarative learning and memory.

Our understanding of learning and memory in cellular terms is very rudimentary. New techniques that enable functional imaging of the brain in ways that reveal changes in the metabolic activity of specific regions and structures are greatly enhancing progress in this area.

Language abilities are localized in the left cerebral hemisphere

No aspect of brain function is as integrally related to human consciousness and intellect as is language. There-fore, studies of the brain mechanisms that underlie the acquisition and use of language are extremely interesting to neuroscientists. A curious observation about language abilities is that they are usually located in only one cerebral hemisphere—which in 97 percent of people is the left hemisphere. This phenomenon is referred to as the **lateralization** of language functions.

Some of the most fascinating research on this subject was conducted by Roger Sperry and his colleagues at the California Institute of Technology; Sperry received the Nobel prize in medicine for this work. The two cerebral hemispheres are connected by a tract of white matter called the **corpus callosum**. In one severe form of epilepsy, bursts of action potentials travel from hemisphere to hemisphere across the corpus callosum. Cutting the tract eliminates the problem, and patients function nearly normally following the surgery. But experiments revealed interesting deficits in the language abilities of these "split-brain" persons. Without the connection between the two hemispheres, the knowledge or experience of the right hemisphere could no longer be expressed in language, nor could language be used to communicate with the right hemisphere.

Another curious feature of our nervous systems is that the left side of the body is served (in both sensory and motor aspects) mostly by the right side of the brain, and the right side of the body is served mostly by the left side of the brain. Thus, sensory input from the right hand goes to the left cerebral hemisphere, and sensory input from the left hand goes to the right cerebral hemisphere. Language abilities reside predominantly in the left hemisphere.

The mechanisms of language in the left hemisphere have been the focus of much research. Again, the experimental subjects are persons who have suffered damage to the left hemisphere and are left with one of many forms of *aphasia*, a deficit in the ability to use or understand words. These studies have identified several language areas in the left hemisphere (Figure 46.16).

Broca's area, located in the frontal lobe just in front of the motor cortex, is essential for speech. Damage to Broca's area results in halting, slow, poorly articulated speech or even complete loss of speech, but the patient can still read and understand language. In the temporal lobe, close to its border with the occipital lobe, is *Wernicke's area*, which is more involved with sensory than with motor aspects of language. Damage to Wernicke's area can cause a person to lose the ability to speak sensibly while retaining the abilities to form the sounds of normal speech and to imitate its cadence. Moreover, such a patient cannot understand spoken or written language. Near Wernicke's area is the *angular gyrus*, which is believed to be essential for integrating spoken and written language.

Normal language ability depends on the flow of information among various areas of the left cerebral cortex. Input from spoken language travels from the primary auditory cortex to Wernicke's area (see Figure 46.16a). Input from written language travels from the primary visual cortex to the angular gyrus to Wernicke's area (see Figure

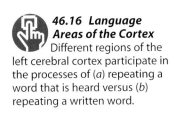

46.16 Language Areas of the Cortex
Different regions of the left cerebral cortex participate in the processes of (a) repeating a word that is heard versus (b) repeating a written word.

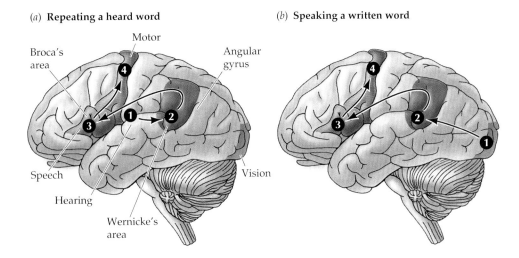

(a) **Repeating a heard word**

(b) **Speaking a written word**

46.16b). Commands to speak are formulated in Wernicke's area and travel to Broca's area and from there to the primary motor cortex. Damage to any one of those areas or the pathways between them can result in aphasia. Using modern methods of functional brain imaging, it is possible to see the metabolic activity in different brain areas when the brain is using language (Figure 46.17).

What is consciousness?

This chapter has only scratched the surface of our knowledge about the organization and functions of the human brain, but it may give you some idea of the incredible challenge that neurobiologists face in trying to understand their own brains. Progress is being aided by powerful new technologies, such as patch clamping (see Chapter 44), functional imaging (see Figure 46.17), and neurochemical and molecular methods. However, even these sophisticated new research tools may not allow us to answer the question "What is consciousness?"

If you look at a black dog, you are conscious of the fact that it is a dog, it is black, and it is a Labrador retriever, and you may remember that its name is Sarina, it is 3 years old, it belongs to your friend Meera, and so on. From what you have learned in this chapter, imagine how many neurons would be active during this experience: neurons in the visual system, the language areas, and in different regions of association cortex. But is being conscious of the black dog simply a result of the fact that all of these neurons are firing at the same time? Your brain is simultaneously processing many other sensory inputs, but you are not necessarily conscious of those inputs. What makes you conscious of the black dog and associated memories and not conscious of other information the brain is processing at the same time?

If we could describe all the neurons and all the synapses involved in the conscious experience of seeing and naming a black dog, and then build a computer with devices that modeled all these neurons and connections, would that computer be conscious? It has been said that the question of consciousness resolves into two types of problems: "easy" and "hard." The easy problems deal with all the cells and circuits that process the information that is involved in conscious experience. The implication of "easy" is that we seem to have the tools to solve these kinds of problems, as complex as they may be. The hard problems involve explaining how properties of cells and networks result in consciousness, and we seem to lack the proper tools or concepts even to begin to solve these problems.

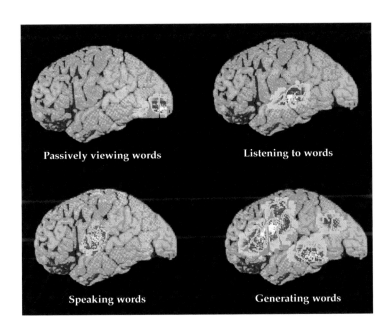

46.17 Imaging Techniques Reveal Active Parts of the Brain
Positron emission tomography (PET) scanning reveals the brain regions that are activated by different aspects of language use. A radioactive form of glucose is given to the subject. Radioactivity accumulates in brain areas in proportion to their metabolic use of glucose. The PET scan visualizes levels of radioactivity in specific brain regions.

47.5 Vertebrate Muscle Tissue

The fibers of cardiac, or heart muscle (*top*), branch and create a meshwork that resists tearing or breaking. Intercalated discs provide strong mechanical adhesions between the cells. In smooth muscle (*center*), the cells are usually arranged in sheets. Skeletal muscle (*bottom*) appears striped, or striated. The individual cells, called muscle fibers, are very large and are multinucleated.

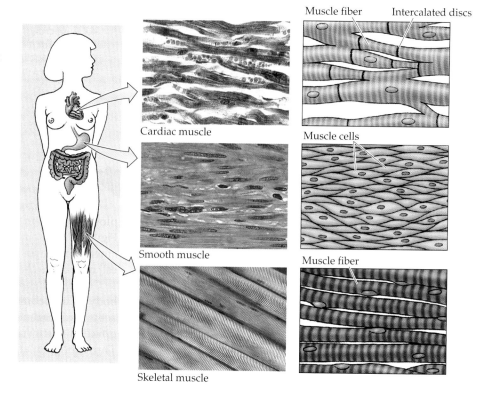

Cardiac muscle

Muscle fiber   Intercalated discs

Smooth muscle

Muscle cells

Skeletal muscle

Muscle fiber

Actin filaments consist of two actin molecules twisting around each other, and myosin filaments are bundles of many myosin molecules. The actin and myosin filaments lie parallel to each other. When contraction is triggered, the actin and myosin filaments slide past each other in a telescoping fashion.

There are three types of vertebrate muscle: smooth muscle, cardiac (heart) muscle, and skeletal muscle (Figure 47.5). Contraction in all three types is triggered by action potentials moving along their membranes. Although they all use the same contractile mechanism, these three muscle types have important differences that adapt them to their particular functions.

Smooth muscle causes slow contractions of many internal organs

Smooth muscle provides the contractile force for most of our internal organs, which are under the control of the autonomic nervous system. Smooth muscle moves food through the digestive tract, controls the flow of blood through blood vessels, and empties the urinary bladder. Structurally, smooth muscle cells are the simplest muscle cells. They are usually long and spindle-shaped, and each cell has a single nucleus. Because the filaments of actin and myosin in smooth muscle are not as regularly arranged as those in the other muscle types, the contractile machinery is not obvious when the cells are viewed under the light microscope (Figure 47.5, center).

If we study smooth muscle tissue from a particular organ, such as the wall of the digestive tract, we find that it has some interesting properties. The cells are arranged in sheets, and individual cells in the sheets are in electrical contact with one another through gap junctions. As a result, an action potential generated in the membrane of one smooth muscle cell can spread to all the cells in the sheet of tissue.

Another interesting property of a smooth muscle cell is that the resting potential of its membrane is sensitive to being stretched. If the wall of the digestive tract is stretched in one location (as by receiving a mouthful of food), the membranes of the stretched cells depolarize, reach threshold, and fire action potentials that cause the cells to contract. Thus smooth muscle contracts after being stretched, and the harder it is stretched, the stronger the contraction. (Later in this chapter we will see how membrane depolarization triggers contraction.)

Other factors that alter the membrane potential of smooth muscle cells are the neurotransmitters of the autonomic nervous system (see Figure 46.11). In the case of the digestive tract, acetylcholine causes smooth muscle cells to depolarize and thus makes them more likely to fire action potentials and contract. Norepinephrine causes these muscle cells to hyperpolarize and therefore makes them less likely to fire action potentials and contract (Figure 47.6).

Cardiac muscle causes the heart to beat

Cardiac muscle looks different from smooth muscle or skeletal muscle when viewed under the microscope (Figure 47.5, top). The cells appear striped, or *striated*, because of the regular arrangement of bundles of actin and myosin filaments within them. Actin and myosin are arranged in a similar way in skeletal muscle, as we'll see below.

The unique feature of cardiac muscle cells is that they branch. The branches of adjoining cells are interdigitated into a meshwork that gives cardiac muscle an ability to resist tearing. As a result, the heart walls can withstand high pressures while pumping blood without the danger of developing leaks. Also adding to the strength of cardiac muscle are *intercalated discs* that provide strong mechanical adhesions between adjacent cells.

As is true of smooth muscle, the individual cells in a sheet of cardiac muscle are in electrical contact with one another. Gap junctions present in the intercalated discs present low resistance to ions or electric currents. Therefore, a

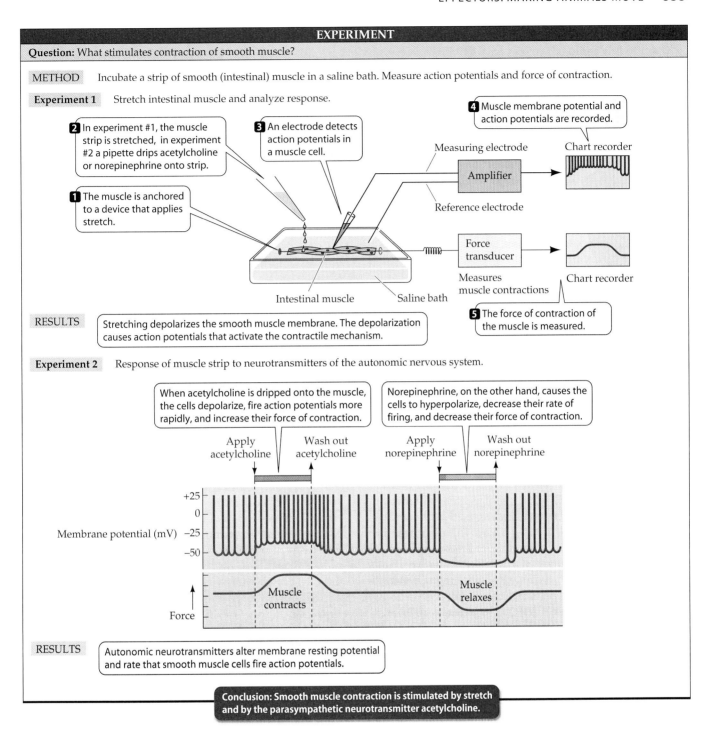

EXPERIMENT

Question: What stimulates contraction of smooth muscle?

METHOD   Incubate a strip of smooth (intestinal) muscle in a saline bath. Measure action potentials and force of contraction.

Experiment 1   Stretch intestinal muscle and analyze response.

2 In experiment #1, the muscle strip is stretched, in experiment #2 a pipette drips acetylcholine or norepinephrine onto strip.

3 An electrode detects action potentials in a muscle cell.

4 Muscle membrane potential and action potentials are recorded.

1 The muscle is anchored to a device that applies stretch.

Measuring electrode

Chart recorder

Amplifier

Reference electrode

Force transducer

Measures muscle contractions   Chart recorder

Intestinal muscle   Saline bath

5 The force of contraction of the muscle is measured.

RESULTS   Stretching depolarizes the smooth muscle membrane. The depolarization causes action potentials that activate the contractile mechanism.

Experiment 2   Response of muscle strip to neurotransmitters of the autonomic nervous system.

When acetylcholine is dripped onto the muscle, the cells depolarize, fire action potentials more rapidly, and increase their force of contraction.

Norepinephrine, on the other hand, causes the cells to hyperpolarize, decrease their rate of firing, and decrease their force of contraction.

Apply acetylcholine   Wash out acetylcholine   Apply norepinephrine   Wash out norepinephrine

Membrane potential (mV)   +25 0 −25 −50

Muscle contracts   Muscle relaxes

Force

RESULTS   Autonomic neurotransmitters alter membrane resting potential and rate that smooth muscle cells fire action potentials.

Conclusion: Smooth muscle contraction is stimulated by stretch and by the parasympathetic neurotransmitter acetylcholine.

47.6 Smooth Muscle Action
Stretching depolarizes the membrane of smooth muscle cells, and this depolarization causes action potentials that activate the contractile mechanism. The neurotransmitters acetylcholine and norepinephrine also alter the membrane potential of smooth muscle, making it more or less likely to contract.

depolarization initiated at one point in the heart spreads rapidly through the mass of cardiac muscle.

An interesting feature of vertebrate cardiac muscle is that certain specialized muscle cells, called **pacemaker cells**, initiate the rhythmic contractions of the heart. We'll learn about the molecular basis for this pacemaking func-

tion in Chapter 49. Because of these specialized pacemaker cells, the heartbeat is *myogenic*—generated by the heart muscle itself. The autonomic nervous system modifies the rate of the pacemaker cells, but is not essential for their continued rhythmic function. A heart removed from an animal continues to beat with no input from the nervous system. The myogenic nature of the heartbeat is a major factor in making heart transplants possible.

Skeletal muscle carries out behavior
Skeletal muscle carries out, or *effects*, all voluntary movements, such as running or playing a piano, and generates the movements of breathing. Skeletal muscle is also called

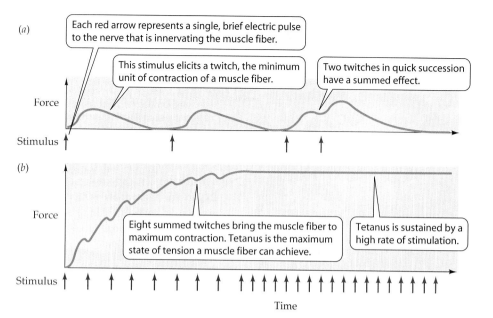

(a)

Each red arrow represents a single, brief electric pulse to the nerve that is innervating the muscle fiber.

This stimulus elicits a twitch, the minimum unit of contraction of a muscle fiber.

Two twitches in quick succession have a summed effect.

Force

Stimulus

(b)

Force

Eight summed twitches bring the muscle fiber to maximum contraction. Tetanus is the maximum state of tension a muscle fiber can achieve.

Tetanus is sustained by a high rate of stimulation.

Stimulus

Time

47.11 Twitches and Tetanus
(*a*) Action potentials from a motor neuron cause a muscle fiber to twitch. Twitches in quick succession can be summed. (*b*) Summation of many twitches can bring the muscle fiber to the maximum level of contraction, known as tetanus.

47.11*b*), and an increase in the number of motor units involved in the contraction causes spatial summation. (Remember that a motor unit consists of all the muscle fibers innervated by a single neuron, and that a single muscle consists of many motor units.)

Many muscles of the body maintain a low level of tension called **tonus** even when the body is at rest. For example, the muscles of the neck, trunk, and limbs that maintain our posture against the pull of gravity are always working, even when we are standing or sitting still. Muscle tonus comes from the activity of a small but changing number of motor units in a muscle; at any one time, some of the muscle's fibers are contracting and others are relaxed. Tonus is constantly being readjusted by the nervous system.

Muscle fiber types determine endurance and strength

Not all skeletal muscle fibers are alike, and a single muscle may contain more than one type of fiber. The two major types of skeletal muscle fibers are slow-twitch fibers and fast-twitch fibers (Figure 47.12*a*). **Slow-twitch fibers** are also called *red muscle* because they have lots of the oxygen-binding molecule myoglobin, they have lots of mitochondria, and they are well supplied with blood vessels. A single twitch of a slow-twitch fiber produces low tension.

The maximum tension a slow-twitch fiber can produce is low and develops slowly, but these fibers are highly resistant to fatigue. Because slow-twitch fibers have substantial reserves of fuel (glycogen and fat), their abundant mitochondria can maintain a steady, prolonged production of ATP if oxygen is available. Muscles with high proportions of slow-twitch fibers are good for long-term aerobic work (that is, work that requires lots of oxygen). Champion long-distance runners, cross-country skiers, swimmers, and bicyclists have leg and arm muscles consisting mostly of slow-twitch fibers (Figure 47.12*b*).

Fast-twitch fibers are also called *white muscle* because, in comparison to slow-twitch fibers, they have fewer mitochondria, little or no myoglobin, and fewer blood vessels. The white meat of domestic chickens is composed of fast-twitch fibers. Fast-twitch fibers can develop maximum tension more rapidly than slow-twitch fibers can, and that maximum tension is greater, but fast-twitch fibers become

fiber increases and becomes more sustained. Thus an individual muscle fiber can show a graded response to increased levels of stimulation by its motor neuron.

At high levels of stimulation, the calcium pumps in the sarcoplasmic reticulum can no longer remove Ca^{2+} ions from the sarcoplasm between action potentials, and the contractile machinery generates maximum tension—a condition known as **tetanus** (Figure 47.11*b*). (Do not confuse this condition with the disease *tetanus*, which is caused by a bacterial toxin and is characterized by spastic contractions of skeletal muscles.)

How long a muscle fiber can maintain a tetanic contraction depends on its supply of ATP. Eventually the fiber will become fatigued. It may seem paradoxical that the *lack* of ATP causes fatigue, since the action of ATP is to break actin–myosin bonds. But remember that the energy released from the hydrolysis of ATP "recocks" the myosin heads, allowing them to cycle through another power stroke. When a muscle is contracting against a load, the cycle of making and breaking actin–myosin bonds must continue to prevent the load from stretching the muscle. The situation is like rowing a boat upstream: You cannot maintain your position relative to the stream bank by just holding the oars out against the current; you have to keep rowing. Likewise, actin–myosin bonds have to keep cycling to maintain tension in the muscle.

The ability of a whole muscle to generate different levels of tension depends on how many fibers in that muscle are activated. Whether a muscle contraction is strong or weak depends both on how many of the motor neurons that synapse with that muscle are firing and on the rate at which those neurons are firing. These two factors can be thought of as *spatial summation* and *temporal summation*, respectively.

Both types of summation increase the strength of contraction of the muscle as a whole. Faster twitching of individual fibers causes temporal summation (see Figure

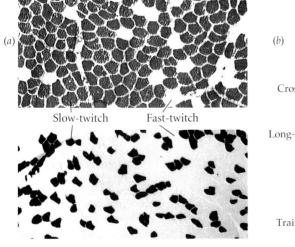

Slow-twitch   Fast-twitch

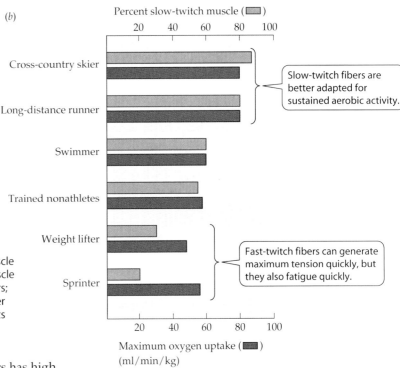

Percent slow-twitch muscle (▢)

Slow-twitch fibers are better adapted for sustained aerobic activity.

Fast-twitch fibers can generate maximum tension quickly, but they also fatigue quickly.

Maximum oxygen uptake (■)
(ml/min/kg)

47.12 Two Types of Muscle Fibers
(*a*) Skeletal muscles stained with a reagent that shows slow-twitch fibers as dark. The upper photo shows muscle from a professional cyclist. The lower photo shows muscle from a nonathlete who has about 75% fast-twitch fibers; this person would probably perform better as a sprinter than as a distance runner. (*b*) Athletes in different sports have different distributions of muscle fiber types.

fatigued rapidly. The myosin of fast-twitch fibers has high ATPase activity, so they can put the energy of ATP to work very rapidly, but they cannot replenish it quickly enough to sustain contraction for a long time. Fast-twitch fibers are especially good for short-term work that requires maximum strength. Champion weight lifters and sprinters have leg and arm muscles with high proportions of fast-twitch fibers.

What determines the proportion of fast- and slow-twitch fibers in your skeletal muscles? The most important factor is your genetic heritage, so there is some truth to the statement that champions are born, not made. To a certain extent, however, you can alter the properties of your muscle fibers through training. With aerobic training, the oxidative capacity of fast-twitch fibers can improve substantially. But a person born with a high proportion of fast-twitch fibers will never become a champion marathon runner, and a person born with a high proportion of slow-twitch fibers will never become a champion sprinter.

Skeletal Systems Provide Support for Muscles

Muscles can only contract and relax. Without something rigid to pull against, a muscle would just be a formless mass that twitches and changes shape. Skeletal systems provide rigid supports against which muscles can pull, creating directed movements. In this section, we'll examine the three types of skeletal systems found in animals: hydrostatic skeletons, exoskeletons, and endoskeletons.

A hydrostatic skeleton consists of fluid in a muscular cavity

The simplest type of skeleton is the **hydrostatic skeleton** of cnidarians, annelids, and many other soft-bodied invertebrates. It consists of a volume of incompressible fluid (water) enclosed in a body cavity surrounded by muscle.

When muscles oriented in a certain direction contract, the fluid-filled body cavity bulges out in the opposite direction.

The sea anemone, a cnidarian (see Figure 31.6*c*), has a hydrostatic skeleton. Its body cavity is filled with seawater. To extend its body and its tentacles, the anemone closes its mouth and constricts muscle fibers that are arranged in circles around its body. Contraction of these circular muscles puts pressure on the water in the body cavity, and that pressure forces the body and tentacles to extend. The anemone retracts its tentacles and body by contracting muscle fibers that are arranged longitudinally in the body wall and along the tentacles.

An earthworm uses its hydrostatic skeleton to crawl. The earthworm's body cavity is divided into many separate, fluid-filled segments. The body wall surrounding each segment has two muscle layers: one in which the muscle fibers are arranged in circles around the body cavity, and another in which the muscle fibers run lengthwise (Figure 47.13*a*). If the circular muscles in a segment contract, the compartment in that segment narrows and elongates. If the lengthwise (longitudinal) muscles of a segment contract, the compartment shortens and bulges outward.

Alternating contractions of the earthworm's circular and longitudinal muscles create waves of narrowing and widening, lengthening and shortening, that travel down the body. The bulging, short segments serve as anchors as the narrowing, expanding segments project forward, and longitudinal contractions pull other segments forward. Bristles help the widest parts of the body to hold firm against the substrate (Figure 47.13*b*).

Another type of locomotion made possible by hydrostatic skeletons is the jet propulsion used by squid and oc-

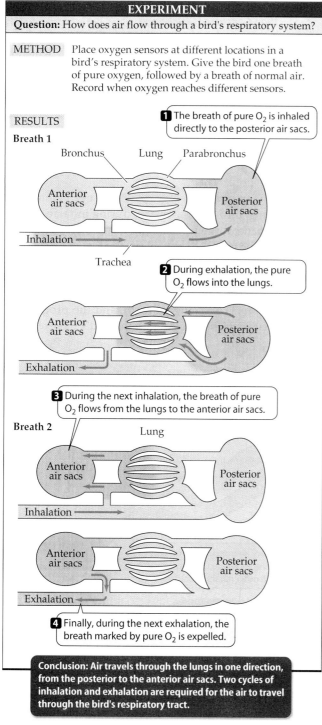

EXPERIMENT

Question: How does air flow through a bird's respiratory system?

METHOD   Place oxygen sensors at different locations in a bird's respiratory system. Give the bird one breath of pure oxygen, followed by a breath of normal air. Record when oxygen reaches different sensors.

RESULTS

Breath 1

1 The breath of pure O_2 is inhaled directly to the posterior air sacs.

Bronchus   Lung   Parabronchus

Anterior air sacs   Posterior air sacs

Inhalation

Trachea

2 During exhalation, the pure O_2 flows into the lungs.

Anterior air sacs   Posterior air sacs

Exhalation

3 During the next inhalation, the breath of pure O_2 flows from the lungs to the anterior air sacs.

Breath 2

Lung

Anterior air sacs   Posterior air sacs

Inhalation

Anterior air sacs   Posterior air sacs

Exhalation

4 Finally, during the next exhalation, the breath marked by pure O_2 is expelled.

Conclusion: Air travels through the lungs in one direction, from the posterior to the anterior air sacs. Two cycles of inhalation and exhalation are required for the air to travel through the bird's respiratory tract.

48.10 The Path of Air Flow through Bird Lungs
The fresh air a bird takes in with one breath (blue) travels through the lungs in one direction, from the posterior to the anterior air sacs. Two cycles of inhalation and exhalation are required for the air to travel the full length of the bird's respiratory system.

system could be followed by the oxygen sensors. This experiment showed that a single breath remains in the bird's gas exchange system for two cycles of inhalation and exhalation, and that the air sacs work as bellows maintaining a

continuous and unidirectional flow of fresh air through the lungs (Figure 48.10).

The advantages of the bird gas exchange system are similar to those of fish gills. Because the air sacs keep fresh air from the outside flowing unidirectionally and practically continuously over the gas exchange surfaces, the P_{O_2} on the environmental side of those surfaces is maximized. Furthermore, the unidirectional flow of air through the system makes possible a pattern of blood flow to minimize the P_{O_2} on the internal side of the exchange surfaces.

It is now clear how birds can fly over Mount Everest. A bird is able to supply its gas exchange surfaces with a continuous flow of fresh air that has a P_{O_2} close to that of the ambient air. Even when the P_{O_2} of the ambient air is only slightly above the P_{O_2} of the blood, O_2 can diffuse from air to blood. Next we will see why humans find it difficult to sustain even low levels of metabolic activity at such high altitudes.

TIDAL BREATHING IN MAMMALS. At the beginning of their evolution, lungs were dead-end sacs, and they remain so today in all air-breathing vertebrates except birds. Because lungs are dead-end sacs, ventilation cannot be constant and unidirectional, but must be **tidal**: Air flows in and out by the same route.

A *spirometer* shows how we use our lung capacity in breathing (Figure 48.11). When we are at rest, the amount of air that our normal breathing cycle moves per breath is called the *tidal volume* (about 500 ml for an average human adult). We can breathe much more deeply and inhale more air than our resting tidal volume; the additional volume of air we can take in above normal tidal volume is our *inspiratory reserve volume*. Conversely, we can forcefully exhale more air than we normally do during a resting exhalation. This additional amount of air that can be forced out of the lungs is the *expiratory reserve volume*. But even after the most extreme exhalation possible, some air remains in the lungs. The lungs and airways cannot be collapsed completely; they always contain a *residual volume*. The *total lung capacity* is the sum of the residual volume, expiratory reserve volume, tidal volume, and inspiratory reserve volume.

Tidal breathing severely limits the partial pressure gradient available to drive the diffusion of O_2 from air into the blood. Fresh air is not moving into the lungs during half of the respiratory cycle; therefore, the average P_{O_2} of air in the lungs is considerably less than it is in the air outside the lungs. Furthermore, the incoming air mixes with the stale air that was not expelled by the previous exhalation. The lung volume that is not ventilated with fresh air is called *dead space*. This dead space consists of the residual volume and, depending on the depth of breathing, some or all of the expiratory reserve volume.

The scale in Figure 48.11 tells us that a tidal volume of 500 ml of fresh air mixes with up to 2,000 ml of stale air before reaching the gas exchange surfaces in our lungs. When the P_{O_2} in the ambient air is 150 mm Hg, the P_{O_2} of the air that reaches our gas exchange surfaces is only about 100

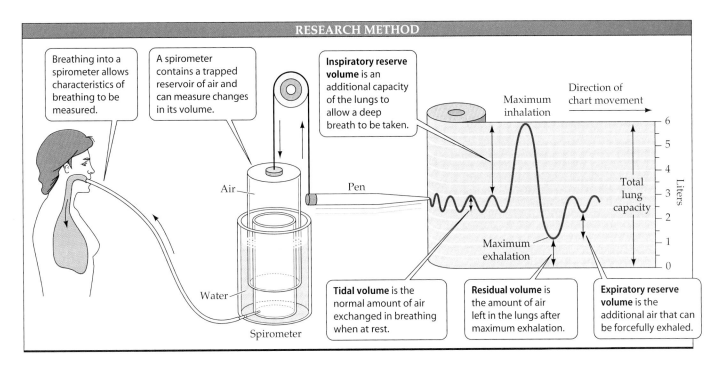

RESEARCH METHOD

Breathing into a spirometer allows characteristics of breathing to be measured.

A spirometer contains a trapped reservoir of air and can measure changes in its volume.

Inspiratory reserve volume is an additional capacity of the lungs to allow a deep breath to be taken.

Air

Pen

Water

Spirometer

Maximum inhalation

Direction of chart movement

Total lung capacity

Liters

Maximum exhalation

Tidal volume is the normal amount of air exchanged in breathing when at rest.

Residual volume is the amount of air left in the lungs after maximum exhalation.

Expiratory reserve volume is the additional air that can be forcefully exhaled.

48.11 Measuring Lung Ventilation with a Spirometer
Breathing from a closed reservoir of air and measuring the changes in the volume of that reservoir demonstrates the characteristics of mammalian tidal breathing.

mm Hg. By contrast, the P_{O_2} in the water that bathes the lamellae of fish gills or in the air that flows through the air capillaries of bird lungs is the same as the P_{O_2} in the outside water or air.

In addition to reducing the partial pressure gradient, tidal breathing reduces the efficiency of gas exchange in another way: It does not allow countercurrent gas exchange between air and blood. Because air enters and leaves the gas exchange structures by the same route, there is no anatomical way that blood can flow countercurrent, or even crosscurrent, to the air flow.

Mammalian Lungs and Gas Exchange

To offset the inefficiencies of tidal breathing, mammalian lungs have some design features that maximize the rate of gas exchange: an enormous surface area, and a very short path length for diffusion. Mammalian lungs serve the respiratory needs of mammals well, considering the ecologies and lifestyles of these animals.

Air enters the lungs through the oral cavity or nasal passage, which join together in the *pharynx* (Figure 48.12). From the pharynx, the esophagus conducts food to the stomach and a single airway leads to the lungs. At the beginning of this airway is the *larynx*, or voice box, which houses the vocal cords. The larynx is the "Adam's apple" that you can see or feel on the front of your neck. The major airway, the *trachea*, is about 2 cm in diameter. The thin walls of the trachea are prevented from collapsing by rings of cartilage that support them as air pressure changes during the breathing cycle. If you run your fingers down the front of

your neck just below your larynx, you can feel a couple of these rings of cartilage.

The trachea branches into two smaller *bronchi*, one leading to each lung. The bronchi branch repeatedly to generate a treelike structure of progressively smaller airways extending to all regions of the lungs (see Figure 48.12). As the branching of the bronchial tree continues to produce still smaller airways, the cartilage supports eventually disappear, marking the transition to **bronchioles**. The branching continues until the bronchioles are smaller than the diameter of a pencil lead, at which point tiny, thin-walled air sacs called **alveoli** begin to appear.

The alveoli are the sites of gas exchange. Because the airways only conduct air to and from the alveoli and do not themselves conduct gas exchange, their volume is physiological dead space. The number of alveoli in human lungs is about 300 million. Even though each alveolus is very small, the combined surface area for diffusion of respiratory gases is about 70 m² — the size of a badminton court.

Each alveolus is made of very thin cells. Between and surrounding the alveoli are networks of the smallest of blood vessels, the *capillaries*, whose walls are also made up of exceedingly thin endothelial cells. Where capillary meets alveolus, very little space separates them (see Figure 48.12), so the length of the diffusion path between air and blood is less than 2 μm. Even the diameter of a red blood cell is greater — about 7 μm.

Respiratory tract secretions aid breathing

Mammalian lungs have two other important adaptations that do not directly influence their gas exchange properties:

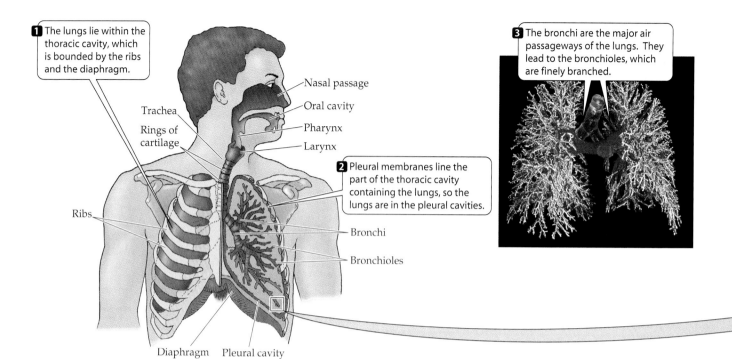

1 The lungs lie within the thoracic cavity, which is bounded by the ribs and the diaphragm.

3 The bronchi are the major air passageways of the lungs. They lead to the bronchioles, which are finely branched.

Nasal passage

Oral cavity

Trachea

Rings of cartilage

Pharynx

Larynx

2 Pleural membranes line the part of the thoracic cavity containing the lungs, so the lungs are in the pleural cavities.

Ribs

Bronchi

Bronchioles

Diaphragm   Pleural cavity

48.12 The Human Respiratory System
The diagram traces the hierarchy of human respiratory structures from the lungs down to the minuscule alveoli.

the production of mucus and the production of surfactant.

A **surfactant** is a substance that reduces the surface tension of a liquid. Lung surfactant reduces the surface tension of the film of fluid lining the insides of the alveoli. What is surface tension and how does it affect lung function? Surface tension is a result of cohesion between the molecules of a liquid. Cohesion gives the surface of the liquid the properties of an elastic membrane. Surface tension is what allows some insects to walk on the surface of water (see Figure 2.16). A surfactant interferes with the cohesive forces that create surface tension. Detergent is a surfactant, and when added to water, it makes walking on water difficult for the water strider.

The thin, aqueous layer that lines the alveoli has surface tension, which must be overcome to inflate the lungs. Surface tension normally is reduced by surfactant molecules produced by certain cells in the alveoli. If a baby is born more than a month prematurely, however, these cells may not yet be producing surfactant. Such a premature baby has great difficulty breathing because an enormous effort is required to stretch the alveoli against the surface tension. This condition, known as *respiratory distress syndrome*, may cause the baby to die from exhaustion and suffocation. Common treatments have been to put the baby on a respirator to assist its breathing and to give the baby hormones to speed its lung development. A new approach, however, is to apply surfactant to the lungs via an aerosol.

Many cells lining the airways produce a sticky *mucus* that captures bits of dirt and microorganisms that are inhaled. This mucus must be continually cleared from the air-

ways. Other cells lining the airways have cilia (see Figure 47.2) whose beating moves the mucus with its trapped debris up toward the pharynx, where it is swallowed. This phenomenon, called the **mucus escalator**, can be adversely affected by inhaled pollutants. Smoking one cigarette can immobilize the cilia of the airways for hours. A smoker's cough results from the need to clear the obstructing mucus from the airways when the mucus escalator is out of order.

Lungs are ventilated by pressure changes in the thoracic cavity

As Figure 48.12 shows, human lungs are suspended in the **thoracic cavity**, which is bounded on the top by the shoulder girdle, on the sides by the rib cage, and on the bottom by a domed sheet of muscle, the **diaphragm**. The thoracic cavity is lined on the inside by the **pleural membranes**, which divide it into right and left **pleural cavities** enclosing each lung. Because the pleural cavities are closed spaces, any effort to increase their volume creates negative pressure—suction—inside them.

Negative pressure within the pleural cavities causes the lungs to expand as air flows into them from the outside. This is the mechanism of inhalation. The diaphragm contracts to begin an inhalation. This contraction pulls the diaphragm down, increasing the volume of the thoracic and pleural cavities (Figure 48.13). As pressure in the pleural cavities becomes more negative, air moves into the lungs. Exhalation begins when the contraction of the diaphragm ceases. The diaphragm relaxes and moves up, and the elastic recoil of the lungs pushes air out through the airways. During tidal breathing, inhalation is an active process and exhalation is a passive process.

The diaphragm is not the only muscle that changes the volume of the pleural cavities. Between the ribs are two sets

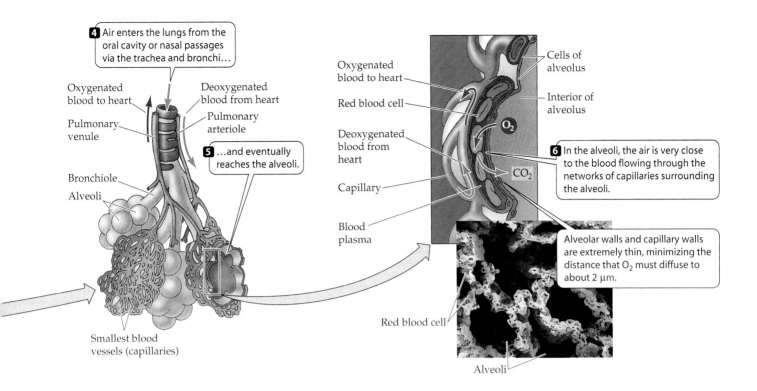

4 Air enters the lungs from the oral cavity or nasal passages via the trachea and bronchi...

Oxygenated blood to heart

Pulmonary venule

Deoxygenated blood from heart

Pulmonary arteriole

5 ...and eventually reaches the alveoli.

Bronchiole

Alveoli

Smallest blood vessels (capillaries)

Oxygenated blood to heart

Red blood cell

Deoxygenated blood from heart

Capillary

Blood plasma

Cells of alveolus

Interior of alveolus

O_2

CO_2

6 In the alveoli, the air is very close to the blood flowing through the networks of capillaries surrounding the alveoli.

Alveolar walls and capillary walls are extremely thin, minimizing the distance that O_2 must diffuse to about 2 μm.

Red blood cell

Alveoli

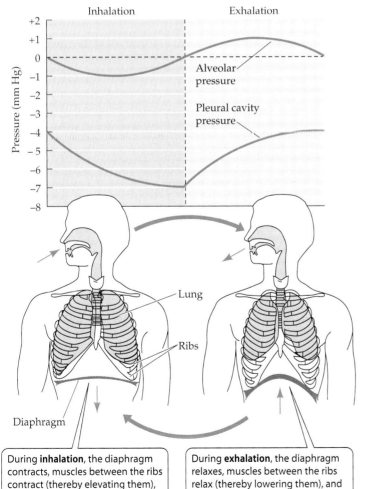

Inhalation   Exhalation

Pressure (mm Hg)

Alveolar pressure

Pleural cavity pressure

Lung

Ribs

Diaphragm

During **inhalation**, the diaphragm contracts, muscles between the ribs contract (thereby elevating them), and the pleural cavities expand, sucking air in.

During **exhalation**, the diaphragm relaxes, muscles between the ribs relax (thereby lowering them), and the pleural cavities contract, pushing air out.

of **intercostal muscles**. The *external intercostal muscles* expand the pleural cavities by lifting the ribs up and outward. The *internal intercostal muscles* decrease the volume of the thoracic cavity by pulling the ribs down and inward. When heavy demands are placed on the gas exchange system, such as during strenuous exercise, the external intercostal muscles increase the volume of air inhaled, making use of the inspiratory reserve volume, and the internal intercostal muscles increase the amount of air exhaled, making use of the expiratory reserve volume.

When the diaphragm is at rest between tidal breaths, the pressure in the pleural cavities is still slightly negative. This slight suction keeps the alveoli partly inflated. If the thoracic wall is punctured—by a knife wound, for example—air leaks into the pleural cavity, and the pressure from this air causes the lung to collapse. If the hole in the thoracic wall is not sealed, the breathing movements of the diaphragm and intercostal muscles pull air into the pleural cavity rather than into the lung, and ventilation of the alveoli in that lung ceases.

48.13 Into the Lungs and Out Again
Inhalation is an active process spurred by the contraction of the diaphragm. Exhalation generally is a passive process as the diaphragm relaxes. During inhalation, the negative pressure in the pleural cavity increases, expanding the elastic lung tissue and sucking air into the lungs. During exhalation, the negative pressure in the pleural cavity decreases, allowing the elastic lung tissue to recoil to create a positive pressure in the lungs and expel air.

Blood Transport of Respiratory Gases

The circulatory system is the subject of the next chapter, but since two of the substances the blood transports are the respiratory gases (O_2 and CO_2), we must discuss blood here. The circulatory system uses a pump (the heart) and a network of blood vessels to transport blood and the substances it carries around the body. As O_2 diffuses across the gas exchange surfaces into the blood vessels, the circulating blood sweeps it away. As we have seen, this internal perfusion of the gas exchange surfaces minimizes the P_{O_2} on the internal side and promotes the diffusion of O_2 across the surface at the highest possible rate. The blood then delivers this O_2 to the cells and tissues of the body.

The liquid part of the blood, the **blood plasma**, carries some O_2 in solution, but its ability to transport O_2 is quite limited. Blood plasma can carry only about 0.3 ml of oxygen per 100 ml, which is inadequate to support the metabolism of a person at rest. Fortunately, the blood also contains **red blood cells**, which are red because they are loaded with the oxygen-binding pigment **hemoglobin**. Hemoglobin increases the capacity of blood to transport oxygen by about 60-fold. There is quite a diversity of O_2-binding pigments among the animals; there is even considerable diversity of hemoglobin molecules. The discussion that follows focuses on human hemoglobin.

Hemoglobin combines reversibly with oxygen

Red blood cells contain enormous numbers of hemoglobin molecules. Hemoglobin is a protein consisting of four polypeptide subunits (see Figure 3.7). Each of these polypeptides surrounds a heme group—an iron-containing ring structure that can reversibly bind a molecule of O_2.

As O_2 diffuses into the red blood cells, it binds to hemoglobin. Once O_2 is bound, it cannot diffuse back across the red cell plasma membrane. By mopping up O_2 molecules as they enter the red blood cells, hemoglobin maximizes the partial pressure gradient driving the diffusion of O_2 into the cells. In addition, it enables the red blood cells to carry a large amount of O_2 to the tissues of the body.

The ability of hemoglobin to pick up or release O_2 depends on the partial pressure of O_2 in its environment. When the P_{O_2} of the blood plasma is high, as it usually is in the lung capillaries, each molecule of hemoglobin can carry its maximum load of four molecules of O_2. As the blood circulates through the rest of the body, it encounters lower P_{O_2} values. At these lower P_{O_2} values, the hemoglobin releases some of the O_2 it is carrying (Figure 48.14).

As you can see from the figure, the relation between P_{O_2} and the amount of O_2 bound to hemoglobin is not linear, but S-shaped (sigmoid). The sigmoid hemoglobin–O_2 binding curve reflects interactions between the four subunits of the hemoglobin molecule, each of which can bind one molecule of O_2. At low P_{O_2} values, only one subunit will bind an O_2 molecule. When it does so, the shape of this subunit changes, causing an alteration in the quaternary structure of the whole hemoglobin molecule (see Chapter 3). This structural change makes it easier for the other subunits to bind a molecule of O_2; that is, their O_2 affinity is increased. Therefore a smaller increase in P_{O_2} is necessary to get most of the hemoglobin molecules to bind two O_2 molecules (that is, to become 50 percent saturated) than it was to get them to bind one molecule of O_2 (25 percent saturated). The influence of the binding of O_2 by one subunit on the O_2 affinity of the other subunits is called **positive cooperativity**, because binding of the first molecule makes binding of the second easier, and so forth.

Once the third molecule of O_2 is bound, however, the relationship seems to change, as a larger increase in P_{O_2} is required to reach 100 percent saturation. This upper bend of the sigmoid curve is due to a probability phenomenon: The closer we get to having all subunits occupied, the less likely it is that a single O_2 molecule will find a place to bind. Therefore it takes a relatively greater P_{O_2} to achieve 100 percent saturation.

This is a good place to mention the danger posed by carbon monoxide (CO), which can come from a faulty furnace or from combusting a fuel such as charcoal or kerosene without adequate ventilation. CO binds to hemoglobin with a higher affinity than does O_2. Thus, CO destroys the ability of hemoglobin to transport and release O_2 to the tissues of the body. The victim loses consciousness and can die because the brain lacks O_2.

1 The normal P_{O_2} of deoxygenated blood returning to the heart is 40 mm Hg.

2 The P_{O_2} of blood leaving the lungs is about 100 mm Hg.

3 Of the O_2 in arterial blood, 25% is released to tissues during normal metabolism.

4 An oxygen reserve of 75% is held by the hemoglobin and can be released to tissues with a low P_{O_2}.

48.14 The Binding of Oxygen to Hemoglobin Depends on the P_{O_2}
Hemoglobin in blood leaving the lungs is 100 percent saturated (four molecules of O_2 are bound to each hemoglobin). Most hemoglobin molecules will drop only one of their four O_2 molecules as they circulate through the body, and are still 75 percent saturated when the blood returns to the lungs. The steep portion of this oxygen-binding curve comes into play when tissue P_{O_2} falls below the normal 40 mm Hg, at which point the hemoglobin will "unload" its oxygen reserves.

Llama guanaco

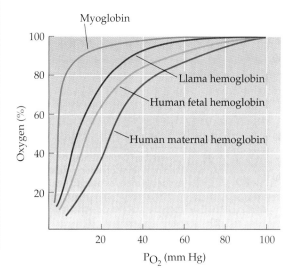

The O_2-binding properties of hemoglobin help get O_2 to the tissues that need it most. In the lungs, where the P_{O_2} is about 100 mm Hg, the hemoglobin is 100 percent saturated. The P_{O_2} in blood returning to the heart from the body is usually about 40 mm Hg. You can see from Figure 48.14 that at this P_{O_2} the hemoglobin is still about 75 percent saturated. This means that as the blood circulates around the body, only about 1 of 4 O_2 molecules it carries is released to the tissues. That seems inefficient, but it is really quite adaptive, because the hemoglobin keeps 75 percent of its oxygen in reserve to meet peak demands.

When a tissue becomes starved of oxygen and its local P_{O_2} falls below 40 mm Hg, the hemoglobin flowing through that tissue is on the steep portion of its sigmoid binding curve. That means that relatively small decreases in P_{O_2} below 40 mm Hg will result in the release of lots of O_2 to the tissue. Thus the cooperativity of O_2 binding by hemoglobin is very effective in making O_2 available to the tissues precisely when and where it is needed most.

Myoglobin holds an oxygen reserve

Muscle cells have their own oxygen-binding molecule, **myoglobin**. Myoglobin consists of just one polypeptide chain associated with an iron-containing ring structure that can bind one molecule of oxygen. Myoglobin has a higher affinity for O_2 than hemoglobin does (see Figure 48.15), so it picks up and holds oxygen at P_{O_2} values at which hemoglobin is releasing its bound O_2.

Myoglobin provides a reserve of oxygen for the muscle cells for times when metabolic demands are high and blood flow is interrupted. Interruption of blood flow in muscles is common because contracting muscles constrict blood vessels. When tissue P_{O_2} values are low and hemoglobin can no longer supply more O_2, myoglobin releases its bound O_2. Diving mammals such as seals have high concentrations of myoglobin in their muscles, which is one reason they can stay underwater for so long. (We will learn more about adaptations for diving in the next chapter.) Even in nondiving animals, muscles called on for extended periods of work frequently have more myoglobin than muscles that

 48.15 Oxygen-Binding Adaptations
Evolution has adapted the oxygen-binding properties of different hemoglobins and of myoglobin. The hemoglobin of llamas, for example, is adapted for binding oxygen at high altitudes, where P_{O_2} is low.

are used for short, intermittent periods. This is one of the reasons for the difference in appearance between fast-twitch and slow-twitch muscle (see Figure 47.12).

The affinity of hemoglobin for oxygen is variable

Various factors influence the oxygen-binding properties of hemoglobin, thereby influencing oxygen delivery to tissues. In this section we examine three of these factors: the chemical composition of the hemoglobin, pH, and the presence of 2,3 diphosphoglyceric acid.

HEMOGLOBIN COMPOSITION. As we noted above, there is more than one type of hemoglobin. The chemical composition of the polypeptide chains that form the hemoglobin molecule varies. The normal hemoglobin of adult humans has two each of two kinds of polypeptide chains—two *alpha* chains and two *beta* chains—and the oxygen-binding characteristics shown in Figure 48.14.

Before birth, the human fetus has a different form of hemoglobin, consisting of two *alpha* chains and two *gamma* chains. The functional difference between these two types of hemoglobin is that the fetal hemoglobin has a higher affinity for O_2. Therefore, the hemoglobin–O_2 binding curve of fetal hemoglobin is shifted to the left in comparison to the curve for adult hemoglobin (Figure 48.15). You can see from these curves that if both types of hemoglobin are at the same P_{O_2}, the fetal hemoglobin will pick up oxygen released by the adult hemoglobin. This difference in O_2 affinities facilitates the transfer of O_2 from the mother's blood to the blood of the fetus in the placenta.

Llamas and vicuñas are mammals native to high altitudes in the Andes Mountains of South America. In the natural habitat of these animals, more than 5,000 m above sea

Reptiles have two aortas instead of one. This simplified representation of reptilian cardiovascular anatomy shows that the right aorta can receive blood from either the right side or the left side of the ventricle:

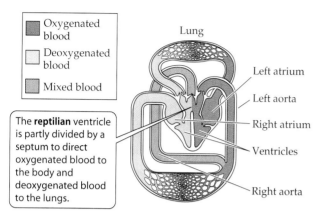

When the animal is breathing air, two factors cause blood from the right side of the ventricle to go preferentially into the pulmonary circuit rather than into the systemic circuit. First, the resistance in the pulmonary circuit is lower than that in the systemic circuit. Second, there is a slight asynchrony in the timing of ventricular contraction, so the blood in the right side of the ventricle tends to be ejected slightly before the blood in the left side. As the ventricle contracts, the deoxygenated blood in the right side of the ventricle moves first into the lung circuit. When the oxygenated blood in the left side of the ventricle starts to move, it encounters resistance in the lung circuit, which is already filled with the deoxygenated blood from the right side. Therefore the blood from the left side tends to flow into the two aortas.

When the reptile stops breathing, blood flow is rerouted by constriction of vessels in the lung circuit. As resistance in the lung circuit increases, the blood from the right side of the ventricle tends to be directed into one of the aortas. As a result, blood from both sides of the ventricle flows through the aortas to the systemic circuit.

The ability of snakes, lizards, and turtles to redirect blood flow from the lung circuit to the systemic circuit depends on the incomplete division of their ventricles. Crocodilians have true four-chambered hearts with completely divided ventricles. Yet the crocodilians have not lost the ability to shunt blood from the lung circuit when they are not breathing. The crocodilians have one aorta originating in the left ventricle and one aorta originating in the right ventricle. However, a short channel connects these two aortas just after they leave the heart.

Because the crocodilians' ventricles are separate, they can generate different pressures when they contract. When the animal is breathing, the pressure in the left ventricle and the left aorta is higher than the pressure in the right ventricle. This higher pressure is communicated through the connecting channel to the right aorta, and this high back pressure prevents right-ventricle blood from entering that aorta. As a result, both aortas carry blood from the left ventricle, and the blood from the right ventricle flows to the lung circuit.

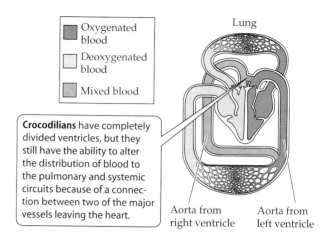

When a crocodilian is not breathing, constriction of vessels in the lung circuit increases the resistance in that circuit. As a result, pressure builds up in the right ventricle to a level that exceeds the back pressure in the right aorta. Under these conditions, blood from both ventricles flows through the two aortas and the systemic circuit, and little blood flows into the lung circuit.

You can now appreciate the fact that reptilian and crocodilian hearts are not primitive. Rather, these hearts and their major vessels are highly adapted to operate efficiently over a wide range of metabolic demands.

Birds and mammals have fully separated pulmonary and systemic circuits

The four-chambered hearts of birds and mammals completely separate their pulmonary and systemic circuits.

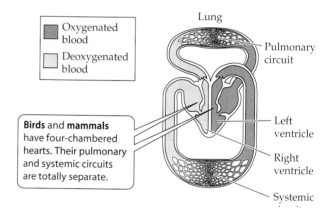

Separate circuits have several advantages:

▶ Oxygenated and deoxygenated blood cannot mix, and therefore, the systemic circuit is always receiving blood with the highest oxygen content.
▶ Respiratory gas exchange is maximized because the blood with the lowest oxygen content and highest CO_2 content is sent to the lungs.
▶ Separate systemic and pulmonary circuits can operate at different pressures.

The tissues of birds and mammals have high nutrient demands and thus a very high density of the smallest vessels, the capillaries. Many small vessels present lots of resistance to the flow of blood. Therefore, high pressure is required in the systemic circuits of birds and mammals. Their pulmonary circuits do not have as many capillaries and as high resistances as their systemic circuits, so the pulmonary circuits of birds and mammals can run at lower pressures.

The Human Heart: Two Pumps in One

Like all other mammalian hearts, the human heart has four chambers: two atria and two ventricles (Figure 49.4). The atrium and ventricle on the right side of your body are called the right atrium and right ventricle. They can be thought of as the right heart. The atrium and ventricle on the left side of your body are called the left atrium and left ventricle. They can be thought of as the left heart. The right heart pumps blood through the pulmonary circuit, and the left heart pumps blood through the systemic circuit.

Valves between the atria and ventricles, the **atrioventricular valves**, prevent backflow of blood into the atria when the ventricles contract. The **pulmonary valve** and the **aortic valve**, positioned between the ventricles and the arteries, prevent the backflow of blood into the ventricles.

In what follows, we'll first focus on the flow of blood through the heart and through the body. Then we'll examine the unique electrical properties of cardiac muscle and see how the heart's electrical activity can be recorded in an EKG (electrocardiogram).

Blood flows from right heart to lungs to left heart to body

Let's follow the circulation of the blood through the heart, starting in the right heart. The right atrium receives deoxygenated blood from the **superior vena cava** and the **infe-**

rior **vena cava** (see Figure 49.4), large veins that collect blood from the upper and lower body, respectively. The veins of the heart itself also drain into the right atrium. From the right atrium, the blood flows into the right ventricle. Most of the filling of the ventricle is due to passive flow while the heart is relaxed between beats. Just at the end of this period of ventricular filling, the atrium contracts and adds a little more blood to the ventricular volume. The right ventricle then contracts, pumping blood into the **pulmonary artery**, which transports the blood to the lungs.

The **pulmonary veins** return the oxygenated blood from the lungs to the left atrium, from which the blood enters the left ventricle. As with the right side of the heart, most left ventricular filling is passive, and ventricular volume is topped off by contraction of the atrium just at the end of the period of filling.

The walls of the left ventricle are powerful muscles that contract around the blood with a wringing motion starting from the bottom. When pressure in the left ventricle is high enough to push open the aortic valve, the blood rushes into the aorta to begin its circulation throughout the body and eventually back to the right atrium. In Figure 49.4, observe that the left ventricle is more massive than the right ventricle. Because there are many more arterioles and capillaries in the systemic circuit than in the pulmonary circuit, resistance is higher in the systemic circuit, and the left ventricle must squeeze with greater force than the right, even though both are pumping the same volume of blood.

49.4 The Human Heart and Circulation
In the human heart, blood flows from right heart to lungs to left heart to body. The atrioventricular valves prevent blood from flowing back into the atria when the ventricles contract. The pulmonary and aortic valves prevent blood from flowing back into ventricles from the arteries when the ventricles relax.

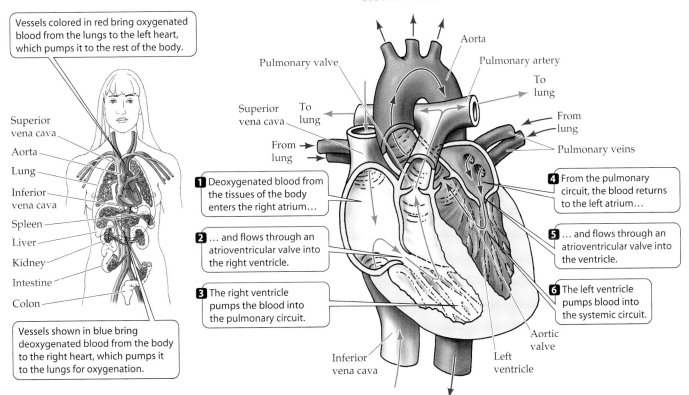

Vessels colored in red bring oxygenated blood from the lungs to the left heart, which pumps it to the rest of the body.

Superior vena cava
Aorta
Lung
Inferior vena cava
Spleen
Liver
Kidney
Intestine
Colon

Vessels shown in blue bring deoxygenated blood from the body to the right heart, which pumps it to the lungs for oxygenation.

Aorta
Pulmonary valve
Pulmonary artery
To lung
Superior vena cava
To lung
From lung
From lung
Pulmonary veins

1 Deoxygenated blood from the tissues of the body enters the right atrium…

2 …and flows through an atrioventricular valve into the right ventricle.

3 The right ventricle pumps the blood into the pulmonary circuit.

4 From the pulmonary circuit, the blood returns to the left atrium…

5 …and flows through an atrioventricular valve into the ventricle.

6 The left ventricle pumps blood into the systemic circuit.

Inferior vena cava
Aortic valve
Left ventricle

The pumping of the heart—the contraction of the two atria followed by the contraction of the two ventricles, and then relaxation—is the **cardiac cycle**. Contraction of the ventricles is called **systole**, and relaxation of the ventricles **diastole** (Figure 49.5). Just at the end of diastole, the atria contract and top off the volume of blood in the ventricles. The sounds of the cardiac cycle, the "lub-dub" heard through a stethoscope placed on the chest, are created by the slamming shut of the heart valves. The shutting and opening of these valves is simply a mechanical event resulting from pressure differences on the two sides of the valves. As the ventricles begin to contract, the pressure in the ventricles rises above the pressure in the atria, and the atrioventricular valves close ("lub"). When the ventricles begin to relax, the high back pressure in the aorta and pulmonary arteries causes the aortic and pulmonary valves to bang shut ("dub"). Defective valves produce the sounds of *heart murmurs*. For example, if an atrioventricular valve is defective, blood will flow back into the atrium with a "whoosh" sound following the "lub."

The cardiac cycle can be felt in the pulsation of arteries such as the one that supplies blood to your hand. You can feel your pulse by placing two fingers from one hand lightly over the wrist of the other hand just below the thumb. During systole, blood surges through the arteries of your arm and hand, and you can feel the surge as a pulsing of the artery in your wrist.

Blood pressure changes associated with the cardiac cycle can be measured in the large artery in your arm by using an inflatable pressure cuff called a sphygmomanometer and a stethoscope (Figure 49.6). This method measures the minimum pressure necessary to compress an artery so that blood does not flow through it at all (the systolic value) and the minimum pressure that permits intermittent flow through the artery (the diastolic value). In a conventional blood pressure reading, the systolic value is placed over the diastolic value. Normal values for a young adult might be 120 mm of mercury (Hg) during systole and 80 mm Hg during diastole, or 120/80.

The heartbeat originates in the cardiac muscle

Cardiac muscle, as we saw in Chapter 47, has some unique properties that allow it to function as an effective pump. First, the cardiac muscle cells are in electrical continuity with each other. Gap junctions enable action potentials to spread rapidly from cell to cell. Because a spreading action potential stimulates contraction, large groups of cardiac muscle cells contract in unison. This coordinated contraction is important for pumping blood.

49.5 The Cardiac Cycle
The rhythmic contraction (systole) and relaxation (diastole) of the ventricles is called the cardiac cycle.

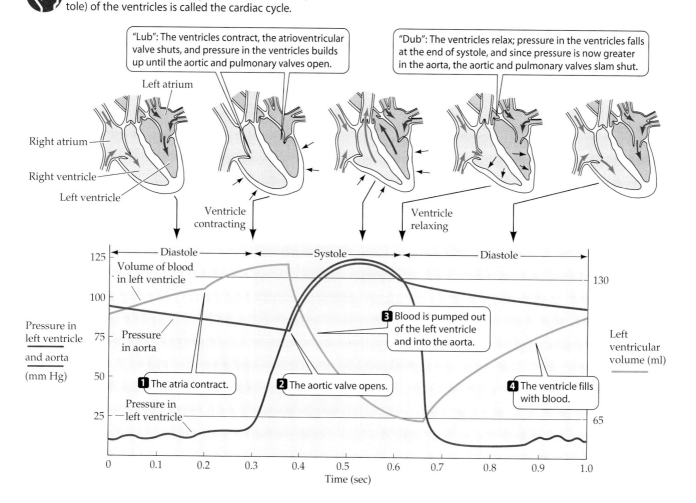

"Lub": The ventricles contract, the atrioventricular valve shuts, and pressure in the ventricles builds up until the aortic and pulmonary valves open.

"Dub": The ventricles relax; pressure in the ventricles falls at the end of systole, and since pressure is now greater in the aorta, the aortic and pulmonary valves slam shut.

Left atrium

Right atrium

Right ventricle

Left ventricle

Ventricle contracting

Ventricle relaxing

Diastole — Systole — Diastole

Volume of blood in left ventricle

Pressure in aorta

Pressure in left ventricle and aorta (mm Hg)

Pressure in left ventricle

Left ventricular volume (ml)

1 The atria contract.

2 The aortic valve opens.

3 Blood is pumped out of the left ventricle and into the aorta.

4 The ventricle fills with blood.

Time (sec)

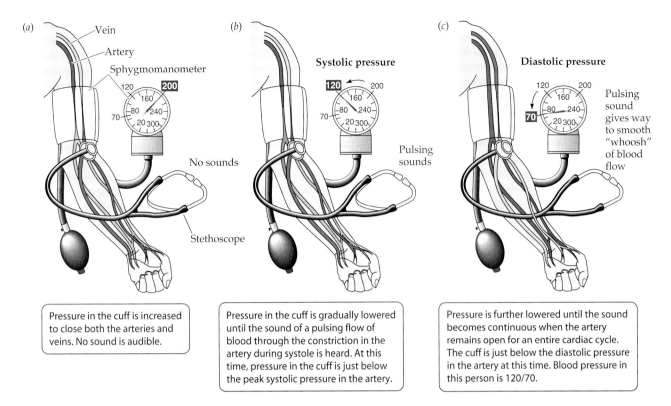

(a) Vein
Artery
Sphygmomanometer

No sounds

Stethoscope

Pressure in the cuff is increased to close both the arteries and veins. No sound is audible.

(b) Systolic pressure

Pulsing sounds

Pressure in the cuff is gradually lowered until the sound of a pulsing flow of blood through the constriction in the artery during systole is heard. At this time, pressure in the cuff is just below the peak systolic pressure in the artery.

(c) Diastolic pressure

Pulsing sound gives way to smooth "whoosh" of blood flow

Pressure is further lowered until the sound becomes continuous when the artery remains open for an entire cardiac cycle. The cuff is just below the diastolic pressure in the artery at this time. Blood pressure in this person is 120/70.

49.6 Measuring Blood Pressure
Blood pressure in the major artery of the arm can be measured with an inflatable pressure cuff called a sphygmomanometer.

Second, some cardiac muscle cells have the ability to initiate action potentials without stimulation from the nervous system. These cells stimulate neighboring cells to contract, thereby acting as pacemakers. The important characteristic of a pacemaker cell is that its resting membrane potential gradually becomes less negative until it reaches the threshold voltage for initiating an action potential (Figure 49.7).

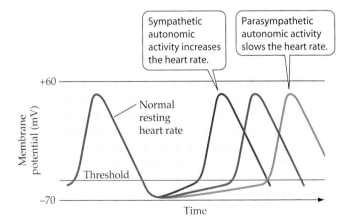

49.7 The Autonomic Nervous System Controls Heart Rate
The resting potentials of the plasma membranes of pacemaker cells spontaneously depolarize to threshold and fire action potentials. Signals from the two divisions of the autonomic nervous system raise and lower the heart rate, respectively.

These action potentials look different from the neuronal action potentials you saw in Chapter 41 because the depolarization is due primarily to the opening of voltage-gated calcium channels rather than voltage-gated sodium channels.

Like neurons, cardiac muscle repolarizes in part by opening potassium channels. The potassium channels in cardiac pacemaker cells, however, are unique. After an action potential, they open, causing the membrane potential to fall to its most negative level. Then they *gradually* close, and as they do so, the membrane potential becomes less negative—it *slowly* depolarizes. Sodium and calcium channels contribute to this gradual depolarization between action potentials. When membrane potential reaches threshold for the voltage-gated calcium channels, another action potential occurs.

The nervous system controls the heartbeat (speeds it up or slows it down) by influencing the rate at which pacemaker cells gradually depolarize between action potentials. Acetylcholine released by parasympathetic nerve endings onto the pacemaker cells slows their rate of depolarization and thereby slows the heart rate. Norepinephrine released by sympathetic nerve endings onto the pacemaker cells increases their rate of depolarization and thereby speeds the heart rate (see Figure 49.7).

Under normal circumstances, the heartbeat originates from pacemaker cells located at the junction of the superior vena cava and right atrium, in the **sinoatrial node** (Figure 49.8). An action potential spreads from the sinoatrial node across the atrial walls, causing the two atria to contract in unison. Since there are no gap junctions between the atria and the ventricles, however, this depolarization does not

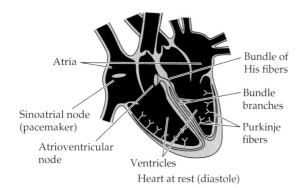

Atria

Bundle of His fibers

Bundle branches

Sinoatrial node (pacemaker)

Purkinje fibers

Atrioventricular node

Ventricles

Heart at rest (diastole)

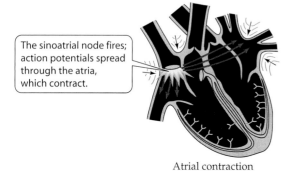

The sinoatrial node fires; action potentials spread through the atria, which contract.

Atrial contraction

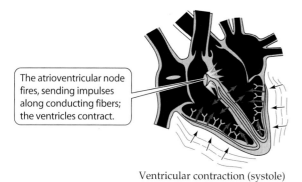

The atrioventricular node fires, sending impulses along conducting fibers; the ventricles contract.

Ventricular contraction (systole)

49.8 The Heartbeat
Pacemaker cells in the sinoatrial node initiate action potentials that spread through the walls of the atria, causing them to contract.

flow directly to the ventricles, and the ventricles do not contract in unison with the atria.

The action potential initiated in the atria passes to the ventricles through another node of modified cardiac muscle cells, the **atrioventricular node**. The atrioventricular node passes the action potential on to the ventricles via modified muscle fibers called the **bundle of His**. The bundle of His divides into right and left branches, which connect with **Purkinje fibers** that branch throughout the ventricular muscle.

The timing of the spread of the action potential from atria to ventricles is important. The atrioventricular node imposes a short delay in the spread of the action potential from atria to ventricles. Then the action potential spreads very rapidly throughout the ventricles, causing them to contract. Thus the atria contract before the ventricles do, so the blood passes progressively from the atria to the ventricles to the arteries.

(a) **A normal EKG**

P corresponds to the depolarization and contraction of the atrial muscle.

Q, R, and S together correspond to the depolarization of the ventricles.

T corresponds to the relaxation and repolarization of the ventricles.

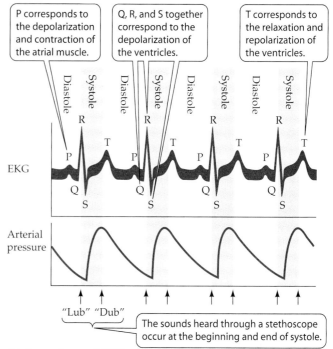

EKG

Arterial pressure

"Lub" "Dub"

The sounds heard through a stethoscope occur at the beginning and end of systole.

(b) **Some abnormal EKGs**

Tachycardia (heart rate of more than 100 beats/minute in a resting person)

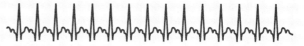

Ventricular fibrillation (uncoordinated contraction of the ventricles)

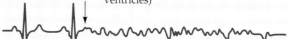

Heart block (failure of stimulation to ventricles following atrial contraction)

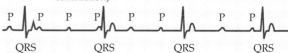

Besides detecting rhythmic irregularities in the heartbeat (arrhythmias), EKGs can detect damage to the heart muscle (infarctions) or decreased blood supply to the heart muscle (ischemias) by changes in the size and shape of the EKG curves.

(c)

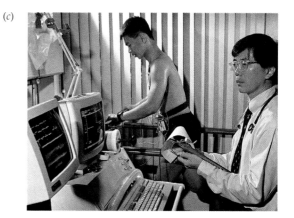

49.9 The Electrocardiogram
An EKG can be used to monitor heart function.

The EKG records the electrical activity of the heart

Electrical events in the cardiac muscle during the cardiac cycle can be recorded by electrodes placed on the surface of the body. Such a recording is called an **electrocardiogram**, or **EKG** ("EKG" because the Greek word for heart is *kardia*, but "ECG" is also used). The EKG is an important tool for diagnosing heart problems.

The action potentials that sweep through the muscles of the atria and the ventricles before they contract are such massive, localized electrical events that they cause electric currents to flow outward from the heart to all parts of the body. Electrodes placed on the surface of the body at different locations—usually on the wrists and ankles—detect those electric currents at different times and therefore register a voltage difference. The appearance of the EKG depends on the exact placement of the electrodes used for the recording. Electrodes placed on the right wrist and left ankle produced the normal EKG shown in Figure 49.9*a*. The waves of the EKG are designated P, Q, R, S, and T, each letter representing a particular event in the cardiac muscle.

The EKG is used by cardiologists (heart specialists) to diagnose heart problems. Figure 49.9*b* shows some abnormal EKGs that result from different problems. For patients who have had heart attacks, it is possible to determine which region of the heart has been damaged by placing electrodes at different locations on the chest. Comparing EKGs from the different electrodes tells the cardiologist which region of the heart is behaving abnormally.

The Vascular System: Arteries, Capillaries, and Veins

Blood circulates throughout the body in a system of blood vessels: arteries, capillaries, and veins. Arteries receive blood from the heart; accordingly, they are built to withstand high pressures. Arteries are important in controlling blood pressure and in the distribution of blood to different organs . Veins return blood to the heart at low pressures and serve as a blood reservoir. The capillaries are the site of all exchanges between the blood and the internal environment. In this section, we see how the structure of each of these vessel types supports its functions. In addition to arteries, capillaries, and veins, we consider another set of vessels, the lymphatic vessels, which return tissue fluid to the blood.

Arteries and arterioles have abundant elastic and muscle fibers

The walls of the large arteries have many elastic fibers that enable them to withstand high pressures (Figure 49.10). These elastic fibers have another important function as

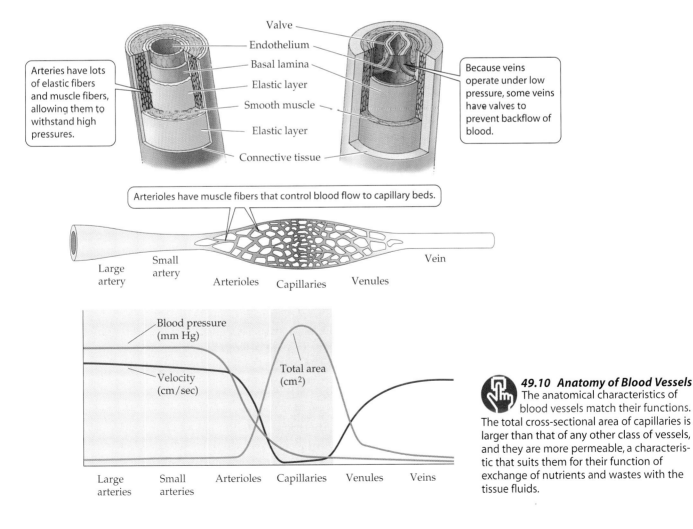

49.10 Anatomy of Blood Vessels The anatomical characteristics of blood vessels match their functions. The total cross-sectional area of capillaries is larger than that of any other class of vessels, and they are more permeable, a characteristic that suits them for their function of exchange of nutrients and wastes with the tissue fluids.

well: During systole, they are stretched, and thereby store some of the energy imparted to the blood by the heart. During diastole, they return this energy by squeezing the blood and pushing it forward. As a result, even though the flow of blood through the arterial system pulsates, it is smoother than it would be through a system of rigid pipes.

Smooth muscle cells in the arteries and arterioles make the diameter of those vessels variable. When their diameter changes, their resistance to blood flow changes as well, and the amount of blood flowing through them changes as a result. By influencing the contraction of the smooth muscle in the walls of arteries and arterioles, neural and hormonal mechanisms can control the distribution of blood to the different tissues of the bodyas well as central blood pressure. The arteries and arterioles are referred to as the *resistance vessels* because their resistance varies.

Materials are exchanged between blood and tissue fluid in the capillaries

Beds of capillaries lie between arterioles and venules. No cell of the body is more than a couple of cell diameters away from a capillary. The needs of the cells are served by the exchange of materials between blood and tissue fluid across the capillary walls. Capillaries have thin, permeable walls, and blood flows through them slowly, facilitating this exchange (Figure 49.11).

To anyone who has played with a garden hose, it may seem strange that blood flows through the large arteries rapidly at high pressures, but when it reaches the small capillaries, the pressure and rate of flow decrease. When you restrict the diameter of a garden hose by placing your thumb over the opening, the pressure in the hose increases, which in turn increases the velocity of the water spraying out of the hose.

This puzzle is solved by two more pieces of information. First, arterioles are highly branched. When one is restricted, blood flows into other branches, so pressure does not build up quickly. Second, each arteriole gives rise to many, many

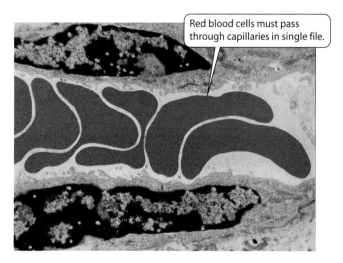

Red blood cells must pass through capillaries in single file.

49.11 A Narrow Lane
Capillaries have a very small diameter, and blood flows through them slowly.

capillaries. Even though each capillary has a diameter so small that red blood cells must pass through in single file (see Figure 49.11), there are so many capillaries that their total cross-sectional area is much greater than that of any other class of vessel. As a result, all of the capillaries together have a much greater capacity for blood than do the arterioles. Returning to our garden hose analogy, if we connect the hose to many junctions leading to small irrigation tubes, the pressure and the flow in each of the irrigation tubes will be quite low.

Materials are exchanged in capillary beds by filtration, osmosis, and diffusion

The walls of capillaries are made of a single layer of thin endothelial cells. In most tissues of the body other than the brain, these tubes of endothelial cells have tiny holes called *fenestrations*. Surrounding the endothelial cells is a very permeable basal lamina. So, capillaries are leaky. They are permeable to water, to some ions, and to some small molecules, but not to large molecules such as proteins. Blood pressure therefore tends to squeeze water and some solutes out of the capillaries and into the surrounding intercellular spaces. This process is called **filtration**. The large molecules that cannot cross the capillary wall create a difference in osmotic potential (also called osmotic pressure) between the plasma and the tissue fluid, which tends to draw water back into the capillary.

Recent research suggests that bicarbonate ions in the capillary plasma are an important contributor to the osmotic attraction of water back into the capillary. As we saw in Chapter 48, CO_2 diffuses into the plasma as the blood flows through the capillary. The conversion of this CO_2 to bicarbonate ions is catalyzed by the enzyme carbonic anhydrase. Therefore, there is a rise in bicarbonate concentration as blood flows through the capillary, and this increased concentration contributes to the resorption of water from the tissue fluid.

Blood pressure is highest on the arterial side of a capillary bed and steadily decreases as the blood flows to the venous side. Therefore, more water is squeezed out of the capillaries on the arterial side of the bed. The osmotic potential pulling water back into the capillary rises as blood flows toward the venous side. Gradually, osmotic potential becomes the dominant force, pulling water back into the plasma. The interaction of these two opposing forces—blood pressure versus osmotic potential—determines the net flow of water between the plasma and the tissue fluid (Figure 49.12).

The balance between blood pressure and osmotic potential changes if the blood pressure in the arterioles or the permeability of the capillary walls changes. Such a change leads to the inflammation that accompanies injuries to the skin or allergic reactions. A major mediator of inflammation is a hormone called *histamine* that is released mainly by white blood cells, called mast cells, that move to the damaged tissue (see Chapter 41). Histamine relaxes the smooth muscle of the arterioles, thus increasing blood flow to the damaged tissue and increasing pressure in the capillaries.

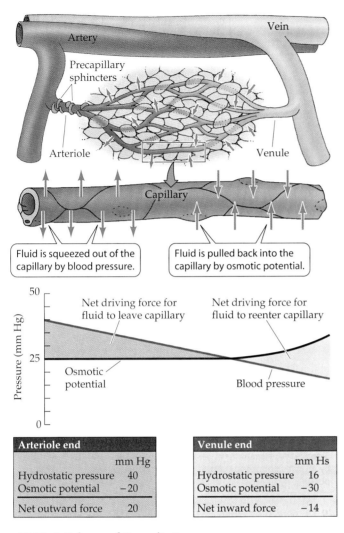

The capillaries in different tissues, however, are differentially selective as to the sizes of molecules that can pass through them. In all capillaries, O_2, CO_2, glucose, lactate, and small ions such as Na^+ and Cl^- can cross. The capillaries of the brain do not have fenestrations, and therefore not much else can pass through them unless it is a lipid-soluble substance such as alcohol. This high selectivity of brain capillaries is known as the *blood–brain barrier* (see Chapter 44).

In other tissues, the capillaries are much less selective. Such capillaries are found in the digestive tract, where nutrients are absorbed, and in the kidneys, where wastes are filtered. Some capillaries have large gaps that permit the movement of even larger substances, such as red blood cells. These capillaries are found in the bone marrow, spleen, and liver. Substances move across many capillary walls by endocytosis (see Chapter 5).

Lymphatic vessels return tissue fluid to the blood

The tissue fluid that accumulates outside the capillaries contains water and small molecules, but no red blood cells, and less protein than there is in blood. A separate system of vessels—the **lymphatic system**—returns tissue fluid to the blood.

After entering the lymphatic vessels, the tissue fluid is called **lymph**. Fine lymphatic capillaries merge progressively into larger and larger vessels and end in a major vessel—the **thoracic duct**—that empties into the superior vena cava, returning blood to the heart (see Figure 19.1). Lymphatic vessels have one-way valves that keep the lymph flowing toward the thoracic duct. The force propelling the lymph is pressure on the lymphatic vessels from the contractions of nearby skeletal muscles.

Mammals and birds have *lymph nodes* along the major lymphatic vessels. Lymph nodes are an important component of the defensive machinery of the body (see Chapter 19). They are a major site of lymphocyte production and of the phagocytic action that removes microorganisms and other foreign materials from the circulation. The lymph nodes also act as mechanical filters. Particles become trapped there and are digested by the phagocytes that are abundant in the nodes.

Lymph nodes swell during infection. Some of them, particularly those on the sides of the neck or in the armpits, become noticeable when they swell. The nodes also trap metastasized cancer cells—that is, those that have broken free of the original tumor. Because such cells may start additional tumors, surgeons often remove the neighboring lymph nodes when they excise a malignant tumor.

49.12 A Balance of Opposite Forces
Blood pressure and osmotic attraction (both expressed as millimeters of mercury, mm HG) control the exchange of fluids between blood vessels and intercellular space.

Histamine also increases the permeability of the capillaries, so that more water leaves the vessels. The accumulation of fluid in the intercellular spaces causes the tissue to swell, a condition known as **edema**. The use of drugs called *antihistamines* can alleviate inflammation and allergic reactions.

The loss of water from capillaries increases if the protein content of the blood decreases, as is seen in cases of liver failure due to alcoholism or liver disease. The liver is the major producer of blood proteins, and when it fails, blood protein levels fall. With a lower protein concentration in the plasma, there is less of an osmotic potential to pull water back into the capillaries. The result is that tissue fluid builds up, swelling the abdomen and the extremities.

Which specific small molecules can cross a capillary wall depends on the architecture of the capillary, the type of substance, and the concentration gradient of the substance between the blood and the tissue fluid. Capillary walls consist of the plasma membranes of endothelial cells and, as mentioned, may have actual holes (fenestrations) in them. Therefore, lipid-soluble substances and many small solute molecules can pass through them from an area of higher concentration to one of lower concentration (see Chapter 5).

Blood flows back to the heart through veins

The pressure of the blood flowing from capillaries to venules is extremely low, and is insufficient to propel blood back to the heart. Blood tends to accumulate in veins, and the walls of veins are more expandable than the walls of arteries. As much as 80 percent of the total blood volume may be in the veins at any one time. Because of their high capacity to store blood, veins are called *capacitance vessels*.

Plasma is very similar to tissue fluid in composition, and most of its components move readily between these two fluid compartments of the body. The main difference between the two fluids is the higher concentration of proteins in the plasma.

Control and Regulation of Circulation

The circulatory system is controlled and regulated by neural and hormonal mechanisms at both the local and systemic levels. Every tissue requires an adequate supply of blood that is saturated withoxygen, carries essential nutrients, and is relatively free of waste products. The nervous system cannot monitor and control every capillary bed in the body. Instead, each tissue regulates its own blood flow through **autoregulatory mechanisms** that cause the arterioles supplying the tissue to constrict or dilate.

The autoregulatory actions of every capillary bed in every tissue influence the pressure and composition of the arterial blood leaving the heart. If many arterioles suddenly dilate, for example, allowing blood to flow through many more capillary beds, arterial blood pressure falls. If all the newly filled capillary beds contribute metabolic waste products to the blood at one time, the concentration of wastes in the blood returning to the heart increases. Thus events in all the capillary beds throughout the body produce combined effects on arterial blood pressure and blood composition. The nervous and endocrine systems respond to these changes by changing breathing, heart rate, and blood distribution to match the metabolic needs of the body.

Autoregulation matches local flow to local need

The autoregulatory mechanisms that adjust the flow of blood to a tissue are part of the tissue itself, but they can be influenced by the nervous system and certain hormones.

The amount of blood that flows through a capillary bed is controlled by the degree of contraction of the smooth muscle of the arteries and arterioles feeding that bed. The flow of blood in a typical capillary bed is diagrammed in Figure 49.18. Blood flows into the bed from an arteriole. Smooth muscle "cuffs," or **precapillary sphincters**, on the arteriole can completely shut off the supply of blood to the capillary bed. When the precapillary sphincters are relaxed and the arteriole is open, the arterial blood pressure pushes blood into the capillaries.

Autoregulation depends on the sensitivity of the smooth muscle to its chemical environment. Low O_2 concentrations and high CO_2 concentrations cause the smooth muscle to relax, thus increasing the supply of blood, which brings in more O_2 and carries away CO_2. Increases in other by-products of metabolism, such as lactate, hydrogen ions, potassium, and adenosine, promote increased blood flow through

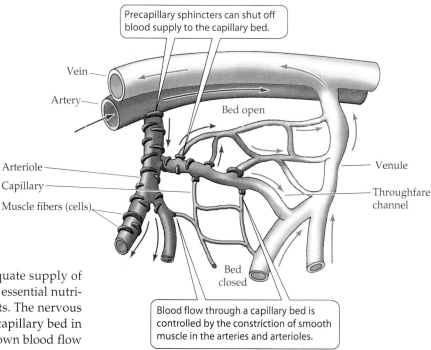

Precapillary sphincters can shut off blood supply to the capillary bed.

Vein

Artery

Bed open

Arteriole

Capillary

Muscle fibers (cells)

Venule

Throughfare channel

Bed closed

Blood flow through a capillary bed is controlled by the constriction of smooth muscle in the arteries and arterioles.

49.18 Local Control of Blood Flow
Low O_2 concentrations or high levels of metabolic by-products cause the smooth muscle of the arteries and arterioles to relax, thus increasing the supply of blood to the capillary bed.

the same mechanism. Hence, activities that increase the metabolism of a tissue also increase blood flow to that tissue.

Arterial pressure is controlled and regulated by hormonal and neural mechanisms

The same smooth muscle of arteries and arterioles that responds to autoregulatory stimuli also responds to signals from the endocrine and central nervous systems. Most arteries and arterioles are innervated by the autonomic nervous system, particularly the sympathetic division. Most sympathetic neurons release norepinephrine, which causes the smooth muscle cells to contract, thus constricting the vessels and reducing blood flow. An exception is found in skeletal muscle, in which specialized sympathetic neurons release acetylcholine, causing the smooth muscle of the arterioles to relax and the vessels to dilate, increasing blood to flow to the muscle.

Hormones also can cause arterioles to constrict. Epinephrine, which has actions similar to those of norepinephrine, is released from the adrenal medulla during massive sympathetic activation—the fight-or-flight response. *Angiotensin*, produced when blood pressure in the kidneys falls, causes arterioles to constrict. *Vasopressin*, released by the posterior pituitary when blood pressure falls, has similar effects (Figure 49.19). These hormones influence arterioles located for the most part in peripheral tissues (extremities) or in tissues whose functions need not be maintained continuously (such as the gut). By reducing blood flow in those arterioles, the hormones increase central blood pressure and blood flow to essential organs such as the heart, brain, and kidneys.

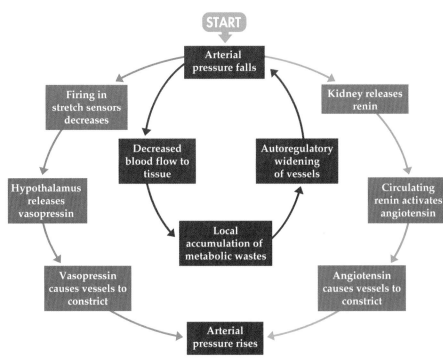

START

Arterial pressure falls

Firing in stretch sensors decreases

Kidney releases renin

Decreased blood flow to tissue

Autoregulatory widening of vessels

Hypothalamus releases vasopressin

Circulating renin activates angiotensin

Local accumulation of metabolic wastes

Vasopressin causes vessels to constrict

Angiotensin causes vessels to constrict

Arterial pressure rises

49.19 Hormonal Control of Blood Pressure through Vascular Resistance
A drop in arterial pressure reduces blood flow to tissues, resulting in local accumulation of metabolic wastes. This change in the extracellular environment stimulates autoregulatory opening of the arteries, which would lead to a further decrease in central blood pressure if this were not prevented by the negative feedback mechanisms shown in this diagram, which work by promoting the constriction of arteries in less essential tissues.

The autonomic nervous system activity that controls heart rate and constriction of blood vessels originates in cardiovascular centers in the medulla of the brain stem. Many inputs converge on this central integrative network and influence the commands it issues via parasympathetic and sympathetic nerves (Figure 49.20). Of special importance is information about changes in blood pressure from stretch receptors in the walls of the great arteries leading to the brain—the aorta and the carotid arteries.

Increased activity in the stretch receptors indicates rising blood pressure and inhibits sympathetic nervous system output. As a result, the heart slows, and arterioles in peripheral tissues dilate. If pressure in the great arteries falls, the activity of the stretch receptors decreases, stimulating sympathetic output. As a result, the heart beats faster, and the arterioles in peripheral tissues constrict. When arterial pressure falls, the change in stretch receptor activity also causes the hypothalamus to release vasopressin, which helps increase blood pressure by stimulating peripheral arterioles to constrict.

You experience the action of the aortic and carotid stretch receptors when you get up each morning. While you are lying down, your blood pressure is rather evenly distributed from head to toe, but when you get up, gravity pulls blood to the lower part of your body. Blood return to the heart decreases, and therefore cardiac output decreases. As a result, the pressure in the aorta and the carotid arteries falls. This change is detected by the carotid and aortic stretch receptors, which stimulate corrective responses within two heartbeats. Now imagine the change in blood pressure detected by the carotid stretch receptors in the giraffe on the cover of this book when it raises and lowers its head. The giraffe has a much larger heart than would be expected for a mammal of its size because of the need to generate blood pressure sufficient to pump blood against gravity to a height more than 3 m above its heart.

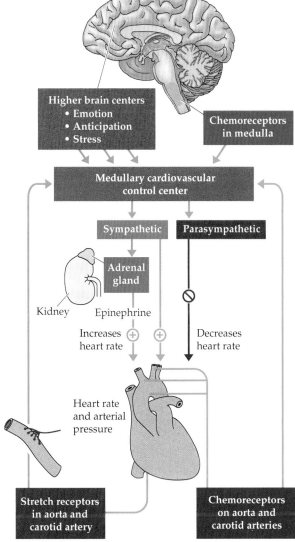

Higher brain centers
• Emotion
• Anticipation
• Stress

Chemoreceptors in medulla

Medullary cardiovascular control center

Sympathetic

Parasympathetic

Kidney

Adrenal gland

Epinephrine

Increases ⊕ heart rate

⊕ Decreases heart rate

Heart rate and arterial pressure

Stretch receptors in aorta and carotid artery

Chemoreceptors on aorta and carotid arteries

49.20 Regulating Blood Pressure
The autonomic nervous system controls heart rate in response to information about blood pressure and blood composition that is integrated by regulatory centers in the medulla.

Other information that causes the medullary regulatory system to increase heart rate and blood pressure comes from chemoreceptors in the carotid and aortic bodies. These nodules of modified smooth muscle tissue respond to inadequate O_2 supply. If arterial blood flow slows or the O_2 content of the arterial blood falls drastically, these receptors are activated and send signals to the regulatory center. The regulatory center also receives input from other brain areas. Emotions or the anticipation of intense activity, as at the start of a race, can cause the center to increase heart rate and blood pressure.

Control and regulation in the cardiovascular system begins with the local autoregulatory mechanisms that cause dilation of local arterioles and precapillary sphincters when a tissue needs more oxygen or has accumulated wastes. As more blood flows into the tissues, the central blood pressure falls, and the composition of the blood returning to the heart reflects the exchanges that are occuring in the tissues. Changes in central blood pressure and blood composition are sensed, and both endocrine and central nervous system responses are activated to return blood pressure and composition to normal. Thus circulatory functions are matched to the regional and overall needs of the body.

Cardiovascular control in diving mammals conserves oxygen

We began this chapter with the observation that when a seal begins underwater activity, its heart rate slows and blood flow to all of its tissues except its brain drops dramatically. This "diving reflex" of marine mammals is in stark contrast to our increase in heart rate and blood flow when we begin exercise. The obvious difference between the situation of the seal and the human is that the human has access to atmospheric oxygen during exercise and the diving seal does not.

The adaptations of the seal that enable it to remain underwater for a long time are several. The seal's oxygen storage capacity is about twice ours due to greater blood volume, the greater oxygen carrying capacity of its blood, and more myoglobin in its muscles. That is not sufficient, however, to explain dives of half an hour or more. The most important adaptation is the *diving reflex*, which is a slowing of the heart (Figure 49.21) and a constriction of major blood vessels going to all tissues except certain critical ones such as the nervous system, the heart, and the eyes. Central blood pressure remains high, but blood flow to the tissues decreases. This reduced blood flow has two effects: one is to switch the tissue to glycolytic (anaerobic) metabolism, and the other is to suppress the metabolism of the tissue.

While diving, the seal accumulates lactic acid in its muscles, which constitutes an "oxygen debt" to be paid back through elevated metabolism after the dive ends. But the total metabolic "payback" is much less than the metabolism that would have occurred over the same period of time had the seal not dived. The diving reflex caused the seal to be

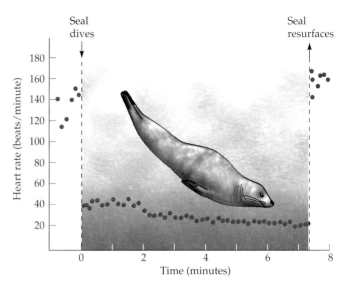

49.21 The Diving Reflex
When a marine mammal dives, its heart rate slows and the arteries to most of its organs constrict, so almost all blood flow and available oxygen goes to the animal's heart and brain. These adaptations enable some seals to remain underwater for up to an hour.

hypometabolic (below the basal metabolic rate) during the dive. Hypometabolism, increased oxygen stores, and a high capacity for anaerobic metabolism make it possible for the seal to perform amazing feats.

Chapter Summary

Circulatory Systems: Pumps, Vessels, and Blood

▶ The metabolic needs of the cells of small aquatic animals are met by direct exchange of materials with the external medium. The metabolic needs of the cells of larger animals are met by a circulatory system that transports nutrients, respiratory gases, and metabolic wastes throughout the body. **Review Figure 49.1**

▶ In open circulatory systems the blood or tissue fluid leaves vessels and percolates through tissues. **Review Figure 49.2**

▶ In closed circulatory systems the blood is contained in a system of vessels. **Review Figure 49.3**

Vertebrate Circulatory Systems

▶ The circulatory systems of vertebrates consist of a heart and a closed system of vessels containing blood that is separate from the tissue fluid. Arteries and arterioles carry blood from the heart; capillaries are the site of exchange between blood and tissue fluid; venules and veins carry blood back to the heart.

▶ The vertebrate heart evolved from two chambers in fishes to three in amphibians and reptiles and four in crocodilians, mammals, and birds. This evolutionary progression has led to an increasing separation of blood flow to the gas exchange organs and blood flow to the rest of the body. **Review Pages 868–870**

▶ In birds and mammals, blood circulates through two circuits: the pulmonary circuit and the systemic circuit.

The Human Heart: Two Pumps in One

▶ The human heart has four chambers. Valves in the heart prevent the backflow of blood. **Review Figure 49.4**

▶ The cardiac cycle has two phases: systole, in which the ventricles contract; and diastole, in which the ventricles relax. The sequential heart sounds ("lub-dub") are made by the closing of the heart valves. **Review Figure 49.5**

▶ Blood pressure can be measured using a sphygmomanometer and a stethoscope. **Review Figure 49.6**

▶ The autonomic nervous system controls heart rate. Sympathetic activity increases heart rate, and parasympathetic activity decreases it. These actions are due to the effects of norepinephrine and acetylcholine on the rate of depolarization of the membranes of pacemaker cells. **Review Figure 49.7**

▶ The sinoatrial node controls the cardiac cycle by initiating a wave of depolarization in the atria, which is conducted to the ventricles through the atrioventricular node. **Review Figure 49.8**

▶ The EKG records electric potentials resulting from the contraction and relaxation of the cardiac muscles. **Review Figure 49.9**

The Vascular System: Arteries, Capillaries, and Veins

▶ Arteries and arterioles have many elastic fibers that enable them to withstand high pressures. Abundant smooth muscle cells allow these vessels to contract and expand, altering their resistance and thus blood flow. **Review Figure 49.10**

▶ Capillary beds are the site of exchange of materials between blood and tissue fluid.

▶ The exchange of fluids between blood and tissues is determined by the balance between blood pressure and osmotic potential in the capillaries. **Review Figure 49.12**

▶ The ability of a specific molecule to cross a capillary wall depends on the architecture of the capillary, the type of substance, and the concentration gradient between the blood and the tissue fluid.

▶ A separate system of vessels, the lymphatic system, returns the tissue fluid to the blood.

▶ Veins have a high capacity for storing blood. Aided by gravity, by contractions of skeletal muscle, and by the actions of breathing, they carry blood back to the heart. **Review Figure 49.13**

▶ Cardiovascular disease is responsible for about half of all deaths in the United States and Europe. Atherosclerosis and thrombus formation can lead to potentially fatal conditions such as heart attack and stroke. Diet and behavior are the keys to good cardiovascular health. **Review Figure 49.14**

Blood: A Fluid Tissue

▶ Blood can be divided into a plasma portion (water, salts, and proteins) and a cellular portion (red blood cells, white blood cells, and platelets). All of the cellular components are produced from stem cells in the bone marrow. **Review Figure 49.15**

▶ Red blood cells transport respiratory gases. Their production in the bone marrow is stimulated by erythropoietin, which is produced in response to hypoxia in the tissues. **Review Figure 49.16**

▶ Platelets, along with circulating proteins, are involved in clotting responses. **Review Figure 49.17**

▶ Plasma is a complex solution that contains gases, ions, nutrient molecules, proteins, and other molecules.

Control and Regulation of Circulation

▶ Blood flow through capillary beds is controlled by local autoregulation mechanisms, hormones, and the autonomic nervous system. **Review Figure 49.18**

▶ Blood pressure is controlled in part by the hormones vasopressin and angiotensin, which stimulate contraction of blood vessels. **Review Figure 49.19**

▶ Heart rate is controlled by the autonomic nervous system, which responds to information about blood pressure and blood composition that is integrated by regulatory centers in the brain. **Review Figure 49.20**

▶ Diving mammals conserve blood oxygen stores by slowing the heart rate during dives. **Review Figure 49.21**

For Discussion

1. At the beginning of a race, cardiac output increases immediately before there is any change in blood oxygen or carbon dioxide concentrations. Explain two factors that contribute to this effect. Include the Frank–Starling law in your answer.

2. Explain how the hearts of crocodilians have the advantages of mammalian hearts during exercise but the efficiency of reptilian hearts during rest.

3. A sudden and massive loss of blood results in a decrease in blood pressure. Describe several mechanisms that help return blood pressure to normal.

4. You can describe the cycle of events in a ventricle of the heart by a graph that plots the pressure in the ventricle on the y axis and the volume of blood in the ventricle on the x axis. What would such a graph look like? Where would the heart sounds be on this graph? How would the graph differ for the left and the right ventricles?

5. If the major arteries become clogged with plaque and become less elastic because of calcification, the left ventricle must work harder and harder to pump an adequate supply of blood to the body. As a result, the left ventricle can become weakened and begin to fail even though the right ventricle is healthy. A heart attack primarily affecting the left ventricle can have the same effect. This condition is known as congestive heart failure, and commonly leads to fatal pulmonary edema. Explain how left ventricular failure can result in pulmonary edema, and why is it said that this condition creates a vicious circle that makes itself worse rapidly.

(a) **Swallowing**

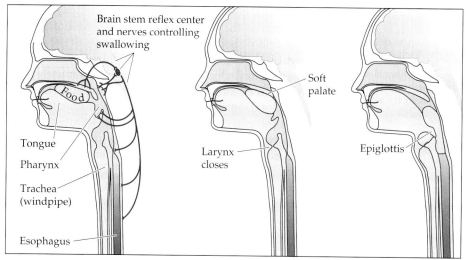

(b) **Peristalsis**

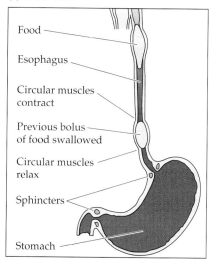

1 Food is chewed and the tongue pushes the bolus of food to the back of the mouth. Sensory nerves initiate the swallowing reflex.

2 The soft palate is pulled up as the vocal cords close the larynx.

3 The larynx is pulled up and forward and is covered by the epiglottis; the bolus of food enters the esophagus.

4 Peristaltic contractions propel the food to the stomach.

50.14 Swallowing and Peristalsis
Food pushed to the back of the mouth triggers the swallowing reflex. Once food enters the esophagus, peristalsis propels it through the gut.

prevents the food from entering the trachea (windpipe) or nasal passages (Figure 50.14).

Once the food is in the esophagus, peristalsis takes over and pushes it toward the stomach. **Peristalsis** is a wave of smooth muscle contraction that moves progressively down the gut from the pharynx toward the anus. The smooth muscle of the gut contracts in response to being stretched. Swallowing a bolus of food stretches the upper end of the esophagus, and this stretching initiates a wave of contraction that slowly pushes the contents of the gut toward the anus.

The movement of food from the stomach into the esophagus is normally prevented by a thick ring of circular smooth muscle at the junction of the esophagus and the stomach. This ring of muscle, the *esophageal sphincter*, is normally constricted. Waves of peristalsis cause it to relax enough to let food pass from the esophagus into the stomach. Sphincter muscles are found elsewhere in the digestive tract as well. The *pyloric sphincter* governs the passage of stomach contents into the intestine. Another important sphincter surrounds the anus.

Digestion begins in the mouth and the stomach

Food is chewed in the mouth, and carbohydrate digestion begins there. The enzyme **amylase** is secreted with saliva and mixed with the food as it is chewed. Amylase hydrolyzes the bonds between the glucose monomers that make up starch molecules. The action of amylase is what makes a piece of bread or cracker taste sweet if you hold it in your mouth long enough.

Most vertebrates can rapidly consume a large volume of food, but digesting that food is a long, slow process. The stomach stores the food consumed during a meal. The secretions of the stomach kill microorganisms that are taken in with the food and begin the digestion of proteins.

The major enzyme produced by the stomach is an endopeptidase called **pepsin**. Pepsin is secreted as a zymogen called **pepsinogen** by cells in the **gastric glands**—deep folds in the stomach lining (Figure 50.15). Other cells in the gastric glands produce hydrochloric acid, and still others near the openings of the gastric glands and throughout the stomach mucosa secrete mucus.

Hydrochloric acid (HCl) maintains the stomach fluid (the gastric juice) at a pH between 1 and 3. This low pH activates the conversion of pepsinogen to pepsin, which is achieved by the cleavage of a masking sequence of 44 amino acids from the N-terminal end of the pepsinogen molecule. The conversion is amplified as the newly formed pepsin activates other pepsinogen molecules, a process called **autocatalysis**. Hydrochloric acid also provides the right pH for the enzymatic action of pepsin. The low pH also helps dissolve the intercellular substances holding the ingested tissues together. Breakdown of the ingested tissues exposes more food surface area to the action of pepsin and eventually other digestive enzymes in the small intestine.

Mucus secreted by the stomach mucosa coats the walls of the stomach and protects them from being eroded and digested by HCl and pepsin. Sometimes, however, the walls of the stomach are exposed to HCl and pepsin; the resulting damage is called an *ulcer*. It was previously thought that ulcers were mostly due to stress and oversecretion of digestive juices. In recent years, however, it has been discovered that the basis for most ulcers is an infectious bacterium called *Helicobacter pylori*, which has the remarkable

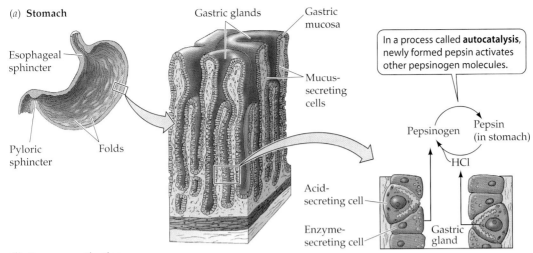

(a) **Stomach**

Esophageal sphincter

Pyloric sphincter

Folds

Gastric glands

Gastric mucosa

Mucus-secreting cells

Acid-secreting cell

Enzyme-secreting cell

Gastric gland

In a process called **autocatalysis**, newly formed pepsin activates other pepsinogen molecules.

Pepsinogen

Pepsin (in stomach)

HCl

(b) **Zymogen activation**

Inactive zymogen: pepsinogen

Active enzyme: pepsin

Low pH

Active site

A masking sequence is cleaved from the pepsinogen molecule…

Masking sequence

…transforming pepsinogen into the active digestive enzyme pepsin.

50.15 *The Stomach*
(a) The human stomach stores and breaks down ingested food. (b) Cells in the gastric glands secrete hydrochloric acid and the proteolytic enzyme pepsin. Both the glands and gastric mucosa secrete mucus that protects the stomach. (c) Pepsin is secreted as an inactive zymogen, pepsinogen, that is activated by low pH through the cleavage of a masking sequence of amino acids. Active pepsin also activates pepsinogen through autocatalysis.

ability to live in the highly acidic environment of the stomach. Lesions started by the bacterial infection are made worse by HCl and pepsin.

Contractions of the muscles in the walls of the stomach churn its contents, thoroughly mixing them with the stomach secretions. The acidic, fluid mixture of gastric juice and partly digested food in the stomach is called **chyme**. A few substances can be absorbed from the chyme across the stomach wall, including alcohol (hence its rapid effects), aspirin, and caffeine, but even these substances are absorbed in rather small quantities in the stomach.

Peristaltic contractions of the stomach walls push the chyme toward the bottom end of the stomach. These waves of peristalsis cause the pyloric sphincter to relax briefly so that little squirts of the chyme can enter the first region of the intestine. The human stomach empties itself gradually over a period of approximately 4 hours. This slow passage of food enables the intestine to work on a little material at a time and extends the digestive and absorptive processes throughout much of the time between meals.

The small intestine is the major site of digestion

In the **small intestine**, the digestion of carbohydrates and proteins continues, and the digestion of fats begins and the absorption of nutrients begins. Although the small intestine takes its name from its diameter, it is a very large organ. The small intestine of an adult human is more than 6 m long; its coils fill much of the lower abdominal cavity (see Figure 50.12). Because of its length, and because of the folds, villi, and microvilli of its lining, its inner surface area is enormous: about 550 m^2, or roughly the size of a tennis court. Across this surface the small intestine absorbs all the nutrient molecules derived from food. The small intestine has three sections. The initial section—the **duodenum**—is the site of most digestion; the **jejunum** and the **ileum** carry out 90 percent of the absorption of nutrients.

Digestion requires many specialized enzymes, as well as several other secretions. Two accessory organs that are not part of the digestive tract—the liver and the pancreas—provide many of these enzymes and secretions.

The liver synthesizes a substance called **bile** from cholesterol. Bile secreted from the liver flows through the **hepatic duct** to the gallbladder and to the duodenum. Bile reaches the gallbladder through a side branch of the hepatic duct (Figure 50.16). It is stored in the gallbladder until it is needed to assist in fat digestion. When fat enters the duodenum, a hormonal signal causes the walls of the gallbladder to contract rhythmically, squeezing bile back out toward the hepatic duct. Below the branch point to the gallbladder, the hepatic duct is called the **common bile duct**. Bile from the gallbladder flows down the common bile duct to the duodenum.

To understand the role of bile in fat digestion, think of the oil in salad dressing: it is not soluble in water (it is hydrophobic), and it tends to aggregate together in large globules. The enzymes that digest fat, the **lipases**, are water-soluble and must do their work in an aqueous medium. Bile stabilizes tiny droplets of fat so that they cannot aggregate into large globules. One end of each bile molecule is soluble in fat (it is lipophilic, or hydrophobic); the other end is soluble in water (it is hydrophilic, or lipophobic). Bile molecules bury their lipophilic ends in fat droplets, leaving their

is controlled so that food passes through the small intestine slowly enough for digestion and absorption to be complete, but quickly enough to ensure an adequate supply of nutrients for the body. Most of the available nutrients have been removed from the material that enters the colon, but the material contains a lot of water and inorganic ions.

The colon absorbs water and ions, producing semisolid feces from the slurry of indigestible materials it receives from the small intestine. Absorption of too much water in the colon can cause constipation. The opposite condition, diarrhea, results if too little water is absorbed or if water is secreted into the colon. (Both constipation and diarrhea can be induced by toxins from certain microorganisms.) Feces are stored in the last segment of the colon and are periodically excreted.

Immense populations of bacteria live within the colon. One of the resident species is *Escherichia coli*, the bacterium that is so popular among researchers in biochemistry, genetics, and molecular biology. This inhabitant of the colon lives on matter indigestible to humans and produces some products useful to its host. Vitamin K and biotin, for example, are synthesized by *E. coli* and absorbed across the wall of the colon. Excessive or prolonged intake of antibiotics can lead to vitamin deficiency because the antibiotics kill the normal intestinal bacteria at the same time they are killing the disease-causing organisms for which they are intended. The intestinal bacteria produce gases such as methane and hydrogen sulfide as by-products of their largely anaerobic metabolism. Humans expel gas after eating beans because the beans contain certain carbohydrates that bacteria—but not humans—can break down.

The large intestine of humans has a small, fingerlike pouch called the **appendix**, which is best known for the trouble it causes when it becomes infected. The human appendix plays no essential role in digestion, but it does contribute to immune system function. It can be surgically removed without serious consequences. The part of the gut that forms the appendix in humans forms the much larger cecum in herbivores (see Figure 50.10), where it functions, as we will see, in cellulose digestion. As our primate ancestors evolved to exploit diets less rich in indigestible cellulose, the cecum no longer served an essential function and gradually became *vestigial* (reduced to a trace).

Herbivores have special adaptations to digest cellulose

Cellulose is the principal organic compound in the diets of herbivores. Most herbivores, however, cannot produce **cellulases**, the enzymes that hydrolyze cellulose. Exceptions include silverfish (well known for eating books and stored papers), earthworms, and shipworms. Other herbivores, from termites to cattle, rely on microorganisms living in their digestive tracts to digest cellulose for them.

The digestive tracts of **ruminants** (cud chewers) such as cattle, goats, and sheep are specialized to maximize the benefits of their endosymbiotic microorganisms. In place of the usual mammalian stomach, ruminants have a large, four-chambered organ (Figure 50.18). The first two chambers, the *rumen* and the *reticulum*, are packed with anaerobic microorganisms that break down cellulose. The ruminant periodically regurgitates the contents of the rumen (the cud) into the mouth for rechewing. When the more thoroughly ground-up vegetable fibers are swallowed again, they present more surface area to the microorganisms for their digestive actions.

The microorganisms in the rumen and reticulum metabolize cellulose and other nutrients to simple fatty acids, which become nutrients for the host. In addition, the microorganisms themselves provide an important source of protein for the host. The plant materials ingested by a ruminant are a poor source of protein, but they contain inorganic nitrogen that the microorganisms use to synthesize their own amino acids. A cow can derive more than 100 g of protein per day from digestion of its endosymbiotic microorganisms.

Carbon dioxide and methane are by-products of the fermentation of cellulose carried out by microorganisms. A single cow can produce and belch 400 liters of methane a day. Methane is the second most abundant of the "greenhouse gases" whose concentration in the atmosphere is increasing,

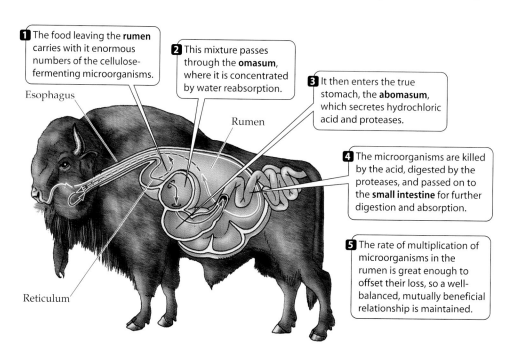

1 The food leaving the **rumen** carries with it enormous numbers of the cellulose-fermenting microorganisms.

2 This mixture passes through the **omasum**, where it is concentrated by water reabsorption.

3 It then enters the true stomach, the **abomasum**, which secretes hydrochloric acid and proteases.

4 The microorganisms are killed by the acid, digested by the proteases, and passed on to the **small intestine** for further digestion and absorption.

5 The rate of multiplication of microorganisms in the rumen is great enough to offset their loss, so a well-balanced, mutually beneficial relationship is maintained.

Esophagus

Rumen

Reticulum

50.18 The Ruminant Stomach
Four specialized stomach compartments enable ruminants to digest and subsist on protein-poor plant material.

and domesticated ruminants are second only to industry as a source of methane emitted into the atmosphere.

The food leaving the rumen carries with it enormous numbers of the cellulose-fermenting microorganisms. This mixture passes through the *omasum*, where it is concentrated by water absorption. It then enters the true stomach, the *abomasum*, which secretes hydrochloric acid and proteases. The microorganisms are killed by the acid, digested by the proteases, and passed on to the small intestine for further digestion and absorption. The rate of multiplication of microorganisms in the rumen is great enough to offset their loss, so a well-balanced, mutually beneficial relationship is maintained.

Some mammalian herbivores other than ruminants have microbial farms and cellulose fermentation vats in a branch off the large intestine called the **cecum**. Rabbits and hares are good examples. Since the cecum empties into the large intestine, the absorption of the nutrients produced by the microorganisms is inefficient and incomplete. Therefore, some of these animals reingest some of their own feces, a behavior known as **coprophagy**. Coprophagous species usually produce two kinds of feces, one consisting of pure waste (which they discard), and one consisting mostly of cecal material, which they reingest directly from the anus. As this cecal material passes through the stomach and small intestine, the nutrients it contains are digested and absorbed.

Control and Regulation of Digestion

The vertebrate gut is an assembly line in reverse—a disassembly line. As with a standard assembly line, control and coordination of sequential processes is critical. Both neural and hormonal controls govern gut functions.

Autonomic reflexes coordinate functions in different regions of the gut

Everyone has experienced salivation stimulated by the sight or smell of food. That response is an autonomic reflex, as is the act of swallowing following tactile stimulation at the back of the mouth. Many autonomic reflexes coordinate activities in different regions of the digestive tract. Loading the stomach with food, for example, stimulates increased activity in the colon, which can lead to a bowel movement.

The digestive tract is unusual in that it has an intrinsic (that is, its own) nervous system. In addition to autonomic reflexes involving the CNS, such as salivation and swallowing, neural messages can travel from one region of the digestive tract to another without being processed by the CNS.

Hormones control many digestive functions

Several hormones control the activities of the digestive tract and its accessory organs (Figure 50.19). The first hormone ever discovered came from the duodenum; it was called **secretin** because it caused the pancreas to secrete digestive juices. We now know that secretin is one of several hormones that control pancreatic secretion; specifically, secretin stimulates the pancreas to secrete a solution rich in bicarbonate ions.

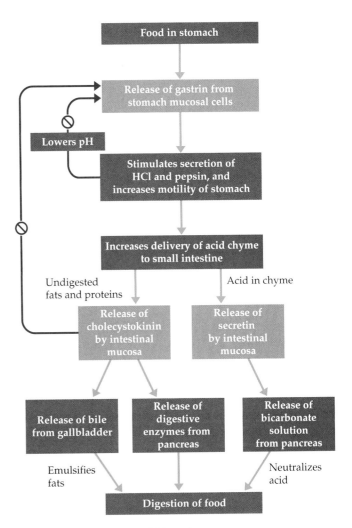

50.19 Hormones Control Digestion
Several hormones are involved in feedback loops that control the sequential processing of food in the digestive tract.

In response to the presence of fats and proteins in the chyme, the mucosa of the small intestine secretes **cholecystokinin**, a hormone that stimulates the gallbladder to release bile and the pancreas to release digestive enzymes. Cholecystokinin and secretin also slow the movements of the stomach, which slows the delivery of chyme into the small intestine.

The stomach secretes a hormone called **gastrin** into the blood. Cells in the lower region of the stomach release gastrin when they are stimulated by the presence of food. Gastrin circulates in the blood until it reaches cells in the upper areas of the stomach wall, where it stimulates the secretions and movements of the stomach. Gastrin release is inhibited when the stomach contents become too acidic—another example of negative feedback.

Control and Regulation of Fuel Metabolism

Most animals do not eat continuously. When they do eat, food is present in the gut and nutrients are being absorbed for some period of time after the meal, called the **absorp-**

tive period. Once the stomach and small intestine are empty, nutrients are no longer being absorbed. During this **postabsorptive period**, the continuous processes of energy metabolism and biosynthesis must run on internal reserves. Nutrient traffic must be controlled so that reserves accumulate during the absorptive period are used appropriately during the postabsorptive period.

The liver directs the traffic of fuel molecules

The liver directs the traffic of the nutrients that fuel metabolism. When nutrients are abundant in the blood, the liver stores them in the forms of glycogen and fats. The liver also synthesizes blood plasma proteins from circulating amino acids. When the availability of fuel molecules in the blood declines, the liver delivers glucose and fats back to the blood.

The liver has an enormous capacity to interconvert fuel molecules. Liver cells can convert monosaccharides into either glycogen or fat, and vice versa. The liver can also convert certain amino acids and some other molecules, such as pyruvate and lactate, into glucose—a process called **gluconeogenesis**. The liver is also the major controller of fat metabolism through its production of lipoproteins.

Lipoproteins: The good, the bad, and the ugly

In the intestine, bile solves the problem of processing hydrophobic fats in an aqueous medium. The transportation of fats in the circulatory system presents the same problem, and lipoproteins are the solution. A **lipoprotein** is a particle made up of a core of fat and cholesterol and a covering of protein that makes it water-soluble. The largest lipoprotein particles are the chylomicrons produced by the mucosal cells of the intestine, which transport dietary fat and cholesterol into the circulation (see Figure 50.17). As the chylomicrons circulate through the liver and to adipose (fat) tissues throughout the body, receptors on the capillary walls recognize their protein coats, and lipases begin to hydrolyze the fats, which are then absorbed into liver or fat cells. Thus the protein coat of the lipoprotein both makes it water-soluble and serves as an "address" that directs it to a specific tissue.

Lipoproteins other than chylomicrons originate in the liver and are classified according to their density. Fat has a low density (it floats in water), so the more fat a lipoprotein contains, the lower its density.

▶ **Very-low-density lipoproteins** (**VLDL**) produced by the liver contain mostly triglyceride fats that are being transported to fat cells in tissues around the body.

▶ **Low-density lipoproteins** (**LDL**) consist of about 50 to 60 percent cholesterol, which they transport to tissues around the body for use in biosynthesis and for storage.

▶ **High-density lipoproteins** (**HDL**) serve as acceptors of cholesterol (they consist of about 25 percent cholesterol) and are believed to remove cholesterol from tissues and carry it to the liver, where it can be used to synthesize bile.

Because of their differing functions in cholesterol regulation, LDL is sometimes called "bad cholesterol" and HDL "good cholesterol"—designations that are somewhat controversial. However, we do know that a high ratio of LDL to HDL in a person's blood is a risk factor for atherosclerotic heart disease. Cigarette smoking lowers HDL levels, and regular exercise increases them.

Fuel metabolism is controlled by hormones

During the absorptive period, blood glucose levels are high as carbohydrates are digested and absorbed. During this time, the liver takes up glucose from the blood and converts it to glycogen and fat, fat cells take up glucose from the blood and convert it to stored fat, and the cells of the body preferentially use glucose as their metabolic fuel.

During the postabsorptive period these processes are reversed. The liver breaks down glycogen to supply glucose to the blood, the liver and the adipose tissues supply fatty acids to the blood, and most of the cells of the body preferentially use fatty acids as their metabolic fuel.

One tissue that does not switch fuel sources during the postabsorptive period is the nervous system. The cells of the nervous system require a constant supply of glucose. Even though the nervous system can use other fuels to a limited extent, its overall dependence on glucose is the reason it is so important for other cells of the body to shift to fat metabolism during the postabsorptive period. This shift preserves the available glucose and glycogen stores for the nervous system for as long as possible.

What directs the traffic in fuel molecules? Insulin and glucagon, two hormones produced and released by the pancreas, are responsible for controlling the metabolic directions that fuel molecules take (Figure 50.20). The most

50.20 Regulating Glucose Levels in the Blood
Insulin and glucagon maintain the homeostasis of circulatory glucose.

(a) **Fuel traffic during the absorptive period**

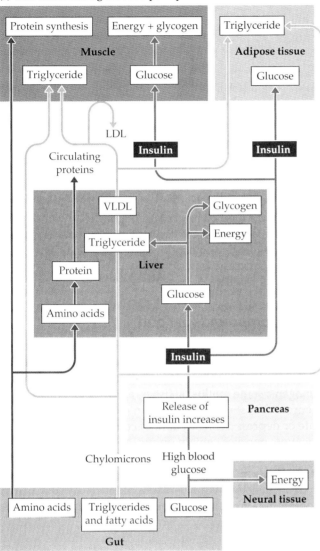

(b) **Fuel traffic during the postabsorptive period**

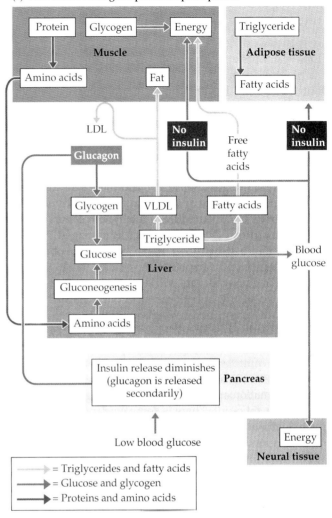

50.21 Fuel Molecule Traffic during the Absorptive and Postabsorptive Periods
Insulin promotes glucose uptake by liver, muscle, and fat cells during the absorptive period. During the postabsorptive period, the lack of insulin blocks glucose uptake by these same tissues and promotes fat and glycogen breakdown to supply metabolic fuel.

important of these hormones is **insulin**, which is produced in response to high blood glucose levels.

The pancreas releases insulin into the circulatory system when blood glucose rises above the normal postabsorptive level. Insulin facilitates the entry of glucose into most cells of the body. When insulin is present, most cells burn glucose as their metabolic fuel, fat cells use glucose to make fat, and liver cells convert glucose to glycogen and fat.

As soon as blood glucose falls back to postabsorptive levels, insulin release diminishes rapidly, and the entry of glucose into cells other than those of the nervous system is inhibited. Without a supply of glucose, cells switch to using glycogen and fat as their metabolic fuels. In the absence of insulin, the liver and fat cells stop synthesizing glycogen and fat and begin breaking them down. As a result, the liver supplies glucose to the blood rather than taking it from the blood, and both the liver and the adipose tissues supply fatty acids to the blood.

The pancreas releases **glucagon** when the blood glucose concentration falls below the normal postabsorptive level. Glucagon has the opposite effect of insulin: it stimulates liver cells to break down glycogen and to carry out gluconeogenesis. Thus, under the influence of glucagon, the liver produces glucose and releases it into the blood.

The traffic of fuel molecules during the absorptive and postabsorptive periods is summarized in Figure 50.21, which indicates the steps controlled by insulin and glucagon. During the absorptive period, all fuel molecules move toward storage, and glucose is the preferred energy source for all cells. During the postabsorptive period, most cells switch to metabolizing fat so that blood glucose reserves are saved for the nervous system. The level of circulating glucose is maintained through glycogen breakdown and gluconeogenesis.

ing swelling and pain. The excretion of ammonia is an important mechanism for regulating the pH of the tissue fluid.

In some species, different developmental forms live in quite different habitats and have different forms of nitrogen excretion. The tadpoles of frogs and toads, for example, excrete ammonia across their gill membranes, but when they develop into adult frogs or toads, they generally excrete urea. Some adult frogs and toads that live in arid habitats excrete uric acid. These examples show the considerable evolutionary flexibility in how nitrogenous wastes are excreted.

The Diverse Excretory Systems of Invertebrates

Most marine invertebrates are osmoconformers and have few adaptations for salt and water balance other than active transport mechanisms for ionic regulation. To excrete nitrogen, they can passively lose ammonia by diffusion to the seawater. Freshwater and terrestrial invertebrates, however, have a wide variety of adaptations for maintaining salt and water balance and excreting nitrogen. All of these adaptations are based on the same set of mechanisms: filtration of body fluids and active secretion and resorption of specific ions.

Protonephridia excrete water and conserve salts

Many flatworms, such as *Planaria*, live in fresh water. These animals excrete water through an elaborate network of tubules running throughout their bodies. The tubules end in *flame cells*, so called because each tubule has a tuft of cilia beating inside it, giving the appearance of a flickering flame (Figure 51.4). A flame cell and a tubule together form a **protonephridium** (plural protonephridia; from the Greek *proto*, "before," and *nephros*, "kidney").

Tissue fluid enters the tubules (how it does so is not entirely clear), and the beating of the cilia causes this fluid to flow through the tubules toward the animal's excretory pore. As it flows, the cells of the tubules modify the fluid. As the modified tubule fluid (urine) leaves the planarian, it is less concentrated than the animal's tissue fluid, so ions are conserved and water is excreted by the protonephridium.

Metanephridia process coelomic fluid

Filtration of body fluids and modification of urine by tubules are highly developed processes in annelid worms, such as the earthworm. Recall that annelids are segmented and have a fluid-filled body cavity, called a coelom, in each segment (see Figure 31.23). Annelids have a closed circulatory system through which blood is pumped under pressure (see Figure 49.3). The pressure causes the blood to be filtered across the thin, permeable capillary walls into the coelom. This process is called *filtration* because the cells and large protein molecules of the blood stay behind in the capillaries while water and small molecules leave them and enter the coelom. In addition, some waste products, such as ammonia, diffuse directly from the tissues into the coelom. But where does this coelomic fluid go?

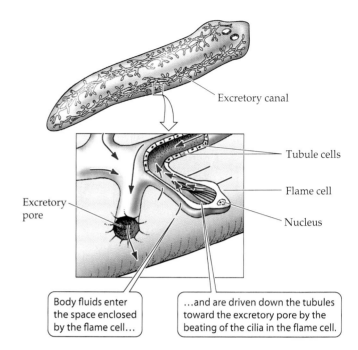

Excretory canal

Tubule cells

Flame cell

Excretory pore

Nucleus

Body fluids enter the space enclosed by the flame cell…

…and are driven down the tubules toward the excretory pore by the beating of the cilia in the flame cell.

51.4 Protonephridia in Flatworms
The protonephridia of the flatworm *Planaria* consist of tubules ending in flame cells. The tubule cells modify the composition of the fluid passing through them.

Each segment of the earthworm contains a pair of **metanephridia** (singular metanephridium; from the Greek *meta*, akin to, and *nephros*, kidney). Each metanephridium begins in one segment as a ciliated, funnel-like opening in the coelom called a *nephrostome*, which leads into a tubule in the next segment. The tubule ends in a pore called the *nephridiopore*, which opens to the outside of the animal (Figure 51.5). Coelomic fluid enters the metanephridia through the nephrostomes. As the fluid passes through the tubules, the cells of the tubules actively resorb certain molecules from it and actively secrete other molecules into it. What leaves the animal through the nephridiopores is a hypotonic (dilute) urine containing nitrogenous wastes, among other solutes.

Malpighian tubules are the excretory organs of insects

Insects can excrete nitrogenous wastes with very little loss of water. Therefore, some species can live in the driest habitats on Earth. The insect excretory system consists of blind tubules called **Malpighian tubules**. An individual insect has from 2 to more than 100 of these tubules attached to the gut between the midgut and hindgut and projecting into the spaces containing tissue fluid (recall that insects have open circulatory systems) (Figure 51.6).

The cells of the Malpighian tubules actively transport uric acid, potassium ions, and sodium ions from tissue fluid into the tubules. As these solutes are secreted into the tubules, water follows because of the difference in solute potential. The walls of the Malpighian tubules have muscle fibers that contract to help move the contents of the tubules toward the hindgut.

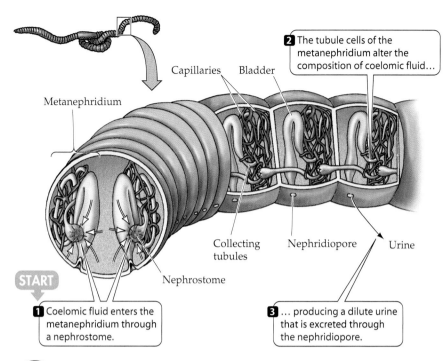

2 The tubule cells of the metanephridium alter the composition of coelomic fluid...

Capillaries   Bladder

Metanephridium

Collecting tubules

Nephridiopore   Urine

Nephrostome

START

1 Coelomic fluid enters the metanephridium through a nephrostome.

3 ... producing a dilute urine that is excreted through the nephridiopore.

51.5 Metanephridia in Earthworms
The metanephridia of annelids are arranged segmentally. The cross section (left) shows a pair of metanephridia. Longitudinal sections (right) show only one metanephridium of the two in each segment.

The tubule fluid changes in composition while it is in the hindgut. The contents of the hindgut are more acidic than the tubule fluid; as a result, uric acid becomes less soluble and precipitates out of solution as it approaches and enters the rectum. The epithelial cells of the hindgut and rectum actively transport sodium and potassium ions from the gut contents back into the tissue fluid. Because the uric acid molecules have precipitated out of solution, water is free to

follow the resorbed salts back into the tissue fluid through osmosis. Remaining in the rectum are crystals of uric acid mixed with undigested food; this dry matter is what the insect eliminates. The Malpighian tubule system is a highly effective mechanism for excreting nitrogenous wastes and some salts without giving up a significant fraction of the animal's precious water supply.

Vertebrate Excretory Systems Are Built of Nephrons

The major excretory organ of vertebrates is the **kidney**. The functional unit of the kidney is the **nephron**. Each human kidney has about a million nephrons. All vertebrate kidneys consist of nephrons, yet the kidneys of different species can serve opposite functions to maintain water and salt balance. The kidneys of freshwater fishes, for example, excrete water, but the kidneys of most mammals conserve water.

To understand how the kidney can fulfill opposite functions in different animals, we need to understand how the different parts of the nephron work and the different ways in which they can work together to influence the composition of the urine. The nephron has three main parts:

▶ A ball of capillaries called the glomerulus that filters the plasma

▶ Renal tubules that receive and modify the filtrate

▶ Peritubular capillaries that serve the tubules

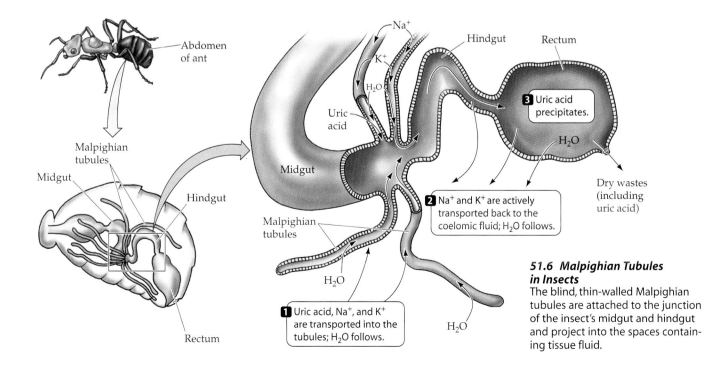

Abdomen of ant

Na⁺

K⁺

H_2O

Hindgut   Rectum

Uric acid

3 Uric acid precipitates.

H_2O

Midgut

Malpighian tubules

Midgut

Hindgut

Malpighian tubules

Dry wastes (including uric acid)

2 Na⁺ and K⁺ are actively transported back to the coelomic fluid; H_2O follows.

H_2O

Rectum

1 Uric acid, Na⁺, and K⁺ are transported into the tubules; H_2O follows.

H_2O

51.6 Malpighian Tubules in Insects
The blind, thin-walled Malpighian tubules are attached to the junction of the insect's midgut and hindgut and project into the spaces containing tissue fluid.

Blood is filtered in the glomerulus

Each nephron has both vascular and tubule components (Figure 51.7). The vascular component is unusual in that it consists of two capillary beds that lie between the arteriole that supplies it and the venule that drains it. The first capillary bed is a dense knot of very permeable vessels called the **glomerulus** (plural glomeruli) (Figure 51.8*a*). Blood enters the glomerulus through an **afferent arteriole** and exits through an **efferent arteriole**. The efferent arteriole gives rise to the second set of capillaries, the **peritubular capillaries**, which surround the tubule component of the nephron (see Figure 51.7).

The tubule component of the nephron, called a **renal tubule**, begins with **Bowman's capsule**, which encloses the glomerulus. The glomerulus appears to be pushed into Bowman's capsule much like a fist pushed into an inflated balloon. Together, the glomerulus and its surrounding Bowman's capsule are called the **renal corpuscle**. The cells of the capsule that come into direct contact with the glomerular capillaries are called **podocytes** (see Figure 51.7). These highly specialized cells have numerous armlike extensions, each with hundreds of fine, fingerlike projections. The podocytes wrap around the capillaries so that their fingerlike projections interdigitate and cover the capillaries completely (Figure 51.8*b*).

The glomerulus filters the blood to produce a tubule fluid that lacks cells and large molecules. The walls of the capillaries, the basal lamina of the capillary endothelium, and the podocytes of Bowman's capsule all participate in filtration. The endothelial walls of the capillaries have pores that allow water and small molecules to leave, but are too small to permit red blood cells to pass through. The meshwork of the basal lamina is even finer than the pores between the endothelial cells, and it prevents large molecules from leaving the capillaries. Also smaller than the pores in the capillaries are the narrow slits between the fingerlike projections of the podocytes. As a result of these anatomical adaptations, water and small molecules pass from the capillary blood and enter the renal tubule of the nephron (Figure 51.8*c*), but red blood cells and proteins remain in the capillaries.

The force that drives filtration in the glomerulus is the pressure of the arterial blood. As in every other capillary bed, the pressure of the blood entering the permeable capillaries causes the filtration of water and small molecules. The glomerular filtration rate is high because glomerular capillary blood pressure is unusually high, and because the capillaries of the glomerulus, along with their covering of podocytes, are much more permeable than other capillary beds in the body.

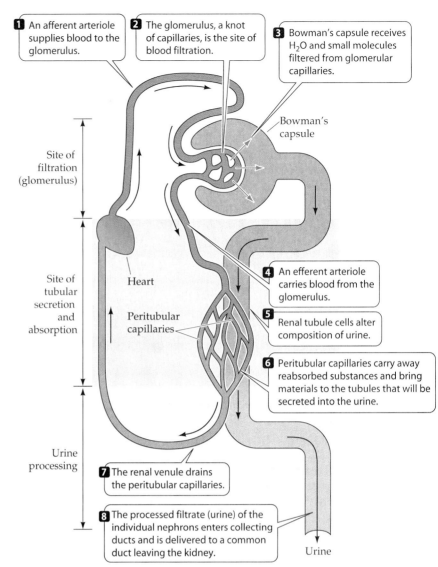

51.7 The Vertebrate Nephron
The vertebrate nephron consists of a renal tubule closely associated with a system of blood vessels. The end of the renal tubule system envelops the glomerulus so that the filtrate from the glomerular capillaries enters the tubules. The tubules change the composition of the filtrate by active absorption and secretion of solutes.

The renal tubules convert glomerular filtrate to urine

The composition of the filtrate that enters the nephron is similar to that of the blood plasma. This filtrate contains glucose, amino acids, ions, and nitrogenous wastes in the same concentrations as in the blood plasma, but it lacks the plasma proteins. As this fluid passes down the renal tubule, its composition changes as the cells of the tubule actively resorb certain molecules from the tubule fluid and secrete other molecules into it. When the tubule fluid leaves the kidney as urine, its composition is very different from that of the original filtrate.

The function of the renal tubules is to control the composition of the urine by actively secreting and resorbing specific molecules. The peritubular capillaries serve the needs of the renal tubules by bringing to them the molecules to be secreted into the tubules and carrying away the molecules that are resorbed from the tubules.

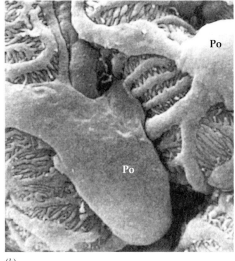
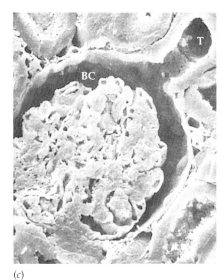

(a)   (b)   (c)

51.8 An SEM Tour of the Nephron

These scanning electron micrographs show the anatomical bases for kidney function.
(a) The blood vessels in the kidney, showing the knots of capillaries that form the glomeruli.
Each glomerulus (Gl) has an afferent and an efferent arteriole (Ar). Peritubular capillaries (Pt)
are looser networks surrounding the tubules of the nephron. (b) Those cells that are in direct
contact with the capillaries are the podocytes (Po). Each podocyte has hundreds of tiny
fingerlike projections that create filtration slits between them. Anything passing from the
glomerular capillaries into the tubule of the nephron must pass through these slits.
(c) A cross section of a glomerulus shows that it is surrounded by the tubule cells that form
Bowman's capsule (BC), which collects the filtrate and funnels it into the tubule (T) of the
nephron. The relationship between the glomerulus and Bowman's capsule is like that of a fist
punched into a balloon. Therefore, some of the renal tubule cells are in direct contact with
the glomerular capillaries.

Both marine and terrestrial vertebrates must conserve water

Since the vertebrate nephron evolved as a structure for excreting water while conserving salts and essential small molecules, how have vertebrates adapted to environments where water must be conserved and salts excreted? The answer to this question differs for each vertebrate group. Even among the marine fishes, the adaptations of bony fishes are different from those of cartilaginous fishes.

MARINE BONY FISHES. Marine bony fishes cannot produce urine more concentrated than their tissue fluid, but unlike most marine animals, they osmoregulate their tissue fluid to only one-fourth to one-third the solute potential of seawater. They prevent excessive loss of water by producing very little urine. Their urine production is low because their kidneys have fewer glomeruli than do the kidneys of freshwater fishes. In some species of marine bony fishes, the kidneys have no glomeruli at all. Even though the glomeruli are reduced or absent, renal tubules with closed ends are retained for active excretion of ions and certain molecules.

Marine bony fishes take in seawater with their food, which results in a large salt load. The fish handle these salt loads by simply not absorbing some ions (such as Mg^{2+} or

SO_4^{2-}) from their guts and by actively excreting others (such as Na^+ and Cl^-) from the gill membranes and from the renal tubules. Nitrogenous wastes are lost as ammonia from the gill membranes.

CARTILAGINOUS FISHES. Cartilaginous fishes are osmoconformers, but not ionic conformers. Unlike marine bony fishes, cartilaginous fishes convert nitrogenous wastes to urea and another compound called tri-methyl amine oxide and retain large amounts of these compounds in their tissue fluids. As a result, their tissue fluids have an osmolarity close to that of seawater. These species have adapted to a concentration of urea in the body fluids that would be fatal to other vertebrates.

Sharks and rays still have the problem of excreting the large amounts of salts they take in with their food. They have several sites of active secretion of NaCl, but the major one is a salt-secreting *rectal gland*.

AMPHIBIANS. Most amphibians live in or near fresh water and stay in humid habitats when they venture from the water. Like freshwater fishes, most amphibian species produce large amounts of dilute urine and conserve salts. Some amphibians, however, have adapted to habitats that require water conservation.

Lymnodynastes dumerilii

51.9 Burrowing Frogs
The banjo frog of the Australian desert survives long droughts by burrowing deep in the sand and entering estivation, a state of low metabolic activity. These frogs store water in the form of dilute urine in their enormous bladders.

Amphibians living in very dry terrestrial environments have reduced the water permeability of their skin. Some secrete a waxy substance that they spread over the skin to waterproof it. Several species of frogs that live in arid regions of Australia burrow deep into the ground, where they remain during long dry periods (Figure 51.9). There they enter *estivation*, a state of very low metabolic activity and therefore low water turnover. When it rains, these frogs come out of estivation, feed, and reproduce. But their most interesting adaptation is that they have enormous urinary bladders. Before entering estivation, they fill their bladders with dilute urine, which can amount to one-third of their body weight. This dilute urine serves as a water reservoir that they use gradually during the long period of estivation. Australian aboriginal peoples dig up estivating frogs as an emergency source of drinking water.

REPTILES.   Reptiles occupy habitats ranging from aquatic to extremely hot and dry. Three major adaptations have freed the reptiles from maintaining the close association with water that is necessary for most amphibians. First, reptiles do not need fresh water to reproduce, because they employ internal fertilization and lay eggs with shells that retard evaporative water loss. Second, they have scaly, dry skins that retard evaporative water loss.

Third, they excrete nitrogenous wastes as uric acid solids, therefore losing little water in the process.

BIRDS.   Birds have the same adaptations for water conservation that reptiles have: internal fertilization, shelled eggs, skin that retards water loss, and uric acid as the nitrogenous waste product. In addition, some birds can produce a urine that is more concentrated than their tissue fluids. This last ability is most developed in mammals.

The Mammalian Excretory System

The adaptations of mammals and birds for producing urine hypertonic to their tissue fluids were an important step in vertebrate evolution. These adaptations enabled the excretory system to conserve water while still excreting excess salts and nitrogenous wastes. Mammals and birds have high body temperatures and high metabolic rates, and therefore have the potential for a high rate of water loss. Being able to minimize water loss from their excretory systems made it possible for these highly active species to occupy arid habitats.

We have seen how the nephron originally evolved to excrete water; now we will see how it can serve as the basic structural unit of an organ that is able to conserve water. To understand this evolutionary change of function, it is necessary to understand the structure and function of the nephron in the context of the overall anatomy of the kidney. First, however, let's look at the mammalian excretory system as a whole.

Kidneys produce urine, which the bladder stores

We will use humans here as our example of the mammalian excretory system. Humans have two kidneys just under the dorsal wall of the abdominal cavity in the mid-back region (Figure 51.10). Each kidney filters blood, processes the fil-

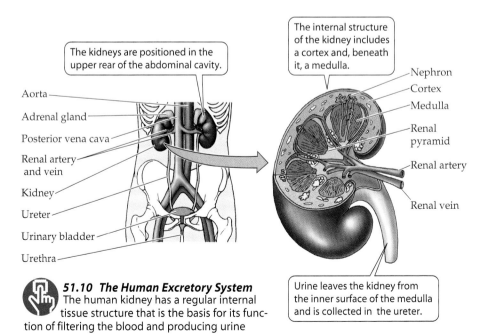

The kidneys are positioned in the upper rear of the abdominal cavity.

The internal structure of the kidney includes a cortex and, beneath it, a medulla.

Aorta
Adrenal gland
Posterior vena cava
Renal artery and vein
Kidney
Ureter
Urinary bladder
Urethra

Nephron
Cortex
Medulla
Renal pyramid
Renal artery
Renal vein

Urine leaves the kidney from the inner surface of the medulla and is collected in the ureter.

51.10 The Human Excretory System
The human kidney has a regular internal tissue structure that is the basis for its function of filtering the blood and producing urine

trate into urine, and releases that urine into a duct called the **ureter**. The ureter of each kidney leads to the **urinary bladder**, where the urine is stored until it is excreted through the urethra. The **urethra** is a short tube that opens to the outside at the end of the penis in males or just anterior to the vaginal opening in females.

Two sphincter muscles surrounding the base of the urethra control the timing of urination. One of these sphincters is a smooth muscle and is controlled by the autonomic nervous system. When the bladder is full, a spinal reflex relaxes this sphincter. This reflex is the only control of urination in infants, but the reflex gradually comes under the influence of higher centers in the nervous system as a child grows older. The other sphincter is a skeletal muscle and is controlled by the voluntary, or conscious, nervous system. When the bladder is *very* full, only serious concentration prevents urination.

Nephrons have a regular arrangement in the kidney

The kidney is shaped like a kidney bean; when cut down its long axis and split open as a bean splits open, its important anatomical features are revealed (see Figure 51.10). The ureter and the **renal artery** and **renal vein** enter the kidney on its concave (punched-in) side. The ureter divides into several branches, the ends of which envelop kidney tissues called **renal pyramids**. The renal pyramids make up the internal core, or **medulla**, of the kidney. The medulla is surrounded by tissue with a different appearance, called the **cortex**. The renal artery and vein give rise to many arterioles and venules in the region between the cortex and the medulla.

Each human kidney contains about a million nephrons, and their organization within the kidney is very regular. All of the glomeruli are located in the cortex. The initial segment of a renal tubule is called the **proximal convoluted tubule**—"proximal" because it is close to its glomerulus and "convoluted" because it is twisted (Figure 51.11). All the proximal convoluted tubules are also located in the cortex.

At a certain point, the renal tubule takes a dive directly down into the medulla. The portion of the tubule in the medulla is called the **loop of Henle**. It is called a loop because it runs straight down into the medulla, makes a hairpin turn, and comes straight back to the cortex. Where the ascending limb of the loop of Henle reaches the cortex, it becomes the distal convoluted tubule—"distal" because it is farther from its glomerulus than the proximal tubule is. The distal convoluted tubules of many nephrons join a common **collecting duct** in the cortex. The collecting ducts then run in parallel with the loops of Henle down through the medulla and empty into the ureter at the tips of the renal pyramids.

Blood vessels also have a regular arrangement in the kidney

The organization of the blood vessels of the kidney closely parallels the organization of the nephrons (see Figure 51.11). Arterioles branch from the renal arteriy and radiate into the cortex. An *afferent* arteriole carries blood to each

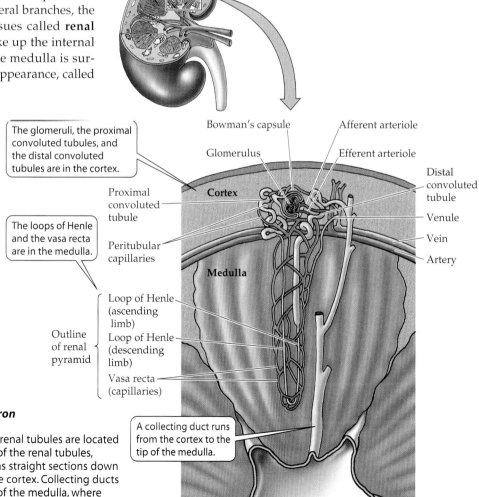

51.11 The Organization of the Nephron within the Mammalian Kidney
The glomeruli and major portions of the renal tubules are located in the cortex of the kidney, but portions of the renal tubules, called the loops of Henle, run in parallel as straight sections down into the renal medulla and back up to the cortex. Collecting ducts run from the cortex to the inner surface of the medulla, where they open into the ureter.

glomerulus. Draining each glomerulus is an *efferent* arteriole that gives rise to the peritubular capillaries, most of which surround the cortical portions of the tubules.

A few peritubular capillaries run into the medulla in parallel with the loops of Henle and the collecting ducts. These capillaries form the **vasa recta**. All the peritubular capillaries from a nephron join back together into a venule that joins with venules from other nephrons and eventually leads to the renal vein, which takes blood from the kidney.

The volume of glomerular filtration is greater than the volume of urine

Most of the water and solutes filtered in the glomerulus are resorbed and do not appear in the urine. We can reach this conclusion by comparing the rate of filtration by the glomeruli with the rate of urine production. The kidneys receive about 1 liter of blood per minute, or more than 1,400 liters of blood per day. How much of this huge volume is filtered in the glomeruli? The answer is about 12 percent. This is still a large volume—180 liters per day! Since we normally urinate 2 to 3 liters per day, about 98 to 99 percent of the fluid volume that is filtered in the glomerulus is resorbed into the blood. Where and how is this enormous fluid volume resorbed?

Most filtrate is resorbed by the proximal convoluted tubule

The proximal convoluted tubule is responsible for most of the resorption of water and solutes from the glomerular filtrate. The cells of this section of the renal tubule are cuboidal, and their surfaces facing into the tubule have thousands of microvilli, which greatly increase their surface area for resorption. These cells have lots of mitochondria—an indication that they are biochemically active. They actively transport Na+ (with Cl− following passively) and other solutes, such as glucose and amino acids, out of the tubule fluid. Almost all glucose and amino acid molecules that are filtered from the blood are actively resorbed by these cells and transported back into the tissue fluid. The active transport of solutes into the tissue fluid causes water to follow osmotically. The water and solutes moved into the tissue fluid are taken up by the peritubular capillaries and returned to the venous blood leaving the kidney.

Despite the large volume of water and solutes resorbed by the proximal convoluted tubule, the overall concentration, or osmolarity, of the fluid that enters the loop of Henle is not different from that of the blood plasma, although its composition is quite different. How, then, does the kidney produce urine that is hypertonic to the blood plasma?

The loop of Henle creates a concentration gradient in the surrounding tissue

Humans can produce urine that is four times more concentrated than their blood plasma. The vampire bat we encountered at the beginning of this chapter can produce a urine that is twenty times more concentrated than its blood plasma, and some desert-dwelling animals, as we will see, can produce even greater concentrations. This concentrat-

ing ability of the mammalian kidney is due to the loops of Henle, which function as a **countercurrent multiplier system**. The term "countercurrent" refers to the fact that tubule fluid in the descending limb of the loop flows in the opposite direction from that in the ascending limb. "Multiplier" refers to the ability of this system to create a concentration gradient in the renal medulla. The loops of Henle do not themselves produce a concentrated urine; rather, they increase the solute potential of the surrounding tissue fluid.

The segments of the loop of Henle differ anatomically and functionally. Cells of the descending limb and the initial cells of the ascending limb are flat, with no microvilli and few mitochondria. They are not specialized for transport. Partway up the ascending limb, the cells become specialized for active transport. They are cuboidal and have lots of mitochondria. Accordingly, the loop of Henle is divided into the *thin descending limb*, the *thin ascending limb*, and the *thick ascending limb*. To understand the countercurrent multiplier mechanism, it is easiest to move backward through the renal tubule, starting with the thick ascending limb (Figure 51.12).

The thick ascending limb actively resorbs Cl− (with Na+ following passively) from the tubule fluid and moves it into the surrounding tissue fluid. The thick ascending limb is not permeable to water, so the resorption of Na+ and Cl− raises the concentration of these solutes in the surrounding tissue fluid.

The thin descending limb, in contrast, is rather permeable to water, but not very permeable to Na+ and Cl−. Since the surrounding tissue fluid has been made more concentrated by the Na+ and Cl− resorbed from the neighboring thick ascending limb, water is withdrawn osmotically from the tubule fluid in the descending limb. Therefore, the fluid in the descending limb becomes more and more concentrated as it flows toward the bottom of the renal medulla.

The thin ascending limb, like the thick ascending limb, is not permeable to water. It is, however, permeable to Na+ and Cl−. As the concentrated tubule fluid flows up the thin ascending limb, it is more concentrated than the surrounding tissue fluid, so Na+ and Cl− diffuse out of it. When the tubule fluid reaches the thick ascending limb, active transport continues to move Na+ and Cl− from the tubule fluid to the tissue fluid, as we saw above.

As a result of the processes described above, the tubule fluid reaching the distal convoluted tubule is less concentrated than the blood plasma, and the solutes that have been left behind in the renal medulla have created a concentration gradient in the surrounding tissue fluid. The tissue fluid of the renal medulla becomes more and more concentrated as we move from the border with the cortex down to the tips of the renal pyramids.

Urine is concentrated in the collecting ducts

As Na+ and Cl− are transported out of the tubule fluid, urea and other waste products make up a greater proportion of its total solute content as it flows toward the collecting duct. Therefore, the tubule fluid entering the collecting duct is at the same *concentration* as the blood plasma, but its *composition* is considerably different from that of the plasma.

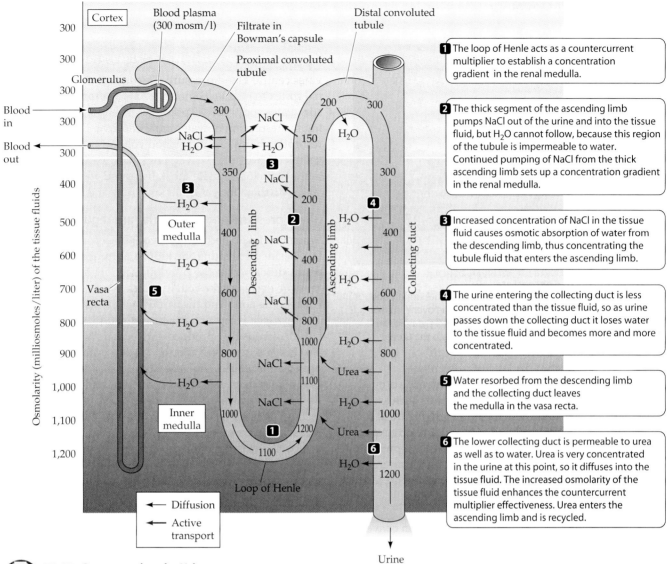

51.12 Concentrating the Urine
The countercurrent multiplier mechanism enables the kidney to produce urine that is far more concentrated than mammalian blood plasma.

The tubule fluid entering the distal convoluted tubule loses water osmotically as it flows toward the collecting duct.

The concentration gradient established in the renal medulla by the loops of Henle enables the urine to be concentrated in the collecting ducts. The collecting ducts begin in the renal cortex and run through the renal medulla before emptying into the ureter at the tips of the renal pyramids. As the solute concentration of the surrounding tissue fluid increases, more and more water is absorbed from the urine in the collecting duct. By the time it reaches the ureter, the urine has been greatly concentrated.

It follows from the process we have just described that the ability of a mammal to concentrate its urine will be determined by the maximum concentration gradient it can establish in its renal medulla. One way to increase the concentration gradient is to increase the lengths of the loops of Henle. That is precisely the adaptation we find in mammals that live in extremely arid habitats. The desert gerbil, for example, has such extremely long loops of Henle that its renal pyramid (each of its kidneys has only one, in contrast to

ours) extends far out of the concave surface of the kidney and into the ureter (Figure 51.13). These animals are so effective in conserving water that they can survive on the water released by the metabolism of their dry food; they do not need to drink!

Control and Regulation of Kidney Functions

Control and regulatory mechanisms act on the kidneys to maintain blood osmolarity and blood pressure. We will discuss these various mechanisms separately, but keep in mind that they are always working together.

The kidneys act to maintain the glomerular filtration rate

If the kidneys stop filtering blood, they cannot accomplish any of their functions. The *glomerular filtration rate (GFR)* depends on an adequate blood supply to the kidneys at an adequate blood pressure. Therefore, the kidneys have mechanisms to maintain their blood supply and blood pressure regardless of what is happening elsewhere in the body. Because these adaptations of the kidney support the maintenance of kidney function, they are called *autoregula-*

▶ Development of male sexual behavior requires the brain of the newborn rat to be exposed to testosterone, but development of female sexual behavior does not require the neonatal brain to be exposed to estradiol.

▶ Neonatal exposure to testosterone masculinizes the nervous systems of both genetic males and females so that they express male sexual behavior as adults.

Thus, the sex steroids that are present during development determine which pattern of sexual behavior develops, and the sex steroids that are present in adulthood determine whether that pattern is expressed.

Testosterone affects the development of the brain regions responsible for song in birds

As we saw above, learning is essential for the acquisition of bird song. Both male and female birds hear their species-specific song as nestlings, but only the males of most song-bird species sing as adults. Male birds use song to claim territory, compete with other males, and declare dominance. They also use song to attract females, which suggests that the females know the song of their species even if they do not sing. Do sex steroids control the learning and expression of song in male and female songbirds?

After leaving the nest where they heard their father's song, young songbirds from temperate and arctic habitats migrate and associate with other species in mixed flocks. During this time they do not sing, and they do not hear their species-specific song again until the following spring. As that spring approaches and the days become longer, the young male's testes begin to grow and mature. As his testosterone level rises, he begins to try to sing. Even if he is isolated at this time from all other males of his species, his song will gradually improve until it is a proper rendition of his species-specific song. At that point the song is **crystallized**— the bird expresses it in similar form every spring thereafter. The young male's brain has learned the pattern of the song by hearing his father. During the subsequent spring, under the influence of testosterone, he learns to express that song—a behavior that then becomes rigidly fixed in his nervous system.

Why don't the females of most songbird species sing? Can't they learn the patterns of their species-specific song? Do they lack the muscular or nervous system capabilities necessary to sing? Or do they simply lack the hormonal stimulus for developing the behavior? To answer these questions, investigators injected female songbirds with testosterone in the spring. In response to these injections, the females developed their species-specific song and sang just as the males did. Apparently females learn the song pattern of their species when they are nestlings and have the capability to express it, but they normally lack the hormonal stimulation.

What does testosterone do to the brain of the songbird? A remarkable discovery revealed that testosterone causes the parts of the brain necessary for learning and expressing song to grow larger (Figure 52.10). Each spring, certain regions of the males' brains grow. Individual neurons in-

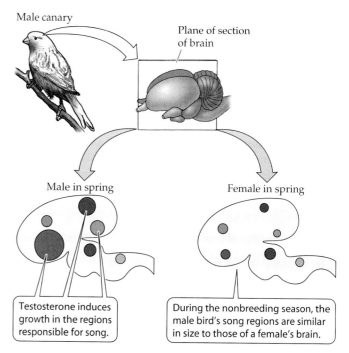

Testosterone induces growth in the regions responsible for song.

During the nonbreeding season, the male bird's song regions are similar in size to those of a female's brain.

52.10 Effects of Testosterone on Bird Brains
In spring, rising testosterone levels in the male cause the song regions of the brain to develop. The size of each circle is proportional to the volume of the brain occupied by that region.

crease in size and grow longer extensions, and the numbers of neurons in those regions of the brain increase. Such research on the neurobiology of bird song has revealed that hormones can control behavior by influencing brain structure as well as brain function, both developmentally and seasonally.

The Genetics of Behavior

To say that behavior is inherited does not mean that specific genes code for specific behaviors. Genes code for proteins, and there are many complex steps between the expression of a gene as a protein product and the expression of a behavior. In no case are all the steps between a gene and its influence on a behavior known. Nevertheless, it is clear that behavior has genetic determinants. In this section we will look at three approaches to investigating how genes affect behavior: hybridization, artificial selection and crossing of the selected strains, and molecular analysis of genes and gene products.

Hybridization experiments show whether a behavior is genetically determined

The effects of hybridization on the courtship displays of duck species were the subject of a classic ethological experiment, as we saw above. A more recent set of hybridization experiments was performed on the songs of crickets. Crickets songs, like bird songs, are species-specific, and as in birds, only male crickets "sing." They do so by rubbing one wing against another that has a serrated edge. These sounds can be recorded and analyzed quantitatively.

When two species of crickets were crossed, their offspring (the F_1 generation) expressed songs that had features of the songs of the two parental species. Backcrosses of F_1 individuals with the parental species produced individuals that had songs closer to the parental species used in the backcross. Clearly the genetic background determined the song pattern. What was amazing, however, was the demonstration that female preferences for male songs were under similar genetic control. Given a choice, females from each parental species preferred the calls of males from their own species, but hybrid females preferred the calls of hybrid males.

These genetic differences between the cricket species and the hybrids were reflected in the properties of their nervous systems. When specific neurons in the crickets' brains were stimulated, songs were expressed that reflected the genotypes of the crickets.

Artificial selection and crossbreeding experiments reveal the genetic complexity of behaviors

Domesticated animals provide abundant evidence that artificial selection of mating pairs on the basis of their behavior can result in strains with distinct behavioral as well as anatomical characteristics. Among dogs, consider retrievers, pointers, and shepherds. Each has a particular behavioral tendency that can be honed to a fine degree by training. However, dogs and other large animals are not the best subjects for genetic studies. Most artificial selection experiments in behavioral genetics have been done on more convenient laboratory animals with short life cycles and large numbers of offspring.

A favorite subject for behavioral genetic studies has been the fruit fly (*Drosophila*). Artificial selection has been successful in shaping a variety of behavior patterns in fruit flies, especially aspects of their courtship and mating behavior. Crossing of these artificially selected strains reveals that most of these behavioral differences are due to multiple genes that probably influence the behavior indirectly by altering general properties of the nervous system. Some single-gene effects, however, can be isolated. One example is the gene *per* (short for "period"), which alters the frequency of the wing vibrations that are part of the male's courtship display. The *per* gene is not a courtship behavior gene, however. It has subsequently been found that this gene codes for a transcription factor that plays an important role in the generation of daily rhythms of rest and activity, as we will see below. How it alters the development of wingbeat frequency is not clear.

Few behavioral genetic studies reveal simple Mendelian segregation of behavioral traits. An exception is nest-cleaning behavior in honeybees. One genetic strain of honeybees practices nest-cleaning, or *hygienic*, be-

havior, which makes them resistant to a bacterium that infects and kills the larvae of honeybees. When a larva dies, workers uncap its brood cell and remove the carcass from the hive. Another strain of honeybees does not show this hygienic behavior and therefore is more susceptible to the spread of the disease (Figure 52.11).

When these two strains of honeybees were crossed, the results indicated that the hygienic behavior was controlled by two recessive genes. All members of the F_1 generation were nonhygienic, indicating that the behavior is controlled by recessive genes. Backcrossing the F_1 with the hygienic strain produced the typical 3:1 ratio expected for a two-gene trait (see Chapter 10). The behavior of the nonhygienic hybrid individuals was very interesting. One-third of them showed no hygienic behavior at all; one-third uncapped the cells of dead larvae but did not remove them; and one-third did not uncap cells, but did remove carcasses if the cells were open.

52.11 Genes and Hygienic Behavior in Honeybees
Some honeybee strains remove the carcasses of dead larvae from their nests. This behavior seems to have two components: uncapping the larval cell (*u*) and removing the carcass (*r*), each of which is under the control of a recessive gene.

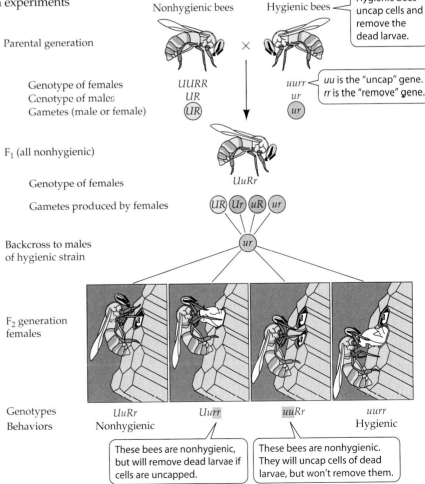

Even though these results appear to indicate a gene for uncapping and a gene for removal, these behavior patterns are complex. They involve sensory mechanisms, orientation movements, and motor patterns, each of which depends on multiple properties of many cells. The genetic deficits of nonhygienic bees could influence very small, specific, yet critical properties of some cells. If a single critical property, such as a crucial synapse or a particular sensory receptor, were lacking, the whole behavior would not be expressed. The responsible gene, then, is not a specific gene that codes for the entire behavior.

Molecular genetics techniques reveal specific genes that influence behavior

Molecular geneticists are investigating specific genes that influence behaviors. Male courtship behavior in fruit flies (*Drosophila*) is a subject of many such studies. This behavior is stereotypic, species-specific, and requires no learning. Males recognize potential mates, follow them, tap the female's body with their forelegs, extend and vibrate one wing, and lick the female's genitals. If the female is receptive, the male copulates with her (Figure 52.12*a*). Research in molecular genetics has now shown that most of this male courtship behavior is controlled by a single gene.

In fruit flies with two X chromosomes (females), a gene called *sex-lethal* (*sxl*) is expressed. This gene is at the top of a genetic hierarchy that determines all aspects of sexual differentiation and behavior (Figure 52.12*b*). The Sxl protein causes another gene called *transformer* (*tra*) to produce the female-specific Tra protein. Fruit flies without the *tra* gene develop into males anatomically and behaviorally, regardless of how many X chromosomes they have. But it is still another gene in the sex determination hierarchy that is responsible for male behavior.

The Tra protein controls two additional genes called *doublesex* (*dsx*) and *fruitless* (*fru*). The *dsx* gene mostly controls the anatomical differentiation of males, and *fru* causes the formation of a nervous system that expresses male courtship behavior. Mutations of the *fru* gene do not affect male body form, but they disrupt male courtship behavior. We don't know all of the actions that the male-specific Fru protein has in the development of the fruit fly nervous system, but this is about as close as we can get at present to identifying a gene that controls a complex behavior.

52.12 The fruitless *Gene Controls Male Courtship Behavior in Fruit Flies*

(*a*) Male fruit flies display stereotypic, species-specific courtship behavior. (*b*) Sexual differentiation in *Drosophila* is controlled by a hierarchy of genes, and in that hierarchy, *fru* controls the branch that leads to male courtship behavior.

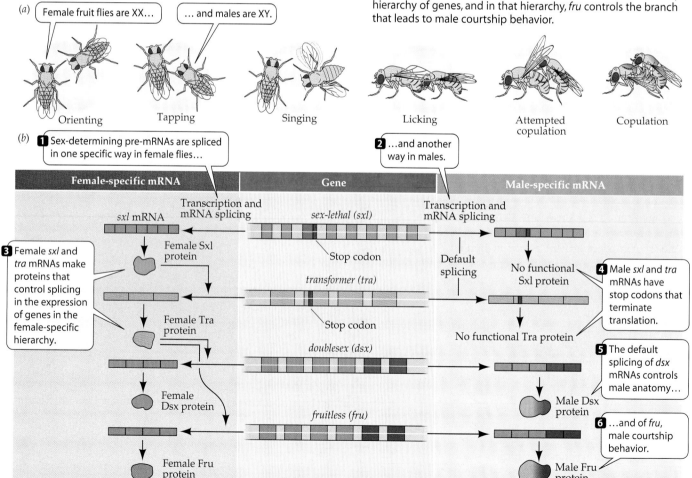

Communication

Communication is behavior that influences the actions of other individuals. It consists of **displays** or **signals** that can be perceived by other individuals and which convey information to them. Natural selection shapes displays or signals into systems of communication if the transmission of information benefits both the sender and the receiver. Thus, the ultimate cause of communication is the selective advantage it gives to individuals that engage in it. The courtship displays of a male, for example, benefit the male if they attract females, and they benefit the female if they allow her to assess whether the male is of the right species and whether he is strong, vigorous, and has other attributes that will make him a good father. A common mutual benefit of communication is the reduction of uncertainty about the status or intentions of the signaler. Even in aggressive interactions, reducing uncertainty helps both sender and receiver to avoid physical harm.

Studies of communication can be complex because they must take into account the sender, the receiver, and the environment. The displays or signals that an animal can generate depend on its physiology and anatomy. Likewise, an animal's ability to perceive displays or signals depends on its sensory physiology and on the environment through which the display or signal must be transmitted.

In Chapter 45, we learned how sensory systems function in chemosensation, tactile sensation, audition, vision, and electrosensation. These are the channels of animal communication. In the discussion that follows, we will explore each of these five channels in turn.

Chemical signals are durable but inflexible

Molecules used for chemical communication between individual animals are called **pheromones**. Because of the diversity of their molecular structures, pheromones can communicate very specific messages that contain a great deal of information. The mate attraction pheromone of the female silkworm moth is a good example (see Figure 45.4). Male moths as far as several kilometers downwind are informed by these molecules that a female of their species is sexually receptive. By orienting to the wind direction and following the concentration gradient of the molecules, they can find her.

Territory marking is another example in which detailed information is conveyed by chemical communication (Figure 52.13). Pheromonal messages left by mammals such as cats and dogs, for example, can reveal a great deal of information about the animal: species, individual identity, reproductive status, size (indicated by the height of the message), and when the animal was last in the area (indicated by the strength of the scent).

An important feature of pheromones is that once they are released, they remain in the environment for a long time. By contrast, vocal or visual displays disappear as soon as the animal stops signaling or displaying. The durability of pheromonal signals enables them to be used to mark trails, as ants do, or to indicate directionality, as in the

Panthera tigris

52.13 Many Animals Communicate with Pheromones
To mark her territory, this female tiger is spraying pheromonal secretions from a scent gland in her hindquarters onto a tree. Other tigers passing the spot will know that the area is "claimed," and they will know something about the animal who claimed it.

case of the moth sex attractant. However, it also means that the message cannot be changed rapidly. This inflexibility makes pheromonal communication unsuitable for a rapid exchange of information.

The chemical nature and the size of the pheromonal molecule determine its speed of diffusion. The greater the speed of diffusion, the more rapidly the message gets out and the farther it will reach, but the sooner it will disappear. Trail-marking and territory-marking pheromones tend to be relatively large molecules that diffuse slowly; sex attractants tend to be small molecules that diffuse rapidly.

Visual signals are rapid and versatile but are limited by directionality

Visual signals are easy to produce, come in an endless variety, can be changed very rapidly, and clearly indicate the position of the signaler. However, the extreme directionality of visual signals means that they are not the best means of getting the attention of a receiver. The receptors of the receiver must be focused on the signaler, or the message will be missed. Most animals are sensitive to light and can therefore receive visual signals, but sharpness of vision limits the detail that can be transmitted. The complexity of the environment also limits visual communication.

Because visual communication requires light, it is not useful at night or in environments that lack light, such as caves and the ocean depths. Some species have surmounted this constraint on visual communication by evolving their own light-emitting mechanisms. Fireflies use a enzymatic mechanism to create flashes of light. By emitting flashes in species-specific patterns, fireflies can advertise for mates at night.

Fireflies also illustrate how some species can exploit the communication systems of other species. There are predatory species of fireflies that mimic the mating flashes of other species. When an eager suitor approaches the signaling individual, it is eaten. Thus, deception can be part of animal communication systems, just as it is part of human use of language.

Auditory signals communicate well over a distance

Compared with visual communication, auditory communication has advantages and disadvantages. Sound can be used at night and in dark environments. It can go around objects that would interfere with visual signals, so it can be used in complex environments like forests. It is better than visual signals at getting the attention of a receiver because the receiver does not have to be focused on the signaler for the message to be received. Like visual signals, sound can provide directional information, as long as the receiver has at least two receptors spaced somewhat apart. By maximizing or minimizing the features of the sounds they emit, animals can make their location easier or more difficult to determine.

Sound is useful for communicating over long distances. Even though the intensity of sound decreases with distance from the source, loud sounds can be used to communicate over distances much greater than those possible with visual signals. An extreme example is the communication of whales. Some whales, such as the humpback, have very complex songs. When these sounds are produced at a certain depth (around 1,000 m), they can be heard hundreds of kilometers away. In this way, humpback whales can locate each other over vast areas of ocean.

Auditory signals cannot convey complex information as rapidly as visual signals can, as the expression "A picture is worth a thousand words" implies. When individuals are in visual contact, an enormous amount of information is exchanged instantaneously (for example, species, sex, individual identity, reproductive status, level of motivation, dominance, vigor, alliances with other individuals, and so on). Coding that amount of information, with all of its subtleties, as auditory signals would take considerable time, thus increasing the possibility that the communicators could be located by predators.

The animal world is relatively silent. Most invertebrates do not produce sound; cicadas and crickets are marvelous exceptions. Many amphibians, most fishes, and most reptiles produce no sound.

Tactile signals can communicate complex messages

Communication by touch is extremely common, although not always obvious. Animals in close contact use tactile interactions extensively, especially under conditions that do not favor visual communication. When eusocial insects such as ants, termites, or bees meet, they contact one another with their antennae and front legs. One of the best-studied uses of tactile communication, beginning with the pioneering work of ethologist Karl von Frisch, is the dance of honeybee. When a forager bee finds food, she returns to

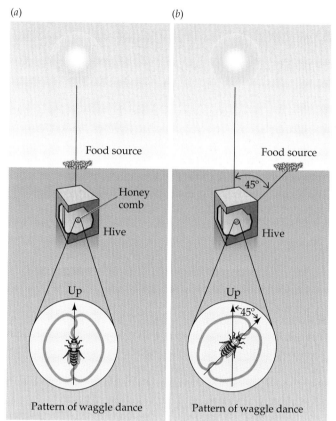

52.14 The Waggle Dance of the Honeybee
(a) By running straight up on the surface of the honeycomb in a dark hive, a honeybee tells her hivemates that there is a food source in the direction of the sun and at least 80 meters from the hive. The intensity of the waggle indicates exactly how far the food source is. If the food source were in the opposite direction from the sun, she would orient her waggle runs straight down. (b) When her waggle runs at an angle from the vertical, the other bees know that the same angle separates the direction of the food source from the direction of the sun.

the hive and communicates her discovery to her hivemates by dancing in the dark on the vertical surface of the honeycomb. The dance is monitored by other bees, who follow and touch the dancer to interpret the message.

If the food is less than 80–100 meters from the hive, the forager performs a *round dance*, running rapidly in a circle and reversing her direction after each circumference. The odor on her body indicates the flower to be looked for, but the dance contains no information about the direction to go—only that it is within 100 meters of the hive.

If the food source is farther than 80–100 meters, the bee performs a *waggle dance*, which conveys information about both the distance and the direction of the food source. The bee repeatedly traces out a figure-eight pattern as she runs on the vertical surface. She alternates half-circles to the left and right with vigorous wagging of her abdomen in the short, straight run between turns. The angle of the straight run indicates the direction of the food source relative to the direction of the sun (Figure 52.14). The speed of the dancing indicates the distance to the food source: The farther away it is, the slower the waggle run.

Electric signals can also communicate messages

Some species of fish have evolved the ability to generate electric fields in the water around them by emitting a series of electric pulses (see Chapter 45). These trains of electric pulses can be used for sensing objects in the immediate surroundings, and they can also be used for communication.

An electrode connected to an amplifier and a speaker can be used to "listen" to the signals generated by glass knife fish in a tank. Each individual fish emits a pulse at a different frequency, and the frequency each fish uses relates to its status in the population. Males emit lower frequencies than females. The most dominant male has the lowest frequency, and the most dominant female has the highest frequency. When a new individual is introduced into the tank, the other individuals adjust their frequencies so that they do not overlap, and the signal of the new individual indicates its position in the hierarchy. In their natural environment—the murky waters of tropical rainforests—these fish can tell the identity, sex, and social position of another fish by its electric signals.

Communication has been a very fruitful area for investigating the ultimate causes of behavior and how the resulting adaptations have been shaped by the environment. Next we will return to some studies of proximate causes of behavior to see some examples of how "how" questions can be addressed.

The Timing of Behavior: Biological Rhythms

Among the important proximate causes of behavior are those that determine its organization through time. The study of biological rhythms has led to major discoveries about brain mechanisms down to the molecular level that enable animals to organize their behavior in time. In the discussion that follows, we will examine two types of biological rhythms: circadian rhythms and circannual rhythms.

Circadian rhythms control the daily cycle of behavior

Our planet turns on its axis once every 24 hours, creating a cycle of environmental conditions that has existed throughout the evolution of life. Daily cycles are characteristic of almost all organisms. What is surprising, however, is that this daily rhythmicity does not depend on the 24-hour cycle of light and dark.

If animals are kept in constant darkness, at a constant temperature with food and water available all the time, they still demonstrate daily cycles of activity, sleeping, eating, drinking, and just about anything else that can be measured. This persistence of the daily cycle in the absence of changes between light and dark suggests that animals have an endogenous (internal) clock. Without time cues from the environment, however, these daily cycles are not exactly 24 hours long. They are therefore called **circadian**

rhythms (from the Latin *circa*, "about," and *dies*, "day").

To discuss biological rhythms, we must introduce some terminology. A rhythm can be thought of as a series of cycles, and the length of one of those cycles is the **period** of the rhythm. Any point on the cycle is a **phase** of that cycle:

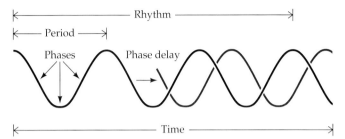

Hence, when two rhythms completely match, they are *in phase*, and if a rhythm is shifted (as in the resetting of a clock), it is *phase-advanced* or *phase-delayed*. Since the period of a circadian rhythm is not exactly 24 hours, it must be phase-advanced or phase-delayed every day to remain in phase with the daily cycle of the environment.

ENTRAINMENT. The process of resetting of the circadian rhythm by environmental cues is called **entrainment**. An animal kept in constant conditions will not be entrained to the 24-hour cycle of the environment, and its circadian clock will run according to its natural period—it will be **free-running**. If its period is less than 24 hours, the animal will begin its activity a little earlier each day (see the middle panel of Figure 52.15).

Animals with free-running circadian rhythms can be used in experiments to investigate the stimuli that phase-shift or entrain the circadian clock. Under natural conditions, environmental cues, such as the onset of light or dark, entrain the free-running rhythm to the 24-hour cycle of the real world. In the laboratory, it is possible to entrain circadian rhythms in free-running animals with short pulses of light or dark administered every 24 hours (bottom panel of Figure 52.15).

When you fly across several time zones, your circadian clock is out of phase with the real world at your destination; the result is jet lag. Gradually your endogenous rhythm synchronizes itself with the real world as it is reentrained by environmental cues. Since your endogenous rhythm cannot be shifted by more than 30 to 60 minutes each day, it takes several days to reentrain your clock to real time in your new location. This period of reentrainment is the time during which you experience jet lag, because your endogenous rhythm is waking you up, making you sleepy, initiating activities in your digestive tract, and stimulating many other physiological functions at inappropriate times of the day.

THE CIRCADIAN CLOCK. Where is the clock that controls the circadian rhythm? In mammals, the master circadian clock is located in two tiny groups of cells just above the optic chiasm, the place where the two optic nerves cross. These structures are called the **suprachiasmatic nuclei** (**SCN**). If

where you want to be. Time and sun position can give information about latitude as well. At a given time of day in the Northern Hemisphere, a sun position higher in the sky than expected indicates you are farther south than you want to be, and if the sun is lower in the sky than expected, you are north of where you want to be. Information about longitude and latitude can also come from sensing Earth's magnetic lines of force and from the positions of the stars.

In spite of the remarkable navigational abilities of animals such as albatrosses, there is currently no evidence that animals use bicoordinate navigation. But, of course, it is not easy to do experiments on animals such as albatrosses!

Human Behavior

As we saw early in this chapter, the behavior of an animal is a mixture of components that are inherited and components that can be molded by learning. However, even some aspects of learned behavior patterns—such as what can be learned and when it can be learned—have genetic determinants. Thus natural selection shapes not only the physiology and morphology of a species, but also its behavior. In some situations natural selection favors inherited behaviors; in others, learned behaviors. In many cases, the optimal adaptation is a mixture of inherited and learned behavioral components. Given these considerations, how would we characterize human behavior?

An important characteristic of human behavior is the extent to which it can be modified by experience. The transmission of learned behavior from generation to generation—**culture**—is the hallmark of humans. Nevertheless, the structures and many functions of our brains are inherited, including drives, limits to and propensities for learning, and even some motor patterns. Biological drives such as hunger, thirst, sexual desire, and sleepiness are inherent in our nervous systems. Is it reasonable, therefore, to expect that emotions such as anger, aggression, fear, love, hate, and jealousy are solely the consequences of learning?

Our sensory systems enable us to use certain subsets of information from the environment; similarly, the structure of our nervous systems makes it more or less possible to process certain types of information. Consider, for example, how basic and simple it is for an infant to learn spoken language, yet how many years that same child must struggle to master reading and writing. Verbal communication is deeply rooted in our evolutionary past, whereas reading and writing are relatively recent products of human culture.

Some motor patterns seem to be programmed into our nervous systems. Studies of diverse human cultures from around the world reveal basic similarities of facial expressions and body language among human populations that have had little or no contact with one another. Infants born blind still smile, frown, and show other facial expressions at appropriate times, even though they have never observed such expressions in others.

Acknowledging that aspects of our behavior have been shaped through evolution in no way detracts from the value we place on our ability to learn and the importance of cultural transmission of information to our species. Even so, we are recognizing that culture, in its simplest form, is not uniquely human. In the introduction to this chapter we saw what has been characterized as pre-cultural behavior in Japanese macaques. Individuals invented new behaviors, and those new behaviors were transmitted by imitative learning through the population.

In a recent study, scientists who have spent years studying chimpanzee behavior in seven widely separated areas of Africa compared their findings on chimpanzee behavior. They were able to identify 39 behaviors, ranging from tool use to courtship behavior, that were common in some populations but absent in others. Moreover, the variation in these behaviors was much greater between populations than within a population, and each population had a distinct repertoire of these behaviors. Just as human societies are characterized by different assemblages of culturally transmitted customs or customary behaviors, so are these chimpanzee populations.

It is increasingly more difficult to draw a line between human behavior and animal behavior, especially that of our closest primate relatives. But why should we expect such a line to exist? We do not expect such a lack of continuity in molecular, biochemical, physiological, or anatomical characteristics. Similarly, human and animal behavior is on a continuum, and the challenge is to understand the common mechanisms and the reasons for quantitative differences.

Chapter Summary

What, How, and Why Questions
▶ Studies of behavior seek to describe behaviors, understand their mechanisms, and understand their evolution.

Behavior Shaped by Inheritance
▶ Many behaviors of many species are stereotypic and species-specific, and are thus largely determined by inheritance. They do not require learning and are minimally modifiable by learning.

▶ Deprivation experiments deprive an animal of opportunities to learn a behavior and can therefore reveal that a behavior is hereditary.

▶ Hybridization experiments can also reveal genetic influences on behavior. **Review Figure 52.2**

▶ Some behaviors are triggered by simple stimuli called releasers. **Review Figure 52.3, 52.4**

▶ Spatial learning enables an animal to learn and use information about its physical environment. **Review Figure 52.5**

▶ Imprinting enables an animal to learn the features of a complex releaser, such as the identity of its parents. **Review Figure 52.6**

▶ The acquisition of bird song is an example in which genetic determinants and learning interact, enabling an animal to learn a behavior by focusing on the correct stimuli at the correct times. **Review Figure 52.7**

▶ Genetically programmed behavior is highly adaptive for species, such as those with nonoverlapping generations, that have little opportunity to learn, for species that might learn the wrong behavior, and in situations in which mistakes are costly or dangerous.

Hormones and Behavior

▶ In rats, the sex steroids present during development determine what sexual behavior patterns develop, and the sex steroids present in the adult control the expression of those patterns. **Review Figure 52.9**

▶ In birds, testosterone determines a bird's ability to sing by causing the brain regions responsible for song to develop. **Review Figure 52.10**

The Genetics of Behavior

▶ There are many complex steps between the expression of a gene as a protein product and the expression of a behavior. Several types of experiments help reveal how genes affect behavior.

▶ Artificial selection and crossbreeding can produce individuals with particular behavioral traits that are inherited. **Review Figure 52.11**

▶ The techniques of molecular genetics can reveal the functions of specific genes that influence behavior. **Review Figure 52.12**

Communication

▶ Communication consists of displays or signals, that can be perceived by other individuals and which influence their behavior. Natural selection favors communication systems when both sender and receiver benefit from the exchange of information.

▶ The evolution of communication signals is constrained by the anatomical and physiological characteristics of a species that are available to be shaped by natural selection.

▶ Many animals communicate by emitting pheromones into the environment and by sensing the pheromones of other animals. Pheromonal messages can last a long time, but they cannot be changed quickly.

▶ Visual communication is easy, versatile, and rapid, but it is limited by its directionality, by the visual acuity of the receiver, and by environmental conditions such as darkness. Many animals communicate via visual signals.

▶ Auditory signals can be used at night, can go around objects that would interfere with visual communication, can easily get the receiver's attention, can provide directional information, and can travel long distances. Compared with visual communication, however, auditory communication is slow. Few animals communicate with auditory signals.

▶ Tactile signals can communicate complex messages, as the dance of the honeybee demonstrates. **Review Figure 52.14**

▶ The electric signals generated by some fishes can be used for communication.

The Timing of Behavior: Biological Rhythms

▶ Animal behaviors are expressed in daily cycles called circadian rhythms. A circadian rhythm is an endogenous rhythm with a period not equal to 24 hours. To remain in phase with the 24-hour daily cycle of the environment, a circadian rhythm must be phase-shifted every day. Phase-shifting cues such as the onset of light and dark entrain circadian rhythms to the natural 24-hour period. **Review Figure 52.15**

▶ In mammals, the clock that controls the circadian rhythm is located in the suprachiasmatic nuclei of the brain. In other animals, different structures function as the circadian clock. **Review Figure 52.16**

▶ Two genes have been identified that are involved in the clock mechanism in a variety of species. **Review Figure 52.17**

▶ Circannual rhythms ensure that animals, such as hibernators and equatorial migrants, that cannot rely on changes in day length as seasonal cues perform the appropriate behaviors at the appropriate times of year.

Finding Their Way: Orientation and Navigation

▶ Piloting animals find their way by orienting to landmarks. **Review Figure 52.18**

▶ Homing animals find their way through unfamiliar territory to specific locations. Migrating animals travel long distances with remarkable accuracy.

▶ Animals that navigate by distance and direction determine distance in part by recognizing landmarks in the vicinity of their destination and in part by biological rhythms timing how far they travel.

▶ Sources of directional information include a time-compensated solar compass and an ability to locate the fixed point in the nocturnal sky. **Review Figures 52.20, 52.21**

▶ The long-distance movements of some species are difficult to explain by distance-and-direction navigation mechanisms. Information for bicoordinate navigation is available from the physical environment, but there is no evidence that any species uses such information.

Human Behavior

▶ Like that of all other animals, human behavior consists of genetically determined and learned components. What sets humans apart from other animals is the extent to which we can modify our behavior on the basis of experience.

For Discussion

1. An oystercatcher is a bird that normally lays a clutch of two eggs. If you place an artificial nest with either three artificial but normal-sized eggs or one very large artificial egg near the oystercatcher's nest, the oystercatcher will abandon its own two eggs and attempt to incubate the artificial eggs. How can you explain this behavior?

2. Cowbirds are nest parasites. A female cowbird lays her eggs in the nest of another species, which then incubates the eggs and raises the young. What do you think would characterize the acquisition of song in cowbirds? In a given area, cowbirds tend to parasitize the nests of particular bird species. How do you think female cowbirds learn this behavior? How would you test your hypothesis?

3. The short-tailed shearwater is a bird that winters in Antarctica and summers in the Arctic. What problems would this species have in using either the sun or the stars for navigation? What is the most likely means it uses to find its way to its summer and its winter feeding grounds?

4. Male dogs lift a hind leg when they urinate; female dogs squat. If a male puppy receives an injection of estrogen when it is a newborn, it will never lift its leg to urinate for the rest of its life; it will always squat. How might this result be explained?

5. If you were able to be the first person to visit a human population that had never been in contact with another culture, how could you use that opportunity to explore whether there were any human behaviors that were genetically determined?

that in an environment stocked with low densities of all three types of prey, the fish would take every water flea that they encountered, but that in an environment with abundant large water fleas, the bluegills would ignore smaller water fleas.

To test their predictions, the investigators put the bluegills in environments containing three different prey densities and observed the proportions of the water fleas of different sizes they actually captured. The proportions of large, medium, and small water fleas taken by the fish were very close to those predicted by the model. Such tests of foraging theory using many different kinds of animals have provided ecologists with a set of rules showing how animals find and choose their prey. They have also provided estimates of the energetic costs and benefits of foraging behavior.

Mating Tactics and Roles

Individual animals choose their associates, how to interact with them, and when to leave them. The most important choice of associates an animal makes is mate selection.

Mating behavior involves only a small set of choices. The most basic mating decision is choosing a partner of the correct species. Once that has been determined, additional decisions can be based on the qualities of a potential mate, on the resources it controls—food, nest sites, escape places—or on a combination of the two. Among those species in which individuals do not control any resources, the traits of the partner are the only criteria for mate selection. Here we will discuss how individuals choose their mating partners and show why males and females approach courtship so differently.

Abundant sperm and scarce eggs drive mating behavior

The reproductive behavior of males and females is often very different. Males usually initiate courtship, and they often fight for opportunities to mate with females. Females seldom fight over males, and they often reject courting males. Why do males and females approach courtship and copulation so differently?

The answer lies in the costs of producing sperm and eggs. Because sperm are small and cheap to produce, one male produces enough to sire a very large number of offspring—usually many more than the number of eggs a female can produce or the number of young she can nourish. Therefore, males of most species can increase their reproductive success by mating with many females

Eggs, on the other hand, are typically much larger than sperm and are expensive to produce. Consequently, a female is unlikely to increase her reproductive output very much by increasing the number of males she mates with. The reproductive success of a female depends primarily upon the the quality of the genes she receives from her mate, the resources he controls, and the amount of assistance he provides in the care of her offspring Thus, females choose among males based on these criteria. By their

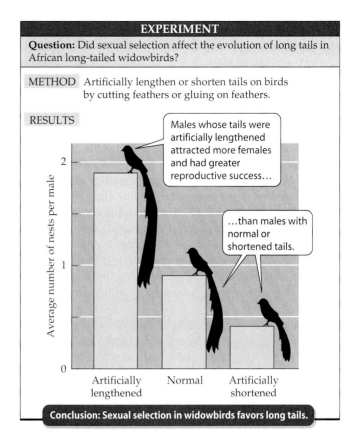

EXPERIMENT

Question: Did sexual selection affect the evolution of long tails in African long-tailed widowbirds?

METHOD Artificially lengthen or shorten tails on birds by cutting feathers or gluing on feathers.

RESULTS

Males whose tails were artificially lengthened attracted more females and had greater reproductive success...

...than males with normal or shortened tails.

Average number of nests per male

Artificially lengthened   Normal   Artificially shortened

Conclusion: Sexual selection in widowbirds favors long tails.

53.3 The Longer the Tail, the Better the Male
Male widowbirds with shortened tails defended their display sites successfully but attracted fewer females than males with long tails.

choices, females may cause the evolution of exaggerated traits that signal male quality.

Sexual selection often leads to exaggerated traits

Traits may evolve among individuals of one sex as a result of **sexual selection**: the selection of traits that confer advantages to their bearers during courtship or when they compete for mates or resources. Successful competitors for resources may gain exclusive access to mates that are attracted to the resources they control. Traits that improve success in courtship may evolve as a result of mating preferences by individuals of the opposite sex.

Sexual selection is responsible for the evolution of the remarkable tails of African long-tailed widowbirds, which are longer than their heads and bodies combined. Male widowbirds compete for display sites, at which they perform courtship displays to attract females. To examine the role of the tail in sexual selection, an ecologist shortened the tails of some males by cutting them, and lengthened the tails of others by gluing on additional feathers. Both short-tailed and long-tailed males successfully defended their display sites, indicating that the long tail does not confer an advantage in male–male competition. However, males with artificially elongated tails attracted about four times more females than males with shortened tails (Figure 53.3).

Why do females prefer males with long tails? Probably because the ability to grow and maintain a long tail, which

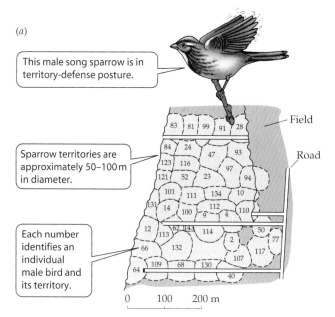

(a)

This male song sparrow is in territory-defense posture.

Sparrow territories are approximately 50–100 m in diameter.

Each number identifies an individual male bird and its territory.

Field

Road

(b) *Morus bassanus*

53.4 Some Territories Provide Everything; Others Provide Only a Nest Site
(a) Male song sparrows defend large breeding territories that contain nesting sites, food resources, and protective cover. (b) The size of a breeding territory among these northern gannets is determined by how far an incubating bird can reach to peck its neighbors without leaving its eggs.

probably carries energetic costs, indicates that the male is vigorous and healthy. Why, then, don't male widowbirds have even longer tails than they do? A likely answer is that the costs of producing a longer tail would exceed the benefits, but the costs of long tails were not measured in these experiments.

Males attract mates in varied ways

Males employ a variety of tactics to induce females to copulate with them. Males of some species defend territories that contain food, nesting sites, or other resources. Some territories are all-purpose: They provide mating sites, nest-

ing sites, and the food necessary to rear offspring (Figure 53.4a). Other territories include a breeding and nesting area, but do not supply all of the food necessary to rear young. The territories of many seabirds, such as gannets, penguins, and cormorants, are very small, consisting of only the area that individuals can defend while sitting on their nests (Figure 53.4b).

If a male controls no resources, he may use courtship behavior that signals in some way that he is in good health, that he is a good provider of parental care, or that he has a good genotype. For example, males of some species of hangingflies court females by offering them dead insects. By capturing an insect and defending it from other males, he demonstrates his ability as a forager and as a competitor. A female hangingfly will mate with a male only if he provides her with food. The bigger the food item, the longer she copulates with him, and the more of her eggs he fertilizes (Figure 53.5).

Whether a male fertilizes the eggs of a female with whom he has copulated depends on when they copulate and whether she copulates with other males. Males have evolved behavior patterns that increase the probability that it will be *their* sperm and no other male's that fertilize a female's eggs. The simplest method is to remain with the female for as long as she is fertile and prevent other males from copulating with her, but this method has high opportunity costs because a male cannot do anything else while he is guarding a female.

Males of many species have evolved behaviors that are more elaborate but take less time. A male black-winged damselfly grabs a female and, using his penis, scrubs out any sperm other males have deposited in her sperm storage chamber. The male removes 90–100 percent of competing sperm before he inserts his own sperm into the chamber. Males of some other insects deposit a plug that effectively seals the opening to the female's genital chamber and prevents other sperm from entering.

53.5 A Male Wins His Mate
The male hangingfly has just presented a moth to his mate, thus demonstrating his foraging skills. She feeds on the moth while they copulate.

Females are the choosier sex

As we have seen, females can improve their reproductive success if they can assess the genetic quality and health of potential mates, the quality of the resources they control, and the quantity of parental care they may provide. But how can females make such assessments when it is to the advantage of males to attempt to signal that they are good in all three of these traits?

By paying particular attention to those signals at which males cannot cheat, females have favored the evolution of "reliable" signals. Possession of a large dead insect indicates that a male hangingfly is a good forager and competitor. Likewise, a male widowbird with a very long tail is likely to be a high-quality mate.

Like tail length, the brightness of the plumage of birds may indicate their health and genetic quality. Male bluethroats (small Eurasian thrushes) have bright blue throat patches (Figure 53.6). Investigators in Norway tested the hypothesis that females use the throat patch as a sign of male quality by blackening the throats of some males and then measuring those males' success in fertilizing eggs. Although most birds form pair bonds, they sometimes engage in extra-pair copulations, so that some of the eggs in a nest may be fertilized by a male other than the one who attends it. A higher proportion of eggs in the nests of males whose throats had been blackened were fertilized by other males than of eggs in the nests of control males.

The investigators also tested the role of the ultraviolet reflectance of the blue feathers on mate choice by female bluethroats. They reduced the UV reflectance of some males by applying to their blue patches a mixture of fat from the glands the birds use to oil their own feathers and UV-absorbing sunblock. Control males received the fat coating with no sunblock. Although the two groups of males looked the same to human observers, female bluethroats could distinguish them. Females started laying eggs sooner on the territories of control males, and they preferred control males as partners for extra-pair copulations. These experiments show that females can use subtle clues to assess the quality of males.

Social and genetic partners may differ

Behavioral ecologists have known for many years that animals nearly always copulate with their mates—the individuals with which they have established pair bonds—but that they sometimes also copulate with other individuals. However, until the recent development of DNA fingerprinting methods, investigators had to assume that mated individuals were the parents of the offspring they raised.

By using the new molecular methods to compare the genomes of offspring with those of their supposed parents and other individuals, ecologists have found that nestling birds are nearly always the offspring of the female attending the nest—that is, females rarely lay eggs in other females' nests. However, the nestlings in a single nest often have different fathers. For example, 34 percent of nestlings

The throat feathers of a male bluethroat reflect ultraviolet light.

Luscinia svecica

53.6 Ultraviolet Reflectance Affects Mate Choice
Female bluethroats are attracted to males whose throat feathers have high reflectance, which signals a healthy, high-quality mate.

in nests of red-winged blackbirds in Washington state were fathered by a male other than the owner of the territory in which the nest was located. All these other fathers were males holding nearby territories; fertile females went to those territories and solicited copulations from the males.

Females that copulated with more than one male raised more offspring than females that remained faithful to their mates. Their reproductive success improved because neighboring males that had copulated with a female were more likely to defend her nest against predators than males that had not copulated with her. Males also let females with whom they had copulated look for food on their territories. Also, there were fewer infertile eggs in nests with multiple fathers than in nests with single fathers. Males try to prevent their mates from copulating with other males, but they must leave them unguarded at times, both to feed and to seek extra-pair copulations of their own.

Costs and Benefits of Social Behavior

Social behavior evolves when individuals that cooperate with others of the same species have, on average, higher rates of survival and reproductive success than those achieved by solitary individuals. Associations for mating may consist of little more than a coming together of eggs and sperm, but individuals of many species associate for longer times to provide care for their offspring. Associating with conspecifics may also improve survival for reasons unrelated to reproduction, such as by reducing the risk of being captured by a predator.

We describe only a few animal social systems, but these examples demonstrate two important concepts. First, social systems are best understood not by asking how they benefit the species as a whole, but by asking how the individuals that join together benefit. Second, social systems are dynamic: Individuals constantly communicate with one another and adjust their relationships.

53.7 Individuals Hunting Together Can Subdue Large Prey
By hunting as a group, lionesses can kill larger animals than a single female could subdue alone.

Group living confers benefits and imposses costs

Living in groups may confer many types of benefits on individuals. It may improve hunting success or expand the range of prey that can be captured. For example, by hunting together, social carnivores improve their efficiency in bringing down prey (Figure 53.7). Such cooperative hunting was a key component of the evolution of human social behavior. By hunting in groups, our ancestors were able to kill large mammals they could not have subdued alone. These social humans could also defend their prey and themselves from other carnivores.

Many small birds forage in flocks. To find out whether flocking provides protection against predators, an investigator released a trained goshawk near wood pigeons in England. The hawk was most successful in capturing a pigeon when it attacked solitary pigeons. Its success decreased as the number of pigeons in the flock increased (Figure 53.8). The larger the flock of pigeons, the sooner some individual in the flock spotted the hawk.

Living in a group typically imposes costs as well as benefits. Individuals in groups may compete for food, interfere with one another's foraging, injure one another's offspring, inhibit one another's reproduction, or transmit diseases to their associates.

The effects of group living on the survival and reproductive success of an individual also depend on its age, sex, size, and physical condition. Individuals may be larger or smaller than the average for their age and sex. Variation in skills, competitive abilities, and attractiveness to potential mates is often associated with these size differences.

An almost universal cost associated with group living is higher exposure to diseases and parasites. Long before the causes of disease were known, people knew that association with sick persons increased their chances of getting sick. Quarantine has been used to combat the spread of illness for as long as we have written records. The diseases of wild animals are not well known, but most of those that have been studied are spread by close contact.

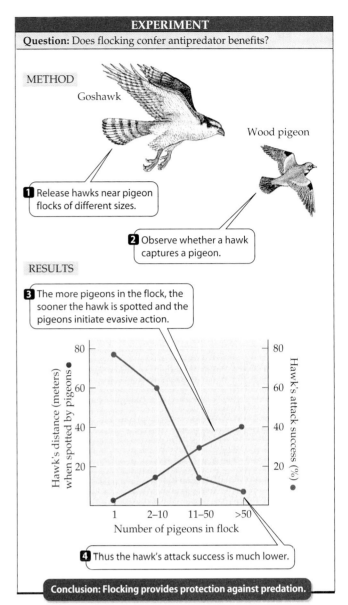

53.8 Flocking Provides Defense against Predators
The larger a flock of pigeons, the greater the distance at which they detect an approaching hawk, and the less likely the hawk is to succeed in capturing a pigeon.

Categories of Social Acts

Individuals living together perform many behaviors that are not performed by solitary animals. These acts can be grouped into four categories according to their effects on the interacting individuals:

▶ An **altruistic act** benefits another individual at a cost to the performer.
▶ A **selfish act** benefits the performer but inflicts a cost on some other individual.
▶ A **cooperative act** benefits both the performer and the recipient.
▶ A **spiteful act** inflicts costs on both the performer and the recipient.

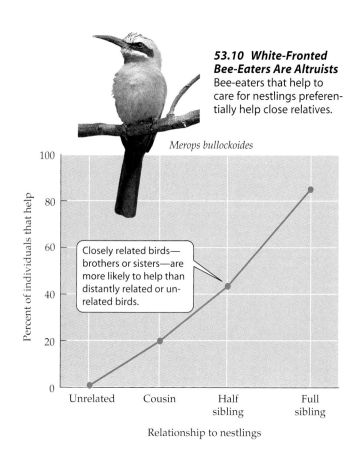

53.9 Types of Social Acts
Social acts can be divided into four categories, based on their effects on the performer and the recipient.

53.10 White-Fronted Bee-Eaters Are Altruists
Bee-eaters that help to care for nestlings preferentially help close relatives.

Merops bullockoides

Closely related birds—brothers or sisters—are more likely to help than distantly related or unrelated birds.

The types of social acts are summarized in Figure 53.9. The terms used are purely descriptive; they do not imply conscious motivation or awareness on the part of the animal. If a genetic basis for a cooperative or selfish act exists, and if performing it increases the fitness of the performer, then the genes governing that behavior will increase in frequency in the lineage. In other words, cooperative or selfish behavior can evolve.

How can behavior that inflicts a cost on the performer evolve? Behavioral ecologists believe spiteful behavior is rare in nature. Altruistic behavior, however, can evolve, both among close relatives and among unrelated individuals.

Altruism can evolve by means of natural selection

Altruistic behaviors evolve most easily when performers and recipients are genetically related. Genetic relatedness is important because an individual can influence its fitness in two different ways. First, it may produce its own offspring, contributing to its individual fitness. Second, it may help its relatives in ways that increase their fitness.

Because relatives are descended from a common ancestor, they are likely to bear some of the same alleles. Two offspring of the same parents, for example, are likely to share 50 percent of the same alleles; an individual is likely to share 25 percent of its alleles with its sibling's offspring. Therefore, by helping its relatives, an individual can increase the representation of some of its own alleles in the population. This process is called **kin selection**. Together, individual fitness and kin selection determine the **inclusive fitness** of an individual. Occasional altruistic acts may eventually evolve into altruistic behavior patterns if the benefits of increasing the reproductive success of relatives exceed the costs of decreasing the altruist's own reproductive success.

Many social groups consist of some individuals that are close relatives and others that are unrelated or distantly related. Individuals of some species recognize their relatives and adjust their behavior accordingly. White-fronted bee-

eaters are African birds that nest colonially (Figure 53.10). Most breeding pairs are assisted by nonbreeding adults that help incubate their eggs and feed nestlings. Nearly all of these helpers assist close relatives. When helpers have a choice of two nests at which to help, about 95 percent of the time they choose the nest with the young most closely related to them.

Several other pieces of evidence suggest that the helping behavior of white-fronted bee-eaters evolved through kin selection. First, both males and females help to care for nestlings, but males help more often than females. Males remain in the social group in which they were born, but females join other social groups when they mature. Therefore, females typically live in social groups composed primarily of nonrelatives.

Second, individuals do not appear to gain anything other than inclusive fitness by helping—helpers are not, for example, more successful when they become breeders themselves than birds that do not help. Finally, nests with helpers produce more fledglings than do nests without helpers, showing that helpers do increase the number of fledglings produced by their close relatives. Notice that all these patterns are consistent with the principle that bee-eaters behave in ways that improve their fitness, not in ways that benefit the species.

Species whose social groups include sterile individuals are said to be **eusocial**. This extreme form of social behavior—the "ultimate altruism"— has evolved in termites and many hymenopterans (ants, bees, and wasps). In these species, worker females defend the group against predators

Eciton burchelli

53.11 Sterile Individuals are Extreme Altruists
Eusocial insect species contain classes of sterile individuals. These soldier ants from Panama protect their nests and nestmates with their large, powerful jaws.

or bring food to the colony, but do not reproduce. Some species have soldiers with large defensive weapons (Figure 53.11). These workers are at risk of being killed while defending the colony.

Both genetic and environmental factors facilitate the evolution of eusociality. The more closely related individuals are to one another, the greater the advantages they can receive by forgoing their own reproduction to help relatives reproduce. The British evolutionary biologist W. D. Hamilton first suggested that eusociality evolved among ants, bees, and wasps because members of the order Hymenoptera have an unusual sex determination system in which males are haploid but females are diploid.

Among the Hymenoptera, a fertilized egg hatches into a female; an unfertilized egg hatches into a male. If a female copulates with only one male, all the sperm she receives are identical because the haploid males have only one set of chromosomes, all of which are transmitted to each sperm cell. Therefore, a female's daughters share all of their father's genes. They also share, on average, half of the genes they receive from their mother. As a result, on average they share 75 percent of their alleles rather than 50 percent, as they would if both parents were diploid. Workers therefore can increase their fitness more by helping their sisters than by reproducing themselves, because they are genetically more similar to their sisters than they would be to their own offspring.

Mating between close relatives, known as *inbreeding*, can also generate close genetic relationships. Even if two mates are unrelated, but each is the product of generations of intense inbreeding, their offspring can be genetically nearly identical. These individuals would also benefit from helping to rear siblings. Genetic similarity generated by inbreeding could explain the evolution of eusociality among the many hymenopteran species in which queens mate with many males and among termites and naked mole-rats—the most extremely eusocial mammals—in which both sexes are diploid.

Eusociality may also be favored if establishment of new colonies is difficult and dangerous. Nearly all eusocial animals construct elaborate nests or burrow systems within which their offspring are reared (Figure 53.12). Naked mole-rats live in underground colonies containing 70 to 80 individuals. The tunnel systems are maintained by sterile workers. Breeding is restricted to a single queen and several kings that live in a nest chamber in the center of the colony. Individuals attempting to found new colonies are at high risk of being captured by predators, and most founding events fail. Thus, high predation rates, which favor cooperation among founding individuals, may facilitate the evolution of eusociality.

Unrelated individuals may behave altruistically toward one another

It is easy to understand how altruistic behavior can evolve among related individuals. It is more difficult to explain the existence of warnings of danger, sharing of food, and grooming among unrelated individuals of the same species, or even between members of different species. How can we explain the evolution of such behavior?

Such behavior among unrelated individuals could evolve through **reciprocal altruism**; that is, if helpers are in turn recipients of beneficial acts by the individuals they have helped. If there is a genetic basis for the acts, natural selection may increase the frequency of alleles governing this behavior. In order for reciprocal altruism to be a force, several social conditions must be present:

▶ Individuals in the group must know one another.
▶ They must associate for long periods.
▶ Individuals must be aware that their altruistic acts are being reciprocated.

53.12 Termite Mounds are Large and Complex
Immense termite mounds such as this one in Kenya are costly to construct and maintain. Elaborate nests or burrows are characteristic of nearly all eusocial animals.

Reciprocal altruism is especially highly developed among humans, in which these conditions prevail. It illustrates the subtle adjustments in behavior among individuals in social groups.

The Evolution of Animal Societies

The decisions animals make about where to live, with whom to mate, and whether and how to care for their offspring all help determine the type of social system they have. Today's social systems are the result of long periods of evolution, but there are few records of past social systems because behavior leaves few traces in the fossil record. Possible routes of the evolution of social systems must therefore be inferred primarily from current patterns of social organization. Fortunately, many stages of social system complexity exist among species, and the simpler systems provide clues about the stages through which the more complex ones may have passed.

Parents of many species care for their offspring

The origins of all animal societies lie in the association of parents with their offspring. Individuals of many species invest time and energy in caring for offspring. Parental care increases the chances of an offspring's survival, but it usually reduces the ability of the parent to produce additional offspring.

Parental care may also lower the chances of survival of the parent itself, because the parent could have used the time and energy to engage in other activities that would improve its own chances of survival. In other words, parents balance trade-offs between the success of their current offspring and their own future survival and reproduction.

Males and females often differ strikingly in the kinds and amounts of parental investment they can and do make. Birds, mammals, and fishes illustrate these differences and why they exist. Only female mammals have functional mammary glands; males cannot produce milk. Therefore parental care among mammals is usually given by females. On the other hand, among birds, all aspects of reproduction except production of eggs and sperm can be performed readily by both males and females. Not surprisingly, both males and females feed their offspring in about 90 percent of bird species.

Sex roles among fishes differ from those of birds and mammals because most fish species do not feed their young. Parental care consists primarily of guarding eggs and young from predators (Figure 53.13). In many fish species, males are the primary guardians. A male can guard a clutch of eggs while attracting additional females to lay eggs in his nest. A female, on the other hand, can produce another clutch of eggs sooner if she resumes foraging immediately after mating than if she spends time guarding her eggs.

The most widespread form of social system is the family, an association of one or more adults and their dependent offspring. If parental care lasts a long time, or if the breeding season is longer than the time it takes for offspring to

The dark area is a large clutch of eggs.

Abudeiduf saxatilis

53.13 A Sergeant Major Guards His Young
This male is defending the eggs a female has deposited. He can court other females while guarding the eggs.

mature, adults may still be caring for younger offspring when older offspring reach the age at which they could help their parents.

Many communal breeding systems, such as the white-fronted bee-eater families described above, most likely evolved via the extended family route. Most mammals evolved social behavior by this route. In simple mammalian social systems, solitary females or male–female pairs care for their young. In species whose young require a long period of parental care, older offspring are still present when the next generation is born, and they often help rear their younger siblings. In most social mammal species, female offspring remain in the group in which they were born, but males tend to leave—or are driven out—and must seek other social groups. Therefore, among mammals, most helpers are females.

53.14 Savanna Weaverbirds Nest Colonially
Many African weaverbirds nest in colonies in isolated trees. Although these nests are highly conspicuous, it is difficult for most predators to get to them at the tips of the small, thorny branches.

53.15 Cooperation among Florida Scrub Jays
These Florida scrub jay helpers, most of which are offspring from the previous breeding season, are helping to feed nestlings and defend the nest against predators, such as the approaching snake. By doing so, they are improving their inclusive fitness.

The environment influences the evolution of animal societies

The type of social organization a species evolves is strongly related to the environment in which it lives. Among the weaverbirds of Africa, species that live in forests eat insects, feed alone, and build well-hidden nests. Most of these species are monogamous, and males and females look alike. In marked contrast, weaverbirds that live in tree-studded grasslands called savannas eat primarily seeds, feed in large flocks, and nest in colonies, usually in isolated *Acacia* trees where their nests are large and conspicuous (Figure 53.14). In most colonial species, males have several mates and are more brightly colored than females.

These striking differences probably evolved because nesting sites and food in forests are common and widely dispersed. Solitary pairs can use these resources more efficiently than animals in groups can. In savannas, however, good nesting trees are scarce and highly clumped. Males compete for these limited nest sites; males that hold the best sites—near the tips of branches where they are safe from predators—attract the most females. Males spend their time attempting to attract additional mates rather than helping to rear the offspring they already have, which explains the evolution of brighter plumage among males.

Florida scrub jays live all year on territories, each of which is home to a breeding pair and up to six helpers (Figure 53.15). Nearly all the helpers are offspring from the previous breeding season that remain with their parents. This social system probably evolved because all suitable territories are occupied, and young individuals have little chance of establishing new territories on their own. By staying in their parents' territory and helping to raise their siblings, they both improve their inclusive fitness and have a chance of taking over the territory if one of their parents dies.

Among the herbivorous hoofed mammals of Africa, social organization and feeding ecology are correlated with the size of the animal. Smaller animals have higher metabolic demands per unit of body weight than do larger ones. Therefore, smaller hoofed mammals feed preferentially on high-protein foods such as buds, young leaves, and fruits. These foods are dispersed throughout forests, which also provide cover in which to hide from predators. Hiding is a tactic that is effective for solitary animals. In contrast, the largest hoofed mammal species are able to eat lower-quality food, but they must process great quantities of it each day. They feed in grasslands with abundant herbaceous vegetation, follow the rains to areas where grass growth is best, and live in large herds (Figure 53.16a).

Among primates, many diurnal species will eat insects and other animal food when they are available, but most of

(b) Papio cynocephalus

(a) Connochaetes taurinus

53.16 Many Mammals of Open Country Live in Large Groups
(a) East African wildebeest live in large herds that follow the rains to places with fresh grass. (b) Baboons are conspicuous as they forage but are seldom attacked because the formidable males cooperate in defending the group.

them eat fruits, seeds, and leaves. In Africa and Asia, primate group sizes are smallest among arboreal forest-dwelling species, whatever their diets, and largest among the ground-dwelling savanna species, such as baboons (Figure 53.16*b*). Troops of foraging baboons are conspicuous to predators, but their large males help protect the other troop members. Baboons have a complex social system. In troops with more than one male, strong dominance hierarchies exist among the males, and one or two of them father most of the offspring. Females may also have dominance relationships, and young females often assume the status of their mothers when they mature.

 Chapter Summary

▶ Ecologists study the nature and consequences of interactions among organisms and their environments.

▶ Behavioral ecology is the study of how animals decide where to carry out different activities, select the resources they need, respond to predators and competitors, and interact with conspecifics.

Balancing Costs and Benefits of Behaviors

▶ Cost–benefit analyses of behavior are based on the principle that animals have only limited amounts of time and energy to devote to their activities.

▶ A cost of defending a territory may be increased risk of mortality. **Review Figure 53.1**

Choosing Where to Live and Forage

▶ Selecting a habitat in which to live is one of the most important decisions an individual makes.

▶ The cues animals use to select habitats are good predictors of conditions suitable for future survival and reproduction.

▶ Foraging theory was developed to understand how animals select prey. **Review Figure 53.2**

Mating Tactics and Roles

▶ Individuals choose their associates, how to interact with them, and when to leave them. The most important choice of associates is the choice of a mate.

▶ Because males produce enough sperm to fertilize many eggs, males typically increase their reproductive success by mating with many females. The reproductive success of females is typically limited by the cost of producing eggs. As a result, males usually initiate courtship and often fight for opportunities to mate with females. Females seldom fight over males and often reject courting males.

▶ Sexual selection often leads to exaggerated traits. **Review Figure 53.3**

▶ While courting, males signal their desirability as mating partners and may perform behaviors that increase the probability that their sperm will fertilize eggs.

▶ Males of some species defend territories that contain food, nesting sites, or other resources. **Review Figure 53.4**

▶ By paying particular attention to those signals at which males cannot cheat, females have favored the evolution of "reliable" signals of mate quality.

▶ DNA fingerprinting methods have shown that social fathers often are not genetic fathers.

Costs and Benefits of Social Behavior

▶ Benefits of social living include better opportunities to capture prey and to avoid predators. **Review Figure 53.8**

▶ Costs of social living include competition for food, interference by conspecifics, and transmission of diseases.

Categories of Social Acts

▶ Acts performed by individuals living together can be grouped into four descriptive categories: altruistic, selfish, cooperative, and spiteful. **Review Figure 53.9**

▶ Altruism among closely related individuals can evolve by means of kin selection because individuals that help close relatives can improve their inclusive fitness. **Review Figure 53.10**

▶ Eusocial systems with sterile individuals have evolved among termites, hymenopterans (ants, bees, and wasps), and in a mammal, the naked mole-rat.

The Evolution of Animal Societies

▶ The origin of most animal societies is the family, an association of one or more adults and their dependent offspring.

▶ The type of social organization a species evolves is strongly related to the environment in which it lives.

For Discussion

1. Most hawks are solitary hunters. Swallows often hunt in groups. What are some plausible explanations for this difference? How could you test your ideas?

2. Because costs and benefits of behaviors can seldom be measured directly, behavioral ecologists often use indirect measures such as correlations between behavior patterns and the presence of predators. What are the strengths and weaknesses of some of these indirect measures?

3. Polyandry is a mating system in which one female has a "harem" of several males. Why is polyandry much rarer among both birds and mammals than polygyny, the situation in which one male forms pair bonds with several females?

4. When frogs mate, a male clasps a gravid female behind her front legs and stays with her until she lays her eggs, at which time he fertilizes them. In most species of frogs, the male remains clasped to the female for a short time, usually no longer than a few hours. However, in some species, pairs may remain together for up to several weeks. In view of the fact that a male cannot court or mate with any other female while clasping one, and that a female lays only a single clutch of eggs, why is it advantageous for males to behave this way? What can you guess about the breeding ecology of frogs that remain clasped for long periods? Why should females permit males to clasp them for so long? (Females do not struggle!)

5. Among vertebrates, helpers are individuals capable of reproducing, and most of them later breed on their own. Among eusocial insects, sterile castes have evolved repeatedly. What differences between vertebrates and insects might explain the failure of sterile castes to evolve in the former?

6. The use of DNA fingerprinting technology has shown that in many species, social partners and genetic partners differ. Under what conditions do individuals benefit from copulating with individuals other than their social mates? Do males and females benefit equally from this behavior?

54 *Population Ecology*

LARGE SAGUARO CACTI ARE CONSPICUOUS features of the Sonoran Desert in southern Arizona. But finding a seedling cactus is difficult—at least, until you learn where to search. All the small cacti are found beneath trees or shrubs.

In the harsh environment of the Sonoran Desert, plants are exposed to intense daytime heat and wide temperature fluctuations. Their roots are in extremely dry soil much of the year. Small plants are most vulnerable to these conditions because they have small root systems, and daytime temperatures are extremely high at the soil surface. Therefore, although seeds of saguaro cacti are dispersed widely over the desert by birds, seedling cacti survive only in the shade of trees and shrubs, called *nurse plants*, where they are protected from the intense daytime heat. Thus, the density and distribution of saguaro cacti are strongly influenced by the number and location of trees and shrubs.

All the individuals of saguaro cacti—or of any other species—within a given area constitute a **population**. The sizes of populations continually change. To understand how and why these changes happen, population ecologists count individuals in different locations and try to determine the factors that influence birth, death, immigration, and emigration rates.

In addition to studying the dynamics of populations in a particular area, population ecologists also investigate changes over the entire ranges of species. They attempt to answer questions such as: What causes a species to be common or rare? Why is a species common in some parts of its geographic range and rare in others? What determines the limits of the ranges of species?

In this chapter we discuss how and why the sizes of populations vary over space and time, and show how this ecological knowledge is used to predict and manage the growth of populations. To set the stage for studies of populations, we present some background information on how the individuals of a population are distributed.

Population Structure: Patterns in Space and Time

At any given moment, an individual organism occupies only one spot and is of one particular age. The members of

a population, however, are distributed over space and differ in age and size. These features are among the components of **population structure**. As we saw in Chapter 21, geneticists and evolutionary biologists also study population structure, but they are interested primarily in distributions of genotypes and their degree of isolation from one another, because that component of population structure influences how populations evolve.

Ecologists study population structure at different spatial scales, ranging from local subpopulations to entire species. They study the numbers and spatial distributions of individuals because these features influence the stability of populations and affect interactions among species.

Density is an important feature of populations

The number of individuals of a species per unit of area (or volume) is its **population density**. Ecologists are interested in population densities because dense populations often

A Cactus Needs Shade
This saguaro cactus (*Cereus giganteus*) grew in the shade of the palo verde tree, which protected it from the intense heat.

54.1 Modular Organisms

(a) Each member of this bryozoan animal colony is called a zooid and is genetically identical to an ancestor zooid from which it has budded. (b) A single modular organism can appear to be a population. This clump of quaking aspen trees looks like a grove of many individuals, but it is actually a single genetic individual.

(a) *Pectinatella magnifica*

(b) *Populus tremuloides*

exert strong influences on their own members as well as on populations of other species. Other scientists—such as those working in agriculture, conservation, or medicine—wish to manage species to raise (in the case of crop plants, aesthetically attractive species, or threatened or endangered species) or lower (in the case of agricultural pests and disease organisms) their densities. To manipulate population densities, we must know what factors make populations increase and decrease in size, and how those factors work.

Because species and their environments differ, population densities are measured in more than one way. Ecologists usually measure the densities of organisms in terrestrial environments by the number of individuals per unit of area, but number per unit of volume is generally a more useful measure for organisms living in water. For species whose members differ markedly in size, as do most plants and some animals (such as mollusks, fishes, and reptiles), the total mass of individuals—their *biomass*—may be a more useful measure of density than the number of individuals.

The fertilized egg of many organisms develops into a unit of construction called **module**, which produces additional modules much like itself. Many plants are modular, and there are many groups of modular protists, fungi, and animals (for example, sponges, corals, and bryozoans; Figure 54.1a). A modular organism may grow to a large size, and it is often difficult to distinguish a modular organism

from a cluster of genetically separate individuals (Figure 54.1b). The effects of modular organisms on their environment often depend primarily on the number and size of the modules. Therefore, ecologists studying modular organisms are often concerned primarily with the number, size, and shape of the modules rather than with the number of genetically distinct individuals.

Under some circumstances, the individuals in a population can be counted directly without missing any of them or counting any of them twice, but this process is usually impossible or too laborious. Ecologists commonly estimate population densities by sampling a population in a representative area and extending their findings to a larger area.

The size of a population can also be estimated by marking and recapturing individuals. For example, if we capture and mark 100 individuals in a population, we can take another sample later and count the individuals in that sample that are already marked. If, say, 10 percent of the individuals in our second sample are already marked, we would conclude that the population contains about 1,000 individuals. This estimate is based on the mathematical assumption that the number of individuals caught the first time is the same proportion of the total population as the propor-

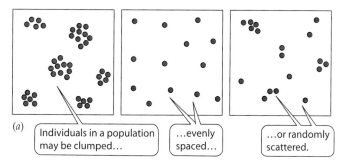

(a) Individuals in a population may be clumped... ...evenly spaced... ...or randomly scattered.

(b)

54.2 Patterns of Spatial Distribution

(a) A diagrammatic representation of clumped, even, and random distribution patterns. (b) The relatively even spacing of these Australian desert plants results because each established plant removes so much water from the surrounding soil that no young plants can grow within its root zone.

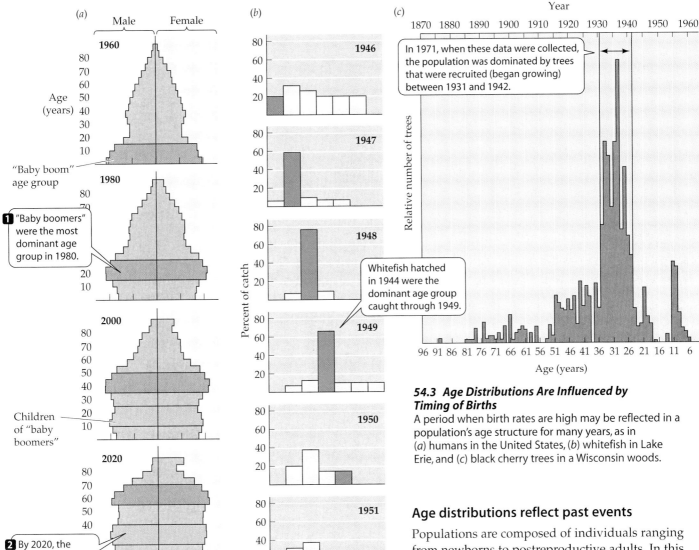

54.3 Age Distributions Are Influenced by Timing of Births

A period when birth rates are high may be reflected in a population's age structure for many years, as in (a) humans in the United States, (b) whitefish in Lake Erie, and (c) black cherry trees in a Wisconsin woods.

Labels within figure:

(a) Male | Female
1960
Age (years)
"Baby boom" age group
1980
1 "Baby boomers" were the most dominant age group in 1980.
2000
Children of "baby boomers"
2020
2 By 2020, the children of "baby boomers" will be as dominant as their parents.
Millions of persons

(b) Percent of catch — 1946, 1947, 1948, 1949, 1950, 1951
Whitefish hatched in 1944 were the dominant age group caught through 1949.
Age group (years)

(c) Year
Relative number of trees
In 1971, when these data were collected, the population was dominated by trees that were recruited (began growing) between 1931 and 1942.
Age (years)

tion of marked individuals to the total number caught the second time

Spacing patterns reflect interactions among individuals

Ecologists studying population structure also look at the way in which the individuals in a population are spaced. Individuals in a population may be tightly clumped together, evenly spaced, or randomly scattered (Figure 54.2a). Distributions can become clumped when young individuals remain close to their birthplaces, when suitable habitat patches are "islands" separated by unsuitable areas, or by chance. The relatively even spacing of many plants is a result of competition for light, water, and soil nutrients (Figure 54.2b). Among animals, defense of space is the most common cause of even distributions (see Figure 53.4b). Random distributions may result when many factors interact to influence where individuals settle and survive. The saguaro cacti in the Sonoran Desert (discussed at the beginning of this chapter) are distributed randomly because that is how suitable nurse plants are distributed.

Age distributions reflect past events

Populations are composed of individuals ranging from newborns to postreproductive adults. In this chapter we consider an individual to be "born" when it leaves its mother's body as a seed, egg, or baby. The proportions of individuals in each age group in a population make up its **age distribution**. The density and spacing of individuals are spatial attributes of a population; age distribution is a *temporal* (time-oriented) attribute.

The timing and rates of births and deaths determine age distributions. If both birth rates and death rates are high, a population will be dominated by young individuals. If birth rates and death rates are low, a relatively even distribution of individuals of different ages results. The age distribution of a population thus reveals much about its past history of births and deaths.

The timing of births and deaths may influence age distributions for many years in populations of long-lived species. The human population of the United States is a good example. Between 1947 and 1964, the United States experienced what is called the post–World War II "baby boom." During these years, average family size grew from 2.5 to 3.8 children; an unprecedented 4.3 million babies were born in 1957. Birth rates declined during the 1960s, but Americans born during the baby boom will constitute the dominant age class into the twenty-first century (Figure 54.3a). "Baby boomers" became parents in the 1980s, producing another bulge in the age distribution—a baby boom

54.1 Life Table of the 1978 Cohort of Darwin's Ground Finch (Geospiza scandens) on Isla Daphne

AGE IN YEARS (X)	NUMBER ALIVE	SURVIVORSHIP[a]	SURVIVAL RATE[b]	MORTALITY RATE[c]
0	210	1.000	0.434	0.566
1	91	0.434	0.855	0.143
2	78	0.371	0.898	0.102
3	70	0.333	0.928	0.072
4	65	0.309	0.955	0.045
5	62	0.295	0.678	0.322
6	42	0.200	0.545	0.455
7	23	0.109	0.651	0.349
8	15	0.071	0.944	0.056
9	14	0.067	0.776	0.224
10	11	0.052	0.923	0.077
11	10	0.048	0.396	0.604
12	4	0.019	0.737	0.263
13	3	0.014	0.714	0.004

[a]Survivorship = the proportion of newborns who survive to age x.

[b]Survival rate = the proportion of individuals of age x who survive to age $x + 1$.

[c]Mortality rate = the proportion of individuals of age x who die before the age of $x + 1$.

The number of individuals in a population at any given time is equal to the number present at some time in the past, plus the number born between then and now, minus the number that died, plus the number that immigrated, minus the number that emigrated. That is, the number of individuals at a given time, N_1, is given by the equation

$$N_1 = N_0 + B - D + I - E$$

where N_1 is the number of individuals at time 1; N_0 is the number of individuals at time 0; B is the number of individuals born, D the number that died, I the number that immigrated, and E the number that emigrated between time 0 and time 1. If we measure these rates over many time intervals, we can determine how a population's density changes.

"echo"—but they had, on average, fewer children than their parents, so the bulge is not as large. Similarly, in Lake Erie, 1944 was such an excellent year for reproduction and survival of whitefish that individuals of that age group dominated whitefish catches in the lake for several years (Figure 54.3b). The population of black cherry trees in a Wisconsin woods is dominated by individuals that began growing between 1931 and 1941 (Figure 54.3c).

Population Dynamics: Changes over Time

At any moment in time, a population has a particular structure determined by the number and distribution of its members in space and their ages. However, as we have just seen, population structure is not static. Changes in the structure of a population influence whether it will increase or decrease; that is, they affect the *dynamics* of the population. We will now examine how ecologists measure birth and death rates and use that information to understand how population densities change. The study of changes in the size and structure of populations is known as **demography**.

Births, deaths, and movements drive population dynamics

Knowledge of when individuals are born and when they die provides a surprising amount of information about a population. Births, deaths, and movements of individuals are demographic events—that is, they determine the numbers of individuals in a population. Ecologists measure the *rates* at which these events take place—the number of such events per unit of time. These rates are influenced by environmental factors, by the life history traits of the species, and by population density.

Life tables summarize patterns of births and deaths

Life tables summarize data about births and deaths that can be used to predict future growth rates of populations. We can construct a life table by determining for a group of individuals born at the same time—called a **cohort**—the number still alive at specific times and the number of offspring they produced during each time interval. An example, based on an intensive study of one population of Darwin's finches carried out on Isla Daphne in the Galápagos archipelago, is shown in Table 54.1.

The data in Table 54.1 are based on a cohort of 210 birds that hatched in 1978 and were followed until 1991, by which time all of them had died. This life table (which presents data only on survival, not on reproduction) shows that the mortality rate was high during the first year of life. It then dropped dramatically for several years, followed by a general increase in later years. Mortality rates fluctuated among years because survival of the birds depends on seed production, which is strongly correlated with rainfall. The Galápagos archipelago experiences both drought years, during which plants produce few seeds, birds do not nest, and adult survival is poor, and years of heavy rainfall, during which seed production is high, most birds breed several times, and adult survival is high. The life table reflects these fluctuations.

Ecologists often use graphs to highlight the most important changes in birth and death rates in populations. Graphs of **survivorship**—the mirror image of mortality—in relation to age show at what ages individuals survive well and at what ages they do not. To interpret survivorship data, ecologists have found it useful to compare real data with several hypothetical curves that illustrate a range of possible survivorship patterns (Figure 54.4*a*). At one extreme, nearly all individuals survive for their entire potential life span and die at about the same age (hypothetical curve I). At the other extreme, the survivorship of young individuals is very low, but survivorship is high for most of the remainder of the life span (hypothetical curve III). An intermediate possibility is that survivorship is the same throughout the life span (hypothetical curve II).

Survivorship data from real populations often resemble one of these hypothetical curves. For example, survivorship of humans in the United States remains high for many decades but then, as in hypothetical curve I, declines significantly in older individuals (Figure 54.4*b*). Many wild birds have survivorship curves similar to hypothetical curve II; the probability of their surviving is about the same over most of the life span once they are a few months old (Figure 54.4*c*).

A widespread survivorship pattern is found among organisms that produce many offspring, each of which receives few energy resources and no parental care. In these species, low survivorship of young individuals is followed by high survivorship during the middle part of the life span, and then low survivorship toward the end of the life span. *Spergula vernalis*, an annual plant that grows on sand dunes in Poland, illustrates this pattern (Figure 54.4*d*).

 Patterns of Population Growth

If a single bacterium selected at random from the surface of this book, and all its descendants, were able to grow and reproduce in an unlimited environment, explosive population growth would result. In a month the bacterial colony would weigh more than the visible universe and would be expanding outward at the speed of light. Similarly, a single pair of Atlantic cod and their descendants, reproducing at the maximum rate of which they are capable, would fill the Atlantic Ocean in 6 years. But, as Darwin observed, this does not happen in nature.

All populations have the potential for explosive growth because as the number of individuals in the population increases, the number of new individuals added per unit of time accelerates, even if the rate per capita of population increase remains constant. This form of explosive increase is called **exponential growth**. If we ignore immigration and emigration and assume that births and deaths occur continuously and at constant rates, such a growth pattern forms a continuous curve (Figure 54.5*a*). This curve can be expressed mathematically in the following way:

Rate of increase in number of individuals

$$= \left(\begin{array}{c} \text{Average per capita birth rate} \\ - \text{ Average per capita death rate} \end{array} \right)$$

$\times$ Number of individuals

or, more concisely,

$$r = \frac{\Delta N}{\Delta t} = (b - d)N$$

where $\Delta N / \Delta t$ is the rate of change in the size of the population (ΔN = change in number of individuals; Δt = change in time). The difference between the average per capita birth rate (b) and the average per capita death rate (d) is the per capita rate of increase (r). In these equations, b includes

(*a*) **Hypothetical curves**

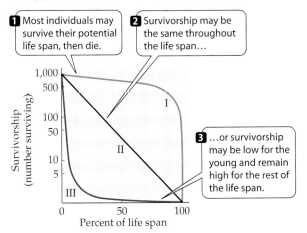

1 Most individuals may survive their potential life span, then die.

2 Survivorship may be the same throughout the life span…

3 …or survivorship may be low for the young and remain high for the rest of the life span.

(*b*)

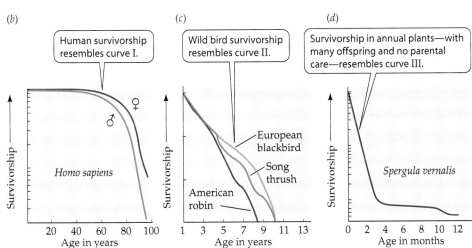

Human survivorship resembles curve I.

(*c*)

Wild bird survivorship resembles curve II.

(*d*)

Survivorship in annual plants—with many offspring and no parental care—resembles curve III.

54.4 Survivorship Curves
Survivorship curves show the number of individuals in a cohort still alive at different times over the life span. (*a*) The range of possible survivorship patterns. Patterns for (*b*) humans in the United States, (*c*) some small wild birds, and (*d*) an annual plant, *Spergula vernalis*, on Polish sand dunes.

(a)

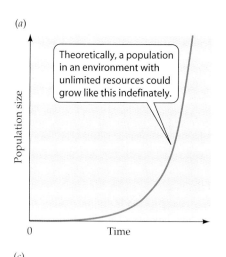

Theoretically, a population in an environment with unlimited resources could grow like this indefinately.

Population size

0   Time

(b)

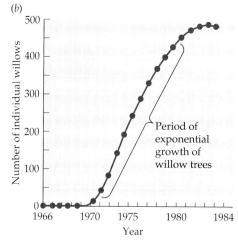

Number of individual willows

Period of exponential growth of willow trees

1966   1970   1975   1980   1984
Year

(c)

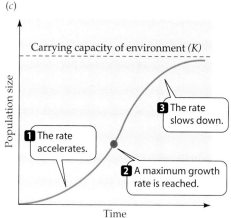

Carrying capacity of environment (K)

Population size

3 The rate slows down.

1 The rate accelerates.

2 A maximum growth rate is reached.

Time

 54.5 Exponential and Logistic Population Growth
(a) A theoretical exponential growth curve. (b) Growth curve of an actual population of willows at Newborough Warren, Wales. The trees experienced a surge of exponential growth when rabbits that fed on their leaves were decimated by disease. (c) A population in an environment with limited resources usually stops growing exponentially long before it reaches the environmental carrying capacity.

both births and immigration and d includes both deaths and emigration.

When there are no limits on population growth, r has its highest value, called r_{max}, the **intrinsic rate of increase**; r_{max} has a characteristic value for each species. Therefore, the rate of growth of a population under optimal conditions is

$$\frac{\Delta N}{\Delta t} = r_{max} N$$

But optimal conditions do not continue indefinitely, and growth rates eventually slow down, as we explore in the following section.

Population growth is influenced by the carrying capacity

Natural populations may experience exponential growth for short periods of time under favorable conditions. For example, one population of willows in Wales increased dramatically in the 1970s after the rabbit population, which had severely nibbled the willows, crashed due to an outbreak of the disease myxomatosis (Figure 54.5b). But no real population can maintain exponential growth for very long because environmental limitations cause birth rates to drop and death rates to rise. In fact, over long time periods, the sizes of most populations fluctuate around a relatively constant number.

The simplest way to picture the limits imposed by the environment is to assume that it can support no more than a certain number of individuals of a particular species. This number, called the **carrying capacity** of the environment, is determined by the availability of resources—food, nest sites, shelter—as well as by disease, predators, and, in some cases, social interactions. Rather than being exponential, population growth slows down as the population approaches the carrying capacity, so that the growth curve has an S shape (Figure 54.5c).

The S-shaped growth pattern, which is characteristic of many populations growing in environments with limited resources, can be represented mathematically by adding to the equation for exponential growth a term, $(K - N)/K$, that slows the population's growth as it approaches the carrying capacity (K). The simplest such equation is that for **logistic growth**:

$$\frac{\Delta N}{\Delta t} = r \left(\frac{K - N}{K} \right) N$$

The biological assumption in this equation is that each individual added to the population makes things slightly worse for the others because it competes with them for available resources, or for other reasons. Population growth stops when $N = K$ because then $(K - N) = 0$, so $(K - N)/K = 0$, and thus $\Delta N/\Delta t = 0$.

The logistic growth equation contains some important simplifications that are not true for most populations. Its most critical assumptions are that (1) each individual exerts its effects immediately at birth; (2) all individuals produce equal effects on the population; and (3) births and deaths are continuous. However, in nature, organisms grow during their lives, and their effects on others normally increase with age, so there may be a delay between the birth of an

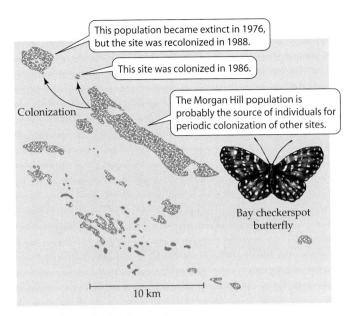

This population became extinct in 1976, but the site was recolonized in 1988.

This site was colonized in 1986.

Colonization

The Morgan Hill population is probably the source of individuals for periodic colonization of other sites.

Bay checkerspot butterfly

10 km

54.6 Subpopulation Dynamics
The population of the bay checkerspot butterfly *Euphydryas editha bayensis* is divided into a number of subpopulations confined to habitat patches that contain the plants its larvae feed on.

individual and the time at which it begins to affect the other members of its population. A seedling tree, for example, exerts a much smaller effect on its neighbors than a large adult tree does, and it does not begin to reproduce until it reaches a relatively large size.

In addition, the logistic equation models a population in a single habitat patch; it does not take immigration and emigration into account. Next we consider the dynamics of an assembly of local populations, which is more complex than the growth of a single population.

Many species are divided into discrete subpopulations

Many populations are divided into discrete *subpopulations* among which some exchange of individuals occurs. Such a pattern is often found where suitable habitat occurs in separated patches, or "habitat islands." Each subpopulation has a probability of "birth" (colonization) and "death" (extinction). Within each subpopulation, growth occurs in the ways we have just discussed, but because subpopulations are typically much smaller than the population as a whole, local disturbances and random fluctuations in numbers of individuals often cause the extinction of a subpopulation. However, if individuals frequently move between subpopulations, immigrants may prevent declining subpopulations from becoming extinct. This process is known as the **rescue effect**.

The bay checkerspot butterfly provides a good illustration of the dynamics of such divided populations. The larvae of this butterfly feed on only a few species of annual plants that are restricted to outcrops of a particular kind of rock on hills south of San Francisco. The bay checkerspot

has been studied for many years by Stanford University biologists. During drought years, most host plants die early in spring, before the butterfly larvae have pupated. At least three butterfly subpopulations became extinct during a severe drought in 1975–1977. The largest patch of suitable habitat, Morgan Hill, typically supported thousands of butterflies (Figure 54.6). It probably served as a source of individuals that dispersed to and colonized small habitat patches where the butterflies had become extinct.

Population Regulation

In a limited environment, population growth slows down as density increases because the members increasingly affect one another adversely. As a result, a population above the environmental carrying capacity is likely to decrease in density, and one that is below the carrying capacity is likely to increase in density. In this section we discuss how populations may be influenced by interactions between their density and the carrying capacity of their environment.

How does population density influence birth and death rates?

If per capita birth or death rates change in response to population density, they are said to be **density-dependent**. Death and birth rates may be density-dependent for several reasons:

▶ As a population increases in abundance, it may deplete its food supply, reducing the amount of food that each individual gets. Poor nutrition may increase death rates and decrease birth rates.

▶ Predators may be attracted to regions where densities of their prey have increased. If predators are able to capture a larger proportion of the prey than they did when prey were scarce, the per capita death rate of the prey rises.

▶ Diseases, which may increase death rates, spread more easily in dense populations than in sparse populations.

A population whose dynamics are influenced primarily by density-dependent factors is said to be *regulated*.

Factors that change per capita birth and death rates in a population independent of its density are said to be **density-independent**. A very cold spell in winter, for example, may kill a large proportion of the individuals in a population regardless of its density. However, even environmental factors whose frequency and severity are unrelated to population density may result indirectly in density-dependent mortality. Cold weather may not kill organisms directly, but may increase the amount of food individuals need to eat each day. Individuals pushed by population density into poorer foraging areas may be more likely to die than those in better foraging areas. Or the death rate may be related to the quality of sleeping places. If population density is high, a larger proportion of individuals may be forced to sleep in places that expose them to the cold.

Various combinations of density-dependent and density-independent factors can influence the density of a popula-

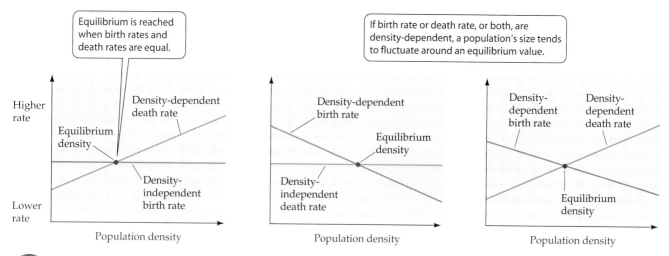

54.7 Density-Dependent Factors Regulate Population Size
The densities of all populations fluctuate, but they tend to return to equilibrium value if either birth rate and/or death rate are density-dependent.

tion. The hypothetical graphs in Figure 54.7 show how birth and death rates can change in relation to population density. When birth and death rates are equal (the point at which the two lines cross) the population neither grows nor shrinks. If birth or death rates, or both, are density-dependent, the population responds to increases or decreases in its density by returning toward the equilibrium density. If neither rate is density-dependent, there is no equilibrium and the population is not regulated.

Fluctuations in the density of a species' population are determined by all the factors and processes, density-dependent and density-independent, acting upon it. The combined action of density-independent and density-dependent factors is illustrated by the dynamics of a population of song sparrows on Mandarte Island, off the coast of British Columbia, Canada.

During recent years, in response to variable winter weather and other physical factors, the number of song sparrows on Mandarte has fluctuated between 4 and 72 breeding females and between 9 and 100 breeding males. Density-independent variation caused by weather is modulated by several density-dependent factors. The number of breeding males is limited by territorial behavior: The larger the number of males, the larger the number that fail to gain territories, so more live as "floaters" (Figure 54.8*a*). Also, the larger the number of breeding females, the fewer offspring each female fledges (Figure 54.8*b*). And, finally, the more offspring are fledged, the more poorly they survive over the winter (Figure 54.8*c*).

Disturbances affect population densities

Disturbances—short-term events that influence populations by changing their environment and, hence, its carrying capacity—regularly affect population densities. Common physical disturbances are fires, hurricanes, ice storms, wind storms, floods, landslides, and lava flows. Biological disturbances include tree falls, disease epidemics, and the burrowing and trampling activities of animals. Disturbances differ in their spatial distribution, frequency, predictability, and severity.

A disturbance typically decreases the environmental carrying capacity for some species while increasing it for others. A landslide, for example, may increase the carrying capacity of the environment for plants that require bare mineral soil for the germination of their seeds and survival of their seedlings. However, it will decrease the carrying capacity for species that require shade and rich organic litter for successful germination.

Populations themselves can influence the frequency of some disturbances. Immediately after a fire, for example, there is not enough combustible organic matter to carry another fire. However, as vegetation grows back, dead wood, branches, and leaves accumulate, gradually increasing the supply of fuel. Thus the frequency of fires may be proportional to the rate at which fuel accumulates through the growth of plant populations, or the rate at which herbivores consume plant materials that would otherwise accumulate. Similarly, as many trees age, their roots and trunks become weakened by fungal infections. Old, large trees are thus susceptible to being toppled by high winds. Therefore, the likelihood of a major blowdown increases as the forest ages.

Organisms cope with environmental changes by dispersing

A common response of animals to environmental changes is **dispersal**—movement to another habitat. If habitat quality declines greatly, individuals may be able to improve their survival and reproductive success by going elsewhere.

If repeated seasonal changes alter a habitat, organisms may evolve life cycles that appear to anticipate the changes. **Migration**—regular seasonal movement from one place to another—is most widespread among birds, but some insects, such as monarch butterflies, and some mammals also

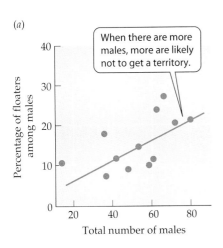

(a)

When there are more males, more are likely not to get a territory.

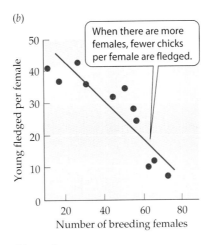

(b)

When there are more females, fewer chicks per female are fledged.

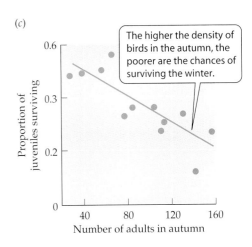

(c)

The higher the density of birds in the autumn, the poorer are the chances of surviving the winter.

54.8 Regulation of an Island Population of Song Sparrows
The number of song sparrows on Mandarte Island is mainly determined by the severity of winter weather, but the weather's effect is modulated by several density-dependent factors, including (a) male territorial behavior, (b) the reproductive success of females, and (c) the survival of juveniles.

migrate (Figure 54.9). Most insectivorous birds leave high latitudes for more favorable wintering grounds in autumn, before conditions seriously deteriorate. In arctic regions, caribou migrate each year between winter and summer ranges.

Life Histories Influence Population Growth

The complete **life history** of an organism consists of its birth, growth to maturity, reproduction, and death. During its life, an individual organism ingests nutrients or food, grows, interacts with other individuals of the same and other species, reproduces, and usually moves or is moved so that it does not die exactly where it was born. Life histories describe how an organism divides its efforts among these activities. In previous discussions we have referred to various components of the life histories of organisms. Now we focus our attention specifically on life history patterns and why they have evolved to be as variable as they are.

The life history traits of organisms have been molded by natural selection acting over many generations. In each lineage, those traits that maximized reproductive success were favored, but natural selection has not produced a single, dominant pattern of reproduction. Some organisms, such as elephants and humans, usually give birth to a single offspring in each reproductive episode; others produce thou-

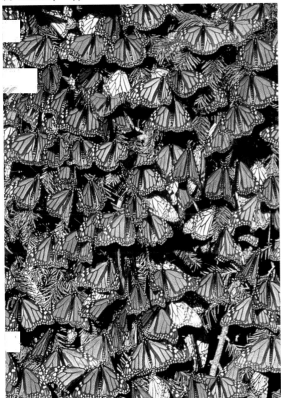

(a) *Danaus plexippus*

(b) *Rangifer tarandus*

54.9 Animals Migrate to Remain in Suitable Environments
(a) Most of North America's monarch butterflies migrate to central Mexico, where they aggregate on conifers in cool mountain valleys. They can survive the winter there without eating because their metabolic rates are low.
(b) These caribou in the American Arctic are migrating from open tundra to their winter foraging grounds at the edge of boreal forest. During the winter they feed on lichens on branches of trees.

sands or millions of eggs or seeds in one bout of reproduction. Some organisms begin to reproduce within days or weeks of being born; others live for many years before reproducing. All life histories are based on a certain set of traits, which includes:

▶ Size and energy supply of the individual at birth
▶ Its rate and pattern of growth and development
▶ How many times individuals disperse
▶ The number and timing of reproductive events
▶ The number, size, and sex ratio of offspring
▶ The ages at which individuals die

Life histories include stages for growth, change in form, and dispersal

For at least part of their lives, all organisms grow by gathering and assimilating energy and nutrients. Some organisms, such as birds and mammals, gather energy and nutrients throughout their lives, even after they reach adult size and stop growing. Energy gathered after growth stops maintains the organism and supports reproduction. In many species, however, energy gathering is confined to a particular life stage. Most moths, for example, feed only when they are larvae. The adults lack mouthparts and digestive tracts, live on the energy they gathered as larvae, and survive only long enough to disperse, mate, and lay eggs (Figure 54.10).

Individuals of many species also change form during their lives. Human babies are unmistakably human, but newborns of many species differ dramatically from adults. Some of the most striking changes are found among insects such as beetles, flies, moths, butterflies, and bees, which undergo metamorphosis from their larval to their adult forms. Many plants have resting stages, such as spores and seeds, that have low metabolic rates and are highly resistant to changes in the physical environment. Growth typically does not take place in these stages.

At some time in their lives, all organisms disperse. Some, such as plants and sessile animals, disperse as eggs, larvae, spores, or seeds. Others, such as insects and birds, disperse primarily as adults. Still others may disperse during several different life stages. Individuals of some species can change their location many times during their lives in response to environmental changes. Others remain in the first place they settle.

Life histories embody trade-offs

Life history evolution is influenced by inevitable trade-offs. These trade-offs exist because changes that improve fitness by means of one life history trait often reduce fitness by means of another. What are the major trade-offs in life history traits?

A universal trade-off exists between number and size of offspring. Every newborn individual begins to grow with energy and nutrients from its maternal parent, but how much energy and nutrients individuals receive from their mothers varies greatly. The larger the amount of energy

Polyphemus sp.

54.10 A Life Stage for Sex and Reproduction Only
This female moth has no mouth or digestive tract. She will live only a few days, just long enough to mate and lay her eggs

provided to each offspring, the larger it can grow before it must gather its own energy, but the fewer offspring a mother can produce for a given amount of energy—a major trade-off.

For animals, a trade-off also exists between the number of offspring produced and the amount of care parents provide to their offspring. The more parental care the parents provide, the fewer offspring they can produce for a given investment in reproduction. Birds and mammals produce few offspring at a time and provide extensive care for each one.

Some organisms invest so much in one reproductive effort that they have no energy left for another—or even for their own survival. If two individuals have the same amount of energy to invest in reproduction, and one reproduces only once while the other reproduces several times, the former can produce more offspring in a single episode than the latter because it reserves no energy for itself. Annual plants invest nearly all of the energy they gain during their single growing season in seed production; they do not survive long after reproducing. Some longer-lived organisms also reproduce once in their lifetimes and die soon afterward. Pacific salmon (genus *Oncorhynchus*) hatch in fresh water, migrate to the sea, spend a number of years at sea, return to fresh water, spawn, and die (Figure 54.11a). Most agaves (century plants) of the American Southwest likewise store up energy for many years before producing a large flowering stalk, forming many seeds, and dying (Figure 54.11b).

Trade-offs also exist between reproduction and growth. Members of many species do not begin to reproduce until they have reached full size, but others, such as most plants, mollusks, fishes, and reptiles, start to reproduce while they are still relatively small and continue to reproduce as they grow. Reproduction usually reduces growth because these two processes compete for the limited amount of energy an individual has at its disposal. Beech trees in Germany, for example, grew more slowly during years when they pro-

(a) *Oncorhyncus nerka*

54.11 A Single Reproductive Effort
(a) These sockeye salmon are ascending Hensen Creek, Alaska. They will lay their eggs in gravel beds in the stream and then die. (b) This century plant has mobilized the energy stored during its long life to produce a large flowering stalk with hundreds of flowers, literally reproducing itself to death.

Flowering
stalk

Nonreproductive
form of *Agave*

duced large crops of nuts than they did during years when their nut crops were small (Figure 54.12).

(b) *Agave* sp.

Offspring are like "money in the bank"

If reproduction compromises future growth and survival, why do some organisms start to reproduce when they are small or young? The potential contribution of an individual's offspring to future generations depends, in part, on when they are produced. A useful analogy compares the production of offspring to earning interest on money deposited in a bank. It pays to deposit money in the bank as soon as possible so that it can begin earning interest. Offspring produced early in an adult's life likewise "yield interest" quickly—that is, they can begin to reproduce sooner than offspring produced later.

Reproductive value is the average number of offspring that remain to be born to individuals of a particular age. A newborn individual does not have the highest reproductive value, even though it has its full reproductive potential ahead of it, because many newborn individuals will die before they have a chance to reproduce. Therefore we must discount the number of offspring an individual could produce if it survived by the chance that it will die before reaching reproductive age, or during reproduction. When we make the appropriate calculations, we find that the reproductive value of an individual steadily increases until it begins to reproduce. Once maturity is reached, reproductive value declines; in most species, it reaches zero when the individual has finished reproducing. However, individuals can still have positive reproductive value after they have stopped reproducing if they continue to assist the survival of their offspring and grandoffspring.

Because reproductive value declines after maturity, the power of natural selection grows increasingly weaker as an individual ages. Once reproductive value has dropped to zero, natural selection cannot act on alleles that first produce

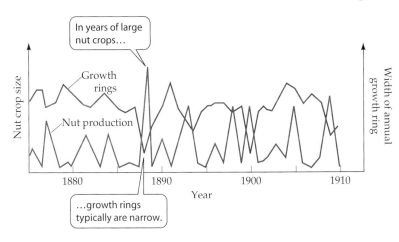

In years of large nut crops…

Growth rings

Nut production

Nut crop size

Width of annual growth ring

…growth rings typically are narrow.

1880   1890   1900   1910
Year

54.12 Reproduction Slows Growth Rates in Beech Trees
The width of annual growth rings shows the growth rate of beech trees in different years. In general, the trees grew slowly in years when they produced large crops of nuts, but during unusually favorable years they grew rapidly and produced many nuts.

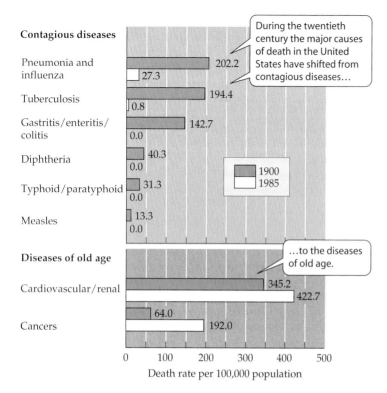

Contagious diseases

During the twentieth century the major causes of death in the United States have shifted from contagious diseases…

Pneumonia and influenza — 202.2 / 27.3

Tuberculosis — 194.4 / 0.8

Gastritis/enteritis/colitis — 142.7 / 0.0

Diphtheria — 40.3 / 0.0

Typhoid/paratyphoid — 31.3 / 0.0

Measles — 13.3 / 0.0

1900
1985

Diseases of old age

…to the diseases of old age.

Cardiovascular/renal — 345.2 / 422.7

Cancers — 64.0 / 192.0

0   100   200   300   400   500
Death rate per 100,000 population

54.13 Causes of Human Death in the United States
Today most people die of diseases of old age because improved sanitation and public water supplies, as well as medical advances such as immunization, have greatly reduced the incidence of contagious diseases that formerly killed many young people.

their phenotypic effects at that age—even those that are highly detrimental to the individual's survival. As a result, increasing numbers of harmful alleles are expressed as individuals age, causing increased mortality rates, especially after reproduction has ceased. In this manner, **senescence**—an increased probability of dying per unit of time with increasing age—has evolved.

As a result of improved hygiene and nutrition, most people in modern industrial societies are now spared the contagious diseases that cause death rates to be high among people of all ages in nonindustrial societies. Most people live to the age when the so-called genetic diseases of old age begin to afflict them. Cancer and heart disease, the main killers in industrialized societies, are much more difficult to cure than the contagious diseases that formerly caused most deaths. For this reason, despite the expenditure of enormous resources to extend life, the average age of death in the United States has changed very little during the past 30 years. As one source of mortality is eliminated, another takes its place (Figure 54.13). Life history theory suggests that this situation is likely to continue indefinitely.

Can Humans Manage Populations?

For many centuries, people have tried to reduce populations of species they consider undesirable and increase populations of desirable species. Strategies for controlling and managing populations of organisms are based on our understanding of how populations grow and are regulated.

A general principle of population dynamics is that the total number of births and the growth rates of individuals tend to be highest when a population is well below its carrying capacity. Therefore, if we want to maximize the number of individuals of a species that we wish to harvest, we should manage the population so that it is far enough below carrying capacity to have high birth and growth rates. Hunting seasons for game birds and mammals are established with this objective in mind.

Life history traits determine how heavily a population can be exploited

Populations with high reproductive capacities can sustain their growth despite a high rate of harvest. In such populations (many species of fish, for example), each female may lay thousands or millions of eggs. Another characteristic of these fast-reproducing populations is that individual growth is often density-dependent. If prereproductive individuals are harvested at a high rate, the remaining individuals may grow faster. Many fish populations can be harvested heavily for many years because only a modest number of females must survive to reproductive age to produce the eggs needed to maintain the population.

Fish can, of course, be overharvested. Many populations have been greatly reduced because so many individuals were harvested that too few reproductive adults survived to maintain the population. The Georges Bank off the coast of New England—a source of cod, halibut, and other prime food fishes—has been exploited so heavily that many fish stocks have been reduced to levels insufficient to support a commercial fishery.

The whaling industry has also engaged in excessive harvests. The blue whale, Earth's largest animal, was hunted nearly to extinction by the middle of the twentieth century. The industry then turned to smaller species of whales that were still numerous enough to support commercially viable whaling operations (Figure 54.14).

Management of whale populations is difficult for two reasons. First, unlike fish, whales reproduce at very low rates. They have long prereproductive periods, produce only one offspring at a time, and have long intervals between births. Thus many whales are needed to produce even a small number of offspring. Second, because whales are distributed widely throughout Earth's oceans, they are an international resource whose conservation and wise management depends upon cooperative action by all whaling nations. This continues to be difficult to achieve.

Life history information is used to control populations

The same principles apply if we wish to reduce the size of populations of undesirable species and keep them at low

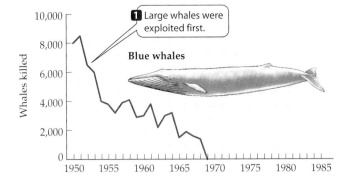

1 Large whales were exploited first.

Blue whales

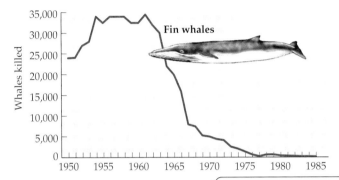

Fin whales

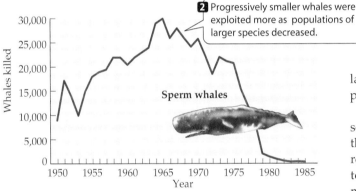

2 Progressively smaller whales were exploited more as populations of larger species decreased.

Sperm whales

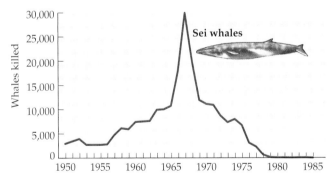

Sei whales

54.14 Overexploitation of Whales
The graphs show the numbers of whales of four species killed each year from 1950 to 1985. As each species reached low population levels, the whaling industry turned to other species. All four species were driven to very low levels by sustained hunting.

Similarly, if we wish to preserve a rare species, the most important step usually is to provide it with suitable habitat. If habitat is available, the species will usually reproduce at rates sufficient to maintain its population. If the habitat is insufficient, preserving the species usually requires expensive and continuing intervention, such as providing extra food.

Humans have introduced many species to new habitats outside their native ranges. When these introduced species undergo population explosions, humans attempt to reduce their numbers by introducing new predators and parasites from the introduced species' original habitat. For example, the cactus *Opuntia*, introduced into Australia from South America, spread rapidly and became a common pest species over vast expanses of valuable sheep-grazing land. The cactus population was controlled by introducing a moth species (*Cactoblastis cactorum*) whose larvae feed on *Opuntia*. Once egg-laying females find a patch of cactus, their larvae completely destroy the patch (Figure 54.15).

But new patches of cactus arise in other places from seeds dispersed by birds. These new patches flourish until they are found and destroyed by *Cactoblastis*. Over a large region, the numbers of both *Opuntia* and *Cactoblastis* are today fairly constant and low, but in the local areas that make up the whole, there are extreme oscillations resulting

densities. At densities well below carrying capacity, populations have high birth rates and can therefore withstand higher death rates than they could closer to carrying capacity. Killing part of a population whose dynamics are influenced primarily by density-dependent factors only reduces it to the density at which it experiences the most rapid rate of growth. A far more effective approach to reducing the population of a species is to remove its resources, thereby lowering the carrying capacity of its environment. We can rid our dumps and cities of rats more easily by making garbage unavailable (reducing the carrying capacity of the rats' environment) than by poisoning rats.

54.15 Biological Control of an Introduced Pest
Cactoblastis caterpillars consume an *Opuntia* cactus in Australia.

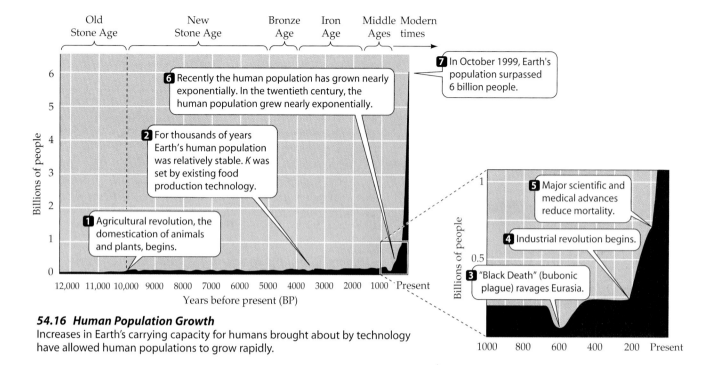

54.16 Human Population Growth
Increases in Earth's carrying capacity for humans brought about by technology have allowed human populations to grow rapidly.

from the extermination of first the plant and then the herbivore. This is another example of a series of subpopulations connected by occasional dispersing individuals.

Can we manage our own population?

Managing our own population has become a matter of great concern. For thousands of years, Earth's carrying capacity for human populations was set at a low level by food and water supplies and disease. Domestication of plants and animals and cultivation of the land enabled our ancestors to increase the resources at their disposal dramatically. These developments stimulated rapid population growth up to the next carrying capacity limit, which was determined by the agricultural productivity possible with only human- and animal-powered tools. Agricultural technology and artificial fertilizers, made possible by the tapping of fossil fuels, greatly increased agricultural productivity, further raising Earth's carrying capacity for humans. The development of modern medicine reduced the effectiveness of disease as a limiting factor on human populations, raising the global carrying capacity still further (Figure 54.16). Medicine and better hygiene have allowed people to live in large numbers in areas where diseases formerly kept numbers very low.

What is Earth's present carrying capacity for people? Today's carrying capacity is set by Earth's ability to absorb the by-products—especially CO_2—of our enormous consumption of fossil fuel energy and by whether we are willing to cause the extinction of millions of other species to accommodate our increasing use of environmental resources. We will explore some of the consequences of high human population densities for the survival of other species in Chapter 58.

Chapter Summary

Population Structure: Patterns in Space and Time

▶ A population consists of all the individuals of a species within a given area.

▶ The number of individuals of a species per unit of area (or volume) is its population density.

▶ Individuals in a population may have uniform, random, or clumped distributions. **Review Figure 54.2**

▶ The age distribution of individuals in a population reveals much about the recent history of births and deaths in the population. The timing of births and deaths may influence age distributions for many years. **Review Figure 54.3**

Population Dynamics: Changes over Time

▶ Births, deaths, immigration, and emigration drive changes in population density and distribution.

▶ Life tables help us visualize patterns of births and deaths in a population. **Review Table 54.1**

▶ Graphs of survivorship in relation to age show when individuals survive well and when they do not. **Review Figure 54.4**

Patterns of Population Growth

▶ All populations have the potential to grow exponentially. However, no population can maintain exponential growth for very long because environmental limitations cause birth rates to drop and death rates to rise.

▶ The number of individuals of a particular species that an environment can support—called the carrying capacity—is determined by the availability of resources and by disease and predators.

▶ A population in a limited environment at first grows rapidly, but growth rates decrease as the carrying capacity is approached. **Review Figure 54.5**

▶ The overall densities of many populations are determined by "births" (colonizations) and "deaths" (extinctions) of local subpopulations. Immigrants may prevent declining subpopulations from becoming extinct, a process known as the rescue effect. **Review Figure 54.6**

Population Regulation

▶ Regulation of a population by changes in per capita birth or death rates in response to density is said to be density-dependent.

▶ If per capita birth and death rates are unrelated to a population's density, the population is not regulated.

▶ The density of a population is determined by the combined effects of all density-dependent and density-independent factors affecting it. **Review Figures 54.7, 54.8**

Life Histories Influence Population Growth

▶ The life history of a species describes how it divides its efforts among growth, dispersal, and reproduction over time.

▶ Trade-offs inevitably exist between number and size of offspring, between number of offspring and parental care, between survival and reproduction, and between growth and reproduction. **Review Figure 54.12**

▶ Reproductive value is the average number of offspring that remain to be born to individuals of a particular age. Reproductive value rises to a peak when individuals first begin to reproduce and declines to zero after reproduction ceases.

▶ Senescence—an increased probability of dying with increasing age—evolves because natural selection cannot act on alleles that first produce their phenotypic effects after a reproductive value drops to zero. **Review Figure 54.13**

Can Humans Manage Populations?

▶ Humans use the principles of population dynamics to control and manage populations of species they consider desirable or undesirable. Nevertheless, many populations have been overexploited. **Review Figure 54.14**

▶ Earth's carrying capacity for humans has been increased several times by technological developments. **Review Figure 54.16**

For Discussion

1. Huntington's disease is a severe disorder of the human nervous system that generally results in death. It is caused by a dominant allele that does not usually express itself phenotypically until its bearer is 35 to 40 years old. How fast is the gene causing Huntington's disease likely to be eliminated from the human population? Would your answer change if the gene expressed itself when its bearer was 20 years old? 10 years old?

2. Many people have improperly formed wisdom teeth and must spend considerable sums of money to have them removed. Assuming, as is probably the case, that the presence or absence of wisdom teeth and their mode of development are partly under genetic control, will we gradually lose our wisdom teeth by evolutionary processes?

3. Some organisms, such as oysters, cod, and elm trees, produce vast quantities of offspring, nearly all of which die before they reach adulthood. What fraction of such deaths are likely to be selective—that is, dependent on the genotypes of the individuals dying? What does your answer imply for the rates of evolution of oysters, cod, and elms?

4. In this chapter we identified a number of trade-offs in life history evolution. Why are these trade-offs inevitable? Why is knowledge about trade-offs important when we attempt to manipulate the life histories of organisms?

5. Ecologists often use the concept of carrying capacity when studying the growth and regulation of populations in nature, even though carrying capacity often changes markedly over time. How can the concept be useful if its value changes so often?

6. Most organisms whose populations we wish to manage for higher densities are long-lived and have low reproductive rates, whereas most organisms whose populations we attempt to reduce are short-lived but have high reproductive rates. What is the significance of this difference for management strategies and the effectiveness of management practices?

7. In the mid-nineteenth century, the human population of Ireland was largely dependent upon a single food crop, the potato. When a disease caused the potato crop to fail, the Irish population declined drastically for three reasons: (1) a large percentage of the population emigrated to the United States and other countries; (2) the average age of a woman at marriage increased from about 20 to about 30 years; and (3) many families starved to death rather than accept food from Britain. None of these social changes was planned at the national level, yet all contributed to adjusting population size to the new carrying capacity. Discuss the ecological strategies involved, using examples from other species. What would you have done had you been in charge of the national population policy for Ireland?

8. From a purely ecological standpoint, can the problem of world hunger ever be overcome by improved agriculture alone? What components must a hunger-control policy include?

55 Community Ecology

SOME PEOPLE LIKE THEIR FOOD SPICY HOT; OTHERS don't. The spices that impart strong flavors to foods are actually antioxidant, antimicrobial, and antiviral chemicals. These chemicals evolved because they protected the plants that produced them from predators and diseases. When we add spices during cooking, we borrow the plants' survival "recipes" and use them to protect our own food. The protection that spices provide was especially important before refrigeration and freezing were widely available, but even today, spices help prevent contamination of foods in most parts of the world.

Organisms interact with one another in a variety of ways. Some of these interactions involve eating and being eaten, but organisms may also interact competitively, or they may benefit one another. All organisms are potentially or actually food for some other organism, and most of them have evolved defenses that make them more difficult to find and capture, or less palatable or nutritious if they are captured. Consumers of those organisms have, in turn, evolved ways of getting around the defenses of their prey.

The organisms that live together in a particular area constitute an **ecological community**. Each species interacts in unique ways with other species in its community and with its physical environment. Some of these interactions are strong and important; others are weak and affect the functioning of the community very little. The study of such interactions, and how they determine which and how many species live in a place, is the focus of **community ecology**.

For several decades, ecologists debated whether which species live together in communities is determined primarily by the interactions of individuals with the physical environment or by their interactions with other organisms. Some ecologists even suggested that a community is a superorganism in which each species plays a particular role, just as each organ plays a role within the body of an individual organism. That view has been abandoned because organisms, unlike organs, do not evolve under the influence of natural selection to serve their community. Nevertheless, determining the roles of species interactions with the biotic and abiotic environment is a major challenge for community ecologists today.

In this chapter we do consider interactions between organisms and their physical environment, but we concentrate on the major biological interactions and show how they influence the structure and functioning of ecological communities.

Types of Ecological Interactions

Organisms interact with one another in five major ways:

▶ Two organisms may mutually harm one another. This type of interaction is common when two organisms use the same resources and those resources are insufficient to supply their combined needs. Such organisms are called *competitors*, and their interactions constitute **competition**.

▶ One organism, by its activities, may benefit itself while harming another, as when individuals of one species eat individuals of another. The eater is called a *predator* or *parasite*, and the eaten is its *prey* or *host*. These interactions are known as **predator–prey** or **parasite–host interactions**.

Some Like It Hot
The peppers on the right of this photo are the seeds of a tropical vine, *Piper nigrum*. The hot chiles (upper left) are the fruits of an unrelated pepper plant, *Capsicum anuum*. The chemicals in peppers and other "hot" spices are often antimicrobial agents.

55.1 *Types of Ecological Interactions*

		EFFECT ON ORGANISM 2		
		HARM	BENEFIT	NO EFFECT
EFFECT ON ORGANISM 1	HARM	Competition (–/–)	Predation or parasitism (–/+)	Amensalism (–/0)
	BENEFIT	Predation or parasitism (+/–)	Mutualism (+/+)	Commensalism (+/0)
	NO EFFECT	Amensalism (0/–)	Commensalism (0/+)	—

▶ If both participants benefit from an interaction, we call them *mutualists*, and their interaction is a **mutualism**.

▶ If one participant benefits but the other is unaffected, the interaction is a **commensalism**.

▶ If one participant is harmed but the other is unaffected, the interaction is an **amensalism.**

These categories of species interactions are summarized in Table 55.1. But they are not clear-cut, both because the strengths of interactions vary and because many cases do not fit the categories neatly. Nevertheless, most interactions fit these categories well enough for us to use them as a guide for exploring interactions among species in this chapter.

Resources and Consumers

Many interactions between organisms within communities center on resources and their consumers. A **resource** is anything directly used by an organism that can potentially lead to the growth of the population and whose availability is reduced when it is used. We usually think first of resources that can be consumed by being eaten, but space—including hiding places, nest sites, and establishment sites for sessile organisms—becomes unavailable if it is occupied, so it, too, is a resource. Factors such as temperature, humidity, salinity, and pH, even though they may strongly affect population size, are not resources because they can be neither consumed nor monopolized.

Some resources, such as nest sites, are not altered by being used and immediately become available for occupancy again when the user leaves. Other resources must regenerate before they are again available to consumers.

Biotic interactions influence the conditions under which species can persist

Each species can persist only under a certain set of environmental conditions, which define its **ecological niche**. If there were no competitors, predators, or disease organisms in its environment, a species would be able to persist under a broader array of physical conditions (its *fundamental niche*) than it can in the presence of other species that negatively affect it (its *realized niche*). On the other hand, the presence of beneficial species may increase the range of physical conditions in which a species can persist.

An experiment performed on two species of barnacles, *Balanus balanoides* and *Chthamalus stellatus*, demonstrated the importance of both abiotic and biotic factors in determining the fundamental and realized niches of these two species. These barnacles live between high tide and low tide levels on rocky North Atlantic shores. Adult *Chthamalus* generally live higher in the intertidal zone than do adult *Balanus*, but young *Chthamalus* settle in large numbers in the *Balanus* zone.

In the absence of *Balanus*, young *Chthamalus* survive and grow well in the *Balanus* zone, but if *Balanus* are present, the *Chthamalus* are smothered, crushed, or undercut by the larger and more rapidly growing *Balanus*. Young *Balanus* settle in the *Chthamalus* zone, but they grow poorly because they lose water rapidly when exposed to air. *Chthamalus* compete successfully with them there, but *Balanus* would persist slightly higher in the intertidal zone in the absence of *Chthamalus*. By experimentally removing one or the other species, researchers have shown that the vertical ranges of adults of both species are greater in the absence of the other species. The result of their interaction is intertidal zonation, with *Chthamalus* growing above *Balanus* (Figure 55.1).

Limiting resources determine the outcomes of interactions

Resources whose supply is less than the demand made upon them by organisms are called **limiting resources**. Resources that are not limiting may have little influence on a species' population dynamics. For example, most terrestrial animals have strict but similar requirements for a certain minimum level of oxygen. However, studying the use of oxygen reveals very little about the structure of terrestrial communities because the concentration of oxygen, which is 21 percent of the atmosphere, is nearly always above that minimum level.

The limiting resources that influence distributions and abundances of terrestrial species are those that are depletable and regenerate slowly, such as food. In freshwater aquatic environments, however, where the maximum concentration of dissolved oxygen is only 0.5 percent, organisms regularly deplete oxygen. Aquatic ecologists, unlike terrestrial ecologists, pay careful attention to oxygen levels.

Which resources are limiting differs among environments, but some kinds of resources, such as food supplies,

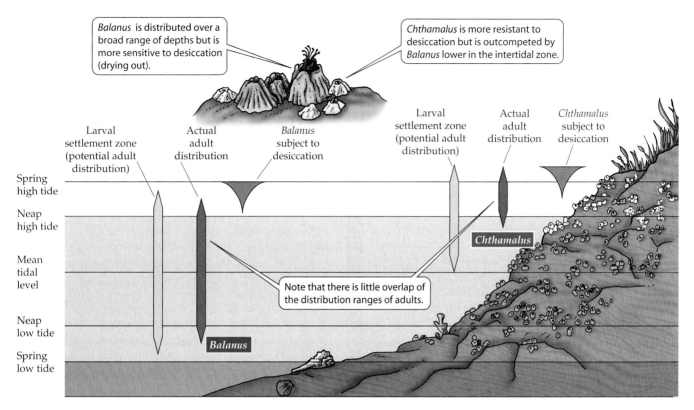

55.1 Potential and Actual Distributions of Two Barnacle Species
Interspecific competition makes the zone each species occupies smaller than the zone it could potentially occupy in the absence of the other species. The width of the red and gold bars is proportional to the density of the populations.

are often limiting. Because of the importance of resources in the lives of all species, we first examine competition that takes place among organisms needing scarce resources, and then consider predation.

Competition: Seeking and Using Scarce Resources

If two or more individuals use the same resources, and those resources are insufficient to meet their demands, the individuals are competitors, whether they are members of the same or a different species. **Intraspecific competition**—competition among individuals of the same species—may result in reduced growth and reproductive rates for some individuals, may exclude some individuals from better habitats, and may cause the deaths of others. **Interspecific competition**—competition among individuals of different species—affects individuals in the same way, but in addition, an entire species may be kept out of habitats where it cannot compete successfully, a phenomenon called **competitive exclusion**. In extreme cases, a competitor may cause the extinction of another species. In this section we will show how ecologists study competition and discuss how it influences species distributions and the composition of ecological communities.

Plants are good subjects for experiments to test the nature and results of competitive interactions because they compete for light, water, and nutrients, all of which can easily be manipulated. For example, the relative importance of root and shoot competition can be assessed by growing plants in shared or separate pots so that either their shoots or roots, or both, compete for resources: nutrients and water in the case of roots, light in the case of shoots. The results of an experiment measuring root and shoot competition between a clover (*Trifolium repens*) and a grass (*Lolium perenne*) are shown in Figure 55.2. The grass outcompeted the clover in both root and shoot competition because a rich supply of nutrients, which the grass was able to use more efficiently than the clover, was provided.

Competition can restrict species' ranges

The role of competition in restricting the ranges of species is illustrated by the interaction of the two barnacle species described in Figure 55.1. In some cases, competition can completely exclude a species from part or all of its range.

Parasitic wasps were introduced into southern California to control outbreaks of scale insects that were seriously damaging citrus orchards. The Mediterranean wasp *Aphytis chrysomphali* was established in southern California by 1900, but it did not effectively control scale insects. Therefore, a close relative from China, *A. lingnanensis*, which has a higher reproductive rate, was introduced in 1948. *A. lingnanensis* increased rapidly, and within a decade it had displaced *A. chrysomphali* from most of its California range (Figure 55.3).

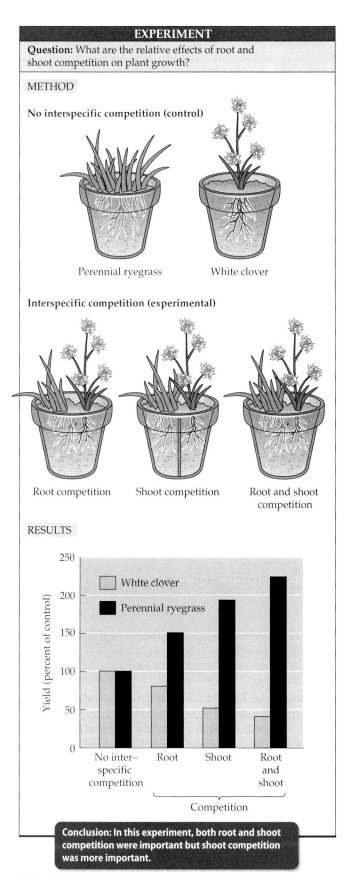

EXPERIMENT

Question: What are the relative effects of root and shoot competition on plant growth?

METHOD

No interspecific competition (control)

Perennial ryegrass   White clover

Interspecific competition (experimental)

Root competition   Shoot competition   Root and shoot competition

RESULTS

Yield (percent of control)

White clover
Perennial ryegrass

No inter-specific competition | Root | Shoot | Root and shoot

Competition

Conclusion: In this experiment, both root and shoot competition were important but shoot competition was more important.

55.2 Plants Compete with their Roots and Shoots
By growing plants in separate pots or together, experimenters can distinguish the influences of root and shoot competition on plant growth. An experiment using white clover (*Trifolium repens*) and perennial ryegrass (*Lolium perenne*) showed that both roots and shoots of these plants are involved in competition.

Competition can reduce species' abundances

Often competition reduces the abundances of competing species rather than eliminating them from an area. Many species of seed-eating ants and rodents live together in the Sonoran Desert of Arizona. To determine whether competition between ants and rodents influences their abundances, ecologists removed ants from some sites, rodents from other sites, and both ants and rodents from a third set of sites. When they removed ants, the density of rodents increased slightly, but seed densities did not change. When they removed rodents, the density of ant colonies nearly doubled, but again, seed densities did not change. When they removed both ants and rodents, seed densities increased to five times their previous value. These results showed that ants and rodents were competing for and influencing their food supply, but that rodents had a much stronger effect on ants than vice versa.

To determine whether different species of rodents also compete with one another, the ecologists erected rodent-proof fences around 50 × 50-m desert plots. The fences around the experimental plots had holes that small rodents could pass through, but were too small to allow the passage of large kangaroo rats. The holes in the fences surrounding control plots were large enough for all rodents to pass through. Within 2 years of the exclusion of kangaroo

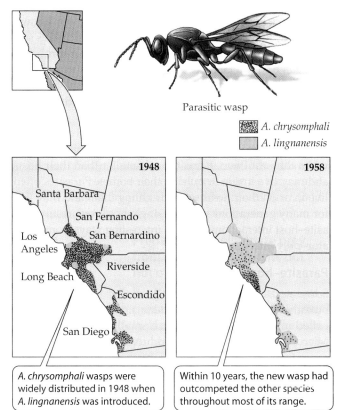

Parasitic wasp

A. chrysomphali
A. lingnanensis

1948

Santa Barbara
San Fernando
Los Angeles
San Bernardino
Long Beach
Riverside
Escondido
San Diego

1958

A. chrysomphali wasps were widely distributed in 1948 when *A. lingnanensis* was introduced.

Within 10 years, the new wasp had outcompeted the other species throughout most of its range.

55.3 A Species May Eliminate a Competitor from Parts of Its Range
Aphytis lingnanensis displaced *A. chrysomphali* over most of its range within a decade.

55.9 A Batesian Mimic Falsely Advertises Danger
By mimicking a wasp, this ctenucid moth is protected from predators.

plants against viruses and bacteria. As we saw at the beginning of this chapter, most of the spices used in human cuisines have antibiotic properties. We can safely include them in our food because they are toxic only to microorganisms.

Digestibility-reducing compounds are secondary compounds that make plant tissues difficult to digest. The most common of these substances are tannins, which are present in the leaves of some herbaceous and most woody species. When an herbivore chews on a leaf, tannins are released from the intracellular compartments in which they are stored. They bind to proteins in the leaf and to the herbivore's digestive enzymes, reducing the ability of the herbivore to extract proteins from the leaves. Tannins may be present in such large quantities that waters draining from forests dominated by tanniferous plants are tea-colored.

Neutral and Beneficial Interspecific Interactions

During predator–prey and competitive interactions, one or both participants in the interaction are harmed. Amensal-

or more unpalatable species. All species in a Müllerian mimicry system, including the predators, benefit when inexperienced predators eat individuals of any of the species because the predators learn rapidly that all species of similar appearance are unpalatable. Some of the most spectacular tropical butterflies are members of Müllerian mimicry systems (Figure 55.10), as are many kinds of bees and wasps.

EVOLVED CHEMICAL DEFENSES ARE WIDE-SPREAD AMONG PLANTS. In addition to physical defenses against herbivores, such as tough leaves, hairs, or spines, most plant tissues also contain defensive chemicals called **secondary compounds**. There are two types of defensive secondary compounds: acute toxins and digestibility-reducing compounds.

Acute toxins disrupt herbivore metabolism. Some of these toxins, such as nicotine, interfere with the transmission of nerve impulses to muscles. Others are hallucinogens, which cause individuals that ingest them to have a seriously distorted view of their environment. Some toxins imitate insect hormones and prevent insects from completing metamorphosis. Still others are unusual amino acids that become incorporated into herbivore proteins and interfere with their functioning. Other acute toxins defend

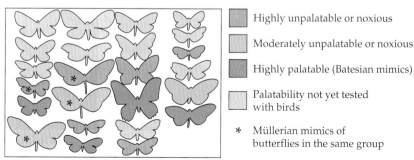

▨ Highly unpalatable or noxious

▧ Moderately unpalatable or noxious

▨ Highly palatable (Batesian mimics)

☐ Palatability not yet tested with birds

✳ Müllerian mimics of butterflies in the same group

55.10 Müllerian and Batesian Mimics
By converging in appearance, the unpalatable Müllerian mimics among these different species of Costa Rican butterflies and moths reinforce each other in deterring predators. The palatable Batesian mimics benefit because predators have learned to associate these color patterns with distasteful or noxious effects.

55.11 Danger Comes From Above
Shrubs and herbaceous plants are often damaged by branches falling from tall trees.

ism causes harm to one of the partners without affecting the other. In the other two types of interspecific interactions—commensalism and mutualism—neither partner is harmed, and one or both may benefit. We examine these interactions in the sections that follow.

In amensalism and commensalism, one participant is unaffected

Amensalisms, in which an individual harms another organism but is unaffected by the species it harms, are widespread and important in nature. Mammals, for example, create bare spaces around water holes. They benefit by drinking water, but not by trampling the plants they kill. Leaves and branches falling from trees damage smaller plants beneath them (Figure 55.11). The trees drop these old structures regardless of whether or not they damage other plants.

Commensalism benefits one partner but has no effect on the other. An example is the relationship between cattle egrets and grazing mammals. Cattle egrets are found throughout the tropics and subtropics. They typically forage on the ground around cattle or other large mammals, concentrating their attention near the heads and feet of the mammals, where they catch insects flushed by their hooves and mouths (Figure 55.12). Cattle egrets that forage close to grazing mammals capture more food for less effort than egrets that do not. The benefit to the egrets is clear; the mammals neither gain nor lose.

Mutualisms benefit both participants

Mutualisms are interactions that benefit both participants. Mutualistic interactions exist between plants and microorganisms, protists and fungi, plants and insects, and among plants. Animals also have mutualistic interactions with pro-

tists and with one another. As you learned in Chapter 27, the evolution of eukaryotic organisms is believed to be the result of mutualistic interactions between previously free-living prokaryotes and the cells they originally infected.

INTERGROUP MUTUALISMS.   Some of the most complex and ecologically important mutualisms are between members of different kingdoms or domains.

Most plants have beneficial associations with soil-inhabiting fungi called mycorrhizae that enhance the plant's ability to extract minerals from the soil (see Figure 30.16). And in the critical mutualistic relationship of some plants with nitrogen-fixing bacteria of the genus *Rhizobium* (discussed at length in Chapter 36), the bacteria receive protection and nutrients from their host plant while providing the host with usable nitrogen.

Lichens are compound organisms consisting of highly modified fungi that harbor cyanobacteria or green algae among their hyphae (see Figure 30.18). The fungi absorb water and nutrients from the environment and provide these as well as a supporting structure for the microorganisms, which in turn provide the fungi with the products of photosynthesis.

Animals have important mutualistic interactions with protists. For example, corals, some anemones, and some tunicates gain most of their energy from photosynthetic protists that live within their tissues. In exchange, they provide the protists with nutrients from the small animals they capture (see Figure 4.16c).

Termites have nitrogen-fixing protists in their guts that help them digest the cellulose in the wood they eat. Young termites must acquire their protists by eating the feces of other termites. If prevented from doing so, they soon die. The protists are provided with a suitable environment in which to live and an abundant supply of cellulose.

55.12 Commensalism Benefits One Partner
Cattle egrets catch more insects with less effort when they forage around large grazing mammals, such as cape buffalos. The buffalos are neither harmed nor helped by the egrets.

55.13 A Plant–Animal Mutualism
Some *Acacia* species have large, swollen, hollow thorns in which ants build their nests.

thorns in which ants of the genus *Pseudomyrmex* construct their nests and raise their young (Figure 55.13). These ants live only on acacias. They feed on nectar that the trees produce at the bases of their leaf petioles and on special nutritive bodies on the leaves. The ants attack and drive off leaf-eating insects, eat the eggs and larvae of herbivorous insects, and even bite and sting browsing mammals. They also cut back the tips of other plants, particularly vines that grow over their host tree. The ants get room and board; the plants get protection against both predators and competitors. Experiments in which acacias are deprived of their ants demonstrate the amount of protection the ants provide (Figure 55.14).

Many angiosperms depend on animals to transport both their pollen and their seeds. The plants benefit from pollination mutualisms by having their pollen carried to other conspecific plants and by receiving pollen to fertilize their ovules. Animals benefit by obtaining food in the form of nectar and pollen (Figure 55.15*a*). Plants also benefit from having their seeds dispersed to sites where they are more likely to germinate and survive than directly under the parent plant (Figure 55.15*b*). Animal dispersers benefit by eating the nutritious fruits surrounding the seeds. The plants pay a price for the benefits they receive: The energy and materials a plant uses to produce nectar, fruits, and other rewards for animals cannot be used for growth or seed production.

ANIMAL–ANIMAL MUTUALISMS. Many species of ants have mutualistic relationships with aphids. Ants "milk" these small, plant-sucking insects by stroking them with their forelegs and antennae. The aphids respond by secreting droplets of partly digested plant sap that has passed through their guts. In return, the ants protect the aphids from predatory wasps, beetles, and other natural enemies. The aphids lose nothing, because plant sap is high in sugar but low in amino acids. Thus, aphids inevitably ingest more sugar than they can use.

PLANT–ANIMAL MUTUALISMS. Terrestrial plants have many mutualistic interactions with animals. A complex mutualism between trees and ants that live in Central America illustrates the benefits of such interactions. Trees of the species *Acacia cornigera* have large, hollow

55.14 An Experiment Demonstrates the Benefits of Housing Ants
Acacia trees that housed ant colonies grew back faster than acacias without ants.

EXPERIMENT

Question: Do ants provide effective protection for acacia trees?

METHOD   Acacia trees were severely pruned. Ants were allowed to recolonize some trees as they regrew, but not others.

RESULTS

With ants   Without ants

Trees grown without ants were heavily attacked by other insects and regained their leaves very slowly.

Conclusion: Ants provide very effective protection for acacia trees.

(a) *Glossophaea* sp.

(b) *Bombycilla garrulus*

55.15 Plants Incur Costs to Attract Mutualists
(a) Some animal pollinators such as this long-tongued bat are attracted by rewards of nectar or pollen. (b) The nutritious pulp of fruits are attractive to many birds, such as this Bohemian waxwing.

Interactions between plants and their pollinators and seed dispersers are clearly mutualistic, but they are not purely mutualistic. Many seed dispersers are also seed predators that destroy some of the seeds they remove from plants. Some organisms that collect these rewards are not mutualists at all. Many animals visit flowers without transferring any pollen. Some of them cut holes to get to the nectar-producing regions at the base of the flowers. On the other hand, some plants exploit their pollinators. The flowers of certain orchids, for example, mimic female insects, enticing male insects to copulate with them (Figure 55.16). The male insects neither sire any offspring nor obtain any reward, but they transfer pollen between flowers, benefiting the orchid.

Coevolution of Interacting Species

The relationships between plants and their pollinators and seed disperers show how the evolution of species traits can be influenced by interactions with other species. Species that have mutually influenced one another's evolution are said to have **coevolved**. In **diffuse coevolution**, species traits are in-

fluenced by interactions with a wide variety of predators, parasites, prey, or mutualists. For example, most flowers are pollinated by a number of animal species, and most pollinators visit many species of flowers.

The traits of the fleshy fruits that surround many seeds are also the result of diffuse coevolution. Few fruits are adapted for dispersal by only a few species of animals. Most bird-dispersed fruits are red, or some combination of red and another color, and have no obvious odor to humans. Fruits dispersed by nonflying mammals, many of which lack color vision, are typically purple, are not highly visible to birds, and usually have a pleasant odor. Bat-dispersed fruits are typically green and have a fruity odor when ripe. They are inconspicuous during the day, but are easy for bats to detect at night. Many bird-dispersed fruits have unpleasant tastes to mammals.

Species-specific coevolution is much rarer than diffuse coevolution, but yucca plants and the moths of the genus *Tegeticula* that pollinate them have this kind of relationship. Female yucca moths lay their eggs only in the ovules of yucca flowers, and yucca flowers are pollinated only by *Tegeticula*. A female *Tegeticula* lays no more than five eggs in any one flower. After she has laid her eggs, she scrapes pollen from the flower's anthers, rolls it into a small ball, flies to another yucca plant, and places the pollen ball on the stigma of a flower before laying another batch of eggs. When the eggs hatch, the larvae burrow into the ovary and feed upon the developing seeds. Each yucca species has a specific moth species associated with it (Figure 55.17).

One feature of the coevolved relationship between *Tegeticula* and *Yucca* is surprising: Why do female moths lay so few eggs per flower? Wouldn't a female moth that laid more than five eggs per flower produce more surviving off-

55.16 Some Orchids Mimic Female Insects
The flowers of this orchid so closely resembles a female wasp that male wasps are tricked into attempting to copulate with them, as this male is doing. The orchid gets pollinated, but the wasp gets no reward.

Ancient events influence current distributions

Early biogeographers believed in an unchanging Earth that was too young to account for the diversity and distribution of life by any means except divine creation. Linnaeus, for example, believed that all organisms had been created in one place—which he called Paradise—from which they later dispersed. Indeed, because most people believed that the continents were fixed in their positions, the only way to account for current distributions was to invoke massive dispersal.

The notion that the continents might have moved was not seriously considered until 1912. Alfred Wegener, the German meteorologist who proposed the idea of continental drift, based his theory on several observations:

- the shapes of continents (the outlines of Africa and South America seemed to fit together like pieces of a puzzle)
- the alignment of mountain chains, rock strata, coal beds, and glacial deposits on different continents
- the distributions of organisms (the distributions of species in Africa and South America were hard to explain if one assumed that the continents had never moved)

When Wegener proposed his ideas, few scientists took them seriously, primarily because there were no known mechanisms to move continents and because no convincing geological evidence of such movements existed. As we learned in Chapter 20, geological evidence and plausible mechanisms were eventually discovered, and the broad pattern of continental movement is now clear.

About 280 million years ago, the continents were united to form a single land mass, Pangaea (see Figure 20.13). By the early Mesozoic era (about 245 million years ago), when the continents were still very close to one another, many groups of nonmarine organisms, including insects, freshwater fishes, frogs, and vascular plants, had already evolved. The ancestors of some organisms that live on widely separated continents today were probably present on those land masses when they were part of Pangaea.

By 100 mya, Pangaea had separated into northern (Laurasia) and southern (Gondwana) land masses, and the southern continents were drifting away from each other (see Figure 20.15). Eventually, continental drift, which continues today, brought India from Africa to southern Asia, Australia closer to Southeast Asia, and South America, which had drifted as an island for 60 million years, into contact with North America. Continental drift has thus influenced the evolution and mixing of species throughout the history of life on Earth.

Modern biogeographic methods

As the great age of Earth and the fact of evolution began to be understood, two groups of investigators developed new methods for generating testable hypotheses about geographic distributions. **Ecological biogeographers** study how current distributions are influenced by interactions among species and by interactions between species and their physical environments. They examine species interactions of the types discussed in Chapter 55 to explain patterns of local and regional species diversity. We will see some examples of their work later in this chapter.

The other group of investigators consists of **historical biogeographers**, who concentrate on longer time frames and larger spatial scales. They ask questions such as

- Where and when did evolutionary lineages originate?
- How did they spread?
- What do the present-day distributions of organisms tell us about their past histories?

An important technique developed by historical biogeographers was the transformation of taxonomic phylogenies into "area phylogenies" by substituting the taxa's geographic distributions for their names. Distribution patterns identified in this manner may suggest routes of dispersal or point to the splitting of biotas due to the appearance of barriers to dispersal. For example, by combining the phylogenetic relationships and current distribution pattern of the horse family, we can better understand why its members are found where they are and where past barriers probably influenced speciation events among them (Figure 57.1).

If we compare the distribution patterns of many evolutionary lineages, we may detect similarities and differences. Similarities suggest common responses to physical events, such as continental drift, mountain building, and sea level changes. Differences suggest that organisms in different lineages responded in different ways, or at different times, to past events, or that they dispersed in different ways and at different times.

The Role of History in Biogeography

Past events influence today's patterns of distribution. We can never know past events with complete certainty, but by using a variety of types of evidence, historical biogeographers can develop and test hypotheses in which they eventually have a high degree of confidence. As we have just seen, biogeographers often base their interpretations on phylogenies, which show the evolutionary relationships among the organisms in a lineage (see Chapter 23). Phylogenies are most useful to biogeographers if the approximate times of evolutionary and geographic separations of lineages can be estimated.

Biogeographers use several approaches to infer the approximate times of separation of taxa within a lineage. First, if a "molecular clock" has been ticking at a relatively constant rate, the degree of difference in the molecules of species will be strongly correlated with the length of time their lineages have been evolving independently (see Chapter 24).

Second, fossils can help to show how long a taxon has been present in an area and whether its members formerly lived in areas where they are no longer found. The fossil record is helpful, but it is always incomplete. The first and

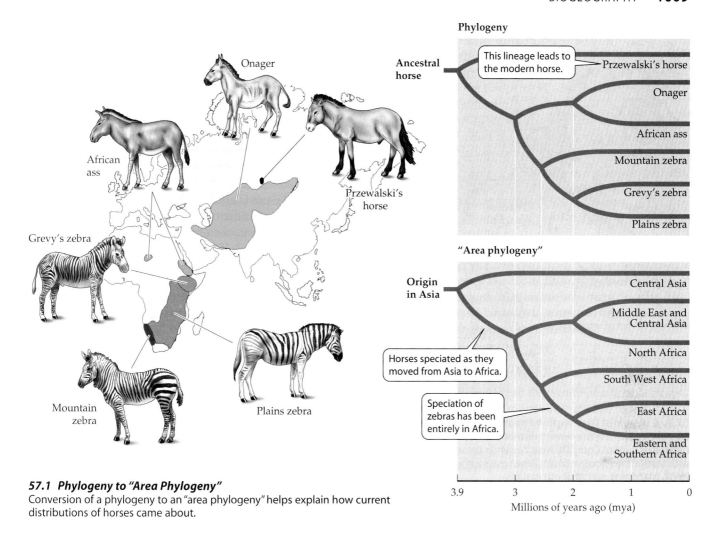

57.1 Phylogeny to "Area Phylogeny"
Conversion of a phylogeny to an "area phylogeny" helps explain how current distributions of horses came about.

last members of a taxon to live in an area are extremely unlikely to have become fossils that are discovered and described. Therefore, dates of colonization and extinction cannot be estimated accurately.

A third valuable source of information is the distributions of living species. Much more complete and extensive information can be gathered on current distributions than will ever be available from fossils. Much can be learned by examining the distribution patterns of many different groups of living organisms. Similarities in their distributions provide clues about past events that affected many of them.

Vicariance and dispersal can both explain distributions

As we have seen, a species may be found in an area either because it evolved there or because it dispersed to that area. But what if a species is found in two or more different places? What accounts for such a split distribution? There are two possibilities.

▶ A barrier may appear that splits a species' distribution. This is called a **vicariant event**, and no dispersal need be postulated to account for a split distribution.

▶ Members of a species may cross an already existing barrier and establish a new population. In this case, the species' split range must be attributed to dispersal.

By studying a single evolutionary lineage, a biogeographer may discover evidence suggesting that the distributions of its ancestors were influenced by a vicariant event, such as a change in sea level, mountain building, or continental movement. If that inference is correct, species in other lineages are likely to have been influenced by the same event and should therefore have similar distribution patterns.

Differences in distribution patterns among lineages indicate either that the lineages responded differently to the same vicariant events, that they separated at different times, or that they have very different dispersal abilities. By analyzing such similarities and differences among lineages, biogeographers can discover the relative roles of vicariant events and dispersal in determining today's distribution patterns.

Species, genera, and families found in only one place are said to be **endemic** to that location. As far as we know, all species are endemic to Earth. Some species are endemic to one continent. Others are restricted to very small areas, such as tiny islands or single mountaintops. Because a species may disperse widely and then die out where it originated,

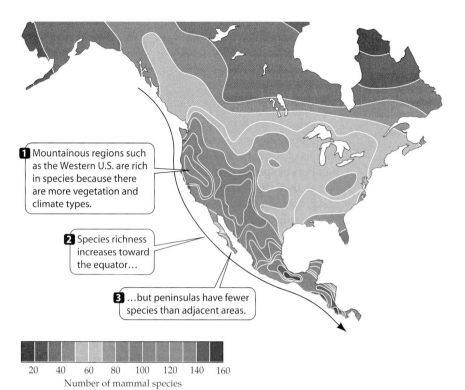

57.8 Latitudinal Gradient of Species Richness of North American Mammals
Lines on the figure connect regions with equal numbers of species. An increase in species richness toward the equator typifies many other taxa, such as birds, amphibians, and trees, as well as mammals.

1 Mountainous regions such as the Western U.S. are rich in species because there are more vegetation and climate types.

2 Species richness increases toward the equator…

3 …but peninsulas have fewer species than adjacent areas.

20   40   60   80   100 120 140 160
Number of mammal species

They found that rates of colonization of the islands by arthropods were very high. Within a year the fumigated islands had about their original number of species, and each census revealed considerable turnover in the number of species present. Both results support the island biogeographic model.

Species richness varies latitudinally

A nearly universal pattern in the distribution of species is that more species live in tropical than in high-latitude regions. Figure 57.8 shows the latitudinal gradient in mammal species richness in North and Central America. Similar patterns exist for birds, frogs, and trees, and for many marine taxa.

The figure also shows two other general patterns of species richness. First, more species are found in mountainous regions than in relatively flat areas because more vegetation types and climates exist within topographically complex regions. Second, species richness declines on peninsulas, such as Florida and Baja California, probably because colonization is possible from only one direction.

Terrestrial Biomes

Another way in which ecologists describe the distribution patterns of organisms is by classifying ecosystems. They apply the name **biome** to a major ecosystem type that differs from other types in the structure of its dominant vegetation. The vegetation of a biome has a similar appearance wherever on Earth that biome is found, but the plant species in these communities, despite their physical similar-

ities, may not be evolutionarily closely related. Although biomes are named for and identified by their characteristic plants, sometimes supplemented by their location or climate, each biome contains many other kinds of organisms. The geographic distribution of biomes is shown in Figure 57.9.

Biomes are identified by their distinctive climates and dominant plants

Because climate plays a key role in determining which types of plants live in a given place, the distribution of biomes on Earth is strongly influenced by annual patterns of temperature and rainfall. In some biomes, such as temperate deciduous forest, precipitation is relatively constant throughout the year, but temperature varies strikingly between summer and winter. In other biomes, both temperature and precipitation change seasonally. In certain biomes, such as tropical rainforest, temperatures are nearly constant, but rainfall varies seasonally.

In the tropics, where seasonal temperature fluctuations are small, annual climatic cycles are dominated by wet and dry seasons. In general, the number of months during which a region is close to the intertropical convergence zone (and hence receives rainfall) increases toward the equator (see Chapter 56). The intertropical convergence zone shifts latitudinally in a seasonally predictable way, resulting in a characteristic latitudinal pattern of distribution of rainy and dry seasons in tropical and subtropical regions (Figure 57.10).

Pictures and graphs capture the essence of terrestrial biomes

It is easiest to grasp the similarities and differences among terrestrial biomes by means of a combination of photographs and graphs of temperature, precipitation, and biological activity, supplemented by a brief description of the species richness and other attributes of those biomes. We use this method in the following pages to describe the major terrestrial biomes of the world.

▶ Each biome is represented by two photographs that show either the biome at different times of year or representatives of the biome in different places on Earth.

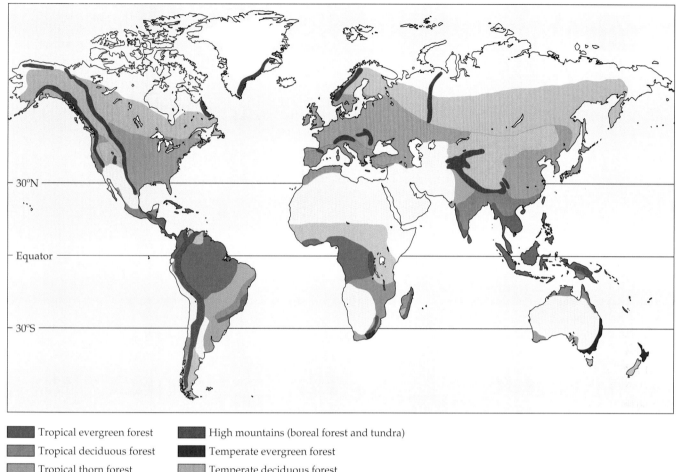

▮ Tropical evergreen forest		▮ High mountains (boreal forest and tundra)	
▮ Tropical deciduous forest		▮ Temperate evergreen forest	
▮ Tropical thorn forest		▮ Temperate deciduous forest	
▮ Savanna		▮ Boreal forest	
▮ Hot desert		▮ Arctic tundra	
▮ Chaparral		▮ Temperate grassland	
▮ Cold desert		▮ Polar ice cap	

57.9 Biomes Have Distinct Geographic Distributions
Compare these biomes with the patterns of net primary production shown in Figure 56.5.

▶ The first set of graphs plots seasonal patterns of temperature and precipitation at a typical site in the biome.

▶ Other graphs show how active different kinds of organisms are during the year. Levels of biological activity (shown by the width of horizontal bars) change either because resident organisms become more active (produce leaves, come out of hibernation, hatch, or reproduce) or because organisms migrate into and out of the biome. (The patterns shown are for the Northern Hemisphere; for high-latitude biomes, patterns in the Southern Hemisphere are 6 months out of phase with those illustrated.)

▶ A small box describes the dominant growth forms of plants in the biome and patterns of species richness there.

These descriptions are very general; they cannot capture the variation that exists within each biome.

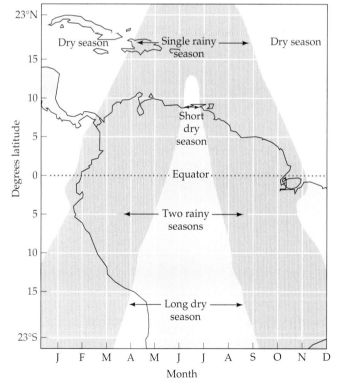

57.10 Rainy and Dry Seasons Change with Latitude
In the tropics and subtropics, which months are rainy and which are dry is highly predictable based on the region's latitude.

TUNDRA

Temperature

20°C is a "comfortable" 68°F.

0°C is the freezing point of water (=32°F).

Upernavik, Greenland 73°N

Winter is very cold and long.

Summer is cool and short.

Range 28°C

Precipitation

5 cm equals just over 2 inches.

Annual total: 23 cm

Biological Activity

Photosynthesis

Flowering

Fruiting

Mammals

Birds

Insects

Soil Biota

Community Composition

Dominant Plants
Perennial herbs and small shrubs

Species Richness
Plants: Low; higher in tropical alpine
Animals: Low; many birds migrate in for summer; a few species of insects abundant in summer

Soil Biota
Few species

Arctic tundra: Northwest Territories, Canada

Tropical alpine tundra: Teleki Valley, Mt. Kenya

Tundra is found at high latitudes and in high mountains

The **tundra** biome is found at high latitudes in the Arctic, Antarctic, and high in mountains at all latitudes (where it is called **alpine tundra**). There are no trees; the vegetation is dominated by short perennial plants.

Permanently frozen soil—**permafrost**—underlies tundra vegetation. The top few centimeters of soil thaw during the short summers, when the sun shines 24 hours a day. Even though there is little precipitation, Arctic tundra is very wet because water cannot drain down through the frozen soil. Plants grow for only a few months each year. Most Arctic tundra animals either migrate into the area only for the summer or are dormant for most of the year.

BOREAL FOREST

Temperature

Winter is very cold and dry.

Summer is mild and humid.

Range 41°C

Ft. Vermillion, Alberta 58°N

°C: 15, 10, 5, 0, −10, −15, −20, −25

Jan — Jul — Dec

Precipitation

Annual total: 31 cm

cm: 5, 0

Jan — Jul — Dec

Biological Activity

Photosynthesis
Flowering
Fruiting
Mammals
Birds
Insects
Soil Biota

Jan — Jul — Dec

Community Composition

Dominant Plants
Trees, shrubs, and perennial herbs

Species Richness
Plants: Low in trees, higher in understory
Animals: Low, but with summer peaks in migratory birds

Soil Biota
Very rich in deep litter layer

Northern conifer forest, Gunnison National Forest, Colorado

Bryophytes and lichens on southern evergreens, Tasmania, southern Australia

Boreal forests are dominated by evergreen trees

The **boreal forest** biome is found equatorward from, or at lower elevations on temperate-zone mountains than, tundra. Winters are long and very cold, and summers are short (although often warm). The short summers favor trees with evergreen leaves because these trees are ready to photosynthesize as soon as temperatures warm in spring.

The boreal forests of the Northern Hemisphere are dominated by coniferous evergreen gymnosperms. In the Southern Hemisphere, the dominant trees are southern beeches (*Nothofagus*). Evergreen forests also grow along the west coasts of continents at middle to high latitudes, where winters are mild but very wet and summers are cool and dry. These forests are home to Earth's tallest trees.

Boreal forests have only a few tree species. The dominant animals—such as insects, moose, and hares—eat leaves. The seeds in the cones of conifers also support a fauna of rodents and birds.

TEMPERATE DECIDUOUS FOREST

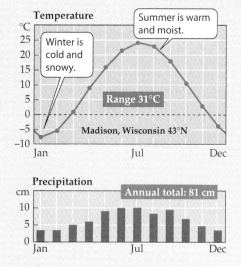

A Rhode Island forest in summer and...

...in winter

Temperature

Winter is cold and snowy.

Summer is warm and moist.

Range 31°C

Madison, Wisconsin 43°N

Precipitation

Annual total: 81 cm

Biological Activity

Photosynthesis

Flowering

Fruiting

Mammals

Birds

Insects

Soil Biota

Community Composition

Dominant Plants
Trees and shrubs

Species Richness
Plants: Many tree species in Southeast USA and East Asia, rich shrub layer
Animals: Rich; many migrant birds, richest amphibian communities on Earth, rich summer insect fauna

Soil Biota
Rich

Temperate deciduous forests change with the seasons

The **temperate deciduous forest** biome is found in eastern North America, eastern Asia, and western Europe. Temperatures in these regions fluctuate dramatically between summer and winter. Precipitation is relatively evenly distributed throughout the year. Deciduous trees, which dominate these forests, produce leaves that photosynthesize rapidly during the warm, moist summers and lose their leaves during the cold winters.

There are many more tree species here than in boreal forests. The temperate forests richest in species are in the southern Appalachian Mountains of the United States and in eastern China and Japan, areas that were not disturbed by Pleistocene glaciers. Many birds migrate into this biome in summer, when insects are abundant.

TEMPERATE GRASSLANDS

Temperature

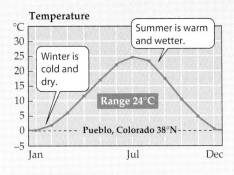

Winter is cold and dry.

Summer is warm and wetter.

Range 24°C

Pueblo, Colorado 38°N

Precipitation

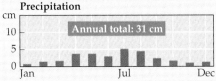

Annual total: 31 cm

Biological Activity

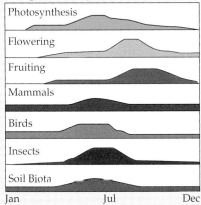

Photosynthesis

Flowering

Fruiting

Mammals

Birds

Insects

Soil Biota

Jan — Jul — Dec

Community Composition

Dominant Plants
Perennial grasses and forbs

Species Richness
Plants: Fairly high
Animals: Relatively few birds because of simple structure; mammals fairly rich

Soil Biota
Rich

Nebraska prairie in spring

The Veldt, Natal, South Africa

Temperate grasslands are widespread

The **temperate grassland** biome is found in many parts of the world, all of which are relatively dry much of the year. Most grasslands have hot summers and relatively cold winters. In some grasslands most of the precipitation falls in winter; in others the majority falls in summer. Such regions as the pampas of Argentina, the veldt of South Africa, and the Great Plains of the United States are components of the temperate grassland biome. Most of this biome has been converted to agriculture.

Grasslands are structurally simple, but they are rich in species of perennial grasses, sedges, and forbs. Grasses are well adapted to grazing and fire because they store much of their energy underground and quickly resprout after they are burned or grazed. As we saw in Chapter 56, many grasslands support large populations of grazing mammals.

COLD DESERT

Temperature

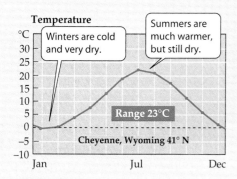

Winters are cold and very dry.

Summers are much warmer, but still dry.

Range 23°C

Cheyenne, Wyoming 41° N

Precipitation

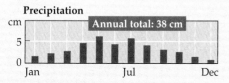

Annual total: 38 cm

Biological Activity

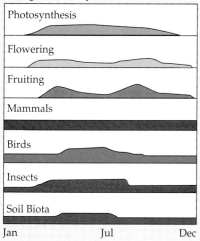

Photosynthesis

Flowering

Fruiting

Mammals

Birds

Insects

Soil Biota

Jan   Jul   Dec

Community Composition

Dominant Plants
Low stature shrubs and herbaceous plants

Species Richness
Plants: Few species
Animals: Rich in small mammals; low in all other taxa

Soil Biota
Poor in species

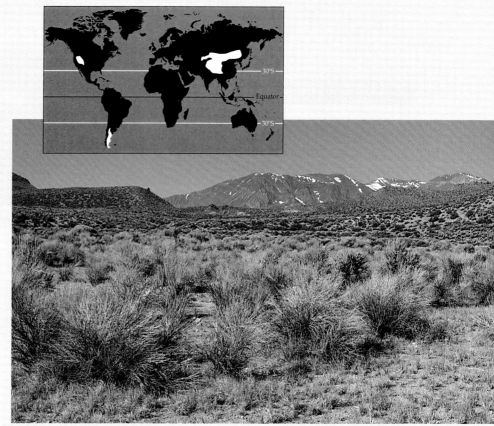

Sagebrush steppe near Mono Lake, California

Los Glacieres National Park, Argentina

Cold deserts are high and dry

The **cold desert** biome is found in dry regions at middle to high latitudes, especially in the interiors of large continents. Cold deserts also are found at fairly high altitudes in the rain shadows of mountain ranges. Seasonal changes in temperature are great.

Cold deserts are dominated by a few species of low-growing shrubs. The surface layers of the soil are recharged with moisture in winter, and plant growth is concentrated in spring. Cold deserts are relatively poor in species in most taxonomic groups, but the plants of this biome tend to produce large numbers of seeds, supporting a rich fauna of seed-eating birds, ants, and rodents.

HOT DESERT

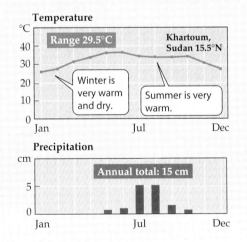

Anzo Borrego Desert, California

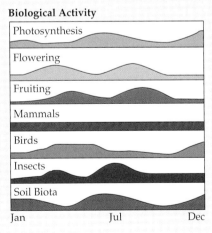

Rainbow Valley in the desert of central Australia

Temperature

°C

Range 29.5°C

Khartoum, Sudan 15.5°N

Winter is very warm and dry.

Summer is very warm.

Jan   Jul   Dec

Precipitation

cm

Annual total: 15 cm

Jan   Jul   Dec

Biological Activity

Photosynthesis

Flowering

Fruiting

Mammals

Birds

Insects

Soil Biota

Jan   Jul   Dec

Community Composition

Dominant Plants
Many different growth forms

Species Richness
Plants: Fairly high; many annuals
Animals: Very rich in rodents; richest bee communities on Earth; very rich in reptiles and butterflies

Soil Biota
Poor in species

Hot deserts form around 30° latitude

The **hot desert** biome is found in two belts, centered around 30°N and 30°S latitudes, respectively. These are the regions where dry air descends, warms, and picks up moisture (see Chapter 56). The driest large regions within this biome are in the center of Australia and the middle of the Sahara Desert of Africa.

Except in these driest regions, hot deserts have a richer and structurally more diverse vegetation than cold deserts. Succulent plants that store large quantities of water in their stems are conspicuous. Annual plants germinate and grow when rain falls. Pollination and seed dispersal by animals are common. Rodents and ants are often remarkably abundant, and lizards and snakes typically are rich in species and common.

CHAPARRAL

Temperature

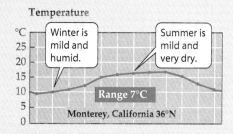

Winter is mild and humid.

Summer is mild and very dry.

Range 7°C

Monterey, California 36°N

Precipitation

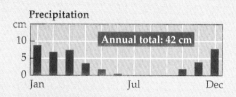

Annual total: 42 cm

Biological Activity

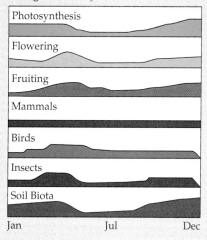

Photosynthesis

Flowering

Fruiting

Mammals

Birds

Insects

Soil Biota

Community Composition

Dominant Plants
Low stature shrubs and herbaceous plants

Species Richness
Plants: Extremely high in South Africa and Australia
Animals: Rich in rodents and reptiles; very rich in insects; especially bees

Soil Biota
Moderately rich

Fynbos vegetation, Cape of Good Hope, South Africa

Mendocino, California

The chaparral climate is dry and pleasant

The **chaparral** biome is found on the west sides of continents at moderate latitudes, where cool ocean waters flow offshore. Winters in this biome are cool and wet; summers are hot and dry. Such climates are found in the Mediterranean region of Europe, coastal California, central Chile, extreme southern Africa, and southwestern Australia.

Chaparral is dominated by low-growing shrubs and trees that have tough, evergreen leaves. The shrubs carry out most of their growth and photosynthesis in early spring, which is when insects are active and birds breed. Annual plants are abundant, producing seeds that are deposited in soil "seed banks." This biome thus supports large populations of small rodents, most of which store seeds in underground burrows.

Chaparral vegetation is naturally adapted to survive periodic fires. Many shrubs of Northern Hemisphere chaparral produce bird-dispersed fruits that ripen in the late fall, when large numbers of migrant birds arrive from the north.

THORN FOREST and TROPICAL SAVANNA

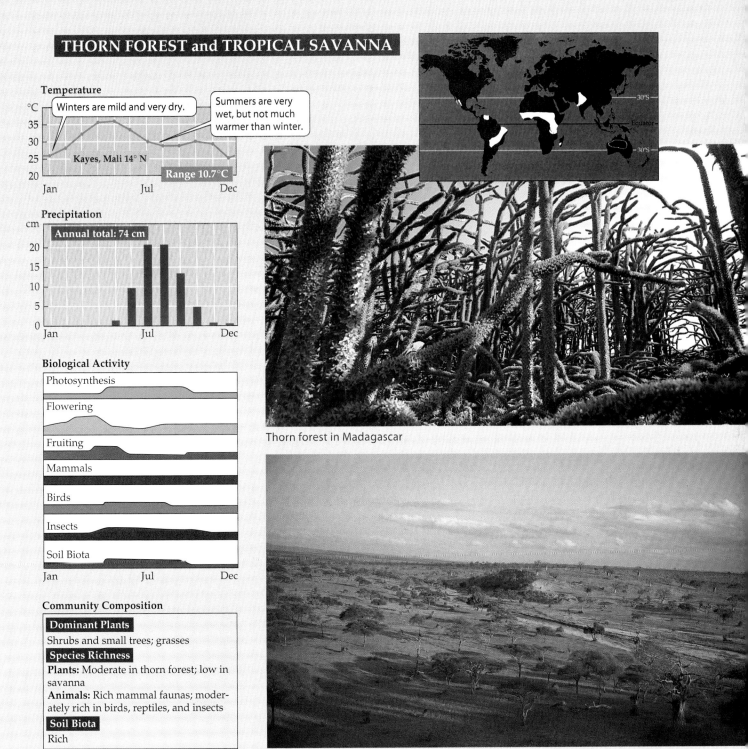

Temperature

°C
- Winters are mild and very dry.
- Summers are very wet, but not much warmer than winter.

35
30
25   Kayes, Mali 14° N
20   Range 10.7°C

Jan   Jul   Dec

Precipitation

cm
Annual total: 74 cm
20
15
10
5
0
Jan   Jul   Dec

Biological Activity

- Photosynthesis
- Flowering
- Fruiting
- Mammals
- Birds
- Insects
- Soil Biota

Jan   Jul   Dec

Community Composition

Dominant Plants
Shrubs and small trees; grasses

Species Richness
Plants: Moderate in thorn forest; low in savanna
Animals: Rich mammal faunas; moderately rich in birds, reptiles, and insects

Soil Biota
Rich

Thorn forest in Madagascar

Savanna in Tanzania

Thorn forests and savannas have similar climates

Thorn forests are found on the equatorial sides of hot deserts. The climate is semiarid; little or no rain falls during the winter, but rainfall may be heavy during the summer. Thorn forests contain many plants similar to those found in hot deserts. The dominant plants are spiny shrubs and small trees. Members of the genus *Acacia* are common in thorn forests worldwide.

The dry tropical and subtropical regions of Africa, South America, and Australia have extensive areas of **savannas**—expanses of grasses and grasslike plants punctuated by scattered trees. The largest savannas are found in central and eastern Africa, where this biome supports huge numbers of grazing and browsing mammals that serve as prey for many large carnivores.

Grazers and browsers maintain the savannas. If savanna vegetation is not grazed, browsed, or burned, it typically reverts to dense thorn forest.

TROPICAL DECIDUOUS FOREST

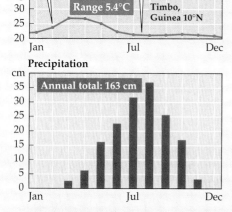

Temperature

Winter is very warm and dry.

Summer is warm and wet.

°C

Range 5.4°C

Timbo, Guinea 10°N

30
25
20

Jan   Jul   Dec

Precipitation

cm

Annual total: 163 cm

35
30
25
20
15
10
5
0

Jan   Jul   Dec

Biological Activity

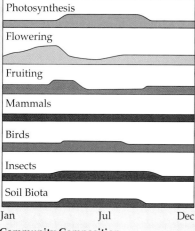

Photosynthesis

Flowering

Fruiting

Mammals

Birds

Insects

Soil Biota

Jan   Jul   Dec

Community Composition

Dominant Plants
Deciduous trees

Species Richness
Plants: Moderately rich in tree species
Animals: Rich mammal, bird, reptile, and amphibian communities; rich in insects

Soil Biota
Rich, but poorly known

Palo Verde National Park, Costa Rica, in the rainy season…

…and in the dry season

Tropical deciduous forests occur in hot lowlands

As the length of the rainy season increases toward the equator, thorn forests are replaced by **tropical deciduous forests**. These forests have taller trees and fewer succulent plants than thorn forests, and they are much richer in species. Most of the trees, except for those growing along rivers, lose their leaves during the long, hot dry season.

Many of them flower while they are leafless. This biome is very rich in species of both plants and animals.

The soils of the tropical deciduous forest biome are some of the best soils in the tropics for agriculture because they are less leached of nutrients than the soils of wetter areas. As a result, most tropical deciduous forests have been cleared for grazing cattle and growing crops.

TROPICAL EVERGREEN FOREST

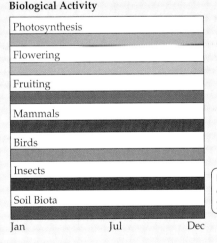

The exterior of lowland wet forest...   ...and its interior, Cocha Cashu, Peru

Temperature

°C

Warm and rainy all year.

Range 2.2°C   Equitos, Peru 3°S

Precipitation

cm

Annual total: 262 cm

Jan   Jul   Dec

Biological Activity

Photosynthesis

Flowering

Fruiting

Mammals

Birds

Insects

Soil Biota

Jan   Jul   Dec

Biological activity is essentially constant year round.

Community Composition

Dominant Plants

Trees and vines

Species Richness

Plants: Extremely high
Animals: Extremely high in mammals, birds, amphibians, and arthropods

Soil Biota

Very rich but poorly known

Tropical evergreen forests are rich in species

Tropical evergreen forests are found in equatorial regions where total rainfall exceeds 250 cm annually. This biome is the richest of all in species of both plants and animals, with up to 500 species of trees per km². Many of these species are rare. Food webs are extremely complex.

Tropical evergreen forests have the highest overall productivity of all terrestrial ecological communities. However, most mineral nutrients are tied up in the vegetation. The soils usually cannot support long-term agriculture.

On the slopes of tropical mountains, temperature decreases about 6° for each 1,000 m of elevation. The trees are shorter than lowland tropical trees. Their leaves are smaller, and there are more *epiphytes*—plants that grow on other plants and derive their nutrients and moisture from air and water rather than soil. Epiphytes thrive in tropical mountain forests where clouds form, bathing the forest in moisture.

Aquatic Biogeography

Three-fourths of Earth's surface is covered by water, most of it in the oceans. Earth's oceans form one large, interconnected water mass with no obvious barriers to dispersal. Fresh waters, in contrast, are divided into river basins and thousands of relatively isolated lakes. For freshwater organisms that cannot survive out of water, terrestrial habitats are barriers to dispersal. However, some aquatic species have flying adults that can disperse widely among water bodies. Others have windborne, desiccation-resistant spores and seeds. Still others are small enough to be transported by means such as mud on the feet of birds. Many freshwater taxa that are capable of dispersing across terrestrial barriers are distributed widely over several continents.

Freshwater ecosystems have little water but many species

Although only about 2.5 percent of Earth's water is found in ponds, lakes, and streams, about 10 percent of all aquatic species live in freshwater habitats. Prominent among these are the more than 25,000 species of insects that have at least one aquatic stage in their life cycle. Most commonly, eggs and larvae are aquatic and adults have wings. Adults of some of these insects, such as dragonflies, are powerful flyers, but adults of mayflies and some other species are weak flyers, desiccate rapidly in air, and live no longer than a few days. As you would expect, oceanic islands have no or very few species of these weak flyers.

Similarly, fishes unable to live in salt water can disperse only within the connected streams and lakes of a river basin. Most families of freshwater fishes are restricted to a single continent. Those families with species distributed on both sides of major saltwater barriers are believed to be ancient lineages whose ancestors were distributed widely in Laurasia or Gondwana.

Marine biogeographic regions are determined primarily by water temperature and nutrients

Ocean water moves in great circular patterns—clockwise in the Northern Hemisphere and counterclockwise in the Southern Hemisphere (see Figure 56.3) These movements disperse organisms with limited swimming abilities. Nevertheless, most marine organisms have restricted ranges, indicating that important environmental limits to their distributions exist in the oceans.

We can divide the oceans into zones based on sharp horizontal and vertical environmental gradients (Figure 57.11). At all depths, the bottom of the ocean is called the **benthic zone**, and the open water column is called the **pelagic zone**. The ocean floor below the level of sunlight penetration is called the **abyssal zone**. The coastal zone from the uppermost limits of tidal action down to the depth where

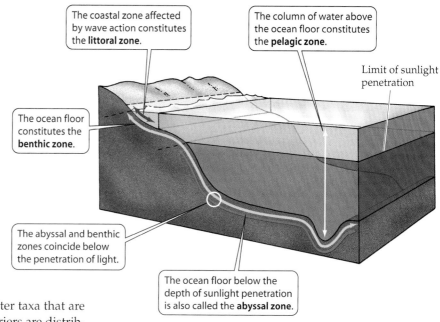

The coastal zone affected by wave action constitutes the **littoral zone**.

The column of water above the ocean floor constitutes the **pelagic zone**.

Limit of sunlight penetration

The ocean floor constitutes the **benthic zone**.

The abyssal and benthic zones coincide below the penetration of light.

The ocean floor below the depth of sunlight penetration is also called the **abyssal zone**.

57.11 Zones of the Ocean
Zones of the ocean are shown schematically in relation to depth and sunlight penetration.

the water is thoroughly stirred by wave action is called the **littoral zone**.

Water temperatures, hydrostatic pressures, and food supplies all change with depth and distance from shore, influencing biotic distributions. Food is scarce, for example, in the permanently dark, cold waters of the deep sea. Living successfully in different zones of the ocean requires different physiological tolerances and morphological attributes. Not surprisingly, even though many organisms can disperse between these zones, organisms from one zone survive poorly if they attempt to live in another.

Ocean temperatures are barriers to colonization because many marine organisms are well adapted to only relatively narrow temperature ranges. The main biogeographic divisions of the pelagic zone coincide with regions where the temperature of surface waters changes relatively abruptly as a result of horizontal and vertical ocean currents (Figure 57.12). These temperature changes, in combination with seasonal changes in the amount of daylight, determine the seasons of maximum primary production. Species of marine algae tend to be adapted to photosynthesize either in summer or in winter, but not during both seasons.

Because nutrients gradually sink to the ocean bottom, high concentrations of nutrients in the pelagic zone are restricted to areas where upwelling currents bring nutrient-rich bottom waters to the surface (see Figure 56.10). Most marine organisms that grow and reproduce well in nutrient-rich waters perform relatively poorly in nutrient-poor waters. Therefore, nutrient-rich waters typically have biotas that differ considerably from those of nutrient-poor waters in the same region.

Deep ocean waters are barriers to the dispersal of marine organisms that live only in shallow water. Eggs and larvae of marine organisms can be carried great distances

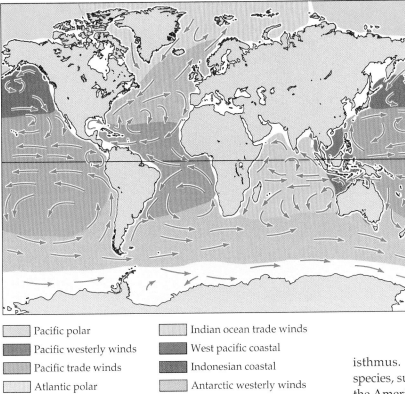

Pacific polar

Pacific westerly winds

Pacific trade winds

Atlantic polar

Atlantic westerly winds

Atlantic trade winds

Indian ocean trade winds

West pacific coastal

Indonesian coastal

Antarctic westerly winds

Antarctic polar

57.12 Pelagic Regions are Determined by Ocean Currents
The arrows represent ocean currents. Regions in which photosynthesis is maximized at different seasons are indicated by different colors.

demonstrated by the richness of reef-building corals in the intertidal and subtidal zones of isolated islands in the Pacific Ocean, which decreases with distance from New Guinea (Figure 57.13).

Marine vicariant events influence species distributions

Ancient vicariant events associated with continental drift do not influence current distributions of marine organisms, but more recent events have left biogeographic traces. An important recent vicariant event was the formation of the Isthmus of Panama about 3 million years ago. The isthmus separated the Pacific Ocean from the Caribbean Sea for the first time in more than 100 million years. Distinct marine biotas are now evolving on opposites sides of the isthmus. It forms a barrier to the dispersal of Pacific species, such as sea snakes, which reached the west coast of the Americas after the isthmus formed (Figure 57.14). Currently the fresh waters of Gatun Lake form a barrier to the dispersal of marine organisms through the Panama Canal. If a sea-level canal were constructed across the isthmus, poisonous sea snakes and other marine organisms would be able to disperse into the Caribbean.

by ocean currents, but the distance they can disperse is determined in large part by the duration of the larval life span. Relatively few species have eggs and larvae that survive long enough to disperse across wide barriers of deep water and settle in new areas. The effect of these barriers is

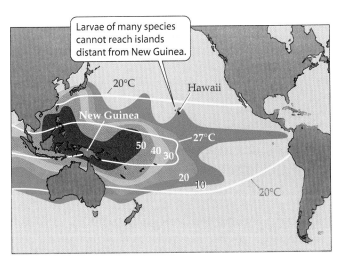

57.13 Generic Richness of Reef-Building Corals Declines with Distance from New Guinea
The lines connect areas with equal numbers of genera. Since temperature also limits the range of these species, the 20°C and 27°C mean annual temperature isotherms are also shown.

Pelamis platurus

57.14 A Block to Dispersal
The existence of the Isthmus of Panama prevents poisonous sea snakes from entering the Caribbean Sea from the Pacific Ocean.

Biogeography and Human History

The distributions of land masses and species on Earth have had a strong influence on human history. The Old World happened to have a large number of species of plants and animals that were suitable for domestication. Eurasia was home to 39 species of large-seeded grasses, many more than were found in Africa or the Americas. Eurasia had 72 species of large mammals, compared with 51 in sub-Saharan Africa and 24 in the New World. Thirteen large mammal species, including pigs, horses, cattle, sheep, goats, and camels, were domesticated in Eurasia. None were domesticated in Africa, and only one, the llama, in the Americas.

To be amenable to domestication, large mammals needed to have three important social characteristics: They needed to live in herds, have well-developed male dominance hierarchies, and be nonterritorial. These traits enabled humans to tame the animals, exert dominance over them, and keep them in herds. All the large mammals of Africa lacked one or more of these traits. They have never been domesticated. The domesticated large mammals found in Africa today all came from Asia.

Domestication of large mammals had other important influences on human history. Many human diseases, such as smallpox and measles, were acquired from domesticated mammals. Eurasian people acquired immunity to these diseases. People on other continents did not. Thus when Europeans colonized the New World, they brought with them diseases that devastated the indigenous people—who transmitted no fatal diseases to the Europeans in turn. In addition, Europeans had horses, a domesticated mammal capable of carrying a person at high speeds. Horses have played a major role in human history, because cultures with horses have readily conquered cultures without them.

In the Old World, most mountain ranges are oriented in an east–west direction. Therefore, dispersal of people and their domesticated plants and animals was relatively easy, and dispersing individuals remained within climates with similar temperatures and day lengths. Humans dispersed into the New World only recently, across the high-latitude Bering land bridge. They brought with them no domesticated plants or animals. North America, as we have seen, had few species of grasses with large seeds. Maize, the grass that came to dominate American agriculture, was difficult to domesticate. Its eventual spread northward from its center of domestication in Mexico was possible only after extensive genetic changes that adapted the plants to the very different day lengths and climates of temperate North America.

Chapter Summary

▶ Biogeography is the science that attempts to explain patterns in the distribution of life on Earth.

Why Are Species Found Where They Are?

▶ If a species occupies an area, either it evolved there, or it evolved elsewhere and dispersed to that area.

▶ If a species is not found in a particular area, either it evolved elsewhere and never dispersed to that area, or it was once present in that area but no longer lives there.

▶ Continental drift has influenced the distributions of organisms throughout Earth's history.

▶ Ecological biogeographers seek to understand how current ecological interactions influence where species are found today.

▶ Historical biogeographers attempt to determine the influence of past events on today's patterns of species distributions.

▶ Historical biogeographers often analyze species distributions by converting phylogenies into "area phylogenies." **Review Figure 57.1**

The Role of History in Biogeography

▶ Biogeographers use the parsimony principle when they attempt to explain distribution patterns. **Review Figure 57.2**

▶ Vicariance and dispersal events have both influenced current distributions. **Review Figure 57.3**

▶ Animal biogeographers divide Earth into six major biogeographic regions. Plant biogeographers recognize two additional regions. **Review Figure 57.4**

Ecology and Biogeography

▶ Ecological biogeographers test theories that explain the numbers of species in different communities, how species disperse, and the effectiveness of barriers to movement.

▶ The island biogeographic model, which predicts the equilibrium species richness on islands, has been tested by examining patterns of distribution and by performing experiments. **Review Figures 57.5, 57.6; Table 57.1**

▶ The number of species in most lineages increases from polar to tropical regions. **Review Figure 57.8**

Terrestrial Biomes

▶ Terrestrial biomes are major ecosystem types that differ from one another in the structure of their dominant vegetation.

▶ The distribution of biomes on Earth is strongly influenced by annual patterns of temperature and rainfall. **Review Figures 57.9, 57.10**

▶ The major terrestrial biomes are tundra, boreal forest, temperate deciduous forest, temperate grassland, cold desert, hot desert, chaparral, thorn forest, savanna, tropical deciduous forest, and tropical evergreen forest.

Aquatic Biogeography

▶ No absolute barriers to the movement of marine organisms exist within the oceans, but most marine organisms have restricted ranges.

▶ Conditions in the oceans change dramatically with depth and sunlight penetration. **Review Figure 57.11**

▶ Boundaries between many pelagic regions are determined by ocean currents. **Review Figure 57.12**

▶ Species that live in shallow waters disperse with difficulty across wide deep-water barriers. **Review Figure 57.13**

Biogeography and Human History

▶ The distributions of plants, animals, and continents have exerted powerful influences on human history.

For Discussion

1. Horses evolved in North America, but subsequently became extinct there. They survived to modern times only

in Africa and Asia. In the absence of a fossil record, we would probably infer that horses originated in the Old World. Today, the Hawaiian Islands have by far the greatest number of species of fruit flies (*Drosophila*). Would you conclude that the genus *Drosophila* originally evolved in Hawaii and spread to other regions? Under what circumstances do you think it is safe to conclude that a group of organisms evolved close to where the greatest number of species live today?

2. The island biogeographic model we described incorporates almost nothing about the biology of the species involved. What traits of species should be incorporated into more realistic models of rates of colonization and extinction of species on islands?

3. Experiments to test theories of species richness are necessarily short-term. What long-term consequences of colonization and extinction are likely to be undetected by these experiments? How could they be studied?

4. In nearly every ecological community, the number of species present is much smaller than the number potentially available to colonize it. What inferences can be drawn from this pattern?

5. A well-known legend states that Saint Patrick drove the snakes out of Ireland. Give some alternative explanations, based on sound biogeographic principles, for the absence of indigenous snakes in that country.

6. Why are there so few species of trees in boreal forests? Why do few species of trees of boreal forests have animal-dispersed seeds?

7. What are some significant present-day human problems whose solutions involve biogeographic considerations? What kinds of biogeographic knowledge are most important for addressing each one?

8. Most of the world's flightless birds are either nocturnal and secretive (such as the kiwi of New Zealand) or large, swift, and well armed (such as the ostrich of Africa). The exceptions are found primarily on islands. Many flightless island species have become extinct with the arrival of humans and their domestic animals. What special biogeographic conditions on islands might permit the survival of flightless birds? Why has human colonization so often resulted in the extinction of such birds? The power of flight has been lost secondarily in representatives of many groups of birds and insects; what are some possible evolutionary advantages of flightlessness that might offset its obvious disadvantages?

58 Conservation Biology

When Polynesian people settled in Hawaii about 2,000 years ago, they exterminated—probably by overhunting—at least 39 species of land birds. Among them were 7 species of geese, 2 species of flightless ibises, a sea eagle, a small hawk, 7 flightless rails, 3 species of owls, 2 large crows, a honeyeater, and at least 15 species of finches.

No people lived in New Zealand until about 1,000 years ago, when the Maori colonized the islands. Hunting by the Maori caused the extinction of 13 species of flightless moas, some of which were larger than ostriches.

When humans arrived in North America over the Bering land bridge, about 20,000 years ago, they encountered a rich fauna of large mammals. Most of those species were exterminated within a few thousand years. A similar extermination of large animals followed the human colonization of Australia, about 40,000 years ago. At that time Australia had 15 genera of marsupials larger than 50 kg, a genus of gigantic lizards, and a genus of heavy, flightless birds. All the species in 13 of those 15 genera had become extinct by 18,000 years ago.

The accelerating pace of human-caused extinctions of species, which raises serious concerns about the future of biological diversity on Earth, has led to the rapid development of the applied discipline of **conservation biology**—the scientific study of how to preserve the diversity of life. Conservation biologists study the causes of endangerment and extinction and develop methods to help preserve genes, species, communities, and ecosystems. The science of conservation biology draws heavily on concepts and knowledge from population genetics, evolution, ecology, biogeography, wildlife management, economics, and sociology. In turn, the needs of conservation are stimulating new research in those fields.

In this chapter we will see how biologists estimate rates of species extinction and the causes of endangerment, and learn how management plans can be used to reduce extinction rates and restore endangered species and communities to states in which they are likely to persist for a long time.

Extinct Flightless Hawaiian Birds
This artist's reconstruction of a flightless Hawaiian goose shows one of the many bird species exterminated by the Polynesian settlers of the islands.

Estimating Current Rates of Extinction

Most human activities that are causing extinctions are not new, but there are many more of us doing those things than ever before (see Figure 54.16). We have also added the results of our advancing technology, such as pesticides and climate change, to the array of pressures created by human activities.

We do not know how many species will become extinct during the next 100 years, first, because we do not know how many species there are on Earth, and second, because the number of extinctions will depend both on what we do and on unexpected events. However, several methods exist for estimating probable rates of extinction resulting from human actions. In this section we will discuss how conservation biologists estimate current rates of extinction and identify species at risk of extinction.

Species–area relationships are used to estimate extinction rates

When we described the island biogeographic model in Chapter 57, we saw that the number of species on an island increases with the size of the island (see Figure 57.6). This **species–area relationship** can be applied to habitat patches

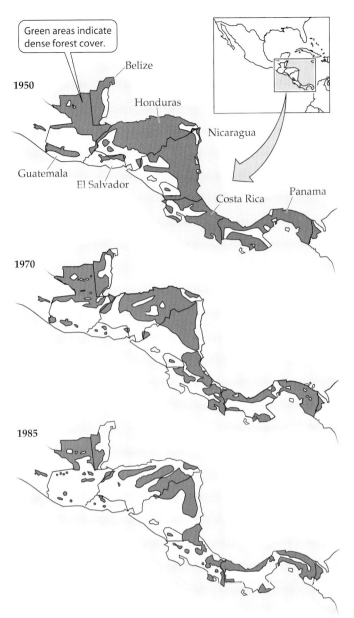

58.1 Deforestation Rates Are High in Tropical Forests
Central America provides an example of the high rate of destruction of tropical forests that has taken place in recent years. As the forests are lost, so are the many species that live in them.

on mainlands as well. Conservation biologists often use the well-established relationship between the size of an area and the number of species present to estimate numbers of species extinctions resulting from habitat destruction.

The rate at which tropical forests are being logged and converted to cropland and pasture is not precisely known, but it is currently very high (Figure 58.1). These forests are Earth's richest biomes, home to perhaps one-half of all the species on the planet. Calculations using the species–area relationship applied to tropical forests are far from exact, but can result in estimates of over 1 million species extinctions in the next few decades.

Even the lowest estimates of current extinction rates predict that at least 10 percent of Earth's species are likely to become extinct during the next two decades. Some estimates predict extinction of 50 percent of Earth's species

during the next 50 years. Extinction rates have been much higher on islands than in mainland areas during the past 400 years (Figure 58.2), but extinction rates on continents are also rising fast.

Population models are used to estimate risks of extinction

To estimate the risk that a population will become extinct, conservation biologists analyze information about interactions between a population's genetic variation, morphology, physiology, and behavior and its environment, both physical and biological. Although rarity itself is not always a cause for concern, species whose populations are shrinking rapidly usually are at risk. Species with only a few individuals confined to a small range are likely to become extinct because they can be eliminated by local disturbances such as fires, unusual weather, disease, and predators.

Both demographic and genetic information were used in assessing the risk of extinction of Furbish's lousewort, a plant that is restricted to the banks of the St. John River in northern Maine. The Furbish's lousewort population is divided into discrete subpopulations growing at separate sites. These subpopulations are found where periodic disturbance, often caused by spring ice scour when blocks of floating ice scrape the stream banks, prevent establishment of trees and shrubs that would outcompete the louseworts (Figure 58.3). Data on annual rates of survival, growth, and reproduction were gathered by following more than 6,000 individually marked plants between 1983 and 1986. Extinctions of subpopulations and foundings of new subpopulations were estimated by counting plants over the entire range of the species.

Although Furbish's lousewort depends on regular disturbance to suppress the growth of shrubs and trees, distur-

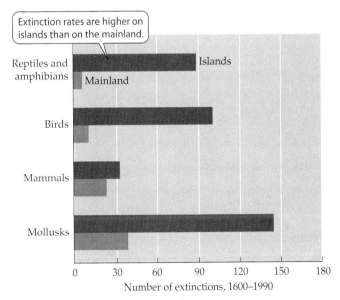

58.2 Extinctions Have Been High on Islands
Between 1600 and 1990, extinctions of terrestrial vertebrates and mollusks were much higher on islands than on continents.

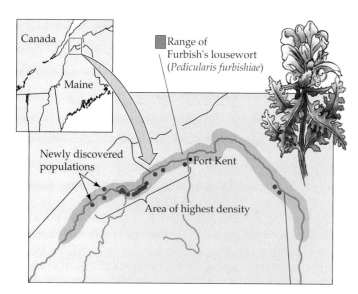

58.3 Furbish's Lousewort Exists as a Metapopulation
Small populations of Furbish's lousewort exist along the St. John River, where ice scour and bank slumping eliminate shrubs and trees. Dispersing individuals colonize newly disturbed sites.

bances can also eliminate subpopulations. Between 1983 and 1984, three of ten study subpopulations were completely destroyed by ice scour and bank slumping, and none of the disturbed sites were recolonized during succeeding years. Thus if disturbance rates were too high, extinction rates of subpopulations are likely to be higher than rates of establishment of new subpopulations. Under current disturbance regimes, the two rates are about equal. Local subpopulations became extinct and new subpopulations were founded at an annual rate of about 3 percent during the study.

The investigators concluded that the persistence of Furbish's lousewort depends on disturbance events at sites currently having subpopulations and also at sites that are currently unoccupied. Clearly a strategy that protected only sites with current subpopulations would not maintain the species for many years. Preserving Furbish's lousewort depends on maintaining disturbance regimes along an entire stretch of the St. John River. If the St. John River continues to flow naturally and spring ice scouring continues, Furbish's lousewort is likely to persist for a long time.

Why Do We Care about Species Extinctions?

We care about species extinctions in part because humans depend on other species in many ways. For example, more than half the medical prescriptions written in the United States contain a natural plant or animal product (Figure 58.4). The search for and use of such products from the living world has hardly begun. Many species may be eliminated by tropical forest destruction before we find out whether they might be sources of useful products.

Extinctions deprive us of opportunities to study and understand ecological relationships among organisms. The

more species are lost, the more difficult it will be to understand the rules that govern the structure and functioning of ecological communities.

We also derive enormous aesthetic pleasure from interacting with other organisms. Many people would consider a world with far fewer species as a less desirable one in which to live. Living in ways that cause the extinction of other species also raises serious moral and ethical issues that are receiving increased attention.

Ecosystem processes, as well as individual species, produce many benefits to humanity. Among them are the generation and maintenance of fertile soils, prevention of soil erosion, detoxification and recycling of waste products, regulation of hydrological cycles and the composition of the atmosphere, control of agricultural pests, pollination, and maintenance of the species richness upon which humanity depends. It is easy to list these ecosystem services, but to justify the allocation of scarce public resources to maintain them, we need quantitative estimates of their value.

A detailed study by economists, ecologists, and land managers in Western Cape Province, South Africa, has shown that an intensive program to eradicate invasive alien plants in the highlands of the region is a cost-effective way of maintaining a reliable regional supply of high-quality water. The native vegetation of these highlands is a species-rich community of shrubs, known as *fynbos* (pronounced "fainbos"). Fynbos can survive regular summer drought, nutrient-poor soils, and the fires that periodically sweep through the highlands (Figure 58.5a).

The fynbos-clad highlands provide about two-thirds of the Western Cape's water requirements. In addition, the flora is harvested for cut and dried flowers and thatching

Catharanthus roseus

58.4 Source of a Life-Saving Drug
A drug derived from the Madagascar rosy periwinkle has greatly increased the survival rate of children with leukemia. Other species that might be sources of drugs are being eliminated by deforestation on Madagascar.

(*a*) The fynbos of South Africa. (*b*) Stream flow from fynbos watersheds is inversely proportional to plant biomass. (*c*) A computer simulation of stream flows from watersheds that have and have not been invaded by trees from outside the region.

(*b*) **Stream flow from fynbos watersheds**

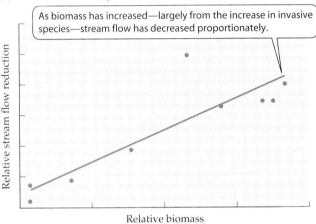

As biomass has increased—largely from the increase in invasive species—stream flow has decreased proportionately.

Relative stream flow reduction

Relative biomass

(*c*) **Computer simulation**

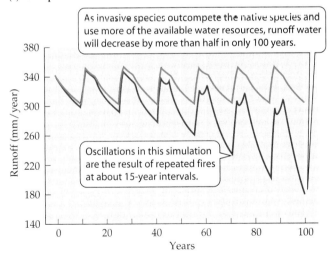

As invasive species outcompete the native species and use more of the available water resources, runoff water will decrease by more than half in only 100 years.

Oscillations in this simulation are the result of repeated fires at about 15-year intervals.

Runoff (mm/year)

Years

cause they are taller and grow faster than the native plants, the exotics increase the intensity and severity of fires. By transpiring larger quantities of water, they decrease stream flows to less than half the amount flowing from mountains covered with native plants (Figure 58.5*b,c*).

Removing the exotic plants by felling and digging out invasive trees and shrubs and managing fire is estimated to cost between $140 and $830 per hectare, depending on the densities of invasive plants. Annual follow-up operations will cost about $8 per hectare. The costs of alternative methods to replace the water lost from watersheds taken over by exotic plants are much higher. A sewage purification plant that would deliver the same volume of water as a well-managed watershed of 10,000 hectares would cost $135 million to build and $2.6 million per year to operate. Desalination of seawater would cost four times as much. Thus, the available alternatives would deliver water at a cost between 1.8 and 6.7 times more than the costs of maintaining natural vegetation in the watershed.

Modern industrial societies often favor technologically sophisticated methods of substituting for lost ecosystem services. The study of water resources in the Western Cape Region shows that simple but labor-intensive methods—cutting and burning—can be cheaper. In addition, they preserve other ecosystem values, such as tourism and commercial plant products.

Some ecosystem values, such as aesthetic benefits, cannot be replaced with technological inventions. Aesthetic benefits may contribute much to a country's economy. One of the largest sources of foreign income in Kenya is nature tourism. The loss of a single species probably would not reduce the flow of tourists to Kenya, but if elephants, rhinoceroses, lions, leopards, and buffalo were all to disappear, fewer people would pay the high price of a Kenyan vacation. Populations of these species can be maintained only if large tracts of the ecosystems in which they live are preserved.

Determining Causes of Endangerment and Extinction

Rare species are more likely to become extinct than common species. Species may be rare for any of several reasons. They may live in a habitat that is rare, such as desert lakes with high salt concentrations or caves (Figure 58.6). Another reason for rarity is trophic level—secondary carnivores are usually rare because so little energy is available to support their populations, as we saw in Chapter 56. Being rare increases a species' chance of becoming extinct, but common species can also become extinct.

grass. The combined value of these harvests in 1993 was about $19 million. Some of the income from tourism in the region comes from people who want to see the fynbos vegetation. About 400,000 people visit the Cape of Good Hope Nature Reserve each year, primarily to see the many unique plants.

During recent decades, a number of plants introduced into South Africa have invaded the fynbos highlands. Be-

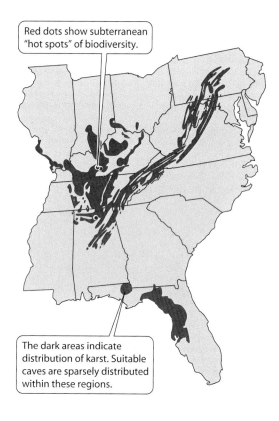

Red dots show subterranean "hot spots" of biodiversity.

The dark areas indicate distribution of karst. Suitable caves are sparsely distributed within these regions.

58.6 The Habitat of Cave Animals Is Patchy and Restricted
The map shows the distribution of subterranean limestone (karst) in which caves with running water are found. The caves that can actually be inhabited by cave animals covers a much smaller area.

An analysis by The Nature Conservancy revealed that habitat loss is the most important cause of endangerment of species in the United States (Figure 58.7), but that other factors, such as exotic species, pollution, overexploitation, and disease, also threaten species. In this section we will examine some of these major threats to species.

Habitat destruction and fragmentation are important causes of extinction today

The 6 billion humans that live on Earth today are fed, clothed, and housed by the agricultural and forestry industries, which convert natural ecological communities containing many species into highly modified communities dominated by one or a few species of plants. Within these communities, humans discourage the presence of other species by killing competing plants, bacteria, fungi, nematodes, insects and other arthropods, and vertebrates.

Although agricultural ecosystems have always harbored fewer species than the complex natural ecosystems they replaced, only recently have farmers planted large tracts of land in single crops. Traditional farmers planted many different crops together, maintaining some of the diversity that is key to the functioning of natural communities. Many species that cannot survive in intensive modern agricultural systems live in traditional agricultural ones. In traditional coffee plantations, for example, coffee bushes are grown in the shade of large trees (Figure 58.8*a*). These structurally rich plantations support populations of many species of birds, and few pesticides need to be applied to them. Some recently developed high-yielding strains of coffee, however, grow best in full sunlight. Pure plantations of these coffee bushes require heavy applications of pesticides and support almost no birds (Figure 58.8*b*).

Agriculture and forestry today are so extensive that more than 30 percent of all net terrestrial primary production is diverted for human use. All other species on Earth must survive on only two-thirds of the total global terrestrial production, and the fraction people divert is steadily increasing.

As natural habitats are progressively destroyed, the remaining patches increasingly become smaller and more isolated. Small habitat patches are qualitatively different from larger patches of the same habitat in ways that affect the

Habitat destruction is the primary cause of extinction and endangerment for freshwater aquatic organisms.

Small animals are vulnerable to habitat destruction because they often have small ranges and highly specialized requirements.

Freshwater mussels
Crayfishes
Amphibians
Freshwater fishes
Flowering plants
Conifers
Ferns
Tiger beetles
Dragonflies, damselflies
Reptiles
Butterflies, skippers
Mammals
Birds

Key:
Vulnerable
Imperiled
Critically imperiled
Presumed extinct

0   10   20   30   40   50   60   70   80
Percent of species

58.7 Proportion of U.S. Species Extinct or at Risk
The groups that are most endangered—mussels, crayfishes, amphibians, and fishes—live in fresh waters, a habitat that has been extensively destroyed and polluted.

(a)

(b)

58.8 The Way Coffee is Grown Affects Biodiversity
(a) A traditional coffee plantation with a canopy of trees supports many species of birds. (b) Few bird species live in plantations of sun-grown coffee.

survival of species. Small patches cannot maintain populations of species that require large areas, and they support only small populations of many of the species that can survive in them.

In addition, the fraction of a patch that is influenced by conditions in adjacent habitats—resulting in **edge effects**—increases rapidly as patch size decreases (Figure 58.9). Close to the edges of forest patches, for example, winds are stronger, temperatures are higher, humidity is lower, and light levels are higher than they are farther inside the forest. Species from surrounding habitats often colonize the edges of patches to compete with or prey upon the species living there.

Usually we do not know which organisms lived in an area before its habitats became fragmented. To address this problem, a major research project near Manaus, Brazil, was launched in an area of tropical forest before logging took place. The landowners agreed to preserve forest patches of certain sizes and locations (Figure 58.10a).

Biologists conducted censuses of those patches while the areas were still part of the continuous forest. Soon after the surrounding forest was cut, species began to disappear from the isolated patches. The first species to be eliminated were monkeys with large home ranges, such as the black spider monkey, the tufted capuchin, and the bearded saki, and antbirds that follow raiding army ant swarms to capture insects flushed by the ants (Figure 58.10b,c).

Species that become extinct in small fragments are unlikely to become reestablished because individuals dispersing from other locations are less likely to find isolated patches. Even if they find them, the patches may be too small to support their populations on a long-term basis. The persistence of species in small patches may be improved if the patches are connected by **corridors** of suitable habitat through which individuals can disperse.

The role of corridors in sustaining populations of species was studied by creating patches of mosses on the surface of a large rock. Eight experimental "landscapes" were established by scraping away the moss to create patches of equal size. Some patches were isolated, others were connected by narrow moss corridors, and still others were connected by pseudocorridors

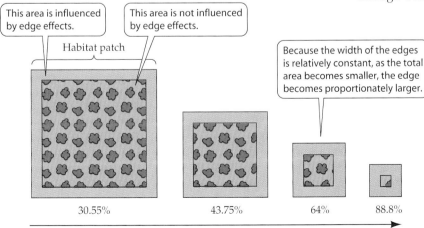

This area is influenced by edge effects.

This area is not influenced by edge effects.

Habitat patch

Because the width of the edges is relatively constant, as the total area becomes smaller, the edge becomes proportionately larger.

30.55%   43.75%   64%   88.8%

Increasing percentage of patch influenced by edge effects

58.9 Edge Effects
The smaller a habitat patch, the greater the proportion that is influenced by conditions in the surrounding environment.

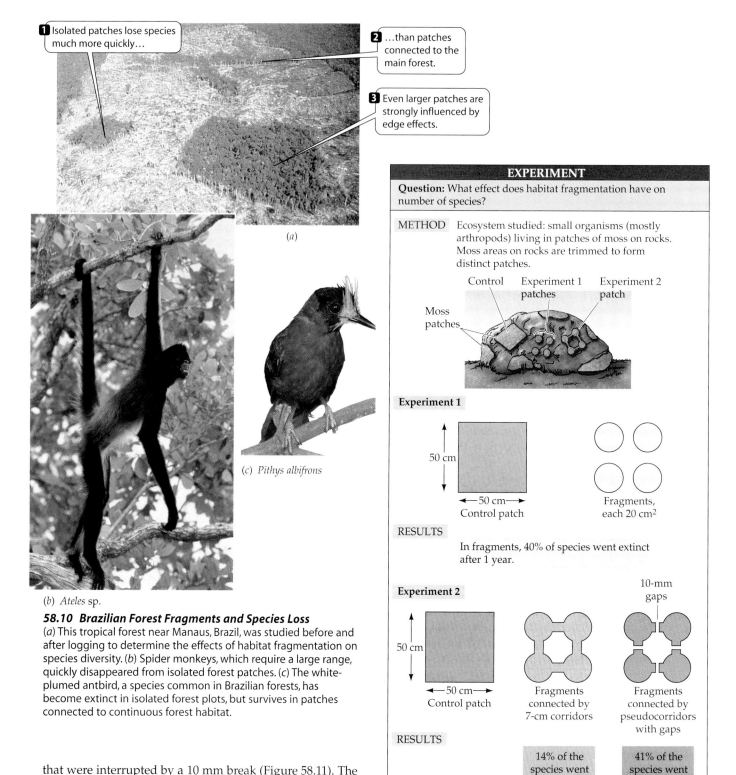

1 Isolated patches lose species much more quickly...

2 ...than patches connected to the main forest.

3 Even larger patches are strongly influenced by edge effects.

(a)

(c) *Pithys albifrons*

(b) *Ateles* sp.

58.10 Brazilian Forest Fragments and Species Loss
(a) This tropical forest near Manaus, Brazil, was studied before and after logging to determine the effects of habitat fragmentation on species diversity. (b) Spider monkeys, which require a large range, quickly disappeared from isolated forest patches. (c) The white-plumed antbird, a species common in Brazilian forests, has become extinct in isolated forest plots, but survives in patches connected to continuous forest habitat.

EXPERIMENT

Question: What effect does habitat fragmentation have on number of species?

METHOD Ecosystem studied: small organisms (mostly arthropods) living in patches of moss on rocks. Moss areas on rocks are trimmed to form distinct patches.

Control   Experiment 1 patches   Experiment 2 patch

Moss patches

Experiment 1

50 cm

←— 50 cm —→
Control patch

Fragments, each 20 cm²

RESULTS

In fragments, 40% of species went extinct after 1 year.

Experiment 2

50 cm

←— 50 cm —→
Control patch

Fragments connected by 7-cm corridors

10-mm gaps

Fragments connected by pseudocorridors with gaps

RESULTS

| 14% of the species went extinct after 6 months. | 41% of the species went extinct after 6 months. |

Conclusion: Even small barriers to dispersal can raise extinction rates.

58.11 An Experiment Demonstrates the Value of Corridors
A small-scale experiment in which patches were created by removing moss from a rock surface demonstrated that even small gaps between suitable habitats can reduce the number of species that persist in patches.

that were interrupted by a 10 mm break (Figure 58.11). The abundance and species richness of tiny arthropods were measured at 2-month intervals over the course of a year—a time period equivalent to many generations for most of the small animals living in the moss.

Remarkably, barriers as narrow as 10 mm were sufficient to greatly reduce the dispersal rates of arthropods between moss patches. Isolated patches and patches connected by pseudocorridors had only about 60 percent as many species of springtails (tiny, wingless insects) and mites as the patches connected by complete corridors.

58.12 Galápagos Tortoises Are Reared in Captivity
Conservationists at the Charles Darwin Research Station remove tortoise eggs from nests. When the eggs hatch, the young are reared in captivity until they are large enough to be invulnerable to predation by introduced pigs and rats. Populations of tortoises on some Galápagos islands would be extinct if they were not propagated in captivity.

cause by removing most of the algae from the water, thereby clarifying it, zebra mussels allow sunlight to penetrate more deeply into the water column. As a result, in areas of high mussel densities, populations of submerged vascular plants and some invertebrates have increased. Not until after many years will we know the full effects of the colonization of North America by zebra mussels.

Some pests proliferated quickly following their introduction to new continents, with destructive consequences. Forest trees in eastern North America, for example, have been attacked by several European diseases. The chestnut blight, caused by a European fungus, virtually eliminated the American chestnut, formerly a dominant tree in forests of the Appalachian Mountains. Nearly all American elms over large areas of the East and Midwest have been killed by Dutch elm disease (caused by a different fungus), which reached North America in 1930. Ecologists suspect that intercontinental movement of disease organisms has caused extinctions throughout life's history, but evidence of disease outbreaks is not usually preserved in the fossil record.

Introduced pests, predators, and competitors have eliminated many species

Deliberately and accidentally, people move many organisms from one continent to another. Pheasants and partridges were introduced into North America for hunting. Europeans introduced rabbits and foxes to Australia for sport. Many plants have been introduced as ornamentals. Weed seeds have been accidentally carried around the world in soil used as ballast in sailing ships and as contaminants in sacks of crop seeds. Despite quarantines, disease organisms have spread widely, carried by infected plants, animals, and people.

A species that has evolved over time in a community with certain predators and competitors may be driven to extinction by newly introduced predators and competitors. Nearly half of the small to medium-sized marsupials and rodents of Australia have been exterminated during the last 100 years by a combination of competition with introduced rabbits and predation by introduced cats and foxes. On the Galápagos archipelago, introduced pigs and rats had exterminated several races of tortoises even before Darwin visited the islands. Populations of tortoises on some islands are maintained today only because conservationists remove eggs and rear the young tortoises in captivity (Figure 58.12).

The zebra mussel, whose larvae were carried in ships' ballast water from Europe, became established in the Great Lakes in about 1985. Zebra mussels dispersed rapidly and today occupy much of the Great Lakes and Mississippi River drainage (Figure 58.13). In some places these mussels have reached densities as high as 400,000 per square meter! Some species of native clams are being covered and smothered by zebra mussels. Zebra mussels have also caused millions of dollars of damage to water intake structures. On the other hand, some native species have benefited, be-

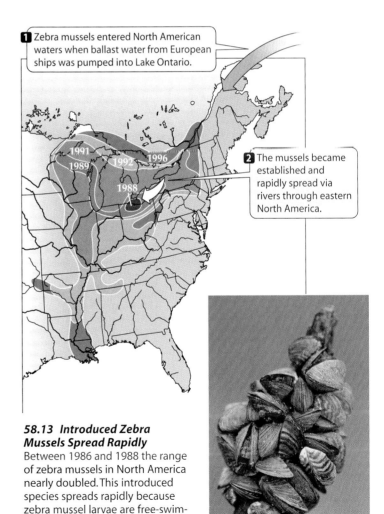

1 Zebra mussels entered North American waters when ballast water from European ships was pumped into Lake Ontario.

2 The mussels became established and rapidly spread via rivers through eastern North America.

58.13 Introduced Zebra Mussels Spread Rapidly
Between 1986 and 1988 the range of zebra mussels in North America nearly doubled. This introduced species spreads rapidly because zebra mussel larvae are free-swimming and adults can attach to moving objects, such as boat hulls.   *Dreissena polymorpha*

The curved bill of the iiwi matches the shape of the *Lobelia* flower.

58.14 Coevolved Mutualists
Declining populations of the iiwi (*Vestiaria coccinea*), a Hawaiian honeycreeper, also threaten the *Lobelia* plant, which has no other pollinator.

Overexploitation has driven many species to extinction

Until recently, humans caused extinctions primarily by overhunting. The passenger pigeon, the most abundant bird in North America in the early 1800s, became extinct by 1914, largely due to overhunting. Russian whalers exterminated the unusual Steller's sea cow of the North Pacific in the late 1800s, just 37 years after it was first described. Such overex-

ploitation continues today. Elephants and rhinoceroses are threatened in Africa because poachers kill them for their tusks and horns. Unfortunately, these animals are not slaughtered for medical or other useful purposes; rather, they are killed for ornaments and because some men believe that powdered horn enhances their sexual potency. Many species of orchids, parrots, reptiles, and tropical reef fishes are currently threatened by lucrative pet and houseplant trades.

Loss of mutualists threatens some species

Many plants have mutualistic relationships with pollinators, but most of these mutualisms are not highly species-specific. On islands, however, where ecological communities contain relatively few species, plant–pollinator interactions often evolve to be highly specific. For example, a single species of the plant *Lobelia* colonized the Hawaiian Islands, where it eventually gave rise to 110 daughter species. A single colonizing species of songbird gave rise to at least 47 species of Hawaiian honeycreepers, some of which have long, slender, curved bills. These nectar-feeding birds were the only pollinators of many species of Hawaiian lobelias (Figure 58.14).

Today, half of the nectar-feeding birds of Hawaii are extinct, leaving many lobelias without pollinators. Many of these lobelias still survive, but populations of some species have been reduced to only a few individuals. A few species survive only because biologists artificially pollinate them.

Global warming may cause species extinctions

Atmospheric scientists predict that, as a result of increasing concentrations of CO_2 and other greenhouse gases in the atmosphere (see Chapter 56), average temperatures in North America will increase 2°–5°C by the end of the twenty-first century. If the climate warms by only 1°C, the average temperature currently found at a certain location will shift 150 km to the north. To remain in the temperature regime to which they are accustomed, species will have to shift their ranges 150 km to the north. Species will need to shift their ranges as much as 500–800 km in a single century if the climate warms 2–5°C. Some habitats, such as alpine tundra, could be eliminated from many areas as forests expand up the mountain slopes.

(a)

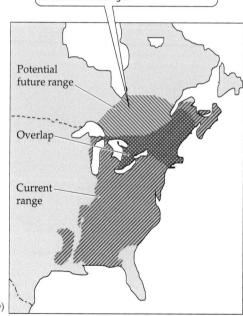

If the climate of eastern North America warms by as little as 4°C, about half the potential future range of beech trees will be beyond the northernmost extent of the current range.

Potential future range

Overlap

Current range

(b)

58.15 Threatened by Global Warming
(a) Seedlings and saplings abound in this beech forest. (b) Beeches would need to migrate 40 times faster than they have in the past to keep up with the anticipated rates of climate change.

Conservation biologists are attempting to predict the effects of global warming on North American species. Trees might be especially vulnerable to climate change because they grow for long periods before they begin to reproduce, and their seeds typically move only very short distances (Figure 58.15).

If Earth warms as predicted, climatic zones will not simply shift northward. New climates will develop, and some existing climates will disappear. New climates are certain to develop at low elevations in the tropics because a warming of even 2°C would result in climates near sea level that are hotter than those found anywhere in the humid tropics today. Adaptation to those climates may prove difficult for many tropical organisms.

Preventing Species Extinctions

Designing recovery plans

Once the causes of endangerment of species have been identified, appropriate remedies can be designed. In the United States, when a species is listed as threatened or endangered under the Endangered Species Act, a recovery plan is typically prepared to guide efforts to improve the status of the species. In this section we will describe how good diagnoses have been used to design management actions to prevent species from becoming extinct.

KIRTLAND'S WARBLER. The Kirtland's warbler is an endangered bird that nests only in 8–18-year-old stands of jack pine growing on sandy soils in Michigan (Figure 58.16). The current population of Kirtland's warblers is less than 1,000 individuals. Field studies determined that the Kirtland's warbler is at risk from both loss of habitat and nest parasitism by brown-headed cowbirds. Fire suppression has reduced the area of young jack pine stands, and cowbirds, which lay their eggs in other birds' nests, have greatly increased in abundance in the area. To prevent further threats to the warblers, conservation biologists ignite controlled fires in jack pine forests to maintain a steady supply of trees of the right age. They are also removing brown-headed cowbirds to reduce nest parasitism rates.

THE CALIFORNIA SEA OTTER. Populations of the California sea otter were hunted nearly to extinction during the nineteenth century. After receiving legal protection in 1911, the species increased steadily to about 2,400 individuals today. The Southern Sea Otter Recovery Team was charged with developing a recovery plan for the sea otter under the U.S. Endangered Species Act. They determined that ample habitat and food supplies are present and that the otters are reproducing and surviving well. They judged that a major oil spill poses the most serious threat to the otters.

A demographic model suggested that the otter population would be endangered if it dropped below 1,850 individuals. A model designed to assess whether a major oil spill could reduce the population to that size suggested that fewer than 800 otters would be killed by 90 percent of the simulated spills. Therefore, the team set the "delisting" criterion at 2,650 individuals (1,850 + 800). Because the California sea otter population has nearly reached this size, it may soon be removed from the list of endangered species.

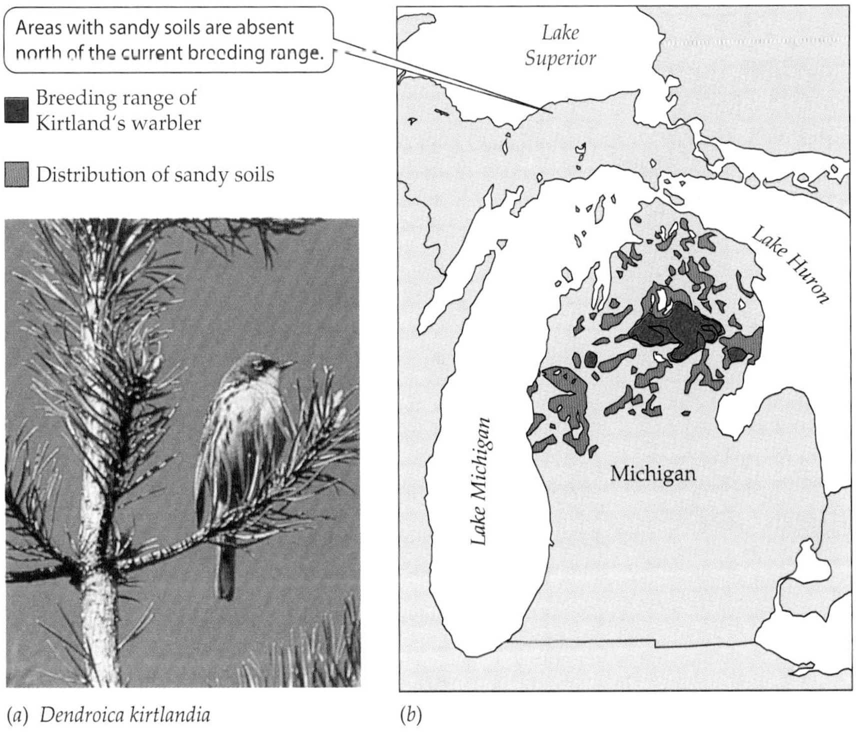

Captive propagation has a role in conservation

Species being threatened by overexploitation, loss of habitat, or environmental degradation through pollution can sometimes be maintained in captivity while the external threats to their existence are reduced or removed. Captive propagation is only a temporary measure that buys time. Existing zoos, aquariums, and botanical gardens do not have enough space to maintain adequate populations of more than a small fraction of Earth's rare and endangered species. Nonetheless, captive propagation can play an important role by maintaining species during critical periods and by providing a source of individuals for reintroduction into the wild. Captive propagation projects in zoos also have raised public awareness of species that are threatened with extinction.

Areas with sandy soils are absent north of the current breeding range.

Breeding range of Kirtland's warbler

Distribution of sandy soils

Lake Superior

Lake Huron

Lake Michigan

Michigan

(a) *Dendroica kirtlandia*

(b)

58.16 Kirtland's Warbler Is Threatened by Habitat Loss
(a) A male Kirtland's warbler in a young jack pine. (b) The warbler's breeding range and the distribution of the sandy soils that support stands of jack pine.

58.17 Peregrine Falcon Populations Have Been Reestablished
(*a*) Peregrine falcons have responded well to captive propagation. Some individuals have adapted to urban life, nesting on the tall buildings of cities and feeding on pigeons. (*b*) Throughout the eastern United States, many pairs of peregrine falcons now attempt to reproduce. Most of them are successful.

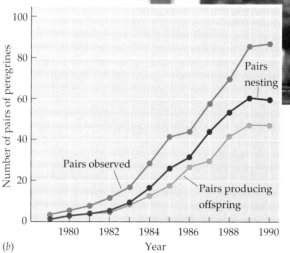

(*a*) *Falco peregrinus*

(*b*)

THE PEREGRINE FALCON. In 1942, about 350 pairs of peregrine falcons bred in the United States east of the Mississippi River. This breeding population disappeared entirely by 1960. The cause of the falcon's disappearance was the widespread use of organochlorine pesticides, such as DDT and dieldrin. These pesticides degrade very slowly in the environment and become concentrated in the falcon's prey. Their accumulation in the peregrines' bodies interfered with the deposition of calcium in eggshells. As a result, most of the falcons' eggs broke before they hatched.

Much of the eastern United States became suitable habitat for peregrines again after the use of DDT in the United States was banned by federal law. Captive breeding of peregrines began at Cornell University in 1970, and by the end of 1986, more than 850 birds reared in captivity had been released in 13 eastern states, with spectacular success (Figure 58.17). Peregrines probably would have recolonized the East by themselves, but they would have done so much more slowly without human assistance.

THE CALIFORNIA CONDOR. With its 9-foot wing span, the California condor is North America's largest bird. Two hundred years ago, condors ranged from southern British Columbia to northern Mexico, but by 1978, the wild population was plunging toward extinction—only 25 to 30 birds remained in southern California. To save the condor from extinction, biologists initiated a captive propagation program in 1983.

The first chick conceived in captivity hatched in 1988. By 1993, nine captive pairs were producing chicks, and the captive population had increased to more than 60 birds. The captive population was large enough that six captive-bred birds could be released in the mountains north of Los Angeles in 1992. These birds are provided with contaminant-free food in remote areas, and they are using the same roosting sites, bathing pools, and mountain ridges as did their predecessors. Captive-reared birds also were released late in 1996 in northern Arizona. It is still too early to pronounce the program a success, but without captive propagation, the California condor would probably be extinct today.

The cost of captive propagation is comparatively low

The California condor rehabilitation program costs about 1 million dollars a year. The Peregrine Fund at Cornell University spent about 3 million dollars over the past 30 years; the expenses of other cooperating agencies add at least another half million to the total. These amounts may seem large, but they are small compared with the costs of other human activities; for example, even a minor Hollywood film costs more than this to produce, and such films often lose money.

Establishing Priorities for Conservation Efforts

Many species and ecosystems are threatened, but the financial and human resources that can be allocated to preservation efforts are limited. How should those resources be spent to achieve the most conservation benefits? Because many species can survive only in the ecological communities in which they evolved, preserving the full array of ecological communities and habitats is vital.

Where should parks be established?

Parks, sanctuaries, and reserves function to maintain species and ecosystems relatively free of human disturbance. Parks are being created in many countries, but where should they be established to achieve the greatest conservation benefits?

High value sites for parks and reserves are those that

▶ Are home to unusually large numbers of different species.
▶ Have many **endemic** species—species that originated in that region and usually are found nowhere else.

Areas of high endemism should receive high conservation priority because if the endemic species are lost there, they often become globally extinct. Madagascar is a good example of such center of endemism: Nearly all the vascular plants and vertebrates of Madagascar are found only on that island (Figure 58.18). Therefore, if the small fragments of tropical and subtropical forests remaining on Madagascar were destroyed, many species would be exterminated.

Andansonia grandieri (giant baobob tree)

Furcifer revocosus
(warty chameleon)

Southern spiny forest

Indri indri (indri)

Enlemur fulvus (brown lemur)

58.18 Madagascar Abounds with Endemic Species

The majority of plant and animal species found on the island of Madagascar, off the eastern coast of Africa, are found nowhere else on Earth.

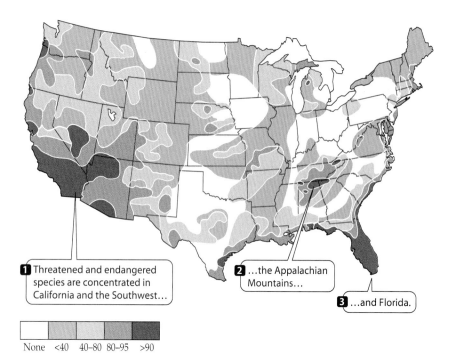

58.19 Geographic Distribution of Threatened and Endangered Species in the United States
Threatened and endangered species are concentrated in just a few locations in the continental United States. These areas are obvious priorities for conservation efforts.

1 Threatened and endangered species are concentrated in California and the Southwest...

2 ...the Appalachian Mountains...

3 ...and Florida.

None   <40   40–80   80–95   >90

Centers of endemism are not the same for all groups of organisms. Nevertheless, some areas have high concentrations of endangered species in many taxa. In the United States, species listed as threatened and endangered under the Endangered Species Act are concentrated in California, the Southwest, Florida, Hawaii, and the Appalachian Mountains (Figure 58.19).

Some economic land uses are compatible with conservation

In most countries, new parks and reserves must be established in already settled areas because few pristine areas remain. The people living there cannot be evicted, nor is it appropriate, in most cases, to prevent hungry people from settling in or hunting in parks. The high rates of human population growth in most tropical countries guarantee that pressures on parks from agricultural settlers will increase rather than decrease.

For these reasons, lands that are exploited for food, medicines, and fiber must play an important role in conservation. These lands are far more extensive than parks and reserves, and they include ecosystems not represented in parks. Fortunately, many species can be preserved on lands that are being used for economic purposes. Only a few species, such as predators on humans and domestic animals or large, destructive herbivores, are incompatible with most human uses of land.

Forest reserves in which economically valuable products are harvested can support both species preservation and economic development. In Belize, people known as *hierbateros* collect medicinal plants in the forests and sell them to the *curanderos* (healers) who provide 75 percent of health care in the country. A botanist and an economist determined that two 1-hectare plots in the second-growth tropi-

cal forest of Belize yielded gross annual revenues of $865 and $4,017, which are greater than the incomes that would result from cultivating squash and corn on those plots.

Conservation requires large-scale planning

Scientists at the World Wildlife Fund have developed a large-scale approach called Ecoregion-Based Conservation (ERBC). An **ecoregion** is an area that has a relatively uniform climate and a biota dominated by a group of widely distributed species. Using all available information on species' distributions and ecological requirements, scientists can identify sites of highest conservation priority within the ecoregion. This information is then used to develop a vision of what successful conservation would look like in 50 years.

An ecoregional conservation project, known as the Yellowstone to Yukon Initiative (Y2Y), is under way in Canada and the northern United States. Y2Y is a binational effort to restore and maintain biological diversity and landscape continuity in an area encompassing 1.2 million square kilometers along the spine of the Rocky Mountains. Information on the distributions of species, human cultures, and soil types in the region is being used to identify the most important areas for maintaining biological diversity. Investigators have also identified threats to biodiversity resulting from mining and its associated toxic waste pollution, unsustainable logging, and conversion of large ranches to suburban developments and sites for summer homes.

An important product of the Y2Y Initiative has been identification of areas that must be preserved or restored as key corridors for the movement of large mammals, such as grizzly bears, wolves, and elk, within the ecosystem. Efforts are now under way to build the base of political and financial support that will be necessary to make the vision a reality.

Restoring Degraded Ecosystems

Many areas that could be incorporated into reserves have been highly altered by human activities. Some of these areas can play their intended roles in biodiversity conservation only if they are restored to their original state. To accomplish this task, a subdiscipline of conservation biology, known as **restoration ecology**, is growing rapidly. Research on methods of restoring populations, communities, and ecosystems is needed because many ecological communities will not recover, or will do so only very slowly, without creative intervention in the recovery process.

The world's largest restoration project is under way in Guanacaste National Park in northwestern Costa Rica. Its goal is to restore a large area of tropical deciduous forest—the most threatened ecosystem in Central America—from small fragments that remain in an area converted mostly to pastures.

The single most important threat to Guanacaste National Park is fires, most of which are started by people. These fires burn the introduced pasture grasses and spread far into surrounding forests. Grazing by domestic livestock lowers the densities of these grasses, and the animals also disperse the seeds of native trees that can invade pastures. Therefore, the restoration program encourages some initial grazing by domestic livestock in the park. When plant succession has progressed to the point where grass no longer poses serious competition to the woody species and is no longer sufficiently dense to carry hot fires, grazing is terminated.

Restoring damaged and degraded habitats is an important activity, but ecologists still have limited ability to restore natural ecosystems. In the United States, the self-serving but false belief that comparable ecosystems can be created somewhere else has made it easy to get building permits for developments that destroy habitats. Developers need only state that they will create substitutes for the ecosystems they are destroying, but promising to do this is much easier than doing so.

Even the most experienced wetland ecologists are having great difficulty creating new wetlands that mimic those being destroyed. Such a "restored" wetland was conceived as part of a compensation agreement that allowed the California Department of Transportation to widen Interstate Highway 5 near San Diego. Despite stringent, court-imposed standards and the involvement of wetland experts, local endangered birds were still not breeding in the "restored" marsh 12 years after it was created. Therefore, as noted by a recent National Research Council committee on wetland restoration: "Wetland restoration should not be used to mitigate avoidable destruction of other wetlands until it can be scientifically demonstrated that the replacement ecosystems are of equal or better functioning."

Markets and Conservation

Most species are common property resources that are "owned" by everyone. Because no individual or group of individuals has strong incentives to use common property resources in a sustainable manner, their preservation usually depends on central governments. Unfortunately, governments generally lack sufficient resources to do the job. Also, governmental actions often are not well attuned to local situations. For these reasons, allowing local people to receive the economic benefits from managing biological resources on their lands can, under proper conditions, assist conservation efforts.

Preserving genes, species, and habitats provides many economic benefits. However, many of these benefits will assist future generations, not the current one. We do not know how future generations will value these benefits or the biodiversity that generates them. We do not even know how the current generation values them. We also cannot predict which species will turn out to be sources of valuable foods, medicines, or drugs.

Between 1951 and 1981, the National Cancer Institute screened extracts from 35,000 different species. To date, only one compound—taxol, derived from the Pacific yew tree—has received regulatory approval as a drug component. That drug is very valuable, but many species had to be searched to find it.

Although it is unlikely that any given species will have market value, extinction is forever. If we purposely or inadvertently exterminate a species, we have irreversibly destroyed a resource of unknown value. Therefore, loss of biodiversity is an especially urgent public issue.

How much should societies invest to preserve biodiversity? This question does not have a scientific answer. Economists and evolutionary biologists can contribute valuable information to the public debate, but the final decision is an ethical and political one that will depend on our beliefs about our responsibilities to the other organisms that share Earth with us.

The preservation of biological diversity and ecosystem services is one of the greatest challenges facing humankind. Many of the scientific tools needed for the task are already available, but appropriate use of these tools requires major changes in people's attitudes toward other species. If species are valued only because they are economically useful, increased losses of species are inevitable. Only when we value biological diversity and ecosystem functioning as the heritage of all humankind, a heritage to be passed on to our descendants as completely as possible, will we begin to reduce the current alarming rates of ecosystem destruction and species extinction.

Chapter Summary

Estimating Current Rates of Extinction

▶ Estimates of current rates of extinction worldwide are based primarily on species–area relationships and rates of tropical deforestation. **Review Figure 58.1**

▶ Rates of extinction are much higher on islands than on the mainland. **Review Figure 58.2**

▶ Demographic and genetic information is used to estimate risks of extinction. **Review Figure 58.3**

Why Do We Care about Species Extinctions?

▶ Diverse species provide the food, fiber, medicines, and aesthetic benefits upon which human life depends.

▶ Ecosystems provide services that can be replaced only by expensive and continuing human effort. **Review Figure 58.5**

Determining Causes of Endangerment and Extinction

▶ Rare species are the most vulnerable to extinction, but common species can also become extinct.

▶ Habitat destruction is the most important cause of species extinction today, but overexploitation, which historically resulted in most human-caused extinctions, is still an important cause of extinctions. **Review Figure 58.7**

▶ The fragmentation of habitats into patches that are too small to support populations is a major cause of extinction.

▶ The proportion of a habitat patch subject to detrimental edge effects increases as patch size decreases. **Review Figure 58.9**

▶ Exotic predators, competitors, and diseases introduced by humans are major causes of extinction. **Review Figure 58.13**

▶ In the future, global warming may be an important cause of extinction. **Review Figure 58.15**

Preventing Species Extinctions

▶ To ensure the recovery of endangered species, human manipulation of their environments and the other species that inhabit them is sometimes needed. **Review Figure 58.16**

▶ Captive propagation plays a useful role in conservation. **Review Figure 58.17**

Establishing Priorities for Conservation Efforts

▶ The best way to maintain populations is to set aside areas in which species and their habitats are protected. High-priority areas for establishing parks and reserves are regions of unusually high species richness and endemism. **Review Figure 58.19**

▶ In most countries, new parks must be created in already settled areas. Management of lands that are exploited for food, medicines, and fiber, must play an important role in conservation.

Restoring Degraded Ecosystems

▶ Restoration is an important component of recovery plans, but restoration of some types of environments, especially aquatic ones, is difficult.

Markets and Conservation

▶ Properly employed, markets can help preserve biodiversity.

▶ The conservation of biodiversity is not just a scientific or economic issue. It raises serious moral and ethical concerns that define what it means to be a human being on Earth.

For Discussion

1. Most species driven to extinction by people in the past were large vertebrates. Do you expect this pattern to persist into the future? If so, why? If not, why not?

2. Species endangered as a result of global warming might be preserved if we could move individuals from areas that are becoming unsuitable for them to those likely to be better for them in the future. What are the major difficulties associated with such interventions? For what types of species might they work well? Poorly? Make no difference?

3. Conservation biologists have debated extensively which is better: many small reserves or a few large ones. What ecological processes should be evaluated in making judgments about size and location of reserves? To what extent should we be concerned with preserving the largest number of species rather than those species judged to be of unusual importance for scientific, aesthetic, or commercial reasons?

4. During World War I, French doctors adopted a "triage" system of dealing with wounded soldiers. The wounded were divided into three categories: those almost certain to die no matter what was done to help them, those likely to recover even if not assisted, and those whose probability of survival was greatly increased if they were given medical attention. The limited resources available to the doctors were directed primarily at the third category. What would be the implications of adopting a similar attitude toward species preservation?

5. Economic arguments dominate discussions about the importance of preserving the biological richness of the planet. In your opinion, what role should moral arguments play?

Appendix:
Some Measurements Used in Biology

QUANTITY	NAME OF UNIT	SYMBOL	DEFINITION
Length	meter (*also* metre)	m	A base unit. 1 m = 100 cm = 39.37 inches
	kilometer	km	1 km = 1000 m = 10^3 m
	centimeter	cm	1 cm = $\frac{1}{100}$ m = 10^{-2} m
	millimeter	mm	1 mm = $\frac{1}{1000}$ m = 10^{-3} m
	micrometer	μm	1 μm = $\frac{1}{1000}$ mm = 10^{-6} m
	nanometer	nm	1 nm = $\frac{1}{1000}$ μm = 10^{-9} m
Area	square meter	m^2	Area encompassed by a square, each side of which is 1 m in length
	hectare	ha	1 ha = 10,000 m^2 = 10^4 m^2 (2.47 acres)
	square centimeter	cm^2	1 cm^2 = $\frac{1}{10,000}$ m^2 = 10^{-4} m^2
Volume	liter (*also* litre)	l	1 l = $\frac{1}{1000}$ m^3 = 10^{-3} m^3 (1.057 qts)
	milliliter	ml	1 ml = $\frac{1}{1000}$ l = 10^{-3} l = 1 cm^3 = 1 cc
	microliter	μl	1 μl − $\frac{1}{1000}$ ml = 10^{-3} ml = 10^{-6} l
Mass	kilogram	kg	A basic unit. 1 kg = 1000 g = 2.20 lbs
	gram	g	1 g = $\frac{1}{1000}$ kg = 10^{-3} kg
	milligram	mg	1 mg = $\frac{1}{1000}$ g = 10^{-3} g = 10^{-6} kg
Time	second	s	A basic unit. 1 s = $\frac{1}{60}$ min
	minute	min	1 min = 60 s
	hour	h	1 h = 60 min = 3,600 s
	day	d	1 d = 24 h = 86,400 s
Temperature	kelvin	K	A basic unit. 0 K = −273.15°C = absolute zero
	degree Celsius	°C	0°C = 273.15 K = melting point of ice
Heat, work	calorie	cal	1 cal = heat necessary to raise 1 gram of pure water from 14.5°C to 15.5°C = 4.184 J
	kilocalorie	kcal	1 kcal = 1000 cal = 10^3 cal = (in nutrition) 1 Calorie
	joule	J	1 J = 0.2389 cal (The joule is now the accepted unit of heat in most sciences.)
Electric potential	volt	V	A unit of potential difference or electromotive force
	millivolt	mV	1 mV = $\frac{1}{1000}$ V = 10^{-3} V

Glossary

Abdomen (ab' duh mun) [L.: belly] • In arthropods, the posterior portion of the body; in mammals, the part of the body containing the intestines and most other internal organs, posterior to the thorax.

Abscisic acid (ab sighs' ik) [L. *abscissio*: breaking off] • A plant growth substance having growth-inhibiting action. Causes stomata to close.

Abscission (ab sizh' un) [L. *abscissio*: breaking off] • The process by which leaves, petals, and fruits separate from a plant.

Absolute temperature scale • Also known as the Kelvin scale. A temperature scale in which zero is the state of no molecular motion. This "absolute zero" is –273° on the Celsius scale.

Absorption • (1) Of light: complete retention, without reflection or transmission. (2) Of liquids: soaking up (taking in through pores or cracks).

Absorption spectrum • A graph of light absorption versus wavelength of light; shows how much light is absorbed at each wavelength.

Abyssal zone (uh biss' ul) [Gr. *abyssos*: bottomless] • That portion of the deep ocean floor where no light penetrates.

Accessory pigments • Pigments that absorb light and transfer energy to chlorophylls for photosynthesis.

Acetylcholine • A neurotransmitter substance that carries information across vertebrate neuromuscular junctions and some other synapses. **Acetylcholinesterase** is an enzyme that breaks down acetylcholine.

Acetyl CoA (acetyl coenzyme A) • Compound that reacts with oxaloacetate to produce citrate at the beginning of the citric acid cycle; a key metabolic intermediate in the formation of many compounds.

Acid [L. *acidus*: sharp, sour] • A substance that can release a proton in solution. (Contrast with base.)

Acid precipitation • Precipitation that has a lower pH than normal as a result of acid-forming precursors introduced into the atmosphere by human activities.

Acidic • Having a pH of less than 7.0 (a hydrogen ion concentration greater than 10^{-7} molar).

Acoelomate • Lacking a coelom.

Acquired Immune Deficiency Syndrome • See AIDS.

Acrosome (a' krow soam) [Gr. *akros*: highest or outermost + *soma*: body] • The structure at the forward tip of an animal sperm which is the first to fuse with the egg membrane and enter the egg cell.

ACTH (adrenocorticotropin) • A pituitary hormone that stimulates the adrenal cortex.

Actin [Gr. *aktis*: a ray] • One of the two major proteins of muscle; it makes up the thin filaments. Forms the microfilaments found in most eukaryotic cells.

Action potential • An impulse in a neuron taking the form of a wave of depolarization or hyperpolarization imposed on a polarized cell surface.

Activating enzymes (also called aminoacyl-tRNA synthetases) • These enzymes catalyze the addition of amino acids to their appropriate tRNAs.

Activation energy (E_a) • The energy barrier that blocks the tendency for a set of chemical substances to react.

Active site • The region on the surface of an enzyme where the substrate binds, and where catalysis occurs.

Active transport • The transport of a substance across a biological membrane against a concentration gradient—that is, from a region of low concentration (of that substance) to a region of high concentration. Active transport requires the expenditure of energy and is a saturable process. (Contrast with facilitated diffusion, free diffusion; see primary active transport, secondary active transport.)

Adaptation (a dap tay' shun) • In evolutionary biology, a particular structure, physiological process, or behavior that makes an organism better able to survive and reproduce. Also, the evolutionary process that leads to the development or persistence of such a trait.

Adenine (a' den een) • A nitrogen-containing base found in nucleic acids, ATP, NAD, etc.

Adenosine triphosphate • See ATP.

Adenylate cyclase • Enzyme catalyzing the formation of cyclic AMP from ATP.

Adrenal (a dree' nal) [L. *ad-*: toward + *renes*: kidneys] • An endocrine gland located near the kidneys of vertebrates, consisting of two glandular parts, the cortex and medulla.

Adrenaline • See epinephrine.

Adrenocorticotropin • See ACTH.

Adsorption • Binding of a gas or a solute to the surface of a solid.

Aerobic (air oh' bic) [Gr. *aer*: air + *bios*: life] • In the presence of oxygen, or requiring oxygen.

Afferent (af' ur unt) [L. *ad*: to + *ferre*: to bear] • To or toward, as in a neuron that carries impulses to the central nervous system, or a blood vessel that carries blood to a structure. (Contrast with efferents.)

AIDS (acquired immune deficiency syndrome) • Condition caused by a virus (HIV) in which the body's helper T lymphocytes are reduced, leaving the victim subject to opportunistic diseases.

Aldehyde (al' duh hide) • A compound with a –CHO functional group. Many sugars are aldehydes. (Contrast with ketone.)

Aldosterone (al dahs' ter own) • A steroid hormone produced in the adrenal cortex of mammals. Promotes secretion of potassium and reabsorption of sodium in the kidney.

Alga (al' gah) (plural: algae) [L.: seaweed] • Any one of a wide diversity of protists belonging to the phyla Pyrrophyta, Chrysophyta, Phaeophyta, Rhodophyta, and Chlorophyta.

Allele (a leel') [Gr. *allos*: other] • The alternate forms of a genetic character found at a given locus on a chromosome.

Allele frequency • The relative proportion of a particular allele in a specific population.

Allergy [Ger. *allergie*: altered reaction] • An overreaction to an antigen in amounts that do not affect most people; often involves IgE antibodies.

Allometric growth • A pattern of growth in which some parts of the body of an organism grow faster than others, resulting in a change in body proportions as the organism grows.

Allopatric speciation (al' lo pat' rick) [Gr. *allos*: other + *patria*: fatherland] • Also called geographical speciation, this is the formation of two species from one when reproductive isolation occurs because of the the interposition of (or crossing of) a physical geographic barrier such as a river. (Contrast with parapatric speciation, sympatric speciation.)

Allopolyploid • A polyploid in which the chromosome sets are derived from more than one species.

Allostery (al′ lo steer′ y) [Gr. *allos*: other + *stereos*: structure] • Regulation of the activity of a protein by the binding of an effector molecule at a site other than the active site.

Alpha helix • Type of protein secondary structure; a right-handed spiral.

Alternation of generations • The succession of haploid and diploid phases in some sexually reproducing organisms, notably plants.

Altruistism • A behavior whose performance harms the actor but benefits other individuals.

Alveolus (al ve′ o lus) (plural: alveoli) [L. *alveus*: cavity] • A small, baglike cavity, especially the blind sacs of the lung.

Amensalism (a men′ sul ism) • Interaction in which one animal is harmed and the other is unaffected. (Contrast with commensalism, mutualism.)

Amine • An organic compound with an amino group (see Amino acid).

Amino acid • An organic compound of the general formula H_2N–CHR–$COOH$, where R can be one of 20 or more different side groups. An amino acid is so named because it has both a basic amine group, $–NH_2$, and an acidic carboxyl group, $–COOH$. Proteins are polymers of amino acids.

Ammonotelic (am moan′ o teel′ ic) [Gr. *telos*: end] • Describes an organism in which the final product of breakdown of nitrogen-containing compounds (primarily proteins) is ammonia. (Contrast with ureotelic, uricotelic.)

Amniocentesis • A medical procedure in which cells from the fetus are obtained from the amniotic fluid. The genetic material of the cells is then examined. (Contrast with chorionic villus sampling.)

Amniote • An organism that lays eggs that can be incubated in air (externally) because the embryo is enclosed by a fluid-filled sac. Birds and reptiles are amniotes.

Amphipathic (am′ fi path′ ic) [Gr. *amphi*: both + *pathos*: emotion] • Of a molecule, having both hydrophilic and hydrophobic regions.

Amylase (am′ ill ase) • Any of a group of enzymes that digest starch.

Anabolism (an ab′ uh liz′ em) [Gr. *ana*: up, throughout + *ballein*: to throw] • Synthetic reactions of metabolism, in which complex molecules are formed from simpler ones. (Contrast with catabolism.)

Anaerobic (an ur row′ bic) [Gr. *an*: not + *aer*: air + *bios*: life] • Occurring without the use of molecular oxygen, O_2.

Anagenesis • Evolutionary change in a single lineage over time.

Analogy (a nal′ o jee) [Gr. *analogia*: resembling] • A resemblance in function, and often appearance as well, between two structures which is due to convergence in evolution rather than to common ancestry. (Contrast with homology.)

Anaphase (an′ a phase) [Gr. *ana*: indicating upward progress] • The stage in nuclear division at which the first separation of sister chromatids (or, in the first meiotic division, of paired homologues) occurs. Anaphase lasts from the moment of first separation to the time at which the moving chromosomes converge at the poles of the spindle.

Anaphylactic shock • A precipitous drop in blood pressure caused by loss of fluid from capillaries because of an increase in their permeability stimulated by an allergic reaction.

Ancestral trait • Trait shared by a group of organisms as a result of descent from a common ancestor.

Androgens (an′ dro jens) • The male sex steroids.

Aneuploidy (an′ you ploy dee) • A condition in which one or more chromosomes or pieces of chromosomes are either lacking or present in excess.

Angiosperm (an′ jee oh spurm) [Gr. *angion*: vessel + *sperma*: seed] • One of the flowering plants; literally, one whose seed is carried in a "vessel," which is the fruit. (See fruit.)

Angiotensin (an′ jee oh ten′ sin) • A peptide hormone that raises blood pressure by causing peripheral vessels to constrict; maintains glomerular filtration by constricting efferent glomerular vessels; stimulates thirst; and stimulates the release of aldosterone.

Animal [L. *animus*: breath, soul] • A member of the kingdom Animalia. In general, a multicellular eukaryote that obtains its food by ingestion.

Animal hemisphere • The metabolically active upper portion of some animal eggs, zygotes, and embryos, which does *not* contain the dense nutrient yolk. The **animal pole** refers to the very top of the egg or embryo. (Contrast with vegetal hemisphere.)

Anion (an′ eye one) • An ion with one or more negative charges. (Contrast with cation.)

Anisogamy (an′ eye sog′ a mee) [Gr. *aniso*: unequal + *gamos*: marriage] • The existence of two dissimilar gametes (egg and sperm).

Annual • Referring to a plant whose life cycle is completed in one growing season. (Contrast with biennial, perennial.)

Anterior pituitary • The portion of the vertebrate pituitary gland that derives from gut epithelium and produces tropic hormones.

Anther (an′ thur) [Gr. *anthos*: flower] • A pollen-bearing portion of the stamen of a flower.

Antheridium (an′ thur id′ ee um) (plural: antheridia) [Gr. *antheros*: blooming] • The multicellular structure that produces the sperm in bryophytes and ferns.

Antibody • One of millions of proteins, produced by the immune system, that specifically recognizes a foreign substance and initiates its removal from the body.

Anticodon • A "triplet" of three nucleotides in transfer RNA that is able to pair with a complementary triplet (a codon) in messenger RNA, thus aligning the transfer RNA on the proper place on the messenger. The codon (and, reciprocally, the anticodon) codes for a specific amino acid.

Antidiuretic hormone • A hormone that controls water reabsorption in the mammalian kidney. Also called vasopressin.

Antigen (an′ ti jun) • Any substance that stimulates the production of an antibody or antibodies in the body of a vertebrate.

Antigen processing • The breakdown of antigenic proteins into smaller fragments, which are then presented on the cell surface, along with MHC proteins, to T cells.

Antigenic determinant • A specific region of an antigen, which is recognized by and binds to a specific antibody.

Antiport • A membrane transport process that carries one substance in one direction and another in the opposite direction. (Contrast with symport.)

Antisense nucleic acid • A single-stranded RNA or DNA complementary to and thus targeted against the mRNA transcribed from a harmful gene such as an oncogene.

Anus (a′ nus) • Opening through which digestive wastes are expelled, located at the posterior end of the gut.

Aorta (a or′ tuh) [Gr. *aorte*: aorta] • The main trunk of the arteries leading to the systemic (as opposed to the pulmonary) circulation.

Apex (a′ pecks) • The tip or highest point of a structure, as the apex of a growing stem or root.

Apical (a′ pi kul) • Pertaining to the apex, or tip, usually in reference to plants.

Apical dominance • Inhibition by the apical bud of the growth of axillary buds.

Apical meristem • The meristem at the tip of a shoot or root; responsible for the plant's primary growth.

Apomixis (ap oh mix′ is) [Gr. *apo*: away from + *mixis*: sexual intercourse] • The asexual production of seeds.

Apoplast (ap′ oh plast) • in plants, the continuous meshwork of cell walls and extracellular spaces through which material can pass without crossing a plasma membrane. (Contrast with symplast.)

Apoptosis (ay′ pu toh sis) • A series of genetically programmed events leading to cell death.

Aquaporin • A transport protein in plant and animals cells through which water passes in osmosis.

Archegonium (ar′ ke go′ nee um) [Gr. *archegonos*: first of a kind] • The multicellular structure that produces eggs in bryophytes, ferns, and gymnosperms.

Archenteron (ark en′ ter on) [Gr. *archos*: beginning + *enteron*: bowel] • The earliest primordial animal digestive tract.

Arteriosclerosis • See atherosclerosis.

Artery • A muscular blood vessel carrying oxygenated blood away from the heart to other parts of the body. (Contrast with vein.)

Ascus (ass' cuss) [Gr. *askos*: bladder] • In fungi belonging to the phylum Ascomycota (the sac fungi), the club-shaped sporangium within which spores (ascospores) are produced by meiosis.

Asexual • Without sex.

Assortative mating • A breeding system in which mates are selected on the basis of a particular trait or group of traits.

Atherosclerosis (ath' er oh sklair oh' sis) • A disease of the lining of the arteries characterized by fatty, cholesterol-rich deposits in the walls of the arteries. When fibroblasts infiltrate these deposits and calcium precipitates in them, the disease become arteriosclerosis, or "hardening of the arteries."

Atmosphere • The gaseous mass surrounding our planet. Also: a unit of pressure, equal to the normal pressure of air at sea level.

Atom [Gr. *atomos*: indivisible] • The smallest unit of a chemical element. Consists of a nucleus and one or more electrons.

Atomic mass (also called atomic weight) • The average mass of an atom of an element on the amu scale. (The average depends upon the relative amounts of different isotopes of an element on Earth.)

Atomic number • The number of protons in the nucleus of an atom, also equal to the number of electrons around the neutral atom. Determines the chemical properties of the atom.

ATP (adenosine triphosphate) • A compound containing adenine, ribose, and three phosphate groups. When it is formed, useful energy is stored; when it is broken down (to ADP or AMP), energy is released to drive endergonic reactions. ATP is an energy storage compound.

ATP synthase • An integral membrane protein that couples the transport of proteins with the formation of ATP.

Atrium (a' tree um) • A body cavity, as in the hearts of vertebrates. The thin-walled chamber(s) entered by blood on its way to the ventricle(s). Also, the outer ear.

Autoimmune disease • A disorder in which the immune system attacks the animal's own antigens.

Autonomic nervous system • The system (which in vertebrates comprises sympathetic and parasympathetic subsystems) that controls such involuntary functions as those of guts and glands.

Autosome • Any chromosome (in a eukaryote) other than a sex chromosome.

Autotroph (au' tow trow' fik) [Gr. *autos*: self + *trophe*: food] • An organism that is capable of living exclusively on inorganic materials, water, and some energy source such as sunlight or chemically reduced matter. (Contrast with heterotroph.)

Auxin (awk' sin) [Gr. *auxein*: increase] • In plants, a substance (indoleacetic acid) that regulates growth and various aspects of development.

Auxotroph (awks' o trofe) [Gr. *auxanein*: to grow + *trophe*: food] • A mutant form of an organism that requires a nutrient or nutrients not required by the wild type, or reference, form of the organism. (Contrast with prototroph.)

Axon [Gr.: axle] • Fiber of a neuron which can carry action potentials. Carries impulses away from the cell body of the neuron; releases a neurotransmitter substance.

Axon hillock • The junction between an axon and its cell body; where action potentials are generated.

Axon terminals • The endings of an axon; they form synapses and release neurotransmitter.

Axoneme (ax' oh neem) • The complex of microtubules and their crossbridges that forms the motile apparatus of a cilium.

Bacillus (buh sil' us) [L.: little rod] • Any of various rod-shaped bacteria.

Bacteriophage (bak teer' ee o fayj) [Gr. *bakterion*: little rod + *phagein*: to eat] • One of a group of viruses that infect bacteria and ultimately cause their disintegration.

Bacteria (bak teer' ee ah) (singular: bacterium) [Gr. *bakterion*: little rod] • Prokaryote in the Domain Bacteria. The chromosomes of bacteria are not contained in nuclear envelopes.

Balanced polymorphism [Gr. *polymorphos*: having many forms] • The maintenance of more than one form, or the maintenance at a given locus of more than one allele, at frequencies of greater than one percent in a population. Often results when heterozygotes are superior to both homozygotes.

Bark • All tissues outside the vascular cambium of a plant.

Baroreceptor [Gr. *baros*: weight] • A pressure-sensing cell or organ.

Barr body • In mammals, an inactivated X chromosome.

Basal body • Centriole found at the base of a eukaryotic flagellum or cilium.

Basal metabolic rate • The minimum rate of energy turnover in an awake (but resting) bird or mammal that is not expending energy for thermoregulation.

Base • (1) A substance which can accept a proton (hydrogen ion; H^+) in solution. (Contrast with acid.) (2) In nucleic acids, a nitrogen-containing molecule that is attached to each sugar in the backbone. (See purine; pyrimidine.)

Base pairing • See complementary base pairing.

Basic • having a pH greater than 7.0 (having a hydrogen ion concentration lower than 10^{-7} molar).

Basidium (bass id' ee yum) • In fungi of the class Basidiomycetes, the characteristic sporangium in which four spores are formed by meiosis and then borne externally before being shed.

Batesian mimicry • Mimicry by a relatively harmless kind of organism of a more dangerous one, by which the mimic enjoys protection from predators that mistake it for the dangerous model. (Contrast with Müllerian mimicry.)

B cell • A type of lymphocyte involved in the humoral immune response of vertebrates. Upon recognizing an antigenic determinant, a B cell develops into a plasma cell, which secretes an antibody. (Contrast with a T cell.)

Benefit • An improvement in survival and reproductive success resulting from a behavior. (Contrast with cost.)

Benign (be nine') • A tumor that grows to a certain size and then stops, uaually with a fibrous capsule surrounding the mass of cells. Benign tumors do not spread (metastasize) to other organs.

Benthic zone [Gr. *benthos*: bottom of the sea] • The bottom of the ocean. (Contrast with pelagic zone.)

Beta-pleated sheet • Type of protein secondary structure; results from hydrogen bonding between polypeptide regions running antiparallel to each other.

Biennial • Referring to a plant whose life cycle includes vegetative growth in the first year and flowering and senescence in the second year. (Contrast with annual, perennial.)

Bilateral symmetry • The condition in which only the right and left sides of an organism, divided exactly down the back, are mirror images of each other. (Contrast with biradial symmetry.)

Bile • A secretion of the liver delivered to the small intestine via the common bile duct. In the intestine, bile emulsifies fats.

Binocular cells • Neurons in the visual cortex that respond to input from both retinas; involved in depth perception.

Binomial (bye nome' ee al) • Consisting of two names; for example, the binomial nomenclature of biology which gives the name of the genus followed by the name of the species.

Biodiversity crisis • The current high rate of loss of species, caused primarily by human activities.

Biogeochemical cycles • Movement of elements through living organisms and the physical environment.

Biogeography • The scientific study of the geographic distribution of organisms.

Biogeographic region • A continental-scale part of Earth that has a biota distinct from that of other such regions.

Biological species concept • The view that a species is most usefully defined as a population or series of populations within which there is a significant amount of gene flow under natural conditions, but which is genetically isolated from other populations.

Bioluminescence • The production of light by biochemical processes in an organism.

Biomass • The total weight of all the living organisms, or some designated group of living organisms, in a given area.

Biome (bye' ome) • A major division of the ecological communities of Earth; characterized by distinctive vegetation.

Biota (bye oh' tah) • All of the organisms, including animals, plants, fungi, and microorganisms, found in a given area.

Biotechnology • The use of cells to make medicines, foods and other products useful to humans.

Biradial symmetry • Radial symmetry modified so that only two planes can divide the animal into similar halves.

Blastocoel (blass' toe seal) [Br. *blastos*: sprout + *koilos*: hollow] • The central, hollow cavity of a blastula.

Blastodisc (blass' toe disk) • A disk of cells forming on the surface of a large yolk mass, comparable to a blastula, but occurring in animals such as birds and reptiles, in which the massive yolk restricts cleavage to one side of the egg only.

Blastomere • A cell produced by the division of a fertilized egg.

Blastopore • The opening from the archenteron to the exterior of a gastrula.

Blastula (blass' chu luh) [Gr. *blastos*: sprout] • An early stage in animal embryology; in many species, a hollow sphere of cells surrounding a central cavity, the blastocoel. (Contrast with blastodisc.)

Blood–brain barrier • A property of the blood vessels of the brain that prevents most chemicals from diffusing from the blood into the brain.

Body plan • A basic structural design that includes an entire animal, its organ systems, and the integrated functioning of its parts. Phylogenetic groups of organisms are classified in part on the basis of a shared body plan.

Bowman's capsule • An elaboration of kidney tubule cells that surrounds a knot of capillaries (the glomerulus). Blood is filtered across the walls of these capillaries and the filtrate is collected into Bowman's capsule.

Brain stem • The portion of the vertebrate brain between the spinal cord and the forebrain.

Brassinosteroids • Plant steroid hormones that promote the elongation of stems and pollen tubes.

Bronchus (plural: bronchi) • The major airway(s) branching off the trachea into the vertebrate lung.

Brown fat • Fat tissue in mammals that is specialized to produce heat. It has many mitochondria and capillaries, and a protein that uncouples oxidative phosphorylation.

Browser • An animal that feeds on the tissues of woody plants.

Bryophyte (bri' uh fite') [Gr. *bruon*: moss + *phyton*: plant] • A moss. Formerly was often used to refer to all the nontracheophyte plants.

Budding • Asexual reproduction in which a more or less complete new organism simply grows from the body of the parent organism and eventually detaches itself.

Buffering • A process by which a system resists change—particularly in pH, in which case added acid or base is partially converted to another form.

C_3 photosynthesis • The form of photosynthesis in which 3-phosphoglycerate is the first stable product, and ribulose bisphosphate is the CO_2 receptor.

C_4 photosynthesis • The form of photosynthesis in which oxaloacetate is the first stable product, and phosphoenolpyruvate is the CO_2 acceptor. C_4 plants also perform the reactions of C_3 photosynthesis.

Calcitonin • A hormone produced by the thyroid gland; it lowers blood calcium and promotes bone formation. (Contrast with parathormone.)

Calmodulin (cal mod' joo lin) • A calcium-binding protein found in all animal and plant cells; mediates many calcium-regulated processes.

calorie [L. *calor*: heat] • The amount of heat required to raise the temperature of one gram of water by one degree Celsius (1°C) from 14.5°C to 15.5°C. In nutrition studies, "Calorie" (spelled with a capital C) refers to the kilocalorie (1 kcal = 1,000 cal).

Calvin–Benson cycle • The stage of photosynthesis in which CO_2 reacts with RuBP to form 3PG, 3PG is reduced to a sugar, and RuBP is regenerated, while other products are released to the rest of the plant.

Calyx (kay' licks) [Gr. *kalyx*: cup] • All of the sepals of a flower, collectively.

CAM • See crassulacean acid metabolism.

Cambium (kam' bee um) [L. *cambiare*: to exchange] • A meristem that gives rise to radial rows of cells in stem and root, increasing them in girth; commonly applied to the vascular cambium which produces wood and phloem, and the cork cambium, which produces bark.

cAMP (cyclic AMP) • A compound, formed from ATP, that mediates the effects of numerous animal hormones. Also needed for the transcription of catabolite-repressible operons in bacteria. Used for communication by cellular slime molds.

Canopy • The leaf-bearing part of a tree. Collectively the aggregate of the leaves and branches of the larger woody plants of an ecological community.

Capillaries [L. *capillaris*: hair] • Very small tubes, especially the smallest blood-carrying vessels of animals between the termination of the arteries and the beginnings of the veins.

Capsid • The protein coat of a virus.

Carbohydrates • Organic compounds with the general formula $C_nH_{2m}O_m$. Common examples are sugars, starch, and cellulose.

Carboxylic acid (kar box sill' ik) • An organic acid containing the carboxyl group, –COOH, which dissociates to the carboxylate ion, –COO$^-$.

Carcinogen (car sin' oh jen) • A substance that causes cancer.

Cardiac (kar' dee ak) [Gr. *kardia*: heart] • Pertaining to the heart and its functions.

Carnivore [L. *carn*: flesh + *vovare*: to devour] • An organism that feeds on animal tissue. (Contrast with detritivore, herbivore, omnivore.)

Carotenoid (ka rah' tuh noid) [L. *carota*: carrot] • A yellow, orange, or red lipid pigment commonly found as an accessory pigment in photosynthesis; also found in fungi.

Carpel (kar' pel) [Gr. *karpos*: fruit] • The organ of the flower that contains one or more ovules.

Carrier • (1) In facilitated diffusion, a membrane protein that binds a specific molecule and transports it through the membrane. (2) In respiratory and photosynthetic electron transport, a participating substance such as NAD that exists in both oxidized and reduced forms. (3) In genetics, a person heterozygous for a recessive trait.

Carrying capacity • In ecology, the largest number of organisms of a particular species that can be maintained indefinitely in a given part of the environment.

Cartilage • In vertebrates, a tough connective tissue found in joints, the outer ear, and elsewhere. Forms the entire skeleton in some animal groups.

Casparian strip • A band of cell wall containing suberin and lignin, found in the endodermis. Restricts the movement of water across the endodermis.

Catabolism [Ge. *kata*: down + *ballein*: to throw] • Degradational reactions of metabolism, in which complex molecules are broken down. (Contrast with anabolism.)

Catalyst (cat' a list) [Gr. *kata*-, implying the breaking down of a compound] • A chemical substance that accelerates a reaction without itself being consumed in the overall course of the reaction. Catalysts lower the activation energy of a reaction. Enzymes are biological catalysts.

Cation (cat' eye on) • An ion with one or more positive charges. (Contrast with anion.)

Caudal [L. *cauda*: tail] • Pertaining to the tail, or to the posterior part of the body.

cDNA • See complementary DNA.

Cecum (see' cum) [L. *caecus*: blind] • A blind branch off the large intestine. In many nonruminant mammals, the cecum contains a colony of microorganisms that contribute to the digestion of food.

Cell adhesion molecules • Molecules on animal cell surfaces that affect the selective association of cells during development of the embryo.

Cell cycle • The stages through which a cell passes between one division and the next. Includes all stages of interphase and mitosis.

Cell division • The reproduction of a cell to produce two new cells. In eukaryotes, this process involves nuclear division (mitosis) and cytoplasmic division (cytokinesis).

Cell theory • The theory, well established, that organisms consist of cells, and that all cells come from preexisting cells.

Cell wall • A relatively rigid structure that encloses cells of plants, fungi, many protists, and most bacteria. The cell wall gives these cells their shape and limits their expansion in hypotonic media.

Cellular immune system • That part of the immune system that is based on the activities of T cells. Directed against parasites, fungi, intracellular viruses, and foreign tissues (grafts). (Contrast with humoral immune system.)

Cellular respiration • See respiration.

Cellulose (sell' you lowss) • A straight-chain polymer of glucose molecules, used by plants as a structural supporting material.

Central dogma • The statement that information flows from DNA to RNA to polypeptide (in retroviruses, there is also information flow from RNA to cDNA).

Central nervous system • That part of the nervous system which is condensed and centrally located, e.g., the brain and spinal cord of vertebrates; the chain of cerebral, thoracic and abdominal ganglia of arthropods.

Centrifuge [L. *fugere*: to flee] • A device in which a sample can be spun around a central axis at high speed, creating a centrifugal force that mimics a very strong gravitational force. Used to separate mixtures of suspended materials.

Centriole (sen' tree ole) • A paired organelle that helps organize the microtubules in animal and protist cells during nuclear division.

Centromere (sen' tro meer) [Gr. *centron*: center + *meros*: part] • The region where sister chromatids join.

Centrosome (sen' tro soam) • The major microtubule organizing center of an animal cell.

Cephalization (sef' uh luh zay' shun) [Gr. *kephale*: head] • The evolutionary trend toward increasing concentration of brain and sensory organs at the anterior end of the animal.

Cerebellum (sair' uh bell' um) [L.: diminutive of *cerebrum*: brain] • The brain region that controls muscular coordination; located at the anterior end of the hindbrain.

Cerebral cortex • The thin layer of gray matter (neuronal cell bodies) that overlays the cerebrum.

Cerebrum (su ree' brum) [L.: brain] • The dorsal anterior portion of the forebrain, making up the largest part of the brain of mammals. In mammals, the chief coordination center of the nervous system; consists of two **cerebral hemispheres**.

Cervix (sir' vix) [L.: neck] • The opening of the uterus into the vagina.

cGMP (cyclic guanosine monophosphate) • An intracellular messenger that is part of signal transmission pathways involving G proteins. (See G protein.)

Channel • A membrane protein that forms an aqueous passageway though which specific solutes may pass by simple diffusion; some channels are gated: they open and close in response to binding of specific molecules.

Chaperone protein • A protein that assists a newly forming protein in adopting its appropriate tertiary structure.

Chemical bond • An attractive force stably linking two atoms.

Chemiosmotic mechanism • The formation of ATP in mitochondria and chloroplasts, resulting from a pumping of protons across a membrane (against a gradient of electrical charge and of pH), followed by the return of the protons through a protein channel with ATPase activity.

Chemoautotroph • An organism that uses carbon dioxide as a carbon source and obtains energy by oxidizing inorganic substances from its environment. (Contrast with chemoheterotroph, photoautotroph, photoheterotroph.)

Chemoheterotroph • An organism that must obtain both carbon and energy from organic substances. (Contrast with chemoautotroph, photoautotroph, photoheterotroph.)

Chemoreceptor • A cell or tissue that senses specific substances in its environment.

Chemosynthesis • Synthesis of food substances, using the oxidation of reduced materials from the environment as a source of energy.

Chiasma (kie az' muh) (plural: chiasmata) [Gr.: cross] • An X-shaped connection between paired homologous chromosomes in prophase I of meiosis. A chiasma is the visible manifestation of crossing over between homologous chromosomes.

Chitin (kye' tin) [Gr. *kiton*: tunic] • The characteristic tough but flexible organic component of the exoskeleton of arthropods, consisting of a complex, nitrogen-containing polysaccharide. Also found in cell walls of fungi.

Chlorophyll (klor' o fill) [Gr. *kloros*: green + *phyllon*: leaf] • Any of a few green pigments associated with chloroplasts or with certain bacterial membranes; responsible for trapping light energy for photosynthesis.

Chloroplast [Gr. *kloros*: green + *plast*: a particle] • An organelle bounded by a double membrane containing the enzymes and pigments that perform photosynthesis. Chloroplasts occur only in eukaryotes.

Choanocyte (cho' an oh cite) • The collared, flagellated feeding cells of sponges.

Cholecystokinin (ko' lee sis to kai nin) • A hormone produced and released by the lining of the duodenum when it is stimulated by undigested fats and proteins. It stimulates the gallbladder to release bile and slows stomach activity.

Chorion (kor' ee on) [Gr. *khorion*: afterbirth] • The outermost of the membranes protecting mammal, bird, and reptile embryos; in mammals it forms part of the placenta.

Chorionic villus sampling • A medical procedure that extracts a portion of the chorion from a pregnant woman to enable genetic and biochemical analysis of the embryo. (Contrast with amniocentesis.)

Chromatid (kro' ma tid) • Each of a pair of new sister chromosomes from the time at which the molecular duplication occurs until the time at which the centromeres separate at the anaphase of nuclear division.

Chromatin • The nucleic acid–protein complex found in eukaryotic chromosomes.

Chromatophore (krow mat' o for) [Gr. *kroma*: color + *phoreus*: carrier] • A pigment-bearing cell that expands or contracts to change the color of the organism.

Chromosome (krome' o sowm) [Gr. *kroma*: color + *soma*: body] • In bacteria and viruses, the DNA molecule that contains most or all of the genetic information of the cell or virus. In eukaryotes, a structure composed of DNA and proteins that bears part of the genetic information of the cell.

Chylomicron (ky low my' cron) • Particles of lipid coated with protein, produced in the gut from dietary fats and secreted into the extracellular fluids.

Chyme (kime) [Gr. *kymus*, juice] • Created in the stomach; a mixture of ingested food with the digestive juices secreted by the salivary glands and the stomach lining.

Cilium (sil' ee um) (plural: cilia) [L. *cilium*: eyelash] • Hairlike organelle used for locomotion by many unicellular organisms and for moving water and mucus by many multicellular organisms. Generally shorter than a flagellum.

Circadian rhythm (sir kade' ee an) [L. *circa*: approximately + *dies*: day] • A rhythm in behavior, growth, or some other activity that recurs about every 24 hours under constant conditions.

Circannual rhythm (sir can' you al) [L. *circa*: approximately + *annus*: year) • A rhythm of behavior, growth, or some other activity that recurs on a yearly basis.

Citric acid cycle • A set of chemical reactions in cellular respiration, in which acetyl CoA reacts with oxaloacetate to form citric acid, and oxaloacetate is regenerated. Acetyl CoA is oxidized to carbon dioxide, and hydrogen atoms are stored as NADH and FADH$_2$. Also called the Krebs cycle.

Class • In taxonomy, the category below the phylum and above the order; a group of related, similar orders.

Class I MHC molecules • These cell surface proteins participate in the cellular immune response directed against virus-infected cells.

Class II MHC molecules • These cell surface proteins participate in the cell-cell interactions (of helper T cells, macrophages, and B cells) of the humoral immune response.

Class switching • The process whereby a plasma cell changes the class of immunoglobulin that it synthesizes. This results from the deletion of part of the constant region of DNA, bringing in a new C segment. The variable region is the same as before, so that the new immunoglbulin has the same antigenic specificity.

Clathrin • A fibrous protein on the inner surfaces of animal cell membranes that strengthens coated vesicles and thus participates in receptor-mediated endocytosis.

Clay • A soil constituent comprising particles smaller than 2 micrometers in diameter.

Cleavages • First divisions of the fertilized egg of an animal.

Cline • A gradual change in the traits of a species over a geographical gradient.

Cloaca (klo ay' kuh) [L. *cloaca*: sewer] • In some invertebrates, the posterior part of the gut; in many vertebrates, a cavity receiving material from the digestive, reproductive, and excretory systems.

Clonal anergy • When a naive T cell encounters a self-antigen, the T cell may bind to the antigen but does not receive signals from an antigen-presenting cell. Instead of being activated, the T cell dies (becomes anergic). In this way, we avoid reacting to our own tissue-specific antigens.

Clonal deletion • In immunology, the inactivation or destruction of lymphocyte clones that would produce immune reactions against the animal's own body.

Clonal selection • The mechanism by which exposure to antigen results in the activation of selected T- or B-cell clones, resulting in an immune response.

Clone [Gr. *klon*: twig, shoot] • Genetically identical cells or organisms produced from a common ancestor by asexual means.

Cnidocytes • The feeding cells of cnidarians, within which nematocysts are housed.

Coacervate (ko as' er vate) [L. *coacervare*: to heap up] • An aggregate of colloidal particles in suspension.

Coacervate drop • Drops formed when a mixture of large proteins and polysaccharides is shaken in water. The interiors of these drops, which are often very stable, contain most of the proteins and polysaccharides.

Coated vesicle • Vesicle, sometimes formed from a coated pit, with characteristic "bristly" surface; its membrane contains distinctive proteins, including clathrin.

Coccus (kock' us) [Gr. *kokkos*: berry, pit] • Any of various spherical or spheroidal bacteria.

Cochlea (kock' lee uh) [Gr. *kokhlos*: a land snail] • A spiral tube in the inner ear of vertebrates; it contains the sensory cells involved in hearing.

Codominance • A condition in which two alleles at a locus produce different phenotypic effects and both effects appear in heterozygotes.

Codon • A "triplet" of three nucleotides in messenger RNA that directs the placement of a particular amino acid into a polypeptide chain. (Contrast with anticodon.)

Coefficient of relatedness • The probability that an allele in one individual is an identical copy, by descent, of an allele in another individual.

Coelom (see' lum) [Gr. *koiloma*: cavity] • The body cavity of certain animals, which is lined with cells of mesodermal origin.

Coelomate • Having a coelom.

Coenocyte (seen' a sight) [Gr.: common cell] • A "cell" bounded by a single plasma membrane, but containing many nuclei.

Coenzyme • A nonprotein molecule that plays a role in catalysis by an enzyme. The coenzyme may be part of the enzyme molecule or free in solution. Some coenzymes are oxidizing or reducing agents.

Coevolution • Concurrent evolution of two or more species that are mutually affecting each other's evolution.

Cohort (co' hort) [L. *cohors*: company of soldiers] • A group of similar-age organisms, considered as it passes through time.

Collagen [Gr. *kolla*: glue] • A fibrous protein found extensively in bone and connective tissue.

Collecting duct • In vertebrates, a tubule that receives urine produced in the nephrons of the kidney and delivers that fluid to the ureter for excretion.

Collenchyma (cull eng' kyma) [Gr. *kolla*: glue + *enchyma*: infusion] • A type of plant cell, living at functional maturity, which lends flexible support by virtue of primary cell walls thickened at the corners. (Contrast with parenchyma, sclerenchyma.)

Colon [Gr. *kolon*: large intestine] • The large intestine.

Commensalism • The form of symbiosis in which one species benefits from the association, while the other is neither harmed nor benefited.

Common bile duct • A single duct that delivers bile from the gallbladder and secretions from the pancreas into the small intestine.

Communication • A signal from one organism (or cell) that alters the pattern of behavior in another organism (or cell) in an adaptive fashion.

Community • Any ecologically integrated group of species of microorganisms, plants, and animals inhabiting a given area.

Companion cell • Specialized cell found adjacent to a sieve tube member in flowering plants.

Comparative analysis • An approach to studying evolution in which hypotheses are tested by measuring the distribution of states among a large number of species.

Comparative genomics • Computer-aided comparison of DNA sequences between different organisms to reveal genes with related functions.

Compensation point • The light intensity at which the rates of photosynthesis and of cellular respiration are equal.

Competitive inhibitor • A substance, similar in structure to an enzyme's substrate, that binds the active site and thus inhibits a reaction.

Competition • In ecology, use of the same resource by two or more species, when the resource is present in insufficient supply for the combined needs of the species.

Competitive exclusion • A result of competition between species for a limiting resource in which one species completely eliminates the other.

Competitive inhibitor • A substance, similar in structure to an enzyme's substrate, that binds the active site and inhibits a reaction.

Complement system • A group of eleven proteins that play a role in some reactions of the immune system. The complement proteins are not immunoglobulins.

Complementary base pairing • The A–T (or A–U), T–A (or U–A), C–G and G–C pairing of bases in double-stranded DNA, in transcription, and between tRNA and mRNA.

Complementary DNA (cDNA) • DNA formed by reverse transcriptase acting with an RNA template; essential intermediate in the reproduction of retroviruses; used as a tool in recombinant DNA technology; lacks introns.

Complete metamorphosis • A change of state during the life cycle of an organism in which the body is almost completely rebuilt to produce an individual with a very different body form. Characteristic of insects such as butterflies, moths, beetles, ants, wasps, and flies.

Compound • (1) A substance made up of atoms of more than one element. (2) Made up of many units, as the compound eyes of arthropods (as opposed to the simple eyes of the same group of organisms).

Condensation reaction • A reaction in which two molecules become connected by a covalent bond and a molecule of water is released. $(AH + BOH \rightarrow AB + H_2O.)$

Cones • (1) In the vertebrate retina: photoreceptors responsible for color vision. (2) In gymnosperms: reproductive structures consisting of many sporophylls packed relatively tightly.

Conidium (ko nid' ee um) [Gr. *konis*: dust] • An asexual fungus spore borne singly or in chains either apically or laterally on a hypha.

Conifer (kahn' e fer) [Gr. *konos*: cone + *phero*: carry] • One of the cone-bearing gymnosperms, mostly trees, such as pines and firs.

Conjugation (kahn' jew gay' shun) [L. *conjugare*: yoke together] • The close approximation of two cells during which they exchange genetic material, as in *Paramecium* and other ciliates, or during which DNA passes from one to the other through a tube, as in bacteria.

Connective tissue • An animal tissue that connects or surrounds other tissues; its cells are embedded in a collagen-containing matrix.

Connexon • In a gap junction, a protein channel linking adjacent animal cells.

Consensus sequences • Short stretches of DNA that appear, with little variation, in many different genes.

Constant region • The constant region in an immunoglobulin is encoded by a single exon and determines the function, but not the specificity, of the molecule. The constant region of the T cell receptor anchors the protein to the plasma membrane.

Constitutive enzyme • An enzyme that is present in approximately constant amounts in a system, whether its substrates are present or absent. (Contrast with inducible enzyme.)

Consumer • An organism that eats the tissues of some other organism.

Continental drift • The gradual drifting apart of the world's continents that has occurred over a period of billions of years.

Convergent evolution • The evolution of similar features independently in unrelated taxa from different ancestral structures.

Cooperative act • Behavior in which two or more individuals interact to their mutual benefit. No conscious awareness by the actors of the effects of their behavior is implied.

Cooption • The act of capturing something for a particular use. In ecology refers to the diversion of ecological production for human use. Such production is said to be coopted.

Copulation • Reproductive behavior that results in a male depositing sperm in the reproductive tract of a female.

Corepressor • A low molecular weight compound that unites with a protein (the repressor) to prevent transcription in a repressible operon.

Cork • A waterproofing tissue in plants, with suberin-containing cell walls. Produced by a cork cambium.

Corolla (ko role' lah) [L.: diminutive of *corona*: wreath, crown] • All of the petals of a flower, collectively.

Coronary (kor' oh nair ee) • Referring to the blood vessels of the heart.

Corpus luteum (kor' pus loo' tee um) [L. *corpus*: body + *luteum*: yellow] A structure formed from a follicle after ovulation; it produces hormones important to the maintenance of pregnancy.

Cortex [L.: bark or rind] • (1) In plants, the tissue between the epidermis and the vascular tissue of a stem or root. (2) In animals, the outer tissue of certain organs, such as the adrenal cortex and cerebral cortex.

Corticosteroids • Steroid hormones produced and released by the cortex of the adrenal gland.

Cost • See energetic cost, opportunity cost, risk cost.

Cotyledon (kot' ul lee' dun) [Gr. *kotyledon*: a hollow space] • A "seed leaf." An embryonic organ which stores and digests reserve materials; may expand when seed germinates.

Countercurrent exchange • An adaptation that promotes maximum exchange of heat or any diffusible substance between two fluids by the fluids flow in opposite directions through parallel tubes in close approximation to each other. An example is countercurrent heat exchange between arterioles and venules in the extremities of some animals.

Covalent bond • A chemical bond that arises from the sharing of electrons between two atoms. Usually a strong bond.

Crassulacean acid metabolism (CAM) • A metabolic pathway enabling the plants that possess it to store carbon dioxide at night and then perform photosynthesis during the day with stomata closed.

Crista (plural: cristae) • A small, shelflike projection of the inner membrane of a mitochondrion; the site of oxidative phosphorylation.

Critical night length • In the photoperiodic flowering response of short-day plants, the length of night above which flowering occurs and below which the plant remains vegetative. (The reverse applies in the case of long-day plants.)

Critical period • The age during which some particular type of learning must take place or during which it occurs much more easily than at other times. Typical of song learning among birds.

Cross section (also called a transverse section) • A section taken perpendicular to the longest axis of a structure.

Crossing over • The mechanism by which linked markers undergo recombination. In general, the term refers to the reciprocal exchange of corresponding segments between two homologous chromatids.

CRP • The cAMP receptor protein that interacts with the promoter to enhance transcription; a lowered cAMP concentration results in catabolite repression.

Crustacean (crus tay' see an) • A member of the phylum Crustacea, such as a crab, shrimp, or sowbug.

Cryptic appearance [Gr. *kryptos*: hidden] • The resemblance of an animal to some part of its environment, which helps it to escape detection by predators.

Cryptochromes [Gr. *kryptos*: hidden + *kroma*: color] • Photoreceptors mediating some blue-light effects in plants and animals.

Culture • (1) A laboratory association of organisms under controlled conditions. (2) The collection of knowledge, tools, values, and rules that characterize a human society.

Cuticle • A waxy layer on the outer surface of a plant or an insect, tending to retard water loss.

Cyanobacteria (sigh an' o bacteria) [Gr. *kuanos*: the color blue] • A division of photosynthetic bacteria, formerly referred to as blue-green algae; they lack sexual reproduction, and they use chlorophyll *a* in their photosynthesis.

Cyclic AMP • See cAMP.

Cyclins • Proteins that activate cyclin-dependent kinases, bringing about transitions in the cell cycle.

Cyclin-dependent kinase (cdk) • A kinase is an enzyme that catalzyes the addition of phosphate groups from ATP to target molecules. Cdk's target proteins involved in transitions in the cell cycle and are active only when complexed to additional protein subunits, cyclins.

Cyst (sist) [Gr. *kystis*: pouch] • (1) A resistant, thick-walled cell formed by some protists and other organisms. (2) An abnormal sac, containing a liquid or semisolid substance, produced in response to injury or illness.

Cytochromes (sy' toe chromes) [Gr. *kytos*: container + *chroma*: color] • Iron-containing red proteins, components of the electron-transfer chains in photophosphorylation and respiration.

Cytokinesis (sy' toe kine ee' sis) [Gr. *kytos*: container + *kinein*: to move] • The division of the cytoplasm of a dividing cell. (Contrast with mitosis.)

Cytokinin (sy' toe kine' in) [Gr. *kytos*: container + *kinein*: to move] • A member of a class of plant growth substances playing roles in senescence, cell division, and other phenomena.

Cytoplasm • The contents of the cell, excluding the nucleus.

Cytoplasmic determinants • In animal development, gene products whose spatial distribution may determine such things as embryonic axes.

Cytosine (site' oh seen) • A nitrogen-containing base found in DNA and RNA.

Cytoskeleton • The network of microtubules and microfilaments that gives a eukaryotic cell its shape and its capacity to arrange its organelles and to move.

Cytosol • The fluid portion of the cytoplasm, excluding organelles and other solids.

Cytotoxic T cells • Cells of the cellular immune system that recognize and directly eliminate virus-infected cells. (Contrast with helper T cells, suppressor T cells.)

Decomposer • See detritivore.

Degeneracy • The situation in which a single amino acid may be represented by any of two or more different codons in messenger RNA. Most of the amino acids can be represented by more than one codon.

Degradative succession • Ecological succession occuring on the dead remains of the bodies of plants and animals, as when leaves or animal bodies rot.

Deletion (genetic) • A mutation resulting from the loss of a continuous segment of a gene or chromosome. Such mutations never revert to wild type. (Contrast with duplication, point mutation.)

Deme (deem) [Gr. *demos*: common people] • Any local population of individuals belonging to the same species that interbreed with one another.

Demographic processes • The events—such as births, deaths, immigration, and emigration—that determine the number of individuals in a population.

Demographic stochasticity • Random variations in the factors influencing the size, density, and distribution of a population.

Demography • The study of dynamical changes in the sizes, densities, and distributions of populations.

Denaturation • Loss of activity of an enzyme or nucleic acid molecule as a result of structural changes induced by heat or other means.

Dendrite [Gr. *dendron*: a tree] • A fiber of a neuron which often cannot carry action potentials. Usually much branched and relatively short compared with the axon, and commonly carries information to the cell body of the neuron.

Denitrification • Metabolic activity by which inorganic nitrogen-containing ions are reduced to form nitrogen gas and other products; carried on by certain soil bacteria.

Density dependence • Change in the severity of action of agents affecting birth and death rates within populations that are directly or inversely related to population density.

Density independence • The state where the severity of action of agents affecting birth and death rates within a population does not change with the density of the population.

Deoxyribonucleic acid • See DNA.

Depolarization • A change in the electric potential across a membrane from a condition in which the inside of the cell is more negative than the outside to a condition in which the inside is less negative, or even positive, with reference to the outside of the cell. (Contrast with hyperpolarization.)

Derived trait • A trait found among members of a lineage that was not present in the ancestors of that lineage.

Dermal tissue system • The outer covering of a plant, consisting of epidermis in the young plant and periderm in a plant with extensive secondary growth. (Contrast with ground tissue system and vascular tissue system.)

Desmosome (dez′ mo sowm) [Gr. *desmos*: bond + *soma*: body] • An adhering junction between animal cells.

Determination • Process whereby an embryonic cell or group of cells becomes fixed into a predictable developmental pathway.

Detritivore (di try′ ti vore) [L. *detritus*: worn away + *vorare*: to devour] • An organism that obtains its energy from the dead bodies and/or waste products of other organisms.

Deuterostome • A major evolutionary lineage in animals, characterized by radial cleavage, enterocoelous development, and other traits. (Compare with protostome.)

Development • Progressive change, as in structure or metabolism; in most kinds of organisms, development continues throughout the life of the organism.

Diaphragm (dye′ uh fram) [Gr. *diaphrassein*, to barricade] • (1) A sheet of muscle that separates the thoracic and abdominal cavities in mammals; responsible for the action of breathing. (2) A method of birth control in which a sheet of rubber is fitted over the woman's cervix, blocking the entry of sperm.

Diastole (dye ahs′ toll ee) [Gr.: dilation] • The portion of the cardiac cycle when the heart muscle relaxes. (Contrast with systole.)

Dicot (short for dicotyledon) [Gr. *di*: two + *kotyledon*: a hollow space] • This term, not used in this book, formerly referred to all angiosperms other than the monocots. (See eudicot, monocot.)

Differentiation • Process whereby originally similar cells follow different developmental pathways. The actual expression of determination.

Diffusion • Random movement of molecules or other particles, resulting in even distribution of the particles when no barriers are present.

Digestibility-reducing chemicals • Defensive chemicals produced by plants that make the plant's tissued difficult to digest.

Digestion • Enzyme-catalyzed process by which large, usually insoluble, molecules (foods) are hydrolyzed to form smaller molecules of soluble substances.

Dihybrid cross • A mating in which the parents differ with respect to the alleles of two loci of interest.

Dikaryon (di care′ ee ahn) [Gr. *dis*: two + *karyon*: kernel] • A cell or organism carrying two genetically distinguishable nuclei. Common in fungi.

Dioecious (die eesh′ us) [Gr.: two houses] • Organisms in which the two sexes are "housed" in two different individuals, so that eggs and sperm are not produced in the same individuals. Examples: humans, fruit flies, oak trees, date palms. (Contrast with monoecious.)

Diploblastic • Having two cell layers. (Contrast with triploblastic.)

Diploid (dip′ loid) [Gr. *diploos*: double] • Having a chromosome complement consisting of two copies (homologues) of each chromosome. A diploid individual (or cell) usually arises as a result of the fusion of two gametes, each with just one copy of each chromosome. Thus, the two homologues in each chromosome pair in a diploid cell are of separate origin, one derived from the female parent and one from the male parent.

Directional selection • Selection in which phenotypes at one extreme of the population distribution are favored. (Contrast with disruptive selection; stabilizing selection.)

Disaccharide • A carbohydrate made up of two monosaccharides (simple sugars).

Dispersal stage • Stage in its life history at which an organism moves from its birthplace to where it will live as an adult.

Displacement activity • Apparently irrelevant behavior performed by an animal under conflict situations, especially when tendencies to attack and escape are closely balanced.

Display • A behavior that has evolved to influence the actions of other individuals.

Disruptive selection • Selection in which phenotypes at both extremes of the population distribution are favored. (Contrast with directional selection; stabilizing selection.)

Distal • Away from the point of attachment or other reference point. (Contrast with proximal.)

Disturbance • A short-term event that disrupts populations, communities, or ecosystems by changing the environment.

Diverticulum (di ver tic′ u lum) [L. *divertere*: turn away] • A small cavity or tube that connects to a major cavity or tube.

Division • A term used by some microbiologists and formerly by botanists, corresponding to the term phylum.

DNA (deoxyribonucleic acid) • The fundamental hereditary material of all living organisms. In eukaryotes, stored primarily in the cell nucleus. A nucleic acid using deoxyribose rather than ribose.

DNA chip • A small glass or plastic square onto which thousands of single-stranded DNA sequences are fixed. Hybridization of cell-derived RNA or DNA to the target sequences can be performed. (See DNA hybridization.)

DNA hybridization • A process by which DNAs from two species are mixed and heated so that interspecific double helixes are formed.

DNA ligase • Enzyme that unites Okazaki fragments of the lagging strand during DNA replication; also mends breaks in DNA strands. It connects pieces of a DNA strand and is used in recombinant DNA technology.

DNA methylation • Addition of methyl groups to DNA; plays role in regulation of gene expression; protects a bacterium's DNA against its restriction endonucleases.

DNA polymerase • Any of a group of enzymes that catalyze the formation of DNA strands from a DNA template.

Domain • The largest unit in the current taxonomic nomenclature. Members of the three domains (Bacteria, Archaea, and Eukarya) are believed to have been evolving independently of each other for at least a billion years.

Dominance • In genetic terminology, the ability of one allelic form of a gene to determine the phenotype of a heterozygous individual, in which the homologous chromosome carries both it and a different allele. For example, if *A* and *a* are two allelic forms of a gene, *A* is said to be dominant to *a* if *AA* diploids and *Aa* diploids are phenotypically identical and are distinguishable from *aa* diploids. The *a* allele is said to be **recessive**.

Dominance hierarchy • In animal behavior, the set of relationships within a group of animals, usually established and maintained by aggression, in which one individual has precedence over all others in eating, mating, and other activities.

Dormancy • A condition in which normal activity is suspended, as in some seeds and buds.

Dorsal [L. *dorsum*: back] • Pertaining to the back or upper surface. (Contrast with ventral.)

Double fertilization • Process virtually unique to angiosperms in which one sperm nucleus combines with the egg to produce a zygote, and the other sperm nucleus combines with the two polar nuclei to produce the first cell of the triploid endosperm.

Double helix • Of DNA: molecular structure in which two complementary polynucleotide strands, antiparallel to each other, form a right-handed spiral.

Duodenum (doo′ uh dee′ num) • The beginning portion of the vertebrate small intestine. (Contrast with ileum, jejunum.)

Duplication (genetic) • A mutation resulting from the introduction into the genome of an extra copy of a segment of a gene or chromosome. (Contrast with deletion, point mutation.)

Dynein [Gr. *dunamis*: power] • A protein that undergoes conformational changes and thus plays a part in the movement of eukaryotic flagella and cilia.

Ecdysone (eck die' sone) [Gr. *ek*: out of + *dyo*: to clothe] • In insects, a hormone that induces molting.

Ecological biogeography • The study of the distributions of organisms from an ecological perspective, usually concentrating on migration, dispersal, and species interactions.

Ecological community • The species living together at a particular site.

Ecological niche (nitch) [L. *nidus*: nest] • The functioning of a species in relation to other species and its physical environment.

Ecological succession • The sequential replacement of one population assemblage by another in a habitat following some disturbance. Succession sometimes ends in a relatively stable ecosystem.

Ecology [Gr. *oikos*: house + *logos*: discourse, study] • The scientific study of the interaction of organisms with their environment, including both the physical environment and the other organisms that live in it.

Ecoregion • A large geographic unit characterized by a typical climate and a widespread assemblage of similar species.

Ecosystem (eek' oh sis tum) • The organisms of a particular habitat, such as a pond or forest, together with the physical environment in which they live.

Ecto- (eck' toh) [Gr.: outer, outside] • A prefix used to designate a structure on the outer surface of the body. For example, ectoderm. (Contrast with endo- and meso-.)

Ectoderm [Gr. *ektos*: outside + *derma*: skin] • The outermost of the three embryonic tissue layers first delineated during gastrulation. Gives rise to the skin, sense organs, nervous system, etc.

Ectotherm [Gr. *ektos*: outside + *thermos*: heat] • An animal unable to control its body temperature. (Contrast with endotherm.)

Edema (i dee' mah) [Gr. *oidema*: swelling] • Tissue swelling caused by the accumulation of fluid.

Edge effect • The changes in ecological processes in a community caused by physical and biological factors originating in an adjacent community.

Effector • Any organ, cell, or organelle that moves the organism through the environment or else alters the environment to the organism's advantage. Examples include muscle, bone, and a wide variety of exocrine glands.

Effector cell • A lymphocyte that performs a role in the immune system without further differentiation.

Effector phase • In this phase of the immune response, effector T cells called cytotoxic T cells attack virus-infected cells, and effector helper T cells assist B cells to

differentiate into plasma cells, which release antibodies.

Efferent [L. *ex*: out + *ferre*: to bear] • Away from, as in neurons that conduct action potentials out from the central nervous system, or arterioles that conduct blood away from a structure. (Contrast with afferent.)

Egg • In all sexually reproducing organisms, the female gamete; in birds, reptiles, and some other vertebrates, a structure witin which early embryonic development occurs.

Elasticity • The property of returning quickly to a former state after a disturbance.

Electrocardiogram (EKG) • A graphic recording of electrical potentials from the heart.

Electroencephalogram (EEG) • A graphic recording of electrical potentials from the brain.

Electromyogram (EMG) • A graphic recording of electrical potentials from muscle.

Electron (e lek' tron) [L. *electrum*: amber (associated with static electricity), from Gr. *slektor*: bright sun (color of amber)] • One of the three most important fundamental particles of matter, with mass approximately 0.00055 amu and charge −1.

Electronegativity • The tendency of an atom to attract electrons when it occurs as part of a compound.

Electrophoresis (e lek' tro fo ree' sis) [L. *electrum*: amber + Gr. *phorein*: to bear] • A separation technique in which substances are separated from one another on the basis of their electric charges and molecular weights.

Electrotonic potential • In neurons, a hyperpolarization or small depolarization of the membrane potential induced by the application of a small electric current. (Contrast with action potential, resting potential.)

Elemental substance • A substance composed of only one type of atom.

Embolus (em' buh lus) [Gr. *embolos*: inserted object; stopper] • A circulating blood clot. Blockage of a blood vessel by an embolus or by a bubble of gas is referred to as an **embolism**. (Contrast with thrombus.)

Embryo [Gr. *en*-: in + *bryein*: to grow] • A young animal, or young plant sporophyte, while it is still contained within a protective structure such as a seed, egg, or uterus.

Embryo sac • In angiosperms, the female gametophyte. Found within the ovule, it consists of eight or fewer cells, membrane bounded, but without cellulose walls between them.

Emergent property • A property of a complex system that is not exhibited by its individual component parts.

Emigration • The deliberate and usually oriented departure of an organism from the habitat in which it has been living.

3′ End (3-prime) • The end of a DNA or RNA strand that has a free hydroxyl group at the 3′-carbon of the sugar (deoxyribose or ribose).

5′ End (5-prime) • The end of a DNA or RNA strand that has a free phosphate group at the 5′-carbon of the sugar (deoxyribose or ribose).

Endemic (en dem' ik) [Gr. *endemos*: dwelling in a place] • Confined to a particular region, thus often having a comparatively restricted distribution.

Endergonic reaction • One for which energy must be supplied. (Contrast with exergonic reaction.)

Endo- [Gr.: within, inside] • A prefix used to designate an innermost structure. For example, endoderm, endocrine. (Contrast with ecto-, meso-.)

Endocrine gland (en' doh krin) [Gr. *endon*: inside + *krinein*: to separate] • Any gland, such as the adrenal or pituitary gland of vertebrates, that secretes certain substances, especially hormones, into the body through the blood.

Endocrinology • The study of hormones and their actions.

Endocytosis • A process by which liquids or solid particles are taken up by a cell through invagination of the plasma membrane. (Contrast with exocytosis.)

Endoderm [Gr. *endon*: within + *derma*: skin] • The innermost of the three embryonic tissue layers first delineated during gastrulation. Gives rise to the digestive and respiratory tracts and structures associated with them.

Endodermis [Gr. *endon*: within + *derma*: skin] • In plants, a specialized cell layer marking the inside of the cortex in roots and some stems. Frequently a barrier to free diffusion of solutes.

Endomembrane system • Endoplasmic reticulum plus Golgi apparatus plus, when present, lysosomes, thus, a system of membranes that exchange material with one another.

Endoplasmic reticulum [Gr. *endon*: within + L. *plasma*: form; L. *reticulum*: little net] • A system of membrane-bounded tubes and flattened sacs found in the cytoplasm of eukaryotes. Exists as rough ER, studded with ribosomes; and smooth ER, lacking ribosomes.

Endorphins • Naturally occurring, opiate-like substances in the mammalian brain.

Endoskeleton [Gr. *endon*: within + *skleros*: hard] • A skeleton covered by other, soft body tissues. (Contrast with exoskeleton.)

Endosperm [Gr. *endon*: within + *sperma*: seed] • A specialized triploid seed tissue found only in angiosperms; contains stored food for the developing embryo.

Endosymbiosis [Gr. *endon*: within + *syn*: together + *bios*: life] • The living together of two species, with one living inside the body (or even the cells) of the other.

Endosymbiotic theory • Theory that the eukaryotic cell evolved from a prokaryote that contained other, endosymbiotic prokaryotes.

Endotherm [Gr. *endon*: within + *thermos*: hot] • An animal that can control its body temperature by the expenditure of its own metabolic energy. (Contrast with ectotherm.)

Endotoxins [Gr. *endon*: within + L. *toxicum*: poison] • Lipopolysaccharides released by the lysis of some Gram-negative bacteria that cause fever and vomiting in a host organism.

Energetic cost • The difference between the energy an animal would have expended had it rested, and that expended in performing a behavior.

Energy • The capacity to do work.

Enhancer • In eukaryotes, a DNA sequence, lying on either side of the gene it regulates, that stimulates a specific promoter.

Enterocoelous development • A pattern of development in which the coelum is formed by an outpocketing of the embryonic gut (enteron).

Enterokinase (ent uh row kine' ase) • An enzyme secreted by the mucosa of the duodenum. It activates the zymogen trypsinogen to create the active digestive enzyme trypsin.

Entrainment • With respect to circadian rhythms, the process whereby the period is adjusted to match the 24-hour environmental cycle.

Entropy (en' tro pee) [Gr. *en*: in + *tropein*: to change] • A measure of the degree of disorder in any system. A perfectly ordered system has zero entropy; increasing disorder is measured by positive entropy. Spontaneous reactions in a closed system are always accompanied by an increase in disorder and entropy.

Environment • An organism's surroundings, both living and nonliving; includes temperature, light intensity, and all other species that influence the focal organism.

Environmental toxicology • The study of the distribution and effects of toxic compounds in the environment.

Enzyme (en' zime) [Gr. *en*: in + *zyme*: yeast] • A protein, on the surface of which are chemical groups so arranged as to make the enzyme a catalyst for a chemical reaction.

Epi- [Gr.: upon, over] • A prefix used to designate a structure located on top of another; for example: epidermis, epiphyte.

Epicotyl (epp' i kot' il) [Gr. *epi*: upon + *kotyle*: something hollow] • That part of a plant embryo or seedling that is above the cotyledons.

Epidermis [Gr. *epi*: upon + *derma*: skin] • In plants and animals, the outermost cell layers. (Only one cell layer thick in plants.)

Epididymis (epuh did' uh mus) [Gr. *epi*: upon + *didymos*: testicle] • Coiled tubules in the testes that store sperm and conduct sperm from the seiminiferous tubules to the vas deferens.

Epinephrine (ep i nef' rin) [Gr. *epi*: upon + *nephros*: a kidney] • The "fight or flight" hormone. Produced by the medulla of the adrenal gland, it also functions as a neurotransmitter. Also known as adrenaline.

Epiphyte (ep' e fyte) [Gr. *epi*: upon + *phyton*: plant] • A specialized plant that grows on the surface of other plants but does not parasitize them.

Episome • A plasmid that may exist either free or integrated into a chromosome. (See plasmid.)

Epistasis • An interaction between genes, in which the presence of a particular allele of one gene determines whether another gene will be expressed.

Epithelium • In animals, a layer of cells covering or lining an external surface or a cavity.

Equilibrium • (1) In biochemistry, a state in which forward and reverse reactions are proceeding at counterbalancing rates, so there is no observable change in the concentrations of reactants and products. (2) In evolutionary genetics, a condition in which allele and genotype frequencies in a population are constant from generation to generation.

Erythrocyte (ur rith' row sight) [Gr. *erythros*: red + *kytos*: hollow vessel] • A red blood cell.

Esophagus (i soff' i gus) [Gr. *oisophagos*: gullet] • That part of the gut between the pharynx and the stomach.

Ester linkage • A condensation (water-releasing) reaction in which the carboxyl group of a fatty acid reacts with the hydroxyl group of an alcohol. Lipids are formed in this way.

Estivation (ess tuh vay' shun) [L. *aestivalis*: summer] • A state of dormancy and hypometabolism that occurs during the summer; usually a means of surviving drought and/or intense heat. Contrast with hibernation.

Estrogen • Any of several steroid sex hormones, produced chiefly by the ovaries in mammals.

Estrus (es' truss) [L. *oestrus*: frenzy] • The period of heat, or maximum sexual receptivity, in some female mammals. Ordinarily, the estrus is also the time of release of eggs in the female.

Ethylene • One of the plant hormones, the gas $H_2C=CH_2$.

Euchromatin • Chromatin that is diffuse and non-staining during interphase; may be transcribed. (Contrast with heterochromatin.)

Eudicots (yew di' kots) [Gr. *eu*: true + *di*: two + *kotyledon*: a cup-shaped hollow] • Members of the angiosperm class Eudicotyledones, flowering plants in which the embryo produces two cotyledons prior to germination. Leaves of most eudicots have major veins arranged in a branched or reticulate pattern.

Eukaryotes (yew car' ry otes) [Gr. *eu*: true + *karyon*: kernel or nucleus] • Organisms whose cells contain their genetic material inside a nucleus. Includes all life other than the viruses, Archaebacteria, and Eubacteria.

Eusocial • Term applied to insects, such as termites, ants, and many bees and wasps, in which individuals cooperate in the care of offspring, there are sterile castes, and generations overlap.

Eutrophication (yoo trofe' ik ay' shun) [Gr. *eu-*: well + *trephein*: to flourish] • The addition of nutrient materials to a body of water, resulting in changes to species composition therein.

Evolution • Any gradual change. Organic evolution, often referred to as evolution, is any genetic and resulting phenotypic change in organisms from generation to generation.

Evolutionary agent • Any factor that influences the direction and rate of evolutionary changes.

Evolutionarily conservative • Traits of organisms that evolve very slowly.

Evolutionary innovations • Major changes in body plans of organisms; these have been very rare during evolutionary history.

Evolutionary radiation • The proliferation of species within a single evolutionary lineage.

Evolutionary reversal • The reappearance of the ancestral state of a trait in a lineage in which that trait had acquired a derived state.

Excision repair • The removal and damaged DNA and its replacement by the appropriate nucleotides.

Excitatory postsynaptic potential (EPSP) • A change in the resting potential of a postsynaptic membrane in a positive (depolarizing) direction. (Contrast with inhibitory postsynaptic potential.)

Excretion • Release of metabolic wastes by an organism.

Exergonic reaction • A reaction in which free energy is released. (Contrast with endergonic reaction.)

Exo- (eks' oh) • Same as ecto-.

Exocrine gland (eks' oh krin) [Gr. *exo*: outside + *krinein*: to separate] • Any gland, such as a salivary gland, that secretes to the outside of the body or into the gut.

Exocytosis • A process by which a vesicle within a cell fuses with the plasma membrane and releases its contents to the outside. (Contrast with endocytosis.)

Exon • A portion of a DNA molecule, in eukaryotes, that codes for part of a polypeptide. (Contrast with intron.)

Exoskeleton (eks' oh skel' e ton) [Gr. *exos*: outside + *skleros*: hard] • A hard covering on the outside of the body to which muscles are attached. (Contrast with endoskeleton.)

Exotoxins • Highly toxic proteins released by living, multiplying bacteria.

Experiment • A scientific method in which particular factors are manipulated while other factors are held constant so that the potential influences of the manipulated factors can be determined.

Exponential growth • Growth, especially in the number of organisms in a population, which is a simple function of the size of the growing entity: the larger the entity, the faster it grows. (Contrast with logistic growth.)

Expression vector • A DNA vector, such as a plasmid, that carries a DNA sequence that includes the adjacent sequences for its expression into mRNA and protein in a host cell.

Expressivity • The degree to which a genotype is expressed in the phenotype— may be affected by the environment.

Extensor • A muscle the extends an appendage.

Extinction • The termination of a lineage of organisms.

Extrinsic protein • A membrane protein found only on the surface of the membrane. (Contrast with intrinsic protein.)

F_1 generation • The immediate progeny of a parental (P) mating; the first filial generation.

F_2 generation • The immediate progeny of a mating between members of the F_1 generation.

Facilitated diffusion • Passive movement through a membrane involving a specific carrier protein; does not proceed against a concentration gradient. (Contrast with active transport, free diffusion.)

Family • In taxonomy, the category below the order and above the genus; a group of related, similar genera.

Fat • A triglyceride that is solid at room temperature. (Contrast with oil.)

Fatty acid • A molecule with a long hydrocarbon tail and a carboxyl group at the other end. Found in many lipids.

Fauna (faw' nah) • All of the animals found in a given area. (Contrast with flora.)

Feces [L. *faeces*: dregs] • Waste excreted from the digestive system.

Feedback control • Control of a particular step of a multistep process, induced by the presence or absence of a product of one of the later steps. A thermostat regulating the flow of heating oil to a furnace in a home is a negative feedback control device.

Fermentation (fur men tay' shun) [L. *fermentum*: yeast] • The degradation of a substance such as glucose to smaller molecules with the extraction of energy, without the use of oxygen (i.e., anaerobically). Involves the glycolytic pathway.

Fertilization • Union of gametes. Also known as syngamy.

Fertilization membrane • A membrane surrounding an animal egg which becomes rapidly raised above the egg surface within seconds after fertilization, serving to prevent entry of a second sperm.

Fetus • The latter stages of an embryo that is still contained in an egg or uterus; in humans, the unborn young from the eighth week of pregnancy to the moment of birth.

Fiber • An elongated and tapering cell of flowering plants, usually with a thick cell wall. Serves a support function.

Fibrin • A protein that polymerizes to form long threads that provide structure to a blood clot.

Filter feeder • An organism that feeds upon much smaller organisms, that are suspend-

ed in water or air, by means of a straining device.

Filtration • In the excretory physiology of some animals, the process by which the initial urine is formed; water and most solutes are transferred into the excretory tract, while proteins are retained in the blood or hemolymph.

First law of thermodynamics • Energy can be neither created nor destroyed.

Fission • Reproduction of a prokaryote by division of a cell into two comparable progeny cells.

Fitness • The contribution of a genotype or phenotype to the composition of subsequent generations, relative to the contribution of other genotypes or phenotypes. (See inclusive fitness.)

Fixed action pattern • A behavior that is genetically programmed.

Flagellum (fla jell' um) (plural: flagella) [L. *flagellum*: whip] • Long, whiplike appendage that propels cells. Prokaryotic flagella differ sharply from those found in eukaryotes.

Flexor • A muscle that flexes an appendage.

Flora (flore' ah) • All of the plants found in a given area. (Contrast with fauna.)

Florigen • A plant hormone (not yet isolated) involved in the conversion of a vegetative shoot apex to a flower.

Flower • The total reproductive structure of an angiosperm; its basic parts include the calyx, corolla, stamens, and carpels.

Fluorescence • The emission of a photon of visible light by an excited atom or molecule.

Follicle [L. *folliculus*: little bag] • In female mammals, an immature egg surrounded by nutritive cells.

Follicle-stimulating hormone • A gonadotropic hormone produced by the anterior pituitary.

Food chain • A portion of a food web, most commonly a simple sequence of prey species and the predators that consume them.

Food web • The complete set of food links between species in a community; a diagram indicating which ones are the eaters and which are consumed.

Forb • Any broad-leaved (dicotyledonous), herbaceous plant. Especially applied to such plants growing in grasslands.

Fossil • Any recognizable structure originating from an organism, or any impression from such a structure, that has been preserved over geological time.

Fossil fuel • A fuel (particularly petroleum products) composed of the remains of organisms that lived in the remote past.

Founder effect • Random changes in allele frequencies resulting from establishment of a population by a very small number of individuals.

Fovea [L. *fovea*; a small pit] • The area, in the vertebrate retina, of most distinct vision.

Frame-shift mutation • A mutation resulting from the addition or deletion of a single base pair in the DNA sequence of a gene. As

a result of this, mRNA transcribed from such a gene is translated normally until the ribosome reaches the point at which the mutation has occurred. From that point on, codons are read out of proper register and the amino acid sequence bears no resemblance to the normal sequence. (Contrast with missense mutation, nonsense mutation, synonymous mutation.)

Free energy • That energy which is available for doing useful work, after allowance has been made for the increase or decrease of disorder. Designated by the symbol G (for Gibbs free energy), and defined by: $G = H - TS$, where H = heat, S = entropy, and T = absolute (Kelvin) temperature.

Frequency-dependent selection • Selection that changes in intensity with the proportion of individuals having the trait.

Fruit • In angiosperms, a ripened and mature ovary (or group of ovaries) containing the seeds. Sometimes applied to reproductive structures of other groups of plants, and includes any adjacent parts which may be fused with the reproductive structures.

Fruiting body • A structure that bears spores.

Fundamental niche • The range of condition under which an organism could survive if it were the only one in the environment. (Contrast with realized niche.)

Fungus (fung' gus) • A member of the kingdom Fungi, a (usually) multicellular eukaryote with absorptive nutrition.

G_1 phase • In the cell cycle, the gap between the end of mitosis and the onset of the S phase.

G_2 phase • In the cell cycle, the gap between the S (synthesis) phase and the onset of mitosis.

G protein • A membrane protein involved in signal transduction; characterized by binding guanyl nucleotides. The activation of certain receptors activates the G protein, which in turn activates adenylate cyclase. G protein activation involves binding a GTP molecule in place of a GDP molecule.

Gametangium (gam i tan' gee um) [Gr. *gamos*: marriage + *angeion*: vessel or reservoir] • Any plant or fungal structure within which a gamete is formed.

Gamete (gam' eet) [Gr. *gamete*: wife, *gametes*: husband] • The mature sexual reproductive cell: the egg or the sperm.

Gametocyte (ga meet' oh site) [Gr. *gamete*: wife, *gametes*: husband + *kytos*: cell] • The cell that gives rise to sex cells, either the eggs or the sperm. (See oocyte and spermatocyte.)

Gametogenesis (ga meet' oh jen' e sis) [Gr. *gamete*: wife, *gametes*: husband + *genesis*: source] • The specialized series of cellular divisions that leads to the production of sex cells (gametes). (Contrast with oogenesis and spermatogenesis.)

Gametophyte (ga meet' oh fyte) • In plants and photosynthetic protists with alternation of generations, the haploid phase that produces the gametes. (Contrast with sporophyte.)

Ganglion (gang' glee un) [Gr.: tumor] • A group or concentration of neuron cell bodies.

Gap junction • A 2.7-nanometer gap between plasma membranes of two animal cells, spanned by protein channels. Gap junctions allow chemical substances or electrical signals to pass from cell to cell.

Gas exchange • In animals, the process of taking up oxygen from the environment and releasing carbon dioxide to the environment.

Gastrovascular cavity • Serving for both digestion (gastro) and circulation (vascular); in particular, the central cavity of the body of jellyfish and other cnidarians.

Gastrula (gas' true luh) [Gr. *gaster*: stomach] • An embryo forming the characteristic three cell layers (ectoderm, endoderm, and mesoderm) which will give rise to all of the major tissue systems of the adult animal.

Gastrulation • Development of a blastula into a gastrula.

Gated channel • A channel (membrane protein) that opens and closes in response to binding of specific molecules or to changes in membrane potential.

Gel electrophoresis (jel ul lec tro for' eesis) • A semisolid matrix suspended in a salty buffer in which molecules can be separated on the basis of their size and change when current is passed through the gel.

Gene [Gr. *gen*: to produce] • A unit of heredity. Used here as the unit of genetic function which carries the information for a single polypeptide.

Gene amplification • Creation of multiple copies of a particular gene, allowing the production of large amounts of the RNA transcript (as in rRNA synthesis in oocytes).

Gene cloning • Formation of a clone of bacteria or yeast cells containing a particular foreign gene.

Gene family • A set of identical, or once-identical, genes, derived from a single parent gene; need not be on the same chromosomes; classic example is the globin family in vertebrates.

Gene flow • The exchange of genes between different species (an extreme case referred to as hybridization) or between different populations of the same species caused by migration following breeding.

Gene pool • All of the genes in a population.

Gene therapy • Treatment of a genetic disease by providing patients with cells containing wild type alleles for the genes that are nonfunctional in their bodies.

Generative nucleus • In a pollen tube, a haploid nucleus that undergoes mitosis to produce the two sperm nuclei that participate in double fertilization. (Contrast with tube nucleus.)

Genet • The genetic individual of a plant that is composed of a number of nearly identical but repeated units.

Genetic drift • Changes in gene frequencies from generation to generation in a small population as a result of random processes.

Genetic stochasticity • Variation in the frequencies of alleles and genotypes in a population over time.

Genetics • The study of heredity.

Genetic structure • The frequencies of alleles and genotypes in a population.

Genome (jee' nome) • The genes in a complete haploid set of chromosomes.

Genotype (jean' oh type) [Gr. *gen*: to produce + *typos*: impression] • An exact description of the genetic constitution of an individual, either with respect to a single trait or with respect to a larger set of traits. (Contrast with phenotype.)

Genus (jean' us) (plural: genera) [Gr. *genos*: stock, kind] • A group of related, similar species.

Geotropism • See gravitropism.

Germ cell • A reproductive cell or gamete of a multicellular organism.

Germination • The sprouting of a seed or spore.

Gestation (jes tay' shun) [L. *gestare*: to bear] • The period during which the embryo of a mammal develops within the uterus. Also known as **pregnancy**.

Gibberellin (jib er el' lin) [L. *gibberella*: hunchback (refers to shape of a reproductive structure of a fungus that produces gibberellins)] • One of a class of plant growth substances playing roles in stem elongation, seed germination, flowering of certain plants, etc. Named for the fungus *Gibberella*.

Gill • An organ for gas exchange in aquatic organisms.

Gill arch • A skeletal structure that supports gill filaments and the blood vessels that supply them.

Gizzard (giz' erd) [L. *gigeria*: cooked chicken parts] • A very muscular port of the stomach of birds that grinds up food, sometimes with the aid of fragments of stone.

Gland • An organ or group of cells that produces and secretes one or more substances.

Glans penis • Sexually sensitive tissue at the tip of the penis.

Glia (glee' uh) [Gr.: glue] • Cells, found only in the nervous system, which do not conduct action potentials.

Glomerulus (glo mare' yew lus) [L. *glomus*: ball] • Sites in the kidney where blood filtration takes place. Each glomerulus consists of a knot of capillaries served by afferent and efferent arterioles.

Glucocorticoids • Steroid hormones produced by the adrenal cortex. Secreted in response to ACTH, they inhibit glucose uptake by many tissues in addition to mediating other stress responses.

Glucagon • A hormone produced and released by cells in the islets of Langerhans of the pancreas. It stimulates the breakdown of glycogen in liver cells.

Gluconeogenesis • The biochemical synthesis of glucose from other substances, such as amino acids, lactate, and glycerol.

Glucose (glue' kose) [Gr. *gleukos*: sweet wine mash for fermentation] • The most common sugar, one of several monosaccharides with the formula $C_6H_{12}O_6$.

Glycerol (gliss' er ole) • A three-carbon alcohol with three hydroxyl groups, the linking component of phospholipids and triglycerides.

Glycogen (gly' ko jen) • A branched-chain polymer of glucose, similar to starch (which is less branched and may be of lower molecular weight). Exists mostly in liver and muscle; the principal storage carbohydrate of most animals and fungi.

Glycolysis (gly kol' li sis) [from glucose + Gr. *lysis*: loosening] • The enzymatic breakdown of glucose to pyruvic acid. One of the oldest energy-yielding machanisms in living organisms.

Glycosidic linkage • The connection in an oligosaccharide or polysaccharide chain, formed by removal of water during the linking of monosaccharides.by root pressure.

Glyoxysome (gly ox' ee soam) • An organelle found in plants, in which stored lipids are converted to carbohydrates.

Golgi apparatus (goal' jee) • A system of concentrically folded membranes found in the cytoplasm of eukaryotic cells. Plays a role in the production and release of secretory materials such as the digestive enzymes manufactured in the pancreas. First described by Camillo Golgi (1844–1926).

Gonad (go' nad) [Gr. *gone*: seed, that which produces seed] • An organ that produces sex cells in animals: either an ovary (female gonad) or testis (male gonad).

Gonadotropin • A hormone that stimulates the gonads.

Gondwana • The large southern land mass that existed from the Cambrian (540 mya) to the Jurassic (138 mya). Present-day South America, Africa, India, Australia, and Antarctica.

Gram stain • A differential stain useful in characterizing bacteria.

Granum • Within a chloroplast, a stack of thylakoids.

Gravitropism • A directed plant growth response to gravity.

Grazer • An animal that eats the vegetative tissues of herbaceous plants.

Green gland • An excretory organ of crustaceans.

Greenhouse effect • The heating of Earth's atmosphere by gases that are transparent to sunlight but opaque to radiated heat.

Gross primary production • The total energy captured by plants growing in a particular area.

Ground meristem • That part of an apical meristem that gives rise to the ground tissue system of the primary plant body.

Ground tissue system • Those parts of the plant body not included in the dermal or vascular tissue systems. Ground tissues function in storage, photosynthesis, and support.

Group transfer • The exchange of atoms between molecules.

Growth • Irreversible increase in volume (probably the most accurate definition, but at best a dangerous oversimplification).

Growth factors • A group of proteins that circulate in the blood and trigger the normal growth of cells. Each growth factor acts only on certain target cells.

Guanine (gwan'een) • A nitrogen-containing base found in DNA, RNA and GTP.

Guard cells • In plants, paired epidermal cells which surround and control the opening of a stoma (pore).

Gut • An animal's digestive tract.

Guttation • The extrusion of liquid water through openings in leaves, caused by root pressure.

Gymnosperm (jim' no sperm) [Gr. *gymnos*: naked + *sperma*: seed] • A plant, such as a pine or other conifer, whose seeds do not develop within an ovary (hence, the seeds are "naked").

Gyrus (plural: gyri) • The raised or ridged portion of the convoluted surface of the brain. (Contrast to sulcus.)

Habit • The form or pattern of growth characteristic of an organism.

Habitat • The environment in which an organism lives.

Habituation (ha bich' oo ay shun) • The simplest form of learning, in which an animal presented with a stimulus without reward or punishment eventually ceases to respond.

Hair cell • A type of mechanoreceptor in animals.

Half-life • The time required for half of a sample of a radioactive isotope to decay to its stable, nonradioactive form.

Halophyte (hal' oh fyte) [Gr. *halos*: salt + *phyton*: plant] • A plant that grows in a saline (salty) environment.

Haploid (hap' loid) [Gr. *haploeides*: single] • Having a chromosome complement consisting of just one copy of each chromosome. This is the normal "ploidy" of gametes or of asexual spores produced by meiosis or of organisms (such as the gametophyte generation of plants) that grow from such spores without fertilization.

Hardy–Weinberg equililbrium • The percentages of diploid combinations expected from a knowledge of the proportions of alleles in the population if no agents of evolution are acting on the population.

Haustorium (haw stor' ee um) [L. *haustus*: draw up] • A specialized hypha or other structure by which fungi and some parasitic plants draw food from a host plant.

Haversian systems • Units of organization in compact bone that reflect the action of intercommunicating osteoblasts.

Heat-shock proteins • Chaperone proteins expressed in cells exposed to high temperatures or other forms of environmental stress.

Helper T cells • T cells that participate in the activation of B cells and of other T cells; targets of the HIV-I virus, the agent of AIDS. (Contrast with cytotoxic T cells, suppressor T cells.)

Hematocrit (heme at o krit) [Gr. *haima*: blood + *krites*: judge] • The proportion of 100 cc of blood that consists of red blood cells.

Hemizygous (hem' ee zie' gus) [Gr. *hemi*: half + *zygotos*: joined] • In a diploid organism, having only one allele for a given trait, typically the case for X-linked genes in male mammals and Z-linked genes in female birds. (Contrast with homozygous, heterozygous.)

Hemoglobin (hee' mo glow' bin) [Gr. *haima*: blood + L. *globus*: globe] • The colored protein of vertebrate blood (and blood of some invertebrates) which transports oxygen.

Hepatic (heh pat' ik) [Gr. *hepar*: liver] • Pertaining to the liver.

Hepatic duct • The duct that conveys bile from the liver to the gallbladder.

Herbicide (ur' bis ide) • A chemical substance that kills plants.

Herbivore [L. *herba*: plant + *vorare*: to devour] • An animal which eats the tissues of plants. (Contrast with carnivore, detritivore, omnivore.)

Heritable • Able to be inherited; in biology usually refers to genetically determined traits.

Hermaphroditism (her maf' row dite' ism) [Gr. *hermaphroditos*: a person with both male and female traits] • The coexistence of both female and male sex organs in the same organism.

Hertz (abbreviated as Hz) • Cycles per second.

Hetero- [Gr.: other, different] • A prefix used in biology to mean that two or more different conditions are involved; for example, heterotroph, heterozygous.

Heterochromatin • Chromatin that retains its coiling during interphase; generally not transcribed. (Contrast with euchromatin.)

Heterocyst • A large, thick-walled cell in the filaments of certain cyanobacteria; performs nitrogen fixation.

Heterogeneous nuclear RNA (hnRNA) • The product of transcription of a eukaryotic gene, including transcripts of introns.

Heteromorphic (het' er oh more' fik) [Gr. *heteros*: different + *morphe*: form] • having a different form or appearance, as two heteromorphic life stages of a plant. (Contrast with isomorphic.)

Heterosporous (het' er os' por us) • Producing two types of spores, one of which gives rise to a female megaspore and the other to a male microspore. Heterosporous plants produce distinct female and male gametophytes. (Contrast with homosporous.)

Heterotherm • An animal that regulates its body temperature at a constant level at some times but not others, such as a hibernator.

Heterotroph (het' er oh trof) [Gr. *heteros*: different + *trophe*: food] • An organism that requires preformed organic molecules as food. (Contrast with autotroph.)

Heterozygous (het' er oh zie' gus) [Gr. *heteros*: different + *zygotos*: joined] • Of a diploid organism having different alleles of a given gene on the pair of homologues carrying that gene. (Contrast with homozygous.)

Hibernation [L. *hibernus*: winter] • The state of inactivity of some animals during winter; marked by a drop in body temperature and metabolic rate.

Highly repetitive DNA • Short DNA sequences present in millions of copies in the genome, next to each other (in tandem). In a In a reassociation experiment, denatured highly repetitive DNA reanneals very quickly.

Hippocampus • A part of the forebrain that takes part in long-term memory formation.

Histamine (hiss; tah meen) • A substance released within a damaged tissue by a type of white blood cell. Histamines are responsible for aspects of allergice reactions, including the increased vascular permeability that leads to edema (swelling).

Histology • The study of tissues.

Histone • Any one of a group of basic proteins forming the core of a nucleosome, the structural unit of a eukaryotic chromosome. (See nucleosome.)

hnRNA • See heterogeneous nuclear RNA.

Homeobox • A 180-base-pair segment of DNA found in a few genes (called **Hox genes**), perhaps regulating the expression of other genes and thus controlling large-scale developmental processes.

Homeostasis (home' ee o sta' sis) [Gr. *homos*: same + *stasis*: position] • The maintenance of a steady state, such as a constant temperature or a stable social structure, by means of physiological or behavioral feedback responses.

Homeotherm (home' ee o therm) [Gr. *homos*: same + *therme*: heat] • An animal which maintains a constant body temperature by virtue of its own heating and cooling mechanisms. (Contrast with heterotherm, poikilotherm.)

Homeotic genes (home' ee ott' ic) • Genes that determine what entire segments of an animal become. Drastic mutations in these genes cause the transformation of body segments in *Drosophila*. Homeotic genes studied in the plant *Arabidopsis* are called organ identity genes.

Homolog (home' o log') [Gr. *homos*: same + *logos*: word] • One of a pair, or larger set, of chromosomes having the same overall genetic composition and sequence. In diploid organisms, each chromosome inherited from one parent is matched by an identical (except for mutational changes) chromosome—its homolog—from the other parent.

Homology (ho mol' o jee) [Gr. *homologi(a)*: agreement] • A similarity between two structures that is due to inheritance from a common ancestor. The structures are said to be homologous. (Contrast with analogy.)

Homoplasy (home' uh play zee) [Gr. *homos*: same + *plastikos*: to mold] • The presence in several species of a trait not present in their most common ancestor. Can result from convergent evolution, reverse evolution, or parallel evolution.

Homosporous • Producing a single type of spore that gives rise to a single type of gametophyte, bearing both female and male reproductive organs. (Contrast with heterosporous.)

Homozygous (home' o zie' gus) [Gr. *homos*: same + *zygotos*: joined] • Of a diploid organism having identical alleles of a given gene on both homologous chromosomes. An organism may be a "homozygote" with respect to one gene and, at the same time, a "heterozygote" with respect to another. (Contrast with heterozygous.)

Hormone (hore' mone) [Gr. *hormon*: excite, stimulate] • A substance produced in one part of a multicellular organism and transported to another part where it exerts its specific effect on the physiology or biochemistry of the target cells.

Host • An organism that harbors a parasite and provides it with nourishment.

Host–parasite interaction • The dynamic interaction between populations of a host and the parasites that attack it.

Hox genes • See homeobox.

Humoral immune system • The part of the immune system mediated by B cells; it is mediated by circulating antibodies and is active against extracellular bacterial and viral infections.

Humus (hew' muss) • The partly decomposed remains of plants and animals on the surface of a soil. Its characteristics depend primarily upon climate and the species of plants growing on the site.

Hyaluronidase (hill yew ron' uh dase) • An enzyme that digests proteoglycans. Found in sperm cells, it helps digest the coatings surrounding an egg so the sperm can penetrate the egg cell membrane.

Hybrid (high' brid) [L. *hybrida*: mongrel] • The offspring of genetically dissimilar parents. In molecular biology, a double helix formed of nucleic acids from different sources.

Hybridoma • A cell produced by the fusion of an antibody-producing cell with a myeloma cell; it produces monoclonal antibodies.

Hybrid zone • A narrow zone where two populations interbreed, producing hybrid individuals.

Hydrocarbon • A compound containing only carbon and hydrogen atoms.

Hydrogen bond • A chemical bond which arises from the attraction between the slight positive charge on a hydrogen atom and a slight negative charge on a nearby fluorine, oxygen, or nitrogen atom. Weak bonds, but found in great quantities in proteins, nucleic acids, and other biological macromolecules.

Hydrological cycle • The sum total of movement of water from the oceans to the atmosphere, to the soil, and back to the oceans. Some water is cycled many times within compartments of the system before completing one full circuit.

Hydrolyze (hi' dro lize) [Gr. *hydro*: water + *lysis*: cleavage] • To break a chemical bond, as in a peptide linkage, with the insertion of the components of water, –H and –OH, at the cleaved ends of a chain. The digestion of proteins is a hydrolysis.

Hydrophilic [Gr. *hydro*: water + *philia*: love] • Having an affinity for water. (Contrast with hydrophobic.)

Hydrophobic [Gr. *hydro*: water + *phobia*: fear] • Molecules and amino acid side chains, which are mainly hydrocarbons (compounds of C and H with no charged groups or polar groups), have a lower energy when they are clustered together than when they are distributed through an aqueous solution. Because of their attraction for one another and their reluctance to mix with water they are called "hydrophobic." Oil is a hydrophobic substance; phenylalanine is a hydrophobic animo acid in a protein. (Contrast with hydrophilic.)

Hydrostatic skeleton • The incompressible internal liquids of some animals that transfer forces from one part of the body to another when acted upon by the surrounding muscles.

Hydroxyl group • The —OH group, characteristic of alcohols.

Hyperpolarization • A change in the resting potential of a membrane so the inside of a cell becomes more electronegative. (Contrast with depolarization.)

Hypersensitive response • A defensive response of plants to microbial infection; it results in a "dead spot."

Hypertension • High blood pressure.

Hypertonic [Gk. *hyper*: above, over] • Having a greater solute concentration. Said of one solution in comparing it to another. (Contrast with hypotonic, isotonic.)

Hypha (high' fuh) (plural: hyphae) [Gr. *hyphe*: web] • In the fungi, any single filament. May be multinucleate (zygomycetes, ascomycetes) or multicellular (basidiomycetes).

Hypocotyl [Gk. *hypo*: beneath, under + *kotyledon*: hollow space] • That part of the embryonic or seedling plant shoot that is below the cotyledons.

Hypothalamus • The part of the brain lying below the thalamus; it coordinates water balance, reproduction, temperature regulation, and metabolism.

Hypothesis • A tentative answer to a question, from which testable predictions can be generated. (Contrast with theory.)

Hypothetico-deductive method • A method of science in which hypotheses are erected, predictions are made from them, and experiments and observations are performed to test the predictions.

Hypotonic [Gk. *hypo*: beneath, under] • Having a lower solute concentration. Said of one solution in comparing it to another. (Contrast with hypertonic, isotonic.)

Imaginal disc • In insect larvae, groups of cells that develop into specific adult organs.

Immune system [L. *immunis*: exempt] • A system in mammals that recognizes and eliminates or neutralizes either foreign substances or self substances that have been altered to appear foreign.

Immunization • The deliberate introduction of antigen to bring about an immune response.

Immunoglobulins • A class of proteins, with a characteristic structure, active as receptors and effectors in the immune system.

Immunological memory • Certain clones of immune system cells made to respond to an antigen persist. This leads to a more rapid and massive response of the immune system to any subsequent exposure to that antigen.

Immunological tolerance • A mechanism by which an animal does not mount an immune response to the antigenic determinants of its own macromolecules.

Imprinting • (1) In genetics, the differential modification of a gene depending on whether it is present in a male or a female. (2) In animal behavior, a rapid form of learning in which an animal comes to make a particular response, which is maintained for life, to some object or other organism.

Inclusive fitness • The sum of an individual's own fitness (the effect of producing its own offspring: the individual selection component) plus its influence on fitness in relatives other than direct descendants (the kin selection component).

Incomplete dominance • Condition in which the heterozygous phenotype is intermediate between the two homozygous phenotypes.

Incomplete metamorphosis • Insect development in which changes between instars are gradual.

Incus (in' kus) [L. *incus*: anvil] • The middle of the three bones that conduct movements of the eardrum to the oval window of the inner ear. (See malleus, stapes.)

Independent assortment • The random separation during meiosis of nonhomologous chromosomes and of genes carried on nonhomologous chromosomes.

Individual fitness • That component of inclusive fitness that results from an organism producing its own offspring. (Contrast with kin selection component.)

Indoleacetic acid • See auxin.

Inducer • (1) In enzyme systems, a small molecule which, when added to a growth medium, causes a large increase in the level of some enzyme. (2) In embryology, a substance that causes a group of target cells to differentiate in a particular way.

Inducible enzyme • An enzyme that is present in much larger amounts when a particular compound (the inducer) has been added to the system. (Contrast with constitutive enzyme.)

Inflammation • A nonspecific defense against pathogens; characterized by redness, swelling, pain, and increased temperature.

Inflorescence • A structure composed of several flowers.

Inhibitor • A substance which binds to the surface of an enzyme and interferes with its action on its substrates.

Inhibitory postsynaptic potential • A change in the resting potential of a postsynaptic membrane in the hyperpolarizing (negative) direction.

Initiation complex • Combination of a ribosomal light subunit, an mRNA molecule, and the tRNA charged with the first amino acid coded for by the mRNA; formed at the onset of translation.

Initiation factors • Proteins that assist in forming the translation initiation complex at the ribosome.

Inositol triphosphate (IP3) • An intracellular second messenger derived from membrane phospholipids.

Instar (in' star) [L.: image, form] • An immature stage of an insect between molts.

Insulin (in' su lin) [L. *insula*: island] • A hormone, synthesized in islet cells of the pancreas, that promotes the conversion of glucose to the storage material, glycogen.

Integrase • An enzyme that integrates retroviral cDNA into the genome of the host cell.

Integrated pest management • A method of control of pests in which natural predators and parasites are used in conjunction with sparing use of chemical methods to achieve control of a pest without causing serious adverse environmental side effects.

Integument [L. *integumentum*: covering] • A protective surface structure. In gymnosperms and angiosperms, a layer of tissue around the ovule which will become the seed coat. Gymnosperm ovules have one integument, angiosperm ovules two.

Intercalary meristem • A meristematic region in plants which occurs not apically, but between two regions of mature tissue. Intercalary meristems occur in the nodes of grass stems, for example.

Intercostal muscles • Muscles between the ribs that can augment breathing movements by elevating and suppressing the rib cage.

Interferon • A glycoprotein produced by virus-infected animal cells; increases the resistance of neighboring cells to the virus.

Interkinesis • The phase between the first and second meiotic divisions.

Interleukins • Regulatory proteins, produced by macrophages and lymphocytes, that act upon other lymphocytes and direct their development.

Intermediate filaments • Fibrous proteins that stabilize cell structure and resist tension.

Internode • Section between two nodes of a plant stem.

Interphase • The period between successive nuclear divisions during which the chromosomes are diffuse and the nuclear envelope is intact. It is during this period that the cell is most active in transcribing and translating genetic information.

Interspecific competition • Competition between members of two or more species.

Intertropical convergence zone • The tropical region where the air rises most strongly; moves north and south with the passage of the sun overhead.

Intraspecific competition • Competition among members of a single species.

Intrinsic protein • A membrane protein that is embedded in the phospholipid bilayer of the membrane. (Contrast with extrinsic protein.)

Intrinsic rate of increase • The rate at which a population can grow when its density is low and environmental conditions are highly favorable.

Intron • A portion of a DNA molecule that, because of RNA splicing, is not involved in coding for part of a polypeptide molecule. (Contrast with exon.)

Invagination • An infolding.

Inversion (genetic) • A rare mutational event that leads to the reversal of the order of genes within a segment of a chromosome, as if that segment had been removed from the chromosome, turned 180°, and then reattached.

Invertebrate • Any animal that is not a vertebrate, that is, whose nerve cord is not enclosed in a backbone of bony segments.

In vitro [L.: in glass] • In a test tube, rather than in a living organism. (Contrast with in vivo.)

In vivo [L.: in the living state] • In a living organism. Many processes that occur in vivo can be reproduced in vitro with the right selection of cellular components. (Contrast with in vitro.)

Ion (eye' on) [Gr.: wanderer] • An atom or group of atoms with electrons added or removed, giving it a negative or positive electrical charge.

Ion channel • A membrane protein that can let ions pass across the membrane. The channel can be ion-selective, and it can be voltage-gated or ligand-gated.

Ionic bond • A chemical bond which arises from the electrostatic attraction between positively and negatively charged ions. Usually a strong bond.

Iris (eye' ris) [Gr. *iris*: rainbow] • The round, pigmented membrane that surrounds the pupil of the eye and adjusts its aperture to regulate the amount of light entering the eye.

Irruption • A rapid increase in the density of a population. Often followed by massive emigration.

Islets of Langerhans • Clusters of hormone-producing cells in the pancreas.

Iso- [Gr.: equal] • Prefix used to denote two separate but similar or identical states of a

characteristic. (See isomers, isomorphic, isotope.)

Isolating mechanism • Geographical, physiological, ecological, or behavioral mechanisms that lead to a reduction in the frequency of hybrid matings.

Isomers • Molecules consisting of the same numbers and kinds of atoms, but differing in the way in which the atoms are combined.

Isomorphic (eye' so more' fik) [Gr. *isos*: equal + *morphe*: form] • having the same form or appearance, as two isomorphic life stages. (Contrast with heteromorphic.)

Isotonic • Having the same solute concentration; said of two solutions. (Contrast with hypertonic, hypotonic.)

Isotope (eye' so tope) [Gr. *isos*: equal + *topos*: place] • Two isotopes of the same chemical element have the same number of protons in their nuclei, but differ in the number of neutrons.

Jasmonates • Plant hormones that trigger defenses against pathogens and herbivores.

Jejunum (jih jew' num) • The middle division of the small intestine, where most absorption of nutrients occurs. (See duodenum, ileum.)

Joule (jool, or jowl) • A unit of energy, equal to 0.24 calories.

Juvenile hormone • In insects, a hormone maintaining larval growth and preventing maturation or pupation.

Karyotype • The number, forms, and types of chromosomes in a cell.

Kelvin temperature scale • See absolute temperature scale.

Keratin (ker' a tin) [Gr. *keras*: horn] • A protein which contains sulfur and is part of such hard tissues as horn, nail, and the outermost cells of the skin.

Ketone (key' tone) • A compound with a C==O group attached to two other groups, neither of which is an H atom. Many sugars are ketones. (Contrast with aldehyde.)

Keystone species • A species that exerts a major influence on the composition and dynamics of the community in which it lives.

Kidneys • A pair of excretory organs in vertebrates.

Kin selection • The component of inclusive fitness resulting from helping the survival of relatives containing the same alleles by descent from a common ancestor.

Kinase (kye' nase) • An enzyme that transfers a phosphate group from ATP to another molecule. Protein kinases transfer phosphate from ATP to specific proteins, playing important roles in cell regulation.

Kinesis (ki nee' sis) [Gr.: movement] • Orientation behavior in which the organism does not move in a particular direction with reference to a stimulus but instead simply moves at an increasing or decreasing rate until it ends up farther from the object or closer to it. (Contrast with taxis.)

Urea • A compound serving as the main excreted form of nitrogen by many animals, including mammals.

Ureotelic • Describes an organism in which the final product of the breakdown of nitrogen-containing compounds (primarily proteins) is urea. (Contrast with ammonotelic, uricotelic.)

Ureter (your′ uh tur) [Gr. *ouron*: urine] • A long duct leading from the vertebrate kidney to the urinary bladder or the cloaca.

Urethra (you ree′ thra) [Gr. *ouron*: urine] • In most mammals, the canal through which urine is discharged from the bladder and which serves as the genital duct in males.

Uric acid • A compound that serves as the main excreted form of nitrogen in some animals, particularly those which must conserve water, such as birds, insects, and reptiles.

Uricotelic • Describes an organism in which the final product of the breakdown of nitrogen-containing compounds (primarily proteins) is uric acid. (Contrast with ammonotelic, ureotelic.)

Urinary bladder • A structure structure that receives urine from the kidneys via the ureter, stores it, and expels it periodically through the urethra.

Urine (you′ rin) [Gk. *ouron*: urine] • In vertebrates, the fluid waste product containing the toxic nitrogenous by-products of protein and amino acid metabolism.

Uterus (yoo′ ter us) [L.: womb] • The uterus or womb is a specialized portion of the female reproductive tract in certain mammals. It receives the fertilized egg and nurtures the embryo in its early development.

Vaccination • Injection of virus or bacteria or their proteins into the body, to induce immunization. The injected material is usually attenuated (weakened) before injection.

Vacuole (vac′ yew ole) [Fr.: small vacuum] • A liquid-filled cavity in a cell, enclosed within a single membrane. Vacuoles play a wide variety of roles in cellular metabolism, some being digestive chambers, some storage chambers, some waste bins, and so forth.

Vagina (vuh jine′ uh) [L.: sheath] • In female mammals, the passage leading from the external genital orifice to the uterus; receives the copulatory organ of the male in mating.

van der Waals interaction • A weak attraction between atoms resulting from the interaction of the electrons of one atom with the nucleus of the other atom. This attraction is about one-fourth as strong as a hydrogen bond.

Variable regions • The part of an immunoglobulin molecule or T-cell receptor that includes the antigen-binding site.

Vascular (vas′ kew lar) • Pertaining to organs and tissues that conduct fluid, such as blood vessels in animals and phloem and xylem in plants.

Vascular bundle • In vascular plants, a strand of vascular tissue, including conducting cells of xylem and phloem as well as thick-walled fibers.

Vascular ray • In vascular plants, radially oriented sheets of cells produced by the vascular cambium, carrying materials laterally between the wood and the phloem.

Vascular tissue system • The conductive system of the plant, consisting primarily of xylem and phloem. (Contrast with dermal tissue system, ground tissue system.)

Vasopressin • See antidiuretic hormone.

Vector • (1) An agent, such as an insect, that carries a pathogen affecting another species. (2) A plasmid or virus that carries an inserted piece of DNA into a bacterium for cloning purposes in recombinant DNA technology.

Vegetal hemisphere • The lower portion of some animal eggs, zygotes, and embryos, in which the dense nutrient yolk settles. The **vegetal pole** refers to the very bottom of the egg or embryo. (Contrast with animal hemisphere.)

Vegetative • Nonreproductive, or nonflowering, or asexual.

Vein [L. *vena*: channel] • A blood vessel that returns blood to the heart. (Contrast with artery.)

Ventral [L. *venter*: belly, womb] • Toward or pertaining to the belly or lower side. (Contrast with dorsal.)

Ventricle • A muscular heart chamber that pumps blood through the body.

Vernalization [L. *vernalis*: belonging to spring] • Events occurring during a required chilling period, leading eventually to flowering.

Vertebral column • The jointed, dorsal column that is the primary support structure of vertebrates.

Vertebrate • An animal whose nerve cord is enclosed in a backbone of bony segments, called vertebrae. The principal groups of vertebrate animals are the fishes, amphibians, reptiles, birds, and mammals.

Vessel [L. *vasculum*: a small vessel] • In botany, a tube-shaped portion of the xylem consisting of hollow cells (vessel elements) placed end to end and connected by perforations. Together with tracheids, vessel elements conduct water and minerals in the plant.

Vestibular apparatus (ves tib′ yew lar) [L. *vestibulum*: an enclosed passage] • Structures associated with the vertebrate ear; these structures sense changes in position or momentum of the head, affecting balance and motor skills.

Vestigial (ves tij′ ee al) [L. *vestigium*: footprint, track] • The remains of body structures that are no longer of adaptive value to the organism and therefore are not maintained by selection.

Vicariance (vye care′ ee unce) [L. *vicus*: change] • The splitting of the range of a taxon by the imposition of some barrier to dispersal of its members.

Vicariant distribution • A distribution resulting from the disruption of a formerly continuous range by a vicariant event.

Villus (vil′ lus) (plural: villi) [L.: shaggy hair] • A hairlike projection from a membrane; for example, from many gut walls.

Virion (veer′ e on) • The virus particle, the minimum unit capable of infecting a cell.

Viroid (vye′ roid) • An infectious agent consisting of a single-stranded RNA molecule with no protein coat; produces diseases in plants.

Virus [L.: poison, slimy liquid] • Any of a group of ultramicroscopic infectious particles constructed of nucleic acid and protein (and, sometimes, lipid) that can reproduce only in living cells.

Visceral mass • The major internal organs of a mollusk.

Vitamin [L. *vita*: life] • Any one of several structurally unrelated organic compounds that an organism cannot synthesize itself, but nevertheless requires in small quantity for normal growth and metabolism.

Viviparous (vye vip′ uh rus) [L. *vivus*: alive] • Reproduction in which fertilization of the egg and development of the embryo occur inside the mother's body. (Contrast with oviparous.)

Waggle dance • The running movement of a working honey bee on the hive, during which the worker traces out a repeated figure eight. The dance contains elements that transmit to other bees the location of the food.

Water potential • In osmosis, the tendency for a system (a cell or solution) to take up water from pure water, through a differentially permeable membrane. Water flows toward the system with a more negative water potential. (Contrast with osmotic potential, turgor pressure.)

Water vascular system • The array of canals and tubelike appendages that serves as the circulatory system, locomotory system, and food-capturing system of many echinoderms; is in direct connection with the surrounding sea water.

Wavelength • The distance between successive peaks of a wave train, such as electromagnetic radiation.

Wild type • Geneticists' term for standard or reference type. Deviants from this standard, even if the deviants are found in the wild, are said to be mutant.

Xanthophyll (zan′ tho fill) [Gr. *xanthos*: yellowish-brown + *phyllon*: leaf] • A yellow or orange pigment commonly found as an accessory pigment in photosynthesis, but found elsewhere as well. An oxygen-containing carotenoid.

X-linked (also called sex-linked) • A character that is coded for by a gene on the X chromosome.

Xerophyte (zee′ row fyte) [Gr. *xerox*: dry + *phyton*: plant] • A plant adapted to an environment with a limited water supply.

Xylem (zy′ lum) [Gr. *xylon*: wood] • In vascular plants, the woody tissue that conducts water and minerals; xylem consists, in various plants, of tracheids, vessel elements, fibers, and other highly specialized cells.

Yeast artificial chromosome • A laboratory-made DNA molecule containing sequences of yeast chromosomes (origin of replication, telomeres, centromere, and selectable markers) so that it can be used as a vector in yeast.

Yolk • The stored food material in animal eggs, usually rich in protein and lipid.

Z-DNA • A form of DNA in which the molecule spirals to the left rather than to the right.

Zooplankton (zoe' o plang ton) [Gr. *zoon*: animal + *planktos*: wandering] • The animal portion of the plankton.

Zoospore (zoe' o spore) [Gr. *zoon*: animal + *spora*: seed] • In algae and fungi, any swimming spore. May be diploid or haploid.

Zygote (zye' gote) [Gr. *zygotos*: yoked] • The cell created by the union of two gametes, in which the gamete nuclei are also fused. The earliest stage of the diploid generation.

Zymogen • An inactive precursor of a digestive enzyme secreted into the lumen of the gut, where a protease cleaves it to form the active enzyme.

26.19: © G. W. Willis/BPS. 26.20: © Science VU/Visuals Unlimited. 26.21: © Michael Gabridge/Visuals Unlimited. 26.23: © Krafft/Hoa-qui/Photo Researchers, Inc. 26.24: © Martin G. Miller/Visuals Unlimited.

Chapter 27 *Opener*: © Mike Abbey/Visuals Unlimited. 27.1a: © David Phillips/Visuals Unlimited. 27.1b: © J. Paulin/Visuals Unlimited. 27.1c: © Randy Morse/Tom Stack & Assoc. 27.7a: © Christian Gautier/Jacana/Photo Researchers, Inc. 27.7b: © Cabisco/Visuals Unlimited. 27.7c: © Alex Rakosy/Dembinsky Photo Assoc. 27.8: © David M. Phillips/Visuals Unlimited. 27.11: © Oliver Meckes/Photo Researchers, Inc. 27.12: © Sanford Berry/Visuals Unlimited. 27.14a: © Mike Abbey/Visuals Unlimited. 27.14b: © Dennis Kunkel, U. Hawaii. 27.14c,d: © Paul W. Johnson/BPS. 27.15b: © M. A. Jakus, NIH. 27.18a: © Manfred Kage/Peter Arnold, Inc. 27.18b: © Biophoto Associates/Photo Researchers, Inc. 27.20a: © Joyce Photographics/The National Audubon Society Collection/Photo Researchers, Inc. 27.20b: © J. Robert Waaland/BPS. 27.21a: © Jeff Foott/Tom Stack & Assoc. 27.21b: © J. N. A. Lott/BPS. 27.23: © James W. Richardson/Visuals Unlimited. 27.24a: © Maria Schefter/BPS. 27.24b: © J. N. A. Lott/BPS. 27.25a: © Cabisco/Visuals Unlimited. 27.25b: © Andrew J. Martinez/Photo Researchers, Inc. 27.25c: © Alex Rakosy/Dembinsky Photo Assoc. 27.31a: © Robert Brons/BPS. 27.31b: © A. M. Siegelman/Visuals Unlimited. 27.32a: © Barbara J. Miller/BPS. 27.32b: © Cabisco/Visuals Unlimited. 27.33a: © D. W. Francis, U. Delaware. 27.33b: © David Scharf/Peter Arnold, Inc.

Chapter 28 *Opener*: © Fred Bruemmer/DRK PHOTO. 28.1a: © Ron Dengler/Visuals Unlimited. 28.1b: © Larry Mellichamp/Visuals Unlimited. 28.4a,b: © J. Robert Waaland/BPS. 28.5a: © Rod Planck/Dembinsky Photo Assoc. 28.5b: © William Harlow/Photo Researchers, Inc. 28.5c: © Science VU/Visuals Unlimited. 28.6: © Dr. David Webb, U. Hawaii. 28.7a: © Brian Enting/Photo Researchers, Inc. 28.7b: © J. H. Troughton. 28.9: Figure information provided by Hermann Pfefferkorn, Dept. of Geology, U. Pennsylvania. Original oil painting by John Woolsey. 28.14a: © Ed Reschke/Peter Arnold, Inc. 28.14b: © Cabisco/Visuals Unlimited. 28.15a: © J. N. A. Lott/BPS. 28.15b: © David Sieren/Visuals Unlimited. 28.16: © W. Ormerod/Visuals Unlimited. 28.17a: © Rod Planck/Dembinsky Photo Assoc. 28.17b: © Nuridsany et Perennou/Photo Researchers, Inc. 28.17c: © Dick Keen/Visuals Unlimited. 28.18: © L. West/Photo Researchers, Inc.

Chapter 29 *Opener*: © Marty Cordano/DRK PHOTO. 29.3: © Phil Gates/BPS. 29.4a: © Roland Seitre/Peter Arnold, Inc. 29.4b: © Bernd Wittich/Visuals Unlimited. 29.4c: © M. Graybill/J. Hodder/BPS. 29.4d: © Louisa Preston/Photo Researchers, Inc. 29.7a: © Dick Poe/Visuals Unlimited.

29.7b: © Richard Shiell. 29.7c: © Richard Shiell/Dembinsky Photo Assoc. 29.8a: © Richard Shiell. 29.8b: © Noboru Komine/Photo Researchers, Inc. 29.11a: © Inga Spence/Tom Stack & Assoc. 29.11b: © Holt Studios/Photo Researchers, Inc. 29.11c: © Catherine M. Pringle/BPS. 29.11d: © Inga Spence/Tom Stack & Assoc. 29.12: © U. California, Santa Cruz, and UCSC Arboretum. 29.12 *inset*: © Sandra K. Floyd, U. Colorado. 29.14a: © Ken Lucas/Visuals Unlimited. 29.14b: © Ed Reschke/Peter Arnold, Inc. 29.14c: © Adam Jones/Dembinsky Photo Assoc. 29.15a: © Richard Shiell. 29.15b: © Adam Jones/Dembinsky Photo Assoc. 29.15c: © Alan & Linda Detrick/The National Audubon Society Collection/Photo Researchers, Inc.

Chapter 30 *Opener*: © S. Nielsen/DRK PHOTO. 30.1a: © Inga Spence/Tom Stack & Assoc. 30.1b: © L. E. Gilbert/BPS. 30.1c: © G. L. Barron/BPS. 30.2: © David M. Phillips/Visuals Unlimited. 30.4: © G. T. Cole/BPS. 30.5: © N. Allin and G. L. Barron/BPS. 30.7: © J. Robert Waaland/BPS. 30.8: © Gary R. Robinson/Visuals Unlimited. 30.9: © Tom Stack/Tom Stack & Assoc. 30.10: © John D. Cunningham/Visuals Unlimited. 30.11a: © Richard Shiell/Dembinsky Photo Assoc. 30.11b: © Matt Meadows/Peter Arnold, Inc. 30.12: © Andrew Syred/Science Photo Library/Photo Researchers, Inc. 30.14a: © Angelina Lax/Photo Researchers, Inc. 30.14b: © Manfred Danegger/Photo Researchers, Inc. 30.14c: © Stan Flegler/Visuals Unlimited. 30.15 *inset*: © Biophoto Associates/Photo Researchers, Inc. 30.16a: © R. L. Peterson/BPS. 30.16b: © Merton F. Brown/Visuals Unlimited. 30.17a: © Ed Reschke/Peter Arnold, Inc. 30.17b: © Gary Meszaros/Dembinsky Photo Assoc. 30.18a: © J. N. A. Lott/BPS.

Chapter 31 *Opener*: © Paolo Curto/The Image Bank. 31.5a: © Don Fawcett/Visuals Unlimited. 31.5b: © Christian Petron/Planet Earth Pictures. 31.5c: © Gillian Lythgoe/Planet Earth Pictures. 31.6a: © Robert Brons/BPS. 31.6b: © Tom & Therisa Stack/Tom Stack & Assoc. 31.6c: © Randy Morse/Tom Stack & Assoc. 31.7, 31.8, 31.9, 31.10: Adapted from Bayerand, F. M., and H. B. Owre, 1968. *The Free-Living Lower Invertebrates*, Macmillan Publishing Co. 31.11a: © G. Carleton Ray/Photo Researchers, Inc. 31.11b: © Fred Bavendam/Minden Pictures. 31.12: © David J. Wrobel/BPS. 31.13: From M. W. Martin, 2000. *Science* 288:841–845. 31.15a: © Fred McConnaughey/Photo Researchers, Inc. 31.17b: © James Solliday/BPS. 31.20a: © Chamberlain, MC/DRK PHOTO. 31.21: © David J. Wrobel/BPS. 31.22: © Jeff Mondragon. 31.24a: © Brian Parker/Tom Stack & Assoc. 31.24b: © Roger K. Burnard/BPS. 31.24c: © Stanley Breeden/DRK PHOTO. 31.24d: © R. R. Hessler, Scripps Institute of Oceanography. 31.26a: © Ken Lucas/Planet Earth Pictures. 31.26b: © Dave Fleetham/Tom Stack & Assoc. 31.26c: © Mike Severns/Tom Stack & Assoc. 31.26d: © Milton

Rand/Tom Stack & Assoc. 31.26e: © Dave Fleetham/Tom Stack & Assoc. 31.26f: © A. Kerstitch/Visuals Unlimited.

Chapter 32 *Opener*: © John Mitchell/The National Audubon Society Collection/Photo Researchers, Inc. 32.2: © Dr. Rick Hochberg, U. New Hampshire. 32.4: © R. Calentine/Visuals Unlimited. 32.5b,c: © James Solliday/BPS. 32.7a: © Doug Wechsler. 32.7b: © Diane R. Nelson/Visuals Unlimited. 32.8: © Ken Lucas/Visuals Unlimited. 32.9a: © Joel Simon. 32.9b: © Fred Bruemmer/DRK PHOTO. 32.10a: © Peter J. Bryant/BPS. 32.10b: © David Maitland/Masterfile. 32.10c: © W. M. Beatty/Visuals Unlimited. 32.10d: © Robert Brons/BPS. 32.11a: © Henry W. Robison/Visuals Unlimited. 32.11b: © Stephen P. Hopkin/Planet Earth Pictures. 32.11c: © Peter David/Planet Earth Pictures. 32.11d: © A. Flowers & L. Newman/The National Audubon Society Collection/Photo Researchers, Inc. 32.13a: © Charles R. Wyttenbach/BPS. 32.13b: © William Leonard/DRK PHOTO. 32.15a: © David P. Maitland/Planet Earth Pictures. 32.15b: © Konrad Wothe/Minden Pictures. 32.15c: © Peter J. Bryant/BPS. 32.15d: © David Maitland/Masterfile. 32.15e: © Steve Nicholls/Planet Earth Pictures. 32.15f: © Brian Kenney/Planet Earth Pictures. 32.15g: © Simon D. Pollard/The National Audubon Society Collection/Photo Researchers, Inc. 32.15h: © L. West/The National Audubon Society Collection/Photo Researchers, Inc.

Chapter 33 *Opener*: © Norbert Wu/DRK PHOTO. 33.3a: © Hal Beral/Visuals Unlimited. 33.3b: © Randy Morse/Tom Stack & Assoc. 33.3c: © Mark J. Thomas/Dembinsky Photo Assoc. 33.3d: © Randy Morse/Tom Stack & Assoc. 33.3e: © John A. Anderson/Animals Animals. 33.4: © C. R. Wyttenbach/BPS. 33.5: © Gary Bell/Masterfile. 33.6b, 33.9: © Norbert Wu/DRK PHOTO. 33.11a: © Dave Fleetham/Tom Stack & Assoc. 33.11b: © Marty Snyderman/Masterfile. 33.12a: © Ken Lucas/Planet Earth Pictures. 33.12b: © Fred Bavendam/Minden Pictures. 33.12c: © Dave Fleetham/Visuals Unlimited. 33.12d: © Dr. Paul A. Zahl/The National Audubon Society Collection/Photo Researchers, Inc. 33.13: © Tom McHugh, Steinhart Aquarium/The National Audubon Society Collection/Photo Researchers, Inc. 33.15a: © Ken Lucas/BPS. 33.15b: © Nick Garbutt/Indri Images. 33.15c: © Art Wolfe. 33.19a: © Michael Fogden/DRK PHOTO. 33.19b: © Joe McDonald/Tom Stack & Assoc. 33.19c: © C. Alan Morgan/Peter Arnold, Inc. 33.19d: © Dave B. Fleetham/Tom Stack & Assoc. 33.19e: © Mark J. Thomas/Dembinsky Photo Assoc. 33.20a: Courtesy of Carnegie Museum of Natural History, Pittsburgh. 33.20b: Fossil from the Natural History Museum of Basel, photographed by Severino Dahint. 33.21a: © Joe McDonald/Tom Stack & Assoc. 33.21b: © John Shaw/Tom Stack & Assoc. 33.21c: © Skip Moody/Dembinsky

Index

INDEX

Turtles, 589–590
Twins, 743, 759
Twin studies, 189–190
Twitches, 839–840
Two-peaked distributions, 406
Tympanic membrane, *802*, 803, 804
Type 1 diabetes. *See* Insulin-dependent diabetes
Typhus, 256
Tyrosine, *37*, *224*, 332
Tyrosine kinase receptors, 283, 287

Ubiquinone (Q), 125, 126, **127**
Ubiquitin, 277
Ulcers, 898–899
Ulna, *843*
Ulothrix, *493*
Ultraviolet radiation, 139
   detection in insects, 812
   mutation effects, 236
Ulva latuca, 492–493
Umbels, **522**
Umbilical cord, 768, 771
Umbilicus, 771
Underground stems, 676
Undernourishment, 132, 888–889
Uniport transporters, 89
Uniramians (Uniramia), 571–574, **578**
United States
   age distribution in, 961–962
   geographic distribution of endangered species, **1042**
   proportion of species extinct or at risk, **1034**
Unsaturated fatty acids, 50
Unsegmented worms, 566–567
Upregulation, of hormone receptors, 729
Upwellings, 999
Uracil, *47*, 48
   deamination, 236, **237**
   genetic code, **224**
   in RNA, 220
   spontaneous formation in DNA, 338
Urbilateria, 552
Urea, 912
Uremic poisoning, 922
Ureotelic animals, 912
Ureter, 919
Urethra, *739*, 740, 919
Uric acid, 912, 913–194, 915
Uricotelic animals, 912
Urinary bladder, 919
Urination, 919
Urine, 911
   of annelids, 914
   concentrating in the kidney, 920–921
   of flatworms, 914
   human, 913–194
   renal tubules and, 916
Urochordata, *578*, 582
Uroctonus mondax, *570*
Urodela, 587
Ursus maritimus, **887**
Urticina lofotensis, **548**
Uterine cycle, 743–745

Uterus, 739
   blastocyst implantation, 742, 745, 758, 769
   egg in, 741
   uterine cycle, 743–745

Vaccination, 359, 360
Vaccine proteins, **323**
Vacuoles
   functions of, 63, 71–72
   in protists, *480*, 481
   See also Plant vacuoles
Vagina, 737, 741, 746
Valine, *37*, *224*, 889, **890**
Vampire bats, 910
Van der Waals forces, *21*, 25, 40
Variable number of tandem repeats (VNTRs), 328
Variable region, of immunoglobulins, *362*, 363, 369
Varicose veins, 878
Vasa recta, 920
Vascular bundles, 615
Vascular cambium, 613, 615–616
Vascular endothelial growth factor, 714
Vascular rays, 615–616
Vascular system (animal)
   arteries and arterioles, 875–876
   capillaries, 876–877
   cardiovascular disease and, 878–879
   lymphatic vessels, 877
   veins, 877–878
   See also Blood vessels; Circulatory systems
Vascular tissue system (plant), 608, **609**, 610–611
   in *Amborella*, 525
   in gymnosperms, 518
   in leaves, 618
   in plant evolution, 502–503, 507
   in roots, 614
   secondary growth, 613, 615–617
   in stems, 615
Vas deferens, *739*, 740, **747**
Vasectomy, *746*, 747
Vasopressin, 718–719, 882, 883. *See also* Antidiuretic hormone
Vectors
   artificial chromosomes, 316
   expression vectors, 322–323
   for plants, 316
   plasmids, 315, 316
   properties of, 315
   viruses and, 242–243, 316
Vegetal hemisphere, *301*, 755, 756
   in cleavage, 757
   developmental potential, 761
Vegetarianism, 889–890
Vegetative organs, 603–606
Vegetative reproduction, 165
   in agriculture, 677
   forms of, 676–677
   See also Asexual reproduction
Veins (animal), 868, 875, 877–878. *See also* Vascular system
Veins (plant), 618

Venter, Craig, 255
Ventilation, 853
   in bird lungs, 854–856
   in fishes, 854
   tidal breathing, 856–857
Ventral horn (spinal cord), 817
Ventricle
   human heart and cardiac cycle, 871–872, 873–874
   in reptilian hearts, 870
   in three-chambered hearts, 869
   in two-chambered hearts, 869
Ventricular fibrillation, **874**
Ventromedial hypothalamus, 906
Venules, 868
Venus flytraps, 634, 644
Vernalization, flowering and, 675–676
Vertebral column, 583, *843*
Vertebrates (Vertebrata), 583
   axis determination in, 309
   body temperature regulation, 707–709
   central nervous system, 774–775
   circulatory systems, 868–871
   colonization of land, 587–591
   dorsal-ventral development, 543
   equilibrium organs in, 801–803
   eye anatomy and function, 807–812
   eye development, 301–302
   generalized body plan, 583
   nephron structure and function, 915–916, **917**
   notochord and, 430, *431*
   origin of, 583–586
   phylogeny, 428–430, *584*
   skeleton and joints, 843–846
   teeth, 894
   tetrapods, **587**
Vertical transmission, 243, 244
Very-low-density lipoproteins (VLDLs), 904, **905**
Vesicles
   from endoplasmic reticulum, 66
   in Golgi apparatus, 65, 66
   in protists, 480–481
Vessels/Vessel elements, 521, 608, **609**
Vestiaria coccinea, **1038**
Vestibular apparatus, 801, *802*, 803
Vestigial organs, 902
Vestimentiferans, 559
Vibrio cholerae, 467
Vicariant distributions, 1009, **1010**, 1027
Victoria (Queen of England), 196
Vicuñas, 861–862
Villi, 896
Vincristine, *347*
Viral retrotransposons, 265
Virginia opossum, 594
Virions, **240**, 241
Viroids, 240, 244
Viruses
   cancer-causing, 343
   characteristics of, 240–241

classifications of, 241
discovery of, 240
as DNA vectors, 316
enveloped, 243
genome, 260
Hershey-Chase experiment, 201–202
regulation of gene expression in, 254, **255**
reproductive life cycles, 241–243
size range, **240**
temperate, 242
transduction and, 247
using in genetic research, 239–240
virulent, 242
Visceral mass, 560
Visual communication, 935–936
Visual cortex, 823–824
Visual disparity, 824
Visual systems
   binocular vision, 824
   detection of ultraviolet light, 812
   in invertebrates, 807
   rhodopsin and photosensitivity, 805–807
   vertebrate eye, 807–812
   visual cortex, 823–824
Vitamin A, 52, **892**
   deficiency, 327
Vitamin B, **892**, 893
   neural tube defects and, 766
Vitamin C, 891, 892
Vitamin D, 52, 891, **892**
Vitamin E, 52, **892**
Vitamin K, 52, **892**, 902
Vitamins, 52, 891–892
Vitelline envelope, 753, 754, **755**
Vitis vinifera, 677
Viviparity, 739
VNTRs, 328
Voice box. *See* Larynx
Volatinia jacarina, **416**
Volcanoes, 383–384, 388, 389, 1004
Voltage, 777
Voltage-gated channels
   in action potentials, 780, 781, **782**, 783
   defined, 778
   in neurotransmitter release, 786, **787**
   in pacemaker cells, 873
   in phototransduction, **806**, 807
   in sensory transduction, **795**, 796
Voluntary nervous system, 815
Volvox, 492
von Frisch, Karl, 928, 936
vp corn mutants, 660
Vries, Hugo de, 178
Vulpes macrotis, **707**
Vulva, in *C. elegans*, 302, *303*

Waggle dance, 936
Walking sticks, 572
Wallace, Alfred Russell, 2, 397
Warblers, 419, **420**
Warty chameleon, **1041**

About the Book

Editor: Andrew D. Sinauer

Project Editor: Carol J. Wigg

Developmental Editor: James Funston

Review Coordinator: Susan McGlew

Copy Editor: Norma Roche

Production Manager: Christopher Small

Book Layout and Production: Janice Holabird, Jefferson Johnson, and Joan Gemme

Art Editing and Illustration Program: J/B Woolsey Associates

Design: Jefferson Johnson

Book Cover Design: Jefferson Johnson

Photo Research: David McIntyre

Index: Grant Hackett

Color Separations: Vision Graphics, Inc. and Burt Russell Litho

Cover Manufacture: Henry N. Sawyer Company, Inc.

Book Manufacture: Courier Companies, Inc.